1 #24
T5-AGX-443

Special Publication No. 2
of the Society for Geology
Applied to Mineral Deposits

Ore Genesis
The State of the Art

A Volume in Honour of Professor Paul Ramdohr
on the Occasion of His 90th Birthday
With Special Reference to His Main Scientific
Interests

Edited by G. C. Amstutz A. El Goresy
G. Frenzel C. Kluth G. Moh
A. Wauschkuhn R. A. Zimmermann

With 398 Figures

Springer-Verlag
Berlin Heidelberg New York 1982

WILLIAM MADISON RANDALL LIBRARY UNC AT WILMINGTON

Prof. Dr. G. C. Amstutz
Prof. Dr. G. Frenzel
Dr. C. Kluth
Prof. Dr. G. Moh
Dr. A. Wauschkuhn
Dr. R. A. Zimmermann
Mineralogisch-Petrographisches Institut der Universität Heidelberg
Im Neuenheimer Feld 236, 6900 Heidelberg, FRG

Prof. Dr. A. El Goresy
Max-Planck-Institut für Kernphysik
Saupfercheckweg, 6900 Heidelberg, FRG

ISBN 3-540-11139-5 Springer-Verlag Berlin Heidelberg New York
ISBN 0-387-11139-5 Springer-Verlag New York Heidelberg Berlin

Library of Congress Cataloging in Publication Data. Main entry under title: Ore genesis, the state of the art. Bibliography: p. Includes index. 1. Ore-deposits – Addresses, essays, lectures. 2. Ramdohr, Paul, 1890– . 3. Mineralogists – Germany (West) – Biography. I. Ramdohr, Paul, 1890– . II. Amstutz, G. C. (Gerhardt Christian), 1922–
QE390.073 549 81-18450 AACR2

This work is subject to copyright. All rights are reserved, whether the whole or part of the material is concerned, specifically those of translation, reprinting, re-use of illustrations, broadcasting, reproduction by photocopying machine or similar means, and storage in data banks. Under § 54 of the German Copyright Law where copies are made for other than private use a fee is payable to 'Verwertungsgesellschaft Wort', Munich.

© by Springer-Verlag Berlin Heidelberg 1982.
Printed in Germany.

The use of registered names, trademarks, etc. in this publication does not imply, even in the absence of a specific statement, that such names are exempt from the relevant protective laws and regulations and therefore free for general use.

Printing and bookbinding: Konrad Triltsch, Graphischer Betrieb, Würzburg.
2132/3130-543210

QE390 .O73 1982

Preface and Outlook

The papers in this volume are all dedicated to Professor Paul Ramdohr, on the occasion of his 90th anniversary. The authors – and many more professional colleagues, friends, and students – owe an immense measure of gratitude to him for his guidance in scientific work and related problems, including principles of teaching and research.

The composition of the present volume would not have been possible without the cooperation of all colleagues mentioned on the front page. But the burden of most of the detailed and tedious editorial work rested with two coworkers to whom we owe cordial thanks for the many hours of work, including much of their free time, during the past year. They are Dr. Cornelia Kluth and Dr. Richard A. Zimmermann. The detailed index is entirely their own work.

Last but not least, we owe a full measure of thanks also to Professor Ramdohr in connection with difficult scientific questions of the editorial work. He received the originals of the papers at the "birthday celebration" on 21 February 1980 and keenly read all the papers. Many comments offered by him were transmitted to the authors and incorporated in the editorial work.

To the publisher we wish to express our thanks for the great amount of work dedicated to maintaining the known Springer quality of composing and printing. In a time of fast changing methods of publishing this is not an easy task, and we hope, together with the publisher, that the elimination of the circulation of galley proofs did not negatively affect the overall quality.

The selection of an appropriate title for a book of this type is not without problems – and this point may well be spelled out in the introductory pages. It is obvious that the tacit subtitle of our "90th Anniversary Volume for Professor Ramdohr" must read: "The state of the art, as seen by some of his colleagues, friends, and students".

Of course not all were able to contribute, and not in every case could the contents of the paper be restricted or directed entirely to "the state of the art", because the flow of new genetic ideas must, in many cases, be allowed to proceed quite independently of set dates of celebrations. Consequently, we did include a few papers referring

242407

rather to mineralogical problems of a more methodological or systematic nature. These will, however, in a broader sense or in the long run also affect the solution of ore genesis problems. We are, however, more concerned about a number of valuable papers which definitely belong to a complete representation of "the state of the art", but which could not be obtained in due time or for other unforeseen reasons. Consequently, every reviewer will find a few holes and gaps, and we wish to state at this time that we are keenly aware of them.

In 1937 Professor Ramdohr himself presented a review paper on "the state of the art" at that time (P. Ramdohr, Fortschritte auf dem Gebiet der Lagerstättenkunde, in Fortschr. d. Min., Krist., Petrogr., Bd. 22, S. 105–184). Characteristically, this paper dealt with ores of magmatic affiliations, ranging down in temperature to volcanic exhalations. Many deposits which now are considered to be formed by sedimentation were, at that time, believed to be hydrothermal replacements. The interest in sedimentary structures was, until recently, at a minimum.

The present volume presents a somewhat contrasting picture. Many of the deposits described are now considered to form a more or less special phase of the rock in which they are found. Not only those deposits in sediments which are now recognized as products of sedimentation and diagenetic crystallization (by some with more, by others with less movement of pore waters, the so-called brines), but also a number of deposits in igneous rocks are now seen as "indigenous" products of at least one phase of the usually complex sequence of magmatic intrusions or extrusions.

The advancement of the science of ore genesis could, in fact, be analyzed and understood quite adequately and drastically by looking into the interplay of various opposites active in its evolution. Some of these opposites occur in pairs, others consist of three extremes, still others are more complex and may be reduced again to interrelated groups of two.

The above-mentioned shift of interpretation from more igneous to more sedimentary origins represents one pair of opposites. But deeper inside this pattern of thought are other pairs of opposites which have undergone a decisive shift, not only in our field of research but also in many neighboring natural sciences, as well as in medicine, philosophy, religion, and art. These are the two pairs,

endogenous – exogenous (space),
syngenetic – epigenetic (time).

Since my 1959 monograph on this topic, the coupled interplay has continued to evolve vigorously and has produced scores of interesting papers. The discussion of whether a metal (or even the cation *and* the anion) has been introduced into a body of rock or whether it belongs to the system is still a much debated problem and always closely linked to the time question as well.

Panepigenesis still has a firm grip on many minds and blocks a free consideration of all possibilities. The tenacity of this mental block I experienced recently when visiting a gold mine in virtually unmetamorphosed ground. A fault cuts off a series with more or less periodic chert layers. These were (and are!) interpreted as formed by an introduction of SiO_2 from below and then by selective replacement. The blind acceptance of this obviously powerful theory inhibits the observation of details such as (1) the lack of any effect of the "hydrothermal SiO_2-solutions" on the other wall of the fault, (2) the fact that the stratigraphic sequence continues below a distinct throw on the fault, and (3) the presence of excellent sedimentary and diagenetic features in the chert and its host rock. The gold is, of course, also believed to be of hydrothermal origin, even though it is basically linked to organic material and pyrite, which is connected with the latter.

The subconscious reasons for the power of the exo-epi-pattern of thought have been described in several earlier papers. At this point I would like merely to recall the fact that the human mind relies heavily (or entirely?) on analogies when proceeding from observation to interpretation. The difficult thing for the scientist is, however, the fact that there are two pools of analogies; one the knowledge of similarities or identities in the outside world, and the other the still larger pool of symbolic, archetypic patterns in our subconscious. As we will see later, some of the "modern" models of ore genesis have certainly returned to old patterns, devoid of observational analogies, dominated instead by powerful subconscious symbols which block access to analogies from the outside world, such as the "common sense" observations described from the gold mine with the chert and the organic matter.

Another potent pair of opposites which emerges in this volume are two extreme origins of ore matter in sediments. This ore matter may be of (volcanic) exhalative origin or it may have been brought in merely from the continents, by the rivers and the winds. Exaggerations and premature interpretations may occur in both directions. Yet again, even the hot sulphide mud on the bottom of a graben or deep sea trough may not necessarily have the same source as the heat. Heat flow is not necessarily identical with diffusional or migrational flow of "hot brines". More details are needed to solve the origin of metal sulphides in such areas. The glamorous and dynamic model is not necessarily the more scientifically correct one.

When following the work on ore genesis of Professor Ramdohr from his comprehensive monograph on the progress of ore genesis work of 1937 till today, one factor is very clear and most impressive: His unwavering striving for better observations to reach a better genetic understanding and conclusion. This is the purest form of the inductive method of scientific investigation, the "process of reasoning from a part to a whole, from particulars to generals, or from the individual to the universal" (Webster). In the pursuit of this true scientific endeavor he visited innumerable mines and was an indefatigable collector of ores and minerals, which means, of observations. When he has found new evidence for a new, perhaps drastically new interpretation, he has, even today, always been prepared to modify a previously held working hypothesis. This attitude and "youthfulness" are also reflected in a number of papers which refer to the change of genetic interpretations and the role of Professor Ramdohr in it. I venture to interpret at least two of the well-known controversies – on the Witwatersrand and on Broken Hill (Australia) – as a typical interplay of the two opposite "types of analogies" mentioned above, the analogy from observations of many analogous outcrops and samples on one hand (on Ramdohr's side), and the analogy with a stern, subconscious, illogical pattern of thought on the opposite side.

But alas! The powerful subconscious fixation on the idea that "ore matter", as a rule, must come from the outside *and later* has produced in recent times two new genetic models. These have not, fortunately, found much support by the authors of this volume – with very few and somewhat hidden exceptions.

One of these theories is the epigenetic accumulation of ore matter by convections of ground- or hydrothermal waters. This mandala-dominated model is applied to both igneous intrusions and volcanic piles around mid-oceanic ridges. Such convection models should certainly be tested. However, the generalization were obviously made much too quickly and without looking at details of wallrock alteration. Since many details cannot be discussed in the frame of this introduction, one example must suffice to characterize the levity of the convection argument. Fossil piles of volcanics interbedded with sediments clearly show alternations of altered and unaltered layers of rocks and even differences in degree of alteration within one and the same lava flow. On such evidence a convection and leaching model does not make sense and an open mind must turn to the (external, not subconscious) analogy of deuteric or diagenetic waters and alterations.

The other theory which is equally based on a subconscious domination of the old panepigenetic dogma, which crept in and blocked the logical, common sense argumentation, is the creation of ore magmas from the subducted plate into and onto the crust of the Earth. Here too, the criteria against such a neo-panepigenetic theory breaks down in view of inductive reasoning. Among the observations speaking against such a creation are (1) the complexity of most porphyry copper intrusions; (2) the fact that, almost always, only one or two of the intrusial magma generations are ore bearing; and (3) the complete absence of any seismic evidence for the creation and movement of such plutons coming up from the plate.

The list of interesting interplays of opposites is still longer, but only two more can be mentioned as being relevant to the papers presented in this volume. A very positive interplay displayed in this book is the one between the experimental and the "naturalist" approaches. Many papers relate both aspects, and this reflects strongly on Professor Ramdohr's life-long striving to incorporate all available experimental evidence into his working hypotheses. Yet, if the observation in nature did not fit the known experimental results, he joined H.H. Hess in assuming that "this must be a case of a missing experiment". It is most unfortunate for the future of economic geology that the gap between naturalists and experimentalists seems to widen. The fact that it is possible to publish a book on experimental ore geology with hardly any link to the observations in nature – at least not to those connected with space – I consider as a warning signal in the development of our science. We absolutely must keep in touch with each other, the naturalist and the experimentalist. I should like to express may satisfaction again that the present book is in most parts an ideal bridge between both approaches.

The other – and last – interplay is in many ways closely linked to the last pair of opposites. These appear to exist, in some schools, an extreme tendency to specialize on analytical results only, leaving aside any geometric, textural evidence. The other extreme I have not yet encountered, but it would be just as detrimental to a true inductive reasoning. Both aspects, the geo-chemical and the geo-metric must be reconciled carefully. Why? Essentially because the compositional, analytical results give us a compositional difference Δc from one place to another and thus a $\Delta c/\Delta s$, where s means space. Too often such a difference is interpreted tacitly as having been introduced at a later time. The expression Δt has sneaked into the concept without any real proof. This is where the geometric, i.e., the textural and structural, observations must come in, to decide whether Δt is x or 0. Fortunately, most authors of the papers in this volume are well aware of the role of geometric observations – after all, most of them have gone through the best ore microscopic school there is, the Ramdohr school, in which textural features and compositional analytical results are inseparable observations; both must be considered when attempting an interpretation.

This leads us to the last and perhaps most important point of introduction to this book, which is at the same time the essential trend of modern ore genesis – as far as it is truly inductive first and deductive only at the very end: This is the reversal of sequence of questioning in ore genetic work. We now start the work with the question "when was the ore formed?", or, "when did the ore minerals crystallize?" This question is more often answered by geometric observations than by analytical data. When the time problem has been reasonably answered, the next problem "where did the ore matter come from?", is already easier, because a few possibilities are normally ruled out automatically if the time of formation is known.

In summary, the sequence of asking the main ore genetic questions has been gradually reversed since about 1950 or 1960, and this fact is clearly visible from the papers of this book.

I should like to close this introduction and outlook with a statement on the subdivision into chapters and sub-chapters. Anyone looking even only at the titles will recognize that almost half the papers could have been entered under one or more headings as well. We tried to strike a happy medium and hope that none of the authors will disagree drastically with our classification.

Summer/Fall 1981 G.C. Amstutz

Contents

Genesis of Ores (Mostly Diagenetic) in Sedimentary and Weakly Metamorphosed Sedimentary Rocks

Detrital Sedimentary Rocks

Carbonates

Evaporites

Genesis of Ores in Volcano-Sedimentary Sequences

Genesis of Deposits in Recent Sediments

Weathered Products of Ore Deposits

Genesis of Ores in Igneous Rocks

Deposits in Acidic Volcanic and Subvolcanic Rocks

Deposits of the Pegmatitic-Pneumatolytic and Hydrothermal Associations

Deposits in Basic Plutonic and Extrusive Rocks

Genesis of Ores in Metamorphic Rocks

Studies of a Regional Nature

Spatial and Temporal Considerations of Ore Genesis

Experimental Studies of Ore Mineral Associations

Mineralogical Studies

Contributors

You will find the addresses at the beginning of the respective contributions

Genesis of Ores (Mostly Diagenetic) in Sedimentary and Weakly Metamorphosed Sedimentary Rocks

Replacement Phenomena of Terrigenous Minerals by Sulphides in Copper-Bearing Permian Sandstones in Poland

M. BANAŚ, W. SALAMON, A. PIESTRZYŃSKI, and W. MAYER[1]

Abstract

Replacement of terrigenous quartz and other minerals by sulphides was observed in the grey sandstone of Lower Permian, SW Poland. The process is connected with diagenetic mobilization of metals in the copper-bearing shale and their migration into the underlying, porous sandstones. The role of clay-organic microlayers acting as semipermeable membranes seems to be valid, controlling the conditions of the environment and the extraction of metals.

1 Introduction

The Fore-Sudetic Monocline is the north-east part of the Sudetic Foreland. The platform sequence of Permo-Mesozoic, Tertiary and Quaternary sediments overlies the crystalline basement. The metal-bearing horizon follows the boundary between Rotliegende and Zechstein (Fig. 1). Extreme concentrations of various metals (Cu, Zn, Pb, Ag, Mo, Ni, Co, V, Au, Pd, Hg, Se, Bi) occur in the so-called copper-bearing shale, lower amounts in both sandstone and dolomite, traces in the boundary dolomite. About 70 ore minerals were identified with Cu and Cu-Fe sulphides prevailing. Vertical zonality of metals and mineral associations is evident.

The grey sandstone, 1–40 m thick, is the uppermost part of the continental Rotliegende clastic sequence. The rock reveals psammitic texture and random structure. Terrigenous minerals are quartz with minor feldspars, rock fragments, titanium oxides and magnetite. The uppermost 20 cm is cemented by carbonates and clay minerals. In the lower parts carbonates disappear. Locally, anhydrite, silica, gypsum and sulphide cement can be observed. Detailed studies on petrology, sedimentology and the origin of the grey sandstone were recently made by Jerzykiewicz et al. (1976), Nemec and Porebski (1977) and Nemec et al. (1978). The grey sandstone is partly mineralized with Cu and Cu-Fe sulphides forming disseminated, sometimes spotty and cementational structures. Depth of mineralization varies from 1 to 20 m.

1 Institute of Geology and Mineral Deposits, University of Mining and Metallurgy, Kraków, Poland

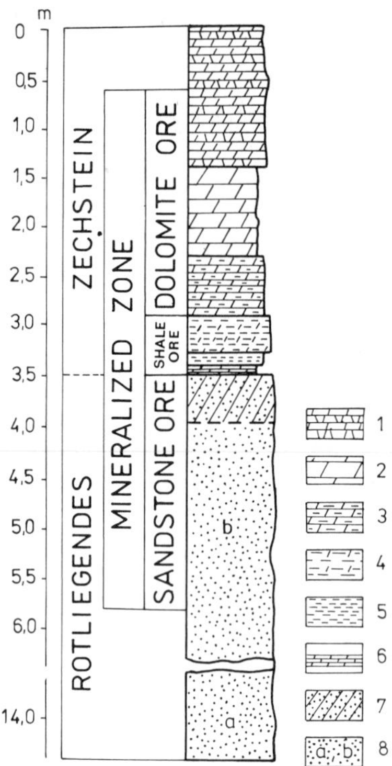

Fig. 1. Schematic profile of the mineralized zone of the Fore-Sudetic copper deposit: *1* dolomites and limestones; *2* streaky dolomite; *3* clayey dolomite; *4* dolomitic shale; *5* clay-organic shale; *6* boundary dolomite; *7* grey, dolomitic sandstone; *8* clayey sandstone: *a* red, *b* grey

2 Replacement Processes in the Mineralized Zone

Detailed microscopic studies of numerous polished sections revealed a common occurrence of metasomatic replacement processes involving ore and gangue minerals in the entire ore-bearing horizon. These processes are closely related to the diagenetic history of the rock sequence, especially to the mobilization of elements and their reprecipitation. Numerous mobilizing reactions can be mentioned: transformation of clay minerals, carbonification and oxidation of organic matter, crystallization of colloids, recrystallization of carbonates and sulphides connected with removal of isomorphic substitutions from crystal lattices, degradation of metastable compounds. All these processes were controlled mainly by pH–Eh conditions. The role of temperature and burial pressure, although less important, cannot be neglected.

Three types of reactions can be distinguished based on the microscope observations:

- mutual replacement of various minerals (sulphides, silicates, carbonates, sulphates, oxides),
- transformations of ore and clay minerals,
- oxidation effects.

The first type is dealt with in local changes of stability of minerals controlled by pH and Eh, resulting in partial or entire dissolution of some rock constituents and precipitation of other compounds. Numerous examples of such a replacement can be mentioned: clay-carbonate cement by sulphides, carbonates by sulphides, terrigenous material by sulphides and carbonates, fossil remains by dolomite, anhydrite and sulphides, early diagenetic sulphides by later generations (e.g.: bornite and chalcocite by native silver, sphalerite by castaingite, chalcopyrite, bornite and tennantite by smaltite-chloantite and/or castaingite, pyrite by galena and/or sphalerite).

The second type is a result of the enrichment of minerals with various ions as an effect of ion exchange between the pore solution and the mineral. The most characteristic illustration of such a process is the formation of bornite by enrichment of digenite with Fe^{+2}. Many other reactions of similar type were observed: replacement of pyrite by chalcopyrite and bornite, chalcocite by stromeyerite, jalpaite and mckinstryite, chalcocite, bornite, djurleite and digenite by castaingite and other minerals of the Cu-Mo-S system. Occasionally, Cu-S minerals were replaced by wittichenite and tennantite. Typical atoll and rim textures are present together with reaction zones between the minerals. Phases of an unusual chemical composition were found.

The oxidation processes are most common among the copper sulphides. Typical examples reveal replacement of bornite by chalcopyrite, chalcocite and bornite by digenite, chalcocite, bornite and digenite by covellite. In many cases relicts of replaced minerals are preserved or pseudomorphs formed.

3 Replacement of Terrigenous Grains

An intense sulphide mineralization is a typical feature of the uppermost layer of the grey sandstone. It occurs in these areas where the copper-bearing shale immediately overlies the sandstone. Various ore structures can be observed in hand specimens and polished sections: spots, nests, cementations and streaks. This mineralization apparently overlaps the earlier phase of the disseminated character. Numerous phenomena were found including replacement of clay cement by carbonates and secondary silica and clay-carbonate cement by sulphides. Locally, a pure chalcocite cement is formed.

A common feature appears to be corrosion of quartz grains by sulphides and carbonates. The main corroding mineral is chalcocite. Reactions with djurleite, pyrite, galena, sphalerite, chalcopyrite and bornite have been found. The process starts from grain boundaries. Intermediate stages are evident (Fig. 2a–d), up to complete replacement and the formation of pure chalcocite mass. In such a "massive" sulphide, relicts of corroded grains are embedded.

Feldspar grains are rare but present. When they exist, they are readily replaced by chalcocite and other sulphides. The process develops from grain boundaries or along the cleavage planes (Fig. 2b). Various stages are also evident up to complete replacement with relicts of grain boundaries.

Detrital magnetite is usually replaced by pyrite, rarely by chalcocite (Fig. 3b), chalcopyrite, galena and sphalerite. Relicts of corroded grains are easily recognizable.

Terrigenous rutile is corroded by galena and sphalerite, occasionally by chalcocite, pyrite and chalcopyrite. The reaction starts from the grain boundaries and spreads usually along microcracks. Relicts of grains are often preserved indicating the initial shape (Fig. 3c). Network textures of microcrystalline rutile/anatase have been

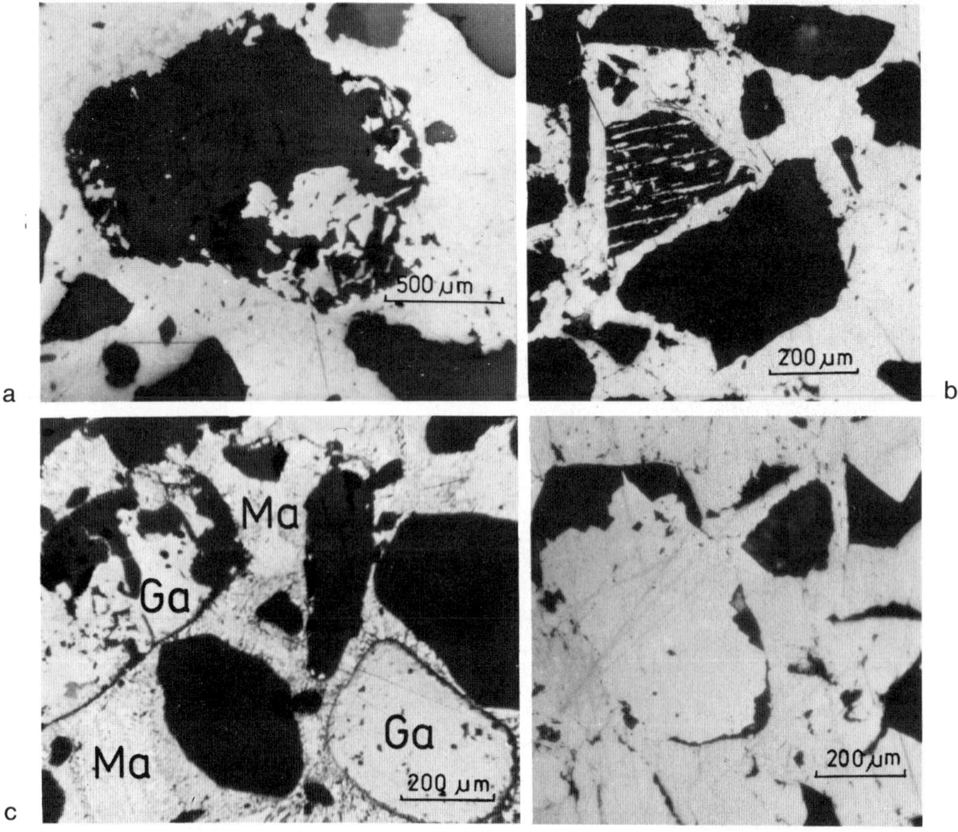

Fig. 2. a Initial stage of quartz *(black)* replacement by chalcocite *(white)*; **b** feldspar replaced by chalcocite along cleavage planes. *Black* quartz grains; **c** advanced replacement of quartz by galena *(Ga)* in marcasite *(Ma)* cemented sandstone; **d** "massive" chalcocite with relicts of quartz grains *(dark)*. Reflected light, 1 N

observed, filled with sulphides. These may result from disintegration of titanomagnetite grains.

4 Conclusion

Three stages of the replacement phenomena in the grey sandstone are visible:

– corrosion of detrital grains by clay minerals and carbonates,
– replacement of clay-carbonate cement by sulphides,
– replacement of detrital grains by sulphides.

The final effect appears to be "massive" sulphide with the relicts of corroded grains (Fig. 2d). It forms typical structures – nests, streaks, lenses, pseudolayers. Their

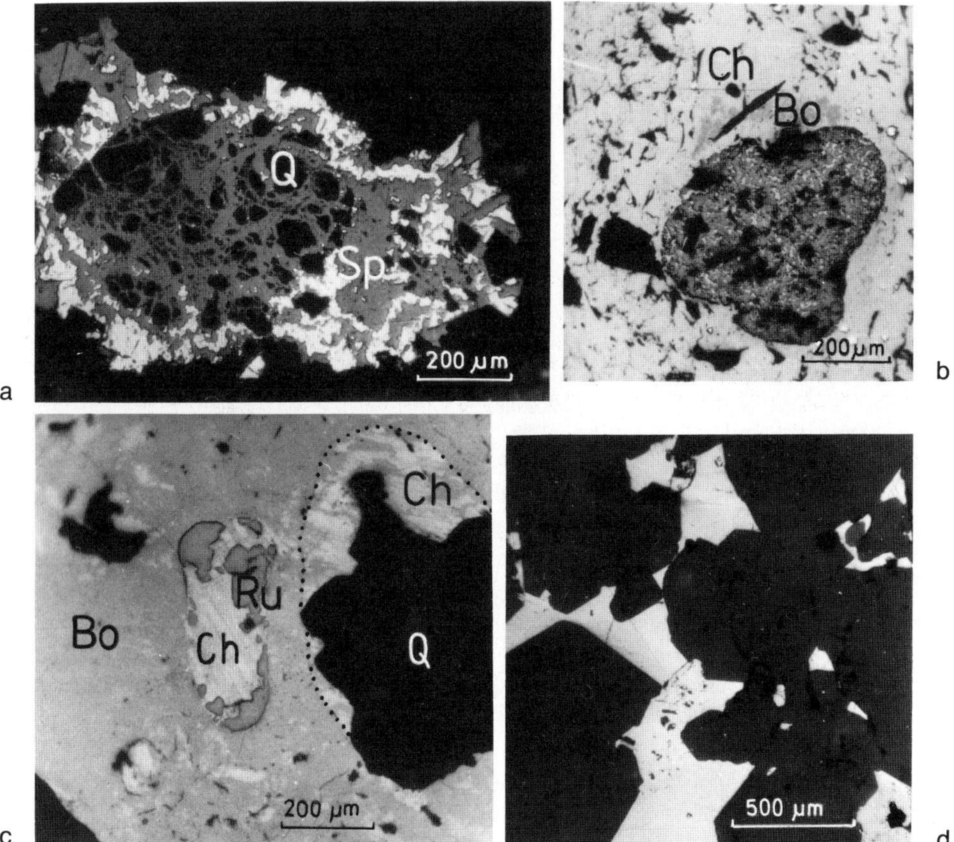

Fig. 3. a Remnants of quartz grains *(Q)* in sphalerite *(Sp)*. Later galena *(white)* replaces sphalerite, which in turn surrounds quartz *(black)*; **b** microcrystalline chalcocite replacing magnetite grain embedded in "massive" chalcocite *(Ch)* with bornite *(Bo)*. *Black spots in magnetite* are defects of polishing; *black spots in chalcocite* relicts of quartz; **c** rutile *(Ru)* and quartz *(Q)* partly replaced by chalcocite *(Ch)*; rim around bornite *(Bo)*, replacing chalcocite; **d** subhedral quartz overgrowth *(dark)* on detrital quartz grains, both cemented by chalcocite *(white)*. Reflected light, 1 N

boundaries are not sharp – commonly gradation is observed from massive structure through sulphide cement up to sulphide disseminations. The relicts preserved are often outlined by clay minerals.

Silica mobilized during the reactions was transported for a distance of several centimeters forming euhedral or subhedral quartz crystals (Fig. 3d), overgrowths or chalcedony cement.

Amorphic silica is a metastable phase at low temperatures and pressures; however, it dissolves under special environmental conditions of pH of about 9 (Krauskopf 1956). Solubility of crystalline SiO_2 phases is much lower. The presence of some ions (Al^{+3}, Mg^{+2}, Iller, see Krauskopf 1956) and organic compounds (Siever 1962) additionally decreases the solubility of SiO_2.

Diagenetic replacement of detrital quartz and other minerals by carbonates (Walker 1957), carbonates and sulphates (Chepikov et al. 1961), carbonates and sulphides (Shenhav 1971), calcite, hydromicas, iron hydroxides and chlorite (Simanovich 1978, p. 43) is well-known (see also Pettijohn et al. 1972, pp. 401–406 and 420–424). The corrosion of terrigenous grains by sulphides was noted in some base-metal deposits in argillaceous and arenaceous rocks (Schüller 1959; Bartura and Würzburger 1974; Jung and Knitzschke 1976; Samama 1976; Sartbaev 1976; Smith 1976).

Stratigraphic position of the replacement zone in the grey sandstone together with its insignificant thickness points to the copper shale as a source-bed for mineralizing solutions. Compaction can be proposed as a possible mechanism of extraction of interstitial fluids from the shale to the porous sandstone. These fluids generated specific pH conditions in which quartz and other minerals could be dissolved and the new phases precipitated. It is possible that corrosion of quartz by clay minerals and carbonates preceded and facilitated the replacement by sulphides; however, microscopic evidence is not convincing. On the other hand, such a sequence of reactions seems to be significant for feldspar replacement. Changes of the environmental conditions enabling dissolution of detrital grains could be caused by changes in the chemistry of interstitial fluids migrating through clay microlayers acting as semipermeable membranes (Siever 1962). Another possibility is local changes of the sandstone environment as a result of ion exchange between interstitial fluid and clay cement (Thomson see Fairbridge 1967, p. 73). A characteristic feature is almost a monometallic composition of the mineralization considered. The reason may be a selective extraction of metal ions from clay-organic microlayers. Weiss and Amstutz (1966) describe such a process leading to the concentration of metals.

The relation of replacement mineralization to the sandstone cement, both primary (clayey) and diagenetic (calcareous) and to the early-diagenetic, disseminated sulphides suggests its late-diagenetic origin.

References

Bartura Y, Würzburger U (1974) The Timna copper deposit. In: Bartholomé P (ed) Gisements stratiformes et provinces cuprifères. Soc Geol Belg, Liège, pp 277–286

Chepikov KR, Yermolova YP, Orlova NA (1961) On the corrosion of quartz grains and the effect of oil on the reservoir properties of sandy rocks. (In Russian). Dokl Akad Nauk SSSR 140: 1167–1169

Fairbridge RW (1967) Phases of diagenesis and authigenesis. In: Larsen G, Chilingar GV (eds) Diagenesis in sediments. Dev Sedimentol 8:19–89

Jerzykiewicz T, Kijewski P, Mroczkowski J, Teisseyre A (1976) Genesis of the Weissliegende sediments from the Fore-Sudetic monocline. (In Polish, English summary). Geol Sudetica 11:57–97

Jung W, Knitzschke G (1976) Kupferschiefer in the German Democratic Republic (GDR) with special reference to the Kupferschiefer deposit in the Southeastern Harz Foreland. In: Wolf KH (ed) Handbook of strata-bound and stratiform ore deposits, vol VI. Elsevier, Amsterdam Oxford New York, pp 353–406

Krauskopf KB (1956) Dissolution and precipitation of silica at low temperatures. Geochim Cosmochim Acta 10:1–26

Nemec W, Porebski S (1977) Weißliegendes sandstone: a transition from fluvial-aeolian to shallow-marine sedimentation (Permian of the Fore-Sudetic monocline). Part 1: Sedimentary structures and textural differentiations. Rocz Pol Tow Geol 47:387–415

Part 2: A study in significance of rock colouration. Rocz Pol Tow Geol 47:513–544

Nemec H, Nemec W, Porebski S (1978) Weißliegendes sandstone: a transition from fluvial-aeolian to shallow marine sedimentation (Permian of the Fore-Sudetic monocline). Part 3: Interpretation in the light of heavy mineral data. Rocz Pol Tow Geol 48:73–95

Pettijohn FJ, Potter PE, Siever R (1972) Sand and sandstone. Springer, Berlin Heidelberg New York, 617 p

Samama JC (1976) Comparative review of the genesis of the copper-lead sandstone-type deposits. In: Wolf KH (ed) Handbook of strata-bound and stratiform ore deposits, vol VI. Elsevier, Amsterdam Oxford New York, pp. 1–20

Sartbaev MK (1976) Mineralogy and structural–textural features of the cretaceous copper sandstones of Alai (South Tan-Shan). (In Russian). Zap Kirg Otd Vses Mineral Ova 9:61–66

Schüller A (1959) Metallisation und Genese des Kupferschiefers von Mansfeld. Abh Dtsch Akad Wiss Berlin, Kl Chem Geol Biol Jg 1958(6):9

Shenhav H (1971) Lower cretaceous sandstone reservoires, Israel: Petrography, porosity, permeability. Bull Amer Assoc Petrol Geol 55:2194–2224

Siever R (1962) Silica solubility 0–200°C and the diagenesis of siliceous sediments. J Geol 70: 127–151

Simanovich JM (1978) Quartz of arenaceous rocks. (In Russian). Trans Acad Sci USSR 314

Smith GE (1976) Sabkha and tidal-flat facies control of stratiform copper deposits in North Texas. In: Wolf KH (ed) Handbook of strata-bound and stratiform ore deposits, vol VI. Elsevier, Amsterdam Oxford New York, pp 407–446

Walker TR (1957) Frosting of quartz grains by carbonate replacement. Geol Soc Am Bull 68:267–268

Weiss A, Amstutz GC (1966) Ion-exchange reactions on clay minerals and cation selective membrane properties as possible mechanisms of economic metal concentration. Miner Deposita 1: 60–66

Problems of Genesis of Copper Sandstones of Central Europe

YU. V. BOGDANOV and V.P. FEOKTISTOV[1]

Abstract

The regional setting of "Kupferschiefer" ("copper slate")-type deposits is discussed, with special reference to the occurrences in Central Europe (specifically those of Lower Silesia, Poland). Localized processes of leaching, bleaching and copper accumulation are discussed.

Copper occurrences in the Permian deposits of Central Europe are widely known as the formation type of "copper slates". The copper mineralization in these deposits is localized in the thin layer (up to 1.0–1.5 m) of mudrock, usually with an increased content of carbonates and coaly matter, occurring at the base of the Zechstein. It is this layer that is called the copper slates. The copper slate layer is traced across extensive areas. The slightly increased copper content is recorded almost ubiquitously; rather high copper content is observed within extensive areas. Due to vast areas of development of the copper slates, the total metal reserves there attain great values. However, since even in the largest copper slate deposits the thickness of various sections of the ore bodies may attain several tens of centimeters, provided that the copper content values are as usual and the reserves are distributed through rather extensive areas, the practical value of such deposits is not great (Jung et al. 1971; Rentzsch 1974).

The discovery and exploitation of very large copper deposits in Lower Silesia (Poland), confined to the Permian and assigned by most investigators to the copper slate type, introduced important amendments into the theory of formation for this deposit type.

There is every reason to regard this formation type of the copper slates as one of the most important commercial types, and the platform deposits as having good potentials for the discovery of large deposits. These conclusions to a certain extent contradicted the confirmed idea that all the large stratiform copper deposits, including unique ores, are localized within the folded regions and belong to the copper sandstone type, whereas no large copper ore accumulations are found within the platform regions (Bogdanov et al. 1972).

Thus, we assume it would be rather interesting to analyze the peculiarities of geological conditions of localization and structure of the copper deposits of Lower Silesia.

1 All-Union Geological Research Institute (VSEGEI), 199026, Leningrad, Sredniy 74, USSR

The Lower Silesian region covers the Polish sector of Hercynides in Central Europe. The major structural elements within the region are (from south to north): the Sudetic Block, The North Sudetic Synclinorium, the cis-Sudetic Block and the cis-Sudetic Monocline.

The Sudetic and the cis-Sudetic Blocks are composed of metamorphic Precambrian rocks, volcanogenic-sedimentary beds of the Lower and Middle Paleozoic, intruded by the Hercynian granitoids. Precambrian rocks and granitoids are mostly developed in the Sudetic Block.

The North Sudetic Synclinorium is the epi-Hercynian orogenic trough where the Paleozoic basement is overlain by a thick sequence of sedimentary, mostly terrigenous rocks of the Upper Carboniferous, Lower (Rotliegende) and Upper (Zechstein) Permian, Triassic and Upper Cretaceous. In places acid volcanites are abundant in the Lower Permian sequence, represented by red sandstones and siltstones. The carbonate and sulphate rocks are widely spread in the Upper Permian and Triassic.

The structure of the cis-Sudetic Monocline is similar to that of the North Sudetic Synclinorium; it represents an epi-Hercynian Foredeep. This foredeep is separated by the pre-Zechstein uplift from the East European Platform, lying to the NNE. Thus, the term "monocline" should be understood only in application to the recent structural framework of this region, but it would be more reasonable to call it the cis-Sudetic Foredeep.

Thus, the deposits occurring in this region and ranging from the Upper Carboniferous to the Triassic might be regarded as the formations of the superimposed trough of the Hercynian folded region. The platform cover proper begins with the Cretaceous deposits. It is only in the N and NE, in the regions of the assumed Caledonian basement, that the Upper Carboniferous, Permian and Triassic sediments are enclosed in the platform cover.

The copper deposits are localized within the eastern and northern limbs of the North Sudetic Synclinorium and the northern limb of the cis-Sudetic Block; strictly speaking, it is this limb that should be called the cis-Sudetic Monocline, i.e., it is controlled by the inner flank of the cis-Sudetic Foredeep.

In the North Sudetic Synclinorium, the copper mineralization is almost exclusively confined to the "copper slate" layer occurring at the base of the Zechstein. The "copper slate" layer is composed of thin-banded grey, dark grey, greenish-grey calcareous mudstone and marl, enriched in organic matter and enclosing fine disseminations, pockets, patches and veinlets of copper and iron sulphides, mostly chalcocite, chalcopyrite and pyrite.

The thickness of the layer varies from 0.1 to 3.0 m. The usual thickness of the layer is about 1 m. Large areas of sulphide mineralization are recorded, but the balance ores are concentrated only within the ore field of the "Conrad" Mine (Konstantinovich 1972; Narkelun et al. 1970).

A distinct association is established between the copper mineralization, the increase of the "copper slate" thickness and the synsedimentary uplift, distinguished in the north and east of the North-Sudetic Anticlinorium on the basis of an abrupt decrease of thickness of the Rotliegende, the increase of the size of sandstone fragments in the Rotliegende and Weißliegende, and the increase of their participation in the conglomerate sequence. To the east and south of the ore field of the "Conrad" Mine, the "copper slate" thickness decreases to the first few tenths of a meter, the mineralization intensity falling abruptly. From the genetic viewpoint the mineralization in the copper slate might be regarded as sedimentary-diagenetic.

The copper deposits of the cis-Sudetic Monocline differ greatly from the copper deposits of the North Sudetic Synclinorium. There, the "copper slate" layer is also distinguished at the base of the Zechstein, bearing the copper mineralization, which is often rather rich. The "copper slate" is represented by black bituminous mudstone, its thickness at the eastern margin of the deposit attaining 1.0 m, but rapidly decreasing westwards to complete wedging out. On pinching out, the "copper slate" is replaced laterally by bituminous limestone or variegated marl. The "copper slate" is not very important in the general balance of the copper ores at the deposits.

Most of the copper mineralization is localized in the upper part of the Weißliegende sandstone unit overlying the Zechstein, and at the western margin of the deposit it is displaced into the overlying limestones. Thus, the mineralization migrates upwards in the section within a thickness interval of 30–40 m. The thickness of the ore bodies varies from 2 to 5–10 m, the average values within the deposit being 2–4 m. The producing rock area is estimated to be about 700 km^2.

The ore-bearing sequence is strongly dislocated. It dips gently, plunging at 5°–10°N. The gentle anticlinal flexures, having southward dips, are distinct against the background of the general southward dip. Numerous tectonic fractures of the fault type with the shift up to several tens of meters are distinguished within the deposits.

The copper content is similar to that recorded at the other copper sandstone deposits, attaining, on the average, several percent. The mineral composition of the ores is almost exclusively bornite-chalcocite. Downdip, where the mineralization pinches out, chalcopyrite ores appear. Considerable lead and zinc concentrations are recorded from the mineralized limestone and dolomite in the upper parts of the mineralized zone.

The ore structure in sandstone is impregnated, in pockets or banded. The ore-bearing sandstone of the Weißliegende represents the upper part of the Lower Permian terrigenous red beds (Rotliegende).

They are quite similar to the underlying red sandstone in their lithological features (composition of the terrigenous material, fragment size and roundness). It is established definitely that the *whitish grey sandstone* is generated due to gleying (decoloration resulting from the ferric iron removal) of the red sandstones. The lower boundary between these units is gradual with numerous red *relics* (patches, lenses). The gleying intensity increases abruptly in the upper part of the unit under the overlying "copper slates". Within the same units sandstone becomes more coherent, and the richest sulphide mineralization is observed.

The carbonate cement prevails, whereas the mineralized varieties are dominated by the sulphide bornite-chalcocite cement.

The sandstones are locally characterized by a distinct cross-bedding. The beds are large, the thickness of the wedge- or trough-shaped series attaining 10–40 cm. The thin layers are straight and slightly convex. The bedding is emphasized by the alternation of thin beds with different grain size. On the basis of their structural features the sandstones can be assigned to the littoral shallow basin sediments, containing separate lenses of the deltaic or eolian deposits.

The areal extent of the copper-bearing rocks in the Lubin-Seroszowice Belt is controlled distinctly by the slope of the cis-Sudetic paleo-uplift, and within the uplift, by local paleo-uplifts, delineated by the occurrence areas of variegated rocks and red beds, replacing the copper-bearing unit (the so-called "Rote Fäule") (Wyżykowski 1971).

The deposits of the cis-Sudetic Monocline are very similar to the Dzhezkazgan copper sandstone deposit. The thickness of the copper-bearing unit of the Lubin-Seroszowice deposits can be correlated with that of ore of the nine ore-bearing units of the Dzhezkazgan deposit, but its area is 7–8 times greater than that of the Dzhezkazgan ore field.

Thus, the largest copper deposits of Poland, no doubt, may be referred to the copper sandstone type, but not to the copper slate type, as is assumed in publications. The copper sandstone deposits are known to be characterized by large metal concentrations in many regions of the world (Bogdanov et al. 1973; Rentzsch 1974).

The genesis of the copper sandstone type deposits is being discussed at present. We assume that the formation of this type of mineralization is mostly influenced by the processes of epigenetic alteration of the ore-bearing red beds.

In general, the following main features of the localization of the copper mineralization in the Permian deposits of Lower Silesia can be distinguished as follows:

1. There are two major types of copper deposits in Lower Silesia:

a) copper slate type;
b) copper sandstone type;

The copper slate deposits are confined to the Zechstein of the North Sudetic Synclinorium. Judging from the amount of the ore reserves and production, they are of minor importance.

The copper sandstone deposits are localized in the cis-Sudetic Monocline. They enclose not only epigenetic ores in sandstones and carbonate rocks, but also ores of the copper slate type, the latter being of minor significance in the general ore balance. The mineralization comprises the upper parts of the Lower Permian terrigenous red beds (Rotliegende) and the lower parts of the Zechstein.

2. The copper deposits are localized within the volcano-sedimentary complex, filling in the inner and outer epi-Hercynian orogenic troughs.

The largest deposits of the Lubin-Seroszowice Belt are controlled by the inner flank of the cis-Sudetic Foredeep.

3. Mineralization is confined to the interval of the section of the terrigenous red formation adjacent to the overlying clay-carbonate grey formation. The copper-bearing formation is closely associated with the sulphate-salt-bearing formations, replacing them up the section, as well as with bituminous and gas-bearing deposits.

4. The copper deposits are controlled by the cis-Sudetic inner paleo-uplift, delineated quite clearly by wedging-out of the Upper Carboniferous and Lower Permian deposits. Within the areas of development of the copper-bearing deposits mineralization is controlled by local paleo-uplifts.

The ore-bearing rocks at the deposits are strongly dislocated.

5. The ore-bearing rocks are characterized by intense epigenetic changes. The intensity of the epigenetic processes probably resulted in high concentration of the copper mineralization at the deposits of Lower Silesia.

The above conclusions concerning the localization features and the structure of the copper deposits in Lower Silesia confirm the assumption that the economic value of the copper slate deposits is rather limited. At the same time, it should be emphasized that the mineralization intensity in the copper slate horizon increases abruptly from the East European Platform to the Late Paleozoic–Early Mesozoic epigeosynclinal orogenic troughs. The increased copper concentrations in the copper slate layer can indirectly point to a possible presence of copper deposits of the copper slate type in the adjacent areas with favourable structural and lithological conditions.

In conclusion it should be emphasized that the genetic models of the copper sandstone deposits, that are being worked out, display extreme differences, varying from sedimentary (terrigenous, etc.) to extreme hydrothermal (telethermal, etc.). Usually, it is most difficult to prove the synchroneity of the ore formation with this or that geological process. This applies, first of all, to the stratigraphic mineralization control, characterized by regional and local features. It is not difficult to establish that mineralization is confined to particular formations, but the lithological control usually does not point to the synchronous generation of ores and ore-bearing layers (Bogdanov et al. 1973).

For instance, new discoveries and the study of the known copper sandstone deposits reveal many features of their distribution which are not given due regard in the course of prospecting and prognostic evaluation of the potential territories.

Lately, it has been established that structural and lithological mineralization control is especially important for the copper sandstones, as well as for many other stratiform deposits (Bogdanov et al. 1973; Feoktistov and Kochin 1970).

All the large copper sandstone deposits are controlled by inner or outer epigeosynclinal orogenic troughs. Within such troughs mineralization is confined to the slopes of synsedimentary paleo-uplifts, especially to those parts where the maximum thickness variations of the ore-bearing strata are observed.

The ore-bearing orogenic structural-formation complexes are represented by siltstone-sandstone red beds, enclosing carbonaceous clay and carbonate basin deposits. Mineralization is confined to large lenses of variegated and grey rocks around which characteristic hydromica-carbonate country rock alterations are observed (Bogdanov et al. 1973).

It should be emphasized that the type and parameters of the copper-bearing strata are to a greater extent determined by the rhythmicity features of red formations. Two morphological types are distinguished: Dzhezkazgan and Lubin-Seroszowice.

The thickness of the copper-bearing strata of the Dzhezkazgan and Udokan deposits is up to 650 and 330 m, respectively, their area attaining several tens of square kilometers (Bogdanov et al. 1973). At the Lubin-Seroszowice deposit the thickness of the coal-bearing strata rarely exceeds 10 m, the area being several hundreds of square kilometers. The specific features of manifestation of these two extreme types of producing strata of the copper sandstone deposits has not yet been adequately studied. However, it is even at present clear that a complete series of morphological types of copper sandstone can be reconstructed between the above extreme types. The position of the copper sandstone deposits in such classification series is determined by $r = \frac{m}{s}$ coefficient (where m is the thickness and s the area of the producing strata), the values of which vary from a few hundredths (Lubin-Seroszowice) to 20–25 (Dzhezkazgan). In a number of cases the ore content coefficient of the producing strata, determined by the ratio of red to grey beds in the rhythms, should also be taken into account. This coefficient is usually equal to 1 or approaches this value.

The above evidence allows us to draw the important practical conclusion that copper sandstone deposits of the Lubin-Seroszowice type will, most probably, be discovered in the variegated foredeep deposits of different ages, occurring within the periphery of the East European, Siberian and other platforms.

References

Bogdanov Yu V, Kochin GG, Feoktistov VP (1972) Potential stratified copper and lead-zinc ore deposits in the sedimentary cover of the Russian platform. Trans VSEGEI New Ser 178:134–150 (in Russian)

Bogdanov Yu V, Burianova EZ, Kutyrev EI, Feoktistov VP, Trifonov NP (1973) Stratified copper deposits of the U.S.S.R. Nedra, Leningrad, 312 pp (in Russian)

Feoktistov VP, Kochin GG (1970) On some peculiar features of localization of stratified copper deposits. Geol Ore Deposits XII 6:80–91 (in Russian)

Jung W, Knitzschke G, Gerlach R (1971) Entwicklungsgeschichte der geologischen Anschauungen über den Mansfelder Kupferschiefer. Geologie 4/5:462–484

Konstantinovich E (1972) Genesis of the Permian copper deposits in Poland. Sov Geol 8:101–117 (in Russian)

Narkelun LF, Filin AM, Bezrodnykh YuP, Trubachev AI (1970) Copper slate deposits of Poland and their correlation with copper manifestations in the east of the U.S.S.R. Sov Geol 10:108–121 (in Russian)

Rentzsch J (1974) The Kupferschiefer in comparison with the deposits of the Zambian Copperbelt. Cent Soc Géol Belg. Gisements Stratiformes et Provinces Cuprifères, pp 395–418

Wyżykowski J (1971) Cechsztyńska formacja miedzionośna w Polsce. Przegl Geol 3:117–122

Hydrothermal Alterations of Conglomeratic Uranium Ores, Pronto Mine, Ontario, Canada

E.WM. HEINRICH[1]

Abstract

Hydrothermal alteration of the conglomeratic uranium ores is intensive and extensive at the Pronto deposit. The sequence of the main alteration is: (1) albitization, (2) chloritization and (3) carbonatization, resulting in the formation of albitized conglomerates, albitites (usually hematitic), chloritic albitites and conglomerates, chloritites and carbonate rocks. In some of the last, dolomite and/or calcite form pseudomorphs after quartz pebbles. The alterations have markedly changed both the distribution of the radioactive minerals, generally concentrating them near footwall contacts, and the habit of both brannerite and uraninite. In both unaltered conglomerate and altered ores none of the pyrite displays any detrital characteristics. Indeed, much of it is euhedral. Sulfur isotope ratios indicate a magmatic-hydrothermal origin for the pyrite.

1 Introduction

The uranium deposits of the Blind River-Elliot Lake area (Algoma district), Ontario, are widely cited as classic examples of disseminated uranium-thorium-pyrite mineralization in weakly metamorphosed conglomerates. The ore beds are generally described as being very uniform not only in structure but also in petrology and as being similar in many respects to the Witwatersrand deposits of South Africa. Although this is the case for most of the deposits in the district, it represents a totally inadequate characterization of the Pronto deposit. In this deposit, which departs radically from this general picture, about 25% of the ore mined was of several non-conglomeratic types (altered ores).

This paper, which is a summary of part of a much larger detailed study on the geology, petrology and mineralogy of the Pronto deposit, describes the mineralogy of the altered ores and compares them to the mineralized conglomerate.

The geology and history of the district and its deposits, which have been summarized in numerous papers, are not recapitulated here. The most recent and complete accounts are those of Robertson (1970, 1978), in which the older literature is cited. The geology of the Pronto deposit itself also is described by Robertson (1970). In his paper the sections on the post-ore dikes, the structural geology, the relations of the non-conglomerate ores to faults and dikes, and the radioactivity of the non-conglomerate ores are extensive quotations from an unpublished company report by P.C. Masterman

1 Department of Geological Sciences, The University of Michigan, Ann Arbor, MI 48109, USA

(1960), chief geologist at Pronto prior to its shutdown in April 1960. At Masterman's invitation this writer began a systematic study of the non-conglomerate ores in 1959. The results of this investigation, which has required dozens of thin and polished sections and scores of autoradiographs, were combined with Masterman's account of the geology in a general paper on the geology and mineralogy of the Pronto non-conglomeratic ores. Unfortunately for several reasons, this paper has not been published; only an abstract has appeared (Heinrich and Masterman 1960a, b).

2 Altered Ores

2.1 General

Recognition of epigenetic alteration of the conglomeratic ores came as early as November 1953 when the first four *deep* drill holes intersected altered ores. F. Bourgeois, one of the geologists who logged the cores, noted both red feldspathic and green chloritic alterations. Unfortunately, inasmuch as no petrographic studies were made, the altered rocks, when encountered underground, were misnamed "radioactive grit" (chloritic) and "radioactive quartzite" (albitic). This led to Holmes' (1957) ill-fated "scarp theory of deposition", which identified the altered rocks as sedimentary facies of the conglomerate. The true nature of these rocks first was determined by the writer in 1956 (Heinrich 1958).

The Pronto ore occurs in a conglomeratic bed of Huronian age formerly called the Lower Mississagi (Bruce Group) but later renamed the Matinenda (Elliot Group) (Robertson et al. 1969). This naming was perhaps unfortunate inasmuch as the Blind River-Elliot Lake ores had come to be universally referred to as "Mississagi conglomerate ores".

The ore horizon, which dips southward into the Pronto thrust fault, is itself broken by two sets of subsidiary faults and cut by two sets of mafic dikes (Fig. 1). The larger dikes are all diabasic, but some of the smaller ones are olivine-augite-biotite lamprophyres. The sequence of faulting and dike intrusions is listed in Table 1. All of the dikes are intensively altered and some are albitized and/or chloritized where they cut the ore bed in contact with altered ores. Thus, it appears that the ore alterations are post-dike intrusion in age. Robertson (1970) reports that altered ores similar to those at the Pronto are present locally in mines of the Quirke and Nordic ore zones "... where it is

Table 1. General sequence of faulting and dike intrusion in the Pronto deposit

7. A curving east-west fault that offsets No. 2 diabase
6. Intrusion of No. 2 diabase
5 Faults that strike northwest
4. Minor faults that strike northeast (offset No. 4 dike)
3. Diabase dikes Nos. 1, 3, 4 and several smaller dikes of lamprophyre
2. No. 3 fault and several other faults that strike west-northwest
1. Normal and reverse faults that strike generally northeast (several phases)

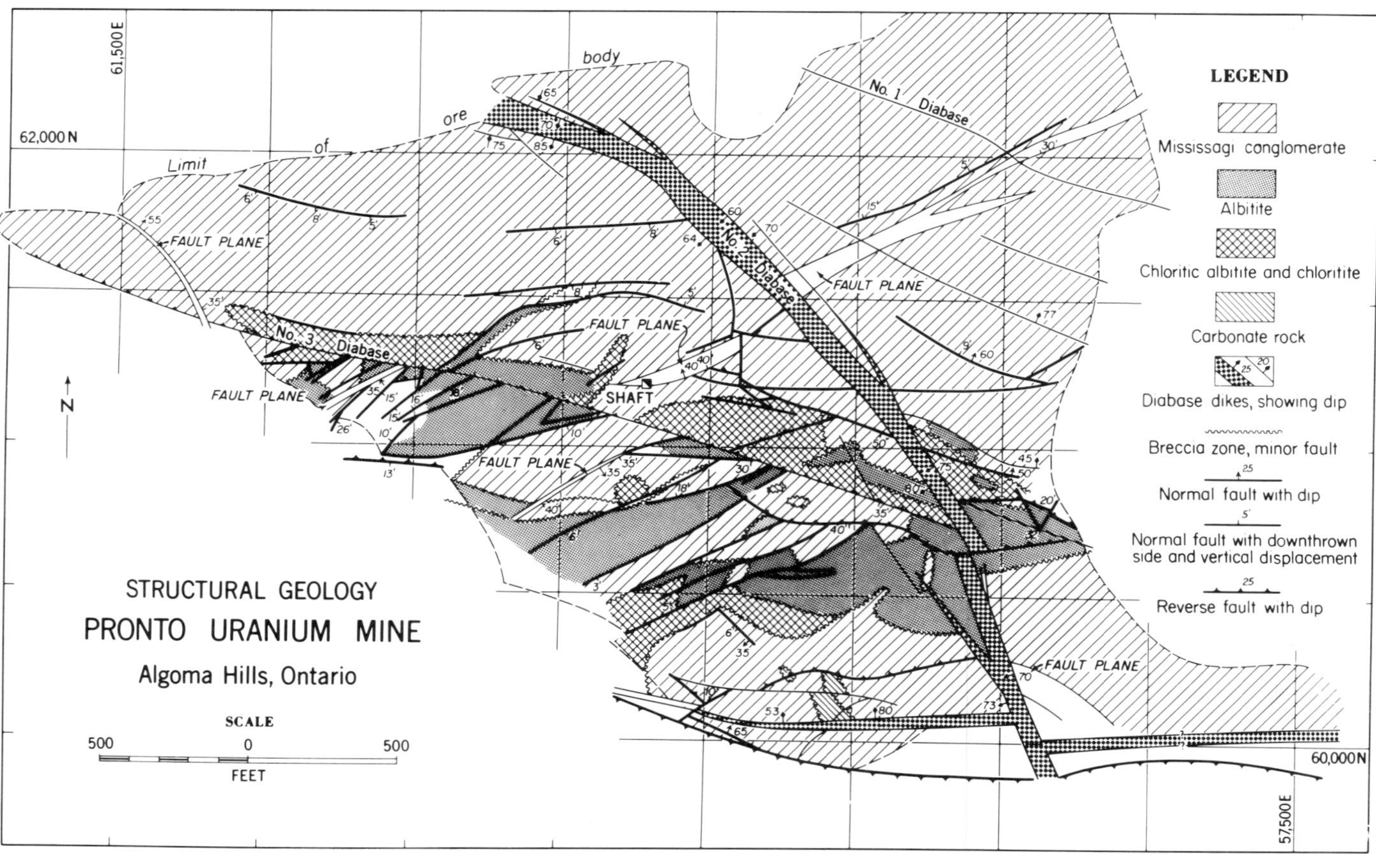

Fig. 1. Petrology and structure of the Pronto deposit (Masterman 1960; Heinrich and Masterman 1960a)

thought that the alteration was related to the end stage magmatic fluids from the Nipissing diabase magma." At Pronto this is unlikely, both from the distributions and enormous volumes of the altered rocks.

2.2 Relations to Conglomerate

Contacts between conglomerate and altered ores are of four types: (1) fault or shear planes; (2) breccia zones; (3) gradational; (4) interlayered (selective replacement).

Fault or shear-plane contacts usually are sharp with very little or no alteration on the conglomeratic side of the fault. Flat-lying shear planes may separate conglomerate from underlying albitite or vice versa.

Breccia zones as much as 10 feet thick may separate conglomeratic ore from altered ore. Usually such zones consist of mixed blocks of altered and unaltered ore, and in these the relationships are usually complicated further by minor faults. Commonly uranium values are highly erratic in brecciated sections.

Gradations from albitite to unaltered conglomerate extend over distances of less than a foot to several feet. The transitional rock commonly consists of subrounded patches of conglomerate set isolate in albitite. Commonly such zones are free of brecciation or multiple fracturing. Directly along contacts between albitite and conglomerate a progressive alteration zoning may be compressed within a distance of from a few inches to a few feet.

Where the ore bed consists of conglomerate interlayered with one or more quartzite bands the conglomeratic layers have been selectively altered, leaving tongues of unaltered quartzite projecting several feet into the albitite. It is noteworthy that uranium values are not redistributed, for the "ghosts" of barren quartzite bands can be traced in albitite for many tens of feet.

2.3 Types and Sequence of Alteration

Although the bulk of the altered rocks occurs within the ore conglomerate bed, the alteration is not entirely restricted to this horizon. Albitization has penetrated the basement where albitized granite is locally present. Similarly, chloritic shear zones also may extend into basement rocks. Zones of albitization also project upward into the quartzite above the conglomerate in a few places. The early mafic dikes also exhibit the results of albitization and chloritization. Such albitites and chloritites external to the conglomerate ore horizon either carry no radioactive minerals or rarely are very weakly radioactive.

The main types of altered rocks are listed in Table 2. Many rocks are composites with several types of alteration, although one type usually is dominant. The following transitions are present:

1. Conglomerate → albitized conglomerate → albitite
2. Albitite → chloritized albitite → chloritite
3. Conglomerate → chloritized conglomerate → chloritite (rare)
4. Albitite → calcitic albite → hematitic calcite rock

5. Conglomerate → dolomitized conglomerate
6. Chloritite → chloritic carbonate rock (either dolomite and/or calcite)

In addition, two other, less widely distributed types of alterations were noted: biotitization and sericitization. The first occurs in association with chloritization, and many chloritites contain variable amounts of biotite. Sericitization occurs in a few carbonate-rich rocks.

The general sequence of alteration is:
1. Albitization
2. Chloritization (± biotite)
3. Carbonatization (dolomite, calcite or both; barite locally abundant)
4. Sericitization

Table 2. Altered ores of the Pronto deposit

Group and map designation	Rock types (most abundant italicized)
Albitite	*Hematitic albitite* (brick red)
	Albitized conglomerate (quartz pebbles relict in albitized matrix)
	Albitite (pink, light brown, gray, green)
	Banded albitite
Chloritic albitite	*Chloritic albitite* (gray, green, mottled red-green)
	Chloritites (dark green)
	Chloritized conglomerate (quartz pebbles relict in chloritized matrix)
Carbonate rocks	*Calcitic albitite*
	Dolomitized conglomerate (dolomite pseudomorphs of quartz pebbles) ± barite
	Hematitic calcite rocks
	Micaceous calcite rocks (chloritic or sericitic)

Of post-altered-rocks age are veins of quartz; massive sulfide pods (as large as 1 × 1.5 ft., with chalcopyrite, galena, pyrite, sphalerite, marcasite, trace bornite); and galena-pyrite veinlets. Chalcopyrite, with sphene, also occurs in skeletal clusters in conglomerate. Much of the sulfide mineralization occurs in close association with calcitic type alterations. One of the most spectacular mineral occurrences was encountered on Level 4, Stope 4–6 south in carbonate rock. Here occur vugs, several feet across, lined with olive to black scalenohedral calcite crystals, 1–2 inches long. The crystals include abundant minute specks of pyrite. In addition the vugs also contain small grains and crystals of sphalerite and highly magnetic pyrrhotite and black spherules of pyrite.

2.4 Unaltered Conglomerate

The unaltered conglomerate has been described in detail many times (e.g., see Robertson 1970, pp. 79–80). It is an oligomictic conglomerate with quartzose pebbles (vein quartz, pegmatite quartz, quartzite, rare chert and even rarer slate) set in a matrix of sericite-pyrite which carried the U-Th values chiefly as "brannerite" aggregates (much is metamict or altered to anatase-uraninite).

The pyrite, which varies greatly in grain size and shape, constitutes 2%–25% of the conglomerate, averaging $\sim$10%. At the Pronto Mine no rounded ("buckshot") pyrite was observed; nor is there any evidence of sorting. The pyrite grains are anhedral to euhedral. In some cases a few inclusions of chalcopyrite and sphalerite are present. Although most of the pyrite is in the matrix, it is not confined to the matrix. Many examples have been noted of:

1. Thin veinlets of pyrite along fractures in pebbles. In some cases the pyrite is flattened parallel with the plane of the vein.
2. Pyrite anhedra and subhedra sparsely disseminated within pebbles.
3. Pyrite grains that lie across pebble-matrix boundaries, with the euhedral faces into the pebble.

Non-conglomeratic quartzite bands that occur within the ore zone rarely carry pyrite. A spectacular exception was encountered in Stope 68-2 from the 6th level where a lens of cross-bedded quartzite (highgrade ore) was very heavily pyritized. Both topset and foreset beds consisted of as much as 75% pyrite. This pyrite is unsorted, with the finer grains in euhedra and the coarser particles in aggregated anhedra.

2.5 Albitized Conglomerates and Albitites

Relict pebbles disappear in the albitites. The albite is usually anhedral and highly variable in grain size; locally some subhedral larger metacrysts are present. Schachbrett-twinning is universal. Pyrite is absent to locally abundant. Apparently much of the iron oxidized and released by pyrite destruction contributed to the formation of dusty hematite. Other new minerals are apatite (some skeletal), rutile, sphene, rare epidote-allanite crystals, and variable amounts of chlorite and calcite.

2.6 Chloritic Albitites and Chloritites

Except in those chloritites derived directly from conglomerates, relict pebbles are absent. Pyrite, exceptionally rare, occurs in a few chloritites as (1) unusally coarse subhedra, (2) fine-grained, post-shearing euhedra, and (3) radial-skeletal pyrite ("wheels"). Variable amounts of albite, sericite, carbonate and secondary quartz are preserved. Biotite occurs in many chloritites, and in a few it exceeds or excludes chlorite. Biotite-chlorite relations are anomalous: in a few specimens biotite appears to be replacing chlorite; in others chlorite replaces biotite; but most chlorite appears to have formed directly without biotite.

2.7 Carbonate Rocks

Two types of carbonate appear: white and pink. The white is dolomite; the pink is either dolomite or calcite. In thin section dolomite may appear as euhedra which are zoned with bands rich in hematite-inclusions. Accompanying calcite is interstitial, anhedral and inclusion-free. Both dolomite and calcite replace and form pseudomorphs after pebbles. Other new minerals are apatite, locally abundant barite (black megascopically) and hematite in coarse tablets. Variable amounts of albite, chlorite, sericite and secondary quartz are present.

Pyrite, present in variable amounts, is fine- to coarse-grained, anhedral to euhedral. In some cases the grain surfaces are highly polished ("fused" pyrite). Chalcopyrite appears as uncommon inclusions. Radial-skeletal pyrites also occur, and some pebble pseudomorphs are pyrite-dolomite aggregates.

3 Changes in Radioactive Mineralization

With increasing alteration of any type the radioactive minerals are increasingly redistributed. The main tendency is to concentrate the values toward the footwall contact of the ore bed. In altered ore values in the lowermost one or two feet commonly are well over mine average, and usually assay 2 to 4 times mine grade ore. In both albitites and chloritites the uranium values are compressed within a narrower width than in adjoining conglomerate. However, inasmuch as the grade increases, the total amount of uranium (width × grade) remains unchanged. Thus it appears that the alterations have neither added nor subtracted uranium.

In addition to redistribution stratigraphically, numerous small-scale changes are evident from thin-section and autoradiograph studies. In albitized conglomerates a high percentage of relict pebbles are cut by radioactive veinlets. In albitites the radioactive components are extremely variable in both form and dissemination patterns.

Brannerite becomes increasingly euhedral and less metamict, thus clear and birefringent. Relatively large boxwork euhedra of brannerite appear. Uraninite in small cubic subhedra and euhedra becomes more common. Veinlets of albite-brannerite, of calcite-uraninite, of uraninite, and of brannerite-uraninite cut the albitites and chloritites. In the carbonate rocks euhedral brannerite and uraninite also become conspicuous, but the radioactive minerals generally avoid the dolomite-barite-rich rock, and are conspicuously absent in the carbonate pseudomorphs after quartz pebbles.

In addition to brannerite and uraninite, small amounts of thucholite and coffinite have been identified. Oxidized uranium minerals are widespread in small amounts from the 400- to the 600-levels. They occur in chloritites and albitites that are strongly sheared, or faulted, particularly along thrust-fault planes. Minerals identified include uranophane, beta-uranophane, boltwoodite and liebigite.

4 Sulfur Isotopes

Sulfur isotope ratios were determined on sulfide samples (14 from the Pronto Mine, 1 from the Panel Mine) by Professor M.L. Jensen. For 8 pyrite samples from conglomerates, chloritites and carbonate rock the $\delta^{34}S$ ‰ range was −1.3 to +0.4. For a composite sample of pyrite from typical conglomerates from several levels in the mine $\delta = +0.4$. It is clear that the pyrite had a primary source that was magmatic-hydrothermal, irrespective of whether it did or did not arrive in the Pronto conglomerate as detrital grains. It also is clear that the various types of alterations did not affect the sulfur isotope ratios, even though in some cases the pyrite was drastically recrystallized. Other sulfides (chalcopyrite, pyrrhotite, galena, sphalerite) have $\delta^{34}S$ ‰ = −2.7 to +3.1 and also must be considered magmatic-hydrothermal.

5 Conclusions

The original conglomerate is a highly siliceous rock (quartzose pebbles) with lesser amounts of K, Al, H_2O (sericite) and Fe and S (pyrite). Albitization introduced large amounts of Na and some Al, removing Si, K, H_2O and S and oxidizing Fe^{2+} to Fe^{3+}. Chloritization required a second major compositional transformation, substituting mainly Mg and Fe^{2+} for Na and Al in albitites and for Si in conglomerates, and reintroducing H_2O. Development of the carbonates required additions of Ca, Mg, CO_2 and Ba to either the conglomerates or albitites. The various alteration stages integrade and appear to be different phases reflecting gradual major changes in the composition of continuously flowing solutions which selectively soaked the more permeable ore conglomerate after ascending along the Pronto fault.

The sequence of the three major introduced elements, Na-Mg-Ca, is strongly reminiscent of the major-element geochemistry of secondary rocks and mineral deposits formed in Alpine ultramafic bodies (albitites, serpentinites, rödingites). However, the source of the solutions remains unknown.

The solutions produced a spectacular variety of replacement rocks and, in so doing, materially redistributed and recrystallized the uranium-thorium minerals, destroyed or replaced the pebbles, and destroyed or recrystallized the pyrite. At the Pronto Mine neither the textures of the pyrite in the unaltered conglomerate nor its sulfur isotope ratios offer any support that the pyrite was first lodged in the conglomerate in detrital form.

References

Heinrich EWm (1958) Mineralogy and geology of radioactive raw materials. McGraw Hill, New York, pp 258–265

Heinrich EWm, Masterman PC (1960a) Nonconglomeratic ores of the Pronto uranium deposit. Unpubl Ms Rep

Heinrich EWm, Masterman PC (1960b) Nonconglomeratic ores of the Pronto uranium deposit (abstr). Geol Soc Am Bull 71:1887

Holmes SW (1957) Pronto Mine. In: Structural geology of Canadian ore deposits, vol 2. Can Inst Min Metall Congr, pp 324–339

Masterman PC (1960) Geology of the non-conglomeratic ores, Pronto mine. Pronto Div, Rio Algom Mines Ltd. Unpubl Rep

Robertson JA (1970) Geology of the Spragge Area. Ontario Dep Mines Geol Rep 76:109

Robertson JA (1978) Uranium deposit in Ontario. In: Kimberley MM (ed) Uranium deposits: Their mineralogy and origin, chap 9. Min Assoc Can, pp 229–280

Robertson JA, Frarey MJ, Card KD (1969) The Federal Provincial Committee on Huronian stratigraphy: Progress report. Can J Earth Sci 6:335–336

The Isotopic Composition of Lead in Galenas in the Uranium Ores at Elliot Lake, Ontario, Canada

W. SCOTT MEDDAUGH[1], H.D. HOLLAND[1], and N. SHIMIZU[2]

Abstract

Ion microprobe isotope analyses of Pb in galenas in uraninite-rich ore from the New Quirke Mine show that the galenas contain large proportions of radiogenic Pb, which was almost certainly generated within the associated uraninite grains and released during two episodes of Pb loss. The isotopic data indicate that the uraninite is at least 2100 m.y. old and that the second episode of Pb loss occurred less than ca. 1350 m.y. ago. The data do not preclude a detrital origin for the uraninite in the Elliot Lake quartz pebble conglomerate ores.

1 Introduction

Several major mid- and early Precambrian uranium deposits in quartz pebble conglomerates contain large tonnages of uraninite that is probably detrital. Arguments based on intergrain chemical heterogeneity (Grandstaff 1974), the high Th, high rare earth element content of the uraninites, textural features, such as grain size and shape (Ramdohr 1958, 1969), and related sedimentological criteria (e.g., Robertson and Steenland 1960; Roscoe 1969; Minter 1976; Theis 1976, 1978) strongly suggest a detrital origin for the uraninite; however, conclusive evidence has been elusive.

An initial, or primary, age for uraninite in such deposits that is greater than the time of deposition of the host conglomerates is strong evidence in favor of a detrital origin of the uraninite. In order to minimize the effects of post-depositional Pb loss on age determinations of uraninites Nicolaysen et al. (1962) obtained "total rock" Pb, U, Th analyses of uranium-bearing minerals in the Witwatersrand conglomerates, and Rundle and Snelling (1977) established the presence of 3050 m.y.-old uranium minerals in the ⩽ 2740 m.y.-old South African conglomerate ores.

Early age determinations of uraninite separates in the Elliot Lake deposits gave Pb-Pb ages of about 1700 m.y. (Mair et al. 1960), considerably less than the 2300 m.y. depositional age of the conglomerates (Roscoe 1969, 1973). Wanless and Traill (1956) reported two even younger U-Pb uraninite ages of 1300 m.y. and 600 m.y. As noted by Roscoe (1969) and many others, the apparent "young" age of the uraninite may reflect

1 Department of Geological Sciences, Harvard University, Cambridge, MA 02138, USA

2 Department of Earth and Planetary Sciences, Massachusetts Institute of Technology, Cambridge, MA 02139, USA

one or more episodes of significant Pb loss. Such Pb loss is suggested by the presence of highly radiogenic lead in galena in the Elliot Lake ores (Mair et al. 1960; Roscoe 1969).

In some uraninite-rich Elliot Lake ore samples numerous small (10–20 μm) blebs of galena occur in the vicinity of many quartz grains, and narrow (5–10 μm) veinlets of galena occur within many uraninite grains. Similar material occurs in the South African deposits (e.g., Burger et al. 1962; Nicolaysen et al. 1962). In the present study an ion microprobe has been used to make in situ Pb isotope analyses of the small galena blebs and veinlets in a sample of high-grade ore from the New Quirke Mine of the Elliot Lake-Blind River district.

2 Geologic Setting and Sample Description

As several discussions of the geology of the Elliot Lake ores have been published (e.g., Derry 1960; Robertson and Steenland 1960; Pienaar 1963; Roscoe 1969; Robertson 1976, 1978), only those features pertinent to the present study will be reviewed. The typical Elliot Lake ore-bearing conglomerate consists of well-rounded, well-sorted, quartz pebbles in a matrix of quartz, feldspar, and sericite; the pyrite content averages 10% to 15% of the rock (Robertson 1978, p. 246). Chert pebbles are also found. In addition to the ore minerals uraninite, brannerite, and monazite, which are invariably restricted to the matrix, the conglomerates contain minor quantities of zircon, other sulfides such as pyrrhotite and galena and several Ti-minerals. Mair et al. (1960) report Pb-Pb ages of 2450 m.y. and 2550 m.y. for zircon and monazite, respectively, in the Elliot Lake ores.

The Elliot Lake ore-bearing conglomerates occur at or near the base of the Matinenda Formation. In the Elliot Lake area the Matinenda Formation marks the base of the Huronian sequence. The ore-bearing conglomerates were probably deposited about 2300 m.y. ago. Fairbairn et al. (1969) report a Rb-Sr isochron age of 2288 ± 87 m.y. for the stratigraphically higher Gowganda Formation. Fairbairn (1965, as reported by Roscoe 1973) reported a Rb-Sr age of ca. 2350 m.y. for the Coppercliff Rhyolite, which underlies the McKim Formation near Sudbury. Roscoe (1969, 1973) considers the McKim Formation stratigraphically equivalent to the Matinenda Formation. The Huronian sequence in the Elliot Lake region unconformably overlies an Archean basement complex dated at about 2550 m.y. (van Schmus 1965).

The Huronian sediments were intruded by the Nipissing Diabase ca. 2155 m.y. ago (van Schmus 1965; Fairbairn et al. 1969; Card and Pattison 1973). The diabase was emplaced during or shortly after an early, major deformation of the Huronian sequence, but before a later, minor deformation associated with regional metamorphism. The latter event occurred ca. 1700 m.y. ago (van Schmus 1965) and resulted in lowermost greenschist facies metamorphism in the Elliot Lake region. Card et al. (1972), however, note that the major metamorphism may be older than about 1950 m.y. South of Elliot Lake, post-Huronian intrusives, the Cutler Granite and the Eagle Granite, were emplaced about 1750 m.y. and 1450 m.y. ago, respectively (van Schmus 1965). According to van Schmus (1965), olivine diabase dikes, dated at about 1225 m.y., cut the Cutler Granite and adjacent formations, and represent the latest intrusive activity in the area.

The isotope data for Pb in galena reported in the present study were obtained from a single high-grade, uraninite-rich sample from the New Quirke Mine. The sample was provided by N.J. Theis. The New Quirke Mine is located on the north limb of the Quirke syncline, approximately 15 km north of the town of Elliot Lake. Uraninite is the most important ore mineral at the New Quirke Mine. The mine ore occurs in the "C" (Denison) reef within the Manfred Member of the Matinenda Formation. Theis (1973, 1976) has described the mineralogy and geology of the New Quirke Mine.

The sample consists of two uraninite-rich seams or bands within typical Elliot Lake conglomerate matrix. The seams, one about 8 mm across and the other about 4 mm across, consist of uraninite (ca. 60% of the seam), 0.15–0.4 mm quartz grains (ca. 12%), and 0.1–0.25 mm pyrite grains (ca. 3%) in a matrix (about 25%) of a very fine-grained quartz-sericite mixture. Minor (ca. 2%) galena, pyrrhotite, and chalcopyrite also occur. The diameter of the uraninite grains ranges from 50 to 300 μm; most grains have a diameter between 75 and 100 μm. The uraninite is sub-angular to sub-rounded; decidedly angular or euhedral grains are less abundant. Grain boundaries vary from smooth to quite ragged, and some extensively corroded or "ghost-like" uraninite similar to that described by Ramdohr (1958) from the Witwatersrand is also present. Most of the uraninite grains are fractured. The fractures are commonly partially or wholly

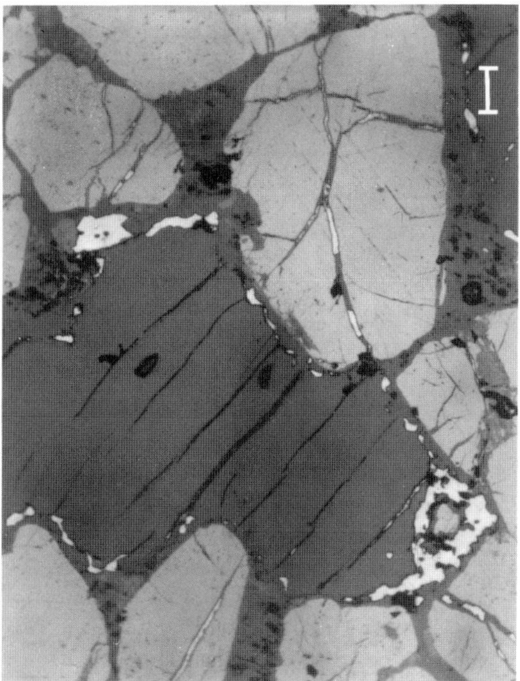

Fig. 1. Photomicrograph of the polished section of uraninite-rich ore from the New Quirke Mine showing small galena blebs *(white)* concentrated near the rim of fractured quartz grain *(dark grey)* and narrow galena veinlets in uraninite grains *(light grey)*. Scale bar is 20 μm

filled with narrow galena veinlets about 5 μm wide. Occasionally, pyrite also occurs in the veinlets. Small blebs of galena, typically about 10–20 μm across, are localized near the rim of quartz grains within or at the boundaries of the seams. Typical galena blebs and veinlets are illustrated in Fig. 1. Less commonly, galena blebs occur within the fine-grained matrix or are enclosed by pyrite. Larger galena grains, generally more than 100 μm across, occur within and between the uraninite-rich seams. Only one or two of these large grains are present per cm^2 of section; they are usually adjacent to grains of pyrite or chalcopyrite.

Partial electron microprobe analyses of uraninite from the larger seam show that their mean Th/U ratio is approximately 0.14; this figure is somewhat higher than the average Th/U ratio of ca. 0.10 reported by Theis (1973, 1976) for other uraninites from the New Quirke Mine.

3 Analytical Procedures

Isotope analyses were acquired with a Cameca IMS 3f ion microprobe installed at the Department of Earth and Planetary Sciences, Massachusetts Institute of Technology. A primary ion beam consisting of negatively charged oxygen, O^-, with an energy of 8.75 KeV was used for all analyses.

A spot size of 5–10 μm was obtained by focussing the 0.7 nA primary beam on a polished Al plate. The beam diameter was estimated in the image mode by comparison with a 25 μm Cu grid superimposed on the Al plate. Optical examination of the samples following analysis confirmed prior estimates of the spot diameter.

The samples selected for ion microprobe analyses were prepared as ordinary polished sections. Following photographic documentation, the sample surface was carefully cleaned to remove surface contamination and was coated with a 200 Å thick gold film. Prior to analysis, each sample spot underwent 3–7 min of sputtering in order to penetrate the gold coating and to reach a steady state rate of secondary ion production.

Secondary ions were accelerated to 4.5 KeV and mass analyzed with a double-focussing mass spectrometer consisting of a 90-degree electrostatic sector followed by a 90-degree magnetic sector. The mass resolving power used for Pb isotope analysis was approximately 600.

Secondary ion intensities were measured with an electron multiplier in a pulse-counting mode. A Hewlett Packard 982A desktop computer was used to change the magnetic field from one peak-top to another and to control data acquisition and reduction. The peak positions were located automatically by scanning across the peak and selecting the magnetic field corresponding to the maximum intensity. During the analysis, the secondary ion signal at each selected mass location was measured for a specified length of time, usually two seconds, first in order of increasing mass number and then in order of decreasing mass number. This cycle was usually repeated fifty times. The background intensity was measured 1/2 mass unit below the highest specified mass number. Isotope analysis of each spot required about 30 min to complete.

Typical peak intensities were $2.5–3.0 \times 10^4$ counts/s (cps) at mass number 206 for the Elliot Lake galena samples and at mass 208 for the Broken Hill galena standard with a primary beam spot of 5–10 μm and 0.7 nA current. The background intensity was normally less than 2 cps. Lower peak intensities were found for the NBS Pb wire standard, about 8×10^3 cps at mass 206. During analysis of each spot the secondary ion intensity remained within 5%–7% of the intensity at the start of analysis. Correction for the dead time of the counting system was negligible with the secondary ion intensities mentioned above.

Precision and Accuracy. Standards with well-characterized isotopic compositions were repeatedly analyzed at 4–6-h intervals. The results for 26 measurements of a Broken Hill galena standard and for 8 measurements of an NBS Equal-Atom Pb wire standard are reported in Table 1. Two corrections have been applied to all measurements. One of these, required by the presence of Pb-hydride ions in the secondary ion mass spectrum, is described below. The second correction is an empirical, mass-related correction factor which was determined by measuring the percent deviation of the isotope ratios after the hydride correction for the Pb standards from their accepted values. The correction factor obtained for both standards has a value of 0.48 ± 0.01 %/mass unit relative to mass 206. We consider this an instrument bias and all data were corrected for this.

The estimated precision of a single isotope ratio analysis is 0.2%–0.5% for ^{207}Pb/^{206}Pb and ^{208}Pb/^{206}Pb and 1%–2% for ^{204}Pb/^{206}Pb for measurements made on the standards. Somewhat lower precision is indicated for the Elliot Lake galena measurements, about 0.5%–1% for ^{207}Pb/^{206}Pb and ^{208}Pb/^{206}Pb and 2%–5% for ^{204}Pb/^{206}Pb. The lower precision of the latter measurements is due to the relatively low abundance of ^{207}Pb, ^{208}Pb, and ^{204}Pb in the Elliot Lake galena samples. The error estimates, confirmed by replicate analyses of the standards and selected Elliot Lake galena grains, can be accounted for by Poisson counting statistics.

In addition to singly charged single ions, the secondary ion mass spectrum typically contains multiple ions, metal-oxide ions, metal-hydride ions and metal-hydroxide ions. The presence of the various "molecular" ions in the secondary ion mass spectrum may produce serious interference problems, as noted, for instance, by Hinthorne et al. (1979) and by Hinton and Long (1979). The galena mass spectrum in the mass range 204–208 is believed to be free of molecular ions except for metal-hydride ions. The presence of Pb-hydride ions is indicated by the occurrence of a peak at mass 209 in

Table 1. Ion microprobe (IMP) analytical data for the Broken Hill galena standard and for the NBS Equal-Atom lead wire standard

Broken Hill	204/206	207/206	208/206
IMP measurements (26)	0.06247 ± 56[a]	0.9614 ± 19	2.2273 ± 85
Accepted standard value[b]	0.062471	0.96187	2.2286
NBS Equal-Atom Pb wire			
IMP measurements (8)	0.02722 ± 23	0.4669 ± 11	1.0012 ± 12
Accepted standard value[c]	0.027219	0.46707	1.00016

[a] Calculated 1σ error. [b] Stacey et al. (1969). [c] Catanzaro et al. (1968)

the Pb wire spectrum. As this standard is considered to contain no ^{209}Bi, the peak is believed to be due to $^{208}PbH^+$ ions. The approximate abundance of Pb-hydride ions can be estimated from the measured 209/208 mass ratio. For the Pb wire standard this ratio was about 2×10^{-3} and about 0.8×10^{-3} for the Broken Hill galena standard. As Bi may be present in the Broken Hill galena, present estimates of Pb-hydrides in the galena mass spectra may be somewhat too large. The Pb-hydride interference was corrected assuming that the abundance of $^{206}PbH^+$ at mass 207 and of $^{207}PbH^+$ at mass 208 was proportional to the measured 209/208 ratios for the respective standards times the 206 and 207 peak intensities, respectively. The following Pb-hydride corrections were applied:

	Broken Hill galena	NBS Pb wire	Elliot Lake galena
207/206	– 0.09%	– 0.64%	– 0.33% to 0.50 %
208/206	– 0.05%	– 0.09%	– 0.16% to 0.21%

No hydride correction is necessary for the 204/206 mass ratio. The variability of the Pb-hydride correction reflects the various proportions of the Pb isotopes in the different samples.

4 Results

Our Elliot Lake Pb isotope data are summarized in Figs. 2 and 3. Figure 2 is a plot of $^{204}Pb/^{206}Pb$ vs. $^{207}Pb/^{206}Pb$. Figure 3 is a plot of $^{208}Pb/^{206}Pb$ vs. $^{207}Pb/^{206}Pb$. Analytical data are represented for a total of 63 galena grains. Each plotted point represents the composition of lead in a single grain; mean values are plotted for 14 grains which were analyzed more than once. As shown in Figs. 2 and 3, there is a strong positive linear correlation between the $^{204}Pb/^{206}Pb$ and $^{207}Pb/^{206}Pb$ ratios and between the $^{208}Pb/^{206}Pb$ and $^{207}Pb/^{206}Pb$ ratios in the galenas. In each case the correlation coefficient for the least squares best fit line is 0.95.

The highest values of the plotted isotope ratios are found in the large galena grains. Lower ratios are characteristic of the galena blebs and galena veinlets. Lead in galena veinlets invariably has low ratios, while lead in the galena blebs shows considerable isotopic variation. The larger galena grains have $^{207}Pb/^{206}Pb$ values between 0.275 and 0.301; the blebs have $^{207}Pb/^{206}Pb$ ratios between 0.194 and 0.273; the galena veinlets have $^{207}Pb/^{206}Pb$ ratios between 0.203 and 0.230. The interpretation of this distribution is considered in a later section. There is no significant difference between the isotopic composition of galena associated with the two uraninite-rich seams.

Several points were analyzed within four of the larger galena blebs and in three of the large galena grains to assess their isotopic homogeneity. Minor variations (about ± 3%–6% of the mean value for each grain) were found in one galena bleb of particularly irregular outline and in the three large galena grains. The variability within single grains is small compared to the total range of the isotopic compositions shown in Figs. 2 and 3.

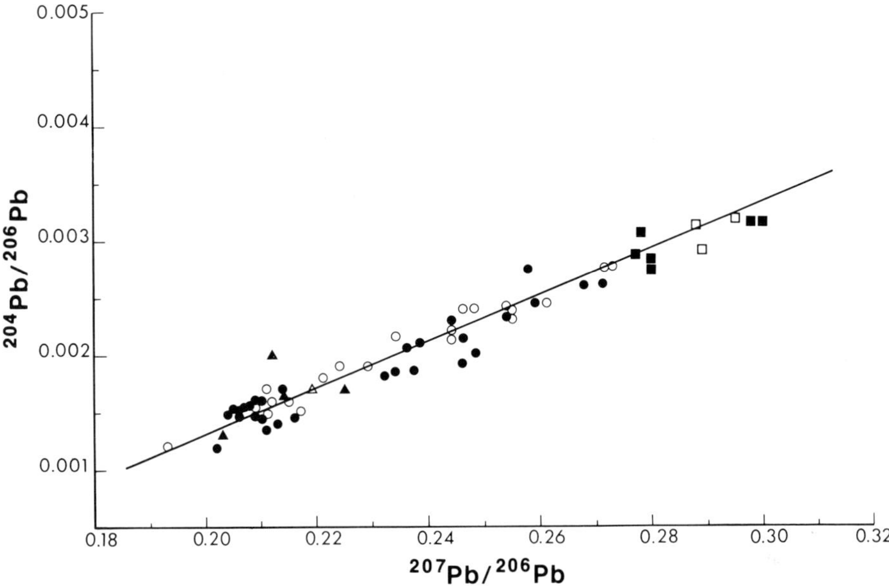

Fig. 2. Plot of $^{204}Pb/^{206}Pb$ vs. $^{207}Pb/^{206}Pb$ in the New Quirke Mine galenas. Isotope data are given for the galena veinlets in uraninite (▲), galena blebs (●), and large galena grains (■) in the large uraninite-rich seam *(solid symbols)* and in the small uraninite-rich seam *(open symbols)*

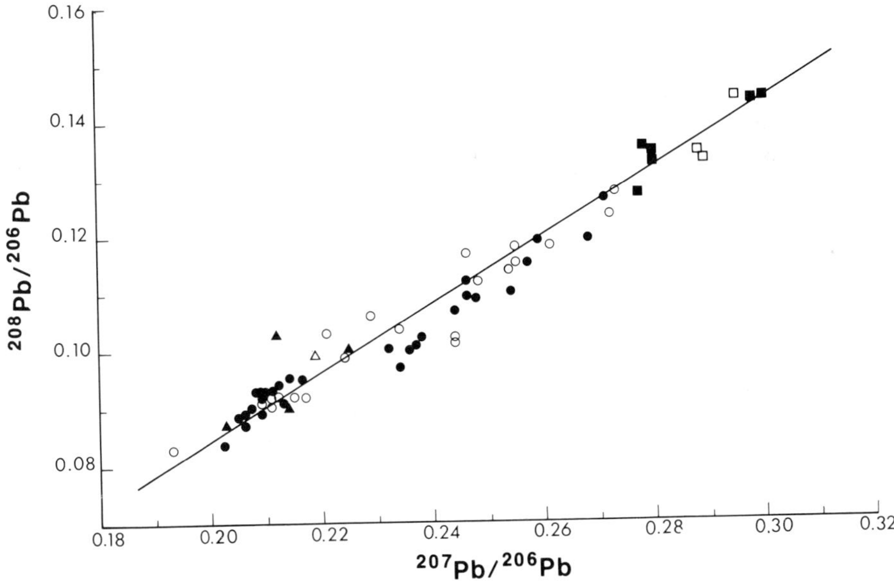

Fig. 3. Plot of $^{208}Pb/^{206}Pb$ vs. $^{207}Pb/^{206}Pb$ in the New Quirke Mine galenas. The symbols are those used in Fig. 2

5 Discussion

The two features of the New Quirke Mine galena Pb isotopic data are particularly striking: the highly radiogenic nature of the Pb in the galenas and the strong covariance of the Pb isotope ratios. The radiogenic component of Pb in the galenas was almost certainly generated within uraninite grains and was released during one or more episodes of Pb loss. Burger et al. (1962) reached the same conclusion regarding the source of the radiogenic Pb in the galena in the South African conglomerates. However, the presence of ^{204}Pb in the galena grains indicates that some non-radiogenic Pb is also present.

The strong correlation between $^{204}Pb/^{206}Pb$ and $^{207}Pb/^{206}Pb$, and between $^{208}Pb/^{206}Pb$ and $^{207}Pb/^{206}Pb$ shown in Figs. 2 and 3 suggests that the Pb in these galenas is the product of mixing leads of distinctly different isotopic composition. As the trends shown in Figs. 2 and 3 do not pass anywhere near the composition of normal crustal Pb (see Figs. 4 and 5), simple mixing of normal crustal Pb with a locally generated radiogenic Pb can not explain the data.

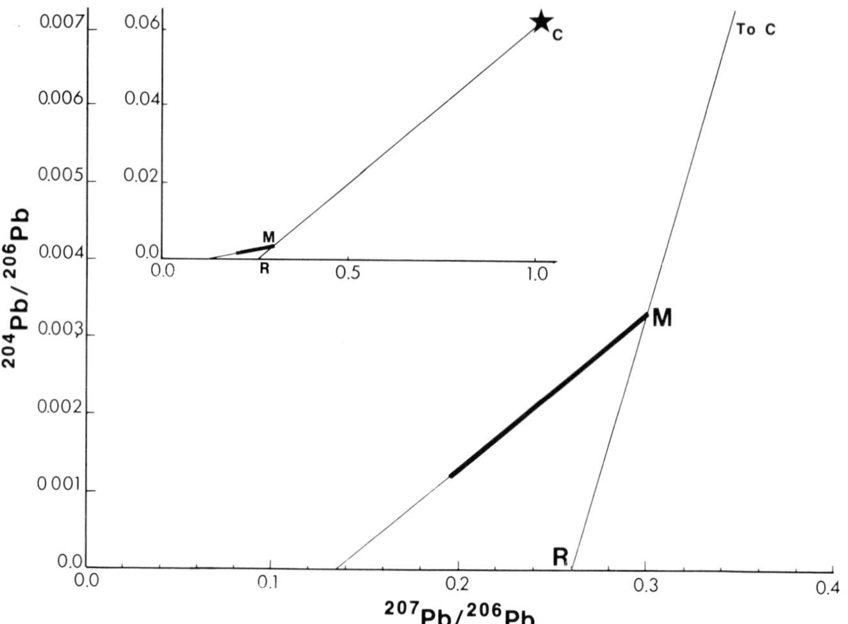

Fig. 4. $^{204}Pb/^{206}Pb$ vs. $^{207}Pb/^{206}Pb$ diagram showing minimum-$^{207}Pb/^{206}Pb$ mixing line *(line CR)*. The *heavy solid line* corresponds to the range of isotope ratio data shown in Fig. 2. *Point C* corresponds to the isotopic composition of crustal Pb ca. 2000 m.y.b.p.

A model involving the introduction of crustal Pb and two periods of Pb loss from the uraninite grains can, however, do justice to the data. In Figs. 4 and 5 point C corresponds to the isotopic composition of single stage crustal Pb at ca. 2000 m.y.b.p.

The position of this point varies only slightly with time during the pertinent time interval (see below). If at a time T_1, Pb is released from the uraninite grains into a solution containing crustal Pb and reduced sulfur, galena is apt to be precipitated. This galena will contain Pb with an isotopic composition intermediate between that of the crustal Pb and the Pb released from the uraninite grains. We propose that this mixed Pb was of nearly constant isotopic composition within a volume at least as large as that of the polished section which we have investigated, and that its composition fell at or beyond the upper end of the mixing lines in Figs. 2 and 3.

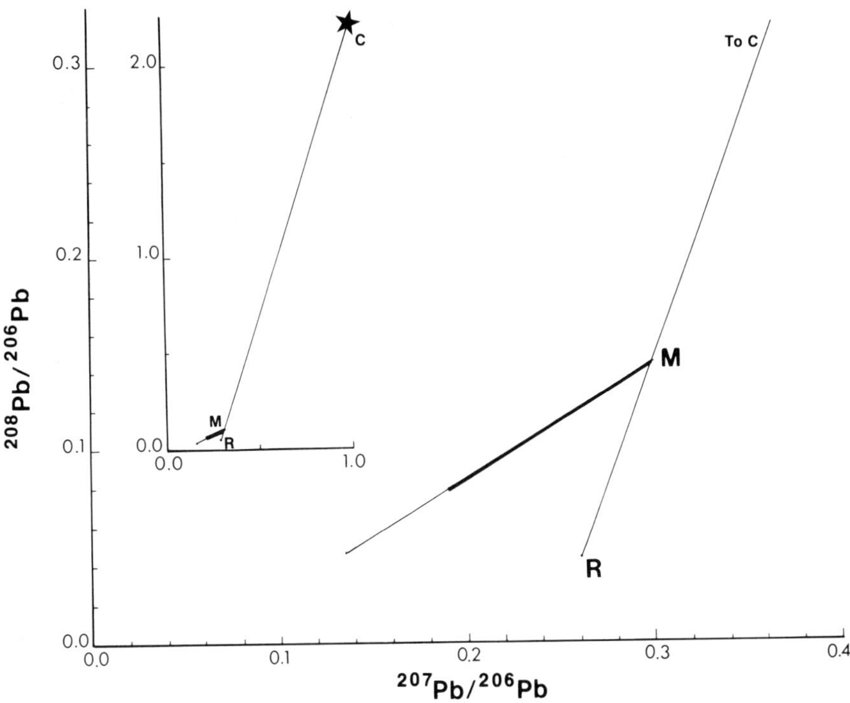

Fig. 5. $^{208}Pb/^{206}Pb$ vs. $^{207}Pb/^{206}Pb$ diagram showing minimum-$^{207}Pb/^{206}Pb$ mixing line *(line CR)*. The *heavy solid line* corresponds to the range of isotope ratio data shown in Fig. 2. *Point C* corresponds to the isotopic composition of crustal Pb at ca. 2000 m.y.b.p.

The mixing lines in Figs. 2 and 3 are then readily explained by a second period of Pb release from uraninite at time T_2, during which isotopic homogeneity was not achieved on the scale of our sample. The isotopic composition of the galena veinlets within the uraninite grains is closest to that of Pb released at time T_2; the Pb in the large galena grains was only slightly affected, and the small galena blebs tend to occupy an isotopically intermediate position.

The isotopic composition of Pb along the lines in Figs. 2 and 3 is, therefore, a function of the isotopic composition of crustal Pb and the isotopic composition of Pb in the uraninites at time T_1, the mixing ratio of these leads, and the isotopic composition of Pb released from the uraninites at time T_2. The composition of Pb released from the uraninite grains at time T_2 is a function of the time of uraninite crystallization, T_0, the times of Pb loss, T_1 and T_2, and the fraction of accumulated Pb lost from the uraninite grains at time T_1.

If T_1 can be estimated reasonably well on the basis of independent geologic data, the isotopic composition of crustal Pb during the period of first lead loss is known. For the Elliot Lake region, T_1 is probably between ca. 1700 m.y., the age of regional metamorphism according to van Schmus (1965), and ca. 2150 m.y., the time of emplacement of the Nipissing Diabase and of the major structural deformation of the Huronian sequence (van Schmus 1965; Fairbairn et al. 1969; Card and Pattison 1973). Fortunately, the isotopic composition of single stage crustal Pb is relatively insensitive to uncertainties in T_1 in this time interval. The least radiogenic isotope analyses tabulated by Roscoe (1969) for Pb in pyrite from the Bruce conglomerate, which lies stratigraphically above the Matinenda Formation, show $^{204}Pb/^{206}Pb = 0.063$, $^{207}Pb/^{206}Pb = 0.97$, and $^{208}Pb/^{206}Pb = 2.19$. These values are quite close to those of single stage crustal Pb between 1700 and 2100 m.y.b.p. If the non-radiogenic Pb in our galenas was derived from feldspar in the conglomerates, and if the age of the feldspar is ca. 2550 m.y., then the isotopic composition of the non-radiogenic or crustal Pb would be about $^{204}Pb/^{206}Pb = 0.070$, $^{207}Pb/^{206}Pb = 1.07$, and $^{208}Pb/^{206}Pb = 2.35$.

Although Pb mixing curves of the type discussed above are non-linear, their departure from linearity in our system is so small that no significant error is introduced into the following discussion by assuming essential linearity. The minimum value of the $^{207}Pb/^{206}Pb$ ratio in the radiogenic Pb released from the uraninites at time T_1 can therefore be obtained by passing a line through the coordinates of the most likely crustal Pb (point "C" in Figs. 4 and 5) and the coordinates of the least radiogenic galena Pb (point "M" in Figs. 4 and 5) and extrapolating this line to $^{204}Pb/^{206}Pb = 0$ (point "R" in Figs. 4 and 5). This minimum $^{207}Pb/^{206}Pb$ mixing curve is shown by the line "CR" in Figs. 4 and 5. The heavy line in Figs. 4 and 5 corresponds to the range of isotope ratios shown in Figs. 2 and 3. Note that if a small amount of ^{204}Pb was present in the uraninite grains and if some or all of this was released at time T_1, the extrapolated minimum value of the $^{207}Pb/^{206}Pb$ ratio of the radiogenic Pb is increased only slightly.

An estimate of the $^{208}Pb/^{206}Pb$ ratio in the Pb released from the uraninites at time T_1 can be obtained from Fig. 5 by drawing a line from point "C" through point "M" to the value of $^{207}Pb/^{206}Pb$ defined by the extrapolation in Fig. 4. A rough cross-check can be obtained by calculating the $^{208}Pb/^{206}Pb$ ratio in lead generated within the uraninite grains on the basis of their Th/U ratio.

As shown in Fig. 4, the minimum $^{207}Pb/^{206}Pb$ mixing line intersects the abscissa at about $^{207}Pb/^{206}Pb = 0.26$. If radiogenic Pb with a $^{207}Pb/^{206}Pb$ ratio equal to or greater than 0.26 was indeed released from the uraninite grains, limits can be placed on the length of time during which Pb accumulated prior to Pb loss, that is, the time between the time of crystallization of the uraninite, T_0, and the time of Pb loss, T_1.

These limits can be inferred from the curves in Fig. 6. All possible combinations of T_0 and T_1 that produce Pb with a $^{207}Pb/^{206}Pb$ ratio equal to or greater than 0.26 are shown by the shaded area in the upper right portion of Fig. 6. The uraninites must therefore be at least 2100 m.y. old, but no unique combination of T_0 and T_1 can be extracted. If, as seems likely, the major period of Pb loss occurred ca. 1700 m.y. ago, the $^{207}Pb/^{206}Pb$ data indicate an age of at least 2450 m.y. for the uraninite grains in our sample.

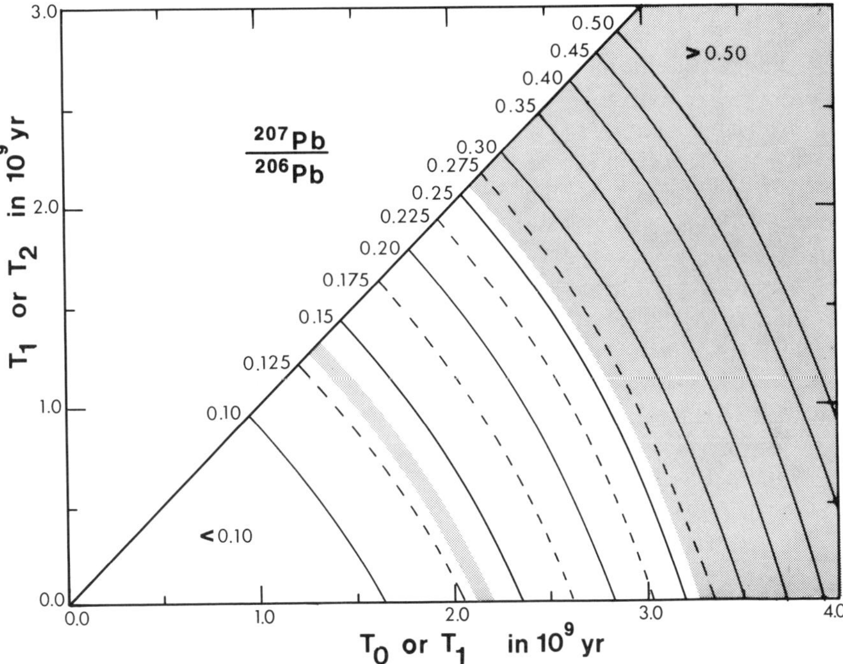

Fig. 6. Combinations of times T_0 and T_1 and of times T_1 and T_2 which yield $^{207}Pb/^{206}Pb$ ratios $\leqslant 0.50$ in galenas produced during lead loss from uraninite grains at times T_1 or T_2 respectively

Additional geochronological constraints on the age and Pb loss history of the uraninite grains can be established by examination of the mixing lines shown in Figs. 2 and 3. The isotopic composition of the Pb released from the uraninite grains during the second Pb loss episode is a function of the time between the time of the first Pb loss, T_1, and the time of the second Pb loss, T_2, and the fraction, F_1, of Pb which was released from the uraninite grains during the first Pb loss episode.

The excellent correlation of the several isotopic ratios in our galenas rules out all but a very small addition of crustal lead at T_2. The isotopic composition of the Pb released during the second Pb loss episode can therefore be obtained by extrapolating the mixing lines shown in Figs. 2 and 3 to $^{204}Pb/^{206}Pb = 0$. As shown in Fig. 4, the secondary mixing line intersects the abscissa at $^{207}Pb/^{206}Pb = 0.135$. If all the Pb that had accumulated in the uraninite grains between T_0 and T_1 was released at T_1, the possible combinations of T_1 and T_2 that produce a $^{207}Pb/^{206}Pb$ ratio of about 0.135 are shown by the heavy line in Fig. 6; the second Pb loss episode must then have occurred more recently than ca. 1350 m.y. ago. No unique combination of T_1 and T_2 can be extracted; however, if it can be shown that the first Pb loss episode occurred ca. 1700 m.y. ago, the second Pb loss episode occurred ca. 950 m.y. ago.

If Pb loss at T_1 was only partial, the heavy arc in Fig. 6 represents only the upper limit of T_2 in various $T_1 - T_2$ combinations. For example, if the age of the uraninite grains is 2500 m.y., the time of first Pb loss 1700 m.y.b.p., and if only 80% of the accumulated Pb was released during the first Pb loss episode, then the second Pb loss episode occurred only about 400 m.y.b.p.

The $^{208}Pb/^{206}Pb$ ratio of the Pb released from the uraninite grains at T_2 can be obtained by extrapolating the mixing line shown in Fig. 5 to the value of the $^{207}Pb/^{206}Pb$ ratio defined by the extrapolation detailed above (Fig. 4). A rough cross-check may again be obtained by calculating the $^{208}Pb/^{206}Pb$ ratio in the uraninite grains based on their Th/U ratio. The extrapolated value of the $^{208}Pb/^{206}Pb$ ratio, about 0.046, requires a Th/U ratio in the uraninite grains of about 0.17, somewhat higher than the value for the Th/U ratio of 0.14 obtained by electron microprobe analysis of the uraninite grains.

A third source of geochronological data can be provided by the present-day isotopic composition of the radiogenic Pb in the uraninite grains. This is a function of the age of the uraninite the time of first Pb loss, the fraction of available Pb released at T_1, the time of second Pb loss, and the fraction of accumulated Pb released at T_2. Meaningful additional constraints on T_0, T_1, T_2, F_1, and F_2 from this source require very precise data for the isotopic composition of lead in the uraninite grains. Such data are not yet available, but will be reported in a later paper.

In the absence of such data we can only state with certainty that the uraninite grains are more than 2100 m.y. old, that they have suffered two periods of lead loss, the second of which occurred more recently than 1350 m.y.b.p., and that they could be of detrital derivation.

Acknowledgements. The authors wish to thank S.R. Hart for many helpful discussions, R. Loucks for assistance with photomicrography, K. Burrhus for skilled technical support in both instrument maintenance and software improvement, and the National Science Foundation for financial support under Grant SR–33659.

References

Burger AJ, Nicolaysen LO, de Villiers JWL (1962) Lead isotopic composition of galenas from the Witwatersrand and Orange Free State, and their relation to the Witwatersrand and Dominion Reef uraninites. Geochim Cosmochim Acta 26:25–59

Card KD, Pattison EF (1973) Nipissing diabase of the Southern Province, Ontario, pp 7–30. In: Huronian stratigraphy and sedimentation. Geol Assoc Can Spec Pap 12:271

Card KD, Church WR, Franklin JM, Frarey MJ, Robertson JA, West GF, Young GM (1972) The Southern Province, pp 335–380. In: Variations in tectonic styles in Canada. Geol Assoc Can Spec Pap 11:688

Catanzaro EJ, Murphy TJ, Shields WR, Garner EL (1968) Absolute isotopic abundance ratios of common, equal atom, and radiogenic lead isotopic standards. J Res Natl Bur Stand Sect A 72: 261–267

Derry DR (1960) Evidence of the origin of the Blind River uranium deposits. Econ Geol 55:906–927

Fairbairn HW, Hurley PM, Card KD, Knight CJ (1969) Correlation of radiometric ages of Nipissing diabase and Huronian metasediments with Proterozoic orogenic events in Ontario. Can J Earth Sci 6:489–497

Grandstaff DE (1974) Microprobe analyses of uranium and thorium in uraninite from the Witwatersrand, South Africa, and Blind River, Ontario, Canada. Geol Soc S Afr Trans 77:291–294

Hinthorne JR, Anderson CA, Conrad RL, Lovering JF (1979) Single grain $^{207}Pb/^{206}Pb$ and U/Pb age determinations with a 10-μm spatial resolution using the ion microprobe mass analyzer (IMMA). Chem Geol 25:271–303

Hinton RW, Long JVP (1979) High resolution ion-microprobe measurement of lead isotopes: variations within single zircons from Lac Seul, Northwestern Ontario. Earth Planet Sci Lett 45: 309–325

Mair JA, Maynes AD, Patchett JE, Russell RD (1960) Isotopic evidence on the origin and age of the Blind River uranium deposits. J Geophys Res 65:341–348

Minter WEI (1976) Detrital gold, uranium, and pyrite concentrations related to sedimentology in the Precambrian Vaal Reef placer, Witwatersrand, South Africa. Econ Geol 71:157–176

Nicolaysen LO, Burger AJ, Liebenberg WR (1962) Evidence for the extreme age of certain minerals from the Dominion Reef conglomerates and the underlying granite in the Western Transvaal. Geochim Cosmochim Acta 26:15–23

Pienaar PJ (1963) Stratigraphy, petrology and genesis of the Elliot Group, Blind River, Ontario, including the uraniferous conglomerate. Geol Surv Can Bull 83:140

Ramdohr P (1958) New observations on the ores of the Witwatersrand in South Africa and their genetic significance. Geol Soc S Afr Trans Annex 61:50

Ramdohr P (1969) The ore minerals and their intergrowths. Pergamon Press, Oxford, 1174 pp

Robertson DS, Steenland NC (1960) On the Blind River uranium ores and their origin. Econ Geol 55:659–694

Robertson JA (1976) The Blind River uranium deposits: the ores and their setting. Ont Div Mines Misc Pap 65:37

Robertson JA (1978) Uranium deposits in Ontario, pp 229–280. In: Kimberley MM (ed) Uranium deposits: their mineralogy and origin. Mineral Assoc Can Short Course Handb 3:523

Roscoe SM (1969) Huronian rocks and uraniferous conglomerates in the Canadian Shield. Geol Surv Can Pap 68–40:205

Roscoe SM (1973) The Huronian supergroup, a Paleoaphebian succession showing evidence of atmospheric evolution, pp 31–47. In: Huronian stratigraphy and sedimentation. Geol Assoc Can Spec Pap 12:271

Rundle CC, Snelling NJ (1977) The geochronology of uraniferous minerals in the Witwatersrand Triad; an interpretation of new and existing U-Pb age data on rocks and minerals from the Dominion Reef, Witwatersrand and Ventersdorp Supergroups. Philos Trans R Soc London Ser A 286:567–583

Stacey JS, Delevaux ME, Ulrych TJ (1969) Some triple filament lead isotope ratio measurements and an absolute growth curve for single-stage leads. Earth Planet Sci Lett 6:15–25

Theis NJ (1973) Comparative mineralogy of the Quirke No. 1 and New Quirke Mines, Rio Algom Mines Limited, Elliot Lake, Ontario. MS Thesis, Queens Univ, Kingston, Canada, 61 pp (unpubl)

Theis NJ (1976) Uranium-bearing and associated minerals in their geochemical and sedimentological context, Elliot Lake, Ontario. PhD Thesis, Queens Univ, Kingston, Ontario, 148 pp (unpubl)

Theis NJ (1978) Mineralogy and setting of Elliot Lake deposits, pp 331–338. In: Kimberley MM (ed) Uranium deposits: Their mineralogy and origin. Mineral Assoc Can Short Course Handb 3:523

van Schmus WR (1965) The geochronology of the Blind River-Bruce Mines area, Ontario, Canada. J Geol 73:775–780

Wanless RK, Traill RJ (1956) Age of uraninites from Blind River, Ontario. Nature (Lond) 178: 249–250

Pre-Witwatersrand and Witwatersrand Conglomerates in South Africa: A Mineralogical Comparison and Bearings on the Genesis of Gold-Uranium Placers

R. SAAGER[1], T. UTTER[2], and M. MEYER[1]

Abstract

Pyritic quartz-pebble conglomerates from pre-Witwatersrand as well as from lower and upper Witwatersrand sediments show far reaching similarities in their ore mineralogy. The erratic and low gold and uranium contents of the investigated pre-Witwatersrand conglomerates, to a large extent, is explained by the inefficiency of the operating sedimentary reworking processes. The detrital chromites of all studied conglomerates have an identical chemical composition which points to an origin from the same provenance areas. "Fly-speck" carbon aggregates have been detected for the first time in many of the pre-Witwatersrand conglomerates indicating the possible presence of advanced, hitherto unknown lifeforms in the late Archaean. The observed extreme scarcity of uraninite in pre-Witwatersrand conglomerates may be due to recent uranium removal in the investigated weathered samples. From this study it is apparent that the Archaean pyritic conglomerates are primitive forerunners of the Witwatersrand gold-uranium placers.

1 Introduction

The Kaapvaal Craton of South Africa (Fig. 1) constitutes the most important gold province on earth, in that more than half of the world's total gold production originates from this craton. The gold mainly is contained in pyritic quartz-pebble conglomerates within early Precambrian sedimentary sequences. These placer deposits are also uraniferous and contain about 10% of today's proven world uranium reserves so that 98% of South Africa's gold and practically all its uranium production originate from conglomerates of the 2500 to 2800 m.y. old Proterozoic Witwatersrand Supergroup (Fig. 2).

A great number of sedimentological, geological, and mineralogical studies have led to the general acceptance of the "modified placer theory" as a metallogenetic concept of ore formation in the sediments of the Witwatersrand Supergroup (Liebenberg 1955; Ramdohr 1955; Brock and Pretorius 1964; Pretorius 1975; and others). The concept envisages that subsequent to their sedimentation some of the heavy minerals, notably the sulphides as well as some of the gold and the uranium phases, were reconstituted and thereby lost their detrital features as a result of low-grade metamorphism.

1 Mineralogisch-Petrographisches Institut der Universität Köln, Zülpicherstr. 49, 5000 Köln, FRG

2 Present address: GEOTEC LTDA, Geologos consultores, Apartado Aéreo 51846, Bogotá, Columbia

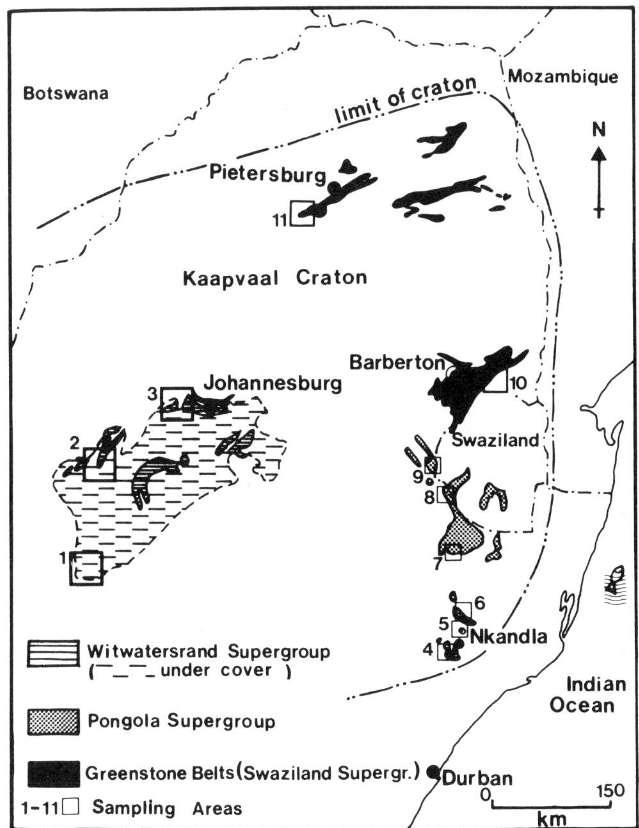

Fig. 1. Sampling areas of early Precambrian conglomerates on the Kaapvaal Craton. *1* Orange Free State Goldfield: *2* Klerksdorp Goldfield; *3* Randfontein area; *4* Cooper's Store; *5* Speedwell and Vuleka; *6* Denny Dalton; *7* Gunsteling; *8* Atalie/Madola; *9* Redcliff; *10* Bearded Man; *11* Mount Robert

On the Kaapvaal Craton pyritic quartz-pebble conglomerates are also found (1) in the approximately 3250 m.y. old Moodies Group of the Swaziland Supergroup in the Barberton Mountain Land, (2) in the Uitkyk Formation of the Pietersburg greenstone belt in the northern Transvaal, and (3) in the 2900 to 3200 m.y. old Pongola Supergroup of the south-eastern Transvaal and of Zululand (Figs. 1 and 2). Based on the geological setting, the Uitkyk Formation has been correlated with the Moodies Group (Grobler 1972; Brandl 1979; Muff and Saager 1979). However, a radiometric age is not yet available on the Uitkyk Formation and the stratigraphic position of this unit must be regarded with some reservation.

The Archaean quartz-pebble conglomerates, only in a few places, are known to be pyritic and to carry erratic gold and uranium values. Thus, these conglomerates so far are of insignificant economic importance. This may be the reason why little information has been published about the distribution, geology, and mineralogy of auriferous and uraniferous pre-Witwatersrand placer occurrences.

Sedimentological investigations on the Moodies Group sediments were carried out by Eriksson (1978a, b). Petrographical and ore mineralogical studies of the gold placers in the Uitkyk Formation were undertaken by Muff and Saager (1979), and Saager and Muff (1980, and in press). Saager and Utter (in press) investigated various old gold workings in conglomerates of the Pongola Supergroup by a geological and radiometrical survey.

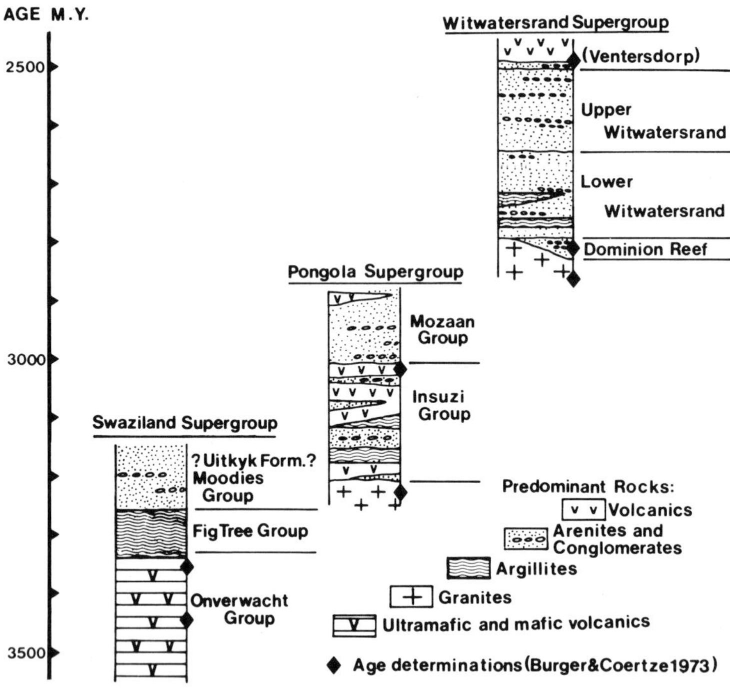

Fig. 2. Schematized stratigraphic columns showing relationship between the Witwatersrand, Pongola, and Swaziland Supergroup sediments

The present paper deals with a mineralogical comparison between the Witwatersrand and pre-Witwatersrand auriferous and uraniferous conglomerates. Apart from a general ore microscopic survey, only some of the results of detailed mineralogical investigations are presented because of the limited space available for this publication. Special attention was given to the chemical composition of the detrital chromites and the occurrence and characteristics of carbonaceous matter and uranium phases. The sampling areas are given in Fig. 1.

The samples from the Witwatersrand Supergroup originate from underground workings of gold mines situated in the Orange Free State Goldfield and from boreholes in the Klerksdorp-Randfontein area (Fig. 1). These samples contain unaltered ore material, whereas all the samples of the pre-Witwatersrand conglomerates had to be taken from surface outcrops and, thus, were more or less intensively affected by supergene alterations.

2 Ore Mineralogy

2.1 General Remarks

The ore mineralogical study was restricted to distinctly mineralized fresh and weathered conglomerate samples. The investigations were carried out by means of standard ore microscopy, by the aid of an electron microprobe, by scanning electron microscopy and by uranium fission-track micromapping.

In practically all fresh or little weathered samples investigated, *pyrite* constitutes 90% of the ore minerals present in the conglomerate matrix (Figs. 3, 4, and 17a). Pyrite, together with rutile/leucoxene form the only macroscopically identifiable ore minerals. Pyrite occurs disseminated throughout the entire widths of individual conglomerate bands and often is concentrated in well-defined layers on footwall contacts or on bedding planes. In all studied pre-Witwatersrand occurrences, pyrite possesses the same morphological features as the pyrite in the Witwatersrand sediments and following the classifications given by Ramdohr (1955), Saager (1970), and Utter (1978a) can be grouped into the following three classes:

1. compact, generally well-rounded pyrite of detrital origin possessing primary inclusions of chalcopyrite, pyrrhotite, and more rarely of sphalerite, galena and in a few cases also of gold,
2. concretionary and porous, occasionally banded pyrite of detrital or of authigenic origin with diameters commonly exceeding those of the compact rounded pyrite, and
3. reconstituted pyrite emplaced during or after the regional metamorphism of the sediments. This pyrite variety often exhibits idiomorphic outlines and forms overgrowths on detrital heavy minerals preponderantly on pyrite.

The genetic significance of the different pyrite varieties to the formation of the Witwatersrand ores has been discussed in detail by Ramdohr (1955) and Saager (1970, 1973).

In the pre-Witwatersrand conglomerates *rutile* and its microcrystalline form, which in the South African literature usually is referred to as *leucoxene,* appear to be more abundant than in many of the Witwatersrand bankets causing a distinct yellowish-brown tinge on polished sections. Rutile/leucoxene in most cases must be regarded as an authigenic constituent which was derived from detrital ilmenite and titaniferous magnetite (Ramdohr 1955). It occurs as irregular earthy masses displaying lamellae or cloth-like textures or as rounded fine-grained aggregates often exhibiting peculiar atoll-like textures whose origin is not yet understood and which are typical for many samples of the Pongola Supergroup (Figs. 5 and 7). The size distribution of the rounded rutile/leucoxene grains, in general, is smaller than that of the compact rounded allogenic pyrite and compares well with that of zircon and chromite. Irregular aggregates of rutile/leucoxene were often found to be closely associated with *brannerite* in samples from the Witwatersrand and the Uitkyk Formation (Thiel et al. 1979).

Chromite is very widely distributed in all investigated samples. This chromite grains vary in shape, but they are usually well rounded and can display a good octahedral cleavage. Most of the chromite grains are shattered (Figs. 3 and 17a). They show remark-

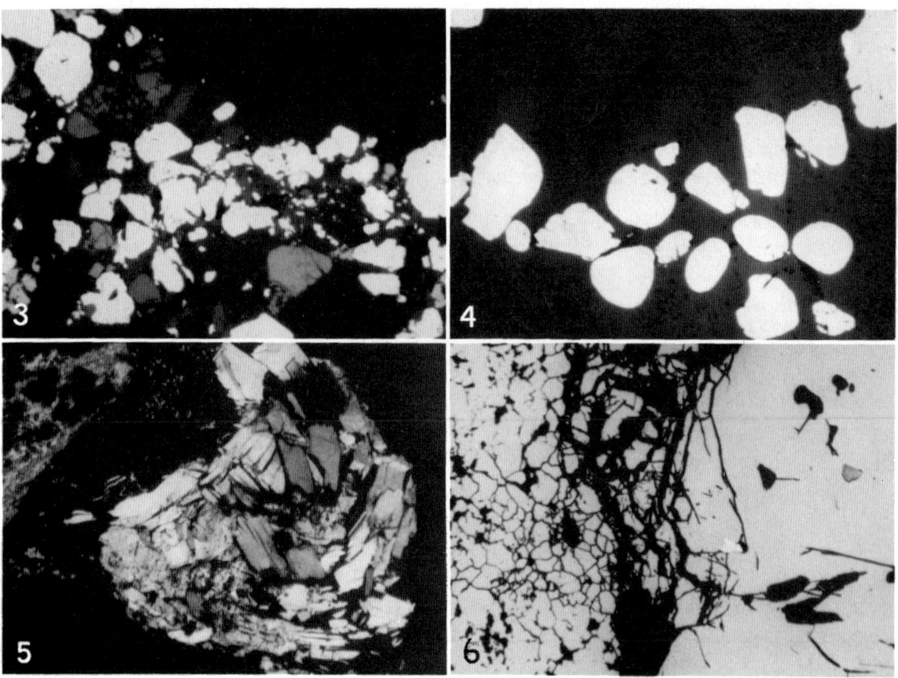

Fig. 3. Moodies Group, Bearded Man. Poorly sorted detrital compact pyrite *(white)* with some authigenic pyrite; shattered detrital chromite *(grey)* and zircon *(dark grey)*; elongated chloritoid porphyroblasts (*darker* than zircon). × 30

Fig. 4. Mozaan Group, Gunsteling. Well-rounded compact pyrite. Note absence of other heavy minerals. × 30

Fig. 5. Uitkyk Formation, Mount Robert. Detrital molybdenite aggregate and irregular masses of leucoxene (*upper left hand corner* of microphotograph). × 170

Fig. 6. Mozaan Group, Gunsteling. Rims of two adjacent large detrital pyrites. Note shattering at the contact point of the two grains which is more pronounced in the more porous pyrite on the *left hand side* of the microphotograph. Gold *(bright white)* and galena *(grey)* form inclusions in the compact pyrite. × 100

able differences in colour and reflectivity from a dark bluish-grey to a distinct yellowish-olive which resembles the colour of ilmenite. Differently coloured chromite varieties were observed within one single polished section, with the olive coloured variety being typical for many of the samples from the Insuzi Group conglomerates (Fig. 10). Zoned chromite grains possessing higher reflecting rims are widely observed features in the investigated material. In a few cases chromite shows oriented exsolution lamellae of ilmenite and of an undetermined mineral phase. The chemical composition of the chromites is elucidated in a separate chapter (p. 48).

Zircon is a consistent heavy mineral in all investigated conglomerates but usually is less abundant than chromite. Zircon shows varying degrees of abrasion from well-rounded to prismatic grains and almost in all cases is zoned and often shattered (Figs. 3 and 17a).

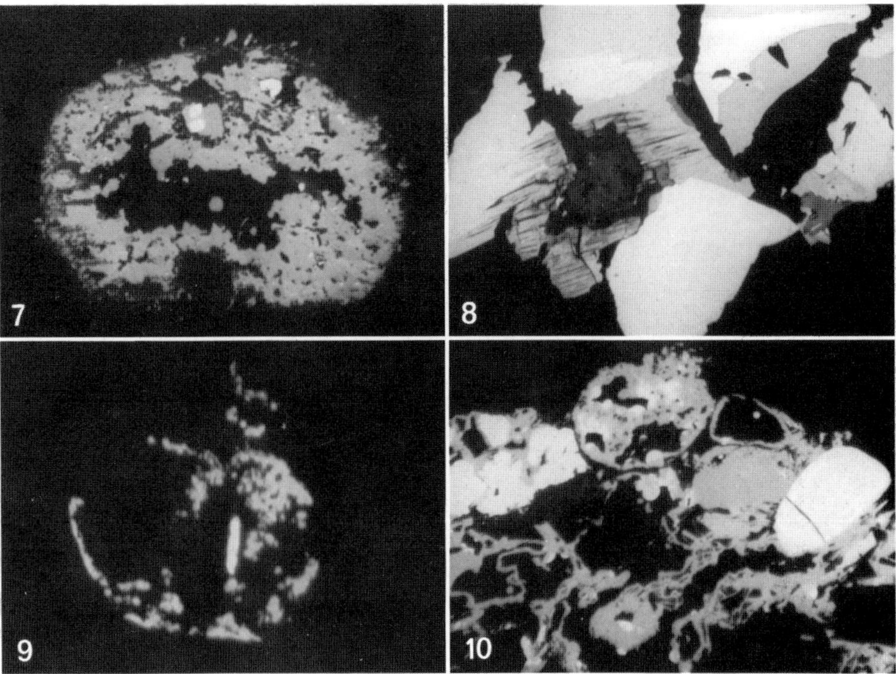

Fig. 7. Insuzi Group, Speedwell. Atoll-like textured rutile/leucoxene aggregate, probably a pseudomorph of a detrital titaniferous black sand mineral. Idiomorphic inclusions of pyrite *(bright white)* and an infiltration of chalcopyrite *(white)* at the rim of the aggregate. × 100

Fig. 8. Promise Reef, Lower Witwatersrand Group, Kafferskraal near Klerksdorp. Complex assemblage of detrital pyrite *(bright white, relief)* being replaced by authigenic pyrrhotite *(light grey)* with pentlandite exsolutions *(white)*. Detrital uraninite *(dark grey)* with white specks of radiogenic galena *(white)*. Note lamellae-type alteration of pyrrhotite / / (0001) to so-called „Zwischenprodukt" (Ramdohr 1975). As the sample originates from an unweathered borehole core, this alteration halo appears to be of radiogenic origin. × 100

Fig. 9. Insuzi Group, Vuleka. Loose pseudomorphs of hematite *(grey)* after detrital pyrite with primary inclusion of gold *(white)*. × 170

Fig. 10. Insuzi Group, Vuleka. Weathered ore showing pseudomorphs of iron-hydroxides *(dark grey)* and hematite *(whitish grey)* after pyrite. Note two juxtaposed grains of rounded chromite on the *right hand side* of the microphotograph *(bright grey* and *dark grey)*. The chromites differ markedly in their colour and reflectivity although they are very similar in chemical composition. × 100

Uraninite as rounded apparently detrital grains is abundant in many of the Witwatersrand bankets, but in pre-Witwatersrand conglomerates it was observed only in samples from the Moodies Group where it occurs as small inclusions in a few grains of "fly-speck" carbon (Fig. 14). The only other uranium phase, so far observed in the pre-Witwatersrand material, are irregular patches of *brannerite* associated with rutile/ leucoxene (Fig. 16). The scarcity of uranium phases, apparently a typical feature of the investigated pre-Witwatersrand conglomerates, is discussed later (p. 52).

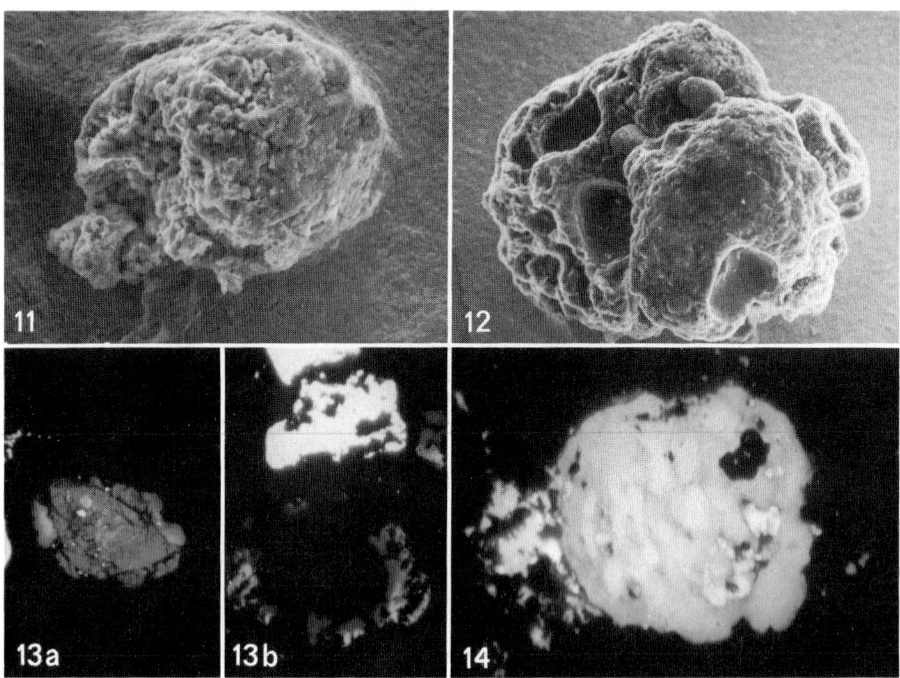

Fig. 11. Moodies Group, Bearded Man. SEM-microphotograph of "fly-speck" carbon. Note the pitted surface of the aggregate. × 100

Fig. 12. Promise Reef, Lower Witwatersrand Group, Kafferskraal near Klerksdorp. SEM-microphotograph of "fly-speck" carbon. Note indentations probably caused by rounded detrital heavy minerals. × 100

Fig. 13a. Basal Reef, Upper Witwatersrand Group, Orange Free State Goldfield. Rounded uraninite grain *(dark grey)* with specks of radiogenic galena *(white)*. The uraninite is rimmed by carbonaceous matter *(dark grey)*. × 100
b Mozaan Group, Gunsteling. Atoll-shaped aggregate of carbonaceous matter *(dark grey)* whose centre part was probably filled by uraninite now removed by surface weathering processes. Irregular authigenic pyrite *(white)* and rutile *(grey)* in *upper part* of microphotograph. × 140

Fig. 14. Moodies Group, Bearded Man. Rounded grain of "fly-speck" carbon *(grey)* with inclusions of uraninite *(lighter grey)* and galena *(white)*. Irregular aggregates of rutile *(grey)* on the *left hand side* of the carbon aggregate. × 190

Gold in the Witwatersrand banket occurs as irregular particles in the fine-grained conglomerate matrix and more seldom as infiltrations in porous pyrites. In the Witwatersrand ores the gold particles, according to Utter (1979), possess average diameters of 70 μm. Seldom gold also occurs as primary inclusions in compact detrital pyrite. In the pre-Witwatersrand conglomerates a few gold particles were observed as primary inclusions in pyrite (Fig. 6), and only in the Uitkyk Formation was gold found to occur in the matrix of the conglomerate.

Carbonaceous matter is a constant component of most of the Witwatersrand banket and its nature and significance have been discussed by Hallbauer (1975) and Hallbauer

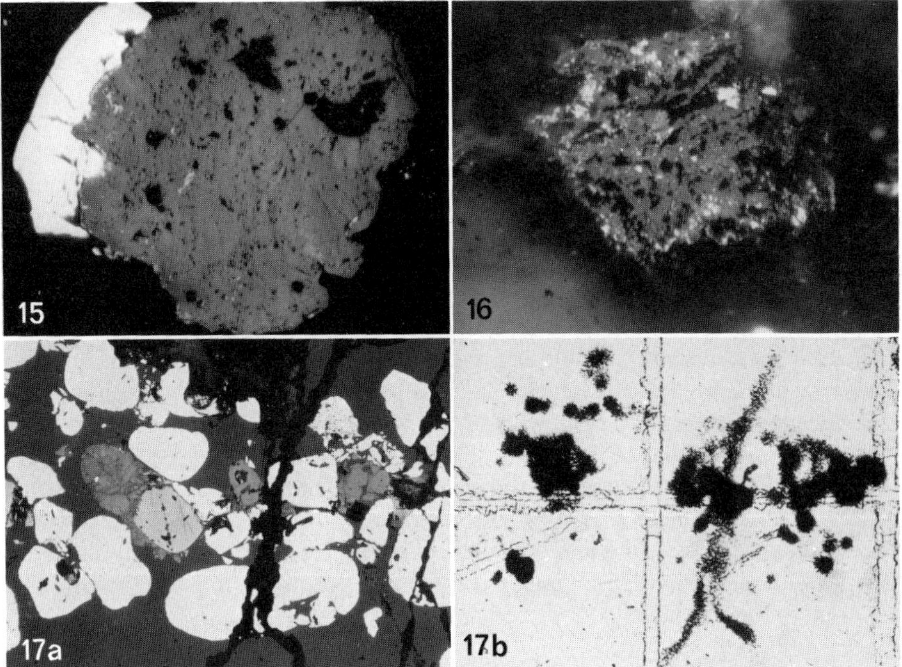

Fig. 15. Promise Reef, Lower Witwatersrand Group, Kafferskraal near Klerkdsdorp. "Fly-speck" carbon with overgrowth of authigenic pyrrhotite *(white)*. An unknown dark grey uranium phase forms the two large inclusions in the *upper right hand corner* of the carbonaceous matter; the irregular small *greyish white* inclusions are rutile. × 100

Fig. 16. Mozaan Group, Denny Dalton. Brannerite *(grey)* intergrown with rutile/leucoxene *(grey)*. The lenticular *bright grey* specks are pyrrhotite. × 100

Fig. 17a. Basal Reef, Upper Witwatersrand Group, Orange Free State Goldfield. Detrital compact pyrite *(white)*, chromite (*light grey* in the *centre* of the microphotograph), strongly shattered zircon *(dark grey)*, small grains of uraninite *(dark grey)* and some aggregates of "fly-speck" carbon *(dark grey)*, a large carbon grain is distinctly visible adjacent to the *left hand edge* of the chromite grain. Note that the sample is transected by recent to subrecent cracks. × 40
b Corresponding uranium distribution on mica detector. Dense track clusters manifest presence of uraninite and carbonaceous matter. Uranium redistribution is indicated by variable uranium concentrations along the crack in the *centre* of the microphotograph. The compact pyrite, the chromite and the quartz are devoid of uranium

et al. (1977). New observations on pre-Witwatersrand carbonaceous matter are given on p. 51.

Fe-hydroxides and hematite are abundant in weathered material and often form pseudomorphs after pyrite (Figs. 9 and 10). All other ore minerals are rare and can be regarded as accessory components of the conglomerates.

2.2 Investigated Areas

The *Moodies conglomerates* from the Bearded Man area in the Barberton Mountain Land (Fig. 1) form a wide zone of immature arenaceous sediments in which the following mineral phases were recognized (order of abundance):

pyrite; rutile/leucoxene, pyrrhotite, chalcopyrite, chromite, zircon, cobaltite, graphite, brannerite, "fly-speck" carbon (Figs. 11 and 14), and uraninite.

Pyrite, chromite, and zircon in comparison with all other occurrences are poorly rounded indicating short transport distances and poor sedimentary reworking (Fig. 3). A few samples show large amounts of authigenic pyrrhotite, chalcopyrite, and cobaltite which together with graphite occur as narrow veinlets transecting conglomerate pebbles and matrix. These epigenetic minerals probably were emplaced as a result of the formation of the gold-quartz veins of the Maid of the Mist occurrence which lies in the immediate vicinity of the investigated sampling locality.

The *Uitkyk Formation* was sampled at the Mount Robert pyritic gold placer occurrence near Pietersburg (Fig. 1). The ore occurs in coarse matrix-supported conglomerates which were deposited in a braided river environment within a rapidly subsiding trough (Muff and Saager 1979; Saager and Muff, in press). The following ore minerals were observed in the Mount Robert ore (order of abundance):

pyrite, rutile/leucoxene, chromite, zircon, chalcopyrite, gold, mackinawite, sphalerite, magnetite, ilmenite, pentlandite, molybdenite, "fly-speck" carbon, brannerite, and galena.

The pyrite grains are frequently shattered by overlying quartz pebbles. As in the Moodies conglomerates, the allogenic pyrite from the Mount Robert occurrence exhibit rather angular outlines reflecting the immature development of the conglomerates. Large porous pyrite forms the bulk of the ore minerals at Mount Robert and is often associated with chloritoid porphyroblasts. Muff and Saager (1979) suggested that these pyrite/chloritoid aggregates represent intraformational clasts originating from iron-sulphide muds which developed in a reducing atmosphere on interchannel flats at the time of sedimentation.

A conspicuous but rare detrital constituent found in the Uitkyk conglomerate is molybdenite forming rounded aggregates of curved occasionally broken laths (Fig. 5). The molybdenite aggregates possess diameters of up to 0.4 mm and are a further indication for very short transport distances of the detrital material.

Conglomerate samples of the *Pongola Supergroup* were investigated from several old gold working and prospecting sites (Fig. 1). Material from the *Insuzi Group* was collected at Speedwell, Cooper's Store, and Vuleka (Fig. 1). With the exception of the Speedwell material, only weathered samples were obtainable. The following ore minerals were determined (order of abundance):

pyrite, rutile/leucoxene, chromite, zircon, cobaltite, pyrrhotite, chalcopyrite, galena, sphalerite, gold, and as secondary minerals Fe-hydroxides, hematite, and covellite.

At Speedwell, a number of broad shear zones irregularly impregnated by white quartz cross-cut the conglomerates. These shear zones probably caused the impregnation of the conglomerates with irregular sulphide grains, mainly pyrite and chalcopyrite, which are clearly distinguishable from the well-rounded allogenic constituents.

Noteworthy is the occurrence of primary gold grains in loose grains of hematite and Fe-hydroxides pseudomorphic after rounded constituents, probably pyrite in the strongly weathered samples from Vuleka (Fig. 9).

The conglomerates of the Pongola Supergroup show a marked lateral inconsistency. They are oligomictic, poorly sorted coarse clastic intercalations in predominantly quartzitic series. According to Saager and Utter (in press) these conglomerates represent fluviatile deposits laid down on alluvial plains which were frequently submerged by shallow marine waters. According to these authors, sedimentary reworking and, thus, concentration of heavy minerals was of minor importance. Only the *Mozaan conglomerates* at Gunsteling, Denny Dalton, and to a minor extent at Redcliff (Fig. 1), where pebble-supported conglomerate bands occur, indicate the presence of reworking processes in a transgressive shallow marine environment. Well-developed layers of often exceptionally rounded pyrite and other heavy minerals were, thus, only encountered in these three localities. The following minerals were observed in the Mozaan conglomerates (order of abundance):

pyrite, leucoxene/rutile, chromite, zircon, arsenopyrite, cobaltite, chalcopyrite, pyrrhotite, galena, brannerite, gold, pentlandite, carbonaceous matter, and as supergene minerals Fe-hydroxides, hematite, covellite, and skorodite.

Compact rounded pyrite typically exceeds the porous pyrite variety (Fig. 4), which is in contrast to the Uitkyk conglomerates and many of the Witwatersrand samples where the porous variety forms the bulk of the allogenic pyrite. Exceptionally large compact well-rounded pyrite with diameters ranging up to 6 mm are characteristic of the basal conglomerate zone at Denny Dalton. Primary inclusions of chalcopyrite, pyrrhotite, pentlandite, sphalerite, and of galena are abundant in the compact detrital pyrite. Other features of the Mozaan conglomerates are the widespread occurrence of detrital cobaltite, which also was found to occur extensively in certain portions of the Basal Reef of the Upper Witwatersrand Group, and at Denny Dalton, the presence of detrital arsenopyrite. Erratic gold particles occur in the Mozaan conglomerates only as rare inclusions in compact pyrite (Fig. 6). Rutile/leucoxene more often forms rounded atoll-like aggregates than the irregular masses which are typical of the Moodies and especially of the Uitkyk conglomerates (Fig. 7). This may indicate that in the Mozaan conglomerates mobilization of rutile/leucoxene was restricted.

The *Witwatersrand conglomerates* were formed as large fluvial fans at the interface between a fluvial system and a shallow-water lacustrine or inland sea environment (Pretorius 1976). Braided stream patterns are present in the sediments which are typically cross-bedded and exhibit a cyclic development with unconformities of varying hiatus usually forming the boundaries between cycles. Longshore currents and movements of the shoreline due to changes in the gradient of the paleoslope caused optimum reworking and winnowing conditions, which led to a concentration of heavy minerals and to the eventual formation of economic placer deposits.

Up to now the conglomerate horizons of the *Lower Witwatersrand Group* are of little economic importance and mining activities ceased some 40 years ago. Thus, fresh samples from these conglomerates can only be examined in borehole cores.

For the present study, eight cores from the Klerksdorp-Randfontein area were studied, and, based on a limited number of polished sections, the following minerals were observed (order of abundance):

> pyrite, leucoxene/rutile, pyrrhotite, chromite, zircon, chalcopyrite, pentlandite, "fly-speck" carbon, cobaltite, arsenopyrite, brannerite, uraninite, sphalerite, and galena.

Rounded compact pyrites form the bulk of the allogenic pyrites, and rounded porous pyrites are less abundant. Some samples show large amounts of authigenic pyrite which often exhibits idiomorphic to hypidiomorphic shapes or forms veinlets cutting through fractured minerals. In a number of cases authigenic pyrrhotite contains exsolved pentlandite (Fig. 8). The relatively large amounts of authigenic pyrrhotite present in the Lower Witwatersrand samples investigated may be caused by mafic dikes present in the vicinity of the drilling sites.

Chromite and zircon are widely distributed in all samples studied. The size of individual grains varies considerably within a particular sample. While zircon occurs almost exclusively in the form of elongated prisms, chromite grains tend to be rounded and often shattered. In a few cases chromite shows orientated exsolution lamellae of ilmenite.

Uraninite was only observed in a few samples where it occurs as rare individual grains, as small inclusions in "fly-speck" carbon, or as conspicuous associations with pyrrhotite (Fig. 8).

Brannerite often associated with rutile/leucoxene aggregates is the most abundant uranium phase in the investigated material of the Lower Witwatersrand Group.

A wealth of mineralogical information is available from the conglomerates of the *Upper Witwatersrand Group.* Up to now more than 70 ore minerals were detected in the Witwatersrand conglomerates and the reader is referred to the papers of Liebenberg (1955), Ramdohr (1955, 1958), Viljoen (1968), Saager (1970, 1973), Schidlowski (1970), and Feather and Koen (1975), which give an excellent insight into the complex ore mineralogy of the reef horizons.

The most conspicuous ore-mineralogical difference between the Upper Witwatersrand conglomerates and all other investigated pyritic conglomerates is the former's widespread content of gold, brannerite, and uraninite, which in the older sediments occur only in minute amounts, although in identical fashion.

3 Composition and Origin of Chromites

The chemical composition of 28 detrital chromites in polished sections from conglomerates of the following stratigraphic units and localities (Fig. 1) was determined by microprobe analyses (EMX-ARL probe, KEVEX analyser):

1. Moodies Group: Bearded Man, Barberton Mountain Land;
2. Uitkyk Formation: Mount Robert;
3. Insuzi Group: Speedwell;
4. Mozaan Group: Gunsteling, and Denny Dalton;
5. Lower Witwatersrand Group: Klerksdorpf, and Randfontein.

The obtained results were compared with analytical data of chromites from the Upper Witwatersrand Group, Klerksdorp Goldfield, published by Utter (1978b).

On each chromite grain at least two measurements were performed, rendering average values of Cr, Fe, Al, Mg, Ti, and Zn. In a few cases, compositional zoning made an individual evaluation of each determination necessary. Assuming stoichiometry and the "normal" spinel formula (Haggerty 1979), the total iron content was recalculated into ferrous and ferric iron and the metal oxides were recalculated arithmetically into weight percents using the method given by Stevens (1944) and Hutchison (1972).

The chromites of the investigated conglomerates including the data of Utter (1978b) exhibit the following compositional ranges:

Cr_2O_3 38 to 58 wt.%, average 52 wt.%
Al_2O_3 2 to 27 wt.%, average 11 wt.%
Fe_2O_3 0 to 27 wt.%, average 8 wt.%
FeO 12 to 34 wt.%, average 25 wt.%
MgO 0 to 12 wt.%, average 5 wt.%

TiO_2 was found to occur in an average concentration of less than 1 wt.%, the range being 0 to 8 wt.%. A few grains contained up to 4 wt.% of ZnO.

Compositional zoning was detected in four chromites. These grains revealed markedly lower MgO and Cr_2O_3, and higher FeO concentrations in their marginal areas (Table 1).

Table 1. Compositional zoning of chromites

Origin	Part of grain	Cr_2O_3	Al_2O_3	Fe_2O_3	FeO	MgO	ZnO	TiO_2
Moodies G.	centre	55.6	12.0	4.0	16.2	11.1	–	0.5
	margin	46.7	12.0	4.1	31.7	0.7	2.7	1.5
Uitkyk F.	centre	53.0	13.2	3.5	18.3	10.3	–	0.6
	margin	45.1	15.1	1.0	32.0	1.0	–	0.7
Mozaan G.	centre	58.9	6.5	4.3	23.8	2.2	0.3	–
	margin	53.4	6.3	5.4	30.1	0.2	0.2	–
Mozaan G.	centre	51.1	11.5	6.3	24.5	6.7	–	1.0
	margin	45.0	14.7	5.4	27.3	5.3	–	1.2

The compositional features of the chromites are demonstrated and plotted in two rectangular projections (Irvine 1965) of a triangular prism (Fig. 18). In the triangular prism, each of the six corners corresponds to one of the major chromian spinel end members (Stevens 1944). The detrital chromites of the Witwatersrand and pre-Witwatersrand auriferous conglomerates plot together in the upper right portion of diagram A (Fig. 18), which marks the compositional field of Cr- and Fe(II)-rich, but Mg-poor chromites. Accordingly in diagram B (Fig. 18) the chromites plot in the lower right portion, i.e. the field of Fe(III)-poor chromites. It is significant that no compositional differences of the chromites are noticeable between the various stratigraphic units and sampling places. The investigated chromites are all similar in their composition.

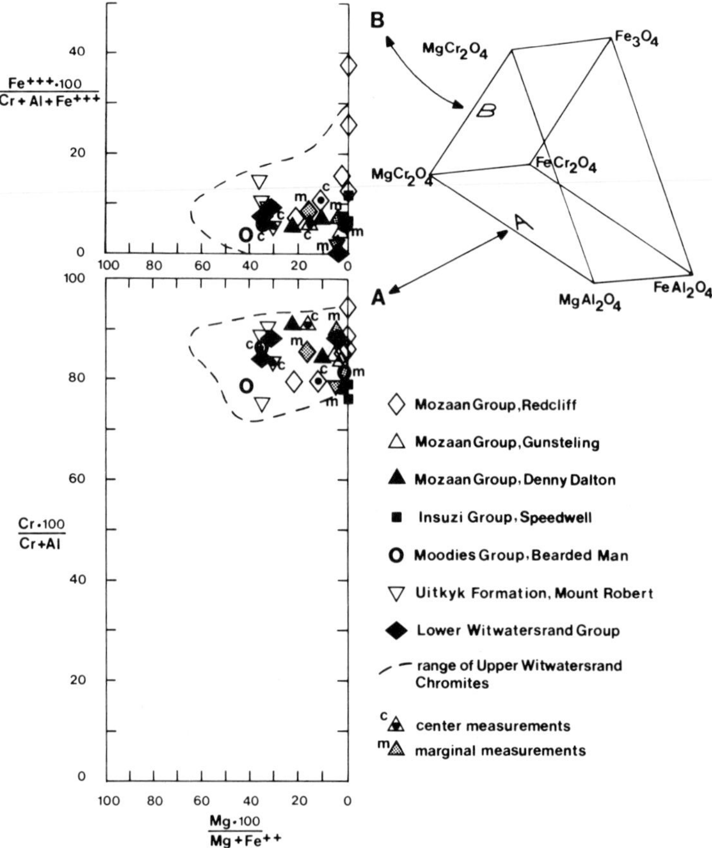

Fig. 18. Composition of investigated detrital chromites, plotted in the projection diagrams after Irvine (1965). The range of Upper Witwatersrand chromites from Utter (1978b)

The origin and probable source area of the investigated chromites can be discussed by a comparison with compositional features of chromites from primary deposits. The detrital Witwatersrand and pre-Witwatersrand chromites differ from chromites of layered stratiform intrusions (Irvine 1967; Thayer 1970) by a higher Cr:Al ratio and a lower Mg:Fe(II) ratio. Chromites from alpine-type ultramafic complexes (Irvine 1967; Thayer 1964) and from "island-arc" basalts (Arculus 1974), in comparison with the chromites from the investigated placer deposits, have a much wider variation in their Cr:Al ratios due to partly reciprocal Al and Cr concentrations. Furthermore, an origin for the Witwatersrand and pre-Witwatersrand chromites from Alaskan-type mafic to ultramafic complexes (Findlay 1969; Jackson and Thayer 1972; Bird and Clark 1974) is unlikely, as the detrital chromites exhibit distinct lower Fe(III)-concentrations than chromites from Alaskan-type magmatic deposits. However, it is possible that the detrital Witwatersrand and pre-Witwatersrand chromites originally derived from Archaean greenstone complexes, because of their compositional convergence to chromites from the Sandrivier Archaean greenstone terrane near Pietersburg where Cr- and Fe(II)-rich and Mg- and Fe(III)-poor chromites have been described to occur either as disseminated or compact ore lenses (Willemse 1948).

A compositional zoning of some of the detrital chromites has previously been described from the Witwatersrand deposit by Mihálik and Saager (1968). According to these authors, it is assumed that radiation from uranium phases can cause a certain amount of lattice distortion in the chromites, which in turn appears to facilitate the in situ leaching of Mg and partly of Cr from the marginal parts of the chromites. Apparently Mg has been replaced by Fe(II).

4 Carbonaceous Matter

Carbonaceous matter is a constant component of most of the *Upper Witwatersrand* auriferous conglomerates and its nature has been discussed by Hallbauer (1975) and Hallbauer et al. (1977). Three main types of carbon have been classified from the Witwatersrand deposit (Hallbauer 1975; Hallbauer et al. 1977): (1) mats of carbon columns 0.5 to 5 mm in length and 2 mm in diameter, arranged perpendicular to the bottom contact of the mats; (2) 0.2 mm to 1 mm large irregularly distributed spherical grains of carbon, so-called "fly-speck" carbon; and (3) apparently amorphous material.

In the present study carbonaceous matter has been studied by polished section microscopy, and with the aid of a scanning electron microscope after the carbon was liberated and isolated from conglomerate samples by hydrofluoric acid treatment (Neuerburg 1975).

In the Archaean pre-Witwatersrand and the Lower Witwatersrand conglomerates, carbonaceous matter occurs as rare aggregates in the form of "fly-speck" carbon and also as irregular jagged particles which occasionally are arranged in an atoll-like structure. In these occurrences the "fly-speck" carbon displays similar morphological and textural features which closely resemble the "fly-speck" carbon described from the Upper Witwatersrand Group. It takes the form of oval-shaped grains and under SEM it shows an irregular, pitted surface texture (Figs. 11 and 12). In pre-Witwatersrand conglomerates the carbon grains possess diameters varying between 100 to 500 μm and in the Lower Witwatersrand conglomerates they are up to 1 mm large (Figs. 14 and 15).

In the Moodies Group the "fly-speck" carbon contains occasionally minute inclusions of uraninite and galena and this represents the only occurrence of uraninite in pre-Witwatersrand conglomerates observed in this study (Fig. 14). Saager and Muff (1980) described "fly-speck" carbon from the Uitkyk Formation where it possesses internal textures of oval-shaped structures which exhibit diameters of about 50 μm. Microprobe analyses showed that the interstitial material between the oval-shaped structures contains up to 40% of uranium, 4% of thorium, and some traces of titanium and lead (Muff and Saager 1979).

Noteworthy in some of the Pongola Supergroup samples is the occurrence of irregular jagged to atoll-shaped aggregates of carbonaceous material having diameters of up to 200 μm (Fig. 13b). These aggregates probably were former overgrowths on detrital uraninite, the latter probably being leached by percolating, oxygen-rich surface waters. Overgrowths of carbon on detrital uraninite grains are an abundant feature in unweathered samples from the Basal Reef, Upper Witwatersrand Group (Fig. 13a). These associations remind one of the earlier suggestion of Liebenberg (1955) and Schidlowski (1970), who proposed the carbon to originate from gaseous or liquid hydrocarbons which were "polymerized" by radioactive radiation of the detrital uraninite.

Hallbauer and van Warmelo (1974), Hallbauer (1975) and Hallbauer et al. (1977) suggest that the columnar-type carbon represents fossilized remains of primitive plants, probably of a symbiotic association of an algal and a fungal partner. The above authors suggested that the "fly-speck" carbon grains are the vegetative diaspores of the columnar lichen-like organisms. However, other workers (Pretorius 1974; Cloud 1976) dispute this interpretation and envisage the "fly-speck" carbon to be detrital carbon fragments. An ultimate biogenic origin of this type of carbonaceous matter is not disputed by these authors.

If the postulation is correct that the "fly-speck" carbon originated from diaspores of the columnar lichen-type organisms, it is remarkable that the columnar-type carbon was not yet detected in pre-Witwatersrand conglomerates. Sedimentary reworking, and thus destruction of mats of columnar-carbon in the pre-Witwatersrand conglomerates, is unlikely as reworking processes played only a minor role in the depositional history of these occurrences (Saager and Muff, in press; Saager and Utter, in press). More probably, the pre-Witwatersrand "fly-speck" carbon should be regarded as detrital fragments of biogenic carbon whose origin is still unknown or, to be speculative, the "fly-speck-type" carbon could reflect life forms situated in their biological evolution between primitive prokaryotes from the Onverwacht and Fig Tree Groups of the Swaziland Supergroup (Knoll and Barghoorn 1977) and the morphologically more complex biogenic structures of Proterozoic age found in the Upper Witwatersrand Group.

5 Uranium Phases

Uranium fission-track micromapping carried out by Simpson and Bowles (1977) on material from the Upper Witwatersrand conglomerates and by Thiel et al. (1979) on material from the Witwatersrand and Uitkyk conglomerates gave evidence that uranium not only occurs in the form of uraninite and brannerite but also as fine dispersions in close association with carbonaceous matter, porous pyrite, leucoxene, and fine-grained phyllosilicates in the conglomerate matrix (Fig. 17a, b). Thiel et al. (1979) explain the nondetrital uranium phases, e.g. the fine uranium dispersions, as a result of synmetamorphic to recent multi-stage redistributions of uranium.

Particularly in samples originating from weathered conglomerate outcrops recent uranium transport by oxidizing meteoric waters was important, causing the removal of uranium. Such a process would explain the absence of uraninite in the investigated pre-Witwatersrand material – only a few inclusions of uraninite in the reducing micro-environment of "fly-speck" carbon grains were detected – and also explains the rare occurrence of the more refractory brannerite. The above explanation of a recent uranium removal is substantiated by high radiometric in situ readings on pre-Witwatersrand conglomerates which have not attained a state of radioactive equilibrium, i.e. conglomerates which were supergenetically altered within the last 1 million years (Saager and Utter, in press).

Outcropping weathered uraniferous conglomerates of the Witwatersrand Supergroup generally have reached a state of radioactive equilibrium as their radiometric readings usually correspond with the low uranium contents (approx. 20 ppm) present. Unweathered underground samples of Witwatersrand conglomerates, however, carry in addition to brannerite widespread detrital uraninite (Liebenberg 1955; Ramdohr 1955; Schidlowski 1966).

6 Discussion and Conclusions

The investigated early Precambrian pyritic quartz-pebble conglomerates form part of sedimentary assemblages which were deposited during a time span ranging from 2500 to 3250 m.y. This period was of great importance for the development and consolidation of the Kaapvaal Craton which, according to van Rendsburg and Pretorius (1977), had reached a state of crustal stability approximately 3000 m.y. ago, so that the cover rocks were laid down and preserved without major modifications by metamorphic or tectonic events.

The Moodies Group sediments – the oldest studied in this work – herald the transition from Archaean to Proterozoic crustal development in that they represent the oldest sediments of marginal marine origin which accumulated along a continental margin on a relatively stable crust (Eriksson 1978a). For the Uitkyk conglomerates, however, a sedimentation in a rapidly subsiding trough is envisaged (Muff and Saager 1979), an environment which would appear to be similar to the depository in which the sediments of the Sioux Lookout greenstone belt in Canada were accumulated (Turner and Walker 1973).

With further maturity of the crust a number of progressively younger sedimentary-volcanic basins were formed on the Kaapvaal Craton. Each of the basins possessed a somewhat similar stratigraphic filling: volcanics at the base and at the top, and fine and coarse clastics in the middle (Pretorius 1976). The oldest of these basins contained the assemblages of the Pongola Supergroup. It was followed by the larger Witwatersrand Basin having its depositional axes to the north-west of that of the Pongola Basin. Whereas the Pongola conglomerates were deposited at the rim of the basin on tectonically inactive plains frequently submerged by shallow marine waters (Saager and Utter, in press), the Witwatersrand conglomerates are regarded as fluvial fan deposits which developed on fault-bounded basin edges (Brock and Pretorius 1964).

This crustal development from Moodies to Witwatersrand times inflicted far reaching changes upon the sedimentary environments of the investigated conglomeratic sediments and, thus, to a certain extent, may explain the compositional and textural differences observed in the ore mineralogy of the investigated conglomerates:

1. *In pre-Witwatersrand strata* sedimentary reworking was little and only locally developed. This explains the observed scarce occurrence of pebble supported conglomerates (Saager and Utter, in press), the poor sorting and rounding of detrital constituents, and the rarity of heavy mineral lag deposits which have very erratic and apparently uneconomic gold values. Only a few conglomeratic bands of the investigated Pongola sediments (Saager and Utter, in press), and locally also of the Uitkyk Formation (Muff and Saager 1979), show signs of reworking. It is significant that in such bands a distinct concentration of heavy minerals does occur and that most of the gold workings and prospects in pre-Witwatersrand strata are situated in sedimentologically reworked conglomerate horizons.
2. In the *Witwatersrand conglomerates* reworking in the fluvial fans was widespread and led to a winnowing concentration of gold and other heavy minerals. Furthermore, basin edge tectonic activities produced unconformities between successive sedimentary cycles, thereby causing an intensification of heavy mineral concentration (Pretorius 1975). In the Witwatersrand conglomerates this is reflected in the prevalence of well-rounded detrital minerals and in the economic importance of the contained gold and uranium deposits, which also carry enormous amounts of pyrite and a host of minor other minerals which are of greater variety than in the pre-Witwatersrand deposits.

From this model, it now can be held that the pre-Witwatersrand pyritic quartz-pebble conglomerates are primitive forerunners of the Witwatersrand banket and that the gold, and most of the pyrite and uraninite, are sedimentary in origin, having been transported as detrital particles to accumulate in the conglomerates. The composition of detrital chromite indicates that greenstone complexes were of importance in the provenance area. These rock types provided also the most favourable source for other heavy minerals, such as gold and detrital sulphides, particularly pyrite, whereas felsic areas, e.g. the granites intruding and surrounding the greenstone belts (see also Viljoen et al. 1970), contributed uraninite and other salic detrital minerals.

The existence of a nonoxidizing atmosphere in the early Precambrian (Schidlowski and Junge 1973; Cloud 1976, and others) is indicated by the occurrence of detrital pyrite in all investigated conglomerates. The apparent absence of detrital uraninite in the investigated pre-Witwatersrand samples, which all originate from surface outcrops, does not contradict a nonoxidizing atmosphere in the early Precambrian as this absence of uraninite may be caused by recent to subrecent weathering.

Carbonaceous matter is abundant in the Witwatersrand sediments and it now has also been discovered in the much older conglomerates of the Pongola Supergroup and the Moodies Group. This may be an indication that in southern Africa hitherto unknown, relatively advanced life forms existed already during the onset of Proterozoic crustal evolution.

Acknowledgments. This work was financed by the Deutsche Forschungsgemeinschaft grant Sa 210/6. Further financial assistance was provided by the Johannesburg Consolidated Investment Company Ltd., Johannesburg, who also kindly provided the borehole cores from the Klerksdorp-Randfontein area. Thanks are also due to Ms. P. Küster for typing the manuscript.

References

Arculus RJ (1974) Solid solution characteristics of spinels: pleonaste-chromite-magnetite compositions in some island-arc basalts. Carnegie Inst Annu Rep Director Geophys Lab 1973-1974:322–327

Bird ML, Clark AL (1974) Microprobe study of olivine chromitites of the Goodnews Bay ultramafic complex, Alaska, and the occurrence of platinum. J Res US Geol Surv 4(6):717–725

Brandl G (1979) Pietersburg group. Unpubl Rep South African Committee Stratigr (SACS), 8 p

Brock BB, Pretorius DA (1964) Rand basin sedimentation and tectonics. In: Haughton SH (ed) The geology of some ore deposits in southern Africa, vol I, Geol Soc S Africa, Johannesburg, pp 549–599

Burger AJ, Coertze FJ (1973) Radiometric age measurements on rocks from southern Africa to the end of 1971. Bull Geol Surv Pretoria 58:46

Cloud P (1976) Major features of crustal evolution: Alex L Du Toit Memorial Lecture No 14, Geol Soc S Africa, Johannesburg, 33 p

Eriksson KA (1978a) Alluvial and destructive beach facies in the Archean Moodies Group of the Barberton Mountain Land. Univ Witwatersrand Econ Geol Res Unit Inf Circ 115:18

Eriksson KA (1978b) Marginal-marine depositional processes in the Archean Moodies Group, Barberton Mountain Land. Univ Witwatersrand Econ Geol Res Unit Inf Circ 122:19

Feather CE, Koen GM (1975) The mineralogy of the Witwatersrand Reefs. Miner Sci Eng 7:189–224

Findlay DC (1969) Origin of the Tulameen ultramafic-gabbro complex, southern British Columbia. Can J Earth Sci 6:399–425

Grobler NJ (1972) The geology of the Pietersburg Greenstone Belt. Unpubl D Sc Thesis, Univ Orange Free State, Bloemfontein, 156 pp

Haggerty SE (1979) Spinels in high pressure regimes. Proc 2nd Int Kimberlite Conf 2:183–197

Hallbauer DK (1975) The plant origin of the Witwatersrand "carbon". Miner Sci Eng 7:111–131

Hallbauer DK, van Warmelo KT (1974) Fossilized plants in thucholite from Precambrian rocks in the Witwatersrand, South Africa. Precambrian Res 1:199–212

Hallbauer DK, Jahns HM, Beltmann HA (1977) Morphological and anatomical observations on some Precambrian plants from the Witwatersrand, South Africa. Geol Rundsch 66:477–491

Hutchison CS (1972) Alpine-type chromite in North Borneo, with special reference to Darvel Bay. Am Mineral 57:835–856

Irvine TN (1965) Chromian spinel as a petrogenetic indicator, Part 1: Theory. Can J Earth Sci 2: 648–672

Irvine TN (1967) Chromian spinel as a petrogenetic indicator, Part 2: Petrologic applications. Can J Earth Sci 4:71–103

Jackson ED, Thayer TP (1972) Some criteria for distinguishing between stratiform, concentric and Alpine peridotite-gabbro complexes. Int Geol Congr 24th, Montreal 1972, Sect 2: Abstr 189–296

Knoll AH, Barghoorn ES (1977) Archean microfossil showing cell division from the Swaziland System of South Africa. Science 198:396–398

Liebenberg WR (1955) The occurrence and origin of gold and radioactive minerals in the Witwatersrand System, the Dominion Reef, the Ventersdorp Contact Reef, and the Black Reef. Geol Soc S Africa Trans 58:101–223

Mihálik P, Saager R (1968) Chromite grains showing altered borders from the Basal Reef, Witwatersrand System. Am Mineral 53:1543–1550

Muff R, Saager R (1979) Petrographic and mineragraphic investigations of the Archaean gold placers at Mount Robert in the Pietersburg Greenstone Belt, Northern Transvaal. Spec Publ Geol Soc S Africa 6:23–31

Neuerburg GJ (1975) A procedure, using hydrofluoric acid, for quantitative mineral separations from silicate rocks. J Res US Geol Surv 3:377–378

Pretorius DA (1974) The nature of the Witwatersrand gold-uranium deposit. Univ Witwatersrand Econ Geol Res Unit Inf Circ 85:50

Pretorius DA (1975) The depositonal environment of the Witwatersrand Goldfields: a chronological review of speculations and observations. Miner Sci Eng 7:18–47

Pretorius DA (1976) Gold in the Proterozoic sediments of South Africa: Systems, paradigms, and models. In: Wolf K (ed) Handbook of stratabound and stratiform ore deposits, vol VII. Elsevier, Amsterdam, pp 1–27

Ramdohr P (1955) Neue Beobachtungen an Erzen des Witwatersrands in Südafrika und ihre genetische Bedeutung. Abh Dtsch Akad Wiss Berlin Kl Math Allg Naturwiss 5:1–43

Ramdohr P (1958) Die Uran- und Goldlagerstätten Witwatersrand – Blind River District – Dominian Reef – Serra de Jacobina: Erzmikroskopische Untersuchungen und ein geologischer Vergleich. Abh Dtsch Akad Wiss Berlin Kl Chem Geol Biol 3:1–35

Ramdohr P (1975) Die Erzmineralien und ihre Verwachsungen, 4. Aufl. Akademie Verlag, Berlin, 1277 pp

Saager R (1970) Structures in pyrite from the Basal Reef in the Orange Free State Goldfield. Geol Soc S Africa Trans 73:29–46

Saager R (1973) Geologische und geochemische Untersuchungen an primären und sekundären Goldvorkommen im frühen Präkambrium Südafrikas: Ein Beitrag zur Deutung der primären Herkunft des Goldes in der Witwatersrand Lagerstätte. Habilitationsschr, Univ Heidelberg 150 pp

Saager R, Muff R (1980) A new discovery of possible primitive lifeforms in conglomerates of the Archean Pietersburg Greenstone Belt, South Africa. Geol Rundsch 69:1391–1397

Saager R, Muff R (in press) The auriferons placer at Mount Robert in the Pietersburg Greenstone Belt northeast of Potgietersrus. Geol Soc S Africa, Ore Deposit Vol

Saager R, Utter T (in press) Geological and radiometrical investigations on gold placer occurrences in the Pongola Supergroup. Geol Soc S Africa, Ore Deposit Vol

Schidlowski M (1966) Beiträge zur Kenntnis der radioaktiven Bestandteile der Witwatersrand Konglomerate I–III. Neues Jahrb Mineral Abh 105/106:183–202, 310–324

Schidlowski M (1970) Untersuchungen zur Metallogenese im südwestlichen Witwatersrand-Becken. Beih Geol Jahrb 85:80

Schidlowski M, Junge C (1973) Die Evolution der Erdatmosphäre. Phys Bl 29:203–212

Simpson PR, Bowles K (1977) Uranium mineralization of the Witwatersrand and Dominion Reef systems. Philos Trans R Soc London, Ser A 286:527–548

Stevens RE (1944) Composition of some chromites of the Western hemisphere. Am Mineral 29: 1–34

Thayer TP (1964) Principal features and origin of podiform chromite deposites, and some observations on the Guleman-Soridag district, Turkey. Econ Geol 59:1497–1524

Thayer TP (1970) Chromite segregation as petrogenetic indicators. Geol Soc S Africa, Spec Publ 1:380–390

Thiel K, Saager R, Muff R (1979) Uranium distribution in early Precambrian gold-bearing conglomerates of the Kaapvaal Craton, South Africa: a case study for the application of U-fission track micromapping. Miner Sci Eng 11:225–245

Turner CC, Walker RG (1973) Sedimentology, stratigraphy and the crustal evolution of the Archean greenstone belt near Sioux Lookout, Ontario. Can J Earth Sci 10:817–845

Utter T (1978a) Morphology and geochemistry of different pyrite types from the Upper Witwatersrand System of the Klerksdorp Goldfield, South Africa. Geol Rundsch 67(2):774–804

Utter T (1978b) The origin of chromites from the Klerksdorp Goldfield, Witwatersrand, South Africa. Neues Jahrb Mineral Abh 133(2):191–209

Utter T (1979) The morphology and silver content of gold from the Upper Witwatersrand and Ventersdorp System of the Klerksdorp Goldfield, South Africa. Econ Geol 74:27–44

van Rendsburg WCJ, Pretorius DA (1977) South Africa's strategic minerals. Valiant Publishers, Johannesburg, 156 pp

Viljoen RP (1968) The quantitative mineralogical properties of the Main Reef and Main Reef leader of the Witwatersrand System. Univ Witwatersrand Econ Geol Res Unit Inf Circ 41:43

Viljoen RP, Saager R, Viljoen MJ (1970) Some thoughts on the origin and processes responsible for the concentration of gold in the early Precambrian of southern Africa. Miner Deposita 5: 164–180

Willemse J (1948) Die Chromiet-Voorkoms op Lemoenfontein 893, Pietersburg Distrik. Geol Soc S Africa Trans 51:195–212

The Tin Deposit of Monte Valerio (Tuscany, Italy): Pneumatolytic-Hydrothermal or Sedimentary-Remobilization Processes?

I. VENERANDI-PIRRI and P. ZUFFARDI[1]

Abstract

All authors which were concerned with Monte Valerio supported pyrometasomatic or pneumatolytic-hydrothermal genetic processes, linked to the Mio-Pliocene acidic magmatic activity. The present authors, on the basis of factual observations directly achieved and/or reported in the existing literature, suggest synsedimentary tin deposition, followed by partial supergene remobilization, and assign a very limited role (if any) to the Mio-Pliocene magmatism and metamorphism on tin distribution.

1 Foreword

Monte Valerio has been the largest tin deposit of Italy. It was known since the Etruscan times, and intensely exploited especially during World War II. The original tonnage of exploited reserves (cutoff grade 0.3% Sn) can be estimated at about 1,000,000 t of crude ore, holding 4000 t of tin. At present, reserves with higher grade than 0.3% Sn are practically exhausted; large reserves with lower grade (0.1%–0.2% minimum) are very probably still available.

The most important paper (with a rich bibliography) on Monte Valerio is the one by Stella (1955). The papers by Giannini (1955) and Bertolani (1958) contain outstanding contributions to the geology and, respectively, to the petrology and ore microscopy of the Campiglia area, in which Monte Valerio is included.

Field work, for the present paper, was carried out by P. Zuffardi; laboratory studies mainly by I. Venerandi-Pirri; both are responsible for the conclusions.

2 Geologic Outline

The geology of the Monte Valerio region is sketched in Fig. 1. The following terranes occur in this area:

1 Cattedra di Giacimenti Minerari, Istituto di Mineralogia, Petrografia, Geochimica, Universita di Milano, Via Botticelli 23, 20133 Milano, Italy

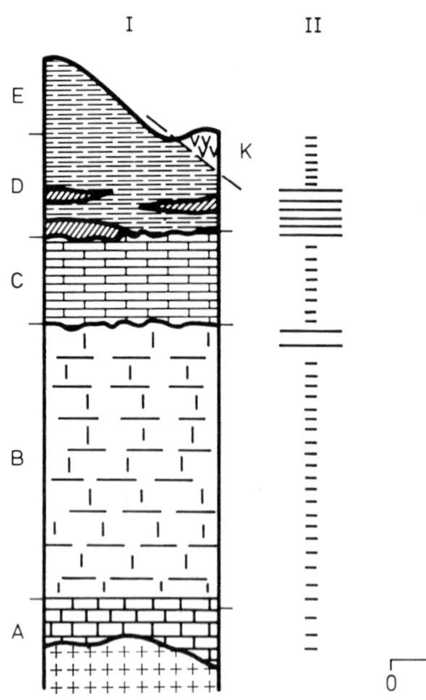

Fig. 1. The stratigraphic column of Monte Valerio *(I)* and the distribution of cassiterite along it *(II)*. See text for explanation of symbols

A. Upper Triassic gray crystalline dolomite, blackish limestones, black shales.

They are not cropping out and have been found in depth by drilling; their base is not visible; proved thickness is about 50 m.

B. Lower Liassic massive, whitish and gray limestones, turning to white massive marbles in depth and in the proximity of Tertiary granites. They are some hundred metres thick (300 or more?).

Their composition is monotonous along their whole thickness except the upper 20 m: pink, somewhat silicic limestone lenses are frequent in the latter section. The pink colour is connected to a diffused presence of hematite-goethite (often pseudomorphous after cubic pyrite) all along the beds (not only at their exposed surfaces). This fact speaks for an early, syndiagenetic, weathering of pyrite and consequent deposition of Fe-oxide.

If this is true, the pink silicic lenses may be considered as evidence of syndiagenetic oxidizing environments possibly related to paleocurrents.

The boundary between formation B and the overlying formation C has the geometric and stratigraphic characters of an emersion surface.

C. Middle (partially Upper?) Liassic thinly bedded marly limestones, disconformably overlying formation B. They are pinkish, in places silicified, with interbedded pinkish silicic lenses close to the base; they are gray and bear flint nodules in the middle and upper section.

The thickness of the whole complex is about 100 m. An evident emersion surface, partially coated with a thick quartzitic crust, separates formation C from the overlying formation D. Features interpretable as connected to (partial) emersion and – in general – to basin instability (slumping, quick heteropic changes) occur also inside the same formation C.

D. Lower Jurassic variegated, more or less silicified slates, with occasional marbles, marly limestones and ferruginous quartzitic beds: the latter are particularly frequent in the lowermost 50 m of the complex.

Their thickness may be estimated about 100 m; they unconformably[2] overlie complex C and are conformably overlain by a thick, Middle Jurassic to Paleogene, variegated slaty complex (E).

A small klippen (K), made of Upper Triassic brecciated dolomitic limestone, covers the Lower Jurassic complex in the S. Barbara region (West of Monte Valerio).

The ferruginous quartzitic lenses of Lower Jurassic deserve some attention: they are lenticular and iso-oriented in plan, with long axes nearly parallel to the bed dips.

They occur at different levels, and pass laterally either to lesser and lesser silicified beds or to silicic iron-manganese hydroxide lenticular concentrations.

Their thickness is about 1 m, the maximum thickness (6–8 m) pertaining to the quartzitic bed occurring at the base of the complex. Short apophyses, especially along their footwalls, are not uncommon.

We interpret them as "hard grounds"; this is in contrast to the existing literature, that explains them either as primary sediments or derived from metasomatic-hydrothermal processes.

3 Cassiterite Occurrences

3.1 Factual Observations on a Large Scale

a) Cassiterite noncommercial occurrences (below 0.3% Sn: most frequently between 0.01% and 0.1%) and/or positive geochemical anomalies are widespread in all the terranes A, B, C, D. It has to be pointed out that also in the Triassic complex of Gavorrano and of Monte Argentario scattered traces of Sn were recently found (Burtet Fabris and Omenetto 1975, and pers. comm. by Omenetto).

b) Cassiterite in commercial concentrations (more than 0.3% Sn) occurs either in stratabound or in cross-cutting deposits.

1. The *stratabound deposits* are located in two well-defined horizons of the stratigraphic sequence described in the foregoing section, namely:
   - the upper 10–20 m of complex B;
   - the lowest 50 m of complex D.
2. The *cross-cutting deposits* occur along faults and fractures: they are fairly frequent in complex B, where they show the typical shapes and compositions of karstic concentrations. A small, limonite-cassiterite stockwork was exploited in complex D.
   It is noteworthy that cassiterite cross-cutting industrial concentrations occur only in connection with the stratabound industrial accumulations.

c) The strata bonds in stratabound concentrations are still more evident if they are investigated at the scale of the deposit.

2 It is difficult to establish if the unconformity is stratigraphic or is tectonic, and connected to differential settling of the two complexes, that have so different plasticity

1. Sn concentrations in section b1 are located in two beds, each one having 1–2 m thickness; the first at the very top of the complex, the second 5–7 m below.
   The pay streaks, in both beds, are iso-oriented, lenticular, with long axes slightly oblique to the bed dips, and therefore they are more or less parallel to the ferruginous quartzitic lenses of complex D.
   It is noteworthy that these pay streaks coincide with the pink, somewhat silicic, limestone lenses described and discussed in point 2B.
   In places, and in very confined portions of the pay streaks (a few cubic metres) grade is very high (up to 70% Sn): miners call them "massello calcareo" (limy small masses) (Fig. 2).
2. Sn concentrations in section b2 are located inside the most silicic interbedded lenses.
   "Massello" type concentrations are present also in them; they are called "massello pietra" (stony small masses) owing to their particular hardness and compactness.

d) Mio-Pliocene granite crops out 3 km north of Monte Valerio; and was reached, at 1130 m depth, by a drill hole beneath the Sn ore deposit.

It is an aplitic, tourmaline-bearing granite in both cases. Some comagmatic porphyries occur north of Monte Valerio; Cu-, Pb-, Zn-bearing skarns are also present; they have been intensely explored and exploited.

The numerous explorations and samplings carried out close to granite and to the porphyries, and in the skarns, clearly demonstrated that there is no change in Sn content approaching them.

On the other hand, tourmaliniferous granites occur also in two other areas of Tuscany: the mining districts of Gavorrano (pyrite) and of Elba Island (complex Fe ores), but no Sn concentrations or, even, positive anomalies, are known in these areas.

One may thus assume that, from the standpoint of Sn distribution, Mio-Pliocene magmatism and allied metamorphism played no important role.

This sentence is evidently in contrast to the ancient genetic hypotheses, which attached particular importance to the presence of tourmaline in the granite and in the Sn deposit.

3.2 Factual Observations on Microscale

Observations on the microscopic scale of the stratabound concentrations seem to have particular bearing on the genetic investigations.

a) The photographs show structural/textural details of a "massello calcareo" sample; they all suggest sedimentary deposition processes.

b) Paragenesis in stratabound concentrations include essentially cassiterite, tourmaline, quartz, calcite, and minor quantities of limonite, hematite, pyrolusite, siderite, pyrite, chalcopyrite, galena, sphalerite as recognized by Bertolani (1958). Stella (1955) quotes the possible presence of acicular transparent actinolite; we have not found it and are inclined to think that, more probably, those acicular crystals were tourmaline, that occurs abundantly, with this same feature, as described also in Bertolani (1958), and it was determined also by us.

The presence of rutile (see Fig. 2D) is herewith reported for the first time.

c) The texture of "massello" is thinly bedded, with thin seams of practically pure cassiterite passing to practically pure tourmaline-quartz seams (see Fig. 2C).

Cassiterite occurs as aggregates of equidimensional well-sorted grains; their average diameter is 0.04 mm, ranging from 0.005 to 1.8 mm; the largest crystals appear to be the product of coalescence and recrystallization.

Tourmaline occurs as small, transparent, prismatic, sometimes zoned, often doubly terminated crystals; the average length of their long axes is 0.02 mm, ranging from 0.003 to 0.1 mm.

Comparing cassiterite and tourmaline grain sizes, and taking into account their specific weights (7 and 3.1 respectively), it follows that they are not "equivalent", in the ore dressing plant meaning.

4 Discussion – Genetic Hypotheses

We incline to attach great significance to the structure and texture of stratabound cassiterite deposits, and to the topographic correlation between stratabound and cross-cutting ore accumulations; the lack of correlation between Mio-Pliocene magmatism and cassiterite occurrences, inferred by us, is – in our opinion – another important argument in the genetic discussion.

On these bases, we are inclined to curtail much the role of Mio-Pliocene magmatism in the formation of cassiterite accumulations and propose the following two-stage

---

Fig. 2A–D (see pp. 62 and 63)

A Thin section of "Massello calcareo"; 1 Nicol, × 6.5 (approx.). Cassiterite is *black;* the *white thin seams/veinlets* and the smallest *white specks* are made of quartz aggregates, holding tourmaline: the latter is not visible at this enlargement. Quartz, in some seams, is evidently pseudomorphous after rhombohedral minerals (probably calcite). The *large white spots* (especially frequent in the *central section of the right side* and *near the left lower corner*) are vugs. These vugs were not produced during the preparation of the thin section (there may be some!): they were present in the rock itself, and were stained or partially filled with goethite; the cubic outline of most of them suggests the presence of an original idioblastic mineral (probably pyrite) formed during diagenesis. Banded structure is the most striking feature; other features which can be explained as produced by sedimentary diagenetic processes, are also evident; slumping *(central part of the right side)*, squeezed diagenetic fractures *(left side)* cross and convolute bedding, load casts *(upper section)*.

B A detail from the upper section of **A**. Thin section – 1 Nicol, × 25 (approx.). The different compositions and textures of the various bands is clearly visible. Convolute bedding in the upper seams is evident. Specks (made of quartz aggregate) with rhombohedral outline (pseudomorphous after calcite?) are frequent in a cassiterite rich seam in the *lower half* of the photo. A cassiterite accumulation with "scour and fill" texture occurs beneath it.

C Thin section of a detail of **A**. 1 Nicol, × 160 (approx.). The parallel texture of the rock is evident: alignments of cassiterite aggregate grains *(dark)* alternate with quartz seams, including colourless crystals of tourmaline, having strong relief and random orientation.

D Polished section of a detail of **A**. 1 Nicol, × 110. The *white spots* are rutile; cassiterite is *pale gray; dark gray* are silicate gangue minerals (quartz + tourmaline). Cassiterite seams are made of aggregates of mainly idiomorphous, slightly pleochroic crystals

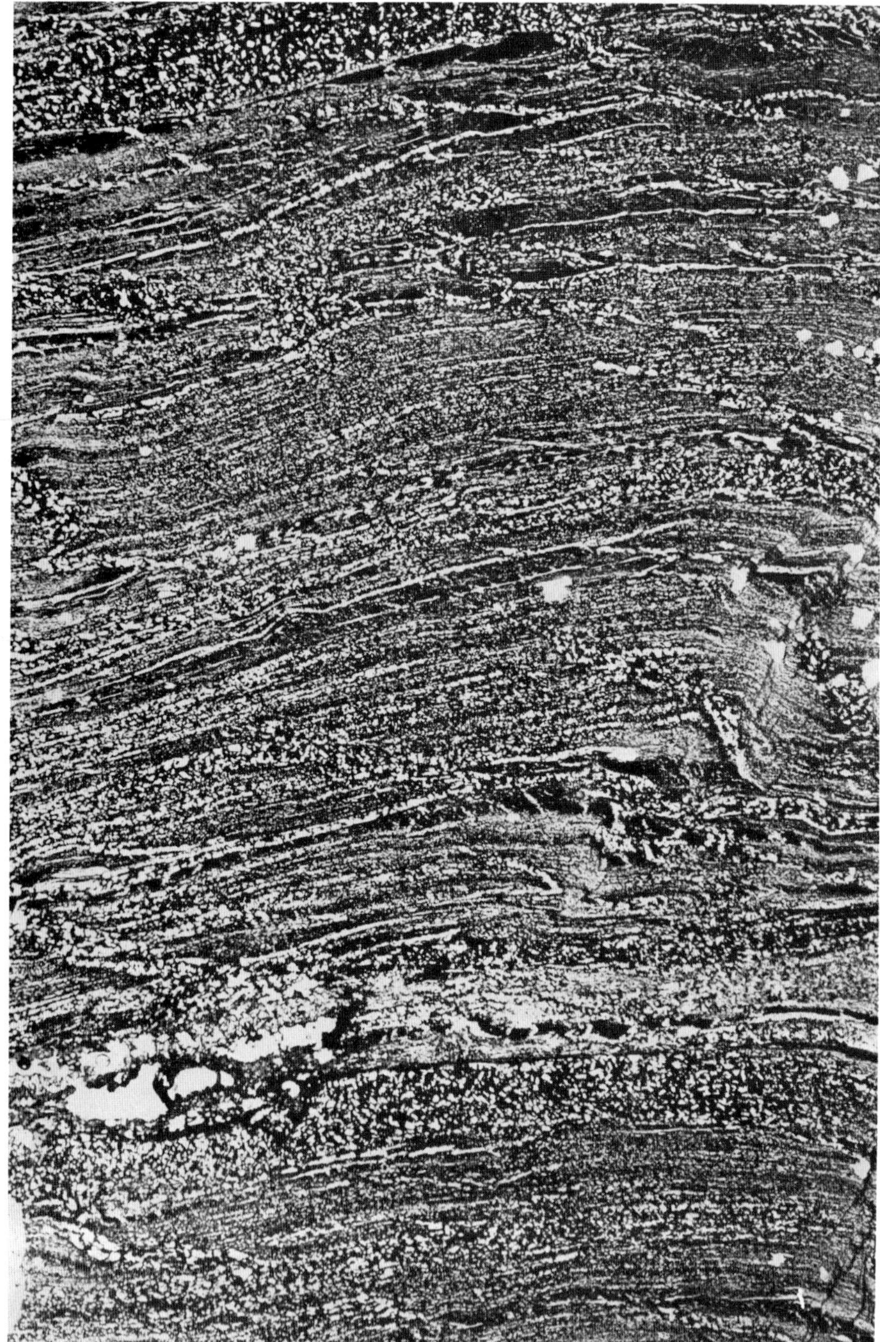

Fig. 2A (legend see p. 61)

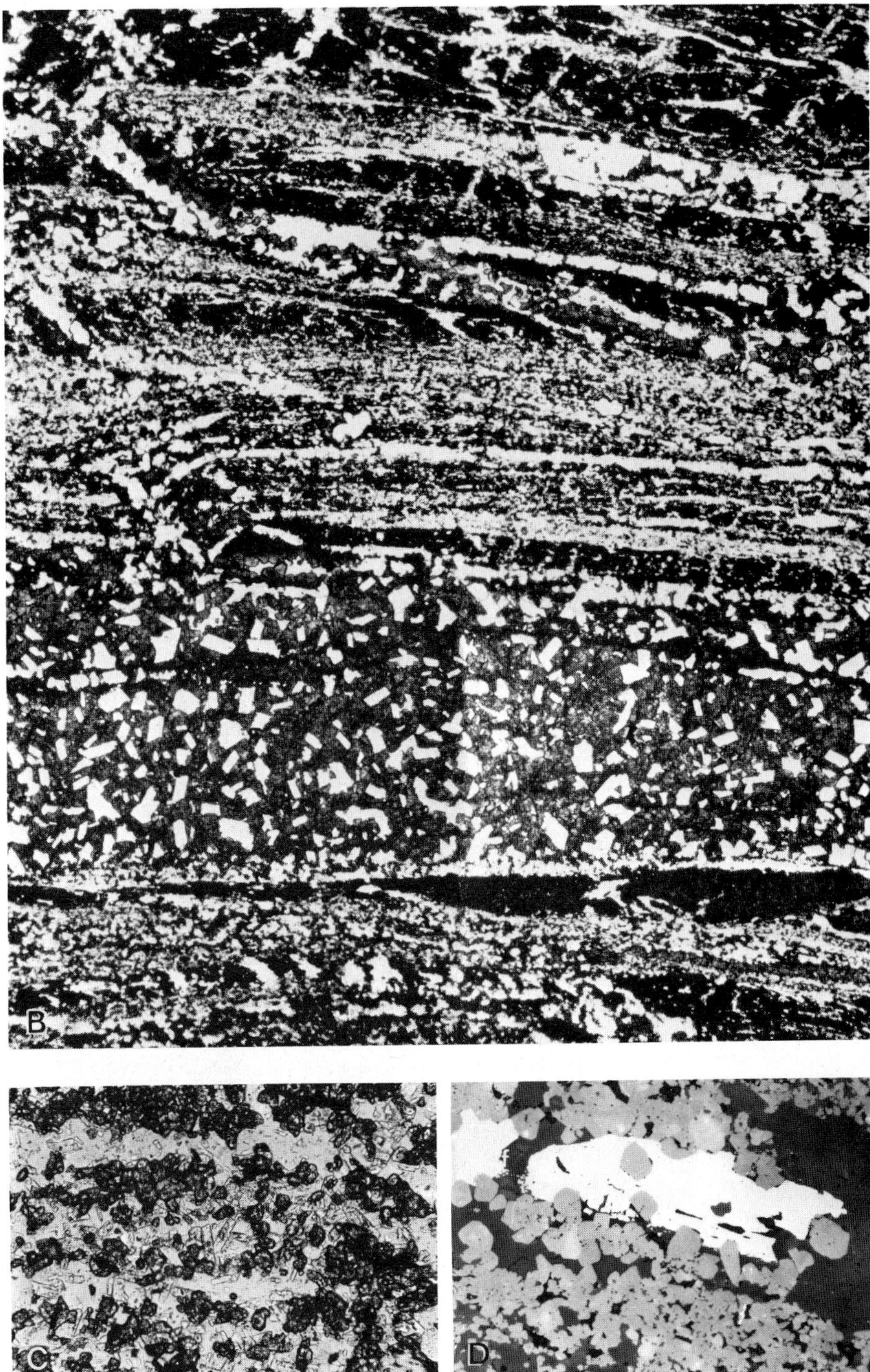

Fig. 2B–D (legend see p. 61)

genetic scheme: (1) synsedimentary deposition of cassiterite-tourmaline in the time span from Triassic to Dogger, with particular intensity in the uppermost part of Lower Liassic and in Lower Dogger; (2) Post-Pliocene supergene partial remobilization, with consequent karstic and/or "sedimentary vein" accumulations.

Recrystallization, well visible in cassiterite, may be diagenetic or may be the (only) result of the (feeble) Mio-Pliocene metamorphism affecting the Monte Valerio area.

The other factual observations listed in the foregoing section fit quite well this scheme: as a matter of fact, we explained the shapes, compositions and orientations of pay streaks in Lower Liassic and in Lower Dogger as sedimentary features (paleocurrent and, respectively, hard grounds).

Also, the thinly bedded texture of "massello" can be interpreted as a sedimentary alluvial feature, in particular, the nonequivalence of cassiterite and of tourmaline particles hindered mixed depositions and helped their separation, according to (slight) changes in current velocity, giving, thus, rise to different, practically monomineralic seams or lenses.

The same accessory presence of rutile grains is in favour of alluvial deposition processes; on the other hand, the (occasional) presence of sulphide suggests the intervention (even if scanty) of chemical metal transport and ore deposition.

The source of cassiterite and tourmaline is a vexing problem, on the basis of the data until now available. The limited roundness of cassiterite and the perfect shape of tourmaline particles suggest, in our genetic scheme, that the source should have been very close to the site of deposition.

It is difficult to go further in trying to unravel this problem. In a merely sedimentary alluvial scheme, the presence of a paleocontinent, just west of Tuscany, undergoing erosion during Mesozoic sedimentation, may be suggested, according to the paleogeographic reconstruction of central-northern Italy by Bosellini (1973), and by Hsü (1971); the Corso-Sardic massive has to be considered a residual part of such a paleocontinent.

The presence of cassiterite, even if in nonconspicuous accumulations, is well known in the Sardinian pre-Hercynian basement, and has been recently emphasized with geochemical prospecting (Marcello et al. 1977); it is not illogical to think that this same situation extends also into Corsica.

A more or less direct volcanic source could also be considered, taking into account that recently Giannelli and Puxeddu (1979), Bagnoli et al. (1978) proved the existence of volcanic activity in the Paleozoic terranes of Tuscany; the presence of hydrothermal sources, of strong hydrothermal alteration phenomena, the same presence of Mio-Pliocene granite and of conspicuous Plio-Pleistocene lava flows close to, or not far from the Monte Valerio area, could be interpreted as the recent execution of a more ancient, till now unknown, volcanism close to the area of Monte Valerio.

The boron content of Larderello hydrothermalism may be also significant, in this volcanic-sedimentary scheme.

5 Conclusion

We do not expect to have solved the genetic problem of Monte Valerio; we think, however, that the factual observations described above are sufficient to suggest to let drop the pneumatolytic-hydrothermal model and to support a synsedimentary genetic scheme, in which alluvial concentration processes of ores/minerals coming from a close source had significant importance.

This working hypothesis should be taken into consideration for future cassiterite prospecting in Tuscany.

Acknowledgments. We are indebted to Prof. M. Bertolani, of Modena University, who kindly supplied us with a sample of the, by now undiscoverable, "massello" and with some thin sections of his collection. We are also indebted to Dr. R. Crespi, G. Liborio and Prof. A. Gregnanin of Milan University for assistance in laboratory work.

References

Bagnoli G, Giannelli G, Puxeddu M, Rau A, Squarci P, Tongiorgi M (1978) The Tuscan Paleozoic: a critical review. In: Report on the Tuscan Paleozoic basement. CNR pp 9–26

Bertolani M (1958) Osservazioni sulle mineralizzazioni metallifere del Campigliese (Livorno). Period Mineral XXVII:311–352

Bosellini A (1973) Modello geodinamico e paleotettonico delle Alpi Meridionali durante il Giurassico-Cretaceo. Sue possibili applicazioni agli Appennini. In: Atti Convegno Moderne Vedute sulla Geologia dell'Appennino. Acc Naz Lincei Quaderno 183, pp 163–205

Burtet Fabris B, Omenetto P (1975) Osservazioni sulle mineralizzazioni ad ossidi e solfuri del Monte Argentario (Grosseto). Rend Soc Ital Mineral Petrogr XXX (3):963–967

Giannelli G, Puxeddu M (1979) An attempt of classifying the Tuscan Paleozoic: chemical data. Mem Soc Geol Ital XX:435–446

Giannini E (1955) Geologia dei Monti di Campiglia Marittima. Boll Soc Geol Ital LXXIV (2): 219–296

Hsü KJ (1971) Origin of the Alps and Western Mediterranean. Nature (Lond) 233:44–47

Marcello A, Pretti S, Salvadori I (1977) Le prospezioni geo-minerarie in Sardegna: la prospezione geo-chimica strategica. Not Ente Mineral Sardo V (2–3):31–59

Stella A (1955) La Miniera di Stagno di Monte Valerio e i giacimenti del Campigliese nel quadro della catena metallifera Toscana. Boll Soc Geol Ital LXXIV (2):109–218

Turbidite Barite in the Ouachita Mountains of Arkansas, U.S.A.

R.A. ZIMMERMANN [1]

Abstract

The problem of the genesis of the stratabound barite deposits in the lower part of the Stanley Shale (Late Mississippian) of Arkansas is investigated. Attention is given to the presence of barite bearing large and small slump structures, turbidites and breccias; disrupted bedded barite, later deposited as fine lenses in turbidites, are especially indicative in proving an early diagenetic age of formation.

1 Introduction

The problem of determining age relations in rocks and, therefore, the genesis of processes, is commonly undertaken by comparing one or more various methods, such as minerals in igneous rocks which crystallize in order (paragenesis) of "decreasing basicity" (Rosenbusch 1882, 1887). To the knowledge of the author, observations on turbidite[2] facies, which may be used as indicators to fix age relations, have not been described for barite.

Turbidite barite beds occur in the barite section (up to 10 m thick) in the lower part of the Stanley Shale (Late Mississippian) in the Ouachita Mountains of Arkansas (Fig. 1). Higher in the stratigraphic section, rhythmic sandstone-shale alternations have been observed and described from the Chamberlain Creek syncline deposit (Zimmermann 1964, 1976a, Fig. 2). The sandstone beds likewise form turbidites (Fig. 3). Comparisons with the literature reveal striking similarities and the occurrence of barite in turbidites indicates an early diagenetic formation.

The present paper is restricted to geometrical features and their meaning; the probable geochemical mechanism of formation of barite in sediments has recently been discussed by Hanor and Baria (1977).

---

1 Mineralogisch-Petrographisches Institut der Universität Heidelberg, Im Neuenheimer Feld 236, 6900 Heidelberg, FRG

2 A turbidite could have been formed from "detritus slumped, slid or otherwise transported into the trough, creating or perpetuating density or turbidity currents" (Glick in McKee et al. 1975, p 160) from which deposited an extensive layer with detrital grains grading upward from coarse to fine

2 Novaculite-Sandstone-Shale-Barite Breccia

At least one breccia bed lies near the upper part of the barite section in the Chamberlain Creek syncline and consists of novaculite[3] (mostly angular fragments of a wide size range, up to several cm in diameter), lumps of sandstone, and baritic sandstone and baritic shale, large subangular to subrounded pyrite fragments (up to 1 cm long) and squeezed and bent barite nodules. A bed of this material forms the top of the barite section, possesses distinct upper and lower contacts and ranges widely in thickness from 30 cm to over 1 m over short distances. The coarse novaculite fragments appear abruptly in the sandstone-shale section. About 120 m up in the Stanley Shale section, above the shale overlying the tuffite, is a thin breccia layer present in the section and contains also tuff fragments. Lumps of sandstone, shale, and barite do not appear to be rounded, lithified materials, but rather mobilized, unlithified sediments. Bent chert and novaculite fragments lie at all angles to the bedding. Barite nodules likewise are squashed. Typically, the nodules and barite lumps do not display radiating or plumose barite from the outer parts inward, but only along a peripheral band bounding the outer parts of the nodules and lumps. Some sandstone with layered parts show the bands to be curved and bent (Fig. 4a, b). The grain sizes are likewise so abrupt and out of congruence with sizes in beds above and below that some origin must account for the presence of the breccia bed different from the sandstone and shale deposits above and below. As the bed marks the upper part of the more massive barite (nodules of barite do, however, lie above this), some connection between its presence and the end of barite deposition may exist.

This breccia bed appears to be the result of a rapid influx of novaculite and sand into a quiet environment of normal barite-shale deposition. The difficulty, however, is to account for the novaculite since the fragments obviously were derived either from the Arkansas Novaculite or from the Big Fork Chert. Folk and McBride (1977) handled the problem of the breccia beds in the Caballos Novaculite (Caballos Novaculite is laterally equivalent to Arkansas Novaculite in the Llanos uplift in Texas) and came up with two possible hypotheses: (1) protonovaculite underwent subaerial exposure, fracturing and development of karst features with subsequent filling, and (2) fractures and breccias formed in unevenly lithified sediment with subsequent filling. Goldstein and Hendricks (1953) considered the siliceous deposits in Oklahoma mainly as volcanic products. Breccias and conglomerates there are attributed to having formed under shallow water conditions in the Ouachita siliceous shales.

On the basis of observations made during the years 1960–1962 (Zimmermann and Amstutz 1961; Zimmermann 1964; etc.), the most feasible mode of formation for the novaculite-sandstone-shale-barite bed appears to be a mudflow, perhaps caused by volcanic and/or seismic activity, and orogenic movements during early phases of the Ouachita orogeny, which could cause extrusion (perhaps along fractures and sedimentary dikes) from openings. The breccias would then represent mobilized fossil mud

3 Novaculite is a very dense, even-textured, light-coloured cryptocrystalline siliceous rock. The term was applied to rocks found in the lower Paleozoic rocks of the Ouachita Mountains of Arkansas (American Geological Institute, Glossary of Geology and Related Sciences, 1957, p 200)

volcano material, a mud volcano flow breccia. The top of the Arkansas Novaculite in the Chamberlain Creek syncline is marked by a boulder breccia, consisting of a blanket of brecciated chert, as well as chert pillows – up to several meters in diameter, having the appearance of pillow lavas – concretions (up to 1 m in diameter, consisting of alternating banded black shale and black chert and flattened parallel to the bedding) with finer novaculite fragments. This horizon could very likely have been tapped as a source for mud volcano and mudflow breccias. A less plausible, alternative hypothesis would be that heaped sand deposits were set in motion and the scouring action produced by some of the first released sands set up turbidity currents which eroded in places the sandstone-shale veneer (over the Arkansas Novaculite) so that the Arkansas Novaculite was exposed and already present breccia fragments were released. These novaculite-breccia mudflows moved into the barite area disturbing the diagenesis so as to produce the peculiar, squeezed barite nodules with radiating barite occurring only on the outer margins of the nodules and filling the spaces with sand and chert between the barite lumps.

Numerous examples may be cited from the literature of present day mud volcanoes, and, less so, from fossil mud volcanoes (Zimmermann and Amstutz 1972; Zimmermann 1976b) wherein the sand-mud facies is almost invariably involved. Sand often deposited rapidly over still loose mud may cause mud volcano activity. In the Hot Springs, Arkansas area, and the Chamberlain Creek syncline, the Hot Springs Sandstone and

Fig. 1. Generalized geologic map of western Arkansas and barite deposits discussed in this report, which include the Chamberlain Creek syncline, Gap Mountain and Fancy Hill. *Crosshatched area,* Cambrian, Ordovician, Silurian and Devonian; *vertically lined,* Carboniferous; *blank,* Cretaceous and Tertiary

Fig. 2. Cross-section of the Chamberlain Creek syncline. The deposits lie at the junction of the relatively thin (1700 m) lower and middle Paleozoic (alternating chert and sandstone with shale) below, and the turbidites and volcanic tuffs above (Carboniferous). The filling of the Chamberlain Creek syncline coincides with the beginning of the Ouachita orogeny. Cross-cutting features are ultra-alkaline dikes of Late Cretaceous age

Fig. 3. Turbidite sandstone-shale section. Note thinning and thickening of beds (due either to climatic or seismic cycles). Volcanic tuffite lies above the black shale band *at the top* of the figure. The barite layers occur about 60 m below this outcrop. The vertical scale of this outcrop is 15 m and it is located to the *right of centre* of Fig. 2 (at the 0) in this figure

Fig. 4a. Chert-sandstone-shale turbidite breccia; a mud flow breccia which originated through an influx of chert and novaculite fragments with sand and clay into barite depositional environment. Chert fragments: *c, cht, i, g; stippled areas:* quartz and barite; barite nodules: *d, h;* barite lumps: *a, b;* barite and quartz: *c, e.*

b Same as **a**. Chamberlain Creek syncline near top of barite section. × 1.5; oriented, top–bottom

Fig. 5. Barite turbidite made up of thin barite lenses bounded by discontinuous shale seams. Probably originally banded, alternating barite and clay. Chamberlain Creek syncline, about 10 m up in barite section: oriented, top–bottom

Fig. 6. Slumped and perhaps partly transported barite. About 6 m above Arkansas Novaculite-Stanley Shale contact (upper part of barite section, Fancy Hill deposit). × 1.5

Fig. 7. Dense, fine-grained, turbidite barite breccia; about 5 m up in barite section, Fancy Hill district. × 1.5; oriented, top–bottom

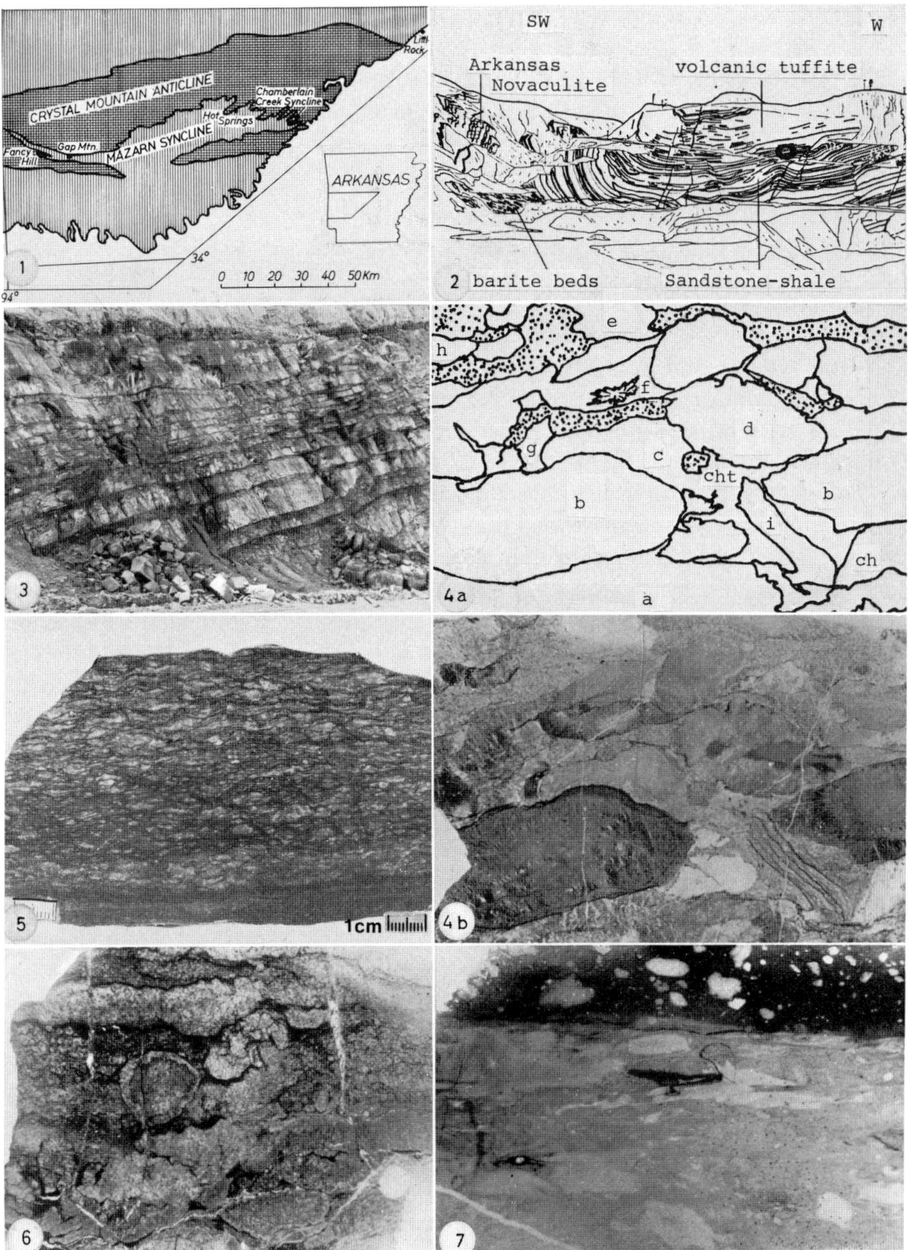
CRYSTAL MOUNTAIN ANTICLINE
MAZARN SYNCLINE
Chamberlain Creek Syncline
Hot Springs
Gap Mtn.
Fancy Hill
Little Rock
ARKANSAS
34°
94°
0 10 20 30 40 50Km
1
SW
W
Arkansas Novaculite
volcanic tuffite
2 barite beds
Sandstone-shale
3
4a
a b c d e f g h i
cht
ch
5
1cm
4b
6
7

Figs. 1–7

Magnet Sandstone, respectively, were quite probably such rapidly deposited sands; such sands overlie the bedded boulder breccia and could well serve as a rapidly deposited material along with the later silts and sands in the barite section, which also appear to have been deposited as turbidites along with the turbidite barite beds (see next section, below).

3 Turbidite (Slumped and Transported) Barite and Undisturbed Barite Bodies

The fact that evaporite beds as well as lenses and beds of carbonate are absent makes it difficult to compare the barite-shale textures with undisturbed features in evaporite and carbonate (in shale) deposits. However, comparisons may be made with the disturbed features of such lithologies (i.e. turbidites and slumps).

To some of the beds in the Chamberlain Creek syncline (Fig. 5) may be compared the gypsum breccias of Schlager and Bolz (1977, Fig. 7) and Meier (1977), interpreted as mass-flow deposits, since they show deposits of apparently only partly consolidated materials put into some type of movement contemporaneous with sedimentation, which again came to rest without the environment first undergoing uplift and erosion. The Zechstein evaporite turbidites (Schlager and Bolz 1977, Fig. 9) extend over on inferred platform and toe-of-slope and would indicate that even locally, within a deposit, turbidite features may develop over short distances of movement. In the Stanley Shale barite, the fine, irregular lenses give the fabric a wavy appearance. It is present in places throughout the section, but is more abundant in the middle and upper parts. A fabric change from layered barite to this fine, wavy lenticular type sometimes occurs when clastics become abundant in the rock; i.e. the greater abundance of clastics over barite and not a decrease in barite precipitation during deposition may result in this fabric. (Flow currents apparently affect the barite before an actual deposit of silt or silty shale is deposited.) (The layered barite gets disturbed and is put in motion and is immediately succeded by sandstone from the flow direction.) The chert breccia in the Chamberlain Creek syncline (over 1 m thick) almost coincides with the end of the barite deposition and may have had something to do with the termination of deposition. But at Gap Mountain, the breccia occurs below the deposit (barite deposited after the breccia). About one third of the barite section is of the wavy, lenticular type in the Chamberlain Creek syncline. At Fancy Hill well over one-third of the barite is lenticular, but apparently mostly undisturbed; the rest is massive and nodular and irregular nodules and lenses (slumped). The black, lensy barite appears as massive beds until one observes the texture closely.

The barite in the Chamberlain Creek syncline is made up essentially of the following main units: (1) banded barite and shale; (2) slumped beds of (1); (3) wavy bedded barite (and with shale) to fine, lenticular barite showing minor slumps to turbidite facies, and finally (4) sandstone which is always accompanied by less to no barite (as nodules) and seems to usher in conditions, or allow conditions to prevail, or accompany conditions whereby barite deposition stops. The higher one goes above the last barite showings, the more pronounced the sandstone lithology becomes in the section.

In the Fancy Hill district, shale, and then sandstone likewise are the lithologies. Fancy Hill deposits are made up mostly of alternating nodular beds (mostly about 10 cm thick, but may form beds up to 1 m thick), dense, gray barite beds (up to 60 cm thick), black lenticular beds in rhythmic fashion, probably reflecting the changing environment resulting in varying abundances of silica in lenses and between the lenses. Slumping has taken place in most of the beds. Some of the textures resemble the deformed (slumped?) laminated to thin bedded anhydrite of Kendall (1979), and one of these is illustrated in Fig. 6. The texture shown in Fig. 7, also in Fancy Hill, is the result of movement within extremely, almost quartz-free, fine-grained, dense barite which looks like novaculite.

Much of the lensy barite at Fancy Hill, however, is not slumped massive barite, since abundant silica occurs in the lenses and cement between the lenses in the barite beds with lenses. Massive barite will show slump structure, but silica is almost absent. In the beds with lenses of barite between silica cement, the particular texture exhibited is a reflection of the mineralogical contents and grain sizes. The barite follows so closely behind the Arkansas Novaculite as to appear connected in origin, and this becomes evident when one takes into consideration the thousands of meters of shale and sandstone to the next appearance of barite in the stratigraphic section (following uplift and erosion of the Ouachitas) in the basal conglomerate of Upper Cretaceous conglomerate and sandstone of the Dierks Sandstone.

4 Conclusions

Disturbed lenses and layers of barite formed through slumping (and, to some extent, turbidity currents) place the age of formation of the barite prior to the movement, which places the age of the deposits as sedimentary formations disrupted during an early diagenetic episode. The presence of turbidite facies reveal additional evidence along with other features previously described which indicate a sedimentary-diagenetic formation for the barite occurring in the Stanley Shale.

Acknowledgments. The author thanks Prof. G.C. Amstutz for reading the manuscript and for marking helpful suggestions; the author likewise thanks Mr. E. Gerike for making some photographs and Mr. H. Lämmler and Mr. O. Stadler for making some thin and polished sections.

References

Folk RL, McBride EF (1977) The Caballos Novaculite revisited. Part I: Origin of novaculite members. J Sediment Petrol 46:659–669

Goldstein A Jr, Hendricks TA (1953) Siliceous sediments of Ouachita facies in Oklahoma. Geol Soc Am Bull 64:421–442

Hanor JS, Baria LR (1977) Controls of the distribution of barite in Arkansas. In: Stone CG (ed) Symposium on the geology of the Ouachita Mountains. Arkansas Geol Comm Bull 2:42–49

Kendall AC (1970) Facies models 14. Subaqueous evaporites. In: Walker RG (ed) Facies models. Geosci Can Reprint series 1. Geol Assoc Can, Toronto, pp 159–174

McKee ED, Crosby EJ et al. (1975) Paleotectonic investigations of the Pennsylvanian system in the United States. US Geol Survey Prof Paper 853:541

Meier R (1977) Turbidite und Olisthostrome; Sedimentationsphänomene des Werra-Sulfates (Zechstein 1) am Osthang der Eichsfeld-Schwelle im Gebiet des Südharzes. Akad Wiss DDR, Zentralinst Phys Erde, Veröff 50:45

Rosenbusch H (1882) Ueber das Wesen der körnigen und porphyrischen Structur bei Massengesteinen. Neues Jahrb Mineral II:1–17

Rosenbusch H (1887) Mikroskopische Physiographie der massigen Gesteine, Bd II, 2. Aufl. Schweizerbart, Stuttgart, 877 S

Schlager W, Bolz H (1977) Clastic accumulation of sulphate evaporites in deep water. J Sediment Petrol 47:600–609

Zimmermann RA (1964) The origin of the bedded Arkansas barite deposits (with special reference to the genetic value of sedimentary features in the ore). Unpubl PhD Dissertation, Univ Missouri at Rolla,367 pp

Zimmermann RA (1976a) Rhythmicity of barite-shale and of Sr in strata-bound deposits of Arkansas. In: Wolf KH (ed) Handbook of strata-bound and stratiform ore deposits, vol III. Elsevier, Amsterdam, pp 339–353

Zimmermann RA (1976b) The Cambro-Ordovician fossil mud volcano of Decaturville, Missouri. Proc Vulcanism, vol III. Int Congr on Thermal Waters, Geothermal Energy and Vulcanism of the Mediterranean area, Athens, Oct 1976, pp 265–279

Zimmermann RA, Amstutz GC (1961) Sedimentary features in the Arkansas-Oklahoma barite district. Geol Soc Am Spec Pap (Abstr) 68:306–307

Zimmermann RA, Amstutz GC (1964) Small scale sedimentary features in the Arkansas barite district. In: Amstutz GC (ed) Sedimentology and ore genesis. Developments in sedimentology, vol II. Elsevier, Amsterdam, pp 157–163

Zimmermann RA, Amstutz GC (1972) The Decaturville sulfide breccia, a Cambro-Ordovician mud volcano. Chem Erde 31:253–274

The Permo-Triassic Paleokarst Ores of Southwest Sardinia (Iglesiente-Sulcis). An Attempt at a Reconstruction of Paleokarst Conditions

MARIA BONI[1, 2] and G.C. AMSTUTZ[2]

Abstract

Karst-type ore deposits are a major new discovery in many countries made by many mining geologists during the past decade. The present paper offers a case study of an area in southern Sardinia where karstification took place at various intervals, producing superimposed karst systems and local sulphide-barite concentrations. The sediments and processes extend from the Cambrian to the Tertiary, with a climax of karstification at the end of Hercynian orogeny, and along structural lines, both of the Hercynian and the rejuvenated directions of the Caledonian orogenesis. The paleokarst ores here described are of Permo-Triassic age.

The paper includes descriptions and genetic discussions of Ba-Pb-Zn-Ag ore bodies in karst cavities and emphasizes the textural evidence for synchronous karst-age deposition.

1 Geological Setting

Sardinia and Corsica resemble more closely the Spanish Meseta and the Montagne Noire of France than the Italian mainland. This fact has been stressed by various authors and the implications were summarized by Cocozza et al. (1974). The main geological traits of the Iglesiente-Sulcis-area (Fig. 1) consist of the abundance of non-metamorphic Paleozoic rocks (excepting the contact aureoles of the Hercynian granites) and of the sparse Mesozoic and Tertiary continental and paralic sediments. According to the present knowledge the base of the Paleozoic still presents some uncertainties as to its age, but it is quite certain that the oldest sediments are Lower- or Infracambrian.

The whole Cambrian sequence of the Lower and Middle-Cambrian consists from bottom to top of the following formations:

- "Nebida Formation" or sandstones
- "Gonnesa Formation" or ore-bearing limestones
- "Cabitza Formation" or shales

The "Gonnesa Formation" is a back-reef facies with carbonate sedimentary rocks. It contains "strata-bound" barite (Sulcis) and Pb-Zn-Fe-sulphides (Iglesiente). In addi-

1 Istituto di Geologia e Geofisica dell' Universitá di Napoli, Italy

2 Mineralogisch-Petrographisches Institut der Universität Heidelberg, 6900 Heidelberg, FRG

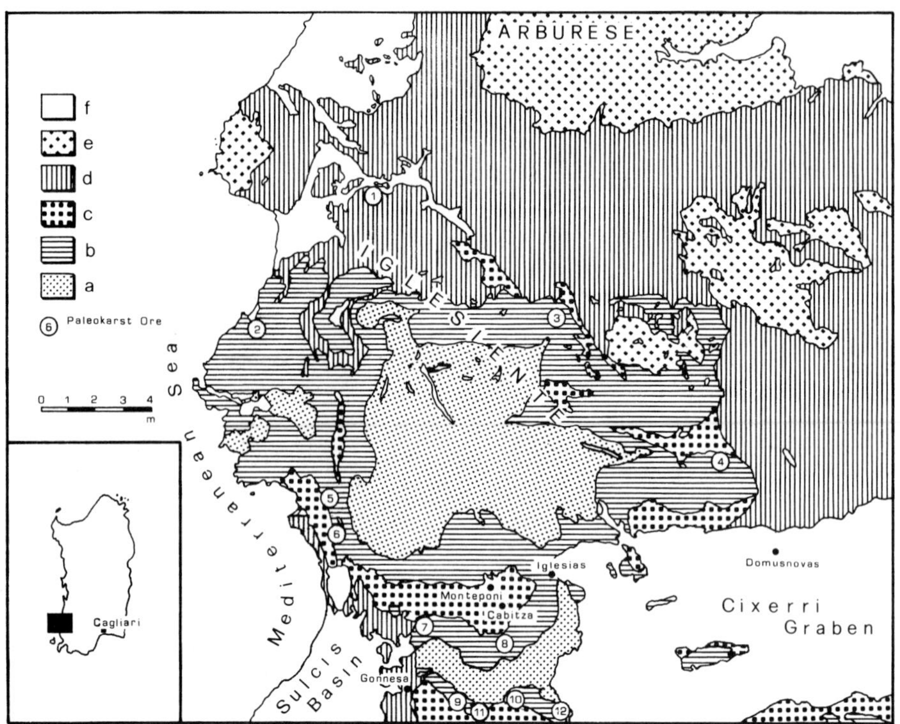

Fig. 1. Geological map of the Iglesiente area, with the most important localities cited in the paper. *a* Nebida Formation (Lower Cambrian); *b* Gonnesa Formation (Lower Cambrian); *c* Cabitza Formation (Middle Cambrian); *d* sedimentary rocks from Ordovician to Lower Carboniferous; *e* late Hercynian batholiths (stocks); *f* post-Paleozoic sedimentary rocks and Tertiary volcanic rocks. Paleokarst ore occurrences: *1* S. Lucia; *2* Buggerru; *3* M. Serrau; *4* M. Marganai (S. Giovanni vein); *5* Masua; *6* Nebida; *7* S. Giovanni; *8* S. Giorgio; *9* M. Onixeddu; *10* Barega; *11* M. Sa Bagattu; *12* M. Arcau

tion the whole formation displays a high average content of these elements. Brusca et al. (1965) recognized the syngenetic sedimentary origin of these ores. The barites are connected with evaporitic phases and a sabkha-environment, and the sulphides with an euxinic lagoonal shelf (Boni and Cocozza 1978; Cocozza and Palmerini 1973; Gandin et al. 1974). The Ordovician rocks rest unconformably on Cambrian sediments folded during the Sardic phase of the Caledonian orogenesis. These Ordovician rocks often occur with the transgressive conglomerates of the "Puddinga" type, followed by finer sediments. During the continental phase before the Ordovician transgression important ore mineral precipitations took place (sulphides, hematite, barite and fluorite), most of which formed in a superficial karst environment, in the course of enrichment processes (Benz 1964; Padalino et al. 1973). The other Paleozoic sediments which contain locally syngenetic ore minerals (Silurian) are insignificant, notwithstanding a certain contribution to the karst-ores.

Since the end of the Carboniferous all of Sardinia has remained tectonically stable. Only a few spotty areas of Permian and Triassic sediments are left over. Even the

Jurassic and Cretaceous sediments show many gaps and are still scarce. Regarding the Triassic, it resembles the germanotype facies in the N of the island, whereas in the Iglesiente-Sulcis area a transitional facies to the Alpidic facies is observed (Baud et al. 1977). In these areas too, special climatic conditions are indicated by sediments of arid, lagoonal and evaporitic sedimentation. Similar observations were made in the Campomá Series (Cocozza and Gandin 1976) where the lithological sequence with many silicification and caliche phenomena as well as pseudomorphs after sulphates indicate a Triassic age. A similar mode of formation was also assumed for the scarce conglomerates resting on the Cambrian limestone containing ore minerals. These conglomerates may thus be contemporaneous (Brusca et al. 1967). Their components consist only of Cambrian rocks in a matrix of quartz and/or barite. Lenses of barite are often seen in the same sequence. Sediments of this type, however, do not seem to occur beyond the Triassic in SW Sardinia. Also, neither the few Jurassic nor the Cretaceous outcrops display rocks which could be assigned to an evaporitic environment.

During Tertiary time the cratonic conditions of SW Sardinia continue unchanged and the island remains to a greater part still above sea level. In the Tertiary sediments too, the characteristic paleogeographic (lithological and mineralogical) features typical for the Permian and the Triassic are absent.

2 Permo-Triassic Karst Ore Deposits

The last phase of the Hercynian orogenesis was followed by a general phase of "rejuvenation" of the Paleozoic Series connected with a long lasting epeirogenic rise and an intensive erosional cycle. Due to the erosion of less resistant rocks (mainly the shales of the Cabitza Formation), the limestones and dolomites of the Gonnesa Formation were uncovered and affected by weathering.

It may be assumed that in these rocks with the abnormal metal contents of over 1000 ppm of Pb + Zn, the weathering processes, the activity of percolating groundwaters in regions of extreme climatic periods produced different and more pronounced effects than in karst systems of more moderate climatic conditions. Yet it appears more likely that the major karst formation set in as early as in Permian time because the climatic conditions to be derived from the sediments of the San Giorgio Formation reflect fluctuations from arid to wet climates (or diurnal or seasonal changes) known to preferentially cause karstifications (Boni 1979).

Over the strongly folded Paleozoic sequences paleosoils rich in iron and barite are often observed (Nuxis, Campanasissa, M. Arcau, Buggerru, Villamassargia) (Fig. 2b). A similar paleoclimatical and paleogeographical significance may probably be assigned to the iron occurrences mined some years ago and named "Ferro dei Tacchi". In East Sardinia these are found in the same position below the Mesozoic transgressive sediments (Salvadori and Zuffardi 1966; Vardabasso 1940).

A complete system of karst caves formed along the major Hercynian and the rejuvenated Caledonian tectonic directions. In addition, karstification also set in along steeply dipping contacts between lithologically different Cambrian formations. The karstic solution process was also enhanced by the elevated sulphide content of the

Fig. 2

a S. Giovanni Mine (Iglesias), Peloggio orebody. Fine-grained karst sediment between the collapse breccias. In the lower and upper parts of the sample Mn-dendrites occur.

b Barega mine (Iglesias), M. Arcau. Red, pisolitic karst sediment with barite fragments.

c Barega mine (Iglesias), Fossa is Bois. Ore bearing karst breccia. In the *centre* of the sample: dolomite fragment with a narrow galena layer visible at the rim.

Cambrian carbonates (mostly pyrite) which led to the formation of sulphuric acid. The karst formation was accompanied by a strong epigenetic dolomitization (yellow dolomite; Table 1) which is visible not only along the tectonic zones and the karstic walls, but also along ore-bearing horizons, marking them down to considerable depths. After Benz (1964) "... la dolomie peut se présenter ici (M. Serrau, near Arenas) comme une sorte de rideau qui descend dans le calcaire et qui se termine par des apophyses verticales en forme de pis de vache: une telle structure évoque bien une phénomène par descensum, suivant des micro-fractures".

Table 1. Analysis of the yellow dolomite of the S. Giovanni mine. (After Brusca and Dessau 1968)

NE Orebody (level +255)		Riccardo Orebody (level +200)
SiO_2	4.36	8.30
Fe_2O_3	1.57	2.57
$CaCO_3$	54.40	53.70
$MgCO_3$	32.36	33.58
$ZnCO_3$	0.67	trace

This type of dolomitization is rather common, also in other areas where Permo-Triassic emersion surfaces developed, in the general areas of ore mineral-bearing limestones (Bogacz et al. 1973; Lagny 1974; Richter 1974). It could have been formed by a slowed-down water percolation with occasional stillstands of flow, during which Mg^{2+} cations can accumulate considerably.

In the course of the Triassic, arid periods became more and more frequent and the precipitations less abundant but more intense. Evaporation increased (i.e., the evaporitic series of Campomá), and so did the salt concentration of circulating waters; consequently, the pH values and the solubility of SiO_2 rose. Ba^{2+} and metal cations became very mobile due to the increased chloride content of the pore solutions and the groundwater in general ("the brines"). The karst erosion is gradually stopped and converted into karst filling by internal sedimentation ("Internsedimente" of Sander 1948). Speleothemes of purely chemical and of detrital origin are formed (Fig. 2a, d–g). Fragments of collapse breccia origin may dominate over normal fine-grained internal sediments (cave or cavity sediments), which could perhaps indicate more arid climatic conditions.

d Barega mine (Iglesias), M. Onixeddu. Finely laminated silicified karst sediment with geopetal texture. Vertical scale are mm. Compare geopetal ore texture of Mississippi Valley deposits in Amstutz et al. (1964) and Amstutz and Bubenicek (1967).
e Nuraghe Paristeris (Carbonia). Finely laminated karst sediment with load casts and local sulphide accumulation *(dark layers)*.
f S. Giovanni mine (Iglesias), Grotta Pisani orebody. Fine-grained karst sediment (precipitated and detrital grains) which are deposited on detrital rosettes; in the transition zone are visible calcite crystals.
g S. Giovanni mine (Iglesias), Grotta Pisani orebody. Fragment of the karst filling material with alternating fine-grained silty and coarse-grained sandy sediments. Galena accumulations occur in the detrital layers

Between the different depositional fabrics ("Anlagerungsgefüge" of Sander 1948) the textures connected with recrystallization during diagenetic consolidation periods prevail. Calcite is present in many different types, i.e., as detrital fragments, as well as idiomorphic zoned grains of variable size, and as chemically precipitated "crystal meadows" ("Kristallrasen"). Barite is found either as tabular large and late crystals (Fig. 3b), which frequently replace carbonates, or it occurs in the matrix as thin needles which are then replaced pseudomorphically by quartz and galena.

The slow but fairly gradual rise of the water table within the karstic network marks the partial submergence of the so-called Hercynian peneplane and the sedimentation of the conglomerate series, both during a period of the Triassic which cannot be determined accurately. The formation of the covering sediments, such as the "covering sediments" of M. Sa Bagattu and the lagoonal, evaporitic Compomá Series represent the closing event of this slow process (Boni 1979).

Barite, as well as the sulphides, occurs mainly in the cement of the "collapse breccia" (Fig. 2c) of the karst cavities, whereas fluorite is encountered only rarely during this karst phase and only in a few areas, as for example within the barite bodies of the S. Lucia (Fluminimaggiore) and Mont'Ega-Conca Arrubia (Narcao) mines.

Whenever the chemical sedimentation prevails, massive sulphate accumulations may form, as for example in the case of the sparry barite of the Litopone and Scraper drifts of the Barega Mine (Tamburrini and Violo 1965). In isolated cases detrital (or broken up) barite was found, whereas no detrital sulphides are known. According to Brusca and Dessau (1968), the sulphides consist mainly of "argentiferous" galena (up to 7 kg Ag/t of Pb-concentrate) (Fig. 3d) with minor amounts of sphalerite, pyrite, covellite and chalcocite. They occur in the matrix of the breccia (Fig. 2c) and more rarely as thin synsedimentary layers within the internal sediments of the same breccia (Fig. 2d). Ag is contained in freibergite, argentite, polybasite, proustite, pyrargyrite and stephanite (Fig. 3f), as already reported by Lauzac (1965).

Several analyses carried out by Dr. V. Stähle in the Mineralogical-Petrological Institute of the University of Heidelberg show the following mean values for some galena inclusions of some specimens: Sb 20%–25%; Ag 15%–16%; Cu 25%–27%; Zn 7%–8%. These values agree with the microprobe analyses in Lauzac's works (1965):

Fahlore of the Burra vein, level +300:

S 26.7%; Sb 21.5%; Cu 40%; Zn 7%; Pb 3.1%; Hg 1%; Total 99.3%

Fig. 3

a S. Giovanni mine (Iglesias), Peloggio orebody. From the karst sediments in the lower part of the sample shown in Fig. 1a. Sparry calcite and quartz most abundant in the light layers, barite in the darker dolomite (thin section, × 10, N //).

b S. Giovanni mine (Iglesias), Grotta Pisani orebody. Partly fragmented barite layers with fluid inclusions in a siliceous, dolomitic groundmass (thin section, × 10, N //).

c Barega mine (Iglesias), M. Sa Bagattu. Idiomorphic, doubly terminated quartz crystals with carbonate inclusions in the carbonate groundmass (thin section, × 80, N //).

d S. Giovanni mine (Iglesias), Grotta grande orebody. General view of a sample of the silver-rich ores: galena *(white)*, sphalerite *(grey)*, quartz *(black)*. Cerussite and fine-grained covellite needles occur at the contact between quartz and sphalerite (pol. section, × 45, air, N //).

e S. Giovanni mine (Iglesias), Grotta Pisani orebody. Tetrahedrite *(light grey)* and galena *(white)* (pol. section, × 200, oil, N //).

f S. Giovanni mine (Iglesias), Grotta Pisani orebody. Freibergite inclusions *(grey)* in galena *(white)* (pol. section, × 900, oil, N //)

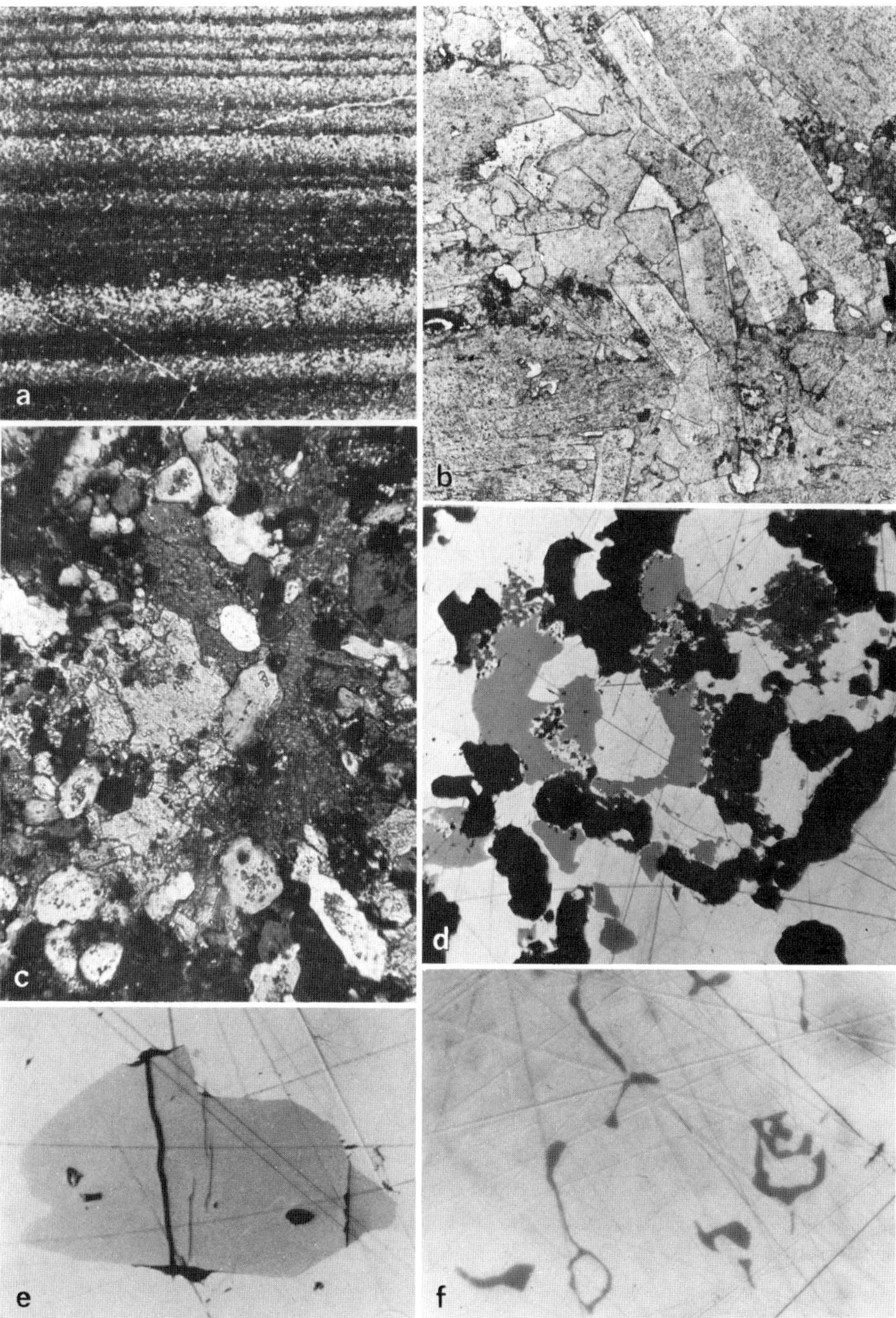

Fig. 3a–f

Calculated formula: $S_{13}\ Sb_3Cu_{10}Zn_2$

Fahlore of the Grotta grande, level +255:
S 19.5%, Sb 24.4%; Cu 29%; Zn 7.2%; Ag 17%; Pb 3.4%; Total 100.5%

Very small amounts of chalcopyrite are also present, but only rarely, whereas many outcrops display spots of copper carbonates (malachite and azurite).

Not considering the breccia fragments, the gangue minerals consist of coarse crystalline quartz and in smaller quantities of calcite. Quartz is also abundant as a microcrystalline constituent in the fine-grained internal sediments of the karst cavities (Fig. 3a) where they are found as idiomorphic bipyramidal crystals with inclusions of carbonates (Fig. 3c). This type of quartz is often linked to a so-called "schizohaline" environment of sedimentation (Folk and Siedlecka 1974; Grimm 1962; Richter 1974).

Quartz also cements the conglomerates which in places rest unconformably on the partly eroded internal karst sediments. This could also indicate the continuation of a limited evaporitic environment which is common to various processes of ore mineral formation. Silicification phenomena or quartz and chert occurrences linked to continental erosion surfaces occur quite frequently and allow the conclusion of conditions similar to those occurring during epigenetic dolomitization (Lagny and Rouvier 1976). They are known in the areas of Triassic ores of Bleiberg (Bechstädt 1975) and with many barite occurrences in karst areas, such as those of the "Hercynian basement" of Morocco (Huvelin 1966), of the old Lead Belt of Missouri, of the lead-zinc belt of Burma etc. Endlicher (1976) describes such occurrences of SiO_2 in sediments of the Permo-Triassic fluorite and barite deposits of West-Asturia, and Richter (1974) such from dolomites of the same age in the Eifel, W. Germany.

Similar features are described not only by Lagny (1974; eastern Calcareous Alps) but also by Baud (1975; Middle Triassic of the Central Prealps) and also by Courel et al. (1977) for early silification and ore mineral formation in Triassic paleokarsts along the border of the French Massif Central. Rouchy reports (1974; Triassic silification and barite contents in the Hérault): "Le dévelopment des quartz bipyramidés dans les lamelles de barytine est un phénomène constant et illustre encore l'association entre cette varieté de quartz et le milieu sulfaté". Bipyramidal idiomorphic quartz crystals were also found in many Triassic Pb-Zn-Ba-F-deposits of the Gorno area (Assereto et al. 1977), where one always finds various faces of ore mineral formation. In the Triassic of the Sierra de Gador Schwerd (1977) reports frequent microcrystalline bipyramidal quartz crystals in primary, xenomorphous, dark-pigmented fluorite.

The paragenetic position of the quartz in the karst of Sardinia during diagenesis appears to be identical with that of the "argentiferous" galena since quartz *and* galena replace the carbonates as well as the barite pseudomorphically, as seen in many polished sections. Among the sulphides the "argentiferous" galena partially replaces the rare sphalerite grains.

The "calamine ores" consist essentially of smithsonite and occur all through the Iglesiente. A minor part of them must be assigned to a karst age because one finds occasional remnants of "Ag-rich" galena and barite, as well as the common, fine, siliceous sediments with the idiomorphic quartz crystals. They may be considered, therefore, to represent at least in part original karst sediments with sphalerite which were oxidized more readily than those with only galena and barite (Boni 1979).

The period of karst formation described in this paper is not the only one in the Iglesiente-Sulcis area, but it is the most important one economically. Two other karst periods show ore mineral occurrences too: these are a Cambrian one and another associated with the Cambro-Ordovician unconformity. Both of them are under investigation in regard to southern Sardinia as an ore province as well as in regard to their paleogeographic position and their genesis.

Acknowledgments. The authors are grateful for the excellent cooperation of the mining companies, as well as to many members of the Institutes of Naples, Cagliari and Heidelberg, with whom numerous topics and methods could be discussed.

References

Amstutz GC, Bubenicek L (1967) Diagenesis in sedimentary mineral deposits. In: Larson G, Chilingar GV (eds) Diagenesis in sediments, vol 8. Elsevier, New York, pp 417–475

Amstutz GC, Ramdohr P, El Baz F, Park WC (1964) Diagenetic behaviour of sulfides. In: Amstutz GC (ed) Sedimentology and ore genesis. Elsevier, Amsterdam, pp 65–90

Assereto R, Jadoul F, Omenetto P (1977) Stratigrafia e metallogenesi del settore occidentale del distretto a Pb, Zn, fluorite e barite di Gorno (Alpi bergamasche). Riv Ital Paleontol 83 (3): 395–532

Baud A (1975) Diagenèse des sédiments carbonatés sous des conditions "hypersalines": quartzine, celestine, fluorine dans les calcaires du Trias moyen des Préalpes medianes (domaine briançonnais, Suisse occidentale). IX Congr Int Sediment Nice, Theme 7:19–24

Baud A, Megard-Galli J, Gandin A, Amadric du Chaffaut S (1977) Le Trias de Corse et de Sardaigne, tentative de corrélation avec le Trias d'Europe sud-occidentale. CR Acad Sci Paris Ser D, t 284: 155–158

Bechstädt T (1975) Lead-Zinc ores dependent on cyclic sedimentation (Wetterstein-limestone of Bleiberg-Kreuth, Carinthia, Austria). Mineral Deposita 10:234–248

Benz JP (1964) Le gisement plombo-zincifère d'Arenas. Atti del Symposium sui problemi minerari della Sardegna. Assoc Mineral Sarda, pp 331–342

Bogacz K, Dzulinski S, Haranczyk C (1973) Caves filled with clastic dolomite and galena mineralization in disaggregated dolomites. Ann Soc Geol Pol XLIII (1):54–72

Boni M (1979) Zur Paläogeographie, Mineralogie und Lagerstättenkunde der Paläokarsterze in Süd-West Sardinien (Iglesiente-Sulcis). Diss Univ Heidelberg, 260 S

Boni M, Cocozza T (1978) Depositi mineralizzati di canale di marea nella Formazione di Gonnesa del Cambrico inf della Sardegna. G Geologia X 411 (1):1–14

Brusca C, Dessau G (1968) I giacimenti piombo-zinciferi di S. Giovanni (Iglesias) nel quadro della geologia del Cambrico sardo. Industria Mineraria 19 (9, 10, 11):470–494, 533–556, 597–609

Brusca C, Dessau G, Jensen ML (1965) L'origine dei giacimenti di zinco e piombo nell'Iglesiente alla luce della composizione isotopica dello zolfo di loro solfuri. Atti del Symposium sui problemi minerari della Sardegna. Assoc Mineraria Sarda, pp 25–31

Brusca C, Pretti S, Tamburrini D (1967) Le mineralizzazioni baritose delle coperture di Monte Sa Bagattu (Iglesiente, Sardegna). Res Assoc Mineraria Sarda 72(7):89–106

Cocozza T, Gandin A (1976) Etá e significato ambientale delle facies detritico-carbonatiche dell' altopiano di Campumari (Sardegna sud-occidentale). Boll Soc Geol Ital 95 (6):1521–1540

Cocozza T, Palmerini V (1973) Ulteriori osservazioni sul "calcare ceroide" del Cambrico sardo al microscopio elettronico, in relazione alle facies coeve della Montaigne Noire e della Spagna. Boll Soc Geol Ital 92 (4):85–104

Cocozza T, Jacobacci A, Nardi R, Salvadori I (1974) Schema stratigrafico-strutturale del massiccio sardo-corso e minerogenesi della Sardegna. Mem Soc Geol Ital XIII(2):85–186

Courel L, Seddoh K, Zoungrana G (1977) Evolution du socle antémésozoique: place de la carbonatation et des minéralisations siliceuses fluorées et baritées. Cas du Charallais et du Bromais (Massif central français). Bull BRGM Sect II 4:259–264

Endlicher G (1976) Die syngenetischen Fluß- und Schwerspatmineralisationen von Arlos und Villabona in Westasturien, Nordspanien. Mineral Deposita 11:329–351

Folk RL, Siedlecka A (1974) The "schizohaline" environment: its sedimentary and genetic fabrics as exemplified by late Paleozoic rocks of Bear Island, Svalbard. Sediment Geol 11:1–15

Gandin A, Padalino G, Violo M (1974) Correlation between sedimentation environment and ore prospecting. Sedimentological and ore genesis studies of Cambrian "arenarie" and "dolomie rigate" formations (Sardinia, Italy): deposition and concentration of barite in an evaporitic environment. Rend Soc Ital Mineral Petrol XXX(1):251–303

Grimm WD (1962) Ausfällung von Kieselsäure in salinar beeinflußten Sedimenten. Z Dtsch Geol Ges 114; 3. Teil, 509–619

Huvelin P (1966) Karst mineralisés en barytine au Jebel Irhoud (Jebilet, Maroc). CR Acad Sci Paris Ser D, t 263:328–331

Lagny Ph (1974) Emersions medio-triassiques et minéralisations dans la région de Sappada (Alpes orientales italiennes). Le gisement de Salafossa: un remplissage paléokarstique plombo-zincifère. Thèse Doct Sci Nat Univ Nancy I:366

Lagny PH, Rouvier H (1976) Les gisements Pb-Zn en roches carbonatées sous inconformité: gisements paléokarstiques ou gisement dans des paléokarsts? Mem Soc Geol Fr 7:57–69

Lauzac F (1965) Il giacimento piombo-zincifero di S. Giovanni (Iglesias). Riassunto. Symposium problemi geo minerari sardi, Cagliari-Iglesias, pp 155–171

Padalino G, Pretti S, Tamburrini D, Tocco S, Uras I, Violo M, Zuffardi P (1973) Ore deposition in karst-formations with examples from Sardinia. In: Amstutz GC, Bernard AJ (eds) Ores in sediments. Springer, Berlin Heidelberg New York, pp 209–220

Richter D (1974) Entstehung und Diagenese der devonischen und permotriassischen Dolomite in der Eifel. Contrib Sediment 2:1–101

Rouchy JM (1974) Etude géologique et métallogénique de la haute vallée de l'Orb (Hérault). Relations socle-couverture. Problème des silicifications et des minéralisations barytiques. Bull Mus Nat Hist Nat 3. Ser, 14; Sci Terre 34, 93 pp

Salvadori I, Zuffardi P (1966) Indizi di mineralizzazioni metallifere nel Mesozoico sardo. Res Assoc Mineraria Sarda LXXI(6):1–29

Sander B (1948) Einführung in die Gefügekunde der geologischen Körper, Teil 1. Springer, Wien Innsbruck, 215 S

Schwerd K (1977) Triassische Karbonatgesteine und schichtgebundene Bleiglanz-Flußspat-Lagerstätten in der westlichen Sierra de Gador (betische Internzone, Provinz Almeria), Spanien. Muenstersche Forsch Geol Palaeontol 43:47–92

Tamburrini D, Violo M (1965) Il giacimento di baritina di Monte Barega – Monte Arcau (Iglesiente, Sardegna). Ric Sci 35 (II-A):814–848

Vardabasso S (1940) Pedogenesi mesozoica e giacimenti limonitici nella Sardegna orientale. Rend Sem Fac Sci Univ Cagliari X:145–151

Observations on Ore Rhythmites of the Trzebionka Mine, Upper Silesian-Cracow Region, Poland

L. FONTBOTÉ and G.C. AMSTUTZ[1]

Abstract

Geochemical analyses and microscopic observations of sphalerite rhythmites of the Trzebionka mine, in the strata-bound Pb-Zn district of Upper Silesia-Cracow, show that they are very similar to diagenetic crystallization rhythmites of other strata-bound deposits in shallow water carbonate facies. The presence of this type of texture can therefore not be used as an argument for an epigenetic origin of this district; rather, some sort of normal diagenetic evolution must have produced these textures, reflecting a clear diagenetic crystallization differentiation.

1 Introduction

In the Trzebionka mine, in the Upper Silesia-Cracow ore district, certain rhythmic ore textures occur, mainly formed by sphalerite, but involving also other sulphides, as well as carbonates and quartz. Among others, Bogacz et al. (1973) offered a description and interpreted the rhythmic ore textures as the result of "epigenetic vein deposits produced by mineralizing solutions spreading along sedimentary interfaces in indurated carbonate rocks". It is assumed by them that this process allows the "preservation of sedimentary patterns" such as bedding, cross-bedding, and slump structures. The rhythmic ore textures were considered by these authors to serve as an argument for an epigenetic origin of the Cracow-Silesia ore deposits.

The present paper presents an alternative hypothesis for the origin of the rhythmic ore textures of the Trzebionka mine. The comparisons with similar textures of many other localities and also with other rhythmic textures of Trzebionka constituted of dolomite, as also with various other diagenetic crystallization features, led to the suggestion that they originated by crystallization differentiation during diagenesis, with the development of several crystal growth generations. The rhythmic textures of Trzebionka would then be another example of diagenetic feature for certain shallow water carbonate facies with and without associated strata-bound ore deposits (Fontboté 1981; Amstutz and Fontboté, in press).

The main characteristics of the Upper Silesia-Cracow ore district are known through numerous papers and monographs (Althaus 1891; Stappenbeck 1928; Assmann 1948; Keil 1956; Ekiert 1959; Gruszczyk 1967; Pawlowska et al. 1979; Dzulynski and Sass-

1 Mineralogisch-Petrographisches Institut der Universität Heidelberg, 6900 Heidelberg, FRG

Gustkiewicz 1980, and others). They can be summarized as follows: (1) a great geometric congruence of the ore textures on all scales with the enclosing rock which is almost exclusively a relatively thin stratigraphic section (about 50 m) of the Lower Muschelkalk over a distance of more than 30 to 40 km. (2) The ores occur in carbonates, mostly dolomites, formed in tidal flat environments in proximity to the coast. (3) The Lower Muschelkalk represents, in this region, the first main transgressive phase, which overlaps evaporitic and clastic sediment of Lower Triassic and Permian age and the Hercynian basement. (4) Karst processes have affected the whole area over at least one time span during Mesozoic; but perhaps some of the karst features are penecontemporaneous with the tidal flats. The ore district of Upper Silesia-Cracow correlates with other strata-bound Pb-Zn-(Ba-F-Cu-) deposits which occur at the borders of the Triassic (and Permian) germanotype basins linked with transgressive phases in central and southern Europe (Amstutz and Fontboté, in press).

Detailed geologic, stratigraphic, and mineralogic descriptions of the Trzebionka mine can be found in Smolarska (1968), Sliwinski (1969), and Pawloska and Szuwarzynski (1979).

2 Description of Examples of Diagenetic Crystallization Rhythmites of Trzebionka

The characteristic banding of the ore rhythmites of Trzebionka is due to the repetition of three geometric elements which correspond to subsequent crystallization generations. Table 1 represents schematically the main features of each crystallization generation.

In example "A" (Fig. 1a–d), generation I is a medium crystalline aggregate of sphalerite with numerous inclusions of dolomite (ϕ 100–200 μ) and sometimes of roundish patches of authigenic quartz. Up to 50% of this first generation can consist of dolomite. Generation II consists of sphalerite without inclusions. Two subgenerations can be distinguished, IIa consisting of schalenblende and IIb of subhedral sphalerite. Generation III is represented in this example by the empty central spaces.

A microprobe profile made in this example "A" (Table 2) reveals that the Fe-content of the sphalerite is extremely low. On the contrary, Cd reaches values around 1% in the sphalerite of generation I and in the schalenblende of subgeneration IIa. The subhedral sphalerite of generation IIb is almost pure ZnS.

Fig. 1a–f. Different aspects of a diagenetic crystallization rhythmite of example "A": it consists almost only of sphalerite; **a** is a handspecimen view; **b** is a thin section (N / /), **c** is a polished section (N / /) view; **d** shows the same polished section with half-crossed nicols to bring out the zoning of generation II (IIa and IIb). The composition of generations, I, IIa, IIb, and III is given in the text. The *black line* indicates the microprobe profile of Table 2. **e** Handspecimen of dolomite of the same ore-bearing horizon showing a very similar texture; **f** is a thin section (N / /) of the same sample. It can be recognized that the *white bands* in the handspecimen correspond to medium to coarsely crystalline subhedral dolomite of generation II. The *dark parts* are constituted of finely to medium crystalline anhedral dolomite of generation I

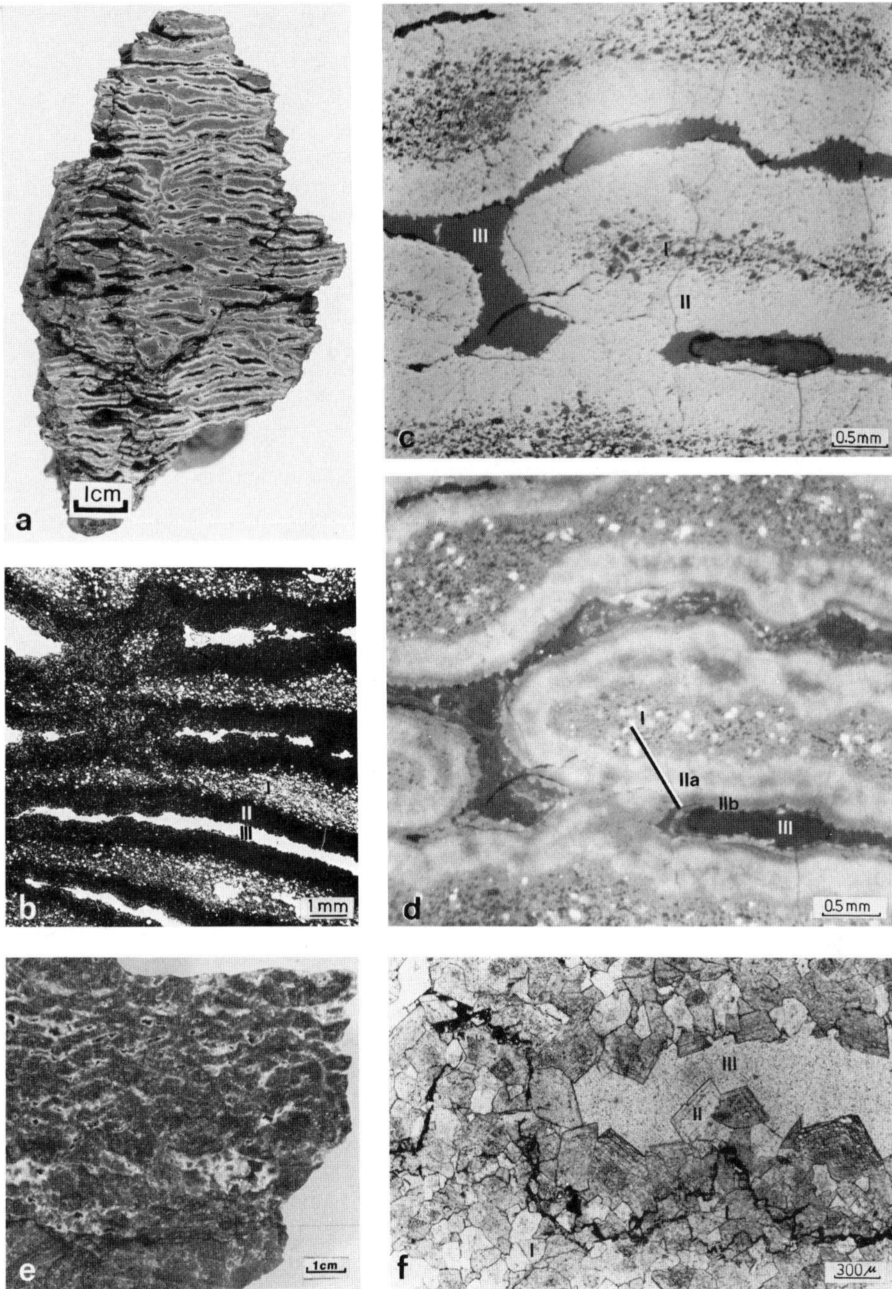
a
1cm
b
I
II
III
1 mm
c
III
I
II
0.5mm
d
I
IIa
IIb
III
0.5 mm
e
1cm
f
III
II
I
I
300 μ

Fig. 1a–f

Table 1. Main features in crystallization generations of diagenetic rhythmites

Generation I ("starting sheet", often with several "subgenerations", i.e., several phases)	Fine to medium-grained Many disseminated inclusions in the grains ("impure grains", therefore dark colors; many of these inclusions are organic in origin) Primary depositional features are often recognizable Dark colors
Generation II (often with several phases or "subgenerations")	Medium to coarse or very coarse subhedral crystals, arranged in a bipolar pattern, i.e., growing above and below the "starting sheets" The limit between generation I and II is gradational in detail Geopetal features are frequent in that the generation II is often better developed above than below the "starting sheets" In contrast to generation I, generation II contains hardly any inclusions (therefore usually light colored) Zonal crystal growth often observed Perfect analogy to the subgenerations of drusy cements Generation II or subgenerations IIa, IIb, etc., respectively, are monomineralic
Generation III (often missing)	The remaining central space or its xenomorphous filling Usually light colors, i.e., no inclusions in crystals The limit between generation II and III is sharp in detail

Table 2. Microprobe analyses of a profile through the sphalerite of generations I, IIa and IIb in one sample of example "A". (For location see Fig. 1d)

	I		IIa		IIb	
Zn	64.5	66.6	65.8	65.2	67.5	67.6
S	32.3	31.2	32.00	31.6	32.1	31.7
Cd	1.04	0.93	1.12	1.44	tr	tr
Fe	0.23	0.12	0.06	0.09	tr	tr
Total	98.2	98.8	99.0	99.3	99.6	99.3

The generations I and II of example "B" (Fig. 2) are very similar to those of example "A". Generation I is also formed mainly by sphalerite with inclusions of dolomite and radial-fibrous, authigenic quartz; this late mineral is, in this example, more abundant than dolomite. Generation II displays three subgenerations. Subgeneration IIa is schalenblende, IIb is formed by elongate sphalerite rods with subhedral terminations into IIc or III. Subgeneration IIc, consisting of pyrite and marcasite, occurs only in certain parts and always over subgeneration IIb (Fig. 2d–f), and, therefore shows a geopetal arrangement. Generation III is composed of calcite. At the bottom of generation III minor inclusions of organic matter and minor amounts of sulphides form a characteristic dark band (Fig. 2d). Some voids are distinguished between the top of generation III and subgeneration IIb.

The fact that these voids are not located in the center of generation III constitutes an exception to all other textures of diagenetic crystallization rhythmites known so far.

Microprobe profiles through one sample of the example "B" (Table 3) show also differences if compared with other diagenetic crystallization rhythmites. The Fe content decreases from generation I to the bottoms of generation IIb, which is probably a normal tendency in diagenetic crystallization. A reversal of this tendency is the increase in the uppermost ends of subgeneration IIb, so that this latest part of the subgeneration of sphalerite is richer in Fe than I and IIa and also the beginning of IIb. The Cd values are very low in all the sphalerite generations of this example "B".

The same measurements were made in generations IIa, and IIb of the bottom layers of a juxtapositional rhythm. The lowermost ends of generation IIb contain clearly less Fe than generation IIa. This fact agrees with the scarcity of pyrite and marcasite layers of generation IIc at the lower part of generation IIb. Only a few isolated pyrite grains were observed.

Table 3. Mean values of two parallel microprobe profiles through sphalerite grains of generation I, IIa, and IIb in example "B" of both, the top (T) layers and the bottom (B) layers of two rhythms in juxtaposition (see Fig. 2c). The sequences of numbering from 1 to 6, and from 7 to 9, respectively, go from "inside out", i.e., from earlier to later generations

	I	IIa (T)			IIb (T)	
	1	2	3	4	5	6
Zn	67.0	66.5	66.2	65.7	65.6	64.7
S	32.2	32.5	32.4	32.5	32.7	32.5
Fe	0.73	0.22	0.47	1.10	1.87	2.25
Cd	tr	tr	tr	tr	tr	tr
Total	99.9	99.2	99.1	99.1	100.2	99.4

	IIa (B)		IIb (B)
	7	8	9
Zn	65.5	65.8	66.2
S	32.6	32.7	32.8
Fe	1.04	0.71	0.36
Cd	tr	tr	tr
Total	99.1	99.2	99.4

From a geometric point of view, some dolomites of the same ore-bearing horizon show a very similar texture (Fig. 1e, f). In this case the dark/light banding is essentially due to changes in the crystallinity grade and not to changes in the mineralogical composition; the dark parts (or generation I) are formed by a fine-grained aggregate of dolomite with some inclusions of sulphides and of organic matter. The light parts (generation II) are clean, subhedral dolomite with polar arrays. Lateral transitions from these dolomites (rhythmitic, mottled to massive) to sphalerite rhythmites of the same type to which example "A" belongs, are observed.

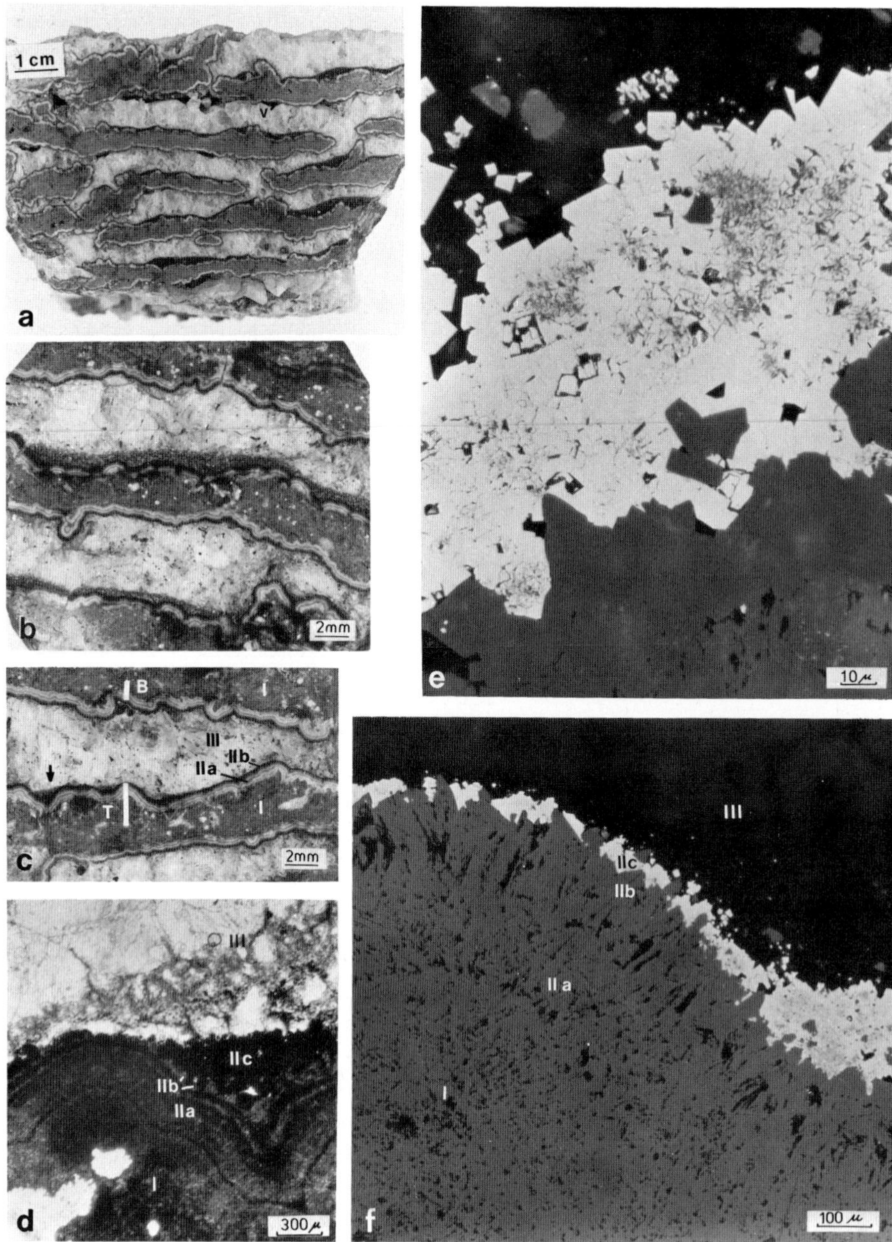

Fig. 2a–f. Different scales of observation of a diagenetic crystallization rhythmite of example "B". **a** is a megascopic view; the *dark bands* correspond to sphalerite of generations I and II (description above and in the text); the *light bands* are calcite of generation III; at the top of the bands some voids *(v)* are recognized. **b** is a polished slab (oblique illumination) showing another cut of the same specimen with more detail, especially in generation II. At the bottom of generation III (calcite), a grey layer formed by inclusions of organic matter and minor amounts of pyrite in the calcite is seen. **c** shows the location of the microprobe profiles of Table 3 (polished slab, oblique illumination). **d** is a thin section (N/ /) of the area marked in **c** with an arrow; generation I consists of

3 Similar Diagenetic Crystallization Rhythmites in Other Localities

Diagenetic crystallization rhythmites as mentioned above are common textures in certain shallow water deposits and also in karst sediments, which, in fact, are also one type of shallow water deposit. A first description of rhythmic diagenetic crystallization was given by Amstutz and Park (1967) and Amstutz and Bubenicek (1967). Levin and Amstutz (1976) offered a geometric classification on an outcrop scale. A review of many more occurrences of diagenetic crystallization rhythmites is contained in Amstutz and Fontboté (in press) and in Fontboté (1981).

Schulz (1978) describes sphalerite rhythmites of Bleiberg, Austria, which are very similar to those of example "A". Ore rhythmites, the texture of which is very similar to that of example "B", are, to mention only two, the barite-silica-galena rhythmites of Sardinia, Italy (Padalino et al. 1973), and sphalerite-barite rhythmites in internal karst sediments in Les Malines, France (Connan and Orgeval 1977).

Dolomites with several diagenetic crystallization generations which often develop rhythmic textures and which under the microscope are completely identical to the dolomites shown in Fig. 1e, f, appear frequently in certain very shallow water facies, often with developments of algal mats.

Descriptions of examples of dolomitic diagenetic crystallization rhythmites of the Triassic of the Alpujarrides in the Betic Cordillera, Southern Spain, and of the strata-bound Zn-Pb-deposit of San Vicente, in the Upper Triassic and Liassic of the Pucará basin, in eastern central Peru, are contained in Fontboté and Amstutz (1980) and in Fontboté et al. (1981) respectively.

4 Discussion

Bogacz et al. (1973, p. 287) recognize also three geometric elements in the sphalerite rhythmites of Trzebionka. They state, "... it appears that the ores under consideration are comprised of three basic constituents: (1) solutional voids, (2) crustified ores, and (3) sphalerite aggregate replacing the host dolomite." According to our description, their "constituent 1" would correspond to generation III, their 2, to generation II, and their 3, to generation I. After Bogacz et al. these ore textures would have formed as "... a record of three processes: (1) cavity making, (2) cavity filling, and (3) metasomatic replacement."

sphalerite *(dark)* and inclusions of roundish patches of authigenic quartz *(light)*; the colloform texture of the schalenblende of subgeneration IIa can be well recognized; on the subhedral sphalerite of generation IIb rests a pyrite and marcasite "layer" which is subgeneration IIc; e and f are polished sections (N//, oil imm.) of equivalent areas. The subhedral sphalerite rods of subgeneration IIb can be distinguished. The pyrite and marcasite layer IIc is formed of single crystals which are not completely cemented.

It must be pointed out that layer IIc occurs only over the sphalerite rhythms. It is therefore a geopetal feature. Probably calcite is exceptionally geopetal also, showing crystal growth pointing upwards only. In addition, the empty spaces occur only at the top, as pointed out above, in the legend to a

Our microscopic observations and geochemical analyses do not support their hypothesis, but show that the crystallization sequence is I–II–III and not the reverse. This is especially evident in the subhedral and polar growth of generation II and in the geopetal features. The lack of evidence of replacement must also be pointed out.

The presence of this type of ore rhythmite cannot serve as an argument for an epigenetic origin of the strata-bound Pb-Zn deposits in Upper Silesia; on the contrary, this type occurs commonly in strata-bound deposits in very shallow carbonate facies and also in karst deposits (Amstutz and Fontboté, in press; Fontboté 1981). In all cases, these textures can be explained by a process of fractional crystallization during diagenesis with development of distinguishable crystallization generations. The identity of the paragenetic sequence in diagenetic crystallization rhythmites and in other common diagenetic textural patterns in shallow water facies, such as pore spaces, in the cementation of spaces between ooliths, and in the cement of diagenetic breccias, is a strong argument for an original diagenetic crystallization sequence.

Acknowledgments. The authors wish to thank Professor Gruszczyk, Professor Banaś, Dr. Maria Sass-Gustkiewicz, Dr. Sliwinsky, and Dr. Szuwarzynski of Cracow and Trzebionka for their generous help in the field, and the mines and the discussions leading to this short communication. Of course, the responsibility for opinions voiced in this paper rests with us, the authors of this paper. The microprobe analyses have been carried out by Dr. V. Stähle. We are grateful for the financial support of the Deutsche Forschungsgemeinschaft and the Deutscher Akademischer Austauschdienst. This work is part of our contribution to the IGPC project no. 6.

References

Althaus R (1891) Die Erzformation des Muschelkalkes in Oberschlesien. Jahrb Preuss Geol Landesanst 12:37–98

Amstutz GC, Bubenicek L (1967) Diagenesis in sedimentary mineral deposits. In: Larsen G, Chilingar GV (eds) Diagenesis in sediments. Elsevier, Amsterdam, pp 417–475

Amstutz GC, Fontboté L (in press) Diagenetic crystallization rhythmites. (A world wide textural pattern in certain shallow water facies, ore-bearing and not ore-bearing)

Amstutz GC, Fontboté L (in press) Correlative observations on the lead-zinc-(barite-fluorite-) occurrences in the "germanotype Triassic" of Central Europe

Amstutz GC, Park WC (1967) Stylolites of diagenetic age and their role in the interpretation of the southern Illinois fluorspar deposits. Miner Deposita 2:44–53

Assmann P (1948) Zur Frage der Entstehung der oberschlesisch-polnischen Blei-Zinkerzlagerstätten. Z Dtsch Geol Ges 98:30–69

Bogacz K, Dzulynski S, Haranczyk C, Sobczynski P (1973) Sphalerite ores reflecting the pattern of primary stratification in the Triassic of the Cracow-Silesian region. Ann Soc Geol Pol 43: 285–300

Connan Jq, Orgeval JJq (1977) Un exemple d'application de la géochemie organique en metallogénie: La mine des Malines (Gard, France). Bull Cent Rech Explor Prod Elf-Aquitaine 1:59–105

Dzulynski S, Sass-Gustkiewicz M (1980) Dominant ore forming processes in the Cracow-Silesian and Eastern-Alpine Zinc-Lead deposits. Proc 5th IAGOD Symp. Schweizerbart, Stuttgart, pp 415–429

Ekiert F (1959) Neue Anschauungen über die Bildung von triassischen Blei-Zinklagerstätten in Oberschlesien. Z Angew Geol Jg 1959:385–392

Fontboté L (1981) Strata-bound Zn-Pb-F-Ba-deposits in carbonate rocks: New aspects of paleogeographic location, facies factors and diagenetic evolution. (Illustrated with a comparison of occurrences from the Triassic of Southern Spain, the Triassic/Liassic of East-Central Perú and other localities). Thesis, Univ Heidelberg, 192 pp

Fontboté L, Amstutz GC (1980) New observations on diagenetic crystallization rhythmites in the carbonate facies of the Triassic of the Alpujarrides (Betic Cordillera, Southern Spain). Comparison with other diagenetic rhythmites. I Symp Diagenesis, Barcelona 1980. Rev Inst Inv Geol Dip Barcelona, 34 pp

Fontboté L, Amstutz GC, Samaniego A (1981) Zur faziellen Stellung und zum diagenetischen Kristallisationsprozeß von Erzmineralien in schichtgebundenen Zn-Pb-Lagerstätten (am Beispiel von San Vicente, im zentralen Ostperu). Proc 7 Geowiss Lateinamerika Koll, Heidelberg 1980. Zbl Geol Paläont Teil I, Stuttgart, Jg 1981:465–477

Gruszczyk H (1967) The genesis of the Silesian-Cracow deposits of lead-zinc ores. Econ Geol Mem 3:169–177

Keil K (1956) Die Genesis der Blei-Zink-Lagerstätten von Oberschlesien (Gorny Slask, Polen). Beih Z Geologie 15:63

Levin P, Amstutz GC (1976) Kristallisation und Bewegung in Erzrhythmiten am Beispiel triassisch-jurassischer Lagerstätten in Ostperu. Muenstersche Forsch Geol Palaeontol 38/39:111–128

Padalino G, Pretti S, Tamburrini D, Tocco S, Uras I, Violo M, Zuffardi P (1973) Ore deposition in karst formations with examples from Sardinia. In: Amstutz GC, Bernard AJ (eds) Ores in sediments. Springer, Berlin Heidelberg New York, pp 209–220

Pawlowska J, Szuwarzynski M (1979) Sedimentary and diagenetic processes in the Zn-Pb host rocks of Trzebionka. Pr Inst Geol Warzawa 95:13–49

Pawlowska J, Chidester AH, Wedow H (1979) Research on the genesis of zinc-lead deposits of Upper Silesia, Poland. Pr Inst Geol Warzawa XCV:151

Schulz O (1978) Kolloforme ZnS-Lagengefüge und ihre Genese in Karbonatgesteinen. Oesterr Akad Wiss, Schriftr Erdwiss Komiss 3:159–168

Sliwinski St (1969) The development of the ore-bearing dolomites in the Cracow-Silesian area (in Polish). Pr Geol Kom Nauk Geol PAN Oddz w Krakowie 57:123

Smolarska I (1968) Textural types of zinc-lead ores in the Trzebionka region. Inst Geol Biol 237: 71–84

Stappenbeck R (1928) Ausbildung und Ursprung der oberschlesischen Blei-Zinkerzlagerstätten. Arch Lagerstaettenforsch 41:143

The Genesis of the Zinc-Lead Ore Deposits of Upper Silesia, Poland

H. GRUSZCZYK[1]

Abstract

In his considerations regarding the genesis of Upper Silesian zinc-lead ore deposits, the author refers to the known facts and the observed regularities in their form and occurrence, presented earlier in a paper published in the Economic Geology Monograph 3, 1967. In this paper, objective and subjective factors affecting the hypothesis of genesis of the deposits are discussed. The hypothesis of sedimentary origin is further substantiated, and the secondary processes which distorted or obliterated the primary structure of the deposits are taken into consideration. From the evidence accumulated to date it appears that at least two deposit-forming stages can be distinguished:

stage I – in which disseminated ores formed, synchronously with dolomites,
stage II – in which further concentration of primary ores and secondary alteration of the primary dolomites took place.

The present picture of the Upper Silesian Zn-Pb ore deposits is the result of several overlapping processes. The deposits can be regarded as polygenetic because both syngenetic and epigenetic processes played a significant role in their formation.

The present paper refers to the hypotheses on the genesis of Upper Silesian zinc-lead ore deposits published in Economic Geology Monograph 3, 1967 as a contribution to the discussion concerning stratiform deposits of lead-zinc, barite and fluorite ores.

The widening scope of research, the use of better methods and consequently, the growing body of knowledge about these deposits have made this evolution inevitable. On the other hand, the information regarding the geotectonic position of the deposits, their stratigraphy, lithology, form and structure, mineralogical and chemical composition, textural and structural habit, and the regularities of their regional distribution is still valid (Econ. Geol. Monograph 3, 1967).

Hypotheses on the genesis of deposits are advanced basing on the existing state of knowledge and specifically, on the data accumulated during the investigations and regarding the regularities in the form and distribution of the deposits. Our knowledge of the zinc-lead ore deposits of Upper Silesia has recently been greatly extended. A great number of data have been accumulated, both of a general nature and on the deposits themselves. Moreover, the scope of geological observation has been widened due to the discovery, surveying, and opening out of the deposits near Chrzanow, Olkusz and Zawiercie. Several new methods have been used to carry out investigations.

1 University of Mining and Metallurgy, Cracow, Poland

As a consequence of this steady inflow of information, several new questions emerge, which complicate the solution of the problem, difficult as it is, of the genesis of the deposits. There are many factors, subjective and objective, which make this problem extremely difficult to solve. The objective factors are, for example, the polygenetic character of geological phenomena, the ambiguity of known facts, and the limited scope of geological observation. The investigator's subjectivity, the fragmentary and unrepresentative character of investigations, the research methods used, the simplification of problems, the overrating of some data and interpretation of disputable features, the reliance on presumed elements, the traditional approach, etc. are some of the subjective factors.

The polygenetic character of geological phenomena is responsible for the fact that the objects of investigations are geological formations which usually owe their origin to a number of successively overlapping processes. Secondary processes, sometimes of minor importance, frequently obscure the primary deposit-forming and alteration processes. The ambiguity of several facts allow various interpretations, which leads in consequence to different hypotheses concerning the genesis of a given deposit. While the objective factors are beyond the investigator's control and generally affect in the same degree the formulation of a hypothesis, the subjective factors depend to a greater extent on his subjective feelings.

Not unimportant for genetic considerations is the scope of investigations and their representativeness in relation to the deposit. This applies particularly to earlier studies, which were confined to certain elements of the deposits, for example, crusty and brecciated ores were investigated and other varieties neglected, or only ores and no barren host rocks were taken into account. Commonly geochemical studies were carried out on a fragmentary collection of samples.

The adopted methods of investigations are of prime importance for the study of deposits, the conditions of their occurrence and the governing regularities. It is imperative that these methods should have a complex character. Equal importance should be attached to more general regional and detailed local investigations, and to macro- and microscopic, chemical and microchemical features.

Over the past decades several hypotheses of sedimentary origin of Upper Silesian zinc-lead ore deposits have been advanced (Gruszczyk 1956, 1967; Smolarska et al. 1972; Smolarska 1974). In view of the growing body of knowledge about the deposits in question, some modifications and extensions to these hypotheses have been inevitable. The most complete theory has been put forward by Smolarska (Smolarska et al. 1972; Smolarska 1974).

The present picture of the zinc and lead ore deposits of Upper Silesia is distorted by secondary processes, which obscure or sometimes even obliterate their primary structure. These are paleokarstic phenomena and the attendant clay accumulations, the rise of brecciated deposits, the filling of fissures, local metasomatic phenomena, changes produced by present-day weathering and karstic processes, etc.

The hypothesis of the sedimentary origin of Upper Silesian zinc-lead deposits rests on the following facts and regularities:

The mineralization of the Triassic rocks has a regional character not only in a Polish but also in European scale.

The deposits and ore seams are associated with the platform cover and subordinately, with the underlying Devonian folded elements of the Cracow branch of the Variscan orogen. They have no connection with magmatic rocks or regional lineaments. Mineralization is confined principally to Muschelkalk and Rhaetian rocks, but recently reports of ore occurrences in the Buntsandstein have begun to appear. The Triassic mineralization is subordinated to stratigraphic factors. The deposits in different areas occur in the defined stratigraphic position, which enables their intercorrelation.

The Triassic zinc-lead ore deposits are as a rule associated with dolomites, particularly with those occurring in the forefield of the stratigraphically corresponding limestones. The latter are not mineralized. It appears, therefore, that the deposits show subordination to lithologic factors.

Ores are related to the rocks that formed in the peripheral parts of the Triassic sedimentary basin and do not occur in zones distant from the former seashore.

The deposits and mineralized zones generally have the form of flat beds or nests, usually discordant with the bedding of the enclosing dolomites. The internal structure of the deposits is more complex due to the superimposed later processes which were responsible for additional local concentration or dissemination of the ore minerals.

A feature deserving note is the multibedded structure of the deposits. This feature is less pronounced in the western areas and fairly distinct in the southern and eastern parts of Upper Silesia. Generally the beds are horizontal, sometimes dislocated with respect to one another, frequently deformed by karstic processes.

The mineralogical composition of the primary zinc-lead ore deposits is simple and stable throughout the district. The principal ore mineralogy is simple Zn, Pb and Fe sulphides with subordinate barite. They appear in varying quantitative ratios, forming macrozones with the prevalence of galena or sphalerite, or containing barite besides galena. From the genetic point of view, it is interesting to note that a certain amount of zinc, lead and iron is combined in dolomite.

The structural and textural habit of ores is constant in the whole area both for the Triassic and Devonian ores. Worth noting is the presence of a variety of structures in the deposits: dispersed, bedded, laminated, porous, pseudo-brecciated, brecciated, crusty, botryoidal, oolitic, sinteritic, concretionary, veined, drusy. Some of them, for example, dispersed, bedded, laminated or concretionary, are indicative of the primary sedimentary origin of the deposits, whilst the others owe their origin to secondary processes, presumably superimposed on the primary deposits.

The assemblage of trace elements present in the ores is quantitatively poor and heterogeneous.

The isotopic composition of lead from the Triassic and Devonian rocks is similar. This feature can be accounted for by infiltration of Triassic ores down to the Devonian rocks.

The isotopic composition of sulphur occurring in sulphides of the zinc-lead deposits is characterized by a high content of the heavy sulphur isotope with the values of δ^{34} S up to 67.1%. This indicates that sulphate-reducing anaerobic sulphur bacteria participated in the formation of the deposits. However, a considerable scatter of results of the determination of sulphur abundance ratios in these deposits suggests that mineralization has been a complex or multistage process.

The Sr content in barite from the Upper Silesian deposits is similar to the concentrations of this element reported from the sedimentary and infiltration deposits of Central Europe.

From the above facts and regularities it appears that the deposits in question show signs of being subordinated to regional, geotectonic, stratigraphic, lithologic and paleogeographic factors. They also exhibit a great many chemical, textural and morphological features which point to their sedimentary origin. All the known tectonic dislocations crosscutting the Mesozoic platform cover are posterior to the zinc-lead deposits. The distribution of deposits is independent of and different from the tectonic directions, the tectonic features being also younger than the primary dolomitization process.

Basing on the analysis of known facts, and particularly on the studies of ore structures and the time relation between the rise of dislocations and Zn-Pb deposits, at least two deposit-forming stages of mineralization can be distinguished:

I. in which disseminated ores formed, synchronously with dolomites,
II. in which further concentration of primary ores and secondary alteration of the primary dolomites took place.

It can be assumed on the available evidence that stage I was syngenetic and stage II epigenetic with the host rocks. Most data published by various authors refer to stage II ores.

The evidence accumulated to date suggests the following sequence of the processes leading to the rise of the deposits and responsible for their alteration (Smolarska et al. 1972):

1. the rise of dolomites and syngenetic mineralization,
2. intraformational brecciation of some fragments of the deposits,
3. the development of paleokarst and epigenetic mineralization,
4. dislocation of the deposits,
5. the rise of katagenetic mineralization,
6. the development of younger karst and hypergenetic processes.

Syngenetic mineralization (stage I) gave rise to ores with dispersed, laminated, bedded, pseudo-brecciated, etc. structures. The formation of zinc, ankeritic and to a lesser extent, lead dolomites is also to be associated with this stage.

Deposits interbedded with breccia occur in the richer parts of the deposits showing typical bedded forms. Breccias presumably formed during diagenesis as a result of their differing progress in the carbonate sediments containing varying amounts of sulphides.

The development of paleokarst phenomena gave rise to dolomitic collapse breccia and caused its epigenetic mineralization (stage II) at the expense of the mobilization of sulphides formed in the first stage of mineralization. The multistage character of these processes was also responsible for the formation of ore breccia in karstic caverns as a result of redeposition of ore-dolomitic grains. The same processes induced changes in the development of the ore-bearing dolomites, involving their secondary dolomitization, recrystallization, dedolomitization etc.

Dislocation of the deposits was attended by the formation of tectonic breccia consisting of fragments of dolomitic and calcareous rocks and ores. They were cemented with typical tectonic detritus made up of crushed dolomites, limestones, ores, clay minerals and coarse-crystalline calcite.

Katagenetic mineralization was responsible for the filling of fissures and caverns in dolomites with ores at the cost of the substance formed both in stage I and II of mineralization. The development of younger karst phenomena and hypergenetic processes brought about further changes in the deposits, such as the rise of oxidized ores, alteration of dolomites, partial dissemination of ores, etc.

The present picture of the Upper Silesian deposits of zinc-lead ores is the result of several overlapping processes. The deposits are polygenetic, as both syngenetic and epigenetic processes played a significant role in their formation.

References

Gruszczyk H (1956) O wykształceniu i genezie śląsko-krakowskich złóż cynkowo-ołowiowych. Biul Inst Geol Warsaw, pp 1–154

Gruszczyk H (1967) The genesis of the Silesian-Cracow deposits of lead-zinc ores. In: Genesis of stratiform lead-zinc-barite-fluorite deposits. Econ Geol Monogr 3:169–177

Gruszczyk H (1978) Hipoteza poligenicznego pochodzenia złoża. Inst Geol Prace Tom LXXXIII

Smolarska I (1974) Studium nad okruszcowaniem triasu w Polsce. Pr Mineral Kom Nauk Miner PAN Oddz, Cracow, nr 37

Smolarska I, Gruszczyk H, Dang Xuan Phong (1972) Breccias in the Silesian-Cracow zinc and lead ore deposits. Bull Acad Pol Sci Ser Terre 20(2) Varsowie

The Influence of Original Host Rock Character Upon Alteration and Mineralization in the Tri-State District of Missouri, Kansas, and Oklahoma, U.S.A.

R.D. HAGNI[1]

Abstract

The genesis and character of the strata-bound ore deposits in the Tri-State district (Missouri, Kansas, and Oklahoma, U.S.A.) depend upon the interrelated sedimentary, tectonic, karst, and mineralizing processes that affected the host rocks.

The depositional environment of the Mississippian Boone Formation in the Tri-State district was that of a warm, shallow sea, which favored the growth of organisms, especially crinoids and bryozoans. The recognition of bioherms or patch reefs in the lower Boone Formation south of the district, where they are unaltered by dolomitization, silicification, and mineralization typical of the mining fields, permits the speculation that these sedimentary features may also be present in the mining district but have not been recognized because they were obliterated by alteration. The sizes, cross-sectional shapes, arcuate map patterns, and other features of the dolomite "cores" in the Tri-State district are suggestive of bioherms or parts of bioherms.

The original and varied characteristics of the Boone limestones influenced the position and intensity of subsequent fracturing, dissolution, and mineralization. Fracturing occurred preferentially either at the margins of biohermal structures or between the dolomite cores and adjacent limestones and formed the arcuate fracture patterns that are exposed in the mines of the Tri-State district. These fractures served as the principal conduits for subsurface ground water, which dissolved parts of the Boone limestones and prepared those host rocks with open spaces and chert breccias that facilitated the introduction of subsequent mineralizing solutions.

1 Introduction

The Tri-State district covers an area of about 5000 km^2 in southwest Missouri, southeast Kansas, and northeast Oklahoma. A total of more than two billion dollars in zinc and lead concentrates has been produced from its ore deposits (Brockie et al. 1968).

Because the district occurs on the northwest flank of the Ozark dome, the sedimentary beds of the district have a gentle northwest regional dip of about 3–5 m/km. Superimposed upon the regional dip is a series of gentle folds, the most prominent of which plunge gently down the flank of the Ozark dome to the northwest. Faults trend northeasterly from Oklahoma into the Tri-State district.

The ore deposits are strata-bound and confined almost entirely to the cherty limestones of the Boone Formation. The individual beds that serve as hosts for the deposits are distinguished by the character of their chert content and are designated by the

1 Department of Geology and Geophysics, University of Missouri-Rolla, Rolla, MO 65401, U.S.A.

letters B through R (Fowler and Lyden 1932). Information about the lithology, thickness, and importance as a host for ore of each bed is given by Hagni (1976, Table 1).

The purpose of this paper is to emphasize the presence of bioherms near the Tri-State district. Because these structures occur in limestones stratigraphically equivalent to those that contain the ore deposits within the district, the implication is that they may have had an important role in the subsequent dolomitization, fracturing, dissolution, silicification (jasperoid), and emplacement of the sulfide ores.

The ore bodies of the Tri-State district occur as "runs", "circles", and "sheet" deposits. Run deposits are elongate tabular bodies of mineralized chert breccias that are confined to a given bed. Runs are especially common in the M, K, and higher beds. Circle deposits, which form partial to nearly complete circular map patterns, are common in the Missouri portion of the Tri-State district. Some of these, such as the Oronogo Circle at the northwest edge of the Webb City field, involve significant slumpage of the overlying Pennsylvanian shale. Circles, 1/2 to 2 1/2 km in diameter, are common in the Picher field. The intense degree of mineralization in the Picher field tends to mask most of the circles, but those on the margins can be distinguished. Sheet ground deposits are flat, extensive ore bodies of relatively uniform grade. They occur in the O, P, and Q beds.

Sphalerite and galena are the economically important minerals in the ore deposits. Additional sulfide minerals include marcasite, pyrite, chalcopyrite, and traces of others. Quartz, dolomite, and calcite are the principal gangue minerals. Chert, the principal quartz gangue, is white to tan and consists of quartz grains 3 to 14 μm across. Brecciated chert was the favored host for sulfide mineralization. Quartz is present also in the form of jasperoid. Jasperoid is a dark gray to nearly black chert-like rock in which the individual grains are 14 to 70 μm and are locally euhedral. Jasperoid generally replaces limestones, but locally replaces other rock types. It was deposited at the time of mineralization and characteristically contains disseminated sulfide grains. Dolomite occurs as gray crystals about 1/2 to 2 mm across and is locally intergrown with jasperoid. It occurs also as pink crystals that coat vugs and fractures. Calcite occurs locally at the margins of the ore bodies as large prismatic and scalenohedral crystals. The depositional sequence of the sulfides and associated gangue minerals is characterized by the repeated deposition of most minerals (Hagni and Grawe 1964).

The writer is indebted to D.C. Brocki, E.H. Hare, Jr., and P.R. Dingess, for many discussions regarding Tri-State geology that he has had with them over a period of 20 years. The writer particularly acknowledges A.C. Spreng, whose experience with the bioherms in Missouri and Oklahoma has contributed much to this paper. The writer has also profited from discussions with G. de Mille.

2 Cherty Limestone and Bioherms

The depositional environment for the carbonate rocks of the Boone Formation in the Tri-State district is presumed to have been that of a warm, shallow, widespread sea, which favored the growth of organisms, especially crinoids and bryozoans. Scotese et al. (1979) have shown that the paleoequator was near the Tri-State district during

Fig. 1. Bioherm developed in lower St. Joe Limestone Member and enclosed by Reeds Spring Member of the Boone Formation. Noel, Missouri

Fig. 2. Two adjacent bioherms with massive limestone cores consisting primarily of fine-grained calcilutite mud. Flank sediments are thin-bedded and consist of coarse crinoidal limestone. Noel, Missouri

Mississippian time. The highly broken character of the fossil debris and the presence of oolites indicate shallow, agitated water.

Bioherms are present near the district and probably within the district itself in the lower part of the Boone Formation. A large number of these structures are located in the vicinity of Noel, Missouri, about 32 km south of the Tri-State district (Troell 1962). They form thick lenses in cross section (Fig. 1) and measure up to 9 m thick and 90 m across. Several occur in close proximity to each other (Fig. 2). The core material is composed principally of what was formerly calcareous mud and contains a few small fragments of bryozoans. The bioherms, which slightly depress the underlying finely laminated limestone beds (Fig. 3), were developed during the deposition of the lower St. Joe Limestone and were built up so that in the case of some reefs the Reeds Spring Limestone was deposited against these structures (A.C. Spreng 1976, pers. comm.). These rock units are the lowest two members of the Boone Formation. In the Picher field, the Reeds Spring contains the lowest ore deposits.

Bioherms have been described by Harbaugh (1957), Huffman (1958), and A.C. Spreng (1976, pers. comm.) from near Salina, Oklahoma, which is about 24 km south of the Tri-State district. These structures differ from those in the Noel area with respect to lithology and stratigraphic position. The limestone cores of the Oklahoma bioherms are relatively massive, and the flank beds are thin-bedded (Fig. 4). The core and flank carbonates are composed of coarse-grained crinoidal rock, which is somewhat similar in lithology to that of the M bed, the principal host for the ore deposits in the Picher field. The core rocks are especially porous or vuggy. The bioherms generally range from 7 1/2 to 15 m in thickness and are 15 m wide. The largest known structure is 45 m thick and 70 m wide (A.C. Spreng 1976, pers. comm.). Harbaugh (1957) noted that some of the bioherms are elongate and have one core overlapping another. Because these bioherms began their growth in upper Reeds Spring time and have upper Osagean strata deposited against them, they occur in precisely the same stratigraphic interval as most of the Tri-State ore deposits.

Mississippian bioherms, which occur west of Spavinaw, Oklahoma, are at a slightly higher stratigraphic level (Harbaugh 1957). In one of these, the mound is 4 1/2 m thick and 90 m wide. Its core consists of very coarse crinoidal limestone.

Larger bioherms occur near the south edge of the Picher field. Harbaugh (1957) found six, and McKnight and Fischer (9170), who described some of them, pictured a single example, which is now covered by the waters of Lake of the Cherokees. The cores of these structures consist of coarse-grained crinoidal limestone and are similar to those in the bioherms near Salina. They are lens-shaped in cross section and have flat bases. Their thicknesses generally range from 9 to 18 m, but some are as much as 30 m thick. Cross-sectional widths reach 335 m. They occur at the base of the Reeds Spring Member.

All of these bioherms near the Tri-State district share many characteristics with reef knolls, or "patch reefs" or "klintar" in Mississippian rocks elsewhere (Cotter 1965; Lees 1961; Morgan and Jackson 1970; Parkinson 1957; Philcox 1963; Schwarzacher 1961; Wilson 1975).

Within the ore fields, the extensive effects of dissolution, dolomitization, and silicification, evidenced by the formation of chert and jasperoid, have obliterated much of the evidence for the presence of biohermal structures. Yet, certain features suggest that

Fig. 3. Detail of bioherm showing depression of the underlying limestone beds. Noel, Missouri

Fig. 4. Bioherm *(right)* with thin-bedded flank sediments *(left)*. Bioherm consists of coarse-grained limestone with abundant large crinoid fragments. Salina, Oklahoma

they were originally present. The arcuate patterns of mineralization and dolomitization that are so common to the district may be largely relicts of the former biohermal structures as discussed below in the section on dolomite cores.

Some of the clays and shales in the mines may represent remnants of the lateral mud facies developed off the flanks of bioherms. These shales contain subangular to subrounded chert fragments, which contrast with the angular chert fragments contained in muds formed as insoluble residues from limestone dissolution in the mining district. Such shales as those shown in Figs. 5 and 6 are significant in that they indicate abrupt lateral facies changes, which are typical of biohermal environments.

Locally, chert may contain relict textures suggesting the former presence of biohermal material. Part of the chert in the L bed exhibits a distinctive mottling that is referred to as "coach dog". McKnight and Fischer (1970) interpret this feature as having resulted from the replacement of fenestrate bryozoan fronds. Chert of this character appears to have replaced biohermal material within the ore deposits themselves.

3 Dolomite Cores and Their Possible Origin as Bioherms

The most significant feature in the Tri-State mining district that supports the postulate of the former presence of bioherms is the character of the dolomitized areas or "cores" in the Boone Formation. This is especially marked by their map patterns, sizes, and cross-sectional shapes. The original and variable characteristics of the Boone Formation influenced the position and intensity of this subsequent dolomitization, fracturing, dissolution, silicification, and mineralization. Most of the run and circle ore bodies in the Tri-State district are located along one side of a dolomitized area. Where ore bodies occur on both flanks of such an area, the dolomite is partly mineralized. The dolomitized areas exhibit elongate to circular map patterns that range from a few to one hundred meters or more in diameter. These areas consist primarily of a coarse-grained gray dolomite and contain lesser amounts of chert and vug-lining pink dolomite. Because the dolomite is very closely associated with the ore bodies, its recognition in drill cuttings and in underground mapping constitutes one of the most important guides to the ore in the district. The subsurface maps of the mining companies and the U.S. Geological Survey record the distribution of many of the dolomitized cores in the Picher field (McKnight et al. 1944). The importance of gray dolomite as an ore guide has been stressed by Lyden (1950), McKnight (1950), Brockie et al. (1968), and McKnight and Fischer (1970).

Although it is obvious that the dolomitized cores were formed by replacement of portions of the Boone limestone, the mechanism that localized the replacement is not known. The most common interpretation is that dolomitization was structurally controlled. McKnight and Fischer (1970, p. 137) state that the dolomitizing and ore-forming fluids were channeled through the fractures located near the centers of what now constitute the dolomitized cores. There are several features, it seems to this writer, that contradict this structural model for the development of Tri-State dolomitized cores. First, the ore fluids would have to move through the central portions of all the dolomitized cores, but some are barren of mineralization and are relatively undisturbed.

Fig. 5. Thin-bedded shale and shaly limestone units and intercalated shaly beds containing subangular to subrounded chert fragments. Massive limestone *(light gray)* is present at the top and lower right. M bed, Kenoyer Mine, Picher field, Oklahoma

Fig. 6. Typical shale and shaly limestone *(dark gray)* containing some chert *(white)* and occurring above L bed massive chert. The shale changes laterally to dolomite. Loose material *(light gray)* covers the mine wall in the lower portion of the photograph. New Chicago No. 1 mine, Picher field, Oklahoma

Second, the model does not accept the fractured and brecciated open ground at the edges of the dolomitized cores as major conduits for the ore fluids (McKnight and Fischer 1970, p. 137), yet these exceedingly permeable zones form some of the main water courses in the M and higher beds today. Although the dolomitized cores in the various beds of the Boone Formation appear to be stacked or superimposed in some mines (McKnight and Fischer 1970, p. 137), in many, the pattern of distribution in one bed does not correspond closely or at all with that in an overlying or underlying bed (Brockie et al. 1968). This suggests that the cores were sedimentary structures that were subsequently dolomitized.

The bioherms described earlier may be the structures that were selectively dolomitized, because they initially possessed a high degree of porosity and permeability, and their size, shape, and distribution are compatible with the present conditions. Dolomitized cores in the M bed have thicknesses of about 5 to 9 m, that are comparable to those of the bioherms in the Noel, Salina, and Spavinaw areas and to those of the smaller bioherms south of the Picher field. The widths of many of the dolomitized cores in the Picher field average about 15 to 75 m and are equivalent to those of the recognizable bioherms. The cross-sectional shapes of some of the cores have margins that appear to slope outward (E.H. Hare Jr. 1973, pers. comm.).

An important structural relationship of the dolomitized cores in the Tri-State district is that they tend to occur over slight positive features in the immediately underlying sedimentary units. Thus, cores in the M bed are often located where the map contours on the top of N bed indicate positive structures. Consequently, structural contouring has long been an important guide to the ore runs in the district (Fowler and Lyden 1932; Brockie et al. 1968). An excellent example of cores that are "perched" upon positive structures in the upper beds can be observed by comparing Figs. 3 and 6 of Lyden (1950) for the Bilharz area in the northwestern part of the Picher field. In this area, the dolomitized cores in the K bed are positioned immediately above positive features in the top of the L bed that are only about 3 to 6 m high. The writer believes that the dolomitized cores have a definite structural relationship to the underlying beds, because their position and location reflect the conditions of bioherms that preferentially grew on slight submarine topographic highs.

The horseshoe type map patterns formed by many of the linked dolomitized cores are interpreted to be similar to horseshoe-shaped reefs known elsewhere (Stafford 1959). The writer believes that the water depth of the Mississippian sea that covered the area of the Picher field was shallower than in the Noel and Salina areas. Consequently, biohermal growth fluorished in much of the area, whereas in the deeper water areas only isolated bioherms grew. The large horseshoe-shaped dolomitized areas in the Picher field are 1/2 to 2 1/2 km in diameter and commonly have their open ends turned away from the center of the field. This arrangement perhaps reflects the presence of deeper water around the perimeter of the field at the time the Boone Formation was deposited. An example of a horseshoe-shaped dolomitized area is depicted by Lyden (1950, Fig. 6). This structure is open to the southwest, where a "pulldrift" has exposed very shaly limestones.

Some of the clays and shales to be found in the mines may represent facies equivalent to nearby bioherms. McKnight and Fischer (1970) believed that the shales of J bed represent lateral facies of K bed limestone (and dolomite) and L bed chert. If the dolomitized cores had been bioherms or parts of bioherms, their original sedimentary features and fossil content have long since been obliterated by intense dolomitization, silicification, and solution thinning.

4 Fracturing, Dissolution, and Mineralization

Minor fractures constituted a major part of the plumbing system for mineralization and played an important part in the localization of the individual ore deposits. The distribution and arcuate pattern of the fractures are readily understood if it is assumed that the fractures developed along the margins of dolomitized bioherms at or near contacts between rocks of different character and response rates to successive periods of mild tectonic activity. The fractures promoted the underground solution of the Boone Formation when it was exposed to erosion at various intervals. Its erosional surface, which forms the post-Mississippian pre-Pennsylvanian unconformity, was particularly amenable to the development of karst topography. Thus, the associated solution channels, caves, and sinks prepared the rocks with open spaces and chert breccias that facilitated the introduction of subsequent mineralizing solutions.

The M bed is 110 feet thick where it is unaffected by dissolution, but in many areas of the Picher field, it is as thin as 20 to 30 feet (Hagni and Desai 1966). Large tracts of the Tri-State district have suffered dissolution but little or no mineralization. These areas are referred to as "boulder ground", because they contain broken residual chert. The Pennsylvanian shales, which were deposited upon the post-Mississippian karst surface, have a very irregular base, the irregularity of which is commonly accentuated by continued dissolution of the underlying Mississippian carbonates.

Areas where the base of the Pennsylvanian shales is unusually low are indicative of structural conditions and dissolution that were favorable to mineralization. Some mineralized shale slumps are circular and conform to the shapes of the sinks. The Oronogo ore deposit in Missouri and the Pleasanton in Kansas are examples.

The mineralizing solutions followed avenues in the Boone Formation that were prepared by the combination of processes outlined above. The replacement of dolomite on one side of a series of fractures took place with some degree of difficulty and commonly resulted in partial replacement. In contrast, the limestone on the other side usually experienced complete replacement by jasperoid and disseminated sulfides. Because the ore bodies occur at the margins of the dolomitized cores, they are related to the flanks of the underlying structures, a relationship particularly emphasized by Fowler and Lyden (1932), and Fowler (1938). Additional dissolution and brecciation followed the emplacement of jasperoid and early mineralization. More coarsely crystallized sulfides, quartz, and pink dolomite formed in open spaces and coated the surfaces of fractures, vugs, and breccias. The age of at least part of the mineralization is post-Pennsylvanian, because locally the Pennsylvanian shales have been mineralized.

5 Conclusions

The effects that fracturing, folding, and dissolution of the Boone Formation have had upon the localization of its ore deposits have received a considerable amount of attention, but the character of the original depositional structures in the formation and the influence that they may have had upon alteration and mineralization have been neglected. Because bioherms are common in lower Boone near the district, the writer believes that they may also have been present in the district but have not been recognized, their former presence having been almost obliterated by dolomitization, silicification, and dissolution. The distribution of coarsely crystalline gray dolomite is a particularly important and puzzling aspect of Tri-State geology. It is important because the margins of dolomitized areas constitute the principal loci for the ore deposits. It is puzzling because the mechanism controlling their distribution is not readily apparent. It is postulated here that much of the rock now composed of gray dolomite originally may have consisted of biohermal limestone. Features that support this conclusion include the similarity in size of the dolomitized cores to known bioherms in the region, the areal distribution of the dolomite, the presence of lateral shaly facies, the shallow character of the depositional environment, and the presence of bioherms in the Boone Formation outside the district where its original character is less altered.

The writer proposes a model for the genesis of the dolomitized cores in the Tri-State zinc-lead district. This model involves early dolomitization of biohermal limestone, because it was more porous and permeable than the surrounding limestone. It also takes into account the fact that the dolomitized areas occur over slight positive structures where former submarine topographic highs favored biohermal growth. The distribution and arcuate pattern of fracturing, which has long been a matter of considerable controversy (Fowler and Lyden 1932), are readily explained by the model as having developed along the margins of the dolomitized bioherms at or near their contacts with the limestones. The fractures in turn controlled the sites of subsequent solution thinning, silicification, and mineralization.

References

Brockie DC, Hare EH Jr, Dingess PR (1968) The geology and ore deposits of the Tri-State district of Missouri, Kansas, and Oklahoma. In: Ridge JD (ed) Ore deposits of the United States, 1933–1967. Am Inst Min Metall Pet Eng, New York, pp 400–430

Cotter E (1965) Waulsortian-type carbonate banks in the Mississippian Lodgepole Formation of Central Montana. J Geol 73:881–888

Fowler GM (1938) Structural control of ore deposits in the Tri-State zinc and lead district. Eng Min J 139:46–51

Fowler GM, Lyden JP (1932) The ore deposits of the Tri-State district. Am Inst Min Metall Eng Trans 102:206–251

Hagni RD (1976) Tri-State ore deposits: The character of their host rocks and their genesis. In: Wolf KH (ed) Handbook of strata-bound and stratiform ore deposits, vol 6. Elsevier, Amsterdam, pp 457–494

Hagni RD, Desai AA (1966) Solution thinning of the M bed host rock limestone in the Tri-State district, Missouri, Kansas, Oklahoma. Econ Geol 61:1436–1442

Hagni RD, Grawe OR (1964) Mineral paragenesis in the Tri-State district, Missouri, Kansas, Oklahoma. Econ Geol 59:449–457

Harbaugh JW (1957) Mississippian bioherms in northeast Oklahoma. Am Assoc Pet Geol Bull 41: 2530–2544

Huffman GG (1958) Geology of the flanks of the Ozark uplift. Okla Geol Surv Bull 77:281

Lees A (1961) The Waulsortian "reefs" of Eire: a carbonate mudbank complex of lower Carboniferous age. J Geol 69:101–109

Lyden JP (1950) Aspects of structure and mineralization used as guides in the development of the Picher field. Am Inst Min Metall Eng Trans 187:1251–1259

McKnight ET (1950) The Tri-State region. In: Behre CH Jr, Heyl AV, McKnight ET (eds) Zinc and lead deposits of the Mississippi valley. Int Geol Congr Rep 18: pt 7, 57–58

McKnight ET, Fischer RP (1970) Geology and ore deposits of the Picher field, Oklahoma and Kansas. US Geol Surv Prof Pap 588:165

McKnight ET et al. (1944) Maps showing structural geology and dolomitized areas in part of the Picher zinc-lead field, Oklahoma and Kansas; based on Field Maps prepared by McKnight ET, Fischer RP, Addison CC, Bowie KR, Thiel JM, Owens MF Jr, Wells FG. In: Tri-state zinc-lead investigations, preliminary maps 1–6. US Geol Surv Washington, DC

Morgan GR, Jackson DE (1970) A probable "Waulsortian" carbonate mound in the Mississippian of northern Alberta. Bull Can Pet Geol 18:104–112

Parkinson D (1957) Lower Carboniferous reefs of northern Ireland. Am Assoc Pet Geol Bul 41: 511–537

Philcox ME (1963) Banded calcite mudstone in the lower Carboniferous "reef" knolls of the Dublin Basin, Ireland. J Sediment Petrol 33:904–913

Schwarzacher W (1961) Petrology and structure of some lower Carboniferous reefs in northwestern Ireland. Am Assoc Pet Geol Bull 45:1481–1503

Scotese C, Bambach RK, Barton C, van der Voo R, Ziegler A (1979) Paleozoic base maps. J Geol 87:217–277

Stafford PT (1959) Geology of part of the horseshoe atoll in Scurry and Kent Counties, Texas. US Geol Surv Prof Pap 315–A:20

Troell AR (1962) Lower Mississippian bioherms of southwestern Missouri and northwestern Arkansas. J Sediment Petrol 32:629–644

Wilson JL (1975) Carbonate facies in geologic history. Springer, Berlin Heidelberg New York, 471 pp

Karst or Thermal Mineralizations Interpreted in the Light of Sedimentary Ore Fabrics

O. SCHULZ [1]

Abstract

In the wake of recently obtained data on Zn-Pb-Fe-Ba-F-enrichments on erosion surfaces and paleo-karsts in carbonate rocks, there is a tendency towards unjustified generalization of ideas, beyond immediately straightforward interpretations. The following features suggest paleo-karst formation: discordant layers exposed by erosion, weathering horizons, rather shallow erosion features, resediments and allothigenic detritus accompanying chemically apposited ore- and accompanying minerals, and the feasibility of their bulk mass being derived from a previous enrichment. Some illustrative examples are given, and critical statements presented, from the mineral deposits in Sardinia, Álgeria, the Karnian Alps, Gailtal Alps, and Kitzbühel Alps. Thermal transport of mass remains a plausible hypothesis from the viewpoint of the study of fabrics for many stratiform and discordantly stratified mineral deposits in carbonate rocks.

1 Statement of the Problem

Mineral enrichments on erosion surfaces or in connection with weathering horizons (clay, kaolin, laterite, bauxite, Fe and Mn ores, magnesite, Ni-silicates, phosphates, Cu, Pb, Zn, V, U minerals, Au, Ag, numerous placer minerals) have been known for a long time; they have their firm place in the science of mineral deposits. Nevertheless, it was only in the last decade that Pb, Zn, Ba, F deposits presumably to be found in carbonate karst rocks increasingly gained attention. There has been a boom of hypotheses leading to the impression that concordant and discordant nonferrous metal enrichments in limestone-dolomite rocks are almost universally indicative or emersion phases and karst phenomena, whereas in crystalline rocks hydrothermal or hydatogenic genesis is still accepted. There is a danger of quick generalizations and partial misinterpretations. The question arises whether the study of fabrics enables us to assess critically the one or the other explanation.

1 Institut für Mineralogie und Petrographie, Abteilung Geochemie und Lagerstättenlehre, Universitätsstr. 4, 6020 Innsbruck, Austria

2 Introduction

The problem has a certain oddity: what is a typical karst mineralization for one school of thought, is characteristic proof of hydrothermal replacement by minerals to other scientists. Such differences are a cause for regret. We note in advance that both explanations could be right at the same time in one deposit, if applied to different strata.

Recent research on mineral enrichments by erosion and karst formation was carried out by Benz (1964), Callahan (1965), Tamburrini and Violo (1965), and Leleu (1966). Valuable contributions were added, e.g., by Bernard et al. (1972), Bernard (1973), Padalino et al. (1973), Perna (1973), and Zuffardi (1976). In a detailed study Zuffardi combines geological, geochemical, and mineralogical considerations in the formation of karst, which can produce deposits of useful minerals. He stresses the action of descendent waters in ore enrichment, and stimulates a new interpretation of so-called telethermal or kryptothermal deposits.

While Boni (1979) was working at detailed descriptions of the karst areas in Sardinia which are so rich in minerals, attempts were being made to adapt some Alpine Pb-Zn deposits of the Middle Triassic to the new hypotheses. Along these lines Lagny (1974) treated the Salafossa deposit in the Sappada Dolomites in emersion or paleokarst cycles, respectively, and Bechstädt (1975) did the same for ore bodies of Bleiberg-Kreuth (Gailtal Alps).

Twenty-seven years ago the discussion was centred around the alternative "post-diagenetic-hydrothermal replacements" or "hydrothermal-sedimentary to diagenetic formations", the almost universally accepted conclusion being that concordant, externally and internally sedimentary deposits can be assigned to a cycle of formation jointly with the rock itself. Starting from the question of where the metallic component of ore bodies came from, the current discussion arose about the "extrusive-sedimentary origin of matter" of the synsedimentary formations in a larger sense, and of the importance of "exogenic" processes towards the enrichment of the "alien" minerals.

3 Considerations on Ambiguous and Unique Fabrics

The conflict arises most easily where ambiguous fabrics exist and no supporting unique evidence can be found. Since we are talking about deposits which are exposed at erosion surfaces, we refer to the three-dimensional shape of the boundary surfaces. *Erosion features* which indicate alternating progressive and regressive events may leave doubts whether the causes are mechanical or chemical, particularly in shallow waters. It is obviously not sufficient to find boundary surfaces which are characteristic of external and internal erosion. Both karst formation, as well as thermal systems, can produce similar contours in caves by solution erosion. Compare, for example, the erosively enlarged joints in karst regions (Fig. 1), and hydrothermal fissures in crystalline rocks (Fig. 2). Deep-reaching fractures enlarged by erosion exclude karst formation, when paleogeography is considered. Quite a number of judgements on apparent cavities and their fillings suffer from the ambiguity of two-dimensional view. The

Fig. 1–6

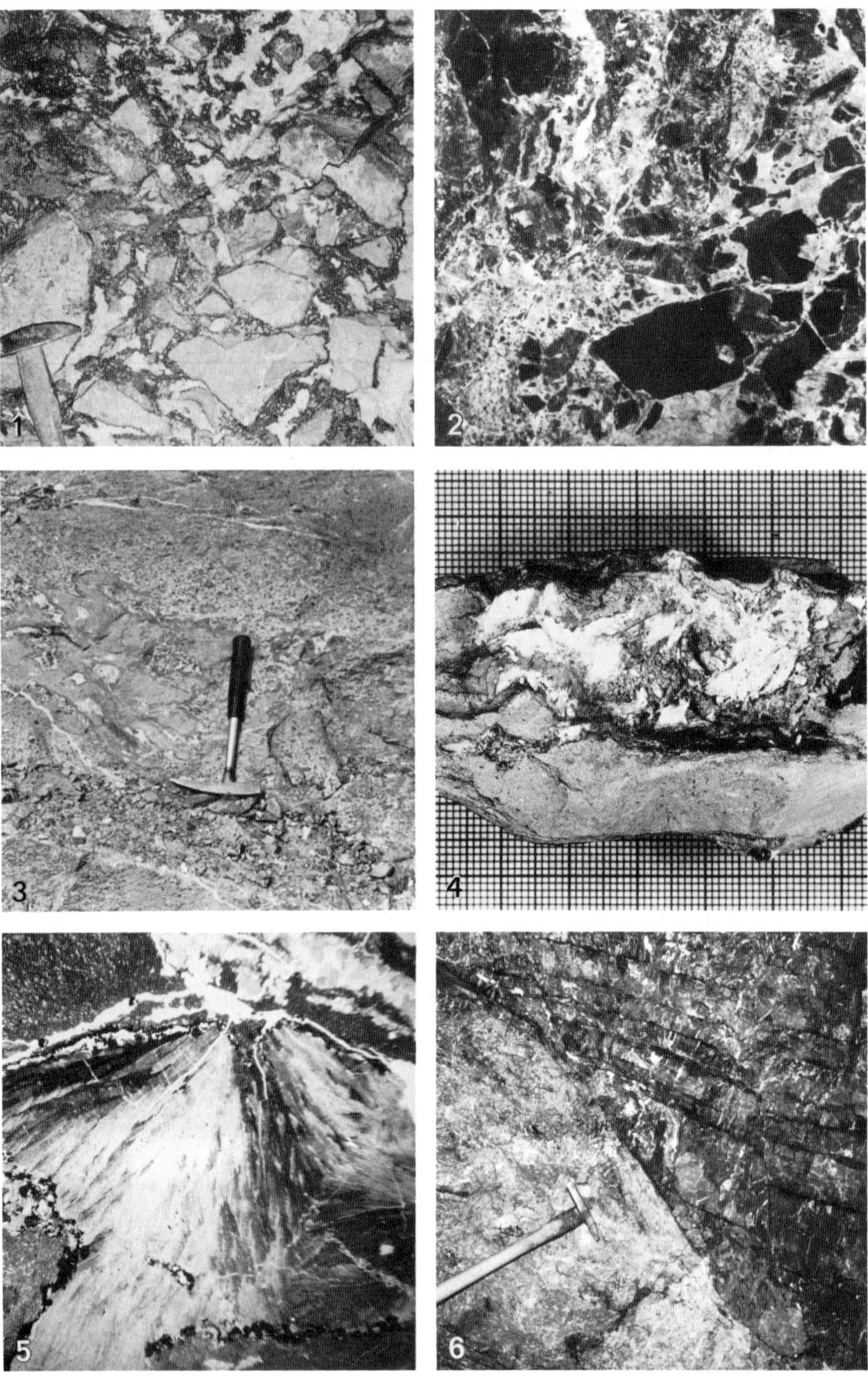

boundary surfaces of the external and internal areas of apposition therefore do not lend themselves ideally to the clarification of such questions.

On the other hand, distinct weathering horizons with detritic sediments, resediments, Fe, Cu, Pb, Zn oxidation minerals, if possible with erosion-discordant new boundary surfaces, do identify the eroding and rebuilding processes.

As misleading, genetic ambiguity exists over *ore-containing breccias,* particularly with *deformation breccias* (Sander 1970) (=intraformational breccias, clastic breccias). Inhomogeneity breccias, syndiagenetic and postdiagenetic-tectonic breccias, solution breccias, collapse breccias are in the same class, as well as the initial stage of breccia formation with ramified–rupturally deformed rocks. The breccia ore bodies may have concordant and a multitude of discordant shapes and sizes. Care is necessary, however, as analogous kataclastic fabrics exist both in sediments and in magmatites and metamorphites! It is, therefore, premature to assign the brecciated ore veins in carbonate rocks to a karst cycle without further proof.

Alternatively, *sedimentary breccias* are much more significant; *resediments,* in particular (Sander 1970), throw light on many a detail of early diagenetic processes which have happened in the young sediment using their fragments.

On the whole, in each size range the *mechanically* apposited component yields the most important clues on origin and transport, and, by its participation, frequently even permits the genetic assessment of the *chemical* apposition. Chemical apposition, by itself (external and internal crystal growths, metasomatites), is an obstacle for the genetic classification of deposits because of its ambiguity, and cannot be taken as proof of sedimentary formations, not even in the case of layered apposition. No doubt, chemical resedimentation has often occurred; but proof of this conjecture is rarely possible because of the difficulty of identifying the material content of a crystal mass as having been previously dissolved.

The presence of *fine-rhythmites, oblique- and cross-beddings, polarity* of sequence with, e.g. vertical grading among the "alien minerals" and *syndiagenetic foldings* are all typical sedimentary features, giving proof of external and internal ore apposition, but they do not solve the problem of metal enrichment by karst formation or thermal springs.

Fig. 1. Wetterstein Limestones – deformation breccia with galena-sphalerite-calcite cementation. Triassic, mining complex Mežica (Karawankes, Yugoslavia). Sample size 55 × 55 cm

Fig. 2. Quartz phyllite – deformation breccia with hydrothermal fluorite-galena-sphalerite cementation. Brixen quartz phyllite, Paleozoic, mining complex Rabenstein (South Tirol, Italy). Sample size 55 × 55 cm

Fig. 3. Ore-containing Carboniferous transgression horizon overlying Devonian limestone. Monte di Val Dolce (Karnian Alps, Italy)

Fig. 4. Quaternary, calcareous-clayey cavity sediment in the Triassic Wetterstein Limestone. Authigenic crystallization of barite *(white),* sphalerite, pyrite, fluorite, and calcite in cave lutite *(light grey).* Mining complex Bleiberg (Gailtal Alps, Austria). Scale: graph paper (mm)

Fig. 5. Thin section – a part of the cave sediment of Fig. 4. Divergent-laminated barite aggregate *(grey white to dark grey),* fringed by sphalerite *(black grains)* and fluorite *(light grey),* calcite *(different light grey);* all around detritic cave lutite *(black grey).* Sample size 11 × 11 mm, Nicols X

Fig. 6. Erosion basin in dolomite of the Raibler strata, Triassic, with filling rich in ZnS. Mining complex Bleiberg (Gailtal Alps, Austria)

Geopetal apposition must, however, be mentioned (as an important "top-bottom criterion") for the cases of discordant layering. Orogenic processes often disrupt sedimentation and cause missing layers, confirmed by discordant layering. Such cases of goepetal continuation of sedimentation are useful for outlining the erosion phases and make such layers rich sources of information. These long- and well-known cases, which, we think, should be well studied sedimentologically and geochemically, should form the basis for all considerations on element concentrations by karst formation in a region.

Karst formation should be excluded as a cause when gradual transitions are present from quasi-horizontal ore-free carbonate fine layers to *fine beddings* with quasi-vertical alien ore material, that is with a gradual change of the sediment supply, without a distinct boundary of interrupted sedimentation. Whenever in such parallel fabrics an admixture of "alien" minerals appears suddenly, we may conclude that the elements are of extrusive origin.

We have so far discussed the information obtainable from mechanical and chemical apposition for our problem; now we need to consider the importance of the sequence of simultaneous *biogenic apposition.* We are not talking here about the presence of organic detritus important to the biofacies description of the region of sedimentation, but rather about the organisms which are immediately involved in the shaping of the regions of sediment build-up through their shell and crust formations, particularly reef-forming organisms. The presence of such organisms in sequences is characteristic of external sediments; mats of algae, e.g. of the type of stromatolites, or cushions and reefs of corals or lamellibranchs will not reside in caves. This self-evident remark may be important for findings where brecciated clastic layers are covered only by a thin crust of stromatolites. These, then, are characteristic of a shallow-water environment, possibly desiccation, but definitely not of cave deposition or deformation breccias.

In my opinion over-interpretations or even wrong interpretations are particularly frequent in the case of *coarse-crystalline parallel fabrics,* which at present exist as chemically apposited aggregates, and which may be derived from different initial fabrics. In these cases there are extensive similarities with metamorphic ore deposits. Already in the sedimentary state of evolution, however, diagenetic recrystallizations can distort the primary fabric or even obliterate it, such that the process of formation and reformation can only occasionally be reconstructed from other evidence. How often have crystalline layered complexes been mistaken for typically sedimentary structures, without evidence for external-sedimentary mineral enrichments, disregarding critical objections which point out another possible explanation as selective replacements! Such objections should be verified by apophysis-like internal ore bodies aligned along veins. Admittedly, stratiform ore deposits extending many miles cannot be explained that way. Why should such deposits, however, be immediately taken for facies-specific emersion horizons? After all, other exogenic and endogenic, extrusive types of metal transport are also feasible.

So far frequently occurring sediment fabrics have been discussed and assessed with regard to their usefulness for deriving the sedimentary causes of ore concentrations. Now we attempt genetic interpretations of some known mining complexes.

4 Examples from Deposits

4.1 Iglesiente-Sulcis (Sardinia)

The paleokarst ores of the Iglesiente-Sulcis region in southwest Sardinia deserve to be the first object of discussion on karst ore deposits. Various typical findings have been described and interpreted in a number of investigations, e.g., by Tamburrini and Violo (1965), Padalino et al. (1973), lately by Zuffardi (1976), and Boni (1979).

The many illustrative exposures with Pb, Zn, Fe (Ag, Cu), Ba, F, Si enrichments, in part almost monomineralic, are associated with genuine karst features (Fig. 1) in carbonate rocks within erosion-discordant layers and their near surroundings. Element concentrations, caused by paleokarst, developed in several cycles: in association with a Cambrian-Ordovician discordant event (Caledonian orogenesis, Sardinian Stage), with the Permo-Triassic "Hercynian Plane" (Hercynian orogenesis), and with Eocene erosions; finally, recent karst phenomena are also evident. The fundamental importance of the Sardinian exposures stems from the well-arranged karst topography with unsurpassed erosion landscapes and their underlying vertical and horizontal cavities and cavity networks. Corrosion was responsible for external and internal deposition cavities, a precondition for the later infillings. Most instructive are the chemically and mechanically apposited sediments which consist of authigenic and allothigenic detritus; first coarsely clastic material, including collapse breccias, then arenitic and lutitic fillings. They are accompanied by mineral enrichments like galena, pyrite, marcasite, sphalerite, barite, fluorite, and quartz, all chemically apposited. To a lesser degree Ag and Cu are also involved (in galena, fahlore, chalcopyrite). Important, however, is the accumulation of residual substances like red clays, clays with kaolinite, halloysite, illite, chlorite, and of oxidation ores like goethite, hematite, cerussite, anglesite, smithsonite, Mn-oxides, and others (Boni 1979).

The importance of the type localities in Sardinia lies furthermore in the obviously belteroporic plan of the karst reliefs and systems, the selective erosion being shown by the preference for certain, sometimes steeply inclined, layers and joint systems of the Caledonian and Hercynian orogenies.

Finally, a precondition for the enrichments of ore parageneses in the various cycles must be stressed emphatically: the previous existence of metal concentrates in the Lower Cambrian Gonnesa Formation, the so-called Metallifero Limestone. Various causes of metal ion supply have been suggested for the process of formation of this strata bound and partly stratiform ore vein with sphalerite, pyrite, marcasite, galena (fahlore, chlacopyrite), barite, quartz, dolomite, calcite and partly fluorite, and also for remarkably high metal trace contents in associated rocks, which were involved in repeated mobilizations with rearrangements into karst systems. One credible process is concentration by chemical resedimentation, e.g. by remobilization under saline conditions from previously existing supplies; other possible synsedimentary formation processes should, however, also be considered, along the lines of Münch (1960), who looks for magmatic connections and, therefore, thermal-sedimentary metal supplies.

Even when paying due attention to the instuctive karst phenomena with mineral concentrations, and respecting their valuable interpretations, one should not go so far as to explain all mineralizations in carbonate rocks solely by erosion phases.

4.2 North Africa, Algeria

Stratabound Pb-Zn-Fe-Ba-F deposits occur frequently in Jurassic and Cretaceous rocks in the Atlas mountains of Morocco–Algeria–Tunisia. In the last decade they have been preferably explained by karst formation processes. The dominating diagenetically recrystallized fabrics are, by themselves, ambivalent. It is no wonder, therefore, that such deposits are still being explained as caused by hydrothermal replacements, especially by Russian scientists.

From my own impressions gained in Algeria I would venture the remark that e.g. in the deposit El Abed, a continuation of Bou Beker-Touissit, Morocco, stratified ore bodies with diagenetically accreted crystallized ZnS-PbS-FeS_2-$BaSO_4$- and CaF_2-layers and mineralized deformation breccias ("solution breccias") do, in fact, dominate by far; the origin of their metal content, however, may be found not only in karst, but also in others like hydrothermal-sedimentary processes. Admittedly, just a few meters above these ore bodies an extensive concordant erosion relief exists with resediments of the dolomitic Jurassic rock as well as arenitic and crystallized ores in synclinal and pocket-line depressions. Such ore accumulations, resedimented on the relief-forming interfaces, confirm in a striking and understandable way the type of ore enrichment occurring on weathering surfaces.

4.3 Karnian Alps, Karawankes

In many places along the Italien-Austrian mountainous borderline, e.g. from the Plöckenpaß across Tarvis into the Karawankes, accumulations of ZnS, fahlore, PbS, FeS_2, $(Co,Ni)As_{3-2}$, $BaSO_4$, CaF_2, SiO_2 may be found at the base of the Carboniferous transgression on Devonian limestone (also dolomite). These are interpreted by Brigo and Di Colbertaldo (1972) as erosional concentrations after removal of older primary metal enrichments. This genetic interpretation is confirmed by the typical denudation relief, the associated taking place of the paragenesis and the participation of ore detritus in the mostly stratified mineral precipitations (Fig. 3).

4.4 Permian Transgression Products, Kitzbühel Alps

A remarkable transgression horizon with mineral precipitation lies in the greywacke zone of North Tirol near Kitzbühel. At the base of the overlapping coarse- to fineclastic Permian sediments above all weak barite precipitations after chemical rearrangement may be found, presumably from the Devonian dolomite rocks mineralized locally with barite (fahlore). They often lie in pocket-like depressions.

Synsedimentary magnesite nodules, also, and U-ore enrichments with Fe-Cu-sulphides enlarge the stratigraphically higher arenitic-lutitic sequence of strata. There are, of course, also modest fahlore, sphalerite, galena concentrations known to exist in the quartz sandstones of the Verrucano complex in the western part of North Tirol, but these latter local metal enrichments of the weathering cycle, which are not the only ones in the Alpine region by far, are not bound to carbonate rocks.

4.5 Bleiberg-Kreuth, Gailtal Alps

The Austrian Pb-Zn-(Fe-Ba-F) deposit in Bad Bleiberg, also contains examples for ore precipitations in cavity networks and external erosion surfaces. Already in 1957 Schulz described mechanical and chemical internal sediments in Quaternary cavities in the Upper Wetterstein Limestone (Middle Triassic). They are interesting not so much for their ruditic, arenitic, and lutitic detritus, which originates from the surrounding sediments and has been geopetally and discordantly sedimentated, but rather for the chemically-internally apposited calcites and the authigenic crystallizates of barite, fluorite, sphalerite, pyrite, galena, and gypsum (Figs. 4 and 5), the latter of which were, however, only found later in an exposure. White, divergent-tabular barite aggregates up to a few centimeters in size are accreted along the fringes and intergrown with fluorite and bright-yellow sphalerite; in this case, the latter prove to be the younger crystallizates. This finding is a telling example of mechanical and chemical rearrangement from the Triassic deposit into Quaternary (Postglacial?) weathering zones. Apart from limonite no oxidation minerals were found here.

In the stratiform ore bodies of the Raibler beds, however, mechanical resedimentation has been known for a long time (Schulz 1960). Lying above the ore deposits, which are contained in shallow-water dolomites, we find ore sediments bedded in a depression-like manner (Fig. 6) which clearly indicate local erosion processes, as shown by analogous examples from the ore-free succession of strata.

Saline lagoonal sedimentation environments have an essential role facies-wise in the genesis of the main ore beds in the Wetterstein Dolomite, in the Wetterstein Limestone, and in the Raibler beds, as well as in the discordant ore veins in the Wetterstein Limestone; but this environment nevertheless cannot explain satisfactorily the origin of the enormous metal content. A survey of this problem is given by Schulz and Schroll (1977); other points of view (Siegl 1956; Dzulynski and Sass-Gustkiewicz 1977) should be taken into account, too, at least for specific questions.

A distinct emersion horizon is visible in the Pb-Zn deposit Mežica (Karawankes) on the Wetterstein (reef) Limestone. The coarse-to fineclastic transgressing resediments contain ores (Štrucl 1970, 1971).

Brigo and Omenetto (1976), too, draw upon the karst phenomenon to explain a series of relief-filling ore deposits in the Wetterstein Dolomite of the Pb-Zn-deposit Raibl (Cave del Predil, Julian Alps). They take diagenetic metal enrichments from the carbonate rocks to play a substantial role in the ore accumulations.

Regarding the tubular ore body of Salafossa (Sappada, Sexten Dolomites), lying somewhat discordantly in the Middle Triassic dolomite complex, Lagny already in 1974 interpreted it to be a deposit formed by karst processes. Future studies will have to show whether this explanation by itself is sufficient for the doubtless sedimentary, Triassic ore body.

5 Concluding Remarks

When trying to interpret genetically sedimentary ore enrichments by way of apposition fabrics, we should not be surprised to be confronted with some uncertainties –

this is a difficult task. For this very reason it is important to search for the few decisive field observations, or otherwise to make only very cautious statements.

The karst model is proven to be a workable model. We must not forget, however, that among the karst erosions and emersions by far the most lack, and only a few have, mineral or ore concentrations. Facies, climatic, and chemical causes can obviously not be the only ones. Preexisting enrichments, which are available for denudation, must therefore be the controlling cause. In my opinion, increased trace contents in carbonates alone are in general not a sufficient explanation. Consider how many weathering horizons do not contain ore enrichments, notwithstanding the fact that in the region of denudation and drainage far more metal ions are available (e.g. in clays) than in the relatively sterile carbonate rocks!

On the other hand, for extrusive material supply of a hydrothermal or secondary- or pseudo-hydrothermal and hydatogenic kind no immediate connection with volcanic rocks or tuffs need be assumed, nor increased metal content of the latter. A magma source can trigger thermal reactions and, together with the oceanic hydrologic cycle, mobilize elements in the earth's crust, and thus stimulate extrusive metal transports. The thermal springs may lead to external and internal erosion in the more or less consolidated sediments, which may topologically resemble the pocket- and doline-shaped karst contours. Purely chemical mineral deposition should remind us of this possibility. In this context Polish researchers use the term "hydrothermal karst" (Dzulynski and Sass-Gustkiewicz 1977). It is well known that under the influence of submarine warm springs there is enormous bacterial activity, which may explain many a difficulty in interpreting S-isotope-ratios.

In conclusion, we may answer the question whether many a Pb-Zn-Fe-Ba-F-Si enrichment in carbonate rocks could have been formed by karst processes by another question: Could not some presumed "karst mineralizations" be of marine-hydrothermal origin, or could at least the primary metal supply be interpreted as being extrusive-sedimentary?

Acknowledgment. The author thanks Ignaz Vergeiner of Innsbruck for the translation of this paper.

References

Bechstädt T (1975) Lead-Zinc-Ores dependent on cyclic sedimentation (Wetterstein-Limestone of Bleiberg-Kreuth, Carinthia, Austria). Miner Deposita 10:234–248

Benz JP (1964) Les gisements plombo-zincifère d'Arenas (Sardaigne). Thèse de doctorat, Fac Sci Univ Nancy, 123 pp

Bernard AJ (1973) Metallogenic processes of intrakarstic sedimentation. In: Amstutz GC, Bernard AJ (eds) Ores in sediments. Springer, Berlin Heidelberg New York, pp 43–57

Bernard AJ, Lagny Ph, Leleu Mg (1972) A propos du rôle métallogénique du Karst. Int Geol Congr 24. Session, Canada 1972, Sect 4 (Gites mineraux), pp 411–423

Boni M (1979) Zur Paläogeographie, Mineralogie und Lagerstättenkunde der Paläokarst-Erze in Süd-West Sardinien (Iglesiente-Sulcis). Dissertation, Univ Heidelberg, 260 S

Brigo L, Di Colbertaldo D (1972) Un nuovo oirzzonte metallifero nel Paleozoico delle Alpi Orientali. 2nd ISMIDA, Bled 1971, Geologija 15, Ljubljana, pp 109–124

Brigo L, Omenetto P (1976) Le mineralizzazioni piombo-zincifere della zone di Raibl. Nuovi aspetti giacimentologici. Ind XXVII Min 49–56

Callahan HW (1965) Paleophysiographic premises for prospecting for stratabound base metal mineral deposits in carbonate rocks. Symp Min Geol Base Metals (Ankara Sept 1964). Cento Treaty Organiz, Ankara, pp 191–248

Dzulynski S, Sass-Gustkiewicz M (1977) Comments on the genesis of the Eastern-Alpine Zn-Pb-deposits. Miner Deposita 12:219–233

Lagny Ph (1974) Emersion medio-triassiques et minéralisations dans la région de Sappada (Alpes orientales italiennes). Le gisement de Salafossa: un remplissage paléokarstique plombo-zincifère. Thèse Doct Sci Nat Nancy – I, 366 pp

Leleu M (1966) Le karst et ses incidentes métallogéniques. Sci Terre 11 (4):385–413

Münch W (1960) Ricerche geo-giacimentologiche in Sardegna. Giornate di studio sulle ricerche geo-giacimentologiche. I, 2, AMMI s.p.a., Roma, 8 pp

Padalino G, Pretti S, Tamburrini D, Tocco S, Uras I, Violo M, Zuffardi P (1973) Ore deposition in karst formations with examples from Sardinia. 8th International sedimentological congress Heidelberg 1971. In: Amstutz GC, Bernard AJ (eds) Ores in sediments. Springer, Berlin Heidelberg New York, pp 209–220

Perna G (1973) Fenomeni carsici e giacimenti minerari. Atti del Seminario di Speleogenesi, Varenna, 1972, Le grotte d'Italia, 4 a, IV, Bologna, pp 1–72

Sander B (1970) An introduction to the study of fabrics of geological bodies. Pergamon Press, London, 641 pp

Schulz O (1957) Über ein Höhlensediment im Bergbau Bleiberg-Kreuth (Kärnten). Mitt Geol Ges Wien 48:297–303

Schulz O (1960) Die Pb-Zn-Vererzung der Raibler Schichten im Bergbau Bleiberg-Kreuth (Grube Max) als Beispiel submariner Lagerstättenbildung. Carinthia 2, Sonderh 22:1–93

Schulz O, Schroll E (1977) Die Pb-Zn-Lagerstätte Bleiberg-Kreuth. Stand der geowissenschaftlichen Forschung 1976 (Projekte 2437, 2776 S). Verh Geol Bundesanst (Austria) 3:375–386

Siegl W (1956) Zur Vererzung der Pb-Zn-Lagerstätten von Bleiberg. Berg- Huettenmaenn Monatsh 101:108–111

Štrucl I (1970) Poseben tip mežiškega svinčevo cinko-vega orudenenja v rudišču Graben. Geologija 13:21–34

Štrucl I (1971) On the Geology of the eastern part of the Northern Karawankes with special regard to the triassic lead-zinc-deposits. Sedimentology of parts of Central Europe. Guidebook 8th Int Sediment Congr. Kramer, Frankfurt/Main, pp 285–301

Tamburrini D, Violo M (1965) Il giacimento di Baritina di Monta Barega-Monte Arcau (Iglesiente, Sardegna). Ric Sci Parte 2 Sez A 35:814–848

Zuffardi P (1976) Karst and economic mineral deposits. "Handbook of strata-bound and stratiform ore deposits, vol III. Supergene and surficial ore deposits; textures and fabrics". Elsevier, Amsterdam, pp 175–212

Atoll Texture in Minerals of the Cobaltite-Gersdorffite Series from the Raipas Mine, Finnmark, Norway

F.M. VOKES[1] and G.S. STRAND[1, 2]

Abstract

Particularly fine examples of "atoll-texture" are shown by Co-Ni-Fe-(Cu) sulpharsenides in specimens from the disused Raipas mine near Alta, Finnmark, northern Norway. Reinvestigations by optical, electron microprobe, and X-ray diffraction techniques have shown that the atolls, and the grains from which they appear to have developed by core replacement, are composed of two mineral phases. One, with a general composition $(Co,Ni,Fe,Cu)_3As_2S_4$ can be shown to be a Co-rich member of the cobaltite-gersdorffite series; the other phase, $(Co,Ni,Fe,Cu)_2AsS_2$, does not appear to correspond to any reported natural mineral having this general composition.

The nature of the often spheroidal grains from which the atolls have developed is not too certain. They bear certain resemblances to the general class of framboidal textures, though their compositions and internal structures are unusual.

1 Introduction

This paper records examples of atoll textures in specimens of copper-cobalt-nickel ore from the long-disused Raipas mine near Alta, Finnmark, northern Norway (latitude 69° 55′ longitude 23° 20′E).

The geology and mineralogy of this small but rich deposit, which the senior author considers to represent infilling of karst-collapse structures in Proterozoic dolomites, has been reported upon previously (Vokes 1955, 1957a). Studies are currently in progress to elucidate the ore genetical problems of the deposit.

The term "atoll texture" is used to describe a texture consisting of a ring or annulus of one mineral, with another mineral or minerals occurring within and without the ring, by analogy with the geographical feature of the same name. An early usage of the term was to describe shells or rings of pyrite forming "atoll-shaped" residuals in galena, sphalerite, and chalcopyrite in specimens of ore from Mt. Isa, Queensland (Grondijs and Schouten 1937). The texture in this case was considered to be the result of hypogene replacement of irregular spherical pyrite concretions starting at their cores and proceeding to an ultimate stage of irregularly rounded shapes of galena or sphalerite. This conclusion was confirmed by laboratory experiments in which bornite was made to replace cores of Mt. Isa pyrite (Schouten 1937, in Edwards 1954).

1 Geologisk Institutt, Universitetet i Trondheim, Norges Tekniske Högskole, Trondheim, Norway

2 Present address: Norges geologiske undersökelse, Trondheim, Norway

Atoll textures also occur in Ag-U-arsenide deposits and have been referred to core replacements by some authors. In particular Furnival (1939) described cobaltite being replaced by native silver, and pitchblende being replaced by sulphides, in an uranium deposit in the Great Bear Lake district of Canada; Bastin (1917), in a study of the silver ores of Cobalt, Ontario, described native silver replacing niccolite at the cores of safflorite-niccolite intergrowths. On the other hand, similar textures between cobalt arsenides and silver in the Kongsberg ores have been interpreted by Neumann (1944) as being due to rim replacement of the silver by the arsenides. Petruk (1971, p. 136) interpreted the textures in the Cobalt ores as reflecting a depositional sequence from the cores outwards, the native silver replacing other minerals (niccolite, cobaltite, etc.) because it continued to be mobile after these had crystallized.

Atoll textures in specimens from the Raipas mine were noted some years ago (Vokes 1957b) but were considered then to be formed by the linnaeite-series mineral, siegenite. A later electron microprobe study indicated that the spheroids and atolls consisted of a cobalt-nickel-arsenic sulphosalt (Vokes 1967, p. 12). Subsequently, the original material, plus new specimens collected at the Raipas mine during the summer of 1971, were restudied in more detail. The work consisted in a careful examination of the textures under the ore microscope and a determination of the chemistry of the phases involved, employing the ARL-EMX electron microprobe housed in the Physics Department of the University of Trondheim, Norwegian Institute of Technology, Trondheim. It revealed a wider range of forms involved in the atoll formation than was previously noted; confirmed, quantitatively, the chemical composition of the phases involved [two Co-Ni-Fe-(Cu) sulpharsenides], and revealed their mineralogical nature.

2 Textural Observations

In polished section the sulpharsenides show a variety of forms, depending on the shape and size of the original grain and on the degree of core replacement which has taken place. The atolls and related features are irregularly distributed in the ore material, occurring typically in rather densely packed swarms, mainly in the abundant bornite of the ore, and in the carbonate gangue, but also to a lesser extent in chalcopyrite (Fig. 1). In the majority of cases observed the mineral forming the cores of the sulpharsenide grains is bornite, although in a small number of cases chalcopyrite appears in this position. Both of the copper-iron-sulphides, especially in the bornite, have been later replaced to varying degrees by digenite and, more rarely, covellite. This latter replacement does not appear to be related to the actual atoll formation.

On the basis of the ore microscopical work, the sulpharsenide phases appear in three textures.

1. Whole, or unreplaced, original grains.
2. Partly replaced grains, or atolls.
3. Residual rim atolls and related "debris".

1. Whole or unreplaced grains are relatively rare in the specimens examined, a fact which has added to the difficulties of determining their original chemical and morpho-

Fig. 1. General view of cobaltite-gersdorffite atolls *(white)* in a groundmass of bornite *(light grey)* and carbonate gangue *(dark grey)*. Plane polarized reflected light, air

logical natures. The most frequent shapes are typically spheroidal, being almost perfectly circular in cross-section, yet exhibiting an indented outer surface, suggestive of the faces of small constituent crystals forming the spheroids. They appear to have had a very uniform diameter, of the order of 35–45 μm. On cursory examination the spheroids appear uniform in optical properties, but high magnifications under oil immersion reveal that they are composed of two mineral phases. Phase A is distinctly white in colour, has a relatively high reflectivity and is slightly the harder of the two; phase B has a lower reflectivity and a distinctly grey tone. Phase A occurs preferentially as a very narrow (1–2 μm), often discontinuous, rim to the spheroids, and as small (1–5 μm), irregularly distributed grains and groups or strings of grains throughout the core of the spheroids, which is composed of phase B material (Fig. 2). Both phases are isotropic under crossed polars. Due to the small grain size and the intricacy of the structure of the spheroids, no quantitative measurements of the reflectivities or microindendation hardnesses of the two phases were possible.

2. The complexly heterogeneous spheroid texture is by far the most abundant of the three types. Typical examples are shown in Fig. 3 from which their complex nature can be seen. The rims to the original spheroids are always preserved intact at this stage and only exceptionally do they appear to be breached by the copper sulphides. The indented nature of the circumferences of the spheroids is accentuated in the topography of the rim in this typical atoll texture. However, in contrast with most of the atoll textures previously reported (see p. 118) the cores of the Raipas atolls do not consist of a homogeneous replacing phase. They show a very intricate structure, which ideally

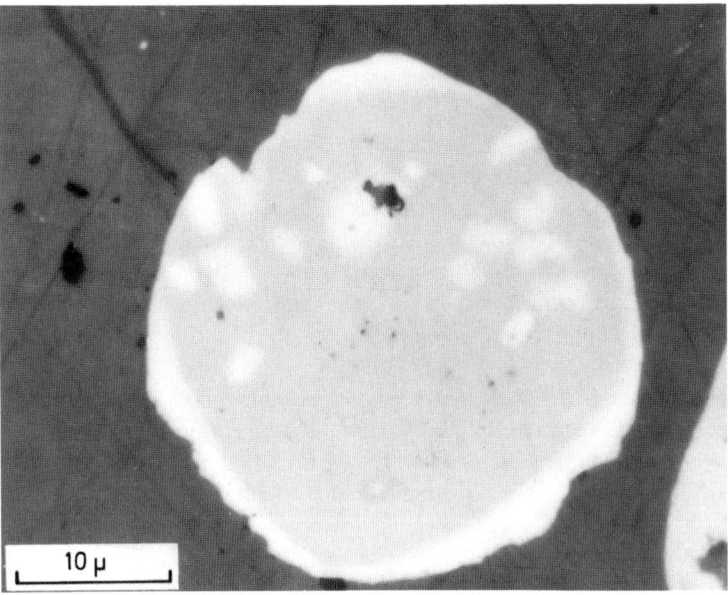

Fig. 2. Close-up view of unreplaced cobaltite-gersdorffite spheroid showing internal texture. Phase A: $Me_2(As,S)_3$, *whiter,* as rim and scattered grains; phase B (cobaltite-gersdorffite): *darker,* as groundmass. Spheroid surrounded by bornite, *dark grey.* Plane polarized reflected light, oil

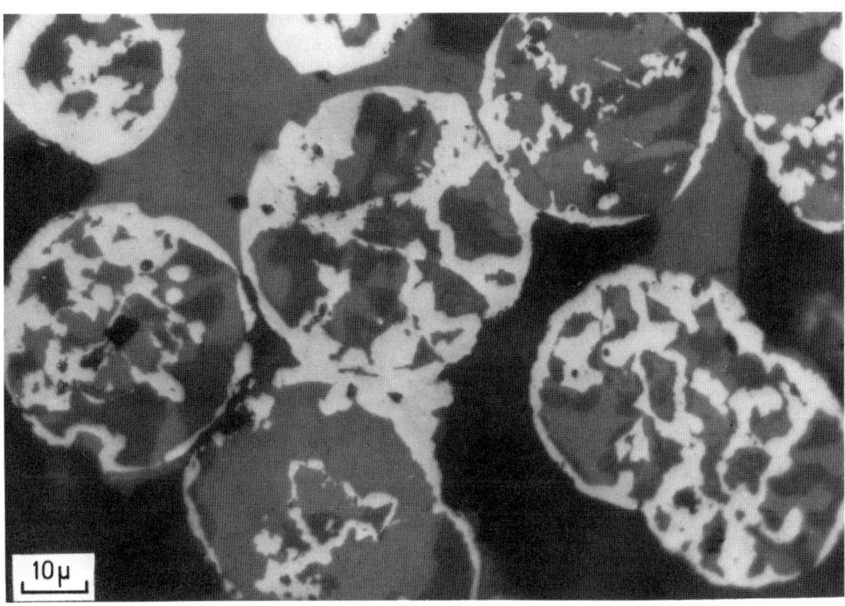

Fig. 3. Complex forms in partially replaced atolls *(white).* Groundmass and replacing media, bornite *(grey)* and digenite *(darkest).* Plane polarized reflected light, oil

seems to consist of an inner atoll of the sulpharsenide phases, concentric to the outer one, and connected to it by several irregular ribs. Variations on this texture are common; for example, the central atoll can take the form of a solid "island", while the connecting ribs can be imperfectly developed or lacking entirely. Others show a more patchy disposition of the copper sulphides and the sulpharsenides in the core of the atolls. Examination of the sulpharsenide patches, islands, or ribs in the interior of the atolls shows that both phases, A and B, are present, though phase A appears to predominate in the outer atolls which represent the rims of the original spheroids. Apparently as the result of increasing core replacement of the sulpharsenides by the copper sulphides, the internal structures of the atolls are destroyed and the rims or atolls become thinned out and breached, so that textures representing transitions to the third type (see below) are produced.

3. Residual atoll type textures appear to represent complete core replacement of the original spheroid by the copper sulphides so that only the extreme outer edges remain (Fig. 4). An extreme expression of this type of texture is occasionally seen in what appears to be an irregular jumble of rim fragments which have lost their annular structure.

Careful examination in these cases does indicate that the last parts of the original spheroid rims to go are composed of phase A.

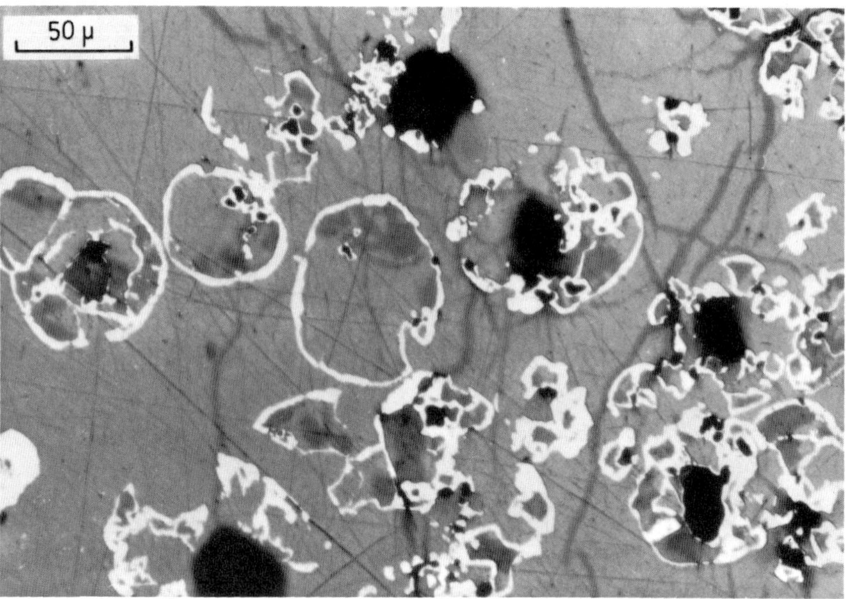

Fig. 4. Residual atoll rims *(white)* in groundmass of bornite *(light grey)* veined by digenite *(darker)*. *Black:* holes. Plane polarized reflected light, oil

3 Composition of the Spheroids

Quantitative electron microprobe, back-scatter X-ray images and linear scans for the elements showed that cobalt was preferentially concentrated around the rims of the spheroids and other forms involved in the atoll textures, as well as at vaguely defined points within their interiors. This distribution of cobalt seemed to parallel that of phase A as seen under the optical microscope. Quantitative electron microprobe determinations were carried out on the two phases for Co,Ni,Fe,Cu,As, and S using pure metals, millerite and chalcopyrite standards. A correction programme, mainly after Springer, was employed to arrive at the compositions of the phases measured. The analysis points were chosen after microscopic examination and photography of the grains. By this means it was possible to analyze phases A and B separately.

However, in some cases it seems that the extremely fine grain size caused the areas being analyzed to overlap slightly from one phase to the other. This is doubtless reflected in the spreads of metal values recorded. Nevertheless, the analyses and the resulting calculations are consistent enough to indicate very close approximations to the chemistries of the two phases. From Table 1 it can be seen that the microprobe analytical results fall into two main groups, corresponding to the phase types distinguished. The four points of phase A analyzed show higher (up to 9%) Co contents, slightly lower (about 1%) Fe contents, lower (2%–3%) As contents and lower (1%–2%) S contents than the six points of phase B. Ni and Cu contents are similar in the two phases, with a tendency to be slightly higher in phase B.

Table 1. Chemical compositions of sulpharsenide phases in spheroids and atolls. Raipas mine material. (Analyst G.S. Strand: ARL microprobe, N.T.H. Trondheim)

Grain No.	Phase	Co	Ni	Fe	Cu	As	S	Total
1	A	26.78	10.09	4.99	1.24	30.52	26.87	100.49
2	A	27.40	10.10	4.71	1.23	30.49	26.12	100.05
3	A	28.83	9.86	4.55	1.05	28.99	26.48	99.76
4	A	29.57	11.06	4.36	0.79	32.78	22.14	100.70
5	B	23.46	9.40	5.40	1.03	32.53	27.67	99.49
6	B	23.41	9.82	5.32	1.07	32.41	27.75	99.78
7	B	21.89	12.02	5.42	1.41	30.84	28.55	100.13
8	B	21.39	10.43	5.78	1.67	32.59	27.62	99.48
9	B	20.81	10.71	5.63	1.48	32.66	27.96	99.25
10	B	19.79	12.82	5.72	1.05	31.99	28.89	100.26

Calculations of the atomic compositions represented by the ten analyses are shown in Table 2. The results seem to be consistent within the two groups of analyses, indicating general formulae:

Phase A: $(Co,Ni,Fe,Cu)_2AsS_2$ or $Me_2(As,S)_3$
Phase B: $(Co,Ni,Fe,Cu)_3As_2S_4$ or $Me(As,S)_2$.

Table 2. Calculated atomic compositions of sulpharsenide phases in spheroids and atolls. Raipas Mine, Finnmark

Analysis points	Co	Fe	Ni	Cu	Sum Me	As	S	Sum As,S	Atom percent			Atomic ratios		Generalized formula
									Me	As	S	As/S	$\frac{Me}{As + S}$	
A.1	1.095	0.212	0.414	0.047	1.768	0.981	2.019	3.0	37.08	20.57	42.34	0.485	0.589	
A.2	1.142	0.207	0.423	0.048	1.820	0.999	2.001	3.0	37.76	20.73	41.51	0.499	0.607	$(Me)_2(As,S)_3$
A.3	1.210	0.201	0.416	0.041	1.868	0.957	2.043	3.0	38.37	19.66	41.95	0.468	0.623	As: S = 0.5
A.4	1.335	0.207	0.501	0.033	2.076	1.164	1.836	3.0	40.90	22.93	36.17	0.634	0.692	Me: As + S = 0.6
B.5	0.921	0.223	0.370	0.038	1.552	1.004	1.996	3.0	34.09	22.06	43.85	0.503	0.517	
B.6	0.918	0.220	0.386	0.039	1.563	1.000	2.000	3.0	34.25	21.91	43.83	0.500	0.521	
B.7	0.856	0.224	0.472	0.051	1.603	0.948	2.052	3.0	34.82	20.59	44.58	0.462	0.534	$(Me)(As,S)_2$
B.8	0.840	0.239	0.411	0.061	1.551	1.007	1.993	3.0	34.08	22.13	43.79	0.505	0.517	As: S = 0.5
B.9	0.810	0.231	0.418	0.053	1.512	1.000	2.000	3.0	33.51	22.16	44.33	0.500	0.504	Me: As + S = 0.5
B.10	0.759	0.232	0.494	0.037	1.522	0.965	2.035	3.0	33.66	21.34	45.00	0.474	0.507	

4 The Identities of the Sulpharsenide Phases

Phase A, with a general formula $Me_2(As,S)_3$, does not seem to be comparable to any natural mineral. Its occurrence as the minor phase in the already minute spheroids and atolls in the Raipas material has made it impossible to investigate its nature by X-ray diffractometry.

Phase B, $Me(As,S)_2$, where Me is Co, Ni, Fe (Cu), would seem to correspond to a mineral in the cobaltite-gersdorffite-arsenopyrite ternary system, in particular the cobaltite-gersdorffite series.

Various authors have investigated minerals in this series by means of electron microprobe analysis, enabling a comparison to be made between their material and that of Raipas.

One of the latest systematic microprobe investigations of naturally occurring gersdorffites is that of Rosner (1970). He analyzed 72 specimens of this mineral from localities around the world. The main difference in metal contents shown by the Raipas analyses from those published by Rosner is in the Co:Ni ratios. The highest Co:Ni ratio found by Rosner was 0.88 in a specimen from a Siegerland locality; the Raipas B phase shows Co:Ni ratios of the order of 2. Figure 5 is a ternary plot of the Fe(As,S), Co(As,S) and Ni(As,S) contents of the Raipas sulpharsenides. Rosner's (1970, Fig. 1) plot of the 72 specimens analyzed by him is shown for comparison. It can easily be seen that none of Rosner's points plot higher than 40% Co(As,S) on the ternary diagram, whereas the Raipas phase B points all plot in the field greater than 60% Co(AsS).

In another recent publication, Petruk et al. (1971, Fig. 109) show an identical ternary plot of Fe-Co-Ni sulpharsenides from the Cobalt and Gowganda, Ontario, mining camps. These authors use the term *cobaltite* to describe material whose compositions extend into the gersdorffite field of Rosner (1970)[3], because the material involved forms a solid solution with the true cobaltite of the deposit, but not with gersdorffite. On a chemical basis, the Raipas phase B sulpharsenide would fall into the cobaltite field of Petruk et al. (1971) (see Fig. 5).

In another microprobe study of naturally occurring sulpharsenides, Klemm and Weiser (1965) described what they term "mixcrystals of cobaltite-gersdorffite" from specimens of chromite-bearing veins in the footwall country rocks of the Outokumpu orebody in Finland. Their sulpharsenide crystals appeared optically homogeneous, but a zoned structure was clearly visible in scans under the electron microprobe. Klemm's and Weiser's analytical results are plotted on Fig. 5, together with other analyses under discussion. It can be seen that they coincide very well with the plotted analyses of the Raipas sulpharsenides.

Thus, from the point of view of metal chemistry and metal: (As + S) ratios, the Raipas B phase, at least, would seem to be a cobaltite or a Co-rich member of the cobaltite-gersdorffite series. However, there is one aspect of the chemistry of the Raipas sulpharsenide phases which distinguishes them from other reported naturally occurring members of the cobaltite-gersdorffite series. This is the As:S ratio. The

---

3 Petruk et al. (1971), for example, plot as cobaltite a specimen with a Co:Ni (at pptn) ratio as low as 0.83.

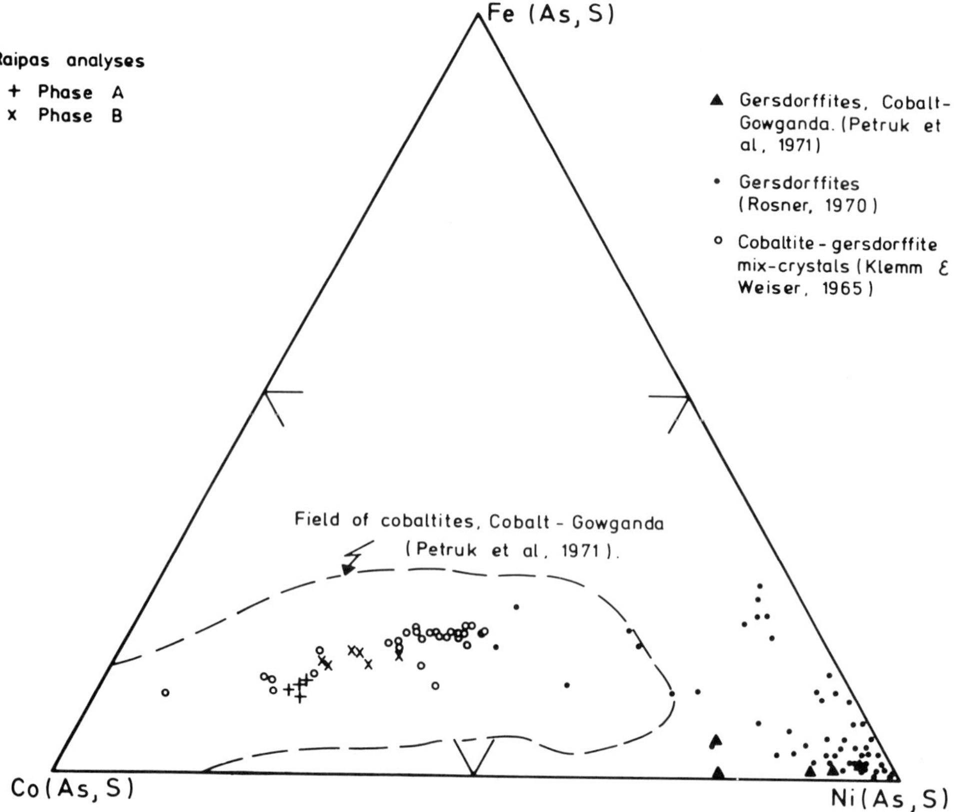

Fig. 5. Triangular Fe(As,S)-Co(As,S)-Ni(As,S) plot of compositions of Raipas sulpharsenide phases. Previously published analyses plotted for comparison

Raipas analyses show As:S ratios of the order of 0.5. The lowest As:S ratio in the specimens analyzed by Rosner (1970) was 0.92. None of the gersdorffites or cobaltites analyzed by Petruk et al. (1971) from the Cobalt-Gowganda ores show an As:S (at pptn) ratio less than 1.0. However, if one turns to synthetical studies of the relevant systems (e.g. Yund 1962; Bayliss 1969) it is readily apparent that the lower As:S ratios are not inconsistent with the structures of the phases produced.

The above comparisons indicate that the Raipas phase B sulpharsenide has a chemical composition not incompatible with those of cobaltites-gersdorffites, though it shows As:S ratios differing considerably from those of hitherto analyzed naturally occurring minerals. Phase A, while being very similar chemically, has a Me:(As + S) ratio which does not correspond to those of the cobaltites. For the time being it must be regarded as an unknown, perhaps new, species.

5 X-Ray Investigations

Attempts to separate out pure spheroid material, let alone the individual phases, for X-ray diffractometry, were unsuccessful due to their extremely fine grain size. The authors are therefore grateful to Professor E.F. Stumpfl, then at Mineralogisches Institut, Universität Hamburg, Federal Republic of Germany, for an investigation employing microdrill techniques on a polished section of Raipas material and the examination of the bornite-spheroid powder so produced. Debye-Scherrer patterns obtained (analyst Miss Cornelisen) showed lines of bornite and a mineral of the cobaltite-gersdorffite series. Subsequent examination of the powder photographs in the X-ray laboratory of Canmet, Ottawa, Canada, confirmed the Raipas material as a mineral of the cobaltite-gersdorffite series. The spacings of the characteristic lines indicated the Raipas material was closer to the pure Co end member than to the Ni end member (Petruk 1976, pers. comm.). It is considered that the X-ray pattern obtained was essentially that of the major phase B; no lines relating to phase A were detected.

6 Discussion

6.1 Formation of the Atoll Textures

The microscopical evidence reviewed in this paper seems to show conclusively that, in the case of the Raipas mineralization, the atoll textures are the results of a continuous process of core replacement acting upon original sulpharsenide spheroids. There seems to be an almost continuous progression of textures from the unreplaced spheroids, through the complexly developed atoll textures to the residual atoll and fragmental textures which represented the outer rims of the spheroids. There is no suggestion that processes involved acted in the other direction, i.e. to form spheroids by replacement of the copper sulphides by the sulpharsenides.

The complex nature of the core replacement textures observed is difficult to explain. It does not seem to be related to the original intricate, internal structure of the sulpharsenide spheroids. Initially, at least, it would seem that both sulpharsenide phases are replaced indiscriminately, though the parts of the spheroids which are the last to be replaced, i.e. the rims, do seem to consist predominantly of phase A.

The replacing phase has been, primarily, bornite, the mineral which also envelopes the spheroids. It seems that at some stage or other, in the history of the Raipas copper deposit, the sulpharsenide spheres became unstable in the environment represented by the bornite. (It must also be remarked that even where the spheroids appear to be totally enclosed in the carbonate or fahlore, the mineral replacing the spheroids is still bornite.)

This instability could have developed under various conditions of formation of the deposit:

1. During the epigenetic introduction of the bornite (by some form of solution) at a stage post-dating the formation of the sulpharsenide spheroids. The replacement

would presumably start almost instantaneously and cease when the activity of the system reached a critical minimum value.

2. During some stage or event later than initial deposition of the sulpharsenides, bornite, and carbonate gangue (which may or may not have been deposited simultaneously).

   The main possibilities here are changes occurring during diagenesis in the case of a sedimentary deposition, or during a later metamorphism, independent of the original mode of formation.

Since there is no evidence whatsoever that the Raipas deposit and its surrounding rocks have been metamorphosed to a level which would initiate readjustments in the original mineralogy, this last possibility may be eliminated.

It is of interest to note that the bornite in the cores of the atolls showed As contents of the order of a few tenths per cent when analyzed by the electron microprobe. Bornite outside the atolls was essentially As-free. This would appear to reinforce the conclusion that the bornite in the atoll cores has actively replaced the sulpharsenides and inherited their chemistry to a certain degree.

The bornite in its turn became unstable at a still later stage and was altered to digenite along cracks and grain boundaries and in irregular patches. This alteration is often accompanied by the development of small, secondary chalcopyrite lamellae and idaite. Such changes appear to be general during the initial stage of the supergene breakdown of bornite (see, e.g. Krause 1965) and are probably related to the recent weathering surface in the area.

6.2 Nature of the Original Spheroids

A consideration of the nature of formation of the original sulpharsenide spheroids is of value in attempting to reconstruct the history of atoll development.

Ramdohr (1969, pp. 100–104) remarks that spheroidal grains "are not common" in ore deposits, and that "great care" must be exercised in their interpretation. Ramdohr cites as examples of spheroidal shapes: trapped droplets of melt; crystals formed under conditions of "inhibited growth" in which the surface tension is still stronger than the energy of idiomorphic development; spheroidal exsolution bodies, secondary fillings of vesicular cavities, grains occurring as inclusions in poikiloblastic intergrowths, corrosion relicts, "and still others".

Through considerations of the geology of the deposit and its surroundings, all possibilities except one may be eliminated. This is, that the spheroids resulted from crystallization under conditions where surface tension determined the shapes of the grains.

Among textures of this type, the one that immediately comes to mind is the framboidal texture displayed by pyrite. Rickard (1970) considers that two main textural features are characteristic of the true framboid:

1. The outer shape is spheroidal or modified spheroidal.
2. The spheroids consist exclusively of "discrete equant microcrysts".

While displaying the first of these textural features, the Raipas spheroids do not display the second one; the disposition shapes and sizes of the crystals or grains of phase A within the body of the spheroid are by no means equant and the outer rims of phase A seem to have no corollary in any undoubted framboids described in the literature. Furthermore, all examples so far described appear to be monomineralic–overwhelmingly FeS_2. Farrand (1970) has succeeded in synthesizing framboids of Cu, Zn, Pb, Ni and As sulphides, but so far none of these compositions have been reported from framboids occurring in natural ores.

Another feature which serves to distinguish the Raipas spheroids from normally accepted framboids is their size. As already stated, the spheroids are fairly evenly sized with an average diameter in section of around 35–45 μm. The size distribution of the measured diameters would indicate that the maximum diameter of the spheroids is of the order of 50 μm. This is very much in excess of the normal size ranges which are quoted for framboids. Rickard (1970, p. 270) cites data of Morissey and Rickard showing a positively skewed unimodal distribution around 4 μm, and with 95% of the spheroids less than 45 μm for framboids from Tynagh, Ireland. Solomon (1965, p. 741) gives the diameters of the Mt. Isa framboids he investigated as 1–5 μm. Love and Zimmermann (1961) quote 1–12 μm as the size range they found for Mt. Isa "microorganisms", as the framboids were then called. On the other hand, Love (1964) notes that pyrite spheroids, showing framboidal textures, found in fine-grained sediments, even those of Recent age, "may be 40 μm in size or even more".

The present authors feel that, while it is not possible to draw a very firm conclusion as to the genesis of the sulpharsenide spheroids in the Raipas deposit, they do show strong resemblances to sulphide spheroids of the framboid type and were probably formed under similar, i.e. low-temperature, conditions, consistent with a possible intrakarstic mode of formation of the deposit.

Acknowledgements. The authors are grateful to Dr. W. Petruk, Canmet, Ottawa, for a critical reading of the manuscript.

References

Bastin ES (1917) Significant mineralogical relations in silver ores of Cobalt, Ontario. Econ Geol 12: 219–236

Bayliss P (1969) X-ray data, optical anisotropism, and thermal stability of cobaltite, gersdorffite and ullmannite. Mineral Mag 31:26–33

Edwards AB (1954) Textures of the ore minerals and their significance. Inst Min Metall, Melbourne, Australia, 242 pp

Farrand M (1970) Framboidal sulphides precipitated synthetically. Miner Deposita 5:237–247

Furnival GM (1939) A silver-pitchblende deposit at Contact Lake, in the Great Bear Lake area. Econ Geol 34:739–776

Grondijs HF, Schouten C (1937) A study of Mt. Isa ores. Econ Geol 32:407–450

Klemm DD, Weiser T (1965) Hochtemperierte Glanzkobalt-Gersdorffit-Mischkristalle von Outokumpu/Finnland. Neues Jahrb Mineral Monatsh 1965:236–241

Krause H (1965) Idaite, Cu_5FeS_6 from Konnerud, near Drammen. Contributions to the mineralogy of Norway, no 33. Nor Geol Tidsskr 45:417–422

Love LG (1964) Early diagenetic pyrite in fine-grained sediments and the genesis of sulphide ores. In: Amstutz GC (ed) Developments in sedimentology, vol II, Sedimentology and ore genesis. Elsevier, Amsterdam, pp 11–17

Love LG, Zimmermann DO (1961) Bedded pyrite and microorganisms from the Mt. Isa shale. Econ Geol 56:873–896

Neumann H (1944) Silver deposits at Kongsberg. Nor Geol Unders Publ 162:133

Petruk W (1971) Mineralogical characteristics of the deposits and textures of the ore minerals. Can Mineral 11:108–139

Petruk W, Harris DC, Stewart JM (1971) Characteristics of the arsenides, sulpharsenides and antimonides. Can Mineral 11:150–186

Ramdohr P (1969) The ore minerals and their intergrowths. Pergamon Press, Oxford, 1124 pp

Rickard DT (1970) The origin of framboids. Lithos 3:269–293

Rosner B (1970) Mikrosonden-Untersuchungen an natürlichen Gersdorffiten. Neues Jahrb Mineral Monatsh 1970:483–498

Schouten C (1937) Metasomatische Probleme, Amsterdam (quoted in Edwards 1954)

Solomon MJ (1965) Investigations into sulfide mineralization at Mount Isa, Queensland. Econ Geol 60:737–765

Vokes FM (1955) Observations at Raipas Mine, Alta, Finnmark. Nor Geol Unders Publ 191 (Årbok 1954):103–114

Vokes FM (1957a) Some copper parageneses from the Raipas formation of northern Norway. Nor Geol Unders Publ 200 (Årbok 1956):74–111

Vokes FM (1957b) On the presence of minerals of the linnaeite series in some copper ores from the Raipas formation of northern Norway. Nor Geol Unders Publ 200 (Årbok 1956):112–120

Vokes FM (1967) Linnaeite from the Precambrian Raipas Group of Finnmark, Norway. An investigation with the electron microprobe. Miner Deposita 2:11–25

Yund RA (1962) The system Ni-As-S; phase relations and mineralogical significance. Am J Sci 260: 761–782

A New Type of Sulphosalt Mineralization in the Myrthengraben Gypsum Deposit, Semmering, Lower Austria

W. TUFAR[1]

Abstract

In the Myrthengraben (Lower Austria), a Carnian slate and dolomite series of the Semmering Mesozoic contains a gypsum-anhydrite deposit characterized by an extremely complex, iron poor, non-ferrous metal mineralization (i.e., Cu, As, Pb, Zn, Fe, S, ± Sb, Sn, Se, U, etc.). Sulphides and sulphosalts are the primary ore minerals. Major constituents of the ore paragenesis are enargite, tennantite, wurtzite, galena, pyrite, and jordanite. Minor and/or accessory ore minerals include luzonite, seligmannite, stibnite, a Cu-Zn-Sn-S-mineral and a new As-S-mineral. Characteristic gangue minerals are gypsum, anhydrite, dolomite, and magnesite. In places, epsomite is relatively abundant. A uranyl-carbonate, andersonite, occurs in the supergene zone. Of great interest is the recognition of several new types of myrmekitic intergrowths of enargite. Based on paragenesis and observed ore mineral intergrowths, the iron poor, complex, non-ferrous metal mineralization in the evaporite deposit of the Myrthengraben represents a totally new type of sulphosalt occurrence.

1 Introduction

The well-known lead-zinc deposits of the type Bleiberg-Kreuth (Carinthia) – Mežica (Slovenia) are a characteristic feature of the East-Alpine Triassic strata. These occurrences are bound to the Anisian, Ladinian, and Carnian rocks of the Upper Austro-Alpine Zone. They are characterized by the predominance of silver-poor galena and sphalerite/schalenblende. Locally, among others, barite, fluorite, and anhydrite are gangue minerals of these mineralizations.

More than 30 years ago, Schwinner (1949) demonstrated that this type of mineralization is of syngenetic-sedimentary origin and constitutes a special sedimentary facies within the Triassic strata. At that time, this clarification was of fundamental importance. The results of subsequent investigations by numerous other workers support these findings.

Galena and sphalerite in association with barite are also found in the Anisian-Ladinian dolomite of the Semmering Mesozoic, which is tectonically part of the Lower Austro-Alpine Zone and shows evidence – if compared with the unmetamorphosed Mesozoic of the Upper Austro-Alpine Zone – of some progressive regional metamorphic overprinting.

The Carnian of the Semmering Triassic consists of a series of "Keuper" slates with dolomite, and is characterized by evaporite occurrences. In the gypsum-anhydrite

1 Fachbereich Geowissenschaften der Philipps-Universität Marburg, Lahnberge, 3550 Marburg/Lahn, FRG

deposit of the Myrthengraben/Semmering (Lower Austria), a non-ferrous metal mineralization of very complex composition, in the most striking cases occurring along with sulphosalts, is found locally.

Brief references concerning the mineral content of this deposit were given by Schroll (1954), and Schroll and Azer Ibrahim (1959). The relatively high selenium content of this particular occurrence has been pointed out by Rockenbauer (1960), and Rockenbauer and Schroll (1960). In galena, selenium concentrations of up to 1 wt.% were determined, the highest selenium content so far known from East-Alpine ore minerals. The first systematic investigations of the mineral paragenesis, by Tufar (1963, 1966, 1967, 1968, 1969, 1980), indicated paragenetic peculiarities in the Myrthengraben deposit, which required further investigation.

A survey of the geology of the area has been given by Kristan and Tollmann (1957), Tollmann (1964, 1977), and Tufar (1963). Geological descriptions of the Myrthengraben gypsum-anhydrite deposit have been advanced by Bauer (1967) and Neuner (1964).

Fig. 1a, b. Crystal aggregate of magnesite with local thin coatings. Magnesite clearly exhibits crystal faces and shows distinct rhombohedral habit. **b** Detail of **a**, crystal faces with coatings. Scanning electron microscope: **a** × 148; **b** × 540

Fig. 2. Crystal aggregate of andersonite exhibiting distinct thin hexagonal platelets. Scanning electron microscope: × 1900

Fig. 3. Gypsum *(black)* with galena aggregate *(light gray)* showing a jordanite replacement rim *(medium gray,* weak bireflectance causing different *shades of gray)* in contact with gypsum. Jordanite, in turn, exhibits a replacement rim of seligmannite *(dark gray)* at the gypsum interface. In the jordanite rim an idioblastic pyrite *(light gray, nearly white)* can be observed. Fractures in ore minerals, especially in jordanite, are healed by mobilized gypsum. Polished section, oil immersion: × 91

Fig. 4. Recrystallized galena (grain boundaries with the 120° "triple junction", typical for recrystallization fabrics, are clearly recognizable from the weak, anomalous anisotropism) with peripheral replacement rim of jordanite (locally brightened up due to anisotropism) in gypsum gangue (illuminated by internal reflections). Polished section: × 65 + Pols

Fig. 5. Gypsum gangue *(black)* with wurtzite inclusion *(dark gray)* clearly being replaced along the {0001} direction mostly by tennantite *(light gray)* and some enargite *(light medium gray)*. Wurtzite encloses abundant primary inclusions of gypsum and some dolomite (also *black*). Later fractures in ore minerals are cemented by mobilized gypsum. Polished section, oil immersion: × 113

Fig. 6. Tennantite (*medium gray,* partly idiomorphic) intergrown with luzonite (also *medium gray to dark gray* due to bireflectance, partly idiomorphic, locally twinning with fine lamellae!), wurtzite *(black)*, and gypsum gangue (also *black*). Polished section, oil immersion: × 245

Fig. 7a, b. Gypsum *(black)* with galena aggregate *(light gray)* showing in contact with gypsum a peripheral replacement rim of jordanite *(light medium gray)*. Jordanite itself shows peripherally along the gypsum interface a narrow replacement rim of seligmannite *(medium gray)*. In the jordanite replacement rim are found relics of wurtzite *(black, right upper half)* and some luzonite and tennantite (both *light dark-gray*). In gypsum and in the jordanite replacement rim an idiomorphic Cu-Zn-Sn-S mineral *(dark gray)*, probably kesterite, and clearly exhibiting cataclasis, is recognizable. Fractures in ore minerals are healed by mobilized gypsum and locally also by mobilized ore minerals (e.g., tennantite, luzonite)

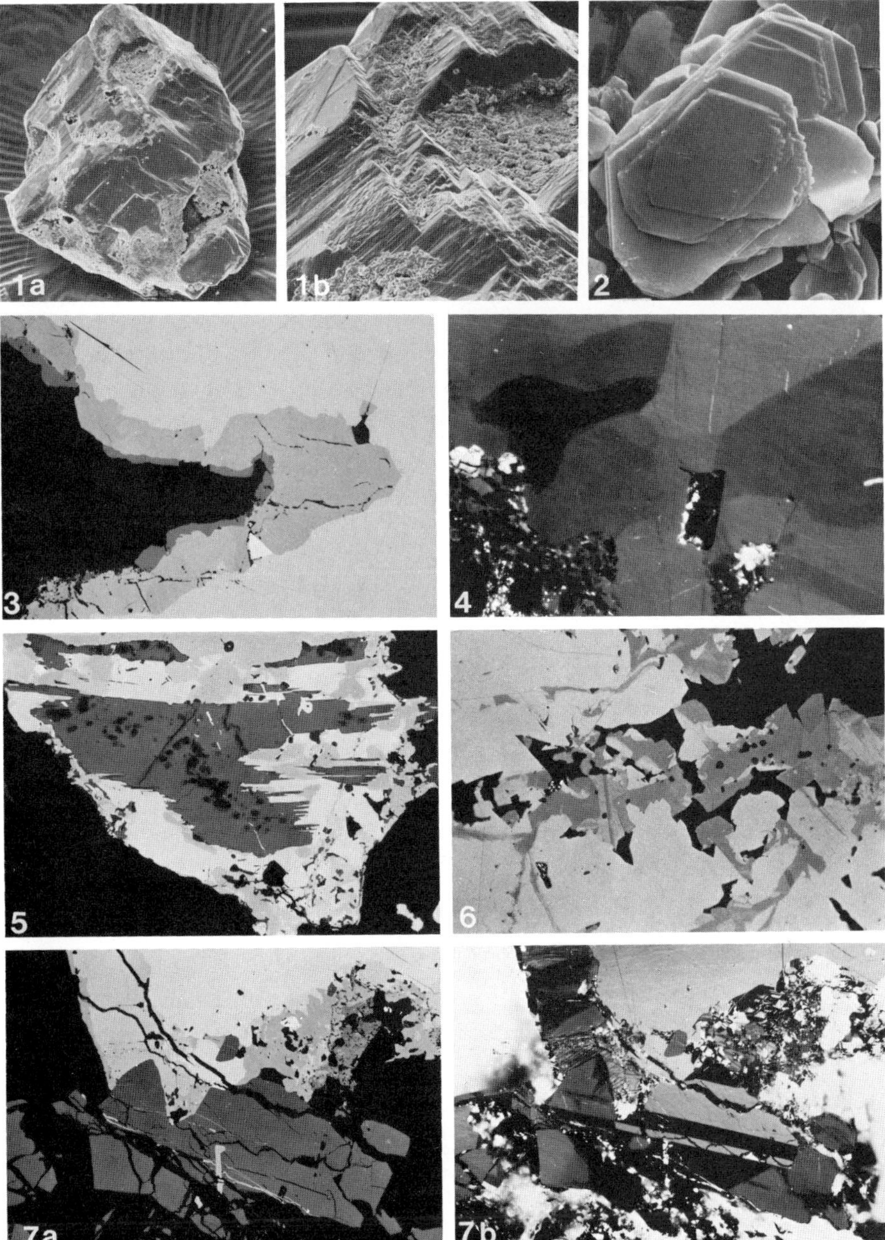

Fig. 7b shows clearly the anisotropism of the Cu-Zn-Sn-S mineral and the fine lamellar twinning of jordanite enhanced through anisotropism *(left upper half)*. Luzonite appears locally brightened up due to anisotropism. Wurtzite and gypsum gangue are both strongly illuminated by internal reflections. Polished section, oil immersion: × 91. **a** 1 Pol.; **b** + Pols

2 On Mineral Paragenesis – Ore and Gangue Minerals

The predominant mineral in the Myrthengraben evaporite deposit is gypsum. Other major gangue constituents are anhydrite and, by way of correlation with the host rocks, dolomite. Magnesite (Fig. 1) was found to be a characteristic gangue mineral, occurring as crystal aggregates. The presence of magnesite in this deposit is of importance in regard to explaining the genesis of other East-Alpine magnesite deposits as well. The geological environment of the Myrthengraben evaporite deposit clearly excludes an epigenetic influx of "hydrothermal" Mg solutions originating from a (hypothetical) deep-seated magma reservoir. On the contrary, it also offers important evidence for a primary, syngenetic accumulation of magnesium during the time of formation of the deposit and its country rocks. In places, relatively large amounts of epsomite are found.

Primary ore mineralization of the Myrthengraben gypsum-anhydrite deposit consists of sulphides and sulphosalts. Major constituents are enargite, tennantite, wurtzite, galena, and pyrite, with enargite and tennantite often predominating. Luzonite and jordanite are mostly found as minor constituents. Other minor and/or trace constituents are seligmannite, stibnite, a Cu-Zn-Sn-S mineral, a new As-S mineral, and bravoite. This mineral paragenesis differs distinctly from all other mineralizations in the Eastern Alps.

Fig. 8. "Idioblast sieves" of pyrite *(light gray)* in "xenoblast sieve" of tennantite *(medium gray)*. Pyrite and tennantite enclose abundant replacement relics of primary inclusions of gypsum and dolomite (both *black*). Polished section: × 48

Fig. 9. Wurtzite *(black)* with some tennantite and luzonite (both *dark gray to medium gray*, hardly distinguishable on photo), as well as with galena and jordanite (both *light to medium gray*, hardly distinguishable on photo), intergrown with idioblastic pyrite *(light gray, nearly white)* in larger amounts. Pyrite shows abundant replacement relics of ore minerals mentioned above and gypsum gangue (also *black*), often in zonal arrangements. Polished section, oil immersion: × 48

Fig. 10. Pyrite xenoblasts, partly "sieves" *(light gray)*, in "xenoblast sieve" of wurtzite *(dark gray)*. Wurtzite and, in places, pyrite contain larger amounts of gypsum and dolomite (both *almost black*) as primary inclusions and replacement relics. Some jordanite *(medium gray)*, already locally replaced oriented by weathering products (also *almost black*), and some subordinate tennantite *(light dark gray)* can be found. Fractures in wurtzite are healed by mobilized tennantite. Polished section: × 96

Fig. 11. Myrmekite of tennantite *(medium gray*, partly internal reflections – *lower left half)* with wurtzite (*nearly black*, internal reflections). Enargite *(light dark-gray to medium gray* – bireflectance!) forms rims locally around tennantite, in some places even myrmekitically intergrown with it *(lower right half)* and with wurtzite. Also small myrmekites of jordanite *(light gray)* with tennantite and enargite. Polished section, oil immersion: × 610

Fig. 12a, b. Portion of enargite *(light dark-gray*, varying shades due to bireflectance! – compare **a** and **b**!) with inclusions of tennantite *(medium gray)*, fine grained wurtzite (*black*, locally base {0001}-developed!), and gypsum (also *black*). Enargite forms fine myrmekites with stibnite *(light gray, nearly white)* and with a new As-S mineral *(medium gray to almost black* due to strong bireflectance! – compare **a** and **b**, e.g., *upper half* of figure as contrast), also in turn in myrmekitic intergrowth with tennantite. Polished section, oil immersion, × 610

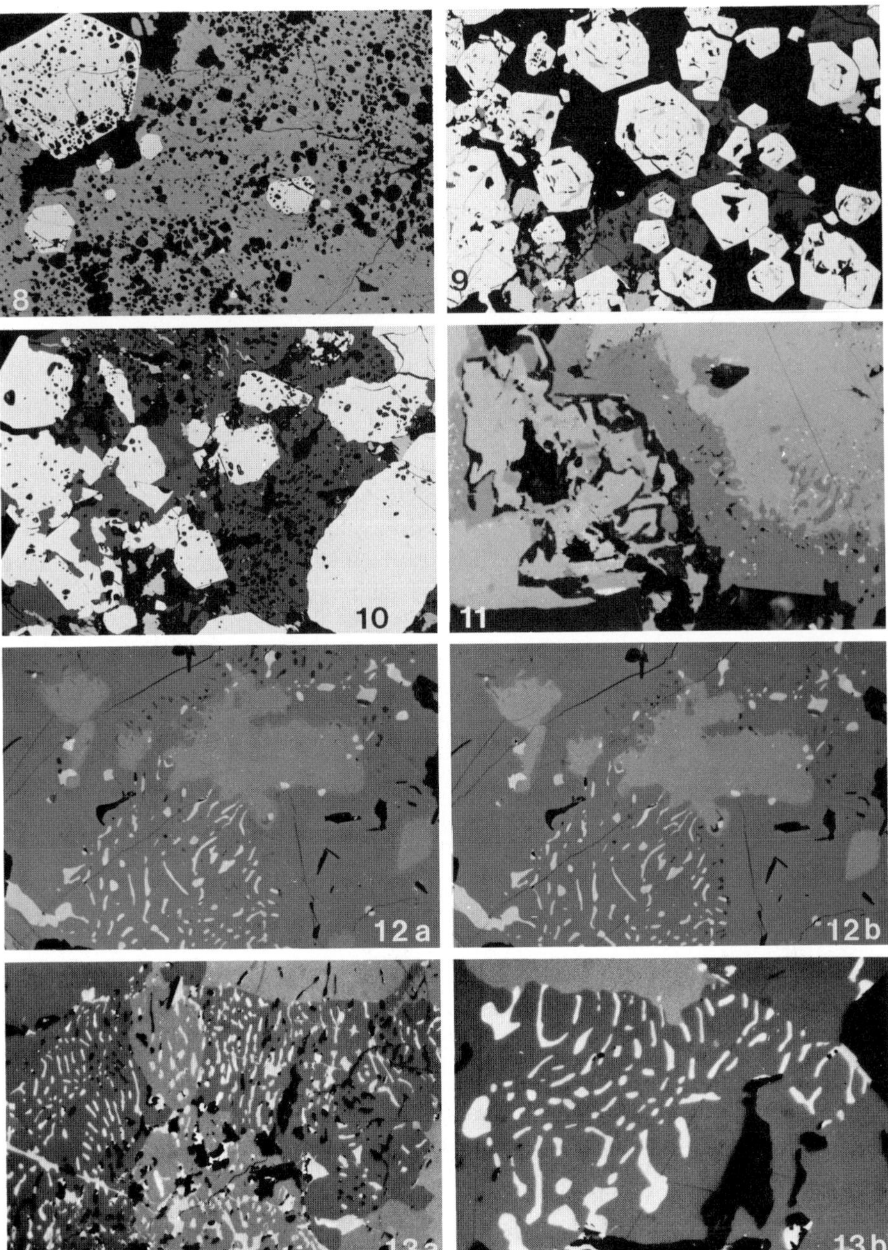

Fig. 13a, b. Myrmekite of enargite (*medium gray to dark gray* due to bireflectance) with jordanite (*light gray*) and some seligmannite (trace darker than jordanite, hardly recognizable in figure), as well as myrmekite of enargite with weathering products (*black*) replacing jordanite and seligmannite. Occasionally some wurtzite and gypsum (both *black*), as well as some tennantite (*medium gray*) can be observed. Polished section, oil immersion: **a** × 610; **b** × 850

The first results of electron microprobe analyses indicated that arsenic is partly substituted by antimony in enargite, tennantite, luzonite, and seligmannite, and also that copper is partly substituted by zinc in tennantite.

Generally, this evaporite deposit displays a very complex, iron-poor, element association (Cu, As, Pb, Zn, Fe, and S, plus, among other elements, Sb, Sn, Se, and U), which is characterized by rare or previously unknown ore minerals and mineral intergrowths. As already indicated by the presence of enargite as one of the major constituents of this ore paragenesis, the iron concentration was rather low, iron evidently having been used up during pyrite formation, causing the absence of chalcopyrite.

Clues concerning the formation temperatures of this mineralization of non-ferrous metals can be deduced from replacement relics and/or primary gypsum inclusions within the "sieve-like" xenoblasts of enargite, wurtzite (Figs. 5 and 10), tennantite (Fig. 8), and pyrite (Fig. 10), as well as in the "sieve-like" idioblasts of pyrite (Figs. 8 and 9). On the basis of these features, formation temperatures of around 100°C and/or lower can be established. This is supported by the occurrence of primary bravoite in association with luzonite and tennantite.

Cu-As(-Sb etc.) mineralization indicates that galena and wurtzite are the oldest components. A special feature displayed by galena is its weak anisotropy, which helps in the recognition of grain boundaries in recrystallized aggregates (Fig. 4). The occurrence of wurtzite (Figs. 5–7 and 9–13) is a further characteristic of this ore paragenesis. Replacement features of zinc sulphide by tennantite and enargite (Fig. 5) along the $\{0001\}$ direction demonstrate that wurtzite and not sphalerite was the primary mineral. This contention is supported by the presence of small crystals of wurtzite idiomorphic along the $\{0001\}$ direction. The iron content of wurtzite is very low, as already indicated by its rather light color ("Honigblende"), as well as the frequently encountered "total illumination" of wurtzite grains due to strong internal reflections. The reflectivities[2] (in %) of galena and wurtzite are:

	Galena		Wurtzite	
λnm	Air	Oil immersion	Air	Oil immersion
464			17.4	5.4
481	46.3	31.3		
546	41.0	27.5	16.1	4.8
584	41.9	27.5		
588			15.8	4.6
624	42.0	27.8	15.6	4.4

Galena and wurtzite frequently occur as replacement relics and inclusions in enargite and tennantite, which predominate among the major constituents, as well as in pyrite. Locally, with a decided preference towards occurring in smaller grains and intergrowths with luzonite, tennantite may show idiomorphism. Enargite and tennantite frequently contain, sometimes as considerable contributions, primary inclusions and replacement relics of gypsum ("sieves") (Fig. 8). The reflectivities (in %) of enargite and tennantite are:

2 Leitz MPV2 Microscope photometer with standardized interference band filter, precision interference line filter and grid monochrometer. See also Bowie et al. (1980), and Picot and Johan (1977) for comparison.

	Enargite				Tennantite	
	Air		Oil immersion		Air	Oil immersion
λ nm	R_p	R_g	R_p	R_g		
464	27.3	29.7	13.3	15.0	30.1	14.2
546	24.7	28.3	10.7	13.3	29.8	14.7
588	24.1	27.9	10.6	13.6	29.1	13.3
624	24.5	28.3	10.8	13.4	27.3	12.3

Luzonite (Fig. 6) typically occurs as fine-grained crystal aggregates, particularly along the grain boundaries of tennantite aggregates. Also, luzonite frequently exhibits idiomorphism which is quite uncommon for this mineral, constituting another peculiarity of this deposit. The reflectivity (in %) of luzonite was determined as follows:

	Air		Oil immersion	
λnm	R_o	R_e	R_o	R_e
464	24.8	26.0	11.0	12.0
546	24.7	26.1	11.5	13.2
588	27.7	29.2	12.8	14.5
624	28.1	30.4	13.4	15.0

Another characteristic feature of this particular ore mineralization is the presence of jordanite, most frequently found in peripheral replacement rims around galena. The rims of jordanite are in turn replaced by seligmannite (Figs. 3 and 7). Jordanite and seligmannite have the following reflectivities (in %):

	Jordanite				Seligmannite			
	Air		Oil immersion		Air		Oil immersion	
λ nm	R_p	R_g	R_p	R_g	R_p	R_g	R_p	R_g
481	38.9	41.9	23.1	26.6	32.1	33.9	18.0	19.4
546	36.3	39.6	21.6	24.1	31.4	32.8	16.5	17.5
584	36.8	39.1	20.9	23.3	30.4	32.1	15.7	16.5
624	35.9	38.3	21.4	23.6	29.5	30.4	15.6	16.4

A partly idiomorphic ore mineral (Fig. 7), which furthermore replaces wurtzite oriented in the $\{0001\}$ direction, is occasionally associated with these replacement rims of jordanite and seligmannite around galena and may also be found within enargite and tennantite. This ore mineral exhibits a reflectivity similar to fahlore, shows weak anisotropism, and rarely displays a very weak bireflectance. Preliminary electron microprobe results indicated a Cu-Zn-Sn-S ore mineral, possibly kesterite (Cu_2ZnSnS_4). The reflectivity (in %) of this tin mineral is:

λnm	Air	Oil immersion
464	24.6	10.5
546	23.3	10.1
588	23.7	9.7
624	23.2	9.7

The presence of a tin mineral is very important, because one of the accepted characteristics of East-Alpine metallogeny is the practically complete absence of tin.

While pyrite is a major constituent of the ore paragenesis, evidence indicates that it was formed relatively late. It occurs in different types; on the one hand in ("sieve-like") idioblasts (Figs. 8 and 9), on the other in ("sieve-like") xenoblasts (Fig. 10). These "pyrite sieves" may be markedly "filled", partly in zonal arrangements. They consist of residual primary inclusions of gangue material (gypsum, dolomite), as well as of replacement relics of older ore minerals (e.g., galena, wurtzite, tennantite, enargite, jordanite).

Another paragenetic phenomenon in the Myrthengraben deposit is the occurrence of a very fine-grained, accessory ore mineral (Fig. 12). It is characterized by a blue-gray color, a moderate reflectivity, strong bireflectance, strong anisotropism, and internal reflections. The reflectivity (in %) of this ore mineral is:

	Air		Oil immersion	
λnm	R_p	R_g	R_p	R_g
548	18.6	28.2	7.0	12.8

The very striking optical properties of this ore mineral are remarkably similar to those given by Ramdohr (1975) for getchellite ($AsSbS_3$), described by Weissberg (1965) from Humboldt County, Nevada. However, preliminary electron microprobe analyses revealed that the Myrthengraben mineral is not getchellite but a new As-S mineral (As_2S_3), with only a small amount of arsenic substituted by antimony.

In addition, stibnite (Fig. 12) is another accessory mineral of the ore paragenesis.

The Myrthengraben ore paragenesis displays peculiarities not at all common even outside of the Eastern Alps. This also applies to characteristic fabric features.

According to Ramdohr (1975), myrmekitic intergrowths of enargite have never before been described. It is therefore remarkable that Myrthengraben samples show a number of different enargite myrmekites, all of which are of a very fine-grained nature. The following myrmekitic intergrowths have been observed: enargite with wurtzite (Fig. 11), enargite with tennantite (Fig. 11), enargite with jordanite (Fig. 13), enargite with seligmannite, enargite with weathering products replacing jordanite and seligmannite (Fig. 13), enargite with stibnite (Fig. 12), as well as enargite with the new As-S mineral (Fig. 12). In addition, there are myrmekites of pyrite with wurtzite, pyrite with galena, tennantite with wurtzite (Fig. 11), tennantite with jordanite, tennantite with seligmannite, and of tennantite with the new As-S mineral.

The ore minerals show evidence of tectonic stress which in part has led to cataclasis. Fractures of the ore minerals are frequently healed by mobilized gypsum and other mobilized minerals such as galena and tennantite. "Ore breccias" can be observed repeatedly, frequently cemented by mobilized gypsum. These breccias are notably of enargite, luzonite, tennantite, wurtzite, the Cu-Zn-Sn-S mineral, and pyrite. Jordanite exhibits bent twinning lamellae and undulatory extinction as a result of postcrystalline deformation.

The deposit also contains numerous supergene ore minerals, among others, a uranylcarbonate, andersonite (Fig. 2). The latter occurs in the form of very small, thin tabular, idiomorphic crystals, another remarkable feature of the Myrthengraben deposit.

3 Conclusions

The Myrthengraben gypsum-anhydrite deposit and its mineral paragenesis presents extraordinary features which have been, up to now, unknown in the Eastern Alps or, for that matter, anywhere else. These include the complex, iron-poor, non-ferrous metal mineralization, the variety of mineral paragenesis, and last but not least, the unusual mineral intergrowths. These peculiarities are of particular interest and permit direct comparison with the well-known sulphosalt occurrence of Lengenbach in the Binn Valley (Ct. Wallis, Switzerland).

Further important results can be expected from future investigations; however, evidence has in any event already been advanced that the Myrthengraben evaporite deposit constitutes a totally new type of sulphosalt occurrence.

Acknowledgments. The author expresses his thanks to Ernst Leitz Wetzlar GmbH, especially to Mr. D. Kristen, for making it possible to carry out extensive photometric measurements. The assistance of Prof. Dr. V. Göbel (Stephen F. Austin State University, Nacogdoches, Texas), Dipl.-Min. W. Baum, M.Sc. (Texasgulf Inc., Golden, Colorado), Dr. J. McMinn and Dipl.-Phys. K. Fecher (Philipps-Universität Marburg) in the critical review of the English manuscript is also gratefully appreciated.

References

Bauer FK (1967) Gipslagerstätten im zentralalpinen Mesozoikum (Semmering, Stanzertal). Verh Geol Bundesanst Wien 1967:70–90

Bowie SHU, Simpson PR, Auld H (1980) An index for the IMA/COM quantitative data file. Int Mineral Assoc, Proc XI Gen Meet IMA, Novosibirsk, 4–10 September, 1978, vol Sulphosalts, platinum minerals and ore microscopy. Akad Sci USSR Mineral Soc, Moscow, pp 287–304

Kristan E, Tollmann A (1957) Zur Geologie des Semmering-Mesozoikums. Mitt Ges Geol Bergbaustud Wien 8:75–90

Neuner KH (1964) Die Gipslagerstätten des Semmerings. Berg- Hüttenmaenn Monatsh 109:319–331

Picot P, Johan Z (1977) Atlas des minéraux métalliques. Mem Bur Réch Geol Min 90:403
Ramdohr P (1975) Die Erzmineralien und ihre Verwachsungen, 4. Aufl. Akademie-Verlag, Berlin, 1277 pp
Rockenbauer W (1960) Zur Geochemie des Selens in ostalpinen Erzen. Tschermaks Mineral Petrogr Mitt 7:149–185
Rockenbauer W, Schroll E (1960) Das Vorkommen von Selen in österreichischen Erzen. Montan-Rundsch 1960:48–52
Schroll E (1954) Ein Beitrag zur geochemischen Analyse ostalpiner Blei-Zink-Erze, Teil I. Mitt Österr Mineral Ges Sonderh 3:1–83
Schroll E, Azer Ibrahim N (1959) Beitrag zur Kenntnis ostalpiner Fahlerze. Tschermaks Mineral Petrogr Mitt 7:70–105
Schwinner R (1949) Ostalpine Vererzung und Metamorphose als Einheit? Verh Geol Bundesanst Wien 1946:52–61
Tollmann A (1964) Exkursion II/6: Semmering-Grauwackenzone. Mitt Geol Ges Wien, 57: Geologischer Führer zu Exkursionen durch die Ostalpen, 193–203
Tollmann A (1977) Geologie von Österreich, Bd I: Die Zentralalpen. Deuticke, Wien, 766 pp
Tufar W (1963) Die Erzlagerstätten des Wechselgebietes. Joanneum, Mineral Mitteilungsbl 1963: 1–60
Tufar W (1966) Bemerkenswerte Myrmekite aus Erzvorkommen vom Alpen-Ostrand. Neues Jahrb Mineral Monatsh 1966:246–252
Tufar W (1967) Andersonit, ein neuer Uranmineralfund aus Österreich. Neues Jahrb Mineral Abh 106:191–199
Tufar W (1968) Der Alpen-Ostrand und seine Erzparagenesen. Freiberger Forschungshefte, C 230 Mineralogie-Lagerstättenlehre: Probleme der Paragenese von Mineralen, Elementen und Isotopen, Teil I, Breithaupt-Kolloquium 1966, Freiberg, pp 275–294
Tufar W (1969) Das Problem der ostalpinen Metallogenese, beleuchtet am Beispiel einiger Erzparagenesen vom Alpenostrand. Sitzungsber Oesterr Akad Wiss, Math-Naturwiss Kl Abt I 177:1–20
Tufar W (1980) Ore mineralization from the Eastern Alps, Austria, as strata-bound-syngenetic formations of pre-Alpine and Alpine Age. Proc 5th Quadrennial IAGOD Symp I:513–544
Weissberg BG (1965) Getchellite, $AsSbS_3$, a new mineral from Humboldt County, Nevada. Am Mineral 50:1817–1826

Genesis of Ores in Volcano-Sedimentary Sequences

Sb-As-Tl-Ba Mineral Assemblage of Hydrothermal-Sedimentary Origin, Gümüsköy Deposit, Kütahya (Turkey)

S. JANKOVIĆ[1]

Abstract

The Sb-As-Tl-Ba mineral assemblage, accompanied by Pb-Zn sulphides and minor amounts of silver and gold, was formed in connection with undifferentiated hydrothermal solutions, which discharged at the bottom of the Neogene basin (syngenetic type of ore mineralization) and/or within volcanic-sedimentary series – tuffs and sinter (epigenetic type of ore mineralization). In the syngenetic type of mineralization, the ore minerals were deposited as disseminated colloids and metastable minerals, while in epigenetic mineralization small massive veins and lenses formed.

The primary sulphides (stibnite, realgar, orpiment, sphalerite, galena, less tetrahedrite, pyrite and other minerals) were mostly transformed into oxides. Barite is occasionally separated from ore sulphides.

Besides the hydrothermal-sedimentary ore bodies, small quartz-stibnite veins have been formed in the Upper Paleozoic schists, under the ore-bearing volcano-sedimentary basin. This almost monometallic ore was derived from highly differentiated hydrothermal solutions.

1 Introduction

In Western Anatolia, Turkey, numerous antimony deposits were formed in genetic association with the Neogene volcanics of a granodiorite magma. These deposits are located in the metallogenic district of Gedis, from Ivrindi to the north of Afion (Fig. 1). Some of these deposits have been intermittently in operation (in the areas of Ivrindi, Simav, Gediz, Afion). Antimony deposits are of the vein type, mostly quartz veins in volcanics and/or schist, and of a replacement type in silicified limestone, close to its contact with impervious rocks in a hanging wall ("jasperoid" type). Ore minerals have been deposited mostly from highly differentiated hydrothermal solutions, forming monometallic associations (locally only minor sulphides of arsenic, mercury; lead/zinc in traces).

The polymetallic deposit at Gümüsköy, located approx. 30 km west from the town of Kütahya, differs regarding other Sb-deposits in the Gediz metallogenic district. It is a hydrothermal-sedimentary type, formed in close spatial and genetic connection with volcanics, and discharge of ore metals from hydrothermal solutions took place in the Neogene sedimentary environment and close to volcanic vents. Ore deposit size indi-

1 Department of Mineral Exploration, Faculty of Mining and Geology, Djušina 7, 11000 Belgrade, Yugoslavia

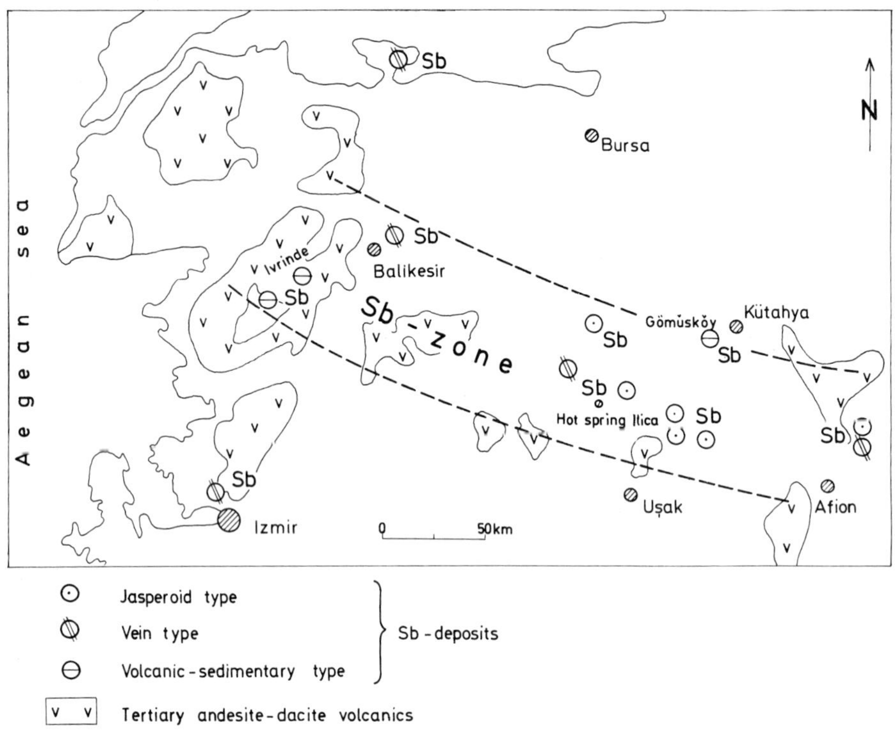

Fig. 1. The location of Sb-deposits in Western Anatolia

cates its economic significance, but ore grade and ore metal form (oxides prevail) are the limiting factors from a techno-economic point of view.

Similar metallogenic environments occur at several places in volcanic regions of the West Anatolia. The travertine, deposited nearby hot spring centres on the north slope of Uludag mountain (Yugitali, a few km south of Bursa), often contains up to 6% antimony, mostly up to 1%, but sulphides have not been recognized. Numerous hot springs at Ilica, between Simav and Gediz, are of particular interest. Water temperature range is 80°–90°C. Evaporated residue from water at the surface of the terrain contains 1.5 ppm Sb and 0.15 ppm Hg. (Analyses performed by N. Ozerova, Moscow.) The precipitates around centres of hot water vents (siliceous sinter and sinter mixtures of calcium carbonate and silica), highly altered (yellow, reddish brown, green) are enriched in many metals (Table 1).

Travertine terraces and siliceous precipitates are of considerable size (thickness of up to 15–20 m), and resemble similar deposits in geyser districts of New Zealand (Weissberg 1969; Browne 1971), the Steamboat Springs in Nevada, the Yellowstone Park (Lindgren 1933), and Kamchatka and Kuril (Ozerova et al. 1971). It is most probable that hot waters of sodium carbonate and/or sodium chloride-silica type yielded abundant amounts of antimony and arsenic at the Gümüsköy deposit and formed in a volcano-sedimentary environment.

Table 1. Analysis of precipitates around centres of hot water vents

		From	To
Sb	%	0.10	2.20
As	%	0.10	2.40
S	%		0.20
Fe	%	0.70	3.80
Pb	ppm	25	600
Zn	ppm	70	200
Cu	ppm	10	15
Ag	ppm		10
Ba	ppm		40

2 Geotectonic Setting

The vicinity of the Gümüsköy deposit is composed of Upper Paleozoic rocks (a series of arkosic sandstones, micaschist, conglomerates, and phyllite) overlain by a series of the Neogene siliceous and carbonate rocks, deposited in a continental basin, and volcanics (dacite-andesite, mostly tuff) and volcano-sedimentary series (travertine, bedded sinter chert, siliceous brecciated sinter, tuffs, volcanic breccia, as well as traces of graphite rocks). The Neogene sediments have been displaced by a system of faults. The thickness of the volcano-sedimentary basin is mostly up to 50–60 m.

Within the Gümüsköy deposit, several ore bodies have been recognized and explored (the Senator mine, Nirengy, Calishilan, Pasha and others). In some parts of ore bodies, like at the Nirengy Hill, several horizons of Sb-mineralization were revealed; barite is not spatially associated with Sb-mineralization (Fig. 2).

Small vertical extent of ore mineralization is a typical feature of all ore bodies.

The volcanic vent in the Senator mine (Bözcukur locality), accompanied by some explosive breccia, is situated in the central part of the ore body, which covers approx. 10,000 m^2. The products both of solfataric and fumarolic activities are in close connection with ore mineral deposition. The wallrocks are highly hydrothermally altered (kaolinization and sericitization prevail). Due to activity of the volcanic gases, wallrocks are often porous, cavernous with incrustations of silica, and/or limonite.

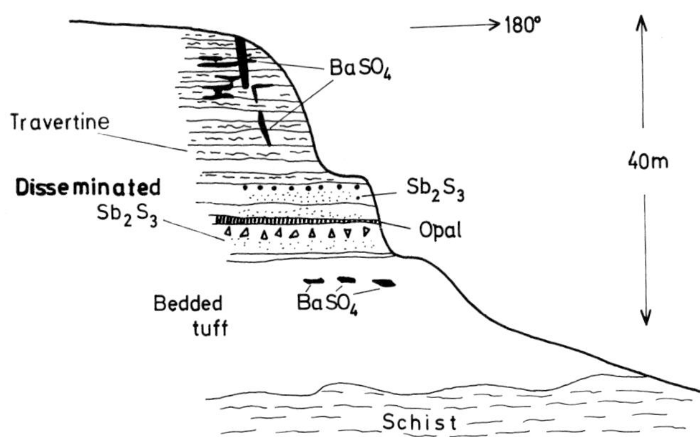

Fig. 2. The cross-section of the Nirengy Hill (schematically only)

3 Ore Deposition Types

There are two principal types of mineral and metal associations, as well as modes of ore deposition:

a) Hydrothermal-sedimentary ore bodies, stratified within bedded volcano-sedimentary series, both of massive and disseminated types, and/or epigenetic veins and lenses, localized in fractures in volcano-sedimentary environments.

Polymetallic mineralization (Sb, As, Tl, Pb-Zn accompanied by barite) is characteristic for this type.

Massive ore bodies are 2–4 m wide. The thickness of individual veins and lenses in the Senator mine is usually 0.5–1.5 m.

The massive and vein types of ore contain 4% Sb, 5% As, mostly as oxides, up to 1% Zn, 0.5% Pb, and approx. 0.15% Tl. The contents of barite ranges from a few percent to over 30% $BaSO_4$. Some single barite veins contain over 80% $BaSO_4$, also occasionally traces of Sb, As, Pb/Zn. Barite occurs either in ore metals, or it forms separate ore bodies, located above an ore mineralization.

The disseminated mineralization, for a width of 50 m, contains 0.1%–2.5% Sb, 0.1%–2% As, 100–3500 ppm Pb, and 100–2500 ppm Zn. In some of the ore bodies, Pb-Zn contains in a single sample even 1%–3% Pb and up to 20% Zn oxide (the same samples contain 0.5% Sb and up to 1.0% As).

b) Epigenetic hydrothermal quartz-stibnite veins in the basement of the Neogene basin (Fig. 3). The ore-bearing solutions originated from the Neogene volcanism.

Antimony is almost the only ore component (the lead-zinc contents are less than 0.02% arsenic, mostly less than 0.3%). The ore veins are 0.1–0.25 m wide.

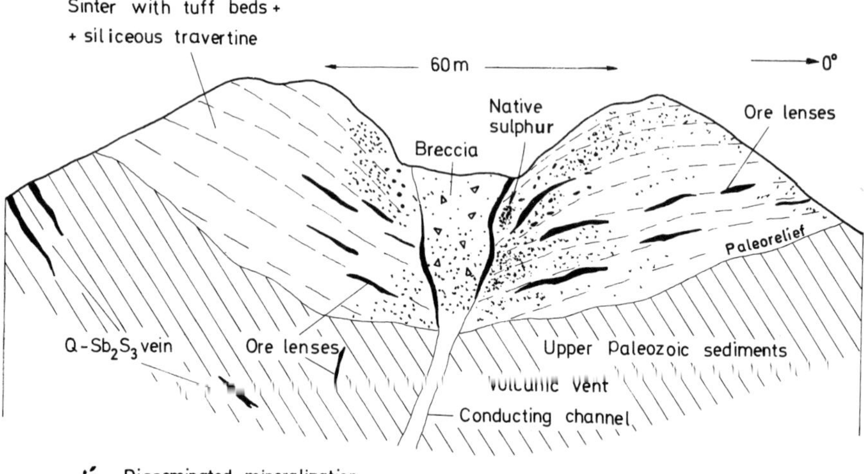

Fig. 3. The cross-section of the Senator mine (schematically only)

4 Mineral Associations

The mineral association of hydrothermal-sedimentary ore is very complex, characterized by widespread oxides, metastable minerals, colloform textures, and overlapping of Pb/Zn minerals by Sb-As-Tl minerals.

The primary sulphides (stibnite, sphalerite, galena, tetrahedrite, realgar, oripiment, rammelsbergite, freibergite, pyrite, melnikovite) were mostly transformed into oxides. The deposits are abundant in barite and gypsum. Native sulphur is restricted to small lenses and pockets. Strengite is widespread in the zone of oxidation.

Stibnite occurs as idiomorphic and collomorphic aggregates, sometimes as fibrous and collomorphic structures. It contains usually 3.2% Tl and 2.2% As, corresponding to the formula $(As,Tl,Sb)_2S_3$ (Janković and Le Bel 1976), and indicating $Sb_2S_3-As_2S_3$ solid solution. It is of interest to note that Tl could be in high percentage associated with stibnite not only with low-temperature As-sulphides (oripiment and realgar).

The spectrographic analysis of stibnite concentrate from the Senator mine display the following trace elements: 0.5% Pb, 55 ppm Ni, 10 ppm Cu, 0.55% Zn, 0.4% Sr, 0.28% Tl, as well as 1% As and 2–12 ppm Hg.

Cervantite and *valentinite* are the principal Sb-minerals, particularly in the disseminated mineralization.

Orpiment and *realgar* are mostly transformed into As-oxides. Only in the opals, xenomorphic development of fine crystalline orpiment can be found.

Skorodite is a very abundant mineral in the Senator mine, interstratified with fine banded siliceous sinters.

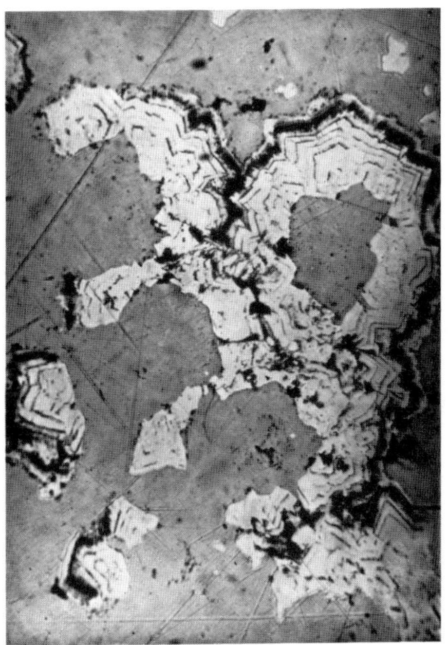

Fig. 4. Replacement structure of galena by rammelsbergite *(white)*. × 420. (Photo: N.N. Mozgova)

Nickel arsenide (rammelsbergite) is of local development, but of particular interest, regarding the source of nickel.

Rammelsbergite is developed by replacement of galena and is characterized by collomorphic and zonal textures, with variable distribution of nickel (0.96%–1.02%) and arsenic (32.1%–28.3%) (Fig. 4). The composition of rammelsbergite in individual zones ranges from 28.3% Ni, 67.2% As, 3.9% S and 0.9% Sb – $Ni_{0.96}(As_{1.78}Sb_{0.01}S_{0.24})_{2.04}$ to 32.1% Ni, 63.5% As, 6.8% S and 0.6% Sb – $Ni_{1.02}(As_{1.58}Sb_{0.1}S_{0.41})_{1.98}$. (Microprobe analyses performed by N.N. Mozgova and Y. Borodaev, Moscow State University.)

Sulphoantimonide of thallium ($TlSb_{11}S_{17}$) occurs in the amorphous state, containing 63.7% Sb, 9.7% Tl, and 25.3% S (Janković and Le Bel 1976). Other Tl-minerals are likely to be found, but detailed investigations are still in progress.

Galena is associated with sphalerite. Very small inclusions of freibergite can be occasionally found in galena.

Sphalerite is mostly transformed into carbonate.

Strengite forms small oval aggregates. Its presence indicates phosphorous in hydrothermal solutions.

Barite occurs mostly in idiomorphic forms, white in colour.

Sulphur isotope composition. The $\delta^{34}S$ values of stibnite range from +3.5‰ to +7.0‰, which indicate slight enrichment in heavy sulphur, and probably volcanic origin.

The mean $\delta^{34}S$ values in barite and gypsum are –17.5‰, and –12‰, respectively, indicating organic origin of sulphur.

5 Some Genetic Considerations

The Sb-As-Tl-Ba mineral assemblage, accompanied by Pb-Zn sulphides, and minor amounts of silver and gold, was formed

a) from hydrothermal solutions, discharged at the bottom of the Neogene shallow-water basin. Ore minerals were deposited together with tuffs and sinter (syngenetic type of mineralization). The ore mineralization was deposited like colloids and metastable minerals.

Due to a sharp drop in pressure and temperature, and changes in $f(O_2)$ and $f(S_2)$ as well as mixing with meteoritic water and multi-stage hydrothermal activity, the vertical interval of ore deposition was short, and ore mineralization does not show any zonality in distribution. Ore mineral deposition took place in the vicinity of volcanic vents and hot springs. Ore was derived from undifferentiated ore-bearing hydrothermal solutions, and it is characterized by very complex composition.

b) Hydrothermal and volcanic activity continues since early sedimentation completed, and ore-bearing solutions penetrated into volcanic-sedimentary series, forming epigenetic veins and lenses. The mineral assemblage is similar to syngenetic mineralization, but barite is much more abundant. Ore is mostly massive and localized within a system of stringers.

c) The quartz-stibnite veins in the basement schists have been formed from hydrothermal solutions derived from subvolcanic intrusives of dacite-andesite. The ore-bearing solutions were highly differentiated, and it is characterized by a simple mineral assemblage, and almost monometallic composition.

Ore mineralization at Gümüsköy proved that small but economically significant deposits, at least with regard to the tonnage of ore mineralization, can be formed from hot springs related to volcanic activity.

Acknowledgment. I would like to thank N.N. Mozgova and V. Borodaev, Moscow, for microprobe analyses.

References

Browne PRL (1971) Mineralization in the Broadlands geothermal field, Tanpo volcanic zone, New-Zealand. Soc Min Geol Jpn Spec Issue 2:64–75 (Proc IMA-IAGOD Meet "70")

Janković S, Le Bel L (1976) Le thallium dans le minérai de Bözcukur près Kitahya, Turquie. Bull Suisse Minéral Pétrogr 56:69–77

Lindgren W (1933) Mineral deposits, 4th edn. McGraw-Hill, New York, 930 pp

Ozerova NA, Naboko SI, Vinogradov VI (1971) Sulfides of mercury, antimony and arsenic, forming from the active thermal springs of Kamchatka and Kuril Islands. Soc Min Geol Jpn Spec Issue 2:164–170 (Proc IMA–IAGOD Meet "70")

Weissberg BG (1969) Gold-silver ore grade precipitates from New-Zealand thermal waters. Econ Geol 64:95–108

A Strata-Bound Ni-Co Arsenide/Sulphide Mineralization in the Paleozoic of the Yauli Dome, Central Peru

H. W. KOBE[1]

Abstract

The pre-Permian Paleozoic of the Yauli Dome in Central Peru contains a strata-bound finely disseminated Ni-Co arsenide/sulphide mineralization in a weakly metamorphosed series of interbedded slates/phyllites, fossiliferous marbles and basic volcanics and volcaniclastics. The intimate interlayering of these formations on both the macro- and micro-scales is taken to indicate a synsedimentary origin for the mineralization, related to submarine deposition of lavas, tuffs and their hydrothermal exhalative derivatives among the seafloor sediments.

This is in stark contrast to other Ni-Co mineralization known from the Eastern Paleozoic Belt in Peru, attributed to either the plutonic ultrabasic or the hydrothermal granite environments.

1 Introduction and General Geology

The Yauli domal complex has a long history of mining activity, but as a structural unit, it was outlined as late as 1940 and 1943 by Harrison. Among the few existing local geological studies only those with relevance to the Paleozoic, i.e., Stone (1928), Kobe (1964), Rivera (1964) and Lyons (1968) have been consulted for the compilation of Fig. 1 (see also Fig. 9).

The Paleozoic, exposed along two major doming anticlines, consists of the Permian Mitu Group subvolcanic intrusives, volcanics and volcaniclastics, underlain discordantly by the pre-Permian Excelsior Group metamorphics which include basic subvolcanic intrusives, volcanics and volcaniclastics, abundant black slates to phyllites, rare quartzites, red cherts and argillites and fossiliferous marbles. The strata-bound Ni-Co mineralization is an integral part of the Excelsior Group. The Paleozoic Mitu and Excelsior Group basement is penetrated by Mesozoic and/or younger basic to acid intrusives (not distinguished on Fig. 1).

1 Geology Department, The University of Auckland, Auckland, New Zealand

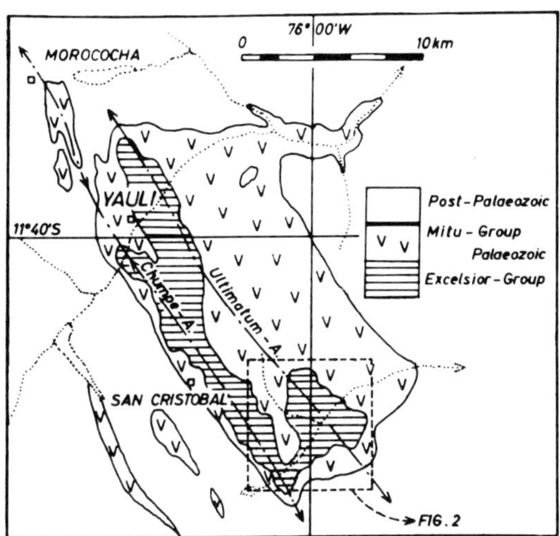

Fig. 1. Simplified geology and structure of Yauli Dome. (After available literature incl. documentation C. de P., Centromin and observations by Kobe 1978)

2 The Country Rocks

The predominant rock type in the Excelsior Group is a black to dark grey *slate to phyllite* occasionally with sandy layers and local *quartzites.* A pronounced schistosity, small-scale folding and boudinage in the quartzitic beds are widespread and pervasive. The intercalated bodies and thick beds of white to light-grey *limestones (marbles)* are strongly distorted and recrystallized but allow usually the recognition of abundant fossils, particularly crinoids. In the region SE of Andaychagua, the limestones are intimately intercalated on all scales (from the outcrop to the microscopic) with *red cherts* and *argillites* and greenish *volcanics* and *volcaniclastics.* This is the characteristic environment for the strata-bound Ni-Co mineralization.

The greater part of the green volcanics/volcaniclastics consists of crystal-lithic *tuffs* with a high content of chlorite, quartz, clear mica, semiopaque titaniferous material, carbonates and with or without disseminated opaques. The fine-layered "fluidal" texture is indicative of accumulation and movements of a fine-grained sediment. The fabric, mineralogy and the high carbonate content allow the interpretation as deposition of volcanic ash in the sea to become intimately mixed with the chemically and biochemically precipitated carbonate sediment. Strongly altered *lavas* with a mineral composition like the tuffs and additional relics of pyroxene and pseudomorphs after mafic constituents and plagioclase, can be interpreted as original andesitic/basaltic flows. Harrison (1940, 1943) recorded occasional pillow structures. The contact with limestones is similar as for the tuffs. A shallow *subvolcanic intrusion* similar in composition to the volcanics has been recognized near Mina Ultimatum. It is a gabbro-diorite with large plagioclase phenocrysts and dendritic ilmenite crystals that has suffered not only intensive movement indicated by the deformation, fracturing and brecciation of the plagioclase phenocrysts, but also a thorough deuteric alteration which affects to various degrees the primary mineral assemblage (now chlorites, micas, carbonates and chalcedonic quartz). The high content of titaniferous minerals, of carbonate, chlorite and the other components in all the greenish derivatives of a basic igneous source in the region, as well as their interaction with the limestone, appears to support the idea of a common origin, although their field relations could not yet be ascertained. In addition, the grade of deformation and metamorphism of these rock types, comparable to that of the associated slates/phyllites is an indication for their contemporaneity and therefore pre-Permian age. In contrast, it appears obvious from the field relations that the stocks and dykes of gabbro/basalt, as well as those intrusives of more acid composition, are of post-Excelsior age (see Fig. 2).

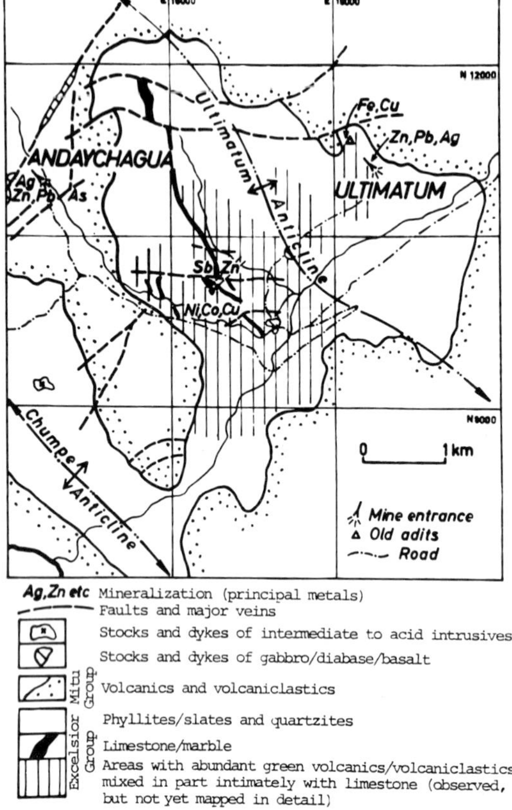

Fig. 2. General geology and mineral deposits in the southern part of Yauli Dome. (Traced from aerophotos by N. Rivera 1964, with modifications from field observations by Kobe)

3 The Mineralizations

Two types of mineralization occur in the Yauli Dome and must be distinguished by their difference in composition, age and mode of emplacement:

a) the *vein mineralization,* obviously related to the structural formation of the Yauli Dome (Kobe 1964; Rivera 1964), displays a zonation pattern around the Chumpe intrusive, demonstrated by the distribution of the metals in the veins, i.e.,

W(Sn) in the centre, followed by Cu, Zn, Pb and Ag and reaching As and Sb in the most distant regions (stibnite 2 1/2 km SE of Andaychagua).

b) the *strata-bound mineralization* is related to the basic volcanism, contemporaneous with the deposition of the Excelsior Group sediments. In the area of Mina Ultimatum, a Zn/Pb sulphide mineralization lies at the contact between underlying green volcanics and overlying Excelsior phyllites. This pyrrhotite-sphalerite-galena association is the possible lateral equivalent of the pyrrhotite-pyrite-chalcopyrite mineralization strata-bound with the volcanics, 1/2 km NW of Mina Ultimatum (Kobe 1978) (see Fig. 2). A concentration of Ni, Co, Cu, As, Fe, Mn, intimately related to the complex of limestone intercalated with volcanics/volcaniclastics has been found in the immediate neighborhood of the vein mineralization of stibnite, sphalerite, calcite and quartz mentioned above (see Fig. 2), but the extent of it is not known.

The mineralized rock has a black manganese oxide coating (derived from the weathering of manganiferous calcite) but when cut is of greenish-grey colour, displaying the finely layered fabric seen in Fig. 3. Under the microscope this can be resolved into three associations of minerals:

a) layers and flat lenses of chlorite, often with mosaic areas of quartz and fine reddish hematite flakes in subparallel layers (Fig. 4), accompanied by layers of semiopaque titaniferous matter, concentrated along the contacts with areas b) (see Fig. 7);
b) irregular patches of carbonate, often with recognizable organic internal structure (fossil remains), as well as spherulitic aggregates of manganiferous calcite (see Fig. 6);
c) the mineralization, disseminated or in discrete patches, elongated along the layering of the rock, though distorting occasionally the layered arrangement of the titaniferous mineral (see Fig. 5). The opaques are usually surrounded by coarse, clear calcite and muscovite blades (Fig. 8).

All this appears in a fine-grained texture of fluidal, contorted character, underlined by the distribution of fine, equidimensional grains of an apparently isotrope mineral, still elusive to determination attempts. It occurs as inclusions throughout all the other minerals (including the opaques) although least in the carbonate spherulites and is not present in the cross-cutting calcite veinlets. This indicates that the opaques cannot be a later addition to a preexisting metamorphosed sedimentary-volcanic rock complex. The mineralization has suffered the same fate is its host, i.e., been subjected to a pre-Permian metamorphic phase.

The *opaque minerals* are the Ni-Co compounds millerite (No. 1), violarite (No. 2), gersdorffite (Nos. 3 and 4), mineral No. 5 (in decreasing abundance) as well as pyrite, chalcopyrite (No. 6) and sphalerite (No. 7) (numbers in brackets refer to Table 1 and Fig. 8).

Either millerite or violarite may form monomineralic or bimineralic patches (for typical intergrowths see Fig. 8). Pyrite, although forming the occasional porphyroblast, occurs usually as deeply corroded inclusions in violarite/millerite. Gersdorffite shows strong tendencies to idiomorphism, particularly when on its own and is then distinctly zoned. It more often forms irregular rims around violarite/millerite. The two varieties, distinguished optically (No. 3 light grey, No. 4 brownish grey, both isotropic) vary considerably in composition (see Table 1). They may be myrmekitically intergrown with chalcopyrite. Violarite and millerite may also be rimmed by a dis-

Table 1. Composition of major Ni-Co compounds

Mineral	wt.% Av. + simplif.		Atom. prop.	Approx. ratios		Bulk formula
No. 1	Ni	65%	1.12	1		
millerite	S	35%	1.09	1		Ni S
No. 2	Ni	35%	0.59	6		
violarite	Co	6%	0.10	1	3	
	Fe	16%	0.29	3		
	S	43%	1.34	13	4	$(Ni,Fe,Co)_3\ S_4$
No. 3	Ni	12%	0.20	2		
gersdorffite	Co	31%	0.53	5	4	
(cobaltian)	Fe	6%	0.10	1		
	As	31%	0.41	4	5	
	S	20%	0.62	6		$Co_5(Ni,Fe)_3As_4S_6$
No. 4	Ni	44%	0.75	7	4	
gersdorffite	Co	3%	0.05	1		
	Fe	4%	0.06			
	As	28%	0.37	10	5	
	S	21%	0.65			$Ni_7(Co,Fe)As_4S_6$
No. 5	Ni	50%	0.85	1		
unnamed	Fe	4%	0.07			
mineral	Sb	20%	0.16	1		
	S	26%	0.81			(see text)

tinctly bireflectant and anisotropic mineral (No. 5). Microprobe analyses indicated Sb and since the ratio Ni:S (resp. Ni:S+Sb) remains approximately the same as for millerite, one has to assume a substitution of some S by Sb to the effect that mineral No. 5 might be an intermediate phase in the system NiS (millerite) – NiSb (breithauptite), i.e., "antimonian millerite". On the other hand by recalculation of atom.prop. to 8 S atoms and rearrangement it is possible ot obtain approximately a formula $(Ni,Fe)_9Sb_2S_8$ akin to that of hauchecornite, $Ni_9(Bi,X)_2S_8$ (where X is As, Sb or Te), but complete substitution of Bi by Sb is said not to be allowable for "antimonian hauchecornite" according to Gait and Harris (1972). However, Feather and Koen

Fig. 3a, b. Polished face of a typical mineralized volcanic sediment under oblique **(a)** and near vertical **(b)** illumination. Note the band with accumulated calcareous fossils *(white)* into which an angular volcanic fragment *(black)* is included. This in turn is draped over by a tuffaceous bed with flat siliceous-chloritic-hematitic-titaniferous lenses *(dark)*. The dissemination of the Ni-Co mineralization *(brilliant white spots* in **b**) follows the layering except where enclosed in the black volcanic fragment. Top is at *right*

Fig. 4. A siliceous-chloritic lens magnified in thin section. Within lens: chlorite *(white)*, titaniferous mineral and/or hematite *(grey layers and schlieren)*, opaques *(black)*. Surrounding lens: titaniferous mineral+chlorite+silica *(dark grey)*, mixture of carbonate *(light grey)* and titaniferous mineral *(grey)*, opaques *(black)*. × 27

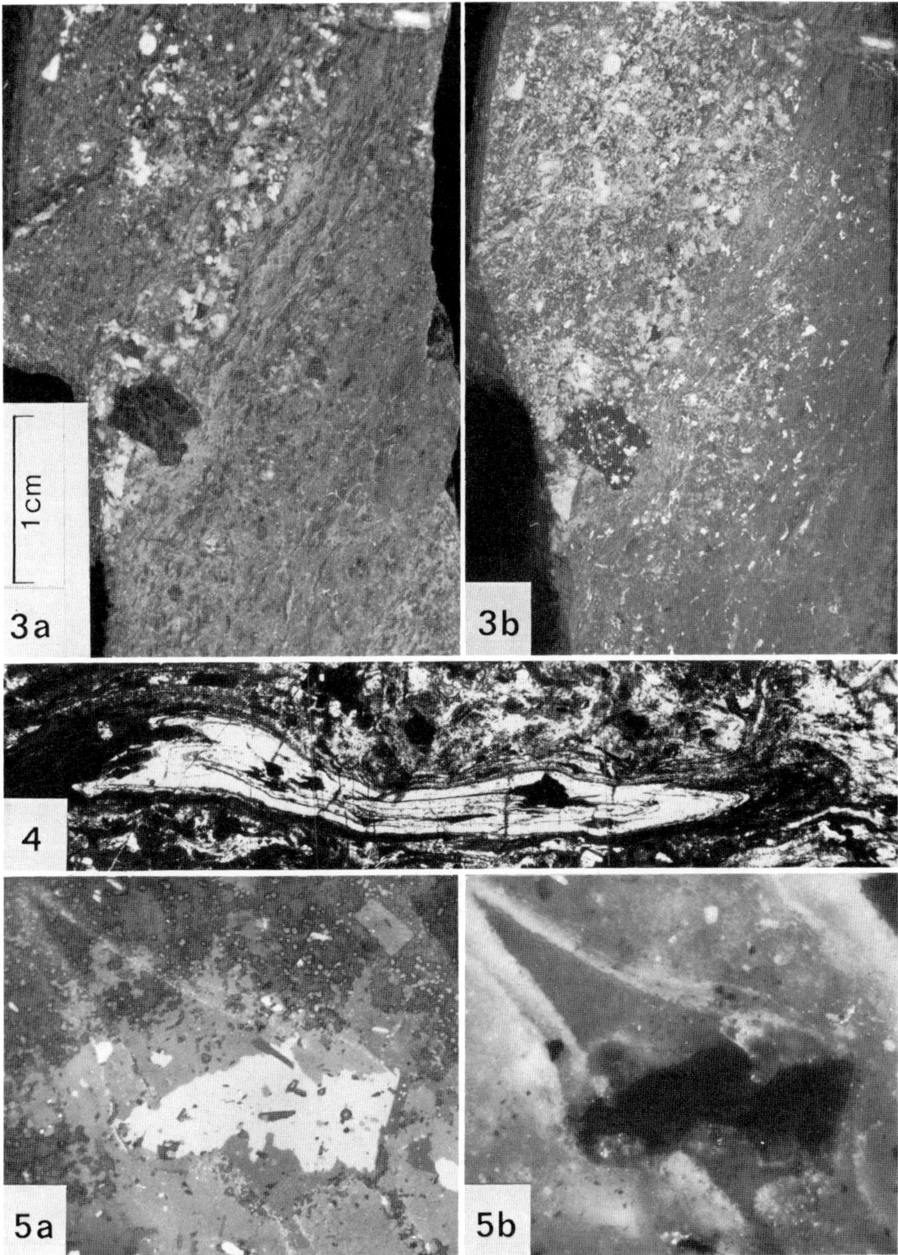

Fig. 5a, b. Polished thin section, plane polarized light **(a)** and with x nicols **(b)** showing a sulphide aggregate pushing aside layers of titaniferous mineral (anatase?) (*light grey* in **a**, *white* in **b**). Siliceous and carbonate matrix (*medium grey* in **a**) and chloritic matrix *(dark grey)* with disseminated equidimensional unknown mineral *(light grey)*, sulphides (*white* in **a**, *black* in **b**). × 260

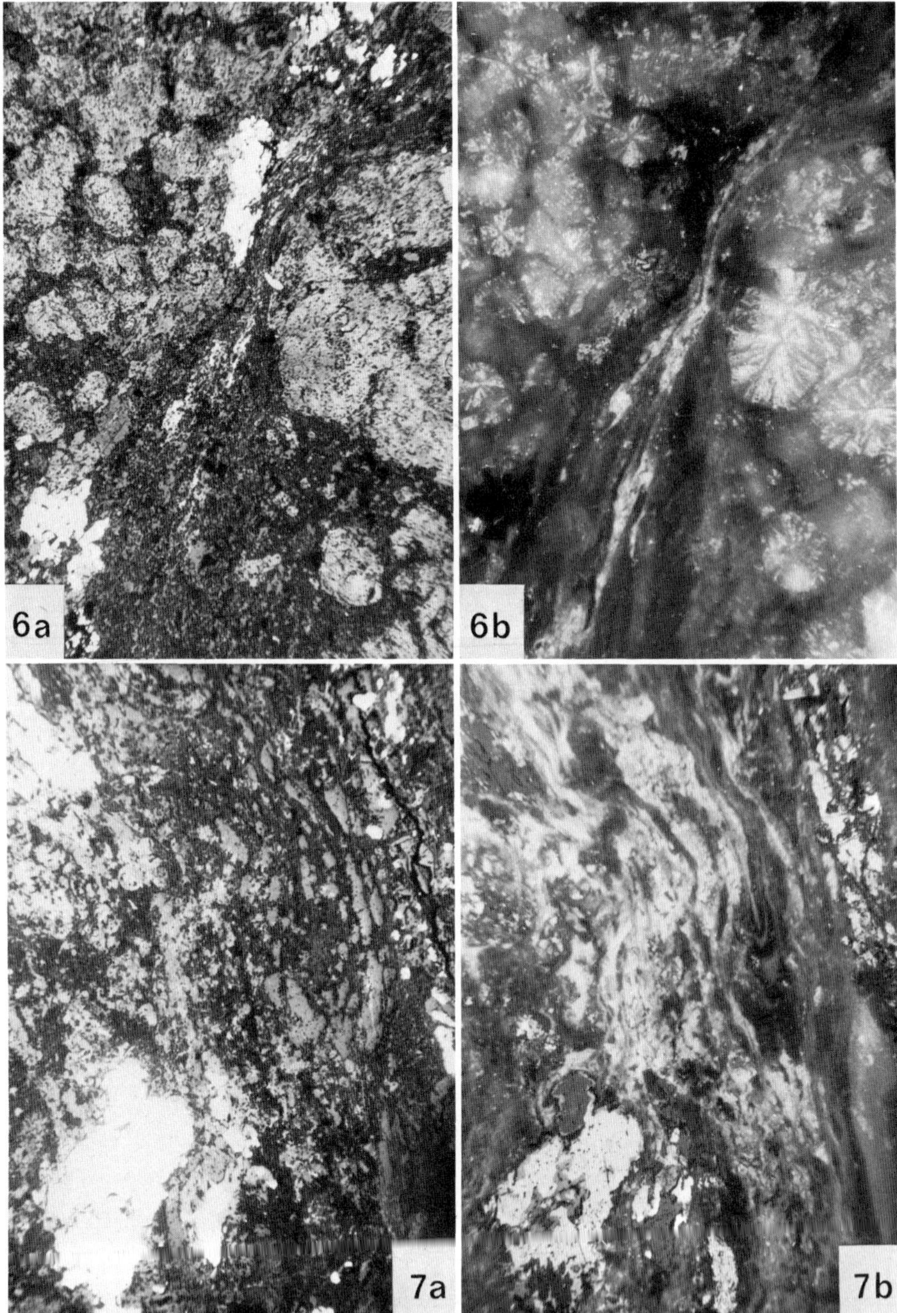

Figs. 6 and 7

Fig. 8. Typical associations of sulphides and arsenides and their immediate surrounding. *Vertical ruling* siliceous–chloritic matrix; *c* clear carbonate; *x* idiomorphic carbonate against millerite; *1* millerite; *2* violarite; *3* gersdorffite (cobaltian); *4* gersdorffite; *5* unnamed mineral; *6* chalcopyrite; *7* sphalerite; *lath-shaped mineral* muscovite; *layers of dashes* hematite; myrmekitically corroded bodies in *1* and *2* pyrite. Camera lucida drawings

Fig. 6a, b. Polished thin section with 1 nicol **(a)** and x nicols **(b)** of mainly spherulitic carbonate bodies (*grey* in **a**) aggregated among siliceous-chloritic matrix (*dark grey* in **a**) including titaniferous mineral schlieren (*light grey* in **b**) and opaques (*white* in **a**). × 58

Fig. 7a, b. Polished thin section with 1 nicol **(a)** and x nicols **(b)**. Chloritic matrix (*dark grey* in **a**) with schlieren and elliptical bodies of siliceous-hematitic material *(grey* in **a**, *right third and left of frame)*, abundant titaniferous matter *(light grey* in **b**, *belt through centre of frame)*, sulphides (*brilliant white* in **a**) (the largest aggregate *bottom left* consists of millerite surrounded by mineral No. 5). × 58

(1975) described a mineral from the Witwatersrand Reefs with the composition Ni 47.80%, Fe 3.75%, Sb 21.87%, Bi 1.02%, As 1.34%, S 25.13%, resulting in a formula $Ni_9Sb_2S_8$. Since its X-ray diffraction pattern is very similar to that of hauchecornite, the authors consider it to be the other end-member in the solid solution series $Ni_9Sb_2S_8 - Ni_9Bi_2S_8$. It appears likely that mineral No. 5 is essentially equivalent to that yet unnamed mineral of Feather and Koen (1975).

Sphalerite is only rarely in direct contact with the Ni-Co compounds, it rather forms its own aggregates with mica and calcite. Chalcopyrite occurs disseminated through silicates and carbonates, rarely associated with the Ni-Co compounds and as fine inclusions in sphalerite. The titanium-bearing compound (anatase?) forms irregular, layered, granular aggregates (see Fig. 5). Occasionally this mineral in some discrete patches is arranged in reticulate lamellae, reminiscent of the "trellis" intergrowth of ilmenite in magnetite. Thin, irregular veinlets with predominant calcite gangue carry sphalerite, chalcopyrite, pyrite and rarer millerite and violarite. The latter may be remobilization products from the surrounding Paleozoic strata-bound mineralization, while the former could be in part components of the epigenetic stibnite-sphalerite vein mineralization. Stibnite has not been found among the strata-bound mineralization.

4 Metallogenetic Considerations

Mineralizations of Ni-Co-Cu-U are known and have in the past been exploited in a number of places in the Peruvian Andes (Rivera Plaza 1953; Ponzoni et al. 1969; Bellido and De Montreuil 1972) (see Fig. 9). They are classified as of the hydrothermal vein type with replacements and disseminations of millerite, violarite, gersdorffite, skutterudite, cobaltite, niccolite, arsenopyrite, loellingite and others, with chalcopyrite, sphalerite, pyrite etc. and uraninite besides quartz and calcite. The above authors suggest that the mineralization was emplaced in Paleozoic metamorphic and sedimentary rocks and genetically related to granitic intrusives of possible Paleozoic age; thus a classical epigenetic situation. However, the characteristics of the Ni-Co arsenide and sulphide mineralization in the Yauli Dome as expressed in the fabric and paragenetic relations of the rock constituents and the metallic mineralization proper, suggest a syngenetic emplacement by hydrothermal solutions related to basic volcanic activity in a submarine environment of chemical/biochemical carbonate precipitation.

The Yauli Dome Ni-Co mineralization compares well with deposits of the "Ag-Co-Ni-Bi-U formations" of Schneiderhoehn (1962, 4th edn.), particularly his account on Kongsberg (Norway), as well as with the summary and interpretation given by Jambor (1971) for Cobalt (Ontario, Canada). Even closer comes the description of the depositional environment and the suggested genesis of the Great Bear Lake silver deposits (N.W.T., Canada) by Badham et al. (1972). It appears that with respect to metallogenetic interpretation the emphasis has shifted from the plutonic to the subvolcanic-volcanic environment and from the acid to the basic igneous rock compositions, so that the basic intrusives-extrusives come to be considered not only as favourable country rocks for "topomineralic" precipitation, but the actual source of the hydrothermal mineral content. The Yauli example extends this shift to the suggestion of an

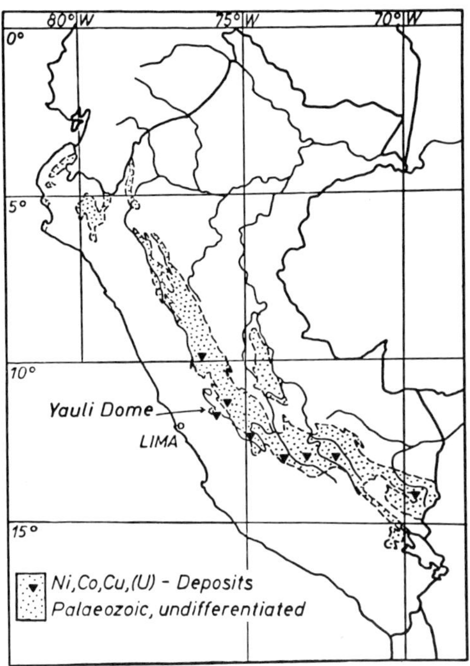

Fig. 9. Ni, Co, Cu, (U) deposits and their relation to the Paleozoic. (After the literature and supplements – Pampas and Yauli Dome – by Kobe)

essentially syngenetic mineralizing process in connection with their volcaniclastic and hydrothermal derivatives in a mixed volcano-sedimentary submarine environment.

The recognition of these relationships calls for a careful revision of the existing reports on the Ni-Co-Cu-U deposits in the Paleozoic of the Eastern Cordillera of Peru; and another attempt to ascertain the actual and not hypothetical field relations appears appropriate in order to establish new guides for exploration.

Acknowledgments. I wish to dedicate the present study to Prof. P. Ramdohr in remembrance of his kind interest in my work and helpful advice when on visit at the Heidelberg Institute 1966 and 1974. I acknowledge gratefully sponsorship and support by the Peruvian Institutions INCITEMI and CENTROMIN respectively and the kind liaison by Prof. G.C. Amstutz to make it possible to carry out metallogenic investigations in Central Peru 1977/78. I thank my colleagues Prof. P.M. Black and Mr. M. Little, Geology Department University of Auckland, for their assistance with electron-microprobe work and the related mineralogical interpretation, but the results as presented are my own responsibility.

References

Badham JPN, Robinson BW, Morton RD (1972) The geology of the Great Bear Lake silver deposits. Abstr 24th Int Geol Conf Sect 4:541–548

Bellido E, De Montreuil L (1972) Aspectos generales de la metalogenía del Perú. Serv Geol Mineral Geol Econ 1:149

Feather CE, Koen GM (1975) The mineralogy of the Witwatersrand reefs. Minerals Sci Eng 7/3: 189–224

Gait RI, Harris DC (1972) Hauchecornite – antimonian, arsenian and tellurian varieties. Can Mineral 11:819–825

Harrison JV (1940) Nota preliminar sobre la geología de los Andes Centrales del Perú. Bol Soc Geol Perú 10:5–29

Harrison JV (1943) Geología de los Andes Centrales en parte del Departamento de Junín, Perú. Bol Soc Geol Perú 16:1–51

Jambor JL (1971) General geology. In: Petruk W et al. (eds) The silver arsenide deposits of the Cobalt-Gowganda Region, Ontario. Can Mineral 11:12–33

Kobe HW (1964) Andaychagua – San Cristobal exploration. Priv Rep

Kobe HW (1978) Mina Ultimatum y alrededores, Geología general. Priv Rep

Lyons WA (1968) The geology of the Carahuacra Mine, Central Peru. Econ Geol 63:247–256

Ponzoni E, Postigo A, Birkbeck J (1969) Guía para el mapa metalogenético del Perú. Soc Nac Mineral Petrol Lima, 128 pp

Rivera N (1964) Ten year program for the San Cristobal Division (1965–1975). Priv Rep

Rivera Plaza G (1953) Métodos de examen mineralógico. El Ingeniero Geólogo, Lima, pp 341

Schneiderhoehn H (1962) Erzlagerstätten, 4. Aufl. Fischer, Stuttgart, 371 S

Stone JB (1928) Geological map of Carahuacra – San Cristobal Region. Unpubl map

Sedimentary and Diagenetic Fabrics in the Cu-Pb-Zn-Ag Deposit of Colquijirca, Central Peru

R.W. LEHNE [1] and G.C. AMSTUTZ [2]

Abstract

Megascopically visible sedimentary and diagenetic textures are very common in and around the ores of the Cu-Pb-Zn-Ag deposit of Colquijirca, Central Peru. The present paper summarizes the most frequent ones of these fabrics, such as various types of stratification, geopetal textures, and slumping.

1 Introduction

The present publication deals with fabrics within the ore as well as textural relations between ore and host rock (country rock). It concentrates on megascopically visible features, because these are the most conspicuous and can already be observed in the field. There is no doubt, however, that macrotextures are nearly always connected with a great variety of equally important microtextures and that the latter are only spatially subordinated to the first. All textural patterns mentioned in this paper lead to one of the main criteria for the new genetic interpretation of the mine as a syn-sedimentary-syndiagenetic deposit, possibly of volcanic-exhalative origin (Amstutz and Lehne 1978; Lehne and Amstutz 1978a,b, 1980).

2 General Aspect of the Deposit

Colquijirca is located in Central Peru, at an altitude of 4300 m (Fig. 1). The ore deposit of Colquijirca is an entirely strata-bound ore deposit. Massive ore bodies lie always concordantly within the host rock sequence (normally sand-, silt-, mudstones, and tuffites) and disseminated ores are always restricted to well-defined horizons. On all scales, from outcrop to microscopic dimensions, there is perfect congruence between ore and country rock. Consequently, the ore-bearing horizons can be regarded as normal members of the stratigraphic sequence.

1 Casilla 131-B, Quito, Ecuador
2 Mineralogisch-Petrographisches Institut der Universität Heidelberg, Im Neuenheimer Feld 236, 6900 Heidelberg, FRG

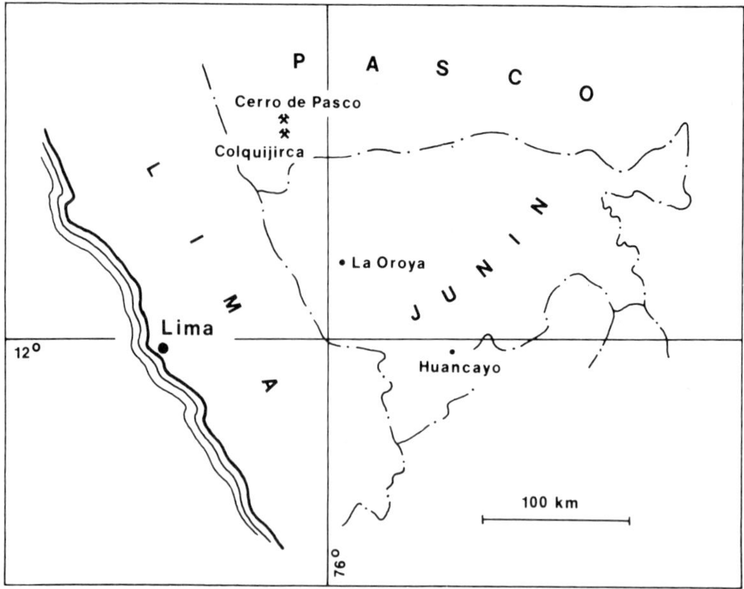

Fig. 1. Location of the mine

3 Macrotextures

The **geometrically** most **simple** patterns of the macrotextures are those that do not show any subordinated megascopic features. They occur in connection with layers, lenses, etc. of massive pyrite or enargite ore. The textural relations between these ores and their host rocks are numerable. Because of their weight massive pyrite or enargite lenses etc. tend to form geopetal fabrics in the underlying strata. Sometimes several strata-bound ore lenses are in line along the strike of their country rock and simulate almost boudinage. Examples for both phenomena are to be found above all among the massive pyrite-enargite ores of the Marcapunta and Smelter mine sections.

Massive pyrite or enargite horizons can often be followed for longer distances without any major thickness changes. They lie concordantly even in tectonically deformed host rock. As an example for this, Ahlfeld (1932) already mentions an ore manto compressed by folding in the area of the Chocayoc/Mercedes anticline in the mine section Sur. Occasionally the ore horizons show slumping caused by synsedimentary movements. Slumping occurs in several places of the mine, and in most cases it is restricted to hand-specimen scale (Figs. 2 and 3). Only in one instance a more extensive area was affected; on a pillar of Crucero 125 in the Marcapunta section an enargite layer with a thickness of about ten cm was found to be torn apart by slumping (Fig. 4). It is so far the most extreme feature of this type in the mine. Ore breccias could not yet be recognized although there are intraformational breccias in the host rocks. A last geometrically simple fabric is represented by massive pyrite horizons which alternate with dark chert layers. Such alternating layers can comprise up to ten single beds with thicknesses of 5 to 30 cm each. They were observed in the Marcapunta section in Galeria Principal between Crucero 225 and 250 and in several places of the mine section Sur.

Fig. 2. Congested fine pyrite layer *(py)* caused by synsedimentary movements

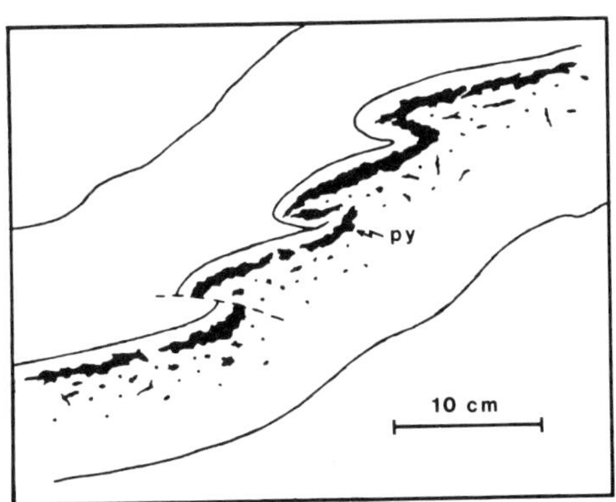

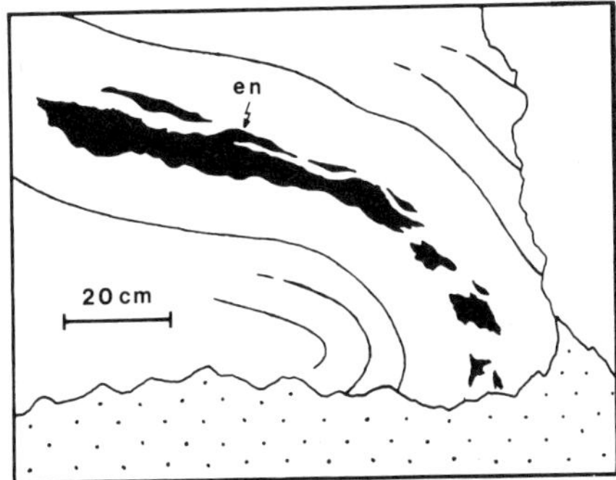

Fig. 3

Fig. 4

Fig. 3. A small pyrite horizon *(py)* shows slumping folds and small thrust

Fig. 4. Enargite layer *(en)* torn apart by slumping. (Bottom of outcrop is covered by detritus)

Fig. 5. Massive enargite *(black)* with stratiform sandy pyrite *(grey areas)*

Geometrically more **complex** textural patterns result from subordinate internal textures within the ore bodies. In most instances these are stratiform textures caused by intercalations of other material. In the Marcapunta section the massive enargite bodies are subdivided by fine pyrite or chert layers of some mm or few cm thickness (Fig. 5) whereas in the northern parts of the mine the same effect is caused by shaly intercalations. McKinstry (1936) writes of these ores:

"The bedding is so well preserved that it is rare to find a layer of ore a foot thick that does not show conspicuous traces of stratification; even in the thickest ore bodies there are generally beds of shale separating the ore into layers."

Further subordinate stratabound textures consist of flaser- and cross-bedding (Fig. 6). Both originate from fine pyrite arranged in the corresponding way and occurring mainly in silty rock, sometimes within massive enargite beds or lenses. Flaser- and cross-bedding can be seen preferably in the Marcapunta section and are restricted to hand-specimen scale. Geopetal features within the ore are distributed with great frequency all over the mine. They appear in connection with pyrite aggregates, chert concretions, and other dense particles and bodies which, by their weight, apparently compressed the underlying ore mud. The crystal-growth orientation which is directed preferably upwards can also be regarded as a geopetal feature (Fig. 7). It is occasionally visible within the massive enargite bodies. According to the observations made so far, all ore-internal geopetal textures can be expected in hand-specimen scale and smaller dimensions.

Between the geometrically **complex** and the **simple** textural patterns exist, of course, many transitions. Both types and their transitions have also the same ore to host rock relations.

Fig. 6 **Fig. 7**

Fig. 6. Cross-bedding of fine pyrite within silty rock (uppermost 2/5 of figure). Bottom shows intercalations of silty material with pyrite seams, locally with geopetal bottom to top growth of pyrite aggregates

Fig. 7. The crystal-growth orientation of the enargite *(en)* is directed preferably upwards. *Black layer* is silty-shaly material, *finely dotted areas* are silty-pyritic material, *white dots* are pyrite *(py)*

4 Microtextures

Many of the microtextures are the microscopic equivalents to the macrotextures. Stratiform textures as well as geopetal features are ubiquitous and photographs of both were already published (Lehne 1977). In addition to these, Colquijirca holds a great variety of partly very complex microtextures which were discussed in many publications (Ahlfeld 1932; Lindgren 1930, 1935; McKinstry 1929, 1936; Orcel and Rivera Plaza 1929). These textures will be summarized and partly reinterpreted in a forthcoming monograph.

References

Ahlfeld F (1932) Die Silberlagerstätte Colquijirca, Peru. Z Prakt Geol 40:81–87

Amstutz GC, Lehne RW (1978) Genese der polymetallischen Lagerstätte Colquijirca, Peru. 6th Geowiss. Lateinam Kolloq Stuttgart, p 6

Lehne RW (1977) Nuevos aspectos acerca del yacimiento de Colquijirca. CITEM Rev Inst Cient Tecnol Min, Lima 3:17–26

Lehne RW, Amstutz GC (1978a) Neue Beobachtungen in der Pb-Cu-Zn-Ag-Lagerstätte Colquijirca, Zentralperu. Münstersche Forsch Geol Palaeontol 44/45:173–178

Lehne RW, Amstutz GC (1978b) Modelos geneticos del yacimiento polimetalico de Colquijirca. 4th Congr Peruano Geol, Lima, abstract vol

Lehne RW, Amstutz GC (1980) The polymetallic ore deposit of Colquijirca, Central Peru: a new genetic model. 26th Int Geol Congr, Paris, abstract, vol III, p 958

Lindgren W (1930) Pseudo-eutectic textures. Econ Geol 25:1–13

Lindgren W (1935) The silver mine of Colquijirca, Peru. Econ Geol 30:331–346

McKinstry HE (1929) Interpretation of concentric textures at Colquijirca, Peru. Am Mineral 14: 431–433

McKinstry HE (1936) Geology of the silver deposit at Colquijirca, Peru. Econ Geol 31:618–635

Orcel J, Rivera Plaza G (1929) Etudes microscopiques de quelques minerais métalliques du Pérou. Bull Soc Fr Min 52:91–107

Sulphide Occurrence Near Tunca, Turkey

V. VUJANOVIĆ [1]

Abstract

Recent field research and laboratory examinations discovered a new sulphide occurrence near Tunca village, northeast Turkey. Examinations have demonstrated that the ore consists of two stages of sulphide assemblages such as: (1) the primary volcanic-sedimentary ore consisting of pyrite, chalcopyrite, sphalerite, tetrahedrite, bornite and galena in the host rhyolite pyroclastics with the lack of the replacement structures; and (2) the subsequently deposited hydrothermal assemblage consisting of the same sulphide minerals and occurring in form of veinlets and metasomatic replacements in the host rock and the primary ore, forming in this way the stockwork type.

1 Introduction

Tunca village is situated about 13 km (air distance) SSE of the Ardešen town, northeast Turkey.

The Tunca occurrence is located about 150 m south of Tunca village, in the Tunca river bed and on both banks. It is actually an ore body which has the exposed length of about 150 m, it is 70 m wide and the known thickness is about 30 m; thinning in any direction was not observed so far.

Host rock is the rhyolite pyroclastics, strongly silicified and argillized, locally also sericitized and limonitized. Its fragments and "phenocrysts" include sanidine-orthoclase, usually altered and replaced by quartz. The matrix is of potash feldspar, is siliceous and contains clay material.

The thickness of the rhyolite pyroclastics and ore body is not known; these are overlain by a volcanic-sedimentary complex of Senonian age (diabase-basalt flows, fossiliferous limestone, tuffaceous sandstones, etc.).

The underlying rocks in the larger Tunca area belong to the same volcanic-sedimentary complex (rhyodacite and its pyroclastics, trachyandesite and its pyroclastics, diabase and basalt and their volcanic breccias and fossiliferous limestone intercalations (Milutinović 1972).

The primary sulphide ore (first stage) forms a bed or lens stratified together with the host rhyolite pyroclastic rock within the described volcanic-sedimentary complex.

---

1 Knez Danilova-36, 1100 Beograd, Yugoslavia

The subsequently deposited sulphide mineralization (second stage) forms the complex veinlets as well as metasomatic replacements in the primary ore body. It can be concluded that the sulphide ore body consists of two stages of sulphide mineralization, i.e.:

1. The primary (first stage) sulphide ore, usually fine-grained, is intergrown mutually and with the host rock, and lacks replacement structures.
2. The second stage, hydrothermal, sulphide assemblage, is mineralogically identical to the first one but formed later, commonly replacing the host rock and the primary sulphide ore.

The preliminary sampling of the sulphide ore body has shown the following average grade: copper 2.60%, lead 0.13%, and zinc 0.81%.

2 First-Stage Sulphide Assemblage (Primary Ore Body)

Pyrite is included as the predominant mineral associated with considerable amounts of chalcopyrite, with frequent sphalerite, tetrahedrite, and bornite and with limited amounts of galena (Fig. 1).

Pyrite is as a rule the chief mineral and can reach up to 80% in some specimens. It is mostly fine or irregularly intergrown with the host rock minerals, but also it forms large patches or grains of megascopic size showing rhythmical textures. The large colloidal pyrite grains sometimes include numerous oval grains of the same mineral or chalcopyrite grains of various sizes. The oval and ring-like grains of pyrite are very common, so diameters of these grains may reach up to 3 mm. Rhythmical, concentrical and ring-like alternations of pyrite and other sulphide minerals are also very frequent, particularly those with chalcopyrite. The multiple alternations of pyrite with the mentioned minerals have also been frequently seen. The oval chalcedony grains from the rock matrix are often rimmed by pyrite or include numerous inclusions of it.

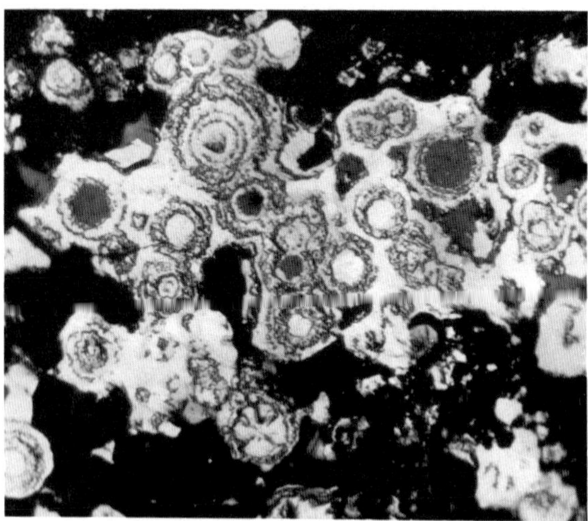

Fig. 1. In the chalcopyrite matrix *(light grey)* sphalerite globules *(dark grey)* are visible as well as numerous ring-like pyrite grains *(light grey, pronounced relief)*. Multiple concentric alternations of pyrite and chalcopyrite are also present occasionally. *Black*, rhyolite pyroclastics. x 220, imm

Chalcopyrite is very common and in places even dominant. However, it is hardly recognizable in the sections rich in the hydrothermal chalcopyrite, as the colloidal textures of the former are in such cases rarely visible.

The first-stage chalcopyrite appears as fine mixtures with pyrite, or the numerous inclusions in the rock. In relatively large chalcopyrite grains numerous pyrite inclusions are also present. Very common is chalcopyrite in form of oval, egg-shaped, ring-like and similar grains, or as rhythmical and concentric alternations with pyrite and sphalerite, less frequently with other sulphides.

Sphalerite is frequent. It occurs in the form of grains of various sizes in the rock or as impregnations, but mostly in the form of oval and ring-like grains. Rhythmical and concentrical alternations with other sulphides, particularly with pyrite and chalcopyrite are rather common.

Tetrahedrite is usually scarce, but occasionally can be rather abundant. It occurs as a rule in the form of small grains and globules, ring-like and egg-shaped grains in the rock or in pyrite and chalcopyrite. Relatively large tetrahedrite patches and specks are not frequent and occur in the rock or in chalcopyrite.

Bornite is frequent but never abundant. It occurs as grains of various sizes, but mostly fine, in host rock or with other sulphides. Very frequently it appears as oval, egg-shaped and ring-like grains in pyrite and sphalerite and rarely in chalcopyrite.

Galena is sparse or lacking and this is the reason why it could not be examined in detail. On the other hand, distinguishing it from the hydrothermal galena is very difficult. Whereas the presence of the primary galena could be proved by the detection of its concentric alternations with primary pyrite and chalcopyrite, as well as of its small globules which are sometimes rimmed by pyrite.

3 Second-Stage Sulphide Assemblage (Hydrothermal Paragenesis)

As stated earlier, the same sulphides occur in this stage as in the previous one, such as pyrite, chalcopyrite, tetrahedrite, sphalerite, bornite, and galena, with quartz and barite as the nonmetallic minerals. The minerals described occur in the form of veinlets or as metasomatic replacements in the primary ore body, forming in this way a type of the stockwork. This assemblage is hydrothermal in origin and later than the primary ore body. The second-stage sulphides frequently enclose the grains of the primary ones and the rock. Thus numerous grains of first-stage minerals are enclosed in hydrothermal pyrite, chalcopyrite, and sphalerite, sometimes also in bornite and tetrahedrite. Frequently enclosed oval, ring-like and egg-shaped grains of the primary sulphides in the hydrothermal ones are visible, but the primary textures are not always preserved.

The widespread extent and the abundance of the hydrothermal sulphide minerals are, in general, similar to the corresponding primary ones. Thus the main mineral in the hydrothermal assemblage is pyrite, associated with frequent or abundant chalcopyrite, frequent sphalerite, bornite and tetrahedrite, and usually sparse galena.

Quartz is abundant, often coarse or crystalline, whereas barite is moderately developed.

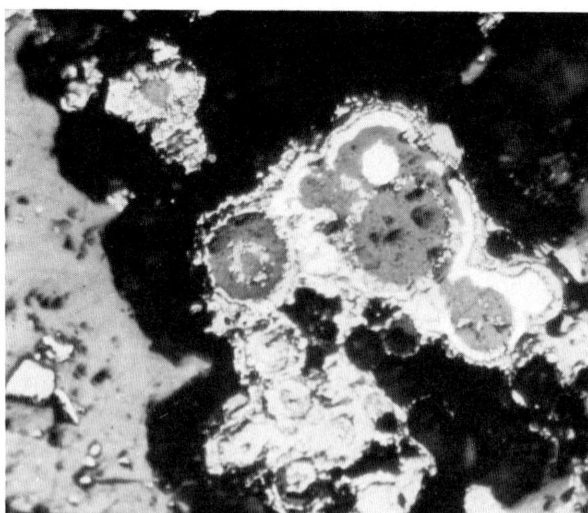

Fig. 2. In the rhyolite pyroclastics *(black)* large irregular grain of chalcopyrite *(light grey)* encloses sphalerite globules *(dark grey)*. Both minerals are partly rimmed by pyrite *(light grey, pronounced relief)*. *Left,* hydrothermal sphalerite replaces the rock. × 220, imm

The mode of occurrence of the second-stage sulphides is as follows (Fig. 2):

Pyrite is in the form of coarse grains and crystals and is the oldest hydrothermal mineral. It replaced the host rock and primary ore body in the form of veinlets or metasomatically, and frequently cemented and enclosed its grains. In some cases numerous oval and egg-shaped chalcedony grains from the host rock are enclosed in pyrite too.

Chalcopyrite is the dominant copper mineral and its contents may reach more than 4%. It occurs in the form of grains of various sizes, as patches and specks, and locally as large grains. Chalcopyrite mostly replaced the rock and the first stage sulphides.

Sphalerite is frequently coarse, sometimes also crystalline. It is younger than pyrite of both stages and than hydrothermal quartz, replacing or cementing these minerals or their fragments. The chalcopyrite *exsolutions* in the sphalerite of the second stage are usually scarce, but occasionally can be frequent.

Bornite and tetrahedrite occur as grains of various sizes replacing the rock, the primary ore body minerals, as well as the second-stage pyrite, sphalerite and quartz, mostly along their grain contacts.

Galena is one of the latest hydrothermal minerals and replaced almost all primary and second-stage sulphides, or enclosed the fragments of the fractured host rock and older sulphides.

Quartz. Strong silification of the rhyolite pyroclastics and of the primary ore body took place mostly before the deposition of the second-stage sulphide assemblage, so the main quantities of the quartz have deposited in this period. During the deposition of later hydrothermal sulphides, considerable amounts of quartz II have also formed, i.e. after pyrite and before the other sulphides.

Barite is not frequent. It is the latest hydrothermal mineral and is fine or coarse-grained.

4 Paragenetic Positions

The paragenetic position of the second-stage sulphide minerals is as follows:

Pyrite-quartz-sphalerite, chalcopyrite exsolutions-bornite, chalcopyrite exsolutions II-tetrahedrite-chalcopyrite-galena-barite.

5 Secondary Minerals

Although sulphide assemblages in both of these stages are mostly preserved, some descending alterations occur sporadically. The most developed is limonitization of the host rhyolite pyroclastics, which has not included the conversion of pyrite. The conversion of bornite into covellite is frequent, occasionally even complete. Bornite transformation into chalcocite is scarce, so the latter mineral occurs always with covellite. Generally, both of the secondary copper minerals are scarce or lacking.

6 Genesis

In the Tunca locality two stages of sulphide deposition can be recognized: the first, volcanic-sedimentary, and the second, hydrothermal.

6.1 Volcano-Sedimentary Stage

The primary sulphide ore body was formed simultaneously with the host rhyolite pyroclastics, which is a member of the volcano-sedimentary Senonian formation of Eastern Pontides.

The primary sulphide minerals are fine to coarse and intergrown with the host rock, with the replacement structures completely lacking. The mutual sulphide intergrowths are also characterized by the lack of these structures. Primary sulphides show dominantly colloidal textures which include oval, egg-shaped and ring-like grains as well as multiple rhythmical and concentric mutual alternations. No sulphide veinlets, crystals, or geologic thermometers (exsolutions) in the primary ore body have been recognized. The host rock consists also partly of oval transparent grains, mostly of chalcedony.

Although microscopic examinations were performed in detail, colloform textures have not been detected in the second-stage sulphide assemblage at all, thus, these textures are the oldest ones in the sulphide ore body (Fig. 3). Finally, the paragenetic position of the primary sulphides cannot differ as all the minerals have deposited contemporaneously. All of these data indicate the volcanic-sedimentary origin of the first-stage sulphide assemblage (Ramdohr 1969).

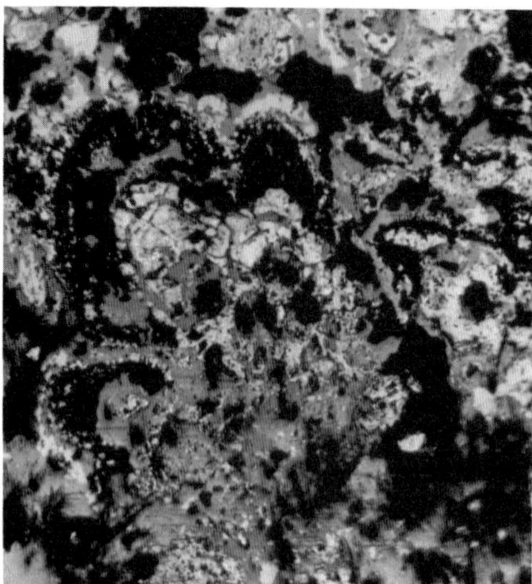

Fig. 3. Colloform intergrowth of pyrite *(light grey, pronounced relief)*, sphalerite *(grey)* and chalcedony *(black)* from the pyroclastic rock. × 180, imm

6.2 Hydrothermal Stage

Throughout the second hydrothermal stage, the same sulphide assemblage has been formed but all of its properties clearly differ from the previous one. The hydrothermal sulphides have been deposited along the fractures and fissures in the pyroclastic rock and the primary ore body, in their void spaces or metasomatically, with commonly developed replacing structures. Therefore, the way of deposition of the hydrothermal assemblages is typically epigenetic, with well-expressed paragenetic positions and relationships of the sulphide minerals deposited in the form of coarse grains and crystals (pyrite, sphalerite, quartz), and with the presence of exsolutions of chalcopyrite in sphalerite and bornite (geological thermometers) (Fig. 4). As for high-temperature minerals in the second-stage sulphide assemblage, they are lacking; thus, all the data presented indicate the moderate hydrothermal deposition temperatures of the second-stage sulphide paragenesis (Ramdohr 1969).

As to the origin of the hydrothermal sulphide assemblage, there are two possible hypotheses:

1. New hydrothermal influx of the same minerals that have formed the volcanic-sedimentary stage, and

2. Remobilization of volcanic-sedimentary sulphide ores by the subsequent thermal waters that have circulated along the fractures, fissures, and void spaces inside the primary deposit, and by redeposition – this time hydrothermal – of dissolved material inside the existing sulphide ore body.

Due to the fact that in the Black Sea coastal region both mechanisms – the juvenile hydrothermal introduction and the regeneration of the existing volcanic-sedimentary deposits – have been observed (Vujanović 1974), it will be very difficult, at the present time and stage of investigations, to reach a final conclusion in this case.

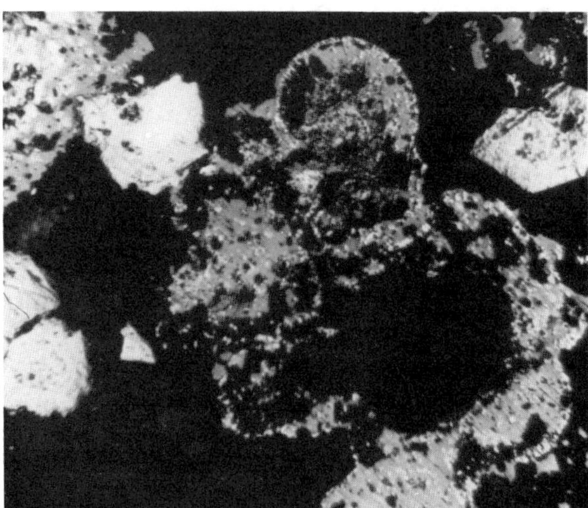

Fig. 4. Globules of sphalerite *(grey)* in rhyolite pyroclastics *(black). Left and right:* hydrothermal pyrite *(light grey, pronounced relief),* partly crystalline. X 180, imm

References

Milutinović D (1972) Report on the geological mapping and prospection in 10,000 scale SE of Pazar and Ardešen. M.T.A. Institute of Turkey, Ankara, pp 13–25

Ramdohr P (1969) The ore minerals and their intergrowths. Pergamon Press, Oxford London Edinburg New York Paris Frankfurt, pp 83–286

Vujanović V (1974) The basic mineralogic, paragenetic and genetic characteristics of the sulphide deposits exposed in the eastern Black-Sea coastal region (Turkey). Bull Miner Res Explor Inst Turkey 82:22–24

The Condestable Mine, a Volcano-exhalative Deposit in the Mesozoic Coastal Belt of Central Peru

A. WAUSCHKUHN and R. THUM [1]

Abstract

The Cu-deposit Condestable in the volcanic rocks of the Mesozoic coastal belt of Central Peru is investigated petrographically, mineralogically and geochemically to distinguish its mode of formation. The deposit consists of stratabound pyrite-chalcopyrite manto ore which is closely associated with andesitic rocks of Lower Cretaceous age. To a lesser degree veinlets are known in the deposit. The ore minerals in the mantos and the country rocks have undergone a low grade metamorphism of the Barrow-type, while the veinlets are unmetamorphosed.

The investigations of Condestable indicate a volcano-exhalative genesis of the deposit. Mineralogical and geochemical evidence points out to a mobilization of the elements Zn, Pb, Co, and Ni from the manto ore by low grade metamorphism and a recrystallization of these elements in veinlets, while the elements Cu and Fe remain in the manto ore and only recrystallize under the given metamorphic conditions.

1 Introduction

The Condestable mine is located about 90 km south of Lima, only a few kilometres from the Pacific coast and the Panamerican highway (Fig. 1). It is situated near the town of Mala at long 76°35′W and lat 12°42′S.

The mine is an Albian stratabound pyritic copper deposit within a sequence of volcanic and sedimentary rocks in the Mesozoic coastal belt. Located in the nearby neighbourhood is the Raúl deposit, which is similar to the Condestable deposit and is geologically related with it. Two other stratabound copper deposits, which show despite some differences a lot of similarities with respect to the country rock, the ore minerals, and its geometrical distribution, are situated south of Condestable and Raúl also in Lower Cretaceous volcanics and sediments of the Mesozoic coastal belt (Wauschkuhn 1979a). In volcanics of Lower Cretaceous time in Chile also stratabound Cu-sulphide deposits are known (Ruiz et al. 1970).

The geological setting, the nature of the country rock, and the ore minerals, and their geometrical distribution suggest a volcano-exhalative and exhalative-sedimentary genesis of all the Peruvian deposits (Ripley and Ohmoto 1977, 1979; Wauschkuhn

1 Mineralogisch-Petrographisches Institut der Universität Heidelberg, Im Neuenheimer Feld 236, 6900 Heidelberg, FRG

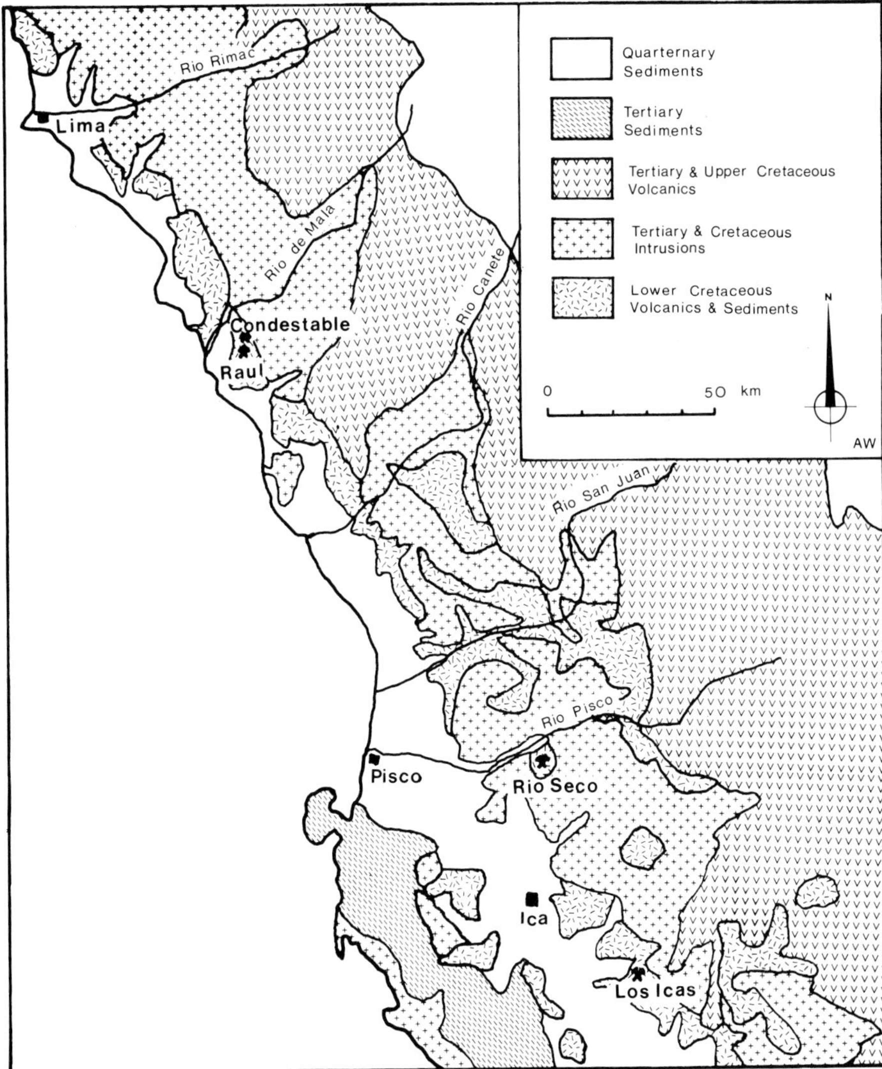

Fig. 1. Geological map of the Western part of Central Peru. (Modified after Bellido 1969)

1979a; Wauschkuhn and Ohnsmann 1980). In this regard they are similar for example to the Kuroko deposits of Japan (Matsukuma and Horikoshi 1970; Sato 1974, 1977), to the Kieslagerstätten of Norway (Vokes 1962), or to different massive sulphide deposits of Canada and the United States (Sangster 1972; Hutchinson 1973).

The country rocks and the ore minerals of Condestable are changed metamorphically. In this paper mineralogical and geochemical data of the country rock and the ore minerals are studied to clear up the formation of the ore and the influence of the metamorphism on it.

2 The Geological Setting

The regional geology of Central Peru is characterized by a number of geological units, which run parallel to the coastline (Petersen 1965). According to Amstutz (1978) many different ore deposits in Peru can be assigned to different geological units. The investigated deposit is situated in the Mesozoic coastal belt (Fig. 1). The rocks of this Mesozoic belt are eugeosynclinal and consist of thick sequences of intermediate volcanic rocks and marine sediments of Jurassic and Cretaceous age.

The structure of these rocks is relatively simple. The area underwent a low-grade metamorphism of the Barrow type. In connection with the rising of the coastal batholith during the Eocene and Paleocene (dated from Giletti and Day 1968; Laughlin et al. 1968; Noble 1973; Noble and McKee 1975) an uplift took place, which effected a tilting of the country rock and intrusions of changing chemical compositions are injected. The beds strike 160°SE and dip 50°SW. The rocks are normally not folded, but block faulting is common. The rocks associated with the deposit consist of volcanic lavas and tuffs, sandy and limy sediments and intrusive rocks (Fig. 2).

The geometrical distribution of the ore minerals is mainly stratabound, as ore layers, the so-called mantos, and as disseminated ore. The thickness of the mantos is up to 3 m. To a lesser degree veinlets are known in the deposit. The country rock consists mainly of tuffs; sediments are less abundant.

The mineral deposit is bordered by a porphyric stock, which was intruded through faults and bedding planes. The intrusion divides the mineral deposit into two sections,

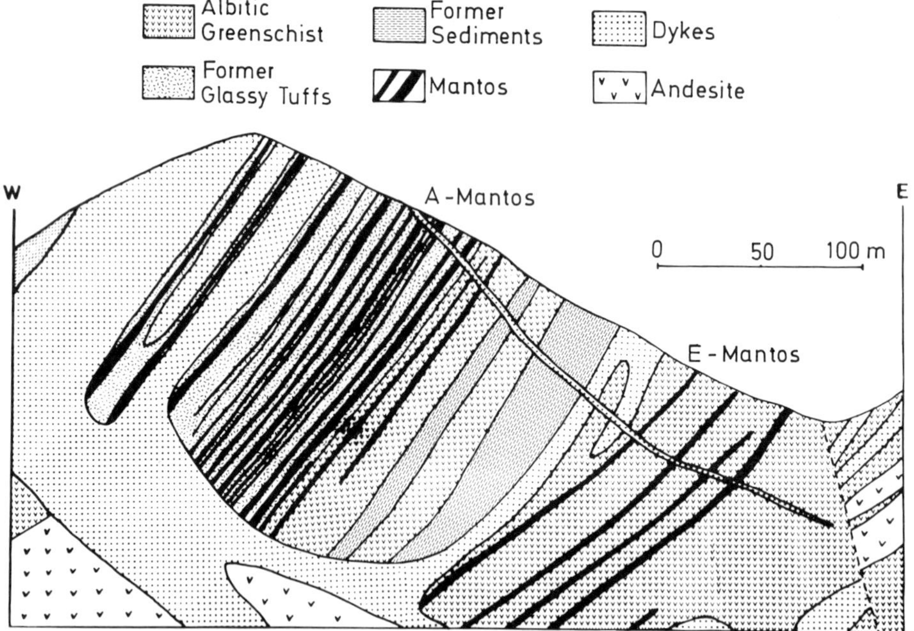

Fig. 2. Geological profile of the Condestable deposit, modified after one from the geological staff of the Condestable mine

the A-deposit (the upper part) and the E-deposit (the lower part) (Fig. 2). Petrographically, A and E form two different units. The deposit is mined for copper which is concentrated in layered and disseminated ore. According to the data provided by the mining company, 400 t of rock with an average content of 1.7% Cu are worked now. The mining of lead and zinc, which are concentrated in the veinlets, is not profitable.

3 Petrography

The country rocks of the Condestable mine are macroscopically very similar. They are rather homogeneous, compact, and hard. The colour varies from green to grey.

Microscopically, four types can be distinguished.

Albitic greenschist appears in the E-part of the mineral deposit. It is mainly built up by plagioclase, calcite, chlorite, grammatite, quartz, prehnite, and titanite; very rarely epidote and piemontite are found. The plagioclase mostly consists of albite (Fig. 3), and sometimes of oligoclase (with 20% An). Anorthoclase could also be located. These feldspars consist of big, almost always decomposed or corroded grains, which are irregularly situated in the fine groundmass of chlorite, quartz, and smaller feldspar. Very often plagioclase is directly intergrown with calcite and is apparently locally displaced by it (Fig. 4).

These appearances point to an autometamorphical change of basic plagioclase to albite and calcite, an autometasomatical albitisation, according to Tröger (1967). A change from plagioclase to sericite is hardly found.

In the lower part of the A-series are sediments, which consist of quartz, plagioclase, calcite, ore minerals (sulphides, limonite, malachite), and some apatite. The plagioclase is intergrown with carbonate similar as in the albitic greenschists, which indicates here also a change of basic plagioclase to albite and calcite. The quartz can possibly be determined as representing an authigene formation; such a new formation can be found in former shaly or carbonatic sediments (Black 1949). According to the mineralogy, the original rocks were shaly and limy sediments.

Former glass tuffs form the upper part of the A-deposit and show typical microfelsic structures which indicate an originally glassy composition. The matrix is made up of extremely fine quartz and plagioclase. In this groundmass, there are intersperses of opaque minerals and of grammatite, prehnite, apatite, leucoxene, and quartz as well as plagioclase. Sometimes plagioclase is nearly changed into chlorite (Fig. 5); also round and egg-formed aggregates of colourless to light green chlorites are found. Sometimes a primary bedding is recognizable in thin sections. In this case an enrichment of leucoxene on the bedding planes is observable.

The actinolite rock is the actual manto rock with 70%–95% actinolite/grammatite, and is only found in close intergrowth with the ore. Two varieties of this rock can be distinguished. The wall rock of the top manto of the A-deposit, which is rich in magnetite, is mainly made up of fine fibrical and often fibroradial grammatite (the Fe-free component of the mixed crystal system). In addition to grammatite also chlorite and orthite are to be found here. The rocks of the mantos in which no iron oxides are apparent are mostly made up of coarse, idiomorphic actinolite and a lot of chlorite, quartz, calcite, and titanite. Sometimes also stilpnomelane, apatite, and epidote/clinozoisite can be found. The two end members of the actinolite/grammatite system can be found together sometimes. Then grammatite rims actinolite, therefore the Fe-free component was formed later (Fig. 6).

4 Ore Mineralogy

The geometrical distribution of the ore in the mineral deposit is concordant with the country rock and to a lesser degree discordant. The discordant ores are not important economically. The stratabound ores can be differentiated into three types:

a) Sulphide manto ore, mainly iron and copper sulphides.
b) Magnetite manto ore, iron, and copper sulphides and magnetite.
c) Disseminated ore, sulphide minerals in the former glass tuffs.

Type a: *Sulphide manto ore.* The main ore minerals are pyrite and chalcopyrite and both build rather big irregular aggregates. Idiomorphic pyrite is rare, but if present it is – in opposition to the xenomorphic pyrite – free from inclusions and can be explained as idioblasts, which have cleaned themselves during the blastese (Anger 1966). Cataclastic pyrite can often be found. The cracks in the cataclastic pyrites are filled by chalcopyrite. Skeleton growth was found in one place, the interstice in this pyrite being mainly filled with stilpnomelane (Fig. 7).

Marcasite is often irregularly intergrown with pyrite. Both minerals seem to have formed at the same time. Pyrrhotite is rarely found as small inclusions in pyrite. A process in which pyrite is formed from pyrrhotite is possible under diagenetic and metamorphic conditions. Possibly the inclusions of pyrrhotite in pyrite are relicts of larger amounts of pyrrhotite.

Fig. 3. Albite (polysynthetic twinning is distinguishable) in a groundmass of chlorite and quartz. Thin section, X nicols, length of the bottom edge of the picture is 2.5 mm

Fig. 4. Plagioclase *(black* and *grey)* partly displaced by calcite *(white).* Thin section, X nicols, length of the bottom edge of the picture is 1.0 mm

Fig. 5. Phenocryst of plagioclase *(black)* nearly totally displaced by chlorite *(white)* in a layer of chlorite, plagioclase, and quartz. Thin section, X nicols, length of the bottom edge of the picture is 1.0 mm

Fig. 6. Actinolite *(dark grey)* overgrown by grammatite *(white)* and ore minerals *(black).* Thin section, // nicols, length of the bottom edge of the picture is 1.0 mm

Fig. 7. Skeleton growth of pyrite *(white)* closely intergrown with stilpnomelane *(black).* Chalcopyrite *(greyish white)* is also present. Polished section, oil immersion, // nicols, length of the bottom edge of the picture is 0.4 mm

Fig. 8. Magnetite *(white)* showing zones, which are displaced by gangue minerals *(black).* Polished section, oil immersion, // nicols, length of the bottom edge of the picture is 0.4 mm

Fig. 9. Relicts of magnetite *(grey)* in pyrite *(white)* and gangue minerals *(black).* Polished section, oil immersion, // nicols, length of the bottom edge of the picture is 0.2 mm

Fig. 10. Sphalerite *(grey)* with colloform textures, surrounded by marcasite *(white).* The inclusions in the sphalerite are galena *(white).* Polished section, oil immersion, // nicols, length of the bottom edge of the picture is 1.8 mm

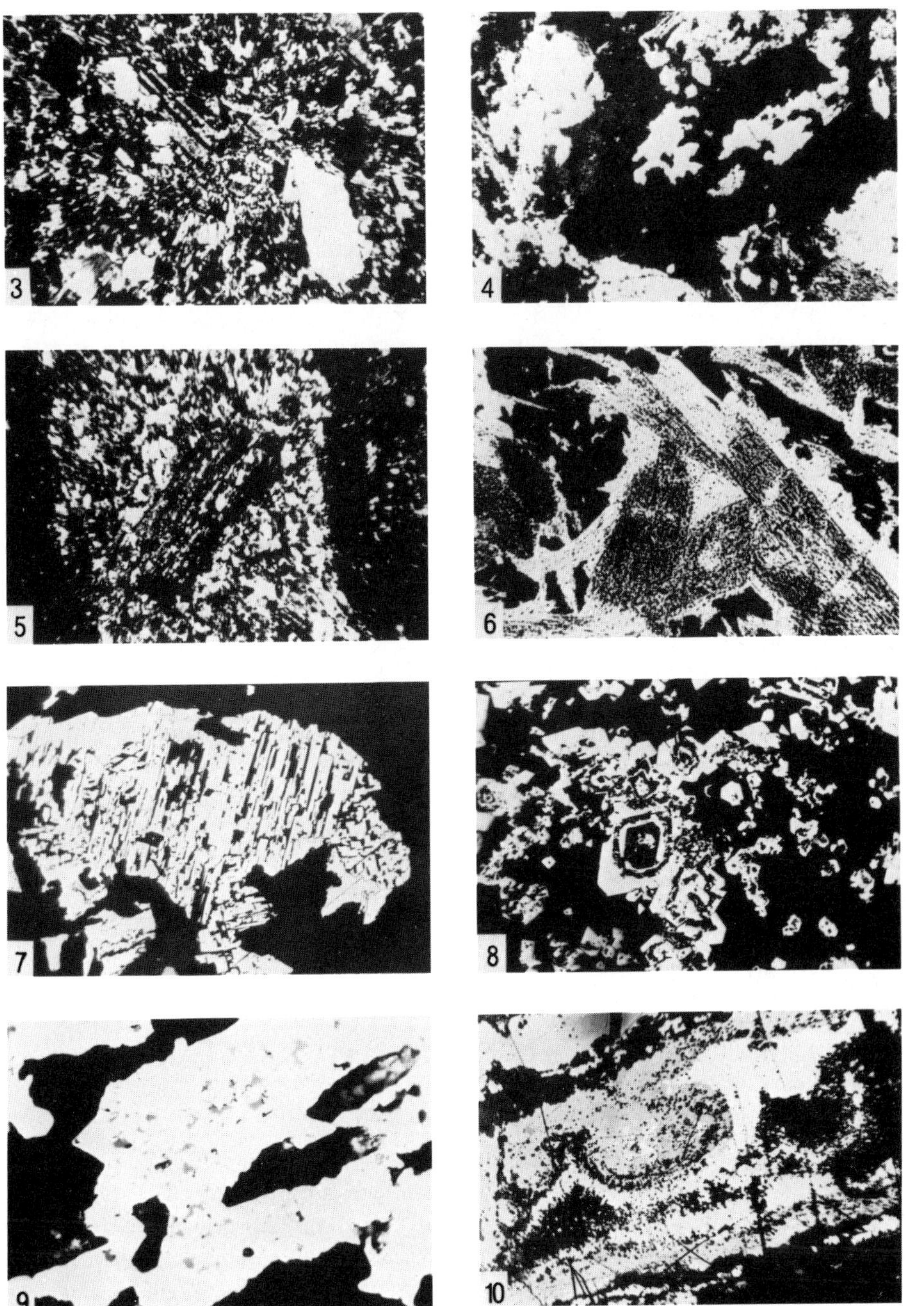

Figs. 3–10

Other ore minerals are mackinawite as exsolution in chalcopyrite, bornite, sphalerite with exsolutions of chalcopyrite and Ti-oxides (leucoxene, titanite). Very seldom galena and molybdenite and traces of gold are found. Covellite and chalcocite occur as alteration products of chalcopyrite.

Textures of sulphidization of gangue minerals to pyrite are often observed. These processes can take place during the sedimentation and ore formation, the diagenesis, and during metamorphism if sulphur is added or is released from sulphides, and react with metal cations from other minerals. A sulphur addition during the metamorphism must be excluded in our case. A release of sulphur out of the existing sulphides during metamorphism – although limited – is possible. But also a mobilization of elemental sulphur during diagenesis and metamorphism can not be excluded. The question whether the existing sulphidization textures are relicts of ore forming, of diagenetic or of metamorphic processes, must remain open.

Type b: *Magnetite manto ore.* The ore minerals of the top layer of the A-series is mainly made up of magnetite next to pyrite and chalcopyrite. Magnetite builds many idiomorphic grains, which are partly grown together in a loose skeleton. Sometimes xenomorphic masses are also present. In one place weakly anisotropic zones are found in the magnetite; such zones are partly replaced by gangue minerals (Fig. 8).

Chalcopyrite and pyrite have similar appearances to type a). Indications of sulphidization of magnetite, especially to pyrite, can often be recognized (Fig. 9).

The formation of the magnetite can hardly be regarded as a metamorphic product, but does not contradict a possible exhalative genesis of the deposit (Zies 1924; Ramdohr 1962). The precipitation of magnetite from an aqueous solution occurs under certain pH-Eh conditions (Garrels 1960). It seems likely that at the end phase of the volcanic processes the chemism and the intensity of the exhalations change, and that the physico-chemical conditions in the sedimentation area vary accordingly so that a magnetite formation is possible.

Type c: *Disseminated ore.* The ore minerals here are pyrite, pyrrhotite, and chalcopyrite. The single grains are spread relatively homogeneously in the groundmass. Idiomorphic pyrite is also present. The texture of the single grains indicates that they are mainly porphyroblasts.

The discordant mineralizations can be distinguished into two types:

1. Small pyrite-calcite veinlets.
2. Veinlets with Pb- and Zn sulphides.

1. The pyrite-calcite veinlets appear nearly stratabound and are only a few centimetres long and a few millimetres thick. The pyrite here is coarse grained, without inclusions and usually idiomorphic. As there are transitions between this pyrite variety and the xenomorphic pyrite in the stratabound ore, it can be concluded that the pyrite was partly mobilized during the metamorphism and recrystallized into small cracks together with the calcite, which originates from altered plagioclase.

2. The veinlets with Pb-Zn sulphides appear mainly in the lower part of the deposit (E-part) and have a length of 1–2 m and a width of 10–20 cm. The ore minerals of the veinlets are closely intergrown with calcite. No metamorphic minerals are observed.

Two of the veinlets were investigated closely. In one there is mostly galena of skeleton growth surrounded by colloform sphalerite (Fig. 10). Also pyrite and sometimes twins of idiomorphic marcasite are present.

In the other veinlet we deal with a rhythmitic intergrowth of galena and sphalerite on the one hand and pyrite with Co-Ni-zonation on the other hand. Together with galena there are arsenopyrite, luzonite, bournonite, and chalcopyrite. This paragenesis is always overgrown by colloform sphalerite. The sphalerite is extremely rich in fine inclusions of the minerals which were mentioned above. The galena-sphalerite zone is surrounded by pyrite. The zonation of pyrite as indicated by microprobe analyses is due to different contents of Co and Ni in the crystal lattice of pyrite (Fig. 11).

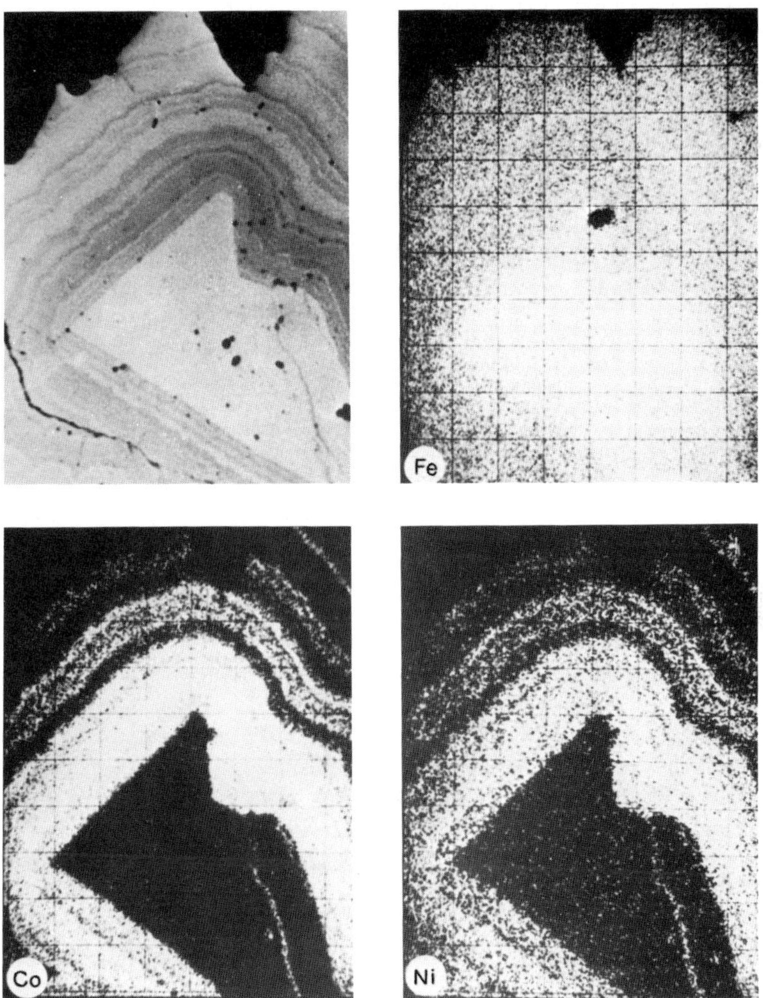

Fig. 11. Pyrite surrounded by brownish zones and the corresponding scanning photographs of Fe, Co, and Ni from a veinlet of the Condestable deposit. Polished section, oil immersion, // nicols, length of the bottom edge of the picture is 0.02 mm

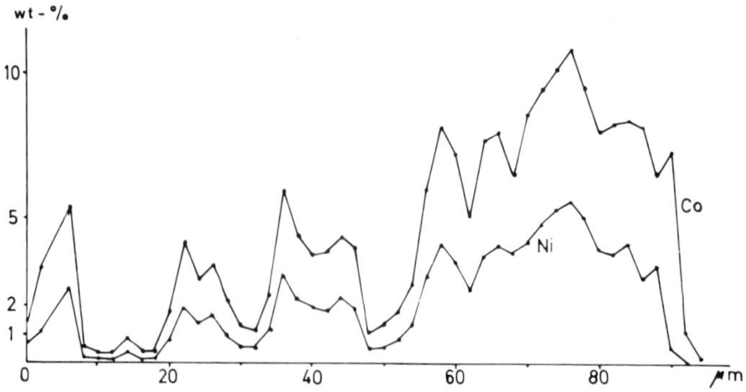

Fig. 12. Microprobe profile across the Co-Ni-containing zone of the grain of Fig. 11

The proportion of Co:Ni is constantly 2:1 (Fig. 12). Besides, the pyrite shows cracks parallel to the direction of its growth. These phenomena suggest that the pyrite originated from melnikovite pyrite and the cracks are shrinking cracks. They are filled with gangue (calcite) and chalcopyrite. On the border of the veinlets to the actinolite rocks sulphidized silicates can be found. This suggests an alteration process of the actinolite rocks caused by the same sulphuric fluids which have created the veinlets.

5 Geochemical Investigations

Thirtynine samples from different parts of the Condestable deposit were analyzed for Cu, Zn, Pb, Co, Ni, Mn, Cr, and Fe (Table 1) to determine relationships between ore mineral-forming elements if present. To clarify the relationships between the analyzed elements, the analyses were treated statistically.

Means ($\bar{x}$), standard deviations (s) and variation coefficients (C) for all elements are listed in Table 2. The variation coefficient C is a measure of the relative variability of the analyzed data and makes a comparison between the different elements possible, independently of their dimensions.

The high C-values of Pb, Zn, Ni, and Co demonstrate the great variation of the analyzed data, and show that the arithmetically found means for this elements are a bad estimate of the true mean of population. Much nearer to the true mean of population is the arithmetically found mean of Cu; for the elements Mn, Cr, and Fe the variation coefficients are distinct < 1.0 and $\bar{x}$ gives a precise estimate for the true mean of population. All distributions were submitted to Chi^2 analyses to determine whether or not they were approximated adequately by either a normal or a lognormal distribution. Using the Chi^2 criterion, it was pointed out that all elements are approximated by lognormal distributions (Table 2).

With the help of a statistical factor analysis, relationship between the elements was established. In this paper only the factor analysis and its applicability to geological problems will be discussed. Studies on factor analysis, their operation and their applica-

Table 1. Content of elements in the ore and country rock of the Condestable deposit (measured by AAS)

Cu ppm	Pb ppm	Zn ppm	Co ppm	Ni ppm	Mn ppm	Cr ppm	Fe %
185	20	100	10	50	450	40	3.12
4740	tr	45	30	40	380	55	3.41
520	tr	115	10	25	440	50	3.12
20580	35	170	90	135	770	50	16.20
17590	15	95	40	80	610	40	20.57
4235	tr	35	20	30	145	40	5.26
5115	tr	30	15	75	220	90	3.11
4170	tr	55	25	35	170	90	3.09
45355	tr	655	60	65	490	40	16.47
12770	10	630	35	45	450	25	12.56
11145	15	360	60	90	520	46	17.38
4950	15	85	25	75	575	50	12.24
40385	20	1210	25	70	460	25	10.67
13590	tr	85	35	60	80	95	3.96
24195	tr	75	30	55	340	20	13.06
1840	80	120	15	95	340	90	2.33
57905	10	45	30	65	145	20	24.91
5685	25	1095	50	60	305	175	30.31
4760	tr	285	35	50	585	35	17.01
21555	20	645	30	35	235	15	40.76
7380	10	550	15	140	235	165	4.34
11980	20	125	50	135	605	35	15.22
2960	15	80	25	35	590	120	13.90
170	tr	55	10	55	400	60	2.53
2605	15	245	25	25	625	80	7.86
72100	20	225	60	55	580	85	26.87
3225	tr	160	15	30	740	65	6.56
14430	40	565	25	70	1030	70	9.85
15235	tr	60	15	20	720	60	4.87
4930	10	2740	25	50	1820	125	8.65
4525	tr	2265	30	40	1250	65	8.22
3350	tr	60	10	40	500	25	4.99
32420	40	725	45	40	330	65	28.22
7020	15	975	25	50	1295	50	11.55
20215	25	2155	35	35	1510	60	9.70
1280	tr	740	15	180	1350	100	10.24
37535	115	1075	50	115	340	40	10.46
9300	4500	32385	255	1250	410	20	12.50
1255	7530	78430	665	1490	535	20	16.94

bility to geological and ore depositional problems were summarized by Imbrie and van Andel (1964), Krumbein and Graybill (1965), Klovan (1968), Nicol et al. (1969), and Saager and Esselaar (1969). The basic assumption for factor analysis is that the measured variables, in our case the eight elements for which analyses were made, are related lineally with each other, and that it must be possible as a consequence to reduce the number of variables. The newly formed variables are then interpreted in terms of geological and geochemical processes.

Table 2. Means, standard deviations (in ppm), correlation coefficients, and Chi^2-tests of the analyzed rocks of the Condestable deposit

Variable	Mean $\bar{x}$	Standard deviation S	Variation coefficient C	Chi^2	Chi^2 (log. transformed)
Cu	14185	16656	1.17	41.60	1.64
Zn	3309	13375	4.04	145.88	12.28
Pb	324	1385	4.28	145.88	42.40
Co	53	108	2.04	110.52	9.28
Ni	130	295	2.27	118.55	11.02
Mn	578	394	0.68	17.06	8.73
Cr	61	37	0.61	10.66	8.48
Fe	121297	88098	0.73	9.41	7.67

Table 3. Varimax factor matrix of 3 factors of the analyzed rocks of the Condestable deposit

Variable	Communalities	Factor 1	Factor 2	Factor 3
Cu	0.82	0.03	*0.90*	0.11
Zn	0.84	*0.80*	0.20	0.41
Pb	0.79	*0.88*	0.14	0.05
Co	0.85	*0.82*	0.42	0.03
Ni	0.85	*0.91*	0.07	0.12
Mn	0.89	0.10	0.03	*0.94*
Cr	0.39	0.34	0.37	0.37
Fe	0.84	0.30	*0.85*	0.17
Variance (%)		39.03	23.70	15.46
Cumulative Variance (%)		39.03	62.72	78.19

The R-Mode factor analysis undertaken here, investigates the relation of all variable pairs, starting from the matrix of the correlation coefficients. The individual factors were determined according to Harmann's (1967) main component method. As the mathematically found main components often are difficult to interpret, they are turned orthogonally around their origin according to the Varimax Method by Kaiser (1958) until they are as near as possible to the variable vectors, and thus make a geological interpretation easier. To carry out the R-Mode factor analysis, a computing program of the Kansas Geological Survey was used (Ondrick and Srivastava 1970). The Varimax factor matrix of the analyzed rocks is given in Table 3. The communalities represent the proportion of the variation in a given variable that is explained by the 3-factor model. The total explained variance of this model is 78.19%. The table shows that each factor consists of significant contributions from certain variables (>0.8) and less important to negligible contributions from others (<0.8).

Factor 1 is made up by the variables Ni (0.91), Pb (0.88), Co (0.82), and Zn (0.80) and explains 39.03% of the analyzed data. This factor contains the elements which build up the postmetamorphic veinlets in the deposit, and comprises the elements which are mobilized from the mantos during metamorphism.

Factor 2 represents the elements Cu (0.90), and Fe (0.85), and explains 23.70% of the analyzed data. It comprises the elements which are not mobilized, i.e., removed over distances of more than a few decimetres, during low-grade metamorphism in the Condestable deposit, but remain in the stratabound ore. These elements are only involved in the recrystallization.

Factor 3 is mainly made up by Mn (0.94) and explains 15.46% of the analyzed data. Mn remains in solution in a volcano-exhalative environment and crystallizes in a greater distance from the volcano-exhalative basin (Wauschkuhn 1979b). This factor reflects the separation of the chalcophile elements Cu, Pb, and Zn and the siderophile elements Fe, Co, and Ni from the lithophile element Mn during the formation of the deposit.

6 Discussion of the Results and Conclusions

The mineralogical association of the country rocks and the gangue minerals of Condestable indicates a low-grade metamorphism of the clinozoisite-albite-actinolite-chlorite zone (greenschist facies, according to Winkler 1976). Winkler gives a temperature range from about 330° to 500°C for the low-grade metamorphism. The presence of prehnite together with actinolite and epidote in Condestable indicates that the metamorphic grade of the rocks is near the border between very low grade to low grade. Regarding this mineralogical association of the country rocks we suggest that the temperature of metamorphism was about 300°C.

The ore microscopical investigations confirm this assumption. Borchert (1934) reports that valleriite exsolutions in chalcopyrite indicate an exsolution temperature of about 200°–250°C, and Ramdohr (1975) points out that Borchert confounded valleriite with mackinawite, but the given temperature range indicates only a more or less qualitative value. The presence of mackinawite exsolutions in chalcopyrite demonstrates a metamorphic temperature higher than 200° to 250°C.

The existence of marcasite on the other hand indicates a temperature not higher than 300°C, as it changes completely to pyrite in temperatures above that. The intergrowth between marcasite and pyrite shows that both have crystallized simultaneously and were apparent before the metamorphism (but an occasional new formation of marcasite during and after metamorphism should not be excluded). Furthermore, the small grain size and fine-grain growth is an indication of low-temperature metamorphism (Stanton 1960).

Regarding the effects of the metamorphism on the ore it must be considered that in spite of low stage of metamorphism the original ore textures were considerably destroyed, as almost only sulphides were found in the ore components, which react rather sensitively when confronted with a metamorphic event (McDonald 1967). No colloform textures in the stratabound ore could be observed, like those known from Norwegian volcano-exhalative "Kieslagerstätten" (Anger 1966), from the Japanese Kuroko deposits (Watanabe 1974), or from the Peruvian Los Icas deposit (Wauschkuhn 1979a). Only the sulphidization textures can partly provide primary structures as they are known from recent volcano-exhalative environments (Wauschkuhn and Gröpper 1975; Wauschkuhn et al. 1977).

To summarize the investigations on the stratabound part of the ore, the following appearances can be classified as clearly or highly probably metamorphic at low temperature and low pressure:

- cataclase of pyrite,
- porphyroblastic formation of pyrite (mainly in tuffs),
- self-purification of pyrite from inclusions,
- mobilization of Zn, Pb, Co, and Ni.

Figure 13 demonstrates the reactions of the ore minerals, which could have happened during low-grade metamorphism.

The veinlets on the other hand are not affected by any metamorphic process and their mineral association indicates that they were formed under low-temperature conditions. The presence of Co-Ni bravoite corroborates the low-temperature formation. Ni entering the crystal lattice of pyrite demands according to Kullerud (1961) and Klemm (1962) either temperatures lower than 135°C or temperatures above 400°C, which can be excluded from the formation of the veinlets. Colloform sphalerite and skeleton-like galena also indicate a very low crystallization temperature. These observations demonstrate that the veinlets are formed after the metamorphic event probably from the residual solutions which are mobilized from the mantos.

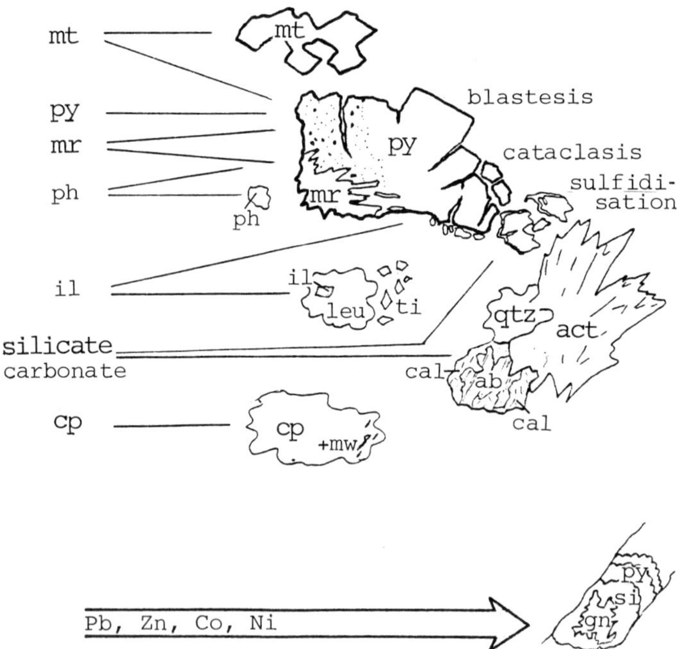

Fig. 13. The possible reactions of the ore minerals during low-grade metamorphism in the Condestable deposit

Mineralogical and geochemical evidence in Condestable indicates a mobilization of the elements Zn, Pb, Co, and Ni by low-grade metamorphism, while Cu and Fe remain in the manto ore and only recrystallize.

Acknowledgments. The authors would like to express their appreciation to Professor Dr. G.C. Amstutz for interesting discussions and useful suggestions regarding this manuscript.

We thank the staff of Minas Condestable for providing geological assistance at the mine and access to company maps.

To Dr. V. Stähle of the Institute of Mineralogy and Petrography of the University of Heidelberg we owe thanks for performing the microprobe analyses.

The research was supported by the "Deutsche Forschungsgemeinschaft".

References

Amstutz GC (1978) Zu einer Metallogenie der Zentralen Anden von Peru. Muenstersche Forsch Geol Palaeontol 44/45:151–158

Anger C (1966) Die genetischen Zusammenhänge zwischen deutschen und norwegischen Schwefelkieslagerstätten unter besonderer Berücksichtigung der Ergebnisse von Schwefelisotopen-Untersuchungen. Clausthaler Hefte Lagerstaettenkd Geochem Miner/Rohst 3:115 S und Anh

Bellido EB (1969) Sinopsis de la geología del Perú. Serv Geol Mineral Peru 22:54 pp

Black WW (1949) An occurrence of authigenic feldspar and quartz in Yoredale limestones. Geol Mag 86:129

Borchert H (1934) Über Entmischungen im System Cu-Fe-S und ihre Bedeutung als geologisches Thermometer. Chem Erde 9:145–172

Garrels RM (1960) Mineral equilibria. Harpers, New York, 254 pp

Giletti BJ, Day HW (1968) Potassium-argon ages of igneous intrusive rocks in Peru. Nature (London) 220:570–572

Harmann HH (1967) Modern factor analysis, 2nd edn. Univ Press, Chicago, 469 pp

Hutchinson RW (1973) Volcanogenic sulfide deposits and their metallogenic significance. Econ Geol 68:1223–1246

Imbrie J, van Andel TH (1964) Vector analysis of heavy-mineral data. Bull Geol Soc Am 75: 1131–1155

Kaiser HF (1958) The varimax criteria for analytical rotation in factor analysis. Psychometrika 23:187–200

Klemm DD (1962) Untersuchungen über die Mischkristallbildung im Dreieckdiagramm FeS_2-CoS_2-NiS_2 und ihre Beziehungen zum Aufbau der natürlichen „Bravoite". Neues Jahrb Mineral Monatsh 1962:76–91

Klovan JE (1968) Selection of target areas by factor analysis. West Miner Feb: 44–52

Krumbein WC, Graybill FA (1965) An introduction to statistical models in geology. McGraw-Hill, New York, 475 pp

Kullerud G (1961) The Fe-Ni-system, the Cu-Ni-system, the Fe-Mo-system. Yearb Carnegie Inst 60:144–152

Laughlin AW, Damon PE, Watson BN (1968) Potassium-argon dates from Toquepala and Michiquillay, Peru. Econ Geol 63:166–168

Matsukuma T, Hirokoshi E (1970) Kuroko deposits in Japan, a review. In: Tatsumi T (ed) Volcanism and ore genesis. Univ Tokyo Press, pp 153–179

McDonald JA (1967) Metamorphism and its effects on sulphide assemblages. Miner Deposita 2: 200–220

Nicol I, Garrett RG, Webb JS (1969) The role of some statistical and mathematical methods in the interpretation of regional geochemical data. Econ Geol 64:204–220

Noble DC (1973) Tertiary pyroclastic rocks of the Peruvian Andes and their relation to lava volcanism, batholith emplacement, and regional tectonism. Geol Surv Am Abstr Prog 5:86–87

Noble DC, McKee EH (1975) Cainozoic stratigraphic and tectonic framework of the Andes of Peru. Geol Surv Am Abstr Prog 7:1214

Ondrick CW, Srivastava GS (1970) Corfan FORTRAN IV computer program, factor analysis (R- and Q-mode) and Varimax rotation. Kans State Geol Surv Comput Contrib 42:97 pp

Petersen U (1965) Regional geology and major ore deposits of Central Peru. Econ Geol 60:407–476

Ramdohr P (1962) Erzmikroskopische Untersuchungen an Magnetiten der Exhalationen im Valley of the 10 000 Smokes. Neues Jahrb Mineral Monatsh 1962:49–59

Ramdohr P (1975) Die Erzmineralien und ihre Verwachsungen. Akademie Verlag, Berlin, 1277 S

Ripley EM, Ohmoto H (1977) Mineralogic, sulfur isotope, and fluid inclusion studies of the stratabound copper deposits at the Raul Mine, Peru. Econ Geol 72:1017–1041

Ripley EM, Ohmoto H (1979) Oxygen and hydrogen isotopic studies of ore deposition and metamorphism at the Raul mine, Peru. Geochim Cosmochim Acta 43:1633–1643

Ruiz C, Aguilar A, Egert E, Espinosa W, Peebles F, Quezada R, Serrano M (1970) Strata-bound copper sulphide deposits of Chile. Soc Min Geol Jpn Spec Issue 3:252–260

Saager R, Esselaar PA (1969) Factor analysis of geochemical data from the Basal Reef, Orange Free State Goldfield, South Africa. Econ Geol 64:445–451

Sangster DF (1972) Precambrian volcanogenic massive sulphide deposits in Canada: a review. Geol Surv Can Pap 72-22:44

Sato T (1974) Distribution and geological setting of the Kuroko deposits. In: Ishihara S (ed) Geology of Kuroko deposits. Min Geol Spec Issue 6:1–9

Sato T (1977) Kuroko deposits: their geology, geochemistry and origin. Spec Publ Geol Soc London 7:153–161

Stanton RL (1960) General features of the conformable "pyritic" ore bodies, part I. Field association. Trans Can Inst Min Metall LXIII:22–27

Tröger WE (1967) Optische Bestimmung der gesteinsbildenden Minerale, Teil 2, Textband. Schweizerbart, Stuttgart, 822 S

Vokes FM (1962) Mineral parageneses of the massive sulfide ore bodies of the Caledonides of Norway. Econ Geol 57:890–903

Watanabe M (1974) On the textures of ores from the Daikoku ore deposit, Ainai mine, Akita Prefecture, Northeast Japan, and their implications on the ore genesis. In: Ishihara S (ed) Geology of Kuroko deposits. Min Geol Spec Issue 6:337–348

Wauschkuhn A (1979a) Geology, mineralogy and geochemistry of the stratabound Cu-deposits in the Mesozoic coastal belt of Peru. Bol Soc Geol Peru 62:161–192

Wauschkuhn A (1979b) Untersuchungen in rezenten vulkanisch-exhalativen Ablagerungsräumen und fossilen vulkanisch-exhalativen Sulfidlagerstätten. Ein Beitrag zur Genese dieses Lagerstättentyps. Eingereicht als Habilitationsschrift, Fak Geowiss, Univ Heidelberg, 174 S

Wauschkuhn A, Gröpper H (1975) Rezente Sulfidbildung auf und bei Vulcano, Äolische Inseln, Italien. Neues Jahrb Mineral Abh 126:87–111

Wauschkuhn A, Ohnsmann M (1980) Geologie, Mineralogie und Geochemie der Sulfidlagerstätte Raul, Peru. Münster Forsch Geol Paläont 51:73–105

Wauschkuhn A, Schwartz W, Amstutz GC, Yagi K (1977) Fumarolic hot lakes on Hokkaido: geochemical, mineralogical and biochemical investigations of their significance for the formation of massive sulfide-deposits. Neues Jahrb Mineral Abh 129:171–200

Winkler HGF (1976) Petrogenesis of metamorphic rocks, 4th edn. Springer, Berlin Heidelberg New York, 334 pp

Zies EG (1924) The fumarolic incrustations of the Valley of Ten Thousand Smokes. Contrib Tech Pap, Natl Geogr Soc (I):159–179

Genesis of Deposits in Recent Sediments

Chemical Forms of Metal Enrichment in Recent Sediments

U. FÖRSTNER [1]

Abstract

Sequential extraction is used to differentiate major chemical forms of metal accumulations in Recent aquatic deposits (examples for limnic sediments, pelagic Fe/Mn-rich deposits, and contaminated coastal marine sediments). Hydrous iron and manganese oxides are found to be the dominant enrichment phases. Organo-metallic associations perform a major function in the transfer of metals between different inorganic phases. Man-induced metal enrichments are predominantly found in labile sediment fractions (reactive organic substances, carbonates, easily reducible phases).

1 Introduction

The type of chemical association between metals and particulate phases in recent sediments has become of major interest both with regard to the research on marine ferromanganese ores and in connection with environmental problems arising from contaminated materials. Chemical extraction procedures used in pedology to ascertain the availability of trace substances important for the fertility of soil (Jackson 1958) are increasingly employed in the differentiation of detrital and authigenic phases in pelagic Mn/Fn concentrations (Chester and Hughes 1967), as well as for the estimation of the toxicity potential, e.g., of polluted dredge sediments or sewage sludges deposited on agricultural land (Keeley and Engler 1974).

Despite differing objectives, the procedures presented here are very similar in both research fields. In general, they serve for the determination of the major accumulative phases for metals in particulates and the assessment of diagenetic changes. Special indications are expected in environmentally relevant investigations as to the relative stability of the various phases of chemical transformations and to the possibility of release of metals into solutions. In the ore-related studies, such phase differentiation can give addition information as to economically favourable processing methods.

In the next section two possibilities of determining the enrichment of metals in specific bonding forms of Recent sediment are presented:

- by measurement of mineral or organic concentrates (examples from lacustrine deposits);
- by sequential chemical extractions (coastal marine and pelagic examples).

1 Institut für Sedimentforschung der Universität Heidelberg, Im Neuenheimer Feld 236, 6900 Heidelberg, FRG

2 Significance of Various Mechanisms and Substrates

There are five major mechanisms of metal accumulation in sedimentary particles: (1) adsorptive bonding on fine-grained substances, (2) precipitation of discrete metal compounds, (3) co-precipitation of metals by hydrous Fe and Mn oxides and by carbonates, (4) associations with organic molecules, and (5) incorporation in crystalline minerals. •

Table 1 gives a compilation of the various mechanisms, products, and substrates of metal enrichment and their estimated significance in lacustrine sediments; this scheme includes all major types of metal associations occurring in both natural and man-affected water systems.

Table 1. Mechanisms of metal enrichment in lacustrine sediments. x–xxxx indicate relative significance of the process. C = substrate for coatings

	Rock debris, solid waste material, etc. incl. organic	Metal – hydroxide – carbonate – sulphide	Organic substance (reactive)	Hydrous Fe/Mn-oxide (autochthonous)	Calcium carbonate
Inert bonding	xxxx		x		
Adsorption	x			x	
Ion exchange	xx	x	xx	xx	x
Precipitation	C	xx		C	C
Co-precipitation				xxxx	xx
Flocculation			xx		

The various organic and inorganic phases are interrelated: clay minerals, carbonates, and other suspended materials form the nucleation centres for the deposition of Fe and Mn hydroxides (Aston and Chester 1973). Significant substrates for coatings are marked by C in Table 1. A generalized sequence of the capacity of solids to sorb heavy metals was found as follows: manganese oxides – humic substances – hydrous iron oxides – clay minerals. According to Jenne (1976), the clay mineral phase can be viewed simply as a mechanical substrate upon which organic and secondary minerals are precipitated. The more important trace metal sinks in sediments are thermodynamically unstable species that tend to be amorphous or poorly crystalline and exhibit extensive isomorphic substitution (Jones and Bowser 1978).

Among such phases the hydroxides and Mn oxides of iron and manganese must be considered as the predominant *inorganic substrates of trace metal* in aquatic environments. Formed at pH/Eh-boundaries such as springs, junction of rivers, thermocline regions in eutrophic lakes and fjords, and mixing zones of acid river water with seawater and mineral surfaces (Lee 1975), these substances can be present in X-ray amorphous, microcrystalline and in more "aged" crystalline forms. It is, however, difficult to associate these formations with an allochthonous influence, e.g., from soils or rocks (Gibbs 1977), or an endogenic (within water column), or authigenic process

(Jones and Bowser 1978). An exception is freshwater Fe/Mn concretions, where the latter mechanisms dominate (Callender and Bowser 1976).

Metal enrichment (or dilution) in other endogenic or authigenic phases is shown from selected lacustrine examples in Table 2. There is a more or less significant dilution of most heavy metals by opaline silica, carbonates, and sulphates. Incorporation into pyrite minerals seems to provide a distinct mechanism of enrichment for transition metals (examples from Black Sea sediments). Unstable iron sulphide of the hydrotroilite type ($FeS \cdot H_2O$) is for the most part finely dispersed (black mud) and is instantly oxidized under aerobic conditions; this usually makes the separate analysis of these components and their sorbed trace metals impossible. Initial investigations of coarse-grained aggregates (separated by sieving from the fine-grained matrix) from Lake Constance surface sediments (Wagner 1971) indicate an approximate sixfold enrichment of iron in the concentrate compared with the composition of the surrounding sediment. When compared to the sediment matrix, Cu, Co, and Cr were 2 times greater, Ni 1.5 times, whereas Pb, Zn, and Mn did not undergo a significant increase (Förstner 1980). It may, therefore, be concluded that coprecipitation with iron sulphides is less effective in concentrating trace metals than is the incorporation into hydrous iron oxides.

Table 2. Effects of autochthonous phases on the enrichment or reduction of heavy metals in lacustrine sediments

	Source (1) Endogenic	Authigenic	Lake examples	Enrichment (+, ++) Reduction (–, – –)
Silicates				
Opaline Silica	x		Lake Malawi (2)	– – HM (heavy metals) – Cu
Nontronite		x	Lake Malawi (2)	± HM; + Co, Mn
Carbonate				
Aragonite	x		Van Gölü (Turkey) (2)	+ Mn; ± Zn; – – HM
Mg-calcite	x		Balaton (Hungary) (2)	± Mn, – Fe, Cu, Zn
Ca-dolomite		x	L. Neusiedl (Austria) (2)	– – Cu, Co, Ni
Siderite		x	Birket Ram (Israel) (3)	no effect
Sulphate				
Gypsum	x		L. Hart (Australia) (2)	– – all HM
Alunite	x		L. Brown (Australia) (2)	– – Fe, Mn, Zn, Cr; – Ni, Cu; ± Pb
Sulphide				
Pyrite		x	Black Sea (4)	+ Cu, Ni; ± Zn, Cr
Hydrotroilite	x	x	L. Constance (5)	+ Co, Cu, Cr, Ni; ± Pb, Zn, Mn
Organic substances	x		Lakes in Norway (6) Perch L./Ontario (7)	+ to ++

(1) Jones and Bowser 1978; (2) Förstner 1978; (3) Singer and Navrot 1978; (4) Volkov and Fomina 1974; (5) Förstner 1980; (6) Tobschall et al. 1978; (7) Jonasson 1976

A high positive correlation between the contents of organic matter and heavy metal concentrations has often been observed in aquatic sediments. This, however, may not always be interpreted as being due to direct bonding of heavy metals to organic substances. In strongly polluted areas, for instance, contamination by heavy metals, such as Zn, Pb, Cu, and Hg, simply coincides with the accumulation of organic substances, both of which are derived from urban, industrial, or agricultural sewage effluent. Table 2 lists two examples of endogenic metal enrichments in organic fractions of lake sediments. In deep-water sediments from Perch Lake in Lanark Country/Ontario, Jonasson (1976) assumed that gelatinous colloidal substances, which are formed from dissolved organic acids, spores, pollen, and decayed leaves, take up the metal ions from water. Similar effects have been observed by Tobschall et al. (1978) from gels and sediments in Gjϕvatn, Fiskevatn, and Sortbrysttjer lakes in subarctic northern Norway. Compared to the organic sediments, the heavy metals Ni, Cu, Zn, Pb, Hg, Th, and U are enriched in the organic gels by factors of 2 to 500.

3 Chemical Extraction Studies on Marine Sediments

The differentiation of the chemical forms of metals in sediments, which is particularly useful for the recognition of diagenetic reactions, has found application on pelagic deposits since the earlier days of manganese nodule research. By chemical leaching with ethylene-diamine-tetraacetic acid (EDTA), dilute hydrochloric and acetic acid, Goldberg and Arrhenius (1958) and Arrhenius and Korkish (1959) determined the distribution of elements in detrital igneous minerals and authigenic phases (mainly the oxide minerals and microcrystalline apatite) in pelagic sediments. Chester and Hughes (1967) introduced a combined acid-reducing agent of 1 *M* hydroxylamine hydrochloride and 25% (v/v) acetic acid for the separation of ferromanganese minerals, carbonate minerals and adsorbed trace elements from marine deposits. The latter technique has been used widely on marine sediment samples, e.g., from the North Atlantic (Chester and Messiha-Hanna 1970; Horowitz 1974; Horowitz and Cronan 1976), the North Pacific (Chester and Hughes 1969) and the East Pacific (Sayles et al. 1975).

According to these studies, the *hydrogeneous character* of trace metals in marine pelagic sediments generally decreases in the order Mn > Co > Cu > Ni > Cr > V > Fe. It seems that the supply of V, Cr, and Fe is governed largely by the input of detrital material from the continents, being higher for the North Atlantic than for Pacific regions.

Further *differentiation of the "non-lithogeneous" metal fraction* was mainly promoted by environmental studies, since these phases of the sediment, dredged materials, or sewage substances constitute "the reservoir for potential subsequent release of contaminants into the water column and into new interstitial waters" (Chen et al. 1976), and are thus predominantly available for biological uptake (Luoma and Jenne 1977). Here sequential extraction techniques were applied including the determination of metal contents in interstitial water and in ion-exchangeable, easily reducible, moderately reducible, organic and residual sediment fractions. It has been established that the surplus of metal contaminants introduced from man's activities into the aquatic systems usually exists in relatively unstable chemical associations (Förstner and Patchineelam 1980). An example of this type of work is shown in Table 3 with data of sediments from the Jade Bay on the western part of the German North Sea coast: as the total concentrations of lead, copper, and zinc increase from Eckwarderhörn via Vareler Tief to Südstrand, there is a significant decrease of the percentage of detrital-bound metals.

Table 3. Chemical fractionation of pelitic sediments (< 2 μm) from Jade Bay, western German Bight (EH: Eckwarderhörn; VT: Vareler Tief; SS: Südstrand)

	Lead			Copper			Zinc		
Fraction %	EH	VT	SS	EH	VT	SS	EH	VT	SS
Easily extractable [a]	–	–	–	8	4	7	–	–	–
Humate fraction [b]	18	7	8	13	10	11	1	1	1
Carbonate	24	22	25	24	25	29	36	40	36
Fe/Mn-oxides	–	37	36	7	4	9	12	22	34
Residual organics	–	6	7	6	13	15	5	7	7
Residual inorganics	58	28	24	42	44	29	46	31	18
Total (ppm)	37	48	65	27	30	38	129	149	196

[a] Extracted by 0.2 *M* $BaCl_2$ triethanolamine (Jackson 1958)

[b] Extracted by 0.1 *N* NaOH (Volkov and Fomina 1974)

Sequential extraction procedures in pelagic deposits have only recently begun to be used more intensely. Bowser et al. (1979) have performed a series of specific extraction experiments on sediments associated with nodule-rich areas of the eastern equatorial Pacific Ocean. It is suggested that Cu is mainly associated with organic matter and Ni with siliceous materials. From our preliminary studies (Förstner, Pfeiffer and Stoffers, in prep.) the following sequence of leaching procedures may be proposed for the differentiation of authigenic metal fractions in marine pelagic sediments (including Fe/Mn-nodules):

1. Exchangeable cations: 1 *M* ammoniumacetate, pH 7 (Jackson 1958); 1:20 solid/solution ratio, 2 h shaking;
2. Carbonate fraction: Acidic cation exchanger (Deurer et al. 1978);
3. Easily reducible phases: (Mn-oxides, amorphous Fe-oxyhydrates) 0.1 *N* $NH_2OH \cdot HCl$ + 0.01 *M* HNO_3 (Chao 1972); 12 h, 10 mg/ml solution (5 mg/ml for nodules);
4. Moderately reducible phases: (amorphous and poorly cryst. FeOOH). 0.2 *M* ammonium oxalate + 0.2 *M* oxalic acid (Schwertmann 1964); 24 h, 5–10 mg/ml solution;
5. Non-silicate iron phases: Citrate-dithionite (Holmgren 1967); 24 h, 5 mg/l solution;
6. Organic fractions: 30% H_2O_2 + NH_4 acetate (Engler et al. 1974);
7. Detrital silicates and others: $HF/HClO_4$-digestion.

It must be clearly pointed out that most of the extraction steps described here are not "selective" as sometimes stated. Repeated treatment often shows a further release of metals, especially in reducible fractions (Heath and Dymond 1977). Readsorption of metals can occur, for example, after extraction with dilute acids and H_2O_2 (Rendell et al. 1980). Formation of soluble basic metal oxides may take place during initial high pH-condition, e.g., for humate extraction using sodium hydroxide or sodium pyrophosphate (Patchineelam and Förstner 1977; Burton 1978).

Table 4 summarizes some data from *chemical fractionation studies of transition elements in Pacific pelagic sediments* (Förstner and Stoffers 1981). Steps (5) and (6) were not employed here; instead of 1 *M* ammoniumacetate in step (1), we used 0.2 *M* $BaCl_2$-triethanolamine (Jackson 1958). The latter agent seems to effect specific interactions with the copper components, possible related to a characteristic association with organic matter (Bowser et al. 1979). The data show that up to 20% of zinc in calcareous ooze is bound to carbonate phases. That the manganese in these sediments does not occur in significant amounts either in cation exchangeable or in carbonate forms – which is the opposite effect of the same metal in limnic or near-coast environments – may be due to the persistent oxygenated conditions in these deep sea areas.

Table 4. Percentage of metal associations in different chemical facies[a] of Pacific pelagic deposits (Förstner and Stoffers 1981)

	Manganese				Nickel				Copper			
Fraction (%)	a	b	c	MN	a	b	c	MN	a	b	c	MN
Easily extractable[b]	–	–	–	–	1	–	–	–	7	8	22	3
Carbonate	–	–	3	–	–	–	–	–	1	–	8	–
Easily reducible	88	73	85	86	77	47	84	55	18	10	24	17
Acid soluble[c]	11	16	11	13	17	19	4	42	45	29	31	76
Residual	1	8	1	1	5	32	12	3	30	53	15	4
Total metal	2.9 %	0.4 %	0.3 %	22.7 %	573 ppm	111 ppm	51 ppm	9780 ppm	738 ppm	418 ppm	113 ppm	7065 ppm

[a] (a) red-brown clay, (b) siliceous material, (c) calcareous ooze; MN: micronodules from areas C and K

[b] 0.2 *M* $BaCl_2$ triethanolamine

[c] 0.3 *M* HCl (Malo 1977)

Considering the various possible pathways for metal accumulation in ferromanganese concretions, there is at least one characteristic that can be derived from the present data (Table 4). Starting either from volcanoclastic material or siliceous ooze composition via pelagic clay (b → a → MN), or from calcareous sediments (c → MN) to the composition of micronodules, a distinct reduction in the percentage of residual-bound transition element occurs. The decrease of residual percentages is as great as the metal enrichment in the nodule material. Simultaneously, the percentages of metals associated with the acid-soluble increase in the same direction, either from carbonate sediments or from volcanoclastics to the nodule material. These effects can be explained either as phase transformations, recrystallization during ageing, or both (Förstner and Stoffers 1981). It seems that in the equatorial northern and southern Pacific pelagic zones, dissolution of siliceous organism is important in controlling metal accumulations. The dissolution of calcareous biological matter may also have some role as a source of metals; it is suggested that organic phases act as the predominant carrier for copper.

As in the freshwater deposits, here too it becomes apparent that the organic substances' occurrence in independent accumulative phases is less important than their major function in the transfer of metals between the different inorganic phases.

Acknowledgments. The support of our investigations by the German Research Society (Deutsche Forschungsgemeinschaft) is greatly appreciated. Thanks are also due to Dr. S.R. Patchineelam and Prof. P. Stoffers for their suggestions and data. D. Godfrey kindly assisted in the preparation of the manuscript.

References

Arrhenius GOS, Korkish J (1959) Uranium and thorium in marine minerals. Int Ocean Congr Am Assoc Adv Sci. Preprints, 497 pp

Aston SR, Chester R (1973) The influence of suspended particles on the precipitation of iron in natural waters. Estuarine Coastal Mar Sci 1:225–231

Bowser CJ, Mills BA, Callender E (1979) Extractive chemistry of equatorial Pacific pelagic sediments and relationship to nodule-forming processes. In: Bischoff J, Piper D (eds) Marine geology and oceanography of the Central Pacific Nodule Province. Plenum Press, New York, pp 587–620

Burton JD (1978) The modes of association of trace metals with certain compounds in the sedimentary cycle. In: Goldberg ED (ed) Biogeochemistry of estuarine sediments. Unesco Press, Paris, pp 33–41

Callender E, Bowser CJ (1978) Freshwater ferromanganese deposits. In: Wolf KH (ed) Handbook of strata-bound and stratiform ore deposits, vol 7. Elsevier, Amsterdam, pp 341–394

Chao LL (1972) Selective dissolution of manganese oxides from soils and sediments with acidified hydroxylamine hydrochloride. Soil Sci Soc Am Proc 36:764–768

Chen KY, Gupta SK, Sycip AZ, Lu JCS, Knezevic M, Choi WW (1976): The effect of dispersion, settling, and resedimentation on migration of chemical constituents during open water disposal of dredged material. Contract Rep US Army Eng Waterways Exp St, Vicksburg, Miss, 221 pp

Chester R, Hughes MJ (1967) A chemical technique for the separation of ferromanganese minerals, carbonate minerals, and adsorbed trace elements from pelagic sediments. Chem Geol 2:249–262

Chester R, Hughes MJ (1969) The trace element geochemistry of a North Pacific pelagic clay core. Deep Sea Res 16:639–654

Chester R, Messiha-Hanna R (1970) Trace-element partition patterns in North Atlantic deep sea sediments. Geochim Cosmochim Acta 34:1121–1128

Deurer R, Förstner U, Schmoll G (1978) Selective chemical extraction of carbonate-associated metals from recent lacustrine sediments. Geochim Cosmochim Acta 42:425–427

Engler RM, Brannon JM, Rose J (1974) A practical selective extraction procedure for sediment characterization. 167th Meet ACS, Atlantic City, 17 pp

Förstner U (1978) Metallanreicherungen in rezenten See-Sedimenten – geochemischer background und zivilisatorische Einflüsse. Mitt Nationalkomm Bundesrep Deutschland Intern Hydrolog Progr UNESCO, Heft 2, Koblenz, 66 S

Förstner U (1980) Recent heavy-metal accumulation in limnic sediments. In: Wolf KH (ed) Handbook of strata-bound and stratiform ore deposits, vol 8. Elsevier, Amsterdam, pp 179–270

Förstner U, Patchineelam SR (1980) Chemical associations of heavy metals in polluted sediments from the lower Rhine River. In: Kavaunagh M, Leckie JO (eds) Particulates in water. Adv Chem Ser 189:177–193

Förstner U, Stoffers P (1981) Chemical fractionation of transition metals in Pacific pelagic sediments. Geochim Cosmochim Acta 45:1141–1146

Gibbs RJ (1977) Transport phases of transition metals in the Amazon and Yukon rivers. Geol Soc Am Bull 88:829–843

Goldberg ED, Arrhenius GOS (1958) Chemistry of Pacific pelagic sediments. Geochim Cosmochim Acta 13:153–212

Gupta SK, Chen KY (1975) Partitioning of trace metals in selective chemical fractions of nearshore sediments. Environ Lett 10:129–158

Heath GR, Dymond J (1977) Genesis and transformation of metalliferous sediments from the East Pacific Rise, Bauer Deep, and Central Basin, northwest Naszca Plate. Geol Soc Am Bull 88: 723–733

Holmgren GS (1967) A rapid citrate-dithionite extractable iron procedure. Soil Sci Soc Am Proc 31:210–211

Horowitz A (1974) The geochemistry of sediments from the northern Reykjanes Ridge and the Iceland-Faroes Ridge. Mar Geol 17:103–122

Horowitz A, Cronan DS (1976) The geochemistry of basal sediments from the north Atlantic Ocean. Mar Geol 20:205–228

Jackson ML (1958) Soil chemical analysis. Prentice Hall, Englewood Cliffs, NJ, 498 pp

Jenne EA (1976) Trace element sorption by sediments and soils – sites and processes. In: Chappel W, Petersen K (eds) Symposium on molybdenum, vol 2. Dekker, New York, pp 425–553

Jonasson IR (1976) Detailed hydrogeochemistry of two small lakes in the Grenville Geological Province. Geol Surv Can Pap 76–13:37

Jones BF, Bowser CJ (1978) The mineralogy and related chemistry of lake sedidments. In: Lerman A (ed) Lakes – chemistry, geology, physics. Springer, Berlin Heidelberg New York, pp 179–235

Keeley JW, Engler RM (1974) Discussion of regulatory criteria for ocean disposal of dredged materials: elutriate test rationale and implementation guidelines. US Army Corps Eng DMRP, Vicksburg, Miss, Rep D-74-14:13

Lee GF (1975) Role of hydrous metal oxides in the transport of heavy metals in the environment. In: Krenkel PA (ed) Heavy metals in the aquatic environment. Pergamon Press, Oxford, pp 137–147

Luoma SN, Jenne EA (1977) Estimating bioavailability of sediment-bound metals with chemical extractants. In: Hemphill DD (ed) Trace substances in environmental health, vol 10. Univ Missouri Press, Columbia, Miss, pp 343–351

Malo BA (1977) Partial extraction of metals from aquatic sediments. Environ Sci Technol 11:277–282

Patchineelam SR, Förstner U (1977) Bindungsformen von Schwermetallen in marinen Sedimenten. Senckenberg Marit 9:75–104

Rendell PS, Batley GE, Cameron AJ (1980) Adsorption as a control of metal concentrations in sediment extracts. Environ Sci Technol 14:314–318

Sayles FL, Ku T-L, Bowker PC (1975) Chemistry of ferromanganoan sediment of the Bauer Deep. Geol Soc Am Bull 86:1423–1431

Schwertmann U (1964) Differenzierung der Eisenoxide des Bodens durch photochemische Extraktion mit saurer Ammoniumoxalat-Lösung. Z Pflanzenernaehr Dueng Bodenkd 105:194–202

Singer A, Navrot J (1978) Siderite in Birket Ram lake sediments. Abstr 10th Int Congr Sediment, Jerusalem, pp 616–617

Tessier A, Campbell PGC, Bisson M (1979) Sequential extraction procedure for the speciation of particulate trace metals. Anal Chem 51:844–851

Tobschall HJ, Göpel C, Rast U (1978) Geochemistry of organic gels and organic sediments from selected lakes of northern Norway. Abstr 10th Int Congr Sediment, Jerusalem, pp 682–683

Volkov II, Fomina LS (1974) Influence of organic material and processes of sulfide formation on distribution of some trace elements in deep-water sediments of the Black Sea. Am Assoc Petrol Geol Mem 20:456–476

Wagner G (1971) FeS-Konkretionen im Bodensee. Int Rev Ges Hydrobiol 56:265–272

Ocean Floor "Metalliferous Sediments" – Two Possibilities for Genesis

H. GUNDLACH and V. MARCHIG[1]

Abstract

There are two possibilities for the genesis of recent "metalliferous sediments":

1. Submarine exhalative, i.e., "hydrothermal", bound to recent tectonic faulting. These sediments contain the metals as sulphides, partly oxidized. The best known examples are the ore brines in the Red Sea.

2. Products of diagenesis and repeated redeposition of siliceous sediments, mainly of radiolarian oozes. These products are of about the same composition as the first type. They contain the metals as oxides. The elements in this type can be subdivided in groups: autochthonous, i.e., elements fixed in compounds formed in the same place as they are sedimented, and terrigenous, i.e., elements fixed in compounds resistant to terrestrial weathering and in clay minerals. The more the diagenesis is advanced, the higher is the content of the autochthonous group of elements (which comprises Mn, Cu, Zn, etc.).

Up to now, no distinctive mark has been found between the two types of metalliferous sediments. The solution of many problems connected with the two types of metalliferous sediments will help in exploration and present new aspects in the knowledge of the genesis of stratiform ore deposits.

1 Introduction

One of the most important targets of exploration for ores are stratiform deposits. Most of them are considered to be of submarine-exhalative origin, i.e., their source is ascribed to volcanic activity in a marine environment. Deposits of this type are found in numerous examples from nearly all geological periods and they are also found in statu nascendi in the present marine environment.

But is the genesis of all stratiform deposits formed in marine environment of magmatic origin? Is the source of material always due to hydrothermal activity? These are questions which have to be answered to find the best way for exploration.

Finlow-Bates (1978) has stated that submarine exhalative deposits are formed when "a relatively concentrated hot hydrothermal fluid enters the seawater through a conduit of limited extent".

Such a deposit has usually two parts: the part beneath the sea floor where veiniform and disseminated sulphides are deposited, and the part at the sea floor formed by direct sedimentation of sulphides.

1 Bundesanstalt für Geowissenschaften und Rohstoffe, Postfach 510153, 3000 Hannover, FRG

The following questions have to be answered before genetic problems for such ore bodies can be solved and before a useful guide for exploration can be made (Finlow-Bates 1978):

1. How significant are the host rocks to the ore-forming process?
2. To what extent is ore localization controlled by tectonic processes?
3. What is the rate of deposition of sulphidic sediment?
4. What are the chemical and physical controls on the transport and deposition of submarine exhalative sulphides?
5. What is the significance of interaction of the ore-forming solution with the sediments, volcanics, and pyroclastics of the depositional site?
6. Do biological processes play any role in ore deposition?
7. Have changes in the atmosphere and seawater composition or processes of crustal evolution been recorded in submarine exhalative deposits?

2 Background

In the last few years several recent (active) submarine exhalative deposits have been found. The biggest, best known, and the only one of economic value is situated in the Red Sea (rift valley). One small deposit was found near the Galapagos Islands (Galapagos Rise) and another in the Gulf of California (East Pacific Rise). All these three presently forming deposits are obviously bound to recent tectonic faulting. In each of these cases, sulphides were the original phase precipitated, although they may have suffered later partial oxidation. All three of the deposits are found together with warm water, i.e., they are hydrothermal (sensu lato) in origin.

Deposits like these promise to give us answers to a sizable portion of Finlow-Bates' questions, which we can hardly answer while investigating only fossil deposits, greatly changed (altered, metamorphosed etc.) since their formation.

The term "metalliferous sediments" is widely used in geochemical literature for recent submarine-exhalative deposits. Although we also use this term, we wish to point out that it is not a good expression. All sediments contain metals, at least Al and Si, or Ca. The difference between "metalliferous sediments" and other sediments is the elevated concentration of transition metals relative to rock-forming metals in the "metalliferous sediments". The expression does not imply this information.

A second group of "metalliferous sediments" besides the recent deposits mentioned has been found widespread in the deep sea: fine-grained, dark brown to reddish-brown sediments whose composition is like that of the deposits described: they have elevated concentrations of transition metals in comparison to rock-forming metals. They do not contain sulphides – the metals are bound as oxides or in silicates. They occur as enormously large fields and in thick layers. They are more or less diluted with other sediments. They occur on active tectonic features as well as at some distance from them (see map). The presence of warm bottom water at the respective sites could not be proved.

These sediments have been found in the "Bauer Deep" (Bischoff and Sayles 1972), in the "Fiji Basin" (Cronan and Thompson 1978), in the "Lan Basin" (Cronan and

Thompson 1978), in the "Tiki Basin" (Hoffert et al. 1979), and underlaying the radiolarian oozes in the radiolarian ooze belt (Beiersdorf and Wolfart 1974). Also a thin layer of "metalliferous sediments" was found directly over the basaltic basement in a number of DSDP (Deep Sea Drilling Program) cores.

The genesis of these sediments is not yet certain. It has been hypothesized in the following way (Bischoff and Sayles 1972): Hydrothermal solutions (again sensu lato) from active tectonic features come in contact with seawater. Metallic compounds are precipitated and transported by bottom-near currents over large areas. Sulphides are oxidized to oxides by the oxygen-rich bottom-near seawater.

We investigated in detail further samples of "metalliferous sediments" underlying radiolarian oozes as described by Beiersdorf and Wolfart (1974). They are to be assigned to the second group of metalliferous sediments as described above: widespread occurrences, lack of sulphides, and possible dilution with other sediments. It could be proved that these "metalliferous sediments" are not of hydrothermal origin. They are residual sediments – products of diagenesis of radiolarian oozes (Marchig 1978).

3 Investigations and Results

Radiolarian ooze samples from two different areas within the radiolarian ooze belt in the Central Pacific were collected during several cruises of the German Research Vessel "Valdivia". The first area was sampled during the cruises VA 08 and VA 13. The sedimentation rates in the cores from this area average 3.8 mm/10^3 years (Heye 1976; Meyer 1977). The second area, sampled during cruise VA 04, has an average sedimentation rate of 2.1 mm/10^3 years (Meyer 1977).

The samples from both areas were analyzed by X-ray fluorescence. Before being analyzed, the grain-size fraction $> 63\ \mu m$ was removed from the sediment. This coarse fraction is composed of radiolarians and manganese micronodules, it does not exceed 5% of the sediment. Correlation matrices for the sediments of both areas are shown in Figs. 1 and 2.

Part of the analyzed elements obviously forms two groups. In each of these groups the concentrations of the elements correlate well with each other, between the groups there are only negative correlations.

Most of the elements typical for the first group are precipitated under deep-sea conditions in the form of highly insoluble compounds such as barite or Mn hydroxide. Some of them, such as apatite, remain as highly insoluble components after the solution of the organogenic phase. Another part of the elements in this group is chemically bound or absorbed on the insoluble components, like La and Y in apatite and Cu and Ni on Mn hydroxide. We called this group of elements the "autochthonous group". That means the compounds containing the elements are formed at the same place and under the same conditions as they are sedimented (Gundlach et al. 1979).

The material for this group may have come from different sources such as the decomposition of the organogenic phase of the sediment or deep-sea weathering or volcanism.

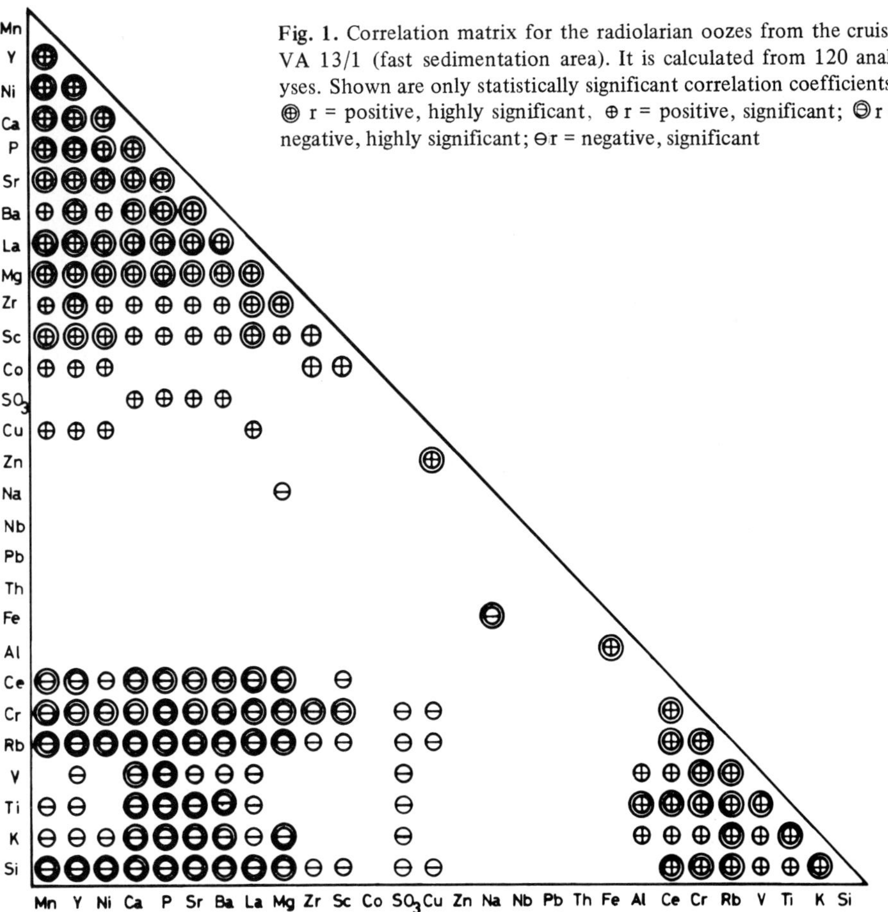

Fig. 1. Correlation matrix for the radiolarian oozes from the cruise VA 13/1 (fast sedimentation area). It is calculated from 120 analyses. Shown are only statistically significant correlation coefficients. ⊕ r = positive, highly significant, ⊕ r = positive, significant; ⊖ r = negative, highly significant; ⊖ r = negative, significant

The other group of elements seems to be of terrigenous origin, since it contains the elements which are known to be resistant to continental weathering as Cr, Ti, and V, and elements typical for clay minerals, such as K and Rb.

The significant difference between the correlation matrices for these two areas is the position of Ca and Sr. In the sediments from the VA 08 and VA 13 area Ca and Sr correlate with the autochthonous group of elements; in the sediments from the VA 04 area, they form a separate group, which correlates with CO_2. That indicates that in this case Ca and Sr are bound predominantly as carbonates. The somewhat higher carbonate content in this field is probably due to the higher primary production by organisms, since the conditions for the dissolution of carbonate are similar in both areas (Gundlach et al. 1979).

If the average composition of the sediments in both areas is compared with the average composition of deep-sea clays (Turekian and Wedepohl 1961), several of the autochthonous elements are obviously elevated. In contrast, most of the elements typical for the terrigenous group are depleted (Figs. 3 and 4).

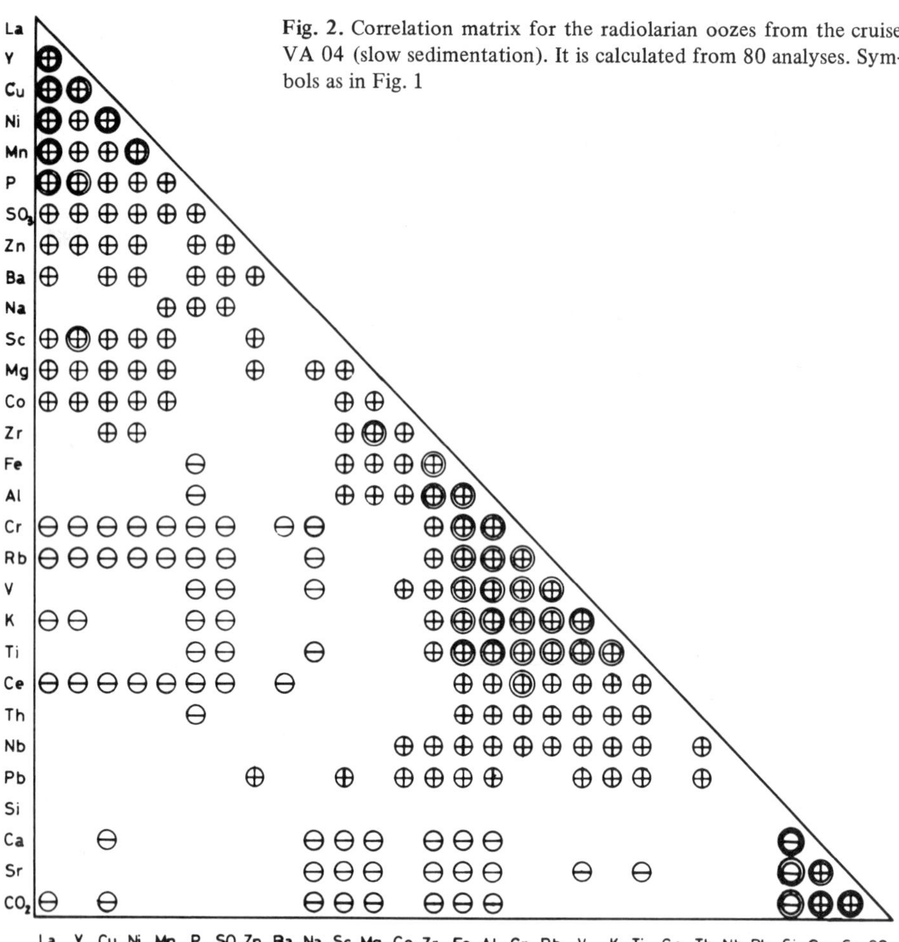

Fig. 2. Correlation matrix for the radiolarian oozes from the cruise VA 04 (slow sedimentation). It is calculated from 80 analyses. Symbols as in Fig. 1

The enrichment of the autochthonous phase is less in the sediments from the VA 08 and VA 13 area (which has higher sedimentation rates), and greater in the sediments from the VA 04 area (which has lower sedimentation rates). Therefore, the sedimentation rate must be the controlling factor in the enrichment of the autochthonous phase.

The origin of this autochthonous phase can be explained by several processes:

1. decomposition of the organogenic compounds and enrichment of the most insoluble components (e.g., barite, apatite, Y and La phosphates);
2. deep-sea weathering (e.g., montmorillonite formation, enrichment of Mg);
3. precipitation from the seawater (e.g., Mn and Fe hydroxides and the ions absorbed on them).

All these processes will be more promoted by lower sedimentation rates as the time of contact with seawater will be longer.

"Metalliferous sediments" found in the VA 08 and VA 13 area can also be classified in that scheme. Figure 5 shows a comparison of the "metalliferous sediments" from

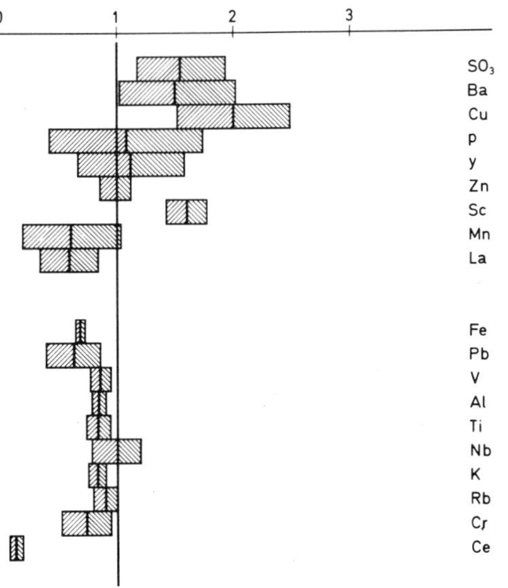

Fig. 3. Comparison of average values of radiolarian oozes from the cruise VA 13/1 (fast sedimentation) with the average composition of deep-sea clays. Shown is the ratio: *average value in radiolarian ooze ± standard deviation/average value in deep-sea clay* (Turekian and Wedepohl 1961)

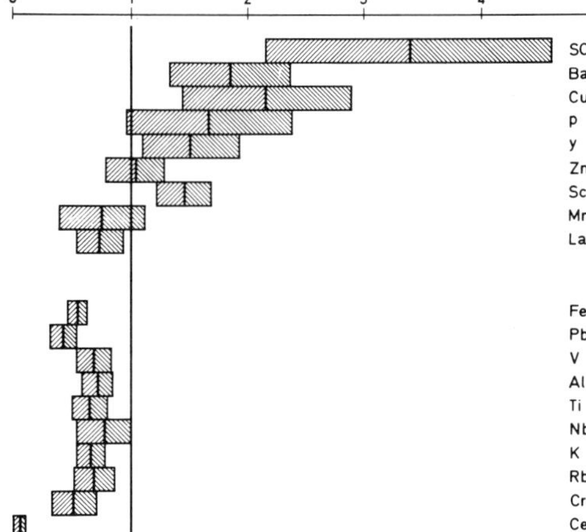

Fig. 4. Comparison of average values of radiolarian oozes from the cruise VA 04 (slow sedimentation) with the average composition of deep-sea clays. Symbols as in Fig. 3

this area and deep-sea clay, as was done for the radiolarian oozes. The enrichment of the autochthonous phase elements in "metalliferous sediments" is distinctly greater compared to deep-sea clays than that in radiolarian oozes.

A comparison of the three groups of investigated sediments (radiolarian oozes with rather high sedimentation rates, radiolarian oozes with rather slow sedimentation rates and "metalliferous sediments") is given in Figs. 6 and 7. Figure 6 shows the autochthonous elements, calculated as a percentage of the contents of these elements in the radiolarian oozes with rather high sedimentation rates.

It is seen that the contents of the autochthonous elements are higher in the radiolarian oozes with lower sedimentation rates; the maximum contents of the autochthonous elements are found in redeposited brown clay ("metalliferous sediments").

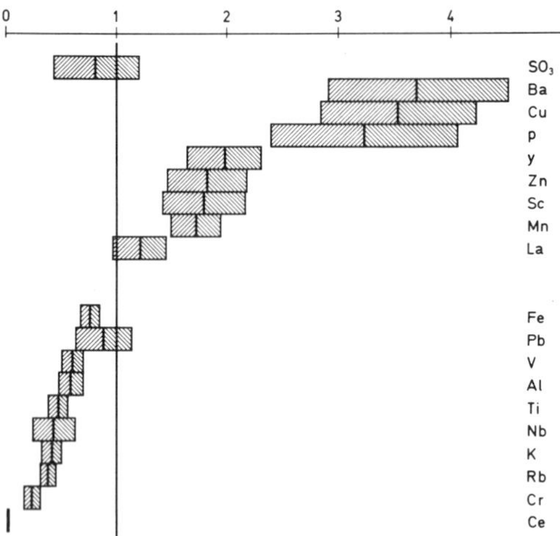

Fig. 5. Comparison of average values of metalliferous sediments (from the cruises VA 08 and VA 13/1) with the average composition of deep-sea clays. Symbols as in Fig. 3

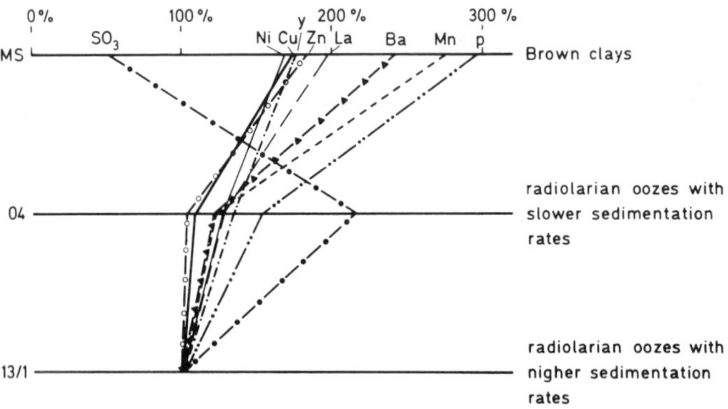

Fig. 6. Changes in the contents of elements which are enriched in the autochthonous phase of the sediment in different sediment types. The contents in the radiolarian oozes from the cruise VA 13/1 are taken as 100%

The opposite trend is obvious with the elements typical for the terrigenous group (Fig. 7). Their contents decrease from the radiolarian oozes with higher sedimentation rates to the radiolarian oozes with lower sedimentation rates and have their lowest concentrations in redeposited brown clay.

The conclusion to be drawn from these observations is that in these sediments there is a succession of diagenesis parallelling the enrichment with autochthonous material. The final product of this diagenetic process is a sediment with a highly enriched autochthonous phase – a "metalliferous sediment".

The conclusions are supported by microscopic and paleontological work:

– A large number of very corroded radiolarian tests within metalliferous sediments (Figs. 8 and 9) were observed with an electron microscope.

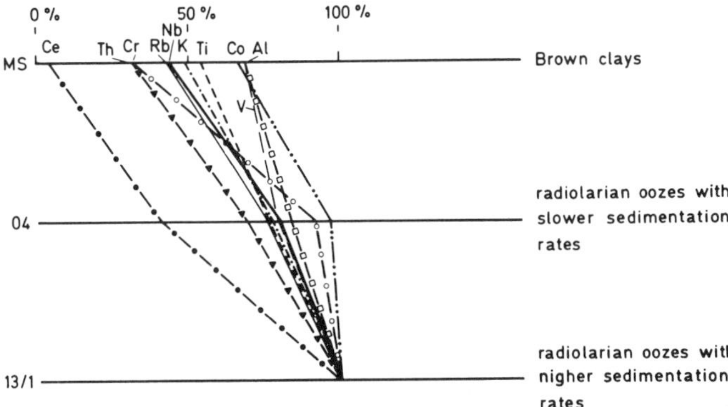

Fig. 7. Changes in the content of elements which are enriched in the terrigenous phase of the sediment in different sediment types. The contents in the radiolarian oozes from the cruise VA 13/1 are taken as 100%

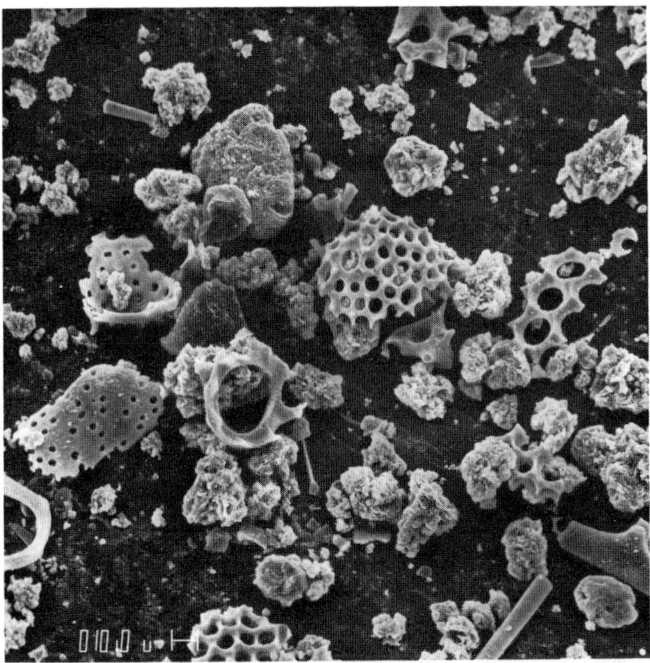

Fig. 8. Size fraction > 63 μm from metalliferous sediments. Clay mineral aggregates and fragments of broken radiolarians. (Photo: Knickrehm)

– Paleontological investigations on these sediments (Beiersdorf and Wolfart 1974) show a mixed fauna of different ages. This accounts for the redeposition of these sediments, and the redeposition explains the advanced progress of diagenesis due to the prolonged contact with seawater.

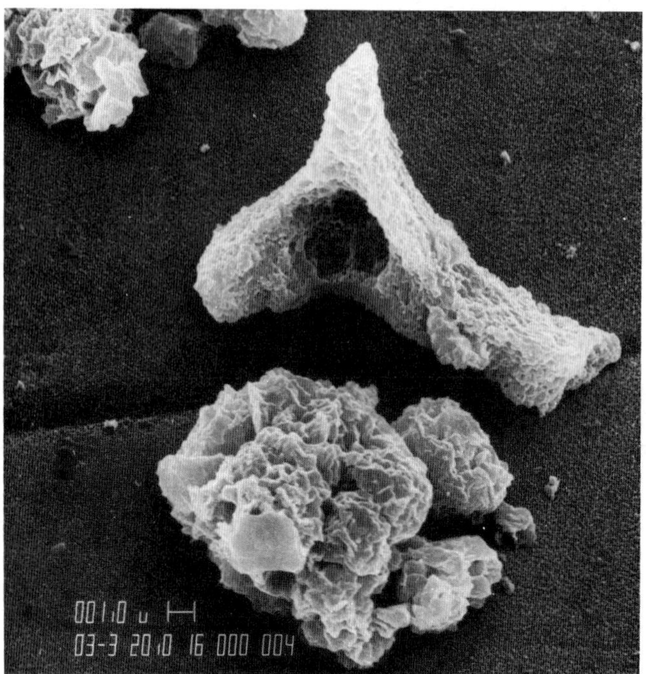

Fig. 9. Detail from the Fig. 8. Clay mineral aggregate and radiolarian fragment. The surface of the radiolarian fragment shows strong corrosion. (Photo: Knickrehm)

4 Outlook for the Future

With the recognition of two possibilities for the genesis of "metalliferous sediments" many questions arise:

1. What is the chemical and mineralogical difference between the two kinds of "metalliferous" sediments? This question includes the finding of reliable indicators for both groups of "metalliferous" sediments. Such indicators would also be useful in exploration work.
2. Where should one explore for one group and where for the other?
3. What is the proportion of each of the groups of "metalliferous sediments" with respect to all of them?

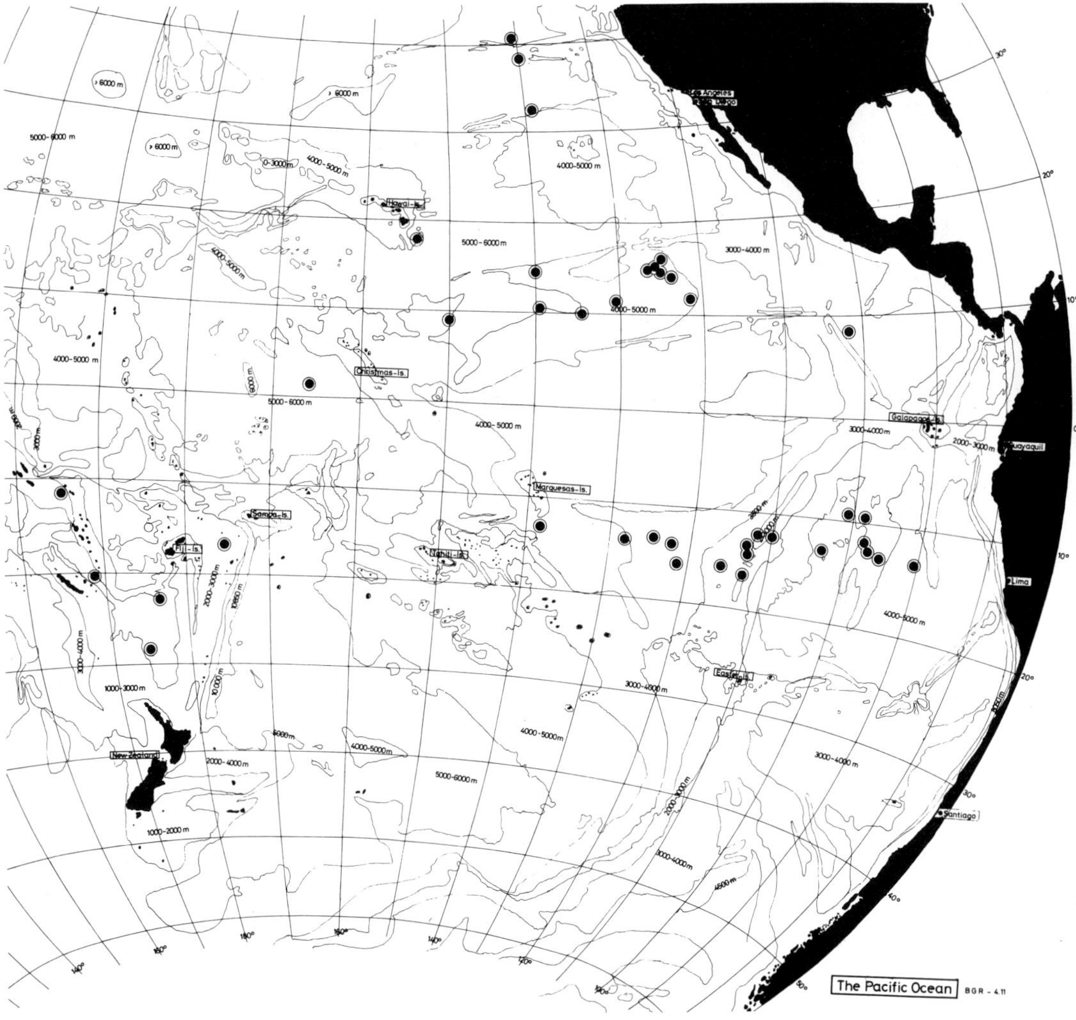

Fig. 10. Map of Pacific Ocean with tectonic features and position where metalliferous sediments have been found and analyzed. The concentration of metalliferous sediments in different zones can be objective, but it can also be the result of patchy distribution of sampling

4. Does the discovery of more "metalliferous sediments" influence calculations of the average composition of deep-sea sediments?
5. If so, would a revised average composition of deep-sea sediments influence calculations of the average residence times of transition metals in seawater?
6. Do the changes described under 4 and 5 influence the concept of the genesis of other marine ores, e.g., manganese nodules?

Several groups have started investigations on the problem of "metalliferous sediments". First sampling will be made possible by the GEOMETEP (*Geo*thermal *Met*allogenesis *E*ast *P*acific) cruises of the German Research Vessel "Sonne" in 1980 in the Pacific (Fig. 10). Thus in the near future progress can be expected.

References

Beiersdorf H, Wolfart R (1974) Sedimentologisch-biostratigraphische Untersuchungen an Sedimenten aus dem zentralen Pazifischen Ozean. Meerestechnik 5:192–198

Bischoff JL, Sayles FL (1972) Pore fluid and mineralogical studies of recent marine sediments: Bauer depression region of East Pacific Rise. J Sediment Petrol 42(3):711–724

Cronan DS, Thompson B (1978) Regional geochemical reconnaissance survey for submarine metalliferous sediments in the southwestern Pacific Ocean – a preliminary note. Appl Earth Sci 87:B87–B89

Finlow-Bates T (1978, unpublished) Controls on the genesis of submarine exhalative ore deposits with reference to Mount Isa Mine Queensland, Australia. Doctoral Diss, Montan-Univ Leoben, Austria

Gundlach H, Marchig V, Schnier C (1979) Woher stammen die Metalle in den Manganknollen? Mar Rohstoffe Meerestech 1 (Marine Rohstoffgewinnung): 46–82

Heye D (1976) Geophysikalische Untersuchungen an Knollen und Sedimenten. Manganknollen. Wissenschaftsfahrt Fahrtber VA 13/1:51–53

Hoffert M, Karpoff A, Schaaf A, Pautot G (1979) The sedimentary deposits of the Tiki basin (South-East Pacific), passage from carbonate oozes to "metalliferous sediments". In: Colloq Int CNRS, no 289. La genèse des nodules de manganèse. CNRS, Paris, pp 101–113

Marchig V (1978) Brown clays from the Central Pacific – Metalliferous sediments or not? Geol Jahrb D 30:3–25

Meyer H (1977) Untersuchungen über Beziehungen zwischen Sedimenten und Manganknollen im zentralen Pazifik SE von Hawaii. Diss TU Braunschweig, 159 S

Turekian KK, Wedepohl KH (1961) Distribution of the elements in some major units of the earth's crust. Bull Geol Soc Am 72:175–191

Textures, Mineralogy and Genesis of Manganese Nodules on the Blake Plateau, Northwestern Atlantic Ocean

SOO JIN KIM [1]

Abstract

The manganese nodules from the Blake Plateau consist mainly of microcrystalline to cryptocrystalline todorokite, with minor quartz, clays, carbonates and phillipsite. The nodules in cross-section show concentric layers, core structure, unconformity and fissure-filling structures megascopically, and colloform, fragmental and diagenetic textures microscopically. A new classification of colloform textures which are applicable to any nodule of any source shows that the colloform textures consist of three basic textural units: banded, cuspate and globular. They occur independently or in combination with each other to form various types of textures. The presence of three predominant textural types suggests that there are three different major modes of nodule growth which are controlled by physical and chemical environments.

1 Introduction

A study has been made of a manganese nodule dredged from the Blake Plateau, northwestern Atlantic Ocean, latitude 31°01,6′ N, longitude 78°18,8′ W at the depth of 849 m. The nodule was supplied by the Wood's Hole Oceanographic Institution with the courtesy of Dr. K.O. Emery and Dr. J. Broda.

According to Manheim (1972), the manganese nodule deposits on the Blake Plateau began forming in Middle Tertiary time concurrently with the establishment of the Gulf Stream. The Gulf Stream has played an essential role in their genesis. It has prevented deposition of carbonates and other sediments on the Blake Plateau, thus permitting accumulation of very slowly accreting metal oxides in an unusually shallow and near-shore environment.

The heterogeneity and cryptocrystallinity of materials in manganese nodules prevent easy access to the resolution of their mineralogy and textures. However, X-ray diffraction analysis has been the principal method for determining the mineralogy of manganese nodules. Microscopic study in polished section under high magnification shows details of the general textural patterns of constituent materials. Analysis of textural patterns of minerals gives important data for interpretation of the mechanism of growth of manganese nodules and thereafter evolution in mineralogy and textures.

---

1 Department of Geological Sciences, College of Natural Sciences, Seoul National University, Seoul 151, Korea

2 Methods of Study

The round manganese nodule having the size of 5.5 × 6.0 cm was cut through the centre in order to study the internal structures and textures, and also the mineralogy. The surfaces of cross-section were polished. The polished surfaces were photographed, and studied under both stereomicroscope and the reflecting microscope. Samples for X-ray analysis were taken under the microscope. A detailed textural study was made under the microscope in order to determine the process of growth of manganese nodules and their diagenetic evolution. Also, a classification of primary colloform textures was undertaken (Fig. 1).

Textural Units	Shape of Base			Textures of Aggregate	Combined Textures	
	Flat	Convex	Concave			
Banded				Banded(B) Laminated(L)	Cuspate banded(Bc)	
Cuspate				Cuspate(C)		
Short				Short cuspate(Cs)		
Medium				Medium cuspate(Cm)	Banded cuspate(Cb)	
Long				Long cuspate(Cl)	Globular cuspate(Cg)	
Globular				Globular(G)	Banded globular(Gb)	

Fig. 1. Classification of primary colloform textures in manganese nodules. Symbol for each type is given in parenthesis

3 Mineralogy

The manganese oxide minerals most commonly occurring in manganese nodules are todorokite and birnessite. In the manganese nodule from the Blake Plateau, todorokite is the only crystalline manganese oxide identified in the X-ray powder diffraction photographs. The manganese nodule studied consists mainly of todorokite with minor quartz, clay, calcite, aragonite and phillipsite.

Todorokite was recognized by two diagnostic lines at 9.6 and 4.8 Å. In addition, it shows diffraction lines at 2.46, 1.42 and 1.14 Å.

Todorokite occurs as two types in the manganese nodule from the Blake Plateau. They are (a) microcrystalline todorokite, and (b) cryptocrystalline todorokite. These two types of todorokite have quite different optical properties in reflected light. The microcrystalline todorokite occurs as very fine-grained flakes, showing distinct anisotropism and white in colour in reflected light. The cryptocrystalline todorokite occurs as a dense homogeneous mass, showing no distinct anisotropism, or even isotropism and light grey in colour in reflected light. Two types of todorokite occur in close association with each other but in different layers or zones. They are easily distinguished by the difference in their reflectivity. Microcrystalline todorokite has higher reflectivity than the cryptocrystalline todorokite.

4 Megascopic Structures

External Textures. The manganese nodule studied is spherical in shape with a slightly wavy surface. It has a very compact surface with a white coating in places.

Internal Structures. The characteristic internal structures of manganese nodules from the Blake Plateau are: concentric layering, core structure, unconformity, and fracture-filling structure.

Concentric Layering. The concentric layering is the most characteristic feature of the manganese nodule which can be easily seen with the naked eye in cross-section. Each individual layer is distinguished from the others by the colour and reflectivity due to the variation in mineralogy and textures. It is characteristic of the manganese nodule from the Blake Plateau that the colloform layers are very thin, laminated and compact. Individual layers have considerable continuity, encrusting the inner layers.

Core Structure. The manganese nodule studied includes an angular nodule fragment of an older generation. The characters of the core fragment are quite different in structure from those of the encrusting materials. The core fragment shows diffusely laminated structure.

Unconformity. The unconformity is also a character of internal structures of manganese nodules from the Blake Plateau. An unconformable contact is seen where the broken structure of an older nodule fragment is overlain at a sharp angle by new layers which encrust the entire fragment. Two different unconformities were found in the different horizons in one cross-section of a nodule. The unconformity is scarcely recognizable where the layers in the older fragment lie parallel to those of the encrusting material. A similar observation was made by Sorem and Forster (1969). Another type of unconformity is found along the fractures. Nodular materials overlie unconformably both walls of fracture, showing well-developed colloform bands.

Fracture-Filling Structure. Fracture-filling structures are abundantly found in the nodule studied. There are two types of fractures: random fractures and concentric fractures. The surfaces of larger fractures are lined with manganese oxides, showing banded and/or cuspate texture. Open spaces are filled with carbonates and clay (Fig. 2). It seems that these fractures are shrinkage cracks resulting from ageing of older oxide materials.

5 Microscopic Textures

Sorem and Forster (1972a, b) classified the textures of manganese nodules into the massive, mottled, compact, columnar and laminated zones.

The manganese nodule from the Blake Plateau also shows massive, mottled, compact, columnar and laminated zones. Among these, massive, laminated and mottled zones are most abundant.

Zones as used by Sorem and Forster are terms indicating the types of aggregations of nodular materials. Therefore, they are simply descriptive terminology without genetic meaning, which can be used to describe the general habit of the nodule materials. The present author proposes a new scheme of classification of nodular textures which can be used without limitation in scale.

Microscopic study shows that the textures of nodules can be classified into two groups: (a) primary growth textures, and (b) secondary diagenetic textures.

Primary Growth Textures. Colloform textures and their classification. The smooth surface is a characteristic feature of colloform growth. Although the concentric colloform layering is the major structure in manganese nodules, other types of colloform textures are also found in the manganese nodules. The general curvature of the concentric layers of nodules is usually governed by the shape of the older mass on which nodule materials are encrusting, but the shape of the surface of each individual layer is considerably controlled by the difference in the growth rate at different sites. The cause of differential growth rate in the same environment is not fully understood. The concentric layering on a megascopic scale is seen under the microscope as either a true continuous layer or a concentric arrangement of discontinuous forms.

Complexity in textures of manganese nodules from the Blake Plateau can be systematized by the systematic classification of colloform shapes. Colloform shapes can be classified into three basic geometrical types on the basis of the variation in the horizontal and vertical dimensions of individual colloform forms; they are (a) banded, (b) cuspate, and (c) globular (Fig. 1).

The banded type of colloform form indicates the parallel layer of colloform material with roughly uniform thickness which is developed parallel to the older surface. This texture suggests the uniform accretion of nodular materials on the whole surface area, resulting in the layer with roughly uniform thickness. Subparallel bands are also abundantly found. Depending upon the shape of the older surface, the following subdivision is possible: flat banded, convex banded, and concave banded (Fig. 1).

The cuspate type of colloform form indicates the concave-convex lens-shaped or bowl-shaped growth of nodular materials. The central part of the cusp is usually thicker than both margins. Continuous successive growth of cusps results in the long cusp resembling the column. The cusp may consist of either homogeneous material or heterogeneous parallel bands depending upon the environment of formation. Dunham and Glasby (1974) described the concentric growth cusps within the nodule. They show that crystallization of these cusps begins at nuclei perhaps 0.05 mm apart. As the cusps grow upward, they also grow sideways but leave delta-shaped areas between cusps. Finally cusps coalesce. When the cusps reach some critical size, two or more new nucleating centres form, causing the cusp to bifurcate. The various types of cusps

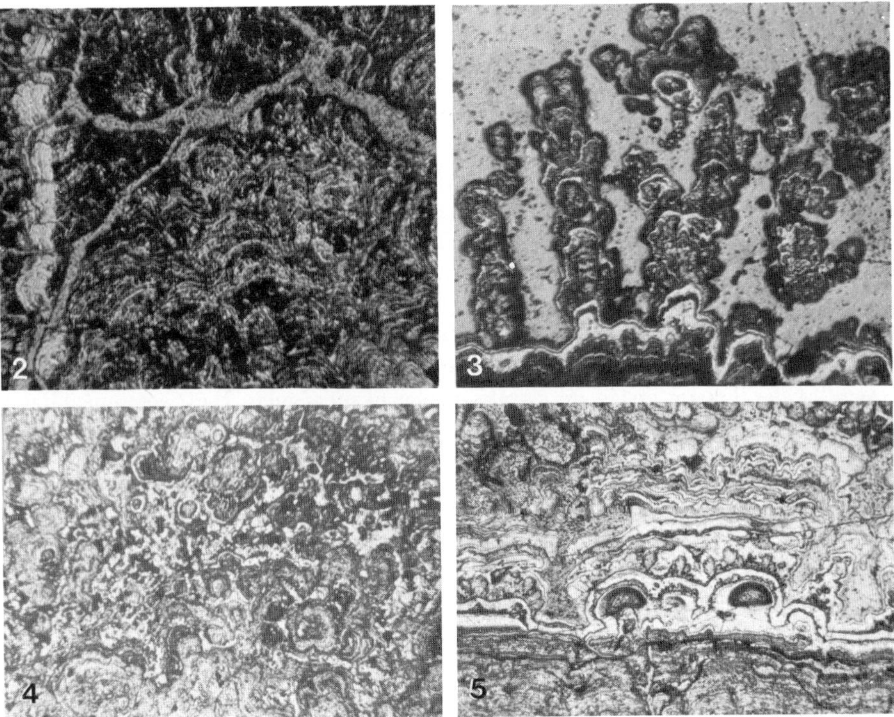

Fig. 2. Banded *(laminated)* type of colloform texture with partial banded cuspate type. Note the lining of microcrystalline todorokite *(whitish)* along the fracture *(left side)*. Carbonates *(grey)* fill the irregular fractures which were probably formed by shrinkage with ageing. × 61

Fig. 3. Typical cuspate and cuspate banded types of colloform texture in the concentric fracture which was probably formed by shrinkage with ageing. The columnar habit of nodule materials results from one-dimensional repetition of the cuspate growth of cryptocrystalline todorokite *(grey)* and the banded growth of microcrystalline todorokite *(white)*. Note the cuspate, cuspate banded, and banded types of colloform texture in the layered structure in the lower part. × 61

Fig. 4. Banded cuspate type of colloform texture with partial globular type. × 37

Fig. 5. Banded and cuspate types of colloform texture. Note the spherical relict structure of microfossils in the carbonate matrix. × 37

visible in the nodule from the Blake Plateau are shown in Fig. 1. Examples are shown in Figs. 3, 4 and 5.

The globular type of colloform formation indicates the globular growth of nodular materials. Globules usually consist of either one homogeneous phase or parallel bands.

From Fig. 1, it can be recognized that the fundamental colloform textural units in manganese nodules are as follows: banded, cuspate and globular, and that important combined textures are as follows: cuspate banded, banded cuspate, globular cuspate, and banded globular. It can be expected that there are wide gradations among these major textural types.

The above new textural classification can be used to describe the textures of manganese nodules systematically. The textural order of growth of nodular materials can be mentioned using the symbols (shown in Fig. 1) for each texture. The manganese nodules from the Blake Plateau have the textural distribution from the centre toward the surface in the following order: Gb-Cb-C-Cb-Bc-(C+B)-L-Cg-L-C-Cb(or G)-Cb-B-Cb-L. The fine alternation of types is within parenthesis.

Since the type of texture within a certain concentric layer is usually considerably uniform, the systematic tabulation of nodular textures in the order of growth as in the above example, makes it easy for one to correlate environments of nodular formation between different places. The above classification of colloform textures are applicable to any nodule of any source.

Fragmental texture. Fine-grained quartz grains are enclosed in the colloform manganese oxides. Quartz grains are usually angular to subangular. They were precipitated together with manganese oxides. They were probably transported from other places by wind or sea water.

Secondary Textures – Diagenetic Textures. Primary growth textures have been partly changed to other textures by diagenetic processes. Diagenetic features in manganese nodules have been studied by Burns and Burns (1978). The diagenetic textures in the manganese nodule from the Blake Plateau are as follows: replacement, solid state crystallization and recrystallization, cementation and deformation textures.

Replacement textures. Replacement took place on a small scale. Quartz grains are replaced by todorokite along the margin of the grains. Microfossils are replaced by manganese oxides, leaving relict structures.

Solid state crystallization and recrystallization texture. Almost all colloform materials which have been originally precipitated as gel ores become more or less crystalline with age. It is uncertain whether the flaky todorokite seen under the microscope is a crystallization product from the amorphous state and subsequent recrystallization, or a direct crystallization product from solution. Carbonate materials and microfossils filling the fractures and open spaces within the nodule are partly recrystallized. Relict structures of microfossils are preserved in the recrystallized carbonate materials.

Cementation (filling) texture. The cavities within the nodule are filled with carbonate and clays after the formation of the nodule.

Deformation texture. Many irregular or concentric fractures have been formed by shrinkage with age. Shrinkage cracks are lined with manganese oxides, resulting in the banded or cuspate texture in places. Remaining open spaces are filled with carbonates and clays. Carbonates consist of calcite and aragonite. This texture is mainly developed in the older part of the nodule.

6 Genetic Implication of Textures

The main problems in the study of manganese nodules are the source of metals and the mechanism of formation. The original source of metals involve hydrogenous, hydrothermal, halmyrolytic, or diagenetic (Bonatti et al. 1972) processes. Since the metals, whatever their source might have been, are precipitated through the medium of the surrounding water, it is hard to find the source for the metals. The adsorption mechanism has been regarded as a dominant means of extracting Mn, Fe, and other trace metals from sea water and incorporation into manganese nodules (Crerar and Barnes 1974; Glasby 1974).

As mentioned above, the colloform textures of manganese nodules are very complex. The complexity in textures suggests the complexity in factors acting on the formation of the nodule. Knowledge of the ancient environments during their historical geology and paleogeochemistry probably play a key role to understanding the general problems of nodule origin as depicted by Sorem and Fewkes (1977). The fact that the complex colloform textures consist of three basic textural units, suggests that there are three different modes of accretion of nodule materials that are probably controlled by the physical and chemical environment. Repeated alternation of these textural units in nodule indicates the repeated change of environment of formation of nodules.

Banded textures may suggest the uniform accretion of nodular materials on the older surface. Cuspate texture may suggest that the number of nuclei of growth was small but the growth rate was different from point to point. Globular texture may indicate the small number of nuclei and concentric growth around them.

Acknowledgments. The author wishes to thank Dr. K.O. Emery, and Dr. J. Broda of the Wood's Hole Oceanographic Institution for their kind supply of nodules for this work.

References

Bonatti E, Kraemer T, Rydell H (1972) Classification and genesis of submarine iron-manganese deposits. In: Horn DR (ed) Ferromanganese deposits on the ocean floor. IDOE Natl Sci Found, Washington, DC, pp 149–169

Burns VM, Burns RG (1978) Diagenetic features observed inside deep-sea manganese nodules from the north equatorial Pacific. In: Johari O (ed) Scanning electron microscopy, vol 1. SEM Inc, AMF O'Hare, Ill, pp 245–252

Crerar DA, Barnes HL (1974) Deposition of deep-sea manganese nodules. Geochim Cosmochim Acta 38:279–300

Dunham AC, Glasby GP (1974) A petrographic and electron microprobe investigation of some deep- and shallow-water manganese nodules. N Z J Geol Geophys 17:925–953

Glasby GP (1974) Mechanism of incorporation of manganese and associated trace elements in marine manganese nodules. Oceanogr Mar Biol Annu Rev 12:11–40

Manheim FT (1972) Composition and origin of manganese-iron nodules and pavements on the Blake Plateau. In: Horn DR (ed) Ferromanganese deposits on the ocean floor. IDOE Natl Sci Found, Washington DC, p 105

Sorem RK, Fewkes RH (1977) Internal characteristics of manganese nodules. In: Glasby GP (ed) Marine manganese deposits. Elsevier, Amsterdam, pp 147–183

Sorem RK, Forster AR (1969) Growth history of manganese nodules west of Baja California, Mexico. Spec Pap Geol Soc Am 121:287 (Abstr)

Sorem RK, Forster AR (1972a) Marine manganese nodules: Importance of structural analysis. 24th IGC, Sect 4:192–200

Sorem RK, Forster AR (1972b) Internal structure of manganese nodules and implication in benefication. In: Horn DR (ed) Ferromanganese deposits on the ocean floor. IDOE Natl Sci Found, Washington DC, pp 167–179

Weathered Products of Ore Deposits

Vanadium Oxides in the Urcal Deposit, Argentina

M.K. DE BRODTKORB[1]

Abstract

The Urcal Mine in the province of La Rioja, Argentina, is a small deposit where tyuyamunite and metatyuyamunite were exploited. These minerals are located in a dense fractured zone developed on the calcareous rocks of the Cabeza de Montero Member of Volcan Formation (Carboniferous age) and in the micrites of San Juan Formation (Ordovician age). Small black pockets up to 20–50 cm in diameter were observed. The polished sections show the following oxidation series of vanadium minerals: karelianite, montroseite, doloresite – häggite?, duttonite, mineral 1 ?, pascoite, associated with mineral 2, coffinite, pyrite, marcasite, galena and tetrahedrite, whose optical and textural characteristics are described. Cation reduction from groundwaters followed by precipitation in a fractured zone are assumed to be the origin of karelianite. The subsequent oxidation of this mineral produced by weathering the other vanadium oxide minerals.

1 Geological Setting of the Urcal Deposit

The Urcal deposit is located in the calcareous conglomerates of the Cabeza de Montero Member which belongs to the Volcan Formation of the Carboniferous, over which the sandstones of the upper member of the Volcan Formation (Brodtkorb and Brodtkorb 1980) lie in angular unconformity. In one shaft, the massive limestones (micrite) of the San Juan Formation of Llanvirnian age, which underlie the Cabeza de Montero Conglomerate, were reached. Andesites as dykes and sills intrude the former, being of a possible Triassic age.

Structurally, the deposit is located on the western Flank of the Sierra de Urcuschún which is a great homoclinal structure with slight flexures that change their strike from N 30° W to N 5° E and with variable dips towards the west. The flexures are conditioned by the basement structure. This delimits local zones of subsidence and gravitational fracturing which reach the Paleozoic cover. A little less than one kilometer from the mine, in the Quebrada de Agüita, there is a direct fault, whose minimum measured throw is of 200 m and on whose upthrown block the Urcal Mine is located. Associated with this fault there is a series of gravitational faults. In the upthrown block there is a conspicuous jointing which diminishes in density and intensity to the north and south of the mine.

1 Paso 258-9A, 1640 Martinez, Argentina

2 Mineralogy

2.1 Basic Concepts

Between the years 1950 and 1960 many new vanadium minerals were found due to the intensive study of the Colorado Plateau type uranium deposits, as uranium is frequently associated with vanadium, forming the uranil-vanadates. These determinations were fundamentally achieved by X-rays, and in general have scarce optical data when they are transparent minerals and practically none when opaque.

In 1964, Geffroy et al. published their first observations on the vanadium minerals of Mounana, with some optical data. Cesbron (1970) returned to the problem of the vanadium minerals but only in the crystallographic aspect.

In 1958, Evans and Garrels published an important paper on the thermodynamic equilibrium of vanadium in aqueous systems applied to the interpretation of the Colorado Plateau deposits. The vanadium minerals are located in zones of stability within the diagram of equilibrium and they propose an alteration sequence that shows the changes that occur in the vanadium ores under weathering conditions. The authors mentioned consider that montroseite is the primary vanadium mineral as, in absence of V_2O_3 in the sandstones of the Colorado Plateau, the VO (OH) would be the most stable phase at the temperature of formation. There would exist two ways in the sequence of weathering: the first would be given by acid conditions, characteristic of zones of concentration of primary vanadium oxides in the sandstones and in the presence of pyrite, following the sequence: montroseite, paramontroseite, doloresite, corvusite, pascoite, hewettite and carnotite. The second would be represented under more basic conditions, where pyrite is lacking and calcite is common, following the sequence: montroseite, paramontroseite, duttonite, simplotite, melanovanadite, rossite and carnotite.

As can be seen, the more reduced vanadium minerals are of valence 3 passing to valence 4 in a first oxidation, both corresponding to optically opaque species. With a further oxidation and above the water table are found the vanadates (valence 5) which correspond to transparent minerals.

In 1960, Evans and Mrose introduced an alternative in the alteration of the montroseite, being able to pass to häggite – phase B or to paramontroseite – doloresite.

2.2 Description of the Species Present in the Urcal Deposit

For descriptive purposes we may form three groups with the minerals present in this deposit:

Vanadium minerals	*Uranium minerals*	*Sulphides*
karelianite	coffinite	pyrite
montroseite	tyuyamunite	marcasite
doloresite	metatyuyamunite	galena
häggite?	metatorbernite	tetrahedrite
duttonite		
mineral 1		
mineral 2		
pascoite		

2.2.1 Vanadium Minerals

Karelianite V_2O_3. It is known in two completely different environments. It was described for the first time as pertaining to a high temperature paragenesis, by Long et al. (1963) who found it in rolled pebbles in the ores of Outokumpu, Finland. Here the karelianite appears in prismatic crystals of up to 0.3 mm and is associated with pyrrhotite, chalcopyrite and pyrite, inclusions of these minerals having been found in the first. From an analysis made with the microprobe, the following formula resulted $(V_{1.82}Fe_{0.08}Cr_{0.07}Mn_{0.03})\,O_3$.

The second occurrence of karelianite was found in the Mounana uranium deposit of Gabon, by Geffroy et al. (1964). This deposit is localized in coarse sandstones of the lower Francevillan (middle Precambrian). The vanadium and uranium minerals are associated to sulphides of iron, lead, zinc and copper, and in the primary zone we find coffinite, pitchblende, karelianite and montroseite.

The karelianite of Urcal is found in compact masses, the crystals being of prismatic habit (Figs. 1 and 2), whose mean sizes are of 0.5 mm. It was determined by X-rays and the spacing is identical to Outokumpu and Mounana.

The optical properties of the Urcal karelianite are as follows: colour dark orange, reflectivity of approx. 20%, low pleochroism, strong anisotropy between reddish and dark grey. Following the diagram established by Evans and Garrels (1958), this karelianite alters on the edges to montroseite.

Montroseite VO (OH) or $V_2O_3 . H_2O$. It was described for the first time by Weeks et al. (1953). Found in several Colorado Plateau deposits, it is an abundant mineral in the Mi Vida Mine, Utah (Gross 1956). It usually occurs as needles and plates cementing sandstones. Its composition was given as VO (OH) and sometimes as (V,Fe) O(OH), as frequently Fe is found substituting V. It is possible to observe a partial oxidation to VO_2, paramontroseite, according to the reaction given by Evans and Mrose (1955):

$$2\,VO(OH) + 1/2\,O_2 \dashrightarrow 2\,VO_2 + H_2O.$$

At Mounana (Geffroy et al. 1964) montroseite occurs in form of plates, sometimes replacing karelianite. Its optical properties in reflected light are: grey with light blue tins, especially on the side of the karelianite, the reflectivity is lower than that of karelianite, pleochroism is weak and anisotropy strong, ranging from yellowish grey to dark brown.

The optical properties of montroseite from the Colorado Plateau are not well described so they may not be compared.

Montroseite from the Huemul deposit (Brodtkorb 1966) occurs well crystallized and is included in asphaltic material and in chalcopyrite. In some crystals an alteration to paramontroseite has been observed.

The montroseite from Urcal is formed by alteration of the karelianite and fills its intergranular spaces (Figs. 1, 2 and 3). The optical properties observed are: light blue colour, ranging with pleochroism from dark greyish blue to light brownish grey, reflectivity lower than 20% and strong anisotropy between yellowish grey to dark grey.

Apparently the metastable phase of paramontroseite is missing here.

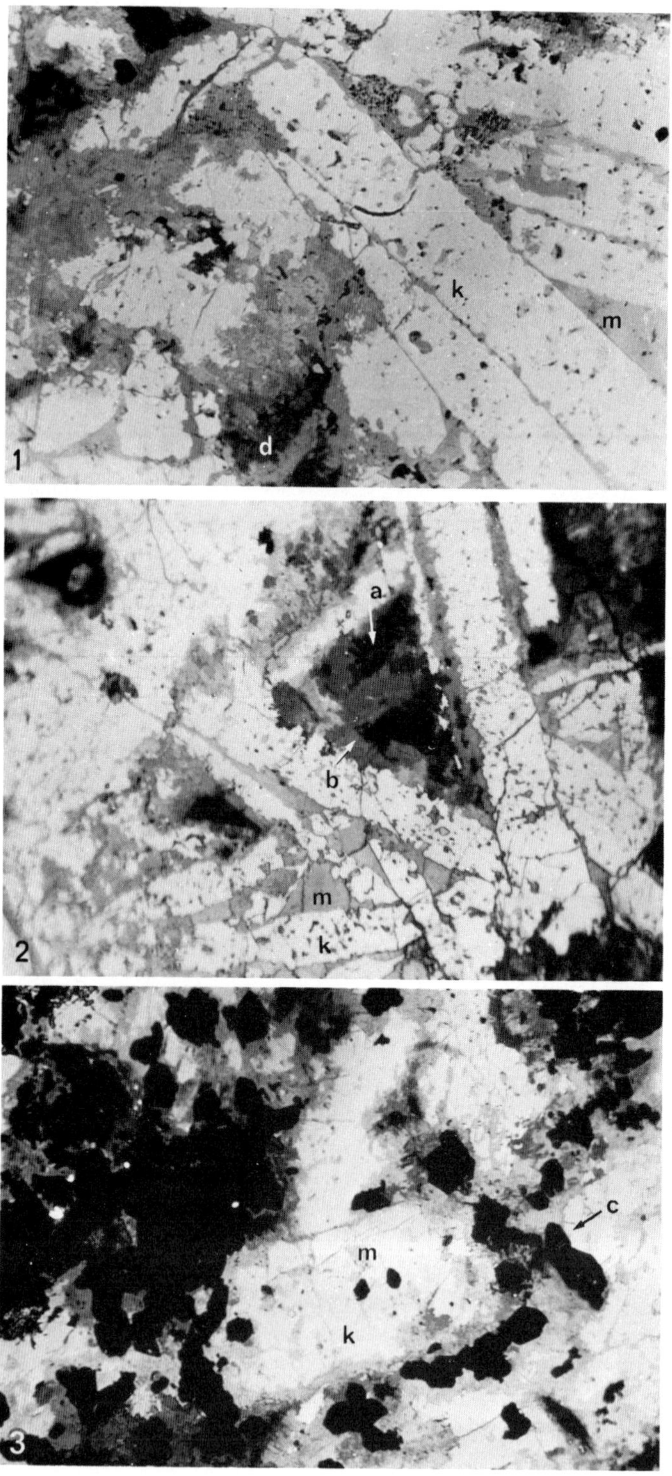
k
m
d
1
a
b
m
k
2
c
m
k
3

Doloresite – Häggite? $3V_2O_4.4H_2O$ or $V_3O_4(OH)_4$ – $V_2O_3.V_2O_4.3H_2O$. Doloresite was described for the first time by Stern et al. (1957) who localized it in numerous deposits of the Colorado Plateau. According to these authors, doloresite is optically opaque, grey, it has pleochroism and is strongly anisotropic. In the Mi Vida Mine, Utah (Gross 1956), doloresite replaces a primary mass of montroseite. It has not been found in Mounana.

Häggite was described for the first time by Evans and Mrose in 1958 from an occurrence in Wyoming, USA.

In a former paper (Brodtkorb 1978) the name doloresite was applied to a mineral that occurs in granular aggregates between tablets of karelianite altered to montroseite. Later studies have determined that they are two minerals which often occur intimately associated, one surrounding the other and also as individuals (Fig. 2). Here a question may arise: could they be doloresite and häggite, and which is which?

The mineral found in the centre of the granular aggregates has the following optical properties: colour grey which depends on the strong pleochroism between medium grey and blackish grey. Anisotropy is strong between medium grey and dark grey. Few reddish internal reflexions. The mineral bordering this nucleus has a higher reflectivity, the colour is greenish grey depending on the pleochroism: light greenish grey–medium grey (colour lying between the positions of extreme pleochroism of the other mineral). Few colourless to yellowish-greenish internal reflexions. The hardness of both minerals is the same, as also that of karelianite and montroseite.

Duttonite $VO(OH)_2$ or $V_2O_4.2H_2O$. It was described for the first time by Thompson et al. (1957). It is not a frequent mineral in the deposits of the Colorado Plateau, but in the Peanut Mine it is considered an oxidation mineral of montroseite.

For Mounana (Geffroy et al. 1964) it is described as a greenish-yellow to dark green mineral. In reflected light, the reflectivity is low and the mineral has a strong pleochroism toward of various shades of grey. Anisotropy is strong but it is superposed by the luminous yellow internal reflexions.

At Urcal, duttonite occurs as radial aggregates, with the same optical properties described for Mounana (Fig. 4).

Both the doloresite-häggite? and the duttonite occur at Urcal as very small grains and interstitially and they have not been confirmed by X-rays, being identified by the optical properties described by Geffroy et al. (1964) and/or confirmed by the oxidation succession given by Evans and Garrels (1958), and Evans and Mrose (1960).

It is to be hoped that in the future we may find these minerals in new deposits in sizes large enough to confirm the scarce optical X-ray data available.

Recently two more minerals have been found in this paragenesis which, due to the small grain size and their intimate intergrowth with other minerals, also could not be identified by X-rays.

Fig. 1. Karelianite *(k)* altered to montroseite *(m)*. × 200, imm

Fig. 2. Karelianite *(k)* altered to montroseite *(m)* and this latter to doloresite – häggite? *(a* and *b)*. × 200, imm

Fig. 3. Karelianite *(k)* altered to montroseite *(m)*. Coffinite *(c)*. × 200, imm

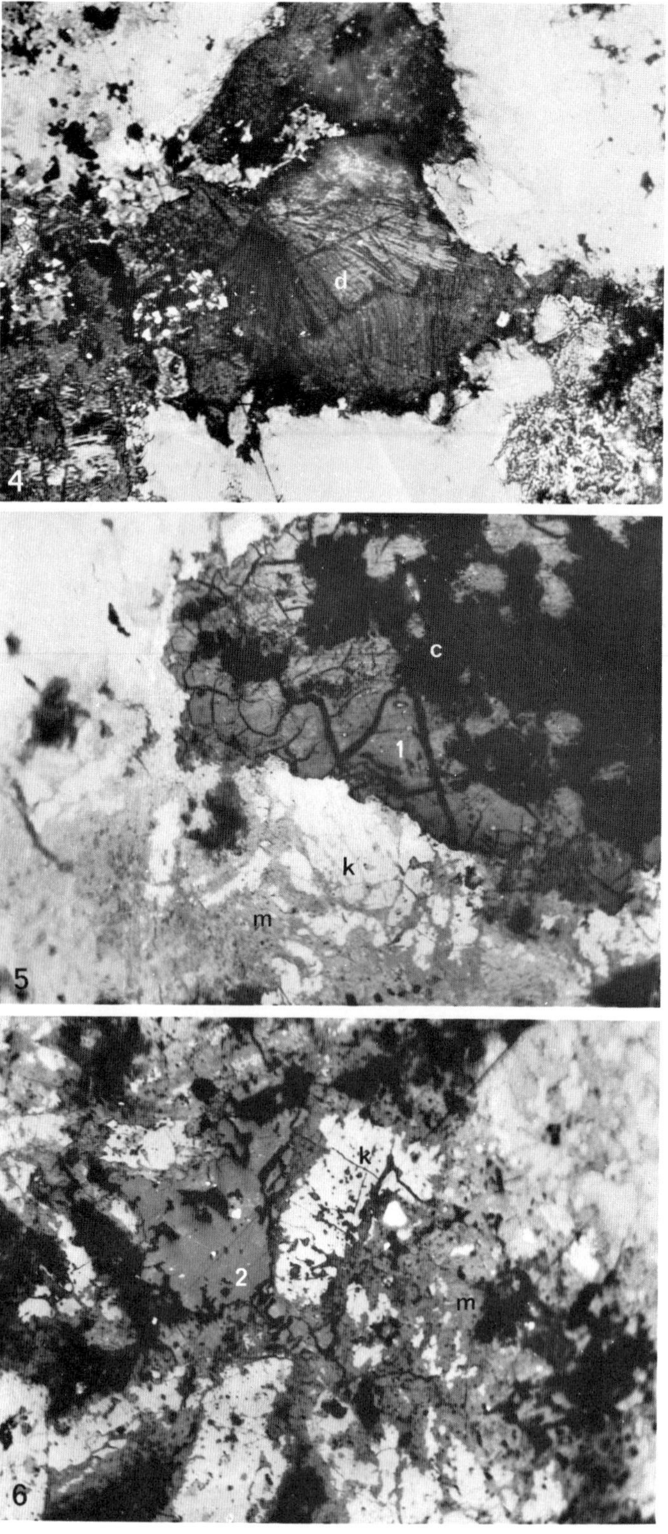
d
4
c
1
k
m
5
k
2
m
6

Mineral 1. It generally occurs on the borders of karelianite – montroseite towards calcite (Fig. 5). It could then be one of the vanadium and calcium oxides (?). Its optical properties are: opaque, low reflectivity, hardness greater than karelianite – montroseite, colour bluish grey. It is slightly anisotropic, in bluish grey tones and apparently has not internal reflexions. It shows contraction cracks and signs of oxidation?

Mineral 2. It occurs as allotriomorphic aggregates, apparently without showing altered edges, which makes it difficult to locate in the oxidation series (Fig. 6). Its optical properties are: opaque, reflectivity somewhat lower than montroseite, colour greenish grey in contrast and hardness greater than in the latter. Good polishing surface. It is isotropic and has orange-yellow internal reflexions.

Pascoite $Ca_3V_{10}O_{28}.16\ H_2O$. It occurs as efflorescences of a characteristic orange colour where oxides of vanadium are found.

2.2.2 Uranium Minerals

Coffinite. It crystallizes in the tetragonal system and its appearance in Urcal is of rather short prisms ending in pyramids (Fig. 3).

Tyuyamunite, Metatyuyamunite and Metatorbernite. They were determined by X-ray due to their powdery character. The first two, of a luminous yellow colour, are what constitutes the ore. The metatorbernite is scarce and occurs as a dark green patina.

2.2.3 Sulphides

Pyrite is very abundant and it is found as idiomorphic and subidiomorphic crystals associated with various vanadium minerals. The other sulphides, marcasite, galena and tetrahedrite are not very frequent and occur in the same paragenesis.

3 Genetic Considerations

In the surroundings of the Urcal Mine two different kinds of mineralizations occur which must be mentioned for the interpretation of the genesis.

On the one hand, some 5 km W of Urcal there is the La Helvecia mining district with a strata-bound mineralization of Pb, Zn and barite in limestones (Brodtkorb and Brodtkorb 1980). On the other hand we find small deposits of uranium minerals, for example the Urcuschún deposit in Carboniferous continental sandstones (Belluco et al. 1973) whose origin is due to circulating waters that precipitated their salts in the sandstones (Colorado Plateau type deposit).

At Urcal, various factors combined. As we have seen before, here we observe a conspicuous jointing in the Cabeza de Montero Formation. Circulating waters coming

Fig. 4. Duttonite *(d)*. × 200, imm

Fig. 5. Mineral 1 *(1)* in the border of karelianite *(k)* to calcite *(c)*. Montroseite *(m)*. × 200, imm

Fig. 6. Mineral 2 *(2)* associated with karelianite *(k)* and montroseite *(m)*. × 200, imm

from the Carboniferous sandstones at a greater altitude, carrying ions of V, U, Cu, etc., have found this jointing as an appropriate place for precipitation. The environmental conditions produced not only the precipitation of the uranil salts (tyuyamunite, metatyuyamunite, metatorbernite) but also in part they were reduced, precipitating karelianite, coffinite and some sulphides. Starting from karelianite, and by weathering and oxidation later, montroseite, doloresite–häggite?, duttonite and pascoite were formed. The presence of galena, sphalerite, barite and their oxidized forms smithsonite, calamine, etc., mentioned for Urcal (Belluco et al. 1973) belong to the mineralization of La Helvecia Mine already mentioned, which are superposed by minerals belonging to the vanadium-uranium paragenesis.

The uranium-vanadium deposits of the Colorado Plateau as also of Mounana, Gabon, occur in sandstones and conglomeratic sandstones, and localitites with U and V in limestones are only mentioned when they are accidentally associated with beds of sandstones. The Tyuya Muyun Mine in Turkestan (Fersmann 1930) occurs in limestones and is related to a karstic relief, that is to say, the mineralization fundamentally of tyuyamunite precipitated together with calcite, barite, gypsum, etc., filling karstic cavities. No trivalent vanadium minerals are mentioned.

This paragenesis, still more in limestones, is possibly not very frequent. The scheme of Evans and Garrels (1958) with respect to two series of weathering, one of them in acid conditions and in the presence of pyrite, and another in absence of this last and in alkaline conditions, would here present an alternative with pyrite and recrystallized calcite in a limestone environment. The presence of karelianite, here as in Mounana, as the most reduced state of the vanadium oxides, would also indicate that the series of Evans and Garrels (1958) could begin with this mineral instead of montroseite.

References

Belluco A, Diez, J, Antonietti C (1973) Los depósitos uraníferos de las provincias de La Rioja y San Juan. J Geol Argent Actas No 5 Vol 2:9–33

Brodtkorb MK de (1966) Mineralogía y consideraciones genéticas del yacimiento Huemul, prov. de Mendoza. Asoc Geol Arg Rev XXI/3:165–179

Brodtkorb MK de (1978) Oxidos de vanadio en calizas: su presencia en el yacimiento Urcal, prov. de La Rioja. Asoc Geol Arg Rev XXXIII/3:97–104

Brodtkorb MK de, Brodtkorb A (1980) La Helvecia: stratabound Pb-Zn-barite deposit, Argentina. V. IAGOD Symposium. Schweizerbart, Stuttgart, pp 785–795

Cesbron F (1970) Minéralogie et cristallochimie du vanadium. Etude du gisement de Mounana, Gabon. Thesis, Fac Sci Paris

Evans HT, Garrels RM (1958) Thermodynamic equilibria of vanadium in aqueous systems as applied to the interpretation of the Colorado Plateau ore deposits. Geochim Cosmochim Acta 15:131–149

Evans HT, Mrose ME (1955) A crystal chemical study of the vanadium oxide minerals montroseite and paramontroseite. Am Mineral 40:861–875

Evans HT, Mrose ME (1958) The crystal structures of three new vanadium oxide minerals. Acta Cryst 11:56–57

Evans HT, Mrose ME (1960) A crystal chemical study of the vanadium oxide minerals häggite and doloresite. Am Mineral 45:1144–1166

Fersmann A (1930) Geochemische Migration der Elemente. Abh Prak Geol Brg 19/II:1–86

Geffroy J, Cesbron F, Lafforgue P (1964) Donnés preliminaires sur les constituants profonds des minerais uranifères et vanadifères de Mounana (Gabon). CR Acad Sci Paris 259:601–603

Gross E (1956) Mineralogy and paragenesis of the uranium ore, Mi Vida Mine, San Juan County, Utah. Econ Geol 51:632–648

Long JVP, Vuorelainen Y, Kouvo O (1963) Karelianite, a new vanadium mineral. Am Mineral 48: 33–41

Stern TW, Stieff LR, Evans HT Jr, Sherwood AM (1957) Doloresite, a new vanadium mineral from the Colorado Plateau. Am Mineral 42:587–593

Thompson ME, Roach CH, Meyrowitz R (1957) Duttonite, a new quadrivalent vanadium oxide from the Peanut Mine, Montrose County, Colorado. Am Mineral 42:455–460

Weeks AD, Cisney EA, Sherwood AM (1953) Montroseite, a new vanadium mineral from the Colorado Plateau. Am Mineral 38:1235–1241

On the Nature of Co-Ni Asbolane; a Component of Some Supergene Ores

F.V. CHUKHROV, A.I. GORSHKOV, I.V. VITOVSKAYA, V.A. DRITS, and A.V. SIVTSOV[1]

Abstract

Co-Ni asbolane specimens from the Lipovsk and Tyulenevsk deposits in the Middle Urals and from Buranovsk deposit in the South Urals were examined by analytical electron microscopy. Chemical analyses are also given. Most suitable for a detailed examination was Co-Ni asbolane from the products of weathering of the Lipovsk serpentinites. Thus, a study of Co-Ni asbolane from the Lipovsk deposit has shown that the mineral has a hybrid structure composed of the layers of Mn^{4+} and Co-Ni octahedra alternating along *c* axis. These layers form hexagonal sublattices I and II with parameters (under vacuum) a = 2.823, c = 9.34 Å (Mn^{4+} sublattice), a = 3.04 Å and c = 9.34 Å (Co-Ni sublattice). In the layers of different types the octahedra differ in azimuth orientation as much as 60°. The chemical formula of the Lipovsk asbolane may be written as

$$[Mn^{4+}O_{2-x}(OH)_x]^{x+}\,[R^{2+}_{1-y}(OH)_{2-2y+x}]^{x-}$$

where: R^{2+} is Ni, Co, Ca; x and y < 1. Both electrostatic and hydrogen bonds exist between the layers of different types.

For a long time, asbolanes were included in a group of wads which, as shown by investigations of the last few decades, combined various earthy minerals from a group of supergene manganese oxides or their mixtures. Besides the earthy structure of aggregates, the asbolanes are characterized by significant to high percentages of cobalt and (or) nickel. Ginsburg and Rukavishnikova (1951) supposed that a carrier of cobalt and nickel in such mixtures should be a definite mineral, asbolane proper. However, the nature of this mineral remained unknown.

In their extensive study of asbolanes, the authors of the present paper used analytical electron microscopy. The mineral was examined in a JEM-100 C electron microscope with attached goniometer and Kevex energy dispersion microanalyzer. The suspension method was used, and for the same microcrystalline sample of asbolane there was successively obtained an electron microscope image, as well as diffraction electron patterns, and spectra of chracteristic X-ray radiation.

The application of the electron microscope technique made it possible to reveal a number of general and rather interesting crystal chemical and structural features of asbolanes. For the first time, it was established (Chukhrov et al. 1979b) that asbolanes are ordered mixed-layer minerals in which, along the *c* axis, the layers of Mn^{4+} octahedra alternate with the layers of Co, Ni, and Co-Ni octahedra. The structure of such a

1 Institute for Ore-Deposit Geology, IGEM, Staromonetnij 35, Moscow Sh-17, USSR

type was suggested by Evans and Allmann (1968) to be hybrid. The layers of different cation composition in asbolanes form the brucite-like structure, but differ in basis parameters, and form two sublattices: sublattice I inherent to Mn layers, and sublattice II inherent to layers of Co, Ni or Co-Ni. It is reasonable to consider asbolanes as hybrid minerals with metrically different layers.

Asbolanes of various composition (Co, Ni, Co-Ni) differ in some structural features, and are worthy of special consideration. The present paper contains the results of the Co-Ni asbolane study. The specimens from the Lipovsk and Tyulenevsk deposits in the Middle Urals and from the Buranovsk deposit in the South Urals were examined.

Distinct under electron microscope, the platy particles of Co-Ni asbolanes have a similar composition (as established by the Kevex energy dispersion analyzer) and yield practically identical diffraction patterns. Most suitable for a detailed examination was Co-Ni asbolane from the products of weathering of the Lipovsk serpentinites. Co-Ni asbolane is confined to high-porous, siliceous rocks, and has an appearance of uniform aggregates or segregations of globules. The size of asbolane plates rarely exceeds 1 mm (0.03–0.05 mm on average).

The colour of the mineral as well as the streak is brownish-black. It has a semi-metallic lustre. By crushing the aggregates, asbolane disintegrates into thin platy fragments with uneven surface, partially covered with goethite coatings.

Data of the chemical analysis of Co-Ni asbolane partly with goethite coatings are given in Table 1. All manganese in the analyzed asbolane is represented by Mn^{4+}. It is possible to distinguish four intervals of water loss in the thermogravimetric curve. The empirical formula of Co-Ni asbolane from the Lipovsk deposit is $0.28\,NiO \cdot 0.17\,CoO \cdot 0.05\,CaO \cdot 1.00\,MnO_2 \cdot 1.00\,H_2O^+ \cdot \sim 0.64\,H_2O^-$ or $(Ni_{0.28}\,Co_{0.17}\,Ca_{0.06})\,Mn(OH)_2\,O_{1.5} \cdot \sim 0.64\,H_2O$.

In the suspension electron microscopic specimens, the asbolane from the Lipovsk deposit (Fig. 1) is represented by platy particles without crystallographic outlines, con-

Table 1. Chemistry of asbolane from the Lipovsk deposit

Components	Asbolane with goethite admixture (%)	Goethite admixture (%)	Asbolane without goethite admixture		
			%	Molecular quantities	Molecular ratio
NiO	11.42	–	11.42	0.159	0.28
CoO	6.91	–	6.91	0.092	0.17
CaO	1.60	–	1.60	0.028	0.05
Fe_2O_3	15.33	15.33	–	–	–
MnO_2	46.91	–	46.91	0.540	1.00
H_2O150°	3.26	–	3.26	0.181	0.33
H_2O150°–300°	4.79	1.73	3.06	0.170	0.31
H_2O300°–600°	7.35	–	7.35	0.408	0.755
H_2O>600°	2.43	–	2.43	0.135	0.248
Total	100.0				

Analyst – R.S. Yashina. Fe_2O_3 is determined from difference to total 100.00%. All iron is assigned to goethite admixture.
Water removed at different temperature ranges is determined by N.I. Shepotchkina from thermogravimetric curve

taining manganese, cobalt and nickel with an insignificant amount of calcium. The plates range in size from a fraction of a μm up to several μm. When the electron beam is perpendicular to the basis plane, the selected area diffraction patterns of the asbolane plates contain two systems of reflexion nets arranged according to the hexagonal law (Fig. 2a, b). The intensities of the reflexions forming the hexagonal net nearest to the pattern centre are several times less than the intensities of the reflexions forming the net situated farther from the pattern centre. Interplanar distances of two six-fold groups of reflexions of these nets nearest to the pattern centre are equal to 2.445 Å and 2.63 Å, respectively.

Fig. 1. SAD pattern of platy particles of Co-Ni asbolane

On the SAD patterns from the bent edges of the plates obtained using Gorshkov's technique (1970), a single set of basal reflexions with $d_{001} = \frac{9.34}{1}$ Å (Fig. 2b) is fixed. The pattern from the plates tilted to the electron beam reveals, in addition to the hk0 and 001 reflexions, some spatial reflexions with d equal to 2.37, 2.17, 1.93 and 1.7 Å (Fig. 2c, d). The SAD data allow us to conclude that the asbolane structure is hybrid, and consists of metric differences in plane (001) and orderly alternating spacing along axis c layers of two types. This structure can be regarded as consisting of two hexagonal sublattices with unit cell parameters (from SAD data) a = 2.823, c = 9.34 Å (sublattice I) and a = 3.04, c = 9.34 Å (sublattice II). The indices of reflexions and corresponding to them d_c-spacings of the two sublattices of asbolane, as well as the experimental values of d_c-spacings determined from the SAD patterns are given in Table 2.

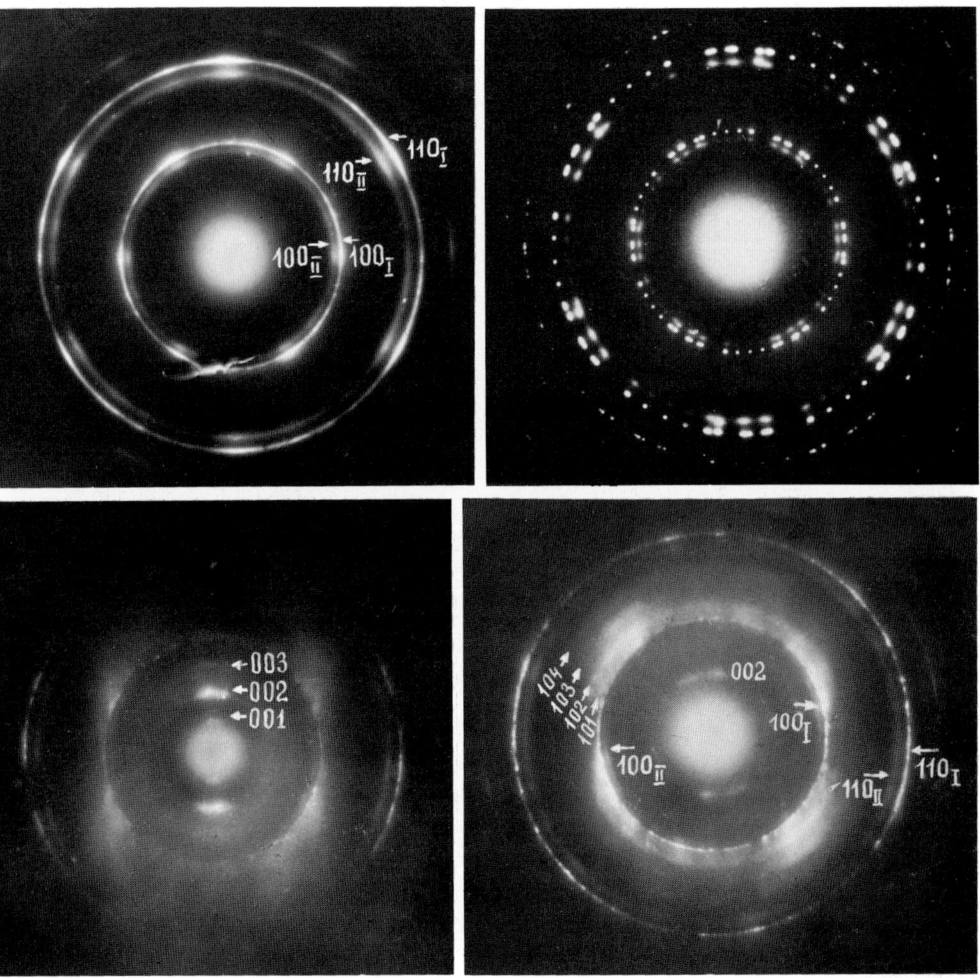

Fig. 2. SAD pattern of Co-Ni asbolane: **a, b** with basis reflexions hk0 of two sublattices (*I* and *II*); **c** from the bent edge of the plate; in addition to basis reflexions basal reflexions 001 are also presented; **d** from the plates with bent edges tilted to the electron beam; fixed reflexions hk0, 001 and hk1

The comparison of calculated and experimental d-values indicates their good accordance in the case of sublattice I; in the case of sublattice II, the accordance of the calculated and experimental d-values is established only for hk0 and 001 reflexions. The absence of spatial reflexions from sublattice II evidently is due to the imperfections in Co-Ni layers, which constitute this sublattice (its island-like appearance).

Using a diffractometer with K radiation, an X-ray diffraction pattern of Co-Ni asbolane was obtained, which shows reflexions with *d* equal to 9.6 (weak), 4.82 (strong), 2.445 (medium strong), 1.7 (very weak), 1.419 Å (very weak). Reflexions with *d* equal to 2.445 and 1.419 Å are identical to the basic hk0 reflexions, 100 and

Table 2. Diffraction characteristic of asbolane from the Lipovsk deposit

SAD data					X-ray data		
d_0Å	Sublattice I		Sublattice II		d,Å	J	hkl
	d_c, Å	hkl	d_c, Å	hkl			
9.33	9.34	001	9.34	001	9.6	weak	001
4.67	4.67	002	4.67	002	4.82	strong	002
3.09	3.11	003	3.11	003			
2.63			2.63	100			
2.45	2.445	100	2.534	101	2.445	medium weak	100
2.37[a]	2.365	101	2.29	102			
2.17[a]	2.166	102					
1.93[a]	1.923	103	2.01	103			
1.7[a]	1.69	104	1.74	104	1.7	very weak	104
1.51			1.518	110			
1.41	1.412	110			1.419	very weak	110
1.22	1.422	200					
	a_0 = 2.823 Å		a_0 = 3.04 Å		a_0 = 2.823 Å		
	c_0 = 9.34		c_0 = 9.34		c_0 = 9.6 Å		

[a] This reflexion has a diffuse character. All X-ray diffraction lines are broad

110 in SAD pattern corresponding to sublattice I. Reflexions 9.6 and 4.82 have 001 and 002 indices, respectively. The only spatial reflexion with index 104 belongs also to sublattice I.

It is likely that a greater value of d_{001} determined from the X-ray diffraction patterns as compared with a value of d_{001} established from SAD patterns may be explained by the presence of molecular water between the layers of the asbolane structure. This water is easily removed in the vacuum of the electron microscope, and is responsible for contraction of the asbolane structure along the normal towards its layers, and as a consequence, for a lower value of d_{001}. Decreased values of d_{001} in the vacuum have been established also for some other manganese minerals with a layer structure, for rancieite, for example (Chukhrov et al. 1979a).

Thus, the X-ray diffraction pattern of asbolane in a natural state contains only the reflexions inherent to sublattice I. The absence of the reflexions typical for sublattice II is evidently due to their extremely weak intensity. It should be noted that the X-ray diffraction patterns of some other hybrid minerals also displayed the reflexions from only one component of their structures. For example, the diffraction patterns of dispersed valleriite contained merely the reflexions referred to the sulphide sublattice (Evans and Allmann 1968; Organova et al. 1975). This fact indicates that in studies of fine-dispersed, poorly crystallized mineral the possiblity of the X-ray analysis is rather limited, so the application of the selected area electron diffraction technique appears to be more effective.

As regards the combination of octahedral layers with different cations, the Co-Ni asbolane is similar to lithiophorite (Wadsley 1952), though the latter structure contains orderly spatial alternations along *c* axis layers of Mn^{4+} and of Al-Li octahedra with the same dimensions in the basis plane, and therefore lithiophorite is characterized by a common monocline lattice. Parameter a = 2.823 Å of the hexagonal sublattice I indicates that the corresponding octahedral layers of asbolane are occupied by cations of Mn^{4+}, as in the case of birnessite (Chukhrov et al. 1979a). A considerably greater value of parameter a = 3.04 Å of the hexagonal sublattice II is consistent with the supposition that the corresponding octahedral layers contain (relative to Mn^{4+}) cations of Co and Ni.

The diffusive character and weak intensity of hk0 reflexions from sublattice II (in comparison with the analogous reflexions of sublattice I) indicate small diffraction areas of the Co-Ni component of the asbolane structure, and their small total volume. As noted above, this fact allows us to regard the Co-Ni layers as imperfect (island-like appearance). This conclusion is supported by a few other facts such as: (a) the absence of spatial hkl reflexions from sublattice II, (b) amorphous state of a specimen heated up to 200°C (X-ray data) most likely due to the decomposition of the structurally imperfect Co-Ni component, (c) essentially weaker intensity of hk0 reflexions from sublattice II relative to the corresponding hk0 reflexions from sublattice I, in spite of the higher scattering power of Co and Ni atoms than of Mn atoms; in the case of two-dimensional continuous octahedral Co-Ni layers within the micromonocrystals of asbolane (like Mn layers), the ratio of the intensities of hk0 reflexions from different sublattices would be inverse relative to those observed experimentally; (d) an island-like character of Co-Ni layers is more favourable for the formation of a stronger bond between Co-Ni and Mn^{4+} layers since a difference between their basis dimensions in the case of two-dimensional continuity of Co-Ni component should result in essential variations and extention of the interlayer bonds.

The assumption that Co-Ni octahedral layers of asbolane are island-like allows us to explain the presence of various types of structurally bound water (thermogravimetric data) in the asbolane under study. The molecular H_2O released in the temperature range of 150°–300°C evidently occupies the positions between the adjacent layers of different types and between the island of the Co-Ni layers. The evidence for this is its easy removal from the structure under vacuum in an electron microscope. The water released at 300°–600°C is likely to correspond to hydroxyl components of an anion part of the Co-Ni layers (islands). The hydroxyl water (released at temperature above 600°C), strongly bound with the structure, forms, together with oxygen atoms, the anion packing of fully filled and apparently perfect Mn layers of the asbolane structure. Calcium, present in minor quantities in each of the asbolane plates, is probably coordinated with H_2O molecules and occurs between the islands of the structure.

Data of IR spectroscopy (Fig. 3) show that the Co-Ni asbolane contains the water of two types: the range of deformation vibrations of the water hydroxyls contains two bands with maxima at 1730 cm^{-1} and 1650 cm^{-1} indicating that H_2O molecules occupy different positions in the mineral structure, and can be removed at different temperatures. In the range of valency vibrations (3600–3200 cm^{-1}) of OH groups, one wide, diffuse band is recognized. With the aim of revealing OH groups in the structure of the mineral at 3700–3100 cm^{-1} in the canal of the device was placed for comparison a specimen whose molecular water was fully removed by heating. Such a specimen allows compensation of the absorption of the mineral itself in the range of OH valency vibrations, and thus makes it possible to reveal more clearly the absorption bands of the structural hydroxyls and hydroxyls of the molecular water. An IR spectrum so measured displays a more intense and isolated band corresponding to the hydroxyl water oscillations with a maximum at 3400 cm^{-1}, and a clearer band in a 3640–3500 cm^{-1} range with a weak maximum, which points to a considerable range of bond energies for the structural OH groups (Fig. 3b). The main absorption band in the IR spectrum of the mineral lies in the range of 700–400 cm^{-1}, the maximum position being at 525 cm^{-1}.

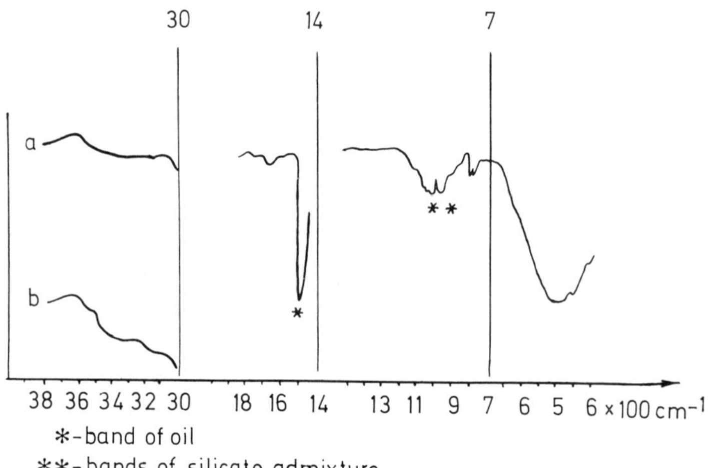

Fig. 3. IR spectra of Co-Ni asbolane. (Data from Rudnitskaya)

The fact that IR spectroscopy has revealed the two types of molecular water variably associated with the structure can be explained to assume the structural model of asbolane, in which Co-Ni layers are divided into island-like separate units. The more feebly bonded molecular water occupies the space between the asbolane layers. It is the water that is easily removable in the vacuum. The band at 1650 cm^{-1} characterizes the molecular water which is strongly bound with the structure, and which is likely to occur between the islands of the Co-Ni layers, and is associated with Ca cations.

Proceeding from the above considerations, the Co-Ni asbolane formula may be written as follows:

$$[MnO_{1.5}(OH)_{0.5}]^{+0.5}\,[(Ni_{0.28}\,Co_{0.17})\,Ca_{0.05}\,(OH)_{1.50}]^{-0.5}\cdot 0.6\,H_2O.$$

As a result of a partial substitution by hydroxyl groups, the Mn^{4+} layers should possess a positive charge, whereas the Co-Ni layers due to cation deficiency should carry a negative charge. Both electrostatic and hydrogen bonds may exist between the adjacent layers of different nature. The positive charge of the Mn^{4+} layers is compensated for by the negative charge of the Co-Ni layers, and vice versa. This is responsible for orderly alternating layers of various types. The oxygen atoms of the Mn^{4+} layers together with hydroxyls of the Co-Ni layers form hydrogen bonds O-H. . .O.

On the basis of the above formula and taking into account the diffraction data, we propose the possible structural models for Mn^{4+} and Co-Ni components of asbolane.

It is obvious that in any possible model the octahedra of layers of every type have a similar azimuth orientation, and their centres lie strictly above each other and project on the (001) plane. As a consequence of this, each of the structural components of asbolane has a primitive hexagonal cell characterized by spatial groups $P\bar{3}m1$. The octahedra of the adjacent layers of different nature have an inverse azimuth orientation (as in gibbsite). The anions in the adjacent layers occur approximately above each other, and consequently the hydrogen bonds should be shorter as compared with the case of other possible azimuth orientation of the layers of different types. The latter is essential because the adjacent layers have different dimensions in the basis plane, and in itself it will ultimately make the bonds longer.

To provide more reliable evidence for the above assumption on the specific structural features of Mn^{4+} and Co-Ni components of the asbolane structure, a comparison was drawn between the intensities of hk0 and 001 reflexions calculated for corresponding structural models, and measured on SAD patterns. It is known that the intensity of spot hk0 reflexions from a thin plate of an ideal micromonocrystal is directly proportional to structural factor $F^2(hk0)$, and to the square number of cells N^2 composing the monocrystal area penetrated by a primary electron beam.

The mosaic structure of the real microcrystals requires an account of the azimuth disorientation of blocks. Then, a value of the local intensity of hk0 reflexions can be written as $I_c\ (hk0) = K^2N^2F_c{}^2(hk0) \cdot d(hk0)$, where K is a constant coefficient for the given experimental conditions, and multiplier d(hk0) equal to a corresponding basis d-spacing is defined by the mosaic structure of the analyzed crystals. In our case, N is equal to the number of octahedra within the average two-dimensional area of coherent dispersion.

On the SAD patterns of the asbolane particles, three reflexions with indices 100, 110 and 200 corresponding to sublattice I, and reflexions 100 and 110 corresponding to sublattice II are fixed. To calculate intensities of considered reflexions, it is necessary to know a dimension ratio of a coherent scattering for Mn and Co-Ni layers in the asbolane microcrystals.

According to the chemical formulae of the asbolane studied, it is natural to assume that $N_I = 2N_{II}$. Then it is convenient to accept $N_I = 1$ for fully cation-occupied layers of the Mn component, and $N_{II} = 0.5$ for imperfect Co-Ni layers. Calculated under this assumption, values of intensities are equal to 12, 20 and 1 for reflexions 100, 110, and 200 of the Mn^{4+} component, 4 and 7 for reflexions 100, 110 of the Co-Ni component. These values agree with the measured ones for several particle values I_{exp} (hk0) for the given reflexions the ratio of which is approximately equal to 11:21:2 and 3,5:6,5 for sublattices I and II, respectively. The agreement between $I_{calc.}$ (hk0) and $I_{exp.}$ (hk0) allows us to conclude that the structural models suggested for Mn^{4+} and Co-Ni components of asbolane correspond to the real structure.

To confirm the conclusion about the alternation of Mn^{4+} and Co-Ni layers along the *c* axis in the structure of asbolane, the comparison of experimental and calculated intensities of reflexions 001 was carried out. Structural factors $F^2(001)$ were calculated for the model shown in Fig. 4; distribution of anions and cations in the layers of this model is given with regard to their average amount in the unit cell of each component. In calculation of F_c^2 (001), the thickness of Mn^{4+} layers was taken approximately equal to the thickness of analogous layers in the lithiophorite structure. A somewhat greater value was taken for the Co-Ni layer thickness in accordance with higher basis dimensions of these layers (relative to the basis dimensions of the Mn^{4+} layers). It is established that values F_c^2 (001) equal to 13, 88, 4.5 and 1 for l = 1, 2, 3, 4, respectively, agree with the experimental intensities of reflexions $I_e(001) = F_e^2$ (001) = 15 (001):88(002):4(003). Due to its weak intensity, reflexion 004 has not been observed in the diffraction patterns.

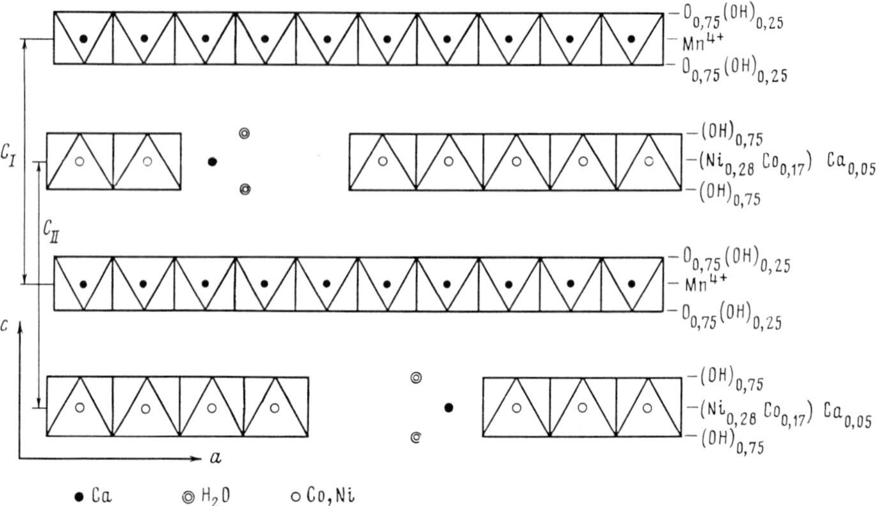

Fig. 4. The scheme of the asbolane structure in projection on plane (010). The distribution of anions and cations in the layers is given with regard to their average amount in a unit cell of each component

Thus, a study of Co-Ni asbolane from the Lipovsk deposit has shown that the mineral has a hybrid structure composed of the layers of Mn^{4+} and Co-Ni octahedra alternating along the c axis. These layers form hexagonal sublattices I and II with parameters (under vacuum) a = 2.823, c = 9.34 Å (Mn^{4+} sublattice), a = 3.04 Å and c = 9.34 Å (Co-Ni sublattice). In the layers of different types, the octahedra differ in azimuth orientation as much as 60°. The chemical formula of the Lipovsk asbolane may be written as

$$[Mn^{4+}O_{2-x}(OH)_x]^{x+}\ [R_{1-y}^{2+}\ (OH)_{2-2y+x}]^{x-}$$

where: R^{2+} is Ni, Co, Ca; x and y < 1. Both electrostatic and hydrogen bonds exist between the layers of different types: a positive charge of Mn^{4+} layers is compensated by a negative charge of Co-Ni layers; oxygen atoms of Mn^{4+} layers, together with hydroxyls of the Co-Ni layers, form hydrogen bonds O-H. . .O. Co-Ni layers are imperfect, and island-like.

The X-ray diffraction pattern of the natural asbolane characterizes only the Mn^{4+} hexagonal sublattice, with the parameters a = 2.823, c = 9.6 Å. No reflexions of the Co-Ni sublattice have been seen, probably because of their weak intensity. An increased value of parameter $c = 9.6$ Å measured on the X-ray diffraction pattern is accounted for by the molecular water present between the layers of the asbolane structure. The removal of water in the electron microscope vacuum results in contraction of the structure along the normal towards its layers, and, as a consequence, in lesser value of parameter c.

References

Chukhrov FV, Gorshkov AI, Rudnitskaya ES, Sivtsov AV (1979a) On the nature of rancieite. Izv Acad Sci USSR Ser Geol 11:71–81 (in Russian)

Chukhrov FV, Gorshkov AI, Vitovskaya IV, Drits VA, Sivtsov AV (1979b) Asbolanes – the minerals with hybrid structure. Thesis VII All-Union Ses on X-ray technique for mineral raw material. Moscow (in Russian)

Evans HT Jr, Allmann R (1968) The crystal structure and crystal chemistry of valleriite. Z Kristallogr 127(1):73–93

Ginsburg II, Rukavishnikova IA (1951) Minerals of the ancient crust of weathering in the Urals. Publ Acad Sci USSR, Moscow (in Russian)

Gorshkov AI (1970) Application of electron microdiffraction for measurement of basal reflexions from platy layer silicates. Izv Acad Sci USSR, Ser Geol 3:133–138 (in Russian)

Organova NI, Genkin AD, Dmitrik AL, Evstigneeva TL, Laputina IP (1975) On structural features and isomorphism of minerals of valleriite group. In: sb. Isomorphism in minerals. Izd Nauka, Moscow, pp 150–161 (in Russian)

Wadsley AD (1952) The structure of lithiophorite, (Al, Li)$MnO_2(OH)_2$. Acta Crystallogr 5:676–680

Genesis of Low-Grade Chromite Ore Deposits in Lateritic Soils from the Philippines

G. FRIEDRICH[1]

Abstract

Deep in-situ weathering of ultramafic rocks of different composition has caused the formation of limonitic soil in various parts of the Philippine Archipelago. The chromitiferous rocks consist of disseminations and minor segregations of chrome-spinels in dunite or harzburgite. The chrome-spinels have been preserved in the in-situ laterites because of their high resistance to weathering. As a result of the removal of Mg and Si from the ultramafics, chromite has been enriched in the lateritic soils by a factor ranging from 2 to more than 10. In some areas, soils of different ultramafic rocks are mixed due to alluvial processes which can cause further enrichments of chromite. These mixed soils often occur as transported soils above non-ultramafic rocks. The abundance of chromite and the content of magnetite, ilmenite and rutile in the limonitic laterite vary considerably according to the occurrence of the accessory opaque minerals of the different types of ultramafics and to local weathering conditions. Depending on the chemical composition of the chrome-spinels in the mafic and ultramafic rocks, chromite ore of metallurgical, chemical and refractory grade occurs in significant reserves in lateritic soils in the Philippines.

1 Introduction

Chromite-bearing lateritic soils occur in various parts of the Philippine Islands. They are found in areas of ultramafic complexes of peridotite-gabbro association within N-S-trending ophiolite belts, as part of the Circum-Pacific mobile belt. The chromitiferous rocks range from disseminations to minor segregations of chrome-spinels in dunites and peridotites. Chromitites, sometimes occurring as small isolated pods and thin layers mainly in serpentinites of dunitic and peridotitic origin, are commercially not viable at present. Dikes of gabbro, pyroxenites and anorthosite cut through the serpentinites. The ultramafic rocks overlie the basement complex consisting of metavolcanics and schists, and are in some places overlain by younger sediments, mainly conglomerates, mudstone and coralline limestone.

The mafic and ultramafic rocks are overlain by lateritic soils which contain grains of chrome-spinels ranging from 0.03 mm to 0.5 mm, and disseminated in a matrix consisting mainly of limonite of size class less than 0.02 mm. The chromite content of the limonitic soil varies from a few kg to more than 100 kg per m^3.

1 Institut für Mineralogie und Lagerstättenlehre, Rheinisch-Westfälische Technische Hochschule Aachen, 5100 Aachen, FRG

This type of "secondary, lateritic chromite deposit" in areas where no minable primary chromite ore bodies occur has not been studied in detail before in the Philippines or in other countries because it was considered to be uneconomic (Bergmann 1941). Detailed exploration was begun in 1977 by different mining companies in various parts of the Philippines, and so far significant geological reserves of lateritic chromite have been found as dispersion of fine-grained chrome-spinels of different chemical composition in the lateritic soils. These projects, in addition to the objective of finding viable ore deposits, provide the possibility of studying the chemistry of chrome-spinels from the underlying highly weathered ultramafic rocks in comparison with those from the soil cover.

A geochemical test sampling programme was carried out in two areas. The topography in the first area of investigation is generally flat, rising gradually towards the foothills of an adjacent ultramafic mountain range. This area has a thick alluvial lateritic soil cover. The terrain of the second area, located on an island several hundred kilometres away, is gentle to moderate but includes gorges and deep valleys in part of the area. Fresh rock is exposed in small streams only. In recent test pits the bedrock is generally deeply weathered and covered by soil that is in places cultivated and elsewhere supports the local forest growth.

One of the areas is outside, the other one inside the typhoon belt of the Philippines, causing significant differences in the average annual rainfall. Heavy rainfalls could occasionally hamper working progress in both areas, but generally exploration work is possible throughout the year.

2 Composition and Genesis of Lateritic Soils

The profiles of limonitic lateritic soils above ultramafics may range from less than 1 m to more than 8 m.

The mineralogical composition of soil samples from different areas is highly variable but in general limonite is the main constituent, especially in the size classes > 0.5 mm and less than 20 microns, where iron occurs as concretionary larger grains or as very small, amorphous particles of iron hydroxides. The chemical analysis of two samples is given in Table 1.

In addition, chromite, gibbsite, quartz, magnetite, ilmenite, anatase, rutile, and organic material (remains of roots) occur as subordinate constituents in the limonitic soil. The frequency distribution of different size classes and the abundance of chromite in each size class are shown in Fig. 1.

Table 1. Chemical composition of soil samples Fr-21 and Fr-22 from two different zones of one exploration area

	Fe_2O_3 %	Cr_2O_3 %	Al_2O_3 %	TiO_2 %	SiO_2 %	MgO %	CaO %	NiO %
Fr-21	42.1	14.8	4.5	1.1	8.7	0.8	0.03	0.14
Fr-22	66.6	6.2	4.2	0.4	0.5	0.7	0.06	0.47

The soil profile has developed by deep weathering of mafic and ultramafic rocks with sufficient removal of aluminium, magnesium and silica to form a thick zone of limonitic material, consisting of amorphous hydroxides and hydrated oxides, mainly goethite. The mobilization and removal of most of the aluminium under pH conditions of 7 to 8 and the redistribution of iron took place due to the disintegration of silicates.

Some of the iron migrated upwards from the weathered bedrock, using the groundwater as transport medium, and has been concentrated in concretionary particles. A fluctuating water table in the lower part of the profile is a very important factor in the genesis of the lateritic soils.

Olivine, serpentine and pyroxene rapidly disintegrate during the process of lateritization and are replaced by various minerals formed by neomineralization. According to Schellmann (1964), the weathering process is mainly restricted to a thin zone of about 25 mm between bedrock and laterite where intensive decrease of SiO_2 and MgO occurs. The higher mobilization of Al_2O_3 compared to Fe_2O_3 in mafic rocks may be due to the higher solubility of aluminate at about pH 8. The zone of enrichment of iron moves downwards during progressive lateritization, forming a thick weathering zone with high iron content. By solution and precipitation during the weathering process, some of the trace elements, such as nickel and cobalt, have been enriched with the main element iron in sol particles of ferric hydroxides.

The chrome-spinels that were originally dispersed throughout the ultramafic rocks and their serpentinized derivates are residually concentrated in the lateritic cover.

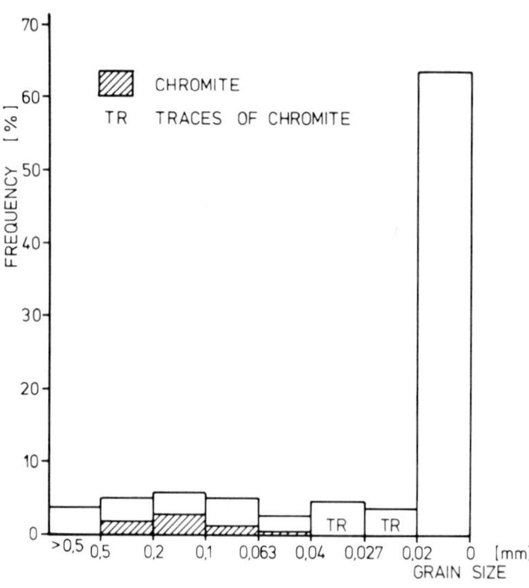

Fig. 1. Frequency distribution of different size classes and chromite content in each size class of the lateritic soil sample Fr-22

3 Chrome-Spinels in Lateritic Soils

One fact of great significance for the evaluation of chromite-bearing lateritic soils in the Philippines is that the chrome-spinels have generally not been destroyed or intensively altered during the lateritic weathering process. In contrast, Seeliger (1961) has noted intensive mobilization of Cr and replacement of chromite by magnetite, maghemite and hematite in lateritic iron ore of Conakry (Guinea) which has been formed during lateritization on top of serpentinite and dunite, respectively. Only small amounts of Cr (less than 1% Cr_2O_3) have been found to occur in the limonitic fraction of size class less than 10 microns, due to the stability of the chrome-spinels in the Philippine soils. It is not known at present whether this chrome content is totally bound to Fe_2O_3-hydrosols or occurs as very fine-grained chromite grains.

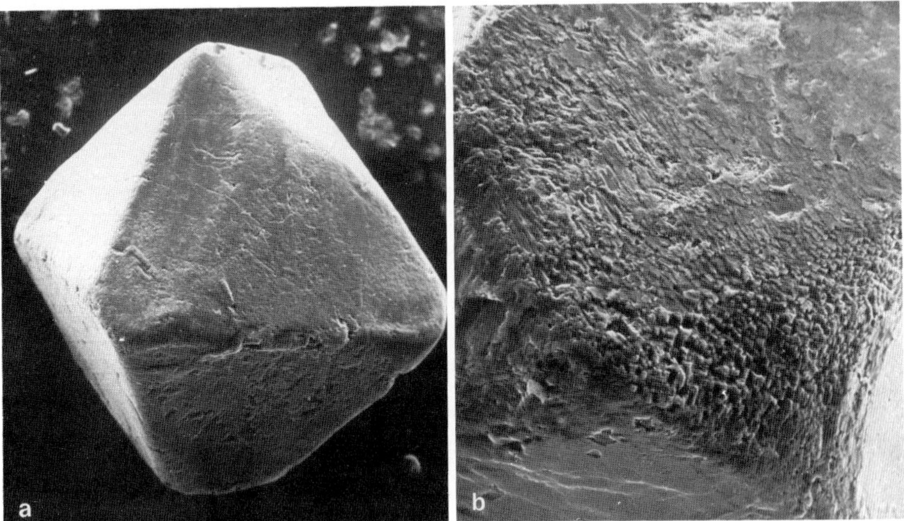

Fig. 2. SEM microphotography of chrome-spinels showing euhedral shape and cubic surface features. Separated from heavy mineral concentrates of lateritic soil. Length of octahedron side 0.5 mm

In the areas of investigation chromite occurring as numerous small grains (mean size 0.15 mm) is predominantly euhedral to subhedral (Fig. 2).

In a recent paper (Friedrich et al. 1980), we have demonstrated the regional variability in chemical composition of chromite concentrates from lateritic soils and a highly variable chromite content in samples taken at close spacing and in different regions.

For studying the local variation of heavy mineral content and the chemical composition of chromite concentrates in the lateritic soils, a special sampling pattern was used. Six auger drill holes were drilled on circles of different radii (5, 10 and 20 m) with different distances between the drill holes (5, 10 and 20 m, respectively). In Tables 2a and 2b, data are given for one location (Fr-20), as an example, where trans-

Table 2a. Local variation of heavy mineral content (kg/m³), percentage of nonmagnetic fraction (= chromite concentrate) and chemical composition of the chromite concentrates separated from lateritic soils of two concentric circles of location Fr-20. Mean depth of each drill hole 4 m, sampling interval 1 m (see Table 2b)

		Average of 6 drill holes from circle I (r = 5 m)	Average of 6 drill holes from circle III (r = 20 m)
Heavy mineral content kg/m³	$\bar{x}$	177.7	216.8
	s	80.7	83.2
Nonmagnetic fraction (%)	$\bar{x}$	98.76	98.78
	s	0.18	0.16
Cr_2O_3 (%)	$\bar{x}$	38.8	38.9
	s	0.91	1.43
FeO (%)	$\bar{x}$	24.52	24.52
	s	0.54	0.82
Cr/Fe	$\bar{x}$	1.40	1.39
	s	0.04	0.04
TiO_2 (%)	$\bar{x}$	6.56	6.36
	s	0.92	1.38

Table 2b. Variation of heavy mineral content in different depth of lateritic soil profiles of location Fr-20 (see Table 2a)

Depth		Heavy mineral content (kg/m³) (circle I)	Heavy mineral content (kg/m³) (circle III)
0–1 m	$\bar{x}$	227.4	281.7
	s	68.1	61.0
1–2 m	$\bar{x}$	185.4	199.9
	s	71.7	29.3
2–3 m	$\bar{x}$	125.1	182.7
	s	56.6	82.4
3–4 m	$\bar{x}$	172.0	203.3
	s	107.6	121.5

$\bar{x}$ = average; s = standard deviation

ported lateritic soils are overlying carbonate rocks. The Cr_2O_3 and FeO values of the total concentrates have very small standard deviations; the heavy mineral content, however, is characterized by a high standard deviation in the same sampling location, especially at different depths (Table 2b). This may be caused by variable alluvial redistribution of the chrome-spinels in the lateritic soil cover.

In other areas with residual laterites overyling serpentinites, the chromite content is also highly variable. This may be due to the variation of chrome-spinel distribution in the bedrock and variable weathering conditions. No distinction can therefore be made between transported lateritic soils and residual lateritic soils by studying the chromite abundances and the bulk compositions of the concentrates only. In the following chapter, first results of microprobe investigations of individual chrome-spinels in concentrates will be discussed. These studies are carried out to distinguish between both types of laterites.

4 Electron Microprobe Analysis of Chrome-Spinels in Lateritic Soils and Rocks

Ore microscopic studies have shown that chromite concentrates from individual sample sites often contain groups of chrome-spinels characterized by different colours and reflectivity values. Microprobe analyses were carried out on selected grains of each concentrate. The range in chemical composition of chrome-spinels from one non-magnetic mineral fraction is given in Table 3: Cr_2O_3 from 45.3% to 54.3%, FeO from 26.6% to 34.6%, Al_2O_3 from 7.3% to 17.5%, and MgO from 6.0% to 10.7%. Grains 1 to 8 are listed according to increasing Cr_2O_3 content. Groups of chrome-spinels with similar chemical composition can also be combined according to decreasing iron contents and increasing Cr/Fe ratios. Grain No. 3, with the highest Cr/Fe ratio, is characterized by low Fe-content and high Al and Mg contents. The average Cr/Fe ratio of about 1.47 is comparable to the Cr/Fe values of 1.40 calculated for the total ore block area, 400 m × 600 m in size. The ore quality of the chromite concentrates is often influenced by small grains of ilmenite and magnetite which could not be magnetically separated. In general, more than 95% of the total heavy mineral concentrate consists of a nonmagnetic chromite fraction.

Table 3. Electron microprobe analysis data of eight characteristic chrome-spinels, listed according to increasing Cr_2O_3 content. Nonmagnetic fraction of a heavy mineral concentrate sample separated from lateritic soil

Grain No.	Cr_2O_3 %	FeO %	Cr/Fe	Al_2O_3 %	MgO %	MnO %	TiO_2 %
1	45.32	34.60	1.15	12.83	7.00	0.42	0.18
2	47.00	32.86	1.26	12.21	6.71	0.42	0.36
3	47.38	22.89	1.82	17.48	10.68	0.23	0.15
4	47.53	29.32	1.43	14.53	6.79	0.41	0.43
5	48.42	29.78	1.43	13.04	5.73	0.42	0.18
6	52.13	28.60	1.60	9.94	8.67	0.40	0.16
7	52.27	32.92	1.40	7.26	6.02	0.47	0.10
8	54.29	26.58	1.80	9.50	8.64	0.46	0.30
$\bar{x}$	49.29	29.69	1.49	12.09	7.53	0.40	0.23
s	3.17	3.81	0.24	3.19	1.67	0.074	0.12

Table 4. Microprobe analyses of 10 chromite grains from serpentinite R-139 and from harzburgite R-52

Grain No.	Cr_2O_3 %	FeO %	Al_2O_3 %	MgO %	MnO %	TiO_2 %	SiO_2 %
(R-139)							
1	53.72	24.98	9.41	8.61	0.46	0.18	0.08
2	54.53	24.66	9.71	9.14	0.43	0.22	–
3	54.61	25.11	9.73	8.67	0.39	0.18	–
4	54.72	25.54	9.66	8.70	0.44	0.25	0.07
5	54.83	25.83	9.78	8.80	0.46	0.22	–
6	54.86	23.99	9.84	9.37	0.41	0.21	–
7	54.95	24.49	9.62	9.01	0.41	0.21	–
8	55.01	25.35	9.04	9.07	0.43	0.20	–
9	55.21	23.67	10.06	9.52	0.42	0.21	–
10	55.25	23.29	9.91	9.79	0.40	0.17	–
$\bar{x}$	54.77	24.69	9.68	9.07	0.43	0.21	
s	0.44	0.83	0.28	0.39	0.024	0.024	
(R-52)							
1	44.12	35.11	12.56	5.98	0.47	0.42	–
2	44.16	34.58	12.34	6.25	0.48	0.47	–
3	45.01	32.31	12.95	6.93	0.44	0.35	–
4	45.17	34.64	12.15	6.03	0.47	0.39	–
5	45.32	33.61	12.63	6.50	0.46	0.41	–
6	45.63	34.12	12.11	6.18	0.48	0.41	–
7	45.65	33.89	11.64	6.25	0.48	0.36	–
8	45.92	33.78	11.94	6.35	0.47	0.38	–
9	45.93	33.25	11.77	6.58	0.44	0.38	–
10	46.08	33.43	11.92	6.43	0.43	0.38	–
$\bar{x}$	45.30	33.87	12.20	6.35	0.46	0.40	
s	0.70	0.80	0.42	0.28	0.019	0.034	

Based on ore microscopic studies, chromite grains of different mafic and ultramafic rocks were studied by the microprobe. It was found that only a very limited variation in chemical composition exists for the ten selected chrome-spinels within each polished section. In Table 4 the microprobe data are given for two rock types, one serpentinite with chrome-spinels of high Cr/Fe ratio, and one harzburgite with chrome-spinels of low Cr/Fe ratio. The Cr_2O_3, FeO, Al_2O_3, and MgO values of both rock types show very small standard deviations.

On the other hand, a high variation in chemistry of chromites from different mafic and ultramafic rocks from all over the area was noted (Table 5). Chrome-spinels from peridotitic rocks have generally lower Cr contents and Cr/Fe ratios than dunitic rocks and serpentinites. The chemical composition of chrome-spinels is shown graphically for average values of ten grains together with the Cr/Fe ratios (Fig. 3) according to Ghisler (1976).

The series of seven rock samples are listed according to decreasing Fe content and increasing Cr content with the exception of rock types 18 and 10 which show deviating Fe and Cr values, respectively.

Table 5. Chemical composition of chrome-spinels (each single value represents the average of microprobe analyses carried out on 10 grains in each rock type) of different types of ultramafic rocks. Philippines

Rock	Cr_2O_3 %	Cr %	FeO %	Fe %	Cr/Fe %	Al_2O_3 %	MgO %	MnO %	TiO_2 %	SiO_2 %
Harzburgite (52)	45.3	31.0	33.9	26.3	1.18	12.2	6.4	0.46	0.40	–
Peridotite (13)	48.7	33.3	34.3	26.7	1.25	9.3	5.8	0.47	0.26	–
Peridotite (18)	50.7	34.7	28.6	22.3	1.56	11.7	8.3	0.45	0.28	0.11
Peridotite (77)	52.9	36.2	29.1	22.6	1.60	9.0	8.0	0.5	0.24	–
Dunite (90)	53.9	36.9	28.0	21.7	1.70	9.5	8.1	0.45	0.28	0.04
Serpentinite (139)	54.8	37.5	24.7	19.2	1.95	9.7	9.1	0.43	0.21	0.01
Serpentinite (10)	53.2	36.4	21.9	17.0	2.14	11.0	10.6	0.35	0.23	0.02
Average	51.4	35.1	28.6	22.3	1.57	10.3	8.0	0.42	0.27	

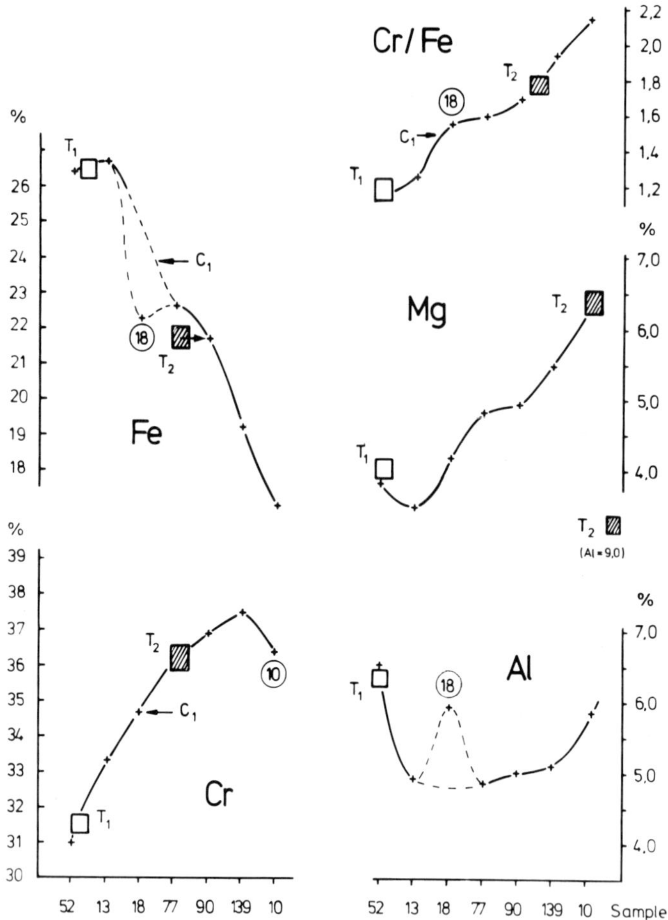

Fig. 3. Variation of main elements in chrome-spinels from different ultramafic rocks (see Table 5) and chromite concentrates (T_1, T_2, C_1) from lateritic soils

In addition, the positions of the two chrome-spinel groups T_1 and T_2 from one chromite concentrate sample are plotted in the curves (Fig. 3). These two groups represent a mixture of chrome-spinels from two different rock types, similar to harzburgite and peridotite. The unusually high Al content of about 9% in group T_2 is not represented by any rock type. The position of the composite concentrate sample C_1 in the curves demonstrates the chemical composition of a mixed product of chrome-spinels in a lateritic soil sample overlying different types of ultramafic rocks.

5 Conclusions

During the last few years, the importance of low-grade and submarginal mineral deposits has continuously increased (Carpenter 1976). It is believed that the chromite-bearing lateritic soils are a typical example of these lower-grade but large-scale resources, occurring in the Philippines and in other countries of the world with ophiolite belts and lateritic weathering conditions.

Depending on the chemical composition of the chrome-spinels in the underlying mafic and ultramafic rocks (varying from harzburgites to dunites) chromite concentrates of metallurgical, chemical or refractory grade can be produced. In addition, nickel and cobalt may be produced as byproducts from the limonitic material of the lateritic soil and limonite may be used as a pigment.

The following genetic interpretation is suggested for the formation of secondary enrichments of chromites of size class 0.03 to 0.5 mm in lateritic soils: deep in-situ weathering of ultramafic rocks of different composition has caused the formation of limonitic soil. The chrome-spinels have been preserved in the in-situ laterites because of their high stability. The chromite grade (in kg/m^3) of the lateritic soils and their content of magnetite, ilmenite and rutile vary highly according to the occurrence of the accessory opaque minerals in the different ultramafics and to local weathering conditions. Each type of mafic and ultramafic rock is characterized by the chemical composition of their different chrome-spinels. As a result of mixing during the formation of the lateritic soil, the chemical composition of the different chromite concentrates separated by panning from the soil varies considerably depending on the types of bedrock occurring in each area.

A more detailed survey of selected ultramafic complexes and studies of the mineralogy and chemistry of the ultramafic rock types and overlying soils will be carried out in the next few years. The research project will involve systematic sampling of mafic and ultramafic rocks in various islands of the Philippine Archipelago following similar studies carried out by Ghisler (1976), Greenbaum (1977) and Malpas and Strong (1975) in other countries. A principal aim will be to establish the economic grade of the lateritic chromite deposits. This requires a study of chromite abundance in areas with lateritic soil, chemical and mineralogical composition of chrome-spinels and matrix, as well as soil thickness and topography.

Acknowledgments. The author would like to express his gratitude to the geologists of the Philippine mining companies for their help in this study and especially to Dr. P. Wallner who provided rock samples for this investigation. The study is part of a project between mining companies and the Institute of Mineralogy and Ore Deposits, RWTH Aachen. The mineralogical and chemical analyses were carried out in the field laboratories and in the laboratories in Aachen. Electron microprobe analyses were partly done in the Institute of Mineralogy of the University of Leoben, Austria. I wish to thank Prof. Dr. E.F. Stumpfl and my co-workers for their help. The research project is continued with the financial support provided by the German Ministry of Research and Technology (BMFT).

References

Bergmann RG (1941) Report on the property of the Zambales Chromite Mining Company, Santa Cruz, Zambales, Luzon, Philippines. Unpubl report

Carpenter RH (1976) A review of low-grade ore potential. In: Recent advances in mining and processing of low-grade and submarginal mineral deposits. Papers of a conference held in New York 1972 (UN Centre Nat Res Eng Transp). Pergamon Press, Oxford, pp 12–23

Friedrich G, Brunemann HG, Thijssen T, Wallner P (1980) Chromit in lateritischen Böden als potentielle Lagerstättenvorräte in den Philippinen. Schr GDMB, Heft 35:65–94

Ghisler M (1976) The geology, mineralogy and geochemistry of the pre-orogenic archaean stratiform chromite deposit at Fiskenaesset, West Greenland. Monogr Ser Mineral Deposits 14:156

Greenbaum D (1977) The chromitiferous rocks of the Troodos ophiolite complex, Cyprus. Econ Geol 72:1175–1194

Malpas J, Strong DF (1975) A comparison of chrome-spinels in ophiolites and mantle diapirs of Newfoundland. Geochim Cosmochim Acta 39:1045–1060

Schellmann W (1964) Zur lateritischen Verwitterung von Serpentinit. Geol Jahrb 81:645–678

Seeliger E (1961) Paragenetische Untersuchungen an C-Erzen (Weicherzen) von Conakry, Guinea. Fortschr Mineral 39:139–141

The Oxidized Zone of the Broken Hill Lode, N.S.W.

J.C. VAN MOORT[1] and C.G. SWENSSON[2]

Abstract

A systematic description is given of the remaining outcrops of the Broken Hill lode together with geochemical data. The oxidized lode is characterized by a coronadite-rich gossan in the upper part followed by a complex heterogeneous cerussitic mineralization at depth. The cerussitic zone is in direct contact with the primary sulphide.

Geochemically, the gossans are characterized by an extreme enrichment in manganese and lead. Significant features of the oxide gossan at the surface are a Pb/Fe ratio of about 1 or more and a Pb/Mn ratio of 1–2 or greater, even when free of lead carbonate.

The high Pb/Mn ratio in the gossan is caused by the presence of a plumboan variety of coronadite containing 35%–58% Pb.

1 Introduction

The aim of this study is to document the uppermost oxide zone of the Broken Hill lode, of which little surface expression remains.

The deposit occurs in high grade metamorphic early Proterozoic rocks (mainly gneisses) described in detail by Andrews (1922) and more recently by Thomson (1975). A recognizable stratigraphic succession known as the "Mine Sequence" can be traced for 24 km.

The Broken Hill ore deposit has an arcuate longitudinal section of which the ends dip 40°–50° to the north-east and 15°–20° to the south-west respectively. Six separate bodies are recognized in the following stratigraphic sequence: the zinc lodes B, A, upper and lower No. 1 lens and the lead lodes No. 2 and No. 3 lens, forming together a large and steep drag fold. The sulphides in all ore lenses consist of sphalerite and galena (with less galena in the zinc lodes) and subordinate amounts of pyrrhotite, chalcopyrite, arsenopyrite and tetrahedrite. Pyrite is an exceedingly minor component of the ore. The gangue mineralogy consists essentially of quartz, the Ca-Mn-Fe silicates rhodonite, bustamite, manganoan hedenbergite, spessartine and, in the zinc lodes, also calcite. Each ore body has a distinct mineralogy. For details see Gustafson et al. (1950) and Johnson and Klingner (1975). The ore bodies are contained within a distinctive unit of the Mine Sequence known as the lode horizon, 50–200 m wide. It

1 Department of Geology, University of Tasmania, Hobart, Australia 7001;
Humboldt Visiting Professor 1979, Heidelberg, FRG

2 C.R.A. Exploration, Melbourne, Australia

comprises the ore body and intervening layers of sillimanite gneiss and quartzite. Garnet quartzite, blue quartz-gahnite lode, garnet sandstone and lode pegmatite are unique to the lode.

The outcrops described in this article are almost entirely within partly retrograde metamorphic sillimanite gneisses and are expressions of the No. 3 and 2 lead lenses, except at the Zinc Corporation in the south-western part of the mining area, where a small outcrop of the zinc lodes was studied.

2 Previous Studies

Reference to the general nature of the oxidized zone is scarce and restricted to publications of Jaquet (1894), Smith (1926), Burrell (1942) and King and O'Driscoll (1953).

The richness and variety of the minerals in the gossan at Broken Hill have been described by Smith (1926), Lawrence (1961, 1968) and Mason (1976) recording eight residual oxides in addition to a total of 52 halides, carbonates, sulphates, tungstates, molybdates, phosphates, arsenates, vanadates and silicates in the oxidized zone, as well as native silver, copper, sulphur and a further six supergene sulphides. Broken Hill is the type locality of some silver halides and has yielded abundant mineable hornsilver, mostly embolite. Most of the minerals mentioned, however, are rare, were collected below the surface and can at present only be seen in collections. The most famous collections are located in Adelaide (University), and Sydney (Australian Museum, Mining Museum, Hugh Dixon collection at the University of Sydney and the private Albert Chapman collection).

Several specific mineralogical papers were published by such authors as Lawrence (1961, 1968) and Barclay and Jones (1971a, b).

3 Description of the Oxidized Zone

3.1 General

Jaquet (1894) described the oxidized zone to consist of quartz, kaolin and manganese oxides, followed at depth by a zone of lead carbonate and kaolin, rich in silver halides and native silver. These oxidized ores were supposedly separated from the primary ore by a thin zone of secondary sulphides. The outcrop was 2.5 km long, 6–23 m wide and up to 15 m higher than the surroundings. The depth of oxidation in general did not exceed 120 m, but small channels of oxidized ore down to 550–610 m in the North Mine and 120–180 m in the BHP area occurred.

The outermost layer of the original outcrop of the main lode horizon has been removed as a result of surface operation in installation. The original state is probably best displayed by pockets of red-black soil on the cliffs of Block 10. This veneer of plumbo-manganiferous limonite would have been discontinuous and never thicker than 1 m. Lawrence (1968) called this uppermost terrigenous zone gossan, in contrast to the underlying oxidized zone. The term gossan in this paper, however, will be used in the broader original meaning of weathered or otherwise decomposed upper part of a lode (Challinor 1973).

The greater part of the outcrop of the main lode is an uncommon gossan: a dense black rock consisting of intergrown aggregates of botryoidal lead manganese oxides and altered rock fragments, with some quartz and goethite, more siliceous towards the top, full of cracks and vughs filled with oxide botryoids and stalactites and towards the base increasingly exhibiting cracks and scattered masses of cerussite and pyromorphite. Lawrence (1968) reports as manganese minerals pyrolusite and psilomelane, some coronadite and rare chalcophanite, in addition to the iron oxides goethite, limonite and lepidocrocite. As shown below most of the manganese minerals prove to be variants of coronadite.

Below this level the gossan gradually becomes less dense, containing less coronadite, which eventually occurs only in the form of veins and patches. Silver halides may also occur. Remnants of the ore are here represented by cerussite and some kaolin, whilst cerussite fillings occur in cracks. At depth, patches of secondary sulphides mark the transition to the primary ore. In most cases the oxidized cerussitic ore is in direct contact with the primary sulphides.

A selection of the few remaining outcrops of the original gossan are described below (see Fig. 1). The grid co-ordinates given are those of the Central Geological Survey, Broken Hill. In addition reference is made to the blue quartz-gahnite lode horizon outcrop at Silver Peak, 7 km south-east, and the shallow gossan at Wolseley, 15 km northwest of Broken Hill.

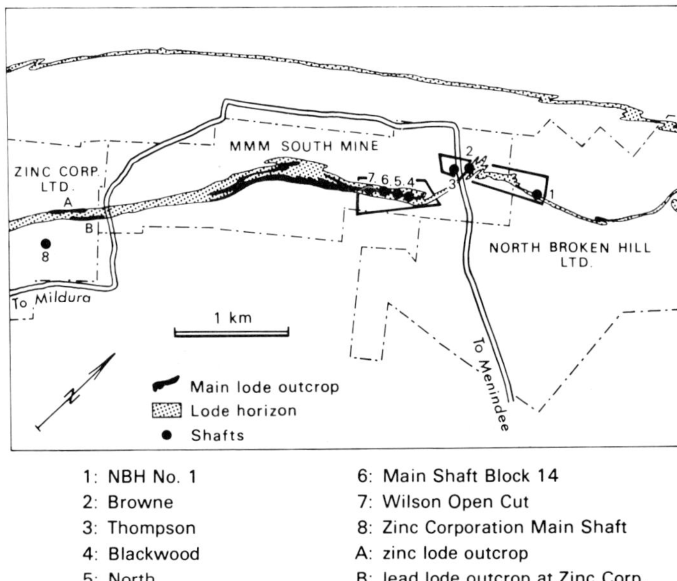

Fig. 1. Locality map of the Broken Hill mine area, with shafts referred to

3.2 Thompson Shaft to North Broken Hill No. 1 Shaft Area

The lode begins to plunge steeply to the north-east just south-west of Thompson Shaft and at the NBH No. 2 Shaft the No. 3 lead ore lens is 450 m under the surface. The rocks at Thompson Shaft (grid ref. 224/2520) represent the extreme top of the No. 2 or No. 3 lens and contained probably originally minor disseminated sulphides and small patches of richer mineralization. This outcrop is the most oxidized and consists of a heterogeneous assemblage of quartz grains and fragments, coronadite, goethite and other Fe-Mn oxides of an amorphous nature. Garnets are totally oxidized, minor boxworks after unidentifiable sulphides occur. Fractures and joints are filled or lined with 5–10 mm layers of botryoidal coronadite.

The garnet quartzites and sillimanite gneisses at the Menindee Road cutting (264/2652), Browne Shaft (264/2680) and NBH No. 1 Shaft (208/2960) exhibit sharp identifiable contacts, whilst the degree of oxidation, occurrence of boxworks, alteration of the garnets and the degree of coating with coronadite are decreasing with increasing depth of the lode. At NBH No. 1 some cryptomelane (electron probe) and ramsdellite (electron probe and XRD) were observed.

3.3 Blackwood Open Pit, Block 14 and Block 15

This area provided in 1977 very good exposures of the oxidized zone, whilst the Main Shaft of Block 14 provided some subsurface access. Figure 2 indicates the approximate levels of the exposures discussed here relative to the original surface of the Wilson Shaft Open Cut discussed below.

1. **Blackwood Crusher Area.** The outcrop of the oxidized zone at the Blackwood Crusher (124/1756) is only 2 m wide on the surface. It consists of an upper zone of coronadite rock, which thins at about 7 m below the surface to a relatively coronadite-poor lode quartzite. The oxidized rocks contact weathered sillimanite gneisses and relatively unleached garnet quartzite. This arrangement can be traced as far as Wilson Shaft Open Cut. The predominant surface outcrop is a dense black rock that consists of highly fractured ragged aggregates of quartz grains (40%), rare spessartine cemented by massive colloform microbanded coronadite (40%) and voids (20%), mostly empty or occasionally filled with amorphous silica. Colloform goethite is minor. Some metres below, the rock is more heterogeneous, with areas of brown jasper and large areas of powdery brown-orange limonite.

2. **Pyromorphite Corner.** Approximately 95 m ENE of the Main Shaft of Block 14 an exposure was mapped at the base of the coronadite zone described above (165/1930). The zone is an exceedingly complex collapsed and oxidized zone with sillimanite gneiss wall rock (see Fig. 3). The base of the coronadite tends to mark the appearance of powdery cerussite. The cerussite increases with depth and becomes associated with a brown oxide forming large boxworks (goethite, jarosite and possibly lepidocrocite). The massive cerussite appears to be a direct replacement product. Etched vitreous surfaces indicate later dissolution. Amorphous goethite and coronadite (as discussed below) coat the cerussite and small pyromorphite crystals have developed on the oxide film.

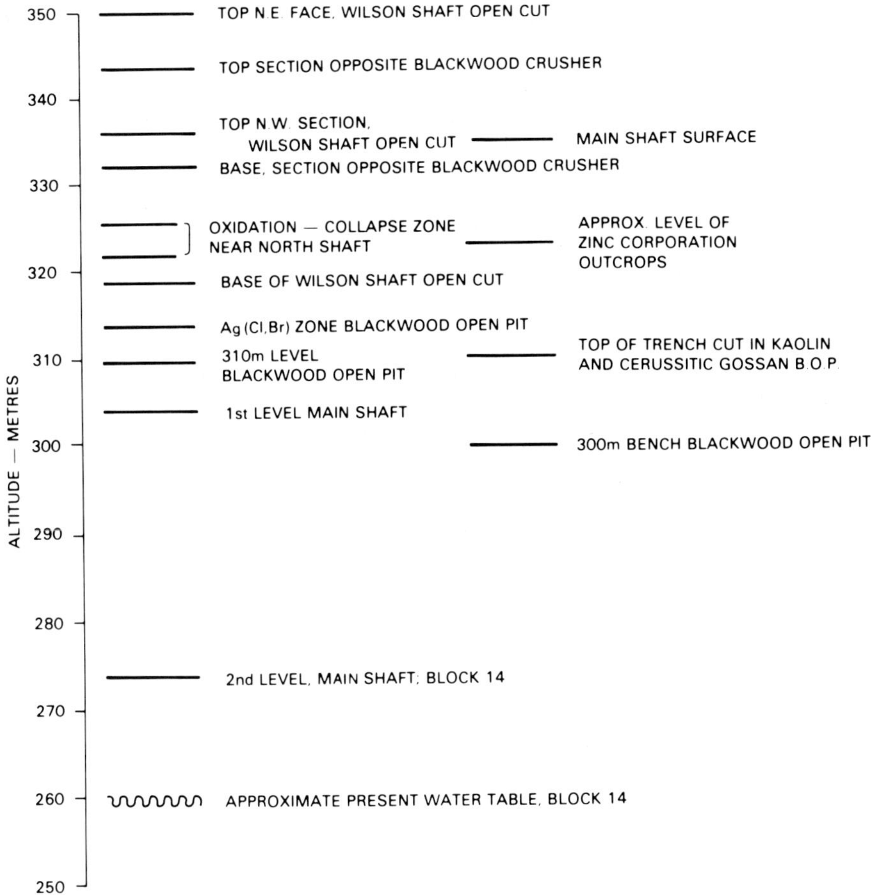

Fig. 2. Approximate levels of various sections and locations at Block 14

In addition to these small crystals considerable pyromorphite development occurs on the surfaces of the quartzite and along fractures in the sillimanite gneiss up to 30 m away and is still evident on the 2nd level of the Main Shaft 50 m beneath the outcrop.

The cinder-textured gossan in the north-western part of the outcrop consists of a more massive cerussite with oxides, unfilled boxworks and remnant aggregates of quartz. This cerussite is also etched, and quite different from the coarsely crystalline cerussite in the fractures of the heavily oxidized garnet quartzite some metres away (see Fig. 3).

The important conclusion which can be drawn from this outcrop is that partial dissolution of the massive cerussite has occurred, suggesting progressive leaching. It is suggested that oxidation collapse was associated with early stages of oxidation. This would account for the isolated lumps of massive cerussite in the zone of powdery oxides, since the first formed massive cerussite might be expected to break up with the substantial ground movement. The oxidation collapse provided channels in the fractured quartzite and gneiss for the passage of further oxidizing solutions.

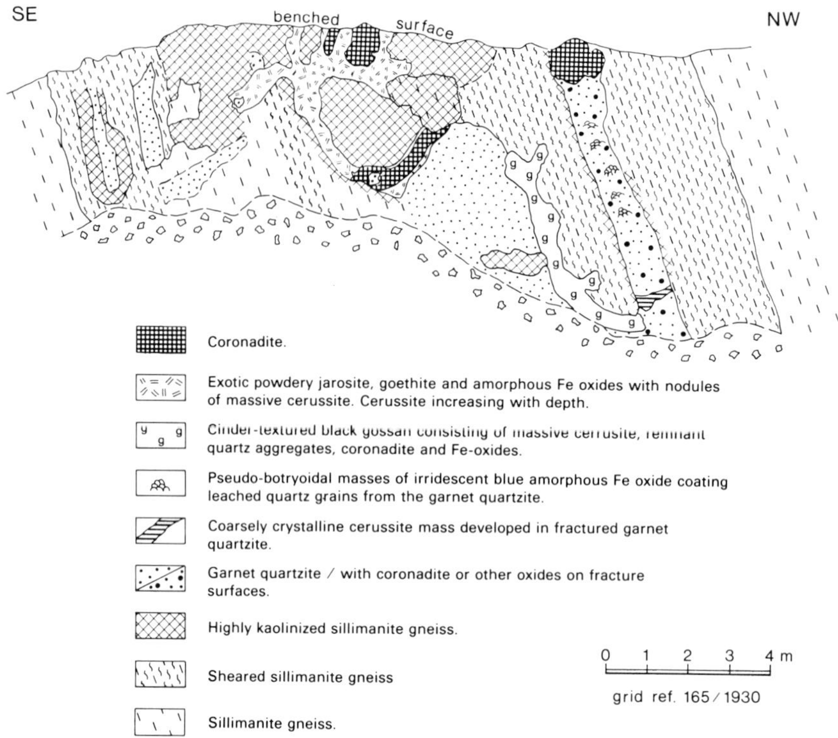

Fig. 3. Section of a collapsed oxidized zone, at Pyromorphite Corner, Blackwoods Open Pit, near North Shaft, Block 14

3. Blackwood's Open Pit. Below the level of the outcrop discussed above, the oxidized zone becomes more cerussite-rich, the sillimanite gneiss tends to be kaolinized, and oxidation collapses are more pronounced. On the 310 m and 300 m bench levels of the Blackwood Open Pit, situation 1977, the oxidized rocks were no longer identifiable as a single unit as they are in the outcrops discussed above. Rather the oxidized lode consisted of an apparently discontinuous, thin and widely scattered cerussitic gossan. Oxidized zones were rarely more than a metre wide and their discontinuous, scattered nature was probably a function of the structure. While the outcrops described so far are almost certainly the expression of the No. 3 lens, some of the oxidized rocks in the Blackwood Open Pit undoubtedly represent the No. 2 lens.

Three different types of oxidized lode could be distinguished: black to brown manganiferous cerussite gossan with a cindery texture and some chlor-bromargyrite; oxide coatings and cerussite developed in the fracture planes of the quartzite and finally cerussite associated with chrysocolla, cuprite and chlor-bromargyrite. The first and second type were developed as replacement of the sulphide ore and as fracture fillings, respectively, of the types described above. The third and least common type was the masses of white cerussite that occur within the quartzite. In places remnant galena grains were preserved. Thin sections showed cerussite replacing ragged galena; galena residuals were often rimmed by covellite. Boxworks and voids were often filled with amorphous silica and rarely with chlor-bromargyrite. Chrysocolla and cuprite occurred in association with these cerussite masses in scattered small areas.

4. Subsurface of the Blackwood Open Pit. On the first and second levels of the Main Shaft of Block 14 the oxidized lode tends to become confined to narrow channels rarely more than 1 m wide. The oxidized material typically consists of small cerussite crystals loosely held by yellow-brown Fe-oxides or vughy leached garnet quartzite. Massive sulphide outcrops are rare above the second mine level. A most notable feature is the absence of a secondary sulphide zone. Anglesite, covellite, chalcocite and tenorite occur on the oxidizing sulphides, but only as spasmodic surface encrustations. The near-absence of secondary sulphides is caused by the original paucity in the lode of pyrite. No significant kaolinisation was observed underground.

3.4 Wilson Shaft Open Cut

The north-east side of the Wilson Shaft Open Cut (188/1650 to 200/1628), which is a preservation area, provides the most continuous vertical exposure of the coronadite-rich rocks consisting of quartz grains and highly fractured garnets in a matrix of colloform coronadite and lepidocrocite (see Fig. 4).

The very top of the section is narrow, no more than 4–5 m wide and consists of highly leached quartzite and siliceous coronadite (60%–80% silica). This leached product consists of quartz grains and highly fractured garnets in a matrix of colloform coronadite and lepidocrocite.

The central portion of the coronadite zone consists of an aggregate of xenomorphic quartz, euhedral and fractured garnets in blue colloform coronadite. The total silica content of this coronadite averages 40%–50%. Large areas consist of almost pure coronadite in horizontal bands or in stalactitic forms. Thin section examination of this coronadite shows it to consist of ragged, turbid quartz grains and many euhedral garnets in a groundmass of coronadite. The coronadite contains many voids and vughs up to several centimetres across. These are unfilled. However, the larger voids and vughs exhibit irridescent botryoidal coronadite and at a depth of about 9 m may contain many blebs of green chlor-bromargyrite and kaolin. Few boxworks are evident. Considering that the garnets in the Thompson Shaft to NBH No. 1 Shaft area tend to break down as discussed earlier, the preservation of the garnets here warrants some explanation, which appears to be due to an early formed coating of coronadite acting as a protective shield.

The base of the coronadite is marked by a transitional but well-defined passage into highly kaolinized sillimanite gneiss, partly consisting of a heterogeneous assemblage of coronadite, jasper and blocks and masses of kaolinized sillimanite gneiss. Near the base of the section a narrow zone of jasper-rich coronadite exists. Both the mine hanging and mine footwall are bordered by quartzite of the kind described as occurring at the Blackwood Crusher, the hanging wall being less leached and altered than the footwall.

Approximately 25 m below the surface on the mine hanging wall there occurs a cavity fill deposit. The deposit consists of near horizontal layers of alternating coarse and fine breccias. The coarser breccia consists of an heterogeneous assemblage of quartz grains, garnets, fragments of quartzite and angular pieces of indurated kaolinized sillimanite gneiss (up to 3 cm long), loosely cemented by yellow amorphous Fe-oxides and jasper. The finer breccias have a similar composition, except for the grain size being in sand to silt range. Colloform coronadite layers mainly subparallel to the bedding permeate the deposit. Chlor-bromargyrite occurs in small vughs of the coronadite. The cavity filling suggests that initial oxidation subsidence occurred early in the oxidation of the lode while coronadite was still being formed.

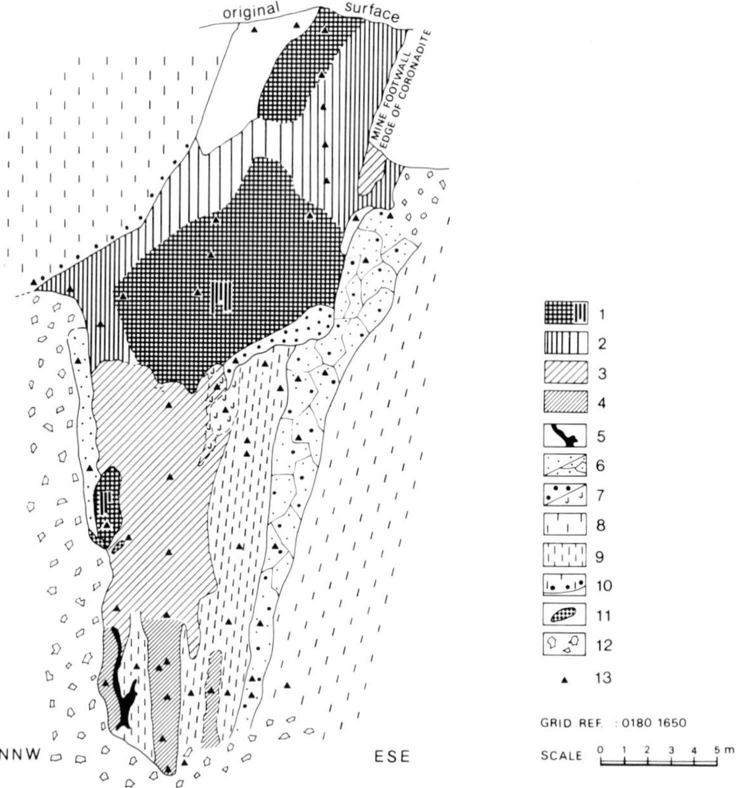

Fig. 4. Section through the coronadite zone, NE face of Wilson Shaft Open Cut, Mineral Lease 14. All coronadite is plumboan, cf. Table 5.
The numbers indicate: *1* aggregate of xenomorphic quartz, euhedral and fractured garnets in blue colloform coronadite; striped parts are stalactitic; typical composition 40%–50% SiO_2, about 22% Mn and 5% Fe, about 25% Pb; *2* siliceous coronadite; increasing line content indicates increasing quartz content; typical composition 60%–80% SiO_2, 5%–15%Mn, 1% Fe and 5%–20% Pb; *3* sillimanite gneiss partially "replaced" by coronadite with local areas of relatively pure coronadite; *4* as above, except quartzite fragments being a common constituent and coronadite development being more pronounced; *5* highly leached lode quartzite, consisting of remnant quartz grains and pegmatite mica flakes cemented by coronadite; *6* lode quartzite, relatively unleached/fractured; *7* lode quartzite permeated with coronadite or other Fe-Mn oxides/ with considerable jarosite *(J)*; *8* sillimanite gneiss; *9* kaolinized sillimanite gneiss; *10* sillimanite gneiss partially invaded by coronadite on the contact interface; *11* zone of cavity fill; *12* scree; *13* centre of sample series

3.5 Lead Lode Outcrop at the Zinc Coproration

In contrast to the rest of the oxidized lode only relatively superficially oxidized parts of the lead and zinc lodes have been naturally preserved at the Zinc Corporation. Galena has been preserved to a certain extent in these outcrops, sphalerite, however, has been leached out. The southwestern-most surface expression, of which part is shown in Fig. 5, occurs 30 m north of the old Zinc Corporation Main Shaft (Zinc Corp.-NBHC metric grid 15N 14W to 68N 24W). The lode here is probably the extreme top of the No. 2 lens.

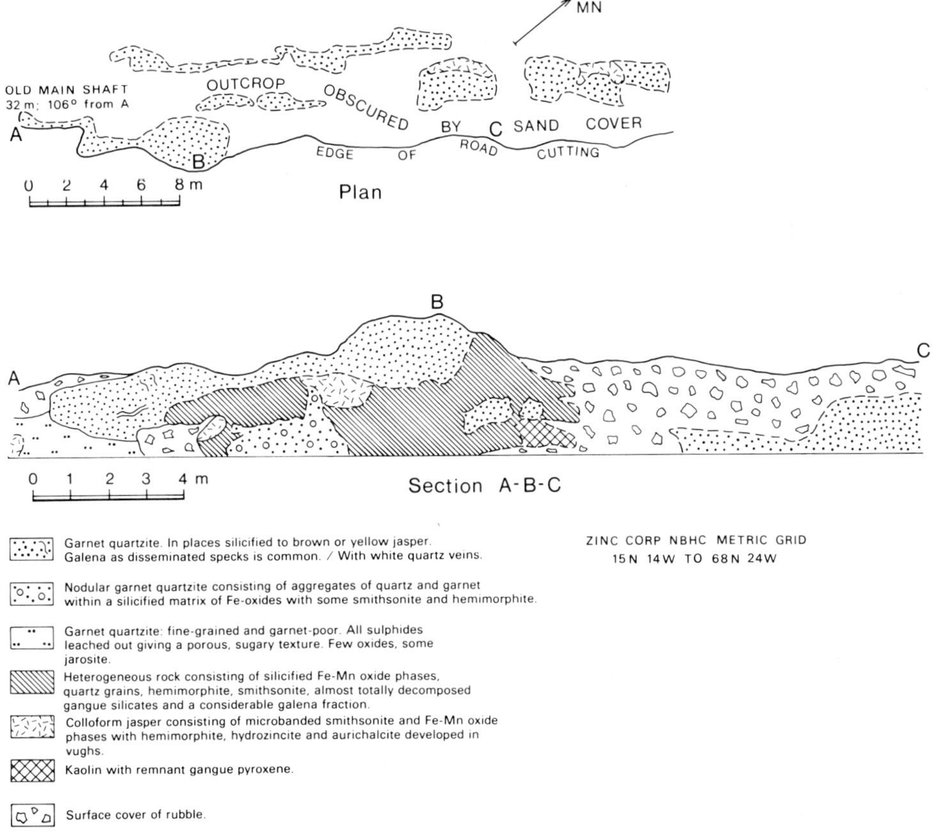

Fig. 5. Plan and section of the lead lode outcrop, Zinc Corporation, Block 6

The predominant rock is vaguely banded garnet quartzite containing some unevenly distributed galena grains surrounded by a narrow zone of anglesite. A portion of the garnet quartzite has been altered into brown jasper. Often a sharp jasper "front" can be observed. The garnet-quartzite-jasper rock passes locally at depth into small masses of silica-rich colloform rock of complex composition. This rock contains up to 6 cm wide vughs filled with hemimorphite, the crystals of which are sometimes coated with druses of hydrozincite and flecks of aurichalcite. The rock itself consists of micro-banded smithsonite, goethite and coronadite.

This colloform smithsonite-bearing rock is closely associated with a dark brown Fe-Mn-rich rock with some galena. The rock is heterogeneous, some parts being a jasper, other areas showing a complex assemblage of partially decomposed gangue silicates, amorphous silica and hemimorphite, and other portions of massive goethite-coronadite with vugh fillings of silica and hemimorphite. Galena in this rock is rimmed by coronadite and/or more rarely by chalcocite.

3.6 Zinc Lode Outcrop, Zinc Corporation

The only surface expression of the zinc lodes is exposed a few tens of metres west of the lead lode outcrop described in Sect. 3.5 (Zinc Corp. NBHC metric mine grid 12S108W to 45N108W). The outcrop consists almost totally of the blue-quartz-gahnite horizon of the lode. Mostly the horizon is little leached and contains some sulphide specks. Local leached areas within the blue-quartz-gahnite horizon have a vughy, nodular appearance due to the disappearance of sulphide specks. Voids are often filled with orange-brown amorphous oxides, with jarosite as a common oxidation product.

A highly quartzose gahnite-bearing rock containing disseminated galena and some chalcopyrite and sphalerite occurs on the north-east side of the outcrop. The sulphides have undoubtedly escaped leaching by their shield of quartz.

4 Geochemical Analyses

4.1 General

All elements in the tables below were analysed by AAS, with the exception of carbon and sulphur, analysed by Leco, and selenium, which was always below the colorimetric detection limit of 5 ppm. XRF scans indicated absence of As, V, W, Br, I, Mo and Cr as major or minor elements.

Two hundred and fifty samples were taken at 3 m intervals on the surface and at 2 m intervals on vertical exposures. Samples were taken by chipping over lengths of 2 m, resultant samples were 5–8 kg, and were prior to analysis carefully crushed, homogenized, quartered, rehomogenized etc. The occurrence of minor minerals with high metal contents like cerussite at 78% Pb may influence the precision of the analyses, although this was not observed in duplicates. A selection of the analyses is shown in Table 1.

4.2 Geochemical Distribution Along Strike, Vertical and Transverse Distribution

The sequence of gossanous outcrops from Thompson Shaft, where the lode is just starting to dip under the surface to NBH No. 1 Shaft shows a decrease of the Pb/Fe and Pb/Mn ratio (Table 2). Even higher values are observed in the central blocks

Table 1. Selection of chemical analyses of coronadite-rich and manganiferous gossans, with some other rock types, of the oxidized lode at Broken Hill

Sample No. Tas. Uni. No.	Fe p.p.m.	Mn p.p.m.	Pb p.p.m.	Zn p.p.m.	Cu p.p.m.	Aq p.p.m.	Ni p.p.m.	Co p.p.m.	Cr p.p.m.	Cd p.p.m.	Sb p.p.m.	Ba p.p.m.	Tot C %	Tot S %	LOCALITY	
46466	125000	123000	165000	3100	2800	54	12	120	90	7	440	BDL	0.18	0.11	Pyromorphite corner.	CR
46475	48000	8300	200000	238000	1430	440	5	110	100	1400	910	70	0.08	14.3	Second level, Main Shaft, Block 14	G/S
46492	72000	157000	177000	4000	1730	31	13	160	40	13	580	150	0.14	n.d.	Thompson Shaft	MG+Q
46503	33000	13600	24000	1400	670	158	6	27	20	3	230	BDL	0.09	0.04	Menindee Road	MG+Q
46508	66000	74000	73000	1370	1540	72	25	200	110	8	800	195	0.12	0.13	Browne Shaft	MG+Q
46517	48000	83000	38000	1270	3300	297	27	63	60	47	220	500	n.d.	n.d.	NBH No.1 Shaft	MG+Q
46524	19000	38000	8000	1140	1170	91	50	39	80	23	208	195	"	"	" " "	M/G+Q
46527	28000	11000	208000	33000	1000	156	10	52	90	49	379	BDL	"	0.39	Lead Lode Outcrop, Zinc Corpn.	G+Q
46528	34000	7000	71000	15200	1780	66	9	34	80	44	682	295	"	0.16	" " " " "	G+Q
46530	26000	13200	22000	4800	1880	22	9	32	90	22	270	80	"	0.07	" " " " "	G+Q
46557	81000	66000	86000	2900	1940	416	8	70	100	8	440	45	0.10	0.17	Zinc Lode Outcrop, Zinc Corpn.	BQG
46589	38000	119000	134000	3000	2080	65	4	95	40	9	440	BDL	0.11	0.13	NE face Wilson Shaft Open Cut	CR
46593	90000	61000	77000	3000	1840	212	6	63	30	7	480	BDL	0.11	0.13	" " " " " "	CR
46594	25000	840	12300	620	310	16	2	4	10	1	125	BDL	0.05	0.52	" " " " " "	CR
46595	57000	65000	75000	1900	1500	205	6	63	30	6	420	BDL	0.07	0.10	" " " " " "	CR
46596	104000	52000	79000	2900	1870	108	9	54	80	14	820	BDL	0.07	0.21	" " " " " "	CR
46597	125000	127000	147000	3500	2500	140	9	81	70	28	1520	BDL	0.09	0.21	" " " " " "	CR
46598	105000	91000	120000	3000	2000	244	8	85	70	9	585	60	0.08	0.15	" " " " " "	CR
46699	43000	194000	215000	2400	1640	232	6	110	70	14	1055	BDL	0.10		" " " " " "	CR
46610	37000	230000	257000	1100	2030	789	6	210	30	12	805	BDL	0.12	0.21	" " " " " "	CR
46611	39000	210000	246000	2000	2620	197	4	130	20	11	635	BDL	0.13	0.21	" " " " " "	CR
46612	33000	260000	297000	2600	2000	133	4	150	40	13	585	BDL	0.13	0.21	" " " " " "	CR
46629	26000	720	4400	320	370	10	8	5	140	1	100	60			Round Hill at Silver Peak	BQG
46633	27000	178000	195000	3300	1380	116	5	100	20	9	175	BDL	0.12	0.11	NE face Wilson Shaft Open Cut	CR
46634	8000	86000	97000	850	840	382	5	59	20	2	205	BDL	0.06	0.13	" " " " " "	CR
46635	9000	45000	52000	690	630	344	6	31	130	2	135	BDL	0.05	0.09	" " " " " "	CR
46636	8000	69000	75000	640	710	210	5	45	80	1	135	40	0.05	0.11	" " " " " "	CR
46637	12000	83000	89000	880	870	219	4	44	80	1	205	BDL	0.06	0.21	" " " " " "	CR
46673	40280	500	642000	4600	1520	150	BDL	15	60	3	30	227	3.78	0.04	Second level, Main Shaft, Block 14	C/MG

Abbreviations Used: BQG = Blue Quartz Gahnite; CR = Coronadite; G+Q = Garnet quartzite; C = Cerussite; G = Galena; MG = Manganiferous gossan

Analyses for Fe, Mn, Pb, Zn, Cu, Ag, Ni, Cr and Cd by C.R.A. Exploration Pty.Ltd. at Zinc Corporation Laboratories, Broken Hill

Analyses for Ba and Sb by C. Swensson

Analyses for total C and total S by J.C. van Moort

Table 2. Longitudinal variation along the strike in the coronadite-rich rocks from the Thompson Shaft area, where the lode is outcropping, to the NBH No. 1 Shaft area where the lode is several hundreds of metres below the surface. Ratios only

	SAMPLE NO.*	Pb/Fe	Pb/Mn	Mn/Fe	Cu/Fe	Zn/Fe	Co/Mn x 10^3	Sb/Fe x 10^3
Thompson Shaft	46492	2.5	1.1	2.2	0.024	0.056	1.0	8.0
	46493	1.3	1.0	1.3	0.046	0.055	1.0	4.9
	46494	1.5	0.9	1.7	0.016	0.054	0.9	4.4
	46495	4.8	0.7	6.6	0.048	0.133	1.1	2.3
	46496	2.2	0.8	2.7	0.027	0.062	1.1	6.7
	46497	1.1	1.0	1.1	0.013	0.044	1.1	33.3
Menindee Road	46499	0.5	1.3	0.4	0.028	0.030	1.4	9.8
	46502	0.7	2.3	0.3	0.017	0.033	1.2	3.8
	46503	0.7	1.8	0.4	0.020	0.042	2.0	1.7
	46504	0.7	2.3	0.3	0.012	0.023	0.6	2.1
Browne Shaft	46507	1.0	0.9	1.1	0.023	0.023	3.2	7.1
	46508	1.1	0.9	1.1	0.023	0.020	2.7	10.8
	46509	2.3	0.9	2.5	0.047	0.024	3.3	7.3
	46510	2.0	0.7	3.0	0.030	0.023	0.2	3.4
	46511	1.6	0.8	2.1	0.028	0.023	0.9	4.3
	46512	0.5	0.6	0.7	0.018	0.019	0.1	1.3
	46513	0.7	0.7	1.0	0.019	0.016	9.4	6.8
North Broken Hill No. 1 Shaft	46515	0.3	0.2	1.3	0.036	0.045	0.3	5.8
	46516	0.01	0.1	1.0	0.052	0.019	0.7	8.9
	46517	0.8	0.2	1.7	0.068	0.026	0.8	2.7
	46520	0.4	0.2	1.9	0.027	0.075	0.1	6.0
	46521	0.7	0.3	2.3	0.048	0.056	0.4	3.7
	46522	0.4	0.2	2.0	0.038	0.046	0.7	4.1
	46523	0.4	0.2	2.0	0.100	0.043	1.0	2.6
	46524	0.4	0.2	2.0	0.062	0.060	1.0	5.5

* Collection Geology Department, University of Tasmania

Table 3. Elemental ratios in the barren Silver Peak Blue Quartz-Gahnite Outcrops and in the ferric gossan at Wolseley. Ratios only

	SAMPLE NO.*	Pb/Fe	Pb/Mn	Mn/Fe	Cu/Fe	Zn/Fe	Co/Mn x 1000	Sb/Fe x 1000
Silver Peak	46616	0.43	63.1	0.007	0.017	0.034	7.6	5.3
	46620	0.12	4.3	0.027	0.009	0.012	2.6	< 7.7
	46621	0.22	7.0	0.032	0.018	0.018	3.9	< 4.2
	46622	0.31	6.5	0.049	0.013	0.024	2.9	15.7
	46623	0.25	11.9	0.021	0.015	0.021	5.0	13.5
	46624	0.14	16.0	0.009	0.014	0.016	10.0	14.1
	46628	0.31	22.1	0.014	0.034	0.018	7.1	< 10
	46629	0.17	6.1	0.028	0.014	0.012	6.9	< 3.8
	46630	0.23	15.1	0.016	0.009	0.035	12.8	< 3.3
Wolseley	46952	0.01	1.1	0.011	0.014	0.157	not determined	
	46953	0.02	2.8	0.008	0.082	0.759		
	46954	0.04	13.1	0.003	0.090	0.415		
	46955	0.01	1.0	0.015	0.012	0.122		
	46956	0.05	18.5	0.003	0.110	0.025		
	46957	0.06	21.4	0.003	0.153	0.032		

* Collection Geology Department, University of Tasmania

(Table 4). Values for the less oxidized Zinc Corporation outcrops are still higher and more erratic. The ratio Co/Mn is constant. The Cu/Fe, Zn/Fe and Sb/Fe ratios are remarkably constant indicating that Cu, Zn and Sb are efficiently scavenged proportional to the Fe available, whatever the Fe values may be (cf. Table 1, cf. Gatehouse et al. 1977). The Pb/Fe values in the NBH No. 1 Shaft area start to approach those of the barren lode horizon outcrops at Silver Peak and the barren gossan at Wolseley (Table 3). The rocks in these areas contain only some hundred of ppm Mn, enhancing the Pb/Mn ratio.

The vertical variations of the metal proportions are given in Table 4. The absolute data show individual peaks for the different elements as a function of depth and will not be discussed.

Table 4. Vertical variation through the coronadite-rich zone of the oxidized lode at the NEface of Wilson Shaft Open Cut. Ratios only

SAMPLE NO.*	DEPTH (in metres)	Pb/Fe	Pb/Mn	Mn/Fe	Cu/Fe	Zn/Fe	Co/Mn x 10^3	Sb/Fe x 10^3
46637	0	7.3	1.0	6.6	0.067	0.070	0.53	16.5
46636	1.5	10.7	5.3	10.7	0.097	0.087	0.65	25.9
46635	3.0	5.3	11.1	4.5	0.064	0.070	0.69	19.4
46634	4.6	11.1	7.2	9.7	0.097	0.098	0.69	23.6
46633	6.1	7.2	8.9	6.5	0.050	0.122	0.56	6.4
46612	7.6	8.9	6.2	7.8	0.062	0.078	0.58	17.5
46611	9.1	6.2	6.9	5.3	0.066	0.051	0.62	16.0
46610	10.7	6.9	5.0	6.2	0.054	0.30	0.91	21.6
46599	13.7	5.0	1.1	4.5	0.038	0.52	0.57	2.4
46598	16.2	1.1	1.1	0.9	0.019	0.028	0.93	5.5
46597	17.7	1.2	1.0	1.0	0.020	0.028	0.64	12.2
46596	19.8	1.0	1.0	0.7	0.023	0.037	1.04	1.3
46595	21.3	1.3	1.2	1.2	0.026	0.033	0.98	0.7
46594	22.9	0.5	0	0	0.012	0.025	undefined	5.1
46593	24.4	0.8	1.3	0.7	0.020	0.033	1.03	5.3
46589	25.9	3.5	1.1	3.1	0.052	0.078	0.80	11.6

* Collection Geology Department, University of Tasmania

Samples were at 25 m intervals collected by Mr. Elliot of the University of N.S.W. along a line running from near the Block 14 Main Shaft in a north-westerly direction, perpendicular to the strike. The absolute and relative values for some of the elements mentioned above are shown in Fig. 6. Note that also here the high Pb/Fe and Pb/Mn ratios near the lode and the low Pb/Fe ratio far away from the lode, not matched by a low Pb/Mn ratio because of the low Mn content.

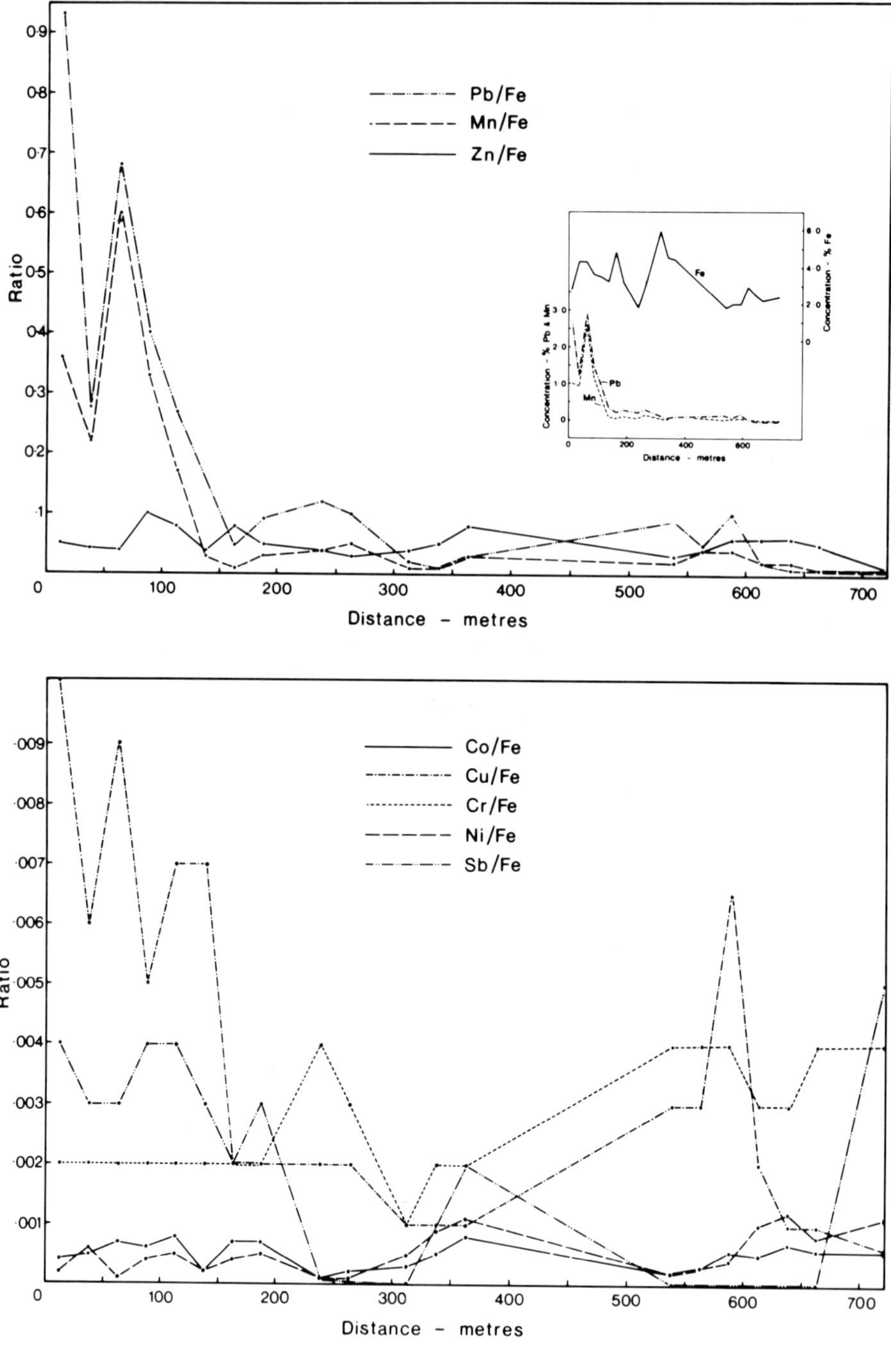

Fig. 6. Transverse dispersion profile from the oxidized lode to the Mine Sequence Rocks, Block 14. Cf. the longitudinal and vertical variations in Tables 2 and 4

4.3 Discussion of Geochemical Trends

The oxidized lode of Broken Hill expresses itself mainly through high Pb values of 10,000 ppm to 100,000 ppm or more, and through associated Mn values of several tens of thousands up to about 100,000 ppm. This unusual enrichment, essentially in the form of coronadite, has been possible because of the original coexistence of galena, spessartine and gangue minerals rich in manganese (bustamite, rhodonite). The width of the Pb-Mn halo is restricted to some hundreds of metres only. Background values for Pb in the surrounding gneisses are several hundred to 1500 ppm.

Iron is sometimes enriched in the gossan to 5% and exceptionally to 10%. These values are lower than usual for oxidized sulphide lodes because of the absence of pyrite (cf. Blanchard 1966). The zinc contents in the oxidized lode and its halo are proportional to the amount of iron present and have as such no particular use as a prospective element. Zinc values of several thousands of ppm occur throughout the Broken Hill district, with a background value of 1500 ppm, and do not indicate as such any mineralization. Usually the Zn/Fe ratio does not exceed 0.05, even not at the outcrop of the zinc lode at the Zinc Corporation. Zinc values appear to increase with depth. Antimony also follows iron.

Copper is only present in the gossan at a level of some hundred ppm and values rarely exceed 3000 ppm. Copper values increase, as expected, somewhat with depth. Essentially, however, the copper contents are a function of the iron content of the gossan. The highest Cu/Fe ratio is encountered laterally near the lode, indicating efficient scavenging.

Cadmium, nickel and chromium distributions do not show specific trends and barium behaves erratically. Silver shows an erratic increase with depth.

5 Occurrence of a Plumboan Coronadite

Unusual high analytical values of up to 40% Pb in the black gossanous material have been known for a long time. Jaquet (1894) mentioned for the cap of the lode a lead content of 20%. Typical areas of high lead values associated with high Pb/Mn ratios are the coronadite outcrops between Browne Shaft and the Zinc Corporation where Pb/Mn ratios of 1.1 are common and sometimes exceed the value of 2, and this in rocks free of lead carbonate, sulphate, plattnerite or minium. The high lead content cannot be attributed to the occurrence of coronadite with a Pb/Mn ratio of 0.45.

Subsequent investigation of the black coronadite-rich rocks gave unidentifiable sharp diffraction peaks, notably at 3.25 Å and not the proper diffraction pattern of coronadite. The rocks consist largely of a coronadite-like mineral in rhythmic colloform bands which may alternate with goethite (once with smithsonite). More often it occurs as homogenous fillings of cavities or cracks and as replacement, whilst the most common form is as massive matrix for fragments of other residual minerals, garnet, quartz and goethite in particular. Featherlike occurrences are rare.

The composition of the mineral, including its banded colloform variety, varies only slightly at any particular locality, but appreciable variations in the Pb/Mn ratio occur

Table 5. Electron microprobe analyses indicating percentage metallic lead and manganese of plumboan coronadite (assuming all lead as PbO, all manganese as MnO_2 and 16 oxygen atoms).

Central Blackwood and Wilson Open Cut Area			NBH No. 1, Browne and Thompson Shaft Area in order of increasing depth of ore		
	Pb	Mn		Pb	Mn
Blackwood Open Pit*	54	26	Thompson Shaft		
	54	27	46495	50	30
Wilson Shaft Open Cut			Menindee Road		
46596	50	29	46501	50	22
46597**	55	26		52	27
	54	27	Browne Shaft		
46598	56	25	46508	51	29
46610	54	26		48	30
	58	24		49	30
44612	56	25		48	37
	40	36		49	36
	44	33		43	34
46633	51	28		43	34
	51	28		41	35
				43	34
			NBH No. 1 Shaft		
			46515	36	39
				35	40

* rel. density of the mineral is 4.81
** Collection Geology Department, University of Tasmania

Table 6. X-ray powder diffraction pattern of plumboan coronadite compared with coronadite from Bou Tazoult, Morocco

Plumboan Coronadite			Coronadite**	
d Å	intensity	hkl*	d Å	I/I_1
≃ 8.1 broad,	strong	110		
5.8 rather broad,	weak	200		
3.8 somewhat broad,	strong	220	3.47	(60)
3.25 sharp,	very strong	310	3.11	(100)
2.46 sharp,	medium	211	2.39	(40)
2.17 sharp,	medium	301	2.15	(20)
1.55 sharp,	medium	521	1.54	(50)

* Indexed according to isostructural equivalent hollandite, JCPDS powder data file 13-115

** Frondel and Heinrich (1942); Ramdohr and Frenzel (1956)

concurrently with whole rock chemical trends discussed above (see Table 5, cf. Table 2).

The powder diffraction pattern of the purest samples was always contaminated to a minor extent with some kaolin, quartz, garnet or ramsdellite. Comparison of 30 diffractograms of different specimens enabled the construction of the Table 6.

The lead content of the plumboan coronadite exceeds the maximum of 27% Pb recorded by Frondel and Heinrich (1942) whilst the d-spacings are also larger than those recorded previously. Detailed study of the mineral is in progress. Common coronadite, as recorded by Lawrence (1961, 1968), was not observed in the present study.

Acknowledgments. The Mining Manager's Association of Broken Hill supported this study through a generous grant in 1977. The study was continued with the assistance of the Humboldt Foundation during 1979 in the Mineralogical Institute Heidelberg.

Valuable assistance was given by Mr. R.J. Ford, Mr. W. Jablonski, Dr. G.D. Klingner, Dr. I. Plimer and Mr. W.G. Widdop in Australia and by Prof. G.C. Amstutz, Dr. H. von der Dick and Dr. E. Krouzek in Germany.

References

Andrews EC (1922) The geology of Broken Hill District. Mem Geol Surv NSW 8:432

Barclay CJ, Jones JB (1971a) The Broken Hill Ag-halides. J Geol Soc Aust 18:149–157

Barclay CJ, Jones JB (1971b) A comparison of the lead phosphates from the oxidized zone, Broken Hill orebody, with those of some other Australian localities. J Geol Soc Aust 18:207–213

Blanchard R (1966) Interpretation of leached outcrops. Nev Bur Mines Bull 66:196

Burrell HC (1942) A statistical and laboratory investigation of ore types at Broken Hill, Australia. Unpubl PhD Thesis, Harvard Univ

Challinor J (1973) A dictionary of geology. Univ Wales Press, Cardiff, 350 pp

Frondel C, Heinrich EWm (1942) New data on hetaerolite, hydrohetaerolite, coronadite and hollandite. Am Mineral 27:48–56

Gatehouse S, Russel DW, van Moort JC (1977) Sequential soil analysis in geochemical exploration. J Geochem Explor 8:487–494

Gustafson JK, Burrell HC, Garretty MD (1950) Geology of the Broken Hill ore deposits. Bull Geol Soc Am 61:1369–1438

Jaquet JB (1894) Geology of the Broken Hill lode and Barrier Ranges Mineral Fields, N.S.W. Mem Geol Surv NSW Geol 5:149

Johnson IR, Klingner GD (1975) The Broken Hill ore deposit and its environment. In: Knight CL (ed) Economic Geology of Australia and Papua New Guinea. Australas Inst Min Metall Monogr Ser 5:476–491

King HF, O'Driscoll ES (1953) The Broken Hill lode. In: Edwards AB (ed) Geology of Australian Ore deposits, pp 578–600

Lawrence LJ (1961) Notes on some additional minerals from the oxidised portion of the Broken Hill lode, N.S.W., with observations on coronadite. J Proc R Soc NSW 95:13–16

Lawrence LJ (1968) Minerals of the Broken Hill District. In: Radmovich and Woodcock (eds) Proc Australas Inst Min Metall Broken Hill Monogr Ser 3:103–136

Mason B (1976) Famous mineral localities. Broken Hill, Australia. Mineral Rec 7:25–33

Ramdohr P, Frenzel G (1956) Die Manganerze. In: Reyna JG (ed) Symposium sobre yaciemientos de manganeso. XX. Congr Geol Int 1:19–73

Smith G (1926) A contribution to the mineralogy of New South Wales. Min Res Geol Surv NSW 34:138

Swensson C (1977) The geology and geochemistry of the oxidised zone of the Broken Hill lode, Broken Hill, N.S.W. Unpubl Hons Thesis, Geol Dep Univ Tasmania 1:162, 2:190

Thomson BP (1975) Tectonics and regional geology of the Willyama, Mount Painter and Denison inlier areas. In: Knight CL (ed) Economic geology of Australia and Papua New Guinea. Australas Inst Min Metall Monogr Ser 5:476–491

Alteration of Radioactive Minerals in Granite and Related Secondary Uranium Mineralizations

J. RIMSAITE[1]

Abstract

Natural losses of uranium and other constituents from partly altered uraninite, uranothorite, pyrochlore and allanite in granitic rocks of Precambrian age (100–1100 Ma) were documented using a scanning electron microscope coupled with an energy dispersive spectrometer and an electron microprobe. Migration and precipitation of the liberated uranium along fractures and reactions between uranium and other ions to form secondary uranium-bearing minerals were studied in polished thin sections using microbeam techniques. Mode of formation of the most common secondary uranium ore minerals: autunite, torbernite, phosphuranylite, thorogummite and uranophane is discussed in relation to supergene enrichment and formation of secondary uranium mineralizations in weathered granites. The purpose of this paper is to describe natural decomposition of primary radioactive minerals in granitic rocks, as well as migration and redeposition of the liberated uranium and other ions on the basis of field and laboratory studies.

1 Introduction

Secondary uranium mineralizations are of economic significance in European, African and in upper portions of weathered Australian and Canadian uranium deposits (Cuney 1978; Leroy 1978; v. Backström and Jacob 1979; Cameron 1979; Fernandes 1979; Rimsaite 1979; Snelling and Dickson 1979). The most common secondary minerals in zone of weathering are uranium phosphates, autunite [Ca $(UO_2)_2 \cdot (PO_4)_2 \cdot 10$–12 H_2O], torbernite [Cu $(UO_2)_2 \cdot (PO_4)_2 \cdot 8$–12 H_2O], and phosphuranylite [Ca $(UO_2)_3$ $(PO_4)_2 \cdot 6\ H_2O$] and their partly dehydrated equivalents; and hydrous uranyl-bearing oxides and silicates, including uranophane [Ca $(UO_2)_2 \cdot Si_2O_7 \cdot 6\ H_2O$] and "thorogummite". All secondary uranium-bearing minerals are associated with the "primary" radioactive minerals, namely uraninite, uranothorite, pyrochlore, allanite and zircon (cyrtolite) in the radioactive granitoid rocks in the Grenville structural province in southern and eastern Canada. The relationship between the primary and secondary minerals was studied in the Canadian granitoid rocks because they are relatively less altered and have better preserved "primary" minerals than the deposits in Australia, South Africa, and southern Europe.

---

1 Geological Survey of Canada, 601 Booth Street, Ottawa, Ontario, KIA 0E8, Canada

Samples for this mineralogical-geochemical study were obtained from the following areas: the Bancroft area (45°03′–45°06′N, 77°55′–78°05′W), Ontario described by Bullis (1965) and Rimsaite (1978, 1980); Mont-Laurier area (46°47′–47°00′N, 74°45′–75°20′W), Quebec (Kish 1977; Rimsaite 1978); and Baie Johan Beetz area (50°17′–50°25′N, 62°49′–62°56′W), including Turgeon Lake and Little Piashti Lake areas, north of St. Lawrence River, Quebec (Baldwin 1970; Fig. 1). Uranium mineralizations in these areas occur in granite, migmatites, and pegmatites which are associated with, and enclose xenoliths of metagrabbro and of metasediments ranging in composition from micaceous gneisses and quartzites to calc-silicate rocks. Uranium concentrations in the samples studied range from 100 ppm to 23,440 ppm (Rimsaite 1978, 1980).

The samples containing partly altered "primary" and secondary uranium-bearing minerals were studied in oil immersion mounts, in polished thin sections and in small chips mounted on a rod using a petrographic microscope, a scanning electron microscope coupled with an energy dispersive spectrometer, and an electron microprobe. Minerals were identified optically and by X-ray diffraction supplemented by mineral spectra obtained using an energy dispersive spectrometer. Radioactive minerals were located in polished thin sections using autoradiographs of these sections. The minerals

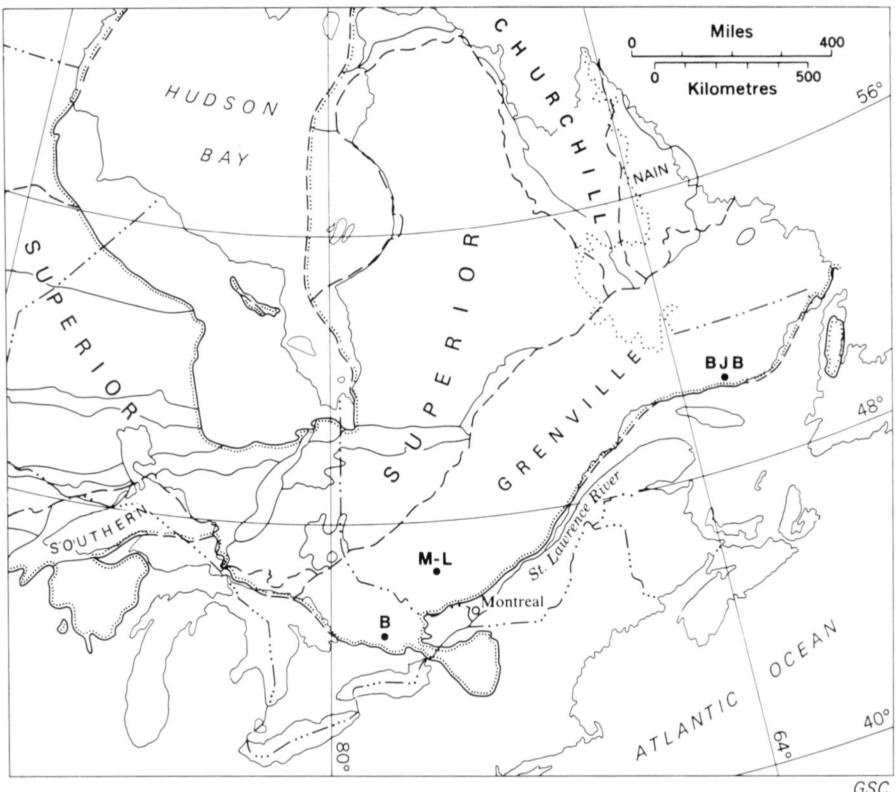

Fig. 1. Geological sketch map showing subdivisions of the Canadian Shield and locations of samples in the Grenville structural province. *B* Bancroft; *M-L* Mont-Laurier; *BJB* Baie Johan Beetz

display zoning and overgrowths, implying more than one episode of crystallization. Thus, the term "primary" is used in quotation marks.

Six modes of occurrence of radioactive minerals were observed in granitic rocks studied: (1) as disseminated grains in biotite, pyroxene, and amphibole in mafic bands and xenoliths; (2) as grains and fracture fillings at the contact between biotite and muscovite in two mica granites; (3) as disseminated crystals along grain boundaries and fractures of rock-forming minerals, including albite, microcline, and quartz in leucocratic portions of a granite; (4) grains in small groups associated with, and enclosed in, zircon (cyrtolite), apatite, xenotime, and titanite crystals: (5) chemically complex radioactive fracture fillings and narrow bands along grain boundaries as well as coatings and rims on radioactive and on nonradioactive grains; (6) as supergene uranyl-bearing mineral aggregates in fractures and vugs in weathered granite. The following features pertaining to the occurrence and decomposition of radioactive minerals, releases of the uranium, and the origin of secondary uranium mineralizations in granitic rocks are discussed:

a) "primary" radioactive minerals and their alterations (Tables 1 and 2, groups I and II; Fig. 2);
b) radioactive fracture fillings and intergranular reactions resulting in the fixation and reprecipitation of uranium from migrating uraniferous aqueous solutions (Fig. 3, Table 2, groups III and IV; Rimsaite 1980); and
c) crystallization of secondary uranium ore minerals and supergene enrichment of uranium in altered granitic rocks (Fig. 3; Table 2, group V).

Table 1. Electron microprobe analyses[a] of relatively fresh "primary" minerals from Madawaska Mine, Bancroft, Ontario

Weight %	Uranothorite			Uraninite			Allanite		Isotropic phases		Pyrochlore	
	Core	Rim	Rim	Edge	Near rim	Rim	Core	Rim	Dark core	Pale rim	Core	Rim
SiO_2	15.8	16.3	15.8	0.0	0.2	29.4	31.6	37.5	31.2	30.9	N.D.[b]	N.D.
TiO_2	N.D.	N.D.	N.D.	N.D.	N.D.	0.3	0.8	0.9	0.3	0.4	2.5	2.3
Al_2O_3	N.D.	N.D.	N.D.	N.D.	N.D.	6.4	11.9	7.3	5.0	4.0	N.D.	N.D.
Fe as FeO	N.D.	N.D.	N.D.	N.D.	N.D.	25.4	16.5	17.9	16.4	4.7	3.1	2.1
MnO	N.D.	N.D.	N.D.	N.D.	N.D.	1.3	0.9	0.9	0.6	0.4	N.D.	N.D.
MgO	N.D.	N.D.	N.D.	N.D.	N.D.	3.6	0.3	1.7	1.1	0.2	N.D.	N.D.
CaO	3.4	3.5	1.6	0.9	1.2	2.4	11.6	5.0	2.6	4.6	5.3	3.4
Y_2O_3	N.D.	N.D.	N.D.	N.D.	N.D.	0.1	0.1	3.9	6.4	9.4	5.4	5.3
La_2O_3	N.D.	N.D.	N.D.	N.D.	N.D.	9.8	6.6	2.3	0.4	0.7	N.D.	N.D.
Ce_2O_3	N.D.	N.D.	N.D.	N.D.	N.D.	10.2	9.8	3.6	1.2	2.0	1.5	1.3
Nd_2O_3	N.D.	N.D.	N.D.	N.D.	N.D.	1.9	2.4	1.5	1.2	1.7	1.4	1.2
Yb_2O_3	N.D.	N.D.	N.D.	N.D.	N.D.	0.3	0.3	1.1	1.8	2.9	2.1	2.0
PbO	1.2	1.9	4.0	10.0	9.8	0.7	0.2	0.2	0.3	0.4	N.D.	N.D.
ThO_2	51.6	47.2	50.0	6.1	4.6	0.3	1.0	1.6	9.0	10.2	2.1	1.9
UO_2	15.8	18.6	16.9	71.1	74.2	0.9	0.1	1.5	4.8	5.6	14.1	17.2
Nb_2O_5	N.D.	N.D.	N.D.	N.D.	N.D.	N.D.	N.D.	N.D.	N.D.	N.D.	44.0	39.4
Ta_2O_5	N.D.	N.D.	N.D.	N.D.	N.D.	N.D.	N.D.	N.D.	N.D.	N.D.	7.6	6.6

[a] Analyst A.G. Plant; [b] not detected

Table 2. Description of uranium-bearing grains, rims, crusts, fracture linings, and secondary mineral aggregates and their spectra obtained using an energy dispersive spectrometer[a]

	Description and mode of occurrence of uranium-bearing minerals and crusts	Element lines listed in order of decreasing intensities in spectra of minerals and crusts
I.	**Uranium-bearing grains in other minerals**	
I-1	Corroded grains of uraninite in thorite	U, Pb phase, replaced by U, Th, Si phase
I-2	Radioactive grains in chloritized mica	Si, U, Fe, Mn, Pb, Y, Ca, Ti, Al, Ce
I-3	Uraninite in zircon (cyrtolite)	U, Pb, Th (Table 1)
I-4	U-, Th-rich grains (7 μm) in xenotime	U, Th, Si, Y, P
I-5	U-, Th-rich spots in allanite	Si, Al, U, Th, Ca, Ti, Fe, Pb
I-6	U-rich bands (< 1 μm) in pyrochlore	U, Nb, Si
I-7	U-, Nb-, Ti-bearing grains in biotite	Ti, Nb, U, Ca, Si, Y, Th, Fe
II.	**Impure U-bearing phases replacing uraninite crystals**	
II-1	Uraninite replaced by Si-, Y-phase	U, Si, Y, Pb
II-2	Uraninite replaced by Si, Th-phase	U, Si, Th, Ca, Pb (Rimsaite 1980)
II-3	Zr-bearing phase in mottled uraninite	U, Th, Zr, Pb, Si, Y, Ti, Fe, Al
II-4	REE-bearing phase replacing uraninite	Si, Ce, La, Ca, U, Th, Pb, Fe, Al
II-5	Pseudomorphous replacement by Fe oxide	Fe, Si, Pb, U (Fig. 2A)
III.	**Uranium-bearing rims on radioactive and nonradioactive minerals**	
III-1	Rims on uraninite (10 μm in width)	U, Th, Si, Y, Pb and specks of REE and $CuFeS_2$
III-2	Uranium-rich rims on uraninite	U, Si, Ca, Pb, Fe (Rimsaite 1980)
III-3	Fe-rich rims on uraninite	Fe, Si, Pb, Al, U, Th
III-4	Fe-rich rims with diverse spots	Fe, Si, Al, U, Pb with spots of U, Fe, Si, Pb, Th
III-5	U-, Y-bearing rims on uraninite	U, Pb, Y
III-6	U-, Ti-bearing rims on biotite/muscovite	Si, U, Ca, Ti, Pb, Fe, Al
III-7	Pb, Y-rich rims on uraninite in mica	Pb, Y, U, Ca, Ce, Cu, Al
III-8	U-, Zr-bearing rims on zircon	U, Zr, Si, Th, Pb
III-9	REE-bearing crusts on titanite	Ce, La, with specks of U-rich phase
IV.	**Uranium-bearing fracture fillings in minerals and at grain boundaries**	
IV-1	Fracture fillings in apatite	U, Nb, Si (Fig. 2E)
IV-2	Fracture fillings in zircon (> 1 μm thick)	U, Th major
IV-3	Colloform fracture linings in uraninite	U, Ti, Pb, Si, Fe (near walls)
IV-4	Central portion enclosed in IV-3	Fe, U, Si, Ti, Th, Pb (1 to 10 μm wide)
IV-5	Between uraninite and apatite	U, P, Ca, Pb
IV-6	Fracture fillings in quartz in granite	Si, U, Th, Y, Pb, Fe with specks of PbS, U, Fe
IV-7	Fracture linings in biotite near zircon	U, Si, Pb, Ti, Zr, Fe, Al
IV-8	Fracture linings between biotite and quartz	U, Si, Ca, Pb (near walls)
IV-9	Core of IV-8	Fe, Si, U, Pb
V.	**Uranium-bearing crusts and secondary mineral aggregates in weathered granite rocks**	
V-1	"Thorogummite" crusts with iron oxides	Th, U, Si, Al, Ca, Pb
V-2	U-, S-rich crusts in radioactive granite	Si, U, Ca, Fe, Al, Pb, Mg
V-3	U-, P-bearing crusts in altered granite	U, Si, P, Al, Ca, Pb, Fe
V-4	Green crusts in argillized granite	U, P, Cu, Pb, Si, Al
V-5	Uranophane crusts	U, Si, Ca (Fig. 3, spectrum 2)

[a] Analyst D.A. Walker

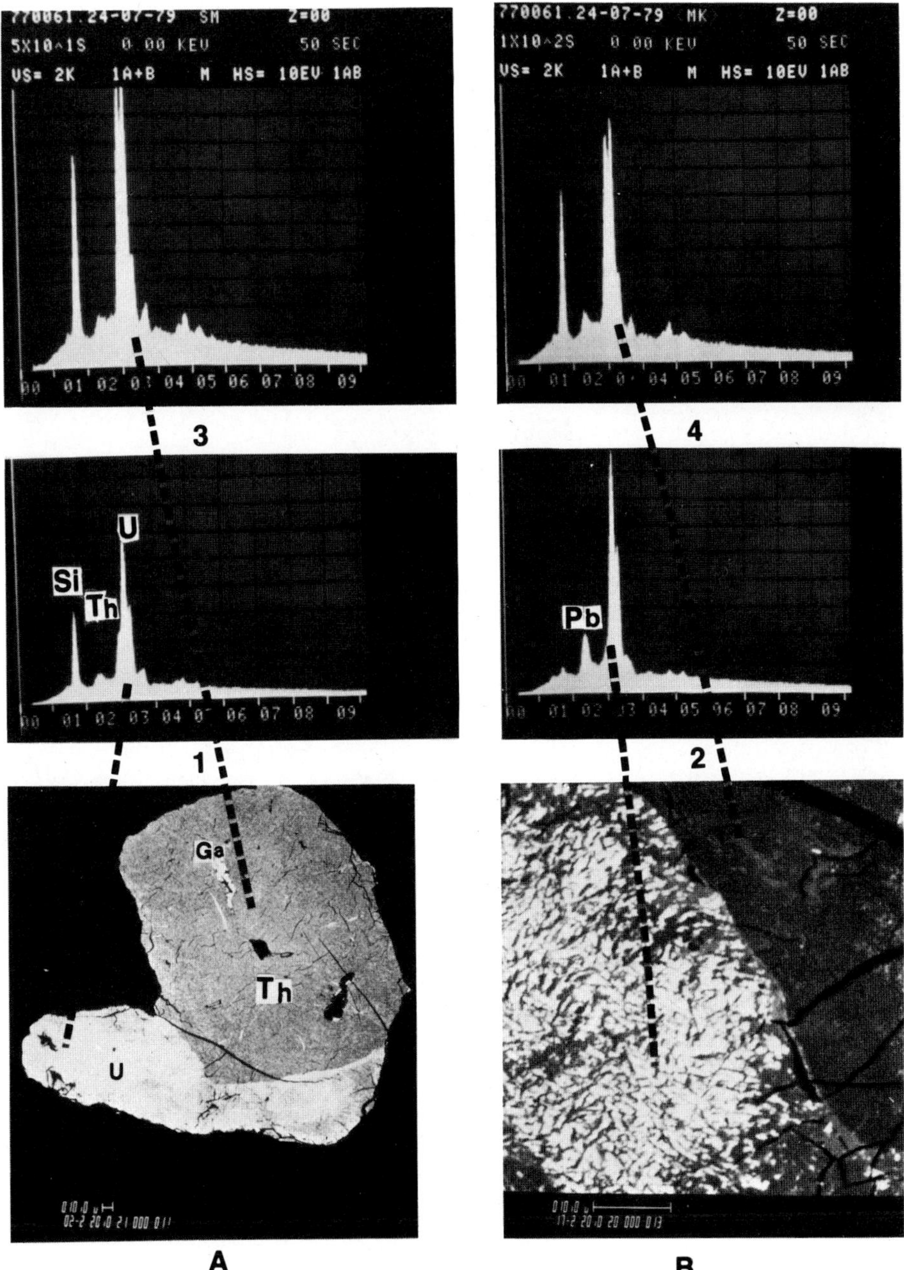

Fig. 2A, B. Back-scattered electron images (**A, B**) and spectra of heterogeneous uraninite and uranothorite from the Bancroft area, Ontario. Analyst D.A. Walker (Negative GSC 203577 A). **A** Uranothorite crystal (*Th*, grey) with white specks of galena *(Ga)* intergrown with heterogeneous uraninite (*U*, pale grey). **B** Uraninite/uranothorite contact, enlarged from **A** showing light grey and darker grey phases in uraninite, and dark grey and lighter grey specks in uranothorite.
Spectrum 1: grey thorian phase replacing white U-Pb phase *(spectrum 2)* in uraninite grain.
Spectrum 3: dark grey Pb-poor phase in uranothorite.
Spectrum 4: Pb-bearing lighter grey specks in uranothorite

TH
Ca
Ce
P
Cu
Ga
UPh
Si
Y
U
Ti
Pb
Fe
T
U
UPh
Th
U-Nb
A
U
Fe
Q
Fe
U
B
C
B
A
6
5
4
3
2
1
7
F
E
D

Fig. 3A–F

2 "Primary" Radioactive Minerals and Their Alterations

The most common "primary" uranium-bearing minerals in the radioactive granitoid rocks studied are uraninite (UO_2); uranothorite [$(Th,U)SiO_4$]; allanite [$(Ce,Ca,Y)_2(Al,Fe)_3(SiO_4)_3(OH)$]; and pyrochlore [$(Na,Ca)_2Nb_2O_6(OH,F)$]. In some radioactive granites, titanite and apatite grains contain variable quantities of uranium that is fixed mainly in radioactive rims and fracture linings. Examples of chemical compositions of the principal uranium-bearing minerals, determined by an electron microprobe using low magnifications and wide beam are given in Table 1, and more detailed variations, using a narrow beam (1/2 μm), in spectra obtained by an energy dispersive spectrometer in Figs. 2 and 3.

Uraninite and uranothorite grains are commonly intergrown with each other and occur in groups associated with zircon (cyrtolite), apatite, and titanite. The relationship between uraninite and uranothorite is very complex. In some granites, uranothorite apparently preceded crystallization of uraninite (Fig. 2). In other granites, corroded and partly decomposed uraninite fragments are enclosed in euhedral crystals of younger uraninite or uranothorite that in turn are surrounded by iron- and rare-earths element (REE)-bearing rims, and outermost uraniferous rims (Rimsaite 1980).

Variations in distribution of U and Th within uraninite grains from Blind River area, Ontario, and in "uranothorianite" grains from Palabora in Northern Transwaal have been studied and illustrated by Ramdohr (1969, Figs. 615 and 616). He attributed the marked alteration of the U-rich core to lesser stability or uranium in the heterogeneous uraninite under conditions of the alteration.

In back-scattered electron images (BEI) under high magnifications, uraninite and uranothorite crystals appear corroded and heterogeneous, consisting of a lighter grey Pb-bearing phase and darker grey Pb-poor and Si-rich phases (Fig. 2). Part of the radiogenic lead and of uranium has been lost from such heterogeneous grains.

Along open fractures, uraninite grains are pseudomorphously replaced by aggregates of siliceous goethite and limonite (Fig. 3A and Fig. 2, spectrum 1 of uraninite and Fig. 3, spectrum 1 of goethite). Depending on the extent of corrosion and replacement, trace amounts to marked quantities of uranium, thorium, and radiogenic lead are liberated from the altered uraninite crystals.

Fig. 3A–F. Back-scattered electron images (**A** to **F**) and mineral spectra (*1* to *7*) obtained using an energy dispersive spectrometer showing pseudomorphous replacements of uraninite grains and reprecipitated U-bearing fracture linings and secondary mineral aggregates in radioactive granites from Mont-Laurier and Baie Johan Beetz areas, Quebec (analyst D.A. Walker) (Negative GSC 203577). **A** Remnant or uraninite crystal *(U)* partly replaced by iron oxide *(Fe)*. **B** Uraninite crystal *(U)* partly replaced by Si-, Th-bearing phase *(Th)*. **C** Crusts of uranophane aggregates *(Uph)* and irregular grains of galena *(Ga)* overgrown on uraninite crystal, enlarged from **B**. **D** Iron-rich fracture fillings *(Fe)* and U-bearing fracture linings *(U)* in a fractured biotite and between biotite *(B)* and quartz *(Q)*. **E** Fractured apatite *(A)* near pyrochlore grain, cemented by unidentified Nb-U fracture fillings. **F** Platy torbernite crystals *(T)* filling fractures and vugs in argillized granite.
Spectra: 1, iron-rich fracture fillings, in central portion (**D**); *2,* uranophane crusts (in **C**) and fracture linings (in **D**); *3,* iron-rich thorian phase replacing uraninite (**B**); *4,* uraniferous Ti-bearing rims on uraninite in **B**; *5,* complex U-, Th-, Zr-, Y-, Ti-, Si-bearing fracture fillings in basal fractures of biotite adjacent to zircon grain; *6,* uraniferous REE-bearing crusts on partly decomposed allanite; *7,* torbernite in **F**

Uranium-rich grains occur also as inclusions in nonradioactive or slightly radioactive minerals such as pyrochlore, allanite, apatite, biotite, chlorite, cyrtolite, and xenotime (Tables 1 and 2, group I; Rimsaite 1978, 1980). Some of the host minerals are relatively resistant to alteration and thus may protect the uraninite inclusion from alteration. Some of the U-rich inclusions are believed to be older than the enclosing host, but some may be products of exsolution or later minerals precipitating from aqueous solutions along fractures in a host grain. Radioactive damage and destruction of crystal structures are the important factors causing fracturing and enhanced alteration of minerals in radioactive granites. The U-bearing pyrochlore and allanite are commonly isotropic (metamict) and very heterogeneous, consisting of various "exsolved" phases, including galena, uraninite, Fe-rich and Fe-poor phases and REE-rich and REE-poor phases (Table 1; Rimsaite and Lachance 1966). The radioactive pyrochlore from the Bancroft area, Ontario, yielded much lower U/Pb, Th/Pb and Pb/Pb isotopic ages than the uraninite (Rimsaite 1980). The pyrochlore breaks down into U-rich and U-poor phases and contains disseminated specks of galena that are visible in high power ($> 1000\times$) back-scattered electron images (BEI). The U-rich phase forms narrow coatings ($< 1/2$ μm in width) on irregular areas that are impoverished in uranium (5–10 μm in size). The heterogeneity increases towards the periphery and along fractures of the decomposing pyrochlore and allanite grains. Some constituents, notably radiogenic lead and uranium, derived from the parent minerals, recrystallize along the edges of the host as "secondary" uraninite and galena, or migrate along fractures. The removal and redistribution of the radioactive and radiogenic elements from the partly decomposed minerals account for their discordant isotopic ages (Rimsaite 1980).

The characteristic features of alteration of granitic rocks are sericitization and argillization of feldspars and chloritization or hydration of biotite. These types of alteration involve liberation of silica, alkalis, and of titanium from the biotite. Under oxidizing conditions, sulphides are replaced by hydroxides (e.g. chalcopyrite and pyrite are replaced by limonite or siliceous goethite). During advanced stages of argillization, the relatively stable accessory minerals, such as tourmaline, apatite, and zircon also alter to secondary phyllosilicates and hydroxides and release boron, phosphoric acid, and other elements (Rimsaite 1979). More detailed data on alteration of radioactive and rock-forming minerals can be obtained from the author upon request.

3 Radioactive Fracture Fillings and Intergranular Reactions Resulting in the Fixation of Uranium from Migrating Uraniferous Aqueous Solutions

Radioactive granites are fractured and contain brightly coloured halos surrounding radioactive grains. Detailed studies of the stained fractures and of mineral interstices using a scanning electron microscope coupled with an energy dispersive spectrometer revealed complex chemical compositions of the fracture fillings and mineral coatings (Fig. 3; Table 2, groups III and IV). As a result of radiation damage, minerals adjacent to uraninite and uranothorite are shattered and altered. Uranium, thorium, and radiogenic lead released from decomposing radioactive minerals react with ions released from adjacent zircon, apatite, biotite, sulphides, and other minerals to form complex

U-Zr, U-Th-Ti, U-Nb, and U-Y bands (Table 2, group III). In addition to these radiation-induced fractures, the host rocks also have fractures induced by cooling and tectonic processes. Autoradiographs of many thin sections of radioactive granites, particularly from the Mont-Laurier area, Quebec, indicated that more uranium occurs in fractures than in the "primary" radioactive grains. In these rocks, uranium has been remobilized, and moved along fractures in aqueous solutions and reprecipitated along fracture walls.

The following nomenclature is used here to describe the precipitates: *mineral coatings and fracture linings* are very narrow bands (< 1 μm) on minerals and on fracture walls (Fig. 3D "U"); *fracture fillings* are wider than fracture linings and commonly fill the central portion of a fracture, thus acting as a plug (Fig. 3D "Fe"). Fracture fillings may impregnate and hold together or cement a shattered mineral (Fig. 3E); *rims and halos* surrounding minerals are wider than coatings. Rims sourrounding uraninite and uranothorite consist of diverse mineral phases, such as calcite, fluorite, pyrite, goethite, uraniferous clays, REE-bearing and U, Th-bearing phases (Rimsaite 1980). The origin of the rims is complex. Some of the rims, such as fluorite, can be deuteric, whereas the others can be metamorphic and supergene. Some of the rims surrounding radioactive grains may exceed in width the enclosed grain and are altered as a result of radiation. Uraninite grains in biotite are enclosed in U, Ti, and Fe-bearing rims; and the pleochroic halo in biotite adjacent to the radioactive inclusion is commonly altered to chlorite and impoverished in Ti and K (Table 2, group III-2; Fig. 3, spectrum 4). Rims surrounding disintegrating allanite are chemically complex and consist of Si, Pb, Th, U, Ca, La, Ce, Fe (Fig. 3, spectrum 6). In fractured areas, outer bands of the rim extend into adjacent fractures.

The most common radioactive fracture fillings consist of U, Fe, Si, and Ca compounds and of uraniferous limonite and phyllosilicates. However, some fracture fillings, adjacent to disintegrating radioactive pyrochlore and to uranothorite, have uncommon chemical compositions, such as U, Nb, Si (in fractured apatite, Fig. 3E) and Fe, U, Si, Ti, Th, Pb (Table 2, group IV-1 and 4).

Radioactive fracture fillings in biotite or at grain boundaries between biotite and quartz commonly contain titanium associated with uranium and iron. Some fracture fillings are zoned and consist of uranophane-like linings adjacent to fracture walls and iron-rich phase filling the central portion of a fracture (Fig. 3D, spectra 1 and 2; Table 2, group IV-3 and 4, and IV-8). Parts of the zoned and colloform fracture fillings are cut by later perpendicular fractures along which the radioactive elements apparently derived from the uraniferous linings at the wall reacted with those from the central portions of the older fracture, thus providing evidences of several episodes of fracturing and of reactions between secondary uraniferous phases (Fig. 3D). Most of the chemically complex uraniferous phases that crystallized as rims and fracture linings are too small or too thin for positive X-ray identification and may comprise several new mineral species.

4 Crystallization of Secondary Uranium Ore Minerals and Supergene Enrichment of Uranium in Altered Granitoid Rocks

Deposition of secondary uraniferous minerals near the surface of some granitic rocks exposed to weathering has caused supergene enrichment of uranium. Thus surface samples are frequently more radioactive than drill core samples of the equivalent fresh rock at depth. Secondary U-bearing minerals are brightly coloured and readily visible in the field. Surfaces of some trenches in radioactive granites are very rich in secondary uranium minerals, as observed at Grandroy radioactive prospect on NW shore of Little Piashti Lake in Baie Johan Beetz area (abundant uraniferous phosphates and uranophane-rich crusts) and Mont-Laurier area ("thorogummite"), Quebec. Uranyl-bearing oxides and silicates commonly occur as fine-grained intergrowths of several phases, uranophane being the ultimate and apparently the most stable phase in the occurrences studied. Uranophane crystallizes in fractures as thin linings and as overgrowths on uraninite, biotite, and other silicates (Figs. 3B–D).

Uranophane fibres either pseudomorphously replace uraninite and uranothorite grains or crystallize as radiating aggregates on surfaces of biotite and feldspar and in vugs of argillized rocks. The uranium is derived from altered "primary" radioactive minerals while calcium and silica are dissolved in aqueous solutions penetrating altered granite and adjacent calc-silicate gneisses. Calcium can also be released from altered apatite, oligoclase and calcite, and silica is liberated during sericitization and kaolinization of feldspars.

$$CaCO_3 + H^+ \rightarrow Ca^{2+} + HCO_3^-; \quad (1)$$
(decomposition of calcite)

$$Ca^{2+} + 2UO_2^{2+} + 2SiO_2 + 6H_2O + 3O^{2-} \rightarrow Ca(UO_2)_2 \cdot Si_2O_7 \cdot 6H_2O; \quad (2)$$
(liberated constituents in an oxidation zone) → (uranophane)

Formation of secondary uranium phosphates, autunite and torbernite, is possible in areas containing decomposed apatite and other P-bearing minerals that can liberate phosphoric acid. Figure 3E provides an evidence that apatite can be replaced along fractures and indeed can liberate phosphoric acid and Ca for crystallization of autunite. Torbernite contains Cu which can be liberated from chalcopyrite that is present along with pyrite and other sulphides in granitic rocks studied (Rimsaite 1980). Some of the reactions pertaining to liberation of ions and formation of supergene radioactive minerals are as follows:

$$CuFeS_2 + 4H_2O \rightarrow Cu^{2+} + Fe^{2+} + 2SO_4^{2-} + 8H^+ + 8e^- \quad (3)$$
(oxidation of chalcopyrite and liberation of Cu and Fe)

$$2[Ca_5(PO_4)_3(F,Cl,OH)] + 5H_2O \rightarrow 6(H_2PO_4)^- + 7\,CaO + 2Ca^{2+} + CaF_2 + 2Cl \quad (4)$$
(decomposition of apatite and formation of phosphoric acid)

$$2(H_2PO_4)^- + Ca^{2+} + 2(UO_2)^{2+} + 8H_2O + 2O^{2-} \rightarrow Ca(UO_2)_2(PO_4)_2 \cdot 10H_2O; \quad (5)$$
(autunite)

$$2(H_2PO_4)^- + Cu^{2+} + 2(UO_2)^{2+} + 10H_2O + 2O^{2-} \rightarrow Cu(UO_2)_2(PO_4)_2 \cdot 12H_2O; \quad (6)$$
(torbernite)

$$2(H_2PO_4)^- + 3(UO_2)^{2+} + Ca^{2+} + 4H_2O + 2O^{2-} \rightarrow Ca(UO_2)_3(PO_4)_2 \cdot 6H_2O; \quad (7)$$
(phosphuranylite)

Thorogummite occurs as orange-yellow crust on radioactive granite exposed to hydration and weathering. Thorogummite crusts are very brittle and are traversed by numerous shrinkage cracks. A fragment of a thorogummite crust analyzed using a scanning electron microscope and an energy dispersive spectrometer showed the presence of Th, U, Si, Al, Ca, and Pb (Table 2, group V-2). Chemical compositions of other spectra of diverse uraniferous crusts on weathered granites are given in Table 2, group V. The supergene uraniferous crusts can penetrate the granite along the fractures and coat "primary" radioactive minerals (Fig. 3C).

Supergene uranium minerals have also been found in the Bancroft area, including uranophane, kasolite, "thorogummite", rutherfordine, and others, Biotite flakes coated with radiating uranophane aggregates contained about ten times more uranium than the host pegmatite.

Some of the reactions involved in alteration of U-bearing minerals and dissolution and transportation of uranium in solution (possibly as bicarbonate) are probably more complex and include metastable phases. Formation and preservation of chemically complex interstitial bands and of supergene radioactive crusts is possible under locally favourable conditions where ions released from adjacent minerals are fixed in secondary compounds instead of being dissolved and removed from the rock by aqueous solutions.

5 Conclusions

Textural relationships between heterogeneous uraninite and uranothorite grains imply at least three episodes of initial crystallization: (1) crystallization of primary uraninite grains that remain as irregular broken-up cores and fragments enclosed in (2) euhedral uraninite crystals that in turn have corroded patches and are replaced by (3) Th, REE, and Si-bearing phases.

Chemically complex rims on radioactive and on nonradioactive minerals imply post-crystallizational reactions between ions released from the adjacent minerals as a result of radiation damage and enhanced alterations in hydrous environment. Uraniferous compounds with spectra containing Ca, Cu, Nb, P, Pb, REE, Fe, Th, Ti, Y, and Zr have been indicated using an energy dispersive spectrometer on less than one micron spots of a rim. Part of the uranium liberated from decomposing "primary" radioactive minerals precipitated as secondary minerals along the margins of the source mineral and as linings and fillings in fractures. The remainder migrated some distance and ultimately formed secondary uranium minerals that produced local supergene enrichment of uranium in weathered zones. Evidences for both types of precipitation have been observed in the composition and textures of the mineral assemblages studied, and can be interpreted in terms of the chemical reactions attendant on hydration and weathering.

Isotope studies of U, Th, and Pb are being made to determine temporal relationships between primary uranium minerals, radioactive fracture fillings, and supergene uraniferous mineral aggregates.

Acknowledgments. The author gratefully acknowledges the following assistance provided by the staff of the Central Laboratories and Administrative Services Division of the Geological Survey of Canada: X-ray identification of mineral aggregates by A.C. Roberts, scanning electron microscope and energy dispersive spectrometric studies by D.A. Walker, electron microprobe analysis by A.G. Plant; and a critical review by S.S. Gandhi of the Geological Survey of Canada.

References

Backström JW von, Jacob RE (1979) Uranium in South Africa and South West Africa (Namibia). Philos Trans R Soc London Ser A 291:307–319

Baldwin AB (1970) Uranium and thorium occurrences on the north shore of the Gulf of St. Lawrence. Bull Can Mining Metall 63:699–707

Bullis AR (1965) Geology of Metal Mines Ltd. (Bancroft Division). Trans Can Inst Mining Metall LXVIII:195–203

Cameron J (1979) Mineralogical aspects and origin of the uranium in the vein deposits of Portugal. IAEA Tech Comm Meet Geol Vein-Type U Depo, Lisbon, TC-295-18, 21 pp

Cuney M (1978) Geologic environment, mineralogy and fluid inclusions of the Bois Noirs – Limouzat uranium vein, Forez, France. Econ Geol 73:1567–1610

Fernandes AP (1979) Alto Alentejo uranium district. Field Excursion. IAEA Tech Comm Meet Geol Vein-Type U Depo, Lisbon, 8 pp, 5 maps

Kish L (1977) Patibre (Axe) Lake area. Preliminary report. Ministère des Richesses Naturelles, Quebec, DPV-487, 13pp + map

Leroy J (1978) The Margnac and Fany uranium deposits of the La Crouzille District (Western Massif Central, France): Geology and fluid inclusions study. Econ Geol 73:1611–1634

Ramdohr P (1969) The ore minerals and their intergrowths. Pergamon Press, Akademie Verlag, New York, 1174 pp

Rimsaite J (1978) Mineralogy of radioactive occurrences in the Grenville structural province Ontario and Quebec. Geol Surv Can Pap 78-1B:49–58

Rimsaite J (1979) Chemical and isotopic evolution of radioactive minerals in remobilized vein-type uranium deposits, Saskatchewan, Canada. IAEA Tech Comm Meet Geol Vein-Type U Depo, Lisbon, TC-295-3, 16 pp

Rimsaite J (1980) Mineralogy of radioactive occurrences in the Grenville structural province, Bancroft area, Ontario. Geol Surv Can Pap 80-1A:253–264

Rimsaite J, Lachance GR (1966) Illustrations of heterogeneity in phlogopite, feldspar, euxenite and associated minerals. In: Int Mineral Assoc IV Gen Meet India, 1964, IMA Volume, pp 209–229

Snelling AA, Dickson BL (1979) Uranium/daughter disequilibrium in the Koongarra uranium deposit, Australia. Miner Deposita 14:109–118

Genesis of Ores in Igneous Rocks

Native Tellurium from Musariu, Brad Region, Metaliferi (Metalici) Mountains, Romania

I. BERBELEAC and MARGARETA DAVID[1]

Abstract

Two veinlets of compact fine-grained and coarse-grained allotriomorphic aggregates of native tellurium were found in the Musariu mine. The visible grains range from 1 mm to 30 mm and show tin-white colour and brilliant lustre. The forms observed are: m $\{10\bar{1}0\}$, R $\{10\bar{1}1\}$ and r $\{01\bar{1}1\}$. D = 6.02–6.16 g/cm^3. Microindentation hardness (VHN$_{10-20}$) = 29–74 kg/mm^2. Main optical properties: white colour with creamy tints; slight reflexion pleochroism; strong anisotropy (grey-white) and reflectance (%) = 63.46 (486), 63.29 (551), 62.48 (589), 59.17 (656) in air, and 56.64 (486), 56.58 (551), 55.0 (589), 54.32 (656) in immersion. The chemical analyses of two specimens gave 99.45% and 96.93% Te and minor amounts of Se, Cu, Fe, Au, SiO_2 and Al_2O_3. The strongest X-ray powder diffraction lines are: 3.23 (100), 2.35 (37), 2.25 (33) 1.824 (20), 3.853 (16), 1.615 (11), 2.083 (10) and 1.475 (10). We note the associations of native tellurium with other tellurides, quartz and clay minerals.

1 Introduction

The main occurrences of native tellurium from Romania belong to the Neogene metallogenic periods and are restricted to the Metaliferi Mountains area. In this part of the Apuseni Mountains, the most characteristic occurrences of native tellurium are to be found in four old mining fields: Săcărîmb, Zlatna-Almaşu Mare, Stănija and Ruda-Barza. Of these, the Ruda-Barza mining field is situated at about 2 km E of the Musariu area. Details concerning these native tellurium occurrences are given by numerous authors (Helke 1934; Giuşcă 1936; Ghiţulescu and Socolescu 1941; Dana and Dana 1961; Rădulescu and Dimitrescu 1966; Vlasov 1966; Ramdohr and Udubaşa 1973; Ianovici et al. 1976; etc.).

As far as the native tellurium occurrence from Musariu is concerned, we underline the fact that it was unknown until recently, and only identified during the summer of 1978. Field and laboratory investigation data furnished the basic information for this paper.

1 Institute of Geology and Geophysics, Str. Caransebeş 1, Bucharest, Romania

2 Geology

The Musariu area is located in the central part of the Metaliferi Mountains, at about 5 km SE of the town of Brad. The regional geology of the Brad region is discussed by Ghiţulescu and Socolescu (1941) and Ianovici et al. (1976, 1977); detailed descriptions of the geology of Musariu gold and prophyry copper mineralization are given by Ianovici et al. (1977) and Borcoş et al. (1978).

In brief, the Musariu area is built up of Mesozoic ophiolites and Neogene sedimentary volcanic rocks (Fig. 1a). These rocks are intruded by the so-called Musariu subvolcanic body which consists of quartz andesite and quartz diorite. It is worth emphasizing the fact that the most important gold, tellurium, gold and silver tellurides and porphyry copper mineralization from the Musariu area are genetically and spatially associated with this subvolcanic body. We also observe that the gold, tellurium and gold and silver tellurides are situated in the upper part of the body, while the porphyry copper mineralization lies in the depth.

3 Native Tellurium Veinlets

The native tellurium described here was found within two veinlets (Fig. 1b, nos. 1 and 2) situated in the neighbourhood of a native gold vein. These veinlets strike NW-SE and E-W and occur at about 20 m below the IInd level from the Musariu mine. They lie within argillized quartz andesites. The thickness of the veinlets is sharply reduced, ranging from 0.5 cm to 3 cm, exceptionally 5 cm. The veins have, as a rule, marginal bands or nets of clay minerals (Fig. 1c, d).

Within the veinlets, the native tellurium and gold and silver tellurides are the principal ore minerals accompanied by some pyrite, chalcopyrite, frohbergite, weissite and tellurite; quartz and clay minerals are the main gangue minerals. Minor amounts of barite also occur. Quartz is the main gangue mineral; it appears as fine grain aggregates or grey-white chalcedony. As a rule, quartz is associated with native tellurium and gold and silver tellurides. The clay minerals, usually kaolinite, lie close to the wall rock and include pyrite, marcasite and alabandite (Fig. 1d). In the gold vein, the clay minerals are predominant.

From these two tellurium veinlets, veinlet no. 2 consists almost exclusively of native tellurium, while veinlet no. 1 contains native tellurium, gold-silver-copper and iron tellurides and common sulphides. In the latter case, the native tellurium predominates in the inner part of the veinlet, whereas the tellurides and sulphides occur in the outer part.

The ore's texture is mostly massive. Less frequently, impregnation, brecciated, drusy and cavernous textures are present.

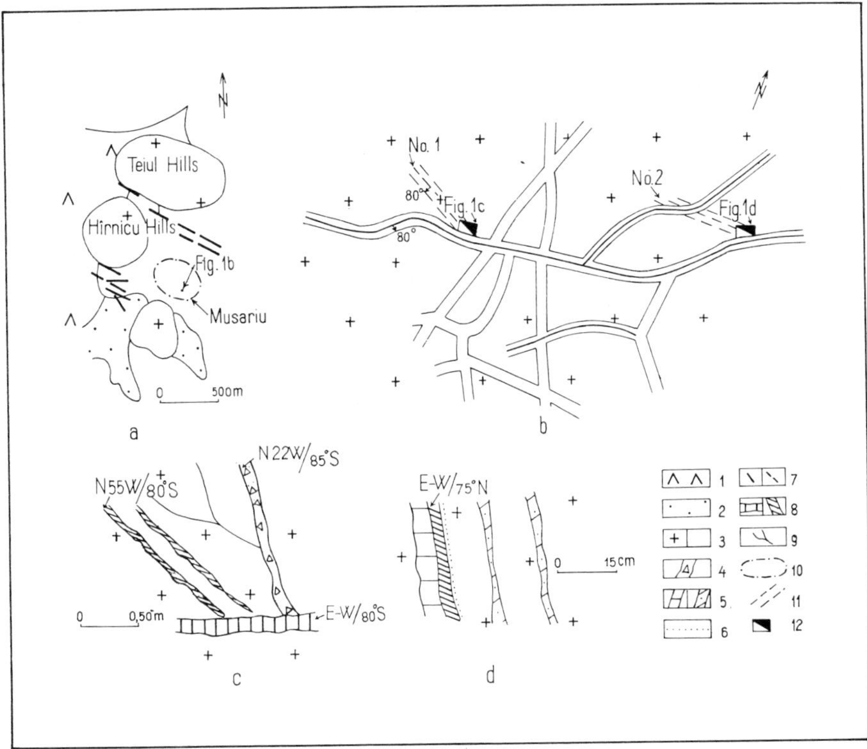

Fig. 1. Geological sketch map of Musariu area **(a)** and some details of native tellurium occurrence **(b–d)**. *1* Mesozoic ophiolites, *2* Tertiary sediments, *3* Tertiary eruptive rocks; **a** quartz andesite intrusion; **b** lavas and pyroclasts of quartz andesite rocks, *4* fracture with breccia filling, *5* kaolinite **(a)** and dark-grey clay minerals **(b)**, *6* pyrite + alabandite, *7* native gold vein at the IInd level **(a)** and native tellurium at the IIIrd level **(b)**, *8* native gold vein in detail **c (a)** and native tellurium in detail **(b)**, *9* clay minerals veinlet, *10* outline of subvolcanic body at 500 m below the surface, *11* the IIIrd level, *12* raise

4 Description of Native Tellurium

Physical and Optical Properties. In the specimens studied, the native tellurium forms visible monomineralic and polymineralic fine or coarse-grained compact allotriomorphic aggregates. The visible individual grains of the coarse-grained aggregates often reach 1–2 cm and exceptionally 3 cm in length. The same large crystals were found in the Zlatna-Almaşu Mare (Romania), Kawatsu and Tein (Japan) regions and John Joy mine (Colorado) (Vlasov 1966). However, many visible grains fall in the range from 2–3 mm to 5 mm in length.

The grains of native tellurium have on the fresh surface a tin-white colour and a brilliant metallic lustre. After a lapse of time, the surface becomes white-grey with black tints. The surface of the grains often conserves the traces of the cleavage planes [$\{10\bar{1}0\}$ perfect and $\{0001\}$ imperfect], while the crushed fragments show the step-like and plain-conchoidal fractures. In these fragments, inequidimensional prismatic and more rarely acicular crystals of different degrees of perfection appear. These crystals have an oblique arrangement. Sometimes, near the wall rock or in geodes, imperfect crystals with diverse and irregular development occur. In this case, the forms which can be observed are: m $\{10\bar{1}0\}$, R $\{10\bar{1}1\}$ and r $\{01\bar{1}1\}$. The best-developed forms were found in veinlet no. 2. These forms are reminiscent of those recognized by Locza (1890, in Dana and Dana 1961) in the Zlatna-Almaşu Mare region and Watanabe (1960, in Vlasov 1966) in the Kawatsu and Tein mine from Japan.

Native tellurium has a grey streak; its microindentation hardness (VHN_{10-20}) ranges between 29 and 74 kg/mm^2. We note that the microhardness is in agreement with literature data (Vlasov 1966; Uytenbogaardt and Burke 1971). The native tellurium polishes well, but in relation with harder minerals it is difficult to obtain a good polished surface. Its hardness is low and smaller than that of calaverite, krennerite, montbrayite, empressite, petzite and hessite, and greater than that of sylvanite and altaite.

The density tests show the following values: 6.02–6.06 g/cm^3 for the native tellurium specimen of the veinlet no. 1 and 6.12–6.16 g/cm^3 for the native tellurium of veinlet no. 2. Each specimen tested by us weighed 0.8–1.2 g and represented an aggregate of crystals in the first case and a single crystal in the latter. The density was measured with the picnometre in 10 ml of water at 22°C. The difference of values between the two densities may reflect the fact that the material collected from veinlet no. 1 consisted of aggregates of crystals which had probably some gangue minerals or some spaces in between. As far as the chemical test is concerned, the literature data are confirmed (Dana and Dana 1961; Vlasov 1966; Ramdohr 1975). We mention the weak etching texture obtained by means of $HCl+CrO_3$ and $KMnO_4+HNO_3$ (1′).

In reflected light, the native tellurium is white with a lightly perceptible creamy tint. Compared with some tellurides with which it may be confused, the colour of native tellurium appears under different aspects, such as: white more lighter than hessite; white lighter than krennerite and calaverite; white less bright than altaite and white with less creamy tints than montbrayite and sylvanite. It has a very high reflectivity. The reflectivity values for the native tellurium from Musariu area are given in Table 1. The measurements were made with the microphotometer ocular, using metallic silicon standard. The material collected from veinlet no. 1 represents a coarse-grained crystal aggregate.

We emphasize that our values, small differences taken into account, agree with literature data (Folinsbee 1949; Ramdohr 1975). These differences may be explained especially by the quality of the polished section.

The reflection pleochroism of native tellurium is slight, especially when seen in isolated crystals. It is considerable at the grain boundaries where two colours following two directions can be noticed: O = white and E = light brown. Often, these two colours range from: O = white to grey-white and E = light brown and creamy brown to brown-grey. These colours are more distinct in oil immersion.

The native tellurium has a strong anisotropy with no pronounced colour effects: grey with bluish and brownish tints. This and the above-mentioned optical properties were used to distinguish it especially from montbrayite, krennerite and calaverite.

Table 1. Reflectance of native tellurium. (Data furnished by N. Pop)

	λmμ			
	486	551	589	656
R_1%	63.46	63.29	62.48	59.17
R_2%	56.64	56.58	55.00	54.32

R_1 – Reflectance in air; R_2 – Reflectance in oil

Forms of Aggregates, Intergrowth and Textures. In the polished sections, three morphostructural types of native tellurium are distinguished: (a) early, drops and minute grain inclusions in sylvanite; (b) euhedral to subhedral crystal aggregates and anhedral grain aggregates deposited after the sulphides and older tellurides (sylvanite, frohbergite, nagyagite, krennerite, calaverite and montbrayite) and before the younger tellurides [petzite, hessite, empressite (stuetzite), altaite] and (c) later, minute grains and films located often along the grain boundaries of sulphides and older tellurides and previous to native tellurium grains and crystals. The last type may be interpreted as a residue rejected from previous crystallizations. Of all these, the second generation is the most representative, appearing in fine-grained and coarse-grained prismatic (Fig. 2a) and allotriomorphic aggregates. The internal geometry of the aggregates was influenced by the competition between the individual grains and the degree of the mixture with mineral inclusions, which led to a more or less inequigranular texture. This fact explains the variable shape and perfection of individual grains and crystals contained in the same volume and of comparable size. Excepting the large grains, macroscopically visible and previously mentioned, many grains of native tellurium range from 0.1 mm to 0.5 mm in length. In the case of partial replacement, the size is measurable in microns.

The outstanding textural features of native tellurium consist in variety and complexity of intergrowths of tellurides and sulphides and the relationships of these minerals with the native tellurium. The investigation of all these permitted us to identify the native tellurium in the following paragenetic associations: sylvanite-nagyagite (Fig. 2b), pyrite-chalcopyrite-weissite (Figs. 2c and 3b), pyrite-chalcopyrite-frohbergite (Figs. 2d and 3a, b), sylvanite-montbrayite, sylvanite-krennerite (Figs. 2f and 3c) and sylvanite-empressite (Fig. 2h). The variety of intergrowths is due to the partial replacement of pyrite (Figs. 3 and 4), alabandite, chalcopyrite (Fig. 3a), weissite (Fig. 2c) and frohbergite (Figs. 2d and 3a). In the intergrowth and replacement zone the sulphides, weissite and frohbergite develop a convex outline and native tellurium adopts complicated forms such as concave splashes with cuspate tentacles within sulphides and tellurides. That is why, in an intimate intergrowth zone in native tellurium, the matrix appears as a series of more or less rounded grains or aggregates of sulphides (Figs. 3 and 4). The same aspect of intergrowth shows, in general, the contact zone between the krennerite (Fig. 3c), calaverite (Fig. 2g), montbrayite (Fig. 2e) and native tellurium. The boundaries of tellurium with these tellurides are, in general, complex and interpenetrating. If in one of the cases mentioned above the replacements are pregnant, sometimes only an incipient replacement was found at the contact between

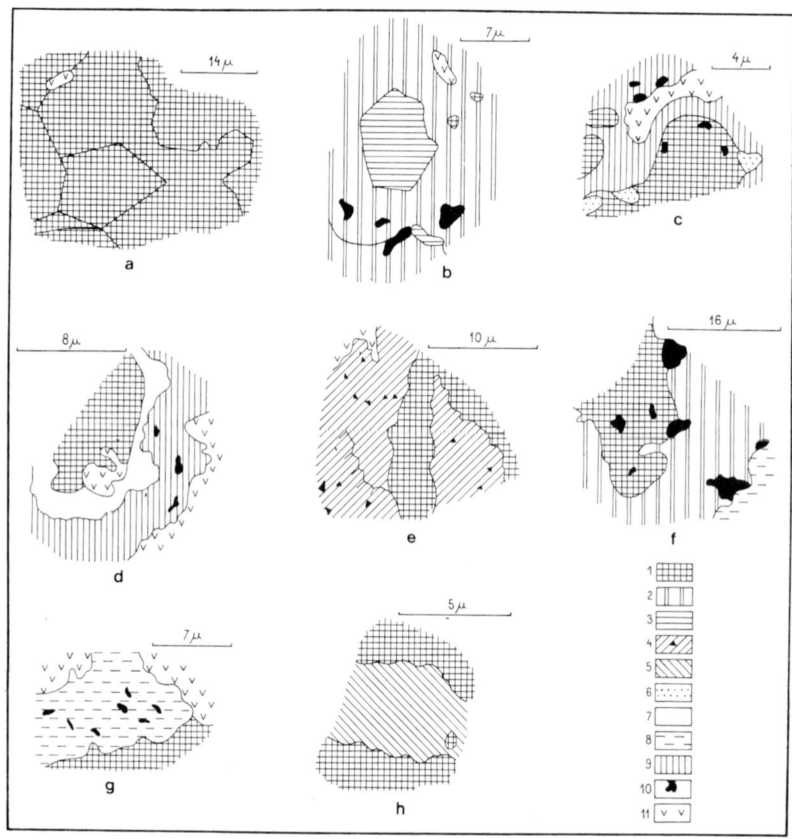

Fig. 2. Paragenetic relations of native tellurium from Musariu. *1* native tellurium, *2* sylvanite, *3* nagyagite, *4* montbrayite, *5* empressite, *6* weissite, *7* frohbergite, *8* calaverite, *9* chalcopyrite, *10* pyrite, *11* gangue minerals (frequently quartz)

Fig. 3

a Chalcopyrite *(grey)* and frohbergite (*white-grey,* down to the *right corner,* between the chalcopyrite and native tellurium) replaced by native tellurium. In black gangue minerals. Veinlet no. 1, × 250, imm

b Pyrite replaced by native tellurium *(white-grey).* The contact zone consists of swarms and minute grains of pyrite inclusions in native tellurium. In centre an inclusion of frohbergite. Veinlet no. 1, × 250, imm

c Intergrowth of native tellurium *(white-grey)* and krennerite *(dark grey).* We note the preference of pyrite inclusions for krennerite. Note the distinct optical contrast between the minerals. In upper corners two pyrite grains. Veinlet no. 1, × 250, imm

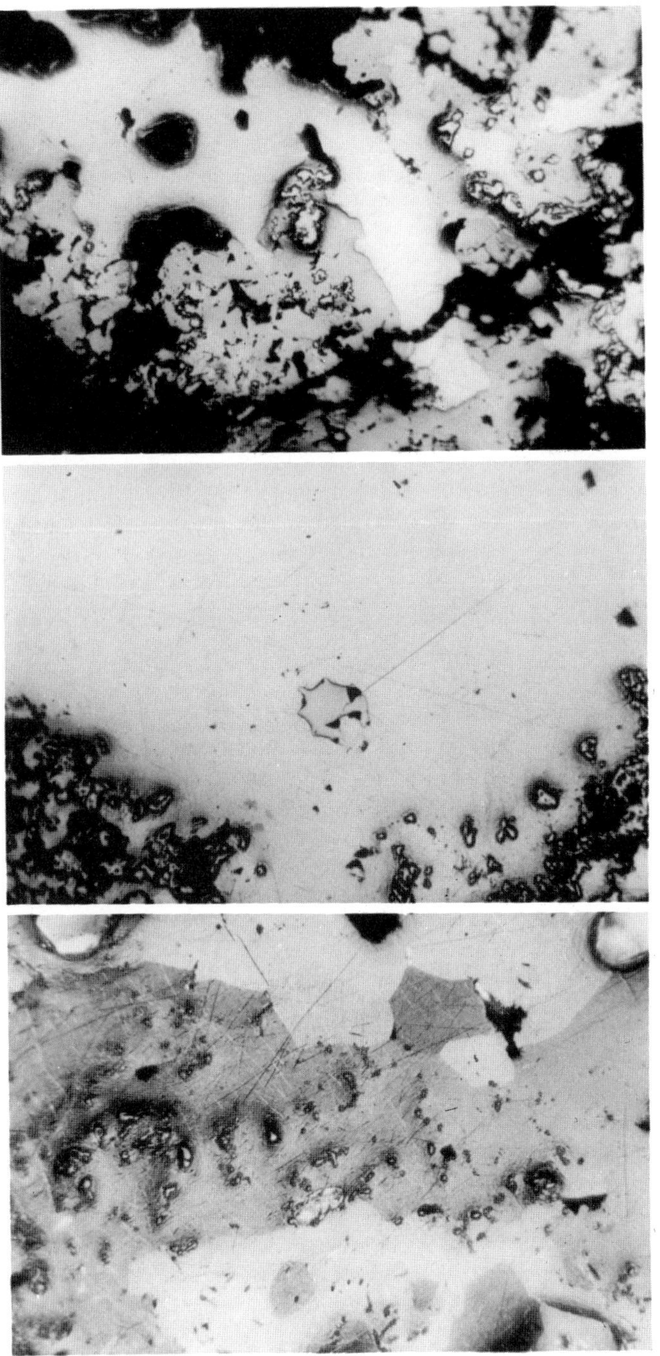

sylvanite and native tellurium. In this case, the outline of the euhedral crystals of sylvanite was not at all or only very little affected (Fig. 4a). For all that, in other cases, the native tellurium deeply penetrates the sylvanite grains (Fig. 2f).

The textural relationships of native tellurium with petzite, hessite, empressite, altaite and two unidentified minerals (grey colour, weak but perceptible pleochroism with greyish to brownish colours, myrmekitic textures, etc.), probably silver and gold tellurides, are in general a very sinuous line. A close examination shows that in many places the above-mentioned tellurides replace tellurium (Fig. 2h) and that tellurium is crossed by telluride veinlets (Fig. 4b).

In many cases, the native tellurium outline grains are wholly subordinated to the crystal faces of quartz. That fact suggests that the native tellurium was formed partially through the filling of ancient geodes. In this case, a typical cemented structure was formed (Fig. 4c).

Taking all these into consideration, it is seen that the native tellurium demonstrates complicated textural relationships with sulphide and tellurides. Under the microscope these relationships helped to specifiy the special ratios among the native tellurium, ore minerals and gangue minerals which can be resumed as follows: (a) native tellurium penetrates the grain boundaries of sulphides and older tellurides and may contain inclusions of these minerals, as well as quartz and clay minerals; (b) native tellurium is penetrated by veinlets of younger tellurides and (c) native tellurium occurs as drops in sylvanite (Fig. 2b) and as irregular grains in younger tellurides and probably, silver-gold unidentified tellurides.

According to the above data, a provisional sequence of the ore minerals from the two described veinlets may be given. The sequence began with common sulphides, continued with copper and iron and gold tellurides and gold and silver tellurides, native tellurium and terminated with petzite, silver tellurides, unidentified tellurides and altaite.

5 Chemical and X-Ray Data

The material selected for chemical analysis was examined both optically and with X-ray in order to obtain, as far as possible, specimens free from impurities.

For analyses, two specimens collected in the mine from veinlet no. 1 (Table 2, no. 1) and veinlet no. 2 (Table 2, no. 2) were used. Specimen no. 1 represents a pure coarse-grained crystal aggregate, while specimen no. 2 is a fine-grained aggregate. A

Fig. 4

a Incipient replacement of sylvanite (some crystals with lamellar twinning) by native tellurium (*grey* and *white-grey* in *half right* and *white* in *left corner* of the picture). We note a strong replacement of pyrite by sylvanite. Veinlet no. 1, × 250, // nicols

b Veinlet with a myrmekitic intergrowth of two unidentified minerals (possibly gold-silver tellurides) in native tellurium *(white-grey)* with minute grains of pyrite *(white)*. Veinlet no. 1, × 250, imm

c Typical cementation structure. Hexagonal and prismatic inclusion crystals of quartz in native tellurium. In *left corner* and in *upper central part* an incipient transformation of native tellurium in tellurite. Veinlet no. 2, × 250, // nicols

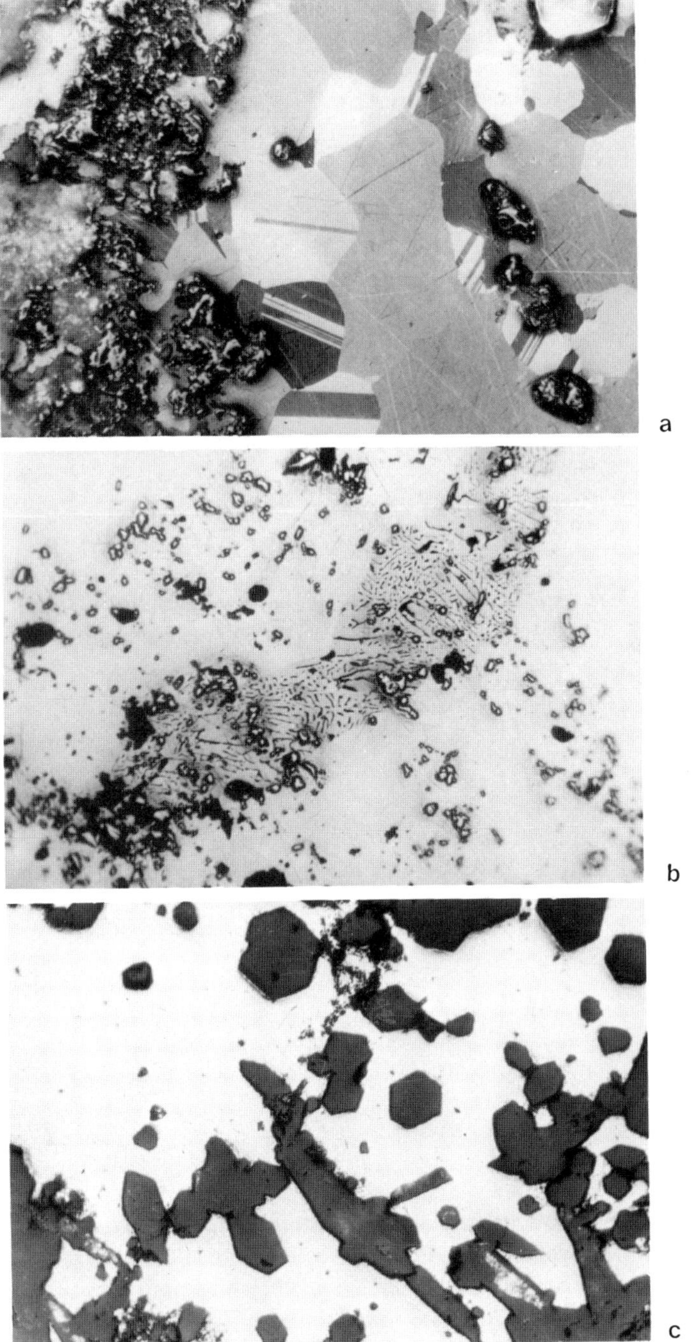
a
b
c

small amount of quartz and clay minerals were present in both specimens. In specimen no. 2, a negligible amount of tellurides could also have been present. The resulting chemical composition is very satisfactory, comparable with the already existing ones given by other authors (Table 2). These analyses show that the native tellurium carries small amounts of Se, Au, Cu and Fe.

Table 2. Chemical analyses of native tellurium

Components	1[a]	2	3	4[b]	5	6
Te	99.45	96.93	97.92	92.29	99.45	96.94
Se	0.025	0.019	traces		0.40	
Au			0.15	3.40		2.40
Ag				1.69		
Cu	0.012	0.027	0.06			
Fe	0.05	0.15	0.53	0.12	0.11	
S	0.002	1.34				
SiO_2	0.31	0.77	1.56			
Al_2O_3	0.12	0.50				
Total	99.96	99.73	100.22	99.94	99.96	99.34

a 1.0235 g/t. Analyst A. Serban. Spectrometric method with ^{198}Au isotope
b 2.44% others

1,2. Musariu mine. Analyst Margareta David. 3. Zlatna-Almaşu Mare. 4. Ballarat District, Colorado. 5. Gunnison Co., Colorado. 6. Hannaus District, Western Australia. 3–6 From Dana and Dana (1961)

X-ray data of native tellurium (Table 3) confirm the hexagonal crystallographic system and represent the measurements made on a pure crystal aggregate from veinlet no. 2. Though some deviations are noted, the results of the X-ray powder pattern are similar to those given by the ASTM card. The strongest lines are as follows: 3.23 (100), 2.35 (37), 2.225 (33), 1.824 (20), 3.853 (16), 1.615 (11), 2.083 (10) and 1.475 (10); Å, $(I/_I)$.

6 Discussion

Native tellurium from the Musariu mine occurrence shows, in contact with other minerals, diverse and complicated relationship textures. However, it seems that the position of native tellurium in paragenetic sequence may be estimated. This paragenetic sequence consists of many ore mineral associations, which formed during two main periods: (1) early, with sulphide (pyrite-alabandite-sphalerite-chalcopyrite) and (2) later, with native tellurium and tellurides. In the later period, two assemblages seem to be distinct: (a) older, in which the weissite, frohbergite, gold tellurides and gold-silver tellurides and native tellurium were formed and (b) younger, which consists of petzite, silver tellurides, unidentified tellurides and altaite. Contrary to the Săcărîmb deposit

Table 3. X-ray powder data for native tellurium

1				2					
dÅ	I	dÅ	I/I_1	dÅ	I/I_1	hkl	dÅ	I/I_1	hkl
3.853	16	1.306	4	3.86	20	100	1.309	6	212
3.230	100	1.255	3	3.230	100	101	1.287	1	300
2.350	37	1.174	9	2.351	37	102	1.257	4	301
2.225	33	1.130	2	2.228	31	110	1.234	1	114
2.083	10	1.093	2	2.087	11	111	1.1802	3	302
1.976	8	1.053	3	1.980	8	003	1.1740	8	{204, 213}
1.927	3			1.930	4	200	1.1334	3	105
1.824	20			1.835	20	201	1.0951	2	221
1.778	5			1.781	7	112	1.0784	< 1	303
1.757	2			1.758	2	103	1.0705	< 1	310
1.615	11			1.616	12	202	1.0535	3	311
1.475	10			1.479	13	113	1.0432	2	115
1.457	8			1.459	8	210	1.0399	3	222
1.45	9			1.457	8	211	1.0104	2	214
1.381	7			1.383	7	{104, 208}	1.0071	3	{205, 312}

1 Native tellurium, Musariu mine, Cu k_α radiations, Ni filter. Diffractometer method. Analyst I. Vanghele

2 Native tellurium, 4.0544 ASTM card (Berry 1974)

(Giuşcă 1936), where the native tellurium appears in the first generation of tellurides (sylvanite → petzite → hessite → Te → altaite), in the Musariu occurrence, the native tellurium seems to have an intermediary position, delimiting the older sequence (weissite → frohbergite → sylvanite → nagyagite → krennerite → calaverite → montbrayite) from the younger sequence [petzite → hessite → empressite (stuetzite) → unidentified tellurides → altaite]. We point out the provisional character of the above order of the two mentioned generations of tellurides. We note also the fact that native tellurium occurs with many typical mineral associations (sylvanite-nagyagite; sylvanite-krennerite; sylvanite-montbrayite and sylvanite-calaverite; etc.), which is in agreement with the phase relations in the Au-Ag-Te ternary system (Markham 1960, in Vlasov 1966).

It should be emphasized that native tellurium occurrence from the Musariu mine belongs to the "young" subvolcanic type (Vatukoula, Fiji Islands; Vlasov 1966) and can be compared with other important and similar occurrences in the world (Kawatsu, Tein, John Joy) (Vlasov 1966; Ramdohr 1975).

The data we dispose of at present are not sufficient to estimate the genetic relationships between the native gold of the vein adjacent to the described veinlets and the native tellurium and tellurides contained by them. However, it seems interesting to note the fact that the vein of native gold does not contain the tellurium nor any of the tellurides mentioned by us. In all events, both mineralization types are low temperature ones, have a primary origin, and were formed at very short time intervals.

Acknowledgments. Thanks are due to the leaders of Mining, Petroleum and Geology Ministry for permission to publish this study and to our colleagues from Brad Mining Enterprise who aided in the sampling of Musariu ore deposit. The first author thanks Dr. G. Udubaşa for suggesting this study and for many helpful discussions. The authors wish to express their gratitude to N. Pop and I. Vanghele for reflectance and X-ray data used in this study.

References

Berry G (ed) (1974) Powder diffraction data for minerals, 1st edn. Philadelphia, 831 pp

Borcoş M, Berbeleac I, Gheorghiţă I, David M, Zămîrcă A, Vanghele I, Şerban A (1978) Raport geologic asupra regiunii Musariu-Valea Talpelor. Unpublished report. Inst Geol Geophys, Bucharest

Dana JD, Dana SE (1961) System of mineralogy. 7th edn, entirely rewritten and enlarged by Palache Ch, Berman H, Frondel C (eds), vol I. New York, London, 834 pp

Folinsbee RE (1949) Determination of reflectivity of the ore minerals. Econ Geol 44:425–436

Ghiţulescu TP, Socolescu M (1941) Etude géologique et minière des Mont Métalifères (quadrilatère aurifère et régions environnantes). An XXI:1–126, Bucureşti

Giuşcă D (1936) Nouvelles observations sur la minéralisation des filons aurifères de Săcărîmb. Bul Acad Roum Sci 18, 3–5: 97–103

Helke A (1934) Die Goldtellurerzlagerstätten von Săcărâmb (Nagyag) in Rumänien. Neues Jahrb Mineral Abh Abt A 68:19–85

Ianovici V, Borcoş M, Bleahu M, Patrulius D, Lupu M, Dimitrescu R, Savu H (1976) Geologia Munţilor Apuseni. Ed Acad, Bucureşti, 631 pp

Ianovici V, Vlad Ş, Borcoş M, Boştinescu S (1977) Alpine porphyry copper mineralization of West Romania. Mineral Deposita 12:307–317

Markham NL (1960) Synthetic and natural phases in the system Au-Ag-Fe. Econ Geol 55/6: 1148–1178

Rădulescu D, Dimitrescu R (1966) Mineralogia topografică a României. Ed. Acad, Bucureşti, 376 pp

Ramdohr P (1975) Die Erzmineralien und ihre Verwachsungen. 4. Aufl. Akademie Verlag, Berlin, 1277 pp

Ramdohr P, Udubasa G (1973) Frohbergit-Vorkommen in den Golderzlagerstätten von Săcărâmb und Fata Băii (Rumänien). Mineral Deposita 8:179–182

Uytenbogaardt W, Burke EAJ (1971) Tables for microscopic identification of ore minerals. 2nd revised edn. Elsevier, Amsterdam London New York, 430 pp

Vlasov AK (ed) (1966) Geochemistry and mineralogy of rare elements and genetic types of their deposits, vol II, 945 pp (English translation, Jerusalem)

Metallogenetic Area of Zletovo-Kratovo, Yugoslavia: Mineralogy and Genesis of Ore Deposits

D. RADUSINOVIĆ[1]

Abstract

The Zletovo-Kratovo metallogenetic area is featured by the deposits of lead, zinc, copper and uranium, formed under hydrothermal conditions related to Tertiary magmatic activity. The repeated tectonic and magmatic activities have influenced the changes in the chemical composition of postmagmatic mineralizing solutions. Consequently, as the result of a given trend in differentiation of mineralizing solutions, both in time and in the space, the mineral parageneses of a group of metals were replaced by the parageneses of other metals. In this way the mineral deposits were formed by this succession: copper–lead/zinc–uranium; the same succession has resulted in the characteristic zonality in their spatial distribution.

1 Introduction

The Zletovo-Kratovo area, within the framework of the geology of Yugoslavia, is located at the boundary of two large geotectonic units – the orogene zone of Dinarides and the internide Serbo-Macedonian mass; actually, this area is located within the axial zone of the Alpine orogene, in this segment of its development.

This area shows a specific contrast in its composition between the formations of mostly Paleozoic age, and those formed in Tertiary. The most widespread are the crystalline rocks of Paleozoic and partly of pre-Paleozoic age, Tertiary volcanites, and the Tertiary sediments.

The crystalline rocks include gneisses, micaschist, chlorite-sericite and amphibole-chlorite schists, phyllite, and quartzite. Here occur locally also the metamorphosed diorite and gabbro, with strong cataclasis; the gabbro is also uralitized as well.

The volcanites consist of andesite and dacite and their pyroclastites – ignimbrite, tuff, and tuffaceous breccia (Divljan 1964). The effusive and extrusive activity in this area was associated with the minor intrusives, solidified within the subvolcanic or even hypabyssal level, such as the quartz latite, latite (trachyandesite), and quartz monzonite-porphyry.

The Eocene sediments, composed of conglomerate, breccia, sandstone, limestone, and marl, were deposited transgressively over the crystalline schists. These sediments contain the tuffaceous sandstone and breccia, which indicate the beginning of the

1 Geoinstitut, Rovinjska 12, 1100 Belgrade, Yugoslavia

volcanic activity in this area. The Miocene-Pliocene sediments consist of conglomerate, sandstone, marl, tuffite, breccia, and limestone.

The Zletovo-Kratovo area is featured by frequent fold- and fault-type structures; in the compressive phases of orogeneses the rock masses were transported in a horizontal direction; during the decompression the deep faults were formed, which served both for the vertical displacements and as the feeders for large masses of magma. The faults are steep, and mostly of NW–SE to N–S, and ENE–WSW to E–W trends. The multiphase characteristics of the magmatic activities are demonstrated by the different composition of volcanic material. The andesitic rocks were followed by dacitic, with pyroclastic extrusive material, mostly of ignimbritic character, dominating in the subsequent phases. Finally, the small subvolcanic and hypabyssal intrusive bodies were formed at their corresponding levels of solidification. Within the different tectonic and magmatic phases, the postmagmatic activities, associated with the mineralizing processes of various composition, have taken place.

2 Mineralogical Features of Ore Deposits

The Zletovo-Kratovo area is featured by the endogene metallogeny of the Alpine epoch; the traces of formations of other metallogenetic epochs were not discovered. Consequently, during the final Tertiary magmatic activity, the numerous and different lead, zinc, copper, and uranium ore deposits and mineral occurrences were formed. From the economic point of view, the most important is the Dobrevo lead-zinc deposit near Zletovo. Among the copper mineralizations the most interesting is Zlatica, and the most important uranium deposit is located at Zletovska Reka.

2.1 Dobrevo Lead-Zinc Deposit

The Dobrevo ore deposit is located within the pyroclastic rocks (ignimbrite) and consists of more than ten ore veins. This mineralization is related to NW–SE fault-type structures, from a few centimetres to more than 10 m wide, and between a few hundred metres to several kilometres long; their depths extend to severel hundred metres.

The mineral composition is highly variable and the following minerals were identified (Ristić and Radusinović 1954; Cissarz and Rakić 1956; Denkovski 1974; Radusinović 1974):

galena, sphalerite, pyrite, pyrrhotite, marcasite, magnetite, hematite, franklinite, jacobsite, alabandite, arsenopyrite, chalcopyrite, valleriite, cubanite, bismuthinite, native bismuth, aikinite, native gold, tetrahedrite, tennantite, proustite, bournonite, boulangerite, enargite, chalcocite, covellite, bornite, famatinite, luzonite, pitchblende, quartz, chalcedony, rhodochrosite, siderite-oligonite, barite, calcite, fluorite, cerussite, anglesite, smithsonite, malachite, azurite, limonite, and psilomelane.

Such a mineral composition consists of several parageneses formed in different tectonic and magmatic phases. In the first place, here are the basic lead and zinc minerals formed in several generations during the main phase of lead-zinc mineralization.

There is also the important high-temperature phase of iron and manganese oxides and sulphides, with copper and iron sulphosalts, and sphalerite with appreciable amounts of isomorphically admixed iron. The minerals of the copper paragenesis are mostly enargite and tennantite. Although these minerals occur frequently in this deposit, particularly in its NW part (less in SE), their concentrations are usually rather insignificant. The uranium parageneses with pitchblende and radioactive chalcedony occur sporadically, but are generally more concentrated in the SE part of this deposit. In the main phase of lead-zinc mineralization, the distribution of galena and sphalerite is such that the galena is rarely more frequent than the sphalerite. The distribution of the gangue minerals gives a vertical zonality in the deposit, with barite and siderite occurring only at the higher levels.

According to mineral composition and the paragenetic relationships among the different minerals, it can be concluded that this deposit was formed in several phases. The first was the high-temperature phase, with minerals of the contact parageneses; its contribution to the deposit as a whole is insignificant. The next were the mesothermal and the epithermal phases, when most of the lead-zinc minerals were deposited.

2.2 Zlatica Copper Deposit

There are numerous copper mineral occurrences within the Zletovo-Kratovo area, frequently associated with lead-zinc mineralization; however, there also exist the important copper concentrations in the deposits consisting exclusively of copper. Among these the most important is Zlatica, which consists of several ore veins, and numerous metasomatic, impregnation- and stockwork-type copper mineralizations. The mineralizations are located inside andesitic and dacitic volcanic rocks and in the redeposited tuffs. These volcanic rocks are strongly altered, with secondary quartzites formed locally. The fault-type structures trend NW–SE. By their morphology, the ore bodies are of stockwork and impregnation type.

Enargite, tennantite, pyrite, chalcocite, covellite, bornite, native gold, luzonite, galena, sphalerite, chalcopyrite, quartz, chalcedony, barite, and siderite were identified. The first three minerals cited form the main components of the ore bodies, and all other minerals, except for quartz, occur in subordinate amounts. It is interesting to note that chalcopyrite, and the gangue minerals barite and siderite, are related to the galena-sphalerite association, with the quartz gangue associated with the enargite-tennantite paragenesis.

According to paragenetic relations among the different minerals and the paleotemperature studies of quartz (Radusinović and Kleut 1963), this deposit was formed within the meso- to epithermal stages of the hydrothermal phase.

2.3 Zletovska Reka Uranium Deposit

The uranium mineralizations are numerous in the Zletovo-Kratovo area, and occur either as the exclusive uranium mineralizations within the propylitized volcanic rocks, in lead-zinc sulphide veins, or in the opal-chalcedony veins with only minor amounts of sulphides or completely without sulphides. The Zletovska Reka is the most important uranium deposit in this area, and belongs to the first mentioned type.

Here the uranium mineralization is localized within the volcanic rocks, with the uranium minerals deposited inside the tectonic zones. These zones trend E–W or ENE–WSW. The ore bodies have the shape of veins, but may have locally the shape of lodes or irregular bands, partially forming the lenses. The ore bodies are always under the control of fault-type structures and follow the direction and dip of these structures.

The pitchblende is the exclusive primary uranium mineral. The other minerals, autunite, torbernite, kasolite, uranophane, and uranospathite are the secondary products formed in the reducing environment due to the alteration of uranium minerals near the surface. The sulphide minerals consist of traces of sphalerite, galena, pyrite, marcasite, bravoite, and tetrahedrite. The gangue minerals also occur in minor amounts, and consist of quartz, chalcedony, barite, siderite, and fluorite.

The pitchblende, which is the dominant mineral, is completely destroyed by the supergene processes in the oxidation zone. The secondary uranium minerals occur only a few tens of metres below the surface.

The bulk of the pitchblende occurs mostly in those parts of the deposit where occur also slightly higher amounts of minerals of the sulphide paragenesis, as the reductive environment, favourable for sulphides, has favoured also the deposition of uranium. The mineralization has been formed within the epithermal to telethermal stages of hydrothermal mineralization.

3 Genesis of Mineral Deposits

The position of the Zletovo-Kratovo area within the activated part of the central massif, which represents the lineament at the contact of Serbo-Macedonian and the Dinaride folded zone, was of particular genetic importance relative to formation of mineral deposits. This lineament was initially formed in the Cambrian, and has been reactivated several times during geological time. During the Tertiary, this area was subjected to intensive magmatic activity, when the volcanic effusive-intrusive/volcanic-plutonic complex of rocks was formed. This complex has originated from a deep-seated granitoid magma reservoir, which has produced not only the volcanites, but also the post-magmatic processes which resulted in formation of lead, zinc, copper, and uranium. The existence of the deep lineament is also of importance for the genesis and the spatial distribution of mineral deposits; also for the movement of magma and formation of various petrogene complexes that could contain various metallogenetic formations, and could have taken place during different geological periods. However, at the present erosional level, only the mineralizations formed within the Alpine metallogenetic epoch are exposed. The detailed studies of mineral composition, structural, geochemical, and geochronological characteristics, all point to the mineralizations in the Zletovo-Kratovo area as having originated from the same magmatic reservoir.

The essential characteristics of all the deposits and mineralizations are as follows:

- All deposits, except for some infrequent lead-zinc occurrences, are spatially related to Tertiary volcanic rocks.
- All deposits were produced by the endogene processes of the hydrothermal stage, but the meso- to epithermal phases were dominant according to results obtained by method of homogenization of gaseous inclusions in quartz and barite.
- The mineralizing solutions have introduced the intensive hydrothermal alterations of surrounding rocks, tracing, in this way, the feeders for mineralizing solutions through the volcanic complex. This hydrothermal alteration has affected the rocks of mineral deposits, whether they contain lead-zinc, copper, or uranium minerals.
- The mineralizations were controlled by the structural features, which resulted in the specific structural and morphological characteristics. Consequently, the ore bodies are mostly of the vein type, less frequently in form of lenses or impregnations; the stockworks were formed in the zones of intensive fracturing. However, the complex veins are dominant.

The mineralized fault-type structures have various spatial orientations. The detailed studies have confirmed that the NW–SE trending structures were favourable for deposition of lead, zinc, and copper; the E–W trending structures were favourable for deposition of uranium minerals.

All the presented data indicate that the mineralizations in the Zletovo-Kratovo area were not formed within a single, continuous phase, but rather were the result of multiphase tectonic and magmatic activity, followed by the corresponding postmagmatic mineralizing processes. The mineralizing solutions, as the result of a pulsating character of tectonic and magmatic activity, have changed their chemical features during geologic time. These changes have resulted in their differentiation within time and space, and in the consequent formation of mineral deposits of different metals. The study of paragenetic relationships of different minerals within the various main phases of major metals points to the fact that each of the phases of the major metals contain, at the beginning, the minerals from the previous phase; at the end of each such mineralizing phase occur the minerals of metals that will represent the main component during the subsequent phase. This conclusion can be confirmed by tracing the mineral composition of a given mineralization according to the order of deposition, which resulted in the characteristic zonality of mineral deposits relative to the lineament. When looking from west to east, copper occurs first, then lead-zinc, and, finally, the uranium mineralization (Fig. 1).

In the Zlatica copper mineralization, some galena and sphalerite are present as well, indicating, in this way, the next phase characterized by the rich lead-zinc mineralizations. The traces of gold in the copper parageneses at Zlatica may indicate that the previous phase, relative to copper, had contained gold as the main metal; however, no gold deposits were discovered so far. The lead-zinc parageneses of Dobrevo contain the minerals from the previous copper rich phase as well, and, at the end of the process, minerals of uranium indicate, in this way, the next phase with uranium as the main metal. The Zletovska Reka uranium deposit contains traces of lead-zinc minerals, formed slightly before the pitchblende, as a paragenesis from the previous phase. However, due to the fact that no younger primary parageneses of other metallic minerals were identified in this deposit, it can be concluded that the uranium mineralizations represent the end of endogene metallogeny of the Alpine phase in this area.

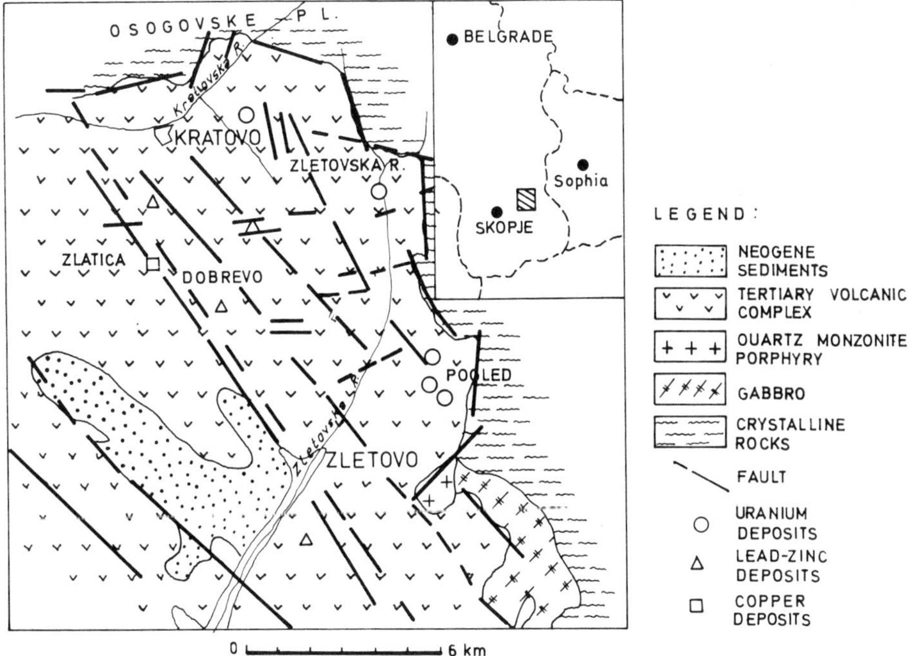

Fig. 1. Metallogenetic sketch map of the Zletovo-Kratovo area

References

Cissarz A, Rakić S (1956) Die Lagerstätte von Zletovo in Mazedonien. Vesn Zavoda Geol Istraz NR Srb XII:224–282

Denkovski DJ (1974) Minerogenesis of vein no. 2 in Dobrevo mine (in Macedonian). 8th Yugoslav Geol Congr, pp 41–48

Divljan S (1964) Some new ideas on the genesis of volcanic rocks in Zletovo-Kratovo area (in Serbo-Croat). Proc Serbian Geol Soc, pp 1–8

Radusinović D (1974) Zletovska Reka Uranium Deposit. IAEA Symposium on the formation of uranium ore deposits. Athens, Greece, pp 593–601

Radusinović D, Kleut DM (1963) Temperature of origin of some minerals (in Serbo-Croat). 2nd Yugoslav-Polish Symp Explor Nuclear Raw Mater, pp 211–221

Ristić M, Radusinović D (1954) Occurrences of ptichblende in Pb-Zn Zletovo mine in Macedonia (in Serbo-Croat). Int Rep Inst Nuklear Raw Mater, Beograd, Yugoslavia, pp 1–34

The Tin Ore Deposit of Orissa, India

S. ACHARYA[1], S.K. SARANGI[2], and N.K. DAS[3]

Abstract

The tin ore deposit of Orissa occurs in the Mathili block (18°36′–18°40′N; 81°54′–82°05′E), Koraput district and is primarily associated with pegmatites intruded into the metapelites of the Bengpal Series (Precambrian). Ore microscopic and X-ray studies reveal the presence of cassiterite, kesterite, varlamoffite, tapiolite, fersmite, sterryite and tantalite which suggest a high temperature hydrothermal origin of the pegmatites. The cassiterite concentration in the pegmatite varies from 0.002% to 0.6% and is noted to be more enriched in the albitised zone. The potential tin deposits in the area are essentially confined to various Quaternary terraces, palaeochannels and river placers where the average cassiterite concentration is more than 0.03% with a maximum of 1.2%. Considering the size of the area and cut-off grade, it may be inferred that the deposit is very promising. The rare minerals may increase the value of the deposit considerably.

1 Introduction

The tin ore deposit in India was first reported by the Madhya Pradesh Mining Directorate in 1973 from Bastar district and was later discovered in Orissa State in 1975, by the second author in the Mathili block of Koraput district. Later field and laboratory investigations revealed the possibility of the occurrence of this deposit as a potential one.

2 Geological Setting

This deposit occurs in the Mathili block of Koraput district (Fig. 1), Orissa in the Survey of India Toposheet No. 65F/14 and J/2 within lat. 18°36′–18°40′N and long. 81°54′–82°05′E and is found to be continuous to the Bastar tin ore field of Madhya Pradesh.

The Precambrian metasediments with metabasics mapped as the Bengpal series (Crookshank 1963; Iyengar and Banerjee 1971) in the area are superimposed by the

1 Utkal University, Bhubaneswar, India
2 Department of Marine Sciences, Berhampur University, India
3 Deputy Director Mines Directorate (South), Orissa, India

Cuddapahs (late Proterozoics). No earlier work exists. The metasediments include andalusite schist, chlorite schist, quartz sericite schist, and quartz-feldspathic gneisses and cover about 70% of the total area. They contain quartz, feldspars, biotite, magnetite, rutile and garnet in varying proportions. Metalavas and their feeders intrude the metapelites; these are overlain by spilitic trap rocks and tuffs. Orthoquartzite overlies the spilites disconformably. Pegmatites of two phases intrude the sequence. The first one is post-metapelite; the second one is younger than the top quartzite, and is known to be stanniferous. Diastrophic features imprinted on the rocks indicate three periods of folding. The first forms the broad E–W folds, the second NNW–SSE and NNE–SSW conjugate set, contains intricate structures and is also possibly responsible for the doming of the Paliam granite and Darbha granite gneiss. Pegmatites have intruded along the axial planes of the folded structures in the metasedimentary rocks. The third fold (E–W direction) is of local significance only. Paliam granite (correlated with stanniferous pegmatite) intrudes Darbha granite gneisses.

The stratigraphic sequence as studied in the area is as follows:

	Quaternary terrace-deposits (1 and 2 around Mundaguda)	
Cuddapah	Limestone Shale Sandstone	
	Pegmatite Orthoquartzite, etc. Spilite and tuffs	
	Pegmatite Metalava with feeder dykes	
Bengpal	Metapelites	andalusite schist, chlorite schist, quartzites
Archaean?	Darbha granite gneiss (out of the area in Fig. 1)	

Attention is confined here to the pegmatites which are the primary source of the tin ore (and other associated ores), and the Quaternary terrace deposits that form the present main source of the ore in the area.

3 Pegmatites and Associated Mineralisation

As already stated, only the younger pegmatite is stanniferous. It contains, in a zoned way, quartz in the core and is followed outward by feldspars (albite, oligoclase, cleavelandite and perthite) as noted after blasting operations of pegmatites. Muscovite,

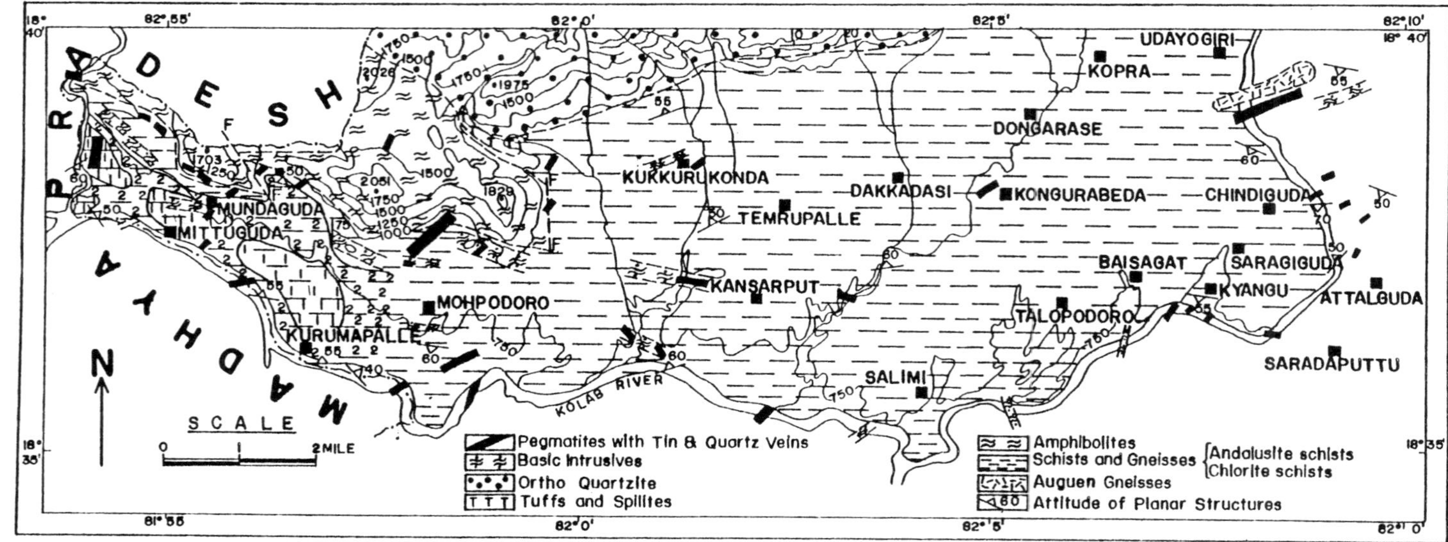

Fig. 1. Geological map of the area of the tin ore deposit in Koraput district, Orissa. The western part of the area was investigated in detail. The deposit continues into Madhya Pradesh, although this is the richest sector. The patches around Mundaguda showing 1 and 2 markings are terraces which continue eastward

biotite and lepidolite are abundant with accessories such as pink and black tourmaline, beryl, garnet, triplite, zinnwaldite, fluorite and apatite, topaz and sphene. Cassiterite occurs as disseminated grains or grain aggregates of micron to 15 cm size. A combined ore microscopic and X-ray study (not fully complete) revealed the presence of cassiterite, kesterite, varlamoffite, tapiolite, fersmite, sterryite and tantalite as rare minerals in pegmatite. Montebrasite and metatriplite occur in Bastar pegmatite (Mookherjee et al. 1979). Cassiterite is sometimes traversed by quartz veinlets. A spectrographic analysis made by the Atomic Minerals Division, Bangalore, has indicated the presence of trace elements as follows:

Ta	1.00%
Nb	0.75%
Ti	0.12%
Mo	0.012%
V	0.007%
W	0.01%

The mineralogy suggests a high temperature hydrothermal origin of pegmatite. High Ta and Nb contents are characteristic of pegmatite (Traub and Moh 1978).

The following mineral groups in pegmatite can be made from the frequency of its broad mineralogical associations.

A – Cassiterite, quartz, muscovite, beryl
B – Cassiterite, cleavelandite, lepidolite and beryl
C – Cassiterite, cleavelandite, wolframite, lepidolite and beryl.

They may be grouped as Type I, II, and IV after Vlassov (cited in Smirnov 1976). Both simple and recrystallised types are common, unlike the neighbouring Bastar tin field where most of the pegmatites are zoned (Mookherjee et al. 1979).

In the greisen zones, quartz, plagioclase, muscovite and minor amounts of disseminated cassiterite are found.

4 Tin Deposit

In the course of the investigations, the genesis of cassiterite was found to be directly associated with pegmatite and greisen zones where the concentration is relatively high. As secondary accumulation patches in terraces and palaeochannels and in the Kolab river deposit, the tin ore is, however, much more abundant with a high tenor.

Based on the frequency of occurrences of pegmatites and delineation of different colluvial-eluvial zones with promising cassiterite content, an area of about 2 km^2 was selected around Mundaguda (lat. 18°37′N; long. 81°53′E) for detailed work.

5 Primary Source

In the case of a primary source, three different varieties of pegmatites in the area were studied in detail by cutting cross-trenches at regular intervals. Keeping in view their zonal variations and the tin mineralisation zone, samples from the pegmatites were collected and then treated by several mechanical processes (screening, jigging and panning) to recover the cassiterite concentrates. The data in Table 1 indicate that a richer concentration is confined to the albitised zone.

Table 1. Variation in tin content with type of pegmatite

Location and nature of the pegmatite	Samples from depth in meters	Percent of cassiterite concentrate
1. Greisenised pegmatite (Kamrupalle)	0 –0.8	0.002
	0.8–1.6	0.0025
	1.6–2.4	0.002
	2.4–3.2	0.0023
	3.2–4	0.016
	4 –4.8	0.019
	4.8–5.9	0.019
2. Albitised pegmatite (Mundaguda)	0 –1	0.3
	1 –2	0.35
	2 –3	0.55
	3 –4	0.60
	4 –5	0.42
3. Zoned pegmatite (Mongerguda)	0 –0.4	0.25
	0.4–1	0.27
	1 –2	0.12
	2 –3	0.01

6 Secondary Source

Similarly samples from Quaternary terrace deposits (1 and 2) and the palaeochannel deposits were collected and treated by the usual mechanical processes to recover the tin concentrates. The average cassiterite concentration computed from the geomorphic terraces varies from 0.01–0.03. However, in certain eluvial zones around Mundaguda (Terraces 1 and 2), the recovery value varies from 0.3 to as high as 1.20 down to a depth of 1 m only, although the depth of terrace and channel deposits is much more than 1 m. From the recovery values of tin, the area can broadly be divided into a richer western sector and a leaner eastern sector, the best concentration zone being around Mundaguda with the dividing line passing near Mohopodar. This preliminary study gives, however, a broad picture as to the potential of the area for tin where the average recovery is computed to be 0.05 down to 1 m depth for an area of 120 km^2. Naturally, tin is to be expected more at depth, and, in the bed of the Kolab river as well. Moreover, the presence of ilmenite, monazite and gold in varying proportions in the river placers is confirmed in addition to many other rare minerals which are likely to go a long way as secondary minerals of commercial importance.

7 Conclusion

The tin ore deposit of Orissa is the first of its type known in the country, the western and northern part of the province being in Madhya Pradesh. The tin ore and other associated rare minerals are primarily confined to the second pegmatite. Distinction between the two pegmatites has not yet been made, although lack of cassiterite in some pegmatites as well as stratigraphic evidence suggests this possibility. The potential ore deposit is, however, essentially a secondary deposit, confined to the Quaternary deposits on the terraces, palaeochannels, and to the Kolab river sands. The tin minerals do not travel far and, as such, the concentration should be, and actually is, very near the source rock – pegmatites whose length varies from 200 to 400 m. In Kudripal in the Bastar area one pegmatite is more than 1 km long. The pegmatites are indistinctly zoned (Despande 1976). In view of the fact that the deposit has the potentialities of being very large, exploitation should begin along with exploration (Moh 1979). Analyses of stream channel sediments in randomly selected places indicate the occurrence of tin in distant localities such as Jayapatna which is more than 100 km from the area. The probability of tin occurring in neighbouring regions is thus high. This is a preliminary study of the mineralogy, mode of occurrence and potentiality of the deposit. Details are being worked out at present in a more systematic manner. In view of the constantly decreasing value of the cut-off grade, this deposit, large as it is, should have good commercial possibilities for the present and also in the future (Gocht 1973).

Acknowledgments. The authors thanks Dr. B.D. Prusti, Director of Mining and Geology, Orissa, for his interest in the work. To Prof. G. Moh of the Mineralogisch-Petrographisches Institut, Heidelberg, W. Germany, the authors wish to convey special thanks for discussion on the genesis and economics of the deposit.

References

Crookshank H (1963) Geology of southern Bastar and Jeypore from the Bailadila range to the Eastern Ghats. Mem Beol Surv India 87:149

Deshpande ML (1976) The cassiterite-lepidolite bearing pegmatites, Bastar district, Madhya Pradesh. Indian Miner 30:67–74

Gocht W (1973) Veränderungen der Bauwürdigkeitsgrenze in Zinnlagerstätten. Dtsch Geol Ges 124:101–109, 4 Abh

Iyengar SVP, Banerjee PK (1971) The iron ore – Bengpal-Dharwar Group. Rec Geol Surv India 101(2):43–89

Moh GH (1979) Pers. comm.

Mookherjee A, Basu K, Sanyal S (1979) Montebrasite and metatriplite from zoned pegmatites of Govindpal, Bastar Dist., M.P. Indian J Earth Sci 6(2):191–199

Smirnov VI (1976) Geology of mineral deposits. Mir Publishers, Moscow, USSR, 520 pp. Translated from Russian

Traub I, Moh GH (1978) Trace elements in tin ores (with special attention in Asian occurrences). Reg Conf Geol Mineral Resour SE Asia, Bangkok, Proc No 3:361–365

The Colquiri Tin Deposit: a Contribution to Its Genesis

D. HANUS[1]

Abstract

The paragenesis of a representative vein in a deeply situated part of Colquiri tin mine (Bolivia) was investigated in detail. The mineral content consists of sulphides (about 80 wt.% pyrrhotite, sphalerite, pyrite and marcasite as main components), cassiterite and gangue minerals. The occurrence of topaz, chalcopyrrhotite and traces of teallite is remarkable. Deposition of the minerals took place during three periods. Topaz is formed possibly katathermally. The kata to mesothermal cassiterite exhibits small grain size and a columnar and needlelike habit. Using experimental sulphide data, the formation temperatures of the primary sulphides were determined to be 300°–400°C. No paragenetic relations to subvolcanic tin mineral assemblages could be observed. Indications of hydrothermal regenerative processes ("hybrid" formations) were not found.

1 Introduction

The Colquiri tin mine is located in the Bolivian Andes relatively far from the nearest tin-bearing granitic plutons in the Cordilleras Quimsa Cruz and Sta. Vera Cruz to the north. The veins of the deposit are situated in Upper Silurian sediments, lutites inlayered with arenites. As yet no intrusion beneath the deposit has been definitely proved. The geological situation in this region, however, makes a relationship between the batholiths and the Colquiri deposit conceivable. The latter could be associated with a deep-seated, southern extension of the aforementioned series of plutons.

There are indications of an intrusion west of Colquiri at the little tin mine of Sta. Rita (Fricke et al. 1964). The occurrence of simple sulphides and cassiterite is common to both the Colquiri and the vein deposits associated with the northern batholiths, however, only small amounts of quartz and a greater abundance of fluorite and siderite have been noted at Colquiri.

On the other hand, the appearance of hydrothermal huebnerite and teallite in the Krupp and San Antonio veins, respectively, belonging to Colquiri area has been reported (Ahlfeld 1932, 1938). Pyrargyrite (Stelzner 1897) and argentite (Mallo et al. 1976) have also been mentioned.

The depth of formation for Colquiri, estimated at about 2000 m by Kelly and Turneaure (1970), coincides with the depth of the subvolcanic type of deposit found in central and southern Bolivia. According to Turneaure (1971), the possibility of subvolcanic influences in the Colquiri area should not be excluded.

1 Mineralogisches Institut der Freien Universität Berlin, Takustraße 6, 1000 Berlin 33, FRG

The origin of these two distinct types of Bolivian tin deposits has been explained differently. According to Schneider-Scherbina (1962), the specific tin-silver parageneses in the subvolcanic deposits in central and southern Bolivia were generated by hybridization of an early Mesozoic plutonic cycle of mineralization and a silver-rich subvolcanic cycle in the late Tertiary.

As a result of their examinations of mineral sequences and formation temperatures in many Bolivian tin deposits based on studies of the fluid inclusions (including samples from Colquiri), Kelly and Turneaure (1970) discovered "a single, prolonged 'event' of mineralization for the individual tin and tungsten deposits throughout Bolivia."

A recent summary of the published concepts were given by Schneider and Lehmann (1977).

2 Scope of Study and Sampling

A renewed detailed mineralogical investigation in the Colquiri deposit with respect to the mineral associations and the conditions of ore formation is intended to aid in solving some of the problems described earlier.

The present investigation is based in part on previous works. Among the publications concerning the Colquiri deposit, Campbell's paper (1947) has provided the most comprehensive geological and mineralogical descriptions. A more recent report by Mallo et al. (1976) is essentially similar to Campbell's study.

Campbell investigated the deposit, as much as had been exposed by mining activities in the forties. After further exploitation, Amstutz had the opportunity to explore the newly opened areas in 1961, during work to recalculate the ore reserves (Amstutz 1964). For a detailed mineralogical study, he selected a vein representative of this area in mineral content and ore texture. This vein was made accessible by means of a raise (situated in the plan square N 2720–2800/W 1600–1640) between the lowest existing levels, 285 and 325. A systematic sampling was carried out by Amstutz, who made the materials available to the author.

3 Mineral Content and Periods of Mineralization

A detailed microscopic investigation (Hanus 1979) yielded a mineral content for the vein similar to the known average of the deposit (Campbell 1947; Mallo et al. 1976) including minerals not described previously. The main components are simple sulphides: pyrrhotite, sphalerite, pyrite and marcasite, which make up about 80 wt.%. Arsenopyrite, chalcopyrite, stannite and galena each amount to less than 1%.

So-called „Zwischenprodukt" (according to Ramdohr 1975), chalcopyrrhotite and mackinawite occur in very small quantities. Traces of teallite were also observed.

The oxide minerals are represented by cassiterite (about 3%), quartz, magnetite, limonite and some rutile. Siderite amounts to about 5%. Fluorite along with the silicates, topaz, kaolin, cronstedtite, serpentine (and quartz) compose about 7%. Wolframite occurs in traces only.

It is possible to distinguish in the vein three distinct regions of mineralization according to their textures. These regions can be correlated with three periods of mineral deposition (Fig. 1):

1. the early period from the beginning of the wall rock alteration and up to the deposition of the early vein minerals (Figs. 2 and 5): topaz, quartz, rutile, cassiterite and wolframite; possibly fluorite and kaolin;
2. the main period, with formation of the primary sulphides: pyrrhotite, sphalerite, arsenopyrite, stannite, chalcopyrite, galena and teallite; and due to exsolution from sphalerite, chalcopyrrhotite and mackinawite;
   at the end of this period the sulphides were crushed by movements;
3. the late period, with alteration of pyrrhotite to marcasite and pyrite and siderite as byproduct; fluorite, serpentine and kaolin deposition in the fractures of the earlier minerals.
   Most recently, limonite has been formed by weathering.

This sequence is summarized in a diagram (Fig. 1). Its main features concur with the sequences established by Campbell (1947) and Mallo et al. (1976); however, the investigated vein noticeably lacks the primary pyrite described by those authors.

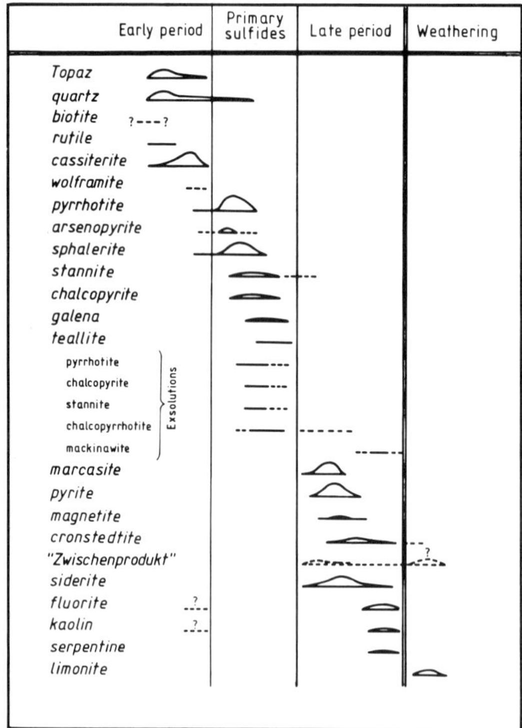

Fig. 1. The mineral sequence in the investigated vein

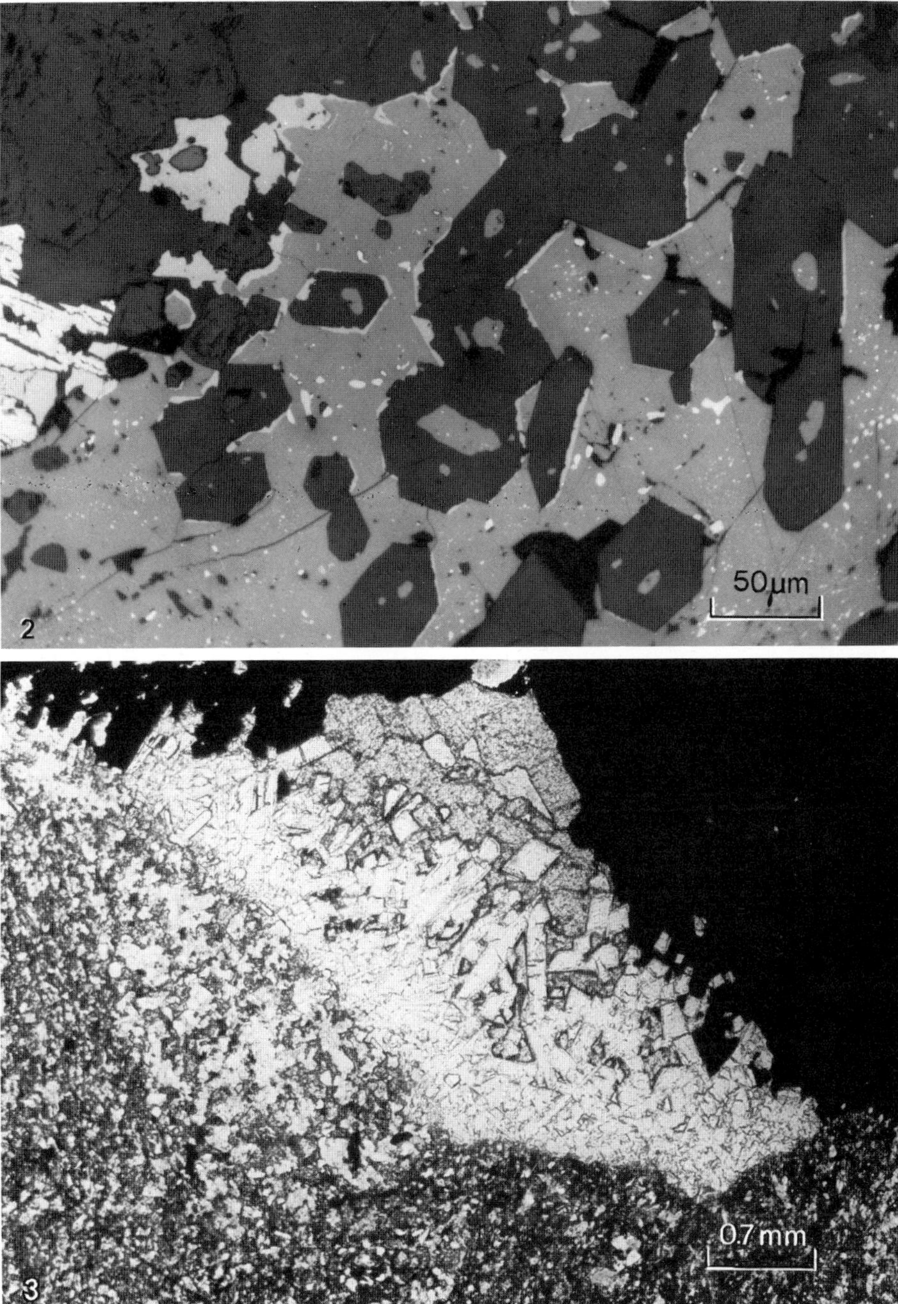

Fig. 2. Idiomorphic quartz *(dark grey)* in sphalerite *(light grey)* near the border of the wall rock. Small rims of stannite *(grey-white)* and exsolution blebs of pyrrhotite *(white)*. Polished section, air, Nic / /

Fig. 3. Columnar topaz crystals *(white)* on the border of altered wall rock (fine-grained topaz-quartz). Sphalerite *(black)* and fluorite *(grey-white)*. Thin section, Nic / /

4 Early Period of Mineralization

Wall rock alteration, as far as it can be observed, has produced mainly topaz and quartz, forming a fine-grained, very hard material with included clay minerals and, rarely, biotite grains. The former bedding is still recognizable.

Tiny grains of rutile occur in the altered wall rock also. According to Schneiderhöhn (1962) these could indicate an acidic and high temperature 350° to 450°C environment.

Up to now, the existence of topaz in Colquiri had been suggested (Campbell 1947), but not proved. Its columnar crystals along the border of the wall rock (Fig. 3) show significant zoning (Fig. 4) and under crossed nicols an indistinct hour-glass structure.

Topaz formation during the early mineralization is unusual for most Bolivian tin deposits. More often tourmaline is formed. It has been observed in Colquiri, also (Mallo et al. 1976).

In the classical sense topaz is an index mineral for "pneumatolytic" conditions, that means, in contrast to "hydrothermal" conditions, mineralization occurs in the temperature range above the critical point of H_2O at 373°C. Supposedly, the mineralization is due to the introduction of so-called fluid solutions; however, because the properties of solutions, e.g., the density, can change continuously from the subcritical to the overcritical stage, the separation of a pneumatolytic stage based *only* on temperature can not be substantiated. This term may be applied to certain types of paragenesis (Schröcke 1973, p. 134).

Considering the depth at which the topaz in this deposit was probably formed, the temperatures during wall rock alteration could be estimated at 373°C or higher, perhaps up to 450°C, indicating katathermal origin. Inclusions of cassiterite, a mesothermal mineral, in the topaz found in the vein indicate that lower formation temperatures may have existed.

The deposition of cassiterite probably began during the wall rock alteration. In the vein, it is most often included in the sphalerite. Two types of idiomorphic grains are common: (1) coarse, squat, columnar crystals up to 1 mm in size; (2) fine-grained, acicular crystals often less than 5 μm long in the form of "needle tin" (Figs. 5 and 6). Ahlfeld (1932, 1958) attempted to relate grain size and habit to formation temperature. According to his morphologic-genetic scale, the larger cassiterite should be mesothermal to katathermal in origin (plutonic and subvolcanic types III and IV; Ahlfeld 1932). The needle tin, on the other hand, would indicate a transition to a lower temperature range.

The problems posed by the coexistence of the latter two essentially incompatible crystal forms plus the higher temperatures required for the formation of the primary sulphides later were the subject of an earlier paper (Hanus 1977).

The deposition of fine-grained cassiterite as a consequence of rapid cooling and precipitation is conceivable.

It is important to note that no redeposition of the coarse cassiterite as a fine-grained mineral took place. Both types of cassiterite definitely belong to the same generation. In the concept of Schneider-Scherbina (1962) an iron-rich cassiterite formed at high temperatures (in a Cretaceous cycle) is re-precipitated as needle tin under the influence of Late Tertiary hydrothermal solutions.

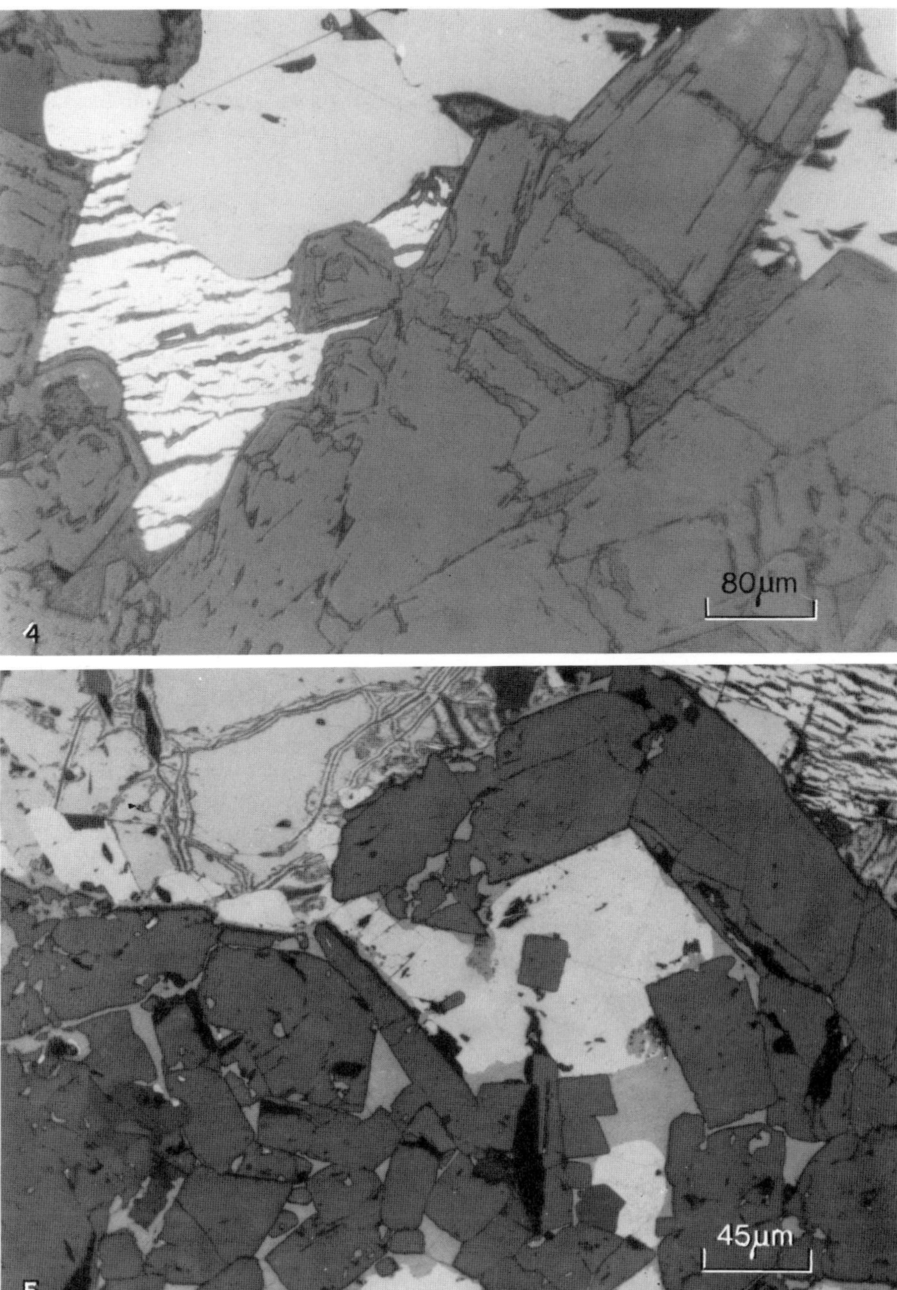

Fig. 4. Zoned topaz crystals *(medium grey)* with pyrrhotite *(grey-white)*, lamellar marcasite *(white)* and kaolin *(dark grey, rough surface)*. Polished section, air, Nic / /

Fig. 5. Coarse-grained cassiterite *(dark grey)* in a groundmass of sulphides: chalcopyrite *(white)*, stannite *(medium grey)*, pyrrhotite *(light grey)* and lamellar marcasite *(white)*. Polished section, air, Nic / /

5 Main Period of Mineralization (Primary Sulphides)

The deposition of the primary sulphides began with small pyrrhotite and sphalerite impregnations in pores in the wall rock. In the vein, the primary sulphides exhibit a mutual boundary texture (Schwartz 1951).

Pyrrhotite and sphalerite (with ca. 22.4 mol.% FeS) are nearly contemporaneous, yet local replacement of pyrrhotite by sphalerite can be presumed, because of a caries texture. Occasionally, idiomorphic sphalerite is included in pyrrhotite. The arsenopyrite is mostly developed idiomorphically. Its intergrowths indicate that it is probably formed during the early sulphide stage.
Generally, the deposition of stannite, chalcopyrite and galena coincides with that of pyrrhotite/sphalerite, but as interstitial fillings in the pyrrhotite/sphalerite groundmass, stannite, chalcopyrite and galena crystallized later.

Frequently, cavities occur lined by idiomorphic sphalerite grains which are coated with stannite exhibiting a similar crystallographic orientation.

The remaining space in these cavities is always filled with siderite (Fig. 7). A later deposition of stannite and chalcopyrite is also demonstrated by small veins of these minerals in sphalerite. Lamellar structures appear in some pyrrhotite grains. These structures are barely visible under crossed nicols, but are revealed better after etching in air. These structures, representing intergrowths of hexagonal and monoclinic pyrrhotite were formed probably by a breakdown of a hexagonal high temperature pyrrhotite and assume a primary pyrrhotite crystallization above 254°C or even above 308°C, according to the experimental results of Kissin (1974). Only the "common" variety of stannite can be observed (Ramdohr 1975). Its fine lamellar twinning is not similar to the parquet twinning usually due to the changes in a high/low temperature. Chalcopyrite apparently has changed from a high temperature form to the tetragonal low temperature form, as evidenced by an "oleander leaf"-twinning. It, too, can be explained as a transformation structure, according to Ramdohr (1975).

Pure chalcopyrite has a transformation temperature of ca. 550°C, while the transformation of chalcopyrite solid solutions with sphalerite and stannite (Bernhardt 1972; Moh 1975) can already take place at an estimated 400°C. The existence of a high-temperature form of chalcopyrite has recently become doubtful (Barton 1973; Cabri 1973). This question will be the subject of future papers (Hanus, in prep.).

Stannite has partially replaced cassiterite. Since chalcopyrite is associated very often with such assemblages a reaction including this sulphide probably took place.

The only mineral related to the sulphostannates observed in the vein is teallite, which occurs as minute grains (30 μm and less) intergrown with stannite and associated with cassiterite. The formation of teallite first began towards the end of the second period of mineralization. No evidence of secondary teallite formation, e.g., reactions of cassiterite or stannite with galena could be found.

Exsolutions. The nearly contemporaneous deposition of the sulphides promoted solid solutions between them. Thus, exsolutions occur within these mixed crystals. Primarily deposited as an extensive, high-temperature solid solution with sphalerite, numerous particles of pyrrhotite, stannite and chalcopyrite were later exsolved. Due to secondary

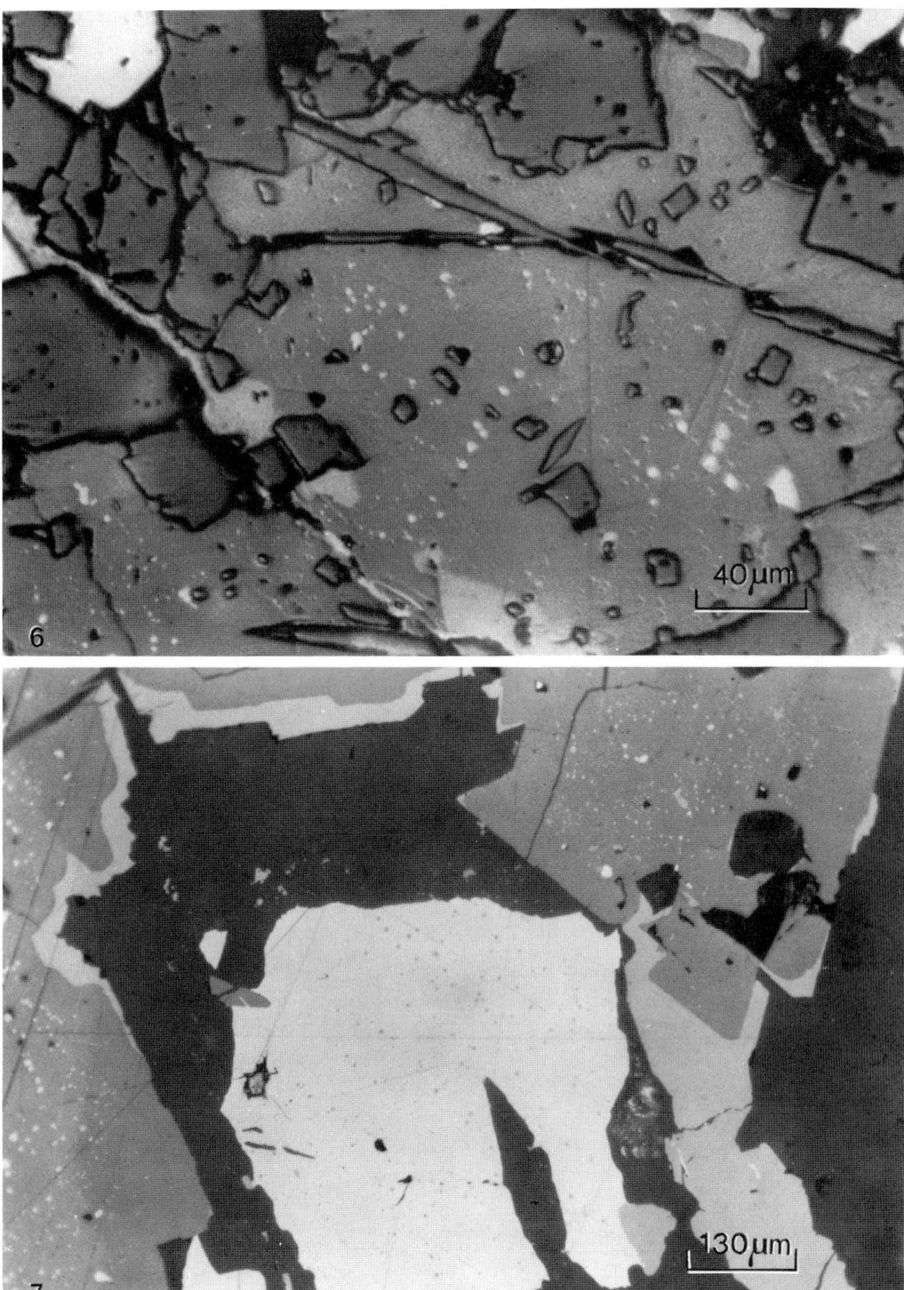

Fig. 6. Squat cassiterite crystals and needle tin *(dark grey)* in sphalerite *(medium grey)*. Exsolutions of chalcopyrite *(white)* and stannite *(light grey)*. Polished section, air, Nic / /

Fig. 7. Idiomorphic sphalerite grains *(medium grey)* coated with stannite *(light grey)* and lining a cavity filled with siderite *(dark grey)* and chalcopyrite *(white)*. Polished section, air, Nic / /

exsolution, clouds of minute sphalerite droplets occur in exsolved stannite, and, in chalcopyrite grains, chalcopyrrhotite and mackinawite have formed.

This secondary unmixing indicates relatively high-temperature primary solid solutions because sufficient kinetic energy must still be available for any succeeding separations. The formation of chalcopyrrhotite (Ramdohr 1975), which corresponds to the synthesized "cubic cubanite" of Yund and Kullerud (1966) and shows structural relation to the so-called intermediate solid solution (Cabri 1973), probably took place between 300° and 400°C.

6 The Late Period of Mineralization

Tectonic activity during the sulphide deposition is documented by the veinlets of stannite and chalcopyrite in fractures in the sphalerite. After the sulphide deposition, greater movements produced the present crushed texture. Beginning in the fractures and along the grain boundaries, alteration of the primary pyrrhotite took place. Monoclinic pyrrhotite was the intermediate form and the end products of this alteration included: (1) lamellar aggregates of fine-grained marcasite, (2) coarse and fine-grained idiomorphic pyrite (with minute magnetite inclusions and cronstedtite grains in the outer zones), (3) idiomorphic marcasite, and (4) "birds-eye" structures of marcasite and pyrite.

This last alteration is characterized by (a) a volume decrease of the disulphides with respect to primary pyrrhotite, indicating a removal of iron, and (b) the intergrowth of the disulphides with siderite and local inclusions of magnetite. The siderite is a by-product of the alteration, fixing in part the free iron.

The so-called "Zwischenprodukt" (Ramdohr 1975) appears as another intermediate phase between monocline pyrrhotite and marcasite, but its formation via oxidation due to weathering is also possible. Among the other late stage minerals, fluorite and kaolin may have been formed as well.

7 Discussion of the Paragenetic Results

Taking into consideration the problems related to the paragenesis in the vein examined here, the following could be concluded. By combining the temperature data described earlier with additional data from other recent experimental investigations of the Cu-Fe-Zn-Sn-S system (Hanus, in prep.), a temperature range from 300° to 400°C could be determined for the formation of the primary sulphides.

The crystallization of the early period minerals most likely took place above these temperatures, which is supported by the presence of rutile (formed at ca. 350° to 450°C). Formation of topaz also suggests a high-temperature environment, but is not sufficient evidence of the existence of a nearby pluton.

The mineral deposition during the late period was probably at lower temperatures than the primary sulphides. One reason for this is that siderite is no longer stable above 300°C (Kullerud 1965).

No clear paragenetic relationship to the subvolcanic type of tin deposits was found, nor were signs of regenerative processes as defined by Schneider-Scherbina (1962) observed. The formation of teallite is conformable to the continuous sequence of the primary sulphides. Although the pyrrhotite alteration to disulphides is surely secondary, it is not caused by the addition of sulphur (as interpreted by Schneider-Scherbina for similar formations in subvolcanic deposits), but due to a removal of iron.

The mineral sequence in the vein and the estimated formation temperatures basically conform to the temperature-time curve of Kelly and Turneaure (1970) for the Bolivian tin deposits, as well as agree with their fluid inclusion temperature measurements for samples from Colquiri. However, there are deviations in the details requiring further examinations.

The results of this study can best be applied to the deeper parts of the Colquiri deposit where the vein investigated was situated.

References

Ahlfeld F (1932) Über Tracht und Genesis des Zinnsteins. Fortschr Mineral 16:47–49

Ahlfeld F (1938) Epithermale Wolframlagerstätten in Bolivien. Neues Jahrb Mineral Abh Beil 74A:1–9

Ahlfeld F (1958) Zinn und Wolfram. Die metallischen Rohstoffe, Bd 11. Enke, Stuttgart, 212 S

Amstutz GC (1964) Paragenetic and statistical study of the minerals in Colquiri Mine, Bolivia. Unpubl rep

Barton PB (1973) Solid solutions in the system Cu-Fe-S. Part I: The Cu-S and CuFe-S joins. Econ Geol 68:455–465

Bernhardt H-J (1972) Untersuchungen im pseudobinären System Stannin-Kupferkies. Neues Jahrb Mineral Monatsh 1972:553–556

Cabri LJ (1973) New data on phase relations in the Cu-Fe-S system. Econ Geol 68:443–454

Campbell DF (1947) Geology of the Colquiri tin mine, Bolivia. Econ Geol 42:1–21

Fricke W, Samtleben Ch, Schmidtkaler H, Uribe H, Voges A (1964) Geologische Untersuchungen im zentralen Teil des bolivianischen Hochlandes nordwestlich Oruro. Geol Jahrb 83:1–29

Hanus D (1977) Observations on Colquiri tin minerals and their interpretation. Neues Jahrb Mineral Abh 131:19–21

Hanus D (1979) Untersuchung über die Bildungsbedingungen einer Gangparagenese in der Lagerstätte Colquiri (Bolivien). Thesis, Univ Heidelberg, 312 S

Hanus D (in preparation) Investigation on the conditions of mineral formation in the Colquiri tin deposit (Bolivia)

Kelly WC, Turneaure FS (1970) Mineralogy, paragenesis and geothermometry of the tin and tungsten deposits of the Eastern Andes, Bolivia. Econ Geol 65:609–680

Kissin SA (1974) Phase relations in a portion of the Fe-S system. PhD Thesis, Univ Toronto, Canada

Kullerud G (1965) Sulfide-carbonate reactions. Carnegie Inst Wash Year Book 64:188–192

Mallo HG, Fuente JC de la, Flores J (1976) Relacion genetico estructural del yacimiento de Colquiri. 1st Congr Geol Boliviano, Combibol, Potosi, 16 p

Moh GH (1975) Tin containing mineral systems. Part II: Phase relations and mineral assemblages in the Cu-Fe-Zn-Sn-S system. Chem Erde 34:1–61

Ramdohr P (1975) Die Erzmineralien und ihre Verwachsungen. Akademie-Verlag, Berlin, 1277 S

Schneider H-J, Lehmann B (1977) Contribution to a new genetic concept on the Bolivian tin province. In: Klemm DD, Schneider H-J (eds) Time- and strata-bound ore deposits. Springer, Berlin Heidelberg New York, pp 153–168

Schneiderhöhn H (1962) Erzlagerstätten. Kurzvorlesungen, 4. Aufl. Fischer, Stuttgart, 371 S
Schneider-Scherbina A (1962) Über metallogenetische Epochen Boliviens und den hybriden Charakter der sogenannten Zinn-Silber-Formation. Geol Jahrb 81:151–170
Schröcke H (1973) Grundlagen der magmatogenen Lagerstättenbildung. Enke, Suttgart, 287 S
Schwartz GM (1951) Classification and definitions of textures and mineral structures in ores. Econ Geol 46:578–591
Stelzner AW (1897) Die Silber-Zinnlagerstätten Boliviens. Z Dtsch Geol Ges 49:51–142
Turneaure FS (1971) The Bolivian tin-silver province. Econ Geol 66:215–225
Yund RA, Kullerud G (1966) Thermal stability of assemblages in the Cu-Fe-S system. J Petrol 7:454–488

Laramide Igneous Rocks and Related Porphyry Copper Occurrences of Southwest Romania

K.A. GUNNESCH[1]

Abstract

The spatial distribution of the Laramide igneous rocks in SW Romania is controlled by three N-S tectonomagmatic lineaments. New data for the subvolcanic bodies and the related copper prospects were obtained by exploration of both the central and the eastern lineament. The Laramide magmatism exhibits a significant multi-stage character. Some principal features of the igneous rocks and the structure of their Cu-pyrite-bearing ores (disseminated and stockwork) provide evidence for the affiliation to the porphyry copper models. Hydrothermal alteration and ore mineral assemblages appear to be zonally distributed. The ore genesis is discussed in terms of the orthomagmatic models. The intrusion of Laramide stocks took place along zones of weakness. The origin of the porphyry copper deposits in the Carpathian-Balkan area interpreted in the light of plate tectonic concepts is discussed. It is still uncertain because of the controversal nature of the different plate tectonic theories.

1 Introduction

In Southwest Romania, the Laramide igneous rocks, widely known as "banatites" (a generic term introduced by v. Cotta 1865) are exposed along three tectonic and, at the same time, petrogenetic lineaments (Giusca et al. 1966) with a NNE–SSW trend (Fig. 1).

The western lineament was recognized long ago not only as a petrogenetic but also as a metallogenic zone in which pyrometasomatic, pyrometasomatic-hydrothermal and hydrothermal ore deposits occur.

By contrast with this area no close studies of either the central or the eastern lineament were carried out until 1972. Unpublished geological reports on specific subjects have been written by Focsa and Radulescu (1956[2]), Constantinoff and Focsa (1959[3]), Hanomolo et al. (1961[4]) and Negut and Popa (1964[5]).

Detailed geological exploration started by Gunnesch et al. (1972[6]) on the central lineament has proven for the first time the presence of a porphyry copper type min-

1 Mineralogisch-Petrographisches Institut, Universität Heidelberg, Im Neuenheimer Feld 236, 6900 Heidelberg, FRG

2–6 2 Focsa, I. and Radulescu, I. (1956); 3 Constantinoff, D. and Focsa, I. (1959); 4 Hanomolo, I., Hanomolo, A., Focsa, I., Hurduzeu, C., Barbu, F., Paraschivescu, C., Cimpeanu, S., and Cimpeanu, N. (1961); 5 Negut, G. and Popa, M. (1964); 6 Gunnesch, K.A., Gunnesch, M., Meitani, I., Caraveteanu, C., Seghedi, I., and Mustetea, E. (1972) – unpublished data

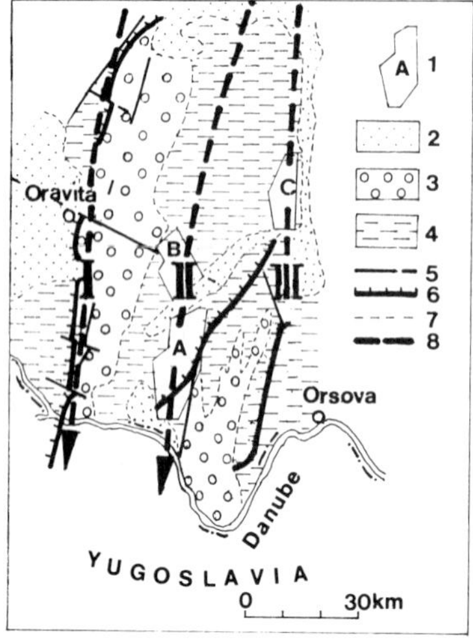

Fig. 1. Sketch map. Distribution of the Laramide igneous rocks in Southwest Romania (Geological and tectonic frame according to 1:1,000,000 geological map published by the Geological and Geophysical Institute, Bucharest). *1* Key to later figures: geological maps of the explored areas, *2* Neogene molasse depressions (intermontane and intramontane basins), *3* Paleozoic and Mesozoic sedimentary rocks, *4* Crystalline schists, *5* Faults, *6* Thrusts, *7* Geological boundaries, *8* Laramide lineaments: *I* western lineament, *II* central lineament, *III* eastern lineament

eral deposit. Exploration was extended on the eastern lineament and succeeded by petrographic, petrochemical and metallogenic studies. So far only a few of these data have been published (Gunnesch et al. 1975, 1978).

The purpose of the present paper is, firstly, to give a concise description of the Laramide igneous rocks both from the central and eastern lineament and, secondly, to present the main features of related porphyry copper prospects. Our work is based on the analyses of data collected from mapping and laboratory studies from 1972 to 1978.

2 Igneous Rocks, Petrography

The spatial distribution of the Laramide igneous rocks is structurally controlled. The emplacement of subvolcanic bodies composed of quartz monzodiorites, quartz diorites, granodiorites, etc. took place along a system of Laramide and reactivated pre-Laramide fractures.

Four major subvolcanic bodies of quartz monzodiorites (Lapusnic, Nasovat, Purcariu, Lilieci, Fig. 2A, B) accompanied by apophyses, dykes and sills extending beyond the intrusive bodies occur on the central lineament, whereas small size igneous bodies and dykes of quartz diorites are to be found in the easternmost area (Fig. 2C).

In the previous paper (Gunnesch et al. 1975) the petrographic uniformity as one of the most important particularities of banatitic rocks in this part of Romania was reported. Thus, using the QAP classification of igneous rocks recommended by the IUGS (Streckeisen 1973), it was established that the "banatites" of the central lineament occupied almost exclusively the compositional field of quartz monzodiorites (Fig. 3). The situation is similar on the eastern lineament. There, the igneous rocks are restricted to the quartz diorite field. In any case, the modal determinations are in accordance with the normative composition established on the basis of the chemical data of the analyzed rocks.

In contrast to both mineralogic and chemical uniformity, the banatitic rocks show a great diversity of textural types. These grade from coarse-grained equigranular to medium-grained phaneritic holocrystalline and further to prophyritic textures. However, most rocks belong to the textural "porphyry" type. The textures of the porphyries again grade from a strong development of phenocrysts over the groundmass, to rocks in which the groundmass predominates over the phenocrysts. The grain size of the groundmasses also varies, but the lowest limit is always "microcrystalline"; a glassy groundmass has never been found. Moreover, no flowage textures could yet be observed.

From earlier preliminary observations and petrochemical data (Gunnesch et al. 1975), we concluded that the Laramide intrusions must have risen from a quartz dioritic magma, and gone through a normal course of differentiation, in a single phase of emplacement. Because of the prevailing porphyritic character of rocks, an emplacement at shallow depth has to be assumed.

Supplementary field work and a recent revision of the data lead us to a different interpretation. Thus, the Laramide magmatism exhibits not a one-stage but a significant multi-stage character. In this way, our interpretation is in accordance with those of Cioflica and Vlad (1973) who found a similar situation in other Laramide areas of Romania.

Magmatism of the central lineament is represented by four phases:

The first phase consists of minor intrusive bodies of dioritic or quartz dioritic composition. Rocks of the early phase are usually encountered as xenoliths in the subvolcanic bodies of the following main phases (phase 2 and phase 3).

Fig. 2A. Geological map of the Nasovat-Purcariu-Lilieci area. *1* Laramide igneous rocks, main subvolcanic bodies: *N* Nasovat, *P* Purcariu, *Li* Lilieci; *2* Cretaceous sedimentary rocks; *3* Crystalline schists of the Danubian autochthon; *4* Crystalline schists of the Getic Unit. *5* Contact metamorphic rocks: *a* hornfels, *b* skarn; *6* Faults; *7* Thrusts; *8* Geological boundaries

B Geological map of the Lapusnicu Mare area. *1* Neogene molasse (intermontane basin); *2* Laramide igneous rocks, *L* subvolcanic body of Lapusnic, *b* contact metamorphic rocks (hornfels); *3* Mesozoic sedimentary rocks; *4* Crystalline schists of the Buceava series; *a* metamorphosed igneous rocks; *5* Paleozoic granites; *6* Crystalline schists of the Sebes-Lotru series; *7* Faults; *8* Geological boundaries

C Geological map of the Teregova-Lapusnicel area. *1* Neogene molasse (intermontane basin); *2* Laramide igneous rocks; *3* Crystalline schists of the Sebes-Lotru series; *4* Trace of synclinal and plunge of axis; *5* Faults; *6* Geological boundaries

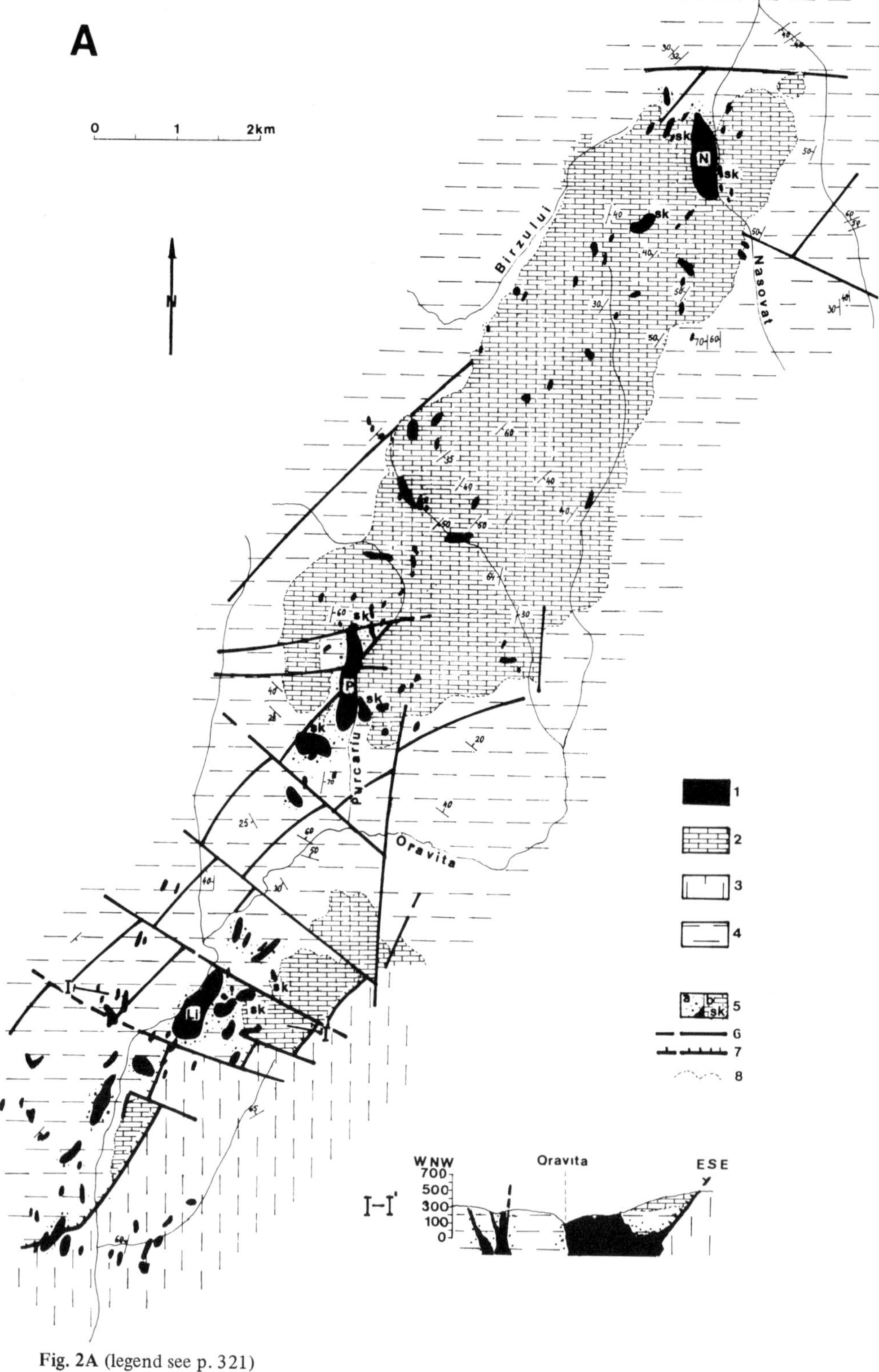

Fig. 2A (legend see p. 321)

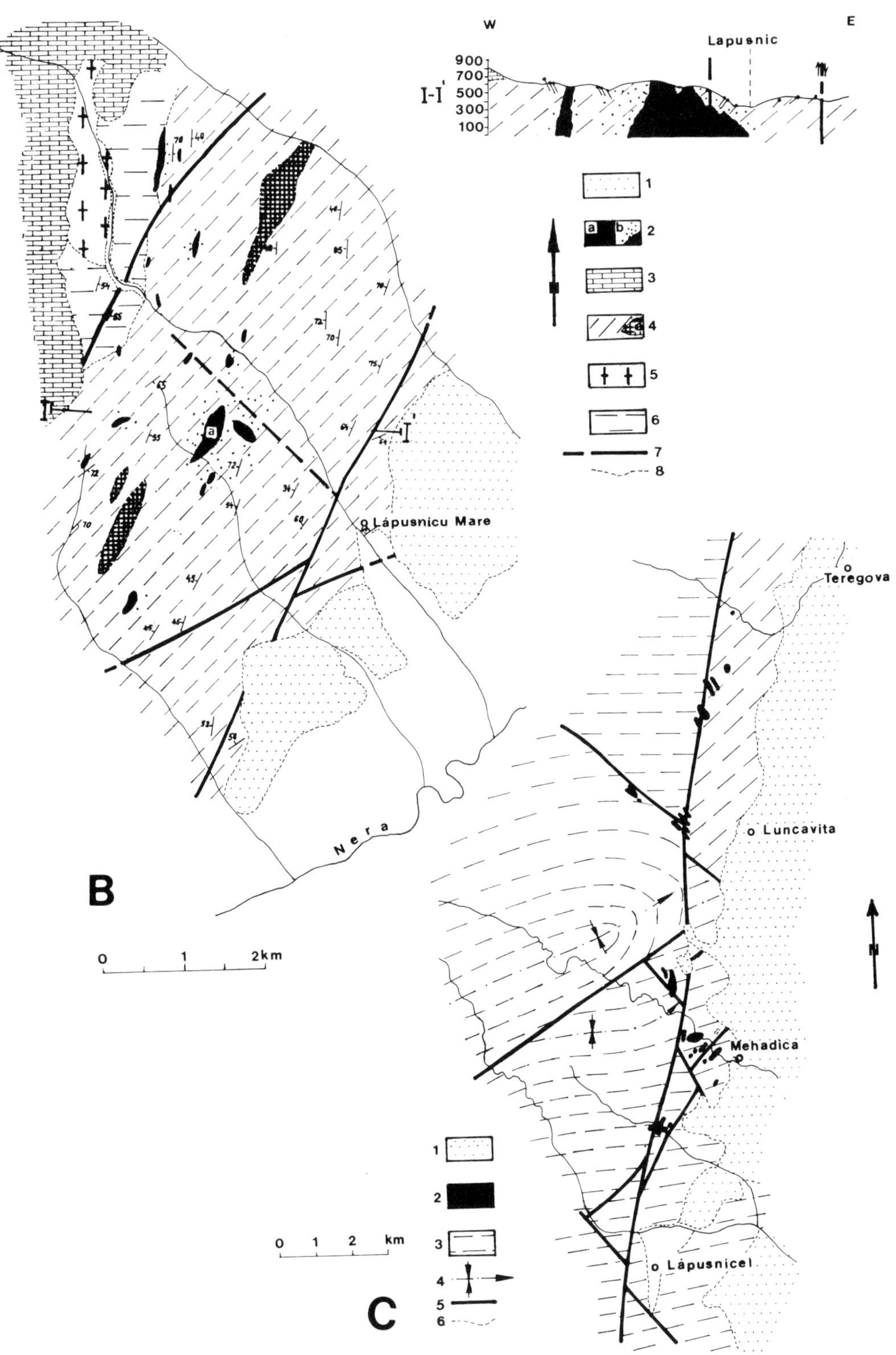

Fig. 2B, C (legend see p. 321)

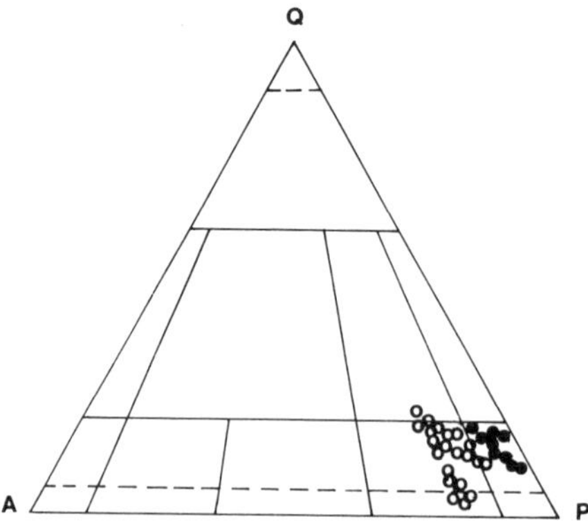

Fig. 3. QAP diagramm of Streckeisen (1973) with modal composition of relatively unaltered Laramide intrusives from the central lineament (○) and from the eastern lineament (●)

The second phase is characterized by emplacement of quartz diorite porphyry stocks and dykes. The rocks are dark grey-coloured, with plagioclase and hornblende phenocrysts in a fine-grained groundmass.

The third phase. The quartz diorite event is followed by the emplacement of quartz monzodiorite stocks and dykes. Unlike the quartz diorite porphyries, quartz monzodiorites are light-coloured and exhibit both equigranular holocrystalline and porphyritic textures. The latter are always predominant.

The fourth phase comprises lamprophyre dykes (spessartites) as the latest products of the Laramide magmatism. They are not shown in Fig. 2A, B because of their small size.

Although the four phases have been established with certainty only in the southern part of the central lineament (Lilieci), there is sufficient evidence for their presence, both at Purcariu and at Lapusnic. No quartz monzodiorite porphyries could yet be found at Nasovat. Since the composition of the igneous rocks is very similar, one may assume that little time lapsed between the earliest and the youngest intrusions.

In the eastern lineament, only two phases of Laramide magmatism were recognized. The emplacement of small subvolcanic bodies and dykes of quartz dioritic composition was preceded by minor dioritic intrusions. Microdiorites always occur as xenoliths in the quartz diorites belonging to the main phase.

It should be emphasized that Laramide magmatism exhibits also short periods of volcanic "explosive" activity. Thus, volcanic breccias and tuffites sporadically occur at Lilieci. Spatial and time relationships are still uncertain, but for the time being we presume their affiliation to the second phase. However, because of the rarity of such rock type occurrences in SW Romania, one may consider the "pyroclastic events" as totally insignificant in the evolution of Laramide magmatism.

3 Hydrothermal Alteration and Ore Mineral Assemblages

The intrusions of the main magmatic phases were accompanied by migration of hydrothermal ore-bearing fluids which gave rise to new mineral assemblages.

At the contact with the Laramide stocks recrystallization and metasomatism took place. Recrystallization resulted in hornfels zones that lack metallogenic importance. However, the carbonatic environment (e.g., the Upper Cretaceous limestones of the Purcariu-Lilieci area) was favourable for significant replacement processes which yielded various skarn deposits. Many skarns are endoskarns developed by skarnificiation of igneous rocks adjacent to limestone. There is almost only one single rock type with widespread occurrence: skarns comprising predominantly calc-silicate minerals (diopside, grossularite, wollastonite) with magnetite, hematite and some chalcopyrite and pyrite. For a long period of time in geological research, attention was paid only to the skarn deposits (e.g., Hanomolo et al. 1961, unpubl. data).

The widespread hydrothermal alteration of Laramide igneous rocks and the structure of their Cu-pyrite-bearing ores (disseminated and stockwork) have been alluded to in 1972 (Gunnesch et al., unpubl. data). These characteristics lead us to suggest that these deposits have similarities with the porphyry copper type. Some other principal features, such as the porphyritic texture of the igneous rocks, the intrusive relationships, the structural style, sequence and zoning of mineral assemblages provide evidence for the affiliation to the porphyry copper models described by Lowell and Guilbert (1970), Hollister (1975), Gustafson (1978).

3.1 Hydrothermal Alteration

Hydrothermal alteration and the related copper ore are associated in space and time with the main phases of Laramide magmatism (phase 2 and phase 3 on the central lineament; quartz diorite emplacement on the eastern lineament). Many of the constituents of hydrothermal alteration and ore mineral assemblages appear to be zonally distributed.

Three main hydrothermal alteration events described subsequently have been recognized: (1) a secondary biotite alteration, (2) a widespread phyllic alteration, surrounded by a halo of (3) propylitic alteration.

Biotite Alteration. Secondary biotite assemblages are the characteristic alteration both at Lapusnic and at Lilieci, but they are to be found at Purcariu and also on the eastern lineament.

Alteration biotite occurs in the following modes: (1) as fine-grained aggregates replacing partly to completely the hornblende phenocrysts leading to hornblende pseudomorphs, (2) as pervasive replacements of groundmass minerals and (3) as veinlet fillings with quartz and chalcopyrite. The primary rock biotite has always a pronounced pale yellow-brown to dark brown pleochroism, whereas alteration biotite is weakly pleochroic (X = greenish-yellow and Y–Z = greenish-brown).

The Phyllic Alteration comprising sericite-quartz-pyrite assemblages is the most strongly developed and to some extent overlaps the biotite alteration zones. The altered Laramide igneous rocks are characterized by the replacement of plagioclase and orthoclase by a fine mat of sericite with fine-grained quartz. Quartz also occurs as vein fillings. Primary biotite is partially sericitized and chloritized; sagenitic rutile which appears along cleavage planes is common (Fig. 4).

The Propylitic Alteration forms a halo several hundred metres wide around the main subvolcanic bodies affected by biotitization and sericitization. Overall, the intensity of the alteration is strong. Plagioclase is variably replaced by calcite ± montmorillonite ± albite ± zeolites. Biotite and hornblende are replaced by chlorite, calcite and epidote, minor sphene, and rutile. Veins and veinlets completely filled with quartz, albite, chlorite, epidote, calcite and/or zeolites are widespread (Fig. 5). Not only the igneous rocks but also the skarns underwent strong subsequent propylitization.

In breccia zones where the spatial relationship between the different vein filling assemblages can be observed better, it is possible to see that the vein quartz has always preceded the chlorite, epidote, calcite ± zeolite vein fillings. Both, calcitization and zeolitization (veins), represent the final stage of hydrothermal activity.

In addition to the above-mentioned alteration zones a sericite-clay alteration is also recognized. This type of alteration is observed only locally and is insignificant. The absence of a hypogene clay-rich alteration zone should be emphasized as one of the most distinct features of porphyry copper prospects in SW Romania.

Because of the extensive cover by vegetation and the lack of mining information, no precise boundaries can yet be given for the extent of each alteration zone.

3.2 Ore Mineral Assemblages

The hypogene copper ore is related to the intrusion of the main Laramide rocks: the quartz diorite and the quartz monzodiorite porphyries on the central lineament, the quartz diorite porphyries on the eastern lineament.

As in the majority of porphyry copper deposits, the principal sulphide minerals in Southwest Romania are chalcopyrite and pyrite. Copper ore minerals occur as discrete grains of chalcopyrite in the matrix of the porphyries (disseminated) and in countless intersecting veinlets accompanied by quartz (stockwork).

An important feature of the ore is the zonal distribution of sulphide minerals within the deposits. The zoning of ore minerals is concurrent with the hydrothermal alteration zones. Both the total sulphide content and the pyrite to chalcopyrite ratio are different in each zone.

As to the data concerning metal contents of the hydrothermally altered host rocks, it is to be noticed that, for the time being, they are entirely provisional. Later exploration work will undoubtedly yield more details.

The primary copper content of unaltered Laramide igneous rocks comprises an average of 60 ppm Cu. On the contrary, Pb values are low (20 ppm). In the hydrothermally altered porphyry intrusions copper ore with values above 0.05% Cu, but rarely above 0.25% Cu (e.g., Lapusnic, Lilieci) are observed.

The highest copper content always occurs in the biotite alteration zone where the pyrite-to-chalcopyrite ratio is about 1:1. Chalcopyrite appears in disseminated grains, whereas pyrite forms mostly veinlets (Fig. 6). Molybdenite has been observed at

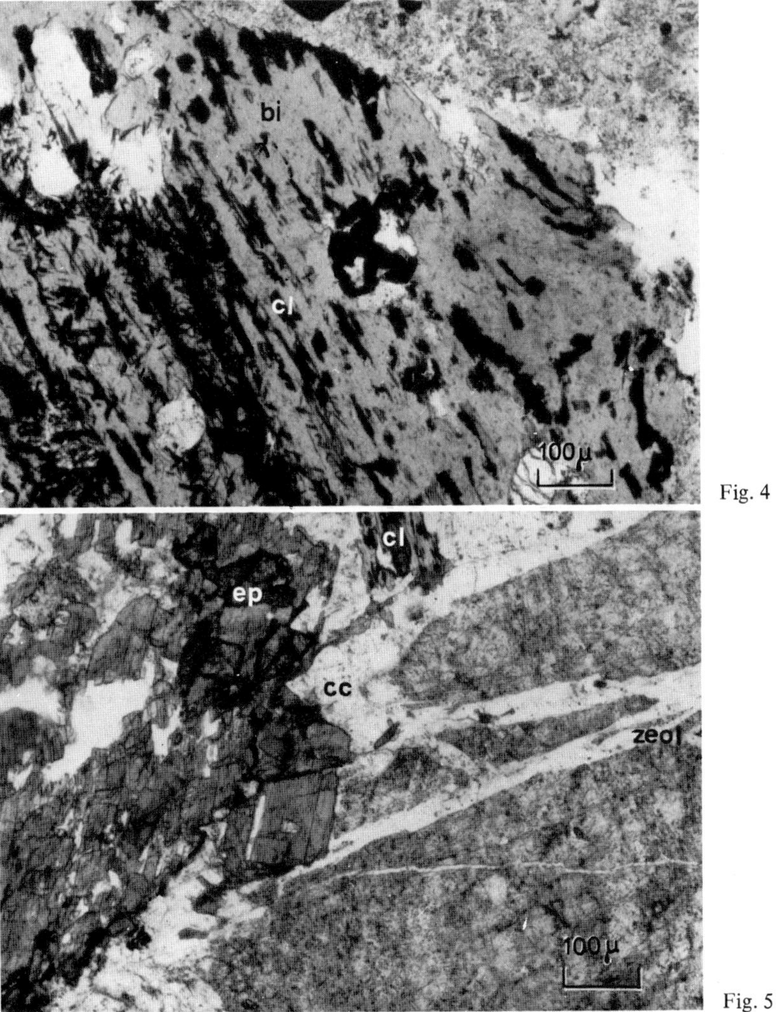

Fig. 4

Fig. 5

Fig. 4. Primary biotite *(bi)* partly replaced by chlorite *(cl)* and acicular rutile along cleavage planes. Plane polarized transmitted light

Fig. 5. Propylitic alteration. Phenocrysts of hornblende and biotite are replaced by epidote *(ep)*, chlorite *(cl)* and calcite *(cc)*. The veinlets are filled with calcite and zeolites *(zeol)*. Plane polarized transmitted light

Lapusnic occurring as fine grains included in veinlet-quartz or associated with chalcopyrite. Elsewhere molybdenite is less common. Minor amounts of magnetite are also present.

By far the highest total sulphide content is present in the phyllic zone, but chalcopyrite is not as well developed as in the biotite zone. In spite of the great tendency to be disseminated, chalcopyrite may also occur in veinlets associated with quartz and

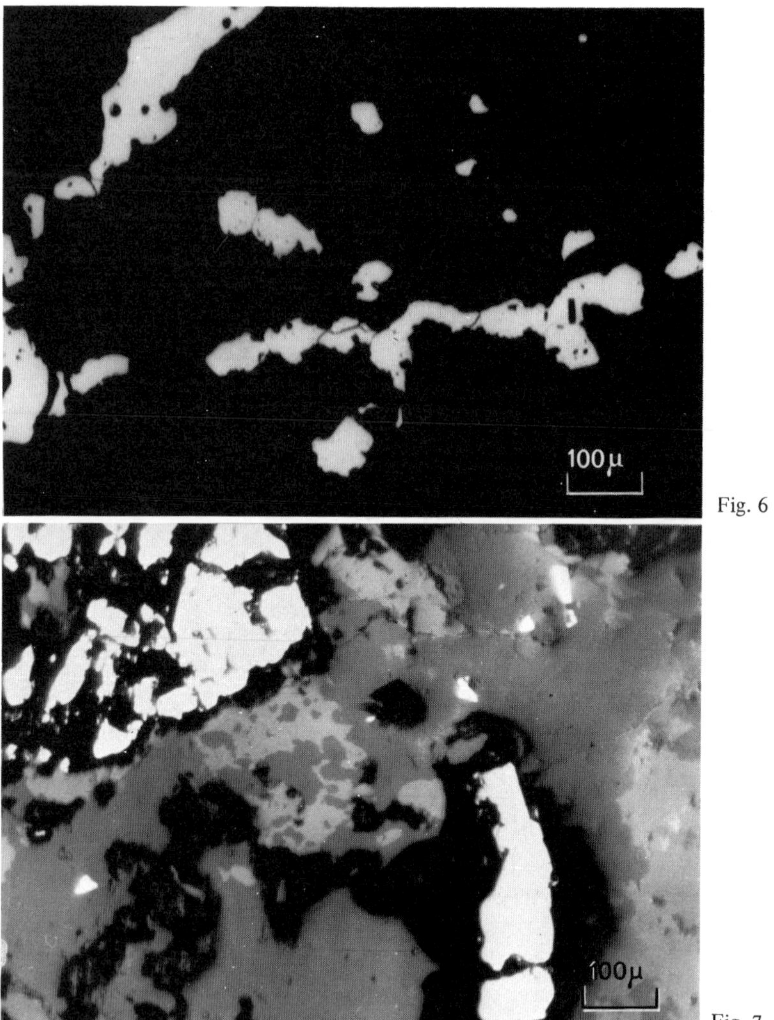

Fig. 6. Veinlet of pyrite in a quartz monzodiorite porphyry of the biotite alteration zone. Polished section

Fig. 7. Pyrite in association with chlorite and epidote in a propylitized quartz monzodiorite. Polished section

pyrite. However, stockworks of quartz-pyrite veins have their greatest development in the phyllic zone. The vein size varies in thickness from paper-thin to several centimetres, spaced from several centimetres to over a metre apart. The pyrite-to-chalcopyrite ratio is about 8:1.

The ore in the propylitic zone consists of pyrite and minor chalcopyrite. Pyrite usually occurs with quartz in veinlets, although chlorite or epidote is also frequently associated with pyrite (Fig. 7).

In marked contrast to the ore mineral assemblage in the Laramide igneous rocks, the primary limestone replacement assemblage is located near the contact of the stocks (Lilieci, Purcariu). Here, a massive magnetite-hematite ore occurs. Though the main ore minerals are magnetite and hematite which partly replace garnets, minor chalcopyrite and pyrite associated with chlorite and epidote are also found. Thus, the propylitic event of hydrothermal activity has affected also the skarn deposits.

The latest hydrothermal fluids gave rise to the deposition of calcite and zeolites, accompanied by pyrite, that lack metallogenic importance. In the supergene zone, the most frequent ore minerals are covellite, malachite and limonite. Significant secondary enrichment is not present. Because of the few data available from former mine workings, there is little to be said about vertical changes in the distribution of the ore minerals.

3.3 Discussion on Ore Genesis

The orthomagmatic models for the formation of porphyry copper deposits are well known (e.g., Schmitt 1954; Amstutz 1965; Lowell and Guilbert 1970; Sutherland Brown et al. 1971; Hollister 1975, and others).

Many recent studies and experimental works on the development of alteration and ore mineral zoning in porphyry systems (e.g., Gustafson and Hunt 1975; Crerar and Barnes 1976; Sheppard and Gustafson 1976; Batchelder 1977; Cathles 1977; Chivas and Wilkins 1977; Ford and Green 1977; Ashley et al. 1978; Cunningham 1978; Eastoe 1978; Ford 1978) suggest a magmatic-hydrothermal origin of the ore-forming fluids. Both the deposition of sulphides and the formation of new mineral assemblages result from the interaction of various factors, e.g., temperature, pressure gradient, Cl^- activity or f_{O_2}, changing pH, mixing of magmatic and meteoric water, reaction with wall rocks, etc. Besides, a structurally favourable environment is important.

The influence of some of these controlling parameters on the evolution of the ore and alteration zoning is briefly discussed below, although neither fluid inclusion nor isotopic studies have been carried out so far in SW Romania. Our discussion is based on the detailed examination of field relations and on an extensive microscopic study.

In the evolution of Laramide copper metallogeny, one may distinguish several stages. The first stage is characterized by the formation of magnetite-hematite massive ore at the contact between the main Laramide intrusions and the carbonate wall rocks. Both magnetite and hematite replace the main skarn minerals (garnets, diopside).

Following this pyrometasomatic event early deposition of copper ore minerals took place as a second stage within the subvolcanic bodies. The stock emplacement into relatively shallow levels of the crust (indicated by the porphyritic texture and the fine to medium-grained groundmass of most of the Laramide intrusions) but also the environmental change from lithostatic to hydrostatic conditions during crystallization may be the main factors responsible for a decrease in pressure and the boiling of the hydrothermal fluid phase (Cunningham 1978). As an effect, the earliest hydrothermal fluids which have the highest temperature and the highest K^+/H^+ ratio introduce Cu, Mo, S and K_2O into the Laramide host rocks, yielding the crystallization of chalcopyrite, molybdenite, quartz, pyrite and secondary biotite. Shrinkage effects in the crystallizing and cooling stocks were probably responsible for the stockwork development. Because of the minor amounts of magnetite in the biotite zone, one may presume that secondary biotite had rapidly equilibrated with the fluid and thus had favoured the precipitation of chalcopyrite. This fact is consistent with thermodynamic evidence (Helgeson 1970; Carson and Jambor 1974).

The next stage is represented by the migration of fluids with high acidity which led to the significant and the extensive silicification of the host rocks in the phyllic zone, but also partly overlapping the biotite zone. The low pH has also enabled the formation of sericite, whereas the lower Cu:Fe ratio (in comparison with the biotite zone) may be responsible for the dominant pyrite precipitation.

At this point, two particular features of the hydrothermal alteration zones and porphyry copper ore in SW Romania may be emphasized: (1) the absence of a clay alteration zone and, (2) the unusually high alkaline character of the late fluids yielding an albite-chlorite-epidote-calcite-zeolite-rich assemblage outward from the phyllic zone. Some of the factors which may be implicated in the appearance of these particularities are referred to below.

The main igneous rocks which underwent hydrothermal alteration are confined to quartz monzodiorite and quartz diorite porphyries. The potassium content in these rock types is relative low; on the contrary, the rocks are rich in Na- and Ca-bearing minerals such as hornblende and plagioclase, which have been partly or completely replaced by hydrothermal pervasive alteration assemblages both in the biotite and in the phyllic zone. The alteration involved K^+ and H^+ ions in the fluid replacing Ca^{2+} and Na^+ ions in the rock. As far as neither anhydrite nor albite was found in the biotite zone, one may suggest that liberated Ca^{2+} and Na^+ were removed in the fluid. Thus, when hydrothermal fluids left the phyllic zone they were characterized by low H^+/Na^+, K^+/Na^+ and H^+/Ca^{2+}, K^+/Ca^{2+} ratios. Decreases in pressure and in temperature (due to wall rock reaction and/or mixing with meteoric waters and/or because of ex-

Table 1. Sequence of ore mineral formation

Mineral	Stages				
	Pyrometasomatic	Hydrothermal			Supergene
		I	II	III	
Magnetite					
Hematite					
Chalcopyrite					
Molybdenite					
Pyrite					
Covellite					
Malachite					
Limonite					

Table 2. Summary of Laramide magmatism and related porphyry copper prospects at Southwest Romania

Prospect	Igneous rock type	Alteration type	Ore mineral assemblages	Structural type of ore
1. Central lineament				
A. Lapusnic	[a]Phase 1: md Phase 2: qd, qdp Phase 3: qmd, qmdp Phase 4: lam	Biot, Phyl, Prop	py+cp±mo	B, C, D
B Nasovat	Phase 1: md Phase 2: qd, qdp Phase 3: possibly absent Phase 4: lam	Phyl, Prop	py+cp	B, (C)
C. Purcariu	Phase 1: md Phase 2: qd, qdp Phase 3: qmd, qmdp Phase 4: lam	Biot (?), Phyl, Prop	py+cp	B, C, E
C. Lilieci	Phase 1: md Phase 2: qd, qdp Phase 3: qmd, qmdp Phase 4: lam	Biot, Phyl ± Arg Prop	py+cp	B, C, E
2. Eastern lineament				
	Phase 1: md Phase 2: qd, qdp	Biot, Phyl, Prop	py+cp±mo	A, B, C

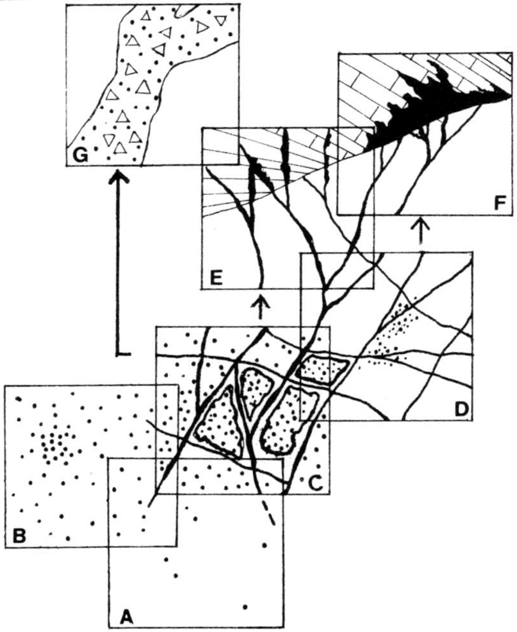

Evolution of patterns in intrusive porphyry coppers (Amstutz 1965 with one more detail)

Abbreviations: *Rocks:* md – microdiorite; qd – quartz diorite; qdp – quartz diorite porphyry; qmd – quartz monzodiorite; qmdp – quartz monzodiorite porphyry; lam – lamprophyre
Alteration type: Biot – biotitic, Phyl – phyllic; Arg – argillic; Prop – propylitic
Minerals: py – pyrite; cp – chalcopyrite; mo – molybdenite

a Phase 1 is in each case the oldest one

pansion within fractures) were presumably the main cause of an inward collapse of the fluid system (Ashley et al. 1978) forming the pervasive chlorite-epidote-carbonate alteration both, in the Laramide igneous rocks, and in the surrounding wall rocks. The last fluids were still enriched in Na^+ and Ca^{2+} ions, so that the excess has been deposited on fissures as calcite and zeolites (laumontite, thomsonite).

The rapid pH-change of the fluids after the phyllic alteration event seems to be also responsible for the lack of argillic alteration, because the conversion of sericite into kaolinite requires an acid environment.

Concerning the ore minerals, it should be noted that chalcopyrite was always confined to the veinlet quartz which presumably belonged to the former phyllic alteration stage. In contrast with chalcopyrite, pyrite occurs also in association with chlorite, epidote and calcite. In any case, one can observe that the vein quartz ± chalcopyrite has been cross-cut by the propylitic vein-filling assemblages.

The sequence of the mineral formation is shown in Table 1.

Laramide magmatism and the related copper from the central and eastern lineament in SW Romania are summarized in Table 2. In order to give a clear idea of the structural type of ores from each zone, we refer to the patterns suggested by Amstutz (1965).

4 The Porphyry Copper Prospects and Their Relation to Regional Tectonics

The structural control of the intrusion of the Laramide stocks and dykes is shown by the diversity of the existing faults.

The most important primary controls are the approximately N–S trending breaks that proceed on south to Yugoslavia (Figs. 1 and 8). It seems probably that certain major dislocations were pre-intrusive and that the Laramide intrusions have obliterated, in some cases, old zones of weakness.

The emplacement of the main subvolcanic bodies in the central lineament occurred along a north-south trending zone of faults. However, the existence of a deep crustal or even subcrustal fracture in this district could only be proved by geophysical data (Cristescu et al. 1975, unpubl. data). Also two other systems of faults and fractures can be discerned: the widespread system of faults that extend from NE to SW and secondly, the less prominent NW trend (Fig. 2A). Some of the main Laramide stocks (e.g., Purcariu, Lilieci) appear to be localized at the intersection of fracture and shear zones. (This connection is visible only on more detailed maps, but not on the generalized map of Fig. 2A.)

A north-south trending major fracture that fed the Laramide magmas to the eastern lineament (Fig. 2C) was recognized recently (Gunnesch et al. 1978). The extension of the break southwards to the Danube (Fig. 8) seems to be unambiguous since Cristescu et al. 1975 (unpubl. data) have proven geophysically the presence of a first-order deep fault in this district. It also seems likely that the very zone of structural weakness has controlled the emplacement of the granodiorite bodies and dykes yielding porphyry Cu-Mo deposits in the Dubova-Mraconia district (Fig. 8). According to these tectonic

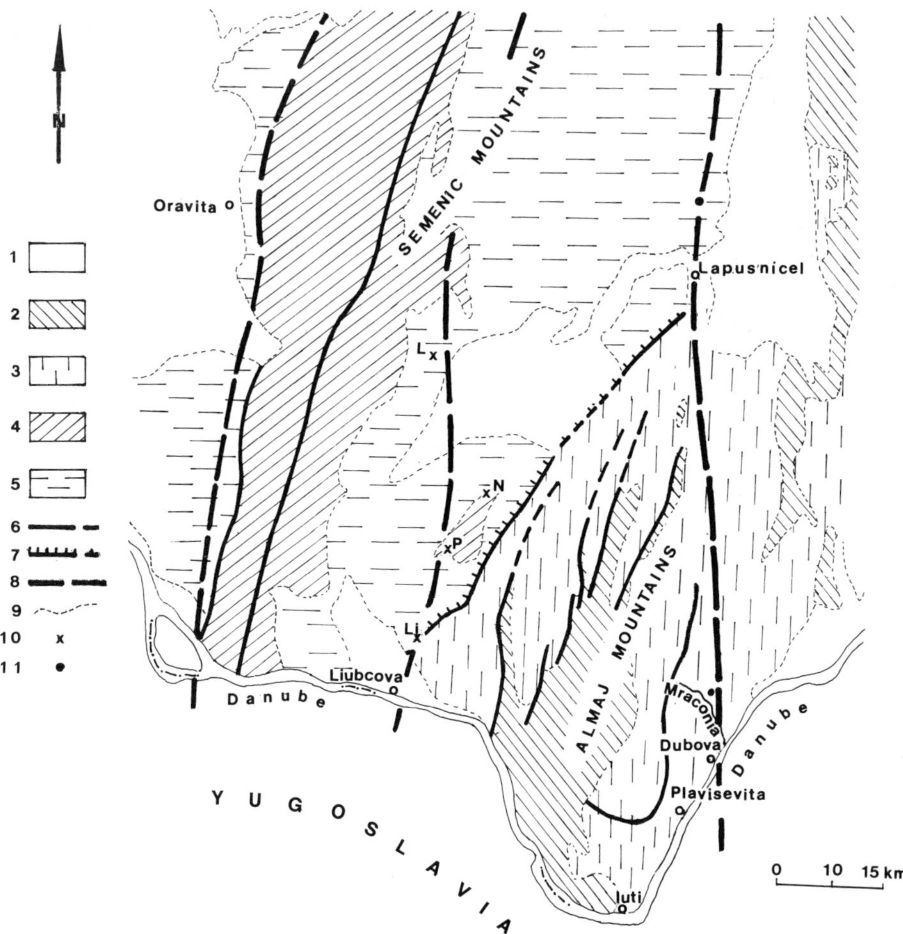

Fig. 8. Tectonic sketch map of the southwestern part of the South Carpathians (Tectonic frame according to 1:1,000,000 tectonic map published by the Geological and Geophysical Institute of Bucharest, with additions made by the author). *1* Intermontane and intramontane neogene basins; *2* Danubian sedimentary cover; *3* Danubian crystalline; *4* Getic sedimentary cover; *5* Getic crystalline; *6* Faults; *7* Thrusts; *8* Deep crustal fractures; *9* Geological boundaries; *10* Main Laramide plutonites from the central lineament; *L* Lapusnic, *N* Nasovat; *P* Purcariu; *Li* Lilieci; *11* Laramide igneous rocks and porphyry copper prospects from the eastern lineament

data, but also providing petrographical, petrochemical and metallogenic evidence, Gunnesch and Gunnesch (1977) have considered the igneous rocks from the Mraconia district as Laramide. Thus, the eastern lineament of the Laramide igneous rocks should be extended southwards from Lapusnicel to Dubova (Fig. 8). It is possible, furthermore, that the north-south tectonic line continues to Eastern Serbia. There, the fracture was active during Early Alpine deformations controlling the emplacement of the Cenomanian volcanic rocks.

The regional structural factors that control the distribution of the porphyry copper deposits are generally interpreted in the light of plate tectonic concepts.

The porphyry copper metallogeny of the Carpathian-Balkan province was also interpreted in terms of plate tectonics by many authors (e.g., Radulescu and Sandulescu 1973; Grubíč 1974; Janković and Petković 1974; Ianovici et al. 1977). Plate tectonic studies in this relatively restricted area are still scarce, and both tectonomagmatic and metallogenic features are interpreted differently by the different authors. One school of thought supported in a recent paper by Cioflica and Vlad (1980) assigns the Laramide magmatism and the related porphyry copper ores from SW Romania to a west-dipping subduction zone along which a micro-ocean was consumed. It is suggested to consider the thrust fault between the Getic and the Danubian Unit (Fig. 8) as representing a paleo-Benioff-zone. In opposition to this concept, Antonijevic et al. (1974) explained the formation of the so-called "Timok magmatic complex" and the related copper deposits of Bor and Majdanpek (situated south of Liubcova) by rift genesis. About the location and inclination of the paleo-Benioff-zones in the Carpathian-Balkan area there is still considerable discussion, as pointed out recently also by Sillitoe (1980).

Because of the contradictions in opinion referring to identical tectonomagmatic environments and metallogeny, no support for any hypothesis may be decisive, however, until many more isotope data are obtained from this area.

Acknowledgments. The author is grateful to Professor Amstutz, for discussions and for critical reading of the manuscript.

References

Amstutz GC (1965) Porfidi Cupriferi, Geologia, Esplorazione, Valorizzazione. Boll Serv Geol Ital 86:3–15

Antonijevic I, Grubic A, Djordevic M (1974) The upper Cretaceous palaeorift in Eastern Serbia. In: Janković S (ed) Metallogeny and concepts of the geotectonic development of Jugoslavia. Belgrade, pp 315–339

Ashley PM, Billington WG, Graham RL, Neale RG (1978) Geology of the Coalstoun porphyry copper prospect, Southeast Queensland, Australia. Econ Geol 73:945–965

Batchelder J (1977) Light stable isotope and fluid inclusion study of the porphyry copper deposit at Copper Canyan, Nevada. Econ Geol 72:60–70

Carson DJT, Jambor JL (1974) Mineralogy, zonal relationships and economic significance of hydrothermal alteration at porphyry copper deposits, Babine Lake area, British Columbia. CIM Bull 67:110–133

Cathles LM (1977) An analysis of the cooling of intrusives by ground-water convection which includes boiling. Econ Geol 72:804–826

Chivas AR, Wilkins RWT (1977) Fluid inclusion studies in relation to hydrothermal alteration and mineralization at the Koloula porphyry copper prospect, Guadalcanal. Econ Geol 72:153–169

Cioflica G, Vlad S (1973) The correlation of Laramian metallogenic events belonging to the Carpatho-Balkan area. Rev Roum Geol Geophys Geogr Ser Geol 17(2):217–224

Cioflica G, Vlad S (1980) Copper sulphide deposits related to Laramian magmatism in Romania. In: Janković S, Sillitoe RH (eds) European copper deposits. Belgrade, pp 67–72

Cotta B von (1865) Erzlagerstätten im Banat und Serbien. Braumüller, Wien

Crerar DA, Barnes HL (1976) Ore solution chemistry. V. Solubilities of chalcopyrite and chalcocite assemblages in hydrothermal solution at 200° to 350°C. Econ Geol 71:772–794

Cunningham CG (1978) Pressure gradients and boiling as mechanisms for localizing ore in porphyry systems. J Res US Geol Surv 6:745–754

Eastoe CJ (1978) A fluid inclusion study of the Panguna porphyry copper deposit, Bougainville, Papua New Guinea. Econ Geol 73:721–748

Ford JH (1978) A chemical study of alteration at the Panguna porphyry copper deposit, Bougainville, Papua New Guinea. Econ Geol 73:703–720

Ford JH, Green DC (1977) An oxygen and hydrogen isotope study of the Panguna porphyry copper deposit, Bougainville. Geol Soc Aust J 24:63–80

Giusca D, Cioflica G, Savu H (1966) The petrologic characterization of the banatitic province. Anu Com Stat Geol 35:1–40 (in Rumanian)

Grubîĉ A (1974) Eastern Serbia in the interpretation of the new global tectonics: consequences of this model for the interpretation of the tectonics of the northern branch of the Alpides. In: Janković S (ed) Metallogeny and concepts of the geotectonic development of Jugoslavia, Belgrade, pp 179–212

Gunnesch KA, Gunnesch M (1977) The banatitic rocks of the Dubova-Ogradena-Mraconia Valley region (South Banat) and the associated metallogeny. DS Inst Geol Geofiz 63:87–98 (in Rumanian)

Gunnesch KA, Gunnesch M, Seghedi I, Popescu C (1975) Contributions to the study of the banatitic rocks from Liubcova-Lapusnicu Mare. DS Inst Geol Geofiz 61:169–189 (in Rumanian)

Gunnesch KA, Gunnesch M, Vlad C (1978) Petrographic and petrochemical study of the banatitic rocks from Teregova-Lapusnicel (Semenic Mountains). St Cerc Geol Geofiz Geogr Ser Geol 23(2):239–248 (in Rumanian)

Gustafson LB (1978) Some major factors of porphyry copper genesis. Econ Geol 73:600–607

Gustafson LB, Hunt JP (1975) The porphyry copper deposit at El Salvador, Chile. Econ Geol 70: 857–912

Helgeson HC (1970) A chemical and thermodynamic model of ore deposition in hydrothermal systems. Mineral Soc Am Spec Pap 3:155–186

Hollister VF (1975) An appraisal of the nature and source of porphyry copper deposits. Miner Sci Eng 7(3):225–233

Ianovici V, Vlad S, Borcos M, Bostinescu S (1977) Alpine porphyry copper mineralization of West Romania. Mineral Deposita 12:307–317

Janković S, Petković M (1974) Metallogeny and concepts of the geotectonic development of Jugoslavia. In: Janković S (ed) Metallogeny and concepts of the geotectonic development of Jugoslavia. Belgrade, pp 443–477

Lowell JD, Guilbert JM (1970) Lateral and vertical alteration mineralization zoning in porphyry ore deposits. Econ Geol 65:373–408

Radulescu D, Sandulescu M (1973) The plate tectonics concepts and the geological structure of the Carpathians. Tectonophysics 16:155–161

Schmitt HA (1954) Origin of the silica of the bedrock hypogene ore deposits. Econ Geol 49:877–890

Sheppard SMF, Gustafson LB (1976) Oxygen and hydrogen isotopes in the porphyry copper deposit at El Salvador, Chile. Econ Geol 71:1549–1559

Sillitoe RH (1980) The Carpathian-Balkan porphyry copper belt – a Cordilleran perspective. In: Janković S, Sillitoe RH (eds) European copper deposits. Belgrade, pp 26–35

Streckeisen AL (1973) IUGS Subcomission on the systematics of igneous rocks: Classification and nomenclature of plutonic rocks. Geotimes 18(10):26–30

Sutherland Brown A, Cathro RJ, Panteleyev A, Ney CS (1971) Metallogeny of the Canadian Cordillera. CIM Bull 64:37–61

Polymorphic Variations of Pyrrhotite as Related to Mineralization in the Pyrometasomatic Copper, Zinc, Iron-Sulphide Deposits at the Kawayama Mine, Japan

T. NAKAMURA[1]

Abstract

The ore deposits of the Kawayama mine are of pyrometasomatic and partly vein-type origin, genetically closely related with acidic magmatism of Late Cretaceous age. The stages of hypogene mineralization in the upper ore deposit are divided into four major stages called I, II, III, and IV from earliest to latest. Major constituent ore minerals are pyrrhotite, chalcopyrite, and sphalerite. The relationship between the polymorphic variations of pyrrhotite and the kinds of the associated characteristic gangue mineral as related to hypogene mineralization has been demonstrated. Pyrrhotites associated with hedenbergite of stage I and actinolite of stage II are completely or predominantly of the sulphur-poor hexagonal phase. Pyrrhotite associated with chlorite of stage III is predominantly of the monoclinic phase. In the stage IV a single phase of monoclinic pyrrhotite associated with hisingerite-vein mineralization is exclusively present as an alteration product from the hexagonal pyrrhotites of stages I, II, and III.

1 Introduction

Pyrrhotite is a common and abundant sulphide mineral, occurring as a major constituent of various types of ore deposits. Because of its common occurrence in nature and complexity of superstructures, phase relations, and magnetic properties, natural and synthetic pyrrhotites have been extensively studied from the viewpoint of ore genesis and mineralogy.

In the course of study of the genesis of pyrometasomatic and partly vein-type deposits at the Kawayama mine (Fig. 1), one of the important copper, zinc, and iron sulphide mines in the Inner Zone of south-west Japan, the author was able to establish the modes of occurrence of the different kinds of structure type of pyrrhotite in the mine. In the present paper the author will demonstrate the relationship between the polymorphic variations of pyrrhotite and the kinds of the associated minerals, chiefly gangue minerals known as skarn minerals, as related to hypogene mineralization in the mine.

1 Osaka City University, Osaka, Japan

2 Geologic Setting and Metallogenic Province of South-West Japan

The Kawayama mining area is mainly composed of Paleozoic formations, acidic volcanics and minor intrusives of Late Cretaceous to Paleogene age. Paleozoic formations consist of the Sangun crystalline schists and the very low-grade metamorphic rocks, formerly called "nonmetamorphic Paleozoic formation". In a wider region including the Kawayama mining area, these rocks are stratigraphically divided into two groups, the Tsuno group and the Nishiki group, in ascending order. In the Kawayama mining area the Tsuno group is composed of green rock, basic, pelitic, and psammitic schists with a subordinate quantity of schistose calcium-rich rock, limestone, and quartz schist. The Tsuno group is characterized by the presence of the epidote-chlorite-actinolite assemblage with or without alkali-amphiboles belonging to glaucophane schist facies or greenschist facies. The Nishiki group is composed of very low-grade pelitic and psammitic rocks, and is characterized by the presence of pumpellyite-chlorite assemblage belonging to the prehnite-pumpellyite metagraywacke facies (Nishimura and Nureki 1966). These two groups are bounded by the thrust zone known as "kitayama" overthrust (Kojima and Sasaki 1950).

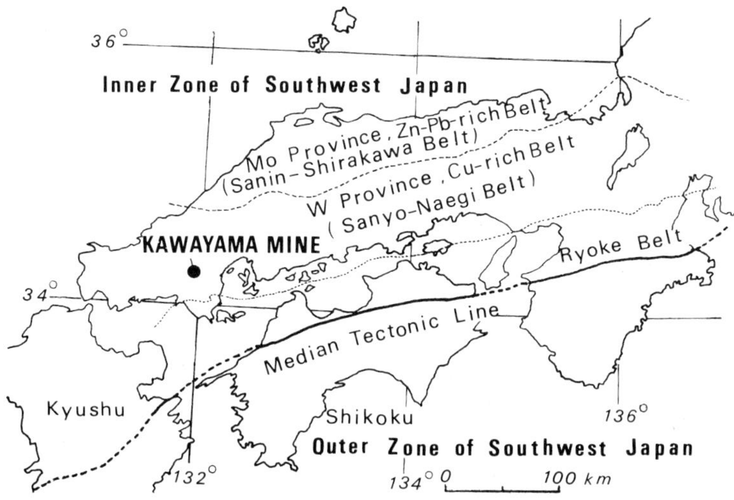

Fig. 1. Location of Kawayama mine, and metallogenic and petrographic provinces in the Inner Zone of south-west Japan. (Modified after Ishihara 1971, 1973; Shimazaki 1975)

The Tsuno group and the Nishiki group are covered by acidic volcanics and are intruded by acidic dike rocks in some limited areas.

The Inner Zone of south-west Japan was a site of intense volcano-plutonism from Late Cretaceous to Paleogene age. Based on chemical and mineralogical composition of granitic rocks, the petrographic provinces in the Inner Zone of south-west Japan are divided into the Ryoke, Sanyo-Naegi, Sanin-Shirakawa belts from south to north (Fig. 1; Kanaya and Ishihara 1973). The acidic rocks of the Kawayama mining area could be interpreted as shallow facies of an acidic magmatism in the Sanyo-Naegi belt.

The K-Ar ages of acidic plutonic and effusive rocks of the Sanyo-Naegi belt are within the ranges of 95–60 m.y. (Kawano and Ueda 1967; Ishihara 1973). In the Inner Zone of south-west Japan tungsten deposits of pyrometasomatic type or plutonic vein-type are genetically associated almost exclusively with the granitic rocks of the Sanyo-Naegi belt, and molybdenum deposits with those of the Sanin-Shirakawa belt (Ishihara 1971). The tungsten and molybdenum provinces are in accord with the Sanyo-Naegi and Sanin-Shirakawa belts respectively. Based on the ratios of Cu/Zn-Pb of pyrometasomatic deposits in the Inner Zone of south-west Japan, the Cu-rich and Zn-Pb-rich belts were proposed by Shimazaki (1975). The Cu-rich and Zn-Pb-rich belts are in accord with the tungsten and molybdenum provinces respectively (Fig. 1). The ore deposit at the Kawayama mine lies in the Cu-rich belt.

The pyrometasomatic or plutonic vein-type tungsten deposits occur in the eastern part of Yamaguchi Prefecture, including the Kawayama mining area. The K-Ar ages of muscovite occurring as gangue minerals are 92.1 m.y. at the Fujigatani mine, and 95.8 m.y. at the Kiwada mine (Shibata and Ishihara 1974). The K-Ar age of muscovite from a two-mica-adamellite, which is a hidden cupola at the Fujigatani mine, is 94.6 m.y. (Shibata and Ishihara 1974). The ore deposits at the Kawayama mine are considered to be genetically closely related to those at the Fujigatani and Kiwada mines. As will be described later, a small amount of scheelite occurs in the copper ore from the Kawayama mine.

3 Outline of Ore Deposits and Mineralization

The ore deposits at the Kawayama mine are of pyrometasomatic and partly vein-type (Nomura and Honda 1952), and occur along a gently dipping thrust zone where schistose calcium-rich rock and a limestone bed of the Tsuno group are observable. The ore deposits are structurally localized by the presence of a thrust zone, limestone bed, and syncline with the axis plunging gently N60°W, and by the intersecting structures of these structural elements.

The ore deposit in the southern part strikes N40°–60°E and dips 5°–30° to NW, while the ore deposit in the northern part strikes NS–N40°W, and dips 5°–30° to W or SW.

The ore deposit, as a whole, is 1200 m along its strike side, and 600 m along its dip side. Its thickness is about 5 m on an average, ranging from 2 to 25 m.

The NE–SW, NW–SE, and NS faults transect the ore body steeply.

Major constituent ore minerals are pyrrhotite, chalcopyrite, and sphalerite. Minor amounts of galena, arsenopyrite, scheelite, and pyrite are observable with them. Microscopically, cubanite, mackinawite, native bismuth, stannite, and wittichenite occur in small amounts (Gohara 1955; Takeno 1963, 1965, 1966; Nakamura and Aikawa 1974). Hedenbergite, ilvaite, actinolite, epidote, quartz, calcite, and hisingerite occur as gangue minerals.

The ore deposits of this mine are divided into two deposits; an upper ore deposit (Honko level to −110 m level) and a lower ore deposit (−110 m level to −240 m level). The modes of occurrence of pyrrhotite and other ore minerals in the lower ore deposit were studied by Takeno (1963, 1965, 1966).

In the present paper the author will deal with the upper ore deposit.

Taking the upper ore deposit as a whole, the stages of hypogene mineralization are divided into four major stages called stage I, stage II, stage III, and stage IV, from earliest to latest. These major stages are separated from one another by a tectonic break.

The characteristic mineral assemblage in each mineralization stage will be briefly given in the following.

Stage I: The mineral assemblage of stage I is represented by a moderate amount of pyrrhotite and hedenbergite with a minor amount of chalcopyrite, sphalerite, quartz, and calcite. Andradite occurs predominantly in the lower ore deposit, but a little andradite occurs in the upper ore deposit. Under the ore microscope, sphalerite contains exsolution bodies of cubanite-chalcopyrite intergrowths.

Stage II: The mineral assemblage of stage II is characterized by dominant massive pyrrhotite associated with actinolite, quartz, calcite, subordinate chalcopyrite and sphalerite, and a little arsenopyrite. Under the ore microscope, sphalerite contains exsolution bodies of cubanite-chalcopyrite intergrowths.

Stage III: The mineral assemblage of stage III is characterized by dominant chalcopyrite and pyrrhotite associated with subordinate sphalerite, chlorite, quartz, and calcite. Small amounts of arsenopyrite, pyrite, and scheelite occur in the chalcopyrite-pyrrhotite ore. Under the ore microscope, chalcopyrite contains "zincblende star."

Stage IV: The mineral assemblage of stage IV is characterized by dominant hisingerite, pyrite, calcite with subordinate quartz.

The upper ore deposit is composed of the ores formed in stages I, II, and III. Especially the ore composed of massive pyrrhotite with subordinate chalcopyrite and sphalerite of stage II is predominant. The copper ore shoot composed of chalcopyrite-pyrrhotite-chlorite-quartz association of stage III occurs along the hanging-wall side of the ore body. The zinc ore shoot composed chiefly of sphalerite and pyrrhotite of stages II and III occurs along the limestone side of the ore body. The ore body composed of the ores of stage I and II is frequently replaced and transected by the chalcopyrite-pyrrhotite-chlorite-quartz ore of stage III. The hisingerite-pyrite-calcite-quartz veins or veinlets of stage IV occur along the steeply dipping faults, fissures, and cracks in the ore body consisting of the ores of stages I, II, and III. Hisingerite occurs frequently as small patches in the pyrrhotite ores formed in stages I, II, and III. As will be described later, monoclinic pyrrhotite of stage IV, formed as an alteration product from hexagonal pyrrhotite of stages I, II, and III, occurs along the borders of the hisingerite-pyrite-calcite-quartz veins or veinlets. There is no relationship between the amount of hisingerite and the depth from the present surface. The occurrence of hisingerite shows that it is not the result of weathering, but is clearly an alteration product from pyrrhotite and other iron minerals which were attacked by hypogene solutions (Sudo and Nakamura 1952).

4 Polymorphic Variations of Pyrrhotite as Related to Mineralization

Polymorphic variations of pyrrhotites of stages I, II, III, and IV were examined by means of X-ray powder diffraction, differential thermal analysis, and structure etching with HI under the ore microscope. The d(102) method of Arnold and Reichen (1962) was applied to determine the iron contents of hexagonal pyrrhotite. Each (102) reflection was measured against the (200) reflection of quartz standard. Scanning and chart speeds were $1°/4$ (2θ) per min and 1 cm per min respectively. Hexagonal pyrrhotite was distinguished from monoclinic pyrrhotite by a single, sharp symmetrical (102) reflection, while monoclinic pyrrhotite was distinguished from hexagonal pyrrhotite by a split reflection of two peaks, (202) and ($20\overline{2}$), of about equal intensity. In examining the structure type of mixtures of hexagonal and monoclinic pyrrhotites, the relative intensity of a split reflection of two peaks, which represent the superposed monoclinic and hexagonal peaks, was referred (Arnold 1966).

The X-ray diffraction pattern of each representative pyrrhotite of different stages is reproduced on a smaller scale in Fig. 2a–d.

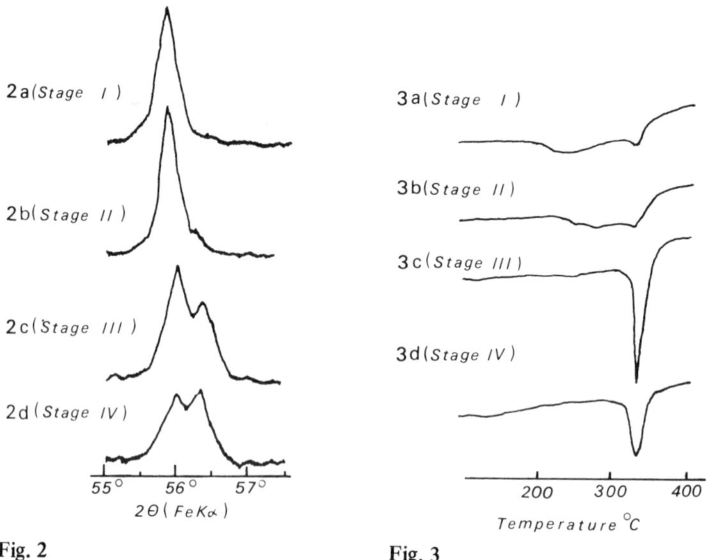

Fig. 2

Fig. 3

Fig. 2a–d. X-ray diffraction patterns of pyrrhotites of stages I, II, III, and IV. **a** H, **b** H > M, **c** H < M, **d** M. "H" denotes the hexagonal phase, "M" the monocline one

Fig. 3a–d. Differential thermal analysis curves of pyrrhotites of stages I, II, III, and IV. **a** H, **b** H > M, **c** H < M, **d** M. "H" denotes the hexagonal phase, "M" the monoclinic one

To correlate the thermal behavior of hexagonal pyrrhotite with monoclinic pyrrhotite, the differential thermal analysis was made on different types of pyrrhotite under vacuum in 10^{-3} mm Hg. The heating rate was 10°C per min. Only a sharp endothermic peak at about 320°C, representing the β-transformation from the low monoclinic form to high hexagonal form, is observable in the differential thermal analysis of monoclinic pyrrhotite. The inversion of monoclinic pyrrhotite to hexagonal pyrrhotite was determined to be 320°C by Grønvold and Haraldsen (1952).

The differential thermal analysis curve of each representative pyrrhotite of different stages is reproduced on a smaller scale in Fig. 3a–d.

To examine the mixed monoclinic-hexagonal character of pyrrhotite in detail, the polished surface of each pyrrhotite of different stages was examined by structure etching with concentrated HI (Kiskyras 1950). The monoclinic pyrrhotite is strongly etched, while the hexagonal pyrrhotite is not etched or only lightly etched by HI. Some etched textures of pyrrhotites of stages I, II, III, and IV are shown in Fig. 4a, b, d, e, f.

Pyrrhotite of stage I: Pyrrhotite associated with hedenbergite of stage I is composed exclusively of the hexagonal phase (structure type: H, Fig. 2a). The composition determined from d(102) is 47.62 atom percent iron. The differential thermal analysis curve (Fig. 3a) shows two endothermic peaks corresponding to γ- and β-transformations. On the etched polished surface, the lightly etched lamellae, 5–50 μ wide, with somewhat haphazard fashion are observable in a host (Fig. 4a).

As to superstructures of pyrrhotite, five different types of pyrrhotite have been confirmed to be stable at room temperature, and are called 4C(Fe_7S_8, monoclinic), 5C(Fe_9S_{10}, hexagonal), 11C($Fe_{10}S_{11}$, orthorhombic), 6C($Fe_{11}S_{12}$, hexagonal), and 2C(FeS, hexagonal) (Desborough and Carpenter 1965; Nakazawa and Morimoto 1970; Morimoto et al. 1971). The compositions of these five types of pyrrhotite are essentially stoichiometric. Recently, it has been clarified that the *n*C pyrrhotites have a composition range approximately from Fe_9S_{10}(5C, orthorhombic) to $Fe_{11}S_{12}$(6C, orthorhombic) and usually show nonintegral types of superstructure (Morimoto et al. 1975). The 5C(orthorhombic) and 6C(orthorhombic) types are considered to be the special cases of the *n*C pyrrhotites, where the *n*-values are generally nonintegral. On the basis of the X-ray diffraction and structure etching with HI, two phases revealed by the structure etching are considered to be a mixture of two types of superstructure with hexagonal (?) symmetry. The question on superstructures of pyrrhotite of stage I remains unsolved.

Pyrrhotite of stage II: Pyrrhotite associated with actinolite of stage II is composed mainly of the hexagonal phase with a small amount of monoclinic phase (structure type: H > M, Fig. 2b). The differential thermal analysis curve (Fig. 3b) shows two endothermic peaks corresponding to γ- and β-transformations. On the etched polished surface a little monoclinic pyrrhotite occurs usually as the strongly etched lamella in a host of hexagonal pyrrhotite. In another case, the lightly etched lamellae of hexagonal pyrrhotite are replaced by the strongly etched monoclinic pyrrhotite (Fig. 4b). The composition of hexagonal pyrrhotite determined from d(102) is 47.37 atom percent iron corresponding to Fe_9S_{10}(5C).

Pyrrhotite of stage III: Pyrrhotite associated with chalcopyrite, chlorite, quartz, and a little sphalerite, arsenopyrite, and scheelite (Fig. 4c) of stage III, is composed mainly of monoclinic phase with a small amount of hexagonal phase (structure type: H < M, Fig. 2c). The differential thermal analysis curve (Fig. 3c) shows a sharp endothermic peak at 320°C corresponding to β-transformation. On the etched polished surface fine

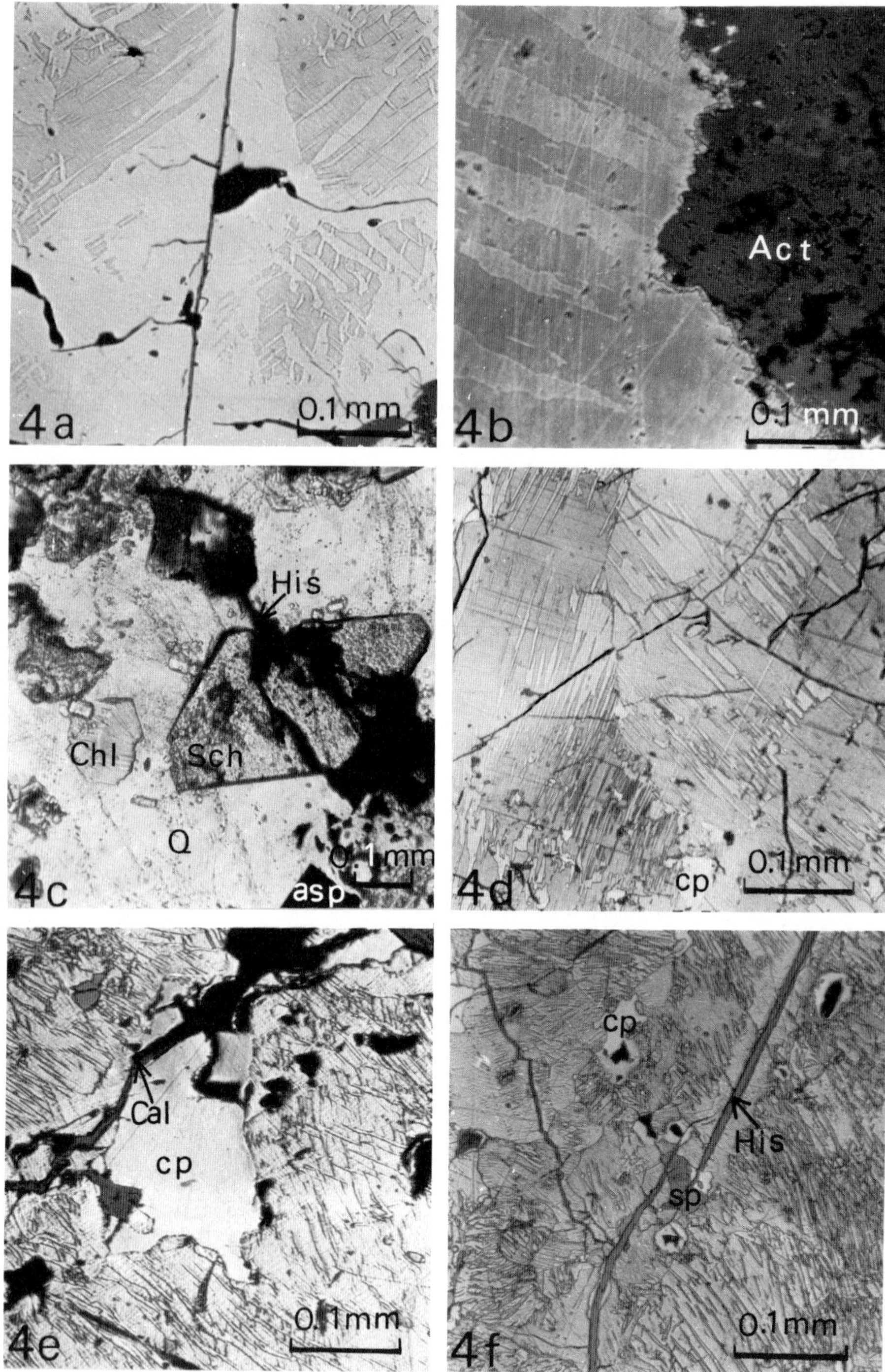
4a
0.1 mm
Act
4b
0.1 mm
His
Chl
Sch
Q
asp
0.1 mm
4c
0.1 mm
cp
4d
Cal
cp
0.1 mm
4e
cp
His
sp
0.1 mm
4f

lamellae, 3–30 μ wide, of hexagonal pyrrhotite occur in a host of the strongly etched monoclinic pyrrhotite (Fig. 4d, e, f). The composition of hexagonal pyrrhotite determined from d(102) is 47.24–47.37 atom percent iron, corresponding approximately to Fe_9S_{10}(5C).

Pyrrhotite of stage IV: A pyrrhotite specimen taken from the borders of a hisingerite-pyrite-calcite-quartz vein shows a single phase of monoclinic pyrrhotite (structure type: M, Fig. 2d) by means of X-ray diffraction, and indicates clearly a sharp endothermic peak at 320°C corresponding to β-transformation. On the etched polished surface, the strongly etched monoclinic pyrrhotite replacing the unetched or lightly etched hexagonal pyrrhotite occurs along the borders of calcite veinlet (Fig. 4e), or hisingerite veinlet (Fig. 4f). The monoclinic pyrrhotite occurring along the borders of hisingerite-pyrite-calcite-quartz vein or veinlet is considered to be an alteration product from the hexagonal pyrrhotites of stages I, II, and III.

5 Conclusions

The relationship between the polymorphic variations of pyrrhotite and the kinds of the associated characteristic gangue minerals, as related to hypogene mineralization at the Kawayama mine, can be summarized from the earliest to latest as follows ("H" denotes the hexagonal phase of pyrrhotite, "M" the monoclinic one): Hexagonal pyrrhotite (H) associated with hedenbergite of stage I → hexagonal pyrrhotite with a small amount of monoclinic pyrrhotite (H > M) associated with actinolite of stage II → monoclinic pyrrhotite with a small amount of hexagonal pyrrhotite (H < M) associated with chlorite of stage III → alteration of hexagonal pyrrhotite to monoclinic pyrrhotite (M) associated with hisingerite-vein mineralization in stage IV.

Pyrrhotites of stages I and II showing little or no alteration by later stage mineralization are completely or predominantly of sulphur-poor hexagonal phase, and the compositions are 47.62 atom percent iron in stage I and 47.37 atom percent iron in stage II. In the stage III the monoclinic pyrrhotite predominates. In the stage IV a single phase of monoclinic pyrrhotite, formed as an alteration product from the hexagonal pyrrhotites of stages I, II, and III, is exclusively present.

Acknowledgements. The author wishes to express his grateful appreciation to Emeritus Professor T. Watanabe, University of Tokyo, Emeritus Professor T. Sudo, Tokyo University of Education, and Professor G. Kojima, Hiroshima University, for their useful suggestions. The author is indebted to geologists of the Nippon Mining Co. Ltd. for their kind instruction in the field, and to Dr.

Fig. 4. a Hexagonal pyrrhotite associated with hedenbergite of stage I. Polished section. Etched with HI. **b** Hexagonal pyrrhotite and monoclinic pyrrhotite *(dark)* associated with actinolite *(Act)* of stage II. Polished section. Etched with HI. **c** Scheelite *(Sch)*, chlorite *(Chl)*, quartz *(Q)*, and arsenopyrite *(asp)* of stage III. *His*, hisingerite. Thin section. Open nicol. **d** Mixtures of hexagonal (lamellae) and monoclinic (host, *dark*) pyrrhotite of stage III. *cp*, chalcopyrite. Polished section. Etched with HI. **e** Mixtures of hexagonal (lamellae) and monoclinic (host) pyrrhotite of stage III. *cp*, chalcopyrite; *Cal*, calcite. Polished section. Etched with HI. **f** Monoclinic pyrrhotite along the borders of hisingerite *(His)* veinlet of stage IV, and mixtures of hexagonal and monoclinic pyrrhotite of stage III. *cp*, chalcopyrite; *sp*, sphalerite. Polished section. Etched with HI

N. Aikawa, Osaka City University, for his help in the laboratory investigation. The study was partly supported financially by the Fund for Scientific Research issued by the Ministry of Education of the Japanese Government.

References

Arnold RG (1966) Mixtures of hexagonal and monoclinic pyrrhotite and the measurement of the metal content of pyrrhotite by X-ray diffraction. Am Mineral 51:1221–1227

Arnold RG, Reichen LE (1962) Measurement of the metal content of naturally occurring, metal-deficient, hexagonal pyrrhotite by an X-ray spacing method. Am Mineral 47:105–111

Desborough GA, Carpenter RH (1965) Phase relations of pyrrhotite. Econ Geol 60:1431–1450

Gohara H (1955) On the ores in Kawayama mine, Yamaguchi Prefecture. Bull Geol Surv Jpn 6: 61–64 (in Japanese)

Grønvold F, Haraldsen H (1952) On the phase relations of synthetic and natural pyrrhotites ($Fe_{1-x}S$). Acta Chem Scand 6:1452–1469

Ishihara S (1971) Modal and chemical composition of the granititc rocks related to the major molybdenum and tungsten deposits in the Inner Zone of Southwest Japan. J Geol Soc Jpn 77:441–452

Ishihara S (1973) The Mo-W metallogenic provinces and the related granitic provinces. Min Geol Jpn 23:13–32 (in Japanese)

Kanaya H, Ishihara S (1973) Regional variation of magnetic susceptibility of the granitic rocks in Japan. J Jpn Assoc Mineral Petrogr Econ Geol 68:211–224

Kawano Y, Ueda Y (1967) Periods of the igneous activities of the granitic rocks in Japan by K-Ar dating methods. Tectonophysics 4:523–530

Kiskyras DA (1950) Untersuchungen der magnetischen Eigenschaften des Magnetkieses bei verschiedenen Temperaturen in besonderem Hinblick auf seine Entstehung. Neues Jahrbuch Mineral Geol Palaeontol Abh Abt A 80:297–342

Kojima G, Sasaki D (1950) Geology of the Kawayama mining district, Yamaguchi Prefecture. J Geol Soc Jpn 56:1–5 (in Japanese)

Morimoto N, Nakazawa H, Tokonami M, Nishiguchi K (1971) Pyrrhotites: Structure type and composition. Soc Min Geol Japn Spec Issue 2:15–21 (Proc IMA-IAGOD Meet 1970, Joint Symp Vol)

Morimoto N, Gyobu A, Tsukuma K, Koto K (1975) Superstructure and nonstoichiometry of intermediate pyrrhotite. Am Mineral 60:240–248

Nakamura T, Aikawa N (1974) Pyrrhotite and its mineral association at the Kawayama mine, Japan. J Mineral Soc Jpn Spec Issue 2:107–115 (in Japanese)

Nakazawa H, Morimoto N (1970) Pyrrhotite phase relations below 320°C. Proc Jpn Acad 46: 678–683

Nishimura Y, Nureki T (1966) Geological relations of the Non-metamorphic Paleozoic formations to the Sangun metamorphic rocks in the Nishiki-cho area, Yamaguchi Prefecture. J Geol Soc Jpn 72:385–398 (in Japanese)

Nomura E, Honda T (1952) Structural features of the ore deposit of the Kawayama mine. Min Geol Jpn 2:17–19 (in Japanese)

Shibata K, Ishihara S (1974) K-Ar ages of the major tungsten and molybdenum deposits in Japan. Econ Geol 69:1207–1214

Shimazaki H (1975) The ratios of Cu/Zn-Pb of pyrometasomatic deposits in Japan and their genetical implications. Econ Geol 70:717–724

Sudo T, Nakamura T (1952) Hisingerite from Japan. Am Mineral 37:618–621

Takeno S (1963) On the ore minerals of the lower ore deposit of the Kawayama mine, Yamaguchi Prefecture. Geol Rep Hiroshima Univ 12:343–359

Takeno S (1965) A note on mackinawite (so-called valleriite) from the Kawayama mine, Japan. Geol Rep Hiroshima Univ 14:59–76

Takeno S (1966) Magnetometric and roentgenometric studies of pyrrhotite from the Kawayama mine, Japan. J Sci Hiroshima Univ Ser C 5:113–156

Plasticity of Cryolite and Brecciation in the Cryolite Deposit, Ivigtut, South Greenland

H. PAULY[1]

Abstract

Two breccia samples from the Ivigtut cryolite deposit are described because they obviously were formed under conditions not involving intensive chemical reactions between the rock fragments and the cryolite which acts as a cementing material in the breccias. In order to explain the breccia formation it seems that it is necessary to expect cryolite to have a rather high mobility as a plastic solid at moderately elevated temperatures. Such a behaviour of cryolite must be taken into account also in the evaluation of the formation of the deposit, because it may have obliterated much of the early formed structures within the deposit.

Breccias play an important role in the cryolite deposit at Ivigtut. Both within the dome-shaped cryolite body (about 70 m high and resting on a nearly circular basis with a diameter of about 170 m) and especially along its contact with the surrounding rocks of the intrusion occurred brecciated materials. Within the cryolite body, brecciation involved the various cryolite types of which siderite-cryolite and fluorite-cryolite (Pauly 1960) were the two main types. Along the margins of the cryolite body the breccias contained xenoliths of the rocks of the intrusion, either the Ivigtut granite or its greisenized equivalents. In most cases cryolite, i.e. pure cryolite, formed the cementing material in the breccias. In the eastern part of the deposit, where a protuberance from the dome-shaped body was exposed, siderite-cryolite formed the matrix between xenoliths of granite and other rocks of the intrusion. Cryolite mining started in this part of the deposit and from the beginning it was mentioned that granite fragments could be found embedded in the cryolite. "Wall rock fragments have presumably become included in the cryolite when it forced its way up through the earth as a molten mass". This summarizes the idea set forth by lecturer (later professor) J. Thomsen (1862) when he read a communication on the Cryolite Industry to the Royal Danish Academy on 10 January 1862.

Bøgvad (1940) published a note on black cryolite, in which he attempted to demonstrate the connection between black cryolite and xenoliths of greisen. He refers in the note to the observation that black cryolite rims xenoliths of granite and greisen "which have fallen down into the cryolite when it was still in a molten state" (translated from the Danish text). In keeping with this viewpoint, he carried out experiments in which he placed fragments of greisen in powdered cryolite in a graphite crucible which was then heated to melting (about 1000°C). In the note he states that

1 Mineralogical Institute, Technical University of Denmark, 2800 Lyngby, Denmark

the greisen fragments were strongly attacked by the melt and if it was kept molten for too long a time the rock fragments were disrupted and settled on the bottom of the crucible as a heap of quartz grains, the mica of the rock having been totally dissolved by the melt. He rightly draws attention to the fact that molten cryolite had earlier been found to have a specific gravity of about 2.2, thus allowing the heavier solids to settle.

As long as one maintained the viewpoint that cryolite had been intruded as a molten mass, it might seem rather straightforward to account for the breccia phenomena. Bøgvad even has some remarks bearing on the aggressive nature of the melt causing the rock fragments to dissolve and to leave only quartz grains, and he mentions observations of clusters of quartz crystals and fragments in an area where also xenoliths have been observed.

What he does not mention, and presumably has not been aware of, is the fact that a great number – in fact most – rock xenoliths show no sign of corrosion. In a very special position within the deposit, rock fragments have been observed showing topazified rims, i.e. their outer part has been transformed into a topaz-rich rock, which over a distance of some cm graded into the original minerals of the rock. It should also be noted that the siderite-cryolite cemented breccia found in the eastern part of the deposit exhibits quartz crystal growth from the surfaces of the rock xenoliths out into the surrounding cryolite-siderite matrix. There have thus been cases where brecciation has been connected with chemical reactions and with conditions allowing quartz to form crystals in continuity with the quartz grains in the xenoliths.

The siderite-cryolite cemented breccia in the eastern part of the deposit constitutes an important part of the deposit because several lines of evidence seem to indicate that this part of the cryolite body was the first to form. Although the breccia here comprises a rather large mass, it is an order of magnitude smaller than the breccias along the margin of the western part of the cryolite body. Rock fragments in this huge brecciated mass, cemented by pure cryolite, exhibit no visible signs of reactions or quartz growth.

Considerations regarding the temperature of formation of the siderite-cryolite (Pauly 1960; Oen and Pauly 1967) pointed to temperatures below the α/β transition of quartz and probably not far above 500°C. This should then also represent the temperature at the time when the breccia in the eastern part of the deposit was formed.

Concerning the temperature for brecciated material along the western part of the cryolite body, it must be noted that fluorite-cryolite occurred as xenoliths cemented by pure cryolite. As fluorite-cryolite formed later than siderite-cryolite, presumably at temperatures around 300°C and below (Karup-Møller 1973), it appears that brecciation in this part of the deposit took place at rather low temperatures, maybe around 200°C.

Deprived of the easily manipulated concept of molten masses cementing the observed breccias in the cryolite deposit, maybe apart from the siderite-cryolite cemented eastern breccia, we have to look for other "mobilization factors" for cryolite. Limitations on such factors may be aptly illustrated by the sample shown in Fig. 1. The sample, which is part of a larger less shattered block, consists of a quartz feldspar rock, a mineralized granite variety from the intrusion in the western part of the quarry. The numerous thin slices of granite are cemented by pure cryolite – here black cryolite.

The relative displacement of the rock slices can be seen to amount to only a few mm. Microscopy of the contacts does not reveal reaction or quartz crystal growth along the edges of the rock fragments. The cement of pure cryolite obviously invaded the shattered rock in such a mobile state that the displacement of the rock pieces was kept to a minimum. Presumably, a reasonably high confining pressure assisted this action though the temperature may be regarded as having been fairly low.

Fig. 1. Polished sample of shattered granite cemented by cryolite. White grains are alkali feldspar, cryolite is black. (Sample is 16 cm long)

Microscopy shows that cryolite is present as grains similar in sizes to the grains of the shattered rock and up to about twice this size, about 0.5 cm in the widest cryolite areas. The twin lamellae of the cryolite grains show them to be oriented in many different directions.

The twin laws present cannot be identified with any degree of certainty but the general pattern seems to be simple and only a few laws seem to be present.

As the sample has not been found in situ, nothing more can be said about its relation to other materials in the deposit.

Figure 2 illustrates another variety of brecciation. Two, maybe three, brecciation stages are present. Apparently this case is fairly representative of breccia samples from along the western wall of the quarry. The triangular greisen xenolith, surrounded on two sides by the cementing black cryolite, is cut by a vein carrying large crystals of quartz which grow from both sides towards the centre of the vein which contains pure cryolite. In the interstices between two quartz crystals is observed a 2 mm large crystal of sphalerite with chalcopyrite inclusions (not visible in the shown polished section; it is found in the corresponding thin section of the sample). The vein is obviously a mineralized vein of a kind which is very common in the deposit. Such veins transect all rocks of the intrusion and similar veins can be found stretching out into the surrounding country rock. Some of these veins carry cassiterite. These veins are regarded as having formed at a stage just preceding the main stage when siderite-cryolite was formed. They thus represent an earlier stage of rock fracturing than the breccias here treated.

Fig. 2. Greisen xenolith with a mineralized quartz vein in a matrix of black cryolite. Note the shattered appearance of the matrix cryolite in *upper left part* of the sample. (Sample is 16 cm across)

When the sample shown in Fig. 2 was collected it was taken as a representative of the main brecciation from the western part of the deposit. On cutting and polishing, however, its peculiar nature was revealed. This is especially well observed in the area occupied by the broad cryolite rim. The cryolite here is itself brecciated and appears as elongated angular xenoliths held together by narrow network-forming veins filled with fine-grained sericitic mica carrying mm-sized grains of cryolite. This "sand fraction" of cryolite also appears with sharply angled contours.

At the contact between this field and the greisen xenolith is a thin film of the same microcrystalline sericitic mica which fills the interstices between the cryolite fragments.

Where the cryolite from this area borders onto the mineralized vein in the greisen fragment one can clearly see the different nature of the cryolite in the vein and in the area outside the xenolith.

The relation between the xenolith and the surrounding cryolite along the other side of the greisen fragment is distinctly different. There is no film of mica at the contact and, as is clear from the photograph, the appearance of the cryolite is more homogeneous. Microscopic observations reveal, however, that the cryolite in this area is also broken up in elongated angular pieces similar to those on the other side of the xenolith, though the sericitic matrix is not present.

Thin section microscopy further reveals that the cryolite fragments are aggregates of more or less differently oriented grains. This is revealed by the twin lamellae of the grains. Some of the cm-sized fragments show twin lamellae structures with a slightly broken pattern indicating that they originally represented a crystallographically homogeneous area. Overprinting by newly formed mechanical twins may account for some of the fine-grained twin structures in the grains. Recrystallization seems not to have taken place to a degree where the clear signs of mechanical breaking have been seriously masked.

In the second sample, it seems rather obvious that cryolite has behaved as a brittle material. As the overall appearance of the sample, however, is similar to breccia samples from the western part of the deposit, this particular sample is thought to represent a piece of material which has been subjected to late-stage mechanical deformation acting at such a low temperature and/or at such a low confining pressure that the cryolite broke up. The presence of the interstitial microcrystalline sericitic mica is an indication of the mineralizing possibilities at that stage. It might be reasonable to consider that water was present to a certain degree but certainly not in amounts which could react with and dissolve the easily soluble cryolite. There is also a possibility that the sericitic mica represents fine-grained debris from the mechanical impact on the micaceous greisen represented in the xenolith.

Leaving aside further chemical speculations, we are left with the problem of brecciation in the deposit and indications for the behaviour of cryolite ranging from a very mobile cement to a brittle substance giving rise to both large and small, hardly recrystallized grains.

The following information appears relevant. Molten cryolite has a specific gravity of about 2.2. Landon and Ubbelohde (1956) found the fractional increase in volume on melting to be 0.25. The expansion from room temperature to just below the melting point is about 14 vol.% (Landon and Ubbelohde 1957). From room temperature to the molten state cryolite thus expands about 39 vol.%.

Cryolite exhibits an α/β transition at 565°C and the β-form has a specific gravity of 2.78 (calculated from cell dimensions in Steward and Rooksby 1953). Construction of the curve volume expansion/temperature shows cryolite to lie between rock salt and quartz. Values for these latter were taken from Skinner (1966) in Handbook of Physical Constants. Apart from the transition at 565°C (and possibly a second transition about 880°C, Landon and Ubbelohde 1957), the curve for cryolite is similar to that of rock salt.

Taken at face value, this constructed curve shows that cryolite on cooling from 500°C contracts much more than quartz and the other minerals usually found in the cryolite. Shearing and breaking of the included minerals would be expected. Thin veins of cryolite invading siderite along cleavage planes may be a result of such processes.

Landon and Ubbelohde (1957) state: "Observations that cryolite behaves as a plastic solid in this range of temperatures (around 880°C) likewise indicates very extensive lattice defect formation".

A further indication of the plasticity of cryolite was obtained in 1960 when Handin (1966) (see Handbook of Phys. Constants) undertook short time triaxial compression/extension tests on cryolite at 5000 bars confining pressure. Deformation up to 40% and 20% was observed after the treatment of the samples.

Plasticity characteristics for cryolite at temperatures relevant to the deposit in Ivigtut are not yet available. Under suitable conditions minerals like ice, rock salt, fluorite, and calcite exhibit pronounced plasticity due to slip, twin gliding, etc. These phenomena are enhanced by increase in temperature. As a working hypothesis, cryolite is assumed to have characteristics between rock salt and fluorite or calcite.

In the near future we hope to explain at least some of the fascinating features connected with this peculiar mineral.

Acknowledgments. I owe sincere thanks to Mr. Ib Nielsen who has prepared the samples, to the photographers at the Mineralogical-Geological Institutes of the University of Copenhagen who have made the photographs, and to John Bailey for suggestions improving the English of the text.

References

Bøgvad R (1940) Sort kryolit. Medd Dansk Geol Foren 9:561–564

Handin J (1966) Strength and ductility. Handbook of Physical Constants, Geol Soc Am Mem 97: 223:289

Karup-Møller S (1973) A gustavite-cosalite-galena-bearing mineral suite from the cryolite deposit at Ivigtut, South Greenland. Medd Grønl 195:1–40

Landon GI, Ubbelohde AR (1956) Volume changes on melting ionic crystals. Trans Faraday Soc 52:647–651

Landon GI, Ubbelohde AR (1957) Melting and crystal structure of cryolite (3NaF, AlF_3). Proc R Soc London Ser A 240:150–172

Oen IS, Pauly H (1967) A sulphide paragenesis with pyrrhotite and marcasite in the siderite-cryolite ore of Ivigtut, South Greenland. Medd Grønl 175:1–55

Pauly H (1960) Paragenetic relations in the main cryolite ore of Ivigtut, South Greenland. Neues Jahrb Miner Abh 94:121–139, Festband Ramdohr

Skinner B (1966) "Thermal expansion". Handbook of Physical Constants, Geol Soc Am Mem 97: 77–96

Steward EG, Rocksby HP (1953) Transitions in crystal structure of cryolite and related fluorides. Acta Crystallogr 6:49–52

Thomsen J (1862) Meddelelser angaaende Kryolithindustrien. Oversigt over det Kgl. danske Vidensk. Selskabs Forhandlinger Aaret 1862 af G. Forchhammer. Nr. 1 og 2, Januar og Februar, pp 1–10

The Formation of Chromite and Titanomagnetite Deposits Within the Bushveld Igneous Complex

D.D. KLEMM, R. SNETHLAGE, R.M. DEHM, J. HENCKEL, and R. SCHMIDT-THOMÉ[1]

During the famous international Bushveld excursion in 1929 some geologists put up a sign: "Here was the Bushveld Complex, before Paul Ramdohr sampled it away". As you may find in the following paper Paul Ramdohr left something for further research.

Abstract

A brief analysis of the early tectonic history of southern Africa reveals that the intrusion of the Bushveld Complex is situated at the intersection of several lineaments. Based on gravity measurements, the Bushveld can be divided into five subcomplexes.

The differentiation of the Bushveld magma can be explained by two model systems: $MgO - SiO_2 - Fe_3O_4 - Cr_2O_3$ and $MgO - SiO_2 - FeO - Fe_2O_3$. However, these model systems can neither explain the formation of the chromitite seams of the Critical Zone nor the massive titanomagnetite layers of the Upper Zone.

f_{O_2}-Measurements on chromites indicate that the formation of chromitite seams is caused by periodical fluctuations of the f_{O_2}. A similar mechanism has been postulated before by various authors for the accumulation of massive titanomagnetite layers.

Detailed geochemical analyses of host rock magnetites are interpreted. Several field evidences indicate the existence of local, but enormous volatile activity and the formation of anomalous magnetite layers by an increase of the f_{O_2}, caused by contamination of the magma with sedimentary fragments.

A new model is presented, which explains the formation of massive chromite and titanomagnetite seams as a function of the f_{O_2}, as well as their extreme lateral consistency within the Bushveld Complex.

1 Introduction

The Bushveld Complex in South Africa is the largest intrusive complex of the earth. Its economic importance regarding the mineral resources of the earth is enormous. Concerning the reserves of chromium, platinum, vanadium and titanium, the Bushveld ranges first in the list of all known mineral resources. Besides this, the production of nickel and copper is also considerable. Abnormal concentrations of apatite in the diorites of the uppermost sequence are not yet sufficiently known in their quantities.

1 Institut für Angewandte Geologie der Universität München, Luisenstraße 37, 8000 München 2, FRG

Therefore, the importance of the Bushveld Complex as a possible reserve of phosphorus in the future is still unknown. Considering these facts, it seems highly justified to concentrate on this extraordinary geological feature and its genetic history.

2 Tectonic Evolution

An intrusion of the dimension of the Bushveld Complex must be the consequence of large-scale tectonic events. Therefore, we draw attention to the main tectonic structures of southern Africa between 2.5 and 1.8 b.y. Already during the formation of the Archean Greenstone belts, at least 3 ENE- and WSW-striking lineaments existed (Fig. 1). These main structures, the Barberton-, the Murchison- and the Soutpansberg-Lineament, date back to about 3.8 b.y. The minimum age of the Limpopo Belt, which connected the Kapvaal and Rhodesian-Craton, is 2.7 b.y. The Limpopo Belt is in fact the oldest mobile belt known. There is evidence in the belt of a significant thermal event at 2.0 b.y., an event which clearly postdates the phases of metamorphism and folding

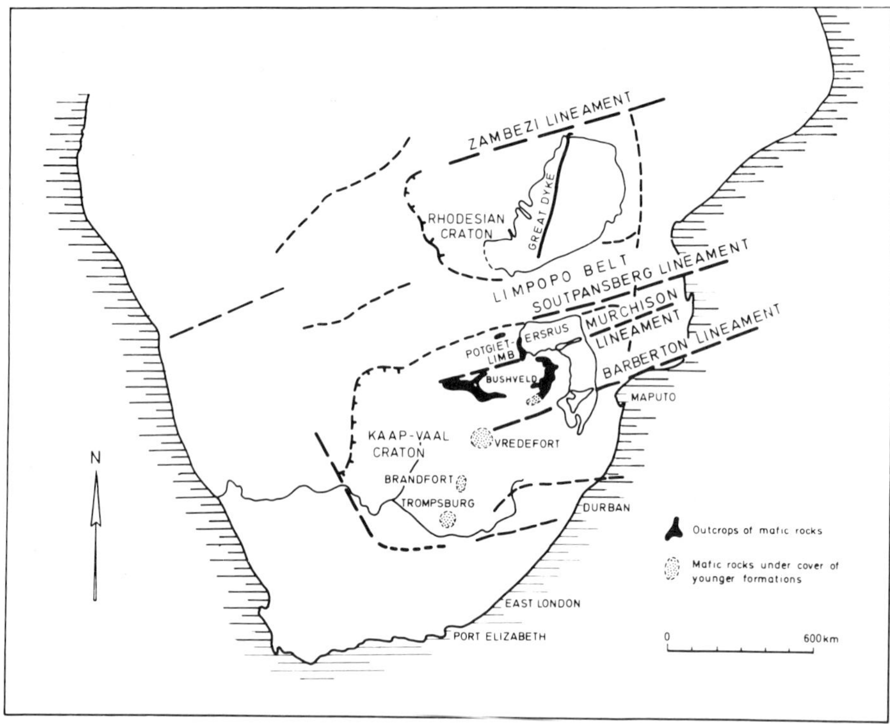

Fig. 1. Generalized tectonical map of Southern Africa (after van Biljon 1976). Several ENE-WSW striking lineaments control the geological evolution from the Precambrian up to now. Basic intrusions are lined up in a NNE-SSW striking direction. The ages of these intrusions (Great Dyke 2200 m.y., Bushveld Complex 2000 m.y., Vredefort Dome 1950 m.y., Brandfort Intrusion unknown, Trompsburg Igneous Complex 1400 m.y.) decrease significantly towards the south

in the belt (Truswell 1977). The strike of this belt is in the same direction as the old lineaments on the Kapvaal-Craton itself. North of the Rhodesia-Craton, the Zambezi-Mobile Belt, which is about 2.0 b.y. old, shows a similar direction. These lineaments played an important role throughout the geological history, not only of the Kapvaal-Craton but also in the whole southern Africa (van Biljon 1976).

Clear evidence indicates the existence of a NNE-SSW striking graben-like structure on the primitive Rhodesia-Craton about 2.2 b.y. ago. This graben-like structure, which caused the intrusion of the Great Dyke, apparently extended down into the upper mantle. A northern continuation of the Great Dyke reaching across the Zambezi Mobile Belt cannot be identified. On the contrary, the very northern part of the Great Dyke was displaced eastwards by the Zambezi Belt.

Roughly 2 b.y. ago, the mafics of the Bushveld Complex intruded. The northern extension, the Potgietersrus-Limb, points in the same direction as the Great Dyke. Obviously, the Bushveld is situated at the intersection of the Great Dyke direction with the Murchison/Barberton lineaments (Fig. 1).

The Vredefort Dome is situated on the southern extension of the Great Dyke–Potgietersrus lineament. Geophysical investigations of the Bouguer anomalies and analyses of the few magmatic rocks indicate that the Vredefort Dome is also a mafic intrusion (Bisschoff 1971, 1973). Further to the south on this lineament, two similar complexes can be found, The Brandfort-Intrusion and the Trompsburg Igneous Complex (Fig. 1). They do not outcrop on the surface. Both were localized by geophysical investigations. The Trompsburg Complex was additionally intersected by boreholes (Coetzee 1976).

In large-scale tectonic considerations, the different ages of these basic intrusions are of special interest. The ages decrease from north to south: Great Dyke 2.5 b.y., Bushveld Complex 2.0 b.y., Vredefort Dome 1.95 b.y., Trompsburg Igneous Complex 1.4 b.y. (Haughton 1969; Truswell 1977; Vail 1977). Considering the age-sequence, genetic hypotheses, like the wandering of a primitive continent, composed of the Kapvaal- and the Rhodesia Craton, across a plume appear suitable. However, they are not the intention of this publication.

2.1 The Setting of the Bushveld Complex

Geophysical investigations of the Bouguer anomalies indicate that the Bushveld is not a single intrusion, but consists at least of five separate complexes: The Potgietersrus-Limb as the most northern part, the Eastern Bushveld Complex, the Bethal Complex in the south-east, the Western Bushveld Complex and a Far-Western Complex north of Zeerust (Fig. 2). The Bethal Complex and parts of the Potgietersrus-Limb are covered by younger Karoo-sediments. Both complexes were intersected by boreholes. The overall extension of these complexes is about 460 km in E-W and approximately 330 m in N-S direction.

Petrographic investigations and searching for ore deposits in all these single complexes showed a remarkable resemblance in their rock-differentiation. Some characteristic layers of the Bushveld could be traced in almost each rock sequence of the five complexes. They can be used as marker horizons and are very helpful for the correlation of the different sequences.

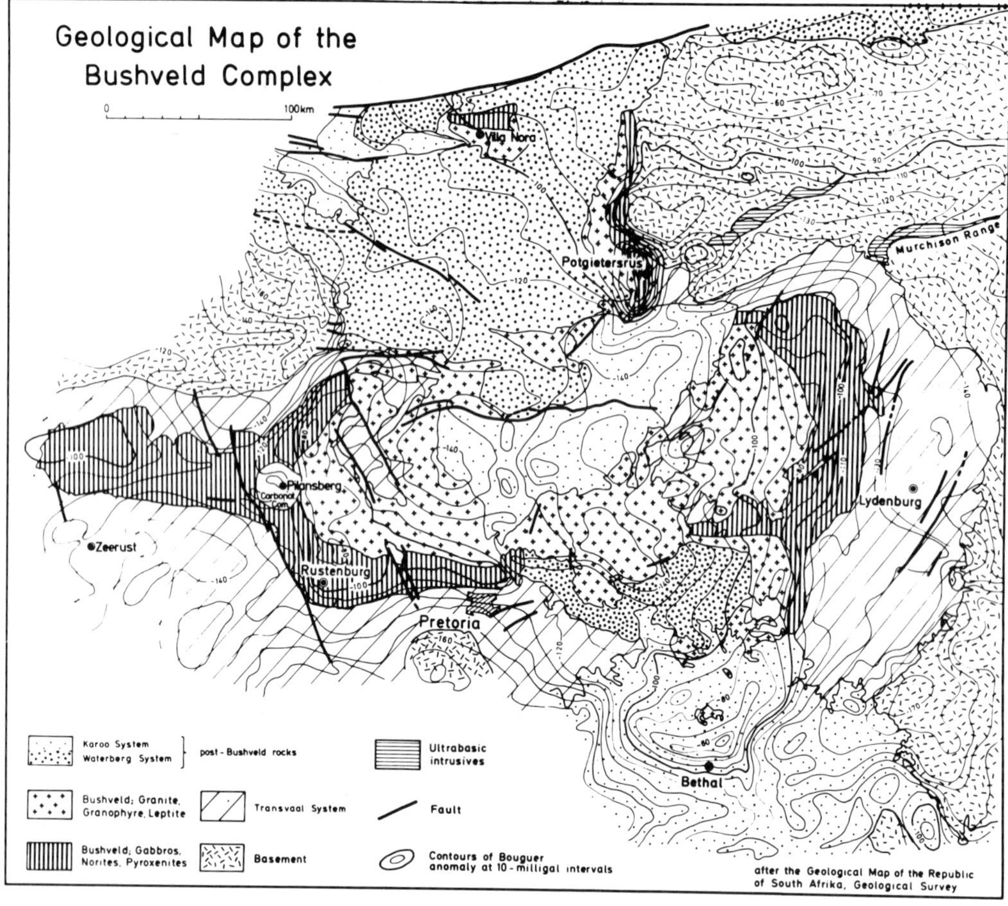

Fig. 2. Generalized geological map of the Bushveld Complex and its surroundings. The Bouguer anomalies clearly indicate that the Bushveld consists at least of five separate complexes: The Potgietersrus Limb in the north, the Eastern Bushveld Complex, the Bethal Complex in the south, the Western Bushveld Complex and the Far-Western Bushveld Complex north of Zeerust. The Potgietersrus Limb is partly, the Bethal Complex completely covered by younger sediments

3 The Layered Sequence

J. Willemse (1969a) compiled a general profile of the whole Bushveld Complex, which was modified by G. v. Gruenewaldt. Along this profile, the differentiation trend of the Bushveld magma can be traced easily.

By its differentiation trend, the Bushveld Complex can be divided into four "stratigraphical" zones (Fig. 3):

The Basal Zone has an average thickness of 1500 m. The main rock types are pyroxenites (bronzitites). Small amounts of harzburgites, anorthositic as well as noritic rocks can also be found. The high anorthite content of their plagioclases (about An 85)

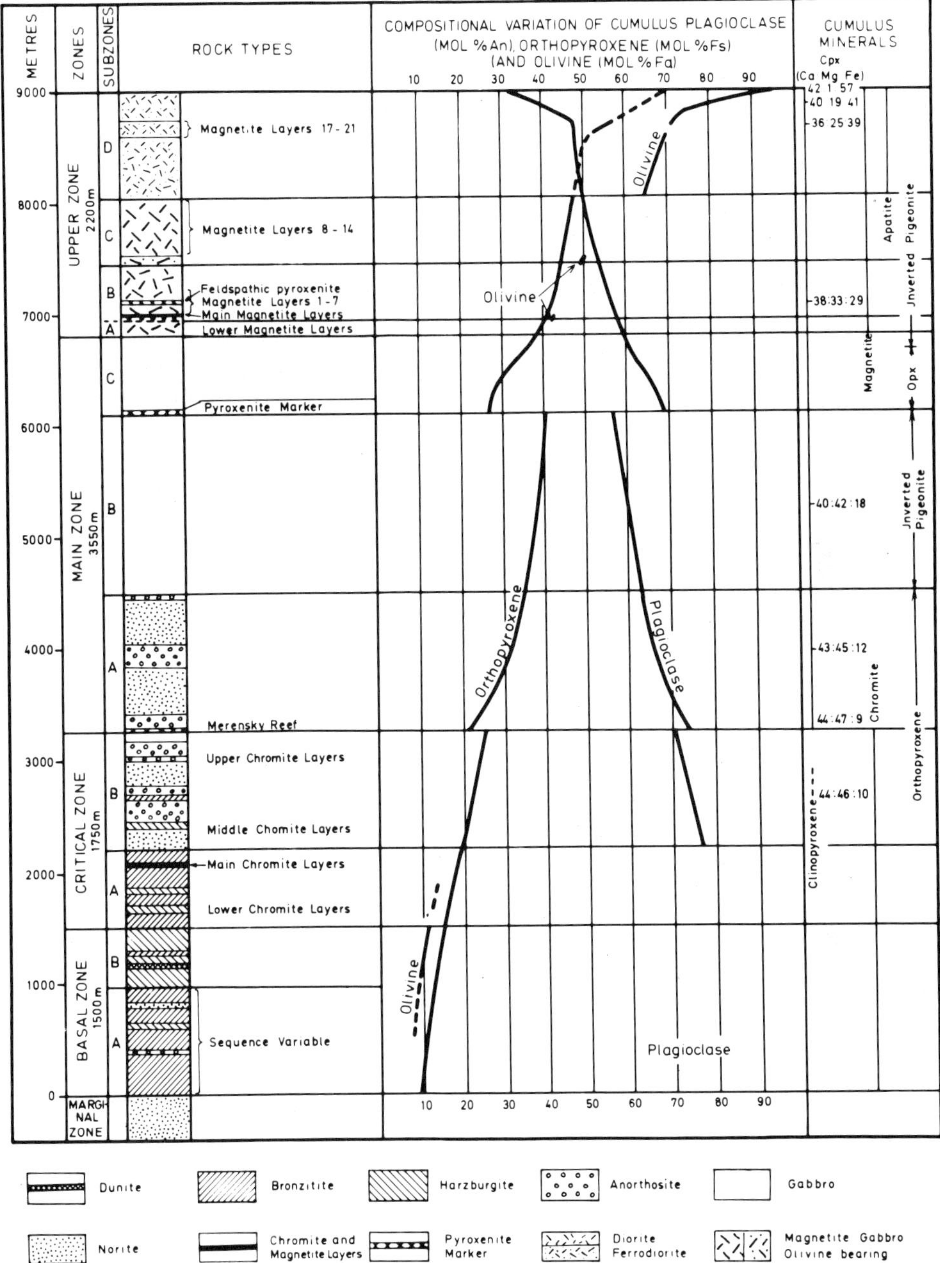

Fig. 3. Generalized columnar section of the Layered Sequence (after Willemse 1969a, modified by G. v. Gruenewaldt). The composition of important cumulus-silicates resembles the differentiation of the Bushveld magma. Remarkable is the compositional break of orthopyroxene and plagioclase at the Merensky Reef at the beginning of the Main Zone and at the Pyroxenite Marker at the base of Subzone C of the Main Zone

fits the differentiation trend. Olivine-rich dunitic rocks are very subordinate in the Basal Zone. The low olivine content is a general feature of the whole Bushveld Complex.

Above the Basal Zone follows the Critical Zone with an average thickness of 1750 m. The lower part of the Critical Zone consists mainly of pyroxenitic rocks, whereas in the upper part noritic and anorthositic rocks are abundant. The Critical Zone is characterized by the appearance of massive chromite seams. These seams are subdivided into three groups: The Lower-, the Middle-, and the Upper group. The occurrence of chromitite layers is restricted to the Critical Zone. Nowadays, the chromite production concentrates on the Main Chromite Seam, which is situated in the lower part of the Critical Zone.

The Main Zone, which is situated above the Critical Zone, has an average thickness of 3550 m. Gabbros and norites are the most abundant rock types. Anorthosites, restricted to the Subzone A within the Main Zone, are very subordinate. A general feature of the Main Zone is that its rocks are very homogeneous. The Merensky Reef, of great economic importance, is defined as the basis of the Main Zone. This horizon can be traced for some 230 km along the western rim and approximately 160 km along the eastern and northeastern segment of the complex (Schwellnus et al. 1976). It is a coarse-grained noritic-pyroxenitic layer. The thickness varies from 25 to 150 cm. Besides the silica content, the Merensky Reef contains considerable concentrations of sulphide minerals, mainly pyrrhotite and pentlandite and some chalcopyrite. Platinum concentrations range from 8 to 15 g/t and are the reason for the mining activities at several places, for example at Rustenburg.

The continuous petrographic progress of the magmatic differentiation in the Main Zone is obviously interrupted at the base of Subzone C. In the field, this interruption is verified by the appearance of a petrologically unexpected pyroxenite layer, the so-called Pyroxenite Marker. It contains an enstatite-rich orthopyroxene. It is remarkable that below this pyroxenite layer, cumulus magnetite appears in the gabbro (v. Gruenewaldt 1973). The significance of this occurrence in connection with the magma differentiation will be discussed in Chapter 4.

The beginning of the Upper Zone is defined with the appearance of the first titanomagnetite seams, respectively with the new appearance of cumulus-magnetite in the silicates some 30 m below the first magnetitite layer. The Upper Zone with an average thickness of 2200 m consists mainly of magnetite gabbros, some olivine gabbros and in its upper part ferrodiorites, which are fayalite-containing diorites. Furthermore several anorthosite layers, fine-grained norites and a feldspathic pyroxenite can be found. Compared with the Main Zone, the rock sequence of the Upper Zone is very heterogeneous.

The economic importance of this zone is related to its larger number of titanomagnetite layers. Besides their Fe and Ti contents, these seams contain up to 2.3% V_2O_5. From the bottom to the top of the Upper Zone the V_2O_5 percentage in the titanomagnetites decreases from 2.3 to 0.3. Consequently, the Bushveld is the world's largest known vanadium resource. At the moment only the Main Magnetite Layer with an average V_2O_5 content of 1.6% is mined.

4 Experimental Investigations

The layered sequence of the Bushveld rock can be sufficiently explained by the model systems $MgO - SiO_2 - Fe_3O_4 - Cr_2O_3$ (Arculus and Osborn 1975) and $MgO - FeO - Fe_2O_3 - SiO_2$ (Muan and Osborn 1956).

Inside the tetrahedron of Fig. 4a, slightly elevated above the basis triangle, there is the stability volume of silicates. Since the chromium content of the primary Bushveld magma is roughly calculated to about 5000 ppm, the composition of the magma would plot near the basis triangle of the tetrahedron. With the estimated 13% Fe_3O_4 and low olivine content of the Bushveld, the composition of the magma must have been close to the stability field of orthopyroxene (point BL in Fig. 4a). Striking is the extreme large stability volume of chromite. For the crystallization of chromite only, however, the Cr_2O_3 concentration of the original melt would have to be considerably higher than it is according to the estimations based on the observed chromium content. This means that the existence of discrete chromitite layers, as seen in the field, cannot be derived from the diagram.

The phase equilibria of this system can be used to explain the evolution of the silicates including the Critical Zone. Starting from point BL (Fig. 4a), the composition of the liquid moves across the stability volumes of olivine and pyroxene until it reaches the cotectical plane of chromite and pyroxene. This part corresponds roughly to the accumulation of the Basal Zone. The cotectical plane reached near the basis of the tetrahedron results in a chromite percentage of about 2%, which is in good accordance with the observed values. In the further course of the differentiation, the composition of the liquid moves along the coexistence of pyroxene and chromite. During this stage, periodical fluctuations of the oxygen-fugacity lead to the formation of massive chromite layers, as will be explained in more detail further on.

During the crystallization of the Critical Zone, the liquid has lost almost all its chromium content. To trace the evolution of the Main and Upper Zone we can therefore use the chromium-free system $MgO - FeO - Fe_2O_3 - SiO_2$ (Fig. 4b). Point BL again represents the estimated composition of the original magma. During the crystallization of the Basal and the Critical Zone the magma was enriched in iron and depleted in Mg. Therefore the composition of the liquid at the beginning of the Main Zone would be situated in the pyroxene volume on the hatched crystallization path.

The first mineral to crystallize is orthopyroxene until the upper plane of the pyroxene volume is reached. This means the formation of pyroxene in spinel-free gabbros and norites, as observed in the whole Main Zone. The crystallization path now runs along this plane while an iron-rich orthopyroxene and magnetite are crystallizing. Again, this agrees well with the field observations in the Upper Zone, Subzone A–C. (In Subzone D orthopyroxene is very rare.)

The already mentioned existence of cumulus magnetite just under the Pyroxenite Marker (basis of Subzone C, Main Zone) indicates that at this point the crystallization had already reached the cotectic plane of magnetite and pyroxene. Consequently, from a petrological point of view, the transition of the Upper Zone had already been performed at the top of Subzone B of the Main Zone. The appearance of the Pyroxenite Marker with subsequent cumulus magnetite-free gabbros and norites means a

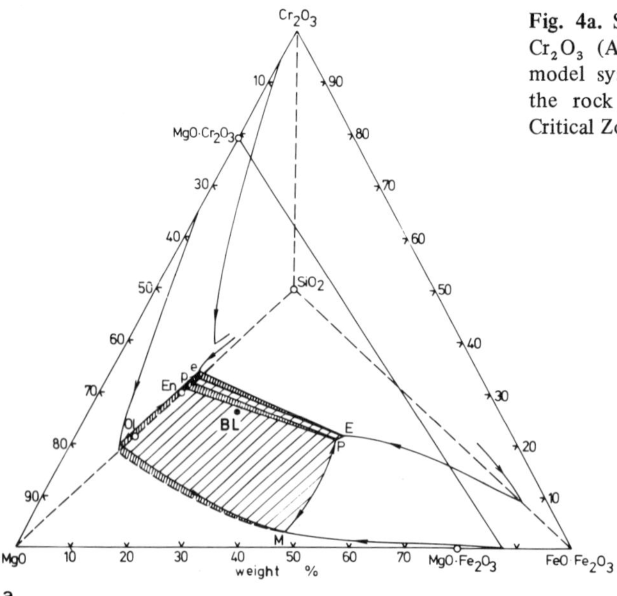

Fig. 4a. System MgO-$FeO \cdot Fe_2O_3$-SiO_2-Cr_2O_3 (Arculus and Osborn 1975) as a model system for the accumulation of the rock sequence of the Basal and Critical Zone of the Bushveld Complex

step back in the differentiation. This is best explained with an injection of undifferentiated magma (v. Gruenewaldt 1970).

At the peritectic point A in the system $MgO - SiO_2 - FeO - Fe_2O_3$, the crystallization of orthopyroxene ceases and is replaced by fayalitic olivine and quartz (together with magnetite). The reliability of this model is proved by the field observations in the Subzone D of the Upper Zone. Fayalitic olivine appears at the basis of Subzone D, whereas cumulus quartz is observed only in the topmost 100 m of Subzone D. Since a gas phase is not involved in the peritectic reaction, the pressure dependence of the peritectic point A is expected to be very small. Considering the system $MgO - SiO_2 - FeO - Fe_2O_3$ is but a model, it fits field observations strikingly well.

Despite the good agreement of the model with the existing rock sequence, it can not explain the formation of pure magnetite seams, which are encountered in the Upper Zone at least 25 times. An additional model, which accounts for the magnetite layers will be presented later.

5 Occurrence of Chromite in the Field

Massive chromitite layers occur in the Critical Zone of the Bushveld Complex. The chromite content of the silicate rocks of these two zones ranges from 2 to 4 %. A vertical profile, cutting the lower part of the Critical Zone from Farm Maandagshoek, Eastern Bushveld Complex, is an instructive example of the situation (for more detailed information see Snethlage and Klemm 1978a, b). The thick chromitite layers are labelled from No. I to XI. The thickest layer is called Steelpoort – or Main

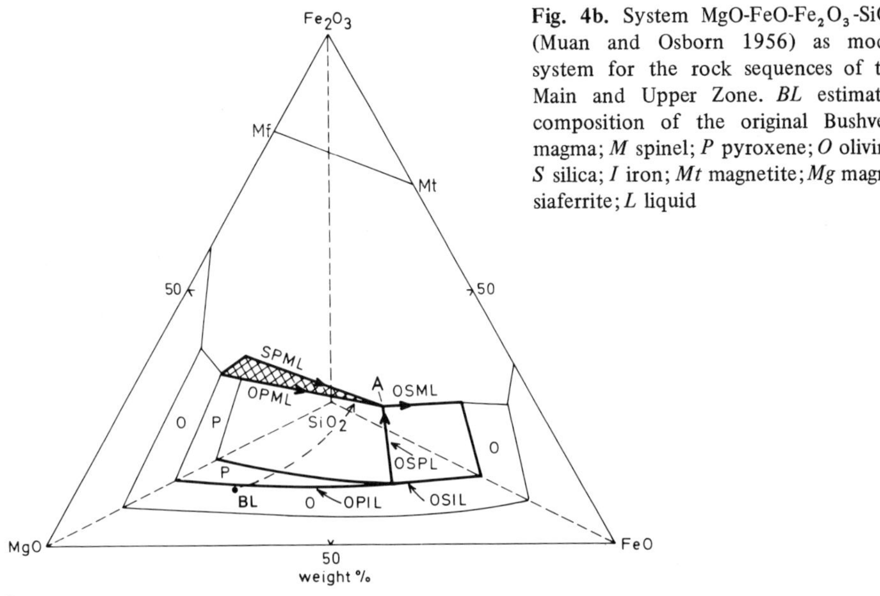

Fig. 4b. System $MgO-FeO-Fe_2O_3-SiO_2$ (Muan and Osborn 1956) as model system for the rock sequences of the Main and Upper Zone. *BL* estimated composition of the original Bushveld magma; *M* spinel; *P* pyroxene; *O* olivine; *S* silica; *I* iron; *Mt* magnetite; *Mg* magnesiaferrite; *L* liquid

Chromite Seam. It is about 1.5 m thick and crops out over about 90 km in the Eastern Bushveld Complex. The main host rock is pyroxenite, but in some cases also anorthosite occurs. Some layers have gradational contacts at the bottom and on the top, some have sharp contacts. Furthermore pyroxenite insertions in the layer can often be observed. At unit "F" (notation after Camaron and Desborough 1969), immediately above a small ca. 2 cm thick chromitite layer, the first cumulus plagioclase occurs in the rock sequence of the Bushveld. This unit subdivides the Critical Zone into a pyroxenitic and an anorthositic unit. Units similar to "F" can be observed in many layered intrusions and seem to be characteristic features. Supposing an annual growth rate of the mining production of 5%, the identified chromium resources would last for ca. 145 years (v. Gruenewaldt 1977).

6 f_{O_2}-Measurements on Chromites and Their Petrologic Application

As shown in Chapter 4, the differentiation of the Bushveld Complex can be understood by means of two model systems, one Cr-bearing, the other Cr-free. More detailed information, however, can be gained through the investigations of Hill and Roeder (1974), who studied among others the liquidus of an olivinetholeiitic basalt containing 500 ppm Cr. Within the liquidus range, they observed a distinct f_{O_2}-T-range in which the crystallization of spinel was extraordinarily enhanced (line E–D, Fig. 5a). As explained by Snethlage and v. Gruenewaldt (1978) and Snethlage and Klemm (1978a) this observation is an excellent basis for the interpretation of f_{O_2}-measurements on chromites. The main idea of this interpretation is to regard the f_{O_2}-T-range of the line E–D as the conditions of formation of massive chromitite layers.

Experimental f_{O_2}-measurements were carried out on chromites from massive layers and on chromites separated from host rocks. Samples from the Eastern (drill core from Farm Maandagshoek) as well as from the Western Bushveld Complex (Zwartkop Chrome Mine) were investigated (detailed information see Snethlage and Klemm 1978a).

f_{O_2}-Measurements by themselves do not offer the possibility of an interpretation with respect to the conditions of formation. A plot of the measured f_{O_2}-T-curves of massive chromites into Hill and Roeder's diagram layers, gives intersections between these chromite curves and the line E–D, which allow to read off the conditions of formation from the diagrams. Since the observation of enhanced chromite crystallization along the line E–D is only qualitative, and since the studied basaltic liquid is certainly not identical with the Bushveld magma, the data derived for the chromite formation cannot be absolutely correct. Their internal ratios, however, are correct. In the following table the derived data are compiled.

Western Bushveld Complex (Zwartkop Chrome Mine)			Eastern Bushveld Complex (Farm Maandagshoek)		
Layer	T (°C)	p_{O_2} (atm)			
			L XI (= Steelpoort Seam)	1115–1150	$10^{-7.80}$–$10^{-8.80}$
LG 3	1160–1237	10^{-5}–$10^{-7.6}$	L X	1125	$10^{-8.25}$
LG 4	1175–1200	$10^{-6.35}$–$10^{-7.20}$	LV	1120	$10^{-8.55}$
LG 6	1162–1207 / mean T 1200°C	$10^{-6.20}$–$10^{-7.50}$	L II	1120–1145 / mean T 1125°C	$10^{-8.0}$–$10^{-8.6}$

The data demonstrate that the chromitite layers of each sequence were obviously formed under nearly identical temperature conditions, 1200°C in the Western and 1125°C in the Eastern Bushveld Complex. This is probably a consequence of the small vertical range covered by the measured samples, which does not reflect detectable temperature differences.

The almost simultaneous assemblage of the massive chromitite and the host rocks implies that they crystallized under isothermal conditions. The temperature data derived from the massive chromitites can consequently also be applied for the host rocks and the included chromites. Therefore, with the aid of the measured f_{O_2}-T-curves of host rock chromites, also the f_{O_2}-conditions of the formation of the host rocks can be derived. In this short representation, the logical evolution of the results and the respective figure cannot be reported; they are, however, explained in detail in Snethlage and Klemm (1978a, b). The data derived for the host rocks are shown in the following table:

Western Bushveld Complex	Eastern Bushveld Complex
T = 1200°C; $P_{O_2} = 10^{-7.25} - 10^{-7.50}$	T = 1125°C; $P_{O_2} = 10^{-8.50} - 10^{-9}$

As the compiled data show, the chromites from the massive layers obviously were formed under higher f_{O_2}-conditions than the host rock chromites. The increase in f_{O_2} is in the order of 0.5–1.0 log unit. This means that cyclic changes of the f_{O_2} have caused the formation of the massive chromitite layers which was already postulated by Ulmer (1969).

From the Western to the Eastern Bushveld Complex, there is an evident slope in f_{O_2} and temperature. Assuming a uniform magma complex, this slope means that the investigated rocks of the Western Bushveld Complex have been formed at an earlier stage of the magma differentiation than the rocks of the Eastern Bushveld Complex. The differences (75°C, ca. 1.5 log units f_{O_2}) agree very well with the conditions of the differentiation at constant total composition of the condensed phases.

The data of the conditions of formation of the host rocks are plotted as black areas I and II into Figs. 5a, b. The line A–B connecting these two areas, could be interpreted as the f_{O_2}-T-decrease of the magma during the differentiation. Since the accumulation of the Bushveld rocks cannot be seen as an equilibrium process, it is reasonable to interpret the line A–B in a Cr-rich (Fig. 5a) and a Cr-poor system (Fig. 5b). At temperatures above 1125°C the magma moves along the line A–B in the chromium-rich system (Fig. 5a), accumulating the Basal- and Critical Zone. As already explained above, here the magma changes from a Cr-rich into a Cr-poor system at the top of the Critical

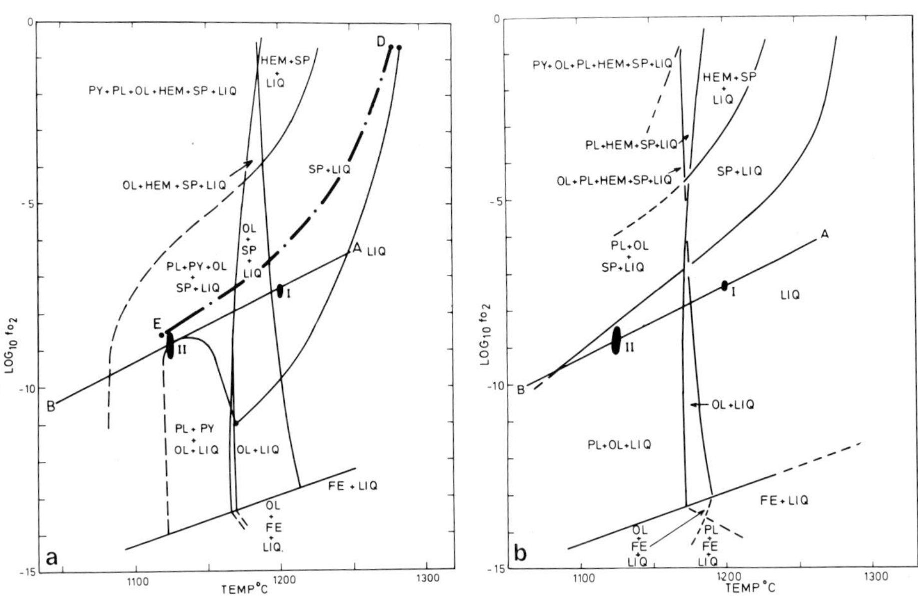

Fig. 5. Phase assemblages in the liquidus of two basaltic liquids, Gl-1921 containing 500 ppm Cr (**a**, simplified after Hill and Roeder 1974) and Gl-RHB containing 80 ppm Cr (**b**, slightly modified after Hill and Roeder 1974). The areas I and II reflect the conditions of formation of the studied host rocks of the Western (I) and the Eastern Bushveld Complex (II). Line *A–B* represents the f_{O_2}-T slope of the magma. The *lower part of the line A–B* between area II and the stability field of plagioclase + olivine + spinel + liquid (**b**) would correspond with the accumulation of the Main Zone

Zone, after the formation of the massive chromitite layers. Discussing further the differentiation of the magma in the Cr-poor system, at temperatures below 1125°C, the line A–B passes through the stability field of plagioclase and olivine. This spinel-free period would correspond with the accumulation of the Main Zone. At its bottom, very Cr-rich clinopyroxenes can be found (Cr_2O_3-content up to 0.88%; Atkins 1969), which collected the Cr-content of the liquid.

Magnetite layers cannot be formed because the f_{O_2} increase necessary to reach the stability field of plagioclase + olivine + spinel is too large. With further decreasing temperature the magma reached the stability field of plagioclase + olivine + spinel. At this stage the magma is sufficiently enriched in ferric iron to form massive magnetite layers by periodic increase of the f_{O_2}. Correlated with the natural rock sequence, this transition corresponds with the beginning of the Upper Zone. Indeed, this zone is characterized by the occurrence of several massive titanomagnetite layers.

7 Occurrence of Titanomagnetite in the Field

The Upper Zone is characterized by the appearance of cumulus-magnetite as a major component in the host rocks, and, as massive, stratiform mineralizations, the magnetite layers. Besides these two types, discordant, pegmatoidal magnetite bodies of various size, the magnetite plugs, occur in the Upper as well as in the Main Zone.

The gabbros, norites, diorites and anorthosites of the Upper Zone contain 4% to 10 vol.% cumulus magnetite. These values are within the range of the magnetite concentration, which is to be expected from the model system $MgO - SiO_2 - FeO - Fe_2O_3$ (Fig. 4b). The average grain size of these magnetites is 0.7 to 1.5 mm.

A good example for the magnetite plugs is the well-known locality of Kennedy's Vale, situated in the Main Zone, with an average V_2O_5 content of 2% (v. Gruenewaldt 1977) (Fig. 6b). Magnetite plugs have been identified in a large number but smaller size, preferentially in two horizons: between Magnetite Layers 7 and 8 in the Upper Zone and in the Subzone C of the Main Zone. The grain size of magnetites in the plugs is distinctly larger than in the host rocks as well as in the magnetite seams. Grain sizes up to 3 cm are observed.

In the Eastern Bushveld Complex in the area of Magnet Heights, Molyneux (1970) distinguished 25 different magnetite layers: 3 Lower Magnetite Layers, the Main Magnetite Layer, and 21 Upper Magnetite Layers. For convenience, the Upper Magnetite Layers are called only Magnetite Layers. The thickness of the seams varies from a few cm (e.g., Magnetite Layer 14) up to 10 m (Magnetite Layer 21). The grain size varies from a few mm to 2.5 cm. During detailed mapping in the scale 1:15,000 (Klemm et al. 1979) in the area north of Roossenekal, Eastern Bushveld Complex, further 10 magnetite layers were found. As their thickness is generally very small, they could not be traced over longer distances. Some magnetite layers actually contain 2 (e.g., Magnetite Layer 14) or even 3 (e.g. Magnetite Layer 7) separate layers, tightly grouped. In the field, all magnetite layers can definitely be identified on the basis of their "stratigraphic" position, thickness, silica-content, grain size, host rocks and contacts. The changes from the host rocks to the magnetite layers are sharp or gradual. Host rocks are gabbros, norites, anorthosites and diorites. Almost every possible combination between host rocks and magnetite seams can be observed. An interdependent system could not be identified. At the time, only the 1.5 m thick Main Magnetite Layer is mined. The reserves of this layer up to a vertical mining-depth of 30 m range about 1030.5×10^6 t of magnetite ore with an average V_2O_5-content of 1.6%. Supposing an annual growth rate of the mining production of 5%, the identified resources would last for more than 130 years (v. Gruenewaldt 1977).

As the mineral assemblages in the Upper Zone have been described in detail by various authors (e.g., Willemse 1969a, b; v. Gruenewaldt 1970, 1973; Molyneux 1970), they are not reviewed in this paper.

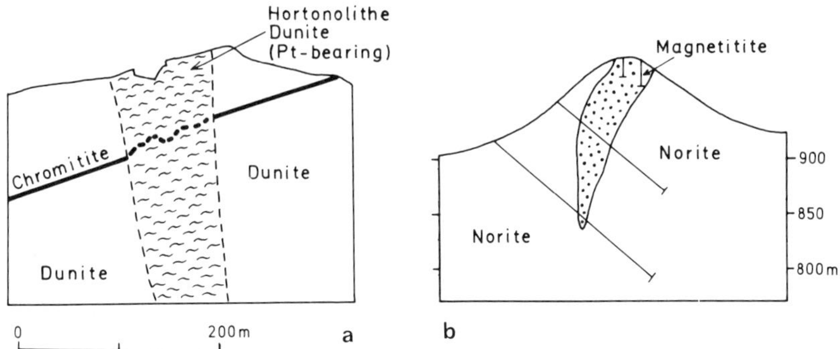

Fig. 6. a Section through the Onverwacht Platinum Pipe. A Fo_{90}-dunite was locally transformed in a hortonolithe-dunite. The released magnesium content formed together with the existing CO_2 magnesite, which is preserved in the joint-system. A chromitite seam was only displaced and brecciated. **b** Section through the Kennedy's Vale Magnetite Plug (after Willemse 1969b). It suggested that this plug was formed by a local but enormous increase of the f_{O_2}

8 Magnetite Crystallization as a Function of f_{O_2}

As already indicated in Chapter 6, periodic fluctuations of the f_{O_2} have caused the formation of massive chromitite layers. Probably, they also control the formation of magnetite layers. This theory can be supported by experimental results. In the system $MgO - FeO - Fe_2O_3 - SiO_2$ (Muan and Osborn 1956) the stability field of magnetite decreases with decreasing f_{O_2}. This applies also in the presence of the anorthite component (Roeder and Osborn 1966). In the pure oxide system, the stability of spinel is also controlled by the f_{O_2}. The comparison of analytical data of natural spinels with the field of stability in the synthetic spinel prism (Ulmer 1969) indicates an optimal f_{O_2} range from 10^{-5} to 10^{-7} atm at 1300°C. At lower f_{O_2} conditions (10^{-9} atm) at 1300°C, however, the spinel field of stability is strongly reduced, because Fe^{3+}-containing spinels are unstable under these conditions.

On the basis of the mentioned experimental results, Ulmer (1969) concluded that the oscillating f_{O_2} could be the controlling factor for the formation of massive spinel ores. Although Ulmer emphasizes the formation of chromitites, this theory may also apply to the formation of titanomagnetites. This is supported by the experimental systems in which the coexisting spinels are solid solutions of $Fe_3O_4 - MgFe_2O_4$. Furthermore, the coexistence of Fe-rich orthopyroxene + Fe-rich olivine + $magnetite_{ss}$ in the experimental systems corresponds very well with the mineral assemblages in the lower parts of the Upper Zone. This indicates that at the beginning of the Upper Zone, the Bushveld magma must have reached a stage, which is very near to the peritectic point A in the system $MgO - FeO - Fe_2O_3 - SiO_2$ (see Fig. 4b). Buchanan (1976) plotted the composition of coexisting magnetite-ilmenite pairs, using the magnetite-ulvöspinel and ilmenite-hematite solid solution curves established by Buddington and Lindsley (1964). The obtained f_{O_2} data range from –log 18 to –log 21, the temperature from 580° to 680°C. Obviously, the estimates of the temperature and the f_{O_2} from the oxide assemblages are extremely low. Buchanan (1976) suggests that substantial subsolidus re-equilibration has taken place. f_{O_2}-Measurements on magnetites of the Upper Zone are not yet available. As shown with the f_{O_2}-measurements on chromites, f_{O_2}-measurements on magnetites would allow important conclusions with respect to the formation of the genesis of the silicate rocks and titanomagnetitite layers of the Upper Zone. The sample material, which is now available, is not appropriate for f_{O_2}-measurements. The magnetite of the Bushveld is oxidized by weathering probably down to 100 m. Fresh material can only be obtained by core drilling which was not carried out so far.

9 Field Evidence for Increased Volatile Activity

Many observations indicate a locally limited volatile activity in the Bushveld Complex. Some of them are briefly discussed in this chapter. However, not all of them can be mentioned.

9.1 Pipes

Indicators for the gas flow of supercritical volatiles through the Bushveld are the well-known platinum pipes of Onverwacht, Mooihoek and Drie Kop. A section through the Onverwacht pipe is shown in Fig. 6a. These jet-like gas flows apparently penetrated already solidified Bushveld rocks. Near Onverwacht, they transformed an originally Fo_{90}-dunite into a hortonolithe-dunite. The released magnesium content formed together with the existing CO_2 magnesite, which is preserved in the joint-system. The platinum content (up to 30 ppm measured in the central part of the pipe) was probably also transported in the gas phase.

For the nickel pipes near Vlakfontein (Western Bushveld), which are described in detail by Vermaak (1976), De Waal (1977, 1978) postulates volatile injections (controlled by faults), which had their origin in the underlying Transvaal Supergroup.

9.2 Magnetite Plugs

Local magnetite concentrations, which correspond to the just explained process, are observed in the Bushveld Complex in the Main and Upper Zone. An example for the already in Chapter 7 mentioned magnetite plugs is given in Fig. 6b (section through the Kennedy's Vale magnetite plug). The geological environment of these plugs, as well as the plugs themselves are of typical pegmatitic texture, which again is an argument for the existence of volatile jets during the formation of the plugs. The intrusive character of these plugs which is sometimes observed in the field may be due to isostatic movements.

It seems probable, that such volatile exhalations also invade the magma above the already crystallized rocks and contribute to a local but enormous increase of the f_{O_2}. In the model of Hill and Roeder (1974) this would cause a transgression from the stability field of plagioclase + pyroxene + olivine + liquid into the stability field of plagioclase + pyroxene + olivine + spinel + liquid, where as a consequence of the increased f_{O_2}, the formation of massive magnetite bodies is possible.

9.3 Potholes

Pothole structures can be observed in the chromitite seams of the Critical Zone, e.g., LG 2 in Rustenberg, Platinum Mine, and in the titanomagnetitite layers of the Upper Zone (Fig. 7). In these potholes the massive ore layers are considerably thicker than usual. They are another indicator for local gas flows of mainly H_2O and CO_2 volatiles through already accumulated rocks. These potholes are normally obliterated again in the crystal mush. Probably, they were more frequent than they can be identified today in the scarce outcrops.

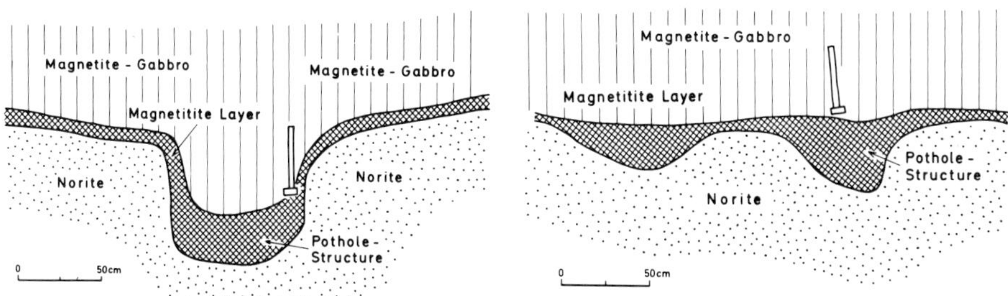

Fig. 7. Potholes in the Magnetite Layer 8 in the Upper Zone near Roossenekal. In these pothole structures the massive magnetite layers are considerably thicker than usual

9.4 Anorthosite Plugs

In the Upper Zone, several anorthosite plugs are encountered in the normal gabbroic and dioritic rocks. Their diameter varies from 1 to 50 m. These small-scale intrusions show pneumatolytic features of alteration: The plagioclases are saussuritizised, pyroxenes are transformed into hornblendes, chlorites, clinozoisite and epidote. The magma of these plugs must have been enriched in volatiles. In one case, in their immediate environment, a thin, but massive magnetite layer was observed.

9.5 Sedimentary Relics

The above-mentioned detailed field study yielded also some details about anomalous anorthosites within the gabbro-diorite sequence in the Upper Zone: areas of almost oxide-free, discordant anorthosites, which cannot be correlated with the normal Bushveld sequence (Fig. 8). They are closely connected with sedimentary fragments which are now enclosed in the Bushveld rocks as relics of hornfelses and calcsilicates. In the same area, two anomalous magnetite layers can be observed (marked × in Fig. 8), which do not correspond to the normal differentiation. A microscopic study of these anomalous anorthosites shows progressive alteration, similar to the one observed in the anorthosite plugs. The plagioclases again are highly saussuritizised, the pyroxenes are

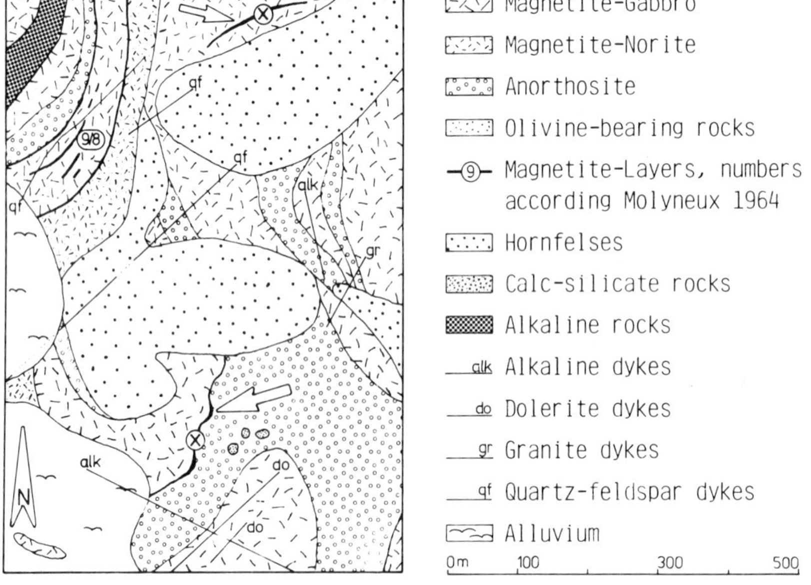

Fig. 8. In the Roossenekal area hornfelses and calcsilicate rocks are encountered in usual Bushveld rocks of the Upper Zone. Together with these inclusions anomalous anorthosites and magnetite layers (marked ×) are recorded. By solution of the sedimentary relics the magma was enriched in Ca and Al and additional volatiles, mainly CO_2 and H_2O were brought into the liquid. The increased concentrations of Ca and Al led to the precipitation of anorthosite; the increased f_{O_2} caused the crystallization of the anomalous magnetite layers

transformed into chlorites, clinozoisite, epidote and Fe-hydroxides. This hydrothermal alteration of the anorthosites is most atypical in normal Bushveld rocks.

The above-mentioned sedimentary fragments are interpreted as collapsed parts of the roof rock: limestones, dolomites and marls of the Transvaal system. They were almost completely dissolved by the magma. This process enriched the melt in Ca and Al, which caused the anomalous anorthosite development. Additionally, the volatiles, mainly H_2O and CO_2, added to the system, caused an increase of f_{O_2}. As a result of their thermal dissociation the f_{O_2} of the magma is buffered at a higher level, which significantly decreases the Fe^{2+}/Fe^{3+} ratio of the magma.

The highly autometamorphosed anorthosites indicate volatile activity. Further evidence for volatile activity is the high sulphide content (occasionally of economic interest) in these anorthosites. An increased f_{O_2} lowers the sulphide solubility of a silica melt (MacLean 1969; Haughton et al. 1974) and favours enhanced sulphide crystallization. A third evidence is given by the above-mentioned anomalous magnetitite seams. They can be taken as a field proof for the postulated formation of massive titanomagnetite layers by sporadic fluctuations of the f_{O_2}.

10 A New Model

The last paragraphs showed the strong correlation of the formation of massive magnetite with the volatile activity, based on field investigations (magnetite plugs, potholes, contamination by sedimentary relics). These massive magnetite concentrations can be explained when the existence of small-scale jet-like volatile injections is postulated. On the other hand, the laterally homogeneous development of the spinel seams indicates a uniform process which causes the changes in f_{O_2}. Experimental research proved that the water vapour diffusion-coefficients in porous solids is 100 to 1000 times greater than in liquids (Klopfer 1974, p. 54).

Since such measurements were not carried out in the temperature range of a mafic magma, but at normal temperature, the assumption is made that this wide range of diffusion velocities can also be applied to crystallized cumulus rocks and basic silica liquids. A contamination of the melt by collapsed roof-rocks and the resulting f_{O_2} increase was used in Section 9.1 to explain the locally limited anomalous magnetite seams. Similar mechanisms Irvine (1977) postulated for the Muskox Intrusion, to cause the formation of massive spinel seams. However, the gas-diffusion coefficients in a melt are significantly too low to explain the enormous lateral extension of some spinel seams in the Bushveld Complex. To explain these large-scale phenomena, a mechanism must be postulated, which is effective over the whole extension of the magma chamber. Additionally, the striking "stratigraphical" similarity of the distinct partial complexes of the Bushveld, especially regarding the ore seams, should be explained by this model. The relatively high rate of gas diffusion in a porous solid renders the possibility for a fast and spatially homogeneous gas flow. This will even be improved by the development of a joint system caused by the solidification of the settled cumulates. The volatile exhalations of the rocks, which underlie the magma chamber (possibly enlarged by the heated underlying crust; Vermaak 1976) cause a laterally well-distributed gas phase. Sporadic changes in this system are largely levelled by diffusion. Consequently, a low but constant gas flow through the accumulated Bushveld-rocks can be expected. The difference of the diffusion coefficients between liquid and crystallized rocks causes a volatile concentration at the contact between these two phases and increases the f_{O_2} in this horizon. It is evident that neither will the f_{O_2} enrichment in this zone be constant, nor will the thickness of this zone be stable. From time to time magma convections may disturb this f_{O_2}-enriched zone so that the f_{O_2} oscillates periodically.

The f_{O_2} measurements on chromites indicated that a f_{O_2} increase of 0.5 to 1.0 lg. units may lead to the formation of massive chromitite seams. The increase of f_{O_2} necessary for the formation of massive magnetite seams should be in the same range.

An essential prerequisite for the formation of massive chromitite or magnetite layers is given by the stage of differentiation of the magma. Favourable conditions for the formation of massive spinel seams are given twice in the differentiation: First, when the magma composition moves along the cotectical plane pyroxene-chromite (Fig. 4a), and second, when the magma has reached the cotectical plane of pyroxene-magnetite (Fig. 4b). At these stages, the enrichment of f_{O_2} at the cumulate-liquid interface is sufficient to cause the formation of massive chromite resp. titanomagnetite in the whole interface. After the precipitation of a seam, only spinel-bearing silicates crystallize until the f_{O_2} is raised again enough to precipitate the next seam.

At the beginning of the Main Zone, the crystallization of chromite has lowered the chrome concentration to about 50 ppm in the remaining melt. The further differentiation process, therefore, can be observed in the Cr-poor basaltic liquid investigated by Hill and Roeder (1974; Fig. 5b). As explained in Chapter 6, at this stage the liquid is in the stability field of plagioclase + olivine + liquid in Hill and Roeder's diagram (Fig. 5b). The f_{O_2} enrichment in the interface is not sufficient to shift the f_{O_2}-T conditions from the normal level of the magma (line A–B) up to the stability field of plagioclase + olivine + spinel + liquid. Therefore the crystallization of spinel ceases and silicate rocks are formed, which built up the Main Zone. On further cooling the magma reaches the stability field of plagioclase + olivine + spinel + liquid. At this point the mechanism of the f_{O_2}-enrichment in the interface plane is again sufficiently effective to favour the formation of massive magnetite layers. Also the presence of sulphides at the top of some anorthosite layers immediately underlying magnetite layers (v. Gruenewaldt 1976) may be a direct indicator for increased oxygen fugacity. So the process of the magnetite seam formation works in the same way as described for the formation of chromitite seams.

References

Arculus RJ, Osborn EF (1975) Phase relations in the system MgO-iron oxide-Cr_2O_3-SiO_2. Annu Rep Geophys Lab Carnegie Inst, Washington, pp 507–512

Atkins FB (1969) Pyroxenes of the Bushveld intrusion, South Africa. Inst Petrol 10:222–249

Biljon WJ van (1976) Goud is nie waar dit gevind word nie! In oorsig van die struktuur, gesteentes en ertsafsettings van die Transvaal. Trans Geol Soc S Afr 79:155–167

Bisschoff AA (1971) Tholeiitic intrusions in the Vredefort Dome. Trans Geol Soc S Afr 75:23–30

Bisschoff AA (1973) The petrology of some mafic and peralkaline intrusions in the Vredefort Dome, South Africa. Trans Geol Soc S Afr 77:27–52

Buchanan DL (1976) The sulphide and oxide assemblages in the Bushveld Complex Rocks of the Bethal Area. Trans Geol Soc S Afr 79:76–80

Buddington AF, Lindsley DH (1964) Iron-titanium oxide minerals and synthetic equivalents. Inst Petrol 54:310–357

Camaron EN, Desborough GA (1969) Occurrence and characteristics of chromite deposits – Eastern Bushveld Complex. In: Wilson HDB (ed) Magmatic ore deposits, a symposium. Econ Geol Monogr 4:23–40

Coetzee CG (ed) (1976) Mineral resources of the Republic of South Africa. Geol Surv Dep Mines S Afr, Pretoria, 462 pp

Gruenewaldt G von (1970) On the phase-change orthopyroxene-pigeonite and the resulting textures in the main and upper Zones of the Bushveld Complex in the Eastern Transvaal. In: Visser DJL, Gruenewald G v (eds) Geol Soc S Afr, Spec Publ 1:67–73

Gruenewaldt G von (1973) The main and upper zones of the Bushveld Complex in the Roossenekal Area, Eastern Transvaal. Trans Geol Soc S Afr 76:207–227

Gruenewaldt G von (1976) Sulfides in the upper zone of the Eastern Bushveld Complex. Econ Geol 71:1324–1336

Gruenewaldt G von (1977) The mineral resources of the Bushveld Complex. Minerals Sci Eng 9:83–95

Haughton SH (1969) Geological history of southern Africa. Geol Soc S Afr 535 pp

Haughton DR, Roeder PL, Skinner BJ (1974) Solubility of sulfur in mafic magmas. Econ Geol 69: 451–467

Hill R, Roeder P (1974) The crystallization of spinel from basaltic liquid as a function of oxygen fugacity. J Geol 82:709–729

Irvine TN (1977) Origin of chromitite layers in the Muskox intrusion and other stratiform intrusions: A new interpretation. Geology 5:273–277

Klemm DD, Dehm RM, Henckel J, Schmidt-Thome R (1979) A detailed geological profile of the upper zone of the Bushveld Complex, South Africa (Resume). 10e Colloq Geol Afr, Montpellier

Klopfer H (1974) Wassertransport durch Diffusion in Feststoffen. Bauverlag, Wiesbaden Berlin, 235 S

MacLean WH (1969) Liquidus phase relations in the FeS-FeO-Fe_3O_4-SiO_2 system, and their application in geology. Econ Geol 64:865–884

Molyneux TG (1970) The geology of the area in the vicinity of Magnet Heights, Eastern Transvaal, with special reference to the magnetic iron ore. Geol Soc S Afr, Spec Publ 1:228–241

Muan A, Osborn EF (1956) Phase equilibria at liquidus temperatures in the system MgO-FeO-Fe_2O_3-SiO_2. J Am Ceram Soc 39:121–140

Roeder PL, Osborn EF (1966) Experimental data for the system MgO-FeO-Fe_2O_3-$CaAl_2SiO_2O_8$-SiO_2 and their petrologic implications. Am J Sci 264:428–480

Schwellnus JSI, Hiemstra SA, Gasparrini E (1976) The Merensky Reef at the Atok Platinum Mine and its environs. Econ Geol 71:249–260

Snethlage R, Gruenewaldt G von (1978) Oxygen fugacity and its bearing on the origin of chromitite layers in the Bushveld Complex. In: Klemm DD, Schneider HJ (eds) Time and stratabound ore deposits. Springer, Berlin Heidelberg New York, pp 352–370

Snethlage R, Klemm DD (1978a) Intrinsic oxygen fugacity measurements on chromites from the Bushveld Complex and their petrogenetic significance. Contrib Mineral Petrol 67:127–138

Snethlage R, Klemm DD (1978b) Über die Theorie zur Messung der "intrinsic oxygen fugacity" und ihre Anwendung auf natürliche Chromite. Neues Jahrb Mineral Monatsh 1978:273–288

Truswell JF (1977) The geological evolution of South Africa. Purnell, Cap Town Johannesburg London, 218 pp

Ulmer GC (1969) Experimental investigations of chromite spinels. In: Wilson HDB (ed) Magmatic ore deposits, a symposium. Econ Geol Monogr 4:114–131

Vail JR (1977) Further data on the alignment of basic igneous intrusive complexes in southern and eastern Africa. Trans Geol Soc S Afr 80:87–92

Vermaak CF (1976) The nickel pipes of Vlakfontein and vicinity, Western Transvaal. Econ Geol 71:261–286

Waal SA De (1967) Carbon dioxide and water from metamorphic reactions as agents for sulphide and spinel precipitation in mafic magmas. Trans Geol Soc S Afr 80:193–196

Waal SA De (1978) Carbon dioxide and water from metamorphic reactions as agents for sulphide and spinel precipitation in mafic magmas (discussion). Trans Geol Soc S Afr 81:227–228

Willemse J (1969a) The geology of the Bushveld Igneous Complex, the largest repository of magmatic ore deposits in the world. In: Wilson HDB (ed) Magmatic ore deposits, a symposium. Econ Geol Monogr 4:1–22

Willemse JA (1969b) The vanadiferous magnetic iron ore of the Bushveld Complex. In: Wilson HDB (ed) Magmatic ore deposits, a symposium. Econ Geol Monogr 4:187–208

Aspects on the Gold Distribution in Some Greek Chromites

A. PANAGOS[1], G. AGIORGITIS[1], and R. WOLF[2]

Abstract

A suite of chromites from different Greek localities was analysed for the element gold by neutron activation analysis. The inhomogeneous distribution points to the presence of submicroscopic atomic gold.

1 Introduction

Occurrences of chromites in Greece have been examined by various authors, e.g. Hiessleitner (1951/52), Zachos (1953), Maratos (1960), Panagos (1967a). These investigations did show that Greek chromites are embedded mainly in dunites and pyroxene peridotites from ultrabasic complexes. The chemical composition of these chromites has been described in detail by Panagos (1967b).

In the present investigation 15 chromite samples from 9 different localities have been selected for neutron activation analysis (Fig. 1). Based on the measured contents, we are trying to calculate the theoretical distribution of atomic and ionic gold.

2 Results

The measurements are part of a systematic investigation of the noble metals in Greek chromites (Agiorgitis and Wolf 1977).

The abundances of gold in the chromites are listed in Table 1. The average value of 2.6 ± 6.4 ppb (± 2s) shows reasonable agreement with observed enrichments in chromites from Mt. Albert pluton (Crocket and Chyi 1972) and the Troodos complex, Cyprus (Agiorgitis and Becker 1979). But it should be pointed out that the above average includes three samples with somewhat higher content. Excluding samples E/4, S/11, and X/15, the average is 1.1 ± 0.6 ppb.

However, the relatively large spread of the values and the lack of any correlation with major elements, such as Cr or other noble metals indicate that gold is not crystallo-

1 Department of Geology, University of Patras, Patras, Greece
2 Mineralogisches Institut der Technischen Universität, 6100 Darmstadt, FRG

Fig. 1. Chromite samples from Greek localities: *1* Soufli, *2* Chalkidiki, *3* Rodiani, *4* Vourinon, *5* Eretria, *6* Domokos, *7* Nezeros, *8* Eubea, *9* Skyros .

Table 1. Au contents and calculated atomic and ionic ratios

Sample	Locality	Au (ppb)	Calculated Au^0/Au^{1+}
B/1	Vourinon	0.48 ± 0.06	$Au^0 = 1 \times 10^5\ Au^{1+}$
B/2	Vourinon	0.40 ± 0.06	$Au^0 = 2.7 \times 10^3\ Au^{1+}$
E/3	Eretria	0.45 ± 0.05	$Au^0 = 3.2 \times 10^3\ Au^{1+}$
E/4	Eretria	10.80 ± 0.40	$Au^0 = 2.3 \times 10^3\ Au^{1+}$
Eu/5	Eubea	1.30 ± 0.08	$Au^0 = 8.5 \times 10^3\ Au^{1+}$
M/6	Domokos	1.35 ± 0.30	$Au^0 = 3.6 \times 10^3\ Au^{1+}$
N/7	Nezeros	0.74 ± 0.06	$Au^0 = 4.1 \times 10^3\ Au^{1+}$
P/8	Rodiani	1.40 ± 0.25	$Au^0 = 7.6 \times 10^3\ Au^{1+}$
P/9	Rodiani	1.60 ± 0.08	$Au^0 = 5.0 \times 10^3\ Au^{1+}$
S/10	Skyros	2.05 ± 0.30	$Au^0 = 4.5 \times 10^3\ Au^{1+}$
S/11	Skyros	6.10 ± 0.20	$Au^0 = 1.1 \times 10^4\ Au^{1+}$
Th/12	Soufli	1.85 ± 0.15	$Au^0 = 5.7 \times 10^3\ Au^{1+}$
Th/13	Soufli	1.55 ± 0.08	$Au^0 = 1.5 \times 10^4\ Au^{1+}$
X/14	Chalkidiki	0.47 ± 0.07	$Au^0 = 1.8 \times 10^4\ Au^{1+}$
X/15	Chalkidiki	8.30 ± 0.30	$Au^0 = 2.3 \times 10^4\ Au^{1+}$

chemically bound in the chromites. The possibility of substitution of Au^{1+} (1.37 Å) and Au^{3+} (0.55 Å) in cations of the chromite lattice must be ruled out because of higher redox potential and electronegativity.

Studies of rocks of the Skaergaard intrusion (Vincent and Crocket 1960) and of rocks and minerals of the Troodos complex (Agiorgitis and Becker 1979) confirm the absence of a significant fractionation of gold.

The question whether gold occurs as ionic or atomic species can be better answered quantitatively. Due to the thermodynamic stability of chromite an oxidation degree of 0.37 can be calculated for the condition of its formation. The formation temperature is assumed about 1150°C. By using the Nernst equation one can calculate the relative distribution of ionic and atomic gold: $E_h = -E^0 + \frac{RT}{nF} \ln \frac{Fe^{2+}}{Fe^{3+}}$

The E^0 value used corresponds to aqueous solution at normal temperature. From the calculated relative abundances (Table 1) it becomes obvious that the largest part of gold in chromites is present as atomic species in finely distributed submicroscopic form. It can be assumed that the uncharged Au occluded during the cooling stage of the chromites. The rest should be associated with sulphides like pyrite, chalcopyrite, etc. which appear in Greek chromites (Panagos 1967a).

The above calculation is performed assuming an undisturbed ratio of the atomic to the ionic species since the cooling stage of the chromites. But we cannot exclude the possibility that this ratio has been disturbed by hydrothermal influence which would remobilize the ionic species preferentially. It has been shown by Hiessleitner (1951/1952) that the above mentioned chromites have experienced hydrothermal alteration at a variable degree.

References

Agiorgitis G, Becker R (1979) The geochemical distribution of gold in some rocks and minerals of the Troodos Complex, Cyprus. Neues Jahrb Mineral Monatsh 7:316–320

Agiorgitis G, Wolf R (1977) Zur Platin-, Palladium- und Goldverteilung in griechischen Chromiten. Chem Erde 36:349–351

Crocket JH, Chyi LL (1972) Abundances of Pd, Ir, Os and Au in an alpine ultramafic pluton. Proc XXIV Int Geol Congr Sect 10:202–209

Hiessleitner G (1951/1952) Serpentin- und Chromerz-Geologie der Balkanhalbinsel und eines Teiles von Kleinasien. Jahrb Geol Bundesanst Wien Sonderbd 1(1, 2):683

Maratos G (1960) Die Ophiolithen des Gebietes Soufli. Inst Geol Min Res Rep T/II

Panagos A (1967a) Beitrag zur Kenntnis der griechischen Chromite. Ann Geol Pays Hell 18:1–42

Panagos A (1967b) Vergleichende Bemerkungen zum Chemismus der griechischen Chromite. Ann Geol Pays Hell 18:471–505

Vincent EA, Crocket JH (1960) Studies in the geochemistry of gold. I. The distribution of gold in rocks and minerals of the Skaergaard intrusion, East Greenland. Geochim Cosmochim Acta 18:130–142

Zachos K (1953) Das Chromerzgebiet von Bourinon, Kozani. Inst Geol Min Res Rep T/III

Ultramafic Rocks and Related Ore Minerals of Lapland, Northern Finland

H. PAPUNEN and H. IDMAN[1]

Abstract

Most of the Precambrian ultramafics of northern Finland underwent metamorphic recrystallization and deformation. The rock types have altered to serpentinites, metaperidotites and carbonate-bearing ultramafics. Also the accessory ore minerals of the ultramafics altered and recrystallized. Hence the present sulphide assemblages include pentlandite, awaruite, heazlewoodite, millerite and mackinawite instead of the common pyrrhotite-pentlandite assemblage of the primary magmatic deposits. Accessory nickel arsenides characterize the carbonatized ultramafics. The oxide minerals are chromite, magnetite and ilmenite. Chromites have two kinds of zoning, called the iron-trend and the aluminium-trend correspondingly. The iron-trend of zoning was formed in deuteric reactions of early chromites or the chromite crystals were mantled by secondary magnetite in serpentinization. The fine-grained intergrowths of chromite and secondary magnetite exist locally between the chromite core and secondary magnetite; also ilmenite exsolution lamellae exist in some chromites and chromite-magnetite lamellae have rarely been met with in primary ilmenite grains.

1 Introduction

The northern Precambrian Baltic Shield is characterized by numerous basic and ultrabasic intrusives. Some of those in the Kola peninsula, U.S.S.R., e.g., Pechenga, Moncegorsk, and Allarecensko (Gorbunov et al. 1973) contain Ni-Cu sulphide ores. Häkli (1971) has pointed out the high tenor of silicate nickel in some of the ultramafites of Finnish Lapland. On the basis of comparable data from southern Finland, they are regarded as potential targets of nickel exploration. In order to collect relevant data and to study the metallogeny of the ultramafites of Lapland, a research project was conducted by the University of Turku in cooperation with Outokumpu Oy in 1974–1979. Some geological and geochemical results have already been given in the progress report by Papunen et al. (1977) and in a comparative study of Finnish nickel deposits by Papunen et al. (1979), but the final detailed report is still being assembled. More than 2000 polished thin sections were made from ultramafites in the course of the project. The following chapters summarize the results of the study of opaque mineral assemblage.

1 Institute of Geology and Mineralogy, University of Turku, 20500 Turku 50, Finland

2 General Geology of the Ultramafites of Lapland

The geological formations of Lapland can be divided into the Archean granitoid complex in the east and north, the Svecokarelian schists in central Lapland and the granitoids of Svecokarelian orogeny (Fig. 1). Some Caledonian schists occupy a minor area in western Lapland. The major part of the Archean complex is composed of granitoids, although remnants of mainly volcanogenic greenstone belts occur in eastern and northern Lapland. The granulite belt yields isotopic ages corresponding to those of Svecokarelian orogeny, ca. 1900 Ma. In contrast to Simonen (1971), Meriläinen (1976) maintains that the schists of the West Inari Schist Belt (called WISB in the following chapters) are Archean in age (Figs. 1 and 2).

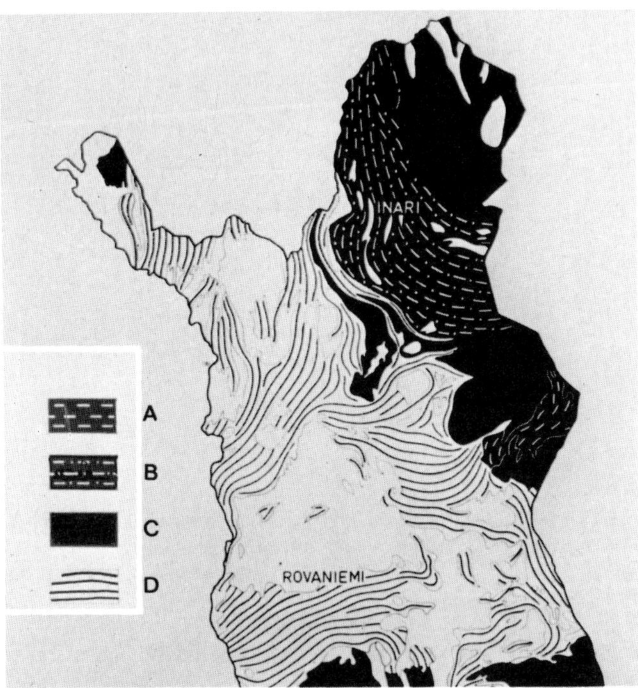

Fig. 1. Simplified geological map of northern Finland (after Simonen 1971). *A* rocks of the Archean greenstone belt association; *B* granulite complex; *C* Archean granitoids; *D* Svecokarelian schists. *White areas:* Postarchean intrusives, mainly Svecokarelian granitoids

The composition of the ultramafic bodies of Lapland varies from dunite and its altered derivative serpentinite to metaperidotite, pyroxenite, cortlandite, and hornblendite. The bodies range in size from a score of metres up to about 2 km.

The distribution of the ultramafic bodies studied is depicted in Fig. 2. Except for the Svecokarelian granitoids and Caledonian schists, the ultramafites occur as single or groups of bodies in all the geological formations mentioned. Ultramafic bodies

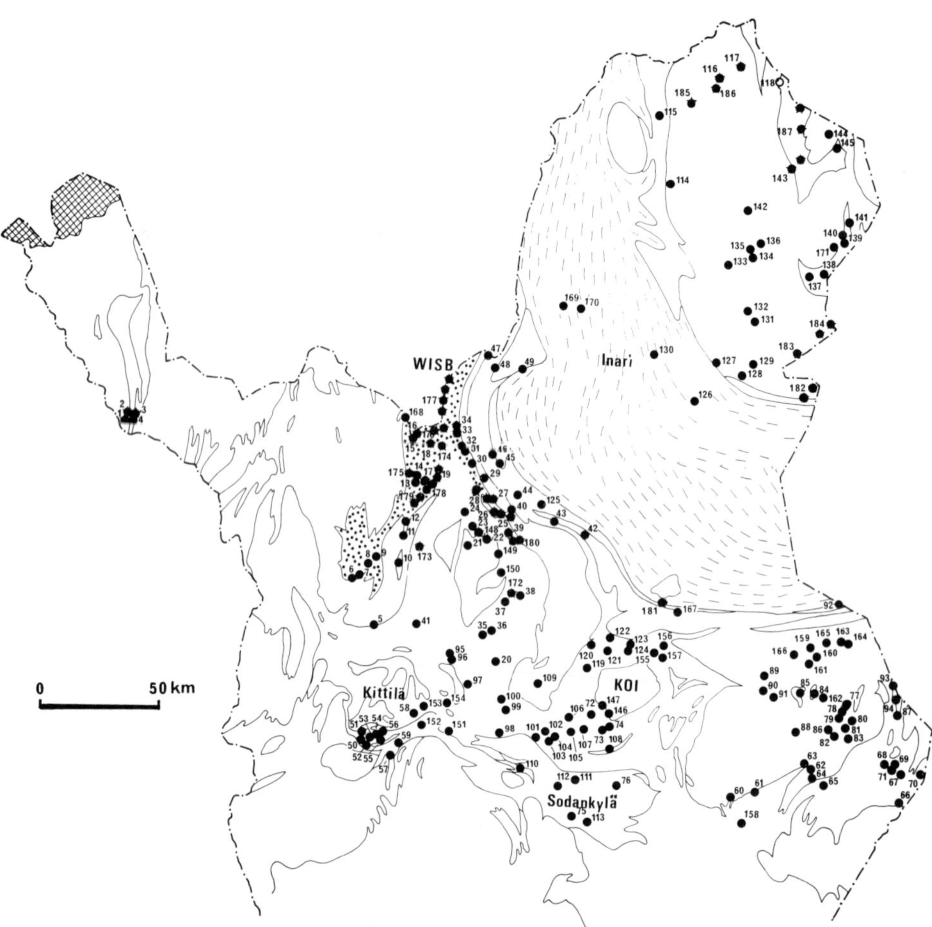

Fig. 2. Location of the ultramafic bodies studied *(dots)* (mainly according to Papunen et al. 1977). *WISB* the West Inari Schist Belt *(stippled); KOI* the gabbro body of Koitelainen

abound in the schist areas, in both the rocks of the Archean greenstone belt association and in Svecokarelian metavolcanic and metasedimentary rocks of central Lapland. Small bodies are locally abundant in the Archean granitoid area, especially in northeastern Inari, where their location is controlled by tectonic shear zones. In the area of WISB, the ultramafites, which are mainly metaperidotitic or dunitic in composition, occur in a lithosome composed of amphibolites, cherty quartzites, iron formations of sulphide and oxide facies, calc-silicate rocks and black schists. These form long, continuous belts, and the association of ultramafites with the predominantly volcanogenic lithosome indicates a stratigraphic relation between them. The metaperidotites and serpentinites in the area are commonly associated with ultramafic amphibolites of extrusive origin. In chemical composition these rocks resemble the komatiitic rock series (Papunen et al. 1979), but for lack of extrusive textures in the metaperidotites and dunites they cannot be considered as true komatiites.

In central Lapland's volcanite areas the metaperidotites constitute lens-shaped bodies interbedded with extrusive rocks. The cumulus textures of the silicates and oxides are still recognizable, and suggest that the ultramafites are shallow intrusives, or sills that intruded close to the surface during extrusion of the mafic members of the series. Some metapyroxenites in the same area likewise display volcanic textures and represent extrusive ultramafites.

In eastern and north-eastern Finnish Lapland the ultramafites are mainly associated with amphibolites that are remnants of the old Archean greenstone belts. These bodies probably represent ultramafic members of the Archean greenstone belt association.

In the Svecokarelian Kittilä greenstone area of central Lapland the dunites and serpentinites form a sinuous zone that runs almost from north to south. These ultramafites also intruded during early stages of geosynclinal development into the pile of volcanics deposited in the geosyncline. In the southern margin of the same area, metaperidotites are encountered in the vicinity of metasediments.

The outer margin of the granulite arch is characterized by a belt of spinel-bearing metacortlandites and by carbonate-bearing orthopyroxenites, or sagvandites (Ohnmacht 1974; Schreyer et al. 1972), which constitute a special ultramafic rock type in the WISB.

Most of the ultramafic bodies intruded or extruded during early stages of the orogeny, partly contemporaneously with Svecokarelian geosynclinal or Archean greenstone-belt volcanism, and hence underwent metamorphic recrystallization and deformation. Some of them, especially those in the greenstone area of Kittilä and in WISB, have experienced premetamorphic serpentinization and carbonatization. Thus, the metamorphic recrystallization of olivine, for example, means that the rock has been deserpentinized.

3 The Ore Mineral Assemblages

3.1 The Oxide Minerals

The most common oxide minerals in the ultramafites of Lapland are the members of the spinel group, viz. chromite, magnetite, and locally the aluminous spinels, although in certain ultramafites ilmenite abounds as well. Hematite exists only in the most oxidized rock types. The mineralogy of the spinel phase in the rocks studied has been discussed in some detail in the review by Papunen et al. (1979), and in a more comprehensive report by Idman (1980).

Chromite commonly occurs as euhedral, zoned crystals in the ultramafites of western Lapland, in WISB, and in the volcanite area surrounding the Koitelainen gabbro body in central Lapland. The cores of the zoned crystals are chromite surrounded by a zone of "ferritechromite" or "gray magnetite" (Onyeagocha 1974), and the whole grain may be mantled by magnetite (Fig. 3a). Only in the ultramafites related to the volcanics of central Lapland does the chemical composition of the chromite core fall in the field of magmatic Mg-Al chromites as described by Evans and Frost (1975). In all of the other chromites the Mg/Fe^{2+}-ratio is lower than in typical magmatic chromites, probably as a result of metamorphism.

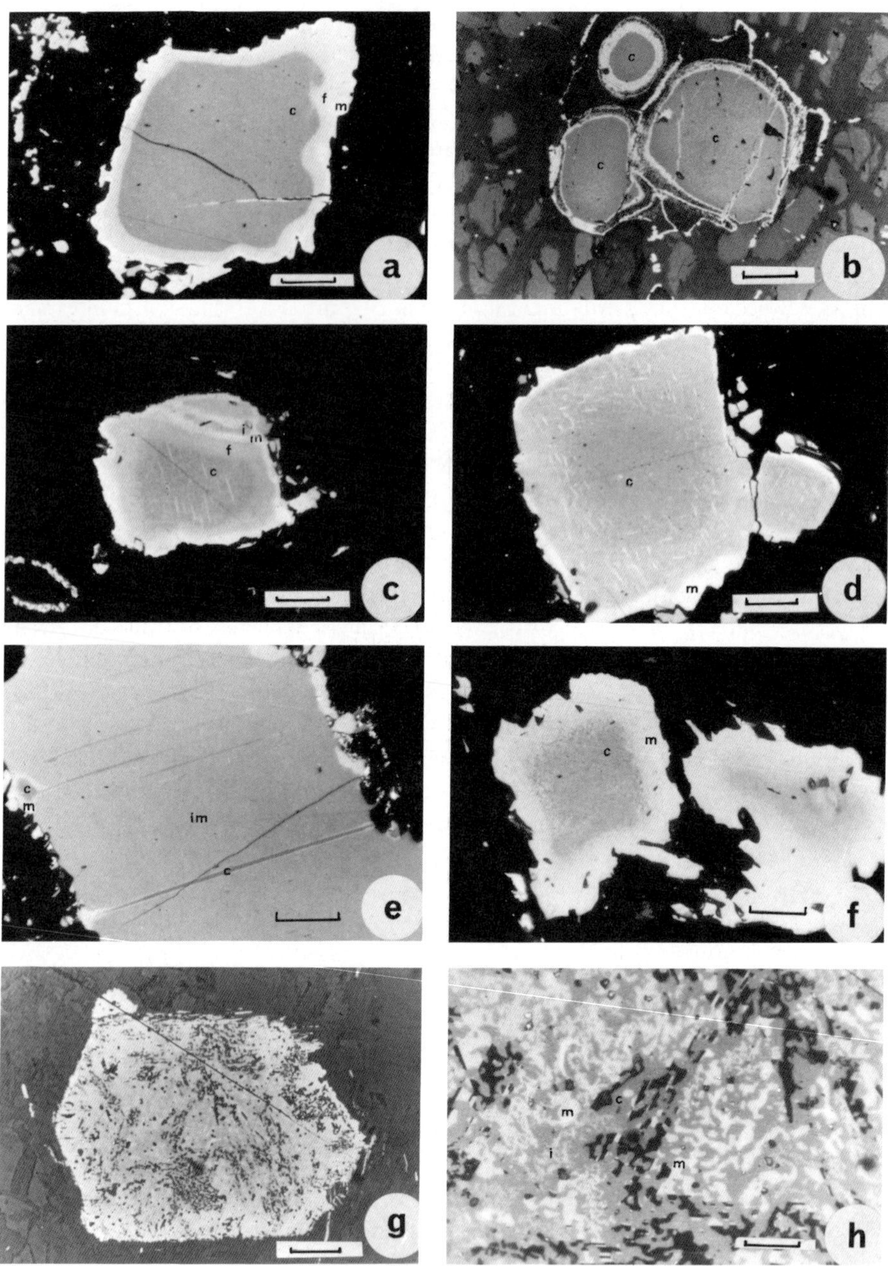

Fig. 3a–h

There are two kinds of zoning in the chromites of Lapland: either the tenor of iron increases towards the margin or the tenor of aluminium increases in the margin. According to Haggerty (1976) the zonal trends of the chromites could be called the iron trend and the aluminium trend. The iron trend is common in the ultramafites, but the aluminium trend has been found only in the cortlandites in the marginal zone of the granulite belt.

The variation of the chemical composition of zoned chromites is presented in Table 1. In the zoned grains of the iron-trend chromites the tenors of Mg, Al, and Cr are highest in the cores, Ti and Ni are enriched in the ferritechromite zone, and Fe and Ni (up to 0.83%) in the chromian or pure magnetite mantle of the crystals. Zn is concentrated in the chromite core and percentages of up to 3 wt.% Zn have been measured in some sulphide-bearing ultramafites of WISB. In the zoned grains of the aluminium-trend chromites the tenor of Cr decreases and those of Al and Fe increase from the core to the margin. Mg seems to increase parallel to the Al content. The tenor of Ti decreases somewhat, but that of Ni is almost constant in both core and rim.

The chromite cores of the crystals are commonly homogeneous, but locally the cumulus chromites in the central Lapland volcanite area display thin ilmenite lamellae along the (111) surface of the chromite (Fig. 3c). The ilmenite lamellae do not extend to the surrounding ferritechromite zone although in some grains the lamellae are concentrated close to the margin of the chromite (Fig. 3d). The ferritechromite zone between the chromite core and the magnetite mantle is locally optically heterogeneous: A very fine-grained intergrowth of chromite and magnetite can be seen at high magnification and under oil immersion (Fig. 3f). The texture resembles ulvite-magnetite exsolution, but the low Ti content in the intergrowth excludes the existence of ulvite. As the bulk chemical composition of the intergrowth corresponds to that of the ferritechromite, it is possible that the ferritechromite zone in general is composed of the submicroscopic intergrowth of chromite and magnetite.

A symplectic intergrowth of chromite and magnetite spinels, ilmenite, and a silicate mineral occurs locally in the peridotites of the central Lapland volcanite area (Fig. 3g, h). The intergrowth is very fine-grained and the species of the anisotropic silicate mineral cannot be identified. A very rare intergrowth of ferritechromite and olivine exists in some ultramafic bodies of the WISB. Ferritechromite occurs as lamellae ca. 0.01 mm wide in olivine. The texture resembles the chromite-olivine exsolution texture described by v.d. Kaaden et al. (1972).

Fig. 3a–h. Textures of oxide minerals. Number of occurrence (cf. Fig. 2) in parentheses: **a** A zoned (deuteric) chromite *(c)* with zones of ferritechromite *(f)* and magnetite *(m)*. Scale 0.02 mm (occurrence No. 147). **b** Chromite crystal mantled by secondary magnetite (*c* = chromite). Scale 0.05 mm (117). **c** Ilmenite lamellae *(light gray)* in the chromite core of the zoned chromite *(c)*, ferritechromite *(f)*, magnetite *(m)* grain. Ilmenite *(i)* exists also between the chromite grains. Scale 0.02 mm (73). **d** Ilmenite lamellae in the margin of the chromite core which is mantled by a thin zone of ferritechromite and magnetite *(m)*. Scale 0.02 mm (73). **e** Chromite lamellae *(c)* in ilmenite *(im)*; magnetite *(m)* exists at the margins of the chromite lamellae. Scale 0.02 mm (73). **f** A "ferritechromite" zone between the chromite core *(c)* and magnetite margin *(m)* is composed of a very fine-grained intergrowth of chromite and magnetite. Scale 0.02 mm (2). **g, h** Symplectic intergrowth of chromite *(c)*, magnetite *(m)*, ilmenite *(i)*, and a silicate mineral *(black)*. **h** is a detail of **g**. Scale in **g** 0.05 mm; in **h** 0.012 mm (73)

Table 1. Chemical composition of zoned chromite crystals (weight per cent)

		Al_2O_3	Cr_2O_3	Fe_2O_3	FeO	MgO	NiO	TiO_2	RAl_2O_4	RCr_2O_4	RFe_2O_4
		a) Iron trend of zoning									
C	x	11.75	48.19	9.40	23.42	6.47	0.10	0.65	23.4	64.6	12.0
	max	20.8	54.6	11.8	30.4	14.0	0.35	3.37			
	min	1.6	4.9	1.3	13.8	0.6	0.02	0.07			
F	x	2.13	33.27	33.54	28.23	2.22	0.27	1.01	4.7	48.7	46.6
	max	5.8	43.6	55.1	31.8	5.4	0.57	3.41			
	min	0.0	14.4	18.2	2.38	0.1	0.05	0.00			
M	x	0.53	8.56	61.56	28.78	1.11	0.47	0.22	1.1	12.6	86.3
	max	1.9	25.0	71.7	31.1	7.6	0.83	0.89			
	min	0.0	1.1	41.6	19.9	0.6	0.10	0.00			
		b) Aluminium trend of zoning									
C	x	29.12	34.32	4.50	22.53	8.70	0.16	0.16	52.9	41:9	5.2
	max	53.4	63.7	9.5	31.1	17.2	0.37	0.36			
	min	7.9	11.7	0.0	16.7	3.4	0.02	0.02			
S	x	40.64	23.99	3.82	19.61	12.03	0.20	0.10	68.7	27.2	4.1
	max	52.8	53.7	8.1	30.8	17.1	0.38	0.36			
	min	16.9	11.2	0.0	13.9	3.4	0.00	0.02			

C = chromite, F = ferritechromite, M = magnetite, S = aluminous spinel
x = arithmetic mean of 18 microprobe analyses in Table a) and 6 analyses in Table b). max and min are the maximum and minimum values of the analyses. The last column indicates calculated percentages of the end members

Magnetite is the predominant oxide mineral in the ultramafites. It is mainly of secondary origin, having formed during serpentinization of olivine. The secondary magnetite exists as fine dust in serpentine, as small veinlets in fissures of former olivine, in replacement of the primary sulphide grains or as chromian magnetite mantling the zoned chromite crystals (Fig. 3b). Magnetite abounds in metaperidotites in certain ultramafic bodies in the central Lapland greenstone area, and in eastern and northeastern Lapland, even though the degree of serpentinization is rather low. In these bodies the olivine is anhedral and locally poikiloblastic; it contains euhedral amphibole needles and magnetite grains as inclusions, indicating that olivine was formed through recrystallization and deserpentinization. Thus, the high abundance of magnetite may also originate from early, premetamorphic serpentinization, while later metamorphism made the fine magnetite dust to recrystallize as medium-grained pure magnetite grains.

Skeletal crystals of primary magnetite occur in the extrusive ultramafites of eastern Lapland (Fig. 4a). Primary magmatic magnetite is also encountered in the intercumulus

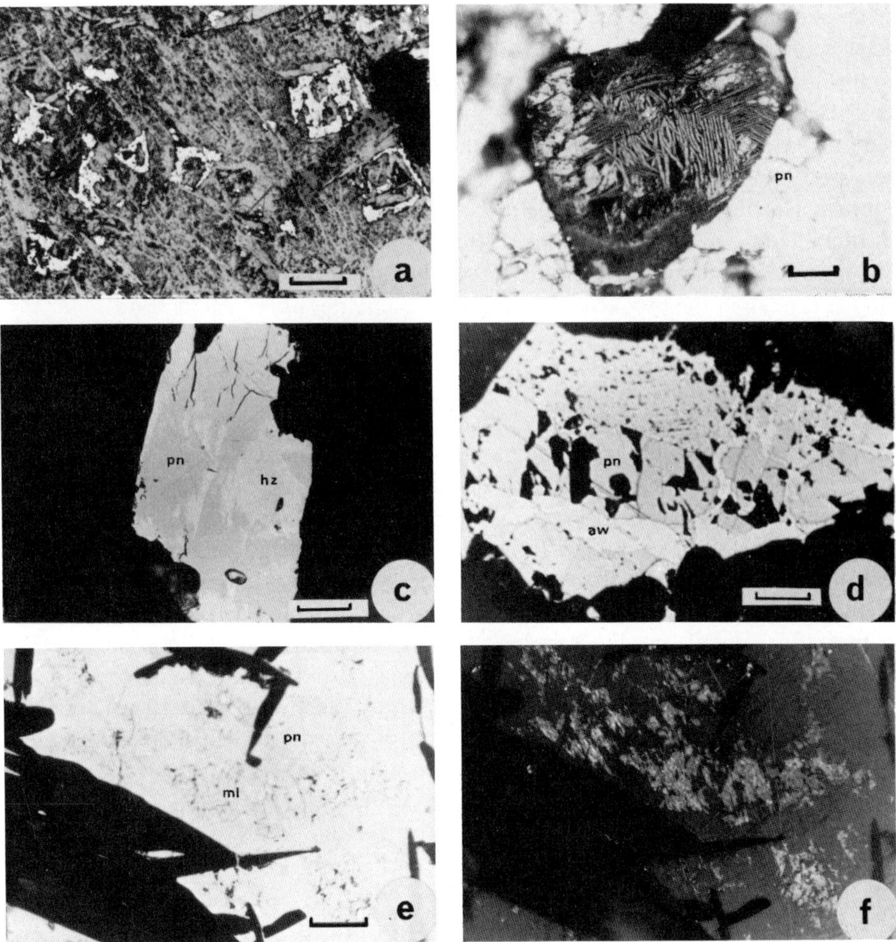

Fig. 4a-f. Textures of oxide and sulphide minerals. **a** Skeletal crystals of magnetite *(light gray)* in metacortlandite. Scale 0.5 mm (84). **b** A valleriite-type mineral replaces magnetite *(black)* which exists as inclusion in pentlandite *(pn)*. Scale 0.02 mm (84). **c** A pentlandite *(pn)* – heazlewoodite *(hz)* grain in serpentinite. Scale 0.02 mm (20). **d** An awaruite *(aw)* – pentlandite *(pn)* grain in serpentinite. Scale 0.02 mm (40). **e** Millerite *(ml)* in pentlandite *(pn)*; euhedral black flakes of antigorite intersect the sulphide grain. Scale 0.02 mm (20). **f** The same as **e** with crossed polarizers

phase of the metaperidotites of central Lapland. This magnetite was originally ilmenomagnetite. The ilmenite lamellae are, however, rare nowadays because ilmenite has accumulated as individual grains along grain boundaries as a result of metamorphic recrystallization.

Transparent green aluminous spinel is characteristic of the ultramafites along the outer marginal arch of the granulite complex. In deformed ultramafites the spinel grains are anhedral but in a lherzolite with cumulus texture the spinel exists in cumulus phase. The grains are commonly zoned with a brown chrome spinel in the core and a green aluminous spinel at the margin. In strongly deformed cortlandites, however, the green spinel grains are homogeneous and anhedral.

Ilmenite is a common but minor component of the oxide phase. It commonly occurs as single, individual grains without exsolution texture and represents a metamorphic, recrystallized variety. Ilmenite abounds only in the metaperidotitic sill in the volcanite area of central Lapland, in which it occurs as an intercumulus phase. Locally ilmenite contains lamellae of exsolved spinel phase (Fig. 3e). The thin lamellae have a ferrite-chromite core and magnetite margins. The ilmenite is obviously of primary magmatic origin and has not been metamorphosed. Primary magmatic ilmenite also exists as tabular crystals in the pyroxenitic, extrusive ultramafites of central Lapland.

3.2 The Sulphide Minerals

In his study of the nickel content of mafic and ultramafic rocks in Finland, Häkli (1971) pointed out the high tenor of silicate nickel and the low sulphide content in the ultramafites of Lapland. Under close microscopic examination, however, the accessory sulphide minerals proved to be rather common. Nevertheless, there are only a few occurrences in which the amount of sulphide sulphur exceeds 1%. The average amount of sulphides is higher in the ultramafites of WISB than in the other ultramafites. As pointed out by Ramdohr (1967), Papunen (1970), Eckstrand (1975), Groves and Keays (1979), the mineral assemblages of accessory sulphides in ultramafites depend on the manner in which the host ultramafites altered. Hence, complex alteration and metamorphic events have left their imprint on the sulphide mineral assemblages in the ultramafites of Lapland.

3.3 Pyrrhotite-Bearing Assemblages

The sulphide mineral assemblage pyrrhotite-pentlandite-chalcopyrite (± cubanite), which is most common in the sulphide Ni-Cu ores of mafic and ultramafic association, is also met within the ultramafites of Lapland. As the copper content is commonly markedly lower than that of nickel, the Ni/Cu ratio being about 10, the ultramafites contain very little chalcopyrite, and the main sulphide minerals in this assemblage are pyrrhotite and pentlandite. Only the mafic gabbroic bodies are chalcopyrite-bearing. Pyrrhotite and pentlandite have been met within weakly altered peridotites and in metapyroxenites and hornblendites but not in moderately to strongly altered metaperidotites or in serpentinites with only small amounts of sulphides. Pyrrhotite-bearing sulphide grains occur interstitially to olivine. In a sulphide-rich metaperidotite of WISB the sulphide grains are locally surrounded by secondary magnetite that partially replaces the sulphides. The combined grains are, in turn, mantled by a zone of mackinawite.

The pyrrhotite is either hexagonal or troilite, or then there are flame-like intergrowths of both phases. The magnetic monoclinic phase has been met with only as thin lamellae in hexagonal pyrrhotite in a supergene environment.

3.4 Pentlandite-Millerite and Pentlandite-Heazlewoodite Assemblages

If the tenor of sulphur does not exceed a few tenths of percent, the prevailing sulphide mineral assemblage in serpentinites and metaperidotites is either pentlandite-heazlewoodite or pentlandite-millerite. Certain ultramafic bodies reveal a gradual change from this assemblage to a pure pentlandite assemblage and further to a pyrrhotite-pentlandite assemblage according to the change in rock type from a serpentinite core to marginal metapyroxenite.

The minerals of pentlandite-heazlewoodite or millerite assemblages, or both, occur as fine-grained disseminations, and the sulphides are commonly intergrown with each other (Fig. 4c, e). Although the tenor of nickel in the sulphide phase of rocks with these minerals may exceed 30%, the total sulphide-bound nickel in the whole rock is less than 0.3%. If the sulphide content increases, the mineral assemblage grades into the pyrrhotite-pentlandite assemblage. In some serpentinites with a high tenor of silicate nickel, awaruite has been noted together with pentlandite (Fig. 4d). Millerite does not seem to be so abundant as heazlewoodite, and the rocks containing millerite have appreciable magnetite as well.

3.5 Minor Sulphides and Arsenides

In a copper-bearing metaperidotite with totally serpentinized olivine in eastern Lapland, the sulphide assemblage includes chalcopyrite, bornite, and covellite. Bornite exists together with chalcopyrite, but lamellar exsolution intergrowths are lacking.

Pyrite is totally absent from fresh ultramafites, but it exists in gabbros and in the ultramafites that have undergone supergene alteration.

The nickel arsenides, maucherite, nickeline and gersdorffite, are common accessory minerals in carbonate-bearing ultramafites. Carbonatization seems to be accompanied by an influx of arsenic, and the arsenides are the result of a reaction between arsenic and nickel liberated from carbonatized olivine or other silicates. The arsenide-bearing opaque mineral assemblage commonly also includes pentlandite and pyrrhotite, or heazlewoodite and millerite. The sagvandite rock type of WISB hosts especially the Ni arsenides.

Mackinawite is a common but minor component of the sulphide-bearing ultramafites. It occurs in serpentinites or metaperidotites as minute flakes in serpentine, as a filling in minute cracks, or as a replacement of pentlandite. Together with minerals of the valleriite type, it was the last sulphide mineral to form during and after serpentinization. In a metaperidotite of WISB, mackinawite exists as rims around the pyrrhotite-pentlandite grains replaced by magnetite, and in a metaperidotite of eastern Lapland a mineral of the valleriite type replaces the magnetite crystals that occur as inclusions in pentlandite (Fig. 4b). Owing to its small grain size, the mineral has not been analyzed, but it resembles the chromian valleriite described from the same type of environment by Nickel and Hudson (1976).

4 Discussion

The primary magmatic silicate mineral associations in the ultramafites of Lapland have undergone alteration, that is, they have been serpentinized and locally also carbonatized. Most of the ultramafites, especially those occurring in the areas of metavolcanites, are metamorphosed, and in some of the occurrences alteration has been followed by metamorphic recrystallization. Thus, olivine, for example, has recrystallized as long, tabular crystals resembling those described by Evans and Trommsdorff (1974). According to Papunen et al. (1977), because of the iron incorporated in secondary magnetite during serpentinization, the metamorphic olivines are poorer in iron than are the primary magmatic olivines. Also the tenor of nickel in the metamorphic olivines is lower than in the magmatic varieties because some nickel entered in sulphide and oxide phases during the first serpentinization.

The chemical composition of the chromites in the Kittilä greenstone area falls in the field of magmatic chromites. Microscopic examination shows that the chromites are unaltered cumulus crystals. The chromites of WISB may also be primary magmatic and unaltered although their Mg/Fe^{2+} ratio is lower than that in the common magmatic chromites. The high Zn content in the WISB chromites (Papunen et al. 1979) is a further indication of primary origin. The ilmenite lamellae, rarely observed in the cores of chromite crystals, are the result of exsolution from a Ti-rich magmatic chromite. Likewise, the chromite lamellae in olivine observed in some metadunitic bodies of WISB are the result of exsolution from magmatic chromian olivine. In contrast, the formation of ferritechromite, either as rims around magmatic chromite crystals or as independent grains, is due to deuteric reaction between early chromites and iron-rich residual melt, or because ferritechromite was formed by secondary growth during alteration and metamorphism. Magnetite, which forms a mantle around ferritechromite, is usually chromian magnetite in composition, and it deposited during serpentinization of the host rock. It is commonly nickel-bearing, the average being 0.75% Ni, owing to the nickel liberated from olivine in serpentinization and which was attached together with iron in the oxide phase. This iron trend in the zoning in chromite is typical of the ultramafic bodies that attained their mineral composition during metamorphism of amphibolite or greenschist facies. An aluminium trend of zoning is typical of the chromites in the ultramafites of the granulite belt and evidently due to crystallization under high stress.

The primary sulphides of the original dunites and peridotites were pentlandite and pyrrhotite. According to Eckstrand (1975), serpentinization followed by carbonatization in ultramafites gives rise to reducing and oxidizing reactions and deposition of the corresponding sulphide mineral assemblages. The mineral associations pentlandite-awaruite, pentlandite-heazlewoodite, or pentlandite-millerite are the result of that reaction in order of increasing f_{O_2}. All of the assemblages mentioned are common in altered ultramafites of Lapland. Since the millerite-bearing assemblages in metaperidotites of Lapland exist without carbonates, the higher fugacity of oxygen required for the deposition of millerite was probably produced by an influx of oxidizing hydrothermal fluids after the main phase of serpentinization.

Mackinawite is a typical mineral of the sulphide assemblage of serpentinites. Compared with the formation of secondary magnetite, its deposition can be interpreted as an alternative or parallel way of attaching iron liberated from olivine during serpentinization. The formation of mackinawite indicates high fugacity of sulphur during serpentinization.

The deposition of nickel arsenides in ultramafites of Lapland is definitely related to the formation of carbonate-bearing ultramafites. These rocks occur in an area that has undergone metamorphic recrystallization. Before metamorphism the ultramafites were altered into talc-magnesite rocks through carbonatization; metamorphic recrystallization accompanied by dehydration then turned them into magnesite-olivine-orthopyroxene rocks. The introduction of As in carbonate-rich alteration fluids was noted by Groves and Keays (1979) in ultramafites of Western Australia.

The sulphide mineral assemblage pentlandite-pyrrhotite in a serpentinized ultramafic body indicates high tenor of sulphur, and the body has been considered as a potential target of nickel exploration. For example, some of the metaperidotites of WISB contain the pyrrhotite-bearing sulphide mineral assemblage and, as a result of intense exploration, some minor sulphide Ni occurrences have been found in the area. The existence of zincian chromite as an explorational indicator of potential primary sulphide-bearing ultramafites is consistent with observations made in the WISB in harmony with data reported from the Ni occurrences of Western Australia by Groves et al. (1977).

The mineral assemblages pentlandite-heazlewoodite or pentlandite-millerite are limited to weakly mineralized parts of the ultramafic bodies, in which the nickel content of the sulphide phase might be very high. Some of the metadunites and serpentinites in central Lapland arouse interest in this respect.

References

Eckstrand OR (1975) The Dumont serpentinite: a model for control of nickeliferous opaque mineral assemblages by alteration reactions in ultramafic rocks. Econ Geol 70:183–201

Evans BW, Frost RB (1975) Chrome-spinel in progressive metamorphism a preliminary analysis. Geochim Cosmochim Acta 39:959–972

Evans BW, Trommsdorff V (1974) On elongate olivin of metamorphic origin. Geology 2:131–132

Gorbunov GI, Jakolev JuU, Astafjev JuA, Goncarov JuV, Bartenev IS, Neradovskii JN (1973) Atlas tekstur i struktur sulfidnyh medno-nikolovyh vud Koskogo poluostrova. Akad Nauk SSSR, Izd NAUKA, Leningrad

Groves DI, Keays RR (1979) Mobilization of ore-forming elements during alteration of dunites, Mt. Keith-Betheno, Western Australia. Can Miner 17:373–390

Groves DI, Barret FM, Binns RA, McQueen KG (1977) Spinel phases associated with metamorphosed volcanic-type iron-nickel sulfide ores from Western Australia. Econ Geol 72:1224–1244

Haggerty SE (1976) Opaque mineral oxides in terrestrial igneous rocks. In: Rumble III D (ed) Oxide Minerals. Miner Soc Am Short Course Not 3:101–300

Häkli TA (1971) Silicate nickel and its applications to the exploration of nickel ores. Geol Soc Fin Bull 43:247–263

Idman H (1980) Lapin ultramafiittien oksidifaasista. M Sc Thesis, Univ Turku, Finland

Kaaden Gvd, Ottemann J, Sahoo KR (1972) Chromite verschiedener Zusammensetzung und Genese aus einem Dunit von Saruabil, Sukinda, Orissa (Indian). Neues Jahrb Mineral Monatsh 1972:256–262

Meriläinen K (1976) The granulite complex and adjacent rocks in Lapland, northern Finland. Geol Surv Fin Bull 281:5–119

Nickel EH, Hudson DR (1976) The replacement of chrome spinel by chromian valleriite in sulphide-bearing ultramafic rocks in Western Australia. Contrib Mineral Petrol 55:265–277

Ohnmacht W (1974) Petrogenesis of carbonate-orthopyroxenites (sagvandites) and related rocks from Troms, Northern Norway. J Petrology 15:303–324

Onyeagocha AC (1974) Alteration of chromite from the Twin Sisters dunite, Washington. Am Mineral 59:608–612

Papunen H (1970) Sulphide mineralogy of the Kotalahti and Hituma nickel-copper ores, Finland. Ann Acad Sci Fenn Ser A III 109:1–74

Papunen H, Idman H, Ilvonen E, Neuvonen KJ, Pihlaja P, Talvitie J (1977) Lapin ultramafiiteista (Summary: The ultramafics of Lapland). Geol Surv Fin Rep Invest 23

Papunen H, Häkli TA, Idman H (1979) Geological geochemical and mineralogical features of sulfide-bearing ultramafic rocks in Finland. Can Mineral 17:217–232

Ramdohr P (1967) A widespread mineral association, connected with serpentinization. Neues Jahrb Mineral Abh 117:241–265

Schreyer W, Ohnmacht W, Mannchen J (1972) Carbonate orthopyroxenites (sagvandites) from Troms northern Norway. Lithos 5:345–364

Simonen A (1971) Das finnische Grundgebirge. Geol Rundsch 60:1406–1421

Chromite and Graphite in the Platiniferous Dunite Pipes of the Eastern Bushveld

E.F. STUMPFL[1] and S.E. TISCHLER[1, 2]

Abstract

Merensky Reef and UG-2 horizon are the economically most important repositories of platinum group metals (PGM) in the Bushveld Complex. The platiniferous dunite pipes of the Eastern Bushveld, although not producing at present, are of considerable interest for our understanding of transport and deposition of PGM. Platinum mineralogy in the pipes and in the Reef differs distinctly; an attempt has now been made at identifying possible compositional differences in associated chromian spinels.

Two types of spinels occur in the pipes: homogeneous spinels plot in the high Fe/high Cr portion of the Mg × 100/Mg + Fe^2 versus Cr × 100/Cr + Al section of the spinel compositional prism. They thus occupy positions well outside the fields of cumulus Bushveld, Muskox and Stillwater spinels. In addition, there occur inhomogeneous spinels with embayment and resorption textures, and antipathetic Fe/Al variations. Clusters of graphite flakes are widespread in some dunites and may have had a catalytic effect on the distribution of PGM; they are attributed to the activity of late magmatic phases enriched in H_2O and CO_2.

1 Introduction

Platinum geology in the Bushveld Complex has, in recent years, been dominated by the considerable advances made in the investigation of the Merensky Reef and the UG-2 horizon. This has largely overshadowed the intriguing problems posed by the platiniferous dunite pipes in the eastern part of the Complex. In view of the present economic insignificance of these occurrences this may be understandable; the last of the once famous mines of Driekop, Onverwacht, and Mooihoek have ceased operations in the early 1960's. They carried some of the highest concentrations of platinum group elements (PGE) ever discovered; more than 2 kg PGE per ton of ore have been recorded in parts of the Onverwacht mine.

Geologically and mineralogically, however, these platiniferous pipes deserve considerable attention as they may well provide additional clues for our understanding of processes responsible for the concentration of PGE in general, and of the important "pothole" features of the Merensky Reef in particular. A comprehensive petrological

1 Institut für Mineralogie und Petrologie, Montan-Universität Leoben, 8700 Leoben, Austria

2 Present address: c/o Carpentaria Exploration Co., P.O.B. 113, Madang, Papua New Guinea

and ore-microscopic investigation of the pipes has been lacking so far; the best account is still to be found in Wagner's (1929) excellent book: there, 49 pages are devoted to the pipe-type deposits ("these are of exceptional interest", p. 50) and 99 pages to the Merensky Reef.

The presence, within the same igneous complex, of two distinctly different types of platinum deposits provides food for thought. The cross-cutting nature of the pipes, in relation to general Bushveld "stratigraphy", has been admirably illustrated by Wagner (1929). The distinct character of platinum group mineral (PGM) associations in the Reef and in the pipes is, by now, firmly established (Stumpfl 1962; Genkin et al. 1966; Tarkian and Stumpfl 1975; Stumpfl and Tarkian 1976; Cabri et al. 1977a–c). Braggite (Pt, Pd, Ni)S, cooperite (PtS), laurite (RuS_2), and sperrylite ($PtAs_2$) with platinum-iron alloy and minor Pt-Pd bismuthotellurides dominate in the Reef of the (economically most important) Rustenburg area. Pt-Fe and Pt-Fe-Cu-Ni alloys, sperrylite ($PtAs_2$), geversite ($PtSb_2$), genkinite $(Pt, Pd)_4Sb_3$ and Rh-Ir-Pt-sulpharsenides are the major PGM in the pipes. Further work may well reveal differences in PGM associations in "reefs" from different parts of the Complex; preliminary results from the UG-2 horizon in the Eastern Transvaal indicate a preponderance of rhodium minerals and arsenides, with Pt-Fe alloys, and tellurides (de Villiers, pers. comm.). This does, however, not affect the unique geological and mineralogical character of the pipes.

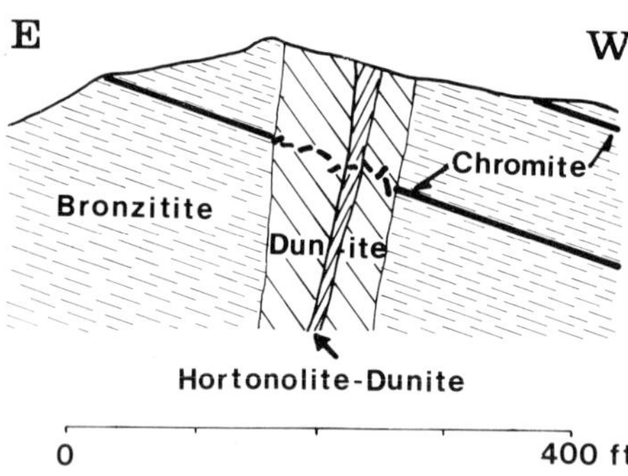

Fig. 1. Section through the Onverwacht pipe, Eastern Bushveld. Olivine in the hortonolite-dunite is Fo_{40-45}; it may be associated with Fe-rich amphibole and phlogopite

The latter are situated within the lower parts of the Critical Zone, with Mooihoek and Onverwacht actually cutting across the Main Chromitite Layer. They are not strictly cylindrical bodies, and do not continue at depth. At Onverwacht (Fig. 1) and Mooihoek, dominant rock types are olivine dunite and hortonolite dunite with hortonolite wehrlite. Mineralization at Driekop occurs in iron-rich olivine dunite and in "normal" olivine dunite. Other, nonplatiniferous pipes, some carrying magnetite or vermiculite, occur throughout much of the stratigraphy of the entire Bushveld Complex.

The Merensky Reef occupies the top of the Critical Zone; the latter has recently been investigated in considerable detail. For the "F" horizon, Flynn et al. (1978) have determined a last equilibrium at an f_{O_2} of $10^{-11.80 \pm 0.40}$ atm at a temperature of 1091 ± 35°C. This is considered a useful guide for crystallization conditions in this part of the Complex.

In view of the wealth of data now available on the petrology of the Merensky Reef, and on the PGM mineralogy of all respective deposits in the Complex, it appears necessary also to consider the mineralogy of oxides and silicates in the pipes. The results presented in this communication are a first step in this direction.

2 The Chromian Spinels

The most widespread oxide minerals in the ultramafic rocks of the pipes are chromian spinels of two major types – homogeneous and inhomogeneous spinels. The former constitute the majority of grains investigated; the latter, however, are considered to be of genetic importance far beyond their quantitative status.

2.1 Homogeneous Spinels

Even at highest magnification and in oil immersion these spinels do not show any sign of intergrowths, exsolution or alteration. They may well be compared, at least optically, with cumulus spinels which are important constituents of the Merensky Reef and of the associated chromite seams in the Critical Zone of the Bushveld Complex. Electron probe microanalyses of these spinels, however, reveal that their composition differs significantly not only from those previously analyzed in other parts of the Bushveld Complex, but also from spinels in the Muskox and Stillwater intrusions of North America. The most significant feature is their low magnesium versus high total iron and their comparatively high chromium content. The results of electron probe analyses of these spinels are summarized in Figs. 2 and 3. It is irrelevant which face of the spinel compositional prism is selected for representing the composition of homogeneous spinels from the pipe deposits – their difference from the other Bushveld spinels remains distinctly obvious. The genetic significance of these findings will be discussed later.

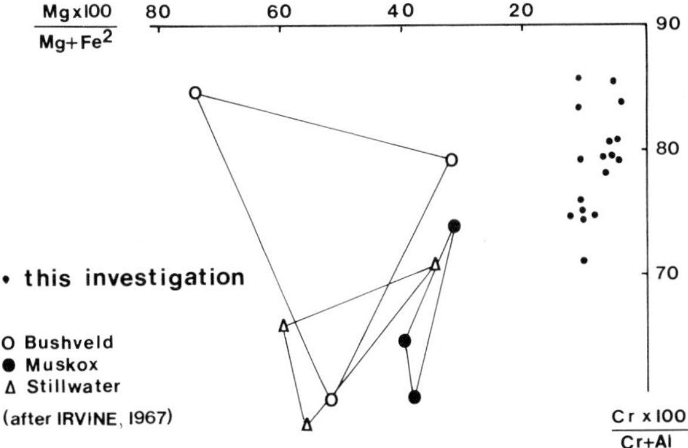

Fig. 2. Microprobe analyses of homogeneous chromian spinels from platiniferous pipes reveal Fe^2 contents which significantly exceed those of cumulate chromites

Irvine (1967) has shown that the MgO/FeO ratios of chromites are generally compatible with the composition of associated olivines and pyroxenes. According to Medaris (1975), spinel compositions from the Seiad Complex, California, can be plotted in a field contoured with respect to Fo-content of olivine.

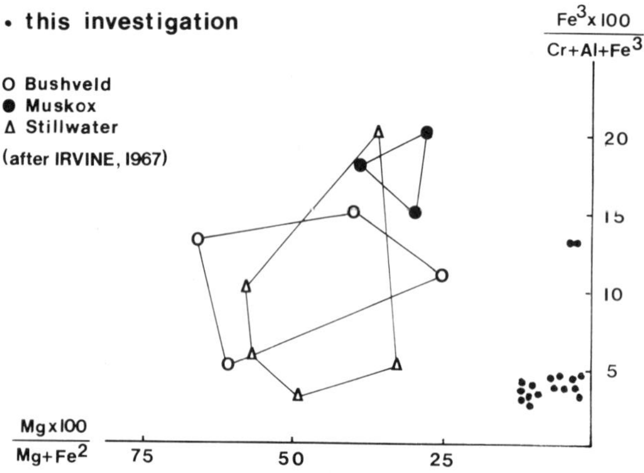

Fig. 3. Microprobe analyses of homogeneous chromian spinels from platiniferous pipes differ distinctly from those of cumulate chromites by higher Fe^2, Cr, and Al values

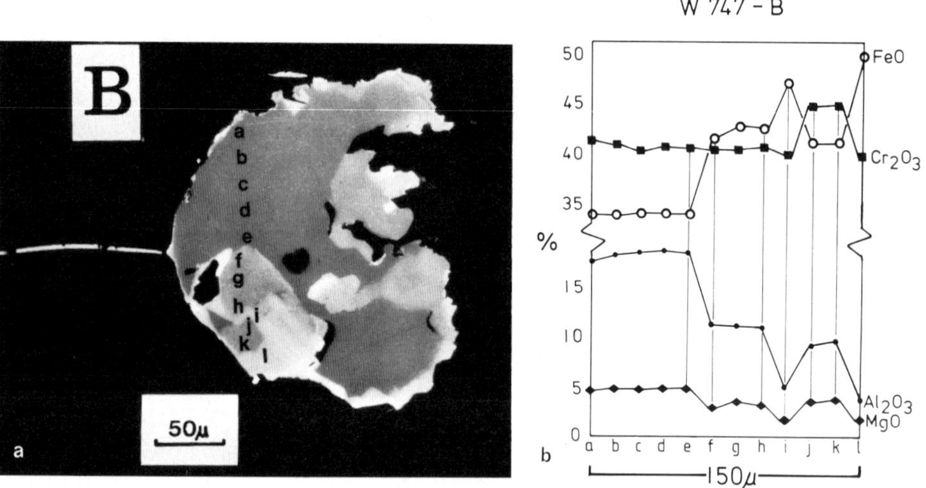

Fig. 4. Mooihoek Mine. Olivine dunite. Inhomogeneous chromite with embayment and resorption textures. Microprobe analysis reveals cryptic compositional variations in addition to those clearly linked to reflectance. Note antipathetic Fe/Al and limited Cr variation

The latter concept is only partially applicable to the pipes. Hortonolite (and, to a smaller extent, wehrlite) zones are widespread and their Fe-rich olivines ($\sim Fo_{40}$), and Fe-rich clinopyroxene (up to 9% FeO, = salite), respectively are correspondingly associated with Fe-rich chromites. However, coexisting forsteritic olivines ($Fo_{88\text{-}92}$) and Fe-rich spinels have also been observed. This is interpreted as the result of nonequilibrium conditions due to a complex interplay of late-magmatic processes. These considerations receive further support from the presence of inhomogeneous polyphase chromites, which are discussed in the next section.

Summarizing the above observations, compositionally homogeneous chromian spinels in the Driekop, Onverwacht, and Mooihoek pipes emerge as useful petrogenetic indicators. Their composition distinguishes them from spinels in the layered sequences of the Bushveld Complex, as well as from those of the Stillwater and Muskox layered intrusions, and even from the Fe-rich chromites of Duke Island, Alaska.

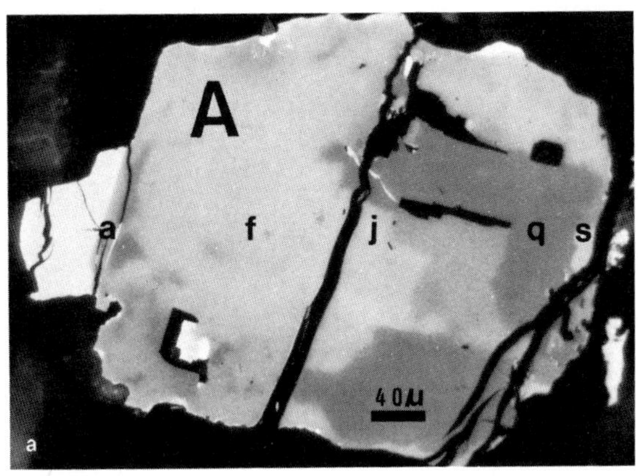

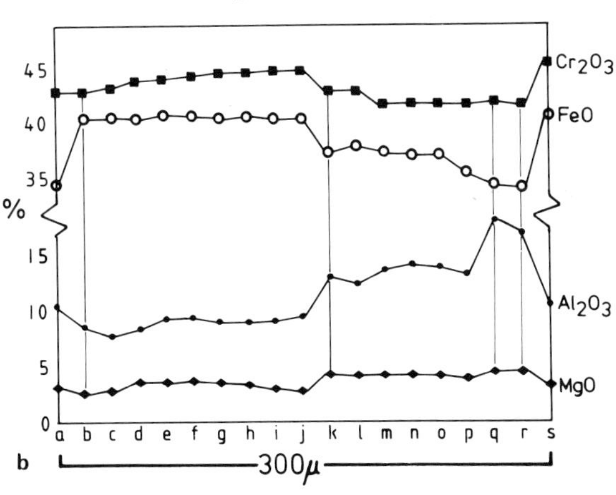

Fig. 5. Mooihoek Mine. Electron microprobe traverse across complex chromite grain. There is a pronounced Fe/Al relationship, which distinctly influences reflectance

2.2 Inhomogeneous Spinels

In addition to the optically and compositionally homogeneous spinels discussed above, complex polyphase spinels also occur. In reflected light and oil immersion up to five phases can be distinguished on the basis of reflectance. Optically, they occupy various positions between chromite (R_{589nm} = 12.1%) and magnetite (R_{589nm} = 20.3%).

These spinels reveal manifold and interesting textures; complex "embayment" and "corrosion" features are widespread. They suggest interaction of primary chromite with late-magmatic liquids (Figs. 4 and 5). It is worthy of note in this context that some chromites from the Merensky Reef also reveal ring-shaped corroded textures (Hallbauer, pers. comm.). This would provide further support for the concept of polyphase evolution of mineralization in the Reef as advocated by Stumpfl (1974).

Microprobe analyses of inhomogeneous chromites from the dunite pipes confirm the microscopic evidence, i.e. the presence of increasingly Fe-rich zones; it also reveals "cryptic" compositional variations (Figs. 4 and 5) which cannot be detected in reflected light.

The chromium content appears to remain more or less constant; major variations are evident in Al_2O_3 and FeO, with MgO confined to the $< 5\%$ level throughout. Quantitative measurements of reflectance show that highest reflectance values correspond with highest Fe/Al ratios (Fig. 6).

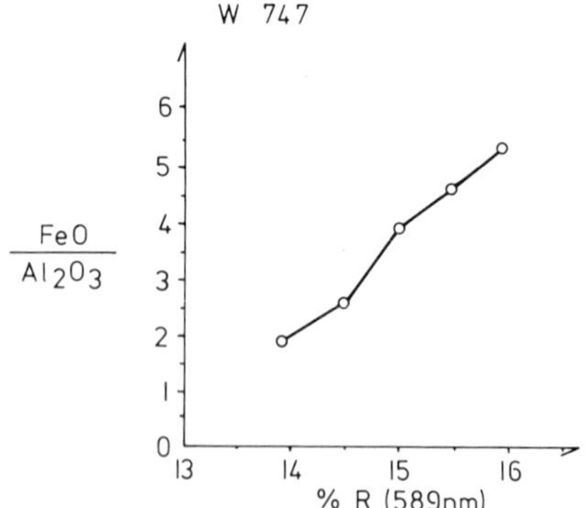

Fig. 6. Mooihoek Mine. Reflectance of phases in complex spinel grains is largely governed by the Fe/Al ratio

Compositional variations in chromian spinels have also been recorded from ultrabasic rocks which have been exposed to metamorphism and/or serpentinization. These are, however, preferentially concentric and show progressive alteration from chromite cores to ferrite-chromites and, finally, magnetite rims (for instance: El Ageed et al. 1980). This contrasts with the complex pattern observed in chromites from the dunite pipes.

3 Graphite

Graphite inclusions are widespread in some dunite pipes from the Eastern Transvaal, especially at Onverwacht. They occur preferentially as clusters and bands of 20–100 μ long flakes; in parts of certain polished sections the total graphite content may reach 10 vol.% of the bulk.

This observation is considered particularly significant for two reasons: firstly, the increasing recognition of carbon as a constituent of the mantle (Wyllie 1979) facilitates the interpretation of graphite in dunites as originally magmatic. Betechtin has recorded carbon-containing vapour phases from ultrabasic massifs in the Urals as long ago as 1961.

Secondly, the mode of transport is envisaged as CO_2, which very likely has been an integral constituent of the late-stage magmatic vapours and liquids instrumental in the formation of the pipes. The presence, in the dunite pipes, of major amounts of phlogopite and amphibole can be attributed to the H_2O content of these phases. The formation of the unusual PGM association of the pipes has previously been linked to processes other than magmatic (Stumpfl 1962). Graphite may well have acted as a catalyst for the precipitation of the PGM. So far, no intergrowths of PGM and graphite have been observed in polished section; such intergrowths, however, are not considered a necessary product of the catalytic activity of carbon.

It is relevant in this context that the occurrence of graphite has also been reported from the "potholes" of the Merensky Reef in the Western Bushveld; the latter are known to carry higher PGE values than the average Reef. A genetic connection between the activity of C-rich vapours and the precipitation of PGM thus seems almost inescapable.

Work performed so far shows that PGM frequently are located in intergranular spaces between the silicates, or even in locations resembling triple points. This also suggests a comparatively late deposition of the PGM in the pipes which, again, are late within the Bushveld Complex as a whole.

4 Conclusions

The position of the pipes as products of complex late-stage magmatic processes within the magmatic evolution of the Bushveld Complex can now be illustrated by five main factors:

1. The geological setting (cross-cutting, rootless "pips") has been outlined succinctly by various authors, including Wagner (1929) and Willemse (1967).
2. Compositions of homogeneous spinels differ distinctly from those encountered in "stratiform" chromites by low 100 Mg/Mg + Fe^{2+} ratios; corroded spinels with complex compositional variations also occur.
3. Iron-rich olivine ($\sim Fo_{40}$) dominates in hortonolite dunite; in "wehrlitic" portions it is associated with clinopyroxene (salite).
4. Graphite is a significant constituent of some dunites.

5. The platinum mineralogy is distinct and differs in both, elements and minerals present, from that of the Merensky Reef.

Cognizance must be taken of the fact that a great number of other pipes occur throughout the stratigraphy of the Bushveld Complex, perhaps with preferential concentration in the Critical Zone. Geological evidence has revealed all pipes to be rootless bodies of limited dimensions, usually not exceeding 200 m in depth, and 30–60 m in diameter. A convincing genetic concept will have to account for the characteristic features mentioned above under 1–5. Potential mantle tapping through major structures during the late stages of magmatic crystallization of the Complex has to be considered in this context. On the other hand, certain components, including H_2O and CO_2, may have been derived from sediments of the Transvaal system in the footwall of the Complex.

The work of Cameron and Desborough (1964) showed that the olivine dunite of the Onverwacht pipe had formed by replacement of pyroxenite by olivine, with water-rich, high-temperature fluids as agents. Evidence from platinum mineral associations in the Driekop pipe (Stumpfl 1974; Tarkian and Stumpfl 1975) points towards deposition from late-magmatic fluids. Von Gruenewaldt (1979) has recently stressed the significance of late solutions "that percolate upward through the pile of semiconsolidated cumulates" for concentrations of platinum-group elements in the iron-rich chromitite layers of the critical zone and in the Onverwacht pipe.

In view of the above evidence, the concept of replacement origin for the pipes, and of late-magmatic ("hydrothermal") solutions as agents for platinum mineralization not only in the pipes, but also in some stratiform chromitites, appears well established.

Acknowledgments. It is a pleasure to dedicate this contribution to Professor Paul Ramdohr on the occasion of his 90th birthday. Professor Ramdohr kindly had arranged for the senior author (E.F.S.) to accompany him on an extensive tour of ore deposits in Southern Africa in 1957, and thus generated a continuing interest in the geology and mineralogy of platinum deposits. This is gratefully acknowledged.

Sincere thanks are due to Prof. A. Reed for the gracious donation of the samples used in this study. The Provincial Government of Styria, Austria, generously financed the acquisition and installation of a fully automated ARL-SEMQ microprobe at the Institute of Mineralogy, Leoben.

References

Betechtin AG (1961) Mikroskopische Untersuchungen an Platinerzen aus dem Ural. Neues Jahrb Min Abh 97:1–34

Cabri LJ, Rosenzweig A, Pinch WW (1977a) Platinum-group minerals from Onverwacht. I. Pt-Fe-Cu-Ni alloys. Can Min 15:380–384

Cabri LJ, Laflamme JHG, Stewart JM (1977b) Platinum-group minerals from Onverwacht. II. Platarsite, a new sulfarsenide of platinum. Can Min 15:385–388

Cabri LJ, Stewart JM, Laflamme JHG, Szymanski JT (1977c) Platinum-group minerals from Onverwacht. III. Genkinite $(Pt, Pd)_4Sb_3$, a new mineral. Can Min 15:389–392

Cameron EN, Desborough GA (1964) Origin of certain magnetite-bearing pegmatites in the eastern part of the Bushveld Complex, South Africa. Econ Geol 59:197–225

El Ageed A, Saager R, Stumpfl EF (1980) Pre-Alpine ultramafic rocks in the Eastern Central Alps, Styria, Austria. In: Panayiotou A (ed) Ophiolites. Proc IAVCEI Ophiolite Symp, Nicosia, Cyprus, pp 601–606

Flynn RT, Ulmer GC, Sutphen CF (1978) Petrogenesis of the Eastern Bushveld Complex: Crystallization of the Middle Critical Zone. J Petrol 19:136–152

Genkin AD, Zhuravlev UN, Troneva NV, Muraveva IV (1966) Irarsite, a new sulfarsenide of iridium, rhodium, ruthenium and platinum. Zap Vses Mineral 95:700–712 (in Russian)

Gruenewaldt G von (1979) A review of some recent concepts of the Bushveld Complex, with particular reference to sulfide mineralization. Canad Mineral 17:233–256

Irvine TN (1967) Chromian spinel as a petrogenetic indicator. Pt. 2 Petrologic applications. Can J Earth Sci 4:71–103

Medaris LG (1975) Coexisting spinels and silicates in alpine peridotites of the granulite facies. Geochim Cosmochim Acta 39:947–958

Stumpfl EF (1962) Some aspects of the genesis of platinum deposits. Econ Geol 57:619–623

Stumpfl EF (1974) The genesis of platinum deposits: further thoughts. Miner Sci Eng 6:120–141

Stumpfl EF, Tarkian M (1976) Platinum genesis: new mineralogical evidence. Econ Geol 71:1451–1460

Tarkian M, Stumpfl EF (1975) Platinum mineralogy of the Driekop Mine, South Africa. Miner Deposita 10:71–85

Wagner PA (1929) Platinum deposits and mines of South Africa. Oliver & Boyd, Edinburgh, 326 pp

Willemse J (1967) The geology of the Bushveld Complex, the largest repository of magmatic ore deposits in the world. In: Wilson HDB (ed) Econ Geol , Monograph 4, Economic Geol Publ Co, pp 1–22

Wyllie PJ (1979) Petrogenesis and the physics of the Earth. In: Yoder HS (ed) The evolution of the igneous rocks: Fiftieth Anniversary Perspectives. Princeton Univ Press, Princeton, pp 483–520

Peridotites and Serpentinites of the East Alps and the Origin of Their Enclosed Ore Minerals

H.H. WEINKE[1] and H. WIESENEDER[2]

Abstract

Geology, petrology and geochemistry of selected samples of Alpine serpentinites and peridotites have been studied. Only serpentinites and related rocks of the Penninic zone and of the Unterostalpine zone, associated with country rocks of young Mesozoic age, can be considered as Alpine ophiolites. During serpentinisation a redistribution of iron, chromium and nickel took place and gave rise to the formation of secondary ore minerals.

1 Introduction

Serpentinites and peridotites occur in different tectonic units of the metamorphic zone of the East Alps. The Penninic zone of the West Alps is dipping below the East Alpine nappes and reappears in the Gargellen-, the Unterengadin-, the Tauern window, as well as in the window group of Rechnitz-Bernstein. Serpentinites, rodingites, metagabbros, greenschists, prasinites and metabasalts are associated with calcareous phyllites and micaschists of Upper Mesozoic age in the Penninic zone and also in the Unterostalpine "frame" of the Tauern window (Matreier zone, Tarntaler Mts.) in Austria. The ultramafititic-mafititic rocks of these zones are considered to be "ophiolites" according to the definition of the Penrose field conference of the G.S.A. 1972. But the term ophiolite must be used critically taking into account the complicated stratigraphy and structures of the metamorphic zones of the Alps. For instance, the Stubachtal peridotite complex and the Ochsner-Rotenkopf peridotite occurrence within the Tauern window do not belong to the Alpine ophiolite suite because the actual implacement of these complexes occurred in Paleozoic time.

1 Institut für Analytische Chemie der Universität Wien, Währinger Straße 38, 1090 Wien, Austria
2 Institut für Petrologie der Universität Wien, Dr. Karl Lueger Ring 1, 1010 Wien, Austria

2 Occurrences

Serpentinites of the Penninic zone and of the Unterostalpine frame.

Antigoritite, Schöneck, Glocknerstraße.

The samples studied are from a lense-shaped occurrence consisting predominantly of antigorite with subordinate actinolite, talc, dolomite, and xenomorphic ore minerals. According to thin section analyses chrysotile was the primary mineral and has been replaced during a second stage of metamorphism by antigorite (see Tables 1 and 2).

Antigoritites from the locality Jungfernsprung near Heiligenblut contain relict intergranular textures of antigorites and metamorphic olivines. The meaning of these textures is now discussed.

Earlier work on Stubachtal peridotite (Hohe Tauern) was done by Becke (1895), Weinschenk (1894), Cornelius and Clar (1939) and others. The largest of these lenses was investigated by Petrakakis (1977, 1978). The peridotite body is NW–SE oriented and has a length of 2 km and a width of 0.5 km. As a result of geochemical studies and mapping, the peridotite and the enveloping amphibolites are genetically related. The peridotite has a primary layering. The individual layers vary between 1 and 30 cm, and consist of dunites, wehrlites and clinopyroxenites. A cumulus process within the upper mantle seems to be the most probable explanation for the origin of the layering of this body.

The mineral associations were transformed during metamorphism into new ones, resulting in coexistence of antigorite and newly formed olivines. The angular interstices between antigorite plates are occupied by olivines. Rocks consisting of this mineral association were named "Stubachite" by Weinschenk (1894). Nickel in metamorphic olivines is lacking.

The occurrence of stubachite textures in Upper Mesozoic antigoritites (Jungfernsprung) indicates alpidic metamorphism already supposed by Cornelius and Clar (1939). During later metamorphic processes the metamorphic olivines became unstable and antigoritite-serpentinites are end products of the Alpine metamorphism.

The Ochsner-Rotenkopf ultramafititic massif forms the eastern end of a row of serpentinite lenses within the Greiner Schiefer of the Untere (Paleozoic) Schieferhülle. Between 2986 m and 3006 m large lenses within the serpentinites consisting of the wehrlitic source rocks were found by Koark (1950). These olivine-clinopyroxene rocks may be compared with the wehrlites of the Stubachtal complex showing a similar geological position. Though we have not yet studied these wehrlites geochemically, a similar origin and metamorphic development as proposed for the Stubachtal peridotite is probable.

Unterostalpine, "Matreier Schuppenzone". A report about petrographical field work in the Lasörling-Gruppe is given by Raith (1977). The rock association is very similar to that of the "Obere Schieferhülle". The original sequence of metasedimentary and metavolcanic rocks is not essentially disturbed by tectonic events. Numerous lense-shaped bodies arranged like a string of pearls occur within the same stratigraphic phyllite horizon. The largest of the serpentinite lenses is the Gößleswand, which has been studied in detail.

Table 1. Primary minerals

	Olivines					opx	cpx
Sample number	1	2	3	4	5	1	5
Number of analyses	3	11	5	6	17	8	15
SiO_2	39.34	41.36	41.49	39.93	40.24	57.43	54.05
TiO_2	–	–	–	–	–	–	0.05
Al_2O_3	–	–	–	–	–	0.42	0.75
Cr_2O_3	–	–	–	–	–	0.14	0.27
FeO	8.25	8.44	2.02	8.46	11.00	8.14	1.92
NiO	0.30	0.33	0.90	0.29	0.32	0.03	–
CoO	–	–	0.21	–	–	–	–
MnO	0.18	0.12	0.21	0.20	0.17	0.19	0.10
MgO	50.72	49.75	54.69	49.86	48.47	33.00	17.42
CaO	–	–	–	–	–	0.24	24.95
Na_2O	–	–	–	–	–	–	0.08
K_2O	–	–	–	–	–	–	–
	98.79	100.00	99.52	98.74	100.20	99.59	99.59

Sample numbers corresponding to localities in Tables 1–4:
1. Kraubath, Wintergraben, Styria. 2. Kraubath, Magnesithalde Wintergraben, Styria; 3. Kraubath, quarry, near Feistritz, Styria. 4. Hochgrößen, Steinkarlalm, Styria. 5. Stubachtal-complex, Tauern, Salzburg. 6. Schöneck, Glocknerstr., Carinthia. 7. Ochsenkogel, Gleinalpe, Styria. 8. Geierspitze, Tarntaler Mts., Tyrol. 9. Gößleswand, 2720 m, Matreier zone, Tyrol. 10. Übelbachgraben, Gleinalpe, Styria. 11. Pölling, Plankogel formation, Carinthia. 12. Lerchkogel, Styria. 13. Gößleswand basis, Tyrol. 14. Kraubath, Gulsen quarry, Styria. 15. Kraubath, Sommergraben, Styria

Table 2. Secondary minerals

	ol	Antigorites			Chrysotiles				Chlorites	
Sample number	5	6	5	7	8	9	3	10	9	8
Number of analyses	5	3	8	1	1	1	3	4	1	2
SiO_2	41.10	40.20	41.09	37.19	41.24	37.19	40.16	42.71	32.07	32.51
TiO_2	–	–	–	–	–	–	0.06	–	–	0.04
Al_2O_3	–	2.93	2.67	3.13	2.58	3.13	0.61	0.49	12.43	15.10
Cr_2O_3	–	0.58	0.56	0.12	0.27	0.12	0.28	0.08	0.66	0.07
FeO	4.76	6.65	5.40	5.69	5.56	5.69	4.68	1.04	5.93	3.74
NiO	–	0.07	0.10	0.16	0.15	0.16	0.20	0.37	0.36	0.19
CoO	–	–	–	–	–	–	0.08	–	0.08	–
MnO	0.26	0.06	0.04	0.08	0.04	0.08	0.06	0.03	0.06	0.08
MgO	52.95	34.21	37.11	38,74	35.35	38.74	39.35	41.10	36.04	32.17
CaO	–	0.04	–	–	0.14	–	0.07	0.01	0.07	–
Na_2O	–	–	–	0.35	–	–	–	–	–	–
K_2O	–	–	–	–	–	–	–	0.01		0.06
	99.07	84.74	86.97	85.46	85.33	85.11	85.55	85.84	87.70	83.96

The WSW–ENE stretched serpentinite body has approx. a length of 2 km and a width of 0.25 km. In contact with carbonate country rocks, the occurrence of tremolite-albite-epidote-actinolite rocks and albite-clinopyroxene rocks is observed.

Unterostalpine, Tarntaler Mts. The serpentinites of this zone are located in the upper Tarntal. The summits of the Lizumer Reckner and of the Narviser Reckner and of the Geier belong to the largest serpentinite body of this zone. It has a length in N–S direction of 1.4 km and a maximal width of 1.0 km. According to Enzenberg (1966) it is underlain and intercalated by Jurassic chertschists, forming a flat syncline. Samples from the serpentinite of the Geier summit and its surroundings are completely transformed to chrysotile mesh serpentinite.

According to Tollmann (1977) the "Mittelostalpin" is divided into two tectonic units of prealpidic origin and metamorphism: the Muriden and the Koriden. Both these units contain ultramafitites. The deeper Muriden nappe consists of lower grade metamorphic rocks like garnet micaschists, staurolite schists, gneisses, and marbles. The largest ultramafic complex of the East Alps, the peridotite mass of Kraubath, Styria, belongs to this unit. The lense shaped body extending NW–SE has a length of 14.5 km and a maximum width of 2.5 km and is fringed by amphibolites. It consists predominantly of dunites and to a lesser amount of orthopyroxenites (Angel 1964; Hiessleitner 1953; Meixner 1938).

The dunite of Hochgrößen, near Oppenberg, Styria, is situated in a different tectonic position (Wieseneder 1969). The folded and partly serpentinized body is fringed by eclogites. The latter is partly transformed to amphibolites. As a result of detailed studies it is supposed that the dunite of Hochgrößen is derived by tectonic movements from the Penninic zone below. It is comparable to the ultramafitites of the Habach formation (Hohe Tauern). The small antigoritite occurrence of Lerchkogel is considered to be in the same tectonic position as the dunite of the Hochgrößen. Stubachite textures are observed in both occurrences.

The Koriden unit consists of kyanite schists, gneisses, micaschists, amphibolites, and eclogites. Peridotites are bound to the "Plankogel Formation" within the Koriden unit. Micaschists and manganese-quartzites are characteristic rocks in this formation.

3 Conclusions

It is a remarkable fact that ultramafitites associated with metasediments of Upper Mesozoic age are more or less completely serpentinized. Ultramafitites of pre-Triassic age occurring in the Tauern window as well as in higher tectonic units, contain many relics of primary minerals, especially olivines, pyroxenes, chromites, and spinels. We suppose that this striking fact is caused by the emplacement of the Penninic ophiolites in water rich sediments. The emplacement of the prealpidic ultramafitites obviously took place in a drier, perhaps metamorphic, environment. Within the Tauern window the Mesozoic ultramafititic ophiolites are represented by antigoritites whose primary textures are completely extinct. In contrast the Unterengadin window and the Rechnitz-Bernstein window contain chrysotile-lizardite serpentinites. The primary textures in these rocks are completely preserved. Higher temperatures in the Tauern area during metamorphism may be the cause for the regional development of antigorite. The preserved primary textures in chrysotile serpentinites point to a metamorphism at constant volume.

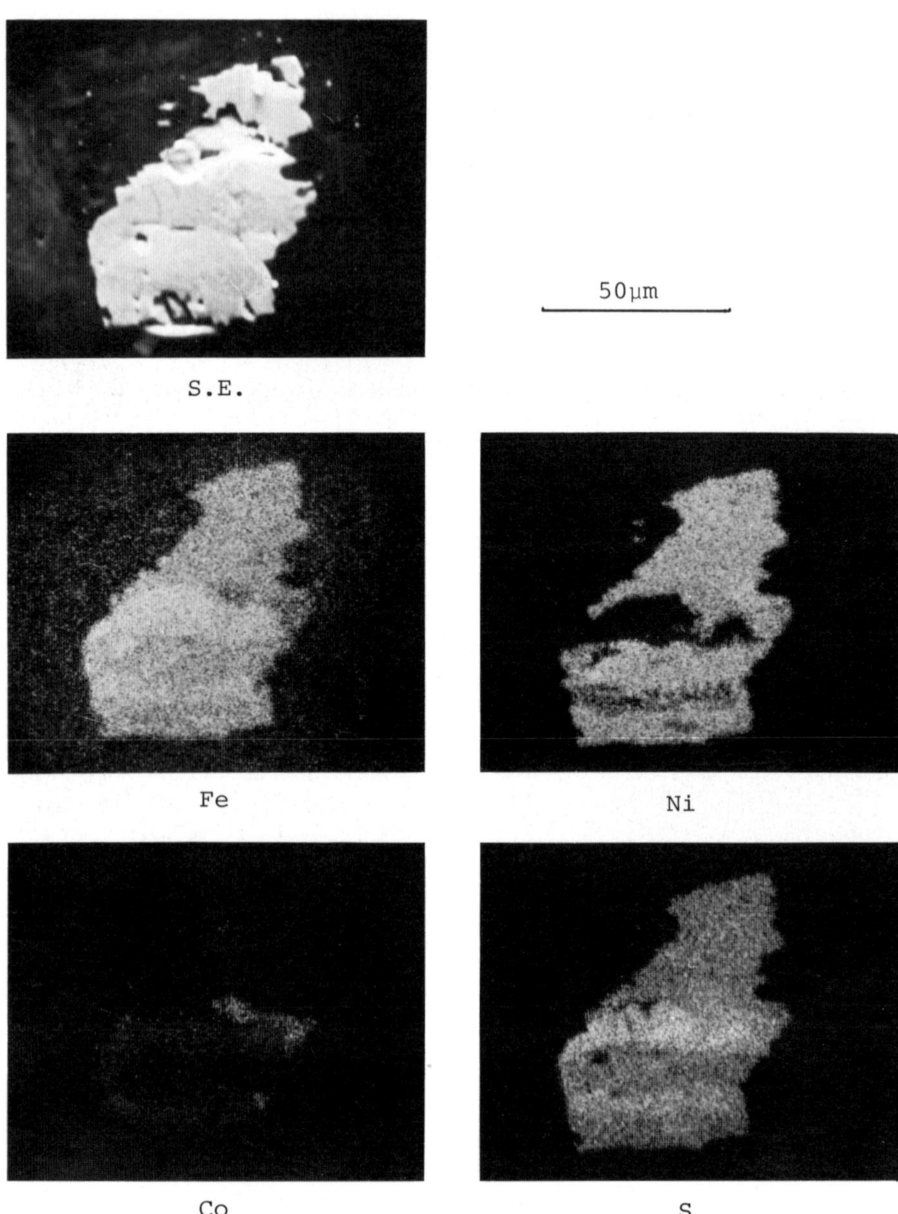

Fig. 1. Shape and elemental distribution of pentlandite from Geierspitze, Tyrol, Austria

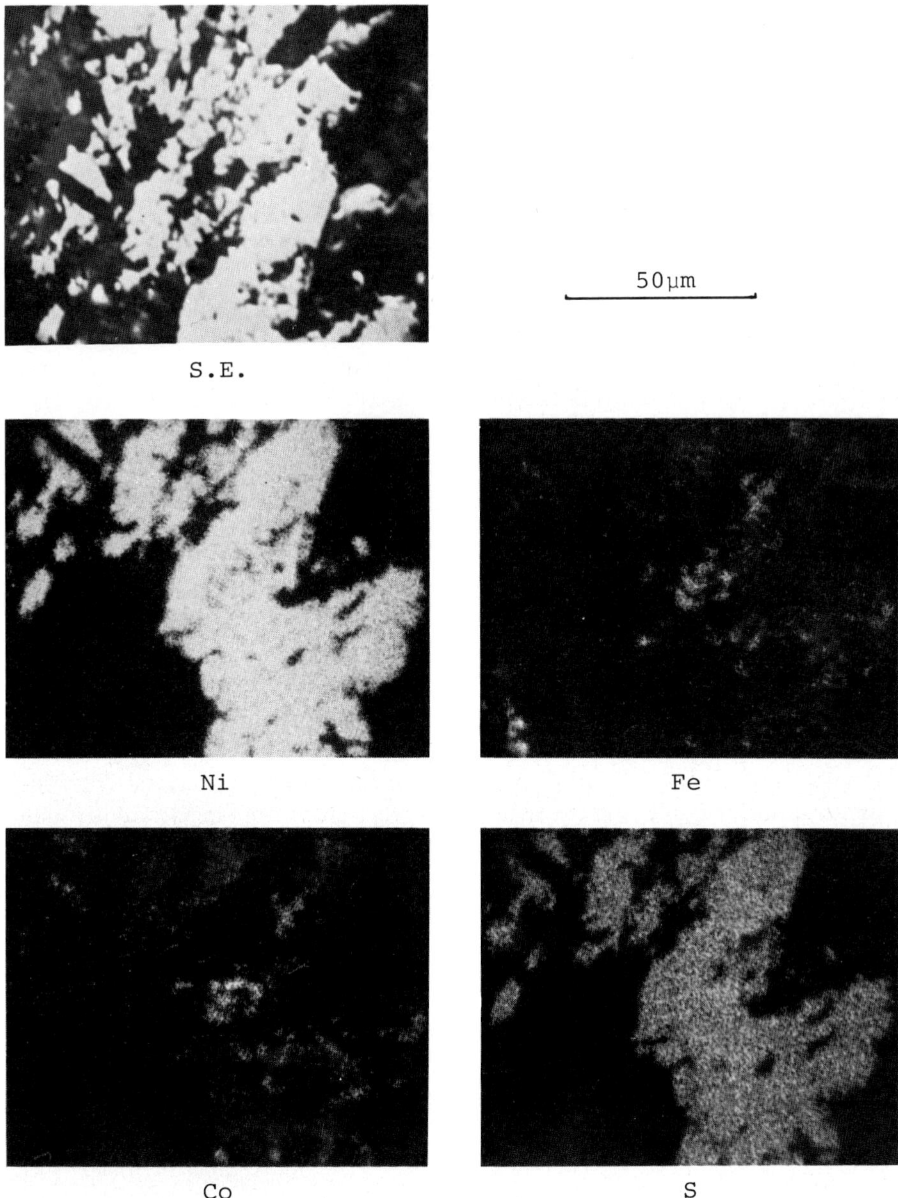

Fig. 2. Shape and elemental distribution of heazlewoodite from Lerchkogel, Styria, Austria

4 Ore Minerals

Microscopic studies (Ramdohr 1952, 1967) and microprobe analyses revealed the fact that nickel sulphides occur widely distributed in serpentinites. However, these minerals are rare in peridotites. Pentlandite is the most abundant sulphide mineral in the rocks studied. Irregular or elongated grains of a length of 5–50 μm are characteristic (Fig. 1). Some of the grains are rich in cobalt. Heazlewoodite, already known from a serpentinite from Friesach, Carinthia, has been confirmed in the Penninic zone and the Muriden nappe. The mineral occurs in groups of crooked rod-shaped or irregular grains up to 100 μm (Fig. 2). Samples of Pölling (Plankogel Formation) contain heazlewoodite with sickle shaped patches of pentlandite reaching a Co content up to 53 weight-%. A mineral of the stoichiometric composition of $(Ni_{3.52}Co_{0.40}Fe_{0.08})$ $(As_{2.93}S_{0.07})$ seems to correspond to orcelite. Irregular grains of violarite with a maximum length of 50–70 μm were found in Gößleswand and Geierspitze. Pyrite with appreciable Ni and Co content occurs in a sample from the Geierspitze, Tarntaler Mts. Chalcopyrite is confirmed from a sample of Geierspitze and Gößleswand. Pyrrhotite in more or less rounded grains of 20–50 μm occurs in serpentinites of Schöneck, Glocknerstraße (Tables 3 and 4).

Magnetite is the most abundant opaque mineral in serpentinites and occurs in irregular shaped grains up to 2 mm in diameter. Grains approximately matching the ideal chemical composition are considered as newly formed minerals during serpentinization. A second group is characterized by Cr_2O_3 14–22 %, MgO 0.3–1.5%, TiO_2 0.2–1.6% and Al_2O_3 0.2–2%. From microprobe studies it seems very probable that these minerals were formed during metamorphism. Pure chromites are observed only in samples from Kraubath and Hochgrößen. Ilmenite has been observed only in a sample from Schöneck, Glocknerstraße. The rounded grains have diameters between 20 and 40 μm.

The frequent distribution of nickel minerals in Alpine serpentinites was an interesting observation during this study. For heazlewoodite, orcelite and violarite we have little doubt that these minerals crystallized during serpentinization. On the other hand, the problem of the origin of pentlandite is more complicated. From general considerations it seems probable that this mineral is a relict phase from the peridotitic source rocks. However the observed distribution does not confirm this opinion. As a consequence of the limited number of samples analyzed further investigations are necessary to solve this problem. The same is true concerning the occurrence of pyrrhotite, chalcopyrite and pyrite. The formation of pure magnetites is clearly to be seen from ore microscopic studies. The iron of the secondary magnetites is derived mainly from serpentinized olivines. Chromium is secondarily concentrated in serpentine minerals and chlorites. Dissolution textures have not been observed in the ore minerals studied.

Acknowledgements. We thank the Fonds zur Förderung der wissenschaftlichen Forschung for making available an electron microprobe ARL-SEMQ.

Table 3. Selected analyses of sulphides and arsenides (in wt.%)

	Pentlandite				Pyrite	Pyr-rhotite	Heazlewoodite			Chalcopyrite		Violarite		Orcelite
Sample number	4	8	9	11	8	6	6	11	12	8	13	8	9	11
Number of analyses	19	3	71	5	4	18	9	10	19	3	3	5	6	23
S	33.10	33.14	32.85	32.49	53.01	38.51	26.88	26.54	26.52	34.62	34.69	41.28	42.70	0.40
Fe	35.20	7.08	25.32	1.97	43.33	58.71	0.65	1.62	0.70	29.75	27.22	14.54	18.12	0.92
Co	1.39	0.02	0.62	52.95	2.43	0.11	0.50	0.11	0.75	0.05	0.18	10.39	4.30	5.17
Ni	30.05	60.69	41.00	12.09	0.86	2.68	72.22	72.05	71.10	0.01	6.49	33.05	35.84	45.17
Cu	–	–	–	–	–	–	–	–	–	34.50	31.07	–	–	–
As	–	–	–	–	–	–	–	–	–	–	–	–	–	48.10
	99.74	100.93	99.79	99.50	99.63	100.01	100.25	100.32	99.07	98.93	99.65	99.26	100.96	99.76

Table 4. Selected analyses of magnetite, spinel, and ilmenite (in wt.%)

	Magnetite			Chromian spinel					Ilmenite
Sample number	4	6	13	9	6	13	14	15	6
Number of analyses	19	5	5	29	3	23	30	22	18
MgO	0.57	0.39	0.43	0.48	0.79	4.41	4.78	3.10	0.79
Al_2O_3	0.23	0.61	1.72	1.96	2.67	8.40	13.36	8.04	0.26
TiO_2	0.40	0.46	0.52	0.50	1.28	0.39	1.02	1.95	52.18
Cr_2O_3	0.21	3.41	16.77	19.99	29.51	52.31	46.07	49.73	0.25
Fe_{tot}	71.00	69.00	56.96	54.88	48.02	25.53	26.18	27.74	35.81
FeO [a]	30.44	29.90	30.22	30.43	30.88	25.53	26.45	26.76	46.05
Fe_2O_3 [a]	67.69	65.42	47.85	44.63	34.34	8.11	8.02	9.92	–
	99.54	100.19	97.51	97.99	99.47	99.15	99.70	99.50	99.53

[a] Stoichiometric calculation

References

Angel F (1964) Petrographische Studien an der Ultramafitit-Masse von Kraubath (Steiermark). Joanneum Mineral Mitteilungsbl 2:1–125

Becke F (1895) Olivinfels und Antigorit-Serpentin aus dem Stubachthal (Hohe Tauern). Tschermaks Mineral Petrogr Mitt 14:271–276

Cornelius HP, Clar E (1939) Geologie des Großglocknergebietes (I. Teil). Abh Reichs Bodenforsch Zweigstelle Wien 25:1–305

Enzenberg M (1966) Die Geologie der Tarntaler Berge. Mitt Ges Geol Bergbaustud Wien 17:5–50

Hiessleitner G (1953) Der magmatische Schichtbau des Kraubather chromerzführenden Peridotitmassives. Fortschr Mineral 32:75–78

Koark HJ (1950) Die Serpentine des Ochsners (Zillertal) und des Reckners (Tarntal) als Beispiele polymetamorpher Fazies verschiedener geologischer Stellung. Neues Jahrb Mineral Abh 81: 399–476

Meixner H (1938) Kraubather Lagerstättenkunde I. Zbl Miner Geol Paläont, Abt A: Mineralogie und Petrographie, p 5–19 (further literature see Angel 1964)

Moody JB (1976) Serpentinisation: a review. Lithos 9:123–138

Petrakakis K (1977) Zur Geologie des Stubachtalultramafitit-Komplexes. Mitt Ges Geol Bergbaustud Oesterr 24:47–57

Petrakakis K (1978) Der Stubachtal-Ultramafitit-Komplex (Salzburg, Österreich). Tschermaks Mineral Petrogr Mitt 25:1–32

Raith M (1977) Feldpetrographischer Kartierungskurs in der Lasörlinggruppe (Neue Reichenberger Hütte), Österreich. Mineral Petrogr Inst, Univ Kiel, BRD, multiplied manuscript

Ramdohr P (1952) Über Josephinit, Awaruit, Souesit, ihre Eigenschaften, Entstehung und Paragenesis. Mineral Mag 29:374–394

Ramdohr P (1967) A widespread mineral association, connected with serpentinization. Neues Jahrb Mineral Abh 107:241–265

Tollmann A (1977) Geologie von Österreich, Bd. 1. Die Zentralalpen. Deuticke, Wien, 766 S

Weinschenk E (1894) Beiträge zur Petrographie der östlichen Centralalpen, speciell des Gross-Venedigerstockes. Bayer Akad Wiss Abh II Cl 18, Abth III, 63 S

Wieseneder H (1969) Der Eklogitamphibolit vom Hochgrößen, Steiermark. Joanneum, Mineral Mitteilungsbl 1/2:12–20

Genesis of Ores in Metamorphic Rocks

Strata-Bound Sulphides in the Early Archaean Isua Supracrustal Belt, West Greenland

P.W.U. APPEL [1]

Abstract

In the ca. 3.8 b.y.-old Isua supracrustal belt, West Greenland, pyrrhotite and chalcopyrite occur as strata-bound mineralizations in layered tuffaceous amphibolites and in an iron-formation. Mineralizations with more complex mineral assemblages comprising galena and sulphides of Ni, Sb, Cd, and Ag are of limited extent only. Field relationships, rare-earth elements and isotopic data indicate that the sulphides are partly of submarine exhalative origin, and partly derived through leaching of basaltic rocks. The sulphides were precipitated on the sea-floor without influence of sulphur bacteria. The mineral zonation was governed by differences in Eh and pH in the depositional environment.

1 Introduction

The Isua supracrustals are exposed in a ca. 30-km-long and up to 4-km-wide arcuate belt, situated ca. 150 km north-east of Godthaab, W. Greenland (Fig. 1). Age determinations on a major iron-formation from within the belt gave an age of 3760 ± 70 m.y. (Moorbath et al. 1973). Field relationships and radiometric work on the surrounding Amitsoq gneisses showed that the supracrustals are older than the gneisses. The Isua rocks are thus the oldest terrestrial rocks known so far.

The geology of the supracrustals has been described by Allaart (1976) and Appel (1979a, 1980). The sulphide mineralizations were discovered in 1976 and further investigated in 1977 and 1978 as part of a research programme carried out at the Max-Planck-Institut für Chemie, Mainz, FRG.

2 Geology and Mineralogy of the Sulphide Mineralizations

The sulphides in the Isua supracrustal belt occur in three different environments, namely: in the banded iron-formation, in layered tuffaceous amphibolites and in cross-cutting veins and veinlets. The three types are presumably cogenetic, but for the sake of clarity their geology and mineralogy will be described individually.

1 Institute of General Geology, Østervoldgade 10, 1350 Copenhagen, Denmark

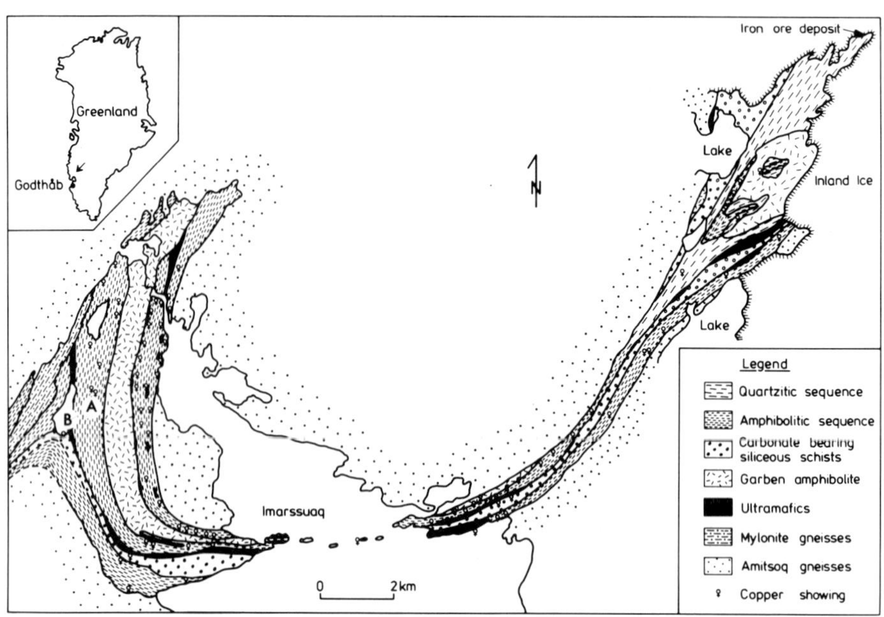

Fig. 1. Simplified geological map of the Isua supracrustal belt. (Slightly modified from Allaart 1976 and pers. comm.)

The sulphides in the iron-formation appear to be facies-dependent (Appel 1979a, 1980), not only as regards to amount of sulphides but also to kind of sulphides. In oxide facies iron-formation pyrite is the sole sulphide. It occurs mainly as subhedral to euhedral grains scattered in quartz and magnetite layers. The size of the crystals ranges up to 1 cm.

Carbonate and silicate facies iron-formation have generally low sulphide content, rarely exceeding 1%–2%. The sulphides are pyrrhotite and chalcopyrite, and the former is dominant. The two sulphides occur mainly as mm-sized anhedral scattered grains. Sulphide facies is not abundant. It consists of up to 60% sulphides, together with varying amounts of grunerite or actinolite and magnetite. The sulphides are chalcopyrite and pyrrhotite, the former usually dominating.

The sulphides in the iron-formation are rarely mobilized during metamorphism, and no replacement of pyrite by pyrrhotite or vice versa has been observed. The preference for pyrite to occur in oxide facies and pyrrhotite and chalcopyrite to occur in the more reducing facies of iron-formation seems therefore to be primary depositional and not a metamorphic phenomenon.

The sulphides in the layered tuffaceous amphibolites occur as mm-sized, mostly anhedral grains, arranged in layers parallel to the bedding of the amphibolites. The individual sulphide layers are normally only a few mm thick, but frequently occur in large numbers across strike. The sulphides are furthermore found as scattered grains throughout the amphibolites. The amount of sulphides can be as high as 10 vol.%, but is normally restricted to a few percent. The layered amphibolites consist of hornblende, intermediate plagioclase and quartz.

The sulphide mineralized amphibolites can be divided into different groups according to mineral assemblage. The most common type characteristically carries iron and copper sulphides only, whereas more complex sulphide parageneses are of limited extent. The iron and copper mineralized amphibolites can be traced throughout the supracrustal belt (Fig. 1). Pyrrhotite and chalcopyrite are always present, but in different proportions. Cubanite is locally abundant.

As a contrast to the rather monotonous iron-copper mineralized layered amphibolites, two horizons in the western Isua belt exhibit a rather complex opaque mineral assemblage (A and B in Fig. 1). The dominating sulphides are still pyrrhotite and chalcopyrite, but in addition galena and a number of complex sulphides occur.

At locality A, two parallel horizons, a couple of metres apart, and traceable for ca. 150 m, are 0.5 to 1 m thick and consist of dense, dark green, slightly layered amphibolite. Sulphides normally make up 1–5 vol.%, but locally more than 60 vol.%. Galena can amount to more than 5%, but is usually less. The fine-grained amphibolite is composed mainly of hornblende, plagioclase and quartz, with minor amounts of biotite and epidote, and occasionally thin carbonate-bearing bands. The sulphides occur as anhedral grains disseminated throughout the rock, with a tendency to form thin, discordant veinlets, rarely exceeding several mm in width and several cm in length.

The opaque minerals from locality A exhibit a rather complex paragenesis. The two dominating sulphides are pyrrhotite and chalcopyrite. Galena occurs as euhedral to anhedral grains. Enclosed or associated with galena are found a number of Ni, Sb, Cd and Ag sulphides of which ullmannite is the most common. Ullmannite occurs up to 0.02 mm as large, mostly euhedral grains, often with tiny inclusions of greenockite, acanthite and a silver-bearing tetraedrite (freibergite?). Euhedral grains of breithauptite and tiny irregular grains of native antimony and dyscrasite are found as inclusions in galena. Exsolutions of a silver-bearing bravoite are found, partly enclosed in chalcopyrite. The galena-bearing amphibolite at loc. B (Fig. 1) is medium-grained, well-layered, with up to 10 vol.% sulphides. The amphibolite is composed of hornblende, quartz and plagioclase. The quartz content is slightly higher in this amphibolite than the amphibolite at locality A. The sulphides at loc. B have a grain size averaging 0.2 mm, and tend to occur as thin parallel layers. The opaque mineral assemblage is simple compared to loc. A. The principal sulphides are pyrrhotite and chalcopyrite, with minor amounts of galena and sphalerite, but with no traces of Sb, Cd and Ag sulphides, instead there is abundant linnaeite. The linnaeite occurs as sub- to anhedral grains, enclosed within or associated with pyrrhotite. Semiquantitative analyses indicate that it is a siegenite, with ca. 10% iron.

The amphibolite at loc. B is cut by a shear zone, up to 5 m wide, characterized by a bright green, closely jointed fuchsite-bearing quartzite. Several joints contain galena with minor sphalerite, with a grain size up to 2 mm. The shear zone further contains several discordant quartz veins, up to 0.5 m wide, carrying small amounts of chalcopyrite, pyrrhotite and galena.

3 Chemistry

The major element composition of the iron-formation and of some layered amphibolites has been published elsewhere (Appel 1979a, 1980). The copper content and the trace element composition of two sulphide-silicate facies iron-formations, two weakly mineralized layered amphibolites and one chalcopyrite vein is shown in Table 1. Sample (1) is sulphide-silicate facies iron-formation from W. Isua consisting of chalcopyrite, magnetite and actinolite. Sample (2) is a sulphide-silicate facies iron-formation from E. Isua, consisting of pyrrhotite, chalcopyrite, grunerite and hornblende with minor amounts of magnetite. Samples (3) and (4) are layered amphibolites with thin sulphide layers, from eastern and western Isua respectively. The chalcopyrite vein (5) is adjacent to sample site (4).

The relatively high gold contents in the iron-formation are interesting. This appears to be a common phenomenon of sulphide facies iron-formations of the early Precambrian (Fripp 1976), which is now extended into the early Archaean. The low zinc contents indicate a very efficient geochemical separation of zinc and copper in the depositional environment. The high nickel content in sample (1) is remarkable. Nickel is not normally found in such quantities in banded iron-formations. Preliminary investigations of nickel contents in the Isua iron-formation also indicate an average level on ca. 50 ppm Ni. The high nickel anomaly displayed by sample (1) is attributed to influx of meteoritic material at the time of deposition (Appel 1979b). The rare-earth element (REE) content in the samples has been normalized to carbonaceous chondrite (Cl) (Appel 1979a), and the results are shown in Fig. 2. Only a few REE could be measured in sample (5), so this sample is omitted from Fig. 2. One of the remarkable features shown in Fig. 2 is that one sample of iron-formation (1) has a pronounced negative

Table 1. Copper contents and trace element composition of two sulphide-silicate facies iron-formations (1) and (2), two weakly mineralized tuffaceous amphibolites (3) and (4), and one discordant chalcopyrite vein (5). Neutron activation analyses by B. Spettel (Max-Planck-Institut, Mainz)

Sample no.	(1)	(2)	(3)	(4)	(5)
Cu %	3.15	2.08	0.30	0.25	26.85
ppm					
Zn	110	230			305
Ni	790			35	
Co	83	52	47	32	6
Ag	13	16		< 5	122
Au	0.076	0.25	0.055	0.005	0.0025
La	19.3	2.68	5.23	21.3	0.20
Ce	36.1	6.6	13.1	60.0	
Nd	20			43	
Sm	2.93	1.08	3.50	11.9	0.23
Eu	0.30	0.32	1.12	2.09	
Tb	0.58		0.78	2.09	
Dy	3.4	1.1	4.9	11.1	
Yb	1.57	0.71	2.46	6.63	0.083
Lu	0.9	0.11	0.34	0.97	0.022

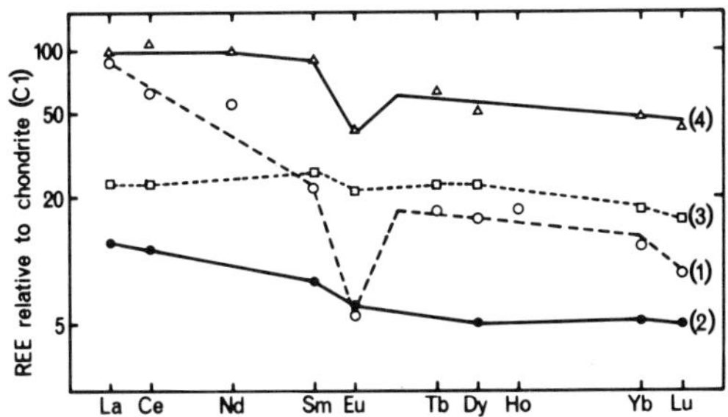

Fig. 2. Chondrite *(C1)* normalized rare-earth elements *(REE)* in two sulphide-silicate facies iron-formation samples and two slightly mineralized amphibolites. Sample numbers are the same as in Table 1

europium (Eu) anomaly, whereas the other (2) has none. The REE are generally trivalent in natural systems, and show very similar geochemical behaviour. Europium, however, can be divalent and will then behave differently, thus being separated from the other REE. Previous investigations of REE distributions in Precambrian iron-formations have shown a tendency towards negative Eu anomalies in pre-2 b.y.-old iron-formations, and none or positive anomalies in younger iron-formations (Fryer 1977a,b; Laajoki 1975). The suggested explanation was that Eu occurred as a divalent ion as a result of the reducing conditions in the atmosphere and oceans prevailing during most of the early Precambrian. After ca. 2 b.y. ago, when free oxygen was available, Eu was oxidized and ceased to be anomalous (Fryer 1977a,b; Laajoki 1975). The REE trend of sample (2) is clearly inconsistent with this theory, whereas the negative Eu anomaly in sample (1) could be taken as a support of the theory. There is, however, a serious problem in explaining negative Eu anomalies as a result of reducing conditions in the oceans. The redox potential (Eh) for the reaction $Eu^{2+} \rightleftharpoons Eu^{3+} + e^-$ is -0.43 volts, which is well below the stability of seawater at pH below 7. A Eu^{2+} ion in seawater (pH < 7) will thus reduce the water and form free hydrogen, itself being oxidized to Eu^{3+}. Furthermore, abundant ferrous and ferric ions were present in the depositional environment. The redox potential for the reaction $Fe^{2+} \rightleftharpoons Fe^{3+} + e^-$ is $+0.77$ volts. This redox potential is of course pH-dependent, but will always be well above Eh = -0.43 volts up to pH 12. As a result, all Eu will be trivalent in aqueous solutions, and will behave like the other REE. We can therefore not expect any fractionation of Eu relative to the other rare-earth elements in a seawater environment. The REE content in iron-formations must thus reflect the REE content in the source of the iron-formation. The different chondrite normalized REE patterns of the two iron-formation patterns [(1) and (2) in Fig. 1] could indicate a different pedigree. The implications will be discussed below.

Lead and sulphur isotopic ratios have been measured in sulphides from the iron-formation and from the layered amphibolites. The results have been published elsewhere (Appel et al. 1978; Monster et al. 1979), so they will only briefly be reported here. The galenas are all very unradiogenic, and a galena from the shear zone has $^{206}Pb/^{204}Pb = 11.151$ and $^{207}Pb/^{204}Pb = 13.037$, which is the least radiogenic terrestrial lead isotopic composition found so far. This gives a model age of 3740 m.y. The sulphur isotopic ratio in iron-formation sulphides as well as in sulphides in layered amphibolites is remarkably constant with an average $\delta\,^{34}S = +\,0.45$ ‰ and $\delta\,^{34}S = +\,0.27$ ‰ respectively (Monster et al. 1979). This is very close to the mantle ratio, and indicates that no bacterial fractionation took place during deposition of the sulphides.

4 Discussion

The genesis of strata-bound sulphide mineralizations in greenstone belts is controversial, whether of epigenetic or syngenetic origin. In the Isua supracrustals field relationships and geochemical characteristics show that the sulphide mineralizations are spatially related and presumably cogenetic with the banded iron-formation. The sulphides in the layered tuffaceous amphibolites and in the iron-formation must have been precipitated contemporaneously with the iron-formation. An epigenetic origin can thus be rejected.

The next question concerns the origin of the sulphides. The most likely sources are (A) from weathering of a continent, (B) from leaching of sea-floor basalts and (C) from submarine exhalations. Chemical weathering on a continent can produce soluble copper compounds, which then can be transported by streams to the sea. This weathering, however, requires the presence of free oxygen, and we have strong indications that the early Precambrian atmosphere was devoid of free oxygen (Schidlowski 1978). We have furthermore indications, that if there was a pre-Isua continent at all, it must have been of very limited extent (Appel 1980). The possibility that the sulphides were derived from weathering of a continent can therefore be dismissed.

The sulphides were thus precipitated on the sea-floor from brines, but the origin of these brines is quite difficult to prove. Seawater, or more likely a seawater hydrothermal water mixture, circulating in a pile of basaltic rocks below the sea-floor is able to leach considerable amounts of metals from the basalts. When the metal-enriched water ascends, most of the dissolved material will be precipitated, either in the uppermost part of the volcanic pile or on the sea-floor. Metal-rich volcanic brines coming directly from submarine exhalations will likewise precipitate the dissolved material on the sea-floor. The precipitates originating from either leaching or exhalations will in most respects be indistinguishable.

However, the rare-earth elements (REE) seem to hold a key to distinguish whether a metalliferous sediment was derived through leaching of basalts or from submarine exhalations. It was shown above that no apparent fractionation of the REE took place in the depositional environment, and that the relative abundances of the REE in the metalliferous sediment are inherited from the source rock or magma. All REE are incompatible elements in basaltic rocks, except for europium (Eu), which is built into plagioclase. A partial leaching of a basalt by circulating hot water will leach all REE in equal proportions apart from Eu. A metalliferous sediment originating from leaching of basalts will thus have a REE pattern resembling that of the basalt, but with a pronounced negative Eu anomaly, as seen in recent metalliferous sediments on the ocean floor (Piper et al. 1975). A metalliferous sediment originating from submarine exhalations can be expected to have a REE pattern similar to that of basalts, maybe with a slight enrichment of light REE. The REE in the Isua metalliferous sediments (Fig. 2), as well as other samples of Isua iron-formation (Appel 1980), show that the samples have a REE trend resembling that of basaltic rocks, but that some have negative Eu anomalies, whereas others have none. Some of the mineralized horizons owe their metal content mainly to leaching [e.g., sample (1) in Fig. 2], whereas others [sample (2) in Fig. 2] mainly originated from submarine exhalations. However, in most cases the origin of the brines is unknown, although they were probably derived from submarine exhalations and/or from leaching. The next problem concerns the sulphide zonation. Here we observe two types. Pyrite occurs in oxide facies and pyrrhotite and chalcopyrite occur in the more reducing facies of the iron-formation, and no indications have been found that pyrite or pyrrhotite are metamorphic replacement products. This distribution pattern of sulphides is in good agreement with geochemical calculations by Garrels (1960), Tischendorf and Ungethüm (1965) and Large (1977) on the stability fields of copper and iron compounds in an aqueous environment. Thus it is likely that the pyrite-pyrrhotite zonation in the Isua iron-formation is primary and not metamorphic.

Another type of sulphide zonation is seen in the tuffaceous amphibolites, where iron and copper sulphides are the only sulphides except for a few horizons where galena and more complex sulphides occur. The main factors controlling the composition of strata-bound sulphide mineralizations are the composition of the brines that supplied the solutions, and secondly the chemistry of the depositional environment. Differences in brine composition could be due to differentiation in the magma that supplied the brines (Spence and de Rosen-Spence 1975). This implies that the copper-iron sulphide mineralizations were precipitated at an early stage of basaltic volcanism. A slight differentiation of the basaltic magma could produce brines slightly enriched in lead and later in antimony, cadmium and silver. A further differentiation would yield metalliferous sediments dominated by galena and sphalerite. These are not found at Isua. The reason is that we either never reached such an advanced state of differentiation or that galena and sphalerite mineralizations were deposited elsewhere in the previously much more extensive Isua belt. At the present time none of these explanations can be excluded.

Another, and perhaps more likely explanation for the metal zonation is based on differences in the depositional environment. The most important factors are Eh, pH, temperature and degree of dilution of the brines. Based on geochemical calculations by Garrels and Christ (1965) and Large (1977) it can be demonstrated that the copper and iron sulphides in the Isua area were deposited from a brine seawater mixture. Deposition of the more complex sulphides would require that the brines did not mix with seawater. Brines can, due to their density, be prevented from mixing with seawater and be concentrated in shallow depressions on the sea-floor. Subsequent cooling of the brine will result in precipitation of the metals carried in solution (Large 1977). This mechanism can adequately explain the occurrence of galena in the tuffaceous amphibolites.

It is concluded that the sulphides are partly of submarine exhalative origin, and partly derived from leaching of basalts by circulating water. The sulphides were deposited on the sea-floor without the influence of sulphur bacteria. The sulphide zonation within the supracrustal belt is attributed to geochemical differences, mainly redox potential and pH, of the depositional environment.

Acknowledgments. The author is indebted to Kira and Lone Appel for making it possible for him to work on the Isua rocks. Preparation of polished thin sections and use of the microprobe in the M.P.I. Dept. Cosmochemie, H. Wänke, is gratefully acknowledged. Part of this work was done while the author held a research stipend at the Max-Planck-Institut für Chemie (M.P.I.), dept. air chemistry. The Isua programme at the M.P.I. was initiated by M. Schidlowski and C.E. Junge, and performed as part of the Sonderforschungsbereich 73 "Atmospheric Trace Components" partially funded by Deutsche Forschungsgemeinschaft.

References

Allaart JH (1976) The pre-3760 m.y. old supracrustal rocks of the Isua area, central West Greenland, and the associated occurrences of quartz banded ironstone. In: Windley BF (ed) The early history of the Earth. John Wiley and Sons, London, pp 177–189

Appel PWU (1979a) Stratabound copper sulphides in a banded iron-formation and in basaltic tuffs in the early Precambrian Isua supracrustal belt, West Greenland. Econ Geol 74:45–52

Appel PWU (1979b) Cosmic grains in an iron-formation from the early Precambrian Isua supracrustal belt, West Greenland. J Geol 87:573–578

Appel PWU (1980) On the early Archaean Isua iron-formation, West Greenland. Precam Res 11: 73–87

Appel PWU, Moorbath S, Taylor PN (1978) Least radiogenic terrestrial lead from Isua, West Greenland. Nature (London) 272:524–526

Fripp REP (1976) Stratabound gold deposits in Archaean banded iron-formations, Rhodesia. Econ Geol 71:58–75

Fryer BJ (1977a) Rare earth evidence in iron-formations for changing Precambrian oxidation states. Geochim Cosmochim Acta 41:361–367

Fryer BJ (1977b) Trace element geochemistry of the Sokoman iron-formation. Can J Earth Sci 14:1598–1610

Garrels RM (1960) Mineral equilibria. Harper, New York, 254 pp

Garrels RM, Christ CL (1965) Solutions, minerals and equilibria. Harper, New York, 450 pp

Laajoki K (1975) Rare-earth elements in Precambrian iron-formations in Värylänkylä, South Poulanka area, Finland. Bull Geol Soc Finl 47:93–107

Large RR (1977) Chemical evolution and zonation of massive sulphide deposits in volcanic terrains. Econ Geol 72:549–572

Monster J, Appel PWU, Thode HG, Schidlowski M, Carmichael CM, Bridgewater D (1979) Sulfur isotope studies in Early Archaean sediments from Isua, West Greenland: Implications for the antiquity of bacterial sulfate reduction. Geochim Cosmochim Acta 43:405–413

Moorbath S, O'Nions RK, Pankhurst RJ (1973) Early Archaean age for the Isua iron-formation, West Greenland. Nature (London) 245:138

Piper DZ, Veh HH, Bertrand WG, Chase RL (1975) An iron-rich deposit from the northeast Pacific. Earth Planet Sci Lett 26:114–120

Schidlowski M (1978) Evolution of the earth's atmosphere: Current state and exploratory concepts. In: Noda H (ed) Origin of life. Center Acad Publ, Tokyo, Japan, pp 3–20

Spence DD, de Rosen-Spence AF (1975) The place of sulfide mineralization in the volcanic sequence at Noranda, Quebec. Econ Geol 70:90–101

Tischendorf G, Ungethüm H (1965) Zur Anwendung von Eh-pH-Beziehungen in der geologischen Praxis. Z Angew Geol 11:55–67

On the Hawleyite-Sphalerite-Wurtzite-Galena Paragenesis from Ragada, Komotini, (Rhodope) North Greece

S.S. AUGUSTITHIS and A. VGENOPOULOS [1]

Abstract

Microscopic and X-ray diffraction studies have determined a complex hydrothermal paragenesis associated with a mylonitized muscovite-gneiss. The main gangue mineral is dolomite which as veinlets invades the mylonitized gneiss. The main ore minerals determined ore-microscopically, by X-ray diffraction, and by microprobe scanning are, sphalerite, wurtzite, galena, chalcopyrite, tetrahedrite (fahlore) and, associated with the sphalerite/wurtzite,hawleyite.

The mylonitized muscovite-gneiss is invaded by veinlets of dolomite. The quartz of the gneiss is often micromylonitized and a fine interlocking quartz texture is produced, often invaded and replaced by dolomite. This fine recrystallized quartz aggregate resembles chalcedony. Often it is difficult to draw the demarcation line between micromylonitized quartz and hydrothermal recrystallized quartz (chalcedonic in derivation).

In places the recrystallized-mylonitized quartz results in orientated quartz patterns, the recrystallized quartz individuals follow two intersecting directions. Comparable textural patterns are described by Sander (1930) due to quartz recrystallization under tectonic influences.

The dolomitic veinlets often extend in the intergranular of the fractured (mylonitized) quartz of the gneiss and as veinlets invade both the quartz of the gneiss and the sphalerite. The intergrowth of the ore minerals with the dolomite and the mylonitized gneiss is illustrated in Fig. 1.

In a preliminary study by Vgenopoulos (1979), hawleyite has been determined, by X-ray diffraction, to be associated with sphalerite. Table 1 shows the comparison of the d-values of the studied hawleyite with the d-values of standard hawleyite from literature. Figure 2 shows hawleyite with strong lemon-yellow-coloured internal reflection associated with sphalerite/wurtzite. A cubic outline of the hawleyite is partly exhibited. The hawleyite shown is marginal to the sphalerite and in contact with a veinlet of gangue minerals (carbonates). Hawleyite is also included in sphalerite/wurtzite and in this case exhibits strong lemon-yellow-coloured internal reflection.

1 National Technical University of Athens, Department of Mineralogy, Petrography and Geology, Athens, Greece

Table 1. d-values of hawleyite from Ragada, Rhodope compared with d-values from literature

Hawleyite from Ragada, Rhodope	Hawleyite literature [a]
d-values	d-values
3.362	3.36
2.902	2.90
1.98	2.058
1.743	1.75
1.676	1.68
–	1.453
1.335	1.337
1.299	1.298
1.183	1.186
1.12	1.12

[a] Joint Committee on Powder Diffraction Standards (1974)

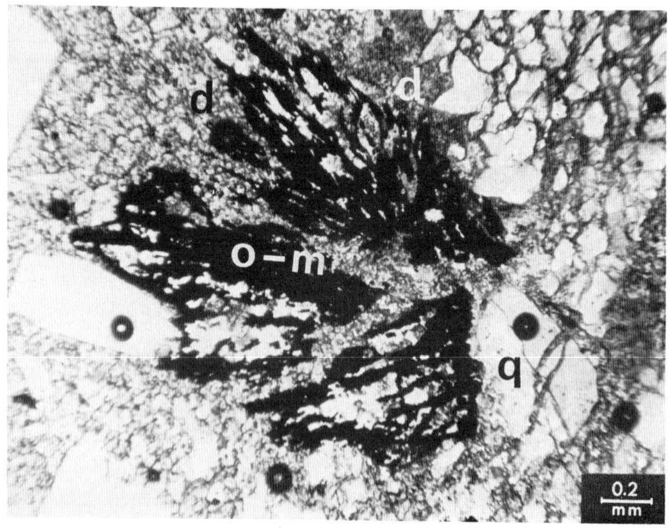

Fig. 1. Ore minerals in intergrowth with the mylonitized quartz of the gneiss and the dolomite. *o-m* ore minerals; *q* quartz; *d* dolomite. Ragada, Rhodope, N. Greece. Without crossed nicols

As a corollary to ore-microscopic observations, semi-quantitative X-ray diffraction studies (Fig. 3) show that a definite crystalline Cd-mineral phase (hawleyite) is present in sphalerite/wurtzite.

The presence of hawleyite as a distinct mineral associated with sphalerite/wurtzite, although not very common, is reported by Berry and Thompson (1962), v. Philipsborn (1967), Ramdohr (1975), Strunz (1957), Trail and Boyle (1955) and Vgenopoulos (1979).

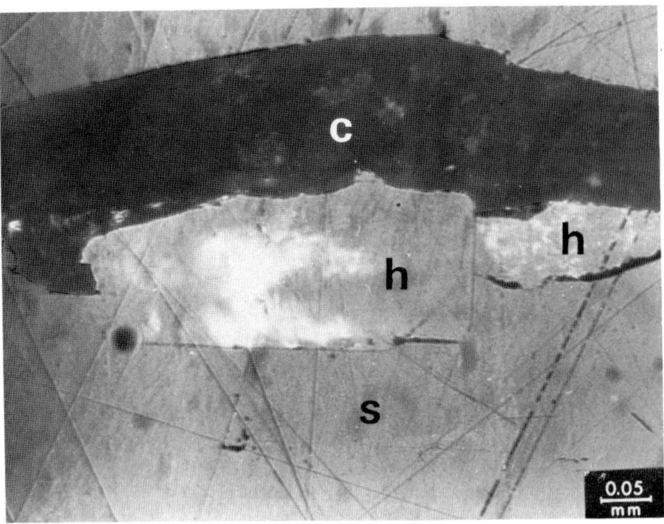

Fig. 2. Hawleyite exhibiting partly cubic outline and strong lemon-yellow internal reflection, marginal to sphalerite/wurtzite. *h* hawleyite; *s* sphalerite/wurtzite; *c* carbonate gangue. Ragada, Rhodope, N. Greece. Polished section (oil immersion, with one nicol)

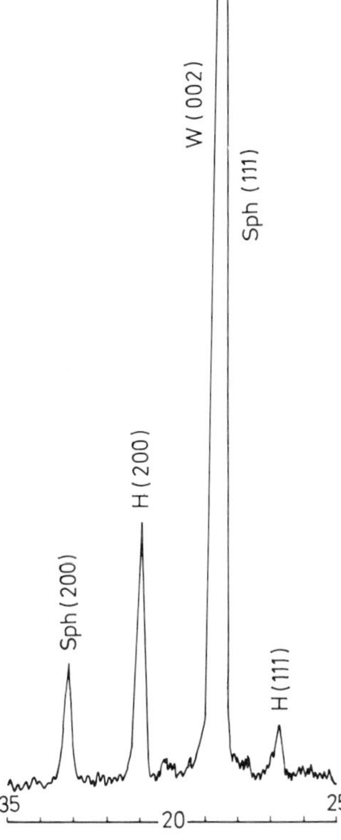

Fig. 3. Debyeogram showing hawleyite in sphalerite/wurtzite from Ragada, Rhodope, N. Greece. (Diffraction diagram, CuKa/Filter Ni 20 mA/40 kV, Philips Generator PW 1130)

Ore-microscopically other sulphides are paragenetically associated with the sphalerite-wurtzite. Sphalerite is associated with galena and tetrahedrite. In contradistinction to the idiomorphic pyrite, spheroids of pyrite (perhaps gel-pyrite) are associated with tetrahedrite, galena and chalcopyrite. Tetrahedrite, often associated with chalcopyrite, is a common and abundant mineral phase in the sulphides paragenesis of Ragada, Rhodope, N. Greece.

Quantitative X-ray fluorescence spectroanalyses show that isolated sphalerite/wurtzite crystal-aggregates contain maximum 1.8% of cadmium. As a corollary semi-quantitative X-ray diffraction studies show maximum 2% of hawleyite in sphalerite/wurtzite. The differentiation of wurtzite from sphalerite was only possible by X-ray analysis.

The association of hawleyite (usually greenockite is present) with sphalerite/wurtzite can be understood by considering the element geochemistry of Cd and Zn (they are related as main elements of the II family of the Periodic Table). Furthermore, the common occurrence in this paragenesis of the elements Cu, Ag, Zn and Cd is again explicable on the basis of the relationships of these elements in accordance to the periodic table. Cu and Zn are related in accordance to the empirical laws of the periodic system, as next to each other main elements of the I and II families of the periodic system and similarly Ag and Cd (horizontal relationship of the main elements). In this connection the presence of Ag is understandable due to its interrelationship with both Cu and Cd.

As the composition of the main sulphides (i.e., chalcopyrite, pyrite, sphalerite, tetrahedrite and hawleyite) of the Ragada paragenesis shows, Cu, Zn, Ag and Cd are the main metallic elements of this paragenesis, in addition to Fe (which is also related to Cu in accordance to the empirical laws of the Periodic Table).

In contradistinction to the group Cu, Zn, Ag and Cd there are also present, in the paragenesis of Ragada, Rhodope, the elements Pb, Sb and as traces As, Bi, Sn, Ge.

Pb is present in the galena and Sb in the tetrahedrite, the other elements occur as traces in the tetrahedrite. The elements As, Sb and Bi are related as main elements of the V family of the Periodic System. The elements Ge, Sn and Pb are similarly interrelated as main elements of the IV family of the Periodic System. Furthermore, the elements Pb, Sb, As, Bi, Sn and Ge are interrelated in accordance to the empirical laws of the Periodic System.

As the element geochemistry of the metallic elements of the Ragada paragenesis shows, two independent groups of elements are present; the elements of each group being interrelated in accordance to the empirical laws of the Periodic System.

References

Berry LG, Thompson RM (1962) X-ray powder data for ore minerals. The Peacock Atlas. Geol Soc Am Mem 85:50

Joint Committee on Powder Diffraction Standards: Search Manual Minerals, printed in Philadelphia (1974)

Philipsborn H von (1967) Tafeln zum Bestimmen der Minerale nach äußeren Kennzeichen. Schweizerbart, Stuttgart, pp 102–103

Ramdohr P (1975) Die Erzmineralien und ihre Verwachsungen, 4. Aufl. Akademie Verlag, Berlin, p 623

Sander B (1930) Gefügekunde der Gesteine. Springer, Wien, p 352

Strunz H (1957) Mineralogische Tabellen, 3. Aufl. Geest and Portig, Leipzig, pp 37–94

Traill RJ, Boyle RW (1955) Hawleyite, isometric cadmium sulphide, a new mineral, Am Mineral 40:555–559

Vgenopoulos A (1979) Geochemical anomalies of Zn, Ni, Cu in Dolomitic limestones from Rhodope Massiv N Greece. International Geological Correlation Programme (IGCP) project No 6. Correlation of diagnostic features in ore occurrences of base metals in Dolomites and limestones, Alger, April 1979

Remobilization of Strata-Bound Scheelite Indicated by Tectonic, Textural and Crystallographic Features in Brejui Mine, NE Brazil

H. BEURLEN, E.B. MELO, and H.S. VILLAROEL [1]

Abstract

Recent literature indicated a polyphasic tectonic evolution of the Seridó Scheelite Province in Northeast Brazil. Structural, textural and crystallographic data obtained during systematic detailed mapping in Brejui Mine, near Currais Novos, Brazil, support the hypothesis that scheelite was originally formed prior to or during the metamorphism and the first isoclinal folding phase. Remobilization and recrystallization took place mainly related with a third, normal folding phase and are responsible for the formation of the richer ore shoots in Brejui Mine.

1 Regional Aspects

More than five hundred scheelite occurrences are spread over the area of about 20,000 km^2, known as the Seridó-Province, in northeast Brazil (Fig. 1).

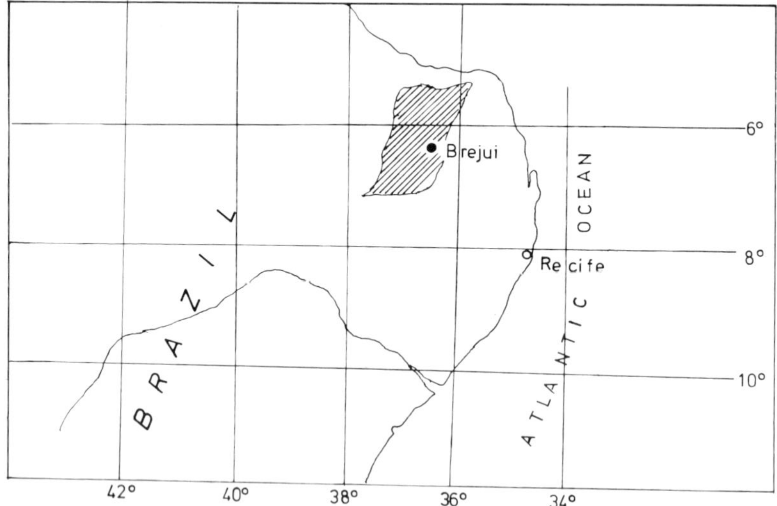

Fig. 1. Location map of the Seridó Province and Brejui Mine

1 Rua Rosa Amélia da Paz 228, 54000 Iaboatão-Piedade-PE, Brazil

This area is made up of basement rocks – mainly gneisses, granites and migmatites – grouped as the Caicó Complex covered by a parametamorphic sequence known as the Ceará (Ebert 1969) or Seridó Group (Ferreira and Albuquerque 1969), supposed to be of Late Precambrian age.

Ebert (1969) distinguished, from base to top, the following formations in the Ceará Group: (1) *Equador-Parelhas* with quartzites and metaconglomerates; (2) *Quixaba* with gneisses, marbles, calcsilicate rocks and amphibolites; (3) *Florânia* with gneisses; (4) *Seridó* with mainly biotite schists. Ferreira and Albuquerque (1969) included the Parelhas, Quixaba, and Florânia units into the Jucurutú Formation, while Maranhão and Siqueira (1977) distinguished the Jucurutú as a discordant, separated group of Middle Algonquian age.

Most scheelite occurrences of the Seridó Province are hosted in the Quixaba (Jucurutú) Formation.

Recent structural analysis by Jardim de Sá (1977) and Ries and Schackleton (1977) showed that a great part of the migmatites supposed to belong to the basement may represent migmatized rocks of the Ceará Group and that no structural disconformity can be evidenced between the Jucurutú and Seridó sequences.

A first isoclinal folding phase with intensive transposition and a later normal folding phase were distinguished by these authors in all formations considered to be of the Ceará Group. Hasui et al. (1979) even recognized four folding phases: the first two with isoclinal style, subparallel axial planes (NNE) and orthogonal axes and the last two phases with subvertical, respectively NNE and NW, axial planes and gently dipping fold axes.

This complicated structural development of the Ceará Group was observed on a regional section cross-cutting the whole Ceará Group and confirmed in the underground survey in several scheelite mines. Thus classical stratigraphic interpretations in the Seridó Province may be the subject of future revisions and the exceptional character of occurrences supposed in the local literature to host in basement rocks remains an open question.

On the genesis of the scheelite, Suszczynski (1975), based on the regional distribution and stratigraphic control of the occurrences, stated a syngenetic-synsedimentary origin. Other authors like Maranhão (1970), Santos (1973), Andritzky and Busch (1975) supposed a hydrothermal origin related to a Late Precambrian migmatization process. A recent lithogeochemical survey in Brejui Mine led Busch (1979) to suppose a volcano-sedimentary origin for the tungsten, based on the orthogenic character of part of the enclosing amphibole-gneisses, gneisses and amphibolites.

A pressure-solution activity during diagenesis and diastrophism should be responsible for the enrichment in the ore shoots. The lack of correlation of tungsten with most of the pneumatophile elements in the host and enclosing rocks also supports his supposition. Based on the amphibolitic envelope, regarded as of volcanic origin, two other occurrences were considered by Moeri and Schnack (1979) to be of volcano-sedimentary origin according to the model of Maucher and Höll (1968).

A preliminary statistical evaluation of the host rock nature on 128 randomly selected occurrences by Beurlen et al. (1979) showed that 67% of the occurrences under consideration are hosted in calcsilicate rocks of very variable mineralogical composition – from simple diopside-epidote-quartz rocks to polymineralic tactites, 18% occur in quartz exudations interbedding or cross-cutting calcsilicate rocks, 17% in amphibolites or silicified and/or epidotized amphibolites, 4% in quartz exudations of schists and gneisses and 1% in pegmatites.

Based on the available geological maps of the province at the scales 1:50,000 and 1:100,000, regarding the proximity of granite bodies, 17% of the considered occurrences are closer than 500 m, 5% occur in a distance between 500 and 1000 m, 10% between 1000 and 2000m, 19% between 2000 and 5000 m and 49% over 5000 m from granitic intrusions.

All these data allow us to classify the majority of the scheelite occurrences as typical strata-bound deposits without direct relation to the granites. Detailed geochemical, petrological and structural analyses are still lacking in almost all occurrences of the scheelite province.

Thus all the different metallogenetic ideas must yet be considered as very preliminary. Much detailed work remains to be done. The present work refers to structural, textural and crystallographic observations made mainly in the Brejui Mine coupled with a systematic mapping regarding a relative datation of the scheelite to the tectonic development of the province.

2 The Brejui Mine

Brejui Mine is the northern portion of the largest scheelite deposit in Brazil, including Boca de Lages, Barra Verde and Zangarelhas Mines producing together over 70% of the Brazilian scheelite production since the 1950's.

The deposit is made up of six main calcsilicate levels, five of them enclosed in marbles and one in biotite gneisses of the Quixaba (Jucurutú) Formation. According to Maranhão (1970) and Ebert (1969) the whole mineralized sequence there corresponds to a large inverse limb of an isoclinal fold related to an overthrust fault at the eastern limit of the mined area. A migmatitic-granitic complex, 200 to 1000 m to the west should correspond to the nucleus of this fold (Fig. 2). Detailed underground mapping in Brejui revealed frequent metric to decametric isoclinal folds with gently dipping axial planes paralleling the schistosity and lithological banding. The axes of these have two preferential directions, NW and SSW respectively. A very intensive axial thickening of the rocks is observed in these folds (F_1 and F_2).

Overprinting these deformation phases, slightly asymmetric folds and/or flexures with steep axial planes and SSW plunges (F_3) paralleling F_2 axes, are sometimes associated with NNE shearing zones and affect the previous schistosity and lithological banding. The structures of this third, regionally most evident phase are observed in metric, decametric and hectometric sizes. Associated to this phase a very conspicuous "boudinage" of the calcsilicate layers is observed, the neck lines being filled by mostly sterile quartz veins with some recrystallized epidote and garnet and, sometimes sulphides.

No important thickening, but intensive recrystallization and formation of porous rich "tactites" are observed in the axial zones of these folds. Most of the ore shoots follow the third-generation fold axes. Minor normal folds of metric size, subvertical

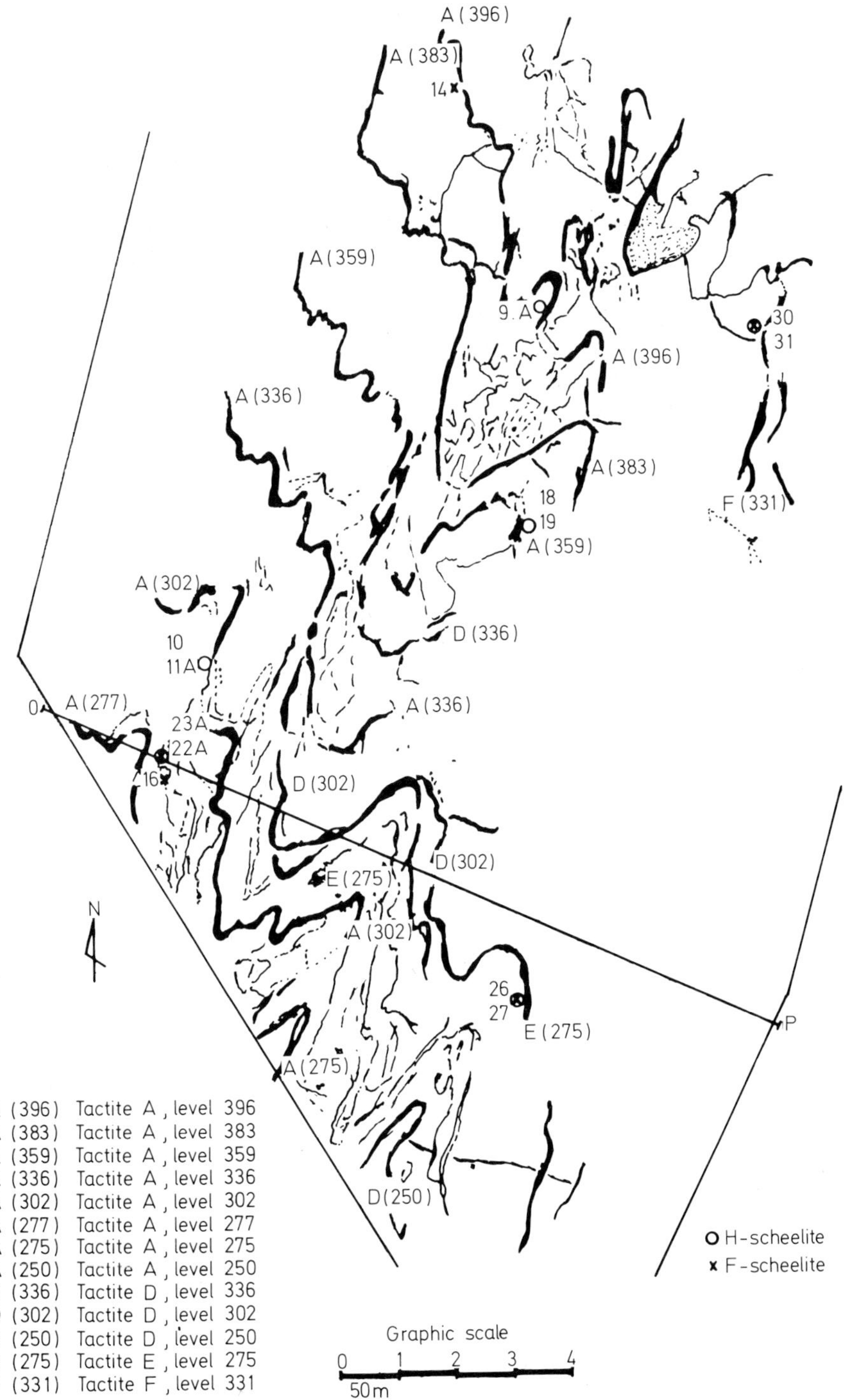

Fig. 2. Structural contours of four different tactite levels of sample location of Brejui Mine indicated by capital letters *A, B, E* and *F*. The numbers in parentheses indicate the topographic level. The total length of the scale is 200 m

axial planes and NW-directed axes are also observed, sometimes again associated with boudinage of the calcsilicate rocks. This phase (F_4) is responsible for the changes of plunge intensity of the F_3 axes and for the typical dome and basin interference pattern observed on the regional geological maps. These data of the Brejui Mine agree completely with the regional observations made by Hasui et al. (1979) and Jardim de Sá (1977) discussed before. Thus the isoclinal foldings observed cannot be understood as a special and local feature, but appear to represent a regionally important structural event.

Two distinct textural patterns of scheelite can be observed in Brejui Mine. A fine-grained scheelite (F) distributed regularly along the few cm-thick contact-zone of "tactites" with marbles and sometimes gneisses. The individual grains, less than 1 mm large, are isolated or may form lenticular aggregates paralleling the contact, lithological banding and schistosity. Mainly close to isoclinal fold hinges "half moon"-like aggregates can also be observed. The scheelite there follows the isoclinal folds and is fractured parallel to schistosity.

The fringes of this F-scheelite are clearly dislocated by fractures and shearing zones and, together with the lithological banding of the tactites follow the minor normal folds with sterile axial plane (Fig. 3).

Contrasting to this regular distribution a very heterogeneously grained scheelite (H) with some individual crystals up to 10 or 20 cm occurs spread erratically in the more porous parts of tactite, preferentially restricted to the hinges of the normal folds (F_3) and to intensively sheared or fractured zones, building up the main ore shoots, plunging 30° SSW.

This ore grades up to 2.5% WO_3 in contrast to the normal tactites with F-scheelite grading 0.0% to 0.3% WO_3.

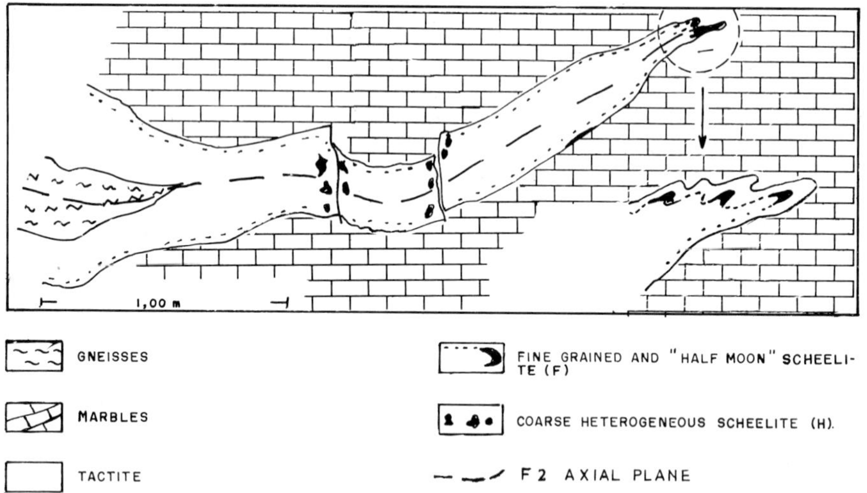

Fig. 3. Schematic section at point 94.98 Brejui II Mine showing the distribution of fine-grained F-scheelite along an isoclinal fold F_2 and coarse H-scheelite along fractures related to a superimposed normal fold F_3

Similar observations were made in several other occurrences of the province, such as Bonsucesso, Diniz, Morada Nova, etc. In Bonfim Mine a very coarse-grained scheelite distributed regularly parallel to the banding and schistosity is clearly oriented and deformed, showing large subidiomorphic sections viewed from the top of the banding and lenticular or halfmoon forms paralleling the schistosity. Under the microscope the "crystals" appear very fractured and with intensive undulating extinction (Fig. 4).

These observations, compared to the recent data on a 4 stage-structural evolution of the area led to the supposition that the F-scheelite and, sometimes, even the coarse-grained scheelite were already submitted to the isoclinal folding F_1 responsible for the schistosity, thus pre-tectonic and pre-metamorphic, whereas the H-scheelite should be a result of a recrystallization and remobilization process related mainly to the third, normal folding phase. In order to test this hypothesis a crystallographic comparison of the two scheelite types was carried out.

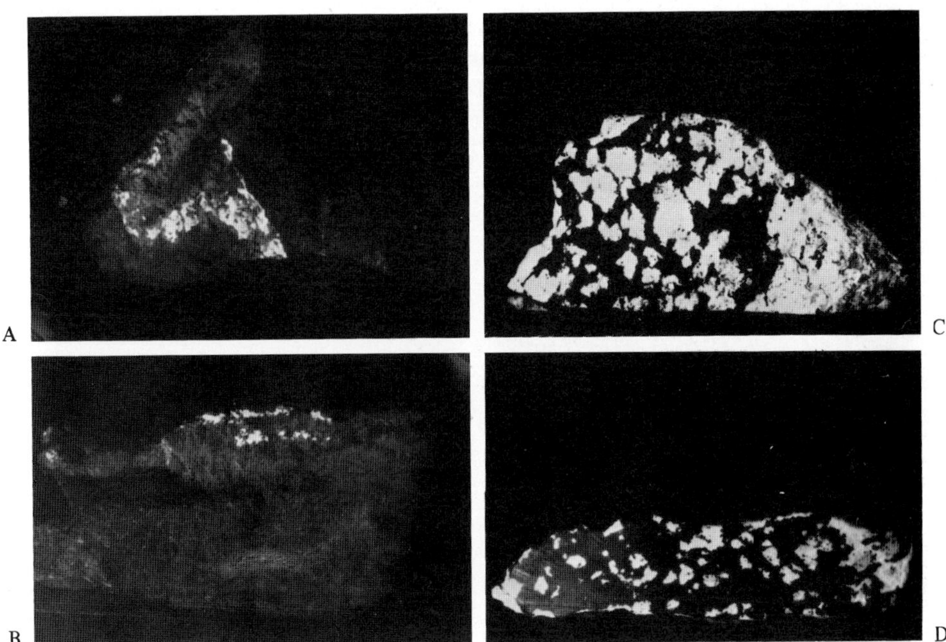

Fig. 4 A–D. Samples from mineralized calcsilicate layers under UV illumination; scale 1/2 natural size. **A, B** sample from Brejui Mine. **A** viewed from top and **B** frontal to banding. The fine-grained scheelite aggregates *(white)* are clearly elongated parallel to the banding/foliation; **C** top, **D** frontal sample from Bonfim Mine; the coarse-grained scheelite is clearly oriented parallel to foliation and banding

3 Crystallographical Data

With the aim of verifying an eventual difference in the crystallographic character of the two texturally distinct scheelite types there were separated with the help of an UV-lamp the scheelite grains of 14 scheelite samples (7 F-scheelites and 7 H-scheelites) and prepared the X-ray diffractograms. The same preparation procedure was observed for the two series of samples (preparation, speed and intensity).

Very suggestive differences could be observed on several peaks, including 008, 305, 323 and 400, as shown in Fig. 5:

1. The intensity of the peaks shown by H-scheelite is systematically higher than the one observed on F-scheelites (about twice in the case of Fig. 5).
2. The peak width is slightly but systematically larger in the case of F-scheelites, making difficult correct readings of the peak maxima. Unfavourable crystallization conditions or later deformation of the lattices could be responsible for this.
3. The peak maxima positions of the fine-grained scheelites are not constant but are normally dislocated to lower values than the ones observed for the H-scheelite. In the case shown by Fig. 5, the dislocation reaches to 1/8 degree. This deviation could be either the result of different deformation intensities of the lattices of F-scheelites, compared with the undeformed H-scheelite, or an apparent result of the scattering effect caused by (1) and (2).

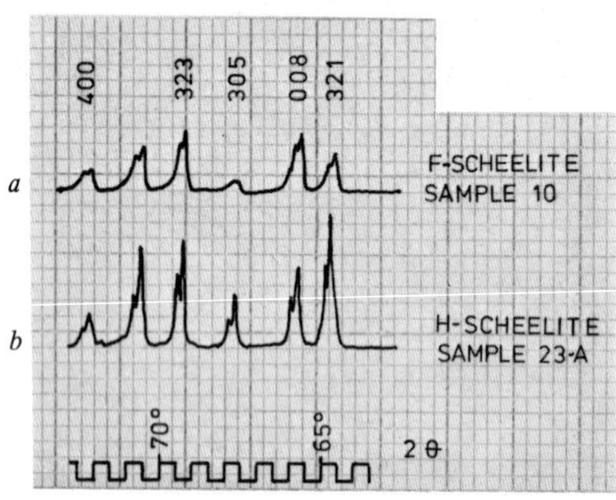

Fig. 5. Partial diffractograms taken under the same operating conditions *a* of fine-grained F-scheelite *(sample 10)* and *b* of heterogenous coarse H-scheelite *(sample 23-A)* from Brejui Mine comparing the intensity and amplitude of several peaks

Preliminary semi-quantitative X-ray fluorescence analyses for trace elements that could substitute for the main ions of the scheelite, such as Mo, Nb, Ti for W and Fe, Cu, Sr, Mn and R.E. for Ca, did not reveal any significant difference between H- and F-scheelites. Monochromatic diffractometer diagrams with very slow velocities will be taken with the aim to define the precise crystallographic parameters and the causes of the differences observed.

Whatever the ultimate cause may be, the differences must result from distinct original crystallization conditions or, according to the tectonic and textural observations, from different histories after crystallization, the F-scheelite being more deformed than the H-scheelite.

Acknowledgments. The authors are indebted to A.N. Sial for reviewing this text. The present results are part of a research program financed by the Conselho Nacional do Desenvolvimento Científico e Tecnológico, CNPq, process no. 700.1.012. 1/79.

References

Andritzky G, Busch K (1975) Erläuterungen zur Geologischen Karte des Scheelit-Gebietes von Santa Luzia, NE-Brasilien, 1:100.000. Bundesanstalt für Geowissenschaften und Rohstoffe, Hannover, 34 pp

Beurlen H et al. (1979) 2nd progess report to SUDENE-CNPq, Projeto Províncias Metalogéneticas. Unpubl Rep, Recife, 49 pp

Busch K (1979) Geochemische Untersuchungen zur Genesis der Wolfram-Lagerstätte in Taktit bei Currais Novos, NE-Brasilien. Unpubl Rep. Bundesanstalt für Geowissenschaften und Rohstoffe, Hannover, 47 pp

Ebert H (1969) Geologia do alto Seridó-Folha de Currais Novos. SUDENE Div Geol Ser Geol Reg, Recife 11:93 pp

Ferreira JA, Albuquerque JTP (1969) Sinopse da Geologia de Folha Seridó. SUDENE Div Geol Ser Geol Reg Recife 18:47 pp

Hasui Y, Silva Filho AF da, Galindo AC, Lima ES, Melo EB, Guimaráes IP, Martins SaI, Rodrigues da Silva MR, Costa MGF (1979) Relatório das atividades desenvolvidas durante o curso de análise estrutural-perfil Currais Novos-Florânia (RN). Unpubl report. Curso de Pós-Graduação em Geociências, Centro de Tecnologia, UFPE, Recife, 13 pp

Maranhão R (1970) Geologia econômica da região de Currais Novos (RN). Thesis, Univ São Paulo, 135 pp

Maranhão R, Siqueira LP (1977) A geossinclinal do Seridó. Bol Mineral Recife 5:51–85

Maucher A, Höll R (1968) Die Bedeutung geochemisch-stratigraphischer Bezugshorizonte für die Alltersstellung der Antimonitlagerstätte von Schlaining im Burgenland, Österreich. Mineral Deposita 3:272–285

Moeri E, Schnack PJ (1979) Mineralizações scheelitíferas na região de Santa Luzia, PB. 9 Simpósio de Geologia do Nordeste. Soc Bras Geol Resumos das Comunica ções, Natal, RN, p 73

Ries AC, Schackleton RM (1977) Preliminary note on structural sequences and magnitude and orientation of finite stains in the Precambrian of Northeast Brazil. Boletim do Núcleo do Nordeste. Soc Bras Geol Campina Grande, PB, pp 397–400

Sá EFJ de (1977) O complexo de embasamento na região do Seridó: desenvolvimento geológico. 8° Simposio de Geologia do Nordeste. Soc Bras Geol. Resumos das Comunicações, Campina Grande, PB, p 60

Santos EJ (1973) Província Scheelitífera do Nordeste. 27 ° Congresso Brasileiro de Geologia. Soc Bras Geol Aracajú, SE, Roteiro das Excursões, pp 31–46

Suszczynski E (1975) Os recursos minerais reais e potenciais do Brasil e sua metalogenia. Interciências, São Paulo, 533 pp

Uranium Occurrences of Bavaria, Germany (With Emphasis on the Bavarian Pfahl of the Altrandsberg District)

H.-W. BÜLTEMANN[1] and R. HOFMANN[2]

Abstract

During surface prospection work carried out systematically and remapping of the area of the Bavarian Pfahl – especially in the region of Altrandsberg, Bayerischer Wald – secondary uranium mineralizations were found near the surface. These were investigated at depth by drilling. A sulphide pitchblende-coffinite mineralization in veinlets up to several mm thickness was found in drillholes in the Pfahl shales at the feather joint endings. First investigation results are presented in this study.

1 Geological Framework

A number of promising uranium anomalies have been discovered on the Bavarian Pfahl between Neukirchen-Balbini and Viechtach as a result of an exploration programme carried out by the company GEWERKSCHAFT BRUNHILDE since 1959 (Bültemann and Hofmann 1960–1962; von Braun 1965) (see Fig. 1). The origin of the Bavarian Pfahl and its mineralization is described below.

In the course of the exploration programme it was observed that radiometric anomalies occur as elongated zones diverging slightly from the general strike of the Pfahl. The prevailing ideas about the Pfahl at that time were mainly based on the mapping results of von Gümbel (1868) and the determination of the hydrothermal character by Ochotzky and Sandkühler (1914) and Hegemann (1936). In their opinion a continuous quartz vein was present. The anomalies which diverge from the general strike of the Pfahl are undoubtedly related to the formation of the Pfahl and could not be explained with the model of a continuous quartz vein.

Remapping revealed the surprising result that the Pfahl is not a uniform continuous vein, and therefore, it is not a filling of a fault fissure but is a vein system (Hofmann 1962). Generally speaking, in the veins of the main Pfahl in the SE portion of the remapped area NW-SE striking feather joint fillings associated with simple shear planes have a NNW-SSE strike. To the NW in addition to NNW-SSE striking shear planes a second shear plane striking E-W is developed, which is frequently vein-filled. In the extreme NW of the mapped area only this second E-W striking shear plane is present.

1 Gewerkschaft Brunhilde, 3162 Uetze/Hannover, FRG

2 Mineralogisches Institut der Universität, Welfengarten 1, 3000 Hannover, FRG

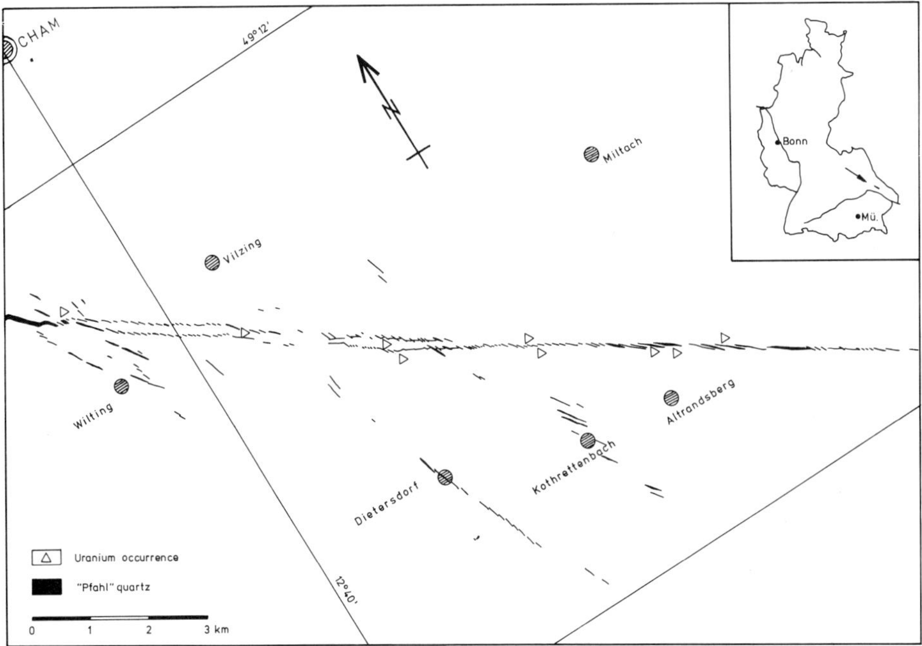

Fig. 1. The Bavarian Pfahl by Hofmann (1962), showing the location of the uranium occurrences in the area of investigation

Investigation of the weaker developed secondary Pfahl feather joint systems has shown that apart from one exception they lie as mirror images to the main Pfahl feather joint system.

The origin of the feather joint systems has been explained fully in Hofmann (1962). According to Hofmann – and alternative prevailing view – there is no connection between the origin of the Pfahl shales (mylonite and blastomylonite) and the Pfahl quartz veins. These findings were contradicted by Steiner (1969), who assumed the existence of a continual quartz vein, which is imbricated by NNW-SSE striking lateral faults. Steiner's finding cannot be convincing, because Hofmann's conclusion has been checked by deep drilling. The relationship between the quartz feather joint veins and the uranium zones was confirmed with these drill holes as mentioned below.

According to the current evidence the main emplacement of quartz in the Pfahl feather joints took place in the Upper Permian. There is no evidence of more recent quartz vein formation. Furthermore, there are no indications of major tectonic events on the Pfahl of Permian or younger age. The post-Cretaceous peripheral fault of the Bodenwöhr Basin only just reaches the Pfahl, which indicates that both structures are independent of each other. Furthermore, quartz deposition has not been observed on the Bodenwöhr peripheral fault.

To return to the previously mentioned observation on the strike of radiometric anomalies, one example from the many observations is given here. At Altrandsberg these radiometric anomalies strike at 130°. The geological mapping has shown that this strike is exactly parallel to the strike of the feather joint quartz veins. The anomalies as a rule lie in the strike continuation of quartz veins, in other words, in the mainly quartz-free extension of the feather joint endings. The uranium mineralization thus surrounds the feather joint quartz veins like an aureole.

The uranium mineralization being peripheral to the quartz veins, must have been deposited first.

The Pfahl feather joints must have been formed by the deformation of a cooling rock mass, which was subsequently mineralized by low temperature hydrothermal solutions. During the Permian these fractures further widened allowing the passage of higher temperature hydrothermal solutions. These solutions undoubtedly caused redeposition of the mineralization, particularly in the cooler extensions of the feather joint endings. Isolated mineralization can sometimes be observed completely enclosed in quartz as a result of later reopening of the fractures.

The above model of the location of the mineralization in relation to the feather joint quartz veins has been confirmed by several deep drill holes (up to 300 m deep). These boreholes were drilled mainly to the NW of Altrandsberg (Fig. 2). Space constraints only allow reference to two vertical drill holes that, as predicted, intersect the ore zone (Fig. 3). The uranium is found as millimetre-wide veinlets occurring swarm-like in the area of the feather joint endings. The mineralized zone has a true thickness of about 3 m in unweathered rock and possesses an average grade of about 0.1% UO_2. Occasional samples yielded grades of up to 1% UO_2.

Independent of the uranium minerals the following sulphides are also present: galena-sphalerite, chalcopyrite and fluorspar. These minerals are currently under investigation (Bültemann and Hofmann 1981).

The primary ore minerals of the deep drill holes A and B is given below.

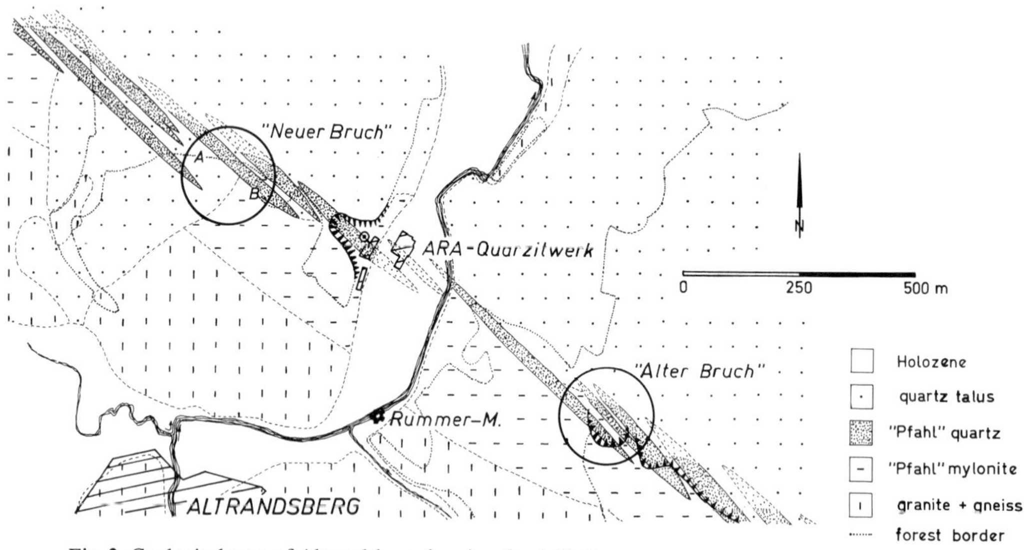

Fig. 2. Geological map of Altrandsberg showing the drill sites

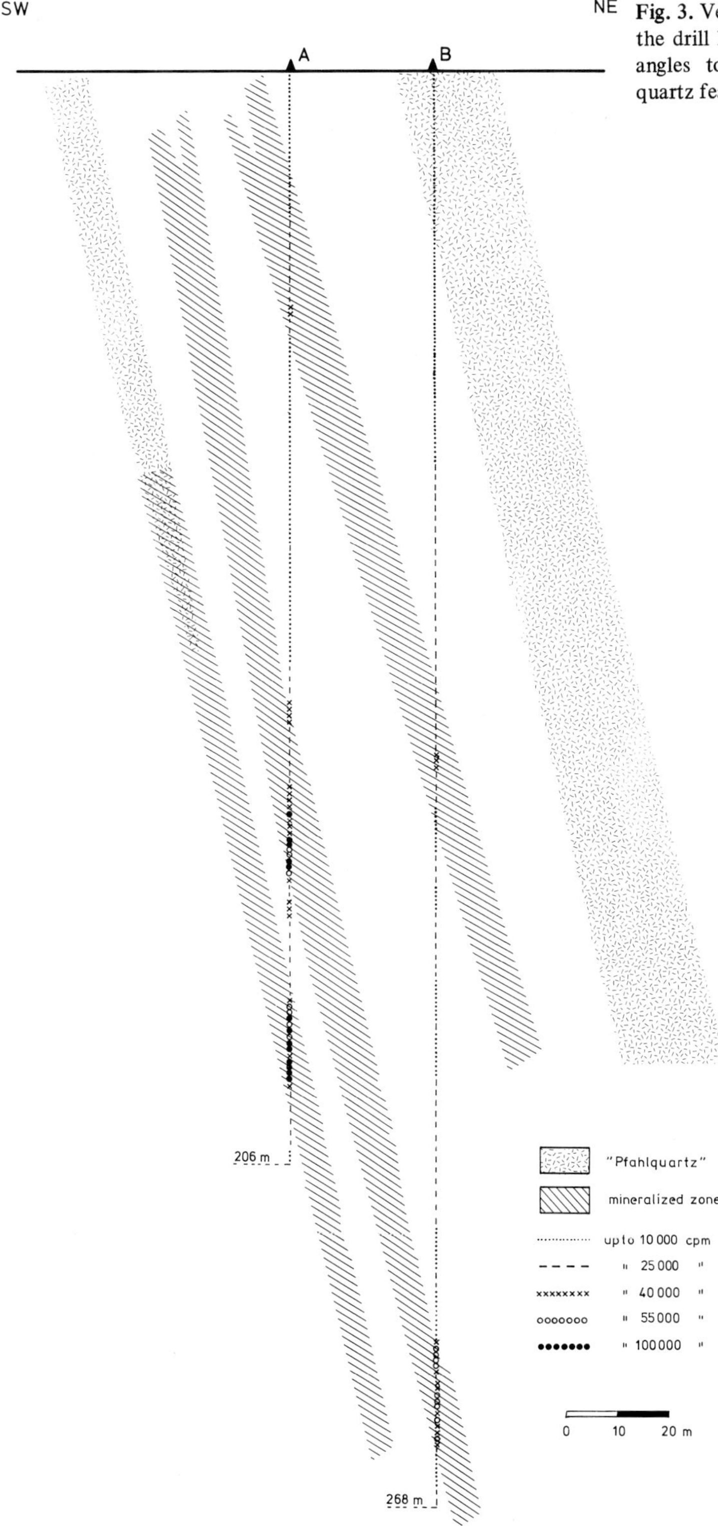

Fig. 3. Vertical profile through the drill holes A and B at right angles to the strike of the quartz feather joints

2 Primary Mineralization

The primary mineralization of the drill holes A and B is made up as follows:

First pitchblende (UO_2) occurs as thick vein wall encrustations, which is not typical for this mineral; however, it can be positively identified by its characteristic reflectance. In addition it occurs occasionally in a reniform, scaly habit, and also as well-developed pseudomorphs after coffinite crystal aggregates. The colour and reflectance suggest in certain cases that initial alteration of the pitchblende to successive minerals (e.g., nasturan) is taking place.

Occasionally sulphides, particularly pyrite and chalcopyrite were observed, replacing the pitchblende outward from contraction cracks.

Coffinite ($USiO_4$) is still commonly present, although most of it has been replaced by pitchblende. Most of the individual crystals can only be recognized under the microscope, although occasionally crystals with edges up to 1 mm in length are present. Crystal aggregates covering an area of up to 0.5 cm^2 are commonly found.

Coffinite is found almost entirely in typical tetragonal habit displaying the faces (110) and (001). The grey to light-grey internal reflection is easily recognized when viewed in oil immersion. The mineral is completely isotropic. Unweathered coffinite is occasionally found on the surface embedded in quartz. The first description of coffinite on the Pfahl was made by Bültemann as referred to in Ramdohr (1961) and Bültemann (1961).

Both uranium minerals are always accompanied by sulphides, particularly pyrite as well as chalcopyrite and galena. The sulphides occur mainly as idioblastic crystals but also as idiomorphic crystals. The coffinite crystals are frequently intergrown with the sulphides.

Galena (PbS) probably of radiogenic origin, is uniformly distributed throughout the coffinite and pitchblende and is easily identifiable. This suggests a relative early age for the uranium mineralization.

Galena-sphalerite mineralization, chalcopyrite-mineralization and occasionally fluorite of a dark violet to an almost black colour were observed in addition to, and independent of the above-mentioned mineralization.

Special attention was paid to the gold occurrence mentioned by Hegemann (1936). Gold, however, could not be detected in any of the numerous samples, by either spectrographic or ore microscopic methods. A paragenesis of this nature is thought to be very unlikely. Native silver, on the other hand, is found in a large number of polished sections, mostly associated with galena.

The above-mentioned "accessory minerals" are described in detail in a separate publication (Bültemann and Hofmann 1981).

The sulphide pitchblende-coffinite mineralization is of hydrothermal origin and seems to have been of relatively high temperature. The deposition of uraninite and coffinite seemed to occur simultaneously with the quartz implacement.

A remobilization and a redeposition of the uranium by the quartz in the feather joint endings was mentioned above. This conclusion is based on field evidence. This separation between quartz and uranium mineralization is the rule, even though occasionally primary uranium minerals are found enclosed in the quartz. It cannot be clearly established whether the mineralization is directly associated with the quartz

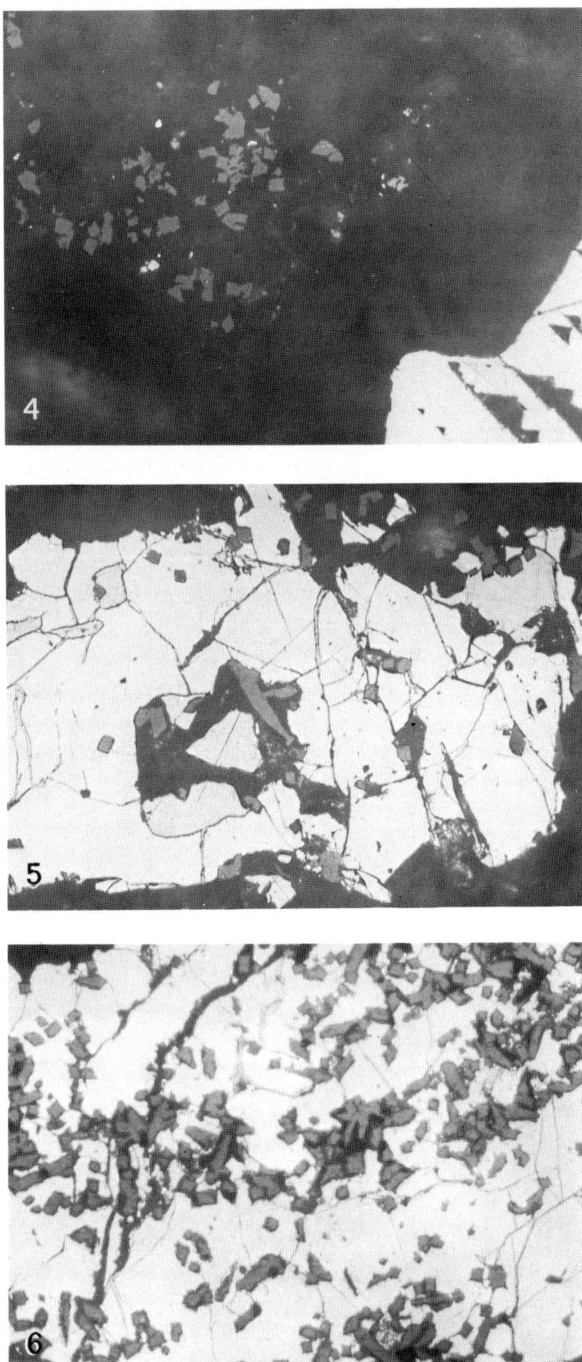

Fig. 4. Coffinite XX *(grey)* in cherty quartz. *Below right* galena displaying typical cleavage controlled fragmentation. Locality: Borehole NW "Neuer Bruch" near Altrandsberg. × 330, oil imm

Fig. 5. Chalcopyrite *(white)* in quartz *(black)* with coffinite XX *(dark grey)*, galena *(light grey with numerous scratches)*. Locality: as Fig. 4. × 280

Fig. 6. Coffinite XX *(grey)* in chalcopyrite *(nearly white)*. Tennantite on fractures *(light grey)*. Locality: as Fig. 4. × 340, oil imm

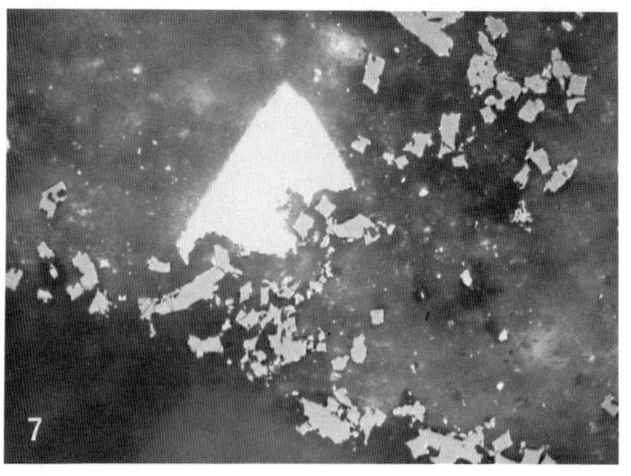

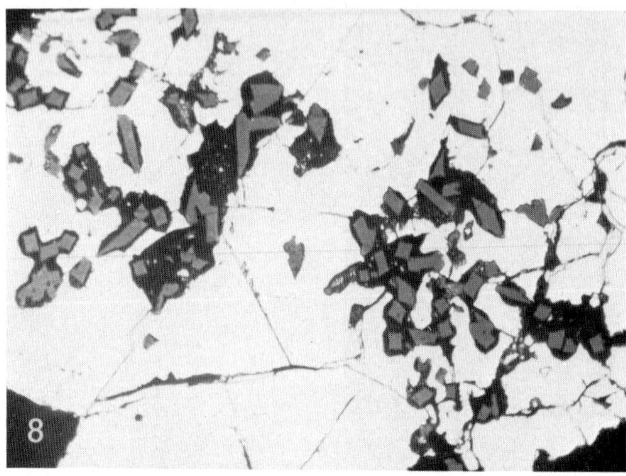

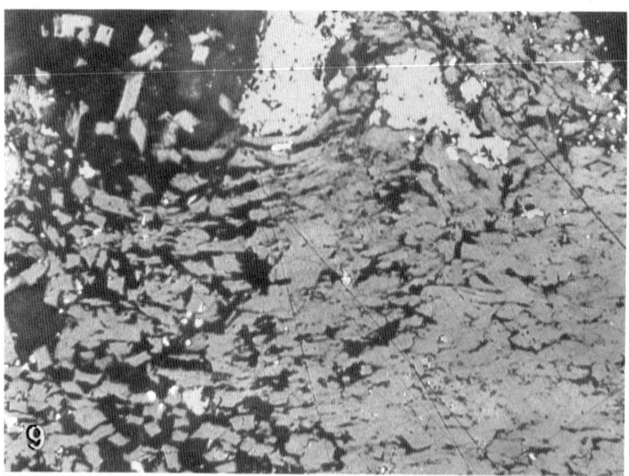

Fig. 7. Pitchblende formerly coffinite XX *(light grey)*, pyrite X *(white)* in quartz *(grey to black)*. Locality: as Fig. 4. × 340, oil imm

Fig. 8. Pyrite *(white)* with coffinite XX *(grey)*, quartz *(black)*. Locality: as Fig. 4. × 410, oil imm

Fig. 9. Pitchblende *(grey)* formerly coffinite XX, traces of galena *(white)* and quartz *(black)*. Locality: as Fig. 4. × 340, oil imm

vein implacement or whether it is rather a remobilization of older, pre-quartz Pfahl-mineralization.

After consideration of the circumstances the latter interpretation is preferred. The uranium-free sulphide mineralization is undoubtedly derived from younger independent hydrothermal activity, because it occasionally intersects the Pfahl quartz.

3 Secondary Uranium Minerals

The following secondary uranium minerals were found mainly in trenches but also in the weathered zone intersected in various drill holes at "Alter Bruch" (see Fig. 2) and "Neuer Bruch".

The main mineral present is torbernite $[Cu(UO_2)_2(PO_4)_2 \cdot 8-12\ H_2O]$(and metatorbernite). This mineral was first mentioned by Hegemann (1936). It occurs as well-developed thin to thick tabular or pyramidal crystals of dark green to light green colour (according to the degree of hydration). Strunz (1961, see also 1962) describes orthotorbernite with 12 H_2O from this locality.

Under UV-light the marginal zones of the torbernite crystals often display a strong yellow-green fluorescence, which is caused by the replacement of Cu by Ca as a result of leaching and thereby forming autunite.

Well-developed tabular, yellow-green crystals of autunite $[Ca\ (UO_2)_2\ (PO_4)_2 \cdot 10\ H_2O]$ were found very rarely with torbernite close to the surface.

Phosphuranylite $[Ca\ (UO_2)_4\ (PO_4)_2\ (OH)_4 \cdot 8\ H_2O]$ is occasionally found with autunite in micro-crystalline encrustations, occasionally as pseudomorphs after autunite.

Parsonsite $[Pb_2\ (UO_2)\ (PO_4)_2 \cdot 2\ H_2O]$ occurs fairly commonly as pale-yellow, needle-like crystal clusters in quartz drusy cavities. In some cases the mineral appears to be an alteration product of a former primary mineral (probably coffinite).

Kasolite $[Pb\ (UO_2)\ SiO_4 \cdot H_2O]$ is found in quartz cavities in form of brownish, wart-like crystal aggregates, occasionally with torbernite.

References

Braun E von (1965) Die mit Bundesmitteln unterstützte Uranprospektion der Jahre 1956–1962. Schriftenr Bundesminist Wiss Forsch 5:111

Bültemann H-W (1961) Zwischenbericht über erzmikroskopische Untersuchungen an Bohrproben aus dem Raum Altrandsberg, Bayerischer Wald. Unpubl Rep Bundesminist Wiss Forsch

Bültemann H-W, Hofmann R (1960–1962) Unpubl Rep GEWERKSCHAFT BRUNHILDE

Bültemann H-W, Hofmann R (1981) Die Mineralisation des Bayerischen Pfahls. In preparation

Gümbel CW von (1868) Geognostische Beschreibung des Ostbayerischen Grenzgebirges, 377 S

Hegemann F (1936) Die Bildungsweise des Quarzes im Bayerischen Pfahl. Chem Erde 10:521–538

Hofmann R (1962) Die Tektonik des Bayerischen Pfahls. Geol Rundsch 52:332–346

Ochotzky H, Sandkühler B (1914) Zur Frage der Entstehung des Pfahls im Bayerischen Wald. Zentralbl Mineral Geol Palaeontol, pp 190–192

Ramdohr P (1961) Das Vorkommen von Coffinit in hydrothermalen Uranerzgängen, besonders vom Co-Ni-Bi-Typ. Neues Jahrb Mineral Abh 95:313–324

Steiner L (1969) Scherflächenpaare in Oberbau und Unterbau. Geotek Forsch 33:62 pp

Strunz H (1961) Orthotorbernit von Altrandsberg. Aufschluß 12:25–27

Strunz H (1962) Die Uranfunde in Bayern von 1804–1962. Acta Albertina Ratisbonensia 24:5–92

Some Features of Ore Fabric, Sasca Montana Skarn Deposit, Romania

E. CONSTANTINESCU [1] and G. UDUBASA [2]

1 Introduction

The Sasca Montana skarn deposit, Banat, West Romania, belongs to the Laramian metallogeny and exhibits copper-rich ore assemblages. It was already known and mined as Cotta was writing his book on the ore deposits in Banat and Serbia (1864).

The most striking feature of this ore deposit is the lack of any zoning both of the skarn and ore mineral assemblages (Constantinescu 1971, 1981). Some new observational data concerning the ore fabric are here presented. Most of the ore minerals form non-equilibrium assemblages. The occurrence of idaite in Romania is reported for the first time with this paper.

2 Ore Mineralogy

About 25 ore- and more than 70 non-metallic minerals have been described to the present from the Sasca Montana ore deposit (Constantinescu 1981). The copper and iron ore minerals, i.e., chalcopyrite, pyrite, chalcocite, bornite, digenite, and idaite are the most common. Some sulphosalts such as jamesonite and boulangerite appear only in places as well as enargite, linnaeite-siegenite, and the near-end members of the fahlore group. Magnetite and molybdenite were observed both within the porphyry-type ore concentrations locally developed in the granodioritic rocks and the skarn bodies. Pyrrhotite and arsenopyrite occur also but only as microscopic constituents of the ore. Fine veinlets and pockets containing sphalerite (iron-poor) and galena appear in places mostly within the recrystallized limestones. The ores hosted by skarns sometimes contain specular hematite. Covellite, cuprite, tenorite, native copper, chrysocolla and aurichalcite occur in the oxidized ores which contain also typically developed anatase grains. A mineral variety of smithsonite, szaszkaite, was also described at Sasca Montana (Cadere 1925).

1 Faculty of Geology and Geography, University of Bucarest, Romania
2 Institute of Geology and Geophysics, Bucarest, Romania

3 Ore Fabric

Exsolution textures are widespread, especially of chalcopyrite in bornite, sphalerite in chalcopyrite, and chalcocite in bornite, well-documented by Superceanu (1969), Constantinescu (1981), and in the *Card Index of Ore Photomicrographs* (Maucher and Ramdohr 1962). The high-temperature ore deposition indicated by such textures correlates well with the association of these exsolution-rich mineral phases with the skarn assemblages.

Substitution textures are often found and their apparition is mainly related to the polyascendent development of the postmagmatic mineral formation. Zoned garnet crystals show beautiful replacements by magnetite with zoning preservation (Fig. 1). Mention should also be made particularly of the lamellar intergrowths of chalcopyrite (cp) and quartz (Fig. 2). The occasional presence of bornite residuals together with the quartz matrix (Fig. 3) is indicative of a preferential substitution of bornite by quartz under preservation of the cp laths.

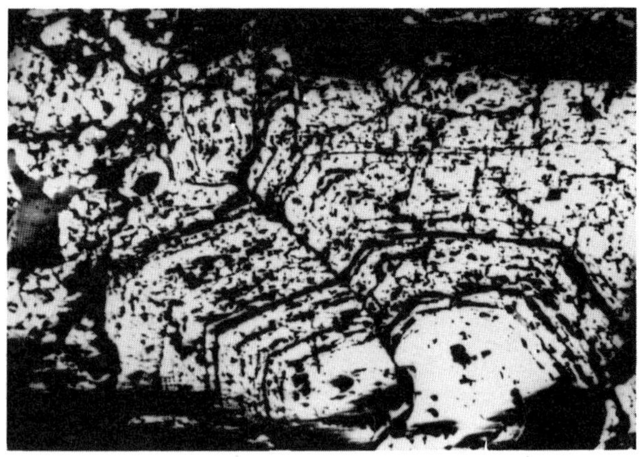

Fig. 1. Zonal replacement of garnet by magnetite. Reflected polarized light (RPL), × 10

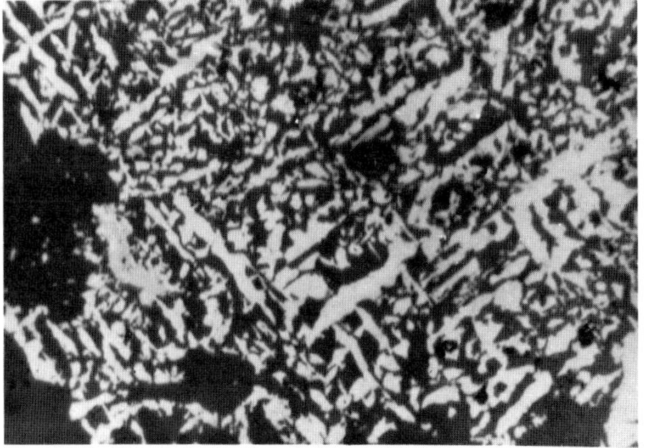

Fig. 2. Lamellar intergrowth of chalcopyrite *(white)* with quartz *(black)*. RPL, oil immersion, × 500

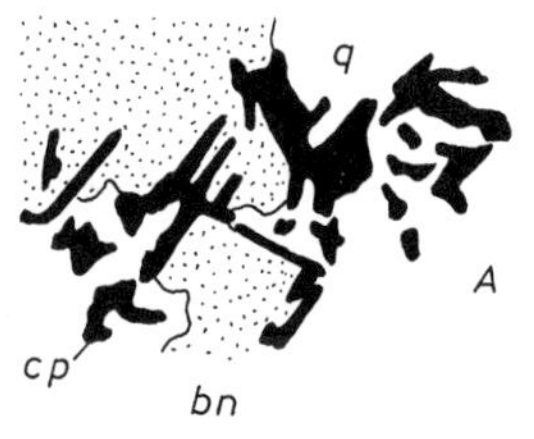

Fig. 3 A,B. Chalcopyrite laths develop both in quartz *(q)* and in bornite *(bn)*. In **B** bornite is completely replaced by quartz. *Scale bars* are 0.01 mm

Intimate intergrowths of enargite-sphalerite, tetrahedrite-cp, and cp-chalcocite (cc) develop sometimes myrmekitically; their mode of formation is, however, very different. The rarely observed myrmekites enargite-sphalerite must probably be regarded as products of a simultaneous crystallization. The very fine intergrowths of cp-fahlore (Fig. 4) probably followed the same way of formation. They are similar in appearence to those mentioned by Ramdohr (1945, Fig. 17; 1969, Fig. 87). Supergene reworking of bornite may perhaps be invoked in the case of such intergrowths where cp and fahlore form irregular aggregates, partly botryoidal in appearance (Fig. 5).

The so-called "disease-cracks" within the monomineralic masses of bornite are related to the development of idaite, locally accompanied by cp and cc during low-temperature decomposition of the primary bornite (Fig. 6). The most frequently observed aspects may perhaps be related to Tufar's (1967) "Vorstufe" (intermediate step) of bornite transformation into idaite.

Incipient alteration of linnaeite-siegenite is a very complex one and commences gradually along the cleavages. Secondary pyrite also forms in places along the cleavages of linnaeite (Fig..7). The end-product of such a transformation is an aggregate of linnaeite grains cemented by cp and bornite (bn). Ramdohr (1945) suggests that the myrmekitic intergrowths of linn+cp+bravoite could represent the breakdown product of the mineral villamaninite. Sporadically occurring millerite lamellae within these intergrowths at Sasca Montana show that a variant of this interpretation may also be accepted for the Sasca ores.

The globular aggregates or apparent single grains of pyrite form a striking feature of the Sasca Montana copper ores (Figs. 8–11). Rounded aggregates of microgranular pyrite are always enclosed in a matrix of cp and bn. The inner part of these globules may locally be replaced by a very fine mixture of cp and bn. Sometimes atoll-textured or shell-like pyrite grains occur and they have nothing to do with the "mineralized bacteria" as Superceanu (1969) states. Such globular pyrites seem to be descendent in nature but their mode of formation is still not understood. Betechtin et al. (1958) and Vujanovic (1974) described similar forms of the pyrite aggregates occurring in a cp-rich matrix and consider their formation to be of colloidal nature under conditions of volcano-sedimentary ore deposition (Vujanovic 1974).

Irrespective of whether the pyrite globules are of volcano-sedimentary origin or whether they formed by supergene reworking of some copper-rich pyrite-bearing assemblages the rounded form with varying size (0.1–0.5 mm) remains to be explained. As in the case of the "Rogenpyrit" and of the framboidal pyrite the nearly spherical form may be regarded as an equilibrium one. It is, however, worth mentioning that the pyrite of the Sasca Montana ores usually transforms into microgranular aggregates by

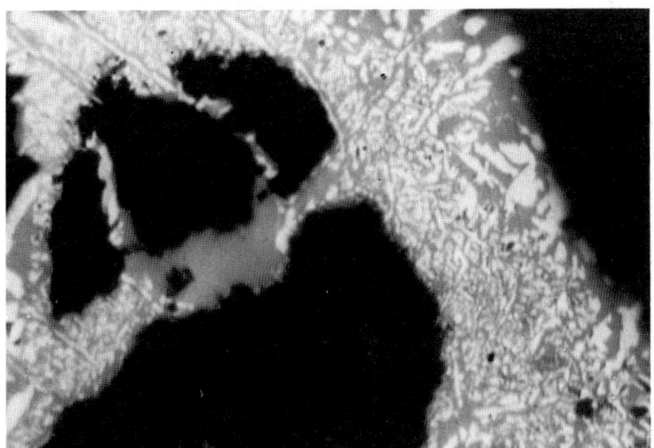

Fig. 4. Myrmekites of chalcopyrite *(white)* and fahlore *(grey)*. RPL, oil immersion, × 400

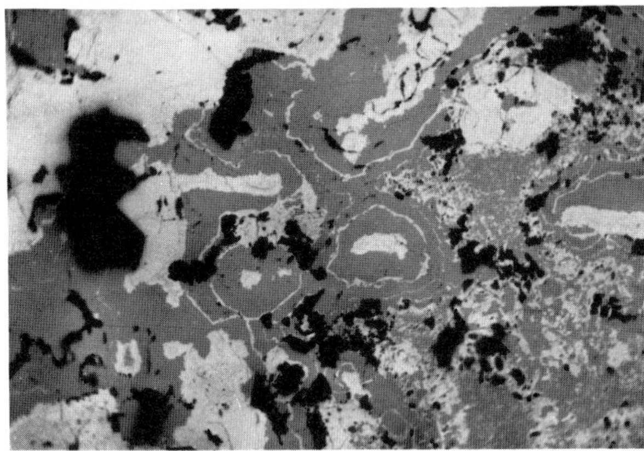

Fig. 5. Intergrowth of chalcopyrite *(white)* with fahlore *(grey)*. *Black* gangue minerals. RPL, oil immersion, × 200

Fig. 6. "disease-crack" in bornite *(dark grey)* with incipient formation of idaite *(light grey)* and minor chalcopyrite and chalcocite (*lighter* than idaite). RPL, oil immersion, × 200

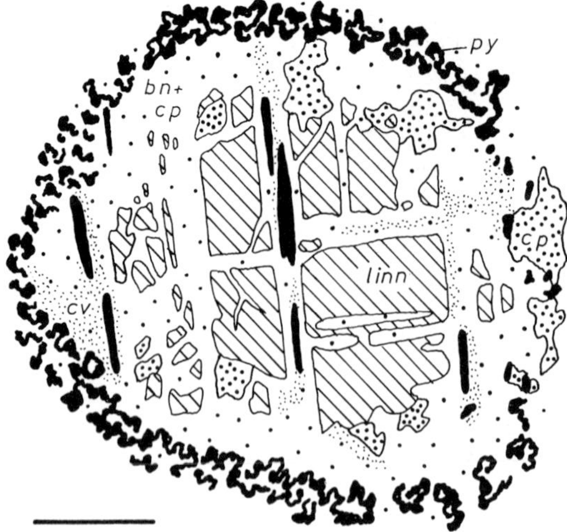

Fig. 7. Transformation of a linnaeite-siegenite grain *(linn)* into a fine-grained aggregate of cp+bn which bears locally irregular masses of cp. Pyrite *(py)* forms both onto linnaeite cleavage and around the linn grain. Irregular patches of covellite *(cv)* appear within the bn+cp. *Scale bar* is 0.15 mm

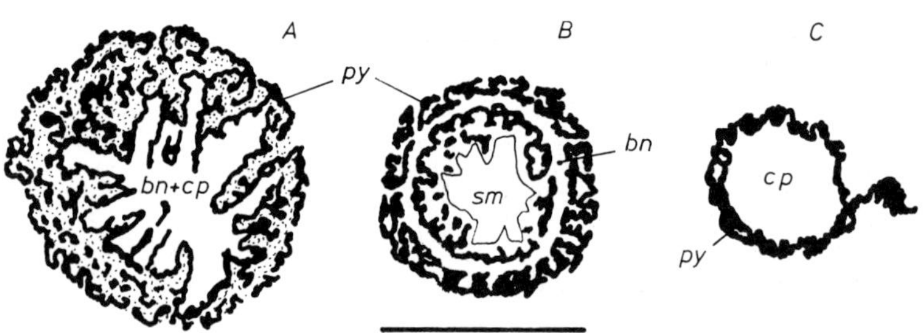

Fig. 8 A–C. Globular pyrite "grains" consisting of very fine pyrite micrograins *(py)* associated with secondary copper minerals. Inside the pyrite aggregate there is a mixture of bn+cp sometimes developed in flower-like pattern **(A)**. Alternating bands of pyrite and bn with a nucleus of secondary copper minerals *(sm)* **(B)**. Simple atoll- or shell-like pyrite grains in cp. *Scale bar* is 0.5 mm

means of a "coating" replacement by cp (Fig. 12). This fact points out that the globular form is here related to a destroying mechanism of the early-formed pyrite grains (by an opposite mechanism of the crystal growth) rather than to a coalescent growth of pyrite micrograins. Reliable evidence of mode of formation could not be obtained, despite its abundance in the copper ores of the Sasca deposit. The prevalence of copper in the ores generally seems to indicate instability of the early- or even simultaneously formed pyrite grains in the copper-rich environment. At Julesti-Valea Fagului deposit, Apuseni Mountains, Romania, the pyrite occurs subordinately within the copper ores and forms scattered grains either in cp or in bn (Udubasa et al. 1981). In other copper-rich ores, e.g., Vorta, Metaliferi Mountains, Ilba and Baiut, Maramures, Romania, occasionally occurring globular pyrite is always confined to the presence of the bornite. Therefore bn seems to induce the instabilization of pyrite.

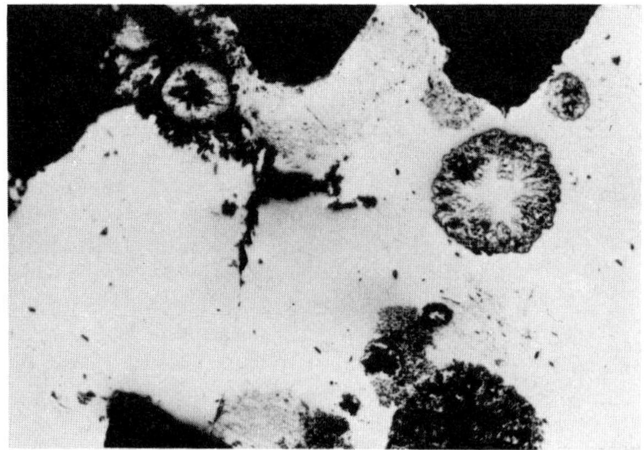

Fig. 9. Globular pyrite in chalcopyrite partly replaced by gangue minerals *(black)*. RPL, × 90

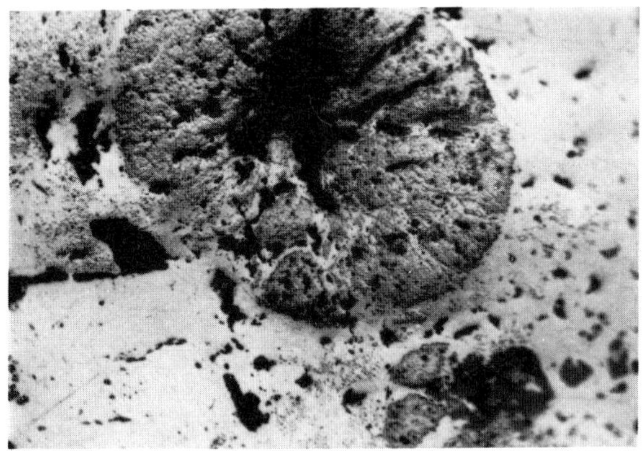

Fig. 10. Globular pyrite (0.5 mm in diameter) in a matrix of chalcopyrite + tetrahedrite. RPL

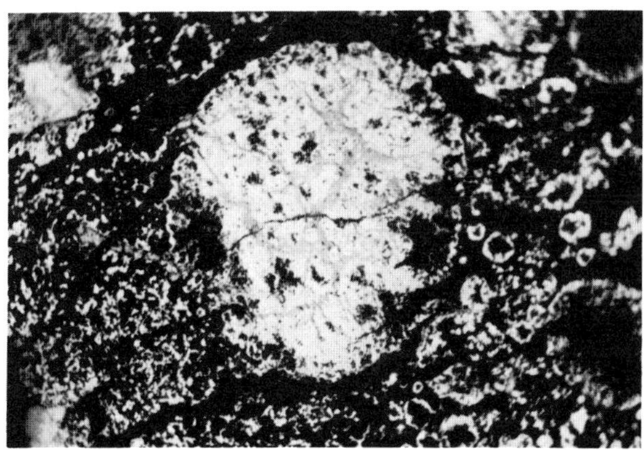

Fig. 11. A globular pyrite grain irregularly replaced by bornite and chalcopyrite, and smaller shell-like pyrite grains in a fine-grained calcite matrix *(black)*. RPL, × 90

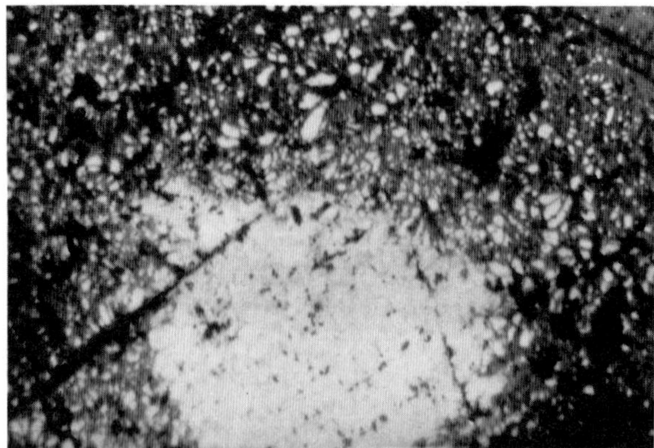

Fig. 12. Pyrite aggregate replaced by chalcopyrite (fine veinlets and patches, indistinguishable on the photograph) which alters further to covellite and secondary copper minerals *(black)*. This matrix contains relict grains of pyrite and chalcopyrite. The sample represents an oxidized ore, but the similar evolution of the pyrite aggregate is presumed to occur at higher temperature. RPL, × 90

It is likely that such "unstable" pyrites represent in fact decomposing products of the high-temperature solid solution series of FeS_2-CuS_2 (Shimazaki 1969). The low-temperature breakdown of this solid solution results in the formation of a metastable copper-bearing pyrite, perhaps also As-bearing (tennantite is locally present). Its further alteration leads to the apparition of the "normal" pyrite cemented both by cp, locally associated with bn, and more rarely fahlore. Under certain circumstances the globular form of pyrite may indicate also the pre-existence of the mineral fukuchilite which typically forms globular aggregates and transforms easily into pyrite and covellite (Kajiwara 1969).

4 Discussion

The ore minerals in the Sasca Montana area are assigned to some major deposition environments, i.e., skarns (garnet-vesuvianite-diopside-wollastonite), altered granodioritic rocks (porphyry-type assemblages), recrystallized limestones, and tectonic breccias. The concurrent ore mineral assemblages differ both in composition and in the intergrowth type, depending on their setting (Constantinescu 1981). The mineralizing process took place discontinuously and a non-equilibrium state prevailed during the entire postmagmatic activity. There is no evidence of a regional zoning of the skarn assemblages in the Sasca Montana area, but zoned and brecciated minerals do frequently occur as well as numerous paramorphs and reverse paragenetic sequences. The non-equilibrium state is mainly due to the repeated re-opening of the faults and fissures which favoured a persistent variation in the CO_2-activity. All this may explain the variety of the ore textures of the Sasca Montana deposit (partly presented in this paper) which characterizes ores formed under varying PT conditions and at varying Fe:Cu ratios of the ore solutions.

References

Betechtin AG, Genkin AD, Filimonova AA, Shadlun TN (1958) Structures and textures of the ores. (Russian). Sci Techn Publ House Geology Environ Preserv, Moscow

Cadere DM (1925) Facts with a view to help the mineralogical description of Romania. (Rumanian). Mem Acad Rom III/3

Constantinescu E (1971) Observations regarding the Laramian skarns and copper mineralizations of the Sasca Montana area. (Rumanian). Anal Univ Bucuresti Geol XX:65–74

Constantinescu E (1981) Skarn genesis at Sasca Montana, Banat. (Rumanian). Academiei, Bucuresti (in press)

Cotta B v (1864) Erzlagerstätten in Banat und Serbien. Wien

Kajiwara Y (1969) Fukuchilite, Cu_3FeS_8, a new mineral from the Hanawa Mine, Akita prefecture, Japan. Mineral J 5:399–416

Maucher A, Ramdohr P (1962) Bildkartei der Erzmikroskopie, Umschau, Frankfurt (M)

Ramdohr P (1945) Myrmekitische Verwachsungen von Erzen. Neues Jahrb Mineral Geol Palaeontol Abh 79A:161–191

Ramdohr P (1969) The ore minerals and their intergrowths. Pergamon Press, Oxford Braunschweig

Shimazaki H (1969) Synthesis of a copper-iron disulfide phase (Abstr). Can Mineral 10:146

Superceanu CI (1969) Die Kupfererzlagerstätte von Sasca Montana im SW-Banat und ihre Stellung in dem alpidisch-ostmittelmeerischen Kupfer-Molybdän-Erzgürtel. Geol Rundsch 58:798–860

Tufar W (1967) Der Bornit von Trattenbach (Niederösterreich). Neues Jahrb Mineral Abh 106: 334–351

Udubasa G, Istrate G, Popa C (1981) Preliminary data on the Julesti-Valea Fagului mineralizations, Apuseni Mountains. (Rumanian, Engl. Summary). Dări Seamă Inst Geol Geofiz Bucuresti LXIV/2 (in press)

Vujanovic V (1974) The basic mineralogic, paragenetic and genetic characteristics of the sulphide deposits exposed in the Eastern Black-Sea coastal region (Turkey). Bull Mineral Res Explor Inst Turk 82:21–36

Diagenetic Crystallisation and Migration in the Banded Iron Formations of Orissa, India

S. ACHARYA[1], G.C. AMSTUTZ[2], and S.K. SARANGI[3]

Abstract

Unmetamorphosed iron formations of the youngest stratigraphic sequence in Orissa, India, were studied for diagenetic features. Attention was concentrated on the oxide and carbonate facies of the iron formation only. Silica (chert), hematite and siderite are primary minerals which show evidences of more than one period of diagenetic crystallisation and migration. The study suggests an early to late diagenetic time during which all the features were produced. It also reveals that carbonate facies is much more important in the study of iron formation than hitherto appreciated.

1 Introduction

The Precambrian banded iron formations of Orissa, India, occur in three separate stratigraphic horizons in three basins (Acharya 1976; Prasadarao et al. 1964). Of these, the youngest and the largest one (in extent) occurs in the thickly forested Keonjhar-Sundergarh (Bonai) districts of the state. The rocks are almost unmetamorphosed and are excellent for different studies of their mineralogical-sedimentological aspects.

The iron formations (known locally as banded hematite jasper/or chert with its variants) and associated rocks form a low NNE plunging synclinorial basin (Jones 1934; Dunn 1940), the western limit of which is bordered by Bolani-Kiriburu-Barsua-Khandadhar range and the eastern boundary is marked by Noamundi-Joda-Mohaparbat range. The closure is represented by the hill range joining Kumritanr, Mankarnacha and Malangtoli to the south. Because of its peculiar shape, Jones named it "Horse-Shoe" (Fig. 1), and it is one of the best producers of iron and manganese ore in the country. Attention, however, has only been focussed in this paper on the banded iron formations and the diagenetic characters they exhibit. Research on a satisfactory definition of the diagenetic history of Precambrian iron formation or associated ore deposits of the unmetamorphosed type is not widely attempted mainly due to the lack of proper understanding of the role of diagenesis in sediments and more so in the case of iron, whose chemical behaviour varies greatly according to its ionic state (Bubenicek in Amstutz and Bubenicek 1967).

1 Department of Geology, Utkal University, Bhubaneswar, India

2 Mineralogisch-Petrographisches Institut der Universität, Heidelberg, FRG

3 Department of Marine Sciences, Berhampur University, Orissa, India (Formerly C.S.I.R. Fellow in Utkal University)

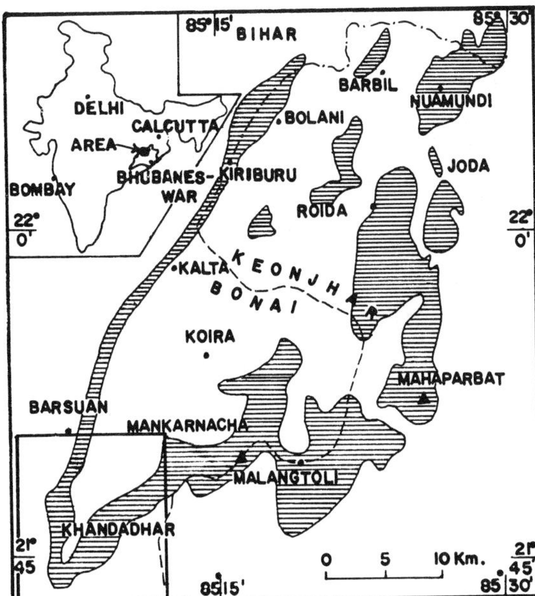

Fig. 1. Location of the area showing along with the generalised map of the youngest iron formation (Horse-Shoe)

2 Regional Setting

Except for Jones (1934), no one so far has worked on any aspect of the entire basin. A few authors have only worked on the petrology, stratigraphy or structure (Murty and Acharya 1975; Misra 1969; Sarangi 1973; Sarkar and Saha 1962, 1966; Sarkar et al. 1969; Misra 1961; Sarangi and Acharya 1975) locally or partly regionally. Acharya et al. (1968) and Acharya (1971) worked on the facies and diagenesis in iron formations. In brief, it may be mentioned that the iron formations (Khandadhar group) disconformably (?) overlie the lower volcanics and are succeeded by younger beds. A younger Kolhan group of rocks overlies the above sequence. The rocks have undergone three periods of folding which have made the correlation of the exposures difficult. The petrographic variations of the iron formations are as follows (Sarangi and Acharya 1975):

1. Banded hematite shale – 15 m approx.
2. Wavy bedded iron formation – 25 m approx.
3. Intercalated banded iron formations and shale – 10 m.
4. Regular bedded iron formation – 100 m.
5. Intercalated shale and banded siderite chert – 20 m.

Their mutual contacts when exposed are found to be gradational.

Samples from these rocks have been taken for the purpose of study. Hematite, siderite, magnetite, martite, goethite, greenalite, minnesotaite and chert/jasper occur in the rocks of iron formations. Pyrite and pyrrhotite occur as additional phases. Diagenetic crystallisation and later migration have changed their original character and in certain ways have enriched iron ore bands. Discussion of silicate and the sulphide facies is excluded from the present investigation.

The iron formations are locally known as BHJ/C or BHQ in India and the oxide facies (James 1954) of iron formation is the most common, as the local term suggests. Sedimentary facies of other minerals mentioned in the previous paragraph have not been frequently reported (Acharya et al. 1968). Nevertheless, the essential units in iron formations are oxides of silica and iron. The former remains fairly constant, while the latter has mineralogical variations.

Both chert and iron oxide of the rock are, however, "original" (Mengel 1973) because their distribution is not controlled by metamorphism, fracturing or folding, depth of burial or proximity to intrusions; the chert in the neighbourhood contains fossils (not shown here). In the same basin associated with cherty shales, stromatolites have also been reported.

Besides the oxides of iron and silica, a portion of the siderite is also primary (Acharya et al. 1968).

Figures 2–13 are given below with explanations.

3 Discussions

From the above figures it is quite clear that the bands of extremely fine chert and hematite are primary sediments which, during their formation, involved also "cut and fill processes" (Fig. 1). These two units are rarely mutually exclusive in bands and one is only larger in proportion to the other in respective bands. Along with them is the siderite which should also be regarded as formed bands similar to the above two mineral crystallites. They are, however, more euhedral peripherally; hematite, chert

Fig. 2. Primary "cut and fill" structure in banded hematite-jasper (black-chert). Note hematite rhombs after siderite. × 20

Fig. 3. Vug-filling from top in hematite bands by chert and siderite changed to hematite after adjusting in the new environment. × 20

Fig. 4. "Eyes" of chert in hematite bands. Note the reflection of depression of the "laces" on the darker band. Unreplaced magnetite *(grey)* is seen between the eyes. × 20

Fig. 5. Banded hematite jasper with vertical diagenetic silica veins *(white)*. Thin veins are deformed more in comparison to the thicker ones. × 20. P.L.

Fig. 6. Bedding plane stylolite bearing banded hematite jasper with vertical diagenetic silica veinlet. Note dissolution of stylolite by widening of the hematite bands. The bedding planes of banded hematite jasper are stylolitic. The veinlets cut the stylolite seams. × 20. P.L.

Fig. 7. The early diagenetic veinlet occurring only between a pair of hematite *(black)* and chert band *(white)* is deformed as shown later by superincumbent load. × 20. P.L.

Fig. 8. Close view of five veinlets, second and fourth are thicker than the rest. They are cut by recrystallised hematite and bent. The veinlets have somewhat protected the thickness of the band during compaction. × 20

Fig. 9. Late diagenetic siderite (euhedral) showing outgrowths of quartz on opposite faces indicating solution along bedding planes. × 20. P.L.

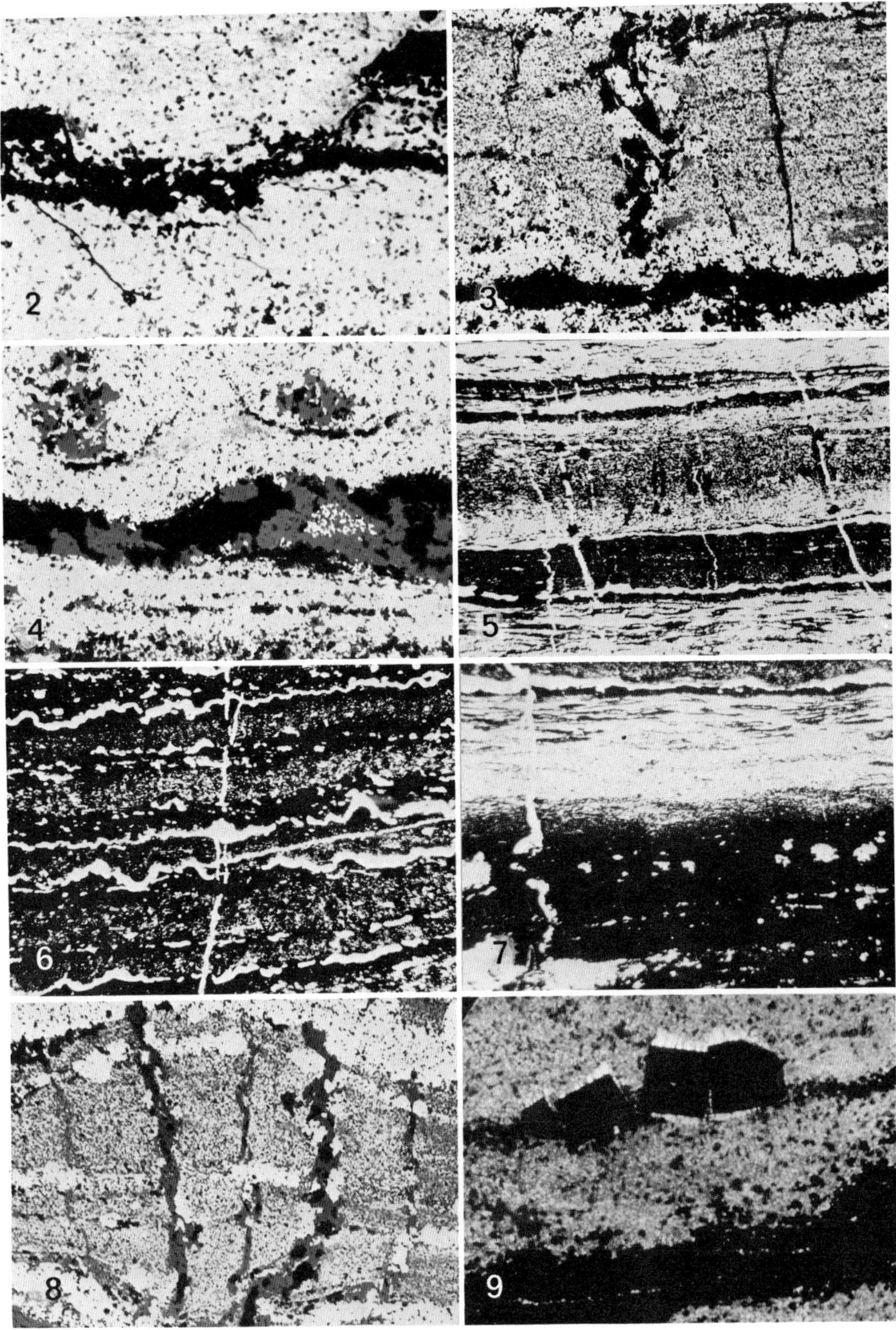
2
3
4
5
6
7
8
9

and siderite are, thus, primary materials. The vug filling and the eyes (Figs. 2 and 3) are criteria of top and bottom and gravitational diagenetic features respectively.

Bedding plane stylolites common in the banded hematite chert are confined to the zones of recrystallised hematite that lies usually above and below them.These seams have possibly formed by the depression on downward migration of the seam due to the weight of the overlying bulk of recrystallised hematite aggregated in the depression. The ridges on the lower sides, thus, display a concentration in hematite. Later it will be shown that the stylolites are diagenetic (Amstutz and Park 1967). Stylolites are responsible for gradual thickening of the iron-oxide bands.

Syneresis has produced cracks that are filled by thin and upright silica veins, which get deformed and bent during the pressure of the overlying rock mass resulting in swelling of bands where they are present (Park and Amstutz 1968). These veins are usually confined to the individual iron oxide bands though wider and longer veins also exist. This reduction of the volume of iron oxide bands might have taken place by compaction; the hematite is probably "returned" to the system to act as recrystallised hematite that cuts the later silica veins which widen downwards. These veins, as shown in Figs. 6 and 11, cut a series of alternating bands of hematite and chert and also cut the stylolite bands. The sharpness of their contact with bands indicates that most of the compaction of the bands took place prior to them. The bands of recrystallised hematite have penetrated these veins, indicating a later period of recrystallisation of hematite (the distinctions of hematite 2 and 3 are at times not clear but inferable).

The mineralogy of the iron formations in this region is simple with only a few minerals; while the chert/jasper and hematite are mutually intimately existing as primary bands consisting of very fine crystallites (dust-like) giving a "pepper and salt" textural pattern (Figs. 4, 7, 9, 11), siderites are also diagenetically recrystallised to rhombohedrons and form bands parallel to those of chert and hematite and in some cases are fully replaced by hematite through magnetite whose vestiges are sometimes present in hematite. While the siderite bands are partly or fully replaced by iron oxide, the grains inside the middle of chert bands are not commonly affected, possibly due to non-availability of iron oxide in the environment. A third type of siderite grains, perfect in shape and of giant size, is found, at times, almost resting on certain bands. These are

Fig. 10. The "pepper and salt" texture of primary hematite and chert. Euhedral siderite grains completely changed to hematite. Usual remnants of magnetite are not noticed. Siderites in the middle part of the bands are less replaced. × 30

Fig. 11. Bands show recrystallised hematite *(white)* after siderite through magnetite. The chert bands *(black)* and fine hematite bands *(grey)* have one in the other. Diagenetic silica veinlet pressed by the overlying rock mass are bent and deformed but have resisted the pressure during compaction. The hematite bands were thicker by about 42%. × 30

Fig. 12. Diagenetic silica vein cuts across the primary chert or hematite bands showing pepper and salt texture *(very fine white and black)* and the stylolite bands are cut across by hematite (recrystallised – *white*). The bedding planes between the recrystallised mini-bands are stylolitic and so also in the chert bands. × 30

Fig. 13. A magnified view of a part of a giant crystal of siderite formed after almost complete compaction (bedding plane is not pressed down). This is partially replaced by hematite obviously at a later diagenetic stage. The hematite bands show intimate mixture with chert. × 60

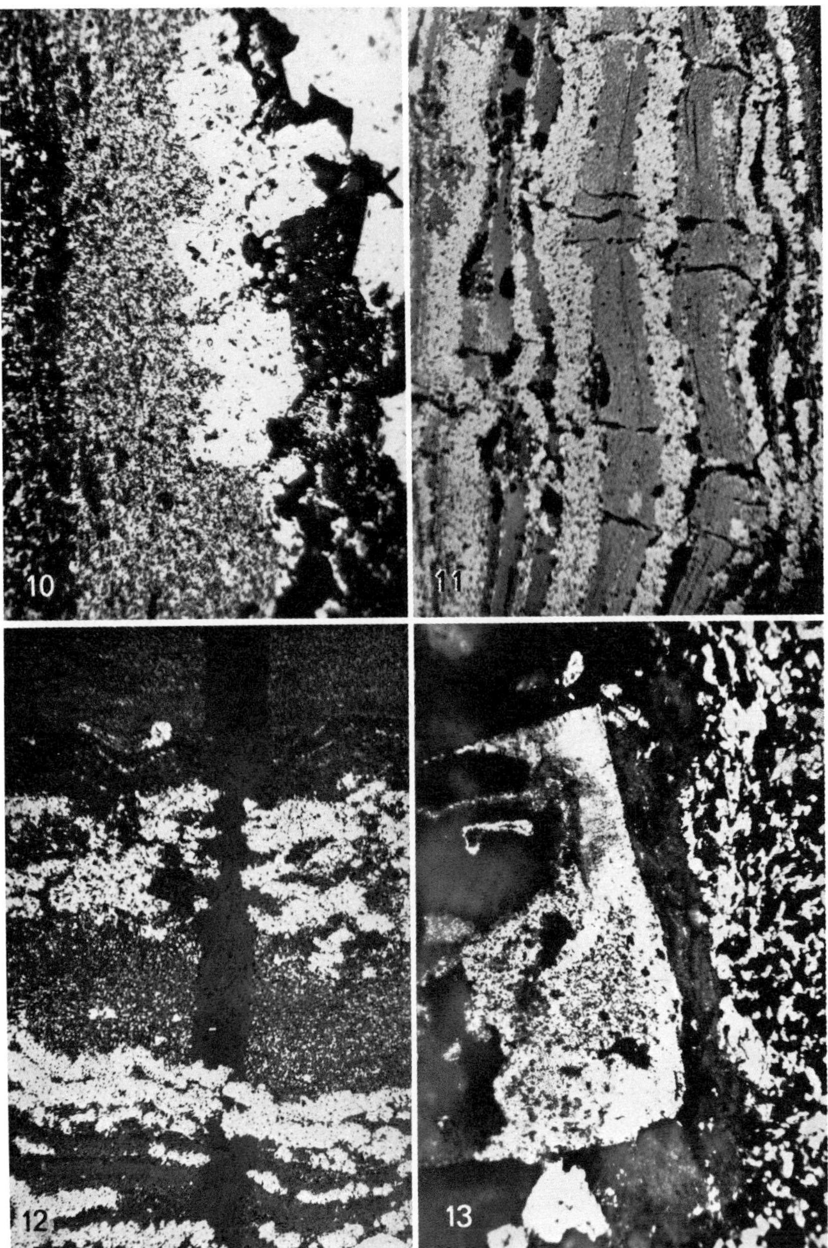
10
11
12
13

also replaced, though only partially, by hematite (Fig. 12). They do not depress the primary bands and are, thus, later than compaction and earlier than the last generation of hematite.

It is apparent from the above discussion that the recrystallisation of hematite is the last in the sequence of events. Effects of metamorphism do not seem to exist at all as chert was nowhere found to be recrystallised to quartzite (or bigger textures). All these features and events can be relegated to a time-period of early diagenesis with the exception of the larger veins which are possibly features of late diagenesis due to their sharp contacts with beds indicating a post compaction time of formation (Dimroth 1979). The confinement of thin silica veinlets to individual bands of only a few bands, their deformations and micro-slips (Figs. 4 and 10) indicate the effect of pressure of overlying sediment. A contrast in thickness of individual bands near the veins and away from them go a long way to support this. The thicker veins are somewhat later and are not deformed by the sediment load but are post-stylolite and pre-recrystallised hematite. In the absence of any regional metamorphic evidence all the features present are ascribed to diagenesis (Dimroth 1979). Figure 14 shows the paragenetic sequence in iron formation in relation to the events.

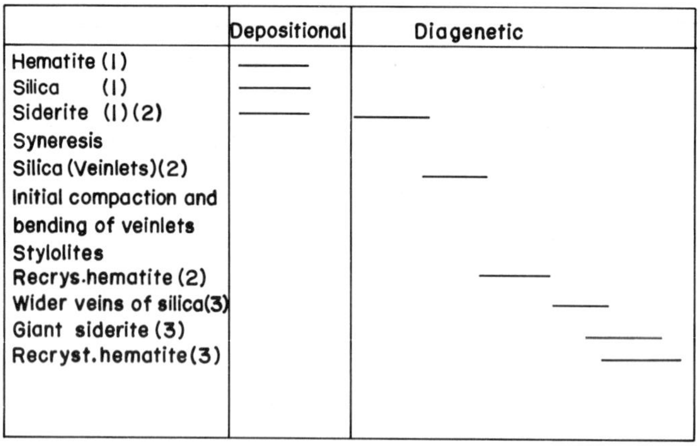

	Depositional	Diagenetic
Hematite (1)		
Silica (1)		
Siderite (1)(2)		
Syneresis		
Silica (Veinlets)(2)		
Initial compaction and bending of veinlets		
Stylolites		
Recrys.hematite (2)		
Wider veins of silica(3)		
Giant siderite (3)		
Recryst.hematite(3)		

Fig. 14. Paragenetic sequence of minerals, along with sequence of events in the iron formation; *(1), (2)* etc. indicate order of crystallisation

The definition of mineral facies proposed by Dimroth and Chauvel (1973) (cited in Dimroth 1979) over James (1954) concept of mineral facies include submicroscopic hematite and fine-grained siderite etc. Zajac (1973) and Klein and Fink (1976) defined it by the assemblage of iron minerals now present. Obviously both Zajac and Klein's works cannot explain diagenesis only and hence the sulphides present in the iron formations have not been discussed to avoid likely confusion.

4 Conclusion

Diagenetic changes play an important role in the Precambrian iron formations which can be well studied from the younger ones undergoing only slight or no metamorphism. The carbonate facies is much more developed than hitherto realised. The oxide facies are mostly diagenetic crystallisation products of the carbonate facies (chert-siderite alternating bands).

Acknowledgments. The award of the Alexander von Humboldt Fellowship to the first author to study at the Mineralogisch-Petrographisches Institut der Universität, Heidelberg, W. Germany, is gratefully acknowledged. The Council of Scientific and Industrial Research, India, financed the field work.

References

Acharya S (1971) Diagenetic features in iron formations. Indian Sci Congr Abstr

Acharya S (1976) Iron formations and iron ores of Orissa – their stratigraphy and correlation. Proc Symp Geol etc. Fer and Fe-alloy Mine, Bangalore, pp 86–100

Acharya S, Ahmed SIS, Sarangi SK (1968) Carbonate facies of iron formations, Orissa. Bull Geol Soc India 3 (2):41–44

Amstutz GC, Bubenicek L (1967) Diagenesis in sedimentary mineral deposits. Dev Sedimentol 8

Amstutz GC, Park WC (1967) Stylolites of diagenetic age and their role in the interpretation of the southern Illinois fluorspar deposits. Miner Deposita 2:44–53

Dimroth (1979) Diagenetic facies of iron formation. Geosci Can 4 (2):83–88

Dunn JA (1940) Stratigraphy of south Singhbhum. Mem G S I 63 (3):303–369

James HL (1954) Sedimentary facies of iron formation. Econ Geol 49 (3):235–293

Jones HC (1934) The iron ore deposits of Bihar and Orissa. Mem G S I 63 (2):162–302

Klein C Jr, Fink RP (1976) Petrology of the Sokomon in the Howells River area. Econ Geol 71: 453–487

Mengel JT (1973) Physical sedimentation in Precambrian cherty iron formation of Lake Superior type. In: Amstutz GC, Bernard AJ (eds) Ores in sediments. Int Union Geol Sc Sr A no 3, pp 179–193

Misra A (1969) The problem of the Precambrian iron formations (the Noamundi iron bearing district of India). Unpubl D Sc thesis, Univ Brussels, p 231

Misra GB (1961) A note on the stratigraphy of the Joda area, Keonjhar. Geol Metall Min Soc India QJ

Murty VN , Acharya S (1975) Lithostratigraphy of the Precambrian rocks around Koira, Sundergarh district, Orissa. J Geol Soc India 16 (1):55–68

Park WC, Amstutz GC (1968) Primary "cut and fill" channels and gravitational diagenetic features. Miner Deposita 3:66–80

Prasadarao GHSV, Murty YGK, Deekshitulu MN (1964) Stratigraphic relations of Precambrian iron formations and associated sedimentary sequences in parts of Keonjhar, Cuttack, Dhenkanal and Sundergarh districts of Orissa, India. 22nd IGC, New Delhi, pt X, pp 72–87

Sarangi SK (1973) Iron formations of Khandadhar-Lusi region, Orissa, India – its structure, sedimentation and chemistry with special reference to the geology of associated iron ores. Unpubl Ph D thesis, Utkal Univ

Sarangi SK, Acharya S (1975) Stratigraphy of the iron group around Kahdadhar, Sundergarh district, Orissa. Indian J Earth Sci 2 (2):182–189

Sarkar SN, Saha AK (1962) A revision of the Precambrian stratigraphy and tectonics of Singhbhum and adjacent regions. Q J G M M Soc India 34:97–136

Sarkar SN, Saha AK (1966) On the stratigraphic relationship of the metamorphic rocks of Singhbhum Gangpur region – Contribution to geology of Singhbhum, Jadavpur Univ, pp 21–31

Sarkar SN, Saha AK, Miller JA (1969) Geochronology of the Precambrian rocks of Singhbhum and adjacent regions, Eastern India. Geol Mag 106 (1):13–45

Zajac (1973) Cited in Dimroth (1979)

Iron Formations of Precambrian Age: Hematite-Magnetite Relationships in Some Proterozoic Iron Deposits – A Microscopic Observation

T.M. HAN[1]

Abstract

Numerous specimens of metamorphosed Proterozoic iron formations in North America were studied microscopically. Textural relations exhibited by magnetite and hematite are described and interpreted.

Hematite-magnetite-quartz is known to be an equilibrium assemblage in oxide iron formations of all metamorphic grades. Most investigators interpret the persistence of this assemblage to indicate the lack of movement of oxygen between layers of iron formation during metamorphism. Microscopic evidence presented here reveals extensive replacement of hematite by magnetite and vice versa. This suggests that oxide iron formations were indeed more open to the movement of oxygen between layers during metamorphism than has often been realized.

The study also shows that textures exhibited by hematite-magnetite depend on the type and metamorphic grade of iron formations.

1 Introduction

Metamorphosed oxide iron formations have been studied by many investigators including James (1955), James and Howland (1955), Klein (1973) and others. Hematite-magnetite-quartz is observed to be a stable assemblage at all metamorphic grades. The persistence of the hematite-magnetite-quartz association has suggested to most that the degree of oxidation of an iron formation is established prior to metamorphism, and that during metamorphism individual layers behave as systems closed to oxygen transfer. However, the present microscopic study suggests that oxidation and reduction during metamorphism are much more common than has been recognized.

This study reveals the replacement textures exhibited by hematite-magnetite in some of the Proterozoic oxide iron formations of varying metamorphic grades in North America.

2 Characteristics of Deposits Related to Metamorphic Grade

Specimens from slightly metamorphosed deposits typically show very fine-grained gangue minerals and many exhibit sedimentary structures. Clastic quartz, if present, is easily distinguished from the recrystallized chert. In such specimens magnetite is

1 The Cleveland-Cliffs Iron Company, 504 Spruce Street, Ishpeming, MI 49849, U.S.A.

considerably coarser than the coexisting hematite, and it frequently contains inclusions of hematite or preexisting hematite (magnetite pseudomorphs after hematite). Greenalite, chlorite, minnesotaite, and stilpnomelane are the most common silicates, whereas magnesian siderite and ankerite are the most frequent carbonates. These specimens were collected from the Empire Mine and vicinity, Marquette District, Michigan; the Gogebic District, Wisconsin and Michigan; the west and central Mesabi District, Minnesota; and the Lake Albanel District, Quebec, Canada.

Specimens from moderately metamorphosed deposits are characterized by the development of grunerite, cummingtonite, actinolite, biotite, epidote, and garnet; the disappearance of greenalite and minnesotaite; the introduction of the exsolution texture of rutile-specularite in the specularite-muscovite-chert assemblage; the obliteration of preexisting hematite crystal outlines in magnetite; the cross-cutting of fine- to medium-grained magnetite by specularite; and the increase of chert grain size along with the decrease of the amount of ore mineral inclusions in chert. Furthermore, the specularite plates are considerably larger than those in the deposits of low metamorphic grade whereas the magnetite grains, in most cases, show little increase in size, except for the presence of some sporadic porphyroblasts closely associated with specularite and fine-grained magnetite. Specimens of this metamorphic grade were collected from the Humboldt, Champion, Republic Mines and their vicinities, Marquette District; Groveland Mine and vicinity, Menominee District, and the Hopes Advance Bay District, Quebec, Canada.

In the deposits of high metamorphic grade, the mineral sizes tend to become nearly equigranular. Chert (quartz) ranges from medium to coarse-grained and contains fewer ore mineral inclusions than that in the lower metamorphic grades. The specularite develops along rather than cutting across the grain boundaries of gangue minerals. Magnetite is commonly coarse-grained, subhedral to euhedral, and usually cuts across the grain boundaries of both specularite and gangue minerals. Specimens from this type of deposit were collected from Quebec Cobalt and O'Keefe Lake of the South Labrador Trough and from the Carter Creek area of Ruby Mountains, Montana, U.S.A.

3 Hematite-Magnetite Relationships

3.1 Slightly Metamorphosed Iron Deposits

In an earlier study (Han 1978), the author found that magnetite formed at the expense of hematite in iron formations of diagenetic and low metamorphic grade. Figs. 1 through 5 show the evidence for this conclusion. The degree of replacement of hematite by magnetite, and the textures exhibited by these minerals are related to the type of iron formations as follows:

1. Magnetite in Jaspilitic Ores. The magnetite in these ores generally occurs as single or coalesced octahedra in hematite or hematitic chert bands. Some of the magnetite octahedra were enlarged by one or more than one stage of overgrowth, as indicated by the presence of concentric dusty rings, rhythmic zoning and core-and-shell arrangement (Han 1978). Some of the magnetite was developed by centripetal replacement of hematite supplemented by extensive overgrowths. This is indicated by the presence of hematite remnants or the outlines of preexisting hematite inclusions in magnetite laminae and coalesced magnetite octahedra. Figure 1 shows mirror images of three hematite-containing magnetite clusters before and after the laboratory-induced oxidation. It clearly demonstrates that the hematite remnants are actually parts of the preexisting hematite rhombs. In some ores, the preexisting hematite inclusions were apparently hallow and nearly spherical bodies, with inner linings made up of wedge-shaped and perfect rhombohedral hematite crystals (Fig. 2). The morphology of these coalesced magnetite octahedra is closely guided by that of the preexisting hematite inclusions.

2. **Magnetite in Banded Carbonate-Rich and Silicate-Rich Ores.** Han (1978) concluded that the magnetite in some of the iron formations referred to as "carbonate", "silicate" or "mixed" facies from the Marquette, Gogebic and Lake Albanel Districts was actually hematite prior to low grade metamorphism. This is evidenced by the presence of preexisting hematite inclusions in bands, laminae, granules, and irregular clusters of fine-grained magnetite. The outlines of these inclusions are lath-shaped with random to subparallel arrangements, and heavily populated. They are exposed either by the induced oxidation procedure employed (Han 1978) or by natural oxidation under certain supergene conditions. The following are some typical examples:

a) *Preexisting Hematite in Elongated Magnetite Grains.* The elongated magnetite grains occur in extremely thin laminae of a finely laminated silicate iron formation from the Lake Albanel District, Quebec, Canada. These grains are in subparallel arrangement and contain preexisting hematite of a similar morphology and orientation (Han 1978).

b) *Preexisting Hematite in Magnetite Bands and Laminae.* The magnetite bands and laminae in the iron formations with a mineral assemblage of magnetite-chert, magnetite-silicate-chert, or magnetite-carbonate-silicate-chert often contain appreciable preexisting hematite inclusions. These inclusions are uniform sized, randomly arranged, typically lath-shaped, and occasionally hexagonal (Fig. 3). In some cases, these bands or laminae exhibit rather complicated microstructures as a result of successive stages of magnetite development (Fig. 4).

3. **Magnetite in Granular Ores.** In the granular ores, the magnetite occurs either as bladed grains or coalesced octahedra in granules with hosts of chert, carbonate-chert, silicate-chert, or carbonate-silicate-chert. The preexisting hematite inclusions are almost always present in these blades and coalesced octahedra, which may become visible upon laboratory-induced oxidation. Those in the coalesced octahedra are commonly lenticular, wedge-shaped, botryoidal, and occasionally in hexagonal or rhombohedral crystals. These inclusions are single or in groups and often display various types of beautiful microstructures. Those in the bladed magnetite are slightly smaller but very similar in morphology as compared to their respective bladed magnetite host. In other words, the bladed magnetite grains are enlarged pseudomorphs after hematite, as a result of pseudomorphic replacement and limited marginal overgrowth. Figure 5 clearly shows that each of the bladed pseudomorphs is rimmed by a thin layer of saw-tooth-like magnetite octahedra. These tiny, well-oriented magnetite octahedra represent the portion of the bladed magnetite formed by the overgrowth.

The granular ores studied were from the Mesabi District, Minnesota; Gogebic District, Wisconsin; and the Lake Albanel District, Quebec, Canada.

3.2 Moderately Metamorphosed Iron Deposits

Magnetite and hematite are, for the most part, intergrown in ores of this metamorphic rank. Nevertheless, the replacement of magnetite by hematite and vice versa are very common. Features observed suggesting the replacements are as follows:

1. **Specularite in Magnetite-Rich Bands.** Specularite embedded in bands or granules of very fine-grained magnetite is not uncommon in the specularite-magnetite-chert ores from the Republic Mine, Marquette District, Michigan. It generally occurs as randomly arranged, lath-shaped crystals much larger than its magnetite host (Fig. 6). These crystals are interpreted as porphyroblasts developed during the period of metamorphism.

2. **Residual Magnetite in Specularite Bands.** Residual magnetite clusters and magnetite-chert islands are frequently observed in the specularite ore bands of the so-called "specular hematite-chert ore" from the Republic Mine, Marquette District, Michigan. The magnetite is, as a rule, rather fine-grained and occurs as partially oxidized clusters distributed in the interspaces of the specularite blades (Fig. 7), or as residual clusters arranged in layers cut into by specularite crystals of the specularite bands (Fig. 8).

3. **Hematite-Containing Magnetite.** The hematite-containing magnetite in the specimens studied displays two different modes of occurrences:

Fig. 1. A Hematite remnants in magnetite prior to induced oxidation. **B** A mirror image of **A** after induced oxidation. × 160

Fig. 2. Coalesced magnetite octahedra with inclusions of rhombohedral and dog-tooth hematite exhibiting a geode type structure. After induced oxidation. Mesabi District. × 325

Fig. 3. Preexisting hexagonal hematite crystals in coalesced magnetite octahedra. After induced oxidation. Mesabi District. × 160

Fig. 4. Note the individuals and clusters of lath-shaped preexisting hematite in enlarged minute magnetite crystals and overgrown coalesced magnetite octahedra of a magnetite lamina. After induced oxidation. Marquette District. × 160

Fig. 5. Sawtooth microstructure of magnetite on preexisting bladed hematite. After induced oxidation. Lake Albanel District. × 160

Fig. 6. Specularite embedded in ultrafine-grained magnetite of a magnetite-chert ore. Republic Mine, Marquette District. × 65

Fig. 7. Specularite-chert ore with residual islands of fine-grained magnetite chert. Republic Mine, Marquette District. × 65

Fig. 8. Coarse-grained specularite layer with magnetite remnants. Note how specularite cuts into clusters of partially oxidized magnetite. Republic Mine, Marquette District. × 65

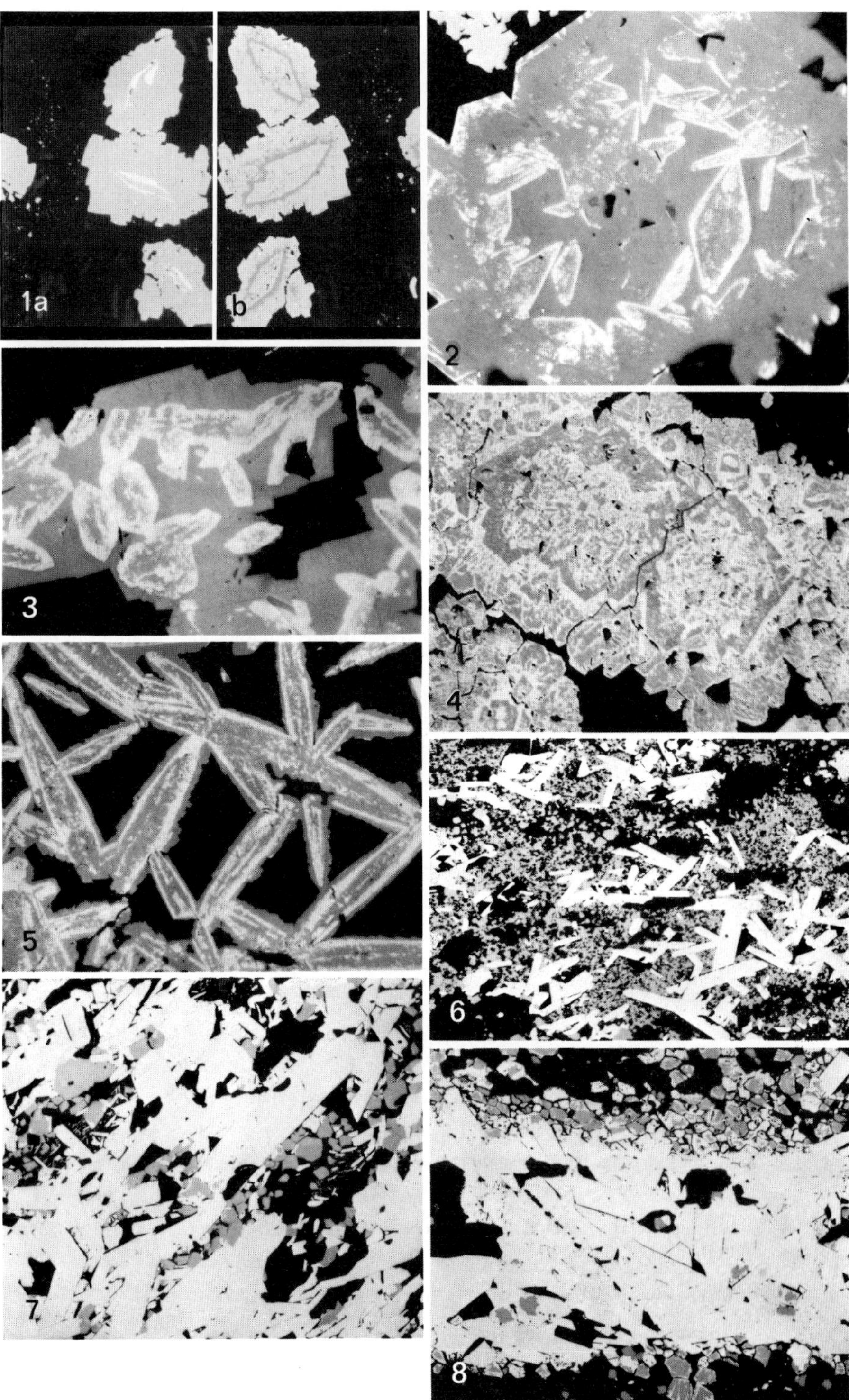
1a
b
2
3
4
5
6
7
8

a) Hematite Inclusions in Magnetite Clusters. Irregular magnetite clusters in the specularite-magnetite-chert ore from the Menominee District, Michigan, often contain numerous, elongate ill-formed hematite blades (Fig. 9). These blades are randomly arranged, often terminated with either featherlike or wedge-shaped ends, and believed to be remnants of preexisting specularite left behind by metamorphic reduction. The hematite-containing magnetite of this type was also observed in some of the ores from the Hopes Advance Bay District, Quebec, Canada.

b) Hematite in Magnetite Octahedra. Some granular hematite grains were seen in magnetite of a garnet-bearing magnetite-specularite-chert ore from the Groveland Mine, Menominee District, Michigan. These grains are irregular and mostly confined within the crystal boundary of magnetite, although some may not be completely enclosed by the magnetite (Fig. 10). Such a textural relationship was apparently established during metamorphism.

4. Magnetite-Containing Garnet in Specularite-Rich Bands. Small octahedra of magnetite coexisting with varying amounts of ultrafine-grained hematite were observed as inclusions in garnet embedded in specularite-rich bands of some magnetite-specularite-chert ores (Fig. 11). Their mode of occurrence is similar to that in some of the slightly metamorphosed jaspilites. These inclusions are believed to be residues of magnetite-jaspilite which survived the metamorphic oxidation process.

5. Magnetite-Containing Specularite. Some specularite blades are characterized by the presence of appreciable magnetite inclusions. These inclusions are irregular, occasionally lath-shaped, and probably remnants left behind from metamorphic oxidation of magnetite pseudomorphs after hematite (Fig. 12). This magnetite-containing specularite is closely associated with both specularite and magnetite in an ore consisting of alternating layers of magnetite-containing specularite-chert and magnetite-specularite-chert from the Hopes Advance Bay District, Quebec, Canada.

Fig. 9. Magnetite with hematite inclusions coexisting with specularite, magnetite, and chert. Menominee District. × 130

Fig. 10. Granular hematite replacing magnetite octahedra in a garnet-bearing specularite-chert ore. Menominee District. × 130

Fig. 11. Garnet containing magnetite enveloped by specularite in a garnet-bearing specularite-magnetite-chert ore. Menominee District. × 65

Fig. 12. Magnetite remnants in specularite of a magnetite-specularite-chert ore. Hopes Advance Bay District, Quebec. × 40

Fig. 13. Magnetite porphyroblasts in a host containing specularite, fine-grained magnetite, and chert. Menominee District. × 160

Fig. 14. Magnetite porphyroblasts in specularite. Note the microstructure of the magnetite. Champion Mine, Marquette District. × 30

Fig. 15. Partially oxidized magnetite cross-cutting specularite. Quebec Cobalt, Labrador Trough, Quebec. × 65

Fig. 16. Specularite with magnetite inclusions coexisting with magnetite in a specularite-magnetite ore. Carter Creek District, Montana. × 80

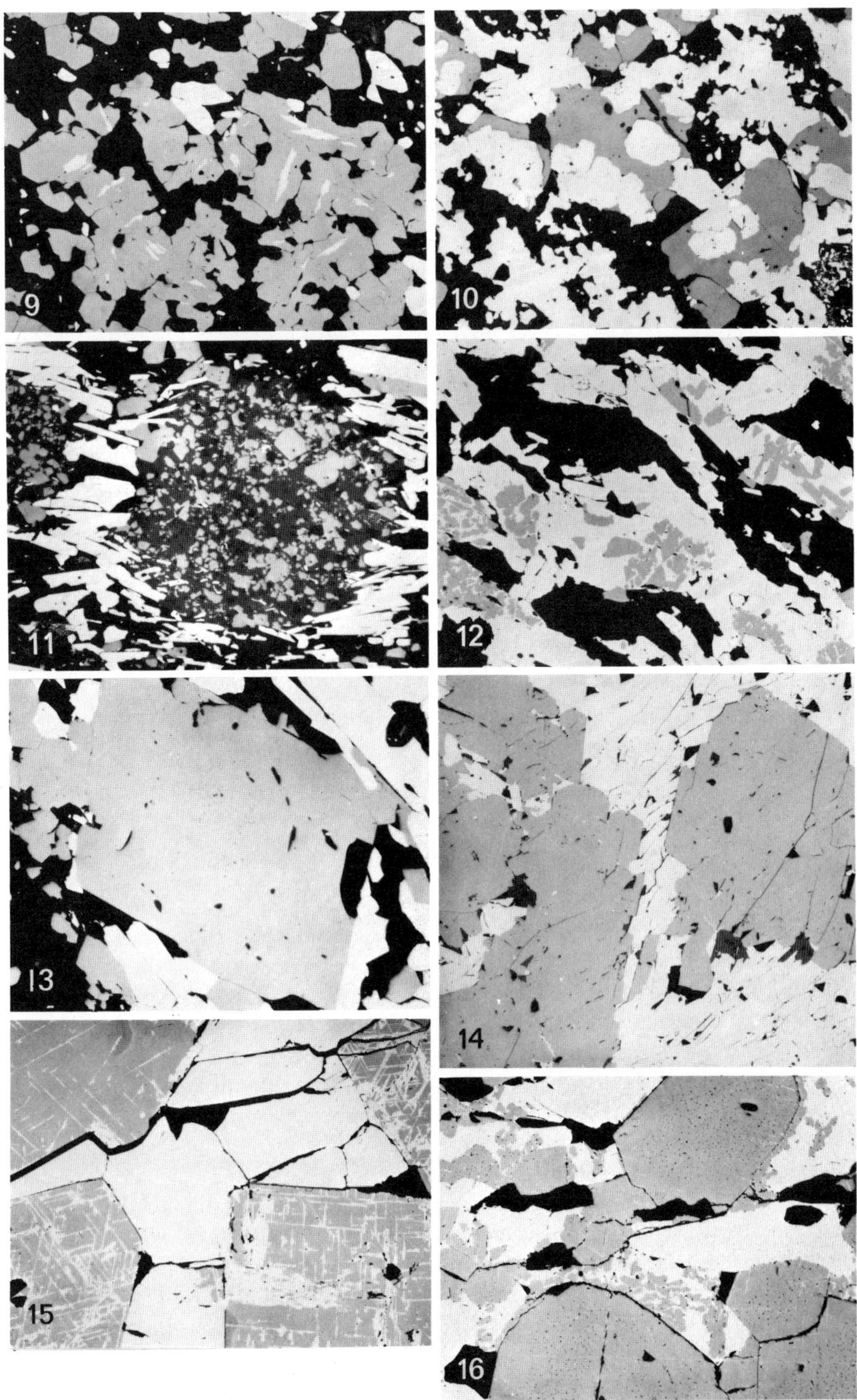
9
10
11
12
13
14
15
16

6. Magnetite in Specularite-Rich Ores. Magnetite grains have often been seen in ore bands of some specularite-chert, magnetite-specularite-chert, and magnetite-silicate-chert ores. These grains are apparently porphyroblasts subsequently developed during metamorphism. Some are in euhedral crystals cross-cutting quartz, magnetite, and specularite of much smaller sizes (Fig. 13); whereas others are anhedral to subhedral frequently containing specularite remnants, some of which are apparently clusters of magnetite pseudomorphs after hematite (Fig. 14).

3.3 Highly Metamorphosed Iron Deposits

The iron deposits of this metamorphic rank in both the Labrador Trough and the Carter Creek area of the Ruby Mountains, Montana, have been investigated by Klein (1973) and Immega and Klein (1976). According to Klein, specularite-magnetite-quartz persists through high-grade metamorphic conditions without reaction. The polished section study on the ores from Quebec Cobalt and O'Keefe Lake of the Labrador Trough shows that the magnetite (or martite) occurs as fractured subrounded grains or euhedral crystals cross-cutting specularite and quartz grains (Fig. 15). Such a textural relationship is interpreted as a result of the development of magnetite porphyroblasts in specularite-quartz ores during metamorphism.

The magnetite and specularite in the ores from the Carter Creek deposits appear to be developed along rather than cutting across the grain boundaries of gangue. They are normally in mutual contact. However, some of the specularite frequently contains some to substantial amounts of magnetite inclusions (Fig. 16). This suggests that there was definitely an exchange in oxygen during the metamorphism.

4 Possible Factors Affecting Hematite-Magnetite Relationships

The hematite-magnetite relationships in the three metamorphic ranks investigated are apparently determined by the primary lithology of the deposits and the local metamorphic conditions to which the deposits were subjected. It is believed that organic carbon and water are probably two key constituents affecting the hematite-magnetite relationships in the deposits during metamorphism.

The effect of primary organic carbon in iron formation on the hematite-magnetite relationship has been emphasized by Han (1978) and will not be discussed further. As to the role of water, Eugster (1959) notes that reduction and oxidation most readily occur during metamorphism where water is present, and whether a mineral assemblage is being reduced or oxidized depends on the composition of the aqueous phase introduced. He further concludes that the limited exchange of oxygen between magnetite and hematite bands in iron formation at the Republic Mine during metamorphism was probably sparked by the presence of water. It is the writer's belief that the development of an intergranular system, microfissures, and microfractures during metamorphism might have functioned as reservoirs and avenues for water, oxygen, hydrogen, etc. to move into and out of the oxide iron formation layers generally referred to as "a closed system with regard to oxygen during metamorphism" (Yoder 1957). Consequently, the presence of water, along with the development of structural weakness in the iron formations, are the driving forces for oxidation and reduction occurring between hematite and magnetite during the period of moderate to high metamorphism.

5 Conclusions

Hematite-magnetite is more than a common mineral assemblage in metamorphosed Proterozoic sedimentary iron deposits. These minerals are more susceptible to interconversions under metamorphic conditions than has generally been realized. In low metamorphic rank deposits, magnetite not only replaces hematite, but is also modified by overgrowth on the preexisting hematite or magnetite of an earlier generation. In moderately to highly metamorphosed deposits, both the specularite and the magnetite are not only the results of recrystallization but also the products of metamorphic oxidation and reduction.

Based on the extent of the replacements, most of the magnetite deposits of the low metamorphic grade were apparently hematite iron formations. Some of the specularite-magnetite deposits of the moderate to high grade metamorphism were probably either magnetite or specularite deposits.

Acknowledgments. The author would like to express his sincere thanks to Professor Henry Lepp for his valuable suggestions and great assistance in the improvement of the manuscript.

This study is a part of the low grade iron ore investigation program carried out by The Cleveland-Cliffs Iron Company. Acknowledgment is therefore due to the management for permission to publish this paper.

References

Eugster HP (1959) Reduction and oxidation in metamorphism. Researches in geochemistry. Wiley, New York, pp 377–426

Han TM (1978) Microstructures of magnetite as guides to its origin in some Precambrian iron-formations. Fortschr Mineral 56:105–142

Immega IP, Klein C (1976) Mineralogy and petrology of some metamorphic Precambrian iron-formations in Southwestern Montana. Am Mineral 61:1117–1144

James HL (1955) Zone of regional metamorphism in the Precambrian of Northern Michigan. Geol Soc Am Bull 66:1455–1488

James HL, Howland AL (1955) Mineral facies in iron and silica-rich rocks (abstract). Geol Soc Am Bull 66:1580

Klein C (1973) Changes in mineral assemblages with metamorphism of some banded Precambrian iron-formations. Econ Geol 68:1075–1088

Yoder HS Jr (1957) Isograde problems in metamorphosed iron-rich sediments. Annu Rep Dir Geophys Lab, Carnegie Inst Washington, Pap 1277, pp 232–237

Thorium Concentrations in the Bodenmais, Bavaria, Sulphide Deposit

E.O. TEUSCHER[1]

Abstract

Concentrations of monazite were observed in the Bodenmais sulphide deposit in a separate layer in the immediate contact with the main ore body; traces of radioactivity in the sulphide ores are possibly products of secondary contamination.

1 Introduction

In the years 1950–1962, just before the old mine Bodenmais in the Bavarian Forest, Germany, was abandoned after an activity period of 500 years, exploration work had been carried out in the ore district. Geophysical investigations were made as well as research work. On occasion, on a visit during exploration, the author found radioactive anomalies at first in a sulphide deposit near Zwiesel (1956) and later on in the Silberberg deposit near Bodenmais (Teuscher 1958). Here, the anomalies were caused by concentrations of heavy minerals, especially of monazite, in a layer accompanying the sulphide ore body.

2 Geological Position

Near Bodenmais, rock series of mainly monotonous metapelitic character range to highly metamorphosed anatexites. The regional structure is determined by a NNE dipping schistosity and by flat-lying fold axes of NW orientation. Generally concordant with these regional structures are intercalated orthogneisses and sulphide ore occurrences, mapped over a distance of ca. 30 km in WNW-ESE direction by Hartmann and Blendinger (see Blendinger and Wolf 1971). Only the sulphide deposit of the Silberberg, Bodenmais, was of local importance, having an extension of nearly 1 km in strike and 150 m in dip direction; it is boudined in lenses with a thickness of sometimes more than 10 m (see ore shoots in Fig. 1), which are generally connected by ore-bearing layers of about 1 m. Forms and mineral assemblages of the sulphide ore occur-

1 Kerschensteiner Str. 25, 8032 Lochham, FRG

rences were already described by Gümbel (1868), Weinschenk (1901), and Münichsdorfer (1908). We have modern petrographic descriptions of the country rock in the vicinity of the Bodenmais deposit by Fischer (1938), Fischer et al. (1968), Blümel (1972), Blümel and Schreyer (1976), and especially by Schröcke (1955, 1956). The main rock types are metatectic garnet-cordierite-sillimanite gneisses with a dark palaeosom (= melanosom = restite) and a light neosom (= leucosom)[2] which is lense- or layer-shaped, ranging in thickness from mm to cm. The separation of dark and light layers is considered to be the result of partial liquefaction and migration of the mobile components; we even find neosoms of granite-like character sometimes with pegmatoid fabric, especially concentrated around the ore body (Schröcke 1955). There is an analogy between the leucocratic neosom and the ore which was relatively mobile at the peak of the regional metamorphism. Both have a less gneissose structure than restites. Former authors had described cordierite-sillimanite gneisses with a few garnets – often corroded – but their sampling was restricted to the mine. Here, the galleries follow the strike of the ore and we find preferably sulphides, neosom and relatively light restites rich in cordierite and sillimanite. But the main rock type of the Silberberg is an anatexite rich in garnet; porphyroblasts had megascopic sizes (some mm to ca. 1 cm). Schreyer et al. (1964) called attention to a replacement following the chemical reaction as:

garnet + biotite + sillimatite + quartz $\rightleftharpoons$ cordiereite + K-feldspar

It seems this reaction proceeds to the right near the sulphide ore body.

3 Mineral Assemblages

According to the investigations of Schröcke (1955) the constituent minerals of the metatectic cordierite-sillimanite gneiss are cordierite (15%–50%), sillimanite (0–50%), biotite (10%–30%), occasional garnet, and spinel; further, ilmenite, pyrrhotite, magnetite, plagioclase, microcline, and quartz. Accessory minerals are zicron and apatite; muscovite and chlorite appear as products of retrograde alterations, being very intense near the sulphide deposit. The ferromagnesian minerals plus sillimanite and feldspar represent the bulk of the restite layers; the neosom consists predominantly of quartz and feldspar, sometimes plus biotite; cordierite is rarely found.

The ore of the Silberberg deposit is coarse-grained. According to the detailed microscopic investigations of Dessau (1933), and Hegemann and Maucher (1938) it has the following mineralogical composition: main constituents are pyrrhotite, pyrite, sphalerite, chalcopyrite, and magnetite, minor constituents are galena, Zn-spinel, ilmenite, native bismuth and silver, and molybdenite. Directly associated with the ore are the silicates cordierite, biotite, amphibole, sillimanite, plagioclase, K-feldspar, and quartz and rutile. Muscovite, chlorite, epidote, valleriite and marcasite are obviously products of later retrograde conditions, limonite and siderite of supergene alteration.

In one slice monazite inclusions in pyrrhotite were observed (Fig. 7).

2 Terminology according to Mehnert , K. R. (1968) Migmatites and the origin of granitic rocks. Elsevier, Amsterdam, 393 pp

4 Radioactive Occurrences and Their Investigation

The prospection had the aim of discovering a continuation of the known ore bodies and was made by geomagnetic, geoelectric and radiometric methods (Paulus 1953; Ostermeier 1957; see Madel et al. 1968). Gamma radiation was only casually measured and no sooner we had found maxima up to 250,000 cpm (while the background in the galleries was of 6000–8000 cpm, caused by K-radiation), than a systematic radioactive survey of all open levels of the Silberberg mine was accomplished. In the Johannesstollen (664 m long), two anomalies of 120 m and 50 m length were observed, which could be confirmed in four other galleries above and below this level. So two radioactive layers of some hundred square metres were constructed (Fig. 1).

In a second stage, short holes of 1,0–1,2 m length were drilled in the anomalous regions at close intervals (2,5 m) with 3–5 holes in each cross-section. Gamma well survey curves and analysis of the drill dust proved radioactive minerals to be concentrated especially in the hanging parts of the sulphide ore body (see Figs. 2 and 3). The highest peaks were observed in the Johannesstollen and, therefore, in a third stage a 13 m inclined shaft and 21 m gallery in dip and strike direction of the deposit were mined and ca. 50 tons of thorium ores produced.

In 100 m cores in the hanging series no remarkable traces of radioactive minerals were discovered at a distance greater than one metre from the deposit.

Other anomalies caused by Th-minerals were noted at distances of ca. 500 m from the Silberberg deposit (Deutsche Schachtbaugesellschaft 1958). The length of outcrop

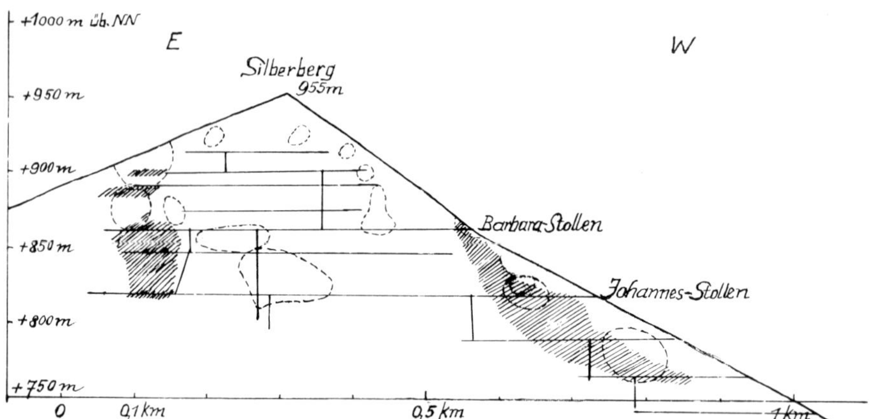

Fig. 1. Vertical section ESE-WSW through the Silberberg mine, Bodenmais. The galleries, following the strike of the sulphide ore deposit, are projected on the plane through the summit of the Silberberg. Dipping of the ore body = 30–40° NNE. *Hatches,* zone of anomalous radioactivity. Ore shoots (thickness of the ore body: 2–10 m) marked by *short lines.* (After Blendinger 1951; see Blendinger and Wolf 1971, Fig. 2)

layers were a few metres to some decametres and these Th ore deposits were not connected with sulphide ores; they produced an impression of eluvial enrichment, the anomalies diminishing in a few metres depth.

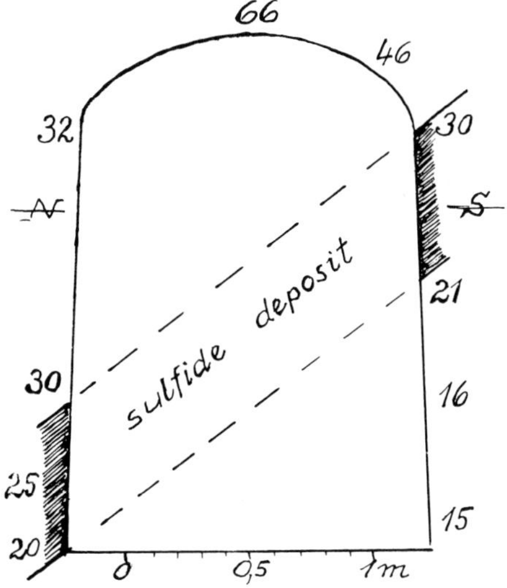

Fig. 2. Cross-section of the Barbara Gallery, Bodenmais. 2,5 m w M.P.4: gamma-wave measurements. Radioactivity *(R)* in thousand cpm (counts per minute). High radioactivity is bound to the hanging wall of the sulphide ore

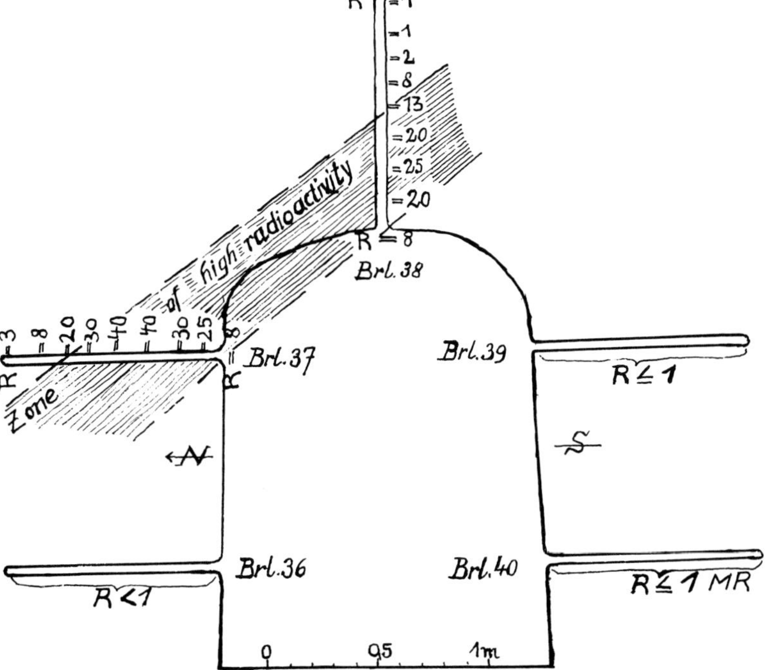

Fig. 3. Cross-section of the Johannes Gallery. Bodenmais, within the zone of anomalous radioactivity. Short boreholes in which the peak of radioactivity was localized in milliroentgen *(MR)*. Rich in monazite is the gneiss layer in the hanging part of the sulphide ore, R > 20 MR

5 The Monazite Layer

The geophysical investigations in the mine were completed by examinations in the laboratories; all samples (about 1000) were tested radiologically, first of all, only U equivalents (e_U) were measured (the e_U even of K in biotite-rich layers is relative small; eU_K max. 10–12 ppm). The richer Th ores were analyzed in detail.

The Th ore from the exploration shaft in the Johannes ore shoot, a Th minerals-bearing gneiss, had the following bulk composition (weight-%):

SiO_2 = 33.82;	Al_2O_3 = 23.36;	TiO_2 = 4.77;	FeO = 9.53;	Fe_2O_3 = 3,54
MnO = 0.06;	MgO = 8.54;	CaO = 0.71;	K_2O = 5.50;	NaO = 1.55
P_2O_5 = 2.21;	SO_3 = 1.27;	H_2O = 2.47;	and	
	CeO_2 = 2.54;	ThO_2 = 0.12;	U = 0.018 [3]	

A high Th concentration was found over a distance of 30 m in the western Johannesstollen. Here the U contents in g per t were determined between (50) 100 and 200 (250); medium content 100 g U/t; the coordinate Th content (235) 500–1100 (2600) and medium content 1050 g Th/t. Rich samples of the ore always have considerably greater Th than U contents; the ratio Th:U varied between 10:1 and 30:1 [4].

Samples of the Johannes anomaly which are much altered as in all old galleries in Bodenmais – the sulphides are intensely oxidized – were mineralogically examined by Dr. Salger of the Bavarian Geological Survey. They had yellow rims of stilbite. The dark brown solid cores, coarse-grained and rich in biotite, were milled and separated by heavy (Clerici-) solution. The fraction G. > 4 consisted mainly of brookite and monazite and a few Zn-spinel (kreittonite) and zircons. Combined X-ray and chemical investigations proved the existence of free Th (thorianite) and Ce-oxides (cerianite). The main radioactive mineral is monazite with 0.007% U_3O_8 (fluorimetric observation) and ca. 0.4% Th; less important are cerianite, thorianite, and zircon. The Th content of kreittonite, described by Strunz (1962), is of subordinate significance; the ratio Th:U of the microlites in the kreittonite was not determined.

The radioactive gneiss, in best developed parts a 10–30 cm layer, is, as a whole, a dark gneiss type of restite character, although in thin sections laminations in millimetre-rhythmites are visible; sometimes more than 20 bands per centimetre perpendicular to schistosity indicating primary, fine bedding. Ca. 50% of the layers are practically pure biotite gneiss; flakes show orientation nearly parallel to schistosity. Subordinate minerals in these layers are monazite, ilmenite, rutile, and some feldspars. The monazite is enriched to 5% or more in small (3–10 mm) melanocratic bands forming irregular rows oriented generally according to the old bedding (Fig. 4); only a small number conforms to boundaries of biotite flakes (Figs. 5 and 6). Growing idioblastic biotite pushed aside the monazite crystals, which surround them in continuous garlands; in the cores we see but few inclusions and radioactive haloes (see basal section of biotites, Fig. 5). A supposed primary content of magnetite and/or ilmenite – in placer deposits often present – may be bound plentifully in the biotites; it reappears in chloritized and muscovitized parts of the rock.

3 anal. Chemisches Laboratorium Fresenius/Wiesbaden, FRG

4 anal. Dr. Abele, Bavarian Geological Survey and Dr. Harre, Bundesanstalt für Bodenforschung/Hannover, FRG

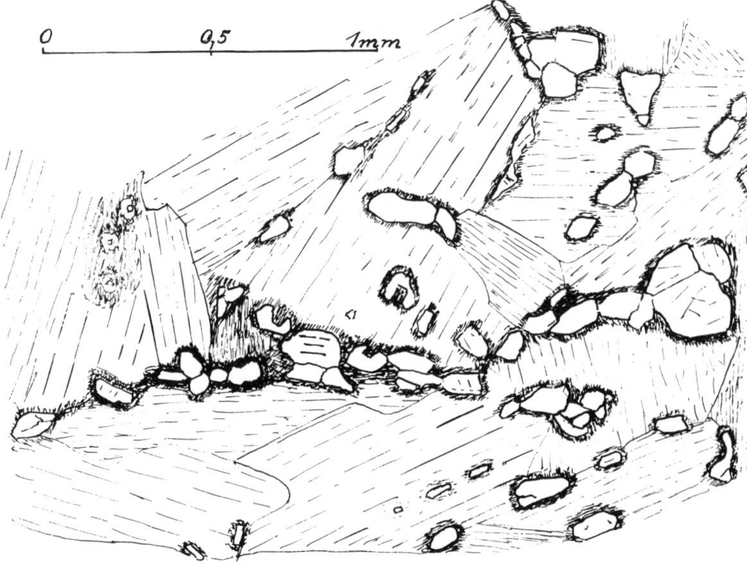

Fig. 4. Thin section. Monazite *(thick lines)* is concentrated in a band parallel to schistosity; other crystals are oriented along the boundaries of biotite flakes. The alignment illustrates an old bedding plane. Most monazite crystals are idioblasts, well-shaped; only a few are rounded grains

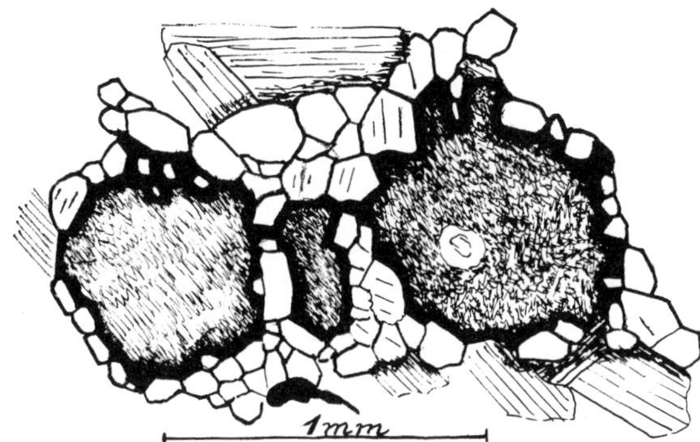

Fig. 5. Tabular biotites with only a few inclusions. The idioblasts are surrounded by a garland of numerous monazites, effecting a continuous radioactive *halo* (*dark rim* of the biotites). *Hatched,* biotite flakes. *Dark grain, below,* ilmenite

The intermediate layers of the radioactive gneiss type are less melanocratic through diminishing of the biotites (to 20% and 10%) forming here short streaks; the mineral assemblages are characterized by cordierite, sillimanite (up to 50%), feldspars, and spinel in varying quantities. Spinel is enriched (to ca. 20%) in the layer next to the ore. Here it replaces often biotite growing in elongated idioblasts within flakes and including small remnants of bleached light-brown biotite. In poikiloblastic structures it encloses numerous small plagioclase crystals which in groups have equal optic orientation. Garnet is rare in this rock type; large crystals are corroded or replaced into little heaps of grains. In intermediate layers of the gneiss, thin lenses of neosom no longer than millimetres consist only of quartz or of quartz and feldspars (mostly microcline).

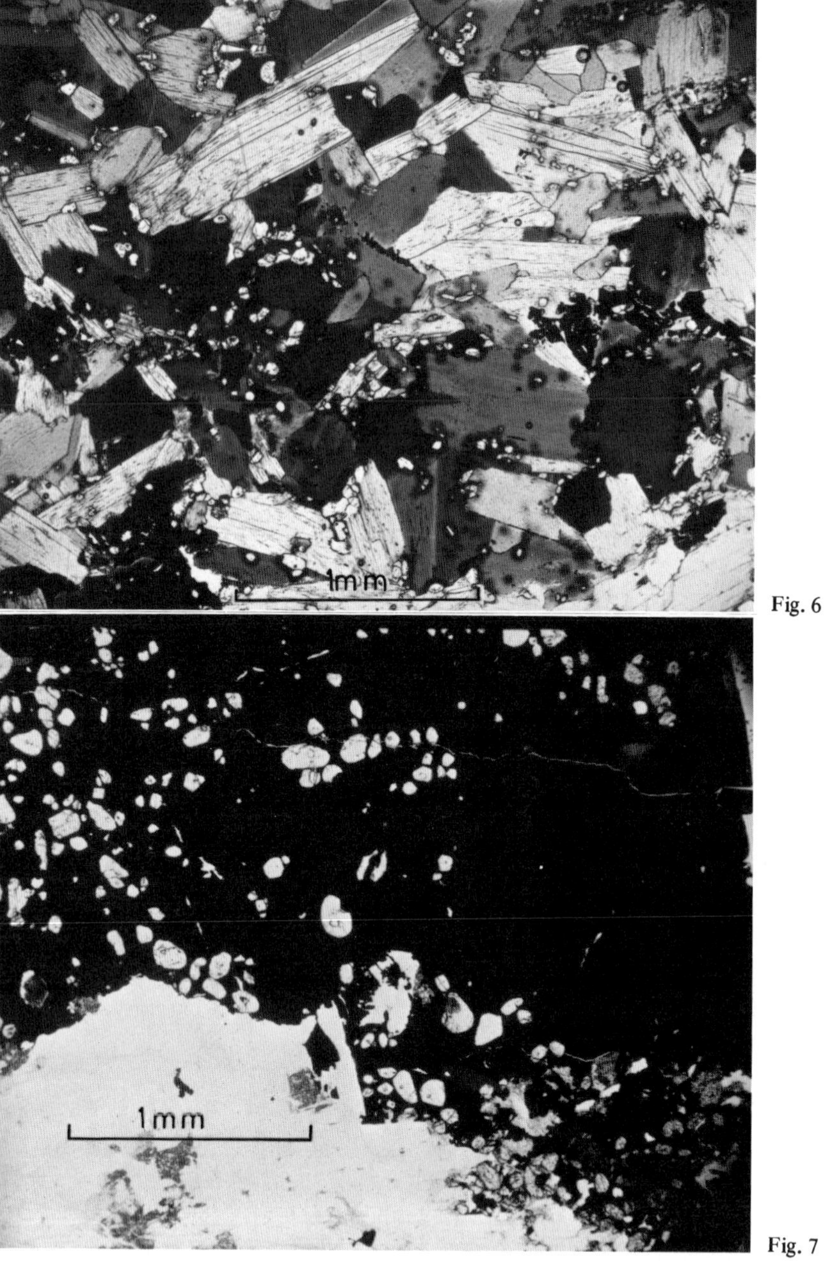

Fig. 6. Dark layer (restite) with ca. 90% biotite; flakes arranged according to schistosity. Monazite, small light crystals, in layers, ca. 5%. Opaque ore not visible in the figure. On the *left side* some tabular biotites (see Fig. 5) surrounded by garlands of monazite crystals. Radioactive anatexite from the exploration shaft in the Johannes anomaly, Bodenmais. Thin section 3054. One Nicol

Fig. 7. Pyrrhotite apophyse into the radioactive biotite gneiss layer. Near the rim of the ore numerous small inclusions of monazite. Exploration shaft in the Johannes anomaly, Bodenmais. Thin section 3059

Sometimes long trains of minute opaque crystals within the quartz – probably ilmenite – mark the general structure. The neosoms are in typical sections encircled by streams of sillimanite needles; in other places by bands of micas.

6 Conclusions

The sulphide ore body of Bodenmais is situated within finely laminated garnet-cordierite-sillimanite gneisses; it is supposed to be an originally marine-sedimentary deposit (type Meggen) within generally monotonous series of metasediments of pelitic character, perhaps connected with initial magmatism (see Schröcke 1955); metabasites were closely related to the deposits (Blümel and Propach 1978). There is no evidence of contact metamorphism from adjacent granites, but the anatectic unit was formed by an equal regional metamorphism under relatively exact determinable PT conditions (with temperatures of about 730°C and pressures of about 3000 bars; see Schreyer et al. 1964).

The changing mineral composition in layers seems to represent old strata. Special rock types like the radioactive gneiss persist over long distances in places where the sulphide layer was relatively uniform in thickness. We regard the radioactive layers in the garnet-cordierite-sillimanite gneiss near Bodenmais as a preserved fossil monazite placer deposit within high-grade metamorphic anatexites.

References

Blendinger H, Wolf H (1971) Die Magnetkieslagerstätte Silberberg bei Bodenmais und weitere Erzvorkommen im hinteren Bayerischen Wald. Aufschluss, Sonderh 21:108–139

Blümel P (1972) Die Analyse von Kristallisation und Deformation einer metamorphen Zonenfolge im Moldanubikum von Lam-Bodenmais. Neues Jahrb Mineral Abh 118:74–96

Blümel P, Propach G (1978) Anthophyllite gneiss from the Silberberg Fe-sulfide deposit Bodenmais, East Bavaria. Neues Jahrb Mineral Abh 134:24–32

Blümel P, Schreyer W (1976) Progressive regional low-pressure metamorphism in Moldanubian metapelites of the northern Bavarian Forest, Germany. Krystalinikum 12:7–30

Dessau G (1933) Einiges über Bodenmais. Neues Jahrb Mineral Beil A 66:381–406

Fischer G (1938) Über das Grundgebirge der Bayerischen Ostmark. Die Gneise nördlich des Pfahls. Jahrb Preuss Geol Landesanst 59:289–352

Fischer G, Schreyer W, Voll G, Hart SR (1968) Hornblendealter aus dem ostbayerischen Grundgebirge. Neues Jahrb Mineral Monatsh 1968:385–404

Gümbel CW v (1868) Geognostische Beschreibung des Ostbayerischen Grenzgebirges oder des Bayerischen und Oberpfälzer Waldgebirges. Perthes, Gotha, 968 S

Hegemann F, Maucher A (1938) Die Bildungsgeschichte der Kieslagerstätte im Silberberg bei Bodenmais. Abh Geol Landesunters Bayer Oberbergamt 11:4–36

Madel J, Propach G, Reich H, mit Beiträgen von Teuscher EO, Vidal H, Ziehr H (1968) Erl Geol Karte Bayern 1:25000, Bl Zwiesel. München, 88 pp

Münichsdorfer F (1908) Mineralogische und petrographische Studien im Silberberg bei Bodenmais. Geogn Jahresh 21:59–91

Paulus R (1953) Die Ursache der erdmagnetischen Anomalien bei Bodenmais und Zwiesel. Unpubl Diss, München

Schreyer W, Kullerud G, Ramdohr P (1964) Metamorphic conditions of ores and country rock of the Bodenmais, Bavaria, sulfide deposit. Neues Jahrb Mineral Abh 101:1–26

Schröcke H (1955) Petrotektonische Untersuchung des Cordieritgneisgebietes um Bodenmais im Bayerischen Wald und der eingelagerten Kieslagerstätte. Heidelb Beitr Mineral Petrogr 4:464–503

Schröcke H (1956) Zur Kenntnis der Metamorphen Kieslagerstätte St. Johanniszeche bei Lam im Bayerischen Wald. Geol Bavarica 25:129–166

Strunz H (1962) Radioaktivität des Zinkspinells von Bodenmais und deren Ursachen. Aufschluss 13:47–52

Teuscher EO (1958) Schlußbericht über die Untersuchungsarbeiten auf Schwefelerze im Raum Bodenmais-Zwiesel-Lam. Arch Bayer Geol Landesamt, 14 pp, 6 Anlagen

Weinschenk E (1901) Die Kieslagerstätte am Silberberg bei Bodenmais. Abh Bayer Akad Wiss II Cl 21:340–410

Petrography, Mineralogy and Genesis of the U-Ni Deposits Key Lake, Sask., Canada

V. VOULTSIDIS [1], E. VON PECHMANN [1], and D. CLASEN [2]

Abstract

The U-Ni deposits at Key Lake, Sask., Canada, were presumably primarily formed under low-thermal near-palaeosurface conditions. The present genetical model is based on investigations of the geometry of the orebodies, weathering decomposition of basement rocks (source) with anomalously high U and Ni clarke values and local synmetamorphic U and Ni mineral enrichments, ore textures and parageneses, geothermometry, age determination of host rock and mineralization and on analogies to U deposits of the same type in Saskatchewan and the East Alligator River uranium district, N.T., Australia. The Key Lake deposits were formed in a fault zone and consist of an older mineralization in the basement, which was probably recrystallized at around 1270 m.y., and a younger dispersion halo with a distinctly different mineral paragenetic assemblage in the overlying Athabasca sandstone. Other genetical concepts (hypogene or diagenetic) are discussed.

1 Introduction

The U-Ni deposit Key Lake is located at the SE rim of the Helikian Athabasca sandstone basin at the intersection with Aphebian metasedimentary rocks of the Wollaston domain of the Churchill structural province (Fig. 1). Both parts of the deposits, in the NE the Deilmann orebody, in the SW the Gärtner orebody, are bound to an ENE-running fault zone which is present and mineralized in both rock units (Fig. 2). The total length of the mineralized zone is over 5 km; the width is 20–40 m in the Gärtner orebody and locally over 100 m in the Deilmann orebody; the maximum depth of the mineralized zone is 150 m. The ore contents are strongly variable, the highest grades being 58% U_3O_8 and 28% Ni over 2 m. The orebodies are covered by up to 80 m Pleistocene glacial sediments.

The deposits were found in 1975 and 1976 by a combination of radiometric boulder tracing, glaciology, geophysics, geochemistry and drilling after 5 years of intensive exploration work (Tan 1976, 1980).

1 Uranerzbergbau-GmbH, 5300 Bonn, FRG
2 Present address: Blieskasteler Str. 11, 6600 Saarbrücken 3, FRG

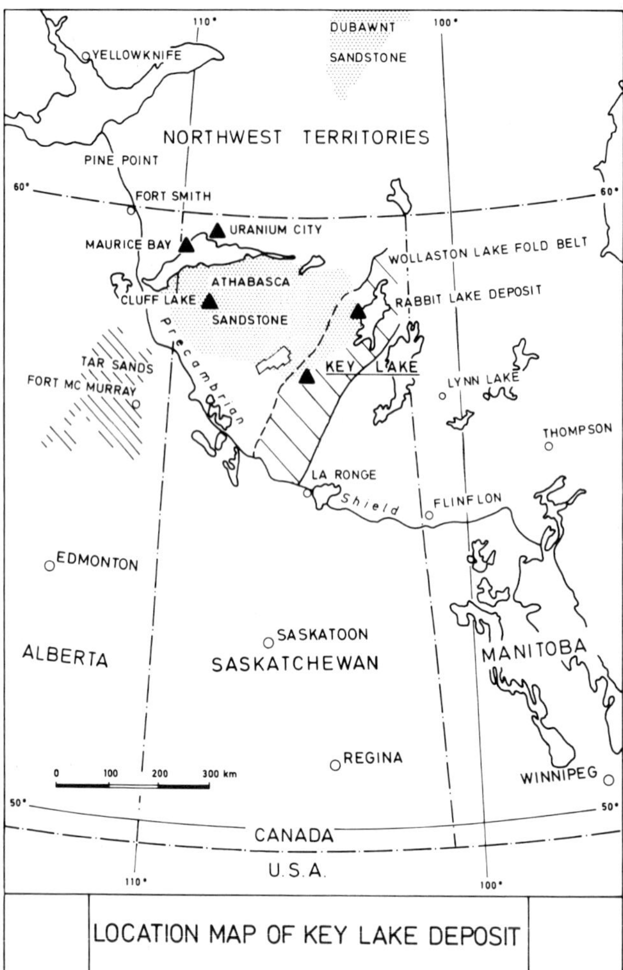

Fig. 1. Location of the Key Lake deposits

2 Petrography and Mineralogy of the Key Lake Deposit

2.1 The Crystalline Basement

The basement rocks were sedimented during Aphebian times and metamorphosed during the Hudsonian orogeny. Archaean rocks form domes within the Aphebian metasediments.

The main rock units of the Aphebian are *biotite-cordierite-feldspar gneisses, graphite gneisses and schists* and *coarse-grained anatexites* (neosome). Of minor importance are amphibolites, biotitites, and biotite-cordierite-feldspar-garnet gneisses. The Archaean unit comprises granite gneisses and granulites.

The predominant biotite-cordierite-feldspar gneisses are of special importance with respect to the metamorphic conditions and the depositional mechanisms of the U-Ni

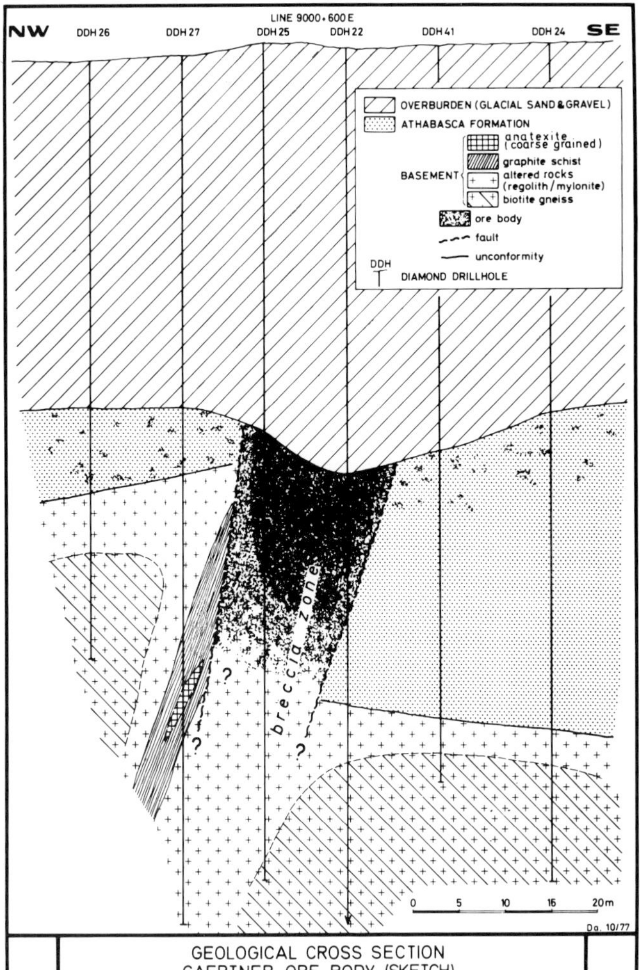

Fig. 2. Schematic cross section of the orezone (from Dahlkamp 1978)

orebodies. Quartz, biotite, cordierite and plagioclase are essential minerals; K-feldspar, graphite, amphibole and almandine are subordinate and partly only locally occurring components. Sillimanite, apatite, zircon, allanite and opaque minerals such as pyrite or chalcopyrite are accessory minerals. Muscovite is not a primary mineral phase of the rock. The paragenesis belongs to the top part of Winkler's (1967) amphibolite facies of the Abukuma type. The rock was probably formed from a pelitic, partly carbonaceous sediment. The graphite gneisses are characterized by the predominance of graphite with varying contents of plagioclase, biotite, cordierite and quartz. The anatexites, consisting of orthoclase, plagioclase, cordierite and quartz with lesser biotite, form up to several feet thick, conformable layers within the metasediments.

Signs of mineral alterations caused by retrograde metamorphism cannot be observed in the A.m. rocks. Several authors from Australia report retrograde metamorphism since garnets are chloritized. The present authors encountered in fresh gneisses completely

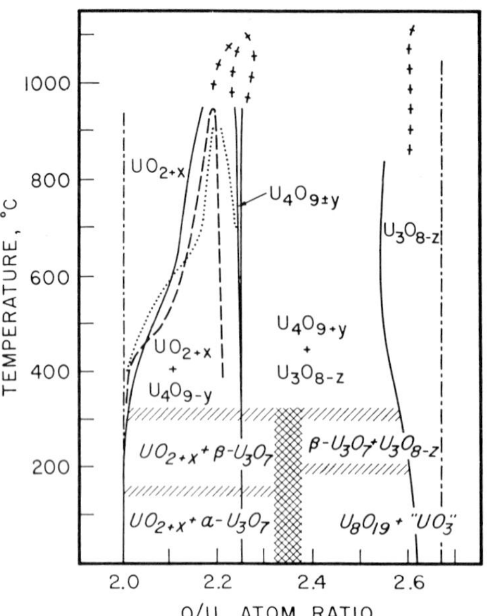

Fig. 3. Temperature stability fields of uranium oxide phases (from Westrum and Grønvold 1962)

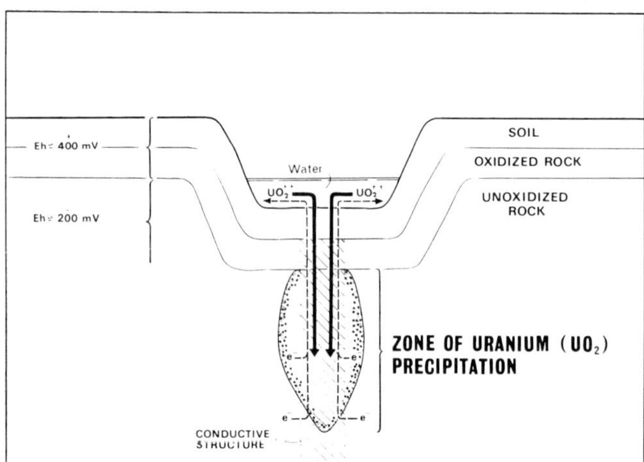

Fig. 4. Principle of an electrolytic cell applied to geology (from Robertson et al. 1978)

unaltered garnets which became chloritized only with the gradual weathering decomposition of the rock. The gneisses are characterized by features of near-surface weathering intensified towards the upper sequences, but reaching depths of 300 m (though weaker at this depth). The mineral decomposition involved all minerals except quartz and graphite so that finally only sericite-chlorite aggregates pseudomorphic after the primary components are recognizable. The stability decreases from K-feldspar over biotite, plagioclase, cordierite, and garnet towards amphibole, the least stable mineral.

The hydration of biotite is of special importance for the later formation of the deposit: after leaching to hydrobiotite all cations are released from the crystal lattice. This so-called baueritization causes opacity by reconcentration of Ti and Fe along the mineral cleavage in form of leucoxenic material and Fe-hydroxides (Fig. 5).

Table 1 shows the chemical change of the rock during increasing decomposition towards the top: it is characterized by leaching of K, Si, Ca and partly Fe. Al, Ti and Mg remain stable and therefore are relatively enriched. This type of weathering suggests a decomposition in a warm-humid climate (beginning palaeolaterization according to Valeton, pers. comm. 1979).

Table 1. DDH H2-example of a chemical variation of the Aphebian metasediments with increasing depth

Sample No.	CN			
	1810	1811	1812–19	1820–31
Depth (ft)	204–11	211–18	212–271	271–349
SiO_2	57.35	71.90	75.75	74.25
TiO_2	0.74	0.32	0.31	0.30
Al_2O_3	21.35	15.12	13.04	12.47
Fe_2O_3 tot.	1.25	1.04	1.88	2.20
CaO	0.08	0.16	0.15	0.18
MgO	9.46	6.14	2.17	4.16
K_2O	0.66	0.75	2.85	3.34
Na_2O	0.40	0.25	0.62	0.20
	← Increasing decomposition			

2.2 Athabasca Formation (AF)

The sequence of the overlying AF starts with a *conglomerate* or *fanglomerate* over the discordance and then grades into a *sandstone (quartzite)*. Both rock units are not metamorphosed. Pelitic horizons (mudstones) are very rare.

The Athabasca sandstone near Key Lake has a matrix-poor texture; the grains show usually a high grain bond number and are not interlocked. The matrix forms generally 5 vol.%, locally reaching up to 20 vol.%. It normally consists of kaolinite and sericite, and minor chlorite. Quartz forms 90–95 vol.%, has grains of 0.2–0.3 mm size (average) and is subrounded; apatite, tourmaline, zircon and opaque grains are accessory minerals. No feldspars have been observed. A second generation of (diagenetic) quartz which results in a quartzitic rock character has clearly been emplaced after the younger U-Ni mineralization (Fig. 6).

Some basal parts of the AF contain varying amounts of rock fragments (quartzites and basement rocks). Inclusions of completely weathered basement fragments of very irregular shape, indicate that they have been deposited after their weathering as relatively soft material (Fig. 7).

This is in contradiction to the hypothesis of Clark and Burrill (1979) that the basement was altered and leached by diagenetic solutions after the sedimentation of the AF.

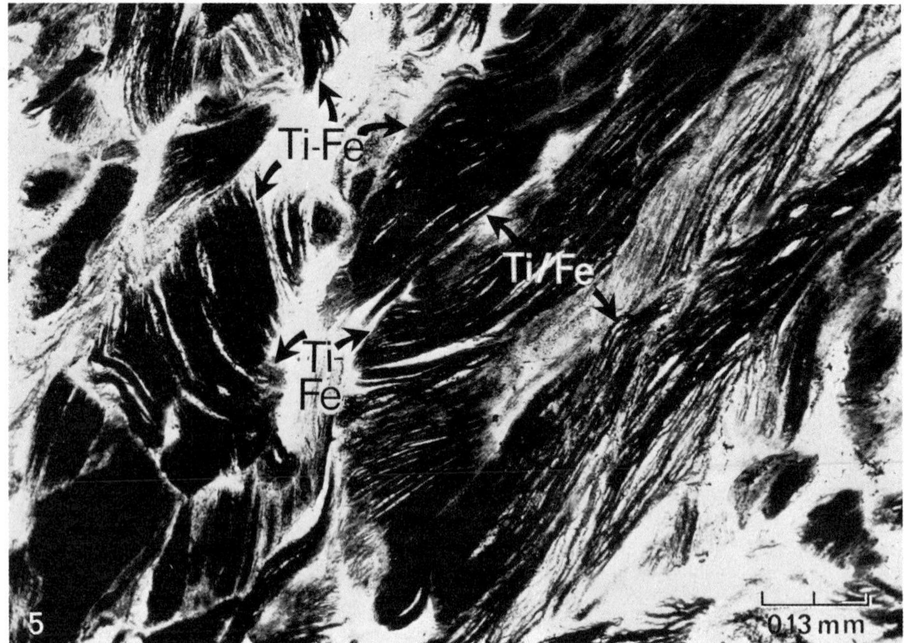

Fig. 5. Microphotograph of a thin section, CN a317 (ddh HG6; 43 ft.); nic. //. Strongly baueritized biotites with abundant opaque-semiopaque Ti/Fe-mineral exsolutions (*black*, Ti/Fe) along their cleavage planes in altered gneiss

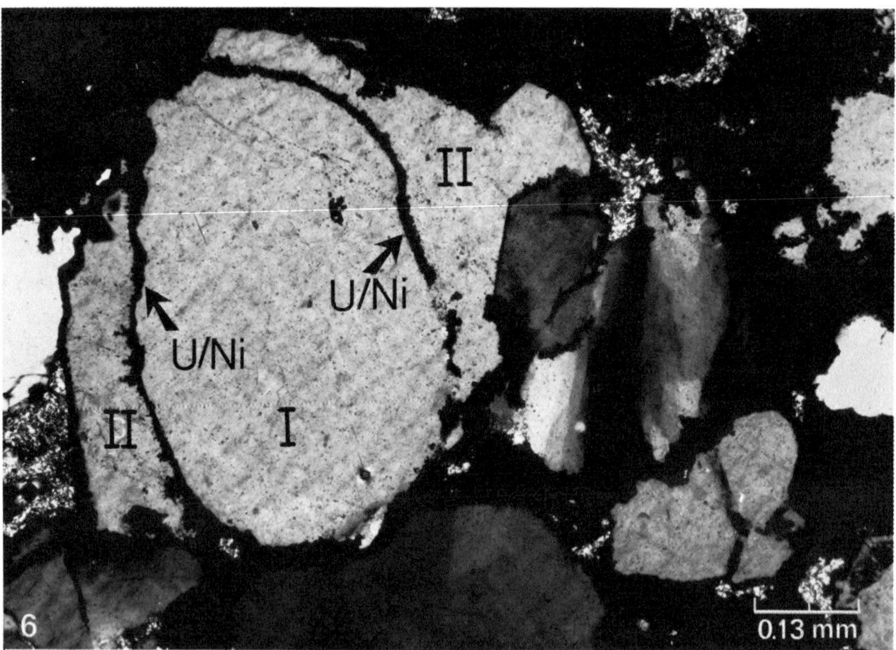

Fig. 6. Microphotograph of a thin section, CN 1787 (ddh H2; 44 m); nic. X. Primarily rounded quartz *(I)* with coating of U/Ni mineralization *(black)* and of a younger *(II)* diagenetic quartz rim in the Athabasca sandstone

Fig. 7. Photograph of a handspecimen, CN 9697 (boulder). Fanglomerate with inclusions of completely weathered basement rock *(wb)*

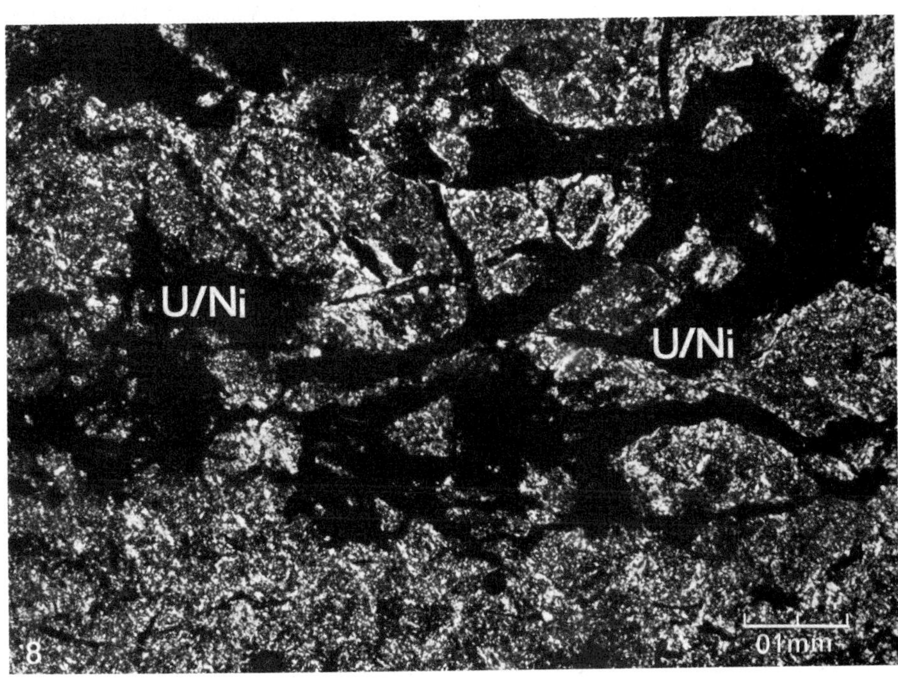

Fig. 8. Microphotograph of a thin section, CN 1312 (ddh 47; 60.3 m); nic. //. Texture of Fe-chlorite mylonite with intensive mineralization (*black*, U/Ni) along shear planes and fissures

The basal parts of the AF that are intersected by the fault are often strongly broken; fissures in the AF components were later filled by the U-Ni ore minerals, carbonates and limonite.

2.3 The Basement Rocks of the Fault Zone

The tectonized rocks of the crystalline basement are subdivided into the mineralized and the non-mineralized units: the mineralized parts are dark green *Fe-chlorite mylonites* and whitish-grey *kaolinite mylonites,* whereas the barren parts contain *sericite, Fe-Mg-chlorite and quartz.* The formation of Fe-chlorite and kaolinite is accompanied by nearly total removal of the primary cations, contrary to the lower non-mineralized chlorite-sericite-quartz mylonites, which have a chemical and mineralogical composition identical to that of the untectonized weathered gneisses. Therefore, the kaolinization and Fe-chloritzation post-dates the regional weathering effects and is caused by the influence of the mineralizing U-Ni solutions.

The microscopic texture of the three mylonite types consists of a fine-felty matrix (max. grain size 0.005 mm; Fig. 8). Locally the mineralized parts are reddened by Fe-hydroxide. Microprobe analysis shows that the Fe-chlorite contains no Mg.

The change from unmineralized tectonized to undisturbed rocks is gradational, whereas the change of the mineralization and its typical Fe-chlorite and kaolinite matrix towards the barren sections within the tectonized parts is abrupt (cm to dm range).

2.4 Ore Mineralogy

The ore mineralogy can be clearly divided into two paragenetical assemblages – an older one in the mylonitized basement domains and a younger one in the overlying Athabasca Formation.

The ore minerals in the basement mylonites consist of α-U_3O_7 ("tetrauraninite", Clasen and Voultsidis 1982, in press), coffinite and sooty pitchblende, gersdorffite, niccolite, millerite, maucherite, rammelsbergite, galena, bravoite, pyrite and marcasite and minor chalcopyrite, covellite and sphalerite. α-U_3O_7, as well as the nickel arsenides niccolite, maucherite and rammelsbergite, do not occur in the sandstone domains, to the authors' present knowledge.

The adjacent graphite schists do not (or at best in places) contain uranium minerals but, to a minor extent, may locally carry gersdorffite, niccolite or millerite. This mineralization might be of (primary) metamorphic formation (cf. Tilsley 1979).

The ore minerals in the Athabasca Formation are predominantly found in the grain interstices of the sandstone and locally between the originally rather well rounded quartz grains and the secondary (younger) quartz rims (Fig. 6). In places late mineralization (above all millerite) is located on fracture planes.

The mineralization in the basement was precipitated along fissures and cracks and on vugs of the mylonitized zones (Fig. 8). Macro- and microtexture of the ore are extremely variable in the cm and even mm range. Massive domains with simple interlocking types occur as well as all the other types of intergrowths described by Amstutz (1961). Of special importance are abundant colloidal concentric-rhythmical intergrowths of ore and gangue minerals (Fig. 9).

The more massive the ore is the more it may be porous. Partly it suffered tectonic deformation resulting in broken ore constituents (Fig. 10).

The essential uranium mineral α-U_3O_7 ("tetrauraninite") is in the form of massive groundmass including all other ore minerals, of filigrane-like pseudomorphs along the cleavage planes of sheet silicates; it occurs also in the colloidal form typical for the uranium oxides (Fig. 10) and locally in the form of idiomorphic grains (Figs. 11 and 12). The latter shows a slight bireflection and a relatively distinct anisotropy (Fig. 11). The α-U_3O_7 is secondarily oxidized to sooty pitchblende at its rims and along numerous shrinkage cracks (Fig. 10). At least part of the α-U_3O_7 is thought to be a product of recrystallization ("Umkristallisation") of former botryoidal pitchblende (Ramdohr, pers. comm. 1980).

Formation of gersdorffite (NiAsS) initiates contemporaneously with that of α-U_3O_7. It occurs as atoll- and ring-like structures and above all in the form of idiomorphic grains (Fig. 12), in places with slight zonality. Noticeable features are concentric intergrowths with uranium and other nickel minerals as well as with gangue minerals (Fig. 9) and pseudomorphs after sheet silicates (Fig. 13). A later gersdorffite phase replaces the other nickel minerals except millerite.

Bravoite – $(Fe,Ni)S_2$ – also belongs to the first ore generation as is clearly visible on Fig. 12 which shows α-U_3O_7 (?) and bravoite included in idiomorphic gersdorffite.

Rammelsbergite ($NiAs_2$) probably the third nickel mineral formed during the first paragenetical stage, is observed in mm to several cm large grains, and is replaced by niccolite along cracks (Figs. 14 and 15).

Niccolite (NiAs) occurs in roundish grains of up to several mm size, similar to maucherite ($Ni_{11}As_8$) which most likely replaces niccolite. Both minerals are also observed along the cleavage planes of sheet silicates, similar to millerite (NiS), the latest essential ore mineral being found as intimate intergrowths with sooty pitchblende, but more often replacing the whole suite of nickel minerals described above. Thus it is the end member of the row: $NiAs_2$ – NiAs – $Ni_{11}As_8$ – NiAsS – NiS (but deviations of this rule are ubiquitous). Moreover, it fills vughs and cracks displaying picturesque aggregates of starlets and needles.

Coffinite ($USiO_4$) and sooty pitchblende (UO_3) (apart from its form as in situ oxidation product of α-U_3O_7) are of a later genetical stage; they are found in the A.m. concentric intergrowths or in the interstices of the other minerals.

Minor, though locally abundant, ore constituents are pyrite (often within gersdorffite), marcasite, chalcopyrite and galena, the latter partly filling cracks in uranium oxide or maucherite. Sphalerite is present as local accumulation or as individual grains, in places as "honigblende" with zonal structures. Accessory covellite, also "blue remaining", is determined, indicating that the temperatures in the deposits did not exceed 157°C (Moh 1971) after the formation of covellite. Besides, there is little hauchecornite – $(Ni,Co,Fe)_9$ $(Bi,As,Sb)_2S_8$ –, bismuthinite – Bi_2S_3 –, clausthalite – PbSe – and probably modderite – CoAs –. The basement ore contains numerous grains of sphene and is cut by locally numerous narrow carbonatic veinlets (calcite and siderite). Iron minerals, essentially iron hydroxides and minor hematite, are abundant particularly in cockade ore zones which are very rich in gangue (matrix) minerals.

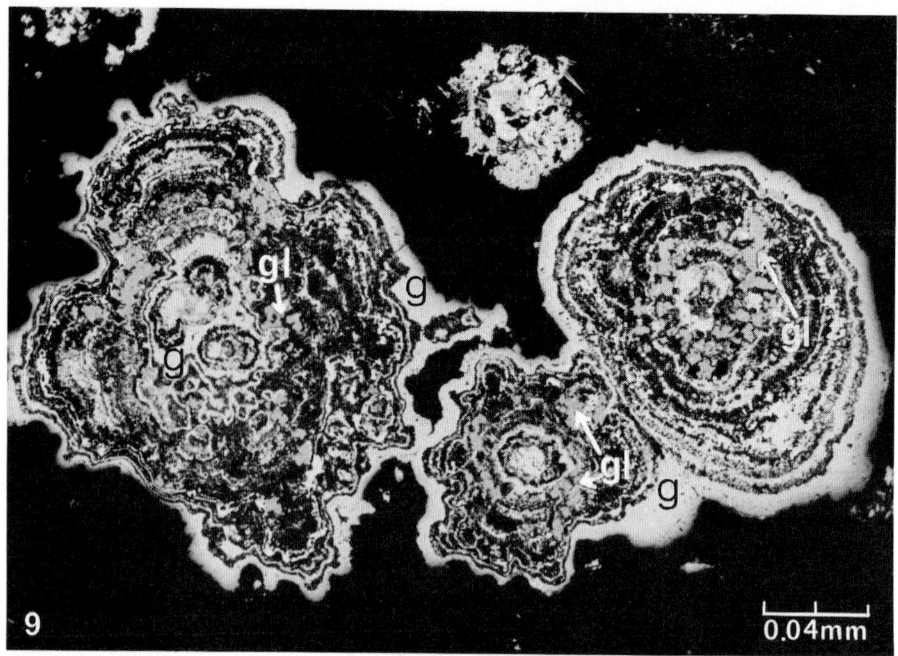

Fig. 9. Microphotograph of a polished section, CN 1342 (ddh 47; 64.8 m); nic. //; × 20 imm. Concentric-colloidal precipitate of gersdorffite (*white*, g), galena (*light grey*, gl) and gangue minerals *(black)*

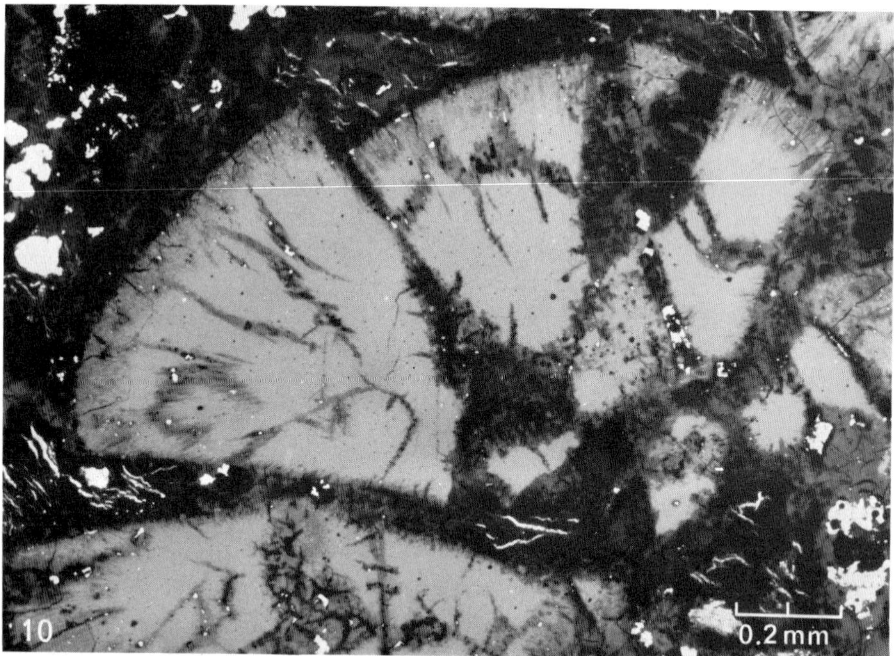

Fig. 10. Microphotograph of a polished section, CN 9980 (ddh 5060); nic. //; × 5 air. Colloidal α-U_3O_7, broken by later tectonism and oxidized along shrinkage cracks to sooty pitchblende

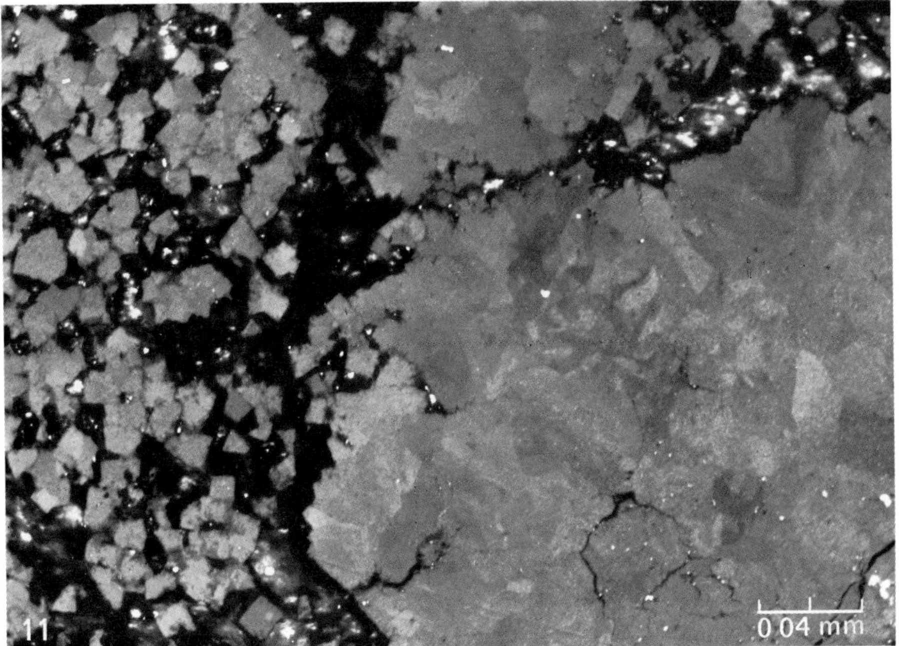

Fig. 11. Microphotograph of a polished section, CN 6408 (ddh 5086); nic. X (-4°); × 20 imm. Idiomorphic and massive α-U_3O_7 displaying its anisotropy

Fig. 12. Microphotograph of a polished section, CN 9979 (ddh 5060); nic. //; × 20 imm. Idiomorphic α-U_3O_7 (?) *(U)* and bravoite *(bv)* in idiomorphic gersdorffite (*white*, g); matrix is *black*

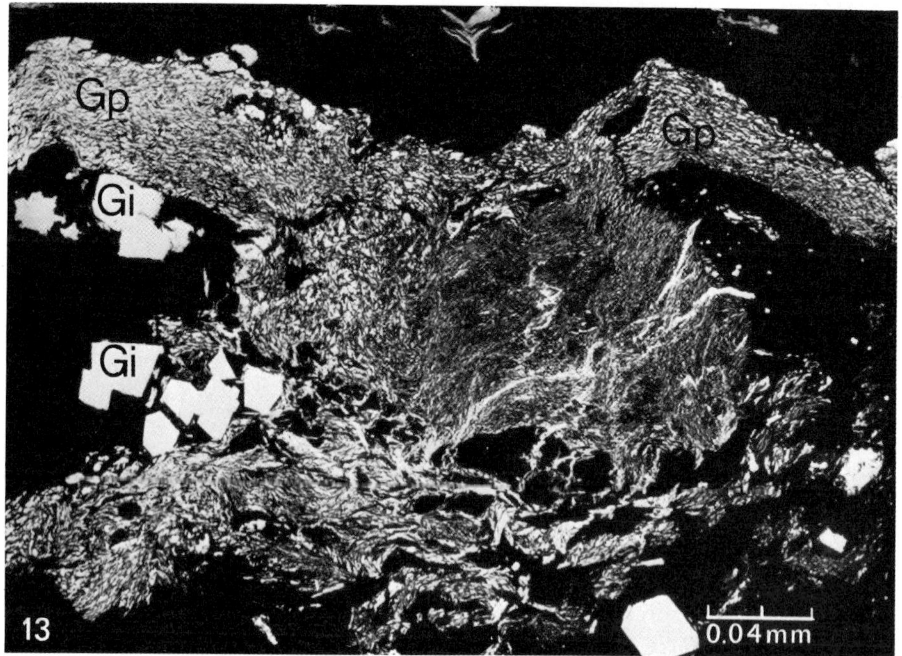

Fig. 13. Microphotograph of a polished section, CN 9987 (ddh 5060); nic. //; × 20 imm. Gersdorffite pseudomorphic *(Gp)* after a sheet silicate; minor idiomorphic gersdorffite *(Gi)*

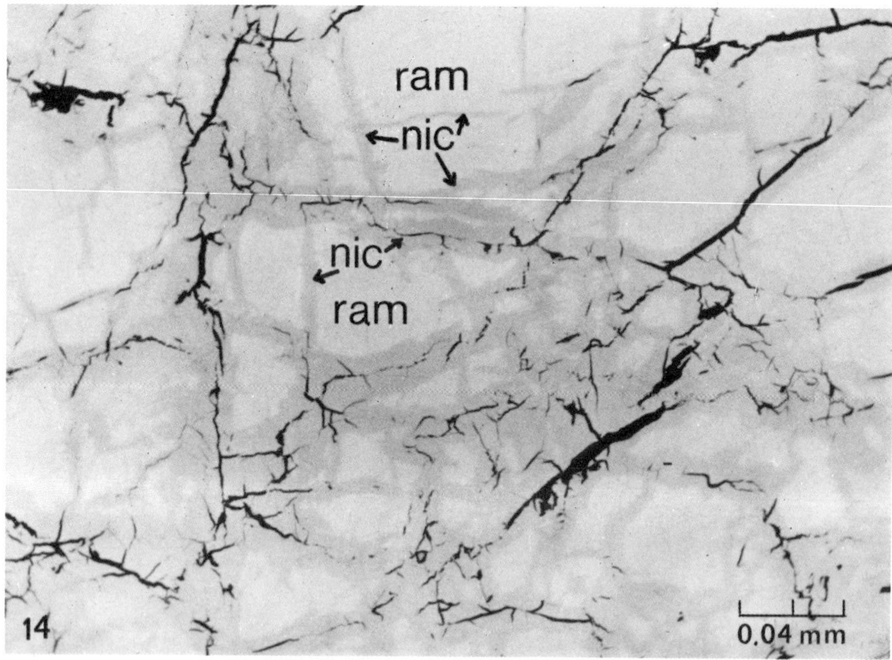

Fig. 14. Microphotograph of a polished section, CN a105 (ddh 70; 58.5 m); nic. //; × 20 imm. Rammelsbergite *(white, ram)* replaced by niccolite *(nic., medium grey)* along cracks

2.5 Age Determinations

The preliminary U/Pb age determinations of the U mineralization carried out by the BGR (German Geological Survey) show a discordia with crossings of the concordia at 1270 m.y. and 270 m.y. (Wendt et al. 1978). This oldest age was obtained from concentrates of α-U_3O_7 (basement mineralization). The sandstone mineralization yielded an age around 300 m.y. so that obviously it represents a secondary halo of the primary basement ore.

A third cumulation of dates lies around 918 m.y. These dates were obtained from basement mineralizations consisting mainly of coffinite.

Basement gneiss samples yielded ages around 1700 m.y. (date of metamorphism).

3 Genetical Considerations

The flood of publications on the genesis of the unconformity-related Proterozoic uranium deposits can roughly be grouped in papers favouring a *hypogene,* a *supergene* or a *diagenetic* ore formation. The points of controversy are:

source of ore elements,
preenrichment/enrichment of ore minerals,
mechanism, age and temperature of ore deposition,
remobilization and redistribution of ore.

These points are discussed in this order in the following. The discussion will refer to data published on other known unconformity related deposits in Saskatchewan (especially Rabbit Lake, Midwest Lake, Cluff Lake, Collins Bay and Maurice Bay) and in the East Alligator River district, N.T., Australia (Jabiluka, Ranger, Koongarra, Nabarlek). Useful reviews are given e.g., by Munday (1979) for the Saskatchewan and Ewers and Ferguson (1979) for the East Alligator deposits.

The authors tried to consider as openly and with as many points pro and contra their own arguments as possible; however, a complete review of all ideas is impossible within the frame of the present paper.

3.1 Primary Source of Ore Elements

Hoeve and Sibbald (1978b) suggest that the primary source of the mineralization are "detrital minerals of the Athabasca Formation" (= AF) (feldspars, mafic and heavy minerals). Feldspars are extremely scarce in the AF, at least around Key Lake, heavy minerals (e.g., zircons, allanites) cannot liberate their U content by the postulated diagenetic solutions, and uraniferous minerals such as uraninites would have been destroyed during sedimentation of the AF because of the oxidzing atmosphere at that time – contrary to the older Witwatersrand uraninites (cf. Ramdohr 1954). Moreover, the probability that uraniferous biotite (see below) has survived the AF sedimentation is extremely low because of the pre-Athabasca decomposition of the gneisses (described above).

Other authors, such as Johns (1970), postulate an old uraniferous roll-front in the AF; however, results of intense exploration did not yield supporting evidence for this theory (e.g., relict redox boundaries).

Our own investigations, partly presented by Dahlkamp (1978) show that part of the Key Lake basement gneisses have elevated U and Ni values [*"protore"* (type 1); Table 2 – average: 12 ppm U_3O_8, 34 ppm Ni], whereas the AF has only around 1.5 ppm U (Hoeve and Sibbald 1978b; Strnad, pers. comm. 1979). U and Ni are predominantly bound to biotite-rich layers, as is proved by long-time autoradiography (fine diffuse distribution of α-tracks in biotitic layers) and by chemical analyses of biotite concentrates (e.g., CN 2918 with 50 vol.% biotite – Table 2). Nickel minerals, such as niccolite and pentlandite (in pyrrhotite), are rare and probably not signs of postmetamorphic igneous activity as postulated by Kirchner et al. (1979) because of their very erratic distribution within the fresh untectonized gneiss. But local thin tourmaline veinlets in the gneiss are of enigmatic genetical implication.

Table 2. Means, standard deviations and errors of U_3O_8-Ni values of Aphebian metasediments from Key Lake [a]

	n	y	$\bar{x}$ (ppm)	s_x (ppm)	$s_{\bar{x}}$ (ppm)	r (ppm)
U_1	31	89	11.19	4.96	0.89	5–24
U_2	16	89	13.87	5.19	1.29	8–22
Ni_1	25	86	34.16	17.67	3.53	12–71
Ni_2	16	89	34.94	14.98	3.75	9–62

[a] Sampling points at least 50 m off the ore zones
U_1 (Ni_1): U-Ni data of fresh biotite-cordierite-plagioclase gneiss
U_2 (Ni_2): U-Ni data of clearly weathered type 1-gneiss
n: number of samples analyzed
y: % of samples used for table (extreme values excluded, e.g., sample CN 2918: 50 ppm U_3O_8)
$\bar{x}$: mean; s_x: standard deviation; $s_{\bar{x}}$: error
r: range of chemical values

Higher clarke values for U (up to 40 ppm U_3O_8) are known also from the Archaean complexes and the metasedimentary rocks adjacent to the East Alligator River deposits (Eupene et al. 1976; Hegge 1977; Ferguson and Rowntree 1979; Ferguson et al. 1979a).

3.2 Preenrichment of Ore Minerals

A preenrichment of ore minerals during metamorphism of the basement rocks is reported by Eupene et al. (1976) and Hegge (1977) for the East Alligator district and by Dahlkamp (1978) and Gatzweiler et al. (1979) from the Key Lake area. Eupene et al. (1976) support their thesis with age determinations suggesting a synmetamorphic age of pitchblende around 1700 m.y. In the Key Lake area a boulder of fresh cordierite-plagioclase-biotite gneiss was found containing considerable amounts of uraninite

Fig. 15. Like Fig. 14, but nic. X (– 3.5°). The lamellae of rammelsbergite are now perfectly visible

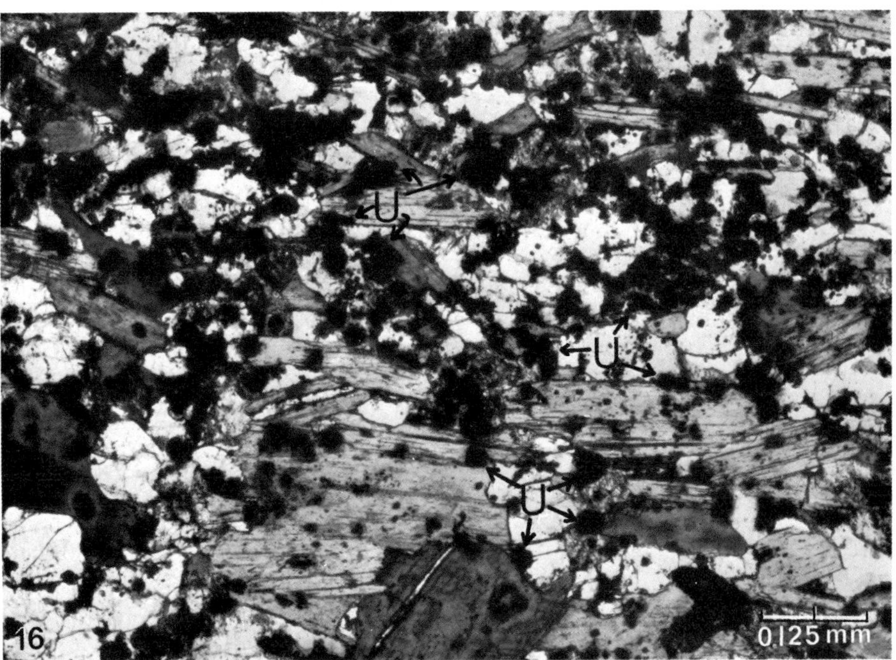

Fig. 16. Microphotograph of a thin section, CN 809 (boulder); nic. //; synmetamorphic enrichment of uraninite *(black dots, U)* in fresh biotite-cordierite-feldspar gneiss

(Fig. 16), strongly suggesting a synmetamorphic accumulation of uranium in mafic domains (protore, type 2). Age dating of this sample is under way. Koeppel (1968) and Dahlkamp (1978) report that parts of the Beaverlodge mineralization (approx. 1700 m.y. old) is of synmetamorphic origin. Tilsley (1979) suggests that carbonaceous pelites, from which graphite schists derived, could have been good electrochemical metal collectors. The local Ni and U anomalies in the graphite schists are the third type of "protores" in the Key Lake area.

3.3 Emplacement of the Uranium Ore

3.3.1 Hypogene Theories: A number of authors postulate orogenic events for the heating of ore-forming solutions. Hoeve and Sibbald (1978a) postulate that diagenetic solutions were heated by magmatic convective activity. Ypma and Fuzikawa (1979) suggest that meteoric waters were heated by orogenic activity, circulating between basement and overlying sandstone.

Hills (1973) presumes that orogenic solutions leached the uranium from basement rocks on their way upwards. Binns et al. (1979) assume a hydraulic fracturing of the host rock at Jabiluka, describe "corroded" graphite and ascribe the strong chloritic gangue matrix to this magmatic event, as does Anthony (1975) for the Nabarlek deposits. Ershov (1976) constructs a connection between Lower and Middle Proterozoic magmatic events during evolution of the Earth's crust (i.e., resulting in mineralizations of Beaverlodge and Key Lake ages). Kirchner et al. (1979) assume an intrusive event of Grenvillian age (1200–950 m.y.) which probably did not take place in the Key Lake area. Finally, Gatzweiler et al. (1979) remark hypogene zonation in the mineralized zone.

At Key Lake, the following features are against the hypogene theories: the presence of two (!) geological thermometers of 135° and 137°C (see below), the geometry of the orebodies which have a limited vertical extent, the ore parageneses, the lack of lateral zonations, and the common absence of typical magmatic-hydrothermal gangue minerals like euhedral quartz, carbonates, barite or distinctly crystalline chlorite. Against a post-Athabasca magmatic emplacement of the ore are the paragenetic and age differences between basement mylonite ore and sandstone ore at Key Lake. Hoeve et al. (1980) for Collins Bay and Lehnert-Thiel et al. (1982) for Maurice Bay remark on the fact that colloidal pitchblende is encountered in the basement ranges as well as in the sandstone domains, thus obliterating part of the paragenetical but not the age differences above and below the unconformity. The presence of colloidal pitchblende in the sandstone could be due to slightly higher temperatures in the Maurice Bay and Collins Bay areas because of a larger cover of Athabasca sandstone above the unconformity.

3.3.2 Diagenetic Theories: The diagenetic hypothesis was first indicated by Hoeve and Sibbald (1976, 1978a). It is supported by Morton and Beck (1978) who also tend to accept a hypogene deposition, by Clark and Burrill (1979), Trow (1979), and Pagel et al. (1979).

From the use of a slightly misleading cross-section presented by Dahlkamp (1978, Fig. 6) which showed ore "roots" in graphitic schist, Hoeve and Sibbald (1978a,

Fig. 9) constructed the theory of the consumption of graphite by "hot diagenetic solutions" producing the reducing agents CH_4 and CO_2. Apart from the thermodynamic fact that graphite would release actually no methane or carbon-dioxide under conditions assumed for the ore formation of only approx. 200°C (Hoeve and Sibbald 1978b), it would be reasonable to expect uranium and nickel enrichments particularly in graphite schists. The author's investigations show however that the graphitic schists have only local and low – probably synmetamorphic– anomalies of U (Ni is higher in places) and that the graphitic schists are not generally and largely coated or soaked with uranium. Thus graphite cannot have acted as a producer for reducing gases. This possibility is also excluded for the East Alligator deposits by Donnelly and Ferguson (1979).

More consistent are Pagel's et al. (1979) considerations; these authors observed CH_4 and CO_2 in fluid inclusions; they do not claim graphite as the source for the gases but rather endogene AF material (Pagel et al., pers. comm. 1980). Pagel et al. (1979) state honestly that "unfortunately, at present, no fluid inclusions have been found in gangue minerals . . . contemporaneous with pitchblende". Fluid inclusions were investigated mainly from secondary quartz rims in the AF which, as shown above, are about 1000 m.y. younger than the oldest dated uranium oxides at Key Lake.

Moreover, diagenetic solutions should also have produced other smaller ore occurrences during their movement along the unconformity (same chemical conditions), but these occurrences have not been detected to date.

The most serious weakness of Hoeve and Sibbald's (1978b) diagenetic theory is the absence of leachable uranium minerals in the AF. Thus both the hypogene and this kind of diagenetic theory have sources of unfortunately "unknown location" (cf. Amstutz 1964), but "our" source is also not yet fully confirmed.

3.3.3 Supergene Theories: The supergene theories can be subdivided into two groups: one preferring a post-Athabasca/post-Kombolgie emplacement of the ore, the other favouring a pre- to syn-sandstone deposition. The post-sandstone theories coincide with the diagenetic ones (already discussed). Therefore, in this chapter only the pre- to syn-sandstone theory is presented.

A pre- to syn-Athabasca emplacement of the uranium ore was first proposed by Knipping (1974) for the Rabbit Lake deposit. Voultsidis and Clasen (1978) gave detailed investigation results from the Key Lake deposits backing this theory. Strong support was given to this hypothesis by Dahlkamp (1978) particularly for the Athabasca region and by Ferguson et al. (1979b) for the East Alligator district.

The supergene pre- to syn-sandstone theory is referring to the above cited high U resp. Ni values in the adjacent (fresh) basement rocks. During the A.m. baueritization of the biotites, the liberated Ni and U concentrate along the cleavages of the biotite skeletons (electron microprobe proof); both elements are thought to be leached during transport and mechanical destruction of the biotite skeletons in a subsequent sedimentation process.

Moreover, the erratically distributed Ni and U minerals, such as niccolite, minor pentlandite and uraninite, are easily chemically dissolved during the erosional weathering (cf. von Pechmann and Voultsidis 1979). The metal content of these solutions is then precipitated under favourable conditions in a fault zone. A condition for the reliability of this theory is that weathered anomalously uraniferous basement rocks were eroded. The erosion at Key Lake is a fact (Fig. 7), but of course the anomalous U/Ni content is assumed by extrapolating the data from basement which is still in situ.

There are several possibilities for the precipitation. Isrusch (1979) found that the fault zone at Key Lake strongly resembles the present day "gley" horizons, i.e., soil sections with little water flow which are undersaturated in oxygen thus being reducing. The little water flow in the fault expected by hydrologists will also have caused an oversaturation of the hydrous solutions by the steady inflow of dissolved salts, so that the ore minerals would have been precipitated on the present clay minerals which are good collectors (Giblin 1979). Robertson et al. (1978) and Tilsley (1978) present the model of a plating (electrolytic) cell (Fig. 4) which is set up by the potential contrast of "high Eh conditions of the oxygenated weathering environment and the low Eh conditions which exist at greater depth" (Robertson et al. 1978) in connection with an underlying conductive structure. The conductive structure could either be a fault zone with dissolved salts or a graphitic layer. The mineralized solutions have oxidized the Fe^{2+} in the Fe-Mg-chlorite to Fe^{3+}, resulting in the formation of Fe^{3+}-chlorite and limonite and in the reduction of U^{6+} to U^{4+} (pitchblende). The fact that not all of the unconformity-related fault-controlled uranium deposits are bound to graphitic horizons as e.g., at Nabarlek, N.T. (Anthony 1975; Eupene et al. 1976) strongly suggests that the presence of graphite is not a necessary requirement for the formation. Nevertheless, a graphite layer adjacent to ore may be very useful as a geophysical guide for the discovery of the deposit, as was proved at Key Lake (Gatzweiler et al. 1981; Tan 1980).

The ore at Key Lake often displays concentric colloidal features (Fig. 9) suggesting not only low temperature conditions of formation but also rapid chemical changes of the ore-bearing solutions (cf. the description of ore of the Rammelsberg-Meggen and Bleischarley types by Ramdohr (1953, 1975).

Two geological thermometers (α-U_3O_7 and bravoite) indicate a maximum temperature of formation of the present ore of around 135°C. Voultsidis and Clasen (1978) have observed the presence of a uranium oxide phase, α-U_3O_7, which is stable in air up to a temperature of 135°C and which is transformed to β-U_3O_7 above 135°C (Westrum and Grønvold 1962) (Fig. 3). This uranium phase at Key Lake is tetragonal and has lattice constants of a_o = 5.462 Å and c_o = 5.398 Å (5.472 Å and 5.397 Å in synthetic crystals). It should be noted that the α-U_3O_7 was synthesized by Westrum and Grønvold in air; formation in hydrous solutions would allow much lower temperatures of formation. The α-U_3O_7 was found in nature for the first time in 1975 and published by Voultsidis and Clasen (1978). The occurrence of (?) primary bravoite with low Co content indicates a very similar low max. temperature of 137 ± 6°C (Clark and Kullerud 1963; Springer et al. 1964). Unfortunately, these facts, published also by Dahlkamp (1978) and Gatzweiler et al. (1979), are often disregarded by authors preferring other genetic concepts.

Higher temperatures obtained from fluid inclusion studies must be handled with great care at present since they may represent (?) diagenetic temperatures and not temperatures of the primary uranium oxide formation. For example, the host minerals of published data of fluid inclusions belong to the younger diagenesis of AF and have nothing to do with the older primary basement mylonite paragenesis!

Very low to low temperatures are also reported from Maurice Bay (bravoite with pitchblende in basement mylonite; Lehnert-Thiel et al. 1981) and from the Jabiluka gold-bearing deposit (paragenesis; Hegge 1977).

Finally: since α-U_3O_7, the oldest dated mineralization at Key Lake, survived later diagenetical and tectonical events it is clear that temperatures in the Key Lake area have never exceeded 135°C for the past approx. 1270 m.y. (cf. Westrum and Grønvold 1962). But this α-U_3O_7 is probably not even of primary origin (cf. above).

Kirchner et al. (1979) argue that an open fault would have been washed out during the following high energy sedimentation (conglomerates and fanglomerates) of the AF.

It cannot be denied that part of the ore will have been eroded before the sandstone sedimentation, but the clayey fault will not have been totally accessible for surface waters.

Another strong argument favouring the supergene emplacement is the abrupt mineralogical change within the fault at the very point where the mineralization stops. The ore mineralization in the basement mylonite is accompanied by contemporaneous Fe-chlorite and predominant kaolinite, whereas the underlying mylonite contains Mg-Fe-chlorite and sericite (and quartz), i.e., the original components of the weathered palaeosurface. In a hypogene deposit the gangue minerals should continue downwards even if unmineralized as it is characteristic for magmatic hydrothermal mineralization.

An argument against the pre- to syn-sandstone deposition are the data obtained from U/Pb age determination. The oldest Key Lake α-U_3O_7 date is 1270 m.y. (Wendt et al. 1978), the oldest pitchblende age for Rabbit Lake is 1281 m.y. (Cumming and Rimsaite 1979), for Cluff Lake 1330 ± 30 m.y. (Gancarz 1979), i.e., about coincident with the end of sedimentation of the AF (1350 ± 50 m.y. – Koeppel 1968; Ramaekers and Dunn 1977; 1428 ± 30 m.y. – Ramaekers 1979). The primary pitchblende ages in the East Alligator district are between 1700 and 1600 m.y. (Ferguson and Rowntree 1979). However, by extrapolating these results to the deposits of Saskatchewan because of the conspicuous geometric and geologic similarities of these mineralizations it can be concluded that the a.m. ages of around 1270–1330 m.y. are not primary and that they are produced by e.g., the final diagenetic processes at the end of the sedimentation of the AF. Moreover, at Key Lake strong mobilization of uranium as well as of lead (probably due to later tectonism and continuous slight water flow etc.) is apparent; radiogenic lead is ubiquitous so that a later falsification or change of the primary age is well imaginable.

3.4 Redistribution – Remobilization

Age dating (Wendt et al. 1978) and the paragenetic differences between the older mylonite ore and the younger sandstone ore, as well as observed mobilization effects (e.g., the radiogenic galena or the young millerite coating joints), show clearly that the Key Lake ore was largely redistributed after its primary emplacement and that therefore the sandstone ore can be termed a secondary halo of the basement ore. The reasons for this redistribution are (still undetermined) orognenic events (Kirchner et al. 1979), or the verified later (epirogenic) rejuvenation of the Key Lake main fault or effects of diagenetic solutions (cf. Fig. 6) as proposed by Dahlkamp (1978) or Hoeve and Sibbald (1978a). Finally, Robertson et al. (1978) suggest that the electrolytic cell was extended after the deposition of the AF so that migrations of the ore became possible. Tilsley (1979) offers the possibility of remobilization by internal heating of the orebody or by influence of solutions which contained components derived from vegetation on the sandstone's surface.

4 Conclusions

From their present knowledge, essentially based on mineralogical data and to a minor extent on analogies to similar U deposits, the authors deny a hypogene or a primarily diagenetic post-Athabasca-sandstone origin of the Key Lake U-Ni orebodies. They favour a supergene pre- to syn-Athabasca formation with strong subsequent redistribution of parts of the ore at around 1300 m.y. and 900 m.y. resulting finally, at about 300 m.y., in a secondary halo in the overlying Athabasca sandstone. Adjacent graphite horizons played no chemical role as producers of reducing gases, but they may have facilitated ore deposition due to their physical properties, possibly in places representing a gliding plane for the tectonic fault or as a conductor in an electrolytic cell; they may also have formed part of a synmetamorphic protore in the area. Favourable exploration areas for the detection of a uranium orebody related to a fault along a Lower/Middle Proterozoic unconformity as represented by the Key Lake deposit are higher U clarke values in the basement host rocks which must have been weathered and finally eroded in order to liberate their uranium which is subsequently precipitated in a mylonitic fault zone, i.e., under near-surface low-thermal conditions. The mineralized fault should have been protected by covering sandstone in later times. Silicification and carbonatization of the sandstone ore may be additional favourable parameters for the ore preservation.

Acknowledgments. The authors wish to thank the Management of Uranerzbergbau-GmbH., Bonn, for the kind permission for the publication of this paper and Dr. R. Gatzweiler, Dr. A.M. Gautier, Dr. J.G. Strnad, and especially Dr. F. Bianconi for their critical review of the manuscript.

References

Amstutz GC (1961) Microscopy applied to mineral dressing. Q Colo Sch Mines 56(3):443–481

Amstutz GC (1964) Introduction to: Sedimentology and ore genesis. Dev Sedimentol 2:1–7

Anthony PJ (1975) Nabarlek uranium deposit. In: Knight CL (ed) Australas Inst Min Metall Monogr Ser 5:304–308

Binns RA, McAndrew J, Sun S-S (1979) Origin of uranium mineralisation at Jabiluka. IUSPCG*, pp 20–25

Clark LA, Burrill GHR (1979) Uranium ore genesis in Saskatchewan and comparisons with Australian deposits. Paper presented at CIM District 4 Annual Meeting, Winnipeg, Sept 1979, 14 pp

* International Uranium Symposium on the Pine Creek Geosyncline, N.T., Australia, Sydney, June 1979 (Extended Abstracts)

Clark LA, Kullerud G (1963) The sulfur-rich portion of the Fe-Ni-S system. Econ Geol 58:853–885

Clasen D, Voultsidis V (1982) Tetrauraninite – a new uranium oxide mineral from the Athabasca uranium province/Canada. Neues Jahrb Mineral (in press)

Cumming CL, Rimsaite J (1979) Isotopic studies of lead-depleted pitchblende, secondary radioactive minerals, and sulphides from the Rabbit Lake uranium deposit, Sask. Can J Earth Sci 16(9):1702–1715

Dahlkamp FJ (1978) Geologic appraisal of the Key Lake U-Ni deposits, Northern Saskatchewan. Econ Geol 73:1430–1449

Donnelly TH, Ferguson J (1979) A stable isotope study of three deposits in the Alligator Rivers Uranium Field, N.T. IUSPCG*, pp 61–64

Ershov AD (1976) Evolution of the conditions of ore formation and of the ore composition in hydrothermal uranium deposits during the evolution of the earth's crust (Russian). Sov Geol 8: 38–47

Eupene GS, Fraser WJ, Ryan GR (1976) The uranium deposits of the top end, 17 pp (unpubl)

Ewers GR, Ferguson J (1979) Mineralogy of the Jabiluka, Ranger, Koongarra and Nabarlek uranium deposits. IUSPCG*, pp 67–70

Ferguson J, Rowntree JC (1979) The uranium cycle and the development of veintype deposits. Tech Committee Meeting on geology of vein and similar types uranium deposits (IAEA), Lisbon, Sept 1979, 11 pp

Ferguson J, Chappell BW, Goleby AB (1979a) Granitoids in the Pine Creek Geosyncline. IUSPCG*, pp 75–78

Ferguson J, Ewers GR, Donnelly TH (1979b) Model for the development of economic uranium mineralisation in the Alligator Rivers Uranium Field. IUSPCG*, pp 79–82

Gancarz AJ (1979) Chronology of the Cluff Lake uranium deposit, Sask., Canada. IUSPCG*, pp 91–94

Gatzweiler R, Lehnert-Thiel K, Clasen D, Tan BH, Voultsidis V, Strnad JG, Rich J (1979) The Key Lake uranium-nickel deposits. Can Min Metall Bull 72(807):73–79

Gatzweiler R, Schmeling BD, Tan BH (1981) History of exploration of the Key Lake uranium deposits. Advisory Group Meeting on case history of uranium exploration. IAEA, Vienna, Nov 1979 (in press)

Giblin A (1979) Experiments on the role of clay adsorption in genesis of uranium ores. IUSPCG*, pp 95–98

Hegge MR (1977) Geologic setting and relevant exploration features of the Jabiluka uranium deposits. CIM Bull 70 (788):50–61

Hegge MR, Rowntree JC (1978) Geologic setting and concepts on the origin of uranium deposits in the East Alligator River Region, N.T., Australia. Econ Geol 73 (8):1420–1429

Hills JH (1973) Lead isotopes and the regional geochemistry of North Australian uranium deposits. Unpubl. Ph D Thesis, Macquarie Univ, Sydney, 192 pp

Hoeve J, Sibbald TII (1976) Rabbit Lake uranium deposit. In: Dunn CE (ed) Uranium in Saskatchewan. Proc Symp Regina (Nov 10, 1976). Sask Geol Soc Spec Publ 3:331–354

Hoeve J, Sibbald TII (1978a) On the genesis of Rabbit Lake and other unconformity-type uranium deposits in Northern Saskatchewan, Canada. Econ Geol 73:1450–1473

Hoeve J, Sibbald TII (1978b) Uranium metallogenesis and its significance to exploration in the Athabasca Basin. In: Parslow GR (ed) Uranium exploration techniques. Proc Symp Regina, pp 161–188

Hoeve J, Sibbald TII, Ramaekers P, Lewry JF (1980) Athabasca Basin unconformity-type uranium deposits: a special class of sandstone-type deposits? In: Ferguson J, Goleby AB (eds) Uranium in the Pine Creek Geosyncline (Proc IUSPCG*), IAEA, Vienna, pp 575–594

Isrusch R (1979) Geochemische und erzmikroskopische Untersuchungen an Proben der Uranlagerstätte Key Lake, Kanada. Unpubl Dipl Thesis, Tech Univ Berlin, 133 pp

Johns RW (1970) The Athabasca sandstone and uranium deposits. West Miner 53 (10):42–52

* International Uranium Symposium on the Pine Creek Geosyncline, N.T., Australia, Sydney, June 1979 (Extended Abstracts)

Kirchner G, Lehnert-Thiel K, Rich J, Strnad JG (1979) Key Lake and Jabiluka: models for Lower Proterozoic Uranium deposition. IUSPCG*, pp 117–120

Knipping HD (1974) The concepts of supergene versus hypogene emplacement of uranium at Rabbit Lake, Saskatchewan, Canada. In: Formation of uranium deposits, Proc Symp, Athens 1974 (IAEA-SM-183/38), pp 531–549

Koeppel V (1968) Age and history of the uranium mineralization of the Beaverlodge area. Sask Geol Surv Can Pap 68-31:111

Lehnert-Thiel K, Kretschmar W, Petura J, v Pechmann E, Voultsidis V (1982) The geology of the Maurice Bay uranium deposit. Can Min Metall Bull (in press)

Moh GH (1971) Blue remaining covellite and its relations to phases in the copper-sulfur system at low temperature. Mineral Soc Jpn Spec Pap 1:226–232

Morton RD, Beck LS (1978) The origins of the uranium deposits of the Athabasca Region, Sask. Canada (abstract). Econ Geol 73 (8):1408

Munday RJ (1979) Uranium mineralization in Northern Saskatchewan. Can Min Metall Bull 72 (804):102–111

Pagel M, Poty B, Sheppard SMF (1979) Contribution to some Saskatchewan uranium deposits mainly from fluid inclusion and isotopic data. IUSPCG*, pp 155–158

Pechmann E von, Voultsidis V (1979) Künstliche Verwitterung von Uran-Nickel-Erz von Key Lake, Sask., Canada – ein Beitrag der Uranmineralogie zum Umweltschutz. Z Dtsch Geol Ges 130(2): 513–526

Ramaekers P (1979) Stratigraphy, sedimentology and structural geology of the Middle Proterozoic basin in the Athabasca area, Sask., Canada and their bearing on uranium mineralization. IUSPCG*, pp 163–165

Ramaekers P, Dunn CE (1977) Geology and geochemistry of the Eastern margin of the Athabasca Basin. In: Dunn CE (ed) Uranium in Saskatchewan. Sask Geol Surv Spec Publ 3:297–322

Ramdohr P (1953) Mineralbestand, Strukturen und Genesis der Rammelsberg-Lagerstätte. Geol Jahrb 67:367–494

Ramdohr P (1954) Neue Beobachtungen an Erzen des Witwatersrandes in Südafrika und ihre genetische Bedeutung. Abh Akad Wiss Berlin 5:43

Ramdohr P (1975) Die Erzmineralien und ihre Verwachsungen, 4th edn. Akademie-Verlag, Berlin, 1277 pp

Robertson DS, Tilsley JE, Hogg JM (1978) The time-bound character of uranium deposits. Econ Geol 73 (8):1409–1419

Springer G, Schachner-Korn D, Long JVP (1964) Metastable solid solution relations in the system FeS_2-CoS_2-NiS_2. Econ Geol 59:475–491

Tan BH (1976) Geochemical case hsitory in the Key Lake area. In: Dunn CE (ed) Uranium in Saskatchewan. Proc Symp Regina (Nov 10, 1976). Sask Geol Soc Spec Publ 3:323–330

Tan BH (1980) Prospektion und Exploration der Uranvorkommen am Key Lake in Saskatchewan (Canada). Unpubl. Ph D Thesis, Tech Univ Berlin, 106 pp

Tilsley JE (1978) Role of electrochemical cells in formation of paleosurface-related vein-type uranium deposits. AAPG-SEPM, Annu Convention April 1978, Oklahoma City (abstract)

Tilsley JE (1979) Continental weathering and development of paleosurface-related uranium deposits: an alternative genesis. IUSPCG*, pp 202–205

Trow J (1979) Unconformity-related Athabasca Formation pitchblende and base-metal sulfides, methane, graphitic conductors, and self potential anomalies. Geol Soc Am – Soc Econ Geol Uranium Deposits Session, San Diego, USA, Nov 1979, 21 pp

Voultsidis V, Clasen D (1978) Probleme und Grenzbereiche der Uranmineralogie. Erzmetall 31 (1): 8–13

Wendt I, Höhndorf A, Lenz H, Voultsidis V (1978) Radiometric age determination on samples of Key Lake uranium deposit. US Geol Surv Open File Rep 78-701:448–449

Westrum EF, Grønvold F (1962) Triuranium heptaoxides: heat capacities and thermodynamic properties of α- and β-U_3O_7 from 5 to 350°. J K Phys Chem Solids 23:39–53

Winkler HGF (1967) Die Genese der metamorphen Gesteine, 2nd edn. Springer, Berlin Heidelberg New York, 237 pp

Ypma P, Fuzikawa K (1979) Fluid inclusion studies as a guide to the composition of uranium ore forming solutions of the Nabarlek and Jabiluka deposits, N.T., Australia. IUSPCG*, pp 212–215

* International Uranium Symposium on the Pine Creek Geosyncline, N.T., Australia, Sydney, June 1979 (Extended Abstracts)

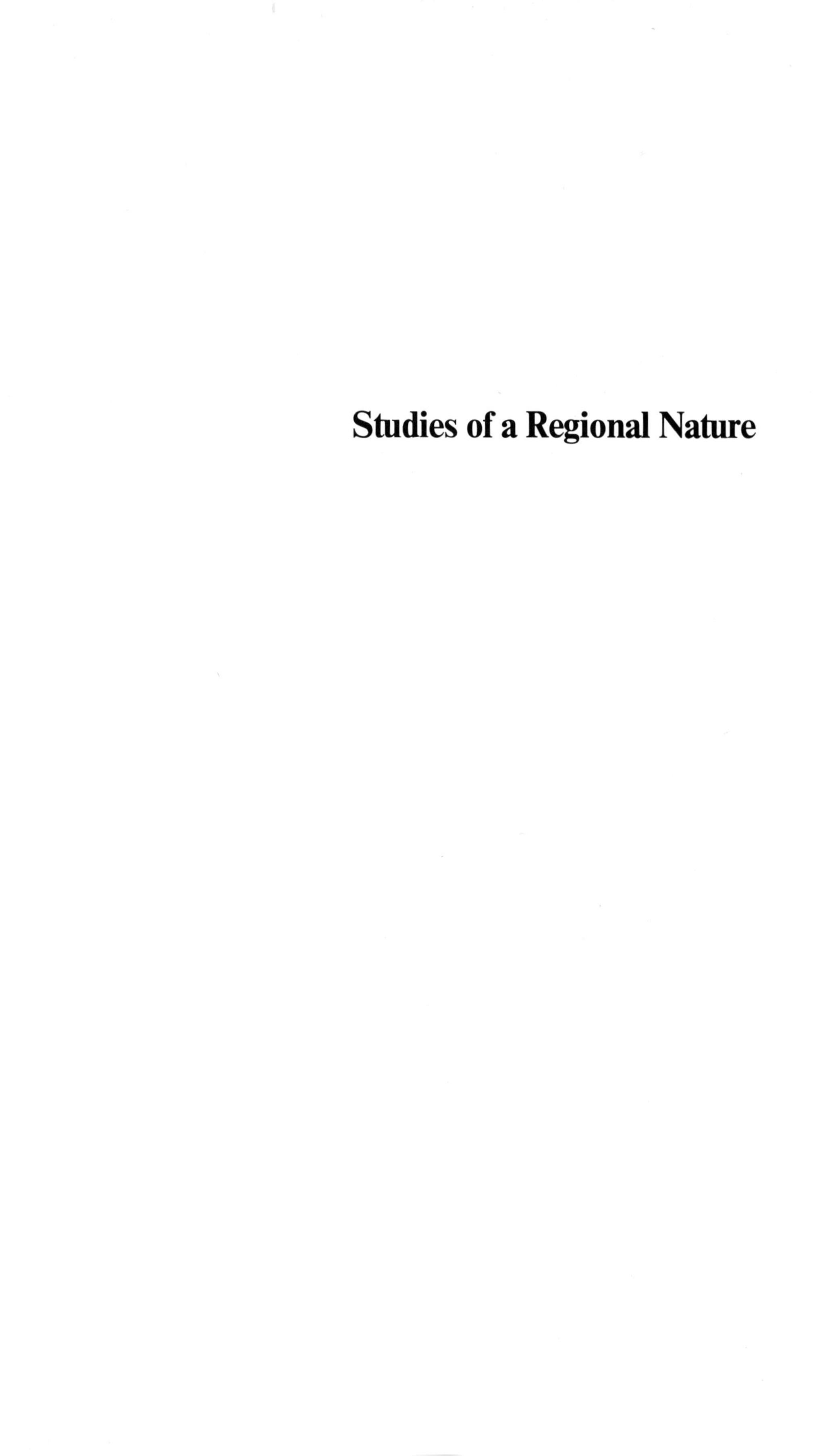

Studies of a Regional Nature

Andean Metallogeny Related to the Tectonic and Petrologic Evolution of the Cordillera. Some Remarkable Points

J. FRUTOS[1]

Abstract

The Central Andes Cordillera shows, according to the distance to the Peru-Chile trench, a metallogenic zonation in sub-parallel bands in which the iron-apatite belt appears, to the west, closest to the trench, followed, to the east, by the Cu-Mo belt, then the Ag-Pb-Zn belt, and finally the Sn-Sb-Bi-W belt. This transversal zonation is closely related to the Mesozoic-Cenozoic evolution of the orogen. Thus, at an early stage of evolution of the geosynclinal system, with volcanic island arcs (Jurassic-Lower Cretaceous) of an incipient rather basic (partly tholeiitic) calc-alkaline magmatic trend, are associated mainly the small atomic number metal deposits (Fe, Ni, Mn, V, Cr, Co) the iron-apatite belt, the Cu-Au porphyries and the Cu-manto type deposits (Coastal Andes of northern Chile and southern Peru). To a more evolved stage of the continental margin (Upper Cretaceous-Middle Tertiary) with a thicker crust and an intermediate calc-alkaline magmatism, the "classical" Cu-Mo porphyries and the polymetallic deposits (Ag-Pb-Zn) are associated. Finally, to the most evolved stage of the orogen (taphrogenic-plateau type of the Central Andes Altiplano) (Middle-Upper Cenozoic) with an extraordinarily thick continental crust and the calc-alkaline magmatic trend reaching the granitic extreme (partly shoshonitic magmatism) mainly the big atomic number metal deposits (Sn-Bi-Sb-W) are associated and also some high Mo/Cu ratio porphyry-type deposits.

Nevertheless a volumetric metal content variation also exists along the Andean metallogenic belts, which seems to be rather in connection with the Precambrian and Paleozoic paleogeographic disposition of ancient basins and magmatic volcanic arcs.

The function played by sedimentary environments and facies, since the Precambrian, in the concentration of metals and their subsequent metallogenic permanence, seems to be of importance at least in the case of Sn, Ag-Pb-Zn, Cu and Mo.

1 Introduction

The present Andean Cordillera has been built from a mixture of rocks of diverse ages, cropping out in belts trending parallel to the recent Cordillera, which shows the effects of the Pliocene-Quaternary tectonism (Andean uplift). In no way do these belts represent the structural orientations of the former episodes. The Andean Belt is the product of the superimposition of successive geosynclinal systems since the Precambrian, each of them with different spatial paleogeographic characteristics.

1 Department Geología-Paleontología, Univ. Concepción, Chile
Mineralogisch-Petrographisches Institut der Universität Heidelberg (Humboldt-Stiftung Stip.)

Subsequent to a Precambrian history, the continental margin of southwestern South America evolved in two Phanerozoic tectonic cycles, the Hercynian (Cambrian to Early Triassic) and the Andean (Late Triassic to Recent), both of them showing eugeosynclinal-miogeosynclinal, orogenic and taphrogenic phases, and the development of successive structural stages.

The genesis of mineral deposits develops in a close relationship with the magmatic-paleogeographic-tectonic features during the evolution of the southwestern continental margin of South America.

2 Paleozoic Hercynian Tectonic Cycle

It has been considered that the Hercynian tectonic cycle in the Andes begins with the tensional period that took place during the Late Upper Precambrian and Lower Cambrian, in the southwestern continental border of the America-Gondwanic plate (Borrello 1969). This event is documented by the presence of extensive post-orogenic molasse deposits that, coupled with a deep erosion, transformed the domain of the former Upper Precambrian orogen (Assyntian cycle) into a platform. Lower to Middle Cambrian conglomerates and red sandstones, demonstrating this stage of evolution, are found in different places, as in Valle de Humahuaca, northwestern Argentina, or in the vicinity of Villazón, southwestern Bolivia. It is probable that, below the volcanic Cenozoic cover of the Chilean Altiplano, some of the coarse sandstones near the Cambro-Ordovician boundary also represent this stage.

After this platform stage, rapid subsidence began in a basin elongated NW-SE, thus making an angle with the preceding WNW-ESE Precambrian orogenic orientations. The eu- and miogeosynclinal phase of this tectonic cycle is well documented by widely distributed Ordovician, Silurian and Devonian rocks. It is possible to distinguish two different sub-parallel marine basins during that period. Close to the Brasilian-Gondwanic platform, over a Precambrian continental crust basement, the Peruvian-Bolivian Basin, with predominating miogeosynclinal characteristics developed. Lutites, limestones, and sandstones were well represented all over this basin (Megard et al. 1971), subordinate to being the volcanic episodes.

Synchronically to the southwest, developed mainly over an oceanic crust basement, the Chilean-Argentinian Basin formed, with predominating eugeosynclinal characteristics. In the southwestern margin of this basin, lavas of probable submarine origin are found interfingered with greywacke-type deposits, cropping out mainly in the coast ranges between the Arauco and Aysén Provinces (37°–47° S) southern Chile. The rocks are metamorphosed from medium-grade quartz-albite-muscovite schists to high-grade glaucophane-lawsonite schists, with serpentinite, in parallel bands oriented NW-SE. The metamorphism decreases in intensity to the northeast. Frutos and Tobar (1973) have interpreted these features, following the idea of Ernst (1971), as produced by the proximity of an active subduction zone migrating slowly with time towards the northeast. From the corresponding volcano-magmatic belt, evolving over a former oceanic crust (Patagonian Arc, Frutos and Tobar 1973), widely distributed submarine tholeiitic volcanic rocks have been described (Godoy 1979). These facies change laterally towards the northeast into a belt of Lower-Middle Paleozoic flysch deposits, oriented NW-SE along the Cuyo Basin (San Juan Province in northwestern Argentina, Copiapo-El Salvador area in central-northern Chile, Coast Ranges of the Aconcagua and Coquimbo Provinces).

Both basins, the Peruvian-Bolivian Basin and the Chilean-Argentinian or Cuyo Basin, were partially separated during the Lower Paleozoic by a sub-positive structure (the Pampean Arc), which seems to have been a volcano-magmatic arc, northwesterly oriented, passing through the northern part of the Antofagasta Province, Chile, and continuing to the SE in coincidence with the Pampean Ranges Massif, northwestern Argentina (Harrington 1962). This structure shows plutonic and volcanic rocks of different Paleozoic ages. Metavolcanic rocks of basic composition, or probable Lower Paleozoic age, in Cerros de Limón Verde, Antofagasta Province, have been recently described. To the southeast, in the same belt, Llano and Escalante (1979) have described marbles, mica schists and basic metavolcanic rocks. Whether this arc is the product of another northeastward-dipping subduction zone during the Lower Paleozoic, remains an open question (Frutos and Tobar 1973).

An important diastrophism, affecting the complete marginal system, took place during the Middle-Upper Devonian producing a wide-spread unconformity that can be recognized all over the region. This diastrophism also produced the emersion of the Pampean Arc, as a positive "dorsal" structure, which became in this way a definitive divide between the Bolivian and the Cuyo Basins, until the Upper Carboniferous orogenic movements that produced the emersion and folding of practically the whole system. This new diastrophism was succeeded by the development of a wide-spread platform during the Middle and Upper Permian to Lower Triassic, characterized by detritic continental post-orogenic sediments and plateau type (taphrogenic) keratophyric-rhyolitic volcanism, which represents both the end of the Paleozoic Hercynian tectonic cycle and the transition to the next (Andean) tectono-magmatic cycle.

3 Paleozoic Ore Deposits

The metallogeny during the Paleozoic cycle shows a close and interesting relationship with the tectonic and paleogeographic scheme before described.

Schneider and Lehmann (1977) studying the Silurian synsedimentary cassiterite enrichments of the Bolivian Basin, have well shown the metallogenic "permanence" of the tin province in that part of the crust, probably since the Precambrian. Even if the Precambrian tin-bearing granites and pegmatites of the Rondônia region in the Brasilian shield do not seem to be directly related to the Andean Bolivian tin belt, they indicate a possible primary source from tin-bearing Precambrian rocks, for the important Lower Paleozoic meta-sedimentary strata-bound tin deposits, forming a belt of more than 600 km, from the NE border of the Titicaca Lake up to Potosi, to the south (Kellhuani and Huallatani mines). It is remarkable that exactly over this belt the Mesozoic and Cenozoic tin deposits are successively situated, mainly related to the calc-alkaline and partly shoshonitic magmatism of these periods.

Ni-Cu-Cr mineralizations in the areas of San Luis, Huánuco and Tapo, near Tarma (central-northern Peru) associated to Lower Paleozoic ultrabasic rocks of possible submarine volcanic origin (Grandin and Zegarra Navarro 1979) have been also reported.

South of the Bolivian Basin, the Pampean Arc, that forms its WSW border, could have been a source area for Paleozoic sedimentary deposits. This seems to be the case of the itabiritic-type iron deposits that appear associated to this structure, from the Arequipa Massif, southwestern Peru, to the northeastern flank of the Pampean Ranges Massif, northwestern Argentina. Over the same Pampean Arc, the Cu and Sn-W deposits related to the magmatism of the arc, form two separated belts, with the same notable metallogenic polarity that the Mesozoic-Cenozoic deposits show, in which the Cu-porphyries are located to the west, in the "oceanic" side of the arc, being the Sn-W deposits to the NE (Fig. 1), in the "continental" side of the arc. This disposition would suggest a relationship with a possible northeastward-dipping paleosubduction zone, if we compare it with the tectonic-metallogenic structure of the Central Andes continental margin during the Mesozoic and Cenozoic.

The Imilac deposit, a hydrothermally altered quartz-diorite stock mineralized with copper and gold in veins and veinlets (Frutos 1978a) has a Rb/Sr age of 240–285 m.y., with an assumed initial $^{87}Sr/^{86}Sr = 0.7045$ (M. Halpern, pers. comm.). It is interesting to note that this figure correlates better with the ratios of the Upper Tertiary island arc porphyry copper deposits (0.702 to 0.706, Kessler et al. 1975) than those corresponding to continental border porphyries as, for instance, the North American deposits whose initial $^{87}Sr/^{86}Sr$ ratios range from 0.705 to 0.710 (Kessler et al. 1975).

Few metallic deposits of Paleozoic age related to the Cuyo Basin volcano-sedimentary series, are known. They are mainly represented by the iron strata-bound (itabiritic type) deposits of Lower Paleozoic age, located rather over the Patagonian Arc (Fig. 2) as the Mahuilque, Relún deposits of the Arauco Province, southern Chile. They consist of magnetite lenses interbedded in mica schists and quartzites in a metamorphic volcano-sedimentary series that also includes amphibolites. Another similar Lower Paleozoic deposit exists in Sierra Grande, southeastern Argentina.

Associated to the ultramafic Paleozoic eruptive rocks of the Patagonian Arc (eugeosynclinal zone) some Cr and Ni mineralizations without economic significance have been also reported (Frutos and Oyarzun 1974). Only one porphyry copper deposit, of Upper Paleozoic age, could be related to this arc (La Voluntad, in the region of Neuquén, Southern Argentina).

Metallogenically compared, the Pampean Arc appears to be of greater importance than the Patagonian Arc, as far as the variety of metals as well as the volume of the mineralizations are concerned. This feature will reflect in the Mesozoic-Cenozoic metallogeny.

The post-orogenic (taphrogenic-plateau type) Permian-Triassic keratophyres and rhyolites, appear to be related to uranium mineralizations and some of these units present extraordinarily high trace contents of this element (Frutos and Oyarzun, unpubl.).

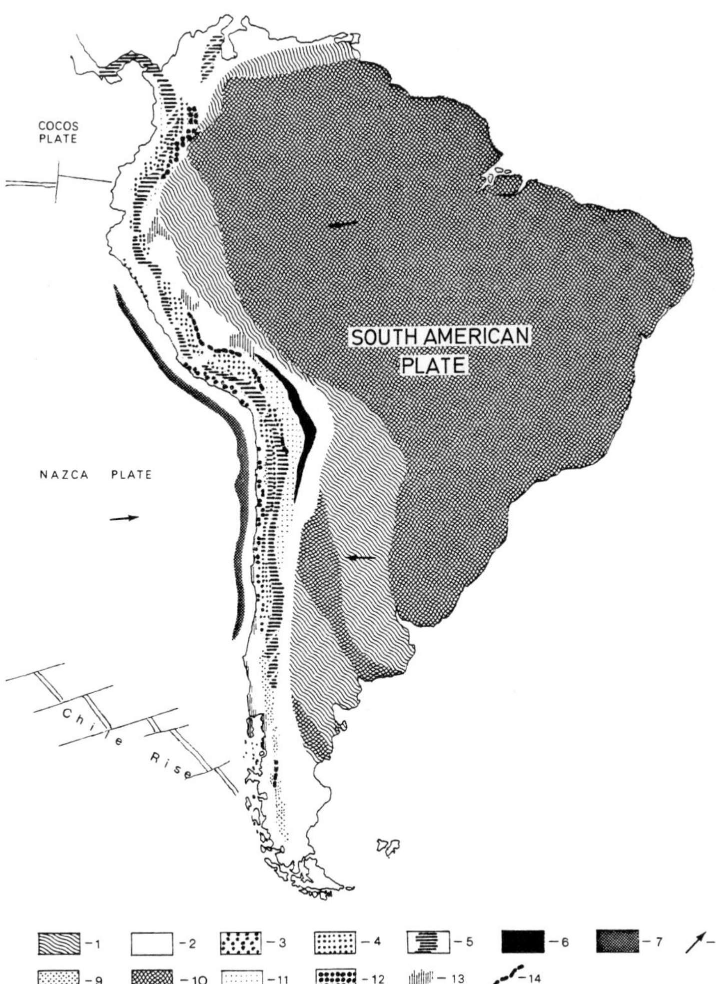

Fig. 1. Mesozoic-Cenozoic Andean metallogenic-paleogeographic scheme: *1* Andean foreland; *2* Andean geosynclinal area; *3* iron-apatite belt; *4* polymetallic deposits; *5* porphyry copper belt; *6* tin belt; *7* Peru-Chile trench; *8* plate movement direction; *9* Patagonian polymetallic belt; *10* Precambrian ranges (Pampean Arc and Patagonian Arc) and shield areas in the Andean foreland; *11* Altiplano polymetallic belt (Cu, Ag, Pb, Zn, and also sedimentary Cu); *12* Fe-P sedimentary deposits, mainly Cretaceous; *13* Au-placer deposits; *14* Mo-(U)-belt (mainly hydrothermal veins)

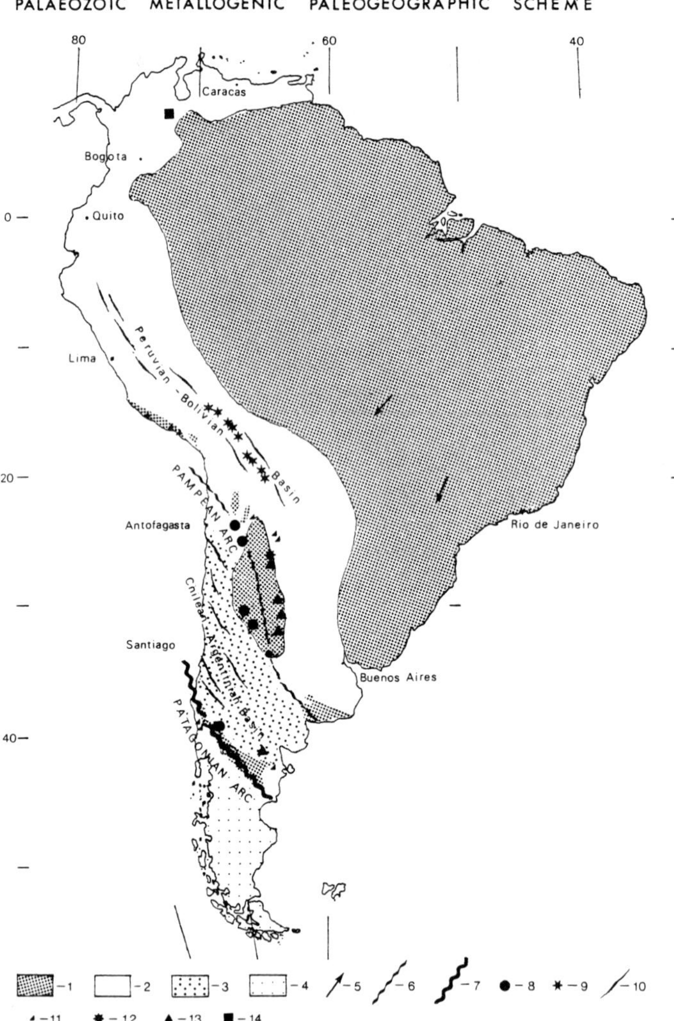

Fig. 2. Paleozoic metallogenic-paleogeographic scheme: *1* Precambrian positive and subpositive zones; *2* Paleozoic Basin: miogeosynclinal type; *3* Paleozoic Basin: eugeosynclinal type; *4* Paleozoic marine series of oceanic platform character; *5* plate movement direction; *6* approximate axis of the Pampean Arc subpositive volcano-magmatic zone; *7* approximate axis of the Patagonian Arc subpositive volcano-magmatic zone; *8* Paleozoic porphyry copper type deposits; *9* tin strata-bound deposits (Ordovician-Silurian marine series); *10* main structural axial directions; *11* iron itabiritic type deposits; *12* tin (magmatic-related type) deposits; *13* wolfram deposits; *14* massive sulphide (Pb-Zn-Cu) deposits

4 Mesozoic-Cenozoic Andean Tectonic Cycle

In broad terms the Andean Geosyncline corresponds to the space and time evolution of a subduction zone (oceanic plate-continental plate convergence type) and the corresponding volcanic arc and marginal basin. From a tectonic point of view, three main sectors could be defined concerning the type of basement: northern, central and southern Andes. Northern Andes (N. Ecuador, Colombia, Panamá) and southern Andes (Magallanes-Fire Land in southern Chile) correspond to a volcanic arc and marginal basin evolving chiefly over an oceanic crust basement near the continental border. There, flysch sediments, tholeiitic volcanism, ophiolites, nappes and high to intermediate pressure metamorphism occur (Aubouin et al. 1973; Aguirre et al. 1974; Goossens et al. 1977). The central Andes (Peru, Bolivia, Argentina and central-northern Chile) correspond to a volcanic arc that has evolved mainly over a continental crust basement along the continental border, except perhaps at the beginning of the cycle (Lower Jurassic) in some places of northern Chile (Antofagasta Prov.) where a tholeiitic volcanism have been reported (Palacios 1976; Frutos 1977). The central Andes, thus defined, lacks ophiolites, nappes, typical flysch sediments and regional metamorphism higher than green schists facies (Aguirre et al. 1974).

The tectonic evolution of the Andes shows a succession of compressive stages (Upper Carboniferous, Late Liassic, Upper Jurassic, Late Neocomian, Late Upper Cretaceous, Lower Eocene, Oligocene, and Upper Miocene) characterized by folds, reverse faulting and a predominance of intrusive magmatism, followed by relatively longer periods of no compression (relaxation phases) generally typified by normal faulting, graben formation and a predominance of extrusive magmatism. Thus, in strict terms, the Andean geosynclinal cycle commenced with a Middle to Upper Triassic extension stage which produced an initial subsidence of the basin and subsequently the eugeosynclinal-miogeosynclinal tectonic phase.

During the Jurassic to Lower Cretaceous span, the central Andes geosyncline consisted of an eugeosynclinal zone (island arc) with predominance of graywacke deposits and important intercalated volcanics, shortly tholeiitic and the beginning (Lower Jurassic) but soon calc-alkaline of rather basic composition (gabbro-dioritic, quartz-dioritic to granodioritic) (Figs. 3 and 4). Towards the foreland a miogeosynclinal zone was built up by marine sediments of platform facies with almost no participation of volcanic materials.

Epeirogenic movements beginning during the Callovian-Oxfordian were followed by a compressive diastrophism during the Kimmeridgian (Araucan tectonic phase). One or both of these processes caused, as the marine domain receded, evaporitic facies to be deposited in the basin. After these events, a narrow marine basin remained during the Titonian and Neocomian, which finally disappeared as a consequence of epeirogenic movements that preceded the Upper Cretaceous (Sub Hercynian) diastrophism. Zn-Pb-Ag-Ba strata-bound deposits related to platform facies of this marine basin, form an important metallotect (Amstutz 1978; Samaniego 1980). After the Neocomian and up to the present the central Andes geosyncline has been mainly an emerged zone, characterized by continental facies.

Eu- and miogeosynclinal domains were separated during the Jurassic and Neocomian, practically along the whole length of the basin, by a subpositive ridge or transitional

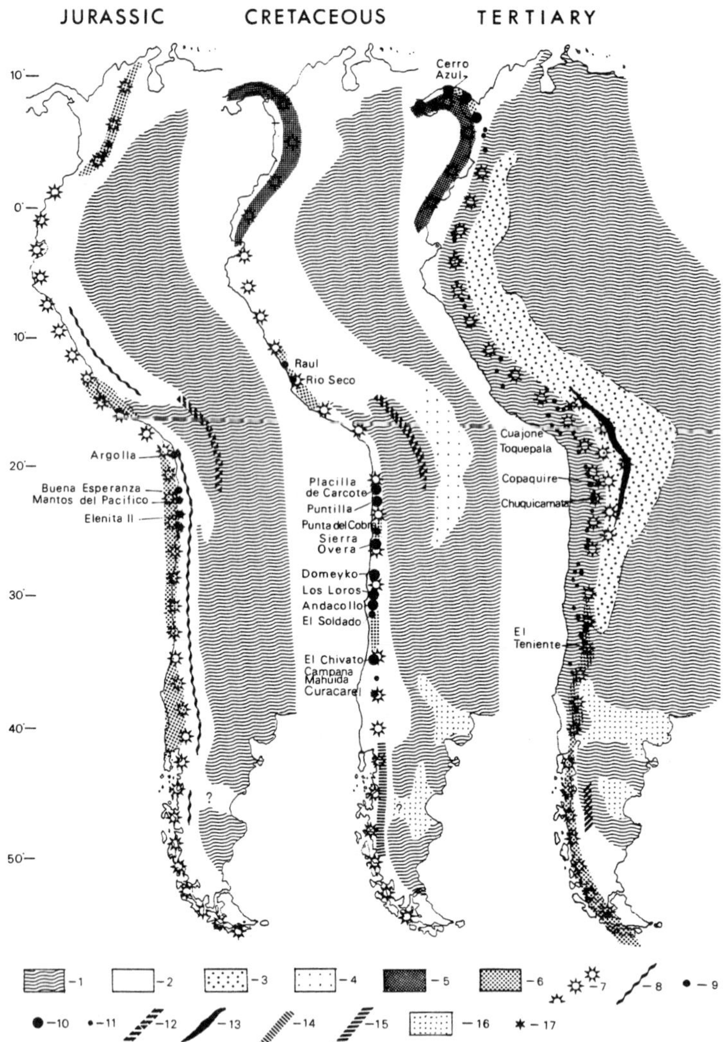

Fig. 3. Mesozoic-Cenozoic Andean metallogenic-paleogeographic evolution scheme: *1* Continental areas; *2* marine basins of active plate-margin type; *3* continental basins; *4* marine-continental active basin; *5* tholeiitic magmatism; *6* calc-alkaline quartz-dioritic magmatism; *7* volcano-magmatic arc; *8* miogeoanticlinal horst ("dorsal mesoliminar"); *9* manto-type Cu deposits; *10* Cu-Au porphyry type deposits; *11* Cu-Mo porphyry type deposits; *12* Mesozoic Au-W-Sb-Sn-Hg belt; *13* Tertiary tin belt (Sn-Bi-Ag-Pb-Sb-W); *14* Pb-Zn (polymetallic) Patagonian belt; *15* Patagonian Mo belt; *16* marine epicontinental areas; *17* Mo-bearing porphyries

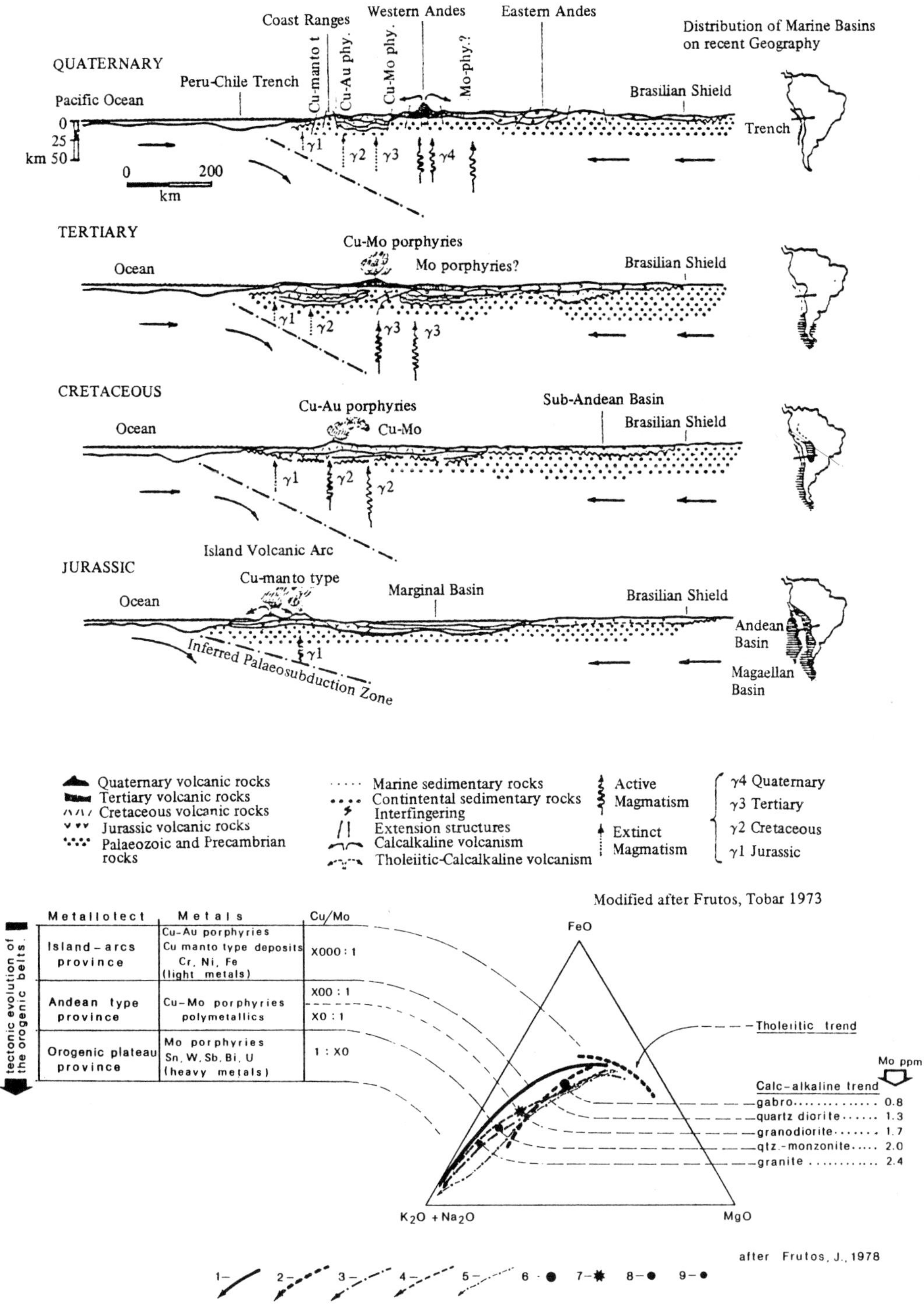

Fig. 4. Relationship between the metallogeny and the magmatic evolution of the American Orogenic Belt: *1* southern Chile Andes; *2* Island Arc (Central America, after Kessler 1975); *3* Canadian Cordillera; *4* western U.S.A. Cordillera; *5* northern Chile Andes; *6* gabbro; *7* granodiorite; *8* quartz-monzonite; *9* granite

zone not always emerged: the "mesoliminar dorsal" (Frutos 1975). This tectonic ridge evolved emerging and migrating with time towards the foreland and finally, during the successive orogenic phases, partially overriding the miogeosynclinal series, along a crust rupture zone. This crustal weakness zone (that importantly controlled the location of the Tertiary hypabyssal intrusives) coupled with the relatively greater block-type uplifting of the "mesoliminar dorsal" during the Upper Tertiary, has been a factor of control of the location of most of the Cu porphyry-type deposits.

The latest tensional period, coinciding with what has been defined as the taphrogenic phase of the Andean cycle, culminated in the Upper Pliocene and is still active; the deep structures (normal faulting) generated in this period appear to be intimately related to the present Andean Chain morphology and its associated volcanism, as well as to the present coastline to which they are parallel.

The calc-alkaline magmatism, which has been evolving since Jurassic to more alkaline-acidic terms and at the same time migrating gradually towards the foreland, reaches, at this phase, the granitic extreme, and seems to be genetically associated with the existence of an intermediate aseismic gape (Hanus and Vanek 1978) interpreted as a partially melted zone, in a well-defined Benioff zone.

Towards the foreland an alkaline volcanic association appears. Notwithstanding, it is interesting to note that a narrow belt of a shoshonitic association (Déruelle 1978) in which the K_2O content is invariably high and almost independent of SiO_2 (K_2O/Na_2O ratio higher than 1) is strictly confined to that zone of the Andes with the thickest crust and deepest seismicity (Altiplano zone: southeastern Peru, central-western Bolivia, northeastern Chile and northwestern Argentina). Lefevre (1973) points out that the shoshonitic suite of southern Peru, which cannot be petrologically related to the alkaline series, is rather associated to the calc-alkaline series, of which it could correspond to an extremely K_2O-enriched differentiation. The shoshonitic belt appears in concordance with the extreme deepness of magma generation and the greater thickness of the crust. As we will see below, that zone is characterized by a highly specialized metallogenic province (Sn-W-Bi-Sb).

The mineralogical and chemical studies carried out at several central Andes Upper Cenozoic andesitic volcanoes (Palacios and Lopez 1979) show that the trace element abundances of these rocks are consistent with either a magma derivation by low degree (3%–5%) of partial melting of uncontaminated garnet peridotite with a mantle trace element composition, or by approximately 15% partial melting of garnet peridotite enriched in large ionic radii lithophile elements, followed in both cases by mafic mineral fractionation. In this latter case, the contamination (3%–5%) could arise from a liquid geochemically similar to a low degree of partial melting of an eclogite of the same composition as the Nazca Plate altered oceanic basalts present (Thorpe et al. 1976; Palacios and Lopez 1979). In any case it seems clear that the continental crust (after Sr isotopic studies) is not acceptable as a source for calc-alkaline Andean cycle magmas. The geological data favour the generation of these magmas from the mantle "wedge" by contamination by liquids and fluids produced by a low degree of partial melting of the subducted oceanic crust.

5 Mesozoic-Cenozoic Metallogeny

The transversal metallogenic zonation in parallel belts in the central Andes appears closely related to the Mesozoic-Cenozoic evolution of the orogen.

During the Jurassic-Lower Cretaceous "youthful" island arc stage, the metallogeny is mainly that of copper veins and stratiform (manto-type) deposits (which gives rise to base metals in some case) associated with a volcanism having tholeiitic affinities in a first phase and later a calc-alkaline andesitic-dioritic phase. These deposits, that appear now closest to the trench in the Coastal Ranges of Peru and central-northern Chile, are well represented by those of Raul, Condestable, Los Icas, southwestern Peru (Wauschkuhn 1979) and Buena Esperanza, Mantos del Pacífico, Mantos de la Luna (Ruiz 1965) in northern Chile.

The iron-apatite deposits, that also appear along the Coastal Ranges of southern Peru and northern Chile, form a narrow and distinct lineament separated in two relatively short and restricted belts. The deposits consist of apatite-amphibolite-bearing iron ores related to the Upper Jurassic-Lower Cretaceous calc-alkaline intermediate magmatism of both plutonic and extrusive type (skarns, injection bodies, veins and manto-type deposits associated with andesitic, partly submarine, volcanism). The close spatial coincidence of these iron deposits with Precambrian and Paleozoic ancient volcanic island arc-itabiritic areas (Arequipa Massif, Pampean Ranges Massif) is rather significant.

The genesis of these apatite-iron deposits, which present field evidence of a magmatic origin but whose geochemistry and paleogeographic control are not consistent with a simple primary magmatic origin, would be in accordance with the idea of a possible magmatic remobilization of iron and phosphorus from the marine ferruginous sedimentary rocks of platform facies that form the pre-Mesozoic basement (Frutos and Oyarzun 1975).

It is interesting to note that, while in the western part of the marginal system "youthful" island arc conditions prevailed, in the eastern part of it, at a distance of about 400 km, a belt of granitic-granodioritic plutons appears (Cordillera Real of Bolivia: Sorata, 180 m.y.; Zongo, 150–211 m.y.; Taquesi, 199 m.y.). To these plutons, which are emplaced in the western border of a relatively thick pre-existent continental crust, are associated early vein-type tin-tungsten mineralizations (Ahlfeld 1967; Grant et al. 1979).

The evolution of the volcanic island arc, in the central Andes (migrating gradually to the east and the magmatism following a calc-alkaline trend) was continued throughout the Jurassic and Lower Cretaceous with a marginal basin (Andean Basin) existing between it and the continent, which definitely emerged during the Late Neocomian. During this stage, clearly associated to both the eruptive processes of the arc and the platform sedimentary conditions of the marginal basin, the polymetallic ore deposit belt appears (located to the east of the iron-apatite belt), which reaches its large volumetric importance in Peru and central-northern Chile.

The older (Late Neocomian-Upper Cretaceous) porphyry copper deposits of the Andean cycle, well represented by those of Almacén and Los Pinos (Coastal Batholith of western Peru), and Placilla de Carcote, Puntilla, Sierra Overa, Domeyko, Andacollo, El Chivato, in central-northern Chile (Frutos 1978b) present some peculiar characteris-

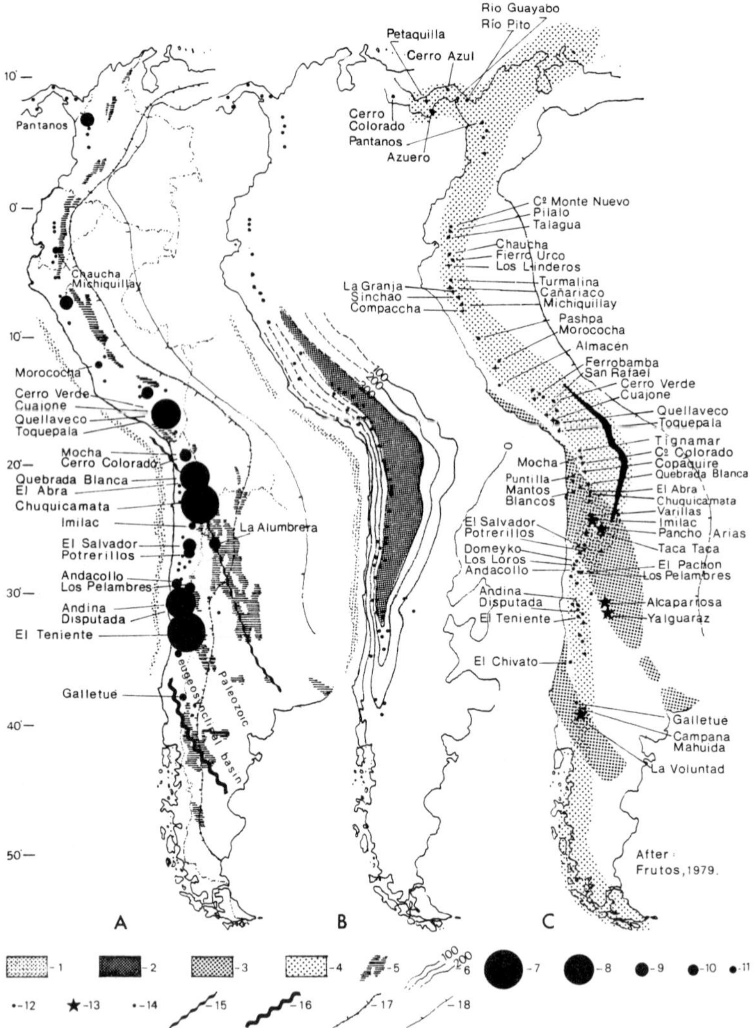

Fig. 5 A–C. Andean porphyry copper deposits related to structural-paleogeographic features: The relationship between the big clusters of metal concentration with: **A** The Precambrian massifs; **B** the thicker crust area; **C** the Paleozoic tectono-magmatic belts. *1* Peru-Chile trench; *2* the thicker crust area; *3* Paleozoic tectono-magmatic belts; *4* Andean (Mesozoic-Cenozoic) magmatism; *5* Precambrian massifs; *6* Bouguer gravity anomal. contour lines; *7* nucleus of Cu-metal concentration: 50 m.t. Cu; *8* 20 m.t. Cu; *9* 5–10 m.t. Cu; *10* 2.5–5 m.t. Cu; *11* 1–2.5 m.t. Cu; 12 less than 1 m.t. Cu; *13* Paleozoic porphyry copper deposits; *14* Mesozoic-Cenozoic porphyry copper deposits; *15* Pampean Arc; *16* Patagonian Arc; *17* Maximum extension of the Andean Mesozoic-Cenozoic geosynclinal system; *18* limit of the Paleozoic geosynclinal area

tics that permit correlating them to the island arc type of porphyries. These generally gold-rich deposits are, nevertheless, relatively poor in molybdenum. They are associated with a quartz-dioritic to granodioritic calc-alkaline magmatism, and are located more to the west (closest to the trench) than the main Middle-Upper Tertiary Cu-Mo porphyry type deposits belt (Frutos 1978b).

During the Cenozoic, the evolution of the orogen was completed with the creation of thicker and thicker crustal conditions, continuing the slow migration of the magmatic arc inside the continental border and the geochemical evolution of the magmatism towards increasingly acid calc-alkaline conditions (granodioritic and granitic; the last of these especially in the margin of the magmatic belt farther away from the trench). A restricted shoshonitic belt has been developed along the easternmost front of the central Andes orogen. With this Cenozoic stage of the mobile border are associated both the main porphyry copper belt (mainly Middle Upper Tertiary deposits of the "classic" Cu-Mo type) (Fig. 5) located chiefly in the western Andes and the tin belt (Ag-Bi-Sb-W) located 300 km towards the east, mainly in Bolivia, approximately coinciding with the shoshonitic magmatism area in the taphrogenic plateau zone of the central Andes.

No Mo-porphyry type deposits have been reported, even though certainly some mineralizations such as those of Compaccha in Peru and Copaquire in northern Chile, would coincide with this model. The geologic-tectonic setting suitable for this type of deposit would be found precisely along the eastern border of the north Chile, southeast Peru, west Bolivia and northwest Argentina. It could be that exploration in these localities, which has been intensified lately, may show some deposits of this type.

In the central Andes, coinciding with an eastward increase in lithophile elements (K, U, Sr) and $^{87}Sr/^{86}Sr$ observed in the intrusives (Zentilli 1975), an increasing Mo/Au ratio as trace contents in the intrusives and also in the mineral deposits, can be documented.

Concerning the paleogeographic-tectonic control of the volumetric metal content variation along the Andean metallogenic belts, a close relationship with the pre-Mesozoic metallogenic provinces and the ancient tectono-magmatic belts disposition can be observed. This is clearly the case of the Andean tin belt, the iron-apatite belt and the three big clusters of copper and molybdenum concentration (Toquepala-Cuajone zone: 20 m.t. Cu; Chuquicamata-El Abra zone: 70 m.t. Cu; El Teniente-Disputada zone: 70 m.t. Cu).

References

Aguirre L, Charrier R, Davidson J, Mpodozis A, Rivano S, Thiele R, Tidy E, Vergara M, Vicente J-C (1974) Andean magmatism: its paleogeographic and structural setting in the Central Part (30°–35°S) of the Southern Andes. Pac Geol 8:1–38

Ahlfeld F (1967) Metal epochs and provinces of Bolivia. Mineral Deposita 2:291–311

Amstutz GC (1978) Zu einer Metallogenie der Zentralen Anden von Peru. Muenstersche Forsch Geol Palaeontol 44/45:151–158

Aubouin J, Borrello AV, Cecioni G, Charrier R, Chotin P, Frutos J, Thiele R, Vicente J-C (1973) Esquisse paléogéographique et structurale des Andes Méridionales. Rev Géogr Phys Geol Dyn II S XV, Fasc 1–2:11–73

Borrello AV (1969) Los Geosinclinales de la Argentina. Min Ec Y Trabajo. Dir Nac Geol Miner Anal XIV, Buenos Aires, Argentina, 188 pp

Déruelle B (1978) Calc-alkaline and shoshonitic lavas from five Andean volcanoes (between Lat. 21°45'–24°30' S) and the zonation of the Plio-Quaternary volcanism of the Southern Andes. J Volcanol Geotherm Res 3:281–298

Ernst WG (1971) Petrologic reconnaissance of Franciscan metagraywackes from the Diablo Range, Central California Coast Ranges. J Petrol 12:413–437

Frutos J (1970) Ciclos tectónicos sucesivos y direcciones estructurales sobreimpuestas en los Andes del Norte Grande de Chile. Upper Mantle Symp (IUMP) Solid Earth Probl Buenos Aures II: 473–483

Frutos J (1975) Porphyry copper-type mineralization and geosynclinal tectonic evolution in the Chilean Andes. Ann Soc Geol Belg 98/1:5–15

Frutos J (1977) Evolution de la marge continentale des Andes du Nord du Chili, le volcanisme et la métallogenèse. Thesis Dr 203, Univ Paris, Fac Sci Orsay, 160 pp

Frutos J (1978a) Informe geológico económico de la Mina La Casualidad, Prov. de Antofagasta, Chile, Inedit Rep Dep Geol Univ Concepción, Chile, 58 pp, 6 map

Frutos J (1978b) World survey of molybdenum deposits. In: Sutulov A (ed) International Molybdenum Encyclopaedia, vol I, pp 375–401

Frutos J, Ferraris F (1973) Mapa tectónico de Chile. Inst Invest Geol, Santiago, Chile

Frutos J, Oyarzun J (1974) Metallogenic belts and tectonic evolution of the Chilean Circum-Pacific continental border. Circum-Pacific energy and mineral resources conference. Honolulu, Hawaii. Am Assoc Petrol Geol

Frutos J, Oyarzun J (1975) Tectonic and geochemical evidence concerning the genesis of El Laco magnetite lava flow deposits, Chile. Econ Geol 70:988–990

Frutos J, Tobar A (1973) Evolution of the southwestern continental margin of South America. III. Int Gondwana Symp Canberra 39:565–578

Godoy E (1979) Metabasitas del basamento metamórfico Chileno, nuevos datos geoquimicos. II. Congr Geol Chileno Inst Invest Geol, pp E133–E148

Goossens PJ, Rose WI, Flores D (1977) Geochemistry of tholeiites of the basic igneous complex of northwestern South America. Geol Soc Am Bull 88:1711–1720

Grandin G, Zegarra Navarro J (1979) Las rocas ultrabásicas en el Perú. Las intrusiones lenticulares y los sills de la región Huánuco-Monzon. Bol Soc Geol Perú 63:99–116

Grant NJ, Halls C, Avila W, Snelling N (1979) K-Ar ages of igneous rocks and mineralization in part of the Bolivian tin belt. Econ Geol 74:838–851

Hanus V, Vanek J (1978) Morphology of the Andean Wadati-Benioff andesitic volcanism and tectonic features of the Nazca Plate. Tectonophysics 44:65–77

Harrington H (1962) Paleogeographic development of South America. Bull Am Assoc Petrol Geol 46 (10):1773–1814

Kessler SE, Jones LM, Walker RL (1975) Intrusive rocks associated with porphyry copper mineralization in island arc areas. Econ Geol 70:515–526

Lefevre C (1973) Les caractêres magmatiques du volcanisme Plio-Quaternaire des Andes dans le Sud tu Pérou. Contrib Mineral Petrol 41:259–272

Llano J, Escalante A (1979) Petrología de las anfibolitas de la Qd. del Gato, Sa. Pié de Palo, Prov. San Juan, Rep. Argentina. II. Congr Geol Chileno, Inst Invest Geol, Chile, pp E39–57

Megard F, Dalmayrac B, Laubacher G, Marocco R, Martinez C, Paredes J, Tomasi P (1971) La chaîne hercynienne au Pérou et en Bolivie: premiers résultats. Cah ORSTOM Sér Géol 1 (3): 5–44

Palacios C (1976) Notes about the Jurassic Paleovolcanism in Northern Chile. Inedit Rep Dep Geol Univ Norte, Chile

Palacios C, Lopez L (1979) Geoquímica y Petrología de andesitas cuaternarias de los Andes Centrales (18°57'–19°28' S). II. Congr Geol Chileno, Inst Invest Geol, Chile, pp E73–E88

Ruiz C (1965) Geología y yacimientos metalíferos de Chile. Inst Invest Geol, Santiago, Chile, 386 pp

Samaniego A (1980) Strata-bound Zn-Pb-(Ag-Cu) ore occurrences in Early Cretaceous sediments in North and Central Peru. Diss Univ Heidelberg, 209 pp

Schneider HJ, Lehmann B (1977) Contribution to a new genetical concept on the Bolivian tin province. In; Klemm DD, Schneider HJ (eds) Time and strata-bound ore deposits. Springer, Berlin Heidelberg New York, pp 153–168

Thorpe RS, Potts PJ, Francis PW (1976) Rare earth data and petrogenesis of andesite from north Chilean Andes. Contrib Mineral Petrol 54:65–78

Wauschkuhn A (1979) Geology, mineralogy and geochemistry of the strata-bound copper deposits in the Mesozoic Coastal Belt of Peru. Soc Geol Peru 62:161–192

Zentilli M (1975) Geological evolution and metallogenic relationships in the Andes of Northern Chile, between 26° and 29° S. II. Congr Ibero-Am Geol Econ, Buenos Aires, Argentina

Correlation of Strata-Bound Mineral Deposits in the Early Cretaceous Santa Metallotect of North and Central Peru

A. SAMANIEGO [1]

Abstract

The metallogenetic investigations carried out in north and central Peru have determined the existence of the "Santa Formation metallotect" based on more than 80 Zn-Pb-(Ag-Cu)-stratabound mineral occurrences related to this formation.

The most important geometric and geologic features of these ore occurrences have been described and correlated between themselves and also with those that characterize the typical locality at the El Extraño mine (Figs. 1 and 2) in Ancash (N. Peru), which represents the most typical and best studied ore occurrence of this metallotect.

The last part of this paper presents a genetic discussion about the origin of these ore occurrences.

1 Introduction

The scope of this paper is to make a correlation of the different stratabound Zn-Pb-(Ag-Cu) mineral occurrences in order to justify its grouping in the Santa Formation metallotect and its genetic interpretation.

These investigations were carried out on five levels including: field geology, stratigraphy, sedimentary petrology, ore microscopy, and geochemistry.

The studied area covers more than 40,000 km^2 in north and central Peru and includes more than 80 mines and prospects (Figs. 1 and 2).

2 Geological Outline of North and Central Peru

The area considered by this study includes part of the Western Cordillera in northern and central Peru and a minor part of the puna surface in central Peru. It consists principally of Mesozoic and Cainozoic sedimentary and volcanic rocks intruded by the Coastal and Cordillera Blanca batholiths. The sedimentary and volcanic rocks were deposited in the Andean geosyncline pair during the liminar period of the Andean orogenesis

1 Mineralogisch-Petrographisches Institut der Universität Heidelberg, Im Neuenheimer Feld 236, 6900 Heidelberg, FRG
Present address: AGESSA-Gerencia de Exploraćiones, Av. Garcilaso de la Vega No. 732-8°, Lima 1, Peru

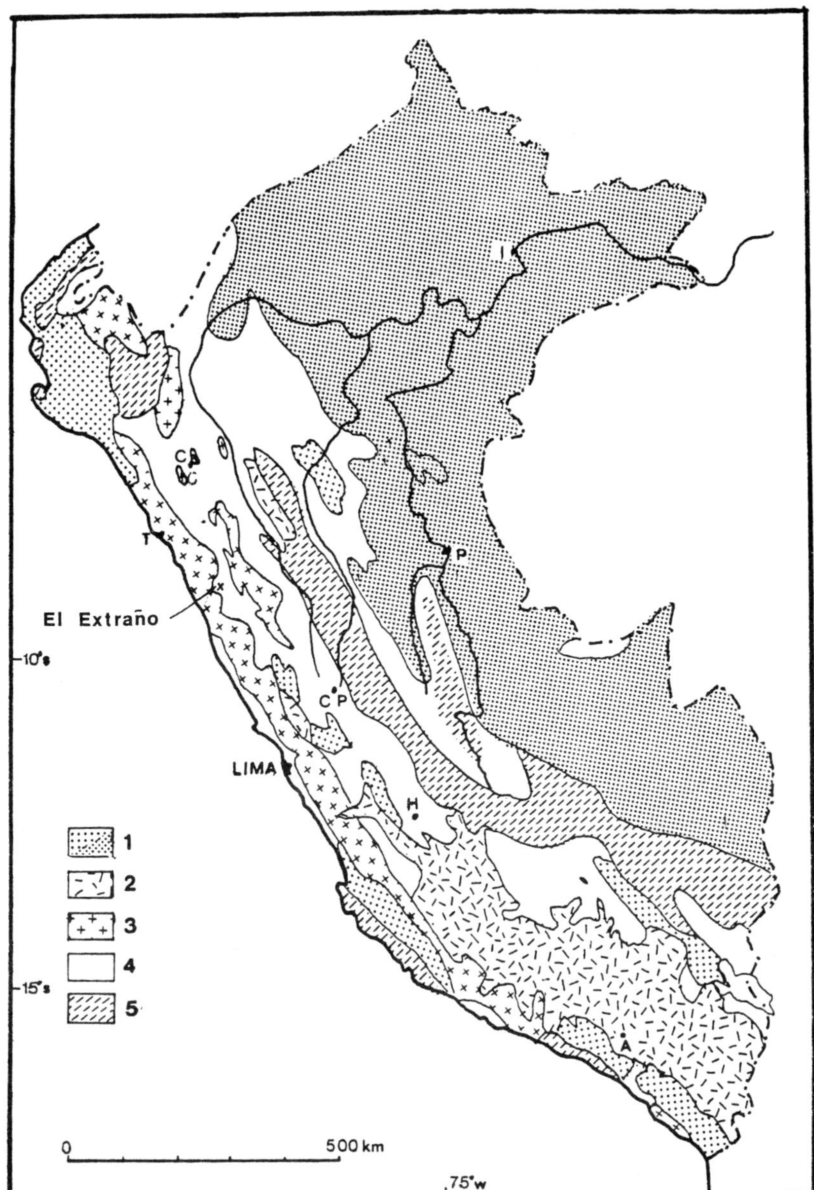

Fig. 1. Major geological elements of Peru. (Simplified after Wilson 1963; Bellido 1969.) *1* Cainozoic rocks; *2* Tertiary volcanic rocks; *3* Coastal batholith; *4* Mesozoic rocks; *5* Paleozoic rocks; *6* Studied area

(terminology after Aubouin 1972). This geosyncline was developed on a block-faulted basement of Precambrian crystalline rocks in which movement occurred along faults parallel to the continental margin (Cobbing and Pitcher 1972; Audebau et al. 1973; Cobbing 1976).

Cobbing (1978) proposed its subdivision into five basins: Pongos, Chavin, Río Santa, Huarmey, and Río Cañete basin which have subsided and were deformed independently

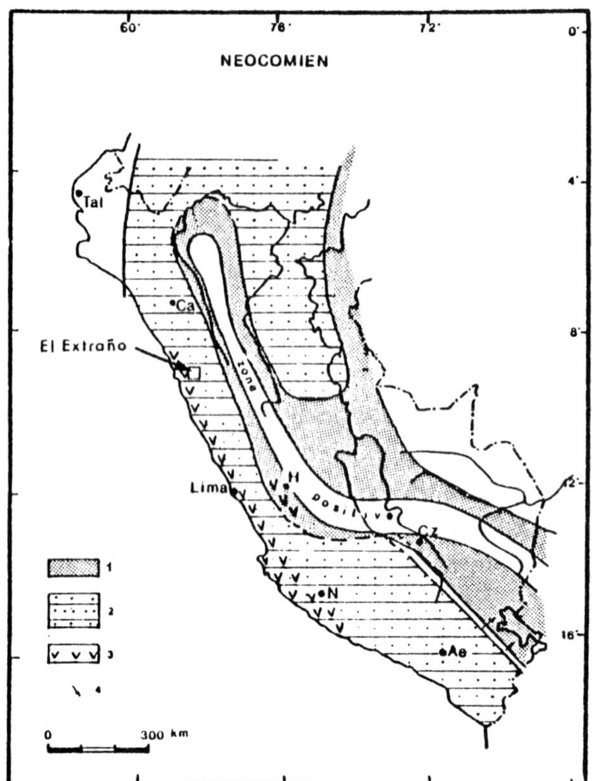

Fig. 2. Paleogeography of Peru at Neocomian times. (After Dalmayrac et al. 1977.) *1* Continental facies; *2* Shallow marine facies; *3* Synsedimentary volcanism; *4* Direction of the detritical provenance. *Tal* Talara; *Ca* Cajamarca; *H* Huancayo; *N* Nazca; *Cz* Cuzco; *Ae* Arequipa

and, consequently, each of them shows a different thickness and type of filling. The Santa Formation was deposited principally in the Chavín basin; other lithological and faunal correlatable sequences were deposited in the Río Santa and Río Cañete basins.

The tardi-liminar and post-liminar periods called by Megard (1978) "tectogenic period" began in the Upper Cretaceous and were characterized by successive, relatively short phases of compression, separated by tectonically quiet periods in which an intense calc-alkaline magmatism of the Coastal Batholith occurred and by the development of subsiding basins filled with continental sediments. After the relatively less important first Upper Cretaceous phase, the main phase occurred at the end of the Eocene, which was followed by three minor phases during the Neogene (Megard 1978).

The Coastal Batholith is, after Pitcher (1978), a multiple intrusion of gabbro-tonalite and granite occupying the core of the Western Cordillera. Its emplacement control and mechanism have been discussed by Cobbing and Pitcher (1972), Myers (1975), Bussell et al. (1976), and Pitcher and Bussell (1977) as an emplacement of magmas along major fault lines in the crystalline basement with close control by transcurrent faults and by smaller scale joint pattern for the emplacement of the individual plutons. The rocks of the batholith were assigned by Cobbing et al. (1977) to distinct plutonic units, superunits, and segments. The studied area is located in the Lima segment and the outcrops of igneous rocks belong principally to the Santa Rosa, Puscao, and

Pativilca superunits. Structurally (after Megard 1978), in central Peru is the crustal shortening of the order of 100 km, and about 50% of the shortening is concentrated in and around the highest parts of the Western Cordillera where there are abundant chevron folds. The eastern side of the Western Cordillera is characterized by steep reverse faults and minor refolded nappes. Pitcher (1978) pointed out that the shortening was not important and that it is unlikely that sufficient shortening during the last 100 m.y. could account for the doubling of the crustal thickness. He remarks on the role played by the intrusion of the Coastal Batholith in the thickening of the crust.

3 The Santa Formation

The Santa Formation at its typical locality near Carhuaz, in the Callejón de Huaylas, was described by Benavides (1956) as a 341 m thick sequence composed of a largely dark gray, fossiliferous, medium-bedded, platy, concretionary limestone which is dolomitic in places and is interbedded with a few thin beds of black splintery shale and chert. At the base of the Santa Formation there are soft, varicoloured, finely splintery shales which rest concordantly on the massive ridge forming top ledge of the Chimú Formation.

The extension of this formation was recognized principally by Benavides (1956) in northern Peru, by Wilson (1963) in northern and central Peru, by v. Hillebrandt (1970), and Megard (1973) in its southern extension in central Peru. The correlation with the eugeosynclinal facies was registered by Myers (1974), Child (1976), and Cobbing (1978).

According to Benavides (1956), the Santa Formation represents a change from the nonmarine conditions that prevailed during the deposition of the Chimú sandstone to a shallow marine brackish water environment; the typical structures observed and measured by them (cross bedding, etc.) record a marine overlap from the west. Wilson (1963) considered that the ripple marks, the sporadically distributed oolithic beds, and the absence of pelagic invertebrates in the Santa sediments of central Peru point to a deposition in a near shore environment. Additional observations made by the author during the measurement of several stratigraphic profiles in the different mining districts, corroborate the findings of both authors.

This formation is considered to be of Late Valanginian age based on the fossil, *Valanginitis broggii* Lissón, together with other fossil records indicating a Neocomian age by Benavides (1956) in the lowest unit of the overlying Carhuaz Formation. Samaniego and Amstutz (1978) reported the finding of the same fossil in the lowest calcareous units of the Santa Formation in the El Extraño mine area.

4 The Extraño Mine, a *Locus Typicus* for the Santa Formation Metallotect

The Extraño mine (Fig. 3) is located in the highest part of the western flank of the Cordillera Negra (Western Cordillera) at altitudes between 3,900 and 4,200 m, at a distance of about 400 km, by road, from Lima.

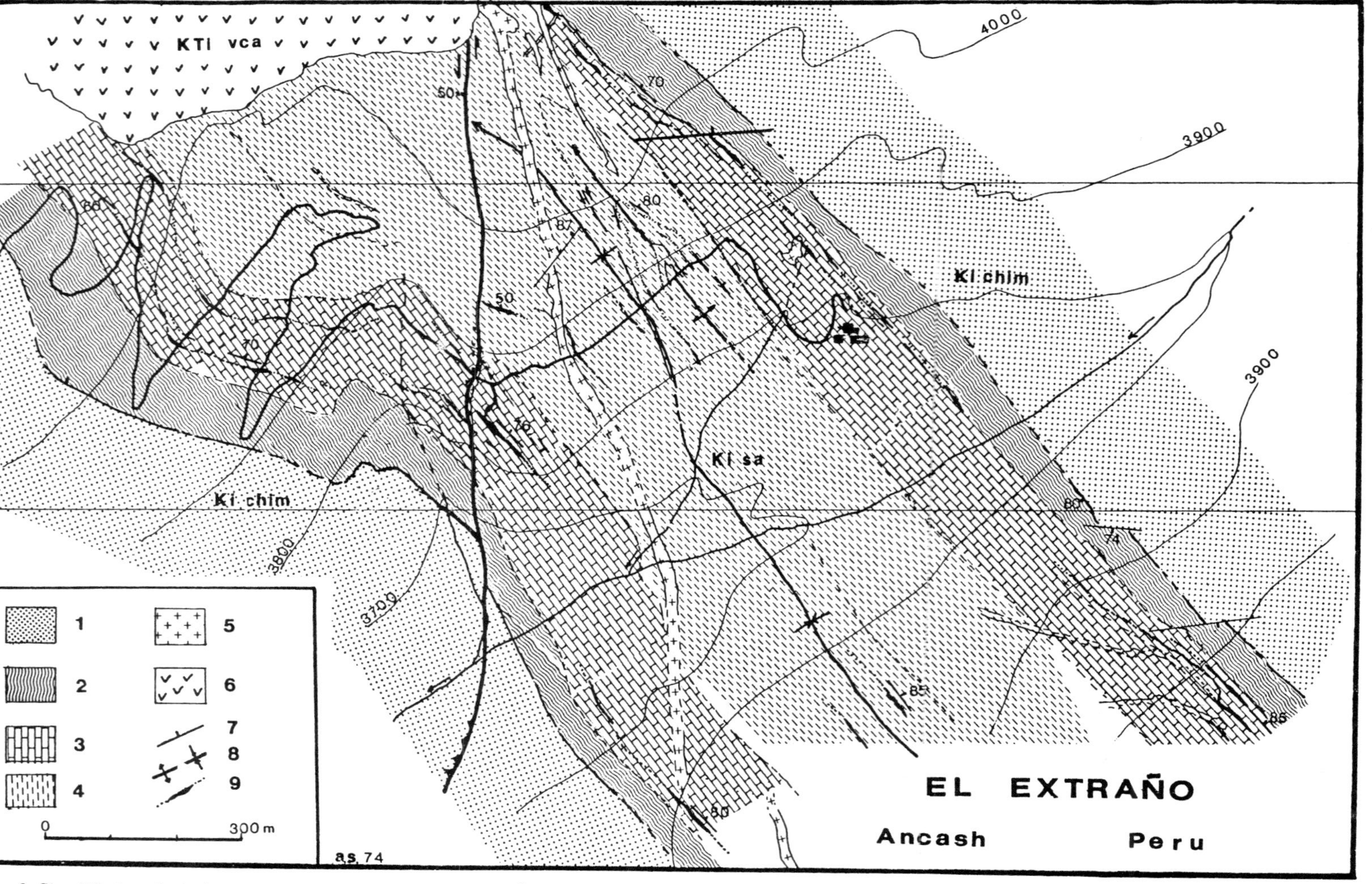

Fig. 3. Simplified geological map of the El Extraño mine. *1* Chimú Formation; *2* Shaly lower member of the Santa Formation; *3* Calcareous member of Santa; *4* Upper shaly member of Santa; *5* Quartz monzonitic dike; *6* Tertiary volcanic rocks; *7* Strike and dip; *8* Axes of folds; *9* Ore mantos outcrops

4.1 Geological Outline

The major structure at El Extraño mine is an asymmetrical, very narrow, elongate syncline (Fig. 4), the axial plane of which strikes NW and dips to the SE. It forms a geologic window and is limited on the north and the west by Tertiary volcanic sequences and, in the south and east, by regional faults.

On the western limb of this syncline, well-bedded, medium-grained quartzites of the Chimú Formation crop out. The eastern limb was thinned tectonically. The core of this structure forms a 282 m sequence of metamorphic assemblages of preexisting shales, limy sandstones, limy tuffitic shales, dolomitic limestones, and dolomites containing some horizons rich in Zn and Pb sulphides. This sequence belongs to the Santa Formation, folded disharmonically into several minor anticlines and synclines.

Two important overthrust faults were recognized on both sides of the major syncline. Several minor joint and fault systems with short displacements were mapped principally in the underground workings of the mines. These faults are postmineral and some remobilization of the ores occurred along them. The metamorphism affecting the ore minerals and the sedimentary rocks of the area shows features of regional character and has mineral assemblages of the greenschist facies. The extremely good preservation of several typical sedimentary and diagenetic textures, such as cross-bedding, load casts, slumping, diagenetic rhythmites, etc., after the metamorphic processes and the absence of a halo-like distribution of the metamorphic paragenesis together with the comparison of their characteristics, results of the investigations of Aguirre et al. (1978), permit the assumption that depth metamorphism affected this area.

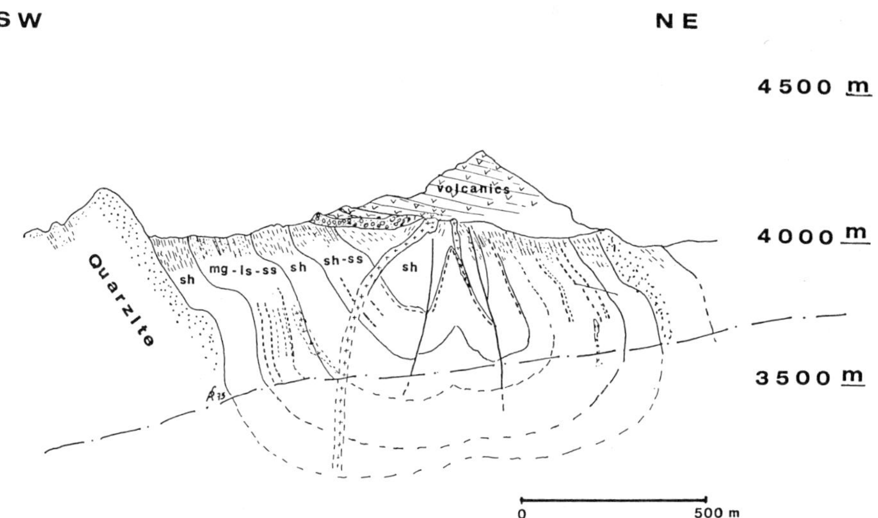

Fig. 4. Schematic geological profile across the syncline at the El Extraño mine. The *dashed lines* represent the position of the ore mantos

Igneous rocks are present in the form of quartz monzonite dikes and rhyolite sills. The intrusive bodies do not show any genetic relationship with the ore mantos.

4.2 Ore Occurrence

The ores at El Extraño mine occur chiefly in stratabound to stratiform mantos in the Santa Formation. The congruence between ore mantos and wall rocks can be observed on several scales: from regional or districtal (10,000 to 1,000 m) to mine scale; also to the local scale (10 to 1 m) and on the microscopic scale as is shown in Figs. 4 and 5.

Mineralogically, the ores are composed mainly by sphalerite and galena (Ag-rich); pyrite, marcasite, and barite are sparse accessories. The Zn/Pb ratio is close to 3/1 and it is, more or less, the same in the various ore mantos.

Secondary minerals as smithsonite, zincite, cerussite, and some greenockite together with Mn-oxydes (chiefly psilomelane) and Fe-oxydes were observed in the poorly developed secondary zone.

Geochemical analyses carried out on Zn and Pb concentrates indicate high Cd and Ge contents in the sphalerite and high As contents in galena.

4.3 Ore Textures

The ore minerals and the wall rocks show a variety of ore textures which are well known in the literature as typical synsedimentary and diagenetic textures (see Fig. 5 and Plate 1).

The following textures affecting both wall rocks and sulphides were observed in many sizes and in many different degrees of development: normal bedding, cross bedding, rhythmic bedding, convolute bedding, and lenticular bedding. Diagenetic textures as load coasts, rhythmites, slumping and others were also observed.

A notable grain size congruency between the sulphides and the metamorphic minerals was observed in several thin and polished sections of the different mantos.

An important observation of a superimposed contact metamorphism over the regional metamorphic assemblages and the destruction of many original textures in a thinly developed metamorphic halo, caused by the intrusion of the quartz monzonite dike, suggest the subordinate to insignificant role played by these intrusions by the mineralization in the area.

Fig. 5 a–f. Stratabound features of the ores at the El Extraño mine at different scales of observation: **a** Stratigraphic column of the Santa Formation at El Extraño mine area showing the distribution of the ore mantos in the reconstructed lithology. **b** Simplified geological map of the Manto 1 at the 830 level West Extraño mine (*dotted area* is the ore manto). **c** Stratiform feature of the outcrop corresponding to the Manto 1a at the access road to El Extraño mine. **d** Stratiform features of hand specimen of the Manto 6; the rhythmites are possibly of diagenetic origin. **e** Stratiform to stratabound features on polished section preparation showing the rhythmic alternation between garnet-diopside-epidote with minor quartz and calcite, and layers containing sphalerite and galena. This sample has been taken at the place shown in **b**

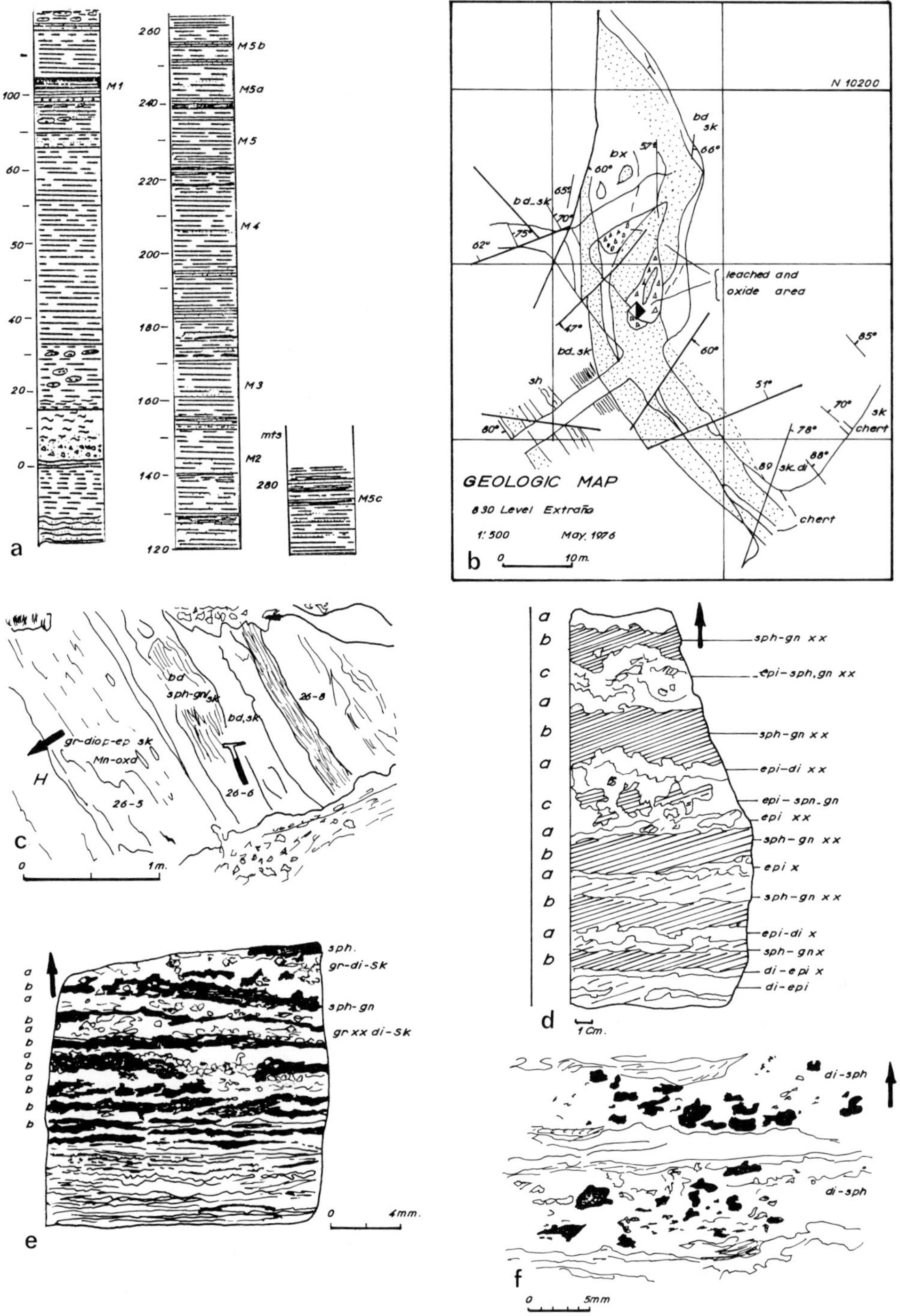
a
M1
M5b
M5a
M5
M4
M3
M2
mts
M5c
b
N 10200
bd
sk
bx
bd_sk
leached and
oxide area
bd_sk
sh
sk
chert
sk.di
chert
GEOLOGIC MAP
830 Level Extraño
1:500
May. 1976
0
10m.
c
bd
sph-gn/sk
bd,sk
26-8
gr-diop-ep sk
Mn-oxd
H
26-5
26-6
0
1m.
d
sph-gn xx
epi-sph,gn xx
sph-gn xx
epi-di xx
epi-spn-gn
epi xx
sph-gn xx
epi x
sph-gn xx
epi-di x
sph-gnx
di-epi x
di-epi
1 Cm.
e
sph.
gr-di-Sk
sph-gn
gr xx di-Sk
0
4mm.
f
di-sph
di-sph
0
5mm

The presence of massive bodies of coarsely grained sulphides is correlatable with remobilizations occurring along faults and folds by the tectonism affecting this area in the Upper Cretaceous and Early Tertiary times.

5 Correlation of the Geologic Features of the Stratabound Ore Occurrences in the Santa Formation Metallotect

In this section is presented a compilation of the correlatable features of the following stratabound mineral deposits having as wall rock the sediments of the Santa Formation in north and central Peru (Figs. 6 and 7). The denomination of some mining areas as mining districts do not have an economic implication but they have been used in order to respect this denomination found in the literature.

1. El Extraño mine area
2. Pueblo Libre mine area
3. Tuco Chira mining district (including the Malaquita and Gumercinda mines)
4. Pachapaqui mining district
5. Huallanca – Huanzalá mining district
6. Pacllón Llamac mining district
7. Oyón area
8. Venturosa mine area
9. Cercapuquio mine area

6 Geometrical Features

In all the cases where ore minerals were seen, their extent could be followed over considerable distances in the same stratigraphic horizon: the ore minerals are stratiform to stratabound into the sediments of the Santa Formation. This geometrical character of the ores has been observed along all the observational ranges (from districtal to microscopic). The behaviour of the ore mineral distribution is exactly the same as that of

Plate 1 a–h. Synsedimentary textures and structures of the ores and wall rocks of the El Extraño mine, the *arrows* indicate the stratigraphic top. **a** Slumping structure at cm scale; the *dark layers* are chiefly composed by sphalerite and minor galena, the *light layers* are lime silicates (epidote and diopside) and quartz. **b** Load cast structure between sphalerite *(dark rounded forms)* and fine crystalline chlorite and diopside layers *(black layers)*. **c** Metarhythmite of sphalerite and galena (probably of diagenetic origin) with layers composed mainly by garnet-diopside and quartz, showing different degrees of crystallinity. **d** Convolute bedding of galena-sphalerite layers and metasediments. **e** Detail of structure described in **b. f** Fine laminations or fissile layers between sphalerite galena *(dark layers)* and diopside-epidote and some garnet and quartz layers. **g** Microscopic detail of **f. h** Stylolitic contact between a metamorphic calcarenite *(lower part)* and recrystallized calcite containing irregular layers of garnet and sphalerite-galena (in the *middle)*. The stylolite is filled with pyrite and remains of organic matter. The upper part is constituted of grains of diopside with minor epidote containing sulphide minerals which form are reminiscent of fossils

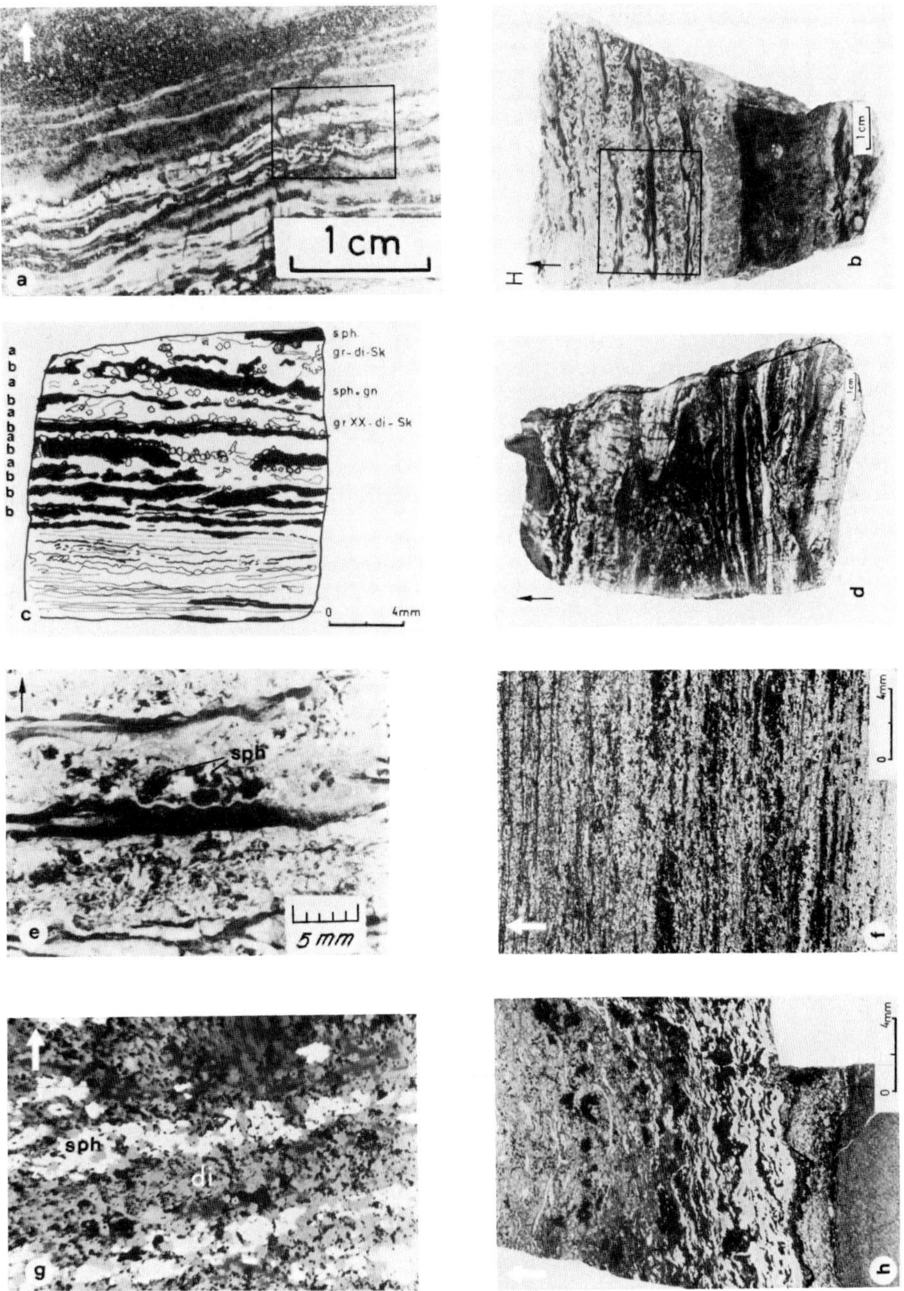
1 cm
a
b
1 cm
H
sph.
gr-di-Sk
sph.gn
grXX-di-Sk
0 4mm
c
1 cm
d
sph
5 mm
e
0 4mm
f
sph
di
g
0 4mm
h

the rock as a whole: where it is lenticular, the ore minerals also show lenticular distribution.

In several places, e.g., at Gumercinda and Tuco mines this congruency is obliterated in some ranges of observation because the appearance of other geological events, as intrusion of folding, were affecting the areas referred to. In other cases, it is obliterated by extensive development of oxidation and secondary zones.

7 Geological Features

Stratigraphy, facies of sedimentation, and paleogeography: Without exception, all these stratabound ore occurrences lie in the sediments of the Santa Formation or others correlatable with them as in the case of the Cercapuquio mine, the ores of which are in the Chauca Limestone. This formation can be correlated with the Santa sequence near Vilca, north of Cercapuquio, described by v. Hillebrandt (1970) (Samaniego and Amstutz 1979).

Looking at the regional scale, all mining districts having stratabound occurrences in this formation occur at or near the borders of the different basins in which the Santa Formation is known (subdivision of basins after Cobbing 1978). The mining districts of Pachapaqui, Pacllón, Tuco Chira, Huanzalá, and Oyón are in the Chimú basin; El Extraño, Pueblo Libre, Malaquita mines are in the Río Santa basin; Cercapuquio and Venturosa mines are in the Río Cañete basin, as is shown in Fig. 8.

The ore mantos are located at different stratigraphic horizons in the different regions; therefore a correlation between them over the studied area is very difficult to make. Otherwise, correlations of the stratigraphic position of the ore mantos are made for very local areas such as the El Extraño, Pachapaqui, and Tuco Chira mining areas. This kind of correlation has been proven as very useful for the exploration of adjacent areas. A "stratigraphical exploration" east of El Extraño gives as a result the localization of four main stratabound structures (ore mantos) at about 6 km from the El Extraño mine.

Petrographical, ore microscopic, and geochemical studies carried out in the samples of several stratigraphic columns (El Extraño, Pachapaqui, Tuco Chira, and Cercapuqio)

Fig. 6 a–c. Stratabound and stratiform features of some mines belonging to the Santa Formation metallotect:

a Stratabound distribution of mineral occurrences in the Santa Formation in the southern part of Ancash Department and the north-western part of Huánuco (after Dunin-Borkowsky (1975). *1* Stratabound mineral occurrences; *2* Santa Formation.

b Generalized geologic map of the Pachapaqui Mining District (after Diaz 1975), showing the location of the stratabound mineral occurrences in the outcrops of the Santa Formation: *Ki-chim* Chimú Formation; *Ki-sa* Santa Formation; *Ki-saca* Santa and Carhuaz Formations (undifferentiated); *Ki-pcp* Pariahuanca-Chulec-Pariatambo Formations.

c Geologic map of the Huanzalá mine area (after Sato and Saito 1977), showing the stratabound character of the ores lying in the Santa Formation: *1* Chimú quartzite with intercalations of shale; *2* Santa Formation; *3* Shales belonging in part to the Santa and Carhuaz formations; *4* Quartz porphyry; *5* Recent formations; *6* Pyrite zone; *7* Cu-Pb-Zn zone; *8* Fault; *9* Overturned syncline axis

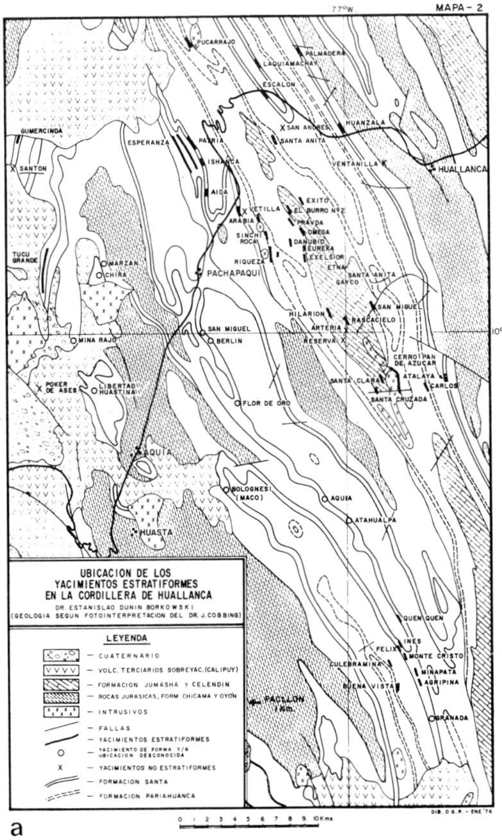

MAPA - 2
77°W
10°S
PUCARRAJO
PALMADERA
LAQUIAMACHAY
ESCALON
SAN ANDRES
HUANZALA
SANTA ANITA
GUMERCINDA
ESPERANZA
PATRIA
ISHANCA
AIDA
SANTON
VENTANILLA
HUALLANCA
EXITO
VETILLA
EL BURRO Nº2
ARABIA
PRAVDA
SINCHI ROCA
OMEGA
DANUBIO
EUREKA
EXELSIOR
RIQUEZA
ETNA
TUCU GRANDE
MARZAN
CHIRA
PACHAPAQUI
SANTA ANITA
GAYCO
SAN MIGUEL
HILARION
RASCACIELO
ARTERIA
MINA RAJO
SAN MIGUEL
BERLIN
RESERVA
CERRO PAN DE AZUCAR
POKER DE ASES
LIBERTAD HUASTINA
SANTA CLARA
ATALAYA
CARLOS
SANTA CRUZADA
FLOR DE ORO
AQUIA
BOLOGNESI (MACO)
AQUIA
ATAHUALPA
HUASTA
UBICACION DE LOS YACIMIENTOS ESTRATIFORMES EN LA CORDILLERA DE HUALLANCA
DR. ESTANISLAO DUNIN BORKOWSKI
(GEOLOGIA SEGUN FOTOINTERPRETACION DEL DR. J. COBBING)
LEYENDA
— CUATERNARIO
— VOLC. TERCIARIOS SOBREYAC.(CALIPUY)
— FORMACION JUMASHA Y CELENDIN
— ROCAS JURASICAS, FORM CHICAMA Y OYON
— INTRUSIVOS
— FALLAS
— YACIMIENTOS ESTRATIFORMES
— YACIMIENTO DE FORMA Y/O UBICACION DESCONOCIDA
— YACIMIENTOS NO ESTRATIFORMES
— FORMACION SANTA
— FORMACION PARIAHUANCA
QUEN QUEN
INES
FELIX
MONTE CRISTO
CULEBRAMINA
MINAPATA
BUENA VISTA
AGRIPINA
PACLLON 1 Km.
GRANADA
0 1 2 3 4 5 6 7 8 9 10Kms
DIB. O.G.R. - ENE '76

a

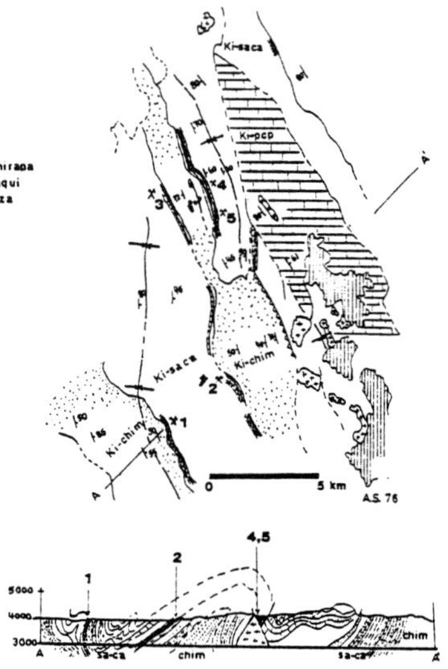

1 Monte Chirapa
2 Pachapaqui
3 Esperanza
4 Patria
5 Otito
Ki-saca
Ki-pcp
Ki-chim
A'
A
0
5 km
A.S. 76
4,5
2
1
5000
4000
3000
sa-ca
chim
sa-ca
chim

b

Q
A'
Recuerdo
Alberto
Julio Cesar
Ki-chim
qda Torres
70
53
55
Ki-sa
Ki-ca
50
A
0 500 1000 m
Qda Torres
3
2
1
4
4200 m
4000 m
3800 m
1
2
3
4
5
6
7
8
9

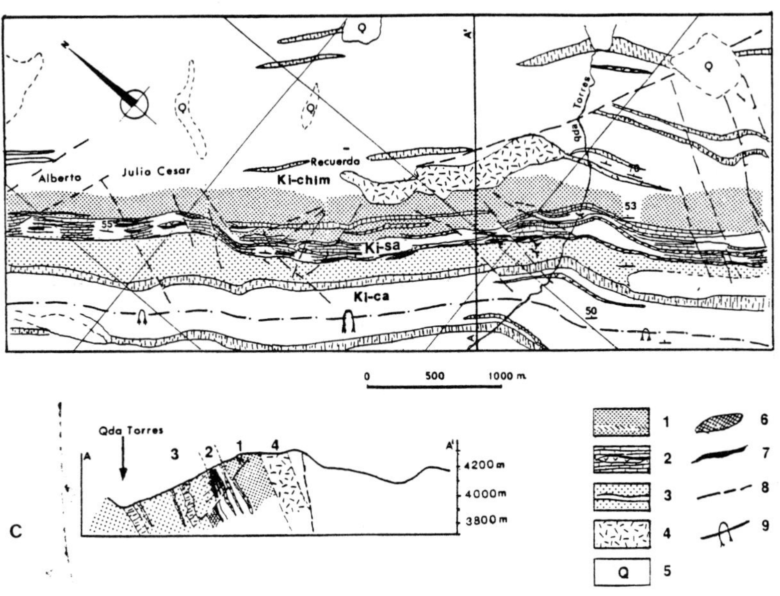

c

have demonstrated the nature of the sedimentary facies in which the ores occur. In all cases, the identified facies correspond to the tidal flats with local development of dolomitic zones and some swamp facies containing organic remains together with the sulphides.

7.1 Mineralogy, Textures, and Geochemistry

The mineralogical composition of the ores in this metallotect is rather homogeneous and consists chiefly of sphalerite and Ag-rich galena; its Zn/Pb ratio varies from zone to zone. Pyrite, marcasite, and chalcopyrite are subordinate in quantity and only in some districts are they abundant. The local variations of the mineralogy occurring in some areas may possibly be correlated with the variations of the geological history of each respective area: the presence of tetrahedrite and pyrrhotite in the Tuco mine is correlatable with the intrusion appearing north of them; the abundance of chalcopyrite at Huanzalá is correlatable with the proximal volcanic activity during the deposition of the Santa Formation in this area.

The typical sedimentary structures which are characteristic for the environments of sedimentation discussed above, embracing the wall rock system and ores (Plate 2) also might be observed in metamorphosed rocks, as in the case of El Extraño mine.

The ore textures appearing at the Huanzalá mine need to be discussed separately. In this mine, in which the volcanogenic character of some part of the sediments and the ores are observed, the ore textures are quite different from all the other deposits forming part of this metallotect. A great number of these ore textures have been described in the literature as typical for volcanic-sedimentary deposits.

Sato and Saito (1977) described a mineral zoning beginning with a pyrite core going outwards to Cu-Zn over a Zn-Pb zone to the unaltered limestone. This zonation is limited to the areas 200 m from the inner pyrite zone.

Preliminary geochemical investigations on the sphalerites of the ore occurrences of this metallotect give a characterization of them; low Fe contents and relatively high Cd and Ge contents. A semiquantitative spectrographical analysis of the fine crystalline sphalerite (in this case called brunckite) of the Cercapuquio mine was published by Miranda (1956) and shows high Cd, Ba, and Ge contents (0.1% to 0.01%) accompanied by low values of Bi, Hg, V (100 to 10 ppm). This result can be correlated with the values published by Haranczyk (1971) for the brunckite of the Olkusz mine in the Upper Silesian mining district.

Fig. 7 a–c. Stratabound features of some stratabound mineral occurrences belonging to the Santa Formation metallotect:

a Cross-section profile along the 200 m line at the Huanzalá mine (after Horita et al. 1973) showing the stratabound character of the ores: *1* Chimú Formation; *2* Santa Formation; *3* Shales belonging in part to the Santa and Carhuaz Formations; *4* Quartz porphyry; *5* Pb-Zn ore; *6* Cu ore; *7* Pyrite zone; *8* Argillaceous breccia zone; *9* Limonite gossan.

b Schematic geological profile along the Ishanca gorge showing the location of the stratabound mineral occurrences of Patria, Ishanca, Otito, and Esperanza. Pachapaqui Mining District. *Ki-chim* Chimú Formation; *Ki-sa* Santa Formation.

c Schematic geological profile across the Culebras ridge showing the stratabound distribution of some ore occurrences of the Pacllón-Llamac Mining District (after Tumialan et al. 1975). *Ki-chim* Chimú Formation; *Ki-sa* Santa Formation; *Ki-ca* Carhuaz Formation

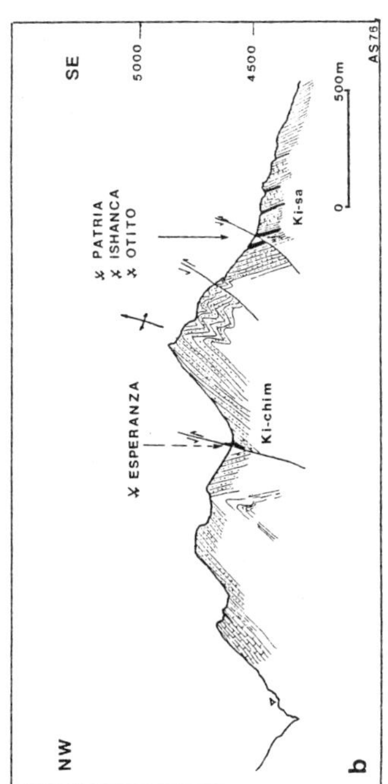
NW
SE
5000
4500
ESPERANZA
PATRIA
ISHANCA
OTITO
Ki-chim
Ki-sa
0
500m
AS78
b

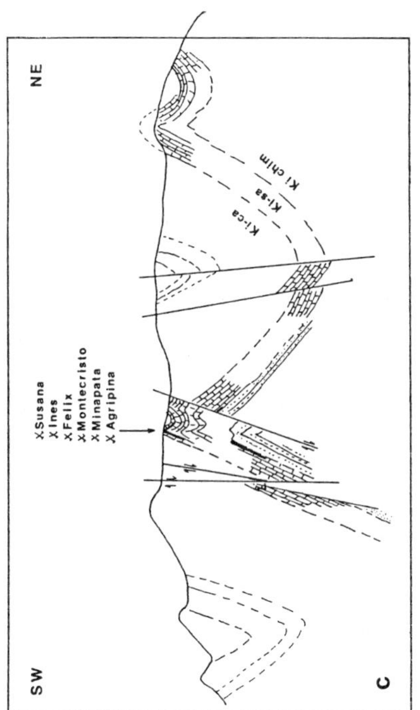
SW
NE
Susana
Ines
Felix
Montecristo
Minapata
Agripina
Ki-ca
Ki-sa
Ki chim
c

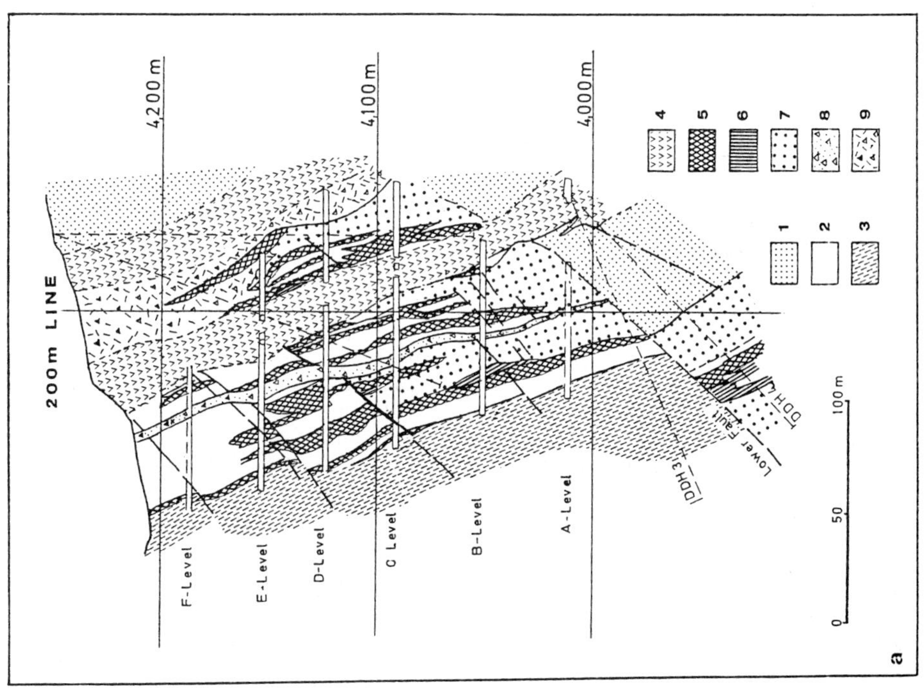
200m LINE
4,200 m
4,100 m
4,000 m
F-Level
E-Level
D-Level
C Level
B-Level
A-Level
DDH 3
Lower Fault
0
50
100 m
1
2
3
4
5
6
7
8
9
a

8 Genetic Implications

With the exception of El Extraño (Kross and Nuñez 1978; Samaniego and Amstutz 1979) and Cercapuquio (Levin 1975), all the other mines are classified as having a selective hydrothermal replacement origin by the different authors who have studied these mineral deposits: on a regional scale by Bodenlos and Ericksen (1955), Bodenlos and Strackzek (1957), Petersen (1965); on a district scale and/or of individual mineral deposits published by Melchori (1955), Diaz (1975), Iberico (1957), Cerro de Pasco Corporation Staff (1970), Tumialan and Vizurraga (1975), Tumialan et al. (1975), Sato and Saito (1977), and many other authors.

The Cercapuquio mine was compared with the Upper Silesian deposits by Melchori (1955); later Petersen (1970) remarked on its stratiform character and the controversy existing over the genetic classification of this deposit.

The possibility of a syngenetic origin for the ores has been pointed out for other mineral occurrences such as Ishcay Cruz (Dueñas 1919), Cercapuquio (Bellido and de Montreuil 1972), and Tuco (Rivera 1979).

After a systematic and inductive analysis of several critical genetic factors, collected over four years of investigation on the Santa Formation metallotect, a genetic interpretation for the ore deposits belonging to this metallotect will be attempted.

The first factor, the congruency, has been postulated by Amstutz (1961, 1968, 1972) as an important and in many cases determinant factor favouring syngenesis. In effect, the ores of the mineral occurrences of this metallotect show a stratiform to stratabound character along most ranges of the observation scales (from district to microscopic). The spatial zonation of the stratabound mineral occurrences is related with the paleogeography of the Early Cretaceous basins: the ores occur almost only in the borders of these basins. The petrological and ore microscopic studies have demonstrated that the ores are principally located in swamps and tidal areas of these basins, and that the ore textures and wall rock structures are characteristic of these geological environments.

Fig. 8 a,b. Stratigraphic correlations of some stratabound ore occurrences of the Santa Formation metallotect:

a Typical locality of the Chimú Formation (Benavides 1956); *b* Typical locality of the Santa Formation, 6 km NW of Carhuaz (Benavides 1956). *1* Typical locality of the Santa Formation metallotect at the El Extraño mine in Ancash (Samaniego and Amstutz 1978); *2* Santa Formation at the Pueblo Libre mine area; *3* Santa Formation at the Malaquita mine south of Huaraz; *4* Santa Formation in the Ishanca gorge, Pachapaqui Mining District; *6* Santa Formation at the Susana and Culebras the Plata mines in the Pacllón-Llamac Mining District; *9* "Unité moyenne calcaire", correlates with the Santa Formation in the Cercapuquio mine area.

b Localization of the studied areas in the schematic isopach map of Cobbing (1978) showing *a*, the main basins of deposition within the geosyncline, *b*, eight numbered measured stratigraphic profiles (*a, b, 1, 2*, etc.).

Nomenclature of the basins: *I* Pongos basin; *II* Chavín basin; *III* Río Santa basin; *IV* Huamey basin; *V* Río Cañete basin.

Studied areas: *1* El Extraño mine area; *2* Pueblo Libre mine area; *3* Tuco-Chira mining district (including the Malaquita and Gumercinda mines); *4* Pachapaqui mining district; *5* Huallanca-Huanzalá mining district; *6* Pacllón-Llamac mining district; *7* Oyón area; *8* Venturosa mine area; *9* Cercapuquio mine area; *C* Cajamarca

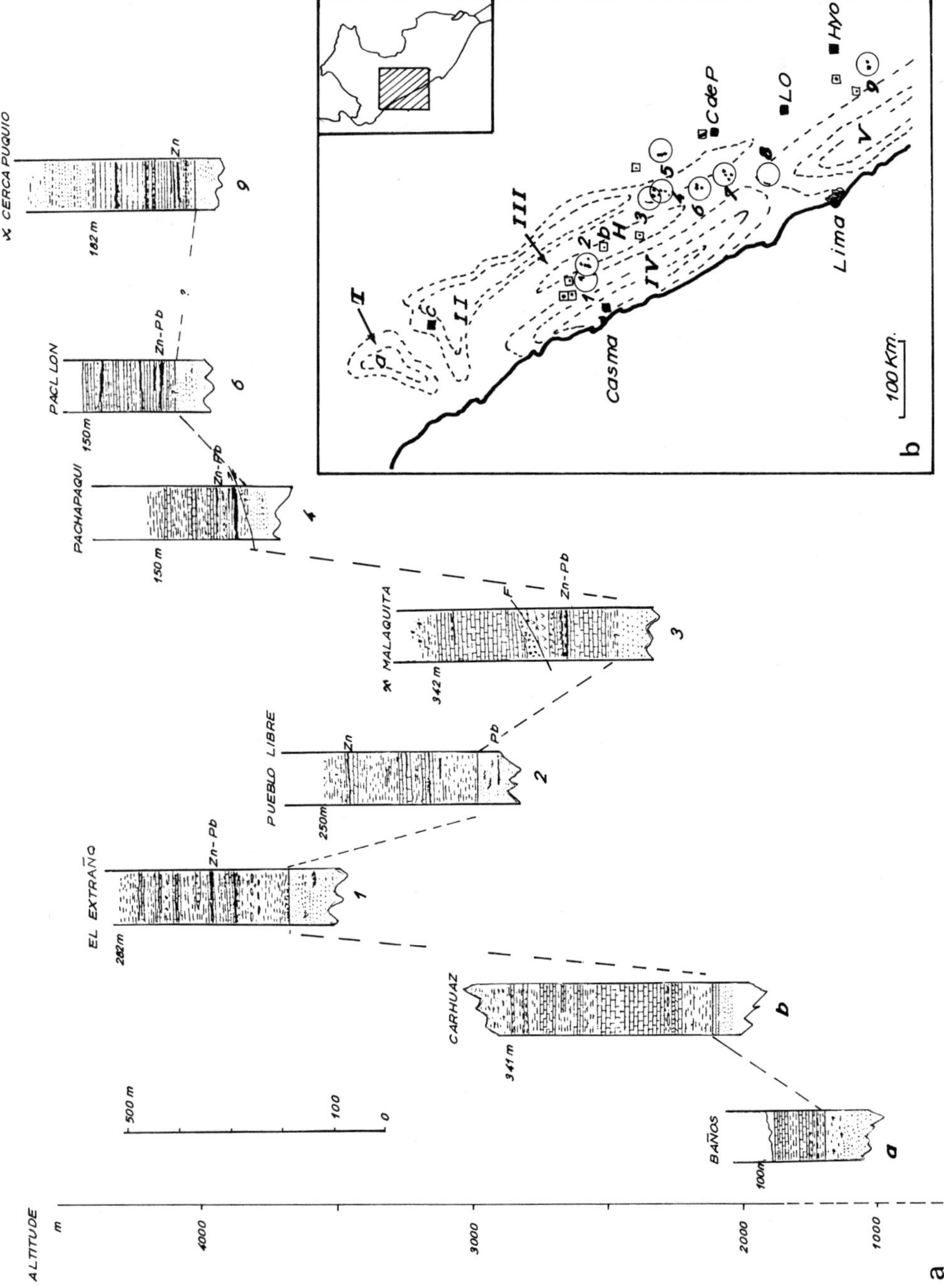
ALTITUDE
m
4000
3000
2000
1000
a
500 m
100
0
BAÑOS
100m
CARHUAZ
341 m
EL EXTRAÑO
282m
Zn-Pb
PUEBLO LIBRE
250m
Zn
Pb
MALAQUITA
342 m
F
Zn-Pb
PACHAPAQUI
150 m
Zn-Pb
PACL LON
150m
Zn-Pb
CERCA PUQUIO
182 m
Zn
b
100 Km
Casma
Lima
C de P
LO
Hyo
I
II
III
IV
V

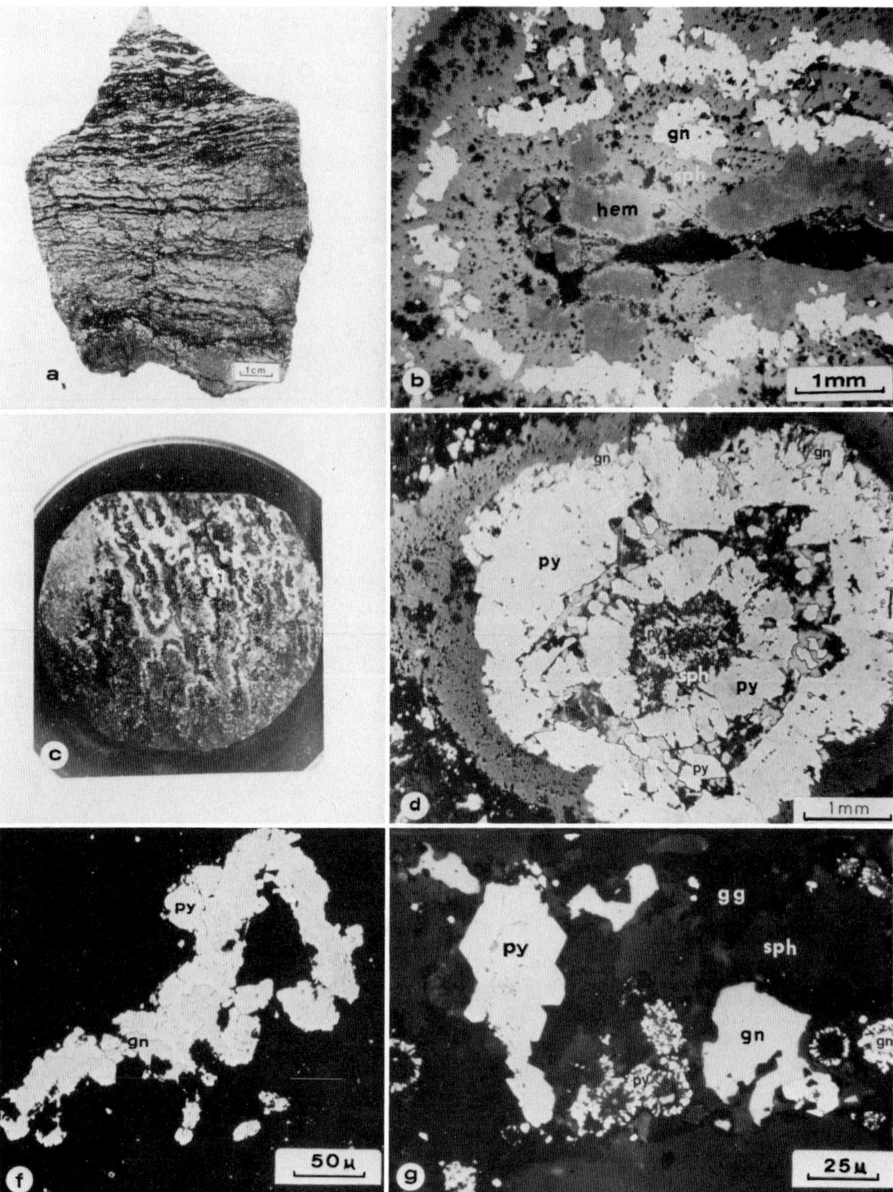

Plate 2 a–g. Structures and ore textures of some ore occurrences in the Santa Formation metallotect: **a** Rhythmic structure between pyrite-marcasite *(light layers)* and minor sphalerite and carbonaceous sandy limestone *(dark layers).* Esperanza mine, Pachapaqui mining district. **b** Rhythmical concentric texture (diagenetic) between galena, sphalerite (brunckite) and hemimorphite. The *dark field* is dolomite aggregate. Cercapuquio mine. **c** Complex rhythmical texture (diagenetic) involving sphalerite, galena, pyrite, siderite, and carbonates. Patria mine, Pachapaqui mining district. **d** Rhythmic concentric texture (diagenetic) or corona-like texture containing pyrite, marcasite, sphalerite, and galena. Ishanca mine, Pachapaqui mining district. **e** Pyritized fossil (probably *Paraglauconia*) filled with galena. Otito mine, Pachapaqui mining district. **f** Microscopic textures of galena, sphalerite, and pyrite in clusters. The gangue is carbonate. Ishanca mine, Pachapaqui mining district

The localization of the ore mantos is in/or near the limits of lithological changes from dolomitic carbonates to shales or sandy carbonates in most of the cases which have been described in the literature as "positive series" (Nicolini 1962, 1964; Routhier 1963) and constitute other important features favouring a syngenetic origin. These phenomena were recorded as a megarhythm in several stratigraphic columns and as microrhythms in nondiagenetic rhythmites in some deposits.

The absence of intrusive masses related to the ore occurrences in most of the studied areas together with the factors presented above made an epigenetic model inconclusive for the origin of these ores.

9 Comments and Needed Research

The grouping of the stratabound mineral occurrences in the Santa Formation metallotect have permitted the investigation of these at regional scale and, consequently, the correlation between its common geological features and the framing of them in the geological history of the studied area. This method of study has permitted the detection of several metallogenetically important features of regional context and they are very useful for the prospection of potentially new areas in north and central Peru.

Because of the extension of the studied area and the relatively short time and the relatively sparse "geological infrastructure", in which these results have been recorded, many important genetic questions can not be answered. As was pointed out by Petersen (1979), some "convincing" evidence to support this theory is not completed and it certainly will be necessary to make more extensive geochemical studies in order to clear up the questions as to source(s) of metal ions and sulphur and their mechanism of precipitation.

Acknowledgments. The author would like to extend his appreciation and thanks to the Compañía Minera Cochas S.A. and the Deutsche Forschungsgemeinschaft for the financial support in all the phases of this work. Further, the author expresses his thanks to many South American and German colleagues for their hospitality and comments on this paper.

Last but not least, the author wishes to thank Prof. G.C. Amstutz for his valuable guidelines, criticism, and comments during the field work and laboratory investigations.

References

Aguirre L, Levi B, Offler R (1978) Unconformities as mineralogical breaks in the burial metamorphism of the Andes. Contrib Mineral Petrol 66:361–366

Amstutz GC (1961) El origen de depósitas minerales congruentes en rocas sedimentarias. Ann I, 2nd Congr Nac Geol Peru. Bol Soc Geol Peru 36:5–30

Amstutz GC (1968) The logic and some relations in the ore genesis. Trans Inter-Univ Geol Congr Leicester, England (Dec 1967), pp 13–30

Amstutz GC (1972) Observational criteria for the classification of Mississippi Valley – Bleiberg – Silesia type of deposits. Proc 2nd Int Symp Miner Deposits of the Alps, Bled, October 1971, Ljubljana, pp 207–215

Aubouin J (1972) Chaines liminaires (Andines) et chaines géosynclinales (Alpines). 24th Int Geol Congr Sect 3:438–461

Audebau E, Capdevilla R, Dalmayrac B, Debelmas J, Laubacher G, Lefevre C, Marocco R, Martinez C, Mattauer M, Megard F, Paredes J, Tomasi P (1973) Les traits géologiques essentiels des Andes centrales (Pérou-Bolivie). Rev Géogr Phys Geól Dyn 1:73–113, pl h–t

Bellido E (1969) Sinopsis de la geología del Perú. Bol Serv Geol Min Peru 22:54 pp

Bellido E, de Montreuil L (1972) Aspectos generales de la metalogenia del Perú. Geol Econ Serv Geol Min Peru 1:149

Benavides VE (1956) Cretaceous system in Northern Peru. Bull Am Mus Nat Hist 108:353–494, Art. 4

Bodenlos AJ, Ericksen GE (1955) Lead-zinc deposits of Cordillera Blanca and northern Cordillera Huayhuash, Peru. US Geol Surv Bull 1017:150 pp

Bodenlos AJ, Strackzek JA (1957) Base metal deposits of the Cordillera Negra, Department Ancash, Peru. US Geol Surv Bull 1040:163

Bussell MA, Pitcher WS, Wilson PA (1976) Ring complexes of the Peruvian Coastal Batholith: a long standing subvolcanic regime. Can J Earth Sci 13:1020–1030

Cerro de Pasco Corporation Staff (1970) Contribución al conocimiento geológico y minero del departamento de Ancash, Perú. Rep Priv Oroya Perú, p 132

Child R (1976) The Coastal Batholith and its envelope in the Casma region of Ancash, Peru. Ph D Thesis, Univ Liverpool, p 328

Cobbing EJ (1976) The geosynclinal pair at the continental margin of Peru. Tectonophysics 36: 157–165

Cobbing EJ (1978) The Andean geosyncline in Peru, and its distinction from Alpine geosynclines. J Geol Soc London 135:207–218

Cobbing EJ, Pitcher WS (1972) The Coastal Batholith of Central Peru. J Geol Soc London 128: 421–460

Cobbing EJ, Pitcher WS, Taylor WP (1977) Segments and super-units in the Coastal Batholith of Peru. J Geol 85:625–631

Dalmayrac B, Laubacher G, Marocco R (1977) Géologie des Andes peruviennes. Caractères généraux de l'evolution géologique des Andes peruviennes. Thèse Univ Sci Tech Languedoc, Montpellier, 361 pp

Diaz N (1975) Geología del distrito minero de Pachapaqui. Bol Soc Geol Peru 50:53–64

Dueñas EI (1919) Reconocimiento geológico-minero de la Cuenca Carbonera Septentrional Lima-Junín (hojas de Oyón, Checras y Pasco). Bol C Ing Minas Peru 97:1–293

Dunin-Borkowsky S (1975) Control litológico y estratigráfico en la Ubicación de los Mantos con súlfuros de metales no ferrosos en las Capas Calcáreas del Perú Central. Bol Soc Geol Peru 50: 25–62

Haranczyk C (1971) Colloidal transport phenomena of zinc sulfide (brunckite) observed in the Olkusz Mine in Poland. Soc Min Geol Jpn, Spec Issue 2:156–162 (Proc IMA–IAGOD Meet '70 Joint Symp Vol)

Hillebrandt A von (1970) Die Kreide in der Zentralkordillere östlich von Lima (Peru, Südamerika). Geol Rundsch 59:1180–1203

Horita A, Oikawa R, Tagami Y (1973) Geological features of the Huanzalá ore deposit, Peru. Min Geol Jpn 23:265–274

Iberico M (1957) Huanzalá mine report. Priv Rep, Cerro de Pasco Co, Lima, 14 pp

Kross G, Nuñez J (1978) Un concepto genético para el Yacimiento de Zinc y Plomo de la Mina El Extraño y su influencia para la Minería. Abstr IV Congr Peruano Geol, Lima, Agusto 1978

Levin P (1975) Der petrographisch-geologische Rahmen des Chanchamayo-Gebietes in Ost-Peru und die Deutung seiner schichtgebundenen Vererzung (Entwurf einer Metallogenese des östlichen Zentralperu). Diss Univ Heidelberg, 242 pp

Megard F (1973) Etude géologique d'une transversale des Andes au niveau dù Pérou Central. Thèse Univ Sciences Techniques du Languedoc, Montpellier, 257 pp

Megard F (1978) Etude géologique des Andes du Pérou Central. Contribution à l'étude géologique des Andes No. 1. Mem ORSTROM 86:303

Melchori J (1955) Exploración geológica y geofísica en las Minas de Cercapuquio. Mineria 10: 25–29; 11:17–21

Miranda LA (1956) El yacimiento mineral de Cercapuquio, Junín y la Brunckita, mineral peruano. Bol Soc Geol Peru 30:243–252 (Ann 1er Congr Geol Perú)

Myers JS (1974) Cretaceous stratigraphy and structure, Western Andes of Peru between latitudes 10° and 10°30′. Bull Am Assoc Petrol Geol 58:474–487

Myers JS (1975) Cauldron subsidence and fluidization: mechanisms of intrusion of the Coastal Batholith of Peru into own volcanic ejecta. Bull Geol Soc Am 86:1209–1220

Nicolini P (1962) L'utilisation des données sédimentologiques dans l'étude et la recherche des gisements stratiformes établissement des «courbes prévisionelles». Chron Mines 30:155–167

Nicolini P (1964) L'application des courbes prévisionales à la recherche des gisements stratiformes du plomb. In: Amstutz GC (ed) Sedimentology and ore genesis. Developments in sedimentology, vol II. Elsevier, Amsterdam, pp 53–64

Petersen U (1965) Regional geology and major ore deposits of Central Peru. Econ Geol 60:407–476

Petersen U (1970) Metallogenetic provinces in South America. Geol Rundsch 59:834–897

Petersen U (1979) Metallogenesis in South America: Progress and problems. Episodes 1979 (4): 3–11

Pitcher WS (1978) The anatomy of a batholith. President's anniversary address 1977. J Geol Soc London 135:157–182

Pitcher WS, Bussell MA (1977) Structural control of batholithic emplacement in Peru: a review. J Geol Soc London 133:249–256

Rivera M (1979) Mineralogische Untersuchungen der Gruben Gumercinda, Nueva Esperanza, Tuco Grande und Magistral im Departamento Ancash, Peru. Dipl Arb, Univ Heidelberg, 95 pp

Routhier P (1963) Les gisement métallifères, 2 vols. Masson, Paris, 1282 pp

Samaniego A (1978) Schichtgebundene Erze der Grube El Extraño (Ancash, Peru) und ihr geologischer Rahmen. Dipl Arb, Univ Heidelberg, 91 pp

Samaniego A, Amstutz GC (1978) Neue Untersuchungen über die Lagerstätte El Extraño, Ancash Peru. Muenstersche Forsch Geol Palaeontol 44/45:159–171

Samaniego A, Amstutz GC (1979) Yacimientos estratoligados de Pb, Zn (Ag, Cu) en en Cretáceo inferior del Perú Central. Bol Soc Geol Peru 62:193–224

Sato H, Saito N (1977) Pyrite zones and zonal distribution of Cu-Pb-Zn ores in Huanzala Mine, Peru. Min Geol Jpn 27:133–141

Tumialan PH, Vizurraga A (1975) Geología económica del distrito minero de Pueblo Libre-Ancash. Ann 2nd Congr Ibero-Am Geol Econ, Buenos Aires III:111–121

Tumialan PH, Sirna C, de la Cruz P (1975) Geología económica del Distrito Minero de Pacllón. Bol Soc Geol Peru 47:123–136

Wilson J (1963) Cretaceous stratigraphy of Central Andes of Peru. Bull Am Assoc Petrol Geol 47: 1–34

A Caldera of Neogene Age and Associated Hydrothermal Ore Formation, Ticapampa-Aija Mining District, Cordillera Negra, Department of Ancash, Peru

P. TRURNIT [1], K. FESEFELDT [1], and S. STEPHAN [2]

Abstract

Most of the hydrothermal Pb-Ag-Zn mineralization in (Upper Cretaceous to) Tertiary volcanics in the Ticapampa-Aija polymetallic mining district in the Cordillera Negra (Cordillera Occidental), Department of Ancash, Peru, is associated with the southeastern quadrant of a Neogene caldera, which has been eroded down to its base. Other controls (structural and rock-mechanic lithological) have also played an important role in the concentration of the ore.

1 Geographical Location, Climate, and Morphology

The Ticapampa-Aija mining district (Fig. 1) is located between 4000 m and 4800 m elevation in the crest region of the Cordillera Negra (part of the Cordillera Occidental), Department of Ancash, Peru, approximately 250 km north of Lima. The central coordinates of the mining district are 77°35′ W; 9°45′ S.

Precipitation in the Cordillera Negra is relatively low. High-mountain climate prevails with rainfall during the summer and sudden spells of cold weather during the winter of the southern hemisphere.

The crestline of the Cordillera Negra is 70 to 80 km from the Pacific coast but only 5 to 15 km from the Río Santa Valley. The Cordillera Negra is an uplifted relic of the Neogene Puna erosion surface, broken and arched along faults with an Andean strike direction, tilted somewhat to the west, and dissected by valleys. Above 4000 m elevation the morphology has been shaped by glacial processes.

2 Mining

Mining activity in this district dates back at least to the early Spanish period. The Collaracra and Hunacapetí mines have been known since 1621 (Purser 1971). Gold and silver were the only metals in demand for a long time. The zones of oxidation and

1 Bundesanstalt für Geowissenschaften und Rohstoffe, Postfach 510153, Stilleweg 2, 3000 Hannover 51, FRG

2 Molderamweg 10, 3352 Einbeck, FRG

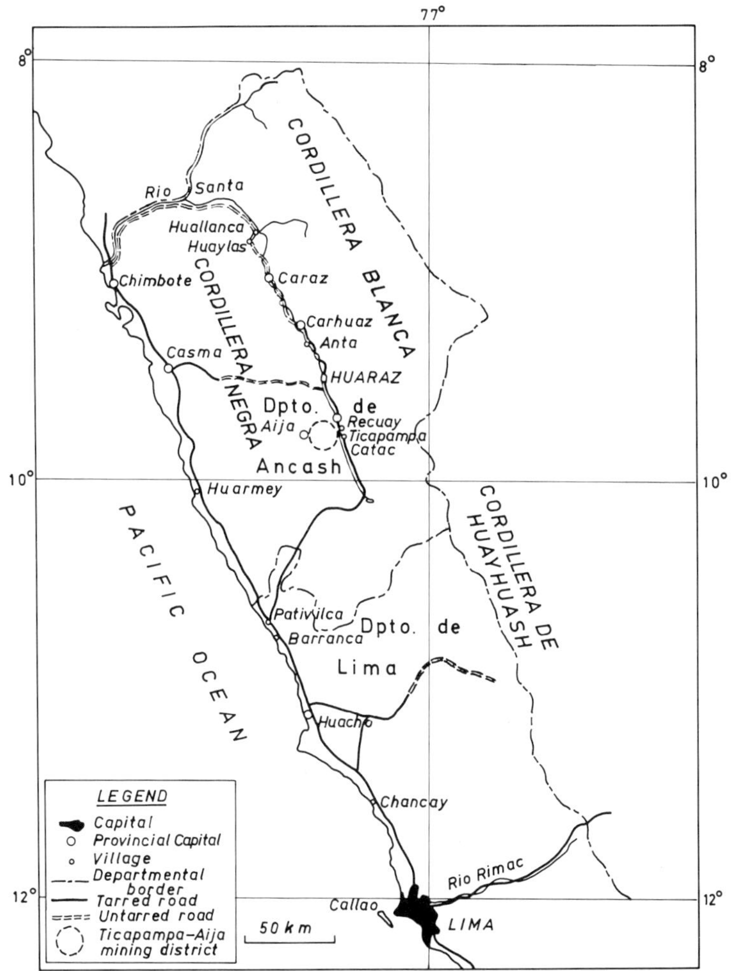

Fig. 1. Geographical location of the Ticapampa-Aija mining district, Cordillera Negra, Ancash, Peru

supergene enrichment (cementation) of many deposits have already been removed by early mining. In terms of quantity, lead and zinc have become the most important metals. Silver is recovered as a by-product from the lead concentrate. Copper plays only a subordinate role. Antimony is unimportant. The presence of arsenic sulphides increases with depth.

Mines are being operated by several small mining companies and a mining venture of the "Mediana Minería", the Cía. Minera Alianza S.A. The small mines are not operated on a continuous basis. Cía. Minera Alianza S.A. puts a total of 1000 t/d through two flotation plants. This ore originates almost exclusively from the Hércules mine.

3 Geology

3.1 Regional Geology

The Andes in central Peru consist of geological units trending approximately NNW, more or less parallel to the morphological features (Petersen 1965; Zeil 1970; Ripley and Ohmoto 1977, Pitcher 1978). Information about pre-Cretaceous sediments, i.e., pre-batholithic, is given by Steinmann (1929), Wilson et al. (1967), Cobbing (1976), Megard (1978), Pitcher (1978), Samaniego (1978) and others.

The intrusive complex of the coastal batholith or plutonite of the Andes forms the "backbone" of the mountain chain. Erosion has freed it from its mantle of sediments and bedded volcanites only at the western margin of the Andes. The coastal batholith is a multiple intrusive consisting of numerous separate intrusive bodies and ring complexes arranged symmetrically and which are increasingly acidic with decreasing age (105 to 30 m.y.) (Hamilton and Myers 1967; Wilson et al. 1967; Cobbing and Pitcher 1972. Knox 1974; Pitcher 1974, 1978; Myers 1975a,b; Pitcher and Taylor 1975; Cobbing 1976).

A cover of bedded volcanic rock, up to 10,000 m thick, accumulated approximately contemporaneously with the emplacement of the Andean pluton (105 to 30 m.y.; Myers 1975a,b). These volcanic rocks originated from the magmas of the complex coastal batholith. The oldest flows were intruded and assimilated locally by younger individual intrusive bodies (Myers 1975a). The volcanics have the same chemical composition as these intrusives (Cobbing and Pitcher 1972; Knox 1974; Pitcher 1974, 1978; Myers 1975a). In the west, up to 7000 m of the Casma volcanoclastics of Lower Cretaceous age (Albian; 105 to 95 m.y.; Myers 1975a,b), which are deposited in the marine environment of a eugeosynclinal island arc facies (basalts to andesites from gabbroic and dioritic magmas), correlate with an up to 5000 m thick calcareous and clastic miogeosynclinal succession of sediments farther to the east (Wilson 1963; Myers 1974; Cobbing 1976, 1978). The bedded, volcanic Calipuy Formation (andesite, dacite, and rhyolite from tonalitic, granodioritic and granitic magmas), first described in detail by Cossio (1964), was deposited in a terrestrial environment. It rests disconformably on Lower Cretaceous sediments or immediately on the multiple intrusive coastal batholith. This formation has an average thickness of 2000 m and a maximum thickness of 3000 to 4000 m (Cobbing 1974, 1976, 1978; Pitcher 1974; Myers 1975a; Samaniego 1978). According to Myers (1975b), the Calipuy Formation is of Upper Cretaceous and Tertiary age (95 to 30 m.y.), according to Pitcher (1974, 1978), it is only Tertiary.

Numerous ore deposits were formed in connection with these magmato-volcanic events. The individual types of ore deposits are arranged in narrow zones parallel to the Andes (Petersen 1970; Sillitoe 1972; Garson and Mitchell 1977; Amstutz 1978).

3.2 Local Geology

3.2.1 Stratigraphy. The geological sequence in the mining district is represented by the following three formations, which are separated by unconformities:

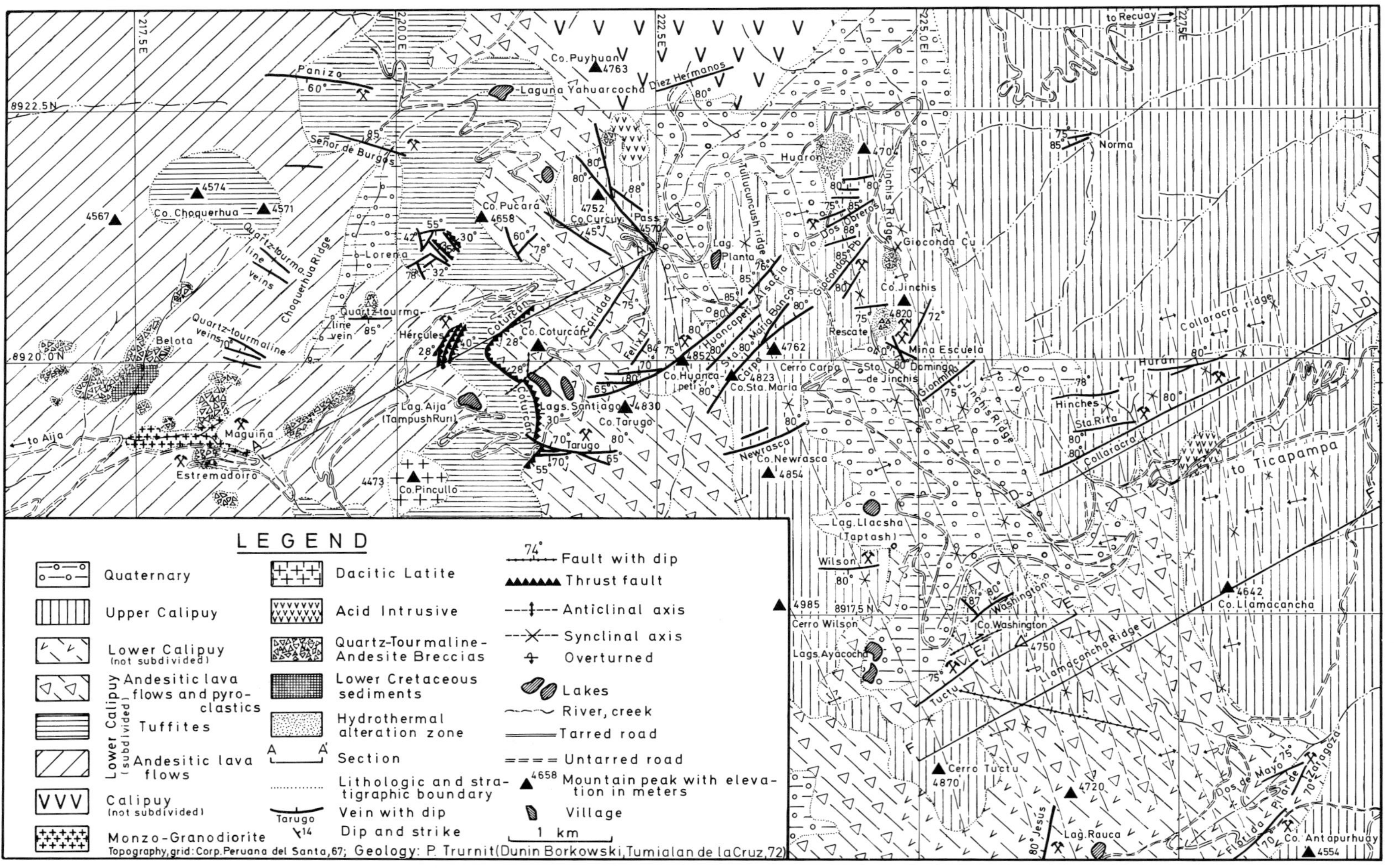

Plan 1. Geological map of the Ticapampa-Aija mining district, Cordillera Negra, Ancash, Peru

- Quaternary; glacial deposits
- Upper Cretaceous (?) to Tertiary (Calipuy); volcanics with sporadic intercalations of thin sediments
- Pre-Albian Lower Cretaceous (Goyllarisquisga); sediments

Lower Cretaceous. Near the pueblo Aija at the eastern margin of the mining district, closely folded, miogeosynclinal sedimentary rocks of pre-Albian Lower Cretaceous age are exposed. They are sandstones, conglomerates, quartzites, calcareous argillites, marls, gypsum, anhydrite, limestones, shales, metamorphic shales, and some thin seams of coal of the Goyllarisquisga Group. The axial planes of the folds have an Andean strike direction and ENE vergence.

Upper Cretaceous (?) and Tertiary (Calipuy). Most of the mining district consists of bedded volcanic rock of the Calipuy Formation. This rock is made up of rather thick flows of lava and pyroclastics, as well as rather thin sheets of ignimbrites and tuffites. A predominance of the former indicates a short distance from the volcanic centre; the predominance of the latter indicates a greater distance from the volcanic centre.

The volcanic rocks lie unconformably on Lower Cretaceous sediments. Together with these sediments or alone, they form the roof of the multiple intrusive coastal batholith. Reduction of the volcanic series took place partly at the bottom by assimilation and at the top by erosion.

The Calipuy Formation is divided into a more intensely folded lower group and a less intensely folded upper group by an angular disconformity that is distinctly visible locally. The lower group is assigned by Myers (1975b) to the Upper Cretaceous and Lower Tertiary, by Pitcher (1978) only to the Lower Tertiary (Paleocene to Oligocene). The upper group, present only as erosion relics, has been assigned mainly to the Upper Tertiary (Oligocene, Miocene, early Pliocene) (Myers 1975b; Pitcher 1978).

In terms of age, genesis, and chemical composition, the predominantly andesitic Lower Calipuy Group correlates with the older, main part of the multiple intrusive coastal batholith; the more acidic, primarily dacitic rhyolitic Upper Calipuy Group correlates with the younger, more acidic, individual intrusive bodies of the coastal batholith and its subvolcanic intrusives (Cobbing and Pitcher 1972; Knox 1974; Pitcher 1974, 1978; Hudson et al. 1975; Myers 1975a). The younger volcanic beds genetically associated with the younger subvolcanic intrusives are frequently already eroded away. The monzonitic neck of the Cerro Pincullo west of the Tarugo mine is an example of these subvolcanic intrusives (Plan 1).

Clastic, partially silicified sediments, mostly only a few metres thick, are locally intercalated in the volcanics, especially in the Upper Calipuy. Such sediments occur more frequently at the bottom of the Upper Calipuy Group, where they are red and have reached greater thicknesses.

Lower Calipuy. In the mining district under consideration, andesites of the Lower Calipuy Group are often encountered on the western flank of the Cordillera Negra (Plans 1 and 2). The lower boundary of the volcanites is exposed in contact with a separate intrusive body of the coastal batholith at an elevation of 3900 m near the Estremadoiro and Maguiña mines. The upper limit of the Lower Calipuy Group can be

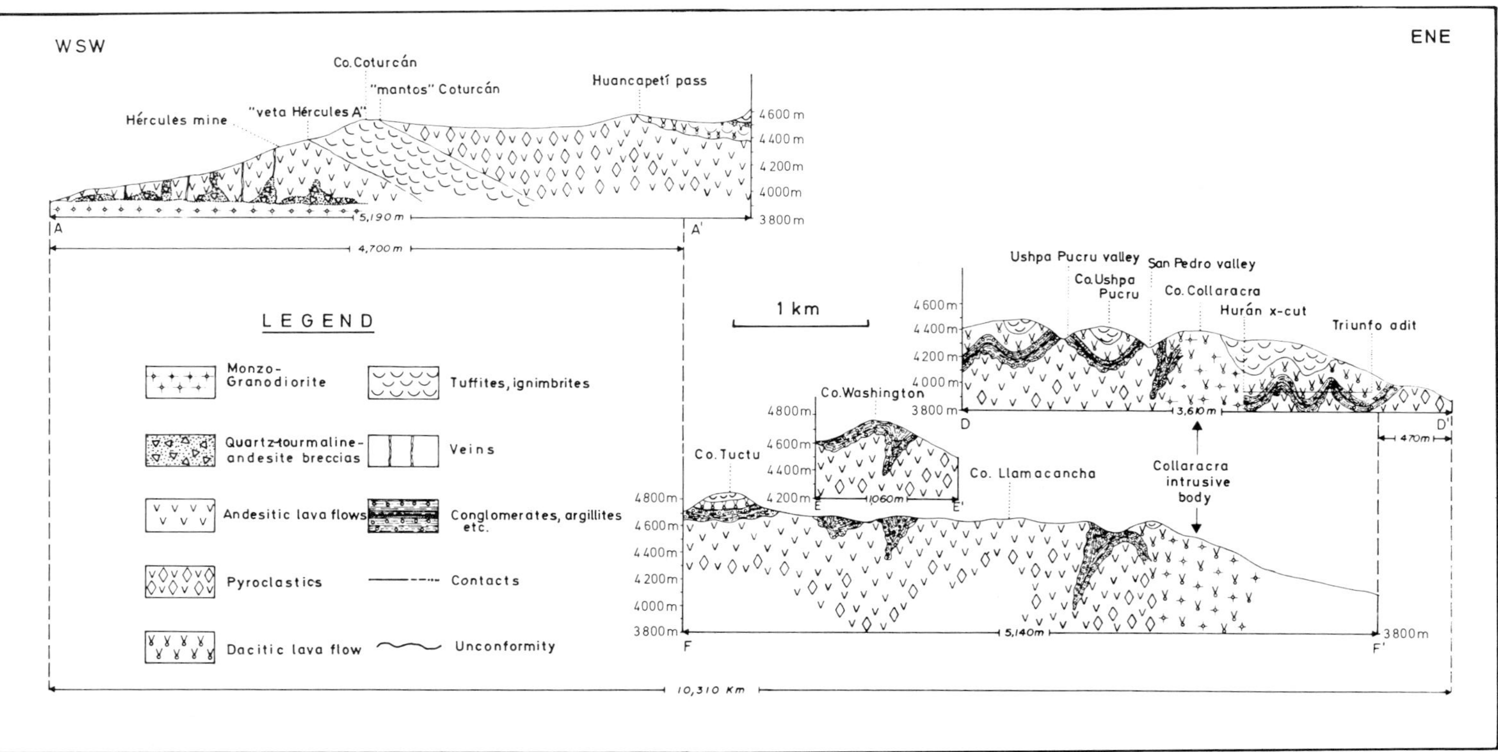

Plan 2. Transverse geological section of the Cordillera Negra summit region, Ticapampa-Aija mining district, Ancash, Peru

identified on the basis of any of several characteristics: either an angular unconformity, distinctly visible locally (e.g., near the Huancapetí mine), a change from the andesites of the Lower Calipuy Group to the dacites and rhyolites of the Upper Calipuy Group or a red, predominantly sedimentary succession deposited only locally at the bottom of the Upper Calipuy.

The Lower Calipuy Group is up to 2600 m thick in the mining district.

Upper Calipuy. The mountain peaks of the Cordillera Negra summit region and the parts of the mountain ridges above 4000 m elevation on the west side of the Río Santa Valley consist predominantly of the Upper Calipuy sequence (Plans 1 and 2). The Upper Calipuy is less thick and more acidic than the Lower Calipuy Group. It is made up of dacitic, rhyolitic, and rarely andesitic lavas, pyroclastics, tuffites and ignimbrites. The beds and flows are the erosional remnants of a once much thicker sequence.

A clastic, limnic lacustrine red series occurs at the bottom of the Upper Calipuy Group in the south-eastern part of the study area. It is best developed and exposed on the south-eastern slope of the Cerro Tuctu (Plan 1). The series consists of argillites, shales, silicified shales, limestones and a considerable amount of conglomerate. There are also such sediments intercalated in the upper volcanic layers of Upper Calipuy remnants east and south-east of the Jinchis Ridge (Plan 1).

The relics of the Upper Calipuy Group are up to 300–400 m thick in the mining district.

Quaternary. Quaternary glacial debris covers valley bottoms and gentle slopes.

3.2.2 Intrusives. The only plutonic intrusive rock exposed at the surface in the mining district is the upper part of a monzonitic-granodioritic separate intrusive complex of the coastal batholith. It crops out in an erosional window at the western slope of the Cordillera Negra at an elevation of 3900 m near the abandoned Maguiña and Estremadoiro mines (Plans 1 and 2). The texture of the rock is granular. It is composed predominantly of plagioclase, considerable orthoclase and hornblende, and only a small percentage of quartz. A network of shrinkage fractures is filled mainly with quartz and tourmaline, some epidote, arsenopyrite and traces of chalcopyrite and pyrite.

Quartz-tourmaline breccia bodies with epidote and altered fragments of andesite are found in the contact zone of the intrusive complex. They extend 200 to 300 m above the contact into the Lower Calipuy Group. Such bodies of andesite breccias with quartz and tourmaline have been encountered on the sixth level of the Hércules mine, 1 to 1.5 km west of the mining area. They are mineralized with pyrite and arsenopyrite. Chalcopyrite, an ore mineral that appears early in the paragenesis of the mining district, occurs only in minor amounts. There are also traces of galena and sphalerite. Such quartz-tourmaline-andesite breccia bodies have been described on a worldwide scale by Sillitoe and Sawkins (1971).

The relationship between the intrusive rock on the one hand and the quartz-tourmaline-andesite breccia bodies together with ore mineralization on the other hand (Estremadoiro, Maguiña, and Hércules mines) suggests a genetic association between ore formation in the district and the intrusive body exposed in the Maguiña mine area.

The time of ore formation is Upper Calipuy (Miocene to early Pliocene).

Presumably, most of the Ticapampa-Aija mining district is underlain by the separate intrusive body of the coastal batholith that crops out at the Maguiña mine. Most probably the volcanic neck of the Cerro Pincullo west of the Tarugo vein is also genetically associated with it (Plan 1).

Subvolcanic bodies, which have intruded rocks higher up, in most cases cut discordantly through the volcanic ejecta. However, they are also observed with a direct transition into volcanic flows.

In many cases it is not easy to distinguish in the field between rather thick lava flows and subvolcanic bodies and stocks that are rather small or have intruded rather high. Their structure and texture are very similar. For instance, the so-called Collaracra intrusive body in the east part of the mining district is exposed only at the deepest level of the Collaracra mine. Closer to the surface there is a transition into a dacite flow, which is locally up to 60 m thick and has a widespread distribution (Upper Calipuy) (Plan 2). Dunin Borkowski (1972) and Tumialan de la Cruz (1972) have interpreted this flow as an intrusive body.

Small apophyses and subvolcanic bodies may be present in the Jinchis district (Plan 1) in the heavily hydrothermally altered zones of the mines and concessions of Gioconda-Cu, Rescate, and Huarón or they are near the surface (Dunin Borkowski 1973; Cabos Yepez 1974). From the intense hydrothermal alteration and the many ore mineral veins present in the Jinchis district, a rather large subvolcanic body might also be expected near the surface. It would be possibly connected with the separate intrusive body of the coastal batholith that crops out at the Maguiña mine.

3.2.3 Structures

Folding and Unconformities. The angular disconformity between the sedimentary rocks of the Lower Cretaceous and the Calipuy Formation is attributed to the "Peruvian Stage" of the Andean orogenesis. Pitcher (1978) who places the beginning of the deposition of the Calipuy volcanics not earlier than the Paleocene, points out a further weak orogenic stage between Late Cretaceous and Lower Tertiary. Accordingly, erosion dominated the whole of the Late Cretaceous.

The angular disconformity between the Lower and Upper Calipuy Groups is caused by the "Inkaic Orogenic Stage" at the beginning of the Oligocene (Steinmann 1929; Pitcher 1978). The "Quechua Orogenic Stage" [Miocene, early Pliocene; according to Petersen (1965), already in the Oligocene] also included the Upper Calipuy volcanics in the folding process. The unconformity caused by this orogenic stage is eroded away in the mining district.

The axial planes of all the folds have an Andean strike direction, i.e., about 330° to 345°. There are local differences in strike direction as far as 20° NNE.

The intensity of folding increases and the distance between folds of the "Quechua Orogenic Stage" decreases in the mining district eastward towards the Colloaracra intrusive body. Some fold limbs are overturned (Cerro Washington) towards ENE (Plans 1 and 2).

Volcanic activity started again after the erosional period following the "Quechua Orogenic Stage". No volcanic deposits of this epoch are preserved in the mining district but only the volcanic vents (dykes) and channelways for the mineralizations (veins).

The development of the Puna erosion surface began during the Miocene or early Pliocene, possibly already during the Oligocene. The uplift of the Andes started during the Pliocene or somewhat earlier, cumulating in several phases (Petersen 1958, 1965). This resulted in the arching of the "Chavin Block" of the Cordillera Blanca (Fig. 1) and the arching and tilting westward of the "Paramonga Block" of the Cordillera Negra (Myers 1975b). Relics of the Puna peneplain today form the summits of the mountain ridges that project westward from the western flank of the Cordillera Negra.

The youngest unconformity is found between the Quaternary, unconsolidated glacial deposits and their Tertiary-Cretaceous base.

Faulting. Several main structural strike directions stand out in the mining district. Most of the veins, faults, fissure planes as well as the drainage system follow this structural pattern. These directions in order of increasing importance are as follows (Fig. 2):

Andean:		330°–345°
Transverse:	"Collaracra System"	60°– 70°
Diagonal:	"Tarugo System"	80°–120°
	"Huancapetí System"	20°– 50°

The most prominent Andean structure is the deep-seated fault of the Río Santa Valley. A series of en échelon faults with Andean strike direction and 70° ENE to vertical dip is exposed in the main adit of the deepest level "El Triunfo" of the Collaracra

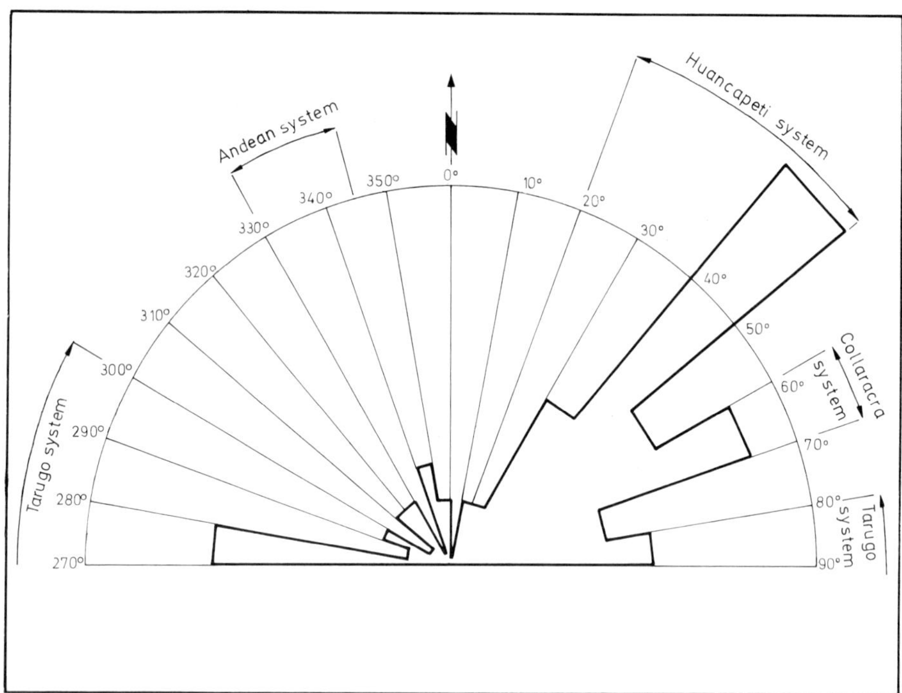

Fig. 2. Structural diagram with frequency distribution of mineralized veins in the Ticapampa-Aija mining district, Cordillera Negra, Ancash, Peru (Cabos Yepez 1974)

mine. Along each of these faults, the west side is lifted with respect to the east side. On a large scale, the arching of the Cordillera Negra must have been caused in this way.

The mineralized "mantos" of the Coturcán mine (2) and the Hércules mine (3–4) are pseudocongruent to the surrounding lava flows, which dip about 28° ENE. These mantos are thrust faults with Andean strike direction; the upper limbs have been thrust westward over the lower ones. Other faults and fissures with Andean strike direction in the mining district are not mineralized with ore.

Almost all of the veins mineralized with ore in the mining district have a transverse or diagonal strike direction.

The mineralized veins with a transverse strike direction (Collaracra System) are approximately normal to the Andean strike direction. In most cases they have a subvertical dip of 80° to 90° NNW. Its main representative in the mining district is the more than 2 km long Collaracra vein.

The diagonal Tarugo System is named after the Tarugo vein. The eastern section of this vein has an almost W-E strike direction; the strike of the western section is between 100° and 120° WNW; it dips between 70° and 90° northward.

The structures of the diagonal Huancapetí System are the most frequently mineralized in the mining district (Fig. 2). It was named after the north-eastern section of the Huancapetí vein. There the vein strikes about 40° NE and dips 75° to 85° NW. The veins Alsacia, Sta. María, and Carpa also belong to this system.

4 Ore Deposits

The Ticapampa-Aija mining district is in the zone of complex polymetallic Pb-Ag-Zn-Cu deposits, which in Peru extends with Andean trend direction through the upper part of the Western and Central Cordillera. The largest concentration of metal in a belt up to 15 km wide and 150 km long with several hundred ore deposits – mainly vein deposits – in the summit region of the Cordillera Negra is in the mining district under consideration. The ore mineralization is associated mainly with veins with a transverse or diagonal strike direction. Thrust faults with Andean strike direction are also mineralized; these are called mantos in the Coturcán and Hércules mines. There are also a few mineralized faults with an Andean strike direction. Locally, ore disseminations are contained in individual lava flows, sediments interbedded with volcanic rocks, breccia bodies, and small, intensely altered subvolcanic bodies.

Numerous authors, e.g., Casadevall and Ohmoto (1977), Harley (1979), Knox (1974), Kouda and Koide (1978), Ohmoto (1977), Smith and Bailey (1968), Spence and de Rosen-Spence (1975), Steven et al. (1974), Tanimura (1973), Thurlow et al. (1975) have described major sulphide ore deposits all over the world that are associated with magmato-volcanic ring structures (calderas). But most of the deposits described belong to the type of massive, stratiform ore deposits of the island arc facies of submarine and syngenetic origin (Sangster 1972, 1980; Mitchell and Bell 1973; Mitchell and Garson 1976; Garson and Mitchell 1977; Large 1977; Sato 1977). The initial to synorogenic volcanism associated with these ore deposits was fed predominantly from gabbro and diorite magmas. Discordant stockwork and vein type mineralization in the roots and vents of the stratiform ore paragenesis deposited at the sea floor play a larger or lesser role, depending on the water depth at the time of ore formation (Finlow-Bates and Large 1978). Some of these deposits may have been remobilized later and epigenetically deposited in hydrothermal veins (Kouda and Koide 1978 – Kuroko type).

This type of deposit is to be distinguished from epigenetic ore formations that were deposited in volcanic rocks of terrestrial origin on an active continental margin and originated in association with a terrestrial subsequent volcanism (plutonic assimilation of stratiform and syngenetic ore deposits in submarine volcanites and sediments included). These mineralizations are pneumatolytic hydrothermal, predominantly discordant vein deposits; the pseudo-concordant mantos were mineralized in the same way as the veins. The volcanism with which these ore formations are associated was fed primarily by monzo-granodioritic magma. This second type of ore deposit can be observed in the polymetallic Ticapampa-Aija mining district in association with a caldera.

4.1 Geometry

Individual veins are up to 2.5 km long (Collaracra).

The veins, including gangue, are locally up to 20 m thick (Huancapetí, Tarugo). Economically minable ore averages 1.2 m thick and in "manto 1" of the Hércules mine reaches a maximum of 4 m. Greater thicknesses have been encountered in mined out sections of the Tarugo vein.

The roots of the ore in the veins reach into the magmatic body. The roof of the mineralizing monzo-granodiorite is exposed near the abandoned Maguiña and Estremadoiro mines at an elevation of 3900 m. The roof of the ore mineralizing source below the veins of the Cordillera Negra crest region (Jinchis district) is probably at a higher elevation than at the Maguiña mine, due to the arching of the Cordillera Negra.

In the summit region, the Tarugo, Huancapetí, and Wilson veins and those of the Jinchis ridge are well mineralized with ore up to an elevation of 4800 m. The mineralization must have reached into still higher levels of the already eroded volcanic beds. According to Myers (1975a,b), individual intrusive bodies were emplaced to within 2.5 to 3 km of the land surface of that time.

The known ore mineralization in the Hércules mine extends about 220 m between 4060 m and 4280 m elevation, in the Tarugo mine ca. 400 m between 4400 m and 4800 m elevation, in the Huancapetí mine about 350 m between ca. 4450 m and 4800 m elevation, in the Collaracra mine about 570 m between ca. 3905 m and 4475 m elevation. The observable range of mineralization does not exceed 600 m. However, it must be kept in mind, especially in the case of the Collaracra vein, that the arching of the Cordillera Negra for the most part postdates ore formation.

The dip of most veins is subvertical between 70° and 85°, predominantly to the northwest. Most sections of the mineralized mantos, together with the volcanic beds, dip about 28° ENE.

The ore mineralizations in the veins frequently occur in lense-like ore shoots, aligned horizontally like a "string of pearls". The ore shoots horizontally have lengths between a few tens of metres to more than 100 m. They are separated from each other by wider,

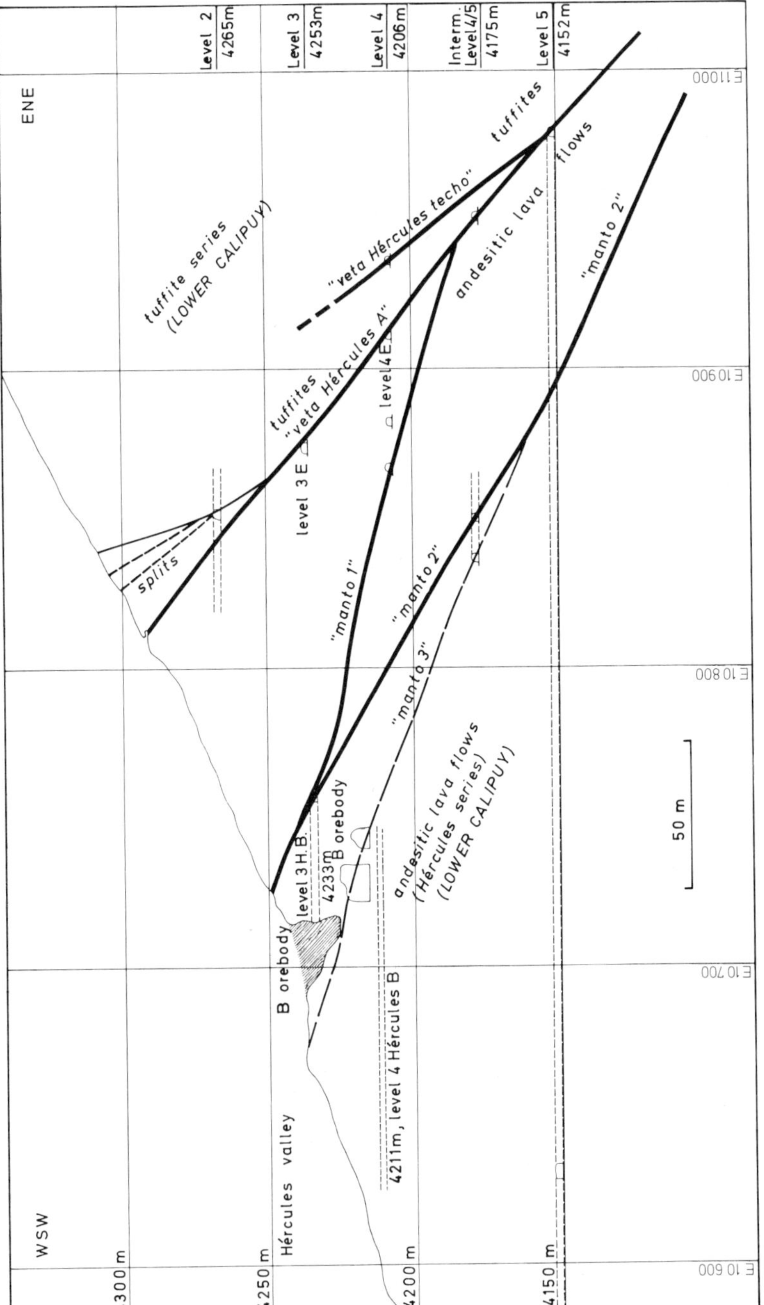

Plan 3. Transverse section of the mineralized structures (predominantly "mantos") of the Hércules mine, Ticapampa-Aija mining district, Cordillera Negra, Ancash, Peru

weakly mineralized or barren sections of the vein, in which even the gangue material frequently thins out.

Viewed on a horizontal plan, the veins towards one end frequently split into horsetail structures (e.g., Huancapetí, Tarugo, Collaracra, Hurán; Plan 1). The "mantos" of the Hércules mine, also viewed on a horizontal plan, unite at the ends in a spindle-like manner.

Vertically, the vein and manto structures (e.g., Hércules mine; Plan 3) tend to split towards the roof and to unite towards depth.

Major veins are frequently composed of many, closely spaced, parallel sigmoidal and stringer structures (e.g., Huancapetí).

In addition to the ore formation in veins and mantos, a disseminated ore mineralization is exposed in shales intercalated with Upper Calipuy volcanic rock in the Gioconda-Cu mine on the Jinchis Ridge.

Non-vein ore bodies are of subordinate importance in the mining district. They occur as

- Quartz-tourmaline breccias with hydrothermally altered wall rock fragments (e.g., Maguiña, Estremadoiro, Belota).
- Stockwork ore bodies (e.g., south-western section of the Huancapetí vein; "Alpha-Beta-Gamma" ore bodies, western section of the Tarugo vein at its intersection with the Coturcán mantos; Hércules-B ore body, Hércules mine).
- Intensely hydrothermally altered zones in the roof region of small intrusive bodies (e.g., Rescate, Huarón, Gioconda-Cu in the Jinchis district).

4.2 Controls on Ore Formation

4.2.1 Structural Controls

Caldera. In the north-western part of the mining district, the arrangement of faults, quartz-tourmaline veins and ore veins in a set of concentric semicircles indicates the presence of the southern half of a caldera. Most of the veins dip between 70° and 85° towards the centre. The mineralization of a vein frequently shifts from one fault to another at their intersection (e.g., Huancapetí, Tarugo). The part of each fault extending outwards from an intersection may have a rich ore shoot in the immediate vicinity of the intersection but becomes lean and thins out within a short distance. The minimum diameter of the caldera ring structure is assumed to be 5 km, but with outer rings may possibly amount to 20 km (Fig. 3; Plan 1). According to MacDonald (1972) caldera diameters average 25 km.

The batholithic foundation of the caldera may be a ring complex composed of separate intrusive bodies of various ages and chemical composition (increasing acidity and decreasing age towards the centre), of the same sort as exposed further to the west and investigated by Cobbing and Pitcher (1972), Knox (1974), Myers (1975a), and Pitcher (1978).

The caldera came into existence during the pneumatolytic stage of mineralization. The cover of bedded volcanic rock was domed up by high vapour pressure. Pressure was released from time to time along a set of concentric ring fractures made up of

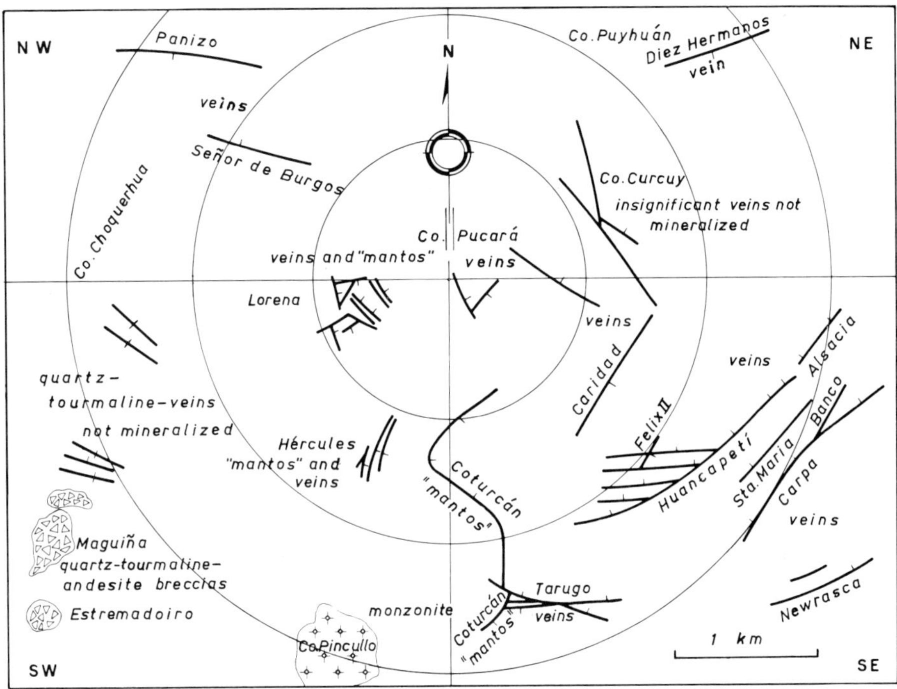

Fig. 3. Schematical plan of the caldera ring structure in the Ticapampa-Aija mining district, Cordillera Negra, Ancash, Peru

intersecting tension faults with Andean, transverse, and diagonal strike directions. Pulsations of emanations deposited primarily quartz and tourmaline, a good amount of arsenopyrite, and minor pyrite and chalcopyrite. After further cooling of the remaining magma and the accompanying decrease of pressure, the roof subsided along pneumatolytically mineralized en échelon ring faults. Intermittently, the feeder channels were closed for some time. Only during a later epoch did horizontal displacement along rejuvenated transverse and diagonal faults create the necessary channelways for the hydrothermal ore-forming solutions. No horizontal displacement took place at this time along veins at right angles to these faults (mainly with an Andean strike direction). These veins contain only pneumatolytic minerals.

According to Myers (1975a,b), some of the younger and more acidic individual intrusive bodies intruded to within 2.5–3 km of the Tertiary land surface. Thus, in the south-western part of the mining district, erosion has left only the lower third of the caldera, i.e., the root zone or base of a volcano.

The south-eastern quadrant of the caldera root zone, with the Huancapetí, Alsacia, Carpa, and Tarugo veins and the mantos of the Hércules and Coturcán mines, is the area with the greatest ore potential of the whole mining district. This quadrant and the geologically little-known north-west quadrant are the most promising areas for prospecting in the mining district.

The veins in the south-western quadrant contain primarily only quartz and tourmaline. Examples are the many barren, north-west striking quartz-tourmaline veins on the Choquerhua mountain ridge between the mines Maguiña/Estremadoiro and Señor de Burgos/Panizo (Plan 1). Accordingly, only weak ore mineralization is to be expected in the north-eastern quadrant of the caldera.

Figure 3 shows the most likely position of the caldera centre in the vicinity of the various veins and mantos of the Lorena concession and the Cerro Pucará. These veins and mantos have various strike directions and are only weakly mineralized with ore. The assignment of the caldera centre (volcanic centre) places the ore veins Señor de Burgos and Panizo in the north-western quadrant (Tarugo System).

Other Structural Controls. Space for the ore shoots in the veins and mantos was created by sliding movements along the faults and overthrust structures and the rather undulating surface of these tectonic structures in most of the different units of the volcanic succession.

Ore shoots are found especially in the more gently dipping sections of the mantos, while the ore mineralization in the more steeply dipping sections thins out and is leaner (e.g., Hércules mine). They are also found at places in veins deviating from the general strike direction (e.g., Hurán vein north of the Collaracra vein).

Ore shoots also occur at intersections, contacts, and junctions or splittings of veins. On closer examination of many of the changes in the strike direction of the veins, it is found that they are better interpreted as intersections of two structures, with the ore mineralization shifting from one structure to the other at the intersection (e.g., Huancapetí and Tarugo). Ore shoots at intersections, contacts, and junctions of mineralized veins and mantos are frequently encountered in the Hércules mine. "Manto 1" meanders between "manto 2" and "Veta Hércules A".

4.2.2 Rock-Mechanic Lithological Control. In the Calipuy volcanic sequence, certain rock units are strongly, others barely mineralized with ore in mantos and cross-cutting veins (Fig. 4). The reason for this is the rock-mechanic property of certain volcanic rocks to form more undulating breaks on a large scale, of others to form more planar fractures. Movements along faults in units broken in an undulating manner result in good channelways for the mineralizing solutions; in units broken in planar fashion, poor channelways or none at all result.

A competent series of tuffite with planar fractures in the Lower Calipuy Group, about 400 m thick, is strinkingly poorly mineralized with ore (Plan 2). In contrast, good to very good ore mineralizations are found in the andesite lavas (containing the mantos of the Hércules mine) underlying these tuffites and in the overlying andesite lavas and pyroclastics (agglomerates, volcanic breccias) containing the Coturcán mantos and the Tarugo, Huancapetí, Carpa, and other veins.

Other poorly mineralized to barren units are the basal beds of the Upper Calipuy (conglomeratic red series containing lutites, limnic-lacustrine sediments and some volcanics). The hanging wall is followed by the Collaracra dacite flow, rhyolitic tuffites and ignimbrites, which are excellently mineralized in the Collaracra and Hurán veins and the veins that cross through the Jinchis Ridge.

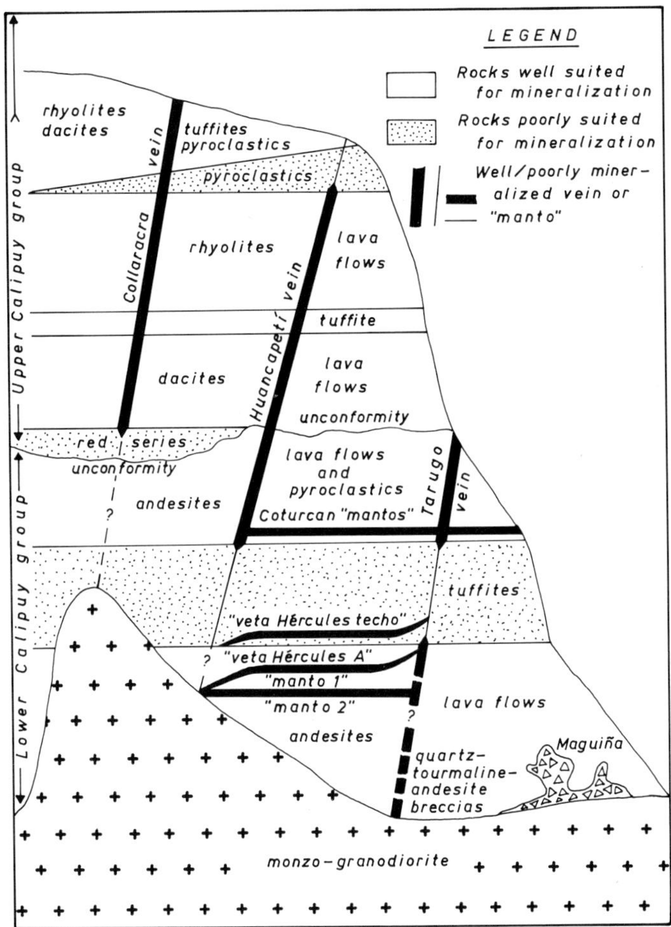

Fig. 4. Schematic section: Rock-mechanic lithologic control of ore formation in the Ticapampa-Aija mining district, Cordillera Negra, Ancash, Peru

The Tarugo vein has been encountered only above the Lower Calipuy tuffite series (Fig. 5). Levels 0 to 8 of the Tarugo mine are in the vein in the andesites and pyroclastics above the tuffite series. The vein had a very good mineralization at these levels and is almost completely mined out. Below level 8, the vein ends at the tuffite series. The Tarugo structure is barely discernible in the section of level 9 that is in the tuffite and only very poorly mineralized with ore. The mineralized Coturcán thrust faults (mantos) are located along or adjacent to the contact between the tuffites and the overlying andesite lavas and pyroclastics. Major ore shoots, most of which have been mined out (Alpha-Beta-Gamma ore bodies), formed at the intersection of the Tarugo vein with the Coturcán mantos. The Tarugo vein above the tuffite series and the Coturcán mantos were mineralized together; they form an integrated mineralization system.

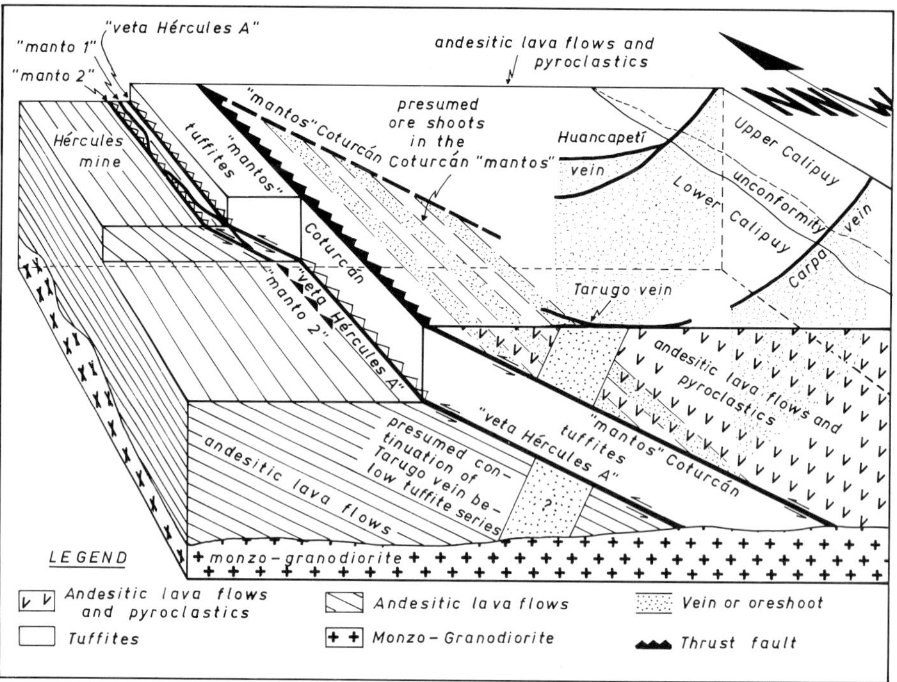

Fig. 5. Block diagram of the veins and "mantos" of Hércules, Coturcán, Tarugo, Huancapeti and Carpa mines, Ticapampa-Aija mining district, Cordillera Negra, Ancash, Peru

A continuation of the ore mineralization in the Tarugo structure below the tuffites in the Hércules series and a connection of the Hércules mantos with the unknown parts of the Tarugo vein to an integrated mineralization system (as above the tuffites with Coturcán) is presumed (Fig. 5).

The known mineralized sections of the Collaracra vein are above the nonmineralized, basal red series, where it is found in the Collaracra dacite flow and in the upper rhyolitic beds of the Upper Calipuy. Below the deepest level ("El Triunfo") of the Collaracra mine, mineralized vein sections are to be expected only there where the Collaracra dacite flow extends deep into the centre of synclinal folds (Plan 2). The ore potential of the vein decreases with depth because of the shortening of the mineralized vein segments in the synclines. There is some possibility for renewed mineralization below the contact between the Upper and Lower Calipuy Groups. The arching of the Cordillera Negra along faults with an Andean strike direction means that the probability of crosscutting mineralized synclinal centres with Collaracra dacite on the lower levels diminishes rapidly towards the west in the direction of the Cordillera Negra crest region.

4.3 Ore Formation

The primary ore mineralization is epigenetic. The paragenetic sequence includes minerals of the pneumatolytic and hydrothermal stages of ore formation.

The comparably thin zones of oxidation and secondary enrichment (cementation) contain minerals of secondary origin.

The major ore minerals of the primary mineralization zone are sphalerite and silver-bearing galena. Jamesonite and plumosite also occur. There are no significant occurrences of copper and silver ores, nor of stibnite. In addition, pyrite and arsenopyrite are present in rather large quantities. The gangue consists predominantly of quartz, tourmaline, rhodochrosite, and calcite. The complete paragenetic sequence is given in Table 1.

Table 1. Paragenetic sequence of minerals in the Ticapampa-Aija mining district, Cordillera Negra, Ancash, Peru; major sulphides are shown in **bold letters**, minor sulphides in *italics;* rare sulphides are given in normal letters

Zone of primary mineralization				
Hydrothermal zone	Telethermal	Alabandite		
		Realgar, auripigment		
	Epithermal	Marcasite		
		Stibnite		Anhydrite
		Bournonite, seligmanite		Calcite
		Jamesonite (Plumosite)		Dolomite
		Primary chalcocite		
		Pyrargyrite – proustite		Barite
	Mesothermal	Tennantite		
		Tetrahedrite		
		Argentite		
		Cubanite		
		Chalcopyrite		
		Galena		Rhodochrosite
		Enargite		Rhodonite
		Luzonite	Pyrite and	Hematite
			quartz in	Siderite
	Katathermal	**Sphalerite**	wide	
		Pyrrhotite	distribution	
		Chalcopyrite		
Pneumatolytic zone		**Arsenopyrite**		Tourmaline

Using the Hércules mine as an example, Cabos Yepez (1974, in Hudson et al. 1975) demonstrated that there are three hydrothermal pulsations after a pneumatolytic stage (Fig. 6).

In addition to those in Table 1, the following ore minerals have been encountered: Gold (e.g., 0.3–12.4 g/t in the Huancapetí mine; Bodenlos and Straczek 1957, 1958), boulangerite, bornite, native silver, silver oxides, secondary chalcocite, covellite, chrysocolla, malachite, azurite, magnetite, iron oxides, manganese oxides, cerussite, angle-

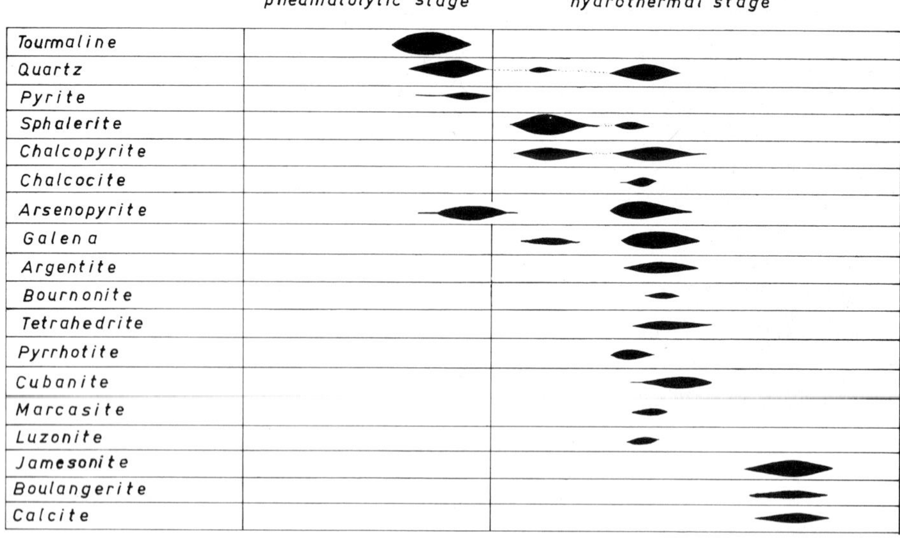

Fig. 6. Paragenesis of ore minerals and stages of ore formation in the Hércules mine, Ticapampa-Aija mining district, Cordillera Negra, Ancash, Peru (slightly changed after Cabos Yepez 1974, in: Hudson et al. 1975)

site, and gypsum predominantly in the oxidation and supergene enrichment zones (Raimondi 1873; Hudson et al. 1975).

The following are primarily from the authors' own microscopic observations (Fig. 7):

Sphalerite sometimes is dark and rich in iron but frequently is also translucent light yellow to red. It almost always contains innumerable dissolution crystals of chalcopyrite throughout the crystal and frequently a cover of small chalcopyrite crystals on the outside. These dissolution phenomena indicate a rather high formation temperature and diffusion.

Chalcopyrite also occurs as free crystals in minor amounts.

Cubanite may occur in chalcopyrite as dissolution crystals (Cabos Yepez 1974, in Hudson et al. 1975).

Pyrite is the most abundant sulphide mineral. Replacement relics of pyrite frequently occur in galena but also in sphalerite.

Galena covers large areas when viewed under the microscope. Silver is probably contained in the galena crystal structure as *argentite.* According to Caldas Vidal (1968) argentite is also found in galena as dissolution crystals. Free argentite only occurs in traces. The primary silver content of the Collaracra mine, which was mined primarily for silver, is supposed to be associated with the frequently occurring fahlore. Secondary silver in the Collaracra mine may also be associated with pyrite in the supergene enrichment zone.

Jamesonite formed by partial replacement of pyrite and galena. It is observed together with pyrite and chalcopyrite. Jamesonite has been found on the lower levels of

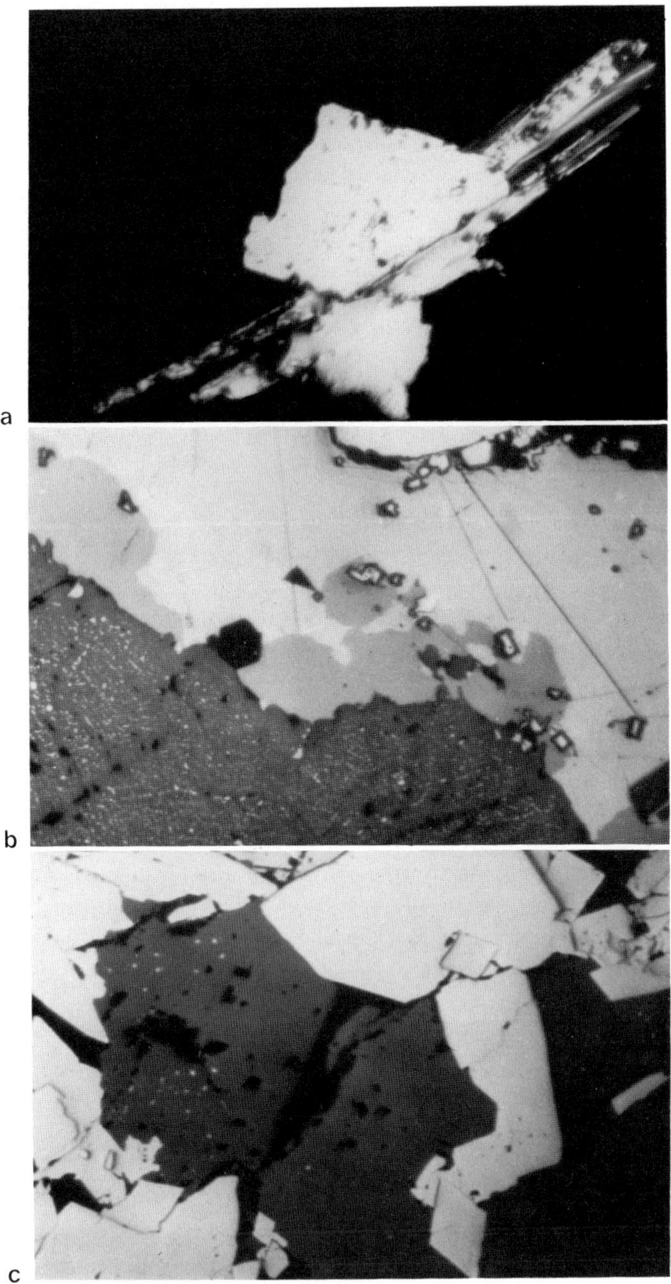

Fig. 7. a Hércules mine; intermediate level 5 1/2; × 750. Idiomorphic needles of jamesonite replace pyrite and galena. b Hércules mine, level 6; × 750. Tetrahedrite is present between galena, which includes replacement relics of pyrite, and sphalerite with innumerable dissolution crystals of chalcopyrite. c Huancapetí mine, level 6; × 375. Galena and sphalerite replace pyrite; at the boundary of the latter, idiomorphic crystals of arsenopyrite. Sphalerite includes some dissolution crystals of chalcopyrite

the Hércules mine in open fractures as a matt of fine needles together with pyrite, chalcopyrite, quartz, and calcite or in huge nests, up to several metres in diameter, of finely felted needles.

Bournonite is found mainly between galena and chalcopyrite crystals. Frequently, relics of tetrahedrite are present in its neighbourhood.

Boulangerite is found exclusively in or at the edge of galena crystals.

The antimony-rich phase ranges from tetrahedrite (prevalent at the deeper levels) to stibnite (closer to the surface).

Tetrahedrite is encountered primarily replacing galena. Its formation usually begins at the edge of a sphalerite crystal or replacement relics of pyrite within the sphalerite.

Tetrahedrite, *tennantite* (containing dissolution crystals of chalcopyrite; Caldas Vidal 1968), *enargite,* and *stibnite* are found more frequently in the Collaracra mine than in the rest of the mining district.

Quartz and tourmaline are normally the main gangue minerals. In the Huancapetí mine, however, rhodochrosite predominates in the gangue.

The greatly increased occurrence of tourmaline and arsenopyrite at the lowest level (6th) of the Hércules mine and of arsenopyrite at the lowest level ("El Triunfo") of the Collaracra mine indicate that mining is approaching the barren roots of the mineralization.

In addition to the primary vertical paragenetic zonation, lateral zonation of ore minerals is found in veins located more to the side of the mineralizing magmatic body. Sphalerite, arsenopyrite, and tourmaline predominate in the WSW section of the Collaracra vein; but in the ENE section, the minerals formed at lower temperatures predominate: stibnite, tetrahedrite, tennantite, and the silver content associated with galena and tetrahedrite (Bodenlos and Straczek 1957, 1958). The magmatic body responsible for the ore formation is to be located, therefore, at depth to the WSW towards the crest region of the Cordillera Negra.

Secondary vertical mineral zonation is not very discernible in the Cordillera Negra because the oxidation and supergene enrichment zones that remain are comparably thin or hardly developed. Bodenlos and Straczek (1957, 1958) give an average thickness of 5 m. This is due to unfavourable climatic conditions and rapid erosion.

4.4 Alteration Processes

Four stages of alteration processes may be distinguished in the mining district according to their age and intensity. The more advanced and intense the alteration, the more limited and restricted its extension. The four stages of alteration in order of decreasing age and increasing intensity are as follows:

- Propylitization (chloritization; epidotization); regionally distributed throughout the mining district.
- Sericitization and pyritization; they accompany many veins from the sidewalls of the veins several metres into the wall rock.
- Koalinization; more restricted to the immediate area of the veins; kaolinized parts of veins are frequently only sparsely mineralized by ore or are completely barren.
- Silicification; frequently accompanied by good ore mineralization.

Strongly sericitized and kaolinized wall rock is soft and shows little rigidity. The soft material is called "panizo". Frequently it creeps, slides or breaks into the drifts and cross-cuts and blocks them, making mining more difficult. Panizos are primarily mylonites. They originated from country rock during sliding movements along faults. Alteration of the panizos was primarily by hydrothermal processes (mainly kaolinization).

Ore formation and alteration processes took place from Miocene to early Pliocene.

5 Results

All over the world calderas are associated with major sulphidic ore deposits. A caldera root zone in Upper Cretaceous (?) to Tertiary volcanic rock of the Calipuy Formation is preserved from erosion in the Ticapampa-Aija polymetallic mining district (Department of Ancash, Peru) in the summit region of the Cordillera Negra. Most of the hydrothermal Pb-Ag-Zn ore mineralization in the mining district is concentrated in the southeastern quadrant of the caldera.

The primary ore formation is associated with subsequent plutonism and volcanism of Neogene age. It occurs in connection with terrestrial or nonmarine volcanism fed by monzo-granodioritic magma on an active continental margin.

Ore formation is primarily associated with faults (veins) and thrust faults (mantos). Ore minerals of economic importance are sphalerite and silver-bearing galena. Jamesonite and plumosite also occur. Minor minerals are chalcopyrite, tetrahedrite and stibnite. Besides these ore minerals, pyrite and arsenopyrite are present in rather large quantities. The main gangue minerals are quartz, tourmaline, rhodochrosite, and calcite.

Primary vertical and lateral zonations are well developed. Zones of oxidation and secondary enrichment are thin. Most of them were mined out long ago for their high silver content.

Besides the role played by the strike direction of the faults (limiting ore formation to specific quadrants of the caldera), other structural and rock-mechanic lithological controls played an important role in the concentration of the ore.

Ore shoots were formed at places in veins with a deviation from their general strike direction, in mantos with a deviation from their general dip to a more gentle dip. Ore shoots also occur at intersections, junctions or splittings of mineralized structures.

In terms of the rock-mechanic and lithological controls on ore formation, the type of rock the vein cuts through is important. Certain units of the Calipuy Formation are fractured with planar surfaces, others with undulating, uneven surfaces. In the latter case, lateral movements along faults or thrust-faulting subparallel to the bedding along the mantos formed good feeder channels for ore-forming solutions; in the former case, poor feeder channels were formed. For this reason, the Tarugo vein is known only above a series of tuffites in the Lower Calipuy Group. The Tarugo structure is barely visible in the tuffites. Below the tuffite series, renewed concentration of the ore as a continuation of the Tarugo vein to depth is to be expected.

An up to 60 m thick dacite flow in the Upper Calipuy Group and rhyolite above it were found to be especially intensely mineralized in cross-cutting veins (e.g., Huancapetí, Collaracra, Hurán). On the other hand, a conglomeratic series of red colour underlying these units at the bottom of the Upper Calipuy Group contains little ore. Below the mined-out parts of the Collaracra and Hurán veins, ore is to be expected only where the Collaracra dacite flow extends to greater depth in the centres of synclinal folds.

Acknowledgments. We are grateful to Cía. Minera Alianza S.A. and to the Ministerio de Energía y Minas, Lima, Peru, for supplying maps, providing lodging quarters and assistance in the field.

We are thankful also to Dipl. Bergingenieur Dr. R. Kónopasek and Ing. geol. C. Hudson for cooperation and useful exchange of views during joint field trips.

We are also indebted to Prof. Dr. M. Kürsten, Dr. K.E. Koch, Dr. W. Hannak, Dr. R.C. Newcomb of the Federal Institute for Geosciences and Natural Resources, Hannover, West Germany, and to Prof. Dr. G.C. Amstutz and Dr. R. Zimmermann of the Mineralogical Institute of Heidelberg, West Germany, for critically reading the manuscript.

References

Amstutz GC (1978) Zu einer Metallogenie der Zentralen Anden von Peru. Muenstersche Forsch Geol Palaeontol 44/45:151–158

Bodenlos AJ, Straczek JA (1957a) Base metal deposits of the Cordillera Negra, Departamento de Ancash, Peru. Geological investigations in the American Republics. US Geol Surv Bull 1040: 160 pp

Bodenlos AJ, Straczek JA (1957b) Depósitos de plomo y zinc de la Cordillera Negra, Departamento de Ancash, Perú. Soc Nac Min Petrol Bol 58:222–235

Cabos Yepez R (1974) Mineralización y análisis espectrométrico de esfaleritas de la mina Hércules, Ticapampa. Tesis para optar el Título de Ingeniero Geólogo. Univ Nac Ingeniería, Lima, Perú.

Caldas Vidal J (1968) Geología de la mina Collaracra, Provincia de Recuay, Departamento de Ancash. Tesis para optar el Grado de Bachiller en Geología. Univ Nac Mayor San Marcos, Fac Ci Esc Geol, Lima, Perú, 55 pp

Casadevall T, Ohmoto H (1977) Sunnyside Mine, Eureka Mining District, San Juan County, Colorado: Geochemistry of gold and base metal ore deposition in a volcanic environment. Econ Geol 72:1285–1320

Cobbing EJ (1974) The tectonic framework of Peru as a setting for batholitic emplacement. Pac Geol 8:63–65

Cobbing EJ (1976) The geosynclinal pair at the continental margin of Peru. Tectonophysics 36: 157–165

Cobbing EJ (1978) The Andean geosyncline in Peru and its distinction from Alpine geosynclines. J Geol Soc London 135:207–218

Cobbing EJ, Pitcher WS (1972) The coastal batholith of Central Peru. J Geol Soc Lodon 128: 421–460

Cossio A (1964) Geología de los Cuadrángulos de Santiago de Chuco y Santa Rosa, Perú. Serv Geol Min Bol 8:60 pp

Dunin Borkowski E (1972) Posibilidades mineras del Distrito Abastecedor de la planta de Catac. Banco Min Perú, Div Ing Mem 178:36 pp

Dunin Borkowski E (1973) Informe preliminar sobre las posibilidades mineras de las minas y prospectos de los pequeños mineros ubicados a menos de 35 kms (in línea recta) de la planta de Catac. Banco Min Perú, Div Sucursales, Mem 351:37 pp

Finlow-Bates T, Large DE (1978) Water depth as a major control on the formation of submarine exhalative ore deposits (Zum geologischen Rahmen schichtgebundener Sulfid-Baryt-Lagerstätten, Nr. 10). Geol Jahrb D 30:27–39

Garson MS, Mitchell AHS (1977) Mineralization at destructive plate boundaries: a brief review. Volcanic processes in ore genesis. Inst Min Metall Geol Soc London, pp 81–97

Hamilton W, Myers WB (1967) The nature of batholiths. US Geol Surv Prof Pap 554-C: 30 pp
Harley DN (1979) A mineralized Ordovician resurgent caldera complex in the Bathurst-Newcastle mining district, New Brunswick, Canada. Econ Geol 74:786–796
Hudson C, Molina J, Gil E (1975) Estudio geológico del distrito minero de Ticapampa y de la mina Hércules. – Informe para la Cía. Minera Alianza S.A., con un anexo de R. Cabos Yepez. Estud Microsc Mina Hércules, Lima, Perú, 88 pp
Knox GI (1974) The structure and emplacement of the Rio Fortaleza centred acid complex, Ancash, Peru. J Geol Soc London 130:295–308
Kouda R, Koide H (1978) Ring structures, resurgent cauldron, and ore deposits in the Hokuroku volcanic field, Northern Akita, Japan. Min Geol 28:233–244
Large RR (1977) Chemical evolution and zonation of massive sulfide deposits in volcanic terrains. Econ Geol 72:549–572
MacDonald GA (1972) Volcanoes. Prentice Hall, New York, 510 pp
Megard F (1978) Etude géologique des Andes du Perou Central. Contrib Etude Geol Andes 1. Mem ORSTROM 86:303
Mitchell AHG, Bell JD (1973) Island arc evolution and related mineral deposits. J Geol 81:381–405
Mitchell AHG, Garson MS (1976) Mineralization at plate boundaries. Miner Sci Eng 8/2:129–169
Myers JS (1974) Cretaceous stratigraphy and structure, Western Andes of Peru between latitudes 19°–10°30′. Am Assoc Petrol Geol Bull 58:474–487
Myers JS (1975a) Cauldron subsidence and fluidization: Mechanism of intrusion of the coastal batholith of Peru into its own volcanic ejecta. Bull Geol Soc Am 86:1209–1220
Myers JS (1975b) Vertical crustal movements of the Andes of Peru. Nature (London) 254:672–674
Ohmoto H (1977) Genetic process of the ore deposits of sea water origin. Min Geol 27:53
Petersen U (1958) Structure and uplift of the Andes of Peru, Bolivia and adjacent Argentina. Bol Soc Geol Peru 33:57–129
Petersen U (1965) Regional geology and major ore deposits of Central Peru. Econ Geol 60/3:407–476
Petersen U (1970) Metallogenic provinces in South America. Geol Rundsch 59:834–897
Pitcher WS (1974) The Mesozoic and Cenozoic batholiths of Peru. Pac Geol 8:51–62
Pitcher WS (1978) The anatomy of a batholith. J Geol Soc London 135:157–182
Pitcher WS, Taylor WP (1975) Los Batolitos del Mesozoico y Cenozoico del Perú. Bol Soc Geol Peru 45:95–98
Purser WFC (1971) Metal mining in Peru, past and present. Praeger Publishers, New York, 340 pp
Raimondi A (1873) El departamento de Ancash y sus riquezas minerales. El Nacional, Lima, Perú, 651 pp
Ripley EM, Ohmoto H (1977) Mineralogic, sulfur isotope, and fluid inclusion studies of the stratabound copper deposits of the Raul Mine, Peru. Econ Geol 72:1017–1041
Samaniego A (1978) Schichtgebundene Erze der Grube El Estraño (Ancash, Peru) und ihr geologischer Rahmen. M Sc Thesis, Mineral Inst Univ Heidelberg, 97 pp
Sangster DF (1972) Precambrian volcanogenic massive sulphide deposits in Canada: A review. Geol Surv Can Pap 72/22: 44 pp
Sangster DR (1980) Quantitative characteristics of volcanogenic massive sulphide deposits. 1. Metal content and size distribution of massive sulphide deposits in volcanic centers (in press)
Sato T (1977) Kuroko deposits: their geology, geochemistry and origin. Inst Min Metall Geol Soc London Spec Publ. Volcanic processes in ore genesis, 7:153–161
Sillitoe RH (1972) Relation of metal provinces in Western America to subduction of oceanic lithosphere. Bull Geol Soc Am 83:813–817
Sillitoe RH, Sawkins FJ (1971) Geologic, mineralogic and fluid inclusion studies relating to the origin of copper bearing tourmaline breccia pipes, Chile. Econ Geol 66:1028–1041
Smith RL, Bailey RA (1968) Resurgent cauldrons. In: Coats et al.(eds) Studies in volcanology. Mem 116:613–662
Spence CD, de Rosen-Spence AF (1975) The place of sulfide mineralization in the volcanic sequence at Noranda, Quebec. Econ Geol 70:90–101
Steinmann G (1929) Geologie von Peru. Carl Winter, Heidelberg, 448 pp

Steven TA, Luedke RG, Lipman PW (1974) Relation of mineralization to calderas in the San Juan volcanic field, Southwestern Colorado. J Res US Geol Surv 2:405–409

Tanimura S (1973) Development of submarine sedimentary basins and environment of Kuroko mineralization in the Hokuroku District. Min Geol 23:237–243

Thurlow JG, Swanson EA, Strong DF (1975) Geology and lithogeochemistry of the Buchans sulfide deposit, Newfoundland. Econ Geol 70:130–144

Tumialan de la Cruz PH (1972) Estudio geológico regional del Distrito Minero de Ticapampa y geología local de las Minas Huancapetí, Collaracra, Florida, Hércules y Coturcán, pertenecientes al mismo distrito minero, Departamento de Ancash. Informe para la Cía. Minera Alianza SA, Lima, Perú, 105 pp

Wilson JJ (1963) Cretaceous stratigraphy of Central Andes of Peru. Bull Am Assoc Petrol Geol 47: 1–34

Wilson JJ, Reyes L, Garayar J (1967) Geología de los Cuadrángulos de Mollebamba, Tayabamba, Huaylas, Pomabamba, Carhuaz y Huari. Serv Geol Min Bol 16:95 pp

Zeil W (1970) Zur Geologie der Anden. Geol Rundsch 59:827–834

The Exhalative Sediments Linked to the Volcanic Exhalative Massive Sulphide Deposits: A Case Study of European Occurrences

A.J. BERNARD, G. DAGALLIER, and E. SOLER [1]

Abstract

Exhalites have been defined as submarine volcano-sedimentary rocks, either detrital or chemical, but deposited by hydrothermal fluids (true solutions or suspensions on the ocean floor. Mineralogical similarities between exhalites, on the one hand, and matter formed by wall rock metasomatic alteration of veins and/or open space filling products (i.e., alterites) on the other hand, is probably the first general observation on which one can rely. Does it imply a simple chemical deposition from a true solution, in both places, or the reworking of previously deposited alterites within the stockwork and further sedimentation on the sea floor? Among the many exhalites generated through one and/or the other processes only a few are metalliferous or signal the vicinity of an ore deposit.

This paper intends to bring into this question some geological information coming from South Spain, Cyprus, and the High Pyrenees (France) where massive sulphide bodies have been described and interpreted as volcano-sedimentary deposits.

Depending on steady or transient convecting cells developing around hot igneous bodies, either back arc acid protrusions or accretion zones, basaltic flows and dykes, a large variety of exhalites could be expected. The question is: which are those of significance for an existing ore-body? Presently, this research field seems to be a very open one!

1 Introduction

Many new studies have been undertaken on exhalative sediments or "exhalites" since Ridler's (1973) review of their definition and use as a new mining exploration tool. Most of these have tried initially to identify exhalative sediments in specific volcano-sedimentary environments and then to test their significance as a guide (control) to ore mineralization. Many of these works have not arrived at definitive conclusions because they are still in progress.

Submarine exhalation is emitted by a vent, stockwork or open fracture network. The wall rocks are metasomatically altered ("alterites") and exhibit the same mineral paragenesis as the exhalites themselves. The exhalites have the appearance of being sedimented particles of alterites: the minerals are identical and it is only their relative proportions which vary.

1 Institut National Polytechnique de Nancy, Ecole Nationale Supérieure de Géologie Appliquée et de Prospection Minière, B.P. No 452, 54001 Nancy Cedex, 94, Avenue de Lattre de Tassigny, 54 Nancy, France

Such mass transfer was first considered (Schneiderhöhn 1941) to be a result of deposition from a true solution which either led to a stratified sediment on the ocean floor or to altered wall rock and/or open space vein filling. Due to hydrodynamic variations in the fissures it is possible that material, either from the vein filling or the altered wall rocks, is carried in suspension by the fluids and later deposited on the sea floor during sedimentation. The relative importance of these two sources of material will depend on the steady or transient state of the convective circulation system (Henley and Thornley 1979) and the nature of the discharge processes.

It would be interesting to be able to distinguish between vein-controlled alterites and sedimentary exhalites in fossil systems. The final distinction is structural: alterite is fracture-controlled and exhalite is related to the sedimentary interface. Chemically the differences are ambiguous, as primary structures are destroyed during metamorphic recrystallization particularly when original features were cross-cutting (e.g., stratoide or pene-concordant stockwork, "Keiko" ore type, Kuroko province, Japan).

Additionally when the submarine magmatic eruption is acidic (Horikoshi 1969), the metalliferous exhalations are definitely sedimented after the last magmatic and/or phreato-magmatic explosive products (Kuroko, Rio Tinto). In contrast, during basic submarine eruptions (e.g., Cyprus) there is no such correlation among basic tuffs, exhalites and sulphide ores. Finally on Cyprus, as at Rio Tinto, there are abundant and well defined exhalites which, although related to basic flows, cannot be correlated with any known sulphide mineralization.

In summary, both mineralized and non-mineralized sedimentary exhalites are derived from the discharge of fluids from a geothermal convective circulation system which is driven by an igneous heat source (Parmentier and Spooner 1978).

Apparently the nature and conditions of acidic or basic magmatism in different geotectonic environments generate a large variety of exhalites including metalliferous ones. From the diverse range of differentiated exhalites resulting from (1) the slow exhalation of true solutions or lightly loaded suspensions from poorly defined vents or (2) powerful plumes of highly mineralized fluids (solution plus suspension) issuing from well defined hydrothermal mounds (chimneys) few of them are metallogenically significant. The problem in prospection is precisely to recognize which exhalite is mineralized and which one is not and then to apply this lesson to mining exploration.

This paper attempts to contribute to the problem of recognising metalliferous exhalites based on studies of three European examples: the metalliferous provinces of Huelva (Spain) and Cyprus, and the mining district of Pierrefitte-Nestalas (High Pyrenees, France). All of them are known to contain massive sulphide deposits.

2 Huelva Province (Spain)

This province is characterized by huge massive sulphide deposits; the whole Rio Tinto district (Atalaya quarry, Filon Sur and Filon Norte, Cerro Colorado, Planes-San Antonio workings) certainly amounts to more than 250 millions metric tons of crude ore in a province whose original reserves exceeded a billion metric tons.

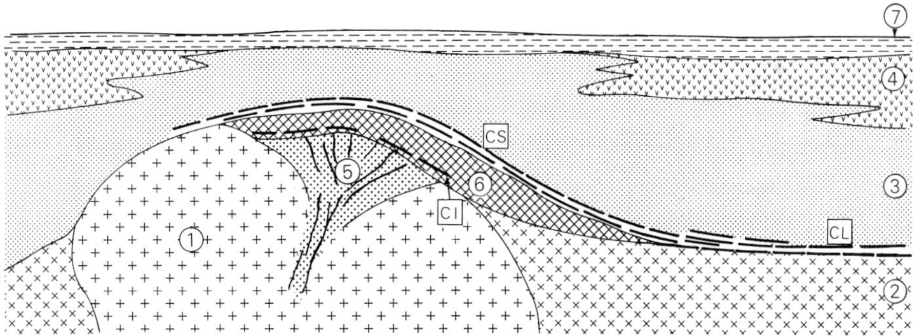

Fig. 1. Exhalite schematic location in the Huelva Province (Soler 1978). *1* Quartz-keratophyre protrusion; *2* Lower tuffs; *3* Upper tuffs; *4* Spilitic flows; *5* Stockwork and hydrothermally altered formations; *6* Pyritic bodies and chloritites: Lower *CI*, upper *CS*, lateral *CL; 7* Polvo hematites and verdes (red and green powders)

None the less, many other ore bodies do not even approach this importance. For example, the South-Iberian province deposits (Fig. 1) belong to the Kuroko type (Bernard and Soler 1974; Soler 1980):

- Close relationship with quartz-keratophyric protrusions, these igneous bodies, being located in a back-arc environment, imply an arc-continent collision.
- Presence of stockworks metasomatically altered by silicification, chloritization, and pyritization.
- Chloritites exhalites associated with massive sulphide ore bodies.

Although notable minor discrepancies do occur: lack of sulphate bodies laterally located relative to the main sulphide masses; no apparent zonal distribution of ores and metals; no explosion breccias above the stockwork . . . On the whole, these deposits should be classified with the Kuroko type.

In a simplified manner Huelva exhalites include:

1. Lower chloritites,
2. pyritites and other sulphides of chalcophilic metals, occasionally ending in a chert seam,
3. upper chloritites.

Thus, the massive pyritic lenses are completely sourrounded, even laterally, by chloritite formations. Above massive bodies and in the underlying stockworks the most common formation is a chloritite which has the composition of the upper chloritites. These rocks are chemically heterogeneous although formed from the same mineral species but in differing proportions:

- Quartz, micro-crystalline,
- Fe and/or Mg chlorites, but belonging to the same ripidolite group,
- ore minerals, mainly iron sulphides.

These compositional variations increase within the same chloritite bed when moving away from the massive sulphide body. Obviously these upper chloritites are the most interesting formation in mining exploration. They can extend for several hundreds meters away from the ore mass and thus provide an excellent guide to a concealed ore body. Presently it is more the petrographic peculiarities of these rocks which lead to

Table 1. Chloritites from Huelva Province (Soler 1980). The first part of the table shows the chemical analyses of the whole rocks (in %). Trace elements composition appears in μg/g in the second part. Analyzed by micro-probe chlorite composition and structural formulae are figured in the third and fourth parts of the table. Besides Fe-Mg variations, the chemical homogeneity of these ripidolites should be noted:

- Stockwork chloritites and chlorites: 1. Rio Tinto, 2. San Miguel
- Upper chloritites and chlorites: 3. Concepcion, 4. La Zarza, 5. Tinto y Santa Rosa
- Lateral chloritites and chlorites: 6. San Platon, 7. La Sorpressa

	1	2	3	4	5	6	7
SiO_2	68.72	26.27	62.47	23.21	52.37	64.78	24.76
Al_2O_3	8.28	17.13	9.24	17.32	14.65	11.73	21.64
Fe_2O_3 total	13.14	25.51	14.88	27.02	16.48	9.72	29.61
Fe_2O_3	2.29	7.62	3.86	5.17	2.79	2.47	7.15
FeO	9.76	16.10	9.92	19.66	12.31	6.52	20.21
MnO	0.08	0.14	0.10	–	0.12	0.13	0.10
MgO	4.00	20.00	6.11	14.96	8.23	7.39	11.97
CaO	–	–	0.05	2.84	–	–	0.04
Na_2O	0.01	–	–	–	0.02	0.02	traces
K_2O	0.11	–	0.02	–	0.77	0.85	traces
TiO_2	0.02	0.28	0.18	0.65	0.64	0.12	0.81
P.F. CO_2	0.14	0.20	0.07	3.89	0.14	0.07	0.10
P.F. H_2O	4.09	10.92	5.63	8.45	6.40	5.07	10.33
Total	98.59	100.45	98.75	98.34	99.82	99.88	99.36
Ba	198	< 10	22	16	113	109	36
Co	15	33	88	67	28	14	66
Cr	< 10	< 10	18	14	30	10	21
Cu	50	45	27	1163	79	10	10
Ni	< 10	19	43	32	16	10	39
Sr	< 10	< 10	10	30	10	10	17
V	< 10	28	10	63	70	10	18
SiO_2	22.50	25.45	23.72	25.39	25.67	26.04	24.31
Al_2O_3	21.52	21.00	21.11	22.13	23.32	21.57	22.56
FeO total	31.25	21.48	30.66	24.01	26.83	24.26	27.61
MgO	10.85	18.10	11.40	17.22	14.67	15.92	13.60
Total	86.12	86.03	86.69	88.75	90.49	87.79	88.08
Si	5.024	5.359	5.216	5.240	5.251	5.427	5.162
Al^{IV}	2.976	2.641	2.784	2.760	2.749	2.573	2.838
Total	8.000	8.000	8.000	8.000	8.000	8.000	8.000
Al^{VI}	2.692	2.569	2.689	2.622	2.873	2.729	2.807
Fer total	5.844	3.783	5.616	4.147	4.591	4.239	4.904
Mg	3.605	5.681	3.742	5.299	4.475	4.954	4.304
Total	12.141	12.033	12.047	12.068	11.939	11.922	12.015

a rapid identification of these exhalites. Their geochemical characteristics are apparently less discriminating and in particular their trace element contents are not distinctive (Table 1).

Spilitic rocks of the Huelva province provide another example of exhalative sediments, i.e., the "polvo", "hematites" or "verdes". The word "polvo", powder in Spanish, emphasizes the fine grain size of these rocks: electron microscope photos reveal an average grain size lower than a few microns. Quartz, colorless phyllites, and hematite (polvo hematites) or chlorite (polvo verdes) are the major components. These formations stratigraphically overlie the spilitic flows of the province. In detail, there is a continuum between "polvo hematites" and the vesicle fillings of the upper part of the spilitic flow on which they lie. Locally and laterally, with regard to the main flow, "polvo hematites" intertongue with Mn-Fe jaspers: starting from a polvo hematites composition, the phyllites decrease and disappear as Mn species appear and increase together with hematite and quartz.

Due to their regular and regional extent these basic exhalites (Table 2) possess a litho-stratigraphic significance like the Cyprus umbers which we are now going to describe.

3 Cyprus Province

The province contains Cu-pyritic mineralizations, which are mainly in veinlets, and their wall rock alterations. The *Symposium on Ophiolites* (Nicosia, 1–8 April 1979) gave a revised review of the ore distribution within the basaltic pile. Contrary to what has been generally accepted, the Skouriotissa strata-bound massive ore body is not the typical example of a Cyprus deposit. The latter are "fracture filled wall replacement" deposits located in the "Lower Pillow Lavas (L.P.L.) – Basal Group (B.G.) – upper part of the Dyke Complex (D.C.)", and never found, except for Skouriotissa, in the Upper Pillow Lavas (U.P.L.) or above.

Guillemot (1979) emphasizes the gap occurring between U.P.L. and the basement (L.P.L., B.G., D.C.). In fact the U.P.L. flowed over a basement already cooled and fractured (Fig. 2). The basalts, extruded from a distant accretion zone, generated their own convective circulation system which was oxidizing, non-steady-state and restricted to formations close to the sea bottom. Nevertheless, they generated exhalative sediments, the numbers (Roberts and Hudson 1973) which are similar to metalliferous deposits sedimented on to the actual ocean floor (Hekinian et al. 1978).

Diffractometric and chemical analyses and electron microscope observations show that these sediments are generally formed of ferric hydroxides, goethite in tiny microspheres (a few tenths of a micron in diameter) and smectites. Although high Mn grades are observed, no Mn-mineral is detectable: apparently this element is adsorbed on Fe-hydroxides. Other metals behave in the same way and are systematically adsorbed on Fe-hydroxides (Cu, Co) whereas others again (Ni, V, Ca) seem only to be occasionally trapped.

It is an illusion to utilize the umbers in mining prospection for Cu-pyritic stockworks from which these umbers are separated by the barren U.P.L. screen. However,

Table 2. "Polvos", umbers and ochres (Guillemot 1979; Soler 1980). Major elements in %, trace elements in μg/g.
Polvos hematites. 1. La Poderosa mine, S of the quarry; 2. Poderosa path to Odiel river; 3. Puebla de Guzman to Paymogo Road, 4.3 km North of Tharsis mine
Polvos verdes. Same location as (1) and (2)
Umbers. 7. Pano Lefkara brown earth; 8. Asgata Umber (above the U.P.L.); 9. Dhrapia Umber (within the U.P.L. flows); 10. Mathiati Umber: note the copper anomaly
Ochres. 11. Mathiati, above mineralization, below U.P.L. (cf. analysis 10); 12. Skouriotissa ochres, stratified abcve U.P.L. (see Fig. 2)
N.B. Total iron is expressed as Fe_2O_3 except for analyses 3 and 4 achieved by traditional chemical method

	1	2	3	4	5	6	7	8	9	10	11	12
SiO_2	77.75	77.66	65.10	56.80	61.84	54.68	60.33	28.44	17.07	12.87	14.73	17.05
Al_2O_3	10.18	9.72	17.10	21.45	16.95	23.58	7.59	5.70	3.23	3.38	4.53	4.88
Fer total	4.33	5.61	6.90	–	7.57	4.43	11.13	36.61	54.17	49.50	56.49	58.26
Fe_2O_3	4.27	5.33	6.75	5.55	6.54	2.28	n.d.	n.d.	n.d.	n.d.	n.d.	n.d.
FeO	0.05	0.25	0.15	2.45	0.93	1.93	n.d.	n.d.	n.d.	n.d.	n.d.	n.d.
MnO	0.10	0.30	0.16	0.11	0.11	0.95	2.42	7.00	5.47	11.71	2.55	tr.
MgO	0.59	1.08	1.15	2.15	1.99	3.45	2.60	1.85	0.86	1.64	3.95	1.44
CaO	–	–	0.25	0.10	–	0.29	0.66	2.63	1.64	2.52	0.79	0.63
Na_2O	0.25	1.72	0.75	0.60	1.44	3.19	0.38	0.34	0.14	0.36	0.23	0.52
K_2O	3.06	1.55	4.40	4.50	2.80	3.17	1.45	0.75	0.20	0.49	0.40	0.77
TiO_2	0.57	0.38	0.68	0.82	0.88	0.87	0.41	0.27	0.23	0.34	0.44	0.50
P_2O_5	–	–	0.09	0.06	–	–	0.39	–	–	–	–	–
P.F. CO_2	0.14	0.07	n.d.	–	0.21	0.14						
P.F. H_2O^-	2.11	1.93	0.15	0.55	5.74	4.59	11.99	15.97	15.29	16.62	14.69	13.83
P.F. H_2O^+			3.25	4.70								
Total	99.80	100.02	99.98	99.84	99.53	99.34	99.35	99.56	98.30	99.43	98.80	97.88
Ba	2256	574	380	750	639	896	210	923	132	1838	76	95
Co	40	32	25	15	27	27	47	131	132	230	159	169
Cr	89	56	40	18	30	30	55	54	87	42	46	67
Cu	10	16	45	14	125	125	340	935	1360	5000	5000	1624
Ni	55	49	50	42	57	57	143	240	216	196	321	406
Sr	78	101	140	80	274	271	288	1202	202	1788	60	82
V	96	91	210	135	159	159	216	415	500	373	254	449

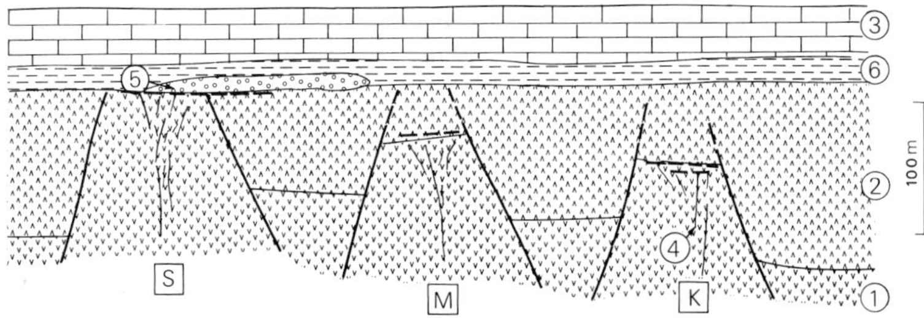

Fig. 2. Schematic section indicating different ore bodies; location in the Cyprus province (see Hutchinson and Searle 1971). *1* Lower pillow-lavas (L.P.L.); *2* Upper pillow-lavas (U.P.L.); *3* Kannaviou Formation (Campano-Maestrichtien); *4* Stockworks and ochres; *5* Skouriotissa reworked ochres and umbers; *6* Perapedhi Formation (umbers and radiolarites: Campanian). *S* Skouriotissa; *M* Mathiati (30 m of U.P.L. above mineralized ochres); *K* Kalavassos (50 m of U.P.L above ochres)

when this screen is thin (Mathiati, Kalavassos deposits) a marked Cu-Zn-anomaly appears in the numbers directly overlying the mineralized stockwork. Penetration depths of the non-steady-state convection cells in the U.P.L. which generated such umber anomalies by leaching the underlying stockwork, did not exceed 50 m. For example in Trulli, umbers lying 50 m above the Cu-pyritic stockwork do not reveal any Cu-Zn anomaly.

Without doubt, the typical geotectonic setting known from Cyprus, with a later layer of sterile basalts, the U.P.L., is not a general one when considering either the present or the past. Presently, as far as we know, metalliferous exhalations are emitted immediately above the sea floor and fossilized or recorded as such. These sediments are similar to hydrothermal sedimented products (Hekinian et al. 1978). They contain sulphides – mainly pyrite – sulphates – anhydrite –, manganese oxide crustifications (todorokite, birnessite) associated with iron hydroxides either crystallized (goethite) or amorphous, and dioctahedral smectites (nontronite). In Cyprus these sediments are always associated with sulphide deposits. From mineralogical arguments, one can consider that these Cypriot ochres are the sedimentary reworked brittle hydrothermal mounds like those recently described on the East Pacific Rise (Francheteau et al. 1978). Cyprus ochres are consistent with this interpretation, but do not provide more information either on the mechanism of hydrothermal fluid dispersal in deep oceanic waters (Solomon and Walshe 1979) or on the hydrothermal mound and stockwork reworking.

This is an open question whose applications we shall not consider. As we have seen basic and cold exhalites of the Huelva province, though widely occurring, are not correlated with any massive sulphide deposit but merely with small Mn concentrations. In fact, the polvos hematites are considered as a litho-stratigraphic marker. In Cyprus, oceanic basic emissions appear to generate massive sulphide deposits in hydrothermal stockworks. The exhalites related to this type of mineralization (ochres), although well characterized by their chemistry and mineralogy, are not extensive in space, at least as far as correlation with sulphide is involved. Obviously Huelva "polvos" are to be compared with Cyprus "umbers".

In conclusion, the Cyprus province seems presently to be unique because of the unusual setting of the ore deposits on the one hand, and of U.P.L. basalt on the other hand. It is difficult to make generalizations from the ochres on Cyprus. In contrast, study of accretion zone exhalites should lead to valuable applications to the exploration of massive sulphides. In that aim, Cyprus does not appear as the ideal field for research.

4 Pierrefitte-Nestalas District (High Pyrenees, France)

These mineralizations of Upper Ordovician age (Bernard and Foglierini 1964) combine the principal problems one may meet when prospecting for volcano-sedimentary deposits in old terrains.

Here metalliferous exhalites are related to acid quartz-keratophyric protrusions intrusive into a thick quartzo-phyllitic sequence stratigraphically ending with a few basic, spilitic flows (cf. Yoder 1973). Hydrothermal circulations generated by the cooling quartz-keratophyric bodies altered both the acid and basic rocks, but in different ways before reaching the ocean floor (Fig. 3). "Alterites" and sedimented "exhalites" are numerous, diversified, ore-bearing or associated with mineralizations: magnetite, pyrrhotite, galena, sphalerite (Besson 1973). The whole assemblage was metamorphosed during the Hercynian orogeny under greenschist P.T. conditions.

We shall consider here only the sedimented exhalites: they are chloritites, biotitites and gruneritites:

- Chloritites: these are associations in different proportions of quartz and chlorites. The latter (Table 3) exhibit a composition placing them either in the thuringite or the ripidolite group. Always Fe-rich, their Mg content is variable.
- Biotitites: generally associated with chloritites they appear as isolated and criss-crossing biotite needles either irregularly distributed or as a dense monomineralic felt.

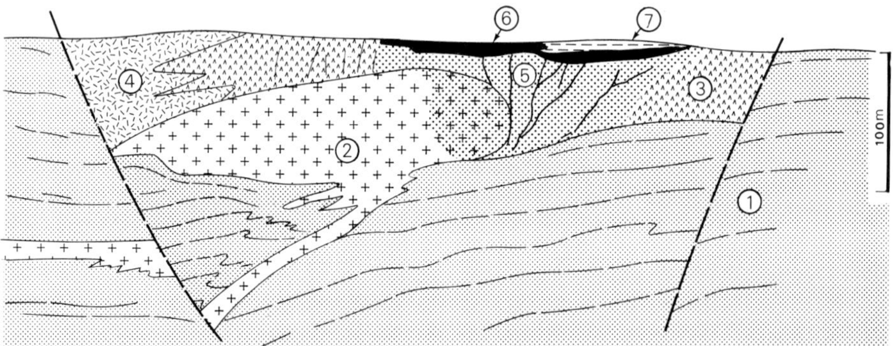

Fig. 3. Old mine of Barbazan (Pierrefitte-Nestalas) according to G. Dagallier (simplified). *1* Schistes and quartzites: Lower Ordovician; *2* Intrusive quartz-keratophyres (protrusions or sills); *3* Spilitic flows, with pillow-lavas; *4* Volcano sedimentary tuffaceous complex; *5* Hydrothermal alteration; *6* Exhalites (chloritites, biotitites, gruneritites); *7* Upper exhalites: sericitites

Table 3. Chlorites, biotites, sericite and amphiboles from Pierrefitte-Nestalas exhalites: major elements micro-probe analysis and corresponding structural formulae (G. Dagallier)
1. Low iron chlorite from acid rocks; 2. Intermediate chlorite; 3. High iron chlorite; 4. Biotite from biotitites lenses; 5. Sericite from the upper sericitites level. 6. and 7. Ferro-siliceous amphiboles (grunerite-cummingtonite)

	1	2	3	4	5	6	7
SiO_2	24.92	23.31	21.08	33.63	47.07	49.56	52.80
Al_2O_3	21.75	21.99	20.30	15.87	34.41	0.42	0.11
FeO	25.01	35.36	40.65	33.16	1.28	39.25	34.82
MnO	0.40	0.32	0.54	0.51	–	2.83	1.56
MgO	15.48	6.54	3.14	2.73	1.11	3.79	9.59
TiO_2	–	–	–	2.08	–	–	–
CaO	–	–	–	–	–	0.23	0.33
K_2O	0.05	–	–	8.12	9.50	–	–
P.F.	12.39	12.48	14.29	3.90	6.66	3.92	0.79
	$Si_{5.26}$ $Al^{IV}_{2.74}$ } 8	$Si_{5.22}$ $Al^{IV}_{2.78}$ } 8	$Si_{5.04}$ $Al^{IV}_{2.96}$ } 8	$Si_{5.51}$ $Al^{IV}_{2.49}$ } 8	$Si_{6.32}$ $Al^{IV}_{1.68}$ } 8	$Si_{8.08}$ $Al_{0.08}$ } 8.16	$Si_{7.97}$ $Al_{0.03}$ } 8
	$Al^{VI}_{2.68}$ $Fe^{2+}_{4.06}$ $Mn_{0.06}$ $Mg_{4.86}$ $K_{0.02}$ } 11.68	$Al^{VI}_{3.02}$ $Fe^{2+}_{6.62}$ $Mn_{0.06}$ $Mg_{2.18}$ } 11.88	$Al^{VI}_{2.76}$ $Fe^{2+}_{8.12}$ $Mn_{0.04}$ $Mg_{1.12}$ } 12.04	$Al^{VI}_{0.58}$ $Ti_{0.25}$ $Fe^{2+}_{4.54}$ $Mg_{0.67}$ $Mn_{0.07}$ } 6.11 $K_{1.70}$	$Al^{VI}_{3.77}$ $Fe^{2+}_{0.01}$ $Mg_{0.02}$ } 3.80 $K_{1.63}$	$Fe^{2+}_{5.35}$ $Mg_{0.92}$ $Mn_{0.39}$ $Ca_{0.04}$ } 6.70	$Fe^{2+}_{4.44}$ $Mg_{2.18}$ $Mn_{0.20}$ $Ca_{0.05}$ } 6.87

These are hydrothermal alteration products primarily developed in the fracture walls, swept away by fluids and finally sedimented with chlorites.
- Gruneritites: associated with the previously described exhalites in sedimentary lenses; these rocks, of high tenacity, are rich in magnetite and "grunerite-cummingtonite" amphiboles. On the whole these amphibolites contain mainly Fe and Si, accessorily Mg and Mn; they are always devoid of Ti.

These exhalites are selectively developed at the top of a volcano-sedimentary complex linked with the setting of a "spilitic-quartz keratophyric" association. In addition to basic flows and acid intrusions this complex exhibits basic and/or acid tuffs, and probably spilitic exhalites. These are carbonated and for this reason are poorly characterized and therefore difficult to recognize as exhalites.

During quartz-keratophyric eruptions, magmatic and/or phreatomagmatic explosions gave acid pyroclastic products on the one hand and on the other hand shattered the already settled basic flow and exhalites, giving epiclastic tuffs. Within mineralized zones it is the whole complex which is transformed by hydrothermal alterations, yielding stockwork alterites which are frequently mineralized. Finally the whole assemblage was metamorphsed under the greenschist P.T. conditions.

It is difficult to untangle such a complex geological situation because hydrothermal alteration has strongly obliterated the primary features of the igneous and sedimentary rocks. Nevertheless, analyses of alterites indicate that there are two successive metasomatic alterations: the first is potassic (sericite, biotite, occasionally adularia), the second is Mg-Fe-Si yielding silicification and chloritisation. These stockwork alterites have been dragged along with the hydrothermal fluids (particle suspension) and finally sedimented at the top of the volcano-sedimentary complex. The gruneritites are more difficult to explain for they could be derived from the metamorphism of an Si-Fe exhalite (polvo hematites type: silica-hematite).

These exhalitic formations, locally called "vein stone" (pierre à filon), have always been used as an exploration guide by the old miners in the Pierrefitte-Nestalas district, although their exact nature was not understood. This example emphasizes exhalite metamorphism. Frequently in old metamorphic terrains some Mg and/or Fe-rich rocks create a difficult problem. They have often been considered to be the result of Fe-Mg metasomatism. One should keep in mind the very characteristic primary chemical composition of exhalites and exhalitic series.

5 Conclusions

Hydrothermal alteration models of fracture wall rock often appeal to stationary fluids where diffusion occurs reducing the existing disequilibrium between the fresh country rock, the pore fluid and the aggressive fluid circulating in the open fracture (Norton 1979). In Cyprus, as elsewhere, the problem of the relationships between moving fluids and wall rocks is raised because of the stockworks (Guillemot 1979). The physicochemical processes may be very variable in space and time: for example, because of the high porosity created during the high temperature leaching of the wall rocks they became crumbly and then may either have been dragged along with the circulating hydrothermal fluids as a suspension, or their open spaces acted as traps of material not necessarily metalliferous for the circulating fluids.

Fluid-rock exchange may have taken place under either steady or transient convective conditions. Acceleration of the circulating fluids due either to temperature increase or to permeability decrease through clogging of the fractures can tear out particles from the walls and form a particle suspension. These fluids clear the choked plumbing and gush out emitting hydrothermal "plumes" which lead to more or less distal sedimentation. This could be the case with exhalative hydrothermal mounds imagined by Solomon and Walshe (1979) and effectively discovered and described by Francheteau et al. (1978).

Alternatively, when a steady convective circulation system discharges in deep water, the permeability increases due to stockwork splaying out underneath the ocean floor. Because of such hydrodynamic changes the fluids slow down without large temperature variations, and deposit the particle load within the vent or just at the outlet.

Temperature decrease, achieved by mixing hydrothermal fluids with deep ocean water, has been frequently called upon (Kajiwara 1973a,b; Seyfried and Bischoff 1977) and certainly remains an effective deposition mechanism. We would like to emphasize the prominent part taken by fluid slackening near discharge outlets for such a mechanism leads equally to sulphide deposition or to other exhalite sedimentation: colloidal iron hydroxides, hematite, chlorite, biotite, sericite, smectites . . . the direct neoformation of which on the sea floor is not always easy to explain. Sedimented exhalites thus appear as a detrital or ultradetrital sediment rather than a chemical one. Their characteristics derive directly from (1) vein deposition (fracture filling and wall rock alterites), (2) the mechanical reworking by fluids which dragged along previously deposited material, (3) sedimentary deposition.

In this concept the morphology of the discharge fracture systems becomes a prominent factor regarding the exhalation condition of:

- A discharge outlet well located on a highly fractured but spatially restricted zone will yield powerful jets emitted from well-formed hydrothermal mounds. Such emissions are easily disseminated in the submarine environment and further reconcentration by distal trapping requires special sedimentological environments. As noted above, the Cyprus province does not provide arguments to answer this question. The associated stockwork below can be extremely rich.
- When phreato-magmatic explosions precede the hydrothermal emission the fracturing thus created prepares the ground for a splayed-out stockwork just below the sea floor (Kuroko, Huelva pro parte, Pierrefitte-Nestalas). There one may expect a proximal exhalite sedimentation immediately above the vent.

The nature of the convective circulation system, whether deep or superficial and its stability and duration are also of importance in determining the outlet geometry and the nature of the hydrothermal emission. For example, the Huelva spilites and Cyprus U.P.L. probably generated unsteady convection cells resulting in poorly defined and varying outlets.

The nature and stability of convective circulations and their metallogenetic capacities also depend on many magmatic factors which we have not considered in this paper. However, we note that magmatic viscosity and in particular, viscosity variations before lava solidification, are critical in determining the morphology of the intrusive or extrusive body and finally the dispersion of the associated heat flux.

References

Bernard AJ, Foglierini F (1964) A propos des "filons hydrothermaux de Pierrefitte-Nestalas". CR Acad Sci Paris 258:274–277

Bernard AJ, Soler E (1974) Aperçu sur la province pyriteuse sud-ibérique. Centenaire Soc Geol Belg, vol 1, Gisements stratiformes et provinces cuprifères (Liège), pp 287–315

Besson M (1973) La formation ferrifère de Pierrefitte. Bull BRGM Sect II, 2:89–114

Francheteau J, Needham HD, Choukroune P, Juteau T, Seguret M, Ballard RD, Fox PJ, Norwark W, Carranea A, Cordoba D, Guerrero J, Rangin C (Equipe Scientifique CYAMEX) et Bougault H, Cambon P, Hekinian R (1978) Découverte par submersible de sulfures polymétalliques massifs sur la dorsale du Pacifique Oriental par 21°N (Projet RITA). C.R. Acad Sci Paris 287: 1365–1368

Guillemot D (1979) Un exemple de métallogénie en milieu océanique: étude des roches témoins des exhalaisons minéralisantes du Troodos (Chypre). Thèse INPL Nancy 1: 160 pp

Hekinian R, Rosendahl BR, Cronan DS, Dimitriev Y, Fodor RV, Goll RM, Hoffert M, Humphris SE, Mattey JP, Natland J, Petersen N, Roggenthen W, Schaader EL, Srivastasa RK, Warren N (1978) Hydrothermal deposits and associated basement rocks from the Galapagos spreading center. Oceanol Acta 1:473–482

Henley RW, Thornley R (1979) Some geothermal aspects of polymetallic massive sulfide formation. Econ Geol 74:1600–1612

Horikoshi E (1969) Volcanic activity related to the formation of the Kuroko-type deposits in the Kosaka district, Japan. Miner Deposita 4:321–345

Hutchinson RW, Searle DL (1971) Stratabound pyrite deposits in Cyprus and relations to other sulphide ores. Soc Min Geol Jpn Spec Issue 3:198–205

Kajiwara Y (1973a) Significance of cyclic seawater as a possible determinant of rock alteration facies in the earth crust. Geochem J 7:23–26

Kajiwara Y (1973b) Chemical composition of ore forming solution responsible for the Kuroko type mineralization in Japan. Geochem J 6:141–149

Norton D (1979) Transport phenomenon in hydrothermal systems: the redistribution of chemical components around cooling magmas. Soc Fr Mineral Bull 102 (5–6):471–486

Parmentier EM, Spooner EJC (1978) A chemical study of hydrothermal convection and the origin of the ophiolitic sulphide ore deposits of Cyprus. Earth Planet Sci Lett 40:33–44

Ridler RH (1973) The exhalite concept: a new mining exploration tool. North Miner 29:59–61

Robertson AHF, Hudson JD (1973) Cyprus umbers: chemical precipitates on a tethyan ocean ridge. Earth Planet Sci Lett 18:93–101

Schneiderhöhn H (1941) Lehrbuch der Erzlagerstättenkunde, vol 1. Die Lagerstätten der magmatischen Abfolge. Gustav Fischer, Jena, 858 pp

Seyfried W, Bischoff JL (1977) Hydrothermal transport of heavy metals by sea water: the role of sea water/basalt ratio. Earth Planet Sci Lett 34:71–77

Soler E (1980) Spilites et métallogénie. La province pyrito-cuprifère de Huelva (SW Espagne). Sc de la Terre, Nancy, Mém 39:1–461

Solomon M, Walshe JL (1979) The formation of massive sulfide deposits on the sea floor. Econ Geol 74:797–813

Yoder HS (1973) Contemporaneous basaltic and rhyolitic magmas. Am Mineral 58:153–171

Strata-Bound Ores of Spain: Newly Discovered Features and Their Interpretation

R. CASTROVIEJO [1]

Abstract

Strata-bound deposits appear frequently in Spain, and some new discoveries were made in the last few years. This brief account cannot summarize all the progress made, but the most outstanding facts are commented upon. The author shows also some results of his work on this type of deposit, bearing mainly on Pb-Zn and on Mn ores: he describes several Pb-Zn deposits of sedimentary origin, most of them – including the Santander Zn province – of the M-B-S type; one of these deposits, in the Sierra Morena Pb province, is older than the Hercynian granites which are currently being related to the Pb ores. Finally, the results are discussed.

1 Introduction

Increasing acceptance of the syngenetic interpretation for the origin of many ore deposits has fostered growing attention to syn-sedimentary and diagenetic ore-forming processes, and consequently the knowledge of strata-bound deposits has increased in Spain.

On the following pages, the author gives a brief review of recent progress in the knowledge of the Spanish strata-bound deposits, showing also some results of his work on this type of ores; some of them are still in a preliminary stage and only a working hypothesis can be offered in this paper.

Figure 1 shows a sketch of the geology of the Iberian peninsula. Two main orogenic areas can be distinguished: the Hesperic massif, due to the Hercynian orogeny, and the Alpine domains. The Alpine orogeny prevails in the Betic Region (SE of Spain), producing complicated tectonic structures characterized by different allochthonous units (Fig. 2). The (Arabic) numbers in Figs. 1–3 show the location of the mentioned deposits, so that they can be related to their geological-geotectonic framework; a geological sketch profile through the Sierra de Dobros Formation is given in Fig. 4.

1 Sección de Metalogenia, ENADIMSA, c/ Serrano, 116, Madrid 6, Spain

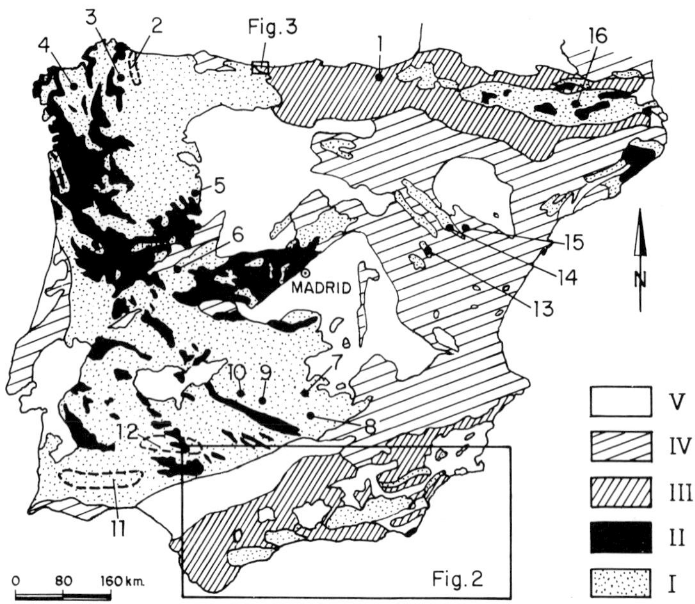

Fig. 1. Geological sketch map of the Iberian peninsula: *I* Folded Paleozoic; *II* Hercynian plutonic rocks; *III* Mesozoic to Paleogene of the Alpine geosyncline, folded and locally metamorphosed; *IV* Mesozoic to Paleogene cover; *V* Neogene to Quaternary. Deposits mentioned: *1* Bilbao; *2* NW oolithic Fe ores; *3* Rubiales; *4* NW massive sulph.; *5* Alcañices; *6* Morille; *7* S. Elena; *8* Linares; *9* Ojailén; *10* Almadén; *11* SW Iberian pyrite belt; *12* SW Fe-prov.; *13* Ojos Negros; *14* N.S. Carmen; *15* Crivillén; *16* Vall d'Arán

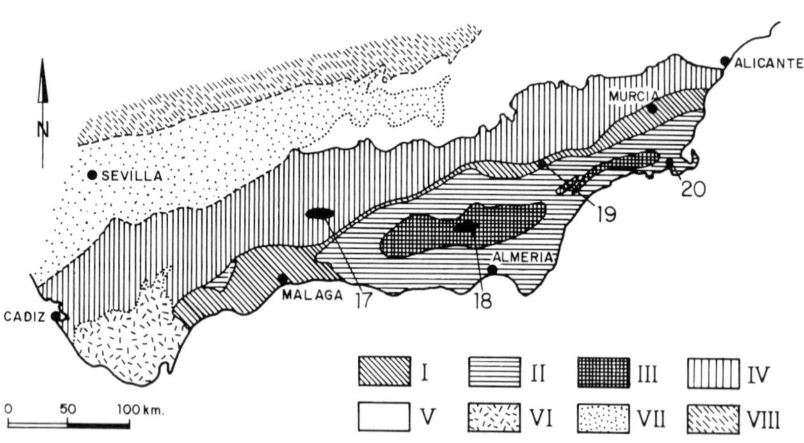

Fig. 2. Geotectonic sketch map of the SE Alpine domain: *I–III* Betic zone (*I* Malaguide, *II* Alpujárride, *III* Nevado-Filábride); *IV* Sub-Betic zone; *V* Pre-Betic zone; *VI* Gibraltar flysch; *VII* Guadalquivir-Neogene; *VIII* Hesperic massif. Deposits mentioned: *17* "Ocres Rojos" (Baena, etc.); *18* Marquesado; *19* Vélez Rubio; *20* Cartagena

2 A Brief Review on Recent Progress

This comment deals only with: (a) occurrences which have a sufficient value to be considered as actual or potential economic ore bodies; (b) the most recent developments: usually only those published in the latter part of the 1970's. Information on some deposits which cannot be discussed here is summarized in Table 1 (industrial minerals are out of the scope of this work).

Table 1. Data on some strata-bound deposits of Spain

Ore	Deposit	Type	Other data
Barite-sulphide	Alcañices (Zamora)	Volcanic-sedimentary [a]	U. Silurian; Meggen-type (Arribas and Moro 1980)
Pb-Zn-F-Ba	Betic province (Almería)	Sedimentary	Triassic, "Complejo Alpujárride" (Espí 1977)
Pb-Zn-(Cu)	Vélez-Rubio (Almería)	Sedimentary	Permo-Triassic, "Complejo Maláguide" (sandstone, dolostone) (Carrasco et al. 1977)
Pb-Zn-(Ag)	Cartagena	Under discussion	NE-end of Betic domain, relation to Tertiary volcanism
Pb	Linares (Jaén)	Sedimentary	Ore as cement in Triassic conglomerates; source: hydrothermal veins of the East S. Morena district
Monazite (Eu, r.e.)	Various: León, Oviedo, etc.	Sedimentary	Aberrant facies of monazite (Vaquero 1979)
Scheelite	Morille (Salamanca)	Sedimentary-metamorphic	Cambro-Ordovician schists (Arribas 1979)
Fe	Nos. 1, 2, 12, 13 (Fig. 1) and Nos. 17, 18 (Fig. 2)	Different types	Controversial interpretations (see Igme 1979)
U	Santa Elena (Jaén) and others	Several stratiform, syngenetic and epigenetic types	Arribas (1975)

[a] The widely known volcano-sedimentary deposits of Almadén (Hg), and of the NW- and SW-massive sulphide provinces (Fig. 1: Nos. 4, 10, 11), will not be discussed here

Sierra Morena Pb (Zn) Province. Strata-bound Ordovician Zn-Pb deposits were found (Rios and Claverias 1979) in the Ojailén district. The ores have been studied by the author (1979a), who found mega-, meso- and microscopic evidence of a syndiagenetic origin. The deposits are bound to the quartz-rich upper levels of the "Caliza Urbana" unit and occur only where the host-rock is dolomitic. Syndiagenetic recrystallization and replacement features, in which the ore minerals are involved, can be seen with the naked eye or with the microscope. Superimposed effects, due to the intensive tectonic activity registered in the area, mask somewhat the syndiagenetic features: e.g., galena is mobilized and appears frequently filling fractures or veins.

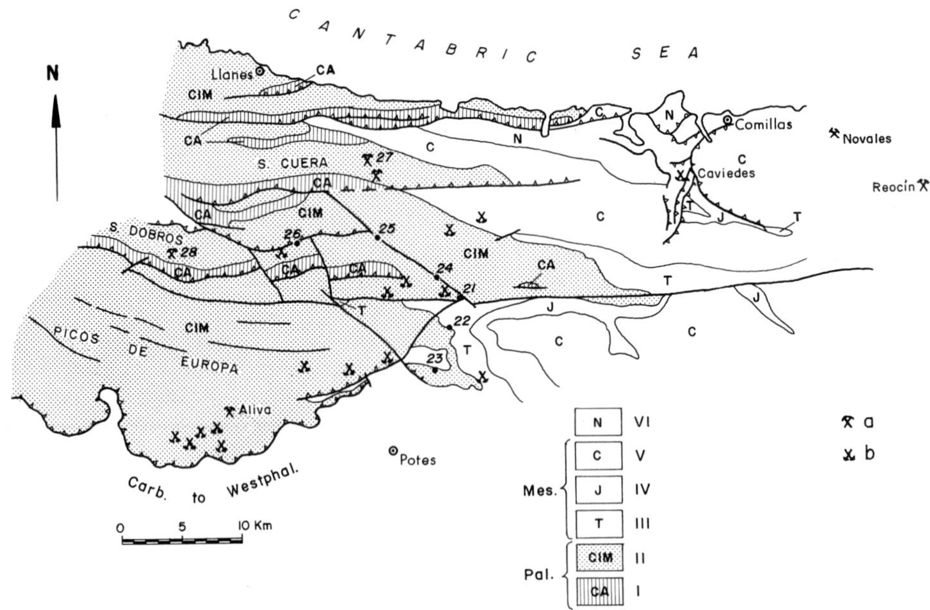

Fig. 3. Geological sketch map of the Santander-Asturias-Cantabric region: *I* Ordovician "Cuarcita Armoricana" Formation; *II* Lower Carboniferous "Caliza de Montaña" Form; *III* Permo-Trias; *IV* Jurassic; *V* Cretaceous; *VI* Tertiary. *a* Mine; *b* prospect. Deposits mentioned: *21* Osina; *22* Jozarco; *23* Lebeña; *24* Coto Rubio; *25* Argallón; *26* Oceño. Aliva, Caviedes, Novales and Reocín: s. map; *27* Pilar mine; *28* S. Dobros mines. (After Rios 1979)

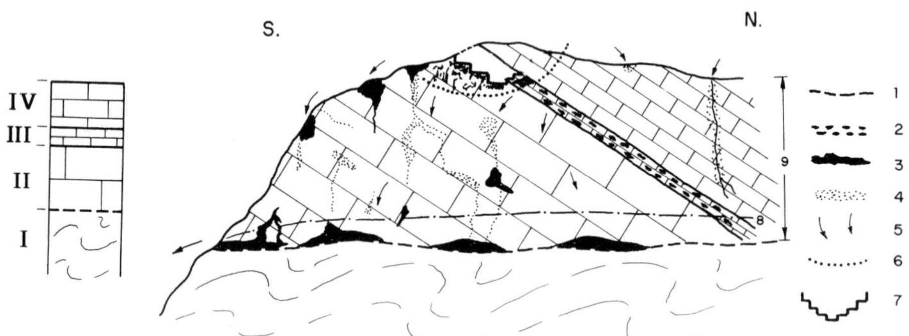

Fig. 4. Geological sketch profile through the Sierra de Dobros formations (after Armengot 1980; not to scale): *I* Ordovician basement: quartzites and schists; *II–IV* Lower Carboniferous limestones; *II* "Caliza de Montaña" Formation; *III* Mn bearing layer; *IV* "Caliza de los Picos" Formation. *1* Unconformity; *2* primary Mn ore; *3* secondarily enriched or karstic Mn (Fe) ore; *4* karst; *5* supergene solutions; *6* local boundary of superficial enrichment; *7* mined zones; *8* water table; *9* percolation zone

Santander Zn Province. Two clearly different zones can be distinguished (Figs. 3 and 4): one is characterized by stratiform deposits bound to Upper Aptian dolostones, and the other by deposits which occur in the Lower Carboniferous "Caliza de Montaña" Formation; the latter have a conspicuous tectonic control and are economically less important.

The author has studied some ores from the Reocín, Novales and Caviedes mines in the first area mentioned. The mineral association consists essentially of sphalerite, galena, marcasite, pyrite, with minor to insignificant amounts of barite, quartz, etc. Banding and colloidal structures, schalenblende, melnikovite-pyrite, etc., frequently appear. Microscopic observations agree largely with those of Monseur (1967), whose sedimentary syngenetic hypothesis is accepted, as is the assumption that the source of the metals might have been the Picos de Europa area.

Nevertheless, further questions remain unanswered. One of them is the origin of the ore occurrences in Paleozoic beds, from which one could proceed onto a comprehensive genetic model holding for the whole province. Evidently, this model cannot be a simple one.

The relations of the ore occurrences to tectonics are generally apparent, but not always clear (some of them, e.g., Jozarco, seem related to a Carboniferous-Permo-Triassic unconformity; Rios 1979).

Samples from Osina, Jozarco, Lebeña, Coto Rubio, Argallón, Oceño and Aliva (Fig. 3), in the "Caliza de Montaña" Formation, were studied (Castroviejo 1979b). The most widespread ore minerals are sphalerite, galena, chalcopyrite, pyrite. Fahlore, covellite and wurtzite also occur. The main mineralogical differences of these ores – except for the Aliva mine – with those of the previous area consist in their relation to sparry calcite instead of dolomite, and to silicif ation (Fig. 5).

Thus, mobilization and silicification are usual features, which could be explained by the prominent role played by tectonic events in the localization of the deposits (Fig. 3): most of them are related to fractures, faults, or even thrusts, not being, however, always located on those weakness planes. Tectonically stressed or fractured minerals are common, although the occurrence of several generations of sulphides and calcite suggest that the relation of tectonics to ore crystallization is not simple.

Information from the main mine of the area (Aliva), which is not far from the "Picos Thrust", contributes some other features to the knowledge of this type of deposits. Microscopic observations indicate that the ore is bound to dolomitized levels, rich in organic matter, whereas the upper and lower limestone beds appear barren. Sphalerite forms "schalenblende" or disseminated grains of variable size, which are recrystallized (Fig. 6), usually without replacement features; minor cracks in sphalerite are welded by calcite or galena, but otherwise no relationship of mineralization to tectonics appears. Geometric evidence points to a simultaneous formation of dolomite and ore, while gel structures and the low-iron composition of sphalerite suggest a low temperature genesis.

Consequently, it is assumed – as a working hypothesis – that a primary, synsedimentary or syndiagenetic ore formation also occurred in the Paleozoic Erathem. Where a tectonic control operates, mobilization of calcite and sulphides, and intensive silicification are usual.

The Santander deposits, as a whole, are considered to be of the Mississippi Valley-Bleiberg-Silesia (M-B-S) type, as defined by Amstutz (1972).

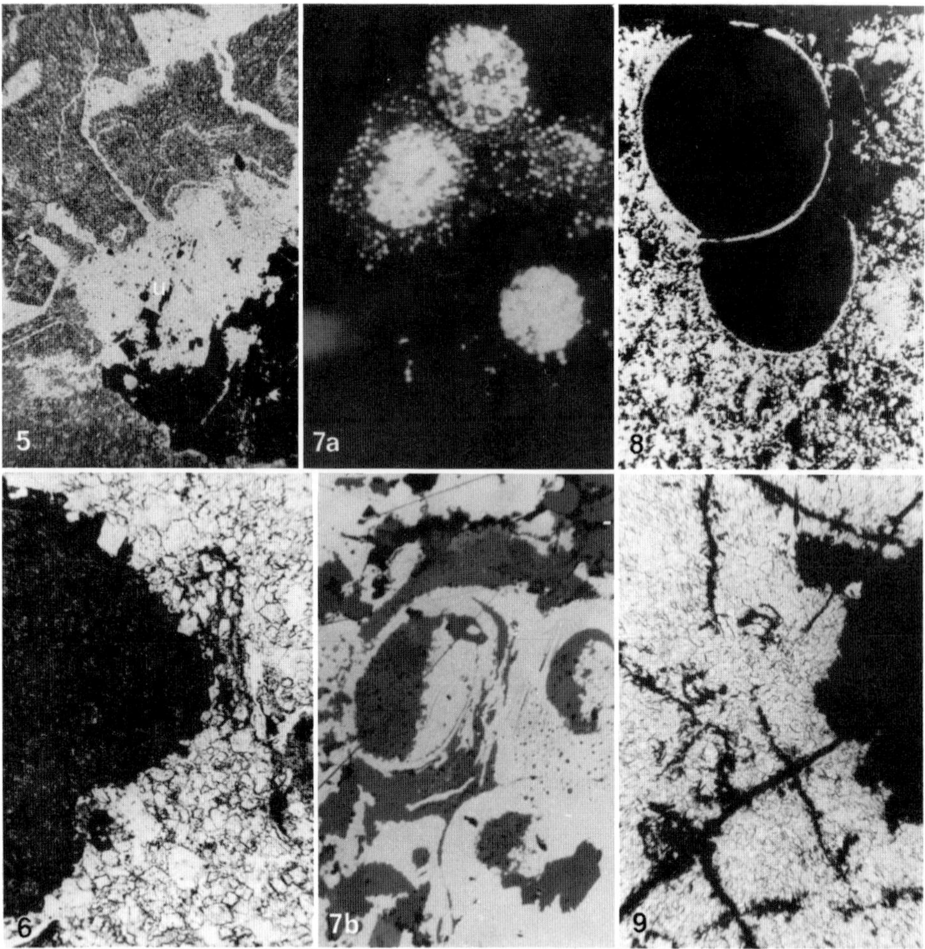

Fig. 5. Argallón, Santander: hypogene replacement of calcite *(grey)* by quartz (*white;* zoning due to minute residual inclusions of calcite), associated to sphalerite *(dark grey)*. Transm.light, // N, × 27

Fig. 6. Aliva, Santander: diagenetic intergrowth of sphalerite *(grey)* and dolomite *(white)*. Transm. light, // N, × 76

Fig. 7a, b. N.S. Carmen, Teruel: **a** framboidal pyrite in dolomite; **b** diagenetic replacement of fossil remnants by sphalerite and dolomite. Both refl. light; // N; **a** = × 1080, oil; **b** = ×63

Fig. 8. Dobros, Asturias: finely intergrown psilomelane *(white)* and calcite *(dark);* fossil remnants replaced by psilomelane, with core of sparry calcite. Reflected light, // N, × 99

Fig. 9. Pilar Mine, Asturias: micritic limestone *(white)*, with incipient replacement by Mn ore (psilomelane, *black*) along fissures. Transm. light, // N, × 95

Other Pb-Zn Deposits. The discovery and development of an impressive sphalerite-galena mine (*Rubiales,* Lugo; EXMINESA), following exploration of an old prospect, has been one of the most prominent events of the last decade. The deposit occurs in the folded Lower Cambrian "Cándana" Formation (mainly limestones and schists). Its genesis is not yet clear; a hypogen epigenetic origin of the ore body is not excluded, since it is closely controlled by Hercynian tectonics and related to massive silicification. In any case, the whole "Caliza de Vegadeo" Formation – latest Lower Cambrian – which occupies a much broader area, appears as an extensive lithostratigraphic metallotect for lead-zinc ores.

The mine "Nuestra Señora del Carmen" (Teruel) develops a stratiform Zn-Pb deposit (M-B-S type), which is bound to Devonian carbonate horizons lying on quartzites. The main minerals are sphalerite, galena and dolomite. Field and microscopic observations (Fig. 7) point – as well for some of the previously mentioned deposits – to a synsedimentary genesis of the ores, probably related to the cycle of sulphur (Castroviejo 1977).

The Strata-Bound Mn Deposits of Crivillén (Teruel) and of the Sierra de Dobros and Sierra de Cuera districts (Asturias, Fig. 3) have been recently studied. The last two, which are similar (Fig. 4), are briefly described. Ores occur in cavities and fracture-fillings, or cementing collapse-breccias and replace carbonates (Figs. 8 and 9). The most complete paragenesis was found in Mina Pilar (S. Cuera): pyrolusite, psilomelane, todorokite, hausmannite, polianite, and calcite; occasional goethite and limonite are due to supergene oxidation of sulphides (e.g., "framboidal pyrite", surrounded by radiating aggregates of marcasite) contained in the biomicritic-limestone (Castroviejo 1980).

The genesis is interpreted as follows: supergene alteration and redistribution of synsedimentary Mn-ores from the III-layer led to local concentrations, controlled either by karstic processes or by the presence of impervious levels (Armengot 1980).

3 Conclusion

Several examples have been discussed showing the development of work on strata-bound deposits in Spain. The results so far obtained bring up new questions that are to be answered by further work:

- The discovery of Ordovician syn-diagenetic ores in the Sierra Morena Pb province leads to the following questions: (1) Is this an isolated occurrence or a more general feature, not observed before due to a prevailing epigenetic view? (2) Did a previously enriched sedimentary cover contribute, through later assimilation and redistribution by Hercynian granites, to the enormous quantities of metal mined in the province?
- The full understanding of the genesis of the Rubiales deposit is needed to help mining exploration.
- A general approach for classification of M-B-S type (Amstutz 1972) Pb-Zn deposits seems useful as a first step aimed at the understanding of their genesis.

– Much work is still to be done to reach a thorough knowledge of the Spanish strata-bound deposits and to achieve a global analysis of all the information about them. Let, then, the conclusion be an urgent call for team work and improved scientific communication, and for a deeper and more detailed geological, sedimentological, paleogeographical, etc., knowledge of the Iberian peninsula.

Acknowledgements. The author acknowledges ENADIMSA/INI for permission to publish results of his work sponsored by them. He is also indebted to Prof. A. Arribas (Univ. Salamanca) for his personal communication, and to all his colleagues in the different research teams of ENADIMSA.

References

Amstutz GC (1972) Observational criteria for the classification of Mississippi Valley-Bleiberg-Silesia type of deposits. Proc 2nd Int Symp Mineral Deposits Alps, Bled, Oct 1971, Ljubljana, pp 207–215

Armengot J (1980) Manganesos de Asturias. Unpubl. Rep ENADIMSA, Madrid

Arribas A (1975) Caracteres geológicos de los yacimientos Españoles de uranio. Su importancia económica e interés en el desarrollo energetico del pais. Stud Geol IX:7–63

Arribas A (1979) Les gisements de tungsténe de la zone de Morille (Salamanca). Chron Rech Min 450:27–34

Arribas A, Moro C (1980) Yacimientos de barita asociados al sinclinorio de Alcañices (Zamora). Univ Salamanca (in press)

Carrasco A, Castroviejo R, Fernandez-Luanco MC, Martin L, Rios S (1979) Estudio de las mineralizaciones de Pb-Zn-Cu del Permotrías del Complejo Maláguide en los alrededores de Vélez Rubio (Almería). Bol Geol Min XC-V:475–488

Castroviejo R (1977) Estudio microscópico de diversas muestras procedentes de la Mina Nuestra Señora del Carmen (Teruel). Unpubl Rep ENADIMSA, Madrid

Castroviejo R (1979a) Génesis de los indicios Zn-Pb en el Ordovícico de la Cuenca del Río Ojailén (Puertollano- C. Real): Aportación del estudio mineralógico-textural. Soc Esp Mineral Extr 1:31–41

Castroviejo R (1979b) Estudio microscópico de diversas muestras del Proyecto "Pb-Zn Cantábrico". Unpubl Rep ENADIMSA, Madrid

Castroviejo R (1980) Estudio microscópico de diversas menas de Asturias. Unpubl Rep ENADIMSA, Madrid

Espí JA (1977) Aspecto metalogénico de los criaderos de flúor-plomo de Sierra de Gádor (Almería). Thesis Doct ETS Ing Minas, Madrid, 184 pp

Igme (1979) I. Reunion de mineralogía y metalogenia del hierro. Temas Geol Miner 3:177 pp

Monseur G (1967) Synthèse des connaissances actuelles sur le gisement stratiforme de Reocín (Santander, Espagne). Int Symp Stratiform Pb, Zn, Ba, F Deposits, New York. Econ Geol Mon 3:278–293

Montoriol J, Campa J, Alvarez A (1979) Estudio del complejo pirenaico de yacimentos de Pb-Zn del Valle de Arán, en relación con los movimientos y las fases de metamorfismo. Fundac J March, Madrid, 323 pp

Rios S (1979) Nota de la visita a various indicios de Pb-Zn en los Picos de Europa. Unpubl Rep, ENADIMSA, Madrid

Rios S, Claverias P (1979) Nota acerca de la existencia de indicious estrato-ligados de Zn-Pb en el Ordoviciense Superior del extremo SE de la Meseta Hercínica Española. Bol Geol Min T XC-I: 1–5

Vaquero C (1979) Descubrimiento por primera vez en España de una monacita de facies aberrante portadora de Europio. Bol Geol Min T XC-IV:374–379

Silver and Associated Minerals in the Northwestern Odenwald, Germany

M. FETTEL [1]

Abstract

Three occurrences of silver and uranium ores in the northwestern Odenwald near Nieder Ramstadt, Nieder Beerbach and Waschenbach are described in a paragenetic, petrographic, and tectonic respect.

The description leads to genetic considerations with the following results:

- All the ore mineralizations belong to one tectonic system which was created in several separated phases in conjunction with the activation of the Rhine Graben lineament.
- Each paragenetic formation is a separate process bound with a specific tectonic phase, such that the Pb-Cu-Zn-As and the Bi-Co-Ni-Ag parageneses are older than the U-Fe-Pb-Zn-Cu one.
- The origin of the ore-bearing "hydrothermal" solutions is discussed, whereby a connection of ascendant and descendant (uranium-bearing) solutions appear probable.

1 Introduction

In 1923, Ramdohr published his first ore microscopic research on *Der Silberkobalterzgang mit Kupfererzen von Nieder Ramstadt bei Darmstadt* (1923, p. 164–192). He came to the conclusion that in the northwestern Odenwald a larger zone of silver ore occurrences must exist besides the small, 1 m^3 silver-cobalt deposit found in the Wingertsberg quarry.

This conclusion was confirmed later when silver was found in 1952 in Waschenbach, in 1977 in neighbouring Nieder Beerbach, and, between these dates, in Nieder Ramstadt as well.

2 The Occurrences

All occurrences are located in the "Gabbrocomplex" of NW Odenwald. This complex of basic intrusive rocks contains as its core the Frankensteinmassif, gabbro which changes towards the metamorphic frame to gabbrodiorite and diorite. In the NW direction, biotitediorites penetrate and dissolve the metamorphites; contrary to this, the

1 Im Kantelacker 5, 6148 Heppenheim-Kirschhausen, FRG

complex to the SE is sharply bound by an Erzgebirge (SW-NE) striking fault and the relatively elevated granodiorite-gneisses of the adjoining southern area (Nickel and Fettel 1979, p. 163).

The occurrences of Nieder Beerbach and Waschenbach are located in the central part of the gabbro complex, Nieder Ramstadt on the northern edge, where diorites, amphibolite and "hornfels" form the boundaries.

2.1 Nieder Ramstadt, Wingertsberg Quarry (TK 25, sheet Nr. 6118 Darmstadt-Ost)

In 1921, Ramdohr found in this quarry the well known Ag-Co ore deposit which he described later (1923, 1975). Until 1974 a diorite complex bound by amphibolite and lime-silicates from the SE and NW was being worked by the quarry company. In some places, the amphibolite is penetrated by diorite and therefore resorbed in its structure. Especially in the northern discovery area, inclusions occur of a quartz-plagioclase-biotite hornfels (see Strecker 1971, p. 287). The entire complex is completely penetrated by an acidic intrusive rock, granite to granodiorites, which forms pegmatitic and aplitic vein fillings.

The tectonics of the ore fissures and veins correlate with the Tertiary activation of the alignment of the Rhine Graben. The location is penetrated by the "rheinisch" orientated fissures to which a very pronounced fissure vein has developed in a 120°–160° direction. These fissure veins of the relatively young tectonic phase are the carrier of the mineralization along with their vein filling.

The Ag-Co mineralization by Ramdohr contained, amazingly, no bismuth or uranium ores and only traces of Ni ore. The paragenesis changes as we go deeper, from silver and silver ores at the top, to galena, sphalerite and copper ores farther down. Striking NW, the mineralized fissures expand to a swarm of parallel striking veins which carry only uranium, nickel and copper ore, but also Ag and Co (found in 1977 and 1979).

The *uranium mineralization* has filled veins up to 7 cm. Together with red chalcedony, calcite, multi-coloured quartz and barite, tennantite, bornite, chalcopyrite and chalcocite also occur, mostly massive, as well as pitchblende, coffinite, nickeline, sphalerite, galena, pyrite and hematite in varying amounts. The hematite causes the red colouring of the chalcedony and parts of the calcite (Rotspat); goethite, erythrite and Cu-sulphates have been identified as byproducts from weathering.

The pitchblende, together with late chalcopyrite and hematite, is the latest ore mineral of the uranium vein. It is black and has a botryoidal, kidney-shaped form with galena phenocrysts, a shelled structure with shrinkage cracks which could be filled with other sulphides. It surrrounds the other ores, mainly copper, pyrites and niccolite, as thin crusts and penetrates the cataclastic cracks. Its formation and the determination of its age lead to the conclusion that it was caused by low temperature ore-bearing solutions.

Here, too, the *Ag-Co paragenesis* presents itself, though still limited to one area which is further proof of the extension of the Ramdohr mineralization. As the oldest mineralization phase within the uranium ore debris, it accompanies the latter partially as its own vein, but will also be enclosed as breccia. The ore minerals are: native silver, argentite, safflorite and smaltite.

Both types occur as mineralization phases of various ages within each other; however, tectonically stressed crack systems more often occur but only when amphibolite or hornfels appear as secondary rocks; in diorite rock they end immediately.

A *second uranium paragenesis,* without silver and cobalt ore, is tied to red aplitic granodiorite. Niccolite, maucherite and gersdorffite are combined with coffinite (first identified by Ramdohr 1960, pers. comm.) and pitchblende as the main ore minerals in comparison to the first paragenesis veins which were quite conspicuous because of the red chalcedony. Annabergite is created here as a product of weathering.

The uranium concentration is strongest in those parts in which either niccolite/maucherite or copper pyrites dominate. The uranium content fluctuates between 600 and 17,900 g/t (Meisl 1975, p. 247).

The following facts could be of genetic importance:

- The Ramdohr Ag-Co mineralization and the present uranium depth belong to the same tectonic system, whereas the Ag-Co belong to an older mineralization phase.
- This system is relatively the youngest and multitudinous in conjunction with the Tertiary activation which originated the Rhine Graben lineament.
- The mineralization occurs only in favourable secondary rock (amphibolite, hornfels, aplitic granodiorite).
- The uranium ore belongs, in the geological timetable, to the youngest mineralization; it is held together by a nickel based compound and copper pyrites.
- The form of these ores indicates that they were formed at low temperatures.

2.2 **Mühltal (Nieder Beerbach) Quarry Glasberg** (TK 25 Bl. Nr. 6218 Neunkirchen)

The silver mineralization (Fettel 1978) is located in the central part of the existing basic complex which consists here of gabbro, gabbrodiorite and amphibolite-inclusions (beerbachite) (Nickel and Fettel 1979, p. 166).

Contrary to the Wingertsberg tectonics, the Hercynian (NW-SE) direction of Glasberg is hardly pronounced. The large vein of the rock formation consists of an Erzgebirgic (SW-NE) system, rare Hercynian (post-Hercynian orogeny) veins and a flat plane parting towards the west which is occasionally outlined by pegmatite-aplite. In connection with several young Rhenic (N-S) faults the Erzgebirgic fissures were so stressed to lead locally to shifting, brecciation and formation of faults. In the mineralization zone the shift of the rock formation is quite pronounced, giving the impression of a SE-striking ladder of ore shoots (Fig. 1).

Here, all the veins in which the ore minerals occur, in particular native silver, which appears as nuggets or wire, are also filled with gangue (quartz, calcite, barite, chlorite, mylonitic rock). Further minerals: argentite, pearceite, cerargyrite, safflorite, smaltite, cinnabar, galena, sphalerite, pyrite, chalcopyrite and erythrite as a weathering product.

Analysis (Dr. Schoembs 1979, written comm.) of moss-like *silver* with gangue shows the following results:

a) 2.24 g weight of sample contained 38.14% Ag; 5.15% Hg; other elements 17.38%; nonsoluble in nitric acid: 44.18%.

b) 2.55 g weight of sample contained 26.69% Ag; 3.24% Hg; other elements 17.38%; nonsoluble in nitric acid: 52.69%.

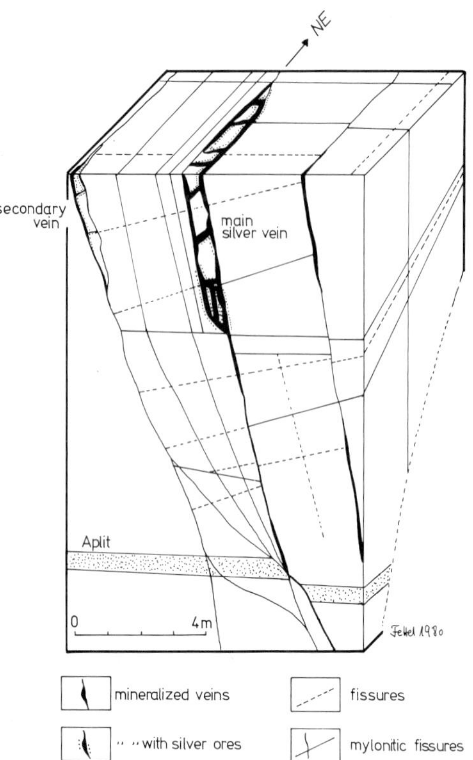

Fig. 1. Silver mineralization, Nieder Beerbach

The ratio of Ag to Hg, indicated by weight of sample, gave the following:

a) 88.09% silver and 11.09% mercury,
b) 89,16% silver and 10.84% mercury.

The silver from this location could also be considered to be "arquerite" due to its high Hg content (Klockmann 1978, p. 391).

Sponge-like porous masses in pockets and wedges of the gangue itself form the largest silver ore mass; in some places they were also the only contents of the veins. Occasionally, it was compressed into massive nuggets or, at the seam, flattened to sheet metal. Crystallized wire or curl-like aggregates are quite rare.

Argentite appears over the entire horizon evenly distributed on the hanging flank. Its form points to a development over 179°C (Ramdohr 1975, p. 505), because the crystals show regular forms with cubic habit.

Pearceite has been discovered here for the first time in the Odenwald and was identified by Dr. Dreyer/Mainz (†).

Cerargyrite occurs quite frequently as crusts on native silver in a light ink-blue colour, perhaps caused by small amounts of Cu (identified by Dr. Hentschel/Wiesbaden).

Cinnabar was found as a few crystals exhibiting the typical colouring together with calcite, also for the first time in Odenwald mineralizations.

Safflorite and rare *smaltite* compare with those from Nieder Ramstadt described by Ramdohr (1923, p. 180).

Galena, sphalerite and chalcopyrite are present but not characteristic for this type of mineralization, this also holds true for the minor traces of digenite.

The following facts probably are of genetic importance:

- The ore mineralization belongs to an Erzgebirgic system which owes its formation to the Tertiary tectonic process in conjunction with the activation of the Rhine Graben lineament.
- Its internal structure, displayed by the positioning of the silver in the geological time structure, shows intense faulting and mineralization was, as at Wingertsberg, not a uniform process, but multiphased process took place.
- Here too the Bi, Ni, and Sb which can be found so abundantly in the southern and middle Odenwald (Fettel 1978) are missing. Instead it indicates a relatively high percentage of Hg (10%) as well as Cl, both of which are missing at Wingertsberg.

2.3 **Mühltal (Waschenbach); Quarry "In der Geberstadt"** (TK 25 Bl. Nr. 6118 Darmstadt-Ost)

The petrographic situation is similar to that at Nieder Beerbach.Tectonically, the 135° direction dominates with the Ag-Cu-Hg mineralization following.

It appears as a narrow, limited ore shoot in an up to 0.5 m-wide barite-calcite vein containing scarce amounts of quartz. Ore minerals are native silver, argentite, jalpaite (identified by Ramdohr 1978), chalcopyrite, galena and cerargyrite.

Native *silver* is the most common ore mineral, it occurs arborescent in veins up to 3 cm in length and also appears as moss and plate forms. The Hg percentage is almost equal to the Glasberg one.

Most noteworthy is the abundance of *cerargyrite.* They are surrounded by argentite-covered silver crystals in drusy barite; violet-brown ore rocks consist primarily of native silver-argentite-jalpaite-cerargyrite. The interior of cerargyrite is completely fresh, consisting of a wax-like colour and consistency along with an adamantine lustre. It occurs as kidney-shaped crusts or cubic crystals which are connected to box- or step-like aggregates.

It is remarkable that the Bi and Co ores are missing as well as any As-Ni-Pb-Zn compounds.

3 Results of the Occurrence Description

In *tectonic respect,* all these ore mineralizations belong to a system of Hercynian (post-Hercynian, post-Variscan orogeny) (NW-SE) and Erzgebirgic (NE-SW) fissure-veins, which were created in different time phases in connection with the activation of the Rhine Graben lineament.

If one extended these directions, the systems would cross in the region of "Silberberg" near Ober-Ramstadt. In this area silver mining took place in the 16th century over 30 m deep, previously stated by Ramdohr (1923, p. 166).

All these ore mineralizations are younger than the tectonic systems and can be oriented in the Tertiary mineralization cycle although a slight time difference can be noted during the formation period.

In a *paragenetic respect* the ore mineralizations belong to the "Bi-Co-Ni-Ag formation" with their creation in the meso- to epithermal temperature zone of the hydrothermal solutions (Fettel 1978).

A discriminatory observation, however, indicates that the separate paragenesis shows distinct development periods. A Pb-Cu-Zn-Fe-As paragenesis is followed by another which contains Ag-Co-Ni-Cu-As (in the middle and southern Odenwald Bi appears occasionally instead of Ag). The youngest link in this mineralization cycle is a Cu-Fe-Pb-Zn-U paragenesis. Result: the so-called Bi-Co-Ni-Ag-U formation does not actually exist in its proposed complexity.

4 Genetic Considerations

4.1 Tectonic

Each type of paragenesis marks a separate, self-contained mineralization cycle in connection with certain phases of tectonic movement.

This differentiation is especially apparent in Mackenheim, southern Odenwald. The Bi-Co-Ni-Cu-As paragenesis (niccolite was found in 1979/12) occurs in a Hercynian system, whereas the U-Fe-Pb-Zn paragenesis occurs in a younger overlaying Erzgebirgic system (Fettel 1978, p. 316). Both systems belong to different movement phases in connection with the Tertiary activation of the Rhine Graben lineament.

According to the results from field and paragenetic observations, there are at least two tectonic stress phases in the "Kristallin-Odenwald" which originated during the Tertiary Period. An *older phase* produced, in particular, faults in the Rhenic direction and also generated a shear plane in the Hercynian direction (120°–160°). In all probability, this phase provided the possibility for the occurrence of the Ag-Co mineralization in Nieder Ramstadt and Waschenbach, the older Pb-Cu-Zn-Fe mineralization in Nieder Ramstadt, the remaining Bi-Co-Ni mineralizations in the Odenwald and the Pb-Cu-Zn-As-Fe paragenesis in the silicified barite veins. The Ag-Hg-Co mineralization of Nieder Beerbach belongs to this phase despite the Erzgebirgic direction; here, the stress was determined by the internal structure of the rock complex in this direction.

A *younger phase* brought new activation of the Hercynian direction, particularly replacing fissure vein systems, preferably fissures of the Erzgebirgic direction.

The U-Fe-Cu mineralization in Nieder Ramstadt and the U-Fe-Pb-Zn paragenesis in Mackenheim could originate in this phase, while the occurrence of Ag-Hg-Co in Nieder Beerbach was brecciated and the mineral content was altered under the influence of solutions.

This picture is in agreement with the tectonics of the Tertiary Period. After the Mesozoic sinking, this area was raised, domed and broken along the Rhine Graben lineament. In the area of this weakened zone, "Rheinische" faults occur with the fractures running parallel to the lineament. At the same time, the area in the eastern part of the lineament (present-day Odenwald) was moved somewhat to the north, shear planes in a Hercynian direction evolved and old faults in this direction were newly activated. This northward movement was accompanied by a slight rotation of the eastern area further east. This again led to more movements of the Hercynian shear plane and to fractures parallel to the bc-axis of the arch along with development of veins and fissures in an Erzgebirgic direction.

4.2 On the Inception of the Hydrothermal Solutions and Their Ore Contents

The paragenesis leaves no question as to the mainly "ascendant-hydrothermal"development of the mineralization and Fazakas (1976) considered this as a probable theory for the development of the silicified barite veins and their Pb-Cu-Zn-As paragenesis. Because they are a part of the Bi-Co-Ni-Ag formation (Fettel 1978), it would not be wrong to apply the development to the entire system. The hydrothermal solutions should be connected, according to Fazakas (1976), with the Tertiary volcanism, although he points out the necessity of the material supply and has in mind the regeneration of foreign melting and assimilation of sialic material.

This assimilation can also occur through lateral secretion, which could explain the genetically hard to believe appearance of Cr and Mo, which have led to the development of vauquelinite and heyite in the silicified veins of Reichenbach. Another possible material supply from original basaltic magmas' final solution would be conceivable through mixing with water which came from the earth's surface and penetrated into deeper crust zones. The possibility of such deep-reaching circulation of surface water stems from the fact that the entire Odenwald complex was sunk far below the present level during the Mesozoic era (Wagner and Storzer 1975, p. 83).

The surface waters then became hydrothermal through increased temperatures and with their lateral secretion enriched through mineralizers.

With this concept "hydrothermal"takes on another meaning. Ramdohr pointed out this change of meaning when he said that a hydrothermal temperature increase was also possible in many other ways, without the "deus ex machina" of a granite (1975, p. 240).

This problem deserves further investigation but regardless of what may be said, the solutions which developed the "Bi-Co-Ni-Ag formation" in the Kristallin-Odenwald, were ascendant and "hydrothermal" in their broadest sense.

This theory is questionable, however, when we consider the origin of the uranium, because it is missing in the mineralization of this formation.

This means that the ascendant-"hydrothermal" solutions were also uranium-free. This eliminates the possibility that the ascendant-"hydrothermal" solutions transported uranium through lateral secretionary extraction of plutonic rock.

This statement leads to the conclusion that a descendant origin of uranium would be very likely.

The process necessary for developing the uranium paragenesis could have been the following: the ascendant-hydrothermal solutions had already released their mineral contents in the primary parageneses of Bi-Co-Ni-Ag and Pb-Cu-Zn-As formations.They continued to circulate as relatively ore-free and cooled-down springs. In zones close to the surface they mix with vadose waters and transport their ore contents, coming out of sediments, in particular the quite migrant uranium, into newly formed fissure systems or in recently fractured older mineralized fissures.

During this process, present mineral parageneses can be descendantly transformed (e.g., apparent cementation by Cu- and Pb ores). Also new epithermal parageneses can originate in older mineralizations; for instance, the vanadium which is always present in the vadose waters creates a rich formation of descloizite, mottramite and vanadinite (Reichenbach); the presence of Cl causes the formation of cerargyrite in Waschenbach and Nieder Beerbach, silver is changed to argentite and minute Hg-percentages will be separated.

This theory requires at least two prerequisites. There had to be sediments at the time of the circulation of juvenile waters and they had to contain the required materials plus uranium.

The first prerequisite did occur. At the beginning of the described process, during the tectonic activity in the Tertiary era, the sediment covering the Kristallin-Odenwald was still present, it was dispersed in younger phases of geological history.

Further prerequisites, adequate substance of the sediments, are probable. Primarily the sediments of the upper Permian come in question. They are locally ± formed sediments of the bedrock. They were the closest to the circulation niveau and were, as the lowest layer of the sediment packages, spared the longest from erosion, thereby allowing for a more complete leaching process. Additionally they contained most probably uranium and other element substances of the covering Permian and Mesozoic layers because of their geochemical environment.

The upper Permian sediments formed from the erosion of the Varistic mountains, in other words from a complex of metamorphites and magmatic intrusive rocks, in which generally ± high percentages of uranium are contained (Dybek 1962, p. 44). During the erosion and sedimentation of this complex, a concentration of the original values of "geochemical falls" and in synclines took place in which the Upper Permian sediments were spared the pre-Permian erosion.

The prerequisites for a descendant supply of uranium and other percentages were also given. The circulating juvenile waters could now deposit their uranium in suitable zones in the newly formed fissure systems.

Such zones in the Odenwald were metamorphites (Nieder Ramstadt, Mackenheim) and Ni, respectively Fe^{2+} containing minerals either in older paragenesis or as associated solutions. In Nieder Ramstadt in particular were catalysts for pitchblende and coffinite chalcopyrite and nickel ore; in Mackenheim the catalysts were pyrite and marcasite; further iron percentages were separated out as hematite along with pitchblende and coffinite which led to the red colouring of the chalcedony in Nieder Ramstadt, and to the formation of a hematite niveau in Mackenheim.

For all practical purposes one could draw the following conclusions:

- If there are any other uranium mineralizations they would be found primarily in the Erzgebirgic-trending fissure system of metamorphic rocks.
- Erzgebirgic-directed anomaly zones make it possible that uranium mineralizations could occur under erosion residue. However, without larger deposits, the danger of misinterpretation exists. It is possible that the uranium percentage of the plutonic

rocks are locally enriched through weathering; such secondary mineralizations can also produce higher percentages than are actually present.

- The uranium mineralization of the Odenwald is connected to the upper parts of the earth surface and should not occur at any greater depths since the mineralization, at the time of its formation, was already "epithermal" – close to the surface, and these surfaces had already been significantly eroded since the formation of the mineralization in the Tertiary period.
- Ascendant or descendant-formed silver mineralization might appear anywhere in the Odenwald, but seems to prefer basic rocks.

Acknowledgment. For the translation I am much obliged to Mr. and Mrs. De Mille, Plankstadt.

References

Dybek J (1962) Zur Geochemie und Lagerstättenkunde des Urans. Clausthaler Hefte zur Lagerstättenkunde und Geochemie der mineralischen Rohstoffe. Borchert (ed), Berlin, H 1, 163 S

Fazakas HJ (1976) Geochemisch-lagerstättenkundliche Untersuchungen an Schwerspatvorkommen des südwestdeutschen Grund- und Deckgebirges. Unpubl Diss, München, 75 pp

Fettel M (1978) Über die Wismut-Kobalt-Nickel-Silber-Uran-Formation im Kristallinen Odenwald. Aufschluss 29:307–320

Klockmann F (1978) Lehrbuch der Mineralogie, 16 edn. (Überarbeitet und erweitert von Paul Ramdohr und Hugo Strunz). Enke, Stuttgart, 876 S

Meisl S (1975) Uranmineralisationen und begleitende Erzparagenesen im Odenwald. Aufschluss, Sonderbd 27:245–248

Nickel E, Fettel M (1979) Odenwald. Samml Geol Führer, vol 65, 202 S

Ramdohr P (1923) Der Silberkobalterzgang mit Kupfererzen von Nieder Ramstadt bei Darmstadt. Notizbl Ver Erdkd, 5 F, vol 6, pp 164–192

Ramdohr P (1975) Der Silberkobalterzgang mit Kupfererzen vom Wingertsberg bei Nieder Ramstadt im Odenwald. Aufschluss, Sonderbd 27:237–243

Strecker G (1971) Die Uranmineralisation am Wingertsberg bei Nieder Ramstadt. Notizbl Hess Landesamt Bodenforsch 99:286–296

Wagner G, Storzer D (1975) Spaltspuren und ihre Bedeutung für die thermische Geschichte des Odenwaldes. Aufschluss, Sonderbd 27:79–85

Late Variscan and Early Alpine Mineralization in the Eastern Alps

J.G. HADITSCH [1] and H. MOSTLER [2]

Abstract

In the space of time between the Upper Westphalian and the Middle Triassic events developed, which hitherto have been associated with the post-Variscan (i.e. Early Alpine) cycle. Contrary to this view, in the southern part of the Eastern Alps (= Southern Alps) the above-mentioned events, controlled by tectonic and magmatic activities, must be assigned to the Late Variscan cycle. In the east of the northern part of the Eastern Alps, the Late Variscan cycle came to an end with the Saalic phase, i.e. in the Lower Permian, and in the west not till the end of the Lower Triassic. Accordingly, the Early Alpine cycle begins at different times, caused by various rift activities.

In connection with the first Early Alpine phase the Late Variscan events lead to five minerogenetic stages, namely:

- Cu-Au-Pb-Zn-U-Ba-F mineralizations related to a subsequent acidic volcanism,
- ore mineralizations in association with acidic intrusions,
- exogenic-sedimentary mineralizations, controlled by the destruction and erosion of the above-mentioned magmatites,
- mineralizations bound to saline depositions and basic magmatism,
- karst ore deposits.

1 Sequence of Geodynamic Events During the Late Variscan and Early Alpine Times

The climax of the Variscan Orogeny in the Eastern Alps is still reached within the Westphalian stage. Subsequent events have been hitherto assigned to the post-Variscan phase as, for example, the sediments deposited within inner and outer basins, including the corresponding volcanites. Contrary to this view, however, it has been recently shown that in the time between the Upper Westphalian and the expiring Lower Permian a highly significant metallogenetic phase, controlled by tectonic and magmatic processes, was in the course of developing. This is unequivocally associated with a Late Variscan cycle.

Initial Late Variscan events within the northern Eastern Alps are marked by the development of inner basins (intermontane basins) during the Upper Westphalian. Conglomerates, sandstones and siltstones with relatively thin coal layers (Grey Molasse) settled into these basins and can be seen as belonging not to the Early Molasse stage, but rather to the Main Molasse stage (after Lützner et al. 1977). The Grey Molasse is totally unimportant in respect to metal enrichments.

1 Mariatroster Straße 193, 8043 Graz, Austria

2 Geologisches Institut der Universität, Universitätsstraße 4/II, 6020 Innsbruck, Austria

Table 1. Correlative chart of the Late Variscan and Early Alpine mineralization in the Eastern Alps

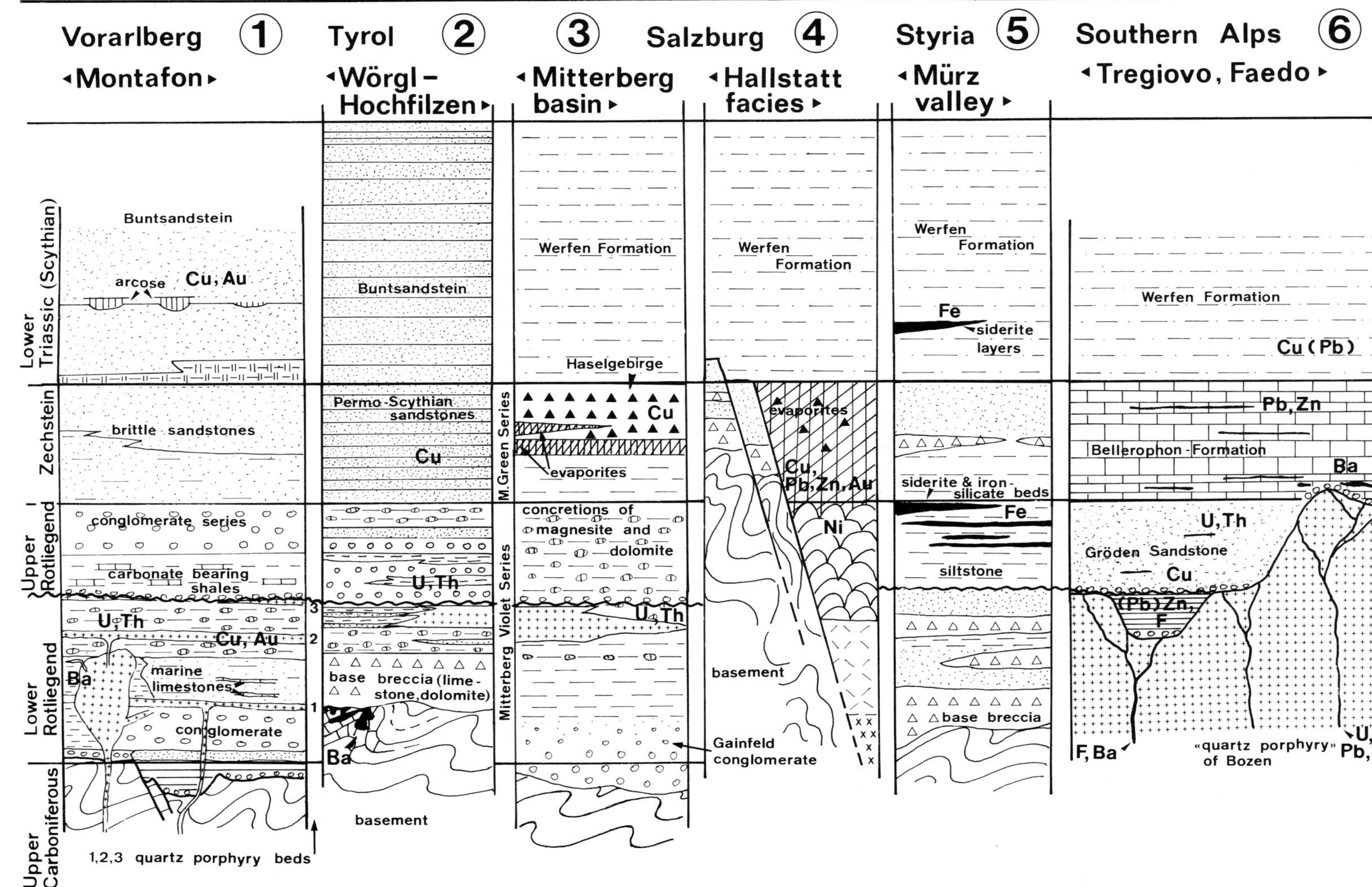

The formation of Early Molasse with sediments analogous to those of the inner basins, except for an interlayering with marine limestones (Auernig-rhythm), occurs within the outer basins of the Carnian Alps (i.e., in the Auernig layers), which lie unconformably above the Flysch deposits. Although the heavy mineral spectra with these sediments typify an erosion product of acidic intrusives, potential ore-bearers and spenders, no significant metal concentrations have been thus far identified here. This also holds true for the Lower Permian Rattendorf Formation, which gives evidence of the initial conditions for the development of a tableland stage ("Tafelstadium"). Coeval with the deposition of the Rattendorf Formation a subsequent primarily acidic volcanism, mineralogically without any effect on the Rattendorf sediments, occurred in the north and west. This first becomes evident with the working of the quartz porphyries during the course of the Gröden Transgression. This event, marking the final stages of the Saalic orogeny, introduced the sedimentation of the Red Molasse (i.e., red sandstones and siltstones along with interlayered conglomerates). This created for the first time the necessary conditions for the formation of interesting metal concentrations within the outer Molasse zone.

The formation of intermontane basins in the southern Eastern Alps west of the outer Molasse (still during the time of magmatic processes), led to the local development of lacustrine 'sapropelitic' sediments with significant metal enrichments (the Tregiovo Formation, Collio Formation and their equivalents; see Table 1, No. 6). Their relationship to the ore mineralization of the quartz porphyries can be conclusively shown.

The previously mentioned inner basins were filled during the time of the diminishing subsequent volcanism (Gröden sandstone), and also proliferate metallizations.

During the Upper Permian a sudden change in the sedimentation occurred, leading to the establishment of a carbonate platform (Bellerophon Formation). The shore areas of the platform developed as evaporites (Sabkha facies), which interfingered with shallow water limestones (of algae limestone facies). These had nothing to do with saline deposition in graben faults of a beginning rift system (as e.g., Dietrich 1976, assured). They are related much more to an already mature tableland stage, which comprised both the outer as well as the inner zones.

The Bellerophon Formation is especially mineralized at the point of interfingering, this being influenced by the continuous erosion of the quartz porphyries. There is direct proof of this in the area of Faedo (Table 1, No. 6), where a transgression of the Bellerophon Formation over a raised block of a quartz porphyry is seen (Mostler 1965).

The development of the tableland stage in the southern part of the Eastern Alps into the metal-poor Werfen Formation continued through the mid-Anisian and partly up to the Lower Ladinian. This phase was ended by the breaking up of the platform, dramatically signalling the beginning of the Alpine Geosyncline in the southern region of the Eastern Alps.

In connection with the dismantling of the platform, land emerged at several localities, leading to the formation of a paleokarst. This was subsequently filled with barite-rich sediments. Basic magmas (of the initial magmatism) penetrated upwards along the borders of several blocks. Only in this stage did it come to the development of narrow basins with pelagic sedimentation. These are directly connected with the Tethys ocean.

Lead-zinc ore deposits, aligned in a pearl-necklace configuration, formed along both sides of a thus created E-W striking graben system (Bleiberg, Gorno, Salafossa etc.). The deposits generally occur along antithetic fault systems. The above-mentioned magmatic and ore deposit-generating events are to be correlated with the Early Alpine geosyncline stage.

After the deposition of Grey Molasse, the formation of Red Molasse occurred in the northern region of the Eastern Alps, also manifested by several beds of subsequent volcanites in the western section. The first ore mineralizations occurred in connection with this volcanism. Also the metallizations occurring in the higher Red Molasse sections were controlled by erosion products of the subsequent volcanism. This can be established even into the short lasting tableland stage (in the Scythian) of this area. Contrary to this, the mineralization in the eastern section, although occurring during the post-Saalic at the earliest, cannot be brought in direct connection with the subsequent volcanism, although this still remains a possibility. This is valid above all for the iron occurrences within the Präbichl and Werfen Formations.

The most significant event is the abrupt alteration in the chemistry of the magmatic phenomena, especially the striking change from acidic subsequent to basic initial magmatism. Although all of the preceding ore mineralizations have been brought in connection with the subsequent cycle, at least a fraction of the post-Saalic mineralizations may have been caused by the basic volcanism, which is interpreted as the early-initial volcanism of the Alpine cycle.

The development of the 'Alpine Geosyncline', having the features of a graben system, immediately following the diminishing of the Saalic phase, began sooner in the northern part of the Eastern Alps than in the southern part.

An Upper Permian saline, generally restricted to an E-W structure, is remarkable in view of its association with basic magmatites (ophiolites in Haselgebirge). The predominance of pillow lavas with characteristic ocean floor chemistry (Kirchner 1977) provides proof that a stripe of ocean crust already existed at this time (Table 1, No. 4). This again verifies the existence of the oldest rift system, approximately corresponding to the Red Sea stage. The mineralizations occurring in the salines can be roughly compared to the brines of the Red Sea. The above occurring Werfen Formation indicates the shortness of the rift. The magmatic activity, however, was not totally interrupted. This is shown by the basic volcanites occasionally occurring within the Werfen Formation and in the Reichenhall Formation which follows.

This relatively ore-barren layer sequence harbours significant mineralizations only in connection with the volcanites.

Contrary to the situation in the western section, a tableland stage never actually developed in the eastern section, although the Werfen and Reichenhall Formations were not affected by any stronger tectogenetic processes.

As a result of powerful tectonic activity during the Middle to Upper Anisian, there occurred an intensive facies differentiation along with a simultaneous diminishing volcanism. These tectonic-magmatic activities represent a further attempt at rifting (Bechstädt et al. 1978). An extensive ore mineralization occurred in connection with this, marking the termination of the relatively long lasting Early Alpine geosyncline stage in the north. This can be divided into two phases as follows:

1. phase in the Upper Permian,
2. phase in the lower Middle Triassic.

2 Minerogenetic Sequence

Corresponding to that which was mentioned in the beginning, one can divide the Late Variscan-Early Alpine time (Upper Carboniferous-Lower Ladinian) into five different minerogenetic stages:

1. Mineralizations in connection with the acidic subsequent volcanism (with Cu, Au, Pb, Zn, U, Ba, F).
2. Ore mineralizations bound on subsequent acidic intrusions.
3. Mineral occurrences originating as a result of the destruction and erosion of the above-named magmatites.
4. Mineralizations existing in relationship to the saline depositions and basic magmatites.
5. Mineralizations existing in connection with the upheaval and karstification of carbonate rocks.

The subsequent volcanism occurred principally as ignimbrites with a rhyolitic composition. On the other hand, lavas with a rhyodacitic and quartz latitic composition occur much more seldom. These rock types, traditionally referred as "quartz porphyries", are more or less broadly distributed in all tectonic stockworks.

Disseminated copper ores in ignimbrite are widely distributed in the west (Montafon, Vorarlberg; see Table 1, No. 1), and are noted for their significant gold content (Angerer et al. 1976; Haditsch et al. 1978). For the first time, a connection could also be made at this locality between a rhyolitic subvolcano and a barium mineralization (Haditsch et al. 1979).

It has been suggested that the barites in the Semmering area originated through lateral secretion, derived from celsian feldspars, without, however, their source being fully explained. The relatively high barium content of the Bozen quartz porphyries (60 samples with an average of 470 ppm, according to an unpublished communication from M. Klau, Hannover) suggests an origin of the barium from feldspars of the quartz porphyries.

In the southern Alpine area, contrary to what we stated previously, there is nevertheless a predominance of Pb-Zn-F mineralizations over Cu, Ba and U mineralizations (di Colbertaldo 1965; Giussani and Leonardelli 1966; Dessau and Duchi 1970; Brusca et al. 1972; Kurat et al. 1974).

The ore mineralizations bound on the subsequent acidic intrusions show not only Cu-Zn-Pb mineralizations, fluorspar and barite, but also an As-characterized hot prephase (Morteani 1966). This is especially true for the mineralizations in the Cima d'Asta pluton. There are also similar ore mineralizations in the granites of Brixen, Iffinger and Kreuzberg.

The predominantly Lower Permian granites (Central Gneiss) of the Tauern assimilated mineralized roof-rock during the course of their formation. This also resulted in the mobilization of W, Mo and Au, etc., from metabasites. In certain cases, as for example at the Alpeiner Scharte (Tyrol), Fe-Cu-Zn mineralizations probably occurred intra-Permian, approximately parallel to the roof. In contrast to this, the Mo metallization remained restricted along steeply inclined veinlets.

Ores which can be derived from the erosional working of the Permian magmatites occur within the 'sapropelitic' sequence of the Tregiovo Formation (Mostler 1966;

see Table 1, No. 6) and also within their equivalents (Brondi et al. 1973). They are also found as detritals in Scythian arkose (Haditsch et al. 1978; Table 1, No. 1). A part of these exogenic sedimentary ores hereupon went into solution and was precipitated as stratiform mm-rhythmites, consisting especially of sphalerite.

Some of the ore mineralizations in the Gröden sandstone (e.g., those of the Oboinig-graben, Friedrich 1968) have still not been explained in respect to the origin of the mineralizing solutions, despite the fact of their relative proximity to quartz porphyries. It can be said in respect to the uranium mineralizations occurring in these sandstones (Mittempergher 1966, 1970) and their metamorphic equivalents (Petrascheck et al. 1976) that the origin of the uranium for that mineralization is clear, the source area consisting exclusively of quartz porphyries. However, the question regarding the origin of the uranium for those mineralizations is much more difficult to answer when metamorphic basement rocks also occur within the source area. The formation of sedimentary uranium deposits in the Eastern Alps occurs quite characteristically only after the termination of the subsequent magmatism, i.e., post-Saalic. In every case the uranium mineralizations (Table 1, No. 2) can be correlated with plant-bearing horizons (Schulz and Lukas 1970).

A decisive change in the mineralization was caused by the Upper Permian taphrogenesis (Table 1, No. 4). Initial basic volcanism in connection with evaporites led to Fe, Zn, (Au) and Pb mineralizations similar to those of the pillow lavas in the northern Apennines (Zuffardi 1977). It appears that the mineralizations can be explained partly by primary hydrothermal events (Haditsch 1968) and partly by a partial leaching of the pillow lavas. Although the parageneses resemble those of the Red Sea types, there is nevertheless an absence of the actual ore muds. This ore paragenesis occurs within the sulphate evaporites (Table 1, No. 4).

Siderite ore deposits of the Hirschwang type (Baumgartner 1976) and siderite-iron-silicate occurrences of the Gollrad-Brandhof type (Weber 1977) can be related to the hydrothermal feeders of the basic volcanism (Table 1, No. 5).

Owing to the structure of a graben fault system with an opening towards the SE to the Tethys ocean, a second climax in the supply of basic volcanites occurred. This culminated in the Lower Ladinian and ended with the Lower Carnian. It has often been maintained (Hegemann 1957) that a relationship exists between this volcanism and with the widely distributed Pb-Zn mineralizations, ranging from the Anisian to the Upper Carnian. The striking fact of the relative poverty of Cu and Fe in Ladinian and Carnian mineralizations (Brigo et al. 1977) can be explained as the result of either the presence of a depth-controlled mineralization (with the consequently resulting zonation features of mineralized stockworks), or as the result of a filtration effect of the migrating ore-bearing solutions (e.g., the fixation of copper in sandstone).

The hydrotherms appearing on the basin floor are, of course, controlled by the facies (Cerny 1977).

The copper content of the Pb-Zn mineralizations in the Northern Tyrolean Calcareous Alps (Reichenhall and Reifling Formations) could have been caused by a weak filtration effect, due to a higher content of silicic clastics, or by an inheriting of the copper ore deposits occurring in the lower stockwork (i.e., in the Graywacke Zone). The occurrence of fahlore in the Middle Triassic Pb-Zn occurrences near Schwaz would provide evidence for this last implication.

Local uplift and the related karstification occurred during both the Late Variscan and Early Alpine times. The oldest recognized karst feature is observed at the Upper Carboniferous/Permian boundary. The Devonian carbonates of the Kitzbühel Alps were thereby subject to an intensive karstification. Some of the karst features were subsequently filled with barite (Emmanuilidis and Mostler 1968). The connection assumed hitherto with the fahlore-barite mineralizations of Kogel (Brixlegg) is therefore not possible because of time factors.

Further emergences exist in connection with the destruction of the lower Middle Triassic carbonate platforms. A part of the karstified carbonate platforms served as a metallotect for the ore spending solutions (Bleiberg also belongs here to some extent).

Acknowledgment. This publication was sponsored partly through the Austrian Funds for the Advancement of Scientific Research (Fonds zur Förderung der wissenschaftlichen Forschung; Projekt 2145). English translation: Mag. rer. nat. L.G. Gould (Leoben, Austria).

References

Angerer H, Haditsch JG, Leichtfried W, Mostler H (1976) Disseminierte Kupfererze im Perm des Montafon (Vorarlberg). Geol Palaeontol Mitt Innsbruck 6 (7/8):1–57

Baumgartner W (1976) Zur Genese der Erzlagerstätten der östlichen Grauwackenzone und der Kalkalpenbasis (Transgressionsserie) zwischen Hirschwang/Rax und Neuberg/Mürz. BHM 121: 51–54

Bechstädt Th, Brandner R, Mostler H, Schmidt K (1978) Middle Triassic block faulting in the Eastern and Southern Alps. Alps, Apennines, Hellenides:98–103

Brigo L, Kostelka L, Omenetto P, Schneider H-J, Schroll E, Schulz O, Strucl I (1977) Comparative reflections on four Alpine Pb-Zn deposits. In: Klemm DD, Schneider H-J (eds) Time- and Strata-Bound Ore Deposits. Springer, Berlin Heidelberg New York, pp 273–293

Brondi A, Carrara C, Polizzano C (1973) Uranium and heavy metals in Permian sandstones near Bolzano (Northern Italy). In: Amstutz GC, Bernard AJ (eds) Ores in sediments, IUGS, Ser A 3: 65–77

Brusca C, Dessau G, Jensen ML, Perna G (1972) The deposits of argentiferous galena within the Bellerophon Formation (Upper Permian) of the Southern Alps. 2nd Int Symp Mineral Dep Alps, pp 159–179

Cerny I (1977) Zur Fazies- und Blei/Zink-Verteilung im „Anis" der Karawanken. Car II 167/87: 59–78

Colbertaldo D di (1965) Il giacimento a barite e fluorite di Zaccon in Valsugana (Trento). Econ Trentino CCIA Trento 5-6:119–133

Dessau G, Duchi G (1970) Mineralizziazioni a solfuri nelle vulcaniti del Montagiù (Trento). Econ Trentina 2-3, Ind Min Nel Trentin-Alto Adige III:115–123

Dietrich V (1976) Plattentektonik in den Ostalpen. Eine Arbeitshypothese. Geotek Forsch 50: 1–84

Emmanuilidis G, Mostler H (1968) Zur Geologie des Kitzbüheler Horns und seiner Umgebung mit einem Beitrag über die Barytvererzung des Spielberg-Dolomites. Festbd Geol Inst Innsbruck, pp 547–569

Friedrich OM (1968) Die Vererzung der Ostalpen, gesehen als Glied des Gebirgsbaues. Arch Lagerstaettenforsch Ostalp 8:1–136

Giussani A, Leonardelli A (1966) Le mineralizzazioni a fluorite della zona tra Cavalese ed il di Lavaze (Trento). Simp Int sui Giacimenti Minerari delle Alpi, Trento-Mendola, 18 pp

Haditsch JG (1968) Bemerkungen zu einigen Mineralien (Devillin, Bleiglanz, Magnesit) aus der Gips-Anhydrit-Lagerstätte Wienern am Grundlsee, Steiermark. Arch Lagerstaettenforsch Ostalp 7:54–76

Haditsch JG, Leichtfried W, Mostler H (1978) Intraskythische, exogen (mechanisch)-sedimentäre Cu-Vererzung im Montafon (Vorarlberg). Geol Palaeontol Mitt Innsbruck 8 (Heissel-FS):183–207

Haditsch JG, Leichtfried W, Mostler H (1979) Über ein stratiformes Schwerspatvorkommen in unterpermischen Schichten des Montafons (Vorarlberg). Geol Palaeontol Mitt Innsbruck 7 (6): 1–14

Hegemann F (1957) Geochemische Untersuchungen zur Entstehung der alpinen Blei-Zink-Erzlagerstätten in triassischen Karbonatgesteinen. BHM 120:233–243

Kirchner E (1977) Vorläufige Mitteilung über eine Pumpellyit-führende Kissenlava vom Grundlsee. Geol Tiefbau Ostalp, Jahresber 1976

Kurat G, Niedermayr G, Korkisch J, Seemann R (1974) Zur Geochemie der postvariszischen Basis-Serien im westlichen Drauzug, Kärnten-Osttirol. Car II 164/84:87–98

Lützner H, Grumbt E, Ellenberg J, Falk F (1977) Sedimentologische Kriterien variszischer Molassen Mitteleuropas. Veroeff Zentralinst Phys Erde, Potsdam 44

Mittempergher M (1966) Le mineralizzazioni ad uranio delle Alpi Italiane. Simp Intern sui Giacimenti Minerari delle Alpi 2:319–333

Mittempergher M (1970) Characteristics of uranium ore genesis in the Permian and Lower Triassic of Italian Alps. In: Uran Explor Geol. IAEA, Wien, Panel Proc Ser, pp 253–264

Morteani G (1966) Petrographisch-geologische und lagerstättenkundliche Untersuchungen im Cima d'Asta-Kristallin. Mem Mus Trid Sci Nat XVI/II:1–136

Mostler H (1965) Bemerkungen zur Genese der sedimentären Blei-Zinkvererzung im südalpinen Perm. Arch Lagerstaettenforsch Ostalp 3:55–70

Mostler H (1966) Sedimentäre Blei-Zink-Vererzung in den mittelpermischen „Schichten von Tregiovo" (Nonsberg, Nord-Italien). Mineral Deposita 2:89–103

Petrascheck WE, Erkan E, Siegl W (1976) Type of uranium deposits in the Austrian Alps. Geology, mining and extraction processing of uranium. IMT London:71–75

Schulz O, Lukas W (1970) Eine Uranerzlagerstätte in permotriadischen Sedimenten Tirols. TMPM 14:213–231

Weber L (1977) Alter und Genese der Eisenspat-Eisensilikatvererzung im Westteil der Gollrader Bucht (Stmk.) BHM 122,2a (Petrascheck-FS):78–80

Zuffardi P (1977) Ore/mineral deposits related to the Mesozoic ophiolites in Italy. In: Klemm DD, Schneider H-J (eds) Time- and strata-bound ore deposits. Springer, Berlin Heidelberg New York, pp 314–323

Radioactive Disequilibrium Studies in Uranium Occurrences of the Odenwald, West Germany

S. MEISL and W. PÖSCHL[1]

Abstract

The radioactive equilibrium of 214 uranium ore samples from two quarries in the Odenwald was measured by means of a gamma ray spectrometer. The uranium ore occurs as pitchblende and coffinite in veins showing Ag-Co-Ni-Bi mineralization in the Nieder-Ramstadt quarry and Bi-Co-As paragenesis in the quarry of Mackenheim. Both occurrences show different radioactive disequilibrium patterns. Whereas the bulk of the samples from Nieder-Ramstadt is almost in a state of radioactive equilibrium, the majority of samples from Mackenheim show high deficiency of most of the radioactive decay products. As by means of radioactive disequilibrium, variations of the ore composition within the last 800,000 years can be determined, the mineralization of the Nieder-Ramstadt occurrence must have happened before this time. The disequilibrium of the Mackenheim samples, however, indicates a recent genesis of the uranium ore, at least in the present level of the outcrop, although a remobilization in the last 800,000 years, of probably older uranium mineralizations within the veins, cannot be totally excluded.

1 Introduction

Like the Black Forest, the Erzgebirge and the Massif Central, the Odenwald is part of the Central European Variscan belt. Whereas uranium mineralizations from the first three regions have long been known, no uranium occurrences were reported from the Odenwald area, until Ramdohr (1961) described pitchblende and coffinite in vein material sampled in the quarry at the Wingertsberg near Nieder-Ramstadt. Later short investigations in this quarry by Bültemann and Wutzler (1968) and Meisl in 1968 were followed by a more detailed study by Strecker (1971).

In 1973 Meisl found pitchblende in a quarry near Mackenheim in the southern Odenwald (Meisl 1974). Both uranium occurrences were studied later in detail by extensive exploration work, carried out by Saarberg-Interplan Corporation and by the Geological Survey of Hessen in a joint venture.

As a third uranium occurrence with very minor pitchblende-mineralization in the Odenwald area, Schriesheim must be mentioned (Levin 1975).

Several hundred samples were taken from the uranium deposits of Nieder-Ramstadt and Mackenheim. Radiometric and colorimetric analyses of uranium showed radioactive disequilibrium of the bulk of the samples, especially in those from the Mackenheim

1 Hessisches Landesamt für Bodenforschung, Leberberg 9, 6200 Wiesbaden, FRG

quarry. As radioactive equilibrium patterns proved to be useful for the interpretation of last stages in uranium ore genesis, the samples were analyzed by gamma ray spectrometry.

2 Geology and Mineralogy

2.1 Nieder-Ramstadt

Nieder-Ramstadt is situated in the northwestern part of the crystalline Odenwald. Diorite is the main rock type in the quarry of Wingertsberg, with some relict bodies of amphibolite and hornfels. There are also numerous dykes and lens-shaped bodies of aplites and pegmatites.

Systematic radiometric measurements in the quarry revealed 78 gamma-ray anomalies. Seventeen of these anomalies showed distinct uranium mineralization in vein structures (Fig. 1). The predominant shear joints show a NNE-SSW direction. These joints are accompanied by a faintly developed system of feather joints, in which the uranium ore occurs. The system of joints consists of veinlets with a strong variation in dip and strike.

Uranium mineralization is linked with a complicated vein structure, such as shear fissures, junctions of veins, and small apophyses. The concentration of ore minerals in the veins is discontinuous and often interrupted. The mineralization is controlled not only by tectonics but also by the chemistry of the wall rock. In most cases, uranium minerals occur in ore veins cutting wall rock like quartz-plagioclase-pyroxene-hornfels or aplitic to aplo-granodioritic bodies. If there is any change in wall rock chemistry only gangue minerals remain.

Two associated, still slightly different, mineral associations are distinguishable: one of them consists of pitchblende, coffinite, niccolite, maucherite, bravoite, chalcopyrite, ± bornite, ± tennantite, ± pyrite, whereas minerals of cobalt and secondary uranium minerals are absent. The gangue minerals are chalcedony and calcite, with hematite staining. This mineral paragenesis favours joints of the granodioritic aplite body.

In the second mineral paragenesis, copper minerals hold a leading position: tennantite, bornite, chalcopyrite, pitchblende, ± niccolite, ± sphalerite, ± pyrite. Coffinite and cobalt minerals are lacking. The gangue minerals are again calcite and chalcedony.

The uranium content varies between 0.06% and 4.48% U with an average of 0.94% U. Small size and discontinuous uranium mineralization make an economic exploitation of this vein system impossible.

Both mineral associations, together with the silver/cobalt mineralization described by Ramdohr (1923), belong to the classical hydrothermal Ag-Co-Ni-Bi-U paragenesis.

2.2 Mackenheim

Mackenheim is situated in the southern part of the crystalline Odenwald. In this quarry, biotite-plagioclase paragneiss occurs with lenses of calc-silicate hornfels, irregu-

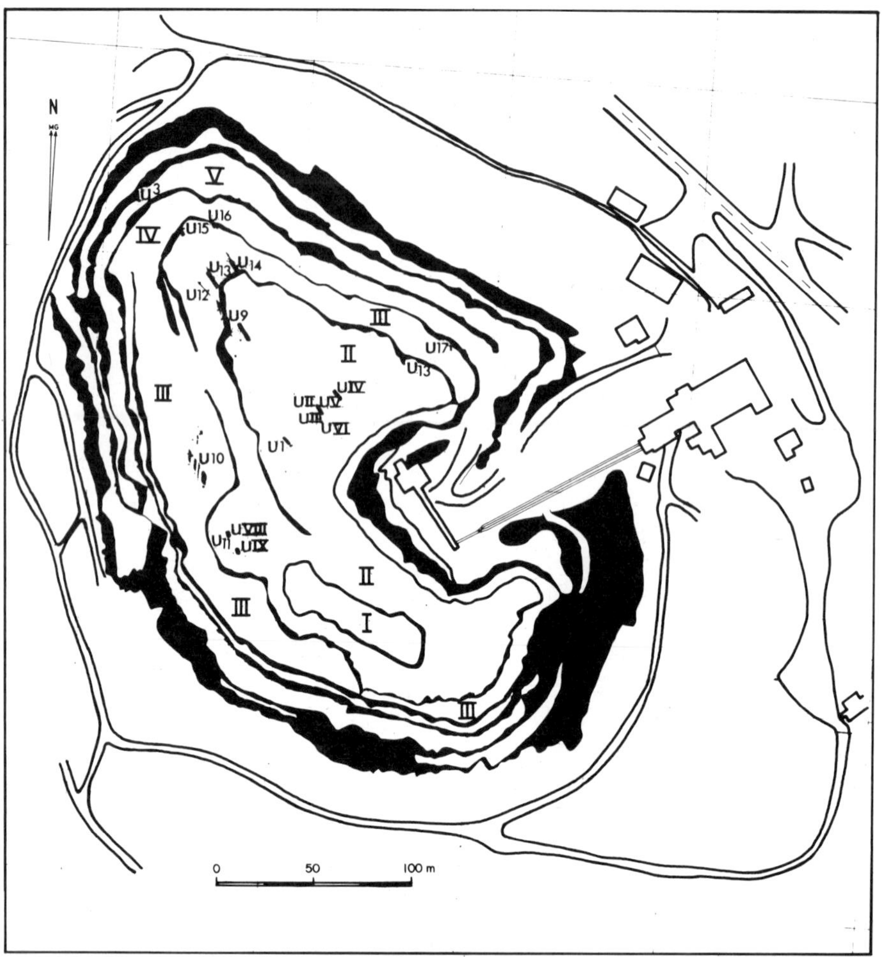

Fig. 1. Sketch map of the Nieder-Ramstadt quarry. *U 1 – U 17,* uranium-bearing veins. *I – V,* working levels

lar-shaped anatectic bodies of dioritic composition and aplitic-pegmatitic dykes. This metamorphic rock series forms a domal anticline dipping NW. The major joints of the quarry show NW-SE trends.

Gamma-ray measurements showed uranium mineralization in five veins (Fig. 2). The veins run parallel to each other and strike NE-SW. They are perpendicular to the b-axis and must be regarded as tension joints. The uranium-bearing ore paragenesis occurs mostly in mylonitic fracture zones. Pitchblende, coffinite, pyrite, arsenopyrite and varying amounts of marcasite, melnikovite-pyrite, sphalerite, galena, chalcopyrite, tennantite, löllingite and safflorite are the ore minerals, accompanied by dark calcite and minor quartz as gangue minerals. In some veins fluorite occurs as well.

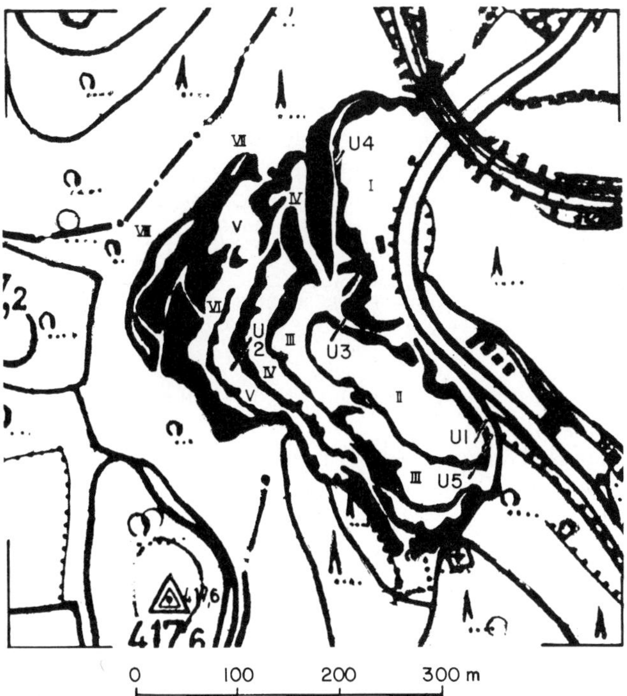

Fig. 2. Sketch map of the Mackenheim quarry. *U 1–U 5,* uranium-bearing veins

These epithermal uranium-bearing veins are superimposed on an older fault system (NE-SW) with higher-temperature pyrite and mesothermal Bi-Co-As mineralization. The gangue minerals in the older veins consist of white calcite, minor quartz and barite.

The average content is 0.45% U, which is less than half of the Nieder-Ramstadt average. As uranium mineralization is very erratic, the deposit is sub-economic at present.

3 Radioactive Disequilibrium

3.1 Analytical Method

All gamma ray energy levels below 620 keV in uranium-bearing vein samples were measured with a planar intrinsic Ge-detector and conventional electronics, following a method described by Szöghy and Kish (1978). The spectra were accumulated in a 2048 channel analyzer and were stored on magnetic tape for further evaluation. Depending on the chemically analyzed U-content, the duration of the runs varied from 15 min, for 4% uranium content, to 2 h for a uranium content of 1000 ppm U in order to obtain statistical errors below 5%.

In nature three radioactive decay series occur: the ^{238}U-series, the ^{235}U-series and the ^{232}Th-series. The last series can be ignored not only because the daughter products of the ^{232}Th-series have too short half-lives but also because the ^{232}Th-content of the samples is very low.

In order to study equilibrium in the radioactive decay series it is sufficient to determine the abundance of some key isotopes, which, by their relative long half-lives, mark a disruption in the isotope series (Rosholt 1958).

Equilibrium in the ^{238}U-series and the ^{235}U-series is established, if, in the 800,000 years since the minerals were formed no uranium or daughter isotopes were lost and no radioactive elements were added to the mineral or host rock. If, on the other hand, gains or losses of radioactive isotopes occurred within the time range represented by ten times the half-life of the longest-lived decay product (^{230}Th), the sample will be in a state of radioactive disequlibrium.

The extent of disequilibrium can be defined if the abundances of the following components are known: ^{234}Th, ^{230}Th, ^{226}Ra, ^{214}Pb, ^{214}Bi (^{238}U-series) and ^{231}Pa, ^{227}Th, ^{223}Ra (^{235}U-series).

^{214}Pb was measured using three different gamma lines to determine the instrumental error. It was found to be below 5%.

Throughout this paper equivalent units are used. All the daughter products in the decay series are expressed as equivalents to the parent nucleides, using the unit percent equivalent. It is defined as the percentage amount of primary parent isotope required to support the amount of daughter products actually present in the sample, always assuming radioactive equilibrium.

3.2 Results and Discussion of Radioactive Disequilibrium

Two hundred and fourteen samples of the Nieder-Ramstadt and Mackenheim quarries were analyzed for radioactive equilibrium. Both localities show different disequilibrium patterns.

Eighty five percent of the Nieder-Ramstadt samples are almost in equilibrium. Figure 3 shows a typical sample of the Nieder-Ramstadt occurrence. With reference to the margins of error, uranium and its daughter products are in radioactive equilibrium. The remaining 15% of the Nieder-Ramstadt samples show a daughter element deficiency in the ^{238}U-decay series, while there is equilibrium in the ^{235}U-decay series (Fig. 4). Disequilibrium starts with the ^{226}Ra daughter element.

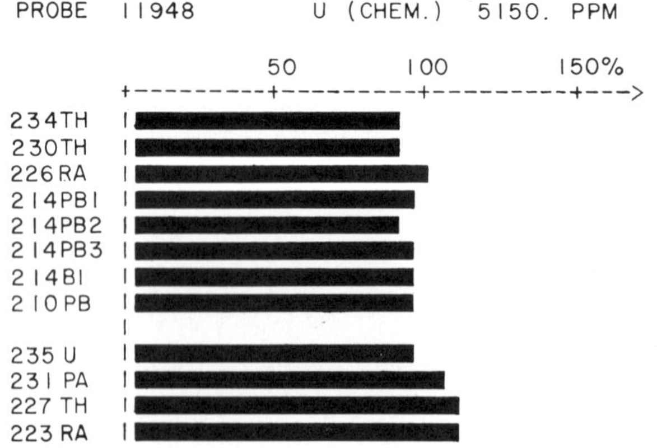

Fig. 3. Typical sample from the Nieder-Ramstadt occurrence, showing radioactive equilibrium

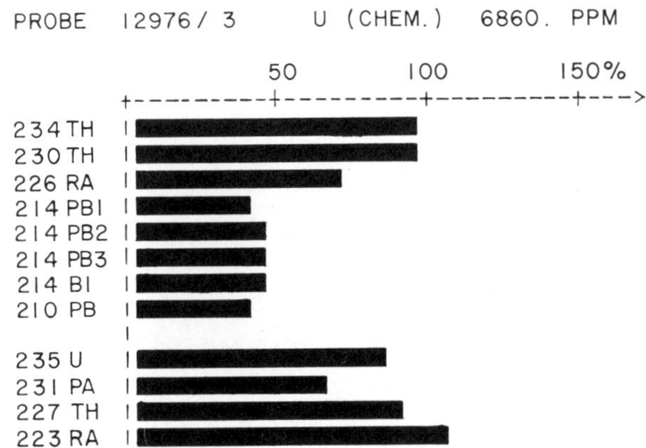

Fig. 4. Typical sample from the small sample group of Nieder-Ramstadt: disequilibrium in the ^{238}U-series

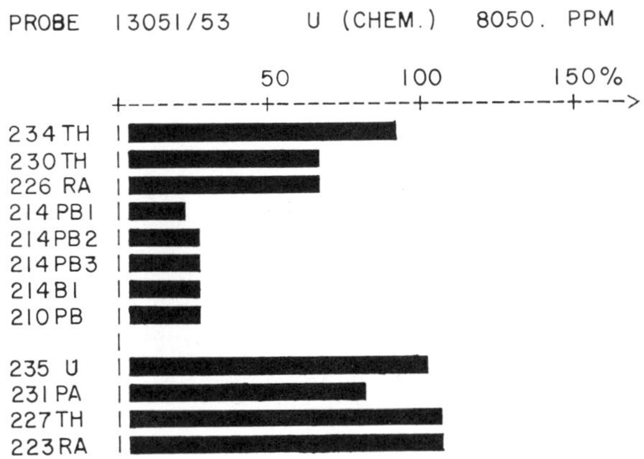

Fig. 5. Typical sample from the Mackenheim quarry (92% group): disequilibrium in the ^{238}U-series

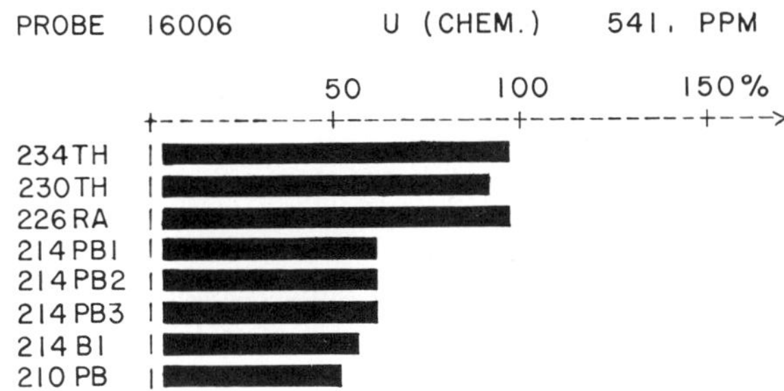

Fig. 6. Sample from altered vein filling of the Mackenheim occurrence: disequilibrium in the radon/lead group (^{238}U-series)

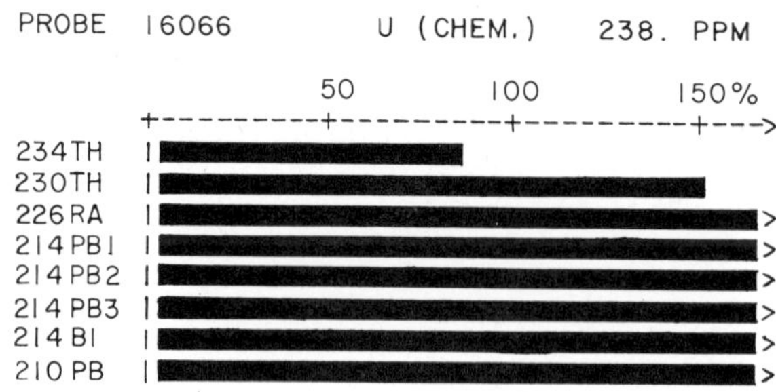

Fig. 7. Vein wall rock sample from the Mackenheim occurrence: daughter product excess

In the Mackenheim quarry 92% of the samples are deficient in daughter isotopes, compared to only 15% of the Nieder-Ramstadt samples (Fig. 5). The extent of the disequilibrium in the radon/lead group is somewhat greater and there exists also equilibrium in the ^{235}U-family. In most samples of this group, disequilibrium starts with ^{230}Th deficiency. The remaining 8% are samples from the host rock and deeply altered vein filling. Specimens from altered vein fillings have disequilibrium patterns similar to those of the 92% group, but the ^{230}Th (and ^{226}Ra) content is somewhat higher (Fig. 6). Host rock samples from the vein wall rock (for example calc-silicate hornfels) are generally low in uranium and show excess of all daughter isotopes (Fig. 7). Isotopes of the ^{235}U-family in such samples were below the detection limit.

Radioactive disequilibrium is a complex phenomenon. It is not easy to find straightforward explanations for various daughter product variations. Disequilibrium in the uranium series can only reveal changes that occurred during the last 800,000 years, ten times the half-life of the longest-lived isotope ^{230}Th in the ^{238}U-series. Events that caused disequilibrium before 800,000 years cannot be detected, for equilibrium has been restored.

Ores with daughter product deficiency can be the result of two primary processes. Uranium may have moved to the place of deposition within 800,000 years, so there has not been enough time to establish radioactive equilibrium.

The other alternative is that there has been preferentially greater leaching of daughter isotopes than of uranium. If the specimen is low in only one long-lived daughter product, then the latter explanation is more probable (Rosholt 1958).

Emanation of radon gas, causing low ^{214}Pb and ^{214}Bi values, is a common source of disequilibrium by weathering. When there is $U > {}^{230}Th \geqslant {}^{226}Ra$, recent deposition of uranium is the likely explanation.

Daughter product excess, found in some wall rock samples of the Mackenheim quarry, may be the result of leaching the uranium from the vein wall. This type of disequilibrium is usually associated with an oxidized environment, typical of radioactive vein outcrops. Through this leaching action, uranium has migrated into the altered vein filling, resulting in disequilibrium patterns showing radon/lead group deficiency (Fig. 6). The slight excess of ^{230}Th indicates a subsequent minor leaching of uranium within these veins (cf. Phair and Levine 1953).

The disequilibrium patterns of the bulk of the Mackenheim samples are not so easily explicable. In line with Rosholt (1958) these samples would indicate a recent deposition of uranium.

The bulk of the samples from Nieder-Ramstadt shows radioactive equilibrium in unaltered material of the veins. The 15% of samples with daughter product deficiency starting with ^{226}Ra, comes from parts of the veins where recent transport of radium (subtraction) or uranium (addition) was possible. A recent addition of ^{238}U without daughter products in these parts of the veins is thought to be geochemically more probable than recent leaching of ^{226}Ra.

4 Conclusion

Pitchblende veins of two quarries from the Odenwald show different radioactive disequilibrium patterns. Most of the samples from Nieder-Ramstadt are close to equilibrium, whereas the bulk of the samples from Mackenheim show a marked daughter product deficiency. This deficiency is explained as the result of recent uranium transport within or into the veins, indicating the young age of the uranium minerals at Mackenheim, at least in the present level of the outcrop.

Daughter product deficiencies in some samples from Nieder-Ramstadt may also be explained with recent differential migration of both uranium and/or radium within the veins.

Uranium deficiency in wall rock samples from Mackenheim indicates the migration of decay products into the wall rock or – more probably – the leaching of uranium from the wall rock by weathering.

Acknowledgment. This study was supported by the „Deutsche Forschungsgemeinschaft".

References

Bültemann H, Wutzler B (1968) Zum Auftreten eines mit Pechblende und Coffinit vererzten Trums im Dioritsteinbruch am Wingert-Berg bei Nieder-Ramstadt südlich Darmstadt. Aufschluss 19: 295–296

Levin P (1975) Über eine gangförmige Vererzung bei Schriesheim im südwestlichen Odenwald. Aufschluss Sonderbd 27:255–262

Meisl S (1974) Neues Pechblende-Vorkommen in Mackenheim/Odenwald. Notizbl Hess Landesamt Bodenforsch Wiesbaden 102:225–228

Phair G, Levine H (1953) Notes on the differential leaching of uranium, radium and lead from pitchblende in H_2SO_4 solutions. Econ Geol 48:358–369

Ramdohr P (1923) Der Silberkobalterzgang mit Kupfererzen von Nieder-Ramstadt bei Darmstadt. Notizbl Ver Erdkd (V) 6:164–192

Ramdohr P (1961) Das Vorkommen von Coffinit in hydrothermalen Uranerzgängen, besonders vom Co-Ni-Bi-Typ. Neues Jahrb Mineral Abh 95:313–324

Rosholt JN (1958) Radioactive disequilibrium studies as an aid in understanding the natural migration of uranium and its decay products. Proceedings of the second United Nations international conference on the peacefull uses of atomic energy, vol 2. UN-Publ, Geneva, pp 230–236

Strecker G (1971) Die Uranmineralisation am Wingertsberg bei Nieder-Ramstadt. Notizbl Hess Landesamt Bodenforsch Wiesbaden 99:286–296

Szöghy JM, Kish L (1978) Determination of radioactive disequilibrium in uranium-bearing rocks. Can J Earth Sci 15:35–44

On the Alpidic Mineralization in Western Central Europe Outside the Alps

H.W. WALTHER [1]

Abstract

Results published on age and genesis of the post-Variscan epigenetic mineral deposits in western Central Europe are rather contradictory. The mineralization is associated with block faulting caused by lateral spreading, usually SW-NE, resulting in the formation of prevailingly NW-SE striking veins. During the development of intraplate tectonics of stable Europe, these conditions were not effective before the Late Mesozoic in the Lower Saxonian Block and eastern Bavaria and during the Tertiary, probably post-Early Miocene in the western and southwestern parts of the Federal Republic of Germany. This structural geological age determination is in agreement with paleomagnetic ones published elsewhere.

Paragenetic and geochemical data could be best explained by precipitation from ascending hydrothermal solutions probably derived from differentiated, deep magmatic bodies. The possibility that this mineralization belongs to a North Atlantic hydrothermal province is discussed.

1 Introduction

It has been well known for a long time that epigenetic ore deposits, mostly veins, were formed in Central Europe in two different metallogenetic epochs at least in some districts. According to present knowledge the first, the Variscan, is characterized in the Federal Republic of Germany mainly by Sn, W, Au and U, with small deposits in the southern part (Moldanubian and Saxothuringian zones) and by Fe and large deposits of Pb-Zn-Ag in the northern part (Rhenohercynian zone and Subvariscan foredeep). The second, the Alpidic, is characterized mainly by Ba, F, Fe-Mn, Cu, Co, Ni and Bi mineralization as well as in some regions by Pb-Zn mineralization, and is well distributed all over the country and beyond its boundaries.

The Freiberg school has carefully investigated the ore deposits of the Erzgebirge, Thuringian Forest, and East Harz districts (Fig. 1) and published the results in a number of excellent papers. Summaries have been given by Werner (1966) and Baumann (1967). Except for some misunderstandings in extrapolating the relationships found to western Central Europe, the results are considered to be essentially correct, with one important exception: The age of the Alpidic (so-called Saxonian) mineralization is considerably younger than early to middle Mesozoic. As many authors have already

1 Bundesanstalt für Geowissenschaften und Rohstoffe, 3000 Hannover, FRG

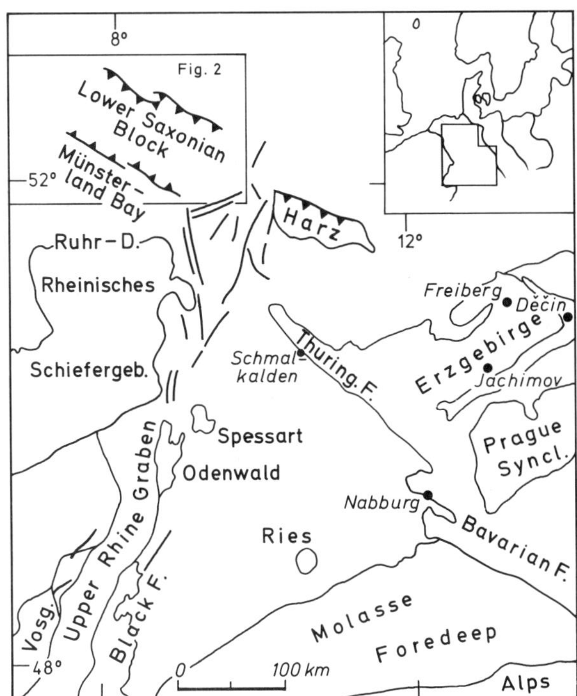

Fig. 1. Location map

pointed out, this mineralization begins during late Mesozoic in NW Germany and during Tertiary in SW Germany. During preparation of the explanatory notes for the Federal Republic part of the "Metallogenic Map of Europe", some further arguments were developed for this age assignment. Before discussing them, a short review of previous papers is given.

2 Previous Papers

In his monograph on German barite deposits, Bärtling (1911) stated that the formation of barite veins reached its maximum during the Tertiary, and in the Central Harz Mountains maybe already during the Late Cretaceous. Barium was thought to have been dissolved from feldspar of the Bunter, and in some places in the Rheinisches Schiefergebirge, where Bunter was not deposited, basalts were discussed as a possible barium source. Cavities for precipitation were created by tectonic stress. v. Engelhardt (1936) also expressed himself in favour of lateral secretion from different types of country rock.

Because the different geological settings of the deposits have not had a significant influence on the mineral paragenesis, Scheiderhöhn (1949) rejected the idea of lateral secretion and concluded that all barite occurrences should be of primary Variscan age and of postmagmatic hydrothermal origin. Later, vein minerals were selectively dissolved at depth by ascending thermal chloride waters and selectively redeposited higher up in the veins. This secondary hydrothermal model assumes that Ba and S are of ascending and O_2 of descending origin. Therefore, barite was deposited close to the surface, an observation repeated by many authors. Because of that, the formation of barite veins was thought to be possible only in an arid climate and not under marine conditions, which would mean Early Permian, Early Triassic, perhaps middle Late Triassic, and middle Tertiary.

According to Murawski (1954), the numerous barite veins in the Spessart Mountains follow the surface and therefore should be of Tertiary age. Ba was brought up by ascending solutions, magmatic influence having probably played an important role.

Gunzert (1961) reviewed the varying regional distribution of barite deposits in Germany and concluded that their appearance depends not only on the presence of barium sources, but on the rock sequence of the subsoil. Only where sulphates and thick carbonates are absent do barite deposits occur, which means (1) in the basement, (2) in the platform cover where Zechstein is not present or present only at the surface, as well as (3) in areas in which the faults have large throw distances.

Werner (1966) attempted a regional synthesis of all the research on barite and fluorite deposits in Central Europe based on detailed work in the Schmalkalden district, Thuringia (Werner 1958). He rejected the secondary hydrothermal model, as well as direct connection with Tertiary basalt volcanism and inferred an Atlantically differentiated simatic magma as subcrustal source. Most of the so-called Saxonian mineralization in Central Europe was thought to be of Late Triassic to Early Jurassic age, and some also of Upper Cretaceous and Tertiary age, according to geophysical dating (see below).

Baumann (1967) summarized the data substantiating two periods of mineralization in the Erzgebirge, discussing geological, petrological, paragenetic, and geochemical factors. He also summarized papers dealing with different methods of geophysical dating: 49 U/Pb determinations by Leutwein (1957) resulted in 18 Permian model ages, which correspond to Variscan mineralization. A smaller number of model ages belonging to the Jurassic, Cretaceous, and Tertiary were interpreted as being from uranium remobilized during different phases of mineralization of the second (Saxonian) metallogenic epoch. U/Pb and Pb/Pb methods were used by Winogradow et al. (1959) with similar results. Rösler and Pilot (1967) carried out 60 K/Ar determinations on samples of hydrothermal paradoxites and sericites from ore veins, their marginal clay selvage, and altered country rock from Saxonia and Thuringia. Post-Variscan model ages, mostly Triassic and Jurassic, were obtained for 29 samples. Paleomagnetic investigations of several hematite-bearing quartz and barite samples from the Freiberg district yielded mainly Tertiary and some Late Paleozoic ages (Baumann and Krs 1967). These authors explain (p. 776): The magnetic poles of samples from the second Freiberg metallogenic epoch point to the area of the Tertiary poles with a range of error on the one hand into the Quaternary, and on the other hand into the Cretaceous. Similar results were obtained earlier for samples from Joachimsthal (Jáchimov) and Tetschen (Děčin) in Czechoslovakia. Thus, the published data practically exclude early and middle Mesozoic ages.

Stadler (1971) described a zonally arranged, ± mesothermal mineralization with indications of telescoping of Late Cretaceous age in the Lower Saxonian Block, which corresponds in space and time with the intrusion of the Bramsche Massif north of Osnabrück and the recently described Massifs of Vlotho and Uchte, west and northwest

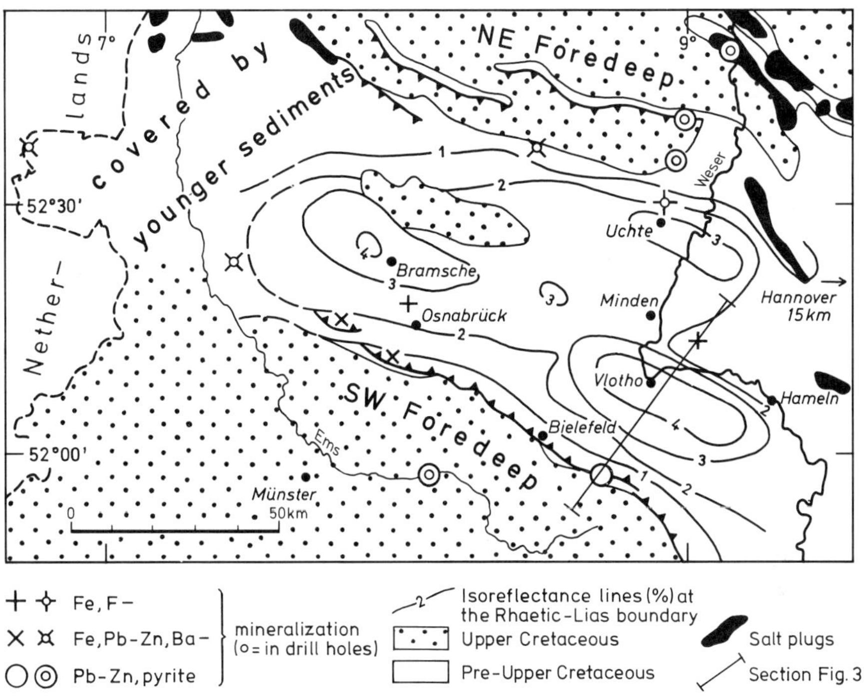

Fig. 2. Coalification and mineralization in the Lower Saxonian Block. (After Stadler 1971; Deutloff et al. 1980)

Table 1. Tectono-magmatic development during the Cretaceous in the Lower Saxonian Block. (After Stadler 1971; Brockamp 1976; Buntebarth and Teichmüller 1979)

Systems and series	Tectonics	Magmatism	Coalification	Minerals	
Tertiary	Weak (Tertiary to Campanian)	Mofettes, following Miocene volcanism			
Cretaceous	Weak				
Maastrichtian + Danian	Upper Campanian sediments not affected by coalification and tectonics				
Campanian	Tangential movements, mostly at the margins		By magmatic heating	PbS, ZnS, FeS_2, $FeCO_3$, PbS, ZnS, $BaSO_4$, $FeCO_3$ (CaF_2)	
Santonian					
Coniacian					
Turonian	Block folding				
Cenomanian	Beginning of inversion				
Albian		Beginning of magmatic intrusion		50%	montmorillonite in the sediments (fraction 2–0.63 μm) with fragments of volcanic glass
Neocomian	Strong subsidence		By subsidence	20%	
Wealden				2%–3%	
Jurassic, Malm				Pyrophyllite up to Oxfordian	

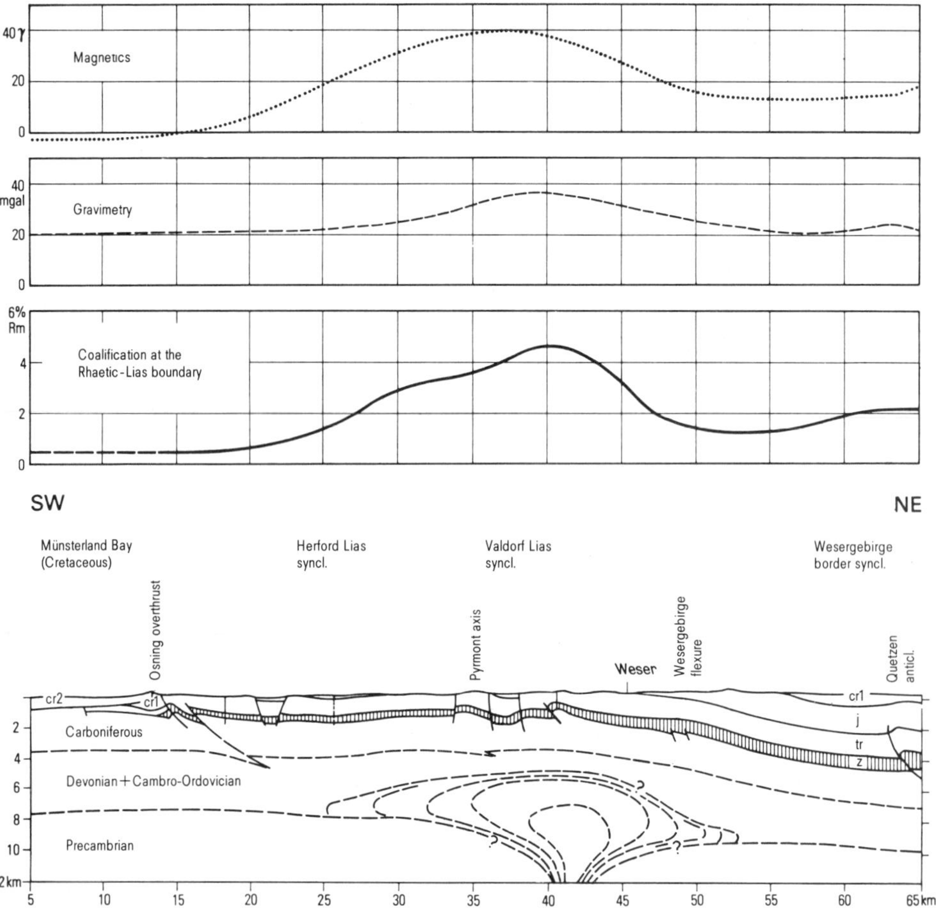

Fig. 3. Cross-section through the Vlotho Massif on the basis of coalification, gravimetry, and magnetics. (After Deutloff et al. 1980)

of Hameln (Deutloff et al. 1980; Fig. 2, Table 1). The mineralization consists of siderite and fluorite veins in the inner zone, metasomatic siderite ore bodies with barite and sulfides in the middle, and epithermal Pb-Zn pyrite veins in the outer zone. The massifs themselves have been identified as laccolithic bodies at a depth of around 6000 m by geophysical, as well as tectonic data and by the very weak metamorphism characterized by the occurrence of pyrophyllite in the sedimentary sequence including the lower part of Upper Jurassic (Fig. 3).

Stoppel and Gundlach (1972) argued for an Upper Cretaceous to Paleogene age of the barite veins in the southwestern Harz Mountains. According to Scherp and Strübel (1974), the NW- to NNW-striking barite veins in the Rheinische Schiefergebirge and the Ruhr district are thought to be related to still migrating chloridic Ba-Sr-bearing Na and Na-Ca solutions. These originated from very deep meteoric waters and from fossil waters migrating in Ba-rich Ordovician slates and ascended after the hanging-wall sequence was tectonically loosened by the uplift of the Rheinische Schiefergebirge during the Upper Cretaceous and Tertiary. Celestite layers and strontianite veins in the Cretaceous rocks of the Münsterland Bay section are explained as later deposits from the same solutions.

Baumann et al. (1975) investigated the relationship between deep fault structures and postmagmatic formation of ore deposits and call special attention to the fact that barite-fluorite veins in Central Europe are bound to NW-SE-striking faults. From tectonic investigations, geostructural considerations, and from petrogenetic geochemical analogue models they conclude that these deposits are genetically associated with an intrusive magmatism with alkalic differentiation. This hypothetical deep magmatism (Baumann and Weber 1977) is thought to have developed during the Mesozoic along deep NNE-SSW striking fault structures in connection with taphrogenic kinematics.

In his just published paper, Hofmann (1979) attempted a correlation of the barite veins and their mineral paragenesis along the Rhenish Lineament between the Black Forest and the Harz Mountains. He assumes a deep magmatic source for hydrothermal solutions, active over a long period of time, and supposed that deposition of barite began during Late Jurassic or Cretaceous but that the bulk of the mineralization is of Tertiary age.

A discussion of the rather numerous papers dealing with individual districts and deposits in the German Federal Republic is reserved for a later more detailed paper.

3 Criteria for an Upper Cretaceous to Tertiary Age of the Post-Variscan Metallogenetic Epoch

Many authors have noted that the majority of the barite-fluorite veins strike NW (135° ± 20°), most of which are 1–15 m thick. This means there must have been special stress conditions over a large area during the formation of these deposits. If it were possible to establish the time range in which this stress regime could have prevailed, it would be an important criterion of the age of the mineralization.

3.1 Southwest Germany

The region of the South German Block seems to be an important key area (Wunderlich 1974). Schönenberg (1975) has ascertained that the platform cover from Lower Triassic to Upper Jurassic may have remained tectonically essentially untouched (unverritzt), so pre-Cretaceous fault tectonics, for the most part, do not need to be taken into account. Carlé (1955) came to the same result. A more precise age indication was given by Illies (1975). He proposed a model in which the counterclockwise rotation of the stress components is ascribed to shear movements of the Afro-Arabian macroplates relative to stable Europe, which had a ball-bearing effect on the intervening microplates in the Mediterranean. These stress components, which during the Paleogene were mainly parallel to the Rhine Graben axis, were rotated until the maximum compressive stress was directed mainly to the northwest. Thus, stress conditions favourable for the opening of the NW-striking faults have been developing in West and Southwest Germany since the Early Miocene.

3.2 Northwest Germany

Woodhall and Knox (1979) investigated the distribution of Mesozoic volcanism in NW Europe with reference to North Sea and North Atlantic rifting. The tectono-magmatic development began in the north during the Early Mesozoic and reached the continent not before the Upper Jurassic (subsidence of the Lower Saxonian Block) and Lower Cretaceous (Zuidval volcano).

The intrusion of the massifs of Bramsche and Vlotho took place during the Upper Cretaceous. Thus, the mineralization in the area of the Lower Saxonian Block, described by Stadler (1971) is considered to be the oldest post-Variscan one in the Federal Republic.

3.3 Geophysical Model Ages

Detailed work has been done by the Freiberg school (see above). The U/Pb determinations have been critically reviewed by Ramdohr (1980) with special consideration of the great mobility of uranium. Ramdohr described single samples in which pitchblende was displaced by coffinite, which was then displaced by younger pitchblende, which is cut by fissures containing a third generation of pitchblende. Thus, he considers that ages as questionable. Also, the K/Ar determinations cited do not seem to be convincing. The great majority of the fractures in the basement are known to be older, mostly of Variscan age, and reworked. Thus, the K/Ar model ages could be explained as mixed ages. Only the paleomagnetic investigations of Baumann and Krs (1967) fit the results obtained from geological and structural considerations.

4 Paragenetic and Genetic Considerations

Werner (1966), Baumann (1967) and others have pointed out that except for local differences, one paragenetic model can be used for all the desposits characterized by quartz, barite, fluorite, Fe-Mn carbonates or oxides, Co, Ni, Bi, As, Sb, Ag, and Hg, as well as varying amounts of Cu, Pb, and Zn sulphides. This is in full agreement with Schneiderhöhn (1949). Only a genetic model applicable for all of Central Europe can therefore be accepted. Such a model has been presented by Baumann et al. (1975) and Baumann and Weber (1977). But the deep magmatism required, which did not exist during the early and middle Mesozoic in Central Europe, is provided by the intrusions in the Lower Saxonian Block during the Upper Cretaceous and by the mantle diapirism underneath the Rhine Graben rift system during the Paleogene (Illies 1975; Baumann et al. 1975). Thus, it seems very probable that the mineralization is not synchronous but that there was a migration of tectono-magmatic development including mineralization from the north, where it began during the early Mesozoic in the northern North Sea, to the south.

Mitchell and Halliday (1976) have demonstrated a single province of hydrothermal ore deposition associated with the early stages of plate separation in the North Atlantic during the early Mesozoic. It is believed that the Central European mineralization could be considered as a younger equivalent of this mineral province.

5 Conclusions

The post-Variscan metallogenic epoch in western Central Europe is of Upper Cretaceous to Tertiary age: Upper Cretaceous in the Lower Saxonian Block, and probably also in the Central Harz Mountains and in the Nabburg district, and Neogene along the Rhine Graben rift system. A magmatic source seems to be absolutely necessary to explain the widespread and uniform mineralization. This mineralization is different in time and paragenesis from that in the Alpine orogenic system during post-Variscan time.

Acknowledgments. The author thanks Dr. Gundlach, Dr. Kockel, and Dr. Stoppel, Hannover, as well as Dr. Stadler, Krefeld, for discussions. Dr. Gundlach and Dr. Stoppel have read the manuscript.

References

Bärtling R (1911) Die Schwerspatlagerstätten Deutschlands. Enke, Stuttgart, 188 pp

Baumann L (1967) Zur Frage der varistischen und postvaristischen Mineralisation im sächsischen Erzgebirge. Freiberg Forschungsh C 209:15–36

Baumann L, Krs M (1967) Paläomagnetische Altersbestimmungen an einigen Mineralparagenesen des Freiberger Lagerstättenbezirks. Geologie 16:765–780

Baumann L, Leeder O, Weber W (1975) Beziehungen zwischen regionalen Bruchstrukturen und postmagmatischen Lagerstättenbildungen und ihre Bedeutung für die Suche und Erkundung von Fluorit-Baryt-Lagerstätten. Z Angew Geol 21:5–17

Baumann L, Weber W (1977) Deep faults, simatic magmatism, and the formation of mineral deposits in the Central Europe outside the Alps. In: Jankovic S (ed) Metallogeny and plate tectonics in the NE Mediterranean. Jugosl Vereinig Wiss Ges, Belgrad, pp 541–551

Brockamp O (1976) Nachweis von Vulkanismus in Sedimenten der Unter- und Oberkreide in Norddeutschland. Geol Rundsch 65:162–174

Buntebarth G, Teichmüller R (1979) Zur Ermittlung der Paläotemperaturen im Dach des Bramscher Intrusivs aufgrund von Inkohlungsdaten. Fortschr Geol Rheinld Westf 27:171–182

Carlé W (1955) Bau und Entwicklung der Südwestdeutschen Großscholle. Beih Geol Jahrb 16:272

Deutloff O, Teichmüller M, Teichmüller R, Wolf M (1980) Inkohlungsuntersuchungen im Mesozoikum des Massivs von Vlotho (Niedersächsisches Tektogen). Neues Jahrb Geol Paläontol Monatsh 1980:321–341

Engelhardt W v (1936) Die Geochemie des Bariums. Chem Erde 10:187–246

Gunzert G (1961) Über das selektive Auftreten der saxonischen Schwerspatvorkommen in Deutschland. Neues Jahrb Mineral Monatsh 1961:25–51

Hofmann R (1979) Die Entwicklung der Abscheidungen in den gangförmigen, hydrothermalen Barytvorkommen Mitteleuropas. Monogr Ser Miner Deposits 17:81–214

Illies H (1975) Intraplate tectonics in stable Europe as related to plate tectonics in the Alpine system. Geol Rundsch 64:677–699

Leutwein F (1957) Alter und paragenetische Stellung der Pechblenden erzgebirgischer Lagerstätten. Geol 6:797–805

Mitchell JG, Halliday AN (1976) Extent of Triassic-Jurassic hydrothermal ore deposits on the North Atlantic margins. Trans Inst Min Metall, Sect B 85:159–161

Murawski H (1954) Bau und Genese von Schwerspatlagerstätten des Spessarts. Neues Jahrb Geol Paläontol Monatsh 1954:145–163

Ramdohr P (1979) Daten und Bilder von Mineralien des U^{4+} verschiedenen Alters und verschiedener Herkunft. Z dt geol Ges 130:439–458

Rösler HJ, Pilot J (1967) Die zeitliche Einstufung der sächsich-thüringischen Ganglagerstätten mit Hilfe der K-Ar-Methode. Freiberg Forschungsh C 209:87–98

Scherp A, Strübel G (1974) Zur Barium-Strontium-Mineralisation. Mineral Deposita 9:155–168

Schneiderhöhn H (1949) Schwerspatgänge und pseudomorphe Quarzgänge in Westdeutschland. Neues Jahrb Mineral Geol Paläontol Monatsh Abt A 1949:191–202

Schönenberg R (1975) Südwest-Deutschland zwischen atlantischer Drift und alpiner Orogenese. Jahresh Ges Naturkde Württemberg 130:54–67

Stadler G (1971) Die Vererzung im Bereich des Bramscher Massivs und seiner Umgebung. Fortschr Geol Rheinld Westf 18:439–500

Stoppel D, Gundlach H (1972) Baryt-Lagerstätten des Südwest-Harzes. Beih Geol Jahrb 124:120

Werner CD (1958) Geochemie und Paragenese der saxonischen Schwerspat-Flußspat-Gänge im Schmalkaldener Revier. Freiberg Forschungsh C 47:117

Werner C-D (1966) Die Spatlagerstätten des Thüringer Waldes und ihre Stellung im Rahmen der saxonischen Metallprovinz Mitteleuropas. Ber dt Ges geol Wiss B 11:5–45

Winogradow AP, Tugarinow AJ, Zhirowa VV, Sykow S, Knorre KG, Lebedew VI (1959) Über das Alter der Granite und Erzvorkommen in Sachsen. Freiberg Forschungsh C 57:73–85

Woodhall D, Knox RWOB (1979) Mesozoic volcanism in the northern North Sea and adjacent areas. Bull Geol Surv Great Britain 70:34–56

Wunderlich HG (1974) Die Bedeutung der Süddeutschen Großscholle in der Geodynamik Westeuropas. Geol Rdsch 63:755–772

Ore Textures[1] as Indicators of Formation Conditions of Mineral Paragenesis in Different Types of Stratiform Lead-Zinc Deposits

T.N. SHADLUN [2]

Abstract

There are three types among stratiform deposits of lead-zinc iron disulphide rich ores occurring in sediments: (1) stratabound hydrothermal-sedimentary in carbonate-sandy-argillaceous strata, (2) hydrothermal in the zones of interlayer brecciation in carbonate rocks, (3) metamorphosed hydrothermal-sedimentary in metamorphic schists. The peculiarities of ore textures observed in polished handspecimens and polished sections are observed using examples of three deposits of the Soviet Union and Cuba. Each of the above mentioned three types of deposits have distinctly different textures of ores. These textures characterize depositional conditions of the main ore mineral assemblages.

1 Introduction

Among the great diversity of stratiform deposits of lead-zinc ores in sediments different types can be distinguished. Not only the occurrence but also the formation of these deposits were observed by Smirnov (1970) to depend on stratification of volcanogenic sedimentary or sedimentary stratified layers. What relates them is both similarity of occurrence and conditions and duration of their formation.

To state the formation conditions for stratiform lead-zinc deposits in sedimentary geosynclinal strata of different ages with different intensity of metamorphism of both enclosing rocks and ores is a very complicated task. To solve it, the whole store of methods that are at the disposal of ore geologists is involved. One of them to be used, in combination with many others but of important value in itself, should be an analysis of texture and structural peculiarities of ores and mineral aggregates. No doubt, this aspect of ore studies can not give a simple answer to questions arising for the solution of the genesis problem if we do not combine it with geological data and geochemical peculiarities of deposits. However, many problems can not be solved if we ignore the ore texture studies.

Deciphering of the ore formation process by relationships of minerals and mineral aggregates is complicated, which is to a large extent caused by ore structure and texture

1 In this article, according to Betekhtin (1958) and Ramdohr (1975) the terms "structure" and "texture" are used to mean: the first – interrelations of mineral grains in one aggregate, the second – interrelations between different mineral aggregates in the ore

2 IGEM Academy of Sciences, Staromonetny 35, 109017 Moscow J-17, USSR

convergence. The appearance of ores, especially their textures connected both with sedimentation and diagenesis and with hydrothermal-metasomatic processes, may often seem to be indiscernible at first sight. Very detailed study of both textures and structures is necessary for their reliable identification. A comparison is needed with ores from other formations for which, in a sufficiently objective way, conditions of their formation are stated by geological and geochemical indicators. Only with certain "standards" available may one consider, with sufficient certainty, different textural types typical for deposits of definite geneses. As such standards, we have chosen and shall analyze texture-structural peculiarities of ores representing deposits of three types: (1) those for which a hydrothermal-sedimentary, syngenetic formation with enclosing sandy-argillaceous-calcareous sediments of Jurassic age is most probable – the Santa Lucia deposit on Cuba, (2) those where a significant part of the ores was deposited from colloidal, hydrothermal solutions in zones of brecciation of carbonate rocks of Middle Paleozoic age – the Utchkulatch deposit in the Uzbek SSR, (3) those where ores jointly with Upper Proterozoic rocks enclosing them were subjected to intensive regional metamorphism – the Kholodninskoe deposit in North Baikal region.

2 Peculiarities of Textures and Structures of Hydrothermal-Sedimentary Ores in Carbonate-Sandy-Argillaceous Strata

Both in geological position and in structure-texture peculiarities of ores, the Santa Lucia deposit in Cuba is of the first type.

Geological position of the Santa Lucia deposit is described in works by Antoneev and Norman (1974) and Zhidkov and Jalturin (1976). The deposit lies on the flanks of the Pinar del Rio anticlinorium. The ore-enclosing stratum of Jurassic age occurs in a monoclinal structure and is complicated by fine folds and disjunctive disturbances. The ore-bearing zone is delimited by faults. The ore-enclosing complex is composed of sandstones, aleurolites, limestones, dolomites and subordinate argillites.

Mineralogical-petrographical study of the composition of the ore-enclosing strata showed all the rocks including sulphide interbeds to be characterized by rhythmic laminations. Both two-component and many-component rhythms are distinguished including argillites-sulphides, carbonate rocks-sulphides, dolomites-sulphides, aleurolites-argillites-carbonates-sulphides. Sulphide layers alternate with carbonaceous argillites, aleurolites containing up to 3% of organic carbon (Zhidkov et al. 1975). Rocks were subjected to folding, jointly with sulphide interbeds and were also subjected to very weak regional metamorphism of the greenschist facies.

Sulphide bed-like deposits consist mainly of fine-globular pyrite in association with sphalerite and galena. For ore-forming sulphides, a low content of trace elements is typical. Silver, copper, antimony, cadmium, and nickel are present in amounts in the region of one-thousandth of a percent. No cobalt was observed in pyrite, although there is arsenic. Data on isotope composition of sulphide sulphur showed a wide range of values from -5.9 to $-34.8‰$ with the mean δ [^{34}S] of $-20.5‰$, which allows us to assume its formation on account of biogenic sulphur in sedimentation processes.

In massive sulphide ore exceptionally fine rhythmic alteration of laminae is typical (Fig. 1), these being composed of a mixture of iron-disulphide with sphalerite, and laminae composed of carbonate (Fig. 2). The lower layer (Fig. 1) shows mainly carbonate-siderite with impregnations of disperse sulphides, increasing in quantity upwards, and alternating with an interbed of massive sulphides with subordinate carbonates, quartz, and argillaceous minerals. Interbedded siderite is crossed perpendicularly to the stratification by diagenetic quartz and pyrite veinlets. In a sulphide mass one sees lenticules of gangue minerals outlining even finer stratification. In sulphide laminae initially marcasite was deposited which turned into pyrite during diagenesis. This is indicated by the presence of elongated lamellae and spear-like forms of disulphide and sometimes of marcasite relics.

G.C. Amstutz and co-authors (Amstutz et al. 1964; Amstutz and Bubenicek 1967) found abundant evidence indicating that the initial disperse silty sediments during the process of diagenetic changes are sometimes subjected to deformation and crystallization. Grain sizes get gradually larger. In the described ores in certain layers and lenticules one can see that along with grains and globules of a few microns in size, the enlarged sphalerite, galena and crystalline grains of iron-disulphide reach tens of microns in diameter and formed as a result of collective crystallization (Figs. 3–5).

The diagenesis process might continue during the period of folding of an entire stratum with only partially consolidated sulphide sediments. Larger grains form from the basic sulphide mixture. Pure sphalerite masses are evidently still in a plastic state and are crossed by an aggregate consisting of iron-disulphide globules, with carbonate and argillaceous minerals (Fig. 3).

Some textural peculiarities testify to plastic deformations. Thus, in stratified carbonate-sulphide ore consisting of rhythmically alternating siderite laminae (1–1.5 mm thick) and also in iron-disulphide laminae with sphalerite (2–2.5 mm thick) a number of small dislocations are observed (Fig. 6) which decline in number very rapidly at a short distance, being evidently formed in a not totally consolidated mass. Under the microscope, the curved siderite laminae are seen to be penetrated by veinlets of iron-disulphide and are boudinaged. At the same time, in sulphide interbeds a fine globular pyrite structure is preserved among sphalerite cement. Siderite laminae display more coarse-grained structures.

In one variety of stratified sulphide ore (Figs. 4 and 5) rhythmic layering alternation is seen. They are fine disperse iron-disulphides with large amounts of the smallest pores and cavities, sphalerite laminae with disperse pyrite (0.5–1 mm thick), almost pure sphalerite with macrolamellar pyrite-marcasite aggregates and carbonate layers with disperse disulphides and sphalerite or galena. Concretionary formations of spheroidal or oval form extend along the stratification (from 2 mm to 2 cm in size) and are of a specific interest. These concretions consist of siderite with disperse sulphides or intimate intergrowth of siderite and sphalerite (Figs. 4 and 6). Formation of these concretions is undoubtedly connected with diagenetic processes. In the largest concretion, stratification relics can be seen marked by broken chains of fine grains of sulphides and at the same time warping of bedding planes around the concretion by its growing. Evidently the mode of formation of diagenetic concretions in our case is very similar to that of carbonate concretions in banded iron ores of Lorraine, described by Amstutz and Bubenicek (1967).

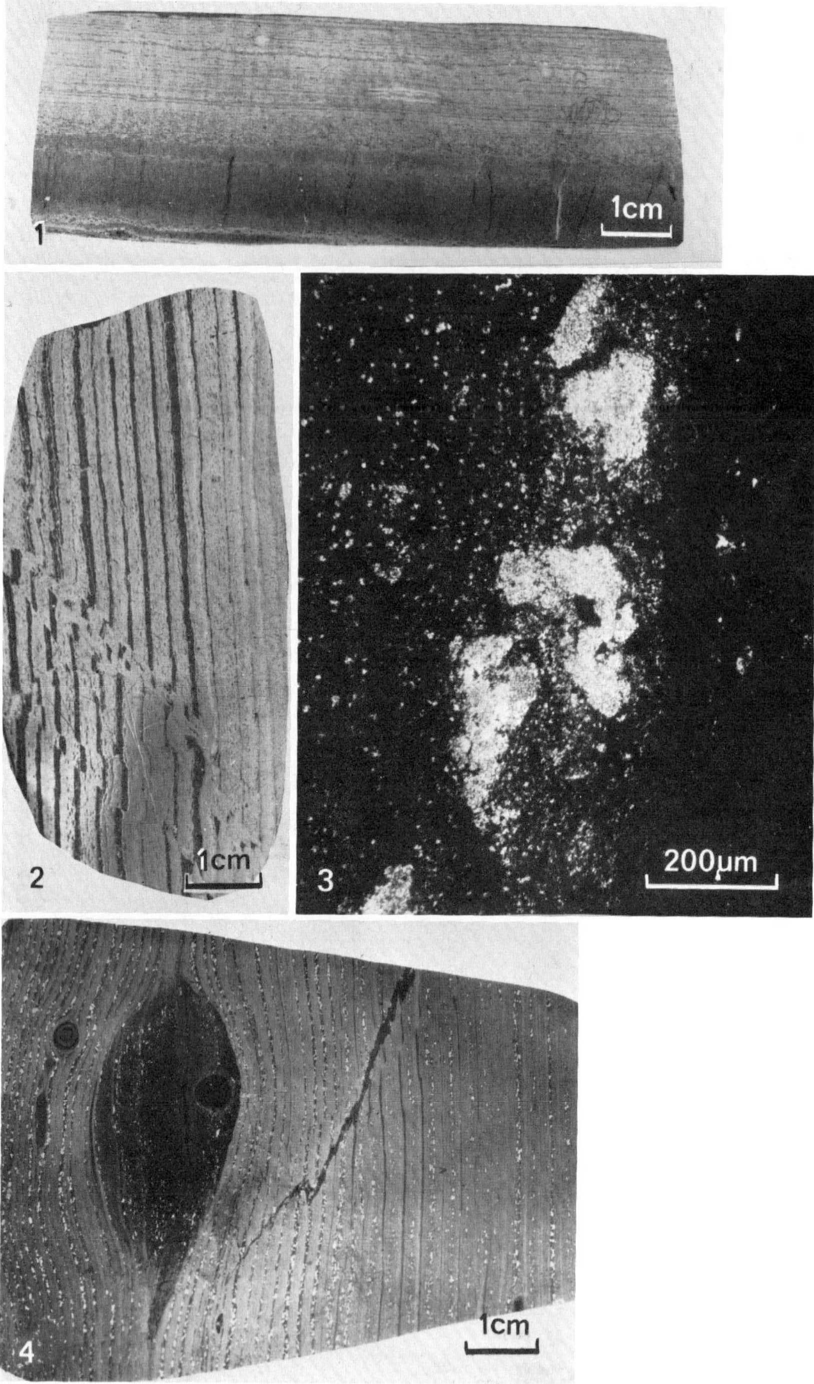

Figs. 1–6. Reflected light. Deposit Santa Lucia, Cuba

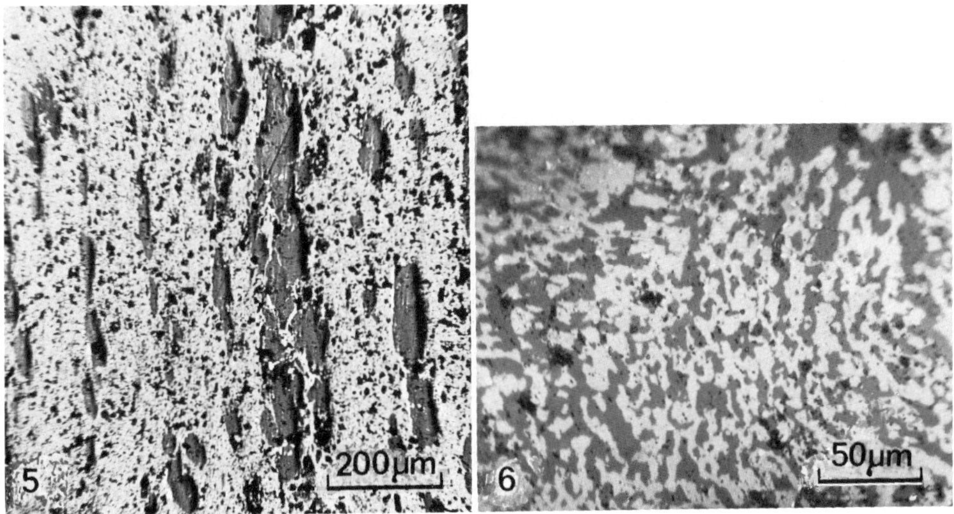

◄ **Fig. 1.** Rhythmically bedded laminated texture of sulphide ore. *Lighter bands,* enriched in iron disulphides; *dark,* rock-forming minerals – argillaceous and carbonate minerals. Polished surface. Natural size

Fig. 2. Rhythmic alternating iron-disulphides-sphalerite laminae *(light)* with laminae of siderite *(black).* Natural size

Fig. 3. Detail similar to Fig. 4 in transmitted light. *Black,* iron-disulphide; *light,* sphalerite. Mag. × 90

Fig. 4. Rhythmically stratified (laminated) carbonate-sphalerite-pyrite-marcasite ore with diagenetic concretions of siderite. Natural size

Fig. 5. Detail of Fig. 4. Lense-like segregation of sphalerite in the mass of iron-disulphides with sphalerite, carbonate. Mag. × 90

Fig. 6. Concretion of siderite from Fig. 4. Fine intergrowth of siderite with sphalerite. Mag. × 320

Detailed study of relationships between sulphides and gangue minerals composing stratified ores does not reveal the phenomena of sphalerite or galena substitution of iron-disulphides. All peculiarities of ore structure testify to syngenetic deposition of sulphides and minerals of enclosing rocks. Different displays of diagenesis and epigenesis create a variability of shapes of sulphide grains (globules, crystalline grains, small lenses, lenticules, and veinlets). One can think though of them forming a single paragenesis, a single mineral association jointly with gangue minerals-carbonates, argillaceous particles or carbonaceous matter. The latter, in all probability, is not preserved in the process of preparation of polished sections, which causes high porosity in sections of certain ore interbeds.

Along with all marked peculiarities of ore structure, the peculiar feature of their composition is the significant prevalence of iron-disulphides and sphalerite over galena, the absence of complex sulphides, and, as was mentioned above, exceptionally low content of trace elements.

Summarizing all of the above-stated peculiarities allows to consider the ores of the described deposit as syngenetic-hydrothermal and, consequently, ore textures are a reflection of sedimentational and diagenetic processes.

Surprisingly similar in textures to Santa Lucia ores are the ores of the early stage of deposits of the Atasuy region in Kazakhstan, in particular of Zhairem, described in a number of works by A.M. Mitryajewa with co-authors (Mitrjajewa and Muratova 1974; Mitrjajewa et al. 1976). They are also characterized by rhythmically laminated textures with alternation of thin layers of globular pyrite or pyrite and sphalerite with argillaceous-siliceous-carbonate layers, often with carbonaceous matter. They differ from Santa Lucia ores by larger dimensions of pyrite framboids, sphalerite and galena grains.

In the Zhairem ores, concretions are developed in a way they were in Santa Lucia ones. In laminated ores, fine nodules of rounded and oval forms are located in sphalerite-pyrite laminae and consist of calcite with sphalerite. The structure and composition of concretions and their correlations with stratification have much in common with what was said earlier. Both deformations of laminated ores in the period of their plastic state and co-sedimentational disturbances are also manifested in the same way. Researchers of these ores point to the collection crystallization leading to formation of idioblastic crystals or metacrystals of pyrite and other minerals. Analysis of the whole complex of textural and geological features led to the conclusion of hydrothermal-sedimentary origin of stratified sphalerite-pyrite ores of Zhairem.

Similar ores were described from the Tekeli deposit (Patalakha 1975) in Kazakhstan as relics among the ores transformed by metamorphism (Shadlun 1958, 1959). To the same type belong the ores of the deposit Ozernoe in the Trans-Baikal region. The texture and structure of these ores were described by Distanov and Kovalev (1975). The ores of marked deposits are similar not only in the presence of finely laminated textures, prevalence of iron-disulphide globular forms, their close association with sphalerite, and the presence of carbonaceous matter, but in geological position, the character of ore-enclosing layers consisting of calcareous-carbonaceous-argillaceous sediments. Indicators of syngenetic accumulation of ore matter are everywhere supported by ore textural peculiarities.

In spite of essential differences in ore and enclosing rock composition there is much in common in textural features of Santa Lucia ores and those of Bleiberg deposit studied in detail by Schulz (1968).

3 Structural-Textural Peculiarities of Ores Deposited from Hydrothermal Solutions in the Zones of Interlayer Brecciation in Carbonate Rocks

Let us dwell upon the following example of the stratiform Utchkulatch deposit in the Uzbek SSR formed in carbonate rocks as a result of circulation of hydrothermal solutions in brecciated and exfoliation zones. According to Pankratyev et al. (1978) the region of the ore field is within the limits of a narrow, extended miogeosynclinal depression, formed in Middle Paleozoic at the pre-Orogenic stage of formation of the southern Tien-Shan folded system. The deposit is located in the preaxial part of a large anticline, limited by faults and composed mainly by carbonate and volcanogenic-sedimentary Devonian rocks. Anticlinal structure is complicated by small brachyanticlines where separate ore deposits are located.

Along with considerable dislocation of sediments and general monoclinal rock occurrence, the rhythmicity is typical. The mineralization enclosing suite consists of dolomites and limestones, alternating with argillites and underlain by tuffs and tuff-lava of acid composition.

In different horizons of stratigraphic cross section, zones of intrastratal fragmentation are widely manifested as causing the appearance of breccia and contributing to ore localization. Mineralization is represented in every part of the deposit by several ore bodies mostly conformable to the bedding of the carbonate enclosing rocks. Bed- and lense-shaped ore bodies are divided by weakly mineralized and ore-free interbeds of rocks. Ore bodies are sometimes dipping at low angle, sometimes as relatively steeply dipping curve-like flexures. Enclosing rocks are poorly metamorphosed. Wall rock hydrothermal alterations appear in weakly recrystallized and redesposited carbonates; baritization, dolomitization and insignificant silicification are present.

In their composition the ores differ depending on the prevalence of one or another mineral association. There are pyritic ores with different amounts of galena, sphalerite, chalcopyrite, sphalerite-galena-baritic ore, bornite-pyrite with chalcocite, chalcopyrite and tennantite. Sulphide-quartz and sulphide-calcite associations have insignificant development, the sulphide-barite association is the main and predominant one. The latter comprises the basic mass of ores.

Trace elements in ore-forming minerals are distributed irregularly. Silver, cobalt, and nickel are present in high amounts more often to one-hundredth of a percent, concentrating sometimes in colloform shaped cryptocrystalline aggregates of iron-disulphide with fine-disperse inclusions of other sulphides.

Typical textures of these ores can be considered breccia-like and crustiform ones with colloform-zonal sulphide crusts around aleurolite or carbonate rock fragments (Fig. 7) with cavity fillings of calcite and barite. Streaky, impregnated textures are developed in those parts where rock brecciation is not sufficiently distinct, but rocks are disturbed and jointed, contributing to solution penetration. The patches of massive sulphide ore composed of iron-disulphide and sphalerite appear only in certain localities.

In breccia-like ores, one can see a thin crust of iron disulphides on every rock fragment. Pyrite and marcasite are either in intergrowths, or they individually compose zones of kidney-shaped colloform aggregates. Iron-disulphide rims become usually covered with a thinner sphalerite rim which in its turn gets covered by an even narrower galena fringe. Sometimes galena is deposited between pyrite and sphalerite or fills in the cavities between kidney-shaped aggregates (Fig. 8).

Sphalerite and chalcopyrite often penetrate along some zones in colloform kidney-shaped cryptocrystalline pyrite aggregates or replace finely disperse cement between iron-disulphide globules.

The phenomena of replacement of sphalerite and chalcopyrite for iron-disulphides are developed very widely and are various in form. Replacement by sphalerite develops in a colloform-zoned kidney-shaped mass of pyrite or along the network of finest syneresis (dehydration) fissures (Fig. 9). Sometimes the whole kidney form of iron-disulphide is nearly totally replaced, the relics of radial structure being preserved.

Different stages of colloform pyrite replacement with bornite are very distinct (Figs. 10 and 11). At the same time, bornite often includes grains of nickel-bearing carrollite. Bornite partly replaces carbonate and corrodes barite. Along the cleavage, barite is partially replaced by sphalerite.

Described relationships of sulphides with each other and with carbonate wall rocks show quite distinctly iron disulphides to be deposited as gel in cavities and on the surface of rock fragments and to a much lesser extent by means of replacement. The features of gel recrystallization structures described by us earlier (Shadlun 1964) in detail here were observed in iron-disulphides. Sulphides of nonferrous metals replaced mainly iron-disulphides which helped to form cryptocrystalline fine-grained structures and high porosity and fracturing of their aggregates.

On the whole, ore textures and structures indicate in a rather simple way sulphide deposition after the occurrence of open space in rocks. They also suggest nonequilibrium relations between earlier iron-disulphides and later solutions delivering nonferrous metals and depositing them either in open spaces or metasomatically at the site of the iron-disulphide which might be made up of finely disperse or cryptocrystalline grains favourable for replacement.

One must not ignore the possibility that sulphides of nonferrous metals in a finely disperse state might be partially contained in the primary gels of iron-disulphide. Solutions that arrived later not only delivered nonferrous metals but redeposited copper and zinc sulphides contained in iron-disulphides. Redeposition as thin veinlets of chalcopyrite in sphalerite, sphalerite in pyrite and galena in sphalerite cross-cutting at the same time iron-disulphides could be a later formation at the same time associated with a local appearance of solutions not supplying metals. Bornite formation is evidently caused by a very local appearance of solutions with very high copper concentration intensively interacting with colloform pyrite and extracting iron from the pyrite.

Thus, the described peculiarities of textures, microtextures, and structures allow to draw a conclusion that the carbonate-barite-sulphide association consists of several consequently deposited parageneses: carbonate-marcasite-pyrite; chalcopyrite-galena-sphalerite and tennantite-bornite.

It would be appropriate to make some comparisons with similar deposits possessing common features both in geological position and in the composition and textures of the ores. One such deposit to our mind is the Sedmotchislennitsi deposit in Bulgaria. Its geology and genesis were frequently described (Mincheva-Stefanova 1965, 1974).

Figs. 7–11. Polished sections in reflected light. Deposit Utshkulatsh, Uzbekistan

Fig. 7. Coating with pyrite and sphalerite on the fragments of rocks

Fig. 8. Colloform pyrite *(light)* and sphalerite *(grey)* are veined by galena *(white).* Reflected light. Mag. × 100

Fig. 9. Veined chalcopyrite *(white)* in very thin dehydration fractures in sphalerite *(dark-grey)* replaced colloform disperse pyrite *(light grey). Black,* carbonate. Reflected light

Fig. 10. Sphalerite and bornite *(grey),* replacement of disperse or fine-grained iron-disulphide encrusted with coarse-grained radial lamellae. Reflected light. Mag. × 160

Fig. 11. Cores and rims of colloform iron-disulphide *(white)* preserved by bornite replacement *(dark grey).* Mag. × 160

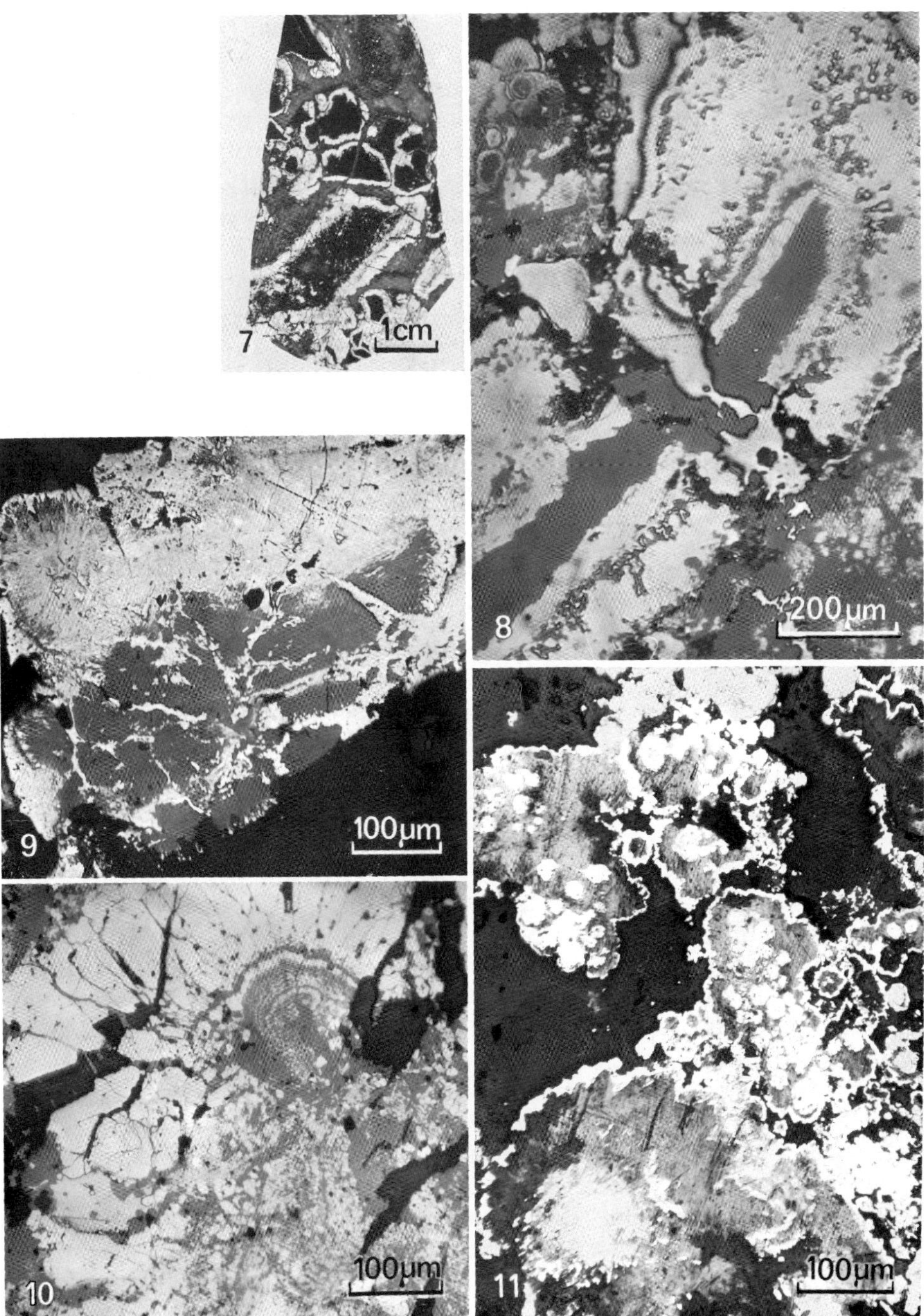
7
1cm
8
200 µm
9
100µm
10
100µm
11
100µm

Recently, peculiarities of composition of its main ore forming minerals were described (Shadlun et al. 1978).

Ores of this deposit are extremely similar to those of Utchkulatch in texture peculiarities. Breccia, and colloform crusts of iron-disulphide are widespread in them around aleurolite or dolomite fragments. Thin-banded colloform microtextures of sphalerite are widely distributed there. The latter replaces dolomite more intensively and forms richer sphalerite ores which in their turn are brecciated. Sphalerite becomes overgrown with later generations of colloform kidney-like iron-disulphide. The presence of silver containing tennantite-bornite association in this deposit is typical. It is separated spatially from the pyrite-galena-sphalerite association. Both these associations are accompanied by calcite, dolomite and barite. Because of their geological and textural features Mincheva-Stefanova (1965, 1974) refers the ores of this deposit to low-temperature hydrothermal metasomatic origin.

Structures and microtextures of Utshkulatsh ores are also similar to any such deposits of the Mississippi-Missouri region and apparently the Magmont Mine, south-east Missouri, recently described by Hagni and Trancynger (1977).

4 Peculiarities of Texture of Metamorphosed Hydrothermal-Sedimentary Ores Occurring in Metamorphic Shales

The most complicated for explanation of the genetic sense of textures and structures are ores of stratiform deposits, occurring in highly metamorphosed schists. As an example, we shall examine the Kholodninskoe deposit in Buryatiya, north Baikal region.

Geological characteristics of the deposit are given in a number of works (Distanov and Kovalev 1977; Rutchkin et al. 1973, 1975). The ore field is within the limits of the Bodaibo synclinal trough of the Baikal folded region. Deeply metamorphosed sediments of Upper Proterozoic age compose a synclinal structure constructed in a complicated way. Along its axis passes the central deep fault. The deposit is located on the flanks of the synclinal fold, complicated by high order plicative structures. Rocks enclosing the ore deposits form a black-shales member consisting of graphite-bearing micaceous-carbonaceous, quartz-mica slates, graphitic quartzite-sandstones, graphite-containing arenaceous marmorized limestones and other intensively metamorphosed varieties of sedimentary rocks. The presence of disthen, staurolite, and garnet in the rocks points to the metamorphic amphibolite facies. High-temperature metasomatic porphyroblastic rocks are located in numerous subparallel faults.

The ore zone includes both bed-like stratiform ore bodies imitating the complex folding of the enclosing rocks, and cross-cutting bodies controlled by longitudinal dislocations with a break in continuity and zones of porphyroblastic and quartz-muscovite slates. Ore bodies form either alternations of banded highly impregnated sulphides and nearly compact pyrite massive ores with interbeds of graphitic, poorly mineralized slates within the limits of bed-like deposits, or streaky-impregnated ores with various intensity of impregnation of sulphides, lenticules, streaks, and segregations imitating or cross-cutting schistosity.

The main ore-forming minerals are pyrite, sphalerite, galena, chalcopyrite, pyrrhotite, quartz, muscovite, and calcite. Arsenopyrite, tetrahedrite-tennantite, magnetite, graphite, plagioclase, disthen, gahnite, and biotite occur sporadically. The following mineral associations are distinguished by researchers who studied the deposit: the early pyrite with subordinate sphalerite and galena, the quartz-chalcopyrite-galena-pyrite-sphalerite composing the main mass of ores of bed-like bodies, the quartz-carbonate-pyrite-sphalerite forming both conformable and cross-cutting bands with peculiar structures, connected with late stages of regressive metamorphism. The pyrite-galena-sphalerite-chalcopyrite-pyrrhotite association composing veinlet-impregnated ores overlies the two first associations.

Combination of geological properties, data of mineragraphic study, thermobarometric and isotope studies led to the conclusion (Rutchkin et al. 1973; Zairi and Kuznetsova et al. 1976) on formation of ores initially in connection with sediments under submarine conditions with participation of magmatic sulphur delivered jointly with metals from the deep source by hydrothermal solutions. The second stage during which modern appearance of ores was created was a complicated and continuous period of regional metamorphism and metasomatism that changed the ores and rocks enclosing them.

Complicated and continuous metamorphism developed during the progressive stage in recrystallization, redeposition of early sulphides and in the regressive stage in deposition by hydrothermal solutions of later associations which caused a variously banded texture and veinlet textures and relatively coarse-grained structures of mineral aggregates.

To the most widespread textures belong those which are banded, relic stratified and impregnated-veinlet-banded of different appearance that form alternation of bands or layers with wall rocks or concordant with the schistosity of irregular veinlets of sulphides (Fig. 12). Areas with spotted or massive texture occur more seldom, mainly in ores composed of the quartz-pyrite-sphalerite association and sometimes in ores of the pyrite-polymetallic type composed of the galena-sphalerite-chalcopyrite-pyrrhotite association. Rarer are veinlet textures of ores of pyrite-pyrrhotite composition and breccia-like textures of ores of sphalerite-pyrite composition distinctly cross-cutting schistosity or relic laminations. The latter ores include those that were conventionally called "ball texture" ores owing to the presence of rounded fragment-like quartz grains in pyrite-sphalerite masses.

The convergence of most banded and banded-veinlet textures of the given type of deposits is that such textures may form both in the process of metamorphism, i.e., redeposition and regrouping of primarily laminated ores with participation of metamorphic solutions and as a result of hydrothermal metasomatic processes due to inheriting of wall rock textures.

The most typical peculiarity of textures and structures of ores (in formation by primary hydrothermal-sedimentary processes) as shown above should be considered the rhythmical laminations and the presence of iron-disulphides as finely disperse (silty) segregations or very thin globules and framboids in the mass of argillaceous or carbonate-sandy-argillaceous and siliceous matter.

Relics of such primary details of structure can be seen in some sections of ores of Kholodninskoe deposit. Thus, pyrite globules 5–10 microns in size are sometimes observed in quartz. In thin-laminated pyrite ores enriched in galena, one can observe pyrite porphyroblasts with pyrite globules contained in them. It is typical that in some pyrite porphyroblasts such globules consist of galena, quartz, or sphalerite. Galena and more seldom sphalerite globules are present in the surrounding mass of quartz along with larger, irregular grains of the same sulphides or in veinlet forms of irregular shape.

There is an assumption (Distanov et al. 1977; Eremin et al. 1977) that galena and sphalerite globules in quartz and pyrite are relics of primary ores preserved the way pyrite globules are. However, it looks hardly probable to us on account of two reasons. First, in the described ores of Santa Lucia, where pyrite globular structures are extremely widely developed, sphalerite or galena globules were never observed though both of these sulphides are most closely connected (similar to pyrite) with sedimentation. Secondly, in the Kholodninskoe ores, sphalerite and galena globules occur only where these sulphides are deposited along fissures in pyrite or along interfaces of their grains corroding pyrite. That is direct evidence of later deposition of these sulphides. As to relics of primary colloform forms of iron-disulphides, we may refer to reniform concentrically zoned, recrystallized aggregates of quartz-pyrite composition that were sometimes observed.

With the examples of massive pyritic ores in volcanogenic layers described (Shadlun 1959, 1964), the processes of ore metamorphism were shown to occur irregularly and not only in different deposits but within the limits of one deposit depending on metamorphic intensity. Owing to this irregularity, relics of primary structures and textures and relics of earlier stages of metamorphism are preserved. Only very strong regional metamorphism occurring at high pressures and temperatures as, for instance, in Mt. Morgan and Broken Hill in Australia or with superposition of high-temperature contact metamorphic effects may transform sulphide ores so as not to preserve any relics of primary textures and structures (Hobbs et al. 1968; Lawrence 1973; Ramdohr 1950).

In spite of the fact that relic, thinly laminated rhythmically stratified textures play rather an essential part especially in the pyritic ore, in ores of the Kholodninskoe deposit, their structures indicate complete recrystallization. Alternation of layers with various degrees of sulphide impregnation and the orientation of grains (in schistosity) coinciding most often with direction of stratification is typical. In banded ores pyrite usually shows granoblastic structures with polygonal boundaries of grains. As is known (Hagni and Trancynger 1977; Ramdohr 1950), one of the indicators of intensive metamorphism of pyritic ores is their relatively coarse-grain nature, and granoblastic and porphyroblastic structures. In comparison with finely disperse or fine-grained structures of aggregates of hydrothermal-sedimentary ores in metamorphosed ores of the Kholodninskoe deposit, the size of grains is larger many times. If we compare them with the size of globules, this size difference is of the order of two; on the average the size of grains in pyrite aggregates is 0.2–0.3 mm, and some porphyroblasts reach 10–20 mm, whereas pyrite globules have the size of 0.005–0.007 mm and less.

The massive and coarsely banded ores composed of the quartz-galena-sphalerite-pyrite or quartz-chalcopyrite-sphalerite-pyrite association possess very similar structures of aggregates. As a rule, pyrite grains which show signs of crushing are fractured and to a different extent subjected to corroding by sulphides cementing them such as chalcopyrite, sphalerite, galena.

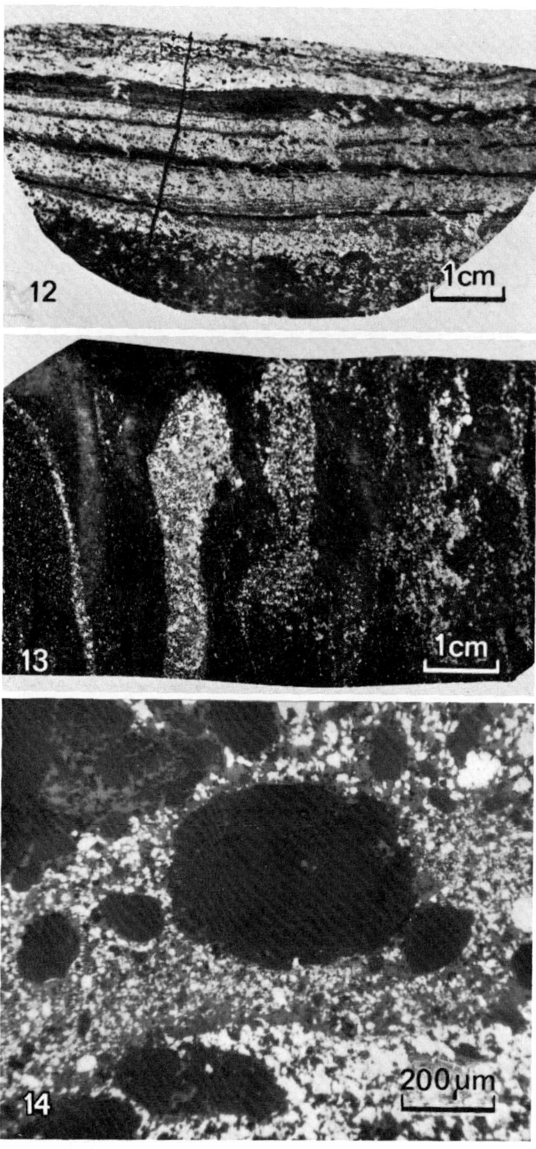

Figs. 12–14. Polished sections, reflected light, Kholodninskoe deposit, Northern Baikal region

Fig. 12. Banded or previous stratified texture of pyrite-pyrrhotite ore. Natural size

Fig. 13. Banded ore; light sulphide bands in schist partially concordant, partially discordant. Natural size

Fig. 14. Detail of Fig. 13. Rounded and oval fragments of quartz *(black)* enclosed in fine-grained mass of sphalerite with pyrite. Mag. × 63

Relationships of the latter with pyrite indicate the late depositional nature of nonferrous metal sulphides. However, the convergence of structures, as has already been said, does not allow to solve, in a simple way, the question as to whether these sulphides were transported by magmatically derived hydrothermal solutions with ensuing progressive metamorphism of pyrite ores or whether they were redeposited at the stage of metamorphic regrouping-redeposition and recrystallization of stratified pyritic ores.

However, we may rather reliably conclude on the nonequilibrium conditions this pair of sulphides underwent at the beginning of sphalerite deposition by viewing the structural correlations of pyrite with sphalerite, for pyrite was subjected to partial replacement with formation of similar skeletal forms of its grains. On sphalerite deposition, pyrite could be partially redeposited and form (crystallizing jointly with them) metacrysts of unusual form. The late quartz also has corrosional boundaries with pyrite, and, along with sphalerite, it is deposited between the grains (interstitially) and along the fractures. Distinct galena substitution for pyrite and sphalerite is developed to a different extent and indicates a certain disturbance of equilibrium during consequent deposition of minerals of one association. Sphalerite crystallization of this association also occurred evidently under somewhat inconstant conditions. To this can be attributed the inhomogeneity of the structure of its aggregates even within the limits of one small section, changing from fine to relatively coarse-grain size. However, zonation in grains is absent.

The relationships of minerals of the quartz-chalcopyrite-pyrite-pyrrhotite association with earlier formed associations of metamorphic minerals of enclosing rocks in combination with data of thermobarometry led to the conclusion (Rutchkin et al. 1973) on the formation of this association during a regressive stage of metamorphism with decreasing temperature. Judging from textures of these ores, deposition of this association was preceded or accompanied by tectonic movements that caused the occurrence of fracturing and crushing of mineral grains.

Arsenopyrite porphyroblasts, 2–5 mm in size, often occur in association with pyrrhotite. Correlation of arsenic and sulphur in the arsenopyrite denotes its formation at relatively low temperatures (about 300°). Judging from the fact that in some idioblasts pyrite relics are preserved one may assume that the arsenic concentration locally increased in metamorphic solutions of the regressive stage and the growth of arsenopyrite porphyroblasts took place at the expense of pyrite, iron, and sulphur. Absolutely even boundaries of these porphyroblasts with pyrrhotite denote their total equilibrium. It is interesting that the late quartz developing along the boundary between these sulphides replaces distinctly only arsenopyrite but does not corrode pyrrhotite at all.

In a number of places in the deposit, quartz-pyrite-sphalerite ores occurred with unusual "ball texture". In them occur a fine-grained aggregate of sphalerite appearing to cement fragmented, sometimes angular, more often rounded grains of pyrite of different shape and value and rounded oval grains of quartz larger than pyrite (Fig. 13 and 14). These ores compose sometimes narrow bands in silicified rocks from several mm to several cm. They cross aggregates of pyrite with pyrrhotite, pyrite being subjected to fracturing and crushing. Texture and structure peculiarities of these aggregates show quite distinctly their formation to be connected with mylonitization phenomena. Sphalerite deposition was evidently accompanied by brecciation and partly by abrasion and solution of earlier pyrite and quartz. Pyrite was partially replaced by sphalerite and was immediately redeposited as fine skeletal grains (Fig. 14). Since the described type of ores cross the intergrowth boundaries of pyrite with pyrrhotite but do not contain pyrrhotite themselves, the latter was evidently deposited earlier than sphalerite. Apparently "ball texture" mylonite-like formations of quartz-pyrite-sphalerite composition are close in time to the period of deposition of the late pyrrhotite-bearing association but their formation was preceded by deformation of aggregates of earlier associations.

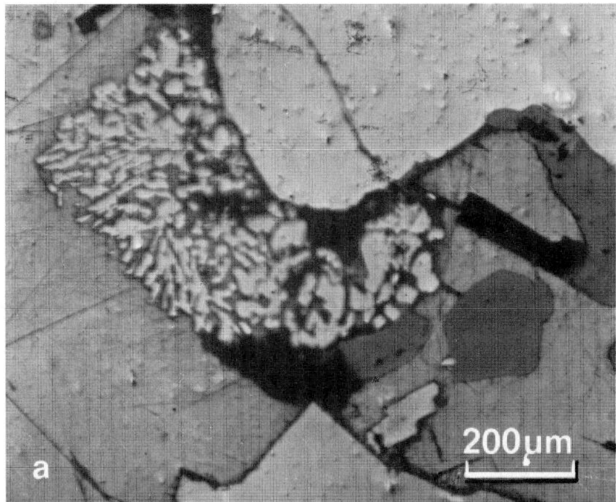

Fig. 15 a,b. Complicated intergrowth of tetrahedrite, sphalerite, arsenopyrite. **a** decomposed tennantite-tetrahedrite body included in sphalerite *(grey)* near the margin with pyrite *(white)*. Mag. × 63. **b** detail of **a**. The shape of individual minerals more precise. Mag. × 250

A very typical detail, specific only for metamorphosed ores, is the structure of tennantite-tetrahedrite decomposition features (Fig. 15 a,b). In this case, owing to tennantite-tetrahedrite of complex composition, an aggregate is formed consisting of a thin intergrowth of isometric grains of silver containing tetrahedrite, sphalerite, and very fine metacrysts of arsenopyrite (the size of mineral grains 10–20 microns). Evidently the stoichiometric tetrahedrite formed in the conditions of high pressures and temperatures becomes unstable and decomposes into the mentioned minerals at the regressive stage of metamorphism. One may observe such formations to occur in pyritic ores of the Urals and Altai deposits subjected to metamorphism. They are not observed in unmetamorphosed ores with colloform structures and textures.

Another peculiarity of metamorphosed rocks is considered to be the occurrence of fine-grained aggregates of graphite or strongly graphitized carbonaceous matter in close association with sulphides along with dust-like impregnations. In the described ores, such aggregates are thin intergrowths of graphitized matter with pyrrhotite enclosed in nonmetalliferous masses or with fine-grained sphalerite located within more coarse-grained sphalerite.

As we can judge from the available literature, ore textures and structural relationships in different mineral aggregates of the Kholodninskoe deposit can be compared with those of Broken Hill, Australia. The mineral composition of ores of the latter is very close to that of Kholodninskoe. The main difference is a negligible role of pyrite and ores much richer with galena in the Broken Hill deposit. The Broken Hill ores went through deeper metamorphism, since the enclosing rocks are represented by the almandine-amphibolite facies and in part the granulite facies. Though there may still exist different viewpoints on the genesis of this deposit, most researchers of the last decade (Pankratyev et al. 1978; Scott et al. 1977) prove, in a convincing way, the ores to have been subjected to regional metamorphism jointly with the enclosing rocks. One must but point out that structures of ores of this deposit caused by metamorphism were described already by P. Ramdohr in 1950.

Unfortunately, there are nearly no data in the literature on Broken Hill ore textures. Lawrence (1972) analyzed in more detail only some structural peculiarities. Similarly to P. Ramdohr, L. Lawrence underlines the homogeneity of the structures, regular coarse-grained nature (the grain size is from 0.5–1.0 to 2–5 mm) of mineral aggregates, and wide development of granoblastic structures of aggregates composed by garnet, orthoclase, quartz, sphalerite, and pyrrhotite. He points out granoblastic structures to be typical for the stage of progressive metamorphism, whereas the brecciated textures and microtextures are formed during the regressive stage, when rocks of the granulite facies were transformed into quartz-sericite slates and brittle minerals were crushed. Especially distinct are quartz and sphalerite fragments cemented by plastic galena, chalcopyrite, and more seldom by pyrrhotite.

A larger part of the mentioned structural peculiarities occur also in ores of the Kholodninskoe deposit where, owing to a lower grade of progressive metamorphism, differences in structures of aggregates of early and late metamorphic stages were not so sharp.

5 Conclusions

1. Peculiarities of ore structure should be considered in combination with all other indicators used to clarify the conditions of ore formation-geological position, structure and composition of ore-enclosing strata, data on substantial composition of ores, trace elements, isotope composition of sulphide sulphur.

2. The appearance of textures and microtextures of ores of stratiform lead-zinc deposits of different type rich in iron-disulphides is not similar and undoubtedly reflects the specific conditions of formation of each of the considered representatives of deposits of three types.

3. The most essential ore peculiarities in hydrothermal-sedimentary deposits occurring in calcareous-sandy-argillaceous sedimentary units are the prevalence of rhythmically-stratified textures, exceptionally thin interstratification of sulphides with layers of sandy-argillaceous material, hydromicaceous minerals or carbonates, the presence of siderite and carbonaceous matter. Quite typical are fine grain and globular structures of iron-disulphides, their closest intergrowth with finely disperse sphalerite and galena.

Typical are signs of diagenetic transformations in the form of concretions, enlarged spots and sulphide lenticules, synsedimentary deformations.

4. The ores deposited by hydrothermal solutions in the zones of tectonic deformations, interbed disturbances in terrigenous-carbonate sediments by means of deposition in open space and partially in a metasomatic way, are characterized by the development of breccia-like, cockade, crustified, veinlet-impregnated textures. Very typical are colloform microtextures with different concentrically zonal kidney-shaped sulphide aggregates and crusts coating the rock fragments. Sulphide aggregates possess the signs of deposition as gels, crypto- and finely crystalline or disperse masses. Concentrically zonal alternations and replacement phenomena of iron-disulphides by copper, zinc and lead sulphides along the zones and fractures of dehydration occur.

5. A typical peculiarity of structure of highly metamorphosed ores of stratiform deposits occurring in graphitic-quartz-micaceous slates, amphibolites, and quartzites are banded textures, alternation of bands of massive sulphide ores and rock interbeds with different amounts of sulphide impregnation. Quite typical features of these ores are their relative coarse-grained character, grano- and porphyroblastic structures, absence of finely disperse mixtures or colloform aggregates of sulphides. Relics of primary textures, signs of depositions as gels and rhythmically stratified shales are absent or play a rather insignificant role.

References

Amstutz GC, Bubenicek L (1971) Diagenesis in sedimentary mineral deposits. In: Larsen G, Chilingar GV (eds) Diagenesis and catagenesis of sedimentary formations. Mir, Moskva, pp 378–425 (translation from English 1967, in Russian)

Amstutz GC, Ramdohr P, El Baz F, Park WC (1964) Diagenetic behaviour of sulphides. In: Amstutz GC (ed) Sedimentology and ore genesis, vol II. Elsevier, Amsterdam, pp 65–90

Antoneev S, Norman A (1974) Estructura geológica, composición, material y genesis de la menas del yacimiento Santa Lucia. Resúmenes Primera Jornada Cientifico-Técnica, Habana, Cuba, vol II

Betekhtin AG (1958) The terms "texture" and "structure" in application to ores. In: Betekhtin AG (ed) Textures and structures of ores. Gosgeoltekhizdat, Moskva, pp 33–62 (in Russian)

Distanov EG, Kovalev KP (1975) Textures and structures of hydrothermal-sedimentary pyrite-polymetallic ores of the Ozernoe deposit. Kusnezov VA (ed) Nauka, Novosibirsk, 174 pp

Distanov EG, Kovalev KP, Shobogorov PC, Bushnev VP, Tsyrenov DT, Buslenko AI (1977) Peculiarities of formation of metamorphosed hydrothermal sedimentary pyrite-polymetallic ores of the Kholodninskoe deposit. In: Kusnezov VA, Distanov EG (eds) Questions of genesis of stratiform lead-zinc deposits of Siberia. Tr Inst Geol Geofiz, Akad Nauk SSSR, Sib Otd 1977, 361: 5–43 (in Russian)

Eremin NI, Kuznetsova T, Konkin VD, Rutchkin GV (1977) Composition of the main ore-forming sulphides of the Kholodninskoe deposit in connection with peculiarities of its genesis. Proc CNIGRI 126:115–130 (in Russian)

Hagni RD, Trancynger TC (1977) Sequence of deposition of the ore minerals at the Magmont Mine, Viburnum Trend, Southeast Missouri. Econ Geol 72:451–464

Hobbs BE, Ransom DM, Vernon RH, Williams PF (1968) The Broken Hill orebody – a review of recent work. Miner Deposita 3:293–316

Lawrence LJ (1972) The thermal metamorphism of a pyritic sulphide ore. Econ Geol 67:487–496

Lawrence LJ (1973) Polymetamorphism of the sulphide ores of Broken Hill, N.S.W. Australia. Miner Deposita 8:211–236

Mincheva-Stefanova J (1965) Über die Formen der Erzkörper und die Textur der Erze aus den Polymetall-Lagerstätten vom Typ "Sedmočislenici" in Bulgarien. Ber Geol Ges DDR, Sonderh 10: 321–334

Mincheva-Stefanova J (1974) Polymetallic deposit of Sedmocislennitsa. In: Kolkovsky B, Dragov P (eds) 12 ore deposits of Bulgaria. IV Symp IAGOD, Sofia, pp 55–76 (in Russian)

Mitrjajewa NM, Muratova DN (1974) Diagenesis phenomena and landslide disturbances in ores of stratiform lead-zinc deposits of Central Kazakhstan. Litol Polezn Iskop 2:101–114

Mitrjajewa NM, Patalakha GB, Muratova DN, Yanulova MK (1976) Textures and structures of ores of lead-zinc deposits. In: Satpaeva TA, Mitrjajewa NM (eds) Atlas textur i struktur rud tsvetnykh metallov Kazakhstana. Nauka, Alma-Ata, pp 7–34

Pankratyev PV, Mikhailova YuB, Vidusov TE (1978) Types of ores and genetic peculiarities of the Utchkulatch deposit (the Dalniy site). Zap Uzb Otd Vses Mineral Ova 31:176–182

Patalakha GB (1975) Hydrothermal-sedimentary concretion ores of the Tekeli deposit. Geol Rudn Mestorozhd 17:96–103

Rutchkin GV, Konkin VD, Kuznetsova TP (1973) Metamorphism of pyrite-polymetallic ores of the Kholodninskoe deposit. Geol Rudn Mestorozhd 15:69–78

Rutchkin GV, Bushuyev KP, Varlamov VA et al (1975) The Kholodninskoe deposit – a representative of Precambrian pyrite-polymetallic deposits. Geol Rudn Mestorozhd 17:3–17

Ramdohr P (1950) Die Lagerstätte von Broken Hill in New South Wales. Heidelberger Beitr Mineral Petrogr 2:291–333

Ramdohr P (1975) Die Erzmineralien und ihre Verwachsungen. Akademie Verlag, Berlin, 1277 pp

Schulz O (1968) Die synsedimentäre Mineralparagenese im oberen Wettersteinkalk der Pb-Zn Lagerstätte Bleiberg-Kreuth (Kärnten). Tscherm Mineral Petrogr Mitt XII (2–3):230–289

Scott SD, Both RA, Kissin SA (1977) Sulphide petrology of the Broken Hill Region, New South Wales. Econ Geol 72:1410–1425

Shadlun TN (1958) Gel recrystallization. Ore recrystallization at metamorphic processes. In: Betekhtin AG (ed) Textury i struktury rud. Gosgeoltekhizdat, Moskva, pp 137–154, 295–328

Shadlun TN (1959) Some regularities of metamorphism in lead-zinc ores of the Tekeli deposit rich with pyrite. Geol Rudn Mestorozhd 1:39–56

Shadlun TN (1964) Pyritic ores. In: Betekhtin AG, Shadlun TN (eds) Strukturno-texturnye osobennosti endogennykh rud. Nedra, Moskva, pp 179–245

Shadlun TN, Mincheva-Stefanova J, Dobrovolskaya MG, Bonev I, Kortman RV (1978) Occurrence conditions and typomorphic peculiarities of sphalerite and galena composition of the most important formations of lead-zinc deposits of Bulgaria and Soviet Union. In: Tchukhrov FV, Petrovskaya NV (eds) Sostav i struktura mineralov kak pokazatel ikh genezisa. Nauka, Moskva, pp 140–157

Smirnov VI (1970) Time factor in formation of stratiform ore deposits. Geol Rudn Mestorozhd 12:3–15

Zairi NM, Kuznetsova TP, Rutchkin GV, Stzizhev WP (1976) Sulphur sources and some questions of the genesis of Kholodninskoe pyrite-polymetallic deposit. Litol Polezn Iskop 2:139–146 (in Russian)

Zhidkov AY, Jalturin NL (1976) Mineralisación estratiforme piritico-polimetálica, zona la Oriental-Baritina. Min Cuba 2:28–39

Zhidkov A, Ovsiannikov V, del Pino J (1975) Papel de la materia orgánica en la formación del yacimiento Santa Lucia. Rev Min Cuba 2:12–18

Constraints on the Genesis of Major Australian Lead-Zinc-Silver Deposits: from Ramdohr to Recent

I.B. LAMBERT [1]

Abstract

Australia is well endowed with large, high-grade, stratiform Pb-Zn-Ag ore bodies: Broken Hill, Mount Isa, Hilton and McArthur. Such deposits were long regarded as having formed by intricate hydrothermal replacement of "favourable" strata. Almost 30 years ago, Professor Ramdohr concluded this was not the case at Broken Hill and, since then, it has become increasingly accepted that these deposits were generated by sedimentary-exhalative processes. However, the past five years have seen a resurgence of the epigenetic versus synsedimentary debate, with particular reference to the virtually unmetamorphosed and little deformed McArthur deposit. The main purpose of this paper is to assess the various lines of information relevant to the origin of the McArthur mineralization.

1 Introduction

Three decades ago, Professor Paul Ramdohr visited Australia and conducted mineragraphic studies on samples of Broken Hill Pb-Zn-Ag ore. At that time, it was generally believed that this fabulously rich deposit in western New South Wales had formed by magmatic-hydrothermal replacement of favourable strata after granulite facies metamorphism and deformation (e.g., Gustafson et al. 1950; King 1976).

However, Ramdohr (1950) concluded that the ore textures were the result of high temperature metamorphism of originally low temperature mineralization. His work did not receive the attention it deserved, partly because it was published in German, but mainly because the replacement school of thought was so firmly entrenched that it tended to ignore contrary opinions. It is now a matter of history that pre-metamorphic deposition of the Broken Hill lodes became increasingly accepted after about 1960, and currently represents the majority view (cf., King 1976).

Broken Hill is the most highly metamorphosed of several large, concordant Pb-Zn-Ag deposits of middle Proterozoic age in Australia. Mount Isa and Hilton (western Queensland) are other examples which have undergone lower greenschist facies metamorphism, whilst McArthur, also known as H.Y.C. (Northern Territory) is relatively

1 Baas Becking Laboratory, P.O. Box 378, Canberra City A.C.T. 2601, Australia

Visiting Alexander von Humboldt Fellow Federal Institute for Geosciences and Natural Resources, Stilleweg 2, 3000 Hannover 51, FRG

undeformed and unmetamorphosed. As the most readily interpreted information on ore genesis should come from detailed studies of the least modified mineralization, this paper considers only the latter deposit.

Since its discovery by Carpentaria Exploration Company in 1955, McArthur has been widely accepted as a classic example of sea-floor exhalative-sedimentary mineralization, though not necessarily "volcanogenic" (e.g., Cotton 1965; Croxford 1968; Croxford and Jephcott 1972; Croxford et al. 1975; Murray 1975; Lambert 1976). However, one worker in the area, Dr. Neil Williams, has strongly advocated that the stratiform ore formed by epigenetic processes (Williams and Rye 1974; Williams 1978a–c, 1979a,b; Rye and Williams 1978). This is not really the exact reverse of the situation which applied some thirty years ago at Broken Hill as Williams (1978a, 1979a) allows that the postulated "epigenetic" replacement could have occurred at depths of $\leqslant$100 m within soft sediments.

We must learn from the past – the replacement mechanism should not be rejected or accepted without extensive discussion of the points it raises. This paper summarizes the epigenetic arguments published to date and endeavours to show that all the features can be accounted for in a relatively uncontrived manner by synsedimentary accumulation of the base metal laminae, followed by predictable post-depositional processes. It is hoped that the points made herein will serve to stimulate further constructive discussion and research on the genesis of this important type of mineralization.

2 General Setting of McArthur Deposit

Figures 1–3 summarize the general setting and some features of the McArthur deposit. For further details, the interested reader is referred to the publications by Cotton (1965), Croxford (1968), Croxford and Jephcott (1972), Murray (1975), Lambert (1976), Walker et al. (1977a), and Brown et al. (1978).

3 Timing and Mechanism of Mineralization

The published arguments for epigenetic formation of the McArthur deposit revolve around interpretations of microscopic textures, sulphide-S/organic-C relationships and isotope data. Before considering these more esoteric features, it is instructive to discuss whether an epigenetic process could adequately account for the perfectly conformable nature of the sulphide bands, and for certain of the soft-sediment structures within the ore.

3.1 Conformable Sulphide Bands

Williams (1978a, 1979a,b) explained the ore banding in terms of influx of hydrothermal fluids along relatively permeable, coarse-grained, inter-ore beds, drawing an analogy

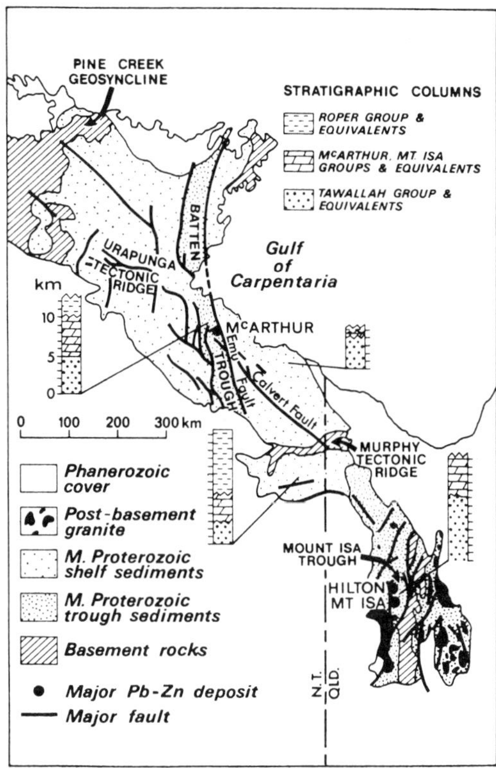

Fig. 1. Regional geology of the central part of northern Australia. (Based on Plumb and Derrick 1975, reproduced from Lambert 1976). The McArthur deposit occurs near the edge of the fault-bounded Batten Trough, which contains a thick sequence of largely shallow-water sedimentary rocks; dolomitic lithologies predominate and these contain abundant evaporite relicts (Walker et al. 1977b; Brown et al. 1978; Muir 1979)

with the formation of alteration zones parallel to the direction of fluid flow during vein formation. However, this is more contrived than synsedimentary precipitation because (1) alteration zones associated with vein deposits are not normally characterized by sulphide laminae and are never perfectly bedded; (2) metalliferous fluids introduced into sediments (i.e., not flowing through fissures) produce banding parallel to the migrating fluid *front* (e.g., Lambert and Bubela 1970), so that solely conformable metal sulphide laminae should not have formed from fluids migrating along permeable strata and thence into adjacent, fine-grained sediments which had been variably slumped and de-watered; and (3) there is no significant mineralization in the inter-ore beds (e.g., Croxford et al. 1975).

3.2 Soft-Sediment Features

A variety of soft-sediment structures occur in the McArthur deposit. The significance of only two of these will be discussed here.

1. There are some examples of mineralized pyritic shale fragments within inter-ore breccias (e.g., Croxford 1968). These are squeezed between dolomite fragments and evidently were derived by dislocation of still-plastic sediment. There are no concentrations of sphalerite and galena around the edges of the fragments as would be expected

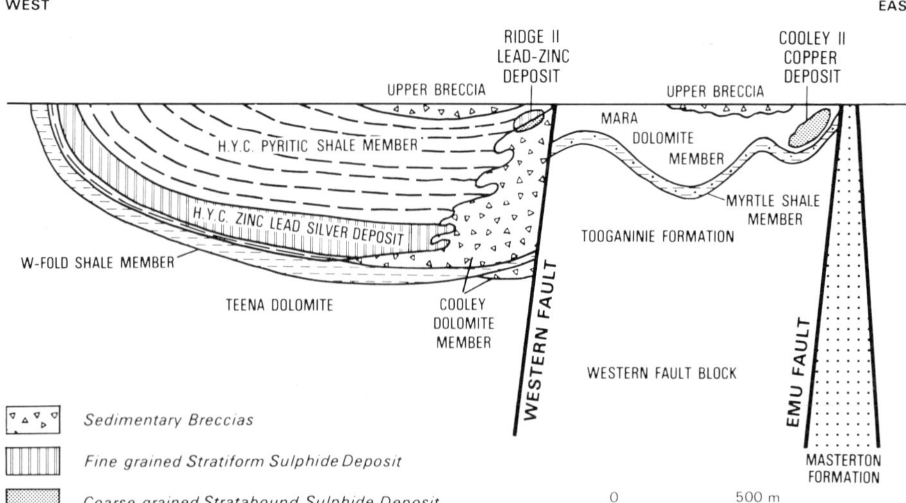

Fig. 2. Schematic east-west cross section through the H.Y.C. sub-basin and Western Fault Block, showing distributions of stratiform and discordant (strata-bound) mineralization (Walker et al. 1977a). The McArthur deposit occurs near the base of the H.Y.C. Pyritic Shale, which accumulated in a local depression. Whilst tuffaceous layers occur sporadically throughout the sedimentary sequence, pyroclastic contribution reached a maximum in the mineralized unit; it is fine-grained, commonly vitroclastic and evidently was transported through the atmosphere from distant (unidentified) terrestrial or shallow water eruptive sites. Widespread discordant mineralization in the Western Fault Block probably precipitated in "feeder channels" for the stratiform mineralization (e.g., Lambert 1976; Williams 1978a, 1979a,b)

if they had been mineralized after incorporation in the breccias. Such mineralized fragments are readily explained in terms of accumulation and eventual disruption of synsedimentary mineralization in the vicinity of the Western Fault, on the unstable eastern flank of the ore basin (Fig. 2). On the other hand, Williams (1979a) explained them in terms of a replacement model. He noted they occur not far beneath the top of the ore, and could have been derived from near the base of the underlying mineralized zone, which now averages 40 m thick, but which he envisaged as being up to 100 m thick at the time. On this basis, he concluded that mineralization could have been introduced at depths of up to 100 m in the sediments. However, with an absence of evidence for extremely localized explosive activity of rapid diapirism, it does not appear that sediments from depths of some tens of metres could have been brought to the surface and incorporated in clastic flows *straight after* their mineralization.

2. There are thin intervals of irregularly folded ore, often with scour features at the base, which are certainly most reasonably interpreted in terms of precipitation and slumping of these sulphide laminae prior to deposition of immediately overlying, undisturbed ore.

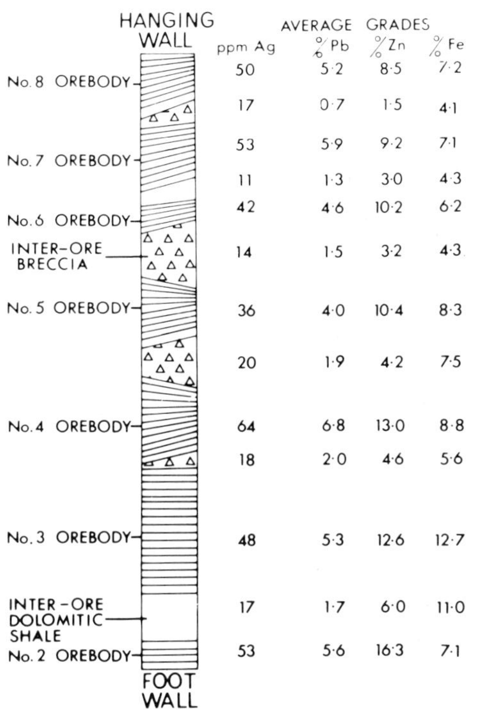

Fig. 3. Sectional column showing the seven ore bodies, inter-ore beds and metal grades in McArthur deposit (based on Murray 1975, reproduced from Lambert 1976). The ore bodies comprise mainly very fine-grained pyrite, sphalerite and galena in perfectly conformable laminae. The sedimentary breccia beds were derived from the Western Fault Block (Fig. 2; Walker et al. 1977a; Brown et al. 1978), presumably mainly as a result of synsedimentary fault movements

3.3 Pyrite Textures

Pyrite is of particular importance in the genetic debate as there were obviously at least two stages of pyrite formation in the ore bodies (Reinhold 1961, quoted by Cotton 1965; Croxford 1968; Croxford and Jephcott 1972; Lambert 1976; Williams 1978a,b, 1979a). The most abundant is the first-formed variety, comprising euhedral to spherical crystals which are normally less than 5 μm across. This form is commonly surrounded by later overgrowth pyrite which occurs as thin rims, concretions, and matrix in the uncommon framboids. Etching with HNO_3 emphasizes the two forms, as later pyrite is relatively readily attacked in the majority of cases.

The outstanding point of contention concerns the timing of formation of later pyrite and of Pb-Zn mineralization. In this regard, Williams (1978a,b, 1979a) emphasized two points about sphalerite and galena associated with pyrite grains comprising both early formed (Py 1) and later-formed overgrowth (Py 2) varieties. The first is that the zinc and lead sulphides always touch the overgrowth pyrite, but rarely the early-formed variety (of course, they can touch Py 1 where this has no Py 2 overgrowth). The second is the absence of inclusions of the other sulphides between the two forms of pyrite. He interpreted these features as indicating that the small Py 1 crystals grew early diagenetically, the pyrite overgrowths grew later, and the sphalerite and galena formed last. However, experimental work (e.g., Stanton 1964; Roberts 1965) has clearly demonstrated that *apparent* paragenetic sequences form readily in mixtures of fine-grained low-temperature sulphide minerals. Bearing this in mind, together with the perfectly concordant sulphide laminae and soft-sediment structures in the McArthur deposit (above), the overgrowth pyrite could well have formed *after* synsedimentary precipitation of the fine-grained zinc and lead minerals. Processes which could have been involved in generation of Py 2 include continuing mobility of iron and reduced sulphur species in the mineralized muds, nucleation on Py 1, and mild recrystallization during burial (cf., Lambert 1973). Microanalyses suggest the presence of trace amounts of base metals within Py 2 (e.g., Williams 1978a,b), which (1) is in accord with Py 2 formation after Pb-Zn mineralization, (2) offers an explanation of its relative reactivity with HNO_3, and (3) means that continued growth of pyrite during burial of the unmineralized sediments should not have produced such base-metal bearing overgrowths.

Another interesting observation is that pyrite commonly encrusts or replaces bacteria preserved in chert blebs within the ore, whilst sphalerite and galena do not (Oehler and Logan 1977). Williams has cited this as evidence that Pb-Zn mineralization had not formed at the time these fossils were pyritized and incorporated in the chert. However, this is again not necessarily the case, for two reasons. Firstly, bacteria preferentially concentrate iron over zinc and lead; secondly, the latter metals were probably virtually completely fixed as sulphides whilst there was still some dissolved iron available (as indicated by the ferroan nature of dolomite in and around the ore). That chert formation and pyritization of bacteria occurred very close to, or at, the sediment-water interface is indicated by the high quality of microfossil preservation (Oehler and Logan 1977).

Pyrite laminae occasionally pass through ferroan dolomite concretions which occur above and below the ore zone (Williams 1978c, 1979a). These laminae are thicker outside the concretions than they are inside, suggesting that a major proportion of this bedded pyrite grew after the concretions. Iron must have continued to migrate within the sediments during their burial to depths that could well have exceeded 10 m. The iron could have been precipitated biogenically as bacterial sulphate reduction can continue in buried sediments, albeit at reduced rates as organic nutrient supplies diminish. This is analogous to the situation observed in the banding experiments referred to above (Lambert and Bubela 1970), whereby the bands thickened with time as metals continued to migrate in the sediments. As similar concretionary structures have not been observed in the ore bodies, they do not provide any control on timing of the Pb-Zn mineralization.

3.4 Geochemical Evidence Against Replacement of Major Components of Host Strata

Chemical analyses are available for a large number of rock samples from the McArthur area (Croxford and Jephcott 1972; Lambert and Scott 1973; Corbett et al. 1975). These do not indicate compositional differences between mineralized and unmineralized H.Y.C. Pyritic Shale of the right order of magnitude for generation of the McArthur ore by replacement of the major alumino-silicate or carbonate components.

Williams (1978a,b, 1979a,b) favours ore formation by replacement of organic matter in the course of abiological sulphate reduction. This is discussed in the next section.

3.5 Sulphide-S and Organic-C Relationships

Williams (1978a, 1979a) noted that poorly mineralized samples from the McArthur deposit (Pb + Zn < 0.2 mol/kg), have a near-symmetrical, unimodal S/C frequency distribution like that of H.Y.C. Pyritic Shale outside the deposit, and unmineralized anoxic sedimentary rocks from other localities. He pointed out that the well-mineralized samples have generally higher, widely varying S/C ratios (Fig. 4), which he assumed formed from the unimodal distribution during deposition of sphalerite, galena and over-growth pyrite. On the basis of mathematical calculations of S/C ratios expected for several potential mineralizing processes, and his interpretation of the textural features (above), he concluded that ore formation involved introduction of metalliferous, sulphate-bearing hydrothermal fluids into the pyritic (Py 1) sediments, where sulphate was reduced by organic matter.

However, Finlow-Bates (1979) strongly criticized Williams' statistical treatment and his conclusions. Furthermore, Scott and Lambert (1979) questioned his fundamental assumption pointing out that he grouped highly pyritic hanging wall shales with the ore samples, even though the former are from up to 250 m above the ore and have insignificant zinc and lead contents. These pyrite-rich hanging wall rocks exhibit a similar S/C distribution to the ore samples (Fig. 4), so it is obvious that such a distribution does *not* indicate sulphate reduction accompanying epigenetic introduction of zinc- and lead-bearing fluids. Williams (1979b) surmised that the bulk of this hanging wall pyrite formed as a result of epigenetic, non-biological sulphate reduction, citing

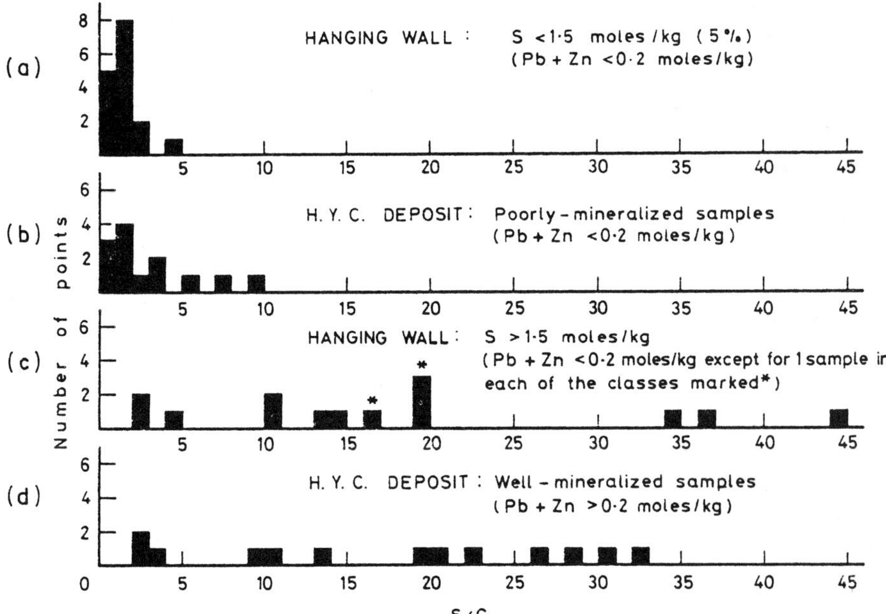

Fig. 4. Sulphide-S/organic-C distributions in and around the McArthur deposit. (From Scott and Lambert 1979)

as evidence the local occurrences of relatively coarse pyrite and growth boundaries in some crystals. These features are not diagnostic of formation of the hanging wall pyrite by such a process; they are adequately explained by formation of the high concentrations of pyrite over a period of time from iron and reduced sulphur migrating within the sediments. The latter would be analogous to the mode of formation discussed above for formation of laminar pyrite associated with dolomite concretions, and it does not require extensive hydrothermal reduction of sulphate by organic matter.

The consideration of S/C ratios masks the important feature that the average organic carbon contents (from Corbett et al. 1975; Lambert and Scott 1975) are similar in the ore (average ~ 0.4 wt.% C) and unmineralized H.Y.C. Pyritic Shale (average ~ 1 wt.% C), particularly after allowance for dilution by sphalerite, galena and clastic dolomite interbeds in the mineralization. Also, organic matter in the ore is compositionally indistinguishable from that outside it (Saxby 1970, 1976). These factors militate against any model which involves a high degree of hydrothermal degradation of organic matter from the mineralized zone, but not from the unmineralized sediments.

The mathematical calculations of Williams are based on the implicit assumption of a "one-for-one" relationship between loss of organic carbon and production of sulphide. This is not sound because any non-biological degradation of organic matter certainly need not have been accompanied by production of quantitatively significant amounts of reduced sulphur. Furthermore, whilst bacterial sulphate reduction was almost certainly an important biological process consuming organic matter, there should have been other microorganisms degrading organic matter in the sediments without producing reduced sulphur. Therefore, S/C relationships should not be used as a key element in assessing ore genesis processes; they do not constitute a cogent argument against the simple introduction of reduced sulphur into the sedimentary basin in zinc- and lead-bearing exhalations.

3.6 Isotopic Features

There is an approach to sulphur isotopic equilibrium coexisting galena and sphalerite, but pyrite incorporated different sulphur which became progressively ^{34}S-enriched from base to top of the ore (Smith and Croxford 1973). Galena has a nearly constant lead isotope composition in the ore and yields a model age of 1600 m.y., which is in the range indicated by other controls (Gulson 1975; Richards 1975). Whilst trace amounts of lead in sphalerite are isotopically identical to the galena, lead in pyrite is variably more radiogenic (Gulson 1975).

Smith and Croxford (1973), Gulson (1975) and Croxford et al. (1975) interpreted their results in terms of sea-floor precipitation of the base metal sulphides. They envisaged that the pyrite incorporated biogenic sulphide, whilst the sphalerite and galena precipitated from reduced zinc-lead-sulphur exhalations derived from a different source to the iron. In contrast, Williams (1978a,b, 1979a) interpreted these isotope data in terms of introduction of the base metals after formation of biogenic pyrite. The fact of the matter is that the sulphur and lead isotope data provide important constraints on the sources of metals and sulphur, but do not provide unequivocal evidence of timing of the Pb-Zn-mineralization; neither do they rule out precipitation of a significant proportion of iron sulphide in the mineralization from the ore fluids.

The $\delta^{34}S$ difference between coexisting sphalerite and galena is not constant through the ore bodies, implying either that there were nonsystematic temperature variations associated with stratiform ore formation, or that there was not complete isotope equilibration. Smith and Croxford (1973) favoured temperature variations from roughly 100° to 260°C, but poor equilibration resulting from rapid precipitation and/or low ore fluid discharge temperatures ($\leqslant$ 150°C) may have been an important factor.

Rye and Williams (1978) and Williams (1979a) have shown that $\delta^{13}C$ and $\delta^{18}O$ values for dolomites associated with discordant and stratiform mineralization exhibit a linear trend which can be interpreted in terms of precipitation at progressively decreasing temperatures from the same hydrothermal system; these data do not distinguish between synsedimentary and epigenetic processes for stratiform ore formation. *Model* temperatures calculated from the results (Williams 1979a) vary from around 300°C near the Emu Fault, down to roughly 150°C in the stratiform mineralization.

4 Summary and Conclusions

1. Sea-floor precipitation of the McArthur deposit is strongly implied by its perfectly conformable nature and a number of soft-sediment structures within it.
2. The evidence for a least two stages of pyrite formation in the ore is compatible with synsedimentary lead-zinc precipitation, followed by predictable post-depositional processes. The conclusion that early pyrite preceded late pyrite, which in turn formed before sphalerite and galena, is much less feasible. The latter would have required several cycles of lateral migration of hydrothermal fluids for distances of up to a kilometre into the H.Y.C. Pyritic Shale, with each cycle comprising an iron-bearing fluid followed by a lead-zinc brine. This is very unlikely on many grounds, not least of which is its inability to explain the intricately conformable nature of the sulphide laminae.
3. Geochemical data do not support formation of the stratiform ore by replacement of any components of the host strata. Furthermore, sulphide-S and organic-C relationships should not be used to decide between potential ore-forming processes because a good deal of oxidation of organic matter could have occurred without production of sulphide minerals. Thus, reduced sulphur for sphalerite and galena formation could well have come in with the lead and zinc.
4. Carbon and oxygen isotope data for dolomite associated with mineralization support ascent of hot ore fluids in the Emu Fault zone and cooling of these fluids as they moved through the Western Fault block, where they precipitated discordant mineralization, before forming the stratiform mineralization. These data are entirely compatible with a sedimentary-exhalative origin for the bedded ore, for which they yield a model temperature of at least 150°C. Brines discharging at 150°C would not boil provided they emanated at water depths of 50–75 m. Had the brines boiled, this would not have had any observable effect on sulphide laminae accumulating

away from the immediate vicinity of the discharge zone, but it could have caused metal sulphide precipitation (e.g., Finlow-Bates and Large 1978).

5. Sulphur isotope data support the importance of bacterial sulphate reduction for precipitation of pyrite within and around the stratiform ore, but not for the base metal sulphides. Furthermore, lead isotope data indicate a significant proportion of the iron was derived from a different source from the lead and zinc. Whilst these results indicate that not all of the iron could have been introduced in the ore fluids, they are consistent with synsedimentary mineralization. For example, base metal sulphides and some of the iron sulphide could have precipitated immediately on discharge of the ore fluids at the eastern margin of the basin. Iron from other sources could have entered the basin and interstitial waters separately and given rise to biogenic pyrite in the near-surface sediments, with trace lead in this pyrite becoming generally more radiogenic away from the influence of the ore fluids.
6. Hydrothermal fluids discharging at temperatures of $\sim$ 150°C are not incompatible with bacterial sulphate reduction, a process only known to occur at reasonable rates at $<$ 75°C. The banded nature of the ore implies that there were multiple distinct pulses of ore fluids, so the small volume of hot fluids emanating at any given time would have cooled rapidly without any long term detrimental effect on microbial activity in the basin-floor muds.

Acknowledgments. The author is honoured to have been invited to present this paper on the occasion of the 90th birthday of Prof. Paul Ramdohr. He is grateful for comments from Duncan Large and Terry Finlow-Bates. The Baas Becking Laboratory is sponsored by the Australian Minerals Industry Research Association Limited, the Bureau of Mineral Resources and the Commonwealth Scientific and Industrial Research Organization. This paper was prepared whilst the author was at the Federal Institute for Geosciences, Hannover, supported by an Alexander von Humboldt Fellowship.

References

Brown MC, Claxton CW, Plumb KA (1978) The Proterozoic Barney Creek Formation and some associated carbonate units of the McArthur Group, Northern Territory. Rec Bur Mineral Resour Geol Geophys Aust 1969/45:59 pp

Corbett JA, Lambert IB, Scott KM (1975) Results of analyses of rocks from the McArthur River area, Northern Territory. CSIRO Minerals Res Labs Tech Commun 57:32 pp

Cotton RE (1965) H.Y.C. lead-zinc-silver ore deposit, McArthur River. In: McAndrew J (ed) Geology of Australian ore deposits. 8th Commonwealth Min Metall Congr. Aust Inst Min Metall Melbourne, pp 197–200

Croxford NJW (1968) A mineralogical study of the McArthur lead-zinc-silver deposit. Proc Aust Inst Min Metall 226:97–108

Croxford NJW, Jephcott S (1972) The McArthur lead-zinc-silver deposit, N.T. Proc Aust Inst Min Metall 243:1–26

Croxford NJW, Gulson BL, Smith JW (1975) The McArthur deposit: a review of the current situation. Mineral Deposita 10:302–304

Finlow-Bates T (1979) Studies of base metal sulfide deposits at McArthur River, Northern Territory, Australia: II. The sulfide-S and organic-C relationships of the concordant deposits and their significance – a discussion. Econ Geol 74:1697–1699

Finlow-Bates T, Large DE (1978) Water depths as a major control on the formation of submarine exhalative ore deposits. Geol Jahrb D 30:27–39

Gulson BL (1975) Differences in lead isotope composition in the stratiform McArthur zinc-lead-silver deposit. Mineral Deposita 10:277–286

Gustafson JK, Burrell HC, Garetty MD (1950) Geology of the Broken Hill deposit, Broken Hill, New South Wales. Bull Geol Soc Am 61:1369–1438

King HF (1976) Development of syngenetic ideas in Australia. In: Wolf KH (ed) Handbook of stratabound and stratiform ore deposits, vol 1. Elsevier, Amsterdam, pp 161–182

Lambert IB (1973) Post-depositional availability of sulphur and metals and formation of secondary textures and structures in stratiform sedimentary sulphide deposits. J Geol Soc Aust 20:205–215

Lambert IB (1976) The McArthur zinc-lead-silver deposit: features, metallogenesis and comparisons with some other stratiform ores. In: Wolf KH (ed) Handbook of stratabound and stratiform ore deposits, vol 6. Elsevier, Amsterdam, pp 535–585

Lambert IB, Bubela B (1970) Banded sulphide ores: the experimental production of monomineralic sulphide bands in sediments. Mineral Deposita 5:97–102

Lambert IB, Scott KM (1973) Implications of geochemical investigations of sedimentary rocks within and around the McArthur zinc-lead-silver deposit, Northern Territory. J Geochem Explor 2:307–330

Lambert IB, Scott KM (1975) Carbon contents of sedimentary rocks within and around the McArthur zinc-lead-silver deposit, Northern Territory. J Geochem Explor 4:365–369

Muir MD (1979) A sabkha model for the deposition of part of the Proterozoic McArthur Group of the Northern Territory, and its implications for mineralization. BMR J Aust Geol Geophys 4: 149–162

Murray WJ (1975) McArthur River H.Y.C. lead-zinc and related deposits. In: Knight CL (ed) Economic geology of Australia and Papua – New Guinea – metals. Aust Inst Min Metall, Melbourne, pp 329–339

Oehler J, Logan RG (1977) Microfossils, cherts, and associated mineralization in the Proterozoic McArthur (H.Y.C.) lead-zinc-silver deposit. Econ Geol 72:1393–1409

Plumb KA, Derrick GM (1975) Geology of the Proterozoic rocks of Northern Australia. In: Knight CL (ed) Economic geology of Australia and Papua – New Guinea – metals. Aust Inst Min Metall, Melbourne, pp 217–252

Ramdohr P (1950) Die Lagerstätte von Broken Hill in New South Wales im Lichte der neuen geologischen Erkenntnisse und erzmikroskopischer Untersuchungen. Heidelb Beitr Mineral Petrogr 2: 291–333

Richards JR (1975) Lead isotope data on three Australian galena localities. Mineral Deposita 10: 287–301

Roberts WMB (1965) Recrystallization and mobilization of sulphide at 2000 atmospheres in the temperature range of 50–140°C. Econ Geol 60:168–180

Rye DM, Williams N (1978) Stable isotope geochemistry of the McArthur River ore deposits: an epigenetic sedimentary-type Pb-Zn deposit. Abstr SEG Meet, Toronto, Oct 1978. Econ Geol 73:1397

Saxby JD (1970) Technique for the isolation of kerogen from sulphide ores. Geochim Cosmochim Acta 34:1317–1326

Saxby JD (1976) The significance of organic matter in ore genesis. In: Wolf KH (ed) Handbook of stratabound and stratiform ore deposits, vol 2. Elsevier, Amsterdam, pp 111–134

Scott KM, Lambert IB (1979) Studies of base metal sulfide deposits at McArthur River, Northern Territory, Australia: II. The sulfide-S and organic-C relationships of the concordant deposits and their significance – a discussion. Econ Geol 74:1693–1694

Smith JW, Croxford NJW (1973) Sulphur isotope ratios in the McArthur lead-zinc-silver deposit. Nature (London) Phys Sci 245:10–12

Stanton RL (1964) Textures of stratiform ores. Nature (London) 202:173–174

Walker RN, Logan RG, Binnekamp JG (1977a) Recent geological advances concerning the H.Y.C. and associated deposits, McArthur River, N.T. J Geol Soc Aust 24:365–380

Walker RN, Muir MD, Diver WL, Williams N, Wilkins N (1977b) Evidence of major sulphate evaporite deposits in the Proterozoic McArthur Group, Northern Territory, Australia. Nature (London) 265:526–529

Williams N (1978a) Studies of base metal sulfide deposits at McArthur River, Northern Territory, Australia: II. The sulfide-S and organic-C relationships of the concordant deposits and their significance. Econ Geol 73:1036–1056

Williams N (1978b) Studies of base metal sulfide deposits at McArthur River, Northern Territory, Australia: I. The Cooley and Ridge deposits. Econ Geol 73:1005–1035

Williams N (1978c) The timing of emplacement of sulphide minerals into the H.Y.C. Pyritic Shale Member at McArthur River, N.T. 3rd Aust Geol Congr Abstr Program, p 32

Williams N (1979a) The timing and mechanisms of formation of the Proterozoic stratiform Pb-Zn and related Mississippi Valley type deposits at McArthur River, N.T., Australia. AIME Preprint No 79-51

Williams N (1979b) Studies of base metal sulfide deposits at McArthur River, Northern Territory, Australia. II. The sulfide-S and organic-C relationships and their significance – a reply. Econ Geol 74:1695–1697

Williams N, Rye DM (1974) Alternative interpretation of sulphur isotope ratios in the McArthur lead-zinc-silver deposits. Nature (London) 247:535–537

Precambrian and Phanerozoic Massive Sulphide Deposits

B. BERCÉ [1]

Abstract

Massive sulphide deposits, which occur throughout the geological history of the Earth, reflect only changes in generational conditions. They retain all basic characteristics of this type regardless of the time of their formation. Consequently, massive sulphide deposits will be regarded as a unique group to which a unitarian classification is to be applied. Therefore, for the description of the deposits in space and time, only the general terms will be used, to avoid any doubts concerning their unity.

1 Introduction

The periods which mark the Earth's evolution are reflected also in the massive sulphide deposits. Are the differences of such significance that they should be taken into account, or can they be neglected, especially in view of the fact that various authors include different deposits into this group? It is true that on the large scale the deposits exhibit gradual passages towards the liquational deposits on one, and/or to the proper stratiform or vein deposits on the other side. As a sharp limit to massive sulphide deposits cannot be drawn, an ambiguity will persist dependent on the approach to the problem. A general classification of the massive sulphide deposits is needed for descriptive purposes. Those proposed (to mention a few of them: Bouladon 1973; Hutchinson 1973; Borodaevskaya et al. 1974; Smirnov 1974; Gilmour 1976; Ripp et al. 1977; Malakhov and Malakhov 1978; Plimer 1978) cannot cover the variability of the deposits, which all occur under the same denomination. It seems that common parameters can serve only as basic coordinates within other subclassifications. However, it will be admitted that the actual treatises on the massive sulphide deposits are so numerous that they largely outstrip over 2000 known massive sulphide deposits of the world. The divergencies of opinions as regards formational conditions, classifications etc., are correspondingly numerous.

1 UNDP Assistance to the East African Mineral Resources Development Centre, P.O. Box 1250, Dodoma, United Republic of Tanzania

2 Precambrian Massive Sulphide Deposits

The deposits of this group may in general be divided into two large parts according to the formational age, i.e., into Archean and Proterozoic. The reason for such separation lies in the fact that some differences have been recorded, although all the basic parameters of the type are preserved.

Archean deposits appear mainly in Canada (Abitibi Belt, Superior Province, etc.), some in Africa (Malagassy Republic, Zimbabwe, Tanzania etc.), Asia (Khetri and Singhbum Belts) and other areas. It might be that some other deposits, actually ascribed to the Proterozoic, are older and pertain to this group. The Archean deposits are located in the mobile belts formed of volcano-sedimentary formation and are metamorphosed to various degrees. In Canada, the metamorphism reaches amphibolite facies, in Africa and Asia, however, up to the gneiss facies. No massive sulphide deposits have been reported in rocks of granulite facies. These belts, which have been formed along sutures, are compatible with the zones of the primordial early parts of geosynclines where the next evolutional stages are lacking. Magmatic rocks here carry signs of hybridism and show an island arc affinity (the sequence is composed of magmatic rock and turbidites). The massive sulphide deposits in Canada, where the prevailing part of Archean deposits occur, are associated with the homodromous contrasting formation, therefore the participation of lead in deposits appears. However, there are cases where the antidromous formation is ore-bearing, as in the Manitouwadge area of Ontario. Whenever the deposits are associated with more basic volcanics, the prevailing role is given to copper and sometimes zinc. The deposits resemble, by volcanic association and ore composition, the younger deposits. In Africa such deposits are associated with dunites, ophiolites, amphibolites, gneisses and quartzites. Also here, when the effect of metamorphism is removed, the same features as mentioned above between volcanic association and ore composition are present. Some of these deposits indicate the link to the deposits of the ophiolite suite. Others indicate that a high degree of metamorphism does not have a significant role in the redeposition of massive ores.

In Archean deposits copper and copper-zinc type ores prevail, which is in full agreement with later but similar evolutional processes in the early part of geosynclines. The admixed elements in the deposits are in accordance with basic ore composition and with the geochemical province in which they occur. Whenever they are associated with more mafic volcanic members the increase of Ni and Co has been noticed. However, a distinctive feature of the Archean deposits is that they lack traces of beryllium and of mercury. In them, only mantle source lead is present (Stacy et al. 1977). These deposits occur mainly as proximal, which is expressed in the shapes of orebodies and with the emplacements whether in volcanics or close to them. Distal deposits are very rare (perhaps some deposits in Zimbabwe can be ranked here).

Proterozoic deposits, in general, frame the zones with Archean mineralization, although some new mineralized zones also appear. The mineralization in places appears in the whole Proterozoic sequence, thus indicating that the process has taken part whenever the depositional conditions have been met, regardless of the time.

In Africa the proximal massive sulphide deposits occur in Proterozoic eugeosynclines (Namibia, South Africa, Zimbabwe, Cameroun, Saudi Arabia). Distal deposits are formed in Proterozoic platform sediments around Archean cratons (Maurętania, Senegal, Congo, Angola, Uganda, Namibia, Malagassy, etc.). In both cases, the deposits are of complex composition, although lead and zinc prevail in distal deposits. They are metamorphosed to various degrees, especially in the southern part of Africa where they occur also in gneisses. The deposits associated with ophiolites are rare (Upper Volta, Morocco). Asian distal deposits are developed as Upper Proterozoic along the Siberian platform (Buryatia, Northern Pribaikal, Enisei Kryazh) in shales and carbonaceous rocks, and as proximal in volcano-sedimentary sequences of Transbaikalia and western Pribalkhash'e, where they grade into Cambrian. Proximal deposits exist also in Iran. Australia carries proximal massive sulphide deposits in the Dugald River, MacArthur, and distal which are metamorphosed to gneiss facies (Broken Hill), or to greenschist facies (Mount Isa), or in shales (Lady Loretta). The northern part of Europe, with more than 600 deposits, contains in Kola and Karelia, Lower and Middle Proterozoic massive sulphide ores, though the Baltic shield was mineralized during Middle and Upper Proterozoic. All deposits here are associated with volcano-sedimentary formations, and to a few of them may be ascribed eventual association with ophiolites. In North America the main part of Proterozoic deposits is concentrated in Canada (Churchill and Slave Provinces, up to some deposits in British Columbia) and on the south-eastern side from Canadian Archean deposits up to the New Jersey area in the U.S.A. Proterozoic deposits occur also in Arizona, northern Idaho, and north-west Montana. These deposits are mainly of proximal and complex type, occasionally some of them are presumably distal. Some Appalachian massive sulphide deposits may appertain to the Proterozoic, however, the data are not in full agreement, as the deposits, together with the host rocks, have been metamorphosed up to the gneiss facies. The North American deposits occur mainly associated with acid volcanics and their frequency drops steeply going to more mafic members (Sangster and Scott 1976).

In places Proterozoic deposits have been formed over a large time span of the stratigraphic sequence, compared to the Archean; therefore they may occur in several well-separated levels. They exhibit a large variation in the sites of emplacement: the variation in ore composition is due to the increase of lead and zinc in the deposits and because distal deposits have been formed at various places and environments. In them, gold and silver are nearly always present, similar to the Archean deposits, though the other trace elements depend on the ore composition and on the geochemical province. The general degree of metamorphism is lower compared to the Archean deposits, although at isolated sites it still reaches up to the gneiss facies. The participation of various iron and manganese components increases in the hanging wall of the mineralization. These massive sulphide deposits may be compared with the lower and the middle sequence deposits developed in the early phase of geosynclinal evolution, although with a slight difference represented by the complete lack of evaporites within the stratigraphical column. Proterozoic volcano-sedimentary formations cannot be distinguished from other similar formations by the accompanying phenomena of the mineralization-like processes of alteration, geochemical halos, admixed elements in pyrites, etc. Consequently, Proterozoic massive sulphide deposits represent a further wide step of the formational evolution in time.

3 Phanerozoic Deposits

Sites where Phanerozoic deposits occur are so numerous that all of them cannot be mentioned. The massive sulphide deposits bring out some new phenomena, of which a repeating mineralized sequence nearly at the same place, and a regeneration of older deposits are the most striking. The evolution of massive sulphide deposits during this time span has abandoned some areas with Precambrian mineralization (Africa) and has settled down on a large scale in others (Asia).

The Euroasian continent is well endowed with massive sulphide deposits. The distal Upper Proterozoic deposits along the Siberian platform extend into the Cambrian, and it seems that at some places they are also younger (NE Tuva). The Proterozoic volcano-sedimentary formation of Transbaikalia extends into the Cambrian, containing massive sulphide deposits. The Paleozoic period, up to the Middle Carboniferous, carries the main part of massive sulphide deposit. The ore composition is dependent in individual zones on whether the oceanic or continental crust prevails in the basement. Uralian deposits formed over the oceanic basement occur within the Silurian/Devonian volcano-sedimentary formation, and tend to the copper-zinc type. They are associated with contrasting and steady homodromous formations. The host rocks have been metamorphosed at places up to the greenschist facies. South Uralian deposits contain platinoids, as they occur in the zone with such mineralization. Caledonide deposits, which extend from Norway to Ireland and further south to the Pyrenees, Portugal, Spain and Sardinia, become younger and reach up to Lower Carboniferous age. They exhibit large variation in ore composition and are associated with the contrasting homodromous formation. When in the basement the continental crust prevails, than complex ores are the most representative type, as in the case of the Devonian deposits of Rudnyi Altai. In the case of Kazakhstan the Devonian-Carboniferous deposits are mostly of Pb-Zn type, showing gradation to the stratiform deposits as they have been formed over the continental basement. In both cases the volcanic activity is subordinated and the mineralization is associated only with the contrasting volcanic formations. The distal deposits are formed in sediments of various ages during this metallogenic cycle in Finland and Germany. In some cases the primary deposits have been redeposited in younger Carboniferous rocks, as exemplified in Tynagh (Ireland) or in Rudnyi Altai. The data indicate that some large areas containing massive sulphide deposits are characterized by a certain regularity which enables prediction of the new discoveries. However, opposite examples may also be cited. Caucasus massive sulphide deposits formed in a eugeosynclinal environment appear along the Devonian/Carboniferous volcano-sedimentary sequence which was metamorphosed up to the amphibolite facies and is by diaphthoresis converted to greenschist facies. The next mineralizing stage occurs in the Lower and part of the Middle Jurassic, in a shaly environment. Both types of mineralization are preorogenic, though the third cycle has been formed in platform secondary geosynclines in the Late Jurassic-Early Cretaceous. The case is similar in Japan where, besides the proximal Upper Paleozoic deposits, also Late Jurassic-Early Cretaceous deposits occur, both with a pronounced trend to copper composition, besides the well-known postorogenic Kuroko deposits. Therefore, the mineralizing process may be repeated in a relatively narrow zone in various intervals and with various formational conditions. Turkey represents in this respect a case where in a relatively small area the deposits of various metallogenic cycles and formational conditions occur. The oldest deposits are distal Silurian/Devonian, which resemble German deposits, then appear Jurassic/Cretaceous deposits (Murgul). However, the Cyprus type is also represented (Ergani Maden). Consequently some areas with massive sulphide deposits show simple regularity of their formation, though others may be very complex. The Cyprus type of deposit is represented by massive sulphide copper mineralization associated with ophiolites. Such deposits occur in Italy and Greece, and most probably extend further east of Turkey along the ophiolitic belt. The Tasman orogenic zone contains Cambrian to Devonian early geosynclinal, mostly proximal deposits associated with contrasting volcanic formation, but Permian deposits here represent a late geosynclinal deposition. The proper copper deposits are rare, although nearly the whole Paleozoic sequence is mineralized. The Tasman Zone may be extended over Sulawesi to the Philippines and Japan. Appalachian deposits in Canada and the U.S.A. are formed within a volcano-sedimentary environment, in association with contrasting formation between Cambrian and Devonian, and similar is the case on the western side of the continent from Alaska to British Columbia. However, Cordilleran deposits are also of Lower Mesozoic age and may be compared with the post-mobile deposits which in Cuba reach Jurassic and Cretaceous age, similar to some deposits in British Columbia.

It appears that the formational environment of the Phanerozoic massive sulphide deposits has been enlarged from the lower to the upper parts of a mobile belt. However, they still preserve all the basic characteristics of its type, regardless of pre- or postmobile belt evolution. In places they also accompany the porphyry copper deposits. In general, geologically younger sequences are characterized by the prevailing role of lead and zinc (Krivtsov et al. 1978), although the differences are not striking. The composition of ore depends namely on two basic factors: the type of the crust, and the course of magmatic differentiation. Copper lodges in younger formations in the porphyry copper type, mainly; thereafter, its significance in massive sulphide deposits diminishes.

4 Final Observations

Massive sulphide deposits exhibit characteristics associated with the Earths evolution, although the variations are subtle and may be registered only through a large number of observations. The deposits grade from the proper pyrite ones to the complex types during the whole of geological history, but the frequency of each type of deposit changed with time. Pyrrhotite occurs at various places. Its presence is due to two reasons. The first is primarily that pyrrhotite has been deposited, instead of pyrite because of the lack of sulphur; but in the second case, its presence is enhanced by the metamorphic process (desulphurization of a deposit). As to size, massive copper deposits lie on the order of 10^7 tons with the total content of the main metals below 10% (Sangster 1977) and no significant difference in size or grade has been registered over time. However, the deposits with a prevailing lead and zinc component reach up to 10^8 tons and exceptionally (in some cases, but not regularly) more.

Phanerozoic deposits may occur in up to five primary mineralized levels (the deposit Gai, Ural), though in the Precambrian deposits up to two primary mineralized levels have been reported until now. A time spread of mineralization at a site, and more frequent repetition of mineralization in a geological sequence, are characteristics mostly expressed in Phanerozoic deposits.

Massive sulphide deposits formed in one stage (about 5%) are rare; two- or three-stage deposits are the most frequent (about 75%), and the remaining deposits were formed in more stages. According to the data of fluid inclusions, the deposits have been formed between 100° and 500°C (Parilov 1978) with more frequent temperatures between 100° and 200°C and over 300°C (Smirnov 1973). The data of sulubility of ore components are 500°C and about 350–500 atm respectively (Ingerson 1977), which indicates the ambiguity with respect to measured data concerning the anticipated formational conditions. In any case, the deposits represent the sequence in the host rocks because the thickness of the volcanic pile increases in zones of ore deposition and diminishes in the sedimentary environment. The depositional process is also reflected by the behaviour of the Kounrad discontinuity, thickness and composition of the crust (Yakovlev and Avdonin 1976), as well as by a regularity in spacing among the deposits in a mineralized belt, which ranges from about 1 km up to 14 km (Solomon 1976), while data from Rudnyi Altai extend the upper limit of spacing up to 35 km. This regularity among the deposits indicates the structural features of a mineralized area.

Consequently, the parameters of the massive sulphide deposits indicate an entity or consistency regardless of their time of formation. Through so many common characteristics, they may be, on the whole, considered as having formed by a uniform process with some deviations in the general evolutional trend.

References

Borodaevskaya MB, Kurbanov NK, Shirai EP, Krivtsov AI, Gadziev TG (1974) Vulkanogennye formatsii bazaltoidnogo ryada i svyaz s nimi kolchedannogo orudeneniya v razlichnykh provintsiyakh SSSR. Evol Vulkan Zemli Moskva, pp 410–420

Bouladon J (1973) Sur les roches de précipitation chimique qui accompagnent le dépôt des minéralisations sulfurées liées au volcanisme. Int Symp Volcano Assoc Metall Bucharest Res, pp 32, 33

Gilmour P (1976) Some transitional types of mineral deposits in volcanic and sedimentary rocks. In: Handbook of stratabound and stratiform ore deposits, vol I. Elsevier, Amsterdam Oxford New York, pp 111–160

Hutchinson RW (1973) Volcanogenic sulfide deposits and their metallogenic significance. Econ Geol 68:1223–1246

Ingerson E (1977) Posledstviya nekotorykh oshibochnykh kontseptisii v teorii rudoobrazovaniya. In: Geokh i probl rudoobr, Moskva, pp 170–177

Krivtsov AI, Samonov IZ, Shabarshov PYa (1978) O produktivnosti razlichnykh metallogenicheskikh epokh dlya kolchedannogo orudeneniya. Geol Rudn Mestorozhd XX:97–102

Malakhov AA, Malakhov DA (1978) Istochniki veshchestva mednokolchedannykh mestorozhdenii Makanskogo rudnogo polya na Yuzhnom Urale. Geol Rudn Mestorozhd XX:40–51

Parilov YuS (1978) Temperaturi, sostavy i svoistva rudoobrazuyushchikh flyuidov stratiformikh mestorozhdenii svintsa i tsinka v Kazakhstane. Geol Rudn Mestorozhd XX:60–71

Plimer IR (1978) Proximal and distal stratabound ore deposits, Mineral Deposita 13:345–354

Ripp GS, Gurulev SA, Kaviladze MSh, Melashvili TA, Truneva MF (1977) Izotopnyi sostav sul'fidnoi sery stratiformykh mestorozhdenii Buryatti. Tr Inst Geol Geofiz, Acad Nauk SSSR, Sib Otd 361:209–261

Sangster DF (1977) Some grade and tonnage relationships among Canadian volcanogenic massive sulphide deposits. Pap Geol Surv Can 77-1A:5–12

Sangster DF, Scott SD (1976) Precambrian strata-bound massive Cu-Zn-Pb sulphide ores of North America. In: Handbook of stratabound and stratiform ore deposits, vol VI. Elsevier, Amsterdam Oxford New York, pp 129–219

Smirnov VI (1973) Ob osobenostyakh formirovaniya nekotorykh kolchedannykh mestorozhdenii po dannym dekripitatsii i variatsiyam izotopov sery slagayushckikh ikh sulfidov. In: Geosinkl Magm Form Ikh Rudn Vyp 102:3–10

Smirnov VI (1974) Mednye i polimetallicheskie mestorozhdeniya SSSR, svyazannye s initsialnym vulkanizmom. In: Int Symp Volcano Assoc Metall Bucharest Res, pp 129–130

Solomon M (1976) "Volcanic" massive sulphide deposits and their host rocks – a review and an explanation. In: Handbook of stratabound and stratiform ore deposits, vol VI. Elsevier, Amsterdam Oxford New York, pp 21–53

Stacy JS, Dow BR, Silver LT, Zartman RE (1977) Plyumbotektonika. IIA. Dokembriiskie kolchedannye mestorozhdeniya. In: Geokh Probl Rudoobr. Nauka, Moskva, pp 93–106

Yakovlev PD, Adonin VV (1978) Neodnorodnost stroeniya kolchedannonosnykh paleovulkanicheskikh zon. Izv Vyssh Uchebn Zaved Geol Razved 9:72–81

Genesis of Uranium Deposits. An Appraisal of the Present State of Knowledge

F.J. DAHLKAMP[1]

Abstract

Based on geometry and formational environment about twenty basic types of uranium occurrences can be distinguished, seven of them of economic importance.

The various types of occurrences can be empirically attributed to three metallogenic categories. The first includes mineralizations derived from magmatic or palingenetic-anatectic origin, the second formed by sedimentary or supergene processes, the third created by metamorphism of the former two generations.

Although geometric and metallotectonic characteristics of most of the types of uranium deposits are relatively well established and theories of their formation abound, their factual metallogenesis is enigmatic. Many open questions still exist, in particular with respect to the provenance of the ore-forming uranium, the conditions of its mobilization, transport, and redeposition, its behavior during metamorphism etc. Extensive and thorough scientific research studies are required to establish reliable data and thus a better understanding of the metallogenesis of the various types of uranium deposits – a challenge to private companies which own most of the discovered deposits and to scientific institutions which employ the professional staff and equipment for such research. Practical exploration for uranium and the geosciences would also benefit from the work and its results.

1 Introduction

Exploration for uranium deposits was rather successful in the past decade. Numerous papers were published on the new and old deposits, and on their formation. Most of the presented genetic models, however, are based widely on academic theories. Apparently very few thorough and comprehensive research programmes on the genesis of the various types of uranium deposits have been initiated, with few exceptions, for example in France.

1.1 Types of Uranium Deposits

As established in the past, about twenty principle types of uranium occurrences exist, of which seven are of economic significance (Dahlkamp 1974, 1978a, 1979). These are, including the subtypes, given in Table 1.

1 Oelbergstraße 10, 5307 Liessem b. Bonn, FRG

Table 1. Classification of uranium occurrences and deposits

Mode of origin	Host rock/Type of Deposit		Example	
Sedimentary	OLIGOMICTIC CONGLOMERATES		Elliot-Quirke Lake (Canada)	1
			Witwatersrand (S-Africa	2
	Blackshales		Ranstad (Sweden)	1
	Phosphates		Florida (USA)	2
			Cabinda (Angola	2
Effusive	ACID VOLCANICS		Peña Blanca (Mexico)	1
			McDermitt (USA)	1
Intrusive	Intraintrusive: Peralkaline Syenites		Ilimaussaq (Greenland)	1 + 2
	Intraintrusive: Carbonatites		Palabora (S-Africa)	2
	Intraintrusive: ALASKITES		Rössing (SW-Africa)	1
	Intraintrusive: Pegmatitic Alkali-Granites		Ross Adams (USA) ⚒	1
	Intraintrusive: Granites		Bingham (USA) (Cu-Porphyry)	2
	Pegmatites		Bancroft (Canada) ⚒	1
	HYDROTHERMAL VEINS		Schwartzwalder (USA)	1
Contact-metasomatic	Calc-Silicates		Mary Kathleen (Australia)	1
Metamorphic	Phyllites		Forstau (Austria)	1
	Schists		Portugal	1
Metamorphic / Supergene	VEINLIKE TYPES	Subtype A	Beaverlodge (Canada	1
			Jabiluka (Australia)	
		Subtype B	Key Lake (Canada)	1
		Subtype C	Massive Central (France)	1
Supergene	SANDSTONES	Subtype A	Grants Mineral Belt (USA)	1
		Subtype B	Wyoming Basins (USA)	1
		Subtype C	Franceville Basin (Gabon)	1
	CALCRETE		Yeelirrie (Australia)	1
	Lignites		N-S Dakota (USA) ⚒	1
	Phosphates		Bakouma (ZAR)	1 + 2
	Karst		Bighorn/Wyo. (USA) ⚒	1

CAPITAL LETTERS (underlined): economic deposits
CAPITAL LETTERS (dotted underline): probably economic deposits
Small letters: subeconomic deposits
Rössing (dashed underline): 1979 in production
⚒ former production

1: main-product
2: by-product

Da. 9/1979

Two additional types will be mentioned: the contact-metasomatic uranium deposit Mary Kathleen in Australia and the Madawaska pegmatitic uranium deposit in the Bancroft district, Canada. Both were mined in the 1960's and after a shutdown for several years they were reopened in the mid 1970's. They are only viable at present due to the existence of the old written-off milling and mining installations.

1.2 Metallogenetic Scheme of Uranium Mineralizations

The various types of uranium occurrences and deposits can be grouped genetically into three categories which, in consequence, can be related to distinct geological processes (see Fig. 1).

These are:

1. the primary, magmatogenic uranium occurrences formed by endogenic processes
2. the secondary uranium occurrences which were formed by subsequent exogenic, i.e., sedimentary or supergene processes from the primary type.
3. the tertiary generation which was formed by endogenic metamorphic processes from the primary as well as from the secondary uranium occurrences.

This cycle closes if by anatexis or palingenesis the secondary and tertiary types are retransformed into the primary type. On the other hand, exogenic processes can leach uranium from the metamorphic environment and reconcentrate it to form secondary deposits.

The formation of uranium deposits, however, should not be understood to be restricted to these principle genetic schemes. Instead, complex polygenetic processes must be considered in the formation of several kinds of deposits.

2 Origin of Uranium and Its Ore-Forming Processes

The genesis of uranium mineralizations and the implications involved in the various inferred metallogenic processes may be summarized as follows:

2.1 Primary Occurrences

The origin of the uranium in the primary occurrences is reasonably well established as either juvenile magmatic, prevailing in the early Precambrian, or palingenetic/anatectic, prevailing in younger times.

To the primary occurrences belong the intra-intrusive and hydrothermal deposits and, with a question mark, the deposits within acid volcanics (Fig. 2).

Examples: At Rössing, Namibia, an *intrusive* alaskite [according to Tröger (1969) nomenclature: aplite granite since it contains 20% placioclase (Adloff, pers. comm.)] is mineralized with irregularly distributed uraninite and betafite. The Upper Proterozoic alaskite is supposed to be derived by anatexis from (meta)sediments of the Middle (?) Proterozoic Nosib Group.

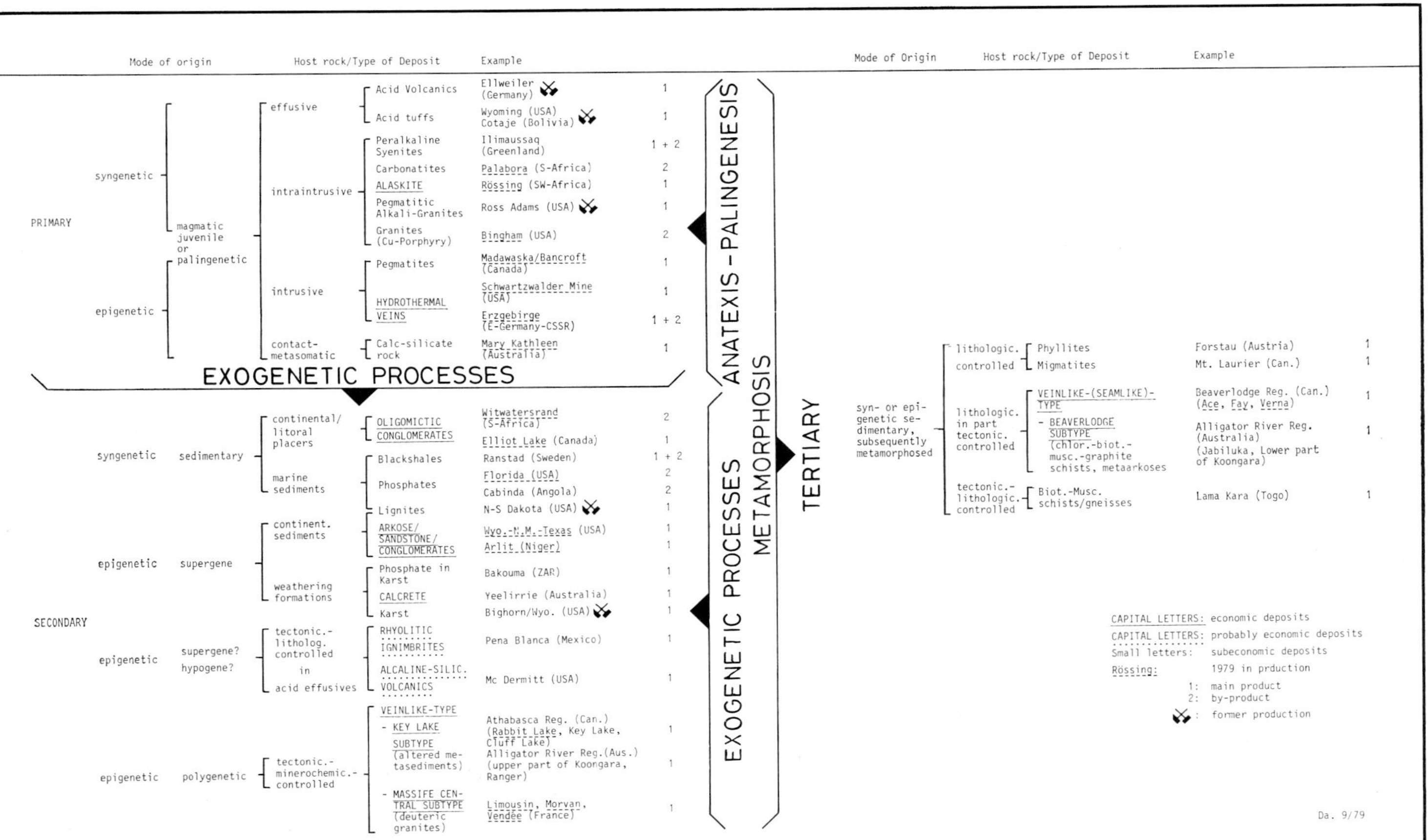

Fig. 1. Classification of uranium occurrences and deposits

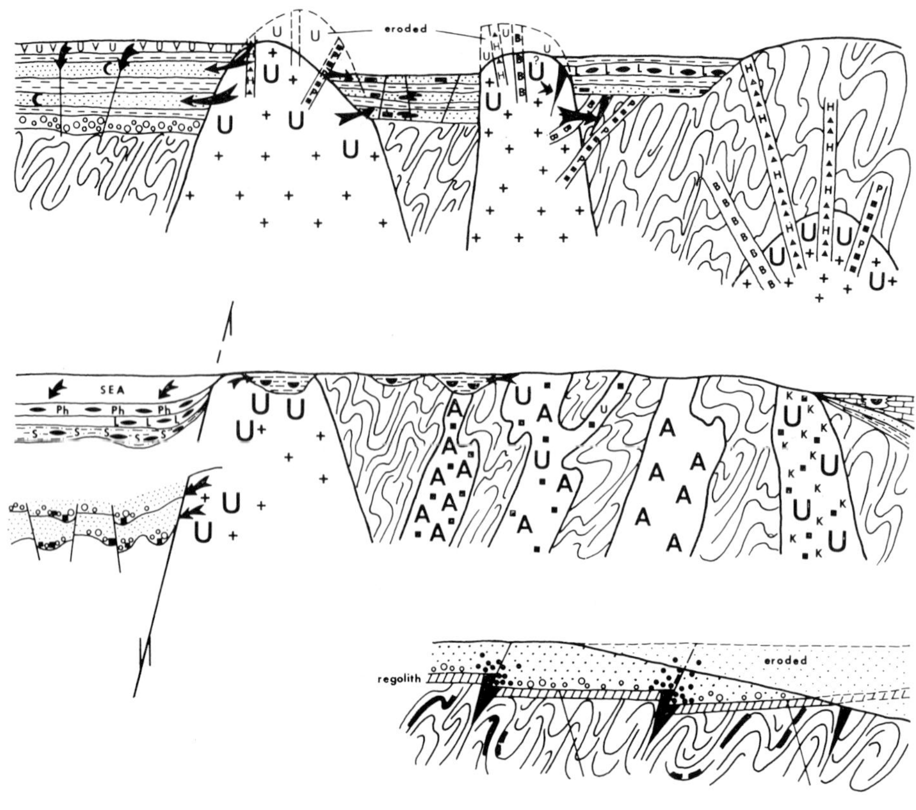

U granitic U-source rocks >5 ppm U
VU uraniferous acid volcanics >5 ppm U
▪ Uraninite, Uranothorianite
• U-impregnation
o▪ oligomictic conglomerates
▬ peneconcordant sandstonetype
(Rollfront-type
tect.-lith. sandstonetype
veinlike type, Key Lake Subtype, Massive Central Subtype
veinlike type, Beaverlodge-Jabiluka Subtype
H▲ hydrothermal veins with U
▪A uraniferous Alaskites
Calcrete type
K▪ U-bearing carbonatite
P▪ U-bearing pegmatite
Ph▬ U-bearing phosphate
L▬ U-bearing lignite
S▬ U-bearing blackshale
B basic dykes

Fig. 2. Diagram of occurrence and formation of U-deposits

Mineable *hydrothermal* deposits are known from the Hercynian orogenic belt in Central (Erzgebirge, Schwarzwald) and Western Europe (Massif Central), from Schwartzwalder Mine, USA; Shinkolobwe, Zaire; Port Radium (?), Canada, and elsewhere. The deposits in Eastern Europe occur overwhelmingly in sediments and metasediments peripheral to granitic batholiths, whereas the deposits in Western Europe concentrate mainly within granites but at their outer contact.

Among the various paragenetic subtypes, the pitchblende-barite-SiO_2 and pitchblende-quartz-carbonate parageneses are the most important ones.

Contrary to former belief and as recently established (Mrna 1963, 1967; Baumann 1967) the various metals in the polymetallic Co, Ni, Ag, Bi, U deposits of, for instance, Joachimsthal in the Erzgebirge, ČSSR, are not paragenetically related. They form independent primary mineralizations, but are superimposed in part on each other by hydrothermal fluids of successive pulsations and ages, thus forming secondary multi-element associations by mineral "telescoping". While it is generally accepted that the source of the uranium is deuteric granite, e.g., of the late Hercynian Eibenstock Massif for the uranium veins of Joachimsthal, the provenance of the Co-Ni-Ag-Bi assemblages remains questionable with regard to the deuteric Eibenstock granite. Some authors (Baumann 1967) propose even an Alpidic age. Others (Ruzicka 1971) question a hydrothermal derivation of all the metals from the Eibenstock granite. They consider a kind of lateral secretion as ore-forming process.

Formerly the intragranitic vein-like deposits in Western Europe were interpreted widely as of supergene origin, e.g., by Geffroy and Sarcia (1958), Moreau et al. (1966), Barbier (1974). Recent studies, particularly on fluid inclusions (Cuney 1978; Leroy 1978) hint more at a complex hydrothermal formation of the initial pitchblende generation, with later supergene processes causing a further concentration or dilution of the primary ore.

Cuney (1978) favours the following evolution for the Bois Noirs deposit in the North-eastern Massif Central, France:

The Bois Noirs granite resulted from the anatexis of uranium-rich sediments near granulite facies conditions. This magma was syntectonically emplaced. A fluid phase formed after a large part of the magma had crystallized. This fluid altered the primary magmatic minerals, resulting in the partial liberation of the uranium of accessory minerals such as sphene, zircon etc. When all the magma had crystallized the fluid phase migrated to the outer zones of the batholith where it precipitated uraninite, with an associated quartz-muscovite alteration. Continuing thermal activity caused hydrothermal convective circulation with water of a supposed meteoric origin which became mixed with carbon dioxide of a deep origin. These CO_2-rich fluids dissolved the uranium contained in uraninite and transported it mainly in form of uranyl-mono-carbonate with a significant amount of H_2S in solution. Deposition of pitchblende in alternation with pyrite and marcasite occurred in breccia zones where the fluid pressure dropped from lithostatic to hydrostatic, i.e., where the CO_2 concentration decreased from a maximum of 3 mol% to less than 1 mol%. Subsequent processes formed five additional stages of mineralization followed finally by supergene remobilization of the primary ore during Oligocene times.

Leroy (1978) concludes from his research results for pitchblende deposits in the Limousin (La Crouzille) district, Western Massif Central, France:

After the emplacement of the St. Sylvestre granite, lamprophyres intruded producing localized hot spots. The heat flux caused the reheating of cold waters which impregnated the plutonic complex. It produced a convective fluid circulation. The fluids were channeled into certain structures causing muscovitic episyenitization of the granite (dissolution of quartz, total muscovitization of plagioclase and biotite etc.). The presence of CO_2, dissolved at depth in these solutions, increased the dissolution of uranium from the granitic host, transporting the uranium as uranyl-carbonate complexes. Where these solutions hit porous rocks (mica episyenite and/or fractures) the pressure dropped, destroying the uranyl-complex. Pitchblende with pyrite were deposited, followed by microcrystalline quartz, marcasite, hematitization of pyrite, and coffinite formation from pitchblende, and in a later stage by fluorite, barite and finally calcite. The primary pitchblende formation took place immediately after the emplacement of the lamprophyres and the muscovitic episyenitization. Younger to recent supergene modifications caused the formation of secondary ore minerals (sooty pitchblende, etc.).

The genesis of uranium deposits within, or associated with *acid volcanics,* as found in the Sierra de Peña Blanca, Mexico, of McDermit, U.S.A., has not yet been investigated in detail. It can be stated only that the mostly rhyolithic host volcanics are generally rich in uranium and form probably the primary source of the deposits, but it has not been established what processes carried the uranium into the host structures and concentrated it to economic magnitudes. Two concepts can be considered either subvolcanic hydrothermal solutions with a uranium source in depth, or supergene fluids leaching uranium from the highly uraniferous tuffs, ignimbrites etc. or a combination of both.

2.2 Secondary Mineralizations

Secondary mineralizations owe their existence to exogenic processes. Three kinds of uranium deposits formed: The oligomictic conglomerates, the various sandstone and in part the vein-like types and, with a question mark, the acid volcanic type (see description "primary occurrences").

In Upper Archean-lowest Lower Proterozoic times when an oxidizing atmosphere was still lacking, mechanical weathering and sedimentation formed placer deposits with detrital uranium as proven by Ramdohr (1955, 1958) in oligomictic conglomerates.

Examples: Blind River-Elliot Lake, Canada; Witwatersrand, South Africa.

After the change to an oxidizing atmosphere which occurred no later than the middle of the Lower Proterozoic (approx. 2200 m.y.) (Roscoe 1969; Frarey and Roscoe 1970; Schidlowski et al. 1974), chemical liberation of uranium led to subsequent mineralization in a suitable lithochemical environment. Dependent on climatic and transport conditions uranium was either:

a) Transported into the sea and concentrated in organic ooze (sapropels, black shales, coal), and in marine phosphates. Examples: Cambrian black shales near Ranstad, Sweden; Tertiary phosphates, Florida, U.S.A.; Lignites, Black Hills, U.S.A. or
b) Concentrated into terrestrial sediments and tensional structures in intracratonic and intramontane basins, occasionally in littoral clastic sediments, and in certain weathering crusts. This was the mode of origin of the sandstone (Examples: Colorado Plateau, Wyoming Basins, U.S.A.; Lake Frome Embayment, Australia; Franceville Basin, Gabon) and calcrete (Yeelirie, Australia; Langer Heinrich, Namibia) types of deposits, and partially of the structurally controlled unconformity subtype of the vein-like types of deposits (Key Lake, Rabbit Lake, Canada) and according to some authors (Barbier 1974) also of the granite endocontact subtype in the Massif Cen-

tral, France (see also section "primary occurrences"). However, for the vein-like subtypes of deposits the supergene concentration process has to be considered as only a part of a polygenetic evolution (see section with hydrothermal deposits).

The local formation of deposits of the secondary category is influenced and controlled by distinct metallogenetic conditions such as provenance and lithology of host rocks, presence of uraniferous source rocks, certain kinds of palaeogeography, palaeoclimate, palaeoweathering and palaeotectonics. Where these factors coincided with suitable sedimentological and/or structural preparation of the host rocks including the required permeability, redox interface etc., uranium deposits formed.

The formation of sandstone-type deposits was extensively studied particularly in the U.S.A. (Grutt 1972; Bayley and Childers 1977; Adams et al. 1978; Davis 1979; see also Dahlkamp 1979; to name a few of many authors).

A model for genesis of the vein-like-unconformity type of deposits was presented by Dahlkamp (1978a, 1981). According to his interpretation of comprehensive studies of the Key Lake ore bodies and other deposits in the Athabasca region, Canada, the uranium was leached from uranium-rich metasediments during a humid climate, transported and concentrated in structural traps, then covered by red bed facies sandstones and altered and further concentrated ± in situ by diagenetic-hydatogenic processes. Subsequent supergene influences modified the mineralization forming new generations of ore minerals.

2.3 Tertiary Mineralizations

The metallogenesis of the tertiary category of uranium mineralizations is still quite vague. Very little is known about the behavior of uranium during metamorphism.

In comparing rock units of different grades or facies of metamorphism, the following can be noted:

a) In phyllites of greenschist facies, such as those from Forstau, Austria, the ore controls are clearly lithologic. Based on all apparent indications, the introduction of uranium was exogenic into lagoonal sediments (Petrascheck et al. 1974, 1977).

Syngenetic or epigenetic introduction has not yet been established. A considerable mobilization of uranium by metamorphic processes does not seem to have taken place. The main ore mineral is pitchblende, but upheating may have caused the start of recrystallization ordering the atoms into the cubic lattice and forming minute uraninite crystals.

b) In chlorite-muscovite-biotite-graphite-schists and -gneisses, meta-arkoses, carbonatic quartzites etc. of amphibolite-granulite facies with retrograde metamorphism, numerous minor mineralizations with uraninite, and the strata-structure bound uranium mineralizations of the Beaverlodge-Jabiluka-subtype of the vein-like type of deposits are observed, e.g., in the Ace-Fay Mine in northern Saskatchewan, Canada, and in Jabiluka in the East Alligator River area, Australia. They still exhibit a distinct strata-bound and lithologic control with a structural overprint.

Certain metallogenetic parameters indicate that the mineralizations originated from uranium-rich pelitic to semipelitic sediments which underwent metamorphism. Apparently the metamorphic effect on the pre-existing uranium was limited to a more or less in situ remobilization and concentration of the uranium without significant transportation, admission or removal, of the ore, as indicated by the commonly almost stratabound appearance of the mineralization.

The main impact of the metamorphosis appears to be a re-crystallization of the uranium as implied from contemporaneous ages for the oldest uranium generation, which comprises uraninite, and its metamorphic host rocks. Later block tectonic activity, probably accompanied by some hydrous processes (water source at least in part supergene, Sassano et al. 1972) caused partial rejuvenation and redistribution into structures of the older uranium generation, thus several uranium generations can be distinguished in these deposits.

c) In biotite-muscovite schists or gneisses of the amphibolite facies, e.g., in the uranium occurrences in Togo, the ore appears structurally and lithologically controlled. The origin of the uranium in this geological environment is ambiguous. It can be explained theoretically to be either of magmatic origin, or by mobilization and reconcentration during or after metamorphism deriving from uraniferous (meta-)sediments. The impact of metamorphism on the formation of these mineralizations, however, remains still unexplained.

3 Conclusion

Physical and metallotectonic characteristics of the various types of uranium deposits are well documented. In contrast, the genesis of most of the deposits remains enigmatical. Extensive and thorough research work was done on the vein-like and/or hydrothermal uranium deposits in France, and also a certain amount on sandstone deposits in the U.S.A. Investigations of varying degree were dedicated to the various vein-like types of deposits in Australia and Canada and likewise to the remaining types. Theoretical and empirical deductions and interpretations of the available data by many knowledgeable geologists have contributed interesting and valid concepts and models; but they cannot replace the irrefutable proof. Many open questions still exist with respect to the provenance of the ore-forming uranium, the conditions of its mobilization, transport and redeposition etc. Comprehensive scientific research programmes have to be undertaken to gather the range of data and parameters, in the field and in the laboratory as well, in order to establish the metallogenesis of the various types of uranium deposits also quantitatively as exactly as possible. Some of the obvious questions are for example: What kind of granites can be considered as fertile source rock for hydrothermal and vein-like type deposits? To what extent can the various uranium oxide phases (uraninite ± Th, pitchblende etc.) be used as indicators for the formational environment (P-T conditions)? Are viable ore bodies of certain types of deposits always surrounded by alteration halos? What are characteristic or indicative alteration- and gangue-minerals, -elements, -isotopes? To what distance do they extend into the wall rock? Can they be indicative for close-by ore bodies? Are diagenetic or metamorphic processes capable of causing lateral secretion of uranium in a magnitude to form deposits? What happens with uranium during the different grades of metamorphism? Why are uranium deposits found only within the Hercynian orogen, not in the Caledonian and Alpidic mobile belts?

This limited selection of problems may hopefully initiate the comprehensive research work on uranium deposits, long neglected in the past, particularly in Germany; a challenge, to scientific institutions, universities and private industry, especially in the present light of the energy crisis and the consequent major and unavoidable demand for uranium in the future.

References

Adams SS, Curtis HS, Hafen PL, Salek-Nejad H (1978) Interpretation of postdepositional processes related to the formation and destruction of the Jackpile-Paguate uranium deposit, Northwest New Mexico. Econ Geol 73:1635–1654

Bailey RV, Childers MO (1977) Applied mineral exploration with special reference to uranium. Westview Press, Golden, pp 1–542

Barbier MM (1974) Continental weathering as a possible origin of vein-type uranium deposits. Mineral Deposita 9:271–288

Baumann L (1967) Zur Frage der varistischen und postvaristischen Mineralisation im sächsischen Erzgebirge. Freiberg Forschungsh C 209:15–38

Cuney M (1978) Geologic environment, mineralogy, and fluid inclusions of the Bois Noirs-Limouzat uranium vein, Forez, France. Econ Geol 73:1567–1610

Dahlkamp FJ (1974) Uranium deposits and reserves – natural uranium supply. Dtsch Atomforum, Mainz, pp 89–125

Dahlkamp FJ (1978a) Classification of uranium deposits. Mineral Deposita 13:83–104

Dahlkamp FJ (1978b) Geologic appraisal of the Key Lake U-Ni deposits, Northern Saskatchewan. Econ Geol 73:1430–1449

Dahlkamp FJ (1979) Uranlagerstätten. Gmelin Handbuch. Springer, Berlin Heidelberg New York, pp 1–280

Dahlkamp FJ, Adams SS (1981) Geology and recognition criteria for veinlike uranium deposits of the Lower to Middle Proterozoic unconformity and strata-related types. U.S. DOE-GJBX-5 (81), Grand Junction, pp 1–254

Davis JF (1979) Uranium in the U.S.A.: genesis and exploration implications. R Soc London: 47–52

Frarey MJ, Roscoe SM (1970) Symposium on basins and geosynclines of the Canadian shield. Geol Surv Can Pap 70-40:151–153

Geffroy J, Sarcia JA (1958) La notion de 'gîte épithermal uranifère' et les problèms qu'elle pose. Bull Soc Geol Fr (6) 8:173–190

Grutt EW (1972) Prospecting criteria for sandstone-type uranium deposits. Uranium prospect. Handbook, London, pp 47–77

Leroy J (1978) The Margnac and Fanay uranium deposits of the La Crouzille District (Western Massif Central, France): Geologic and fluid inclusion studies. Econ Geol 73:1611–1634

Moreau M, Poughon A, Puibaraud Y, Sanselme H (1966) L'uranium et les granites. Chron Mines Rech Min 34:47–51

Mrna F (1963) On the genesis of ore veins in Jachymov. Symp problems of postmagmatic ore deposition, Prag, vol I, pp 446–449

Mrna F (1967) Zum Alter der Bi-Co-Ni-U-Ag formation in Jachymov. Freiberg Forschungsh C 209: 81–86

Petrascheck WE, Erkan E, Neuwirth K (1974) Permo-triassic uranium ore in the Austrian Alps – Paleogeographic control as a guide for prospecting. IAEA-SM-183/25:291–298

Petrascheck WE, Erkan E, Siegl W (1977) Type of uranium deposits in the Austrian Alps. Geol Extr Process Uranium Inst Mineral Metall, London, pp 71–75
Ramdohr P (1955) Neue Beobachtungen an Erzen des Witwatersrands in Südafrika and ihre genetische Bedeutung. Abh Dtsch Akad Wiss Berlin, Kl Math Allg Naturwiss 5:1–43
Ramdohr P (1958) New observations on the ores of the Witwatersrand in South Africa and their genetic significance. Trans Geol Soc S Afr 61: Anhang 1–50
Roscoe SM (1969) Huronian rocks and uraniferous conglomerates in the Canadian shield. Geol Surv Can Pap 68-40:205
Ruzicka V (1971) Geological comparison between East European and Canadian uranium deposits. Geol Surv Can Pap 70-48:1–196
Sassano GP, Fritz P, Morton RD (1972) Paragenesis and isotopic composition of some gangue minerals from the uranium deposits of Eldorado, Saskatchewan, Canada. J Earth Sci 9:141–157
Schidlowski M, Eichmann R, Junge ChE (1974) Evolution des irdischen Sauerstoff-Budgets und Entwicklung der Erdatmosphäre. Umsch Wiss Tech 74:703–707
Tröger WE (1969) Spezielle Petrographie der Eruptivgesteine. Schweizerbart, Stuttgart, in Komm, 360 S

Spatial and Temporal Considerations of Ore Genesis

A Topology of Mineralization and Its Meaning for Prospecting

G.J. NEUERBURG [1]

Abstract

Epigenetic mineral deposits are universal members of an orderly spatial and temporal arrangement of igneous rocks, endomorphic rocks, and hydrothermally altered rocks. The association and sequence of these rocks is invariant whereas the metric relations and configurations of the properties of these rocks are unlimited in variety. This characterization satisfies the doctrines of topology. Metric relations are statistical, and their modes are among the better guides to optimal areas for exploration. Metric configurations are graphically irregular and unpredictable mathematical surfaces like mountain topography. Each mineral edifice must be mapped to locate its mineral deposits. All measurements and observations are only positive or neutral for the occurrence of a mineral deposit. Effective prospecting is based on an increasing density of positive data with proximity to the mineral deposit. This means sampling for maximal numbers of positive data, pragmatically the highest ore-element assays at each site, by selecting rock showing maximal development of lode attributes.

1 Introduction

Mineral deposits are an intrinsic part of a widely distributed and locally repeated petrographic-structural association, conceptually an edifice. Mineral deposits are associated with igneous rocks by proximity, especially superposition, and by a common focus on a major structural discontinuity in the Earth's crust. A single edifice consists of a prism of the crust, alternately and repeatedly opened and sealed (Marler and White 1975; Sibson et al. 1975). The openings are filled by an evolutionary progression of widely varied igneous rocks, recrystallized igneous rocks, secondary rock minerals, and lode (ore and gangue) minerals. Mineralized segments of the crust normally are composed of clusters of mineral edifices, in part interpenetrating, but not necessarily closely related in time. The constructive progression of crystalline components culminates in or converges with the disaggregations and dispersions by weathering onto the crust.

Heuristically, a mineral edifice is composed of four classes of rocks, each of multihued character, located on notably broken segments of the Earth's crust. An analogy with crystallography and its constancy of interfacial angles but wide diversity of crystal

1 U.S. Department of the Interior Geological Survey, Denver Federal Center, Denver, Co 80225, U.S.A.

habit is pertinent. Relations in both ore petrology (Stanton 1972) and crystallography are contained in topology, the branch of mathematics dealing with the properties of position that are unaffected by changes in size or shape. Topology affords a rigorous and objective ordering of the observations and measurements that serve to characterize the phenomenon of mineralization. Neither deterministic nor genetic attributes are in any way predicated by a topology. A mineral edifice is the *set* (igneous rocks, recrystallized igneous rocks, hydrothermally altered rocks, mineralized rocks), and the *operation* addition, as modulated by repeated opening of rock; the topology is four-dimensional. Erosion, corresponding to the tearing operation of topology, and deposition of the pieces restructure the components of the mineral edifice into different arrangements to form the *set* (sedimentary rocks).

The subject of this discourse is (1) a description of the phenomena of mineralization in the framework of topology, and (2) an analysis of the significance of this kind of description to methods of prospecting for ore deposits. The topology described herein is the culmination of a search for rationale and system in dealing with the undeniable uniqueness (Bunge 1962; Tischendorf 1969) of each mineral edifice and the heterogeneity manifest in all spatial properties of mineralized rocks. The problems are those of scientific method and philosophy, and are difficult of expression as well as of agreement (Moss 1979; Stegmüller 1979).

2 Topology of Mineralization

Two principles of topology govern the present description: (1) "structural stability" (Woodcock and Davis 1978) – the invariance of the qualitative spatial and temporal arrangement of the four intrinsic structural components (the building blocks, or *modules*) that make up the edifice and which are ultimately subsumed into the sedimentary *set;* and (2) distortion – the functionally independent and theoretically infinite number of shapes, sizes, and compositions that describe individuals of each module class. Structural stability is the essence of taxonomy; patterns of distortion are the subdivisions of a classification.

Each of the four modules, here called *igneous, endomorphic, alteration,* and *lode,* appears repeatedly in the edifice, apparently in unvarying cycles of pyrogenic crystallizations, igneous recrystallizations, metasomatic recrystallizations, and solute precipitations. Each cycle and subcycle is separated by episodes of rock opening. Compositional properties of the modules change discontinuously between cycles. The changes are broadly unidirectional, that is evolutionary with time, but reversals are numerous along the time axis.

Mappings of modules are metric configurations, statistical or geometrical (March and Steadman 1971) as appropriate, of the physical and chemical properties of each module and of the relations among modules within and between edifices. Examples of such mappings are (1) the mathematical (trend) surfaces that describe metal halos around veins, (2) the sizes and shapes of igneous bodies, (3) the frequency distribution of igneous rock types in a cluster of edifices, and (4) metal specialties (Tischendorf 1970) of the mineral deposits of an edifice. The mappings are unpredictable, mathematically unworkable functions (Burger and Skala 1978) and scale-constrained, which means that the unit of measurement, such as sample size, is arbitrary but must be held constant for any one mapping. Each geometrical configuration is unique, and can be determined only by mapping. An analogy with topographic maps is appropriate and complete.

Empirically, the variety of statistical mappings of module relations – for example, the frequency distribution of tin lode modules among igneous rock types over all edifices – is a limited set and the members of the set are statistically distributed (Taylor 1979). The modality of peakedness of a statistical mapping is greatest within a cluster of edifices – the *parochial* arena, somewhat less in mineralized segment of the crust – the *province* arena, and least pronounced in the full Earth set.

2.1 Igneous Modules

An igneous module is geometrically characterized by continuity of crystallized material within the space it fills, which material has the further characteristic of having congealed from an equally continuous melt. The chemistry and texture of an igneous module are remarkably homogeneous. Observed chemical structuring is of low amplitude and is either without detected system or functionally related to module shape or to the Earth's gravitational field.

The igneous component of mineral edifices is identical with the universal Earth set of igneous rocks. However, some modal associations are evident; thus brecciaform (Gilmour 1977) and porphyritic hypabyssal intrusives are prominent modules of mineral edifices, and felsic igneous rocks are markedly dominant over other igneous rock types in close spatial association with metallic ores. Three major compositional subsets are manifest, each modally associated with a singular part of the crust: (1) mafic rocks in oceanic crust, (2) alkalic rocks in cratonic crust, and (3) calcalkalic rocks in orogenic crust.

Compositions of igneous modules range widely in most mineral edifices, but the aggregated modules are distributed among empirically few compositions. Each composition occurs in several modules, such as dikes and sills, that are repeated and alternated along the time axis. A general progression of compositional changes, paced by an increase in SiO_2 content, is evident. This progression is described by a *fractal* curve (Mandelbrot 1977), like the path of a pebble cascading down a talus slope.

2.2 Endomorphic Modules

An endomorphic module is made up of numerous small bodies of partly space-filling, mainly replacive, and dominantly coarse-crystalline rock. This module includes such diverse rocks as (1) pegmatite intrusives, patches and "veins", (2) miaroles and their linings, (3) internally recrystallized parts of igneous bodies, with or without compositional change (Lutton 1959; Hollister and Baumann 1978), and (4) symplectitic intergrowths of rock minerals and of lode minerals such as tourmaline and topaz. The unifying characteristics are (1) the morphology of clusters of discrete to irregularly merging objects, and (2) the generally radial-structured compositions and (or) textures. These modules are mainly made up of the more felsic minerals of the enclosing rocks.

The endomorphic module is virtually restricted to igneous rocks and to the adjacent metamorphic wall rocks. Modal occurrence is in the late felsic intrusive modules of an edifice. The endomorphic module is morphologically and phenomenologically similar to the lode module; the two are not everywhere separable.

2.3 Alteration Modules

In the present context, alteration is the exchange reaction between rock minerals and aqueous fluids resulting in different mineral compounds and in different rock compositions, with the specific constraint that the secondary minerals retain the spatial coordinates of their progenitors. Alteration petrography varies in concert with the original petrography; thus, tactite with limestone, propylite with andesite, and greisen with granite. Rock mineral species are separately and successively altered to produce a progression of mineralogically distinctive assemblages, which are called *alteration zones.* The number of species altered is a measure of alteration intensity. Extent of alteration is equated with the proportion of a mineral species that has been altered, and is distance-dependent within a zone.

The alteration module is a diffuse body of dispersed secondary minerals – a body with irregular and gradational outline, described by a fractal surface. Its morphology maps a permeating tenuous fluid, in contrast to the displacive viscous fluid of an igneous module. The extent and intensity of alteration decay outward from surfaces of origin (contacts, fractures, faults). The sequence in which rock mineral species reacted, and thus the identity of a zone, differs from cluster to cluster of mineral edifices, and is a parochial property. Zonal mappings in all mineral edifices are the same for all sizes of alteration module, from the thin selvedge on a joint to the huge masses of pervasively altered rocks of a porphyry copper deposit – a property of topology called *self-similarity* (Mandelbrot 1977). Self-similarity is the generalization of Euclidean *similitude;* the general trace of a fractal curve, such as the outline of the British Isles, persists through changing scales of mapping.

The alteration module is prominently localized by extensively fractured crust along with numerous and varied igneous modules such as make up volcanic centers, calderas, and batholith cupolas. Alteration modules are commonly centered on intrusive contacts and are zone-structured outward along a plexus of fractures and intergranular separations in the adjacent rocks.

2.4 Lode Modules

In topologic context, lode minerals comprise a set of chemical compounds that are distinguished by having been precipitated in once open textures and structures. Most lode minerals differ compositionally from the rock minerals of their matrix; all differ in crystal habit and (or) fabric. Chalcophilic elements, like the heavy metals and minor lithophilic elements such as B, Be, and F, occur most commonly in lode minerals –

and these are the majority of the chemical elements of commerce. A lode module is a variously interspersed to dendriform configuration of lode-mineral fillings. The topologic invariants are the physical-chemical integrity and transgressive relationship of lode minerals with respect to a matrix of rock minerals. Mineral deposits of pyrogenic minerals, such as chromite disseminations and some pegmetites, are not lode modules but compositional varieties of igneous modules. These varieties may aptly be called *lode differentiates.* The lode module contains the classical epigenetic mineral deposit(s), but is much larger and is a topologically discrete object of the same taxonomic rank as an igneous dike.

Metric compositions of lode samples are parameters of nearness to ore, both in space and in quality. The lode sample is compounded from four unrelated variables: (1) permeability, or accessible open space, (2) geometry of the lode filling, as fracture-filling versus disseminated, (3) extent of space-filling which may even exceed permeability by way of metasomatism, and (4) chemical composition of the filling matter. Weathering changes of uneven degree are unavoidably a fifth independent – and wild – variable. Spatially dependent but unpredictable differences in each of these variables are characteristic at all scales of sample size, although their magnitude diminishes with increasing sample size. This topologic property is the "nugget effect" of geostatistics (David 1977). Compositional mappings of lode modules are arbitrarily shape- and scale-constrained for internal consistency. The mappings comprise distance-dependent mathematical surfaces that are self-similar fractal sets, like topographic maps of volcanic islands. Compositional magnitudes decay exponentially from distributary structures ("mineralization centers"), which structures map as mathematical singularities – graphically, peaks on the trend surfaces (Neuerburg et al. 1974; Botbol et al. 1978).

Metal specialties, quantity gradients ("halos"), and metal zonings of lode modules are varied, numerous, and complex. Their modalities are dominantly geography-related; that is, provincial-distinctive or parochial-distinctive. A few modal metal-rock associations are apparent, such as molybdenum with granitic rocks, rare earths with alkalic igneous rocks, and diamond with kimberlite.

Lode modules infuse limited segments of endomorphic and alteration modules. As distinct from lode differentiates and the analogous lode facies of sedimentary rocks, lode modules are universally enclosed in altered rock. Structural styles of lode fillings relate to the geometry of openings characteristic of the host module: (1) intergranular moldings in massive igneous modules and interspersions among particles of clastiform igneous modules, (2) central cavity fillings and linings in endomorphic modules, and (3) fillings and linings along the complex of fractures on which alteration modules are focused. The crustal distribution of lode modules is more restricted than that of the other modules; volumetrically, it is the least part of the edifice.

2.5 Sedimentary Set

The mineral edifice decomposes on exposure at the Earth's surface into chemical (dissolved) and physical (clastic) fragments. The fragments are dispersed in myriad sortings onto the crust. The four intrinsic structural components of the mineral edifice are combined to form one infinitely varied structural component of the *set* (sedimentary rocks). Geopetal accumulations of rock fragments and chemical precipitates – and their metamorphic transforms – comprise this sole module of the sedimentary set. Mineral deposits are in the set but are not topologically discrete. They grade without discontinuity into other sedimentary rock types; they are *lode facies,* not lode modules.

The outcrop arena of a mineral edifice is an area of ongoing physical and chemical destruction of the exposed modules and of mechanical dispersion of the pieces. Mappings of the debris ("landscape geochemistry" – Bradshaw 1975) within the outcrop arena are recognizable deformations (Maranzana 1972; Neuerburg et al. 1978; Siems et al. 1979) of the corresponding mappings of the mineral edifice. Overall, modules of the mineral edifice lose their integrity with distance from the outcrop. The fragmented modules are blended and the pieces are variously sorted. The fragments become smaller and their physical-chemical properties are modified so that at some distance from its source a fragment is no longer recognized as being derived from a mineral edifice.

The decay of identity of fragments with distance and (or) time from the materials source (provenance) is the pertinent topologic invariant of the sedimentary set. As used here, "identity" is pragmatically ranked by the certainty of recognizing debris from a mineral edifice despite (1) chemical modifications, as in oxidation of sulfides, (2) extreme reduction of grain size, as with brittle minerals (Mannard 1968), (3) infinite dilutions, as expressed by the particle sparsity effect (Clifton et al. 1967), or (4) chance intersections with exclusive sortings, such as heavy mineral concentrations. The curve of identity decay is inevitably fractal and unpredictable.

3 Topology and Prospecting

The application of topology to prospecting is treated purely from the perspective of how best to find an object of specified characteristics: a mineral deposit, a mapping singularity, a chemical extremum. The object exists and is manifest in the landscape by inherent properties that can be observed and measured, and (or) by known patterns of association (Callahan 1977). Economic aspects of prospecting (Parker 1974; Vannier and Woodtli 1979) are excluded from this analysis because they are ephemeral value judgments.

Prospecting is divisible into three phases, each guided by appropriate rules for its methodology. The phases are: (1) determining an optimal search arena on or in the crust, (2) locating lode modules within that search arena, and (3) mapping the lode module to locate and identify its singularities (metal concentrations; mineral deposits). Testing the singularities for qualification as ore is a separate discipline and is not treated here.

The intrinsic and orderly position of the lode module and lode differentiates in the intracrustal mineral edifice together with the inclusion of lode deposits as nonsingular facies in the supracrustal sedimentary set means that mineral deposits are a normal although unpredictably disposed component of all parts of the crust. With this observation comes a recurring dictum: no spatial or temporal relations, quantitative or qualitative, exclude the possibility of a nearby mineral deposit. Metric data are only positive or neutral (Botbol et al. 1978). Likewise, empirical modes derived from published mappings of mineral edifices serve to outline optimal search areas (Tischendorf 1969), never to exclude areas. Obviously, the positive data and optimal modes are the particular attributes of significance to prospecting.

Apart from known exposures and geologic inferences, lode modules, lode differentiates, and lode facies of a search arena are located by following recognized lode debris upstream to its source, the ancient practice of "shoading" (Thrush 1968). This phase of prospecting is governed by the parameters of identity decay in the sense that the lode identity of fragments is more easily seen and more certainly recognized with proximity to the lode module or lode rock, whether by better preservation, larger number of fragments, or both.

The lode module is all the materials emplaced by mineralization; quantities of these materials vary irregularly in space. The quantity distribution of each lode element is spatially focused on one or more geologic structures, which appear as singularities on a composition mapping. Thus, mapping is the topological method of choice for locating element singularities that may pinpoint an ore deposit.

The quantity of an element at each point in the lode module is measured from a sample and is unique to the particular piece of rock as a function of the size and shape of the rock sample. The scale of sample size must be constant for any mapping in order to preclude artificial singularities or gradients, but may differ between mappings. Ideally, sample size is the bulk from which the subsample for analysis is taken. Sample size is a spatial attribute and must be in units of volume (Whitten 1975). Sample shape must be uniform and preferably isotropic over all styles of lode geometry in order to assure objectivity. Except for homogeneous rocks of massive structure, an isotropic sample shape is ideally spherical (impractical!).

Site heterogeneity and the fractal configurations of lode mappings bring to mind the identity parameter for useful and pragmatic methods of mapping singularities of a lode module. Recognition of a singularity is based on a gradient of increasingly higher assays and on an increasing density of visible signs of lode filling toward the singularity. Thus, samples with maximal attributes of lode filling are chosen, such as multi-fractured sulfidic rocks as little weathered as possible. Several samples are taken from each collecting site as an additional gambling strategy for finding and measuring maximal assays.

To further aid recognition of lode fillings, lode-selective methods of analysis (Olade and Fletcher 1974; Neuerburg 1975) are used. The highest assays from the several samples of an outcrop, or of a specified area, are used for the mapping points of an identity-trend surface of lode modules (Neuerburg et al. 1974, 1978).

4 Topology, Origin, and Prospecting

Theories of origin of mineral deposits address questions of sources and timing in relation to the host rocks, of processes of formation, and of the intensive properties of lode-depositing environment. Each of these questions relates to some aspect of the topology of mineralization. Questions of source and timing relate to invariants of morphology and of position. Theories of processes of formation are explanations of associations, sequences and the parochial distortions. Intensive parameters (such as temperatures of deposition) are equivalent to a conversion and generalization of empirical properties into dynamic properties.

Questions of source and timing are fundamental dichotomies of geology (Amstutz 1960); the dichotomies are inherent in topology. The morphology produced by exogenesis (permeation from a center) and that produced by endogenesis (secretion into a center) are identical; the configuration is a mathematical property of particle distributions about a center (Mandelbrot 1977). Lode facies of the sedimentary module may be judged as epigenetic or syngenetic (Lovering 1963), depending upon whether the lode minerals are clastic particles or interstitial chemical precipitates (cement) in the rock, but this judgment is only a secondary mapping. Parenthetically, a lode facies is part of an identity decay pattern, as illustrated by ancient epigenetic uranium deposits in Wyoming (Stuckless et al. 1977) and by modern syngenetic gold placer deposits in general.

Process models, descriptive of the mechanics of construction of the mineral edifice (such as plate tectonics and igneous differentiation schemes) summarize empirical relations in Ernst Mach's imagery of scientific economy, but add no new road signs to mineral deposits. Likewise, substitution of intensive parameters for the objective measurements of a mapping is conceptually no more than a change of scale – a secondary mapping. Both processes and intensive parameters depend upon empirical mappings, so they have no direct practical application to prospecting, although often suggest some useful relation or parameter that has been overlooked. Finally, "answers" of origin are largely irrelevant for prospecting; the mineral deposit exists however it may have formed (Bridgman 1956; Hosking 1967).

5 Conclusions

The ordering of the known, objective spatial and metric properties of mineralized rocks onto topology makes clear the universality of mineralization as a phenomenon in crustal evolution, and is a rationale for the unpredictability of the dimensional attributes of mineralization as well as of its geographic coordinates. The strictures of this topology are that each metric configuration must be mapped and that scale integrity is mandatory for each mapping. Position is the paramount parameter of mineralization for prospecting. Position is quantified as the degree with which recognized properties of the materials at a sampling site match the properties, the identity of the object being sought. Finally, although topology excludes precise prediction of ore, no measured property or observed relationship excludes the possibility of a mineral deposit.

References

Amstutz GC (1960) Some basic concepts and thoughts on the space-time analysis of rocks and mineral deposits in orogenic belts. Geol Rundsch 50:165–189

Botbol JM, Sinding-Larsen R, McCammon RB, Gott GB (1978) A regionalized multivariate approach to target selection in geochemical exploration. Econ Geol 73:534–546

Bradshaw PBD (ed) (1975) Conceptual models in exploration geochemistry. The Canadian Cordillera and Canadian Shield. J Geochem Explor 4:1–214

Bridgman PW (1956) Probability, logic, and ESP. Science 123:15–17

Bunge W (1962) Theoretical geography. Gleerup, Lund, 208 pp

Burger H, Skala W (1978) Die Untersuchung ortsabhängiger Variablen. Modelle, Methoden und Probleme. Geol Rundsch 67:823–839

Callahan WH (1977) The history of the discovery of the zinc deposit at Elmwood, Tennessee, concept and consequence. Econ Geol 72:1382–1392

Clifton HE, Hubert A, Phillips RL (1967) Marine sediment sample preparation for analysis for low concentrations of fine detrital gold. US Geol Surv Circ 545:1–11

David M (1977) Geostatistical ore reserve estimation. Elsevier, Amsterdam, 364 pp

Gilmour P (1977) Mineralized intrusive breccias as guides to concealed porphyry copper systems. Econ Geol 72:290–303

Hollister VF, Baumann FW (1978) The North Fork porphyry copper deposit of the Washington Cascades. Miner Deposita 13:191–199

Hosking KFG (1967) The relationship between primary deposits and granitic rocks. Tech Conf Tin London 1:269–311

Lovering TS (1963) Epigenetic, diplogenetic, syngenetic, and lithogene deposits. Econ Geol 58: 315–331

Lutton RJ (1959) Pegmatites as a link between magma and copper-molybdenum ore. Mines Mag 49:15–24

Mandelbrot BB (1977) Fractals, form, chance, and dimension. Freeman, San Francisco, 365 pp

Mannard GW (1968) The surface expression of kimberlite pipes. Geol Assoc Can Proc 19:15–21

Maranzana F (1972) Application of talus sampling to geochemical exploration in arid areas: Los Pelambres hydrothermal alteration area, Chile. Trans Inst Min Metall (Sect B) Appl Earth Sci 81:B26–B33

March L, Steadman P (1971) The geometry of environment. RIBA, London, 360 pp

Marler GD, White DE (1975) Seismic geyser and its bearing on the origin and evolution of geysers and hot springs of Yellowstone National Park. Geol Soc Am Bull 86:749–759

Moss RP (1979) On geography as science. Geoforum 10:223–233

Neuerburg GJ (1975) A procedure, using hydrofluoric acid, for quantitative mineral separations from silicate rocks. J Res US Geol Surv 3:377–378

Neuerburg GJ, Botinelly T, Watterson JR (1974) Molybdenite in the Montezuma district of central Colorado. US Geol Surv Circ 704:1–21

Neuerburg GJ, Barton HN, Watterson JR, Welsch EP (1978) A comparison of rock and soil samples for geochemical mapping of two porphyry-metal systems in Colorado. US Geol Surv Open-File Rep 78-383:1–26

Olade M, Fletcher K (1974) Potassium chlorate-hydrochloric acid: a sulphide selective leach for bedrock geochemistry. J Geochem Explor 3:337–344

Parker PD (1974) The nature of unsuccessful projects (Abstr). Min Eng 26:77

Sibson RH, Moore J McM, Rankin AH (1975) Seismic pumping – a hydrothermal fluid transport mechanism. J Geol Soc London 131:653–659

Siems DF, Berger BR, Welsch EP (1979) A comparison of sample media in developing a geochemical exploration model for stockwork molybdenum occurrences, Dillon 1° × 2° quadrangle, Montana-Idaho. Geol Soc Am Abstr Progr 11-6:302

Stanton RL (1972) Ore petrology. McGraw-Hill, New York, 713 pp

Stegmüller W (1979) Moderne Wissenschaftstheorie. Ein Überblick. Naturwissenschaften 66:377–380, 438–445

Stuckless JS, Bunker CM, Busch CA, Doering WP, Scott JH (1977) Geochemical and petrological studies of a uraniferous granite from the Granite Mountains, Wyoming. J Res US Geol Surv 5: 61–81

Taylor RG (1979) Geology of tin deposits. Elsevier, Amsterdam, 543 pp

Thrush PW (comp & ed) (1968) A dictionary of mining, mineral, and related terms. US Bur Mines, Washington, 1269 pp

Tischendorf G (1969) Grundsätze und Methodik der Lagerstättenprognose. Ber Dtsch Ges Geol Wiss A 14:65–79

Tischendorf G (1970) Zur geochemischen Spezialisierung der Granite des westerzgebirgischen Teilplutons. Geologie 19:25–40

Vannier M, Woodtli R (1979) Teaching mineral prospecting by computer-assisted simulation techniques. Comput Geosci 5:369–374

Whitten EHT (1975) Appropriate units for expressing chemical composition of igneous rocks. Geol Soc Am Mem 142:283–308

Woodcock A, Davis M (1978) Catastrophe theory. Dutton, New York, 152 pp

Uniformitarianism and Ore Genesis

D.T. RICKARD [1]

Abstract

A large variety of on-going ore-forming systems have been discovered in recent years. These cover the whole spectrum of ore geology from synsedimentary through volcanic to even orthomagmatic processes. It appears that a large number of those systems which are thought to have been responsible for ore deposition throughout geologic time are observable at present. This has caused ore geology to develop from being one of the least uniformitarian of the natural sciences to becoming one of the most actualistic. Ore deposits are not the result of exceptional processes occurring rarely in earth's history but are seen as normal consequences of geologic evolution.

1 Introduction

Geology exists as a deductive science since it may be assumed that the laws of physics and chemistry are time-independent. Observations of the four billion year record of the rocks and ores of the earth (e.g., Ramdohr 1979) and associated planetary bodies (e.g., Ramdohr 1973) indicate that this is a reasonable assumption. Analyses of the even older spectra of distant galaxies do not, the red shift notwithstanding, provide contradictory evidence. In this petty sense, few would debate the Principle of Uniformitarianism.

In contrast, Hutton's (1788) original idea of an essential uniformity and continuity of process throughout geologic time is still debated. Recent examples include Stacey and Kramers' (1975) proposal of a global resetting of radiogenic clocks at around 3.7 Ga and Green's (1972) suggestion that Archaean greenstone belts were formed by a rash of meteorite impacts. Ore geology is one branch of the science which has especially suffered in the past from nonactualistic reasoning. A major cause of this has been the lack, until recently, of modern examples of ore-forming processes.

The famous and overworked adage that "the present is the key to the past" is important not only in its significance for the interpretation of geologic features but also because of its implications for the development of geology as a true science. It enables geologic hypotheses to be tested.

1 Ore Research Group, Geological Institute, Stockholm University, Box 6801, 113 86 Stockholm, Sweden

Ore geology was described in the 1960s as the "retarded child" of natural science (Amstutz 1969). It can be seen in retrospect that the scarcity, at that time, of known, on-going ore-forming processes meant that the hypotheses of ore geology were not, in fact, testable. Since then, a large variety of ore-forming environments have been discovered and investigated that are geologically very recent or still operating. It is the intention of this essay to summarize those various processes and discuss their implications for ore geology.

2 Synsedimentary Processes

2.1 Detrital Processes

Detrital gold, platinum, and tin deposits have historically been important sources of these metals. As such they are well known and have been thoroughly investigated. In contrast, modern detrital occurrences of sulphide minerals have no direct economic importance and are consequently not well known nor well investigated. Hosking (1969, p. 57) stated, for example, that

"In Cornwall, a number of sulphides including pyrite, arsenopyrite, and chalcopyrite are known to be transported by way of the rivers out to sea, where they are washed back on to the beaches and thus concentrated. In some instances pyrite can be shown to have been present within a few inches of the surface of a river bank for 50 or 60 years without having suffered oxidation. On the beaches themselves, sulphides are preserved between the high and low watermarks, and pebbles containing sulphide are commonplace."

2.2 Chemical Processes

It might be expected that chemical synsedimentary hypotheses for ore formation were more inspired by uniformitarian principles than by any other. In fact, this is not so. Modern chemical precipitation of metals, other than iron and manganese, in any significant concentration in nonvolcanic environments has not been reported. Jedwab (1979) observed for the first time precipitated copper minerals (e.g., malachite) in sea water. Rickard (1973) delineated the limitations of the process by means of a diffusion model. It is theoretically possible for low grade deposits, such as the Kupferschiefer, to be formed in this way. However, similar areas with such high concentrations of metals have not as yet been discovered at the present time. Holland (1979) pointed out the considerable similarity between trace metal patterns in modern Black Sea sediments in the Eocene Suzak Formation, and in the Pennsylvanian Tacket and Mecca quarry shales. He noted that this similarity might extend back to the base of the Proterozoic. He proposed that this suggested that the minor and trace element composition of sea water has remained unchanged throughout this period.

3 Episedimentary Processes

It is beginning to appear that the formation of lead-zinc deposits in permeable sediments (so-called Mississippi Valley-type deposits) is a normal consequence of basin development. Modern lead- and zinc-rich brines, not associated with igneous activity, have been discovered in a number of areas. The best known of these are the Cheleken brines (Lebedev and Nikitina 1968) and the brines of the Mississippi region itself (Carpenter et al. 1974). Other modern brine-generating environments, such as the Sabkha of the Persian Gulf, have been implicated in ore formation but are not as yet known to contain exceptional metal contents.

Bernard (1958) implicated residual, pedological processes in the genesis of certain Pb-Zn-Cu-Ba deposits, based on the ideas of Taupitz (1954) and Erhardt (1956). Certainly, residual concentrations of metals, such as iron, manganese, aluminium, and nickel, are well known. Less economically significant are residual concentrations of cobalt as black oxides associated with Zairan copper ores and zinc carbonate deposits resulting from the oxidation of zinc sulphide ores, as in southern France. However, Barbier (1979) has pointed out that there is no evidence from the hundreds of thousands of analyses of modern soils and alluvia that pedological mechanisms may lead to important Pb, Zn, Cu, and Ba concentrations either in the soils or in the fine fraction of the alluvia.

The large secondary body at Tynagh, Ireland, was produced by Tertiary deep weathering of a primary Alpine-type, limestone-lead-zinc deposit (Morrissey and Whitehead 1970). Bernard's (1973) paleokarstic model for Pb-Zn-Ba deposits in limestone receives considerable actualistic support from this type of observation. Furthermore, modern karsts in southern France, associated with ore-bearing limestones, may contain unusual concentrations of these metals.

It seems probable that the formation of ores of the uranium-vanadium-copper association in fluviatile sedimentary rocks is related to similar episedimentary processes. Doi et al. (1975) described uranium mineralization in Neogene sediments in Japan being formed through the action of ground-water circulation.

4 Igneous Processes

4.1 Volcanic Processes

The publication of the Watanabe volume (Tatsumi 1970) and the Kuroko volume (Ishihara 1974) made available for the first time in English the considerable Japanese work on their volcanogenic sulphide deposits. Although mainly Tertiary age, these deposits have been so little disturbed relative to their more ancient analogues as to be interpretated as actualistic examples of the volcanogenic ore-forming process.

Oftedahl's (1958) revival of the volcanic-exhalative concept of ore formation was not in fact based primarily on an actualistic concept. At that time, Oftedahl only knew of a few, rather poor, examples of exhalative metal formation. He quoted the filling of a meter-wide fissure formed during the 1817 activity of Vesuvius, with hematite resulting from the reaction between ferric chloride and water. He also noted Hjelmqvist's (1951) observation of obsidian blocks in the Eolian islands being nearly completely replaced by pyrite. Since this date a variety of metal concentrations, precipitated from volcanic exhalations, have been discovered.

Skornyakova (1964), Arrhenius and Bonatti (1965), and Boström and Peterson (1966) described iron and manganese oxide deposits from the East Pacific Rise. Bonatti et al. (1972) described an iron-manganese deposit from the Afar region of Ethiopia, which was less than 200,000 years old. Ferguson and Lambert (1972) reported zinc-rich iron and manganese oxides being precipitated from volcanic springs in Matupi Harbor, New Britain. But probably the most spectacular discovery of this type was the metalliferous material precipitating from hot saline brines at depth on the floor of the Red Sea. Well-stratified oxide and sulphide deposits are being formed with high concentrations of Fe, Mn, Cu, Zn, and Pb (Degens and Ross 1969).

Sulphide deposits are less widespread. Bonatti et al. (1976) described copper iron sulphide mineralization in disseminated and stockwork deposits in mafic volcanics from the Mid-Atlantic Ridge. Francheteau et al. (1979) have described zinc-rich cupriferous massive sulphide deposits forming on the East Pacific Rise. These deposits are analogous to Cyprus-type deposits in ophiolite complexes. Puchelt (1973) described iron-rich sediments containing both sulphides and oxides in sediments from Santorini. Similar deposits have been described by Honnorez et al. (1973) from Vulcano. Müller (1979) reported chalcopyrite, pyrite, bornite, covellite, and niccolite in manganese nodules from the northern Central Pacific.

The concept of ore magmas (cf. Spurr 1923) also has a concrete modern analogue. Park (1961) described a shallow intrusive and lava flow consisting of almost pure magnetite-hematite with minor amounts of apatite from Laco Sur in northern Chile.

4.2 Intrusive Processes

The environment of orthomagmatic ore-forming processes is normally inaccessible at the present time. Skinner and Peck (1969) did, however, find an immiscible sulphide melt in the late stage of a cooling lava in Hawaii. The melt was saturated with sulphide-sulphur at a temperature of 1065°C. Two sulphide-rich phases separated – an immiscible sulphide-rich liquid (which quenched to a mixture of pyrrhotite, chalcopyrite and magnetite) followed by a copper-rich pyrrhotite solid solution. Although this discovery is not strictly a modern analogue of the formation of nickel-copper deposits associated with basic and ultrabasic rocks, it is sufficiently close to demonstrate the process that actually occurs.

The major ore-forming processes associated with intrusive activity observed at the present time are related to the secondary development of geothermal systems as a result of intrusion. Although recent vein deposits of mercury and antimony minerals were well known to Lindgren (1933), the potential of these types of processes in ore genesis was really appreciated in the discovery of the Salton Sea brines (cf. White et al. 1963; Skinner et al. 1967), which contained up to 970 ppm Zn and 104 ppm Pb and deposited a variety of minerals such as bornite, chalcocite, digenite, pyrite, chalcopyrite and arsenopyrite.

White noted in 1967 (p. 615): "As recently as twelve years ago surprisingly few data relating base metal ore deposits to hot springs were available'."

In 1978, Browne could review details of hydrothermal alteration in geothermal fields in Iceland, U.S.A., Japan, East Indies, New Zealand, Chile, Italy, and U.S.S.R. and note that pyrite was precipitating in all of them. Nonferrous base metal sulphides were depositing in the Imperial Valley (California), Tongonan, Philippines, and New Zealand geothermal fields.

Concomitant with the enormous expansion in our knowledge concerning metal deposition from geothermal systems has been the discovery of large numbers of geologically recent porphyry copper deposits. Gustafson and Titley (1978) reported over 40 porphyry copper deposits, many less than 5 million years old, in the preface to the Economic Geology special issue on the porphyry copper deposits of the south-western Pacific Islands and Australia. Most of those deposits have been discovered since 1964. It may be noted in addition that part of the mineralization in some of these deposits is believed to be orthomagmatic in origin (e.g., Hine et al. 1978).

5 Uniformitarianism and Ore Genesis

Modern ore-forming processes are summarized in Table 1. This listing reveals that modern ore-depositing systems cover the whole spectrum of ore deposits. Notable exceptions include processes which occur in environments which are inaccessible at the present time. For example, metamorphic processes are completely absent on this listing. Even so, the list is remarkably comprehensive.

Geologically, one of the most striking aspects about ore deposits is the apparent variation in predominant ore types with time and place. For example, copper-cobalt deposits in arenites are mainly limited to the Precambrian of central Africa; few, if any, Precambrian Mississippi Valley-type deposits are known; banded iron formations appear to be concentrated around 2 Ga. The list is long. In this sense, uniformitarianism is not relevant to ore geology, since many of these deposits do not have an exact modern analogue. On the other hand, although the availability of a particular element may vary with time and place, the processes remain the same. After all, the uniformitarianism principle is stating that *the present is the key to the past,* not that *the present and past are identical.* In other words, the discovery of a variety of ore-forming systems occurring at the present day has permitted the uniformitarian doctrine to be applied to ore geology.

Ore formation is no longer a process that happened long ago under unusual circumstances. Processes of ore deposition can be studied in nature; hypotheses may be tested.

Most of the processes responsible for ore deposition in the past are happening, to varying degrees, at the present time. Ore formation can be seen to be a natural consequence of geologic evolution and not the result of rare, but beneficial, catastrophes.

The discovery of numerous on-going, ore-depositing systems has had an important corollary in the importance of ore geology to basic geological research. Studies of ancient deposits can now be used to elucidate ancient geotectonic regimes or to help construct stratigraphies in nonfossiliferous sections. Furthermore, the divergences of ancient ore deposits and ore provinces from their modern counterparts provide an insight into the variations in past environments in the history of the earth.

Table 1. Summary of on-going modern processes of ore deposition and examples of ancient analogues

Classification		Modern Process	Ancient analogue
Sedimentary	Synsedimentary	Mechanical concentration of Au, Pt, Sn and sulphides	Witwatersrand
		Chemical precipitation of metals in black muds	Swedish alum shale
	Episedimentary	Pb-Zn concentrations in basinal brines	Mississippi Valley-type deposits
		Pb-Zn concentrations in karst	Les Malines, France
		Uranium precipitation from ground-waters	Sandstone U-V-Cu association
Igneous	Volcanic	Precipitation of zinc-rich Fe-Mn oxides in association with submarine volcanism	Långban, Sweden; Franklin Furnace, U.S.A.
		Formation of Zn-Cu sulphide deposits at ocean ridges	Cyprus-type deposits
		Deposition of pyrite deposits in association with submarine volcanism	Massive pyrite deposits
	Intrusive	Separation of immiscible sulphide globules from mafic volcanics	Ni-Cu deposits associated with basic and ultrabasic rocks
		Hydrothermal brines	Vein deposits associated with acid intrusives
		Porphyry copper deposits	Porphyry copper deposits

References

Amstutz GC (1969) The logic of some relations in ore genesis. In: James CH (ed) Sedimentary ores. Spec Publ 1, Univ Leicester, pp 13–30

Arrhenius GOS, Bonatti E (1965) Neptunism and volcanism in the ocean. In: Sears M (ed) Progr Oceanogr 3:7–22

Barbier J (1979) Sur la notion de concentration pédologique dans la genèse des gîtes à Pb-Zn-Cu-Ba. Miner Deposita 14:311–321

Bernard A (1958) Contribution à l'étude de la province métallifère sous-cévénole. Thèse, Nancy, 2 vols, 640 pp

Bernard A (1973) Metallogenic processes of intra-karstic sedimentation. In: Amstutz GC, Bernard AJ (eds) Ores in sediments. Springer, Berlin Heidelberg New York, pp 43–57

Bonatti E, Fisher DE, Joensuu O, Rydell HS, Beyth M (1972) Iron-manganese-barium deposit from the northern Afar Rift. Econ Geol 67:717–730

Bonatti E, Guerstein-Honnorez BM, Honnorez J (1976) Copper-iron-sulfide mineralzations from the equatorial Mid-Atlantic Ridge. Econ Geol 71:1515–1525

Boström K, Peterson MNA (1966) Precipitates from hydrothermal exhalations on the East Pacific Rise. Econ Geol 61:1258–1265

Browne PRL (1978) Hydrothermal alteration in active geothermal fields. Annu Rev Earth Planet Sci 6:229–250

Carpenter AB, Trout ML, Picket EE (1974) Preliminary report on the origin and chemical evolution of lead- and zinc-rich oil field brines in central Mississippi. Econ Geol 69:1191–1206

Degens ET, Ross DA (eds) (1969) Hot brines and recent heavy metal deposits in the Red Sea. Springer, Berlin Heidelberg New York, 600 pp

Doi K, Hirono S, Sakamaki Y (1975) Uranium mineralization by groundwater in sedimentary rocks, Japan. Econ Geol 70:628–646

Erhardt H (1956) La genèse de sols en tant que phénomène géologique. Masson, Paris, 90 pp

Ferguson J, Lambert IB (1972) Volcanic exhalations and metal enrichments at Matupi Harbor, New Britain, T.P.N.G. Econ Geol 67:25–37

Francheteau J, and members of the CYAMEX team (1979) Massive deep-sea sulphide ore deposits discovered on the East Pacific Rise. Nature (London) 277:523–528

Green DH (1972) Archaean greenstone belts may include terrestrial equivalents of lunar materia. Earth Planet Sci Lett 15:263–270

Gustafson LB, Titley SR (1978) Preface. Econ Geol 73:597–599

Hine R, Bye SM, Cook FW, Leckie JF, Torr GL (1978) The Esis porphyry copper deposit, east New Britain, Papua New Guinea. Econ Geol 73:761–767

Hjelmqvist S (1951) Resa till Lipariska öarna. Geol Fören Stockhol Förh 73:473–491

Holland HD (1979) Metals in black shales – a reassessment. Econ Geol 74:1676–1680

Honnorez J, Honnorez-Guerstein B, Valette J, Wauschkuhn A (1973) Present day formation of an exhalative sulfide deposit at Vulcano (Thyrrhenian Sea), Part II: Active crystallization of fumarolic sulfides in the volcanic sediments of the Baia di Levante. In: Amstutz GC, Bernard AJ (eds) Ores in sediments. Springer, Berlin Heidelberg New York, pp 139–166

Hosking KFG (1969) Discussion: In: James CH (ed) Sedimentary ores. Spec Publ No 1, Univ Leicester, pp 57–58

Hutton J (1788) Theory of the Earth. Trans R Soc Edinburgh 1:209–304

Ishihara S (ed) (1974) Geology of Kuroko deposits. Soc Min Geol Jpn Spec Iss 6:465 pp

Jedwab J (1979) Copper, zinc and lead minerals suspended in ocean waters. Geochim Cosmochim Acta 43:101–110

Lebedev LM, Nikitina IB (1968) Chemical properties and ore content of hydrothermal solutions at Cheleken. Dokl Acad Sci USSR 183:180–182

Lindgren W (1933) Mineral deposits. McGraw Hill, New York, 930 pp

Morrissey CJ, Whitehead D (1970) Origin of the Tynagh ore-body, Ireland. Proc 9th Commonwealth Mining Metall Cong 2:131–145

Müller D (1979) Sulphide inclusions in manganese nodules of the Northern Pacific. Miner Deposita 14:375–380

Oftedahl Ch (1958) A theory of exhalative-sedimentary ores. Geol Fören Stockholm Förh 80: 1–19

Park CF (1961) A magnetite "flow" in northern Chile. Econ Geol 56:431–436

Puchelt H (1973) Recent iron sediment formation at the Kameni islands, Santorini. In: Amstutz GC, Bernard AJ (eds) Ores in sediments. Springer, Berlin Heidelberg New York, pp 227–245

Ramdohr P (1973) The opaque minerals in stony meteorites. Elsevier, Amsterdam, 245 pp

Ramdohr P (1979) The ore minerals and their intergrowths. Pergamon Press, London, 1300 pp

Rickard DT (1973) Limiting conditions for synsedimentary ore formation. Econ Geol 68:605–617

Skinner BJ, Peck DL (1969) An immiscible sulfide melt from Hawaii. Econ Geol Monogr 4:310–322

Skinner BJ, White DE, Rose HJ, Mays RE (1967) Sulfides associated with the Salton Sea geothermal brine. Econ Geol 62:316–320

Skornyakova IS (1964) Dispersed iron and manganese in Pacific Ocean sediments. Int Geol Rev 7: 2161–2174

Spurr JE (1923) The ore magmas. McGraw Hill, New York, 2 vols, 915 pp

Stacey JS, Kramers JD (1975) Approximation of terrestrial lead isotope evolution by a two-stage model. Earth Planet Sci Lett 26:207–221

Tatsumi T (1970) Volcanism and ore genesis. Tokyo Univ Press, 448 pp

Taupitz K-C (1954) Über Sedimentation, Diagenese, Metamorphose, Magmatismus und die Entstehung der Erzlagerstätten. Chemie Erde 17:104–164

White DE (1967) Mercury and base metal deposits with associated thermal and mineral waters. In: Barnes HL (ed) Geochemistry of hydrothermal ore deposits. Holt Rinehart and Wilson, New York, pp 575–631

White DE, Hem JD, Waring GA (1963) Chemical composition of subsurface waters. USGS Prof Pap 440-F:67 pp

Experimental Studies of Ore Mineral Associations

The Cu-Ni-S System and Low Temperature Mineral Assemblages

G.H. MOH[1] and G. KULLERUD[2]

Abstract

The phase relations in the copper-nickel-sulfur system were studied utilizing silica tubes, in which vapor is an inherent phase, as reaction vessels. Experiments were carried out at 200°C on mixtures of the elements and mixtures of presynthesized nickel and copper sulfides. The experiments at 100°C in addition contained small amounts of carbondisulfide or an aqueous phase to increase reaction rates. Some experiments lasted as long as two years. Even after such extended periods many experiments produced certain phases and phase assemblages which are metastable. Thus, the phase diagram, presented at 100°C does not represent equilibrium in all its parts. However, these phase relations coincide very closely with those observed in many ores. The assemblages found in ores of various types are shown in photomicrographs.

1 Introduction

The copper-nickel-sulfur system has in the past been studied at high temperatures by metallurgists who sought knowledge about the thermodynamic properties of the system's phases in order to improve the methods of extraction of the metals from Cu-Ni-S mattes. It has also been studied at elevated temperatures by geochemists, as a part of the more complicated copper-iron-nickel-sulfur system, in efforts to clarify the conditions under which ores such as those of the Sudbury type were deposited.

The only known ternary compound and mineral in the pure Cu-Ni-S system, villamaninite, is rare in nature as are assemblages of copper sulfide and nickel sulfide minerals. The reason for this rarity is provided by the almost ubiquitous occurrence in the earth's crust of iron in concentrations generally much in excess of those of copper and nickel together. The nickel and copper sulfides cannot stably coexist when iron is available to produce sulfides, and copper iron and nickel iron sulfides such as chalcopyrite and pentlandite form instead. Copper sulfide and nickel sulfide assemblages therefore are confined to isolated and unusual geochemical environments where iron is essentially absent or, when iron is present, to localities where the temperature was so low (probably below 200°C) that equilibrium assemblages could not be established due to sluggish reaction rates.

1 Mineralogisch-Petrographisches Institut der Universität, Im Neuenheimer Feld 236, 6900 Heidelberg, FRG

2 Department of Geosciences, Purdue University, West-Lafayette, IN 47907, USA

This paper discusses the phase relations in the Cu-Ni-S system at low temperatures (200°C and below) and presents examples of coexisting copper sulfides and nickel sulfides under the conditions outlined.

2 Previous Studies

The copper-sulfur system has been studied extensively at temperatures ranging from 1200° to 100°C by a considerable number of researchers. The most recent study covering the 0° to 250°C range was performed by Potter (1977) who also reviews the literature pertaining to the previous experimental investigations of this system.

The nickel-sulfur system was studied in detail over the temperature range 1100° to 150°C by Kullerud and Yund (1959, 1962).

The phase relations in the Cu-Ni system were reviewed by Hansen and Anderko (1958). Complete solid solution exists between copper and nickel at elevated temperatures.

The Cu-Ni-S system has been studied at elevated temperatures by a number of researchers. Kullerud et al. (1969) reviewed the literature and presented new experimental data to clarify the phase relations over the 1200° to 500°C temperature range. The phase relations down to 300°C were investigated by Moh and Kullerud (1963) and preliminary data on the phase relations at 200° and 100°C were given by Moh and Kullerud (1964).

3 Experimental

3.1 Starting Materials

High purity nickel described by Kullerud and Yund (1962) and high purity copper and sulfur described by Yund and Kullerud (1966) were used in all experiments. In some early runs mixtures of elements were directly used for experimentation of the ternary phase relations. However, because of slow reaction rates at 200°C and lower temperatures, it was found advantageous to first synthesize binary and ternary compounds such as chalcocite, digenite, covellite, heazlewoodite, Ni_7S_6, $\alpha Ni_{1-x}S$, polydymite, vaesite, and villamaninite from the elements and/or other presynthesized phases and then use mixtures of these compounds in experiments on the ternary system.

Reaction rates at 100°C are very sluggish in the dry system and experiments at 100°C for that reason were performed in the presence of an aqueous phase.

3.2 Methods

Silica tube experiments, employing the procedures described by Kullerud (1971), with dry mixtures of the elements and of various presynthesized binary compounds and the ternary compound (villamaninite) of the system were performed at 200°C.

The tubes were opened at regular intervals and the materials were ground to a fine powder under pure acetone in order to provide new surface areas for reactions to proceed. The reactions are very sluggish at this temperature and some runs lasted as long as two years. Equilibrium was established or at least approached in most runs except those which contained metal as a stable phase.

In order to obtain, or even approach, equilibrium at 100°C wet methods were utilized. Experiments in aqueous solutions and partly in organic solvents were prepared in different ways to achieve precipitation in some experiments and recrystallization in others which were performed on presynthesized and finely ground solid sulfide assemblages. For instance, the disulfides vaesite and villamaninite were precipitated from aqueous solutions; H_2S and sulfur (polysulfide) or thioacetamide solutions were mixed with precalculated amounts of acidic $NiSO_4$ and $CuSO_4$ solutions in 0.1 mol/l concentrations and kept in 350 ml size sealed pressure bottles for extended periods of time at approx. 100°C on a constantly boiling water bath. For wet syntheses, conducted at temperatures of 100°C or slightly above, different experimental procedures were successfully used employing modified or similar techniques as described by Kullerud (1962), Springer et al. (1964), Moh (1971), and Moh and Taylor (1971).

3.3 Results

The phase relations in the Cu-Ni-S system as obtained in the present study at 200°C are shown in Fig. 1. In this diagram vapor coexists with all phases and phase assemblages. It is noted that villamaninite, the only ternary phase in the system, even at this low temperature displays considerable solid solution, ranging from approx. $CuNiS_4$ to at least $CuNi_2S_6$ composition. The phase is essentially stoichiometric; its metal to sulfur ratio never deviates measurably from the 1:2 cation to anion ratio. The villamaninite phase in this respect behaves very similar to vaesite (Kullerud and Yund 1962) and pyrite (Kullerud and Yoder 1959) which both are essentially stoichiometric. These three phases in addition all have the pyrite crystal structure. Villamaninite of approx. $CuNi_2S_6$ composition coexists with vaesite which contains about 2 wt.% Cu in solid solution. The phases villamaninite + vaesite + liquid of essentially sulfur composition form a univariant assemblage in the presence of vapor. Villamaninite of, or very near, $CuNiS_4$ composition coexists with covellite which is essentially stoichiometric CuS (Kullerud 1965), and which contains less than 0.1 wt.% Ni in solid solution. Villamaninite + covellite + liquid sulfur form a univariant assemblage in the presence of vapor. At this temperature villamaninite + covellite + digenite also form a univariant assemblage. The digenite phase at this temperature displays solid solution which extends over the 1.76 to 1.90 Cu:S atomic ratio. The digenite which coexists with covellite and villamaninite (of $CuNiS_4$ composition) has $Cu_{1.76}S$ composition at 200°C. The divariant region containing villamaninite + digenite + vapor is fairly wide, stretching from $CuNiS_4$ to $CuNi_2S_6$ composition on the villamaninite side, and from $Cu_{1.76}S$ to about $Cu_{1.8}S$ on the digenite side. Villamaninite ($CuNi_2S_6$ composition) + vaesite (containing approx. 2 wt.% Cu) and digenite (of near $Cu_{1.8}S$ composition) in the presence of vapor form a univariant assemblage. The digenite + vaesite + vapor divariant field, which is quite narrow, effectively prevents villamaninite from coexisting stably with any of the phases which lie on the metal side of this field.

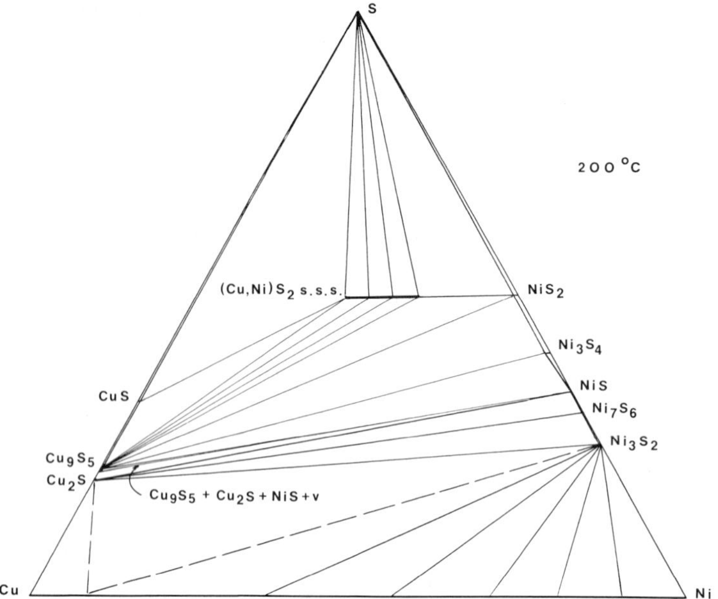

Fig. 1. Phase relations in the dry Cu-Ni-S system at 200°C in weight per cent. Sulfur occurs as a liquid, the other phases are solids; all phases and phase assemblages coexist with vapor. The stable mineral phases are described in the text. Digenite which at 200°C forms solid solution from $Cu_{1.76}S$ to $Cu_{1.90}S$ composition is simply indicated as Cu_9S_5

Digenite + vaesite + polydymite + vapor form a univariant assemblage. The polydymite phase is stoichiometric (Kullerud and Yund 1962) in the presence of digenite it takes 3 wt.%, or more Cu, solid solution. The digenite solid solution when coexisting with polydymite contains less than 0.1 wt.% Ni. A number of experiments with bulk compositions lying inside the digenite-vaesite-polydymite field were undertaken at 200°C to clarify whether a ternary $(Cu,Ni)_3S_4$ type phase exists in the system. Confirmation of such a phase was not obtained even after experimental duration of two years.

Polydymite + millerite (βNiS) + digenite also form a univariant assemblage in the presence of vapor. The digenite in this assemblage has composition of, or very near, $Cu_{1.85}S$. The millerite phase contains less than 0.2 wt.% Cu in solid solution. Digenite (of $Cu_{1.9}S$ composition) + Cu_2S (of the hexagonal intermediate crystal structure) and millerite form a univariant assemblage in the presence of vapor. In this assemblage the hexagonal Cu_2S phase contains less than 0.1 wt.% Ni in solid solution. This phase coexists stably with the low temperature polymorph (βNi_7S_6) of the Ni_7S_6 compound reported by Kullerud and Yund (1962). This Ni_7S_6 phase occurs in nature as a mineral called godlevskite. Godlevskite takes less than detectable (less than 0.1 wt.%) Cu in solid solution. The hexagonal Cu_2S polymorph also coexists with Ni_3S_2 (heazlewoodite) which is stoichiometric (Kullerud and Yund 1962), and which takes no detectable Cu in solid solution.

Experiments with bulk compositions lying on the metal side of the Cu_2S+Ni_3S_2+ vapor divariant region did not produce equilibrium. Mixtures of homogeneous Cu-Ni alloys and Ni_3S_2 or Cu_2S even when heated at 200°C for as long as two years did not produce results to indicate that the Cu-Ni solid solution series is intersected by a solvus at this temperature, nor did mixtures of Ni+Cu when heated with Cu_2S or Ni_3S_2 produce homogeneous Cu-Ni alloys. Figure 1 indicates complete solid solution between Cu and Ni. This is not certain, nor is the composition accurately known of the Cu-Ni alloy which forms a univariant assemblage with hexagonal Cu_2S and Ni_3S_2 at this temperature. The indicated composition is extrapolated from the composition trend observed at elevated temperatures (Moh and Kullerud 1963) and compares well with that determined by analysis of natural mineral assemblages as discussed below.

Most of the experiments carried out at 100°C contained also water in addition to the various starting materials utilized at 200°C. This water was introduced to provide a 1 atm H_2O vapor phase which might serve as a carrier, particularly of sulfur and, thus, promote reaction. This approach has been employed widely in the past by many researchers and has been found to improve greatly reaction rates in many systems. The resulting phase relations at 100°C as seen when projected into the Cu-Ni-S boundary plane of the Cu-Ni-S-H_2O composition tetrahedron are shown in Fig. 2. At this temperature solid monoclinic sulfur coexists with villamaninite, with vaesite, and with covellite. Covellite + villamaninite + monoclinic sulfur + vapor form a univariant assemblage as do vaesite + villamaninite + monoclinic sulfur + vapor. The villamaninite phase remains stoichiometric, but its range of solid solution is slightly more extensive than in

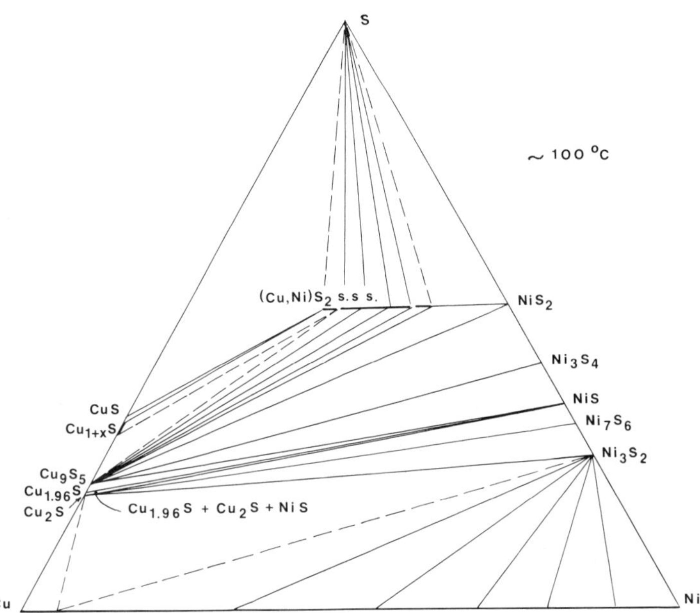

Fig. 2. Phase relations in the Cu-Ni-S system at approx. 100°C expressed in weight per cent. The diagrammatic presentation is based upon experiments performed mainly in aqueous solutions as described in the text

the pure dry system at elevated temperatures. Hydrothermal syntheses performed at approx. 150°C revealed that villamaninite forms a more extensive solid solution at this temperature than at 200°C. The villamaninite solid solution series displays certain similarities to the fukuchilite, FeS_2–CuS_2, series as described by Shimazaki and Clark (1970). The limits of the villamaninite solid solution series could not be accurately determined at 100°C. The covellite + villamaninite + digenite + vapor univariant assemblage which is stable at 200°C (cf. Figs. 1 and 2) does not appear in our experiments at 100°C because a new phase, blaubleibender covellite, containing about 68 wt.% Cu, forms at 157° ± 3°C (Moh 1964, 1971) between covellite and digenite on the Cu-S boundary. Tie lines between this new phase and villamaninite develop a few degrees below 157°C, and at lower temperatures (see Fig. 2) two univariant assemblages, covellite + villamaninite + blaubleibender covellite + vapor and digenite + villamaninite + blaubleibender covellite + vapor, and one divariant assemblage, blaubleibender covellite + villamaninite + vapor, exist where at 200°C only one univariant assemblage (covellite + villamaninite + digenite + vapor) occurred. The blaubleibender covellite phase along the Cu-S binary boundary displays measurable solid solution from about 67.5 wt.% to 68.5 wt.% Cu. The entire blaubleibender covellite solid solution coexists with villamaninite. Moh (1971) by performing experiments containing carbon disulfide in addition to Cu or copper sulfides found that this solid solution series actually consists of two blaubleibender covellite phases. Moh (1964) pointed out that blaubleibender covellite is metastable. However, since first described in ores by Frenzel (1959), blaubleibender covellites have been observed in numerous ore deposits. It is included in Fig. 2 for that reason. The digenite–vaesite tie lines persist at 100°C and so do those between digenite and polydymite and between digenite and millerite. Experiments at about 100°C produced djurleite ($Cu_{1.96}S$) as a phase when mixtures of digenite + Cu_2S were reacted in the presence of NiS. The djurleite phase coexists with millerite, and djurleite + digenite + millerite + vapor as well as djurleite + chalcocite + millerite + vapor form univariant assemblages (see Fig. 2). Various studies of the stability of djurleite (Roseboom 1962; Cook 1971, 1972; Luquet et al. 1972; Potter 1977) agree that this phase is stable below 93° ± 2°C. It is possible that small amounts of Ni in solid solution may stabilize this phase to about 100°C. However, the furnace containing critical experiments designed to clarify the djurleite stability in the presence of millerite unfortunately went down several degrees in temperature during the last few months of the long heating period.

It is noted from Fig. 2 that millerite + chalcocite + godlevskite + vapor as well as godlevskite + chalcocite + heazlewoodite + vapor form univariant assemblages. The inversion between chalcocite of stoichiometric Cu_2S composition and the hexagonal modification takes place at about 104°C. In the presence of pure djurleite the inversion temperature is depressed to 90° ± 2°C (Potter 1977). However, as our experiments would indicate, solid solution of nickel in djurleite may influence the phase relations so that in the ternary system at 100°C chalcocite coexists with heazlewoodite and with godlevskite.

4 Stable Phases at 25° C

Stable nickel sulfide phases at room temperature are vaesite, polydymite, millerite, godlevskite, and heazlewoodite. The only known ternary phase, and mineral, villamaninite, likewise is stable at room temperature. The low temperature stability relations of the copper sulfides are more uncertain. Covellite unquestionably is stable at 25°C and so is chalcocite. Blaubleibender covellite was found to be metastable by Moh (1964). Potter (1977) presents convincing evidence to indicate that both blaubleibender covellite phases first synthesized and described by Moh (1971) are metastable. However, blaubleibender covellites corresponding to the two phases synthesized by Moh (1971) and Potter (1977) do occur in nature as definable minerals described from numerous localities for instance by Frenzel (1959), Sillitoe and Clark (1969), and Globe and Smith (1973). Morimoto et al. (1969) identified a new copper sulfide which was named anilite. Morimoto and Koto (1970) reported this new mineral to have $Cu_{1.75}S$ composition and to be stable below 70° ± 3°C. Digenite was formerly believed to invert from a high- to a low-temperature polymorphic form at 75°C, but Morimoto and Gyobu (1971) found that high digenite instead of inverting to a low-temperature polymorph actually decomposes in the pure Cu-S system to a mixture of anilite and djurleite at 72° ± 3°C. They also were able to show that digenite is stabilized to lower temperatures when it takes some iron into solid solution. Digenite, therefore, may be

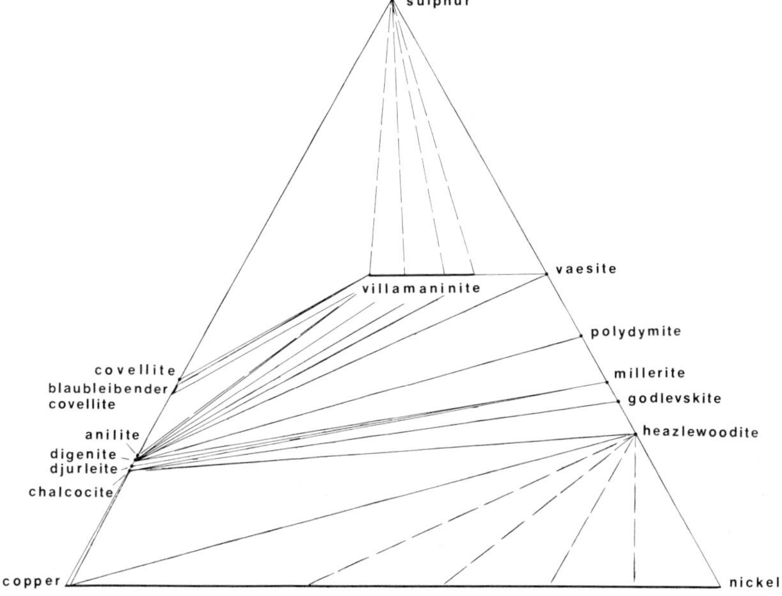

Fig. 3. Phase relations and mineral assemblages in the Cu-Ni-S system at room temperature in weight per cent. The stable and metastable assemblages are described in the text. Anilite, Cu_7S_4, is a stable phase in the pure Cu-S system but was not encountered in experiments having bulk compositions within the ternary Cu-Ni-S system. Minerals of the system and their naturally occurring intergrowths are discussed in the text and illustrated in the photomicrographs of Figs. 4 and 5

stable even at room temperature when it coexists with copper-iron or iron sulfides and contains a few percent iron in solid solution. A phase having hexagonal or tetragonal structure and composition of $Cu_{1.75}S$ to $Cu_{1.93}S$ has been synthesized by various workers (see for instance Flamini et al. 1973). Cook (1971) concluded that this phase is an oxidation product of djurleite or chalcocite. Potter (1977) concludes that it is either a metastable copper sulfide or a phase in the Cu-O-S system. Clark (1972) reported this phase is an ore specimen. Yet another phase reported to have $Cu_{1.90}S$ composition has been described by Eliseev et al. (1964) and others. Potter (1977) shows that $Cu_{1.90}S$ bulk composition consists of a mixture of anilite and djurleite at temperatures below 72° ± 3°C. Mulder (1973) described a phase which is very similar to djurleite. This protodjurleite phase according to Mulder is a metastable copper-rich form of djurleite. Djurleite as mentioned above of approx. $Cu_{1.96}S$ composition is stable below 93° ± 2°C in the pure Cu-S system. This discussion indicates that the stable copper sulfide phases at 25°C are covellite, anilite, djurleite, and chalcocite.

Figure 3 shows minerals and phase assemblages that, from the above discussion, apparently are stable at 25°C in the Cu-Ni-S system. It is possible that sufficient nickel may occur in solution in digenite to stabilize this phase even at room temperature, but, judging from the low degree of such solubility even at 300° and 400°C, this possibility does not seem likely. The stable phase relations as noted in Fig. 3 contrast with the actual experimental results as outlined above, partly because our experiments were conducted at temperatures where anilite does not form, and partly because our experiments even after 650 days produced certain metastable phases. It is interesting that the phases and phase assemblages produced in our experiments correspond more closely, as will be shown below, with those occurring in some ore deposits than with those occurring in the phase equilibrium diagram of Fig. 3. This comparison illustrates the commonly voiced suspicion that metastable phases and phase assemblages frequently occur in nature particularly at very low temperatures.

5 Cu-Ni-S Mineral Associations in Ores

Villamaninite described in detail by Ramdohr (1960) was observed in polished sections of ores from Mina Providencia Carmenes, near Villamanin, Leon, Spain, and recently in copper ores from the Lubin Mine, northwest of Legnica, Lower Silesia, Poland (Moh and Kucha 1980). Figure 4 shows five villamaninite containing mineral assemblages from the type locality and one vaesite-polydymite-digenite assemblage from Shinkolobwe, Katanga, Zaire. These and the other photomicrographs of minerals described below illustrate the phase relations shown in Fig. 3. Figure 4a shows fragments of villamaninite surrounded by coarsely lamellar covellite; Fig. 4b villamaninite and blaubleibender covellite. Figure 4c shows coarsely crystalline villamaninite, partly with clear cubic cleavage fissures caused by fragmentation, healed with digenite and partly replaced by blaubleibender covellite. In Fig. 4d villamaninite is in the process of being replaced by digenite. Figure 4e illustrates zoned crystal growth of vaesite, villamaninite, and cuprian nickel-bravoite. In another polished section of material from the type locality the vaesite + villamaninite + digenite mineral assemblage was observed, but that specimen also contained intergrowths of the minerals with bornite and marcasite.

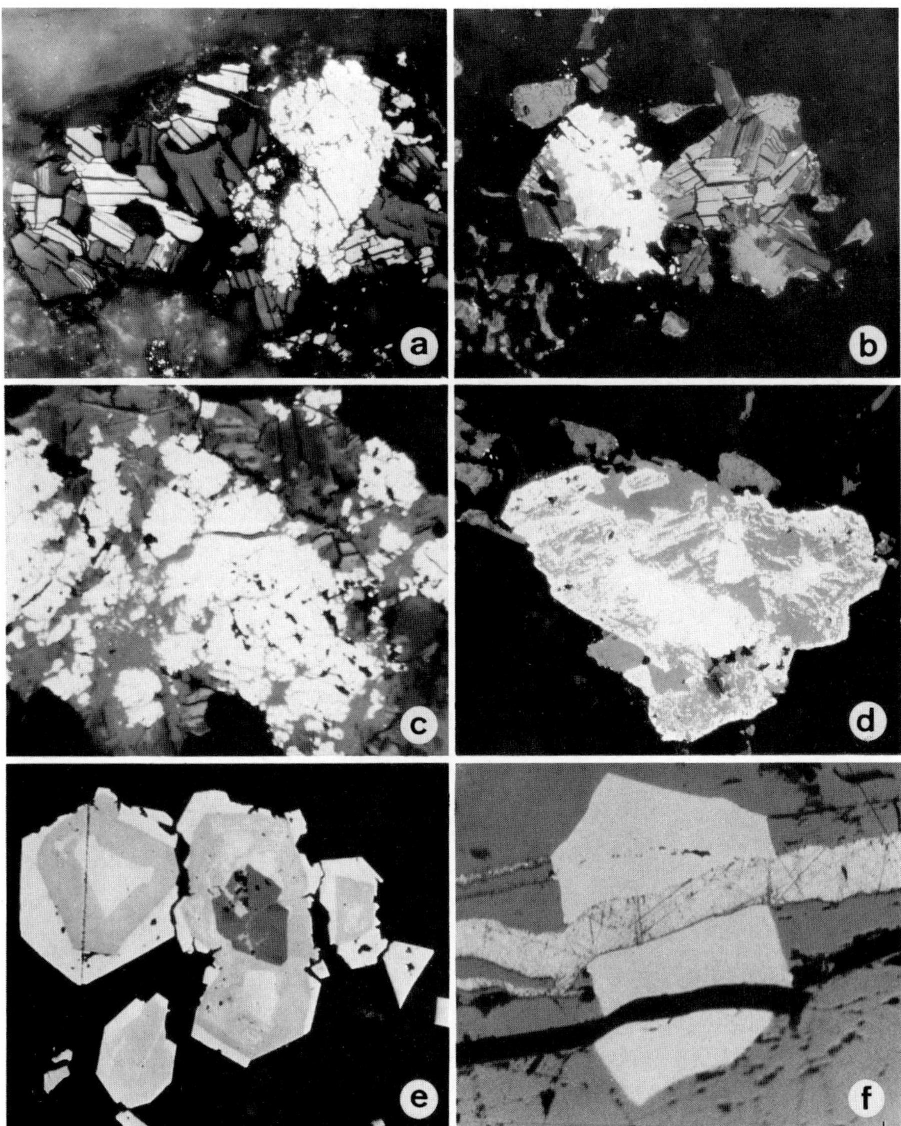

Fig. 4. **a** Villamaninite *(white)* intergrown with or replaced by platy covellite *(white, partly light* or *dark grey)*. **b** Villamaninite *(white)* replaced by blaubleibender covellite *(grey)*. **c** Villamaninite *(white)* replaced by digenite and in turn replaced by blaubleibender covellite *(grey)*. **d** Villamaninite *(white)* replaced by digenite *(grey)*. **e** Vaesite *(dark grey)* sourrounded by villamaninite and cuprian nickel-bravoite. **f** Idiomorphic polydymite crystal in a groundmass of vaesite *(dark grey)*, fractured and partly healed with digenite. Localities: **a–e** Mina Providencia Carmenes, Villamanin, Spain, **f** Shinkolobwe, Katanga, Zaire. Oil imm, × 110

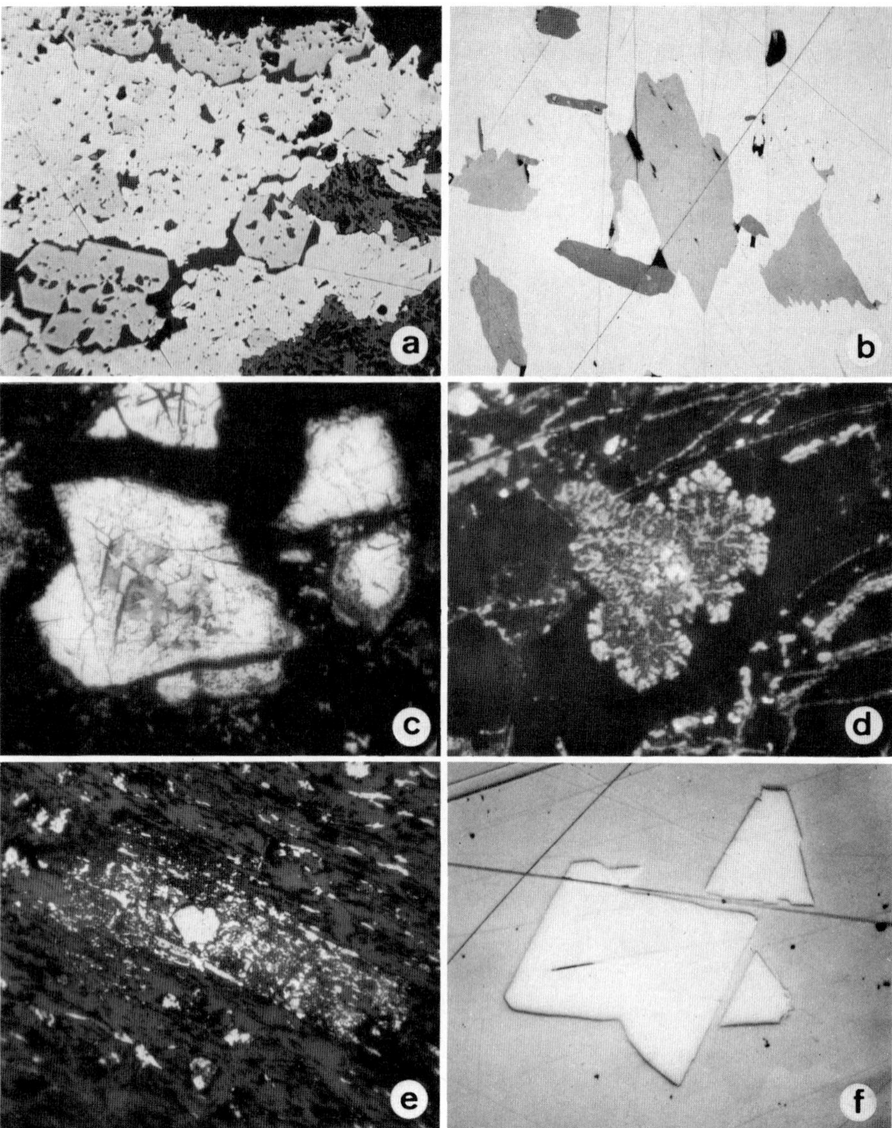

Fig. 5. a Idiomorphic polydymite crystals *(intermediate grey)* in millerite *(light grey)* with digenite *(dark grey)*; blue filter. **b** Djurleite flakes *(dark)* in millerite *(light)*. **c** Heazlewoodite *(white)* – chalcocite *(grey)* intergrowths. **d** Small idiomorphic heazlewoodite crystals *(almost white)* surrounded by copper metal which is marginally replaced by chalcocite *(light grey)*. **e** Coarse heazlewoodite crystals with metallic copper; red filter, less contrast. **f** Idiomorphic nickel metal *(white)* in heazlewoodite. Localities: **a** Königszug mine, near Dillenburg, Germany, **b** Siegen, Germany, **c–e** Kraubath, Austria (specimens from O.M. Friedrich), **f** Bogota, Canala, Nouvelle Calédonie. Oil imm, × 110

In polished sections of ores from Shinkolobwe, Katanga, vaesite coexists with idiomorphic polydymite and fissures caused by fragmentation have been healed, at least partly healed, by digenite as noted in Fig. 4f.

Idiomorphic crystals of polydymite were observed intergrown with millerite and digenite in ore specimens from the Königszug mine near Dillenburg, Germany. This assemblage is shown in Fig. 5a. The same mineral assemblage was also observed in specimens from Evje nickel mine, Norway. One specimen from Siegen, Germany, was found to contain the millerite + djurleite assemblage as shown in Fig. 5b. Millerite + chalcocite mineral assemblages were observed in polished sections made on specimens from Corner de Clerk & Joubert, Transvaal; Bleloch, Lydenburg District, Transvaal, and Siegen, Germany. The minerals godlevskite + heazlewoodite + chalcocite occur together in serpentinized specimens from Allalinhorn and in samples from Selva, Poschiavo, Graubünden; both localities are in Switzerland. Heazlewoodite–chalcocite intergrowths were observed in a pikrite, Tiefenbach, near Braunfels/Lahn, Germany, and in Kraubath, Austria, as shown in Fig. 5c. A number of polished sections made on specimens from the latter locality display also heazlewoodite + chalcocite + copper intergrowths (see Fig. 5d); the same mineral assemblage was observed in one section from Selva, Poschiavo, Switzerland. Figure 5e is a photomicrograph of a polished section of a sample from the Kraubath locality in Austria. It shows coarse heazlewoodite crystals in coexistence with metallic copper; microprobe analyses performed on the copper phase of this assemblage show the Ni content to be only 0.06%. Copper and heazlewoodite intergrowths were observed also in a polished section of a specimen from Linhorka, Bohemia, Č.S.S.R.

As a rarity, idiomorphic nickel metal was observed intergrown with heazlewoodite in a specimen of Bogota, Canala, Nouvelle Calédonie. This assemblage is shown in Fig. 5f; intergrowths of heazlewoodite with intermediate members of the metallic nickel-copper solid solution series are unknown.

All photomicrographs were taken in oil immersion. With the exception of those from Kraubath, Austria, the polished sections utilized in this study were borrowed from the Ramdohr collection in the Mineralogisch-Petrographisches Institut, Heidelberg University. The magnification in all photomicrographs is × 110.

Acknowledgments. We are grateful to Drs. P. Ramdohr, H. Meixner, and O.M. Friedrich for some of the specimens and polished sections used in this study and owe thanks to Dr. A. El Goresy for performing microprobe analyses of copper in one of the Kraubath specimens.

References

Clark AH (1972) Natural occurrence of hexagonal $Cu_{1.83}S$ in Chile. Nature (Lond) 238:123–124

Cook WR Jr (1971) The copper-sulfur phase diagram. Unpubl PhD Thesis, Case Western Reserve Univ Cleveland, Ohio, Gould Labs

Cook WR Jr (1972) Phase changes in Cu_2S as a junction of temperature. US Natl Bur Stand Spec Publ 364:703–712

Eliseev EN, Rudenko LE, Sineo LA, Koshurinikov BL, Solovov NI (1964) Polymorphism of copper sulfides in the system Cu_2S-$Cu_{1.81}S$. Mineral Sb Gosudarst Univ 18:385–400 (in Russian)

Flamini A, Graziani G, Grubessi O (1973) A new synthetic phase in the Cu-S system. Period Mineral 42:257–265

Frenzel G (1959) Idait und Blaubleibender Covellin. Neues Jahrb Mineral Abh 93:87–132

Globe RJ, Smith DGW (1973) Electron microscope investigation of copper sulfides in Precambrian Lewis Series of S.W. Alberta, Canada. Can Mineral 12:95–103

Hansen M, Anderko K (1958) Constitution of binary alloys. McGraw Hill, New York, 1305 pp

Kullerud G (1962) Method for mixing liquids at controlled temperatures. Carnegie Inst Wash Yearb 61:160–161

Kullerud G (1965) Covellite stability relations in the Cu-S system. Freiberg Forschungsh Beitr Lagerstättenkd C 186:145–160

Kullerud G (1971) Experimental techniques in dry sulfide research. In: Ulmer GC (ed) Research techniques for high pressure and high temperature. Springer, Berlin Heidelberg New York, pp 289–315

Kullerud G, Yoder HS (1959) Pyrite stability relations in the Fe-S system. Econ Geol 54(4):533–572

Kullerud G, Yund RA (1959) The Ni-S system. Carnegie Inst Wash Yearb 58:139–142

Kullerud G, Yund RA (1962) The Ni-S system and related minerals. J Petrol 3:126–175

Kullerud G, Yund RA, Moh GH (1969) Phase relations in the Cu-Fe-S, Cu-Ni-S, and Fe-Ni-S systems In: Magmatic ore deposits, a symposium. Econ Geol Monogr 4:323–343

Luquet H, Guastavino F, Bougnot J, Vaissière JC (1972) Etude du système Cu-S dans le domaine $Cu_{1.75}S$-$Cu_{2.1}S$ par analyse thermique différentielle. Mater Res Bull 7:955–962

Moh GH (1964) Blaubleibender covellite. Carnegie Inst Wash Yearb 63:208–209

Moh GH (1971) Blue remaining covellite and its relations to phases in the sulfur rich portion of the copper-sulfur system at low temperatures. Mineral Soc Japn Spec Pap 1:226–232 (Proc IMA-IAGOD Meet '70 IMA Vol)

Moh GH, Kucha H (1980) Villamaninite. In: Moh GH (ed) Ore syntheses, phase equilibria studies and applications, Report 1978/79. Neues Jahrb Mineral Abh 139(2):113–154

Moh GH, Kullerud G (1963) The Cu-Ni-S system. Carnegie Inst Wash Yearb 62:189–192

Moh GH, Kullerud G (1964) The Cu-Ni-S system at 200° and 100°C. Carnegie Inst Wash Yearb 63: 209–211

Moh GH, Taylor LA (1971) Laboratory techniques in experimental sulfide petrology. Neues Jahrb Mineral Monatsh 1971:450–459

Morimoto N, Gyobu A (1971) The composition and stability of digenite. Am Mineral 56:1889–1909

Morimoto N, Koto K (1970) Phase relations of the Cu-S system at low temperatures: stability of anilite. Am Mineral 55:106–117

Morimoto N, Koto K, Shimazaki Y (1969) Anilite, Cu_7S_4, a new mineral. Am Mineral 54:1256–1268

Mulder BJ (1973) Proto-djurleite, a metastable form of cuprous sulfide. Krist Techn 8:828–832

Potter RW II (1977) An electrochemical investigation of the system copper-sulfur. Econ Geol 72:1524–1542

Ramdohr P (1960) Die Erzmineralien und ihre Verwachsungen, 3. Aufl. Akademie Verlag, Berlin, 1089 pp

Roseboom EH (1962) Djurleite, $Cu_{1.96}S$, a new mineral. Am Mineral 47:1181–1184

Shimazaki H, Clark LA (1970) Synthetic FeS_2-CuS_2 solid solution and fukuchilite-like minerals. Can Mineral 10:648–664

Sillitoe RH, Clark AH (1969) Copper and copper-iron sulfides as the initial products of supergene oxidation, Copiapá mining district, Northern Chile. Am Mineral 54:1684–1710

Springer G, Schachner-Korn D, Long JVP (1964) Metastable solid solution relations in the system FeS_2-CoS_2-NiS_2. Econ Geol 59:475–491

Yund RA, Kullerud G (1966) Thermal stability of assemblages in the Cu-Fe-S system. J Petrol 7: 454–488

On the Crystal Chemistry of Some Thallium Sulphides and Sulphosalts

W. NOWACKI, A. EDENHARTER, P. ENGEL, M. GOSTOJIĆ, and A. NAGL[1]

Abstract

In the Introduction a short review on the geochemistry of thallium is given. It is followed by an exposure on hydrothermal syntheses of Tl-As.Sb-sulphosalts. The last chapter yields mainly a table with the coordination numbers (c.n.'s) and distances of Tl against S together with Tl-Tl and other Tl-metal (semi metal) distances of some Tl-sulphides and all natural or synthetic Tl-sulphosalts, known up to present. All c.n.'s between 1 and 12 (except 11) occur. New data are given for natural $Tl_8Pb_4Sb_{21}As_{19}S_{68}$ and for the 5 synthetic compounds ellisite (Tl_3AsS_3), $TlPbAs_3S_6$, parapierrotite ($TlSb_5S_8$), Tl_3SbS_4 and $Tl_2MnAs_2S_5$.

1 Introduction (W.N.)

About 2 years ago we suggested to start a programme on the hydrothermal synthesis of sulphosalts in order to obtain knowledge of the conditions of formation of these synthetical compounds in comparison with the natural minerals found in the Lengenbach quarry (Binntal, Switzerland) and to do crystal structure determinations in connection with these syntheses, i.e. to give a complete structural-chemical description (Figs. 1–10). The ideal situation would be if such investigations could be correlated with the determination of as many physical properties as possible (Table 1).

The geochemistry of Tl is outlined in several books (Rankama and Sahama 1950; Goldschmidt 1958; Ivanov et al. 1960; Vlasov 1966; Wedepohl 1969–78) and review articles (e.g. Rankama 1949; Shaw 1952); the chemistry, especially the structural chemistry of Tl is treated in Lee (1971) and Songina (1964).

At present over 500 carrier minerals of Tl, containing variable quantities of this element, are known (Ivanov et al. 1960; Vlasov 1966, p. 494). Because of the radii [K^+ 1.33, Tl^+ 1.49, Rb^+ 1.49, In^{+3} 0.92, Tl^{+3} 1.05, Y^{+3} 1.05; metallic Zn 1.37, In 1.57, Pb 1.75, Tl 1.71; tetrahedral covalent Cu 1.35, Zn 1.31, In 1.44, Pb 1.46, Tl 1.47 (Goldschmidt 1958)] two main groups may be considered: (1) Tl associated with K, Rb, and Cs [feldspars, micas; certain sulphates (jarosite, alunite, polyhalite) and halides (sylvine, carnallite, vulcanogenic halite)] and (2) Tl-bearing sulphides (galena, pyrite, marcasite, sphalerite, chalcopyrite), sulphosalts, certain hydroxides and sulphates

1 Abteilung für Kristallographie und Strukturlehre, Universität Bern, Sahlistraße 6, CH-3012 Bern, Switzerland

Table 1. Crystallo-chemical data of some thallium sulphides and sulphosalts

Ref.	Name	Formula	Latt. const.	Sp.G.	Z	C.N.	Coor.pol.	Tl-S	Tl-Tl	Fig.
1	Synth.	α-Tl_4S_3	7.97 7.76 13.03 γ 103.99°	$P2_1/a$	4	Tl III (1) 4 Tl I (2) 5 Tl I (3,4) 4	Reg. tetrah. Irreg. Irreg. Irreg. pol.	Tl(1) 2.51–2.59 Tl(2) 2.94–3.36 Tl(3) 2.90–3.16 Tl(4) 3.06–3.31	3.46–3.77	–
2	Synth.	β-Tl_4S_3	7.72 12.98 7.96 103.5°	$P2_1/c$	4	–	–	–	–	–
3	Synth.	Tl S	7.787 6.807	I 4/mcm	8	Tl III 4 Tl I 8	Dist. tetrah. Irreg.	Tl III 2.59 Tl I 3.33	–	–
4	Synth.	red Tl_2S_5	6.660 16.70 6.538	$P2_12_12_1$	4	Tl(1) 3(+3) Tl(2) 2(+3)	Trig.pyr. Irreg.	3.04–3.26 (3.43–3.49) 3.10, 3.12 (3.30–3.47)	–	–
5	Carlinite	Tl_2S	12.12 18.175	R3	27	–	–	–	–	–
6	Synth.	Tl_2S (XR) Tl_2S (ED) (dist.anti-CdI_2 type, $STL_{6/3}$)	12.20 18.17 12.20 18.17	R3 R3	27 27	3 + 3 3 + 3	Trig.pyr. Trig.pyr.	2.61–3.15 2.51–2.70 2.66–3.28	3.50–3.73 3.58–4.29 3.52–4.35	–
7	Chabourneite	$(Tl,Pb)_5(Sb,As)_{21}S_{34}$	16.33 8.53 21.24 93°51′ 95°24′ 104°29′	P1,P$\bar{1}$	2	–	–	–	–	–
8	Nat.	$Tl_8Pb_4Sb_{21}As_{19}S_{68}$	16.32 42.64 8.54 84.0° 89.1° 83.2°	P1	1	6 + 3	Trig.prism.(d)	Tl(1–8) 3.40 3.37 3.34 3.41 3.37 3.38 3.38 3.38 (mean value)	4.08–4.15 4.40–4.47	1
9	Nat.	$Tl(Sb,As)_7S_{11}$	–	–	–	–	–	–	–	–
10	Nat.	$Tl(As,Sb)_{10}S_{16}$	Amorphous	–	–	–	–	–	–	–
11	Chalcothallite	$(Cu,Fe)_6Tl_2SbS_4$	3.87 13.16	Tetr.	–	–	–	–	–	–

12	Christite	Hg Tl As S_3	6.11 16.19 6.11 96.7°	$P2_1/n$	4	–	–	–	–	–
13	Synth.christite	Hg Tl As S_3	6.11 16.19 6.11 96.71°	$P2_1/n$	4	5 + 2	5 to one sheet 2 to other sheet irreg.	3.11–3.31 +3.51, 3.52	3.39 Tl-As 3.50	–
14	Dufrenoysite	$(Pb,Tl)_2 As_2 S_5$	7.90 25.74 8.37 90°21′	$P2_1$	8	9 6 (+1)	Trig.prism.(d) dist.oct.	2.88–3.47 $\underline{2.82}$–3.26 (3.69)		–
15	Ellisite	Tl_3 AsS_3	12.32 9.65	Trig.	7	–	–	–	–	–
16	Synth.ellisite	Tl_3 AsS_3	9.57 6.99	R3m	3	3 (+2)	Trig.pyr. (+ other side)	3.05 3.10 (+3.44)	3.70 3.77 Tl-As = 3.64	3, 5, 6
17	Galkhaite	$[Hg_{0.76}(Cu,Zn)_{0.24}]_{12}$ $Tl_{0.96} \cdot (AsS_3)_8$	10.379	$I\bar{4}3m$	1	12	Trunc.tetrah.	3.863	–	9, 10
18	Hatchite	PbTl Ag $As_2 S_5$	9.22 7.84 8.06 66°25′ 65°17′ 74°54′	$P\bar{1}$	2	(Pb,Tl) 8 (Tl,Pb) 2 + 5 + . . .	Irreg. Irreg.	2.81–3.31, $\overline{3.10}$ $\underline{3.05}$, 3.12, 3.46–3.65 $\overline{\underline{3.41}}$) + . . .	4.06 3.98 Tl-As 3.48 Tl-Ag 3.89	–
19	Hutchinsonite	$(Tl,Pb)_2$ As_5 S_9	10.81 35.36 8.16	Pbca	8	(Pb,Tl) 7 (Tl,Pb) 2 + 5 (+ 1)	Split octah. irreg.	2.77–3.43, $\overline{3.05}$ $\underline{3.12}$, 3.15, $\underline{3.30}$–3.43, $\overline{3.29}$ + (3.79), $\overline{3.35}$	–	–
19a	Synth.	Tl Pb As_3 S_6	47.39 15.46 5.84	–	16	–	–	–	–	–
20	Imhofite	$Tl_{5.6} As_{15} S_{25.3}$	8.76 24.43 5.74 108.3° (sub cell)	$P2_1/n$	1	Tl(1) 8 Tl(2) 7	Dist.trig.prism.(b) split oct.	3.23–3.65, $\overline{3.36}$ 3.08–3.78, $\overline{3.37}$	Tl-As 3.49	–
20a	Jordanit	Pb_{28} As_{12} S_{46} (with up to 1.56% Tl)	8.92 31.90 8.46 117.8°	$P2_1/m$	1	–	–	–	–	–

Table 1 (continued)

Ref.	Name	Formula	Latt. const.	Sp.G.	Z	C.N.	Coor. pol.	Tl-S	Tl-Tl	Fig.
21	Lorandite	Tl As S_2	12.28 11.30 6.10 104°5′	$P2_1/a$	8	2 + 3 (+2)	Irreg.	Tl (1) 2.96, 3.07, 3.19–3.36 (+ 3.69, 3.48) Tl (2) 2.97, 3.014 3.19–3.39 (+ 3.63, 3.89)	(1)–(2) 4.03 (2)–(2) 3.54 (1)–As 3.50 –3.83, (2) As 3.85–3.99	–
22	Orpiment	As_2S_3 (with 0.2% Tl)	11.47 9.57 4.22 90.27°	$P2_1/n$	4	–	–	–	–	–
23	Parapierrotite (≡ Clinopierrotite)	Tl Sb_5S_8	8.02 19.35 9.03 91°58′	$P2_1/n$	4	–	–	–	–	–
24	Synth. Parapierrotite	Tl Sb_5S_8	8.098 19.415 9.059 91.96°	Pn	4	6 + 3 6 + 2	Trig.prism.(d) Trig.prism.(c)	3.23_5–3.69_5, $\overline{3.39}$ 3.13–3.71, $\overline{3.31}$	–	2, 4
25	Picotpaulite	Tl Fe_2S_3	5.40 10.72 9.04	mmmC. . .	4	–	–	–	–	–
26	Pierrotite	$Tl_2(Sb,As)_{10}S_{17}$(?)	8.77 38.8 8.03	$Pbn2_1$, Pbnm	4	–	–	–	–	–
27	Raguinite	Tl Fe S_2	65.25 12.40 10.44 5.26	orth.or mcl. orth.	 –	– –	– –	– –	– –	– –
27a	Synth.	Tl Fe S_2	11.64 5.31 10.51 144.6°	C2/m	4	6 + 3	Irreg.	3.09–3.49, $\overline{3.31}$ +3.73–3.76, $\overline{3.75}$	3.66 Tl-Fe = 3.70	–
27b	Synth.	$Tl_2Fe_xS_4$ (2.6<x<3.6)	∿3.75 ∿13.41(13.31) ∿3.76 3.77 13.38 91°	I4/mmm –	1 –	8 –	Cube –	3.305 –	– –	– –

28	Rathite – I	$(Pb,Tl)_3\ As_4(As,Ag)S_{10}$ (3.1, 3.6% Tl)	25.16 7.94 8.47 100°28′	$P2_1/a$ $(P\bar{1})$	4	9 9 7	Trig.prism.(d) Trig.prism.(d) Trig.prism. (b)	3.00–3.47, $\overline{3.19}$ 2.98–3.42, $\overline{3.19}$ 2.80–3.39, $\overline{3.01}$	–	–
29	Routhierite	$Tl\ Hg\ As\ S_3$	9.977 11.290	I, tetr.	8	–	–	–	–	–
30	Scleroclase	$(Pb,Tl)As_2\ S_4$ (3.9, 2.6%Tl)	19.62 7.89 4.19 90°0′	$P2_1/n$ (pseudostr.)	4	9	Trig.prism.(d)	2.95–3.33, $\overline{3.08}$	–	–
31	Stibnite with 3.2% Tl	$(As,Tl,Sb)_2\ S_3$	–	–	–	–	–	–	–	–
32	"Metastibnite"	Amorphous sulphide with Sb, Hg, Tl (up to 0.5%), As; disch.prec. Sb, Tl (100° ppm), As, Ag, Au	– –	– –	– –	– –	– –	– –	– –	– –
33	Vrbaite	$Hg_3\ Tl_4\ As_8\ Sb_2\ S_{20}$	13.40 23.39 11.29	C2ca	4	Tl(1) 2 + 5 + 1 + 2 Tl(2) 3 + 4 + 1 + 2 As	Dist.cube + Dist.cube +	3.09, 3.15, 3.35 –3.43, $\overline{3.32}$, ∿3.5 (2×) 3.15–3.26, $\overline{3.20}$ 3.34–3.62, $\overline{3.50}$; $\overline{3.37}$	3.75 Tl-As = 3.42, 3.53	–
34	Wallisite	$Pb\ Tl\ Cu\ As_2\ S_5$	8.98 7.76 7.98 65°33′ 65°30′ 73°55′	$P\bar{1}$	2	Pb(Tl) 8 Tl(Pb) 2 + 5 + . . .	Irreg. Irreg.	2.73–3.30, $\overline{3.08}$ 2.99, 3.14, 3.35–3.64, $\overline{3.38}$ + . . .	Pb,Tl-Pb,Tl = 4.01, Tl,Pb-Tl,Pb = 3.78 Tl-Cu = 3.87 Tl-As 3.37, 3.73	–
35	Weissbergite	$Tl\ Sb\ S_2$	11.8 6.4 6.1 109.9° 81.8° 105.4° (cell not reduced!)	P1	4	–	–	–	–	–
36	Synth.	$Tl\ Sb\ S_2$ (ED) $Tl\ Sb\ S_2$ (XR)	5.87–5.94 –	Fm3m –	2 –	Tl, Sb 6 –	Octah. –	2.93_5–2.97 –	– –	– –
37	Synth.	$Tl_3\ Sb\ S_3$ (α)	–	–	–	–	–	–	–	–

Table 1 (continued)

Ref.	Name	Formula	Latt.const.	Sp.G.	Z	C.N.	Coor.pol.	Tl-S	Tl-Tl	Fig.
38	Synth.	Tl_3SbS_4	6.285 6.364 11.647 94.62 98.51 103.93	P1	2	Tl (1) 5 + 1	Irreg.	3.02–3.25, $\overline{3.18}$ + 3.64, $\overline{3.26}$	3.49–3.94	7, 8
						Tl (3) 5 + 1	Irreg.	3.06–3.30, $\overline{3.17_5}$ + 3.68, $\overline{3.26}$	Tl-Sb 3.60–3.91	
						Tl (2) 6 + 1	Irreg.	3.13–3.47, $\overline{3.28}$ + 3.60, $\overline{3.32}$		
						Tl (4) 8	Irreg.	3.09–3.64, $\overline{3.63}$		
38a	Synth.	Tl_3AsS_4	–	–	–	–	–	–	–	–
39	Synth.	$Tl_2MnAs_2S_5$	15.34 7.61 16.65	Cmca or C2ca	8	–	–	–	–	–
40	Nat.	$TlSb_{14}S_{17}$ (amorphous)	–	–	–	–	–	–	–	–
41	Synth.	$TlBiS_2$	$NaFeO_2$ type	–	–	–	–	–	–	–
42	Synth.	$Tl_4Bi_2S_5$	16.76 17.40 4.09	Pnam	4	(1) 1 + 2 + 2 (2) 1 + 2 + 2 (3) 2 + 1 + 1 (4) 2 + 1 + 1	Outside of a tetr.pyr., pol. Outside of a dist.tetrah.,pol.	2.88, 3.12, 3.27 2.98, 3.08, 3.09 2.90, 3.06, 3.06 2.93, 3.18, 3.28	(1)–(3) 3.77 × 2 (1)–(4) 3.83 × 2 (2)–(3) 3.77 × 2 (2)–(4) 3.87 × 2	–

Abbreviations: Latt.const. = lattice constants a, b, c, α, β, γ (tetr., trig., hex.: a, c); Sp.G. = space group; C.N. = coordination number, Coor. pol. = coordination polyhedra; Reg. = regular; Tetrah. = tetrahedron; Irreg. = irregular; Pol. = polar; Dist. = distorted; Trig. = trigonal; Pyr. = pyramidal; Prism. = prismatic; Oct. = octahedron; Trig.prism. (a) = c.n. 6, (b) = c.n. 6 + 1 (6 corners + 1 normal to a side face), (c) = c.n. 6 = 2 (+ normals to 2 side faces), (d) = c.n. 6 + 3 (+ normals to 3 side faces); Split. oct. = c.n. 7 (1 corner is split into 2); Trunc. = truncated; Dist. cube + = c.n. 8 + 2 (8 corners + normals to 2 opposite faces); ED = electron diffraction; XR = X-ray diffraction; Disch.prec. = discharge precipitation; Synth. = synthetic; Nat. = natural; $\overline{\quad}$ = mean value; all distances in Å

References to Table 1

1, 2. Leclerc B, Bailly M (1973) Structure cristalline du sulfure de thallium: Tl_4S_3. Acta Crystallogr B29:2334–2336

Soulard M, Tournoux M (1971) Sur les trois phases du système thallium-soufre: $Tl_4S_3\alpha$, $Tl_4S_3\beta$ et TlS_2. Bull Soc Chim Fr 3:791–793

3. Hahn H, Klinger W (1949) Röntgenographische Beiträge zu den Systemen Thallium/Schwefel, Thallium/Selen und Thallium/Tellur. Z Allg Anorg Chem 260:110–119

Lee AG (1971) The chemistry of thallium. Elsevier, Amsterdam

Scatturin V, Frasson E (1956) Solfuri di tallio da polisolfuro ammonico: Formazione e struttura del monosulfuro di tallio TlS. Ric Sci 26:3382–3386

4. Frasson E, Scatturin V (1955-56) Su alcuni solfuri di tallio. Nota I. – Gruppo spaziale e constanti reticolare di due pentasolfuri di tallio. Atti Ist Veneto Sci Lett Arti Cl Sci Mat Nat 114: 61–66 [Lee, p 102/3, ref. 61, p 110]

LeClerc B, Kabré TS (1975) Structure cristalline du sulfure de thallium Tl_2S_5. Acta Crystallogr B31:1675–1677

Lee AG (197) The chemistry of thallium. Elsevier, New York, pp 366

5. Radtke AS, Dickson FW (1975) Carlinite, Tl_2S, a new mineral from Nevada. Am Mineral 60: 559–565

6. Ketelaar JAA, Gorter EW (1939) Die Kristallstruktur von Thallosulfid (Tl_2S). Z Kristallogr 101:367–375

Man LI (1970) Determination of the structure of Tl_2S by the electron diffraction method. Sov Phys Crystallogr 15:399–403 [Kristallographiya (1970) 15:471–476]

7. Mantienne J (1974) La minéralisation thallifère de Jas Roux (Hautes-Alpes). Thèse, Univ Paris

Picot P, Johan Z (1977) Atlas des minéraux métalliques. Mém BRGM 90:1–402

8. Nagl A (1979) The crystal structure of a Tl-sulfosalt $Tl_8Pb_4Sb_{21}As_{19}S_{68}$. Z Kristallogr 150: 85–106

9. Mantienne J (1974) La minéralisation thallifère de Jas Roux (Hautes-Alpes). Thèse, Univ Paris

10. Guillemin C, Johan Z, Laforêt C, Picot P (1970) La pierrotite $Tl_2(Sb,As)_{10}S_{17}$. Bull Soc Fr Mineral Cristallogr 93:66–71

Mantienne J (1974) La minéralisation thallifère de Jas Roux (Hautes-Alpes). Thèse, Univ Paris

11. Kovalenker VA, Laputina IP, Semenov EI, Evstigneeva TL (1978) Potassium bearing thalcusite from the Ilimaussaq massif and new data on chalcothallite. Dokl Akad Nauk SSSR 239:1203–1206 (in Russian; see Am Mineral (1979) 64:658)

12. Radtke AS, Dickson FW, Slack JF, Brown KL (1977) Christite, a new thallium mineral from the Carlin gold deposit, Nevada. Am Mineral 62:421–425

13. Brown KL, Dickson FW (1976) The crystal structure of synthetitic christite, $HgTlAsS_3$. Z Kristallogr 144:367–376

14. Marumo F, Nowacki W (1967a) The crystal structure of dufrenoysite, $Pb_{16}As_{16}S_{40}$. Z Kristallogr 124:409–419

15. Dickson FW, Radtke AS, Peterson JA (1979) Ellisite, Tl_3AsS_3, a new mineral from the Carlin gold deposit, Nevada, and associated sulfide and sulfosalt minerals. Am Mineral 64:701–707

16. Edenharter A, Gostojić M, Engel P (1978/9) Hydrothermal synthesis of ellisite, Tl_3AsS_3 and parapierrotite, $TlSb_5S_8$ and their crystal structures. Coll Abstracts 78–P3–5a, pp 367. 5th ECM. Copenhagen. Schweiz Ges Kristallogr Jahrestag Lausanne, 5 Okt 1979, Abstr Poster Session

Edenharter A, Peters T (1979) Hydrothermalsynthese von Tl-haltigen Sulfosalzen. Z Kristallogr 150:169–180

Gostojic M (1979) Die Kristallstruktur von synthetischem Ellisit, Tl_3AsS_3. Z Kristallogr 151: 249–254

17. Divjakovic V, Nowacki W (1975) Die Kristallstruktur von Galchait $[Hg_{0,76}(Cu,Zn)_{0,24}]_{12}Tl_{0,96}AsS_3)_8$. Neues Jahrb Mineral Monatsh 1975:291–293; Z Kristallogr 142:262–270, 454

18. Marumo F, Nowacki W (1967b) The crystal structure of hatchite, $PbTlAgAs_2S_5$. Z Kristallogr 125:249–265

19. Dunn PJ (1977) Hutchinsonite from Quiruvilca, Peru. Mineral Rec 8:394

Takéuchi Y, Ghose S, Nowacki W (1965) The crystal structure of hutchinsonite, $(Tl,Pb)_2-As_5S_9$. Z Kristallogr 121:321–348

References to Table 1 (continued)

19a. Edenharter A, Nagl A (1980) Synthesis and crystal structure of $TlPbAs_3S_6$. In prep.

20. Divjaković V, Nowacki W (1976) Die Kristallstruktur von Imhofit, $Tl_{5,6}As_{15}S_{25}$. Z Kristallogr 144:323–333, 418

20a. Harańczyk CO (1956/1957) In: Vlasov KA (ed) (1966) Geochemistry of rare elements. Israel Program of Scientific Translations, Jerusalem, pp 498, 680

Ito T, Nowacki W (1974) The crystal structure of jordanite, $Pb_{28}As_{12}S_{46}$. Z Kristallogr 139: 161–185

Ivanov VV, Volgin VJu, Krasnov AA, Lizunov VN (1960) Thallium. Principal geochemical and mineralogical features, genetic types of deposits and geochemical provinces. Izd Acad Nauk SSSR, Moskva, 156 pp (in Russian; see Vlasov KA (1966) pp 499, 678)

21. Fleet ME (1973) The crystal structure and bonding of lorandite, $Tl_2As_2S_4$. Z Kristallogr 138: 147–160

Radtke AS, Taylor CM, Erd RC, Dickson FW (1974b) Occurrence of lorandite, $TlAsS_2$, at the Carlin gold deposit, Nevada. Econ Geol 69:121–124

22. Radtke AS, Heropoulos C, Fabbi BP, Scheiner BJ, Essington M (1972) Data on major and minor elements in host rocks and ores, Carlin gold deposit, Nevada. Econ Geol 67:975–978

Radtke AS, Taylor CM, Dickson FW, Heropoulos C (1974a) Thallium-bearing orpiment, Carlin gold deposit, Nevada. J Res US Geol Surv 2:341–342

23. Johan Z, Picot P, Hak J, Kvaček M (1975) La parapierrotite, un nouveau minéral thallifère d'Allchar (Yougoslavie). Tschermaks Mineral Petrogr Mitt 22:200–210

Mantienne J (1974) La minéralisation thallifère de Jas Roux (Hautes-Alpes). Thèse, Univ Paris

24. Bohač P, Brönnimann E, Gäumann A (1974) On the ternary compound $TlSbS_2$ and its photoelectric properties. Mat Res Bull 9:1033–1040

Edenharter A, Gostojić M, Engel P (1978/79) Hydrothermal synthesis of ellisite, Tl_3AsS_3 and parapierrotite, $TlSb_5S_8$ and their crystal structures. Coll Abstr 78–P3–5a, pp 367, 5th ECM, Copenhagen. Schweiz Ges Kristallogr Jahrestag, Lausanne, 5 Okt 1979, Abstr Poster Session

Edenharter A, Peters T (1979) Hydrothermalsynthese von Tl-haltigen Sulfosalzen. Z Kristallogr 150:169–180

Engel P (1979) Zur Eindeutigkeit der Kristallstrukturanalyse. Chimia 33:317–324

Engel P (1980) Die Kristallstruktur von synthetischem Parapierrotit, $TlSb_5S_8$. Z Kristallogr 151:203–216

25. Johan Z, Picot P, Schubnel H-J, Permingeat F (1970) La picotpaulite $TlFe_2S_3$, une nouvelle espèce minérale. Bull Soc Fr Mineral Cristallogr 93:545–549

26. Guillemin C, Johan Z, Laforêt C, Picot P (1970) La pierrotite $Tl_2(Sb,As)_{10}S_{17}$. Bull Soc Fr Mineral Cristallogr 93:66–71

27. Johan Z, Picot P, Pierrot R (1969) Nouvelles données sur la raguinite. Bull Soc Fr Mineral Cristallogr 92:237

Laurent Y, Picot R, Permingeat F, Ivanov T (1969) La raguinite, $TlFeS_2$, une nouvelle espèce minérale et le problème de l'allcharite. Bull Soc Fr Mineral Cristallogr 92:38–48

27a. Kutoglu A (1974) Synthese von Kristallstrukturen von $TlFeS_2$ und $TlFeSe_2$. Naturwissensch 61:125–126

Wandji R, Kamsu Kom J (1972) Préparation et étude de la raguinite et de son homologue sélénié: $TlFeS_2$ et $TlFeSe_2$. CR Acad Sci Ser C 275:813–816

Zabel M Range K-J (1979) Ternäre Phasen im System Eisen-Thallium-Schwefel. Z Naturforsch 34b:1–6

28. Marumo F, Nowacki W (1965) The crystal structure of rathite-I. Z Kristallogr 122:433–456

29. Johan Z, Mantienne J, Picot P (1974) La routhiérite, $TlHgAsS_3$, et la laffitite, $AgHgAsS_3$, deux nouvelles espèces minérales. Bull Soc Fr Mineral Cristallogr 97:48–53

30. Iitaka Y, Nowacki W (1961) A refinement of the pseudo crystal structure of scleroclase $PbAs_2S_4$. Acta Crystallogr 14:1291–1292

Nowacki W, Iitaka Y, Bürki H, Kunz V (1961) Structural investigations on sulfosalts from Lengenbach, Binn Valley (Ct. Wallis). Part 2. Schweiz Mineral Petrogr Mitt 41:103–116 (Scleroclase et al.)

References to Table 1 (continued)

Nowacki W, Marumo F, Takéuchi Y (1964) Untersuchungen an Sulfiden aus dem Binnatal (Kt. Wallis, Schweiz). Schweiz Mineral Petrogr Mitt 44:5–9

31. Janković S, Le Bel L (1976) Le thallium dans le mineral de Bözcukur près Kitahya, Turquie. Schweiz Mineral Petrogr Mitt 56:69–77

32. Browne PRL (1971) Mineralisation in the Broadlands geothermal field, Taupo volcanic zone, New Zealand. Proc. IMA-IAGOD Meet '70, Joint Symposium Vol. Soc Min Geol Jpn Spec Issue 2:64–75

Weissberg G (1969) Gold-silver ore-grade precipitates from New Zealand thermal waters. Econ Geol 64:95–108

33. Ohmasa M, Nowacki W (1971) The crystal structure of vrbaite, $Hg_3Tl_4As_8Sb_2S_{20}$. Z Kristallogr 134:360–380

34. Takéuchi Y, Ohmasa M, Nowacki W (1968) The crystal structure of wallisite, $PbTlCuAs_2S_5$, the Cu analog of hatchite, $PbTlAgAs_2S_5$. Z Kristallogr 127:349–365

35. Dickson FW, Radtke AS (1978) Weissbergite, $TlSbS_2$, a new mineral from the Carlin gold deposit, Nevada. Am Mineral 63:720–724

36. Bohać P, Brönnimann E, Gäumann A (1974) On the ternary compound $TlSbS_2$ and its photoelectric properties. Mat Res Bull 9:1033–1040

Edenharter A, Gostojić M (1980) Synthese und Kristallstruktur von $TlSbS_2$, synthetischem Weissbergit. (In prep.)

Semiletov SA, Man LI (1960) Electron diffraction study of the structure of the films of $TlBiSe_2$ and $TlSbS_2$. Sov Phys Crystallogr 4:385–388 [Kristallographiya (1959) 4:414–417]

37. Bohać P, Brönnimann E, Gäumann A (1974) On the ternary compound $TlSbS_2$ and its photoelectric properties. Mat Res Bull 9:1033–1040

Dickson FW, Radtke AS (1978) Weissbergite, $TlSbS_2$, a new mineral from the Carlin gold deposit, Nevada. Am Mineral 63:720–724

38. Gostojić M, Nowacki W, Engel P (1981) The crystal structure of synthetic Tl_3SbS_4. Z Kristallogr (in press)

38a. Engel P, Gostojić M, Nowacki W (1981) The crystal structure of synthetic Tl_3AsS_4. Z Kristallogr (in press)

39. Gostojić M, Edenharter A, Nowacki W, Engel P (1981) The crystal structure of synthetic Tl_2-$MnAs_2S_5$. Z Kristallogr (in press)

40. Janković S, Le Bel L (1976) Le thallium dans le mineral de Bözcukur près Kitahya, Turquie. Schweiz Mineral Petrogr Mitt 56:69–77

41. Gitsu DV, Kantser CT, Kretsu JV, Chumak GD (1970) Nizkotemp Termoelek Mater 17–21. (In: Julien-Pouzol et al. 1979, p 1315

42. Julien-Pouzol M, Jaulmes S, Laruelle P (1979) Structure cristalline du sulfure de bismuth et thallium $Tl_4Bi_2S_5$. Acta Crystallogr B35:1313–1315

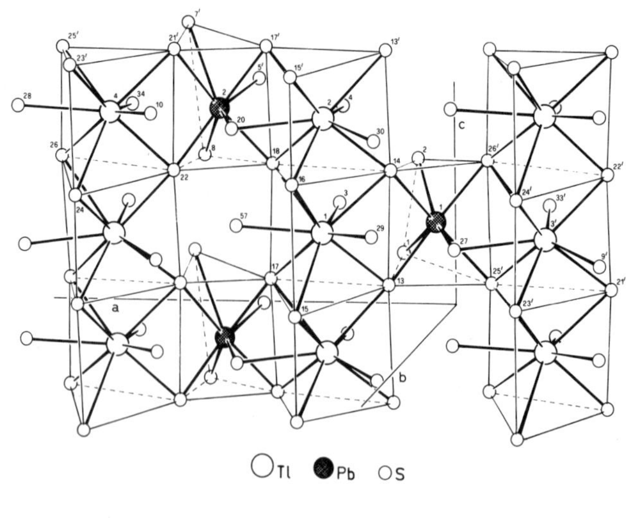

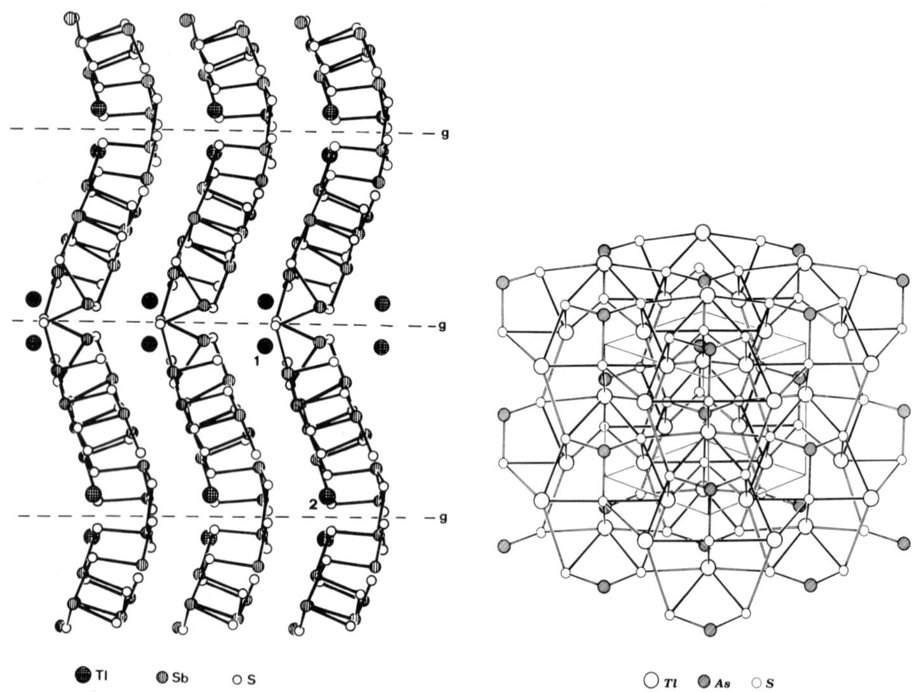

Fig. 1. Structure of the natural sulphosalt $Tl_8 Pb_4 Sb_{21} As_{19} S_{68}$

Fig. 2. Structure of synthetic parapierrotite, $TlSb_5 S_8$

Fig. 3. Structure of synthetic ellisite, $Tl_3 AsS_3$

(Ivanov et al. 1960; Vlasov 1966, p. 499), where Tl is associated with As, Sb, occasionally Bi, Ag, Au, or Hg.

In the sulphosalts of the Lengenbach quarry the Tl content was detected in 1964 by microprobe analyses (Nowacki et al. 1964), Tl substituting Pb, beside the "pure" Tl-mineral hatchite, imhofite, lorandite, and wallisite. The deposit Jas Roux (Hautes Alpes, France) and the Carlin-type gold deposits (Nevada U.S.A.) yielded additional new Tl-, and Tl,Hg-sulphides and sulphosalts (Mantienne 1974; resp. Radtke 1973, 1980; Radtke et al. 1973, 1974; Dickson et al. 1975, 1977, 1979). An interesting question is where the Sb, Tl, Au, . . . atoms occurring in small, but definite quantities in the host structure, are located (e.g. Sb, As, Au, Tl in stibnite, orpiment, and realgar). Bismuth sulphosalts seem to bear no Tl (Godovikov 1965, 1972, 1977) or extremely small amounts. On the other hand, the compounds $TlBiS_2$ and $Tl_4Bi_2S_5$ were synthesized and structurally analyzed (see Chap. 3).

2 The Synthesis of Thallium Sulphosalts (A.E.)

From the mineralogical and geochemical point of view the knowledge of the composition and the relationships of mineral phases found in deposits are very important. By systematic investigations of many sulphide and sulphosalt systems during the last decades much information about the thermodynamic stabilities of minerals and mineral assemblages and of the solid solutions they form together could be obtained.

Several experimental methods were developed to study phase relationships and/or to grow single crystals for detailed studies. The most important methods are: (1) differential thermal analysis (DTA), (2) evacuated silica tube experiments, (3) salt flux experiments, (4) hydrothermal syntheses. The DTA-method (1), used not only for sulphide/sulphosalt research, needs no description. The equipment and procedures for synthesizing sulphides and sulphosalts by dry reaction in evacuated silica tubes (2) were described by Kullerud (1971). Most of the sulphide phase equilibria have been determined by methods (1) and (2). Because solid-state diffusion in some sulphides and sulphosalts is very or too slow to obtain stable equilibrium within a reasonable time, salt flux experiments (3) were done to overcome the reaction kinetics in the evacuated silica tube runs by adding a catalyst or flux to the charge in order to increase the reactivity among the sulphides without shifting their equilibria (Boorman 1967; Kretschmar 1973). Till now only the four binary mixtures, NaCl/KCl, KCl/LiCl, NH_4Cl/LiCl, KCl/$AlCl_3$, have been tested (Moh and Taylor 1971), but there is no reason why other mixtures or more components could not be used. Hydrothermal syntheses (4). The main disadvantages of the methods discussed till now are their inability to produce single crystals for detailed investigations and to allow independent variation of the pressure. Moreover, the sulphosalts of the Lengenbach quarry, synthesized to find out the condition of growth, were formed during a regional metamorphism, by remobilization of As, Cu, Ag, and Tl from the gneisses nearby and by reaction with the base metals in the dolomite of the Lengenbach deposit (Graeser 1965). Because water is very important during the remobilization, our experiments constitute transport reactions in water or aqueous solution using the temperature gradient method. This method can be characterized by the scheme: dissolution – transport – crystallization. The finally crushed sulphide mixture is placed in a silica tube of known volume. The length of the tube may differ due to the desired temperature gradient. Then it is filled with an amount of water to produce the pressure wanted, cooled down with dry ice, and sealed off on vacuum. The ampoule in a steel pressure vessel is placed in a horizontal heater, while the counter-pressure produced by Ar gas is controlled by a manometer. The water or aqueous solutions in which the sulphides should be sufficiently soluble act as a flux from the hotter to the colder part of the ampoule to effect dissolution of the nutrient, transportation and growth of the crystals. In our experiments we have used temperature gradients between 30° and 100°C. Silica tubes were used because gold has been found to be unstable in systems containing sulphides of Pb, Ag, Hg, and As. The silica tubes recrystallize, however, very easily by slightly alkaline solutions.

The hydrothermal synthesis technique described above has produced single crystals of lorandite $TlAsS_2$, ellisite Tl_3AsS_3 (Tl-As-S-system), weissbergite $TlSbS_2$, parapierrotite $TlSb_5S_8$ (Tl-Sb-S-system), and hutchinsonite $(Tl,Pb)_2As_5S_9$ (Edenharter and Peters 1979). Further experiments yielded orange-red orthorhombic crystals of Tl_3AsS_4 and red triclinic Tl_3SbS_4 (Table 1). In the quarternary system Tl-Ms-As-S dark-brown orthorhombic crystals $Tl_2MnAs_2S_5$ (Table 1) could be obtained. Some work was done also in the system Tl-Pb-As-S, where a new dark-red platy sulphosalt $TlPbAs_3S_6$ (Table 1) was found.

3 Crystal Chemistry of Thallium (W.N.)

In Table 1 the data for some Tl-sulphides and Tl-sulphosalts are given (in alphabetical order). The sum of the ionic radii Tl^{+1} + S^{-2} resp. Tl^{+3} – S^{-2} is about 3.28 resp. 2.79 Å, that of the covalent radii 2.59; the metallic radius of Tl for the C.N. 12 is 1.595, that for a single bond 1.549 (Pauling 1960, p. 256). As can be seen from Table 1, all coordination numbers (C.N.) between 1 and 12 (except 11) are realized. Tl^{+1} has the electron structure $4f^{14}$ $6d^{10}$ $6s^2$, i.e. with a lone electron pair; this may yield a polar coordination scheme, which indeed sometimes is present.

The following C.N.'s and coordination schemes for Tl-S are observed (see also Figs. 1–10):

C.N. 1. $Tl_4Bi_2S_5$, Tl(1) and Tl(2) have 1 nearest neighbour (2.88, 2.93), followed by 2+2 at 3.12, 3.27 resp. 3.08, 3.09;

Fig. 4. Coordination of the Tl by the S atoms in synthetic parapierrotite

Fig. 5. Synthetic ellisite crystals (length of the aggregate *lower left* = 2 mm)

Fig. 6. Coordination of the Tl by the S atoms in synthetic ellisite

Fig. 7. Structure of synthetic Tl_3SbS_4

Fig. 8. Coordination of the Tl by the S atoms in Tl_3SbS_4

Fig. 9. Structure of galkhaite $[Hg_{0.76}(Cu,Zn)_{0.24}]_{12}Tl_{0.96}(AsS_3)_8$

Fig. 10. Coordination polyhedra (corners = *S*) around Tl, Hg and As in galkhaite

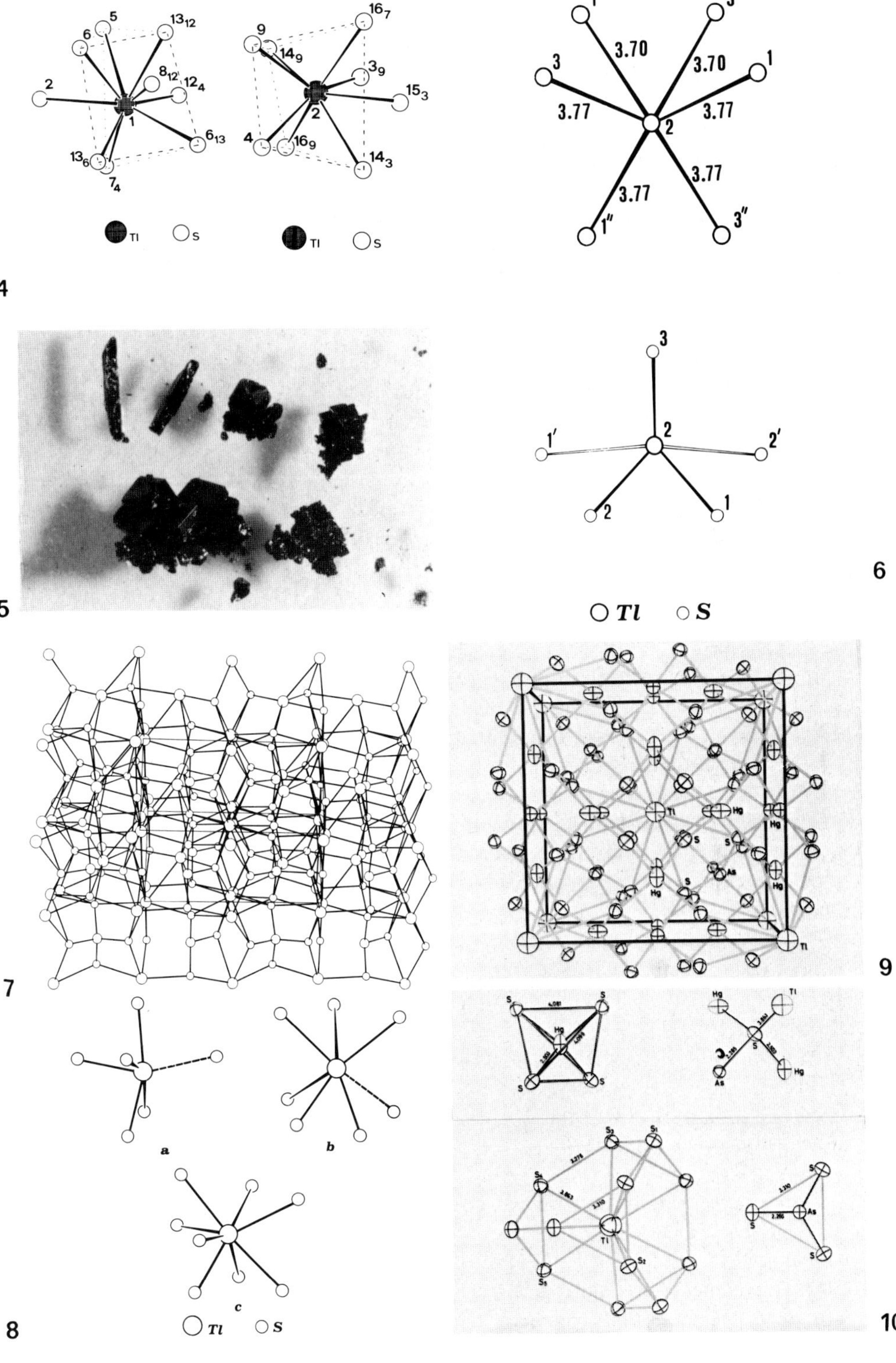
5
13₁₂
6
8₁₂
12₄
2
1
6₁₃
13₆
7₄
Tl
S
16₇
9
14₉
3₉
15₃
2
4
16₉
14₃
Tl
S
1′
3′
3.70
3.70
3
1
3.77
3.77
2
3.77
3.77
1″
3″
3
2
1′
2′
2
1
○ Tl ○ S
a
b
c
○ Tl ○ S
Hg
Tl
S
As

4

5

6

7

8

9

10

C.N. 2. red Tl_2S_5, hatchite, lorandite, vrbaite, wallisite. In many cases it is very difficult to give an "exact" C.N., the distances varying to a large extent;

3. red Tl_2S_5, Tl_2S, synth. ellisite; trigonal-pyramidal;

4. α-Tl_4S_3, TlS; ± regular tetrahedral or irregular; $Tl_4Bi_2S_5$: Tl(3) and Tl(4) are outside of a distorted tetrahedron (polar coordination!) (ideal case: trig. bipyram. with Tl in the middle of the base and an electron pair at one vertex);

5. α-Tl_4S_3, synth. christite, Tl_3SbS_4; irregular; $Tl_4Bi_2S_5$: Tl(1) and Tl(2) are outside of a tetragonal pyramid (polar coordination) (ideal case: octahedron with Tl at the centre and an electron pair at one corner);

6. dufrenoysite (Pb,Tl), $TlSbS_2$ (ED), Tl_3SbS_4; ± regular octahedron or irregular;

7. hatchite (irregular), hutchinsonite (Pb,Tl: split octahedron, i.e. one corner of an octahedron is split into two; Tl, Pb: irreg.), rathite-I [Pb,Tl: trigonal prismatic plus an atom on a normal to one side face, type (b)], Tl_3SbS_4 (irregular);

8. TlS, hatchite, wallisite, Tl_3SbS_4 (irregular), synth. parapierrotite [trigonal prismatic plus normals to two side faces, type (c)], vrbaite (distorted cube);

9. nat. $Tl_8Pb_4Sb_{21}As_{19}S_{68}$, synth. parapierrotite [trigonal prismatic plus normals to all three side faces, type (d); well known for Pb (with some Tl): dufrenoysite, rathite-I, scleroclase; but now also observed for pure Tl], $TlFeS_2$ (irregular);

10. vrbaite (10 = 8+2, i.e. the eight corners of a distorted cube, plus atoms on the normals to two opposite faces of the cube);

12. galkhaite (truncated tetrahedron, "Laves polyhedron"). This is a special case, the Tl atoms being situated statistically in the centre of these holes of the structure.

The observed distances Tl-S are all in accordance with a pure ionic (Tl^{+1}, Tl^{+3}) or covalent-metallic or a mixed (hybridization of several valence states) bond type. A peculiarity of about ten compounds is the existence of relatively short metal-metal or semi-metal distances:

Tl-Tl (2 $R_{met.} \approx 3.19$):	α-Tl_4S_3, synth. carlinite, synth. ellisite, hatchite, lorandite, $TlFeS_2$, wallisite, Tl_3SbS_4, $Tl_4Bi_2S_5$;
Tl-Ag:	hatchite
Tl-As:	synth. ellisite, hatchite, imhofite, lorandite, vrbaite
Tl-Cu:	wallisite
Tl-Fe:	$TlFeS_2$
Tl-Sb:	Tl_3SbS_4

A short review on the crystal chemistry of Tl was given at the meeting of the GDCh in Berlin (Engel et al. 1979); the crystal chemistry and a classification of sulphosalts in general are explained in Nowacki (1969), and Edenharter (1976). In the book by Lee (1971) a comprehensive and critical review of the chemistry of Tl is outlined, taking into account all kinds of Tl-compounds. In a further article on the crystal chemistry of

Tl in general (in preparation) the subject will be discussed in detail, whereas we restricted ourselves in this article mainly to minerals, as it is appropriate to this Festschrift.

In summary, it may be said that Pauling's (1960, p. 448) statement, "The general structure theory of the sulphide minerals still awaits formulation", is still true.

Acknowledgment. We are grateful to the "Stiftung Entwicklungsfonds Seltene Metalle" for financial help.

References

Chapter 1

Dickson FW, Radtke AS (1977) The unique mineralogy of Hg-As-Sb-Tl sulfides at the Carlin gold deposit, Nevada, and implications as to the origin of the deposit. Min Soc Am Friend of Min, 3rd Joint Symp, Crystal growth and habit, Tucson, Ariz, pp 13–14

Dickson FW, Radtke AS, Weissberg BG, Heropoulos C (1975) Solid solutions of antimony, arsenic, and gold in stibnite (Sb_2S_3), orpiment (As_2S_3), and realgar (As_2S_2). Econ Geol 70:591–594

Dickson FW, Rye RO, Radtke AS (1979) The Carlin gold deposit as a product of rock-water interactions. Pap Min Deposits of Western N America, Nev Bur Min Geol Proc, 5th Quadrennial Symp. Int Symp Genesis Ore Deposits, II:101–108

Godovikov AA (1965) Mineraly rjad vismutin-galenit. Nauka, Novosibirsk, Sib Otd, 188 pp

Godovikov AA (1972) Vismutovyye sulfosoli. Nauka, Moskva, 303 pp

Godovikov AA (1977) Orbital'ny radiusy i svoistva elementov. Nauka, Novosibirsk, Sib Otd, 155 pp

Goldschmidt VM (1958) In: Muir A (ed) Geochemistry. Oxford Univ Press, London, 730 pp

Nowacki W, Bahezre C (1963) Die Bestimmung der chemischen Zusammensetzung einiger Sulfosalze aus dem Lengenbach (Binnatal, Kt. Wallis) mit Hilfe der elektronischen Mikrosonde. Schweiz Mineral Petrogr Mitt 43:407–411 (Parker-Festschrift)

Radtke AS (1973) Preliminary geologic map of the Carlin gold mine, Eureka County, Nevada. US Geol Surv Misc Field Stud Map MF–537

Radtke AS (1980) Geology of the Carlin gold deposit, Nevada. US Geol Surv Prof Pap, (in press)

Radtke AS, Dickson FW (1974) Genesis and vertical position of fine-grained disseminated replacement-type gold deposits in Nevada and Utah, U.S.A. Problems of ore deposition, volcanic ore deposits, 4th IAGOD Symp Varna, Bulgaria I:71–78

Radtke AS, Taylor CM, Heropoulos C (1973) Antimony-bearing orpiment, Carlin gold deposit, Nevada. J Res USA Geol Surv 1:85–87

Rankama K (1949) A note on the geochemistry of thallium. J Geol 57:608–613

Rankama K, Sahama ThG (1950) Geochemistry. Univ Chicago Press, Chicago, 912 pp

Shaw DM (1952) The geochemistry of thallium. Geochim Cosmochim Acta 2:118–154

Songina OA (1964) Rare metals. Izd Metallurgija, Moskva, pp 291–319 (transl from Russian)

Vlasov KA (ed) (1966) Geochemistry of rare elements. Israel Program of Scientific Translations, Jerusalem, 494 pp

Wedepohl KH (ed) (1969/1978) Handbook of geochemistry, vol II. Thallium 81. Springer, Berlin Heidelberg New York

Chapter 2

Boorman RS (1967) Subsolidus studies in the ZnS-FeS-FeS_2 system. Econ Geol 62:614–631

Canneri G, Fernandes L (1925) Contributo allo studio di alcuni minerali contenento tallio. Analisi termica dei sistemi: Tl_2S–As_2S_3, Tl_2S–PbS. Atti Accad Naz Lincei Rend 6 Cl Sci Fis Mat Nat 1:671–676

Edenharter A, Peters T (1979) Hydrothermalsynthese von Tl-haltigen Sulfosalzen. Z Kristallogr 150:169–180

Graeser St (1965) Die Mineralfundstellen im Dolomit des Binnatales. Schweiz Mineral Petrogr Mitt 45:597–795

Kretschmar U (1973) Phase relations involving arsenopyrite in the system Fe-As-S and their application. PhD Thesis, Univ Toronto, Canada

Kullerud G (1971) Experimental techniques in dry sulfide research. In: Ulmer GC (ed) Research techniques for high pressure and high temperature. Springer, Berlin Heidelberg New York, pp 288–315

Moh GH, Taylor LA (1971) Laboratory techniques in experimental sulfide petrology. Neues Jahrb Mineral Monatsh 450–459

Peterson JC (1976) Phase relations in the system $Tl_2S{-}As_2S_3$. Thesis, Master of Science, Stanford Univ

Chapter 3

Edenharter A (1976) Fortschritte auf dem Gebiete der Kristallchemie der Sulfosalze. Schweiz Mineral Petrogr Mitt 56:195–217

Engel P, Gostojić M, Nagl A, Nowacki W (1979) Zur Kristallchemie des Thalliums. Ref. 18. GDCh-Hauptversammlung, Berlin 10.–14. Sept, 2. Zirk, S 53, Sekt 9, Festkörperchemie

Lee AG (1971) The chemistry of thallium. Elsevier, New York, 366 pp

Nowacki W (1969) Zur Klassifikation und Kristallchemie der Sulfosalze. Schweiz Mineral Petrogr Mitt 49:109–156

Pauling L (1960) The nature of chemical bond, 3rd edn. Cornell Univ Press, Ithaca, 644 pp

Microbiological Leaching of Tin Minerals by Thiobacillus Ferrooxidans and Organic Agents

G.H. TEH[1], W. SCHWARTZ[2], and G.C. AMSTUTZ[3]

Abstract

Leaching tin from synthetitic minerals like stannite, kesterite, stannoidite, herzenbergite, ottemannite, and berndtite, and natural tin minerals which include stannite, cassiterite, and varlamoffite, in the presence of *Thiobacillus ferrooxidans* and organic agents, especially oxalic acid, EDTA and an oxalic-citric acid mixture has been shown to be possible in the laboratory.

Over a leach period of 35 days with 0.5% pulp density, initial pH of 2.5, using minus 0.16 mm size fraction at 32°C, as much as 54.45, 72.66, 97.13, and 31.30% Sn were extracted from synthetic stannite, kesterite, stannoidite, and natural stannite, respectively.

Varlamoffite found in the dried, leached residues of the tin sulphides provides evidence that bacterial action can be responsible for the genesis of supergene varlamoffite.

1 Introduction

Natural weathering systems, organic agents (such as fulvic and humic acids from humus), which are normally produced during biochemical weathering of rocks by soil microorganisms, decompose minerals dominantly by reacting with polyvalent cations in minerals to form metallo-organic complexes or chelates (Huang and Keller 1972).

Muntz (1890) was the first to suggest that bacteria are involved in rock decomposition. Recently, however, attention has been increasingly called to the activity of bacteria in weathering processes (Ehrlich 1964). A possible aspect of the oxidation and dissolution of stannite and related tin minerals can be attributed to the biochemical weathering processes by *Thiobacillus ferrooxidans.*

The availability of high-grade synthetic specimens, coupled with the research interests in the possible oxidation pathways of stannite and related tin minerals in nature and the applicability of the microbiological leaching technique in the recovery of tin from these minerals, low-grade tin deposits or too complex intergrown grains inseparable by mechanical or other sophisticated beneficiation techniques, have prompted a

1 Department of Geology, University of Malaya, Kuala Lumpur, Malaysia

2 Lehrstuhl für Mikrobiologie der Technischen Universität, Gaußstraße 7, 3300 Braunschweig, FRG

3 Mineralogisch-Petrographisches Institut, Universität Heidelberg, Im Neuenheimer Feld 236, 6900 Heidelberg, FRG

comparative study of the application and rates of microbiological leaching of these minerals in the presence of *Thiobacillus ferrooxidans* and of organic agents.

1.1 Oxidation of Tin Minerals by *Thiobacillus*

It was as early as 1922 that Rudolfs et al. suggested the possible economical extraction of metals from low-grade sulphide ores by microbiological leaching of metal sulphides.

However, initial scientific interest in this field crystallized only after the original isolation of *Thiobacillus ferrooxidans* by Colmer and Hinkle (1947). This was followed by extensive studies on microbiological leaching and processes applicable to the leaching of metals, especially the sulphides, from ores. The literature list on microbiological leaching has in the last 20 years been very extensive and some of the latest review articles applicable to this present study include Malouf and Prater (1961), Duncan et al. (1966), Pings (1968), Torma (1971), Corrans et al. (1972), Tuovinen and Kelly (1974).

At the present, the commercial application of microbiological leaching technique is limited, generally, to the extraction of copper and uranium.

Duncan (1967) demonstrated the sulphide oxidation ability of *Thiobacillus ferrooxidans* on stannite, however not too significant tin solubilization, 12% in 40 days, was reported.

1.2 Leaching by Organic Agents

Organic agents, for instance organic acids which occur in soil and other organic acids of geological and pedologic systems (Huang and Keller 1972), can involve release of cations from minerals and rocks to form soluble metal complexes and to affect chemical weathering of these minerals and rocks (Iskandar and Syers 1972).

The weathering or leaching effects of organic agents (Ginzburg et al. 1963; Huang and Keller 1971; Brockamp 1974) and of acid producing, heterotrophic microorganisms (Wagner and Schwartz 1967; Kiel 1977), have been widely reported but are mainly still in the laboratory stage and have previously been directed to studies of soil weathering and decomposition of common rock-forming minerals.

2 Materials and Methods

2.1 Minerals and Samples

The natural samples were collected mainly from Eu Tong Seng Mine, Tekka and Hock Leong Mine, Ampang, and, in addition, various other tin deposits in Peninsular Malaysia (Teh 1979). For comparison with Malaysian occurrences, cassiterite from Rooiberg Mine, South Africa, and varlamoffite from San Pedro Mine, Tazna, Bolivia, were also used for the microbiological investigations.

The synthetic tin sulphides, herzenbergite (SnS), ottemannite (Sn_2S_3), berndtite (SnS_2), stannite (Cu_2FeSnS_4), kesterite (Cu_2ZnSnS_4) and stannoidite ($Cu_8Fe_3Sn_2S_{12}$), were derived from synthesis in rigid, evacuated, sealed silica glass tubes as described by Kullerud (1971) and confirmed by X-ray diffraction analysis.

2.2 Bacteria

A pure strain of *Thiobacillus ferrooxidans* used in this study was originally isolated from the "Kambrischer Alaunschiefer", Ranstad, Sweden. It has been routinely maintained on medium 9K of Silverman and Lundgren (1959). This acidophilic, chemoautotrophic bacterium oxidizes inorganic compounds of sulphur for energy, assimilates carbon dioxide as the sole source of carbon and requires only inorganic nutrients; nitrogen (as NH_4^+), phosphate, sulphate, magnesium and potassium (Temple and Colmer 1951; Tuovinen and Kelly 1972).

Thiobacillus ferrooxidans is capable of ferrous ion oxidation in acid solutions. The ferric ion resulting from this oxidation is an efficient chemical oxidizing agent at acid pH values.

The catalytic effect of iron in the oxidation of metal sulphides (MS) can be expressed as follows (Ehrlich and Fox 1967):

$$MS + Fe_2(SO_4)_3 \dashrightarrow MSO_4 + 2\ FeSO_4 + S^0 \qquad (1)$$

where M is a bivalent metal. The importance of acid ferric liquors as oxidants of nonferrous metal sulphides is now well established (Woodcock 1967).

In reaction (1) the ferric iron is reduced to the ferrous form while the sulphur content of the sulphide has reacted to the elemental state.

Further, in the presence of sulphuric acid and oxygen, the following oxidation reactions may occur (Silverman and Lundgren 1959; Scott and Dyson 1968; Beck 1969; Habashi 1970):

$$MS + 1/2\ O_2 + H_2SO_4 \dashrightarrow MSO_4 + S^0 + H_2O \qquad (2)$$

$$MS + 2\ O_2 \dashrightarrow MSO_4 \qquad (3)$$

$$2\ FeSO_4 + 1/2\ O_2 + H_2SO_4 \dashrightarrow Fe_2(SO_4)_3 + H_2O \qquad (4)$$

Reactions (2) and (4) predominate with increasing hydrogen ion concentrations. Under the acid conditions of the leaching process the ferrous iron is stable and regeneration of the leaching liquor would be difficult without the agency of bacterial oxidation of the ferrous iron according to Eq. (3).

In the presence of bacteria, reactions (3) and (4) and consequently reaction (1) are catalyzed and accelerated by the enzymic action of the microorganisms.

Acid required for the chemical oxidation of minerals by ferric iron and for the growth of the bacteria, is generated by biological oxidation of elemental sulphur formed in reactions (1) and (2) according to the equation (Silver 1970):

$$S^0 + 3/2\ O_2 + H_2O \xrightarrow{\text{bacteria}} H_2SO_4 \quad (5)$$

Additional acid may be generated by chemical hydrolysis of ferric iron as follows:

$$Fe^{3+} + 3\ H_2O \dashrightarrow Fe(OH)_3 + 3\ H^+ \quad (6)$$

Hence, *Thiobacillus ferrooxidans* acts in the leaching of sulphide minerals in two principal ways: to oxidize solid sulphides directly to soluble sulphates and to oxidize ferrous ion produced by chemical reactions during leaching (Silverman 1967). Direct attack on sulphides by the bacteria has been reported (Bryner and Jameson 1958; Duncan et al. 1967a; Beck and Brown 1968).

While the addition of ferrous or ferric sulphate was reported as not necessary for microbiological leaching, its addition, however, may be beneficial for the maintenance of pH in the case of minerals or ores low in iron (Duncan and Trussell 1964; Duncan and Walden 1972).

3 Culture Technique

3.1 Bacteria

Apart from the preliminary investigations, for the main leaching experiments with *Thiobacillus ferrooxidans,* 0.0375 g of minus 0.16 mm material under study were placed in a flask containing 7.5 ml of medium 9K. The pH was adjusted with 1 N sulphuric acid to pH 2.5, 1 ml of inoculum added and the flask incubated at 32°C on a gyratory shaker.

3.2 Organic Agents

In the experiments with organic agents, 0.1 g of the minus 0.16 mm samples were placed in 7.5 ml of 0.05 M solutions of oxalic acid, salicylic acid, glycollic acid, tannic acid, β-alanine, tartaric acid, acetic acid, citric acid, lactic acid, EDTA, and a mixture of 0.1 M oxalic and citric acids.

The specimen and leaching solution were previously sterilized separately by autoclaving. pH values of the reagents at the start and end of the runs were recorded.

3.3 Metal Extractions

1 ml aliquots of liquor were removed at periodic intervals and analyzed in an Atomic Absorption Spectrophotometer, Perkin-Elmer Model 303, and also confirmed colorimetrically (Stanton and McDonald 1961).

4 Optimum Conditions Determination

Determination of optimum conditions for microbiological leaching of tin was carried out prior to the main leaching experiments involving natural and synthetic tin minerals. The parameters investigated were decided upon after a careful review of previous techniques by various researchers.

The natural "stannite" sample (TL 25) from Hock Leong Mine, Ampang, Malaysia, assaying 30.0% Sn, was used for the experiments which include investigating the effects of pulp density, pH, particle size, temperature, and Fe^{2+}.

The various factors affecting the rate of leaching by *Thiobacillus ferrooxidans* on other substrates have previously been quite extensively investigated (Bryner and Anderson 1957; Malouf and Prater 1961; Razzell and Trussell 1963; Duncan et al. 1967; Torma et al. 1972).

5 Results and Discussion

Throughout the presentation and discussion of results, particle dimensions and pulp densities refer only to their initial values.

As techniques are not developed to measure these changes with time, for the purpose of the current investigations the importance of the variation of these parameters are appreciated.

5.1 Effect of Pulp Density

The effect of pulp density (solid to liquid ratio in %) on the yield of dissolved tin was studied for initial pulp densitites of 0.1%, 0.5%, and 2.0%.

Figure 1 is a plot of tin extraction against pulp density over a period of bacterial leaching by *Thiobacillus ferrooxidans*.

In these experiments the minus 0.16 mm size fraction of a Hock Leong "stannite" ore was used with initial pH 2.5 and at a temperature of 25°C.

The optimum density was found to be 0.1%.

A significant trend is the pronounced decrease in tin extraction by the bacteria with pulp densities of 0.5% and above. This is perhaps because the high solids concentration interferes with the rate of transfer of oxygen and carbon dioxide to the organism (Torma et al. 1972), i.e. cell growth (McGoran et al. 1969), or that the high concentration of tin inhibited the iron-oxidizing activity of the bacteria (Imai et al. 1975).

5.2 The Influence of pH

The influence of pH on dissolution of tin from the natural stannite sample is shown in Fig. 2. pH values of 1.5, 2.0, 2.5, and 3.0 were investigated.

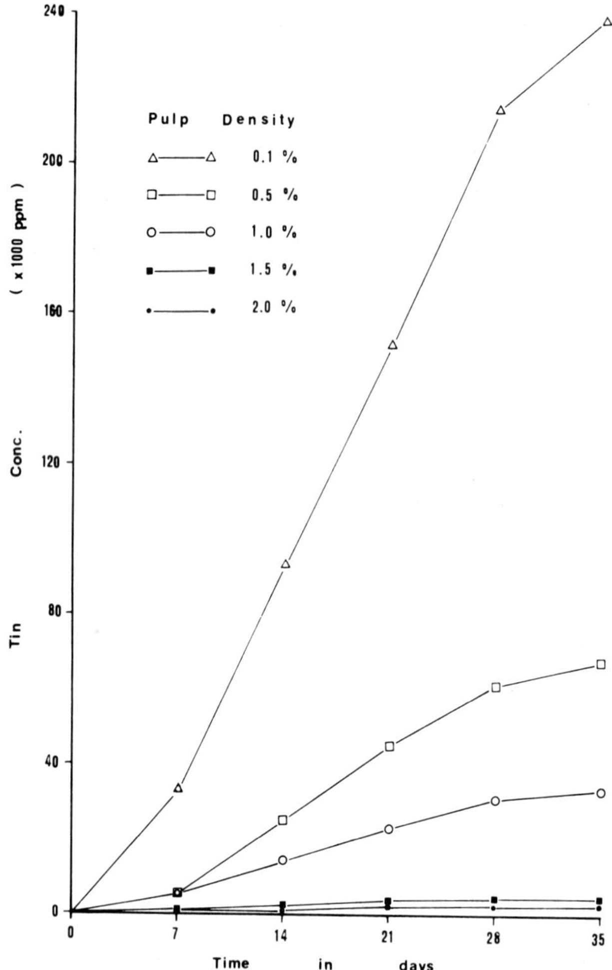

Fig. 1. The effect of pulp density on tin extraction by *Thiobacillus ferrooxidans,* using minus 0.16 mm size fraction stannite from Hock Leong Mine, Malaysia, with initial pH 2.5 and at 25°C

During the lag phase the pH rose and H_2SO_4 had to be added to keep the pH constant. However, during logarithmic growth the pH fell and NaOH had to be added. The optimum pH was found to be 1.5.

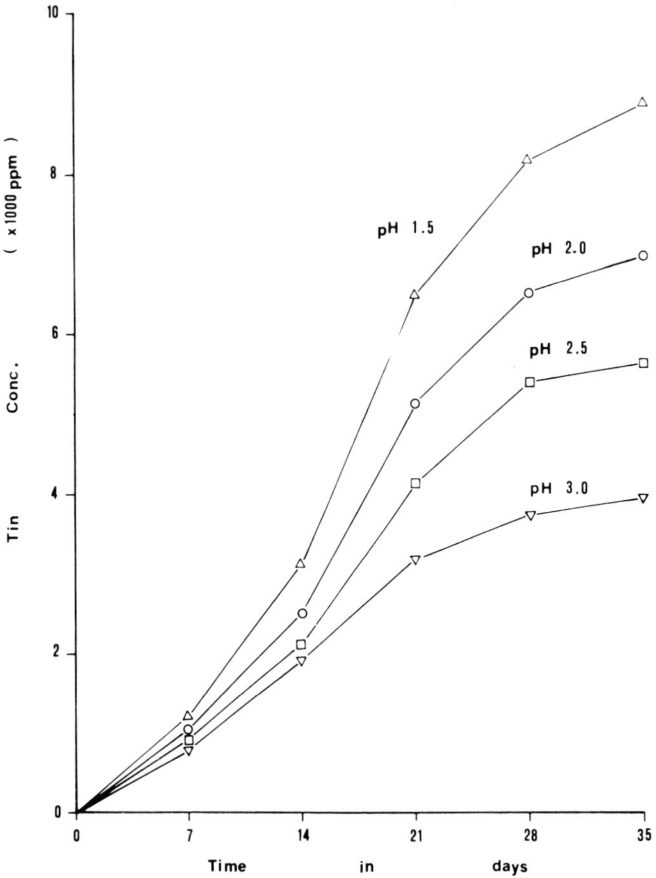

Fig. 2. The effect of pH on tin extraction of *Thiobacillus ferrooxidans* using minus 0.16 mm size fraction stannite from Hock Leong Mine, Malaysia, at 1% pulp density and 25°C

6 Effect of Particle Size

The effect of particle size, hence the surface area of particles, on the rate of leaching of the stannite sample is shown in Fig. 3.

Studies of Razzell and Trussell (1963) and Duncan et al. (1966) indicated that the leaching rate might correlate with the surface area of particles for microbiological oxidation of solid sulphides. Duncan et al. (1966) believe that there seems to be a limit for the minimal particle size below which the bacterial oxidation is not further enhanced.

The optimum particle size for the stannite samples from Hock Leong Mine, Malaysia, was 0.1 to 0.16 mm. Physical variables of the ore, like the preferred and extensive intergrowths with other mineral species and exsolution textures, may have greatly influenced the total surface area exposed to the bacteria.

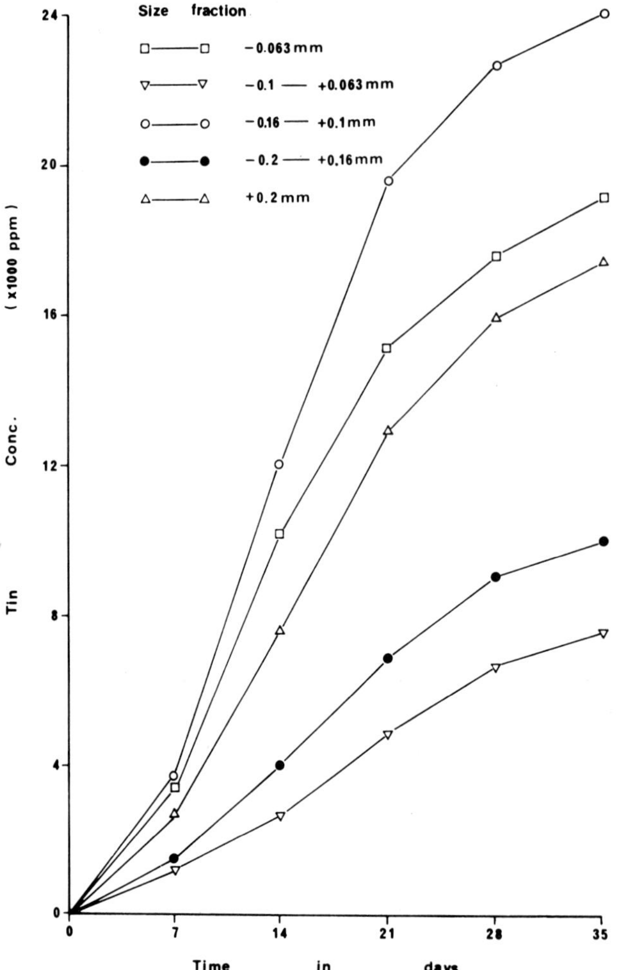

Fig. 3. The effect of particle size on tin extraction by *Thiobacillus ferrooxidans,* using stannite from Hock Leong Mine, Malaysia, with 1% pulp density at initial pH 2.5 and 25° C

6.1 Effect of Temperature

The effect of temperature on the extraction of tin by *Thiobacillus ferrooxidans* is shown in Fig. 4.

The temperatures studied were 25°, 32°, and 37°C.
The optimum temperature was found to be 32°C.

It has been reported that the maximum rate of growth and oxidation processes of *Thiobacillus ferrooxidans* occur between 30° and 40°C, with an optimum usually close to 35°C (Marchlewitz and Schwartz 1961).

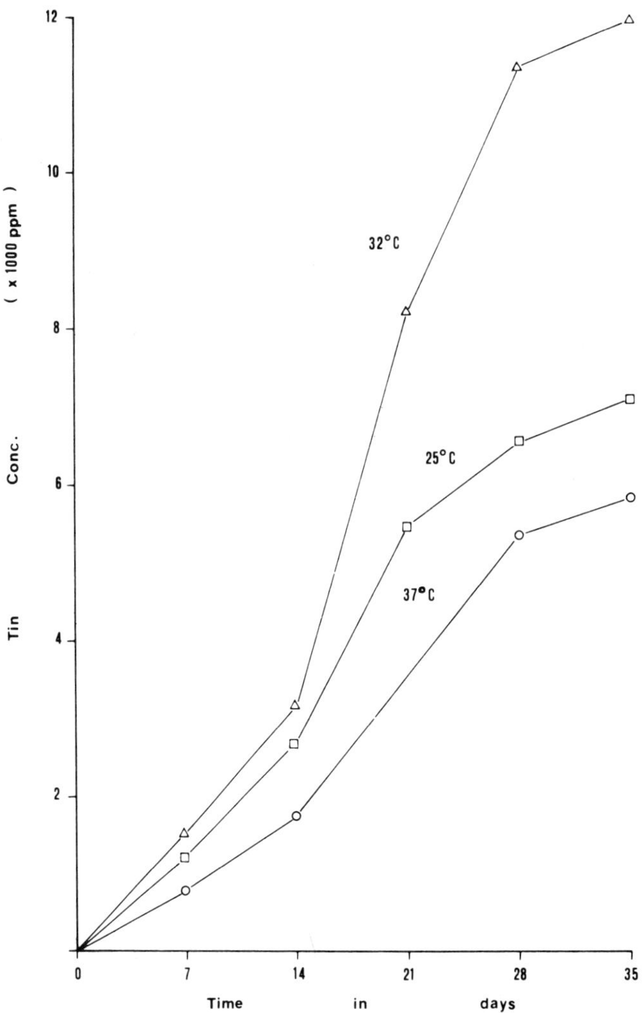

Fig. 4. The effect of temperature on tin extraction by *Thiobacillus ferrooxidans,* using minus 0.16 mm size fraction stannite from Hock Leong Mine, Malaysia, with 1% pulp density at initial pH 2.5

6.2 Influence of Ferrous Iron

The influence of ferrous iron on the extent of tin extraction on the natural stannite sample from Hock Leong Mine and synthetic stannite is displayed in Fig. 5.

The results demonstrate that the presence of ferrous sulphate in the nutrient medium greatly assists in leaching tin from both natural and synthetic stannite. The extent of tin extraction is increased 5-fold in natural stannite and 11-fold in synthetic stannite.

The sterile controls, containing the 9K medium with ferrous sulphate, show slightly higher tin extractions than the medium with bacteria but no ferrous iron.

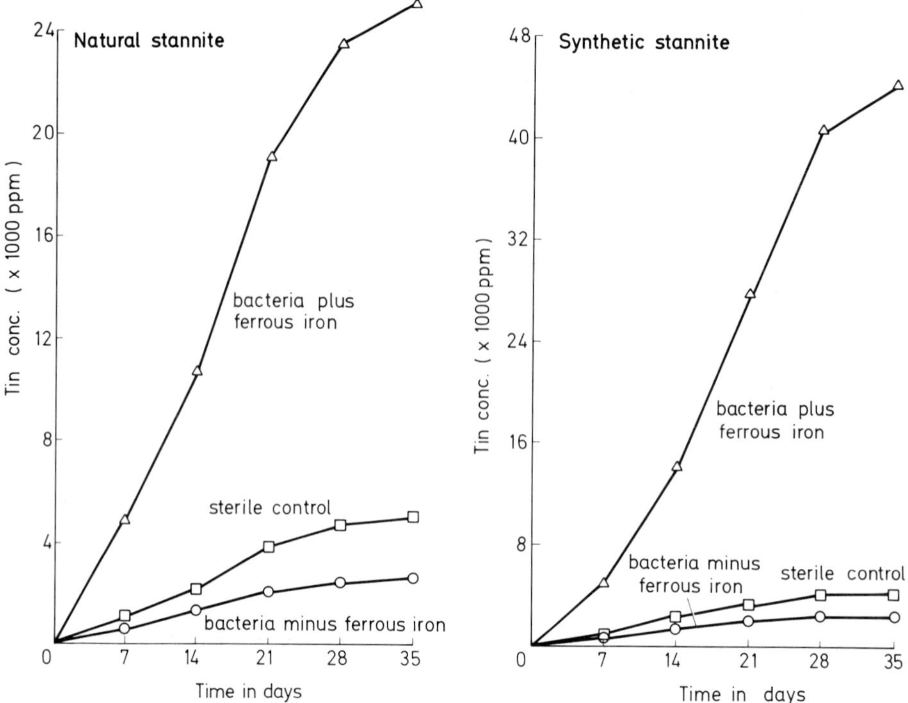

Fig. 5. The influence of ferrous iron on tin extraction by *Thiobacillus ferrooxidans* on natural stannite from Hock Leong Mine, Malaysia, and synthetic stannite; at 25°C, 1% pulp density, initial pH 2.5, using minus 0.16 mm size fraction

The experiments confirm that ferrous iron was necessary in the bacteria nutrient medium to assist in microbiological leaching of tin from stannite. The bacteria oxidize it to ferric iron and this appears to influence the rate and extent of tin extraction.

At 25°C, 1% pulp density, initial pH 2.5, using the minus 0.16 mm size fraction, and after 35 days leaching, bacteria plus ferrous iron in medium yielded 16.17% Sn from synthetic stannite and 8.40% Sn from natural stannite, while those without ferrous iron yielded only 0.98% Sn and 0.86% Sn respectively, and the sterile controls 1.58% Sn and 1.65% Sn respectively.

7 Results of Leaching by *Thiobacillus ferrooxidans*

The resulting curves for the various experiments were plotted as tin concentration (in ppm) against time (in days).

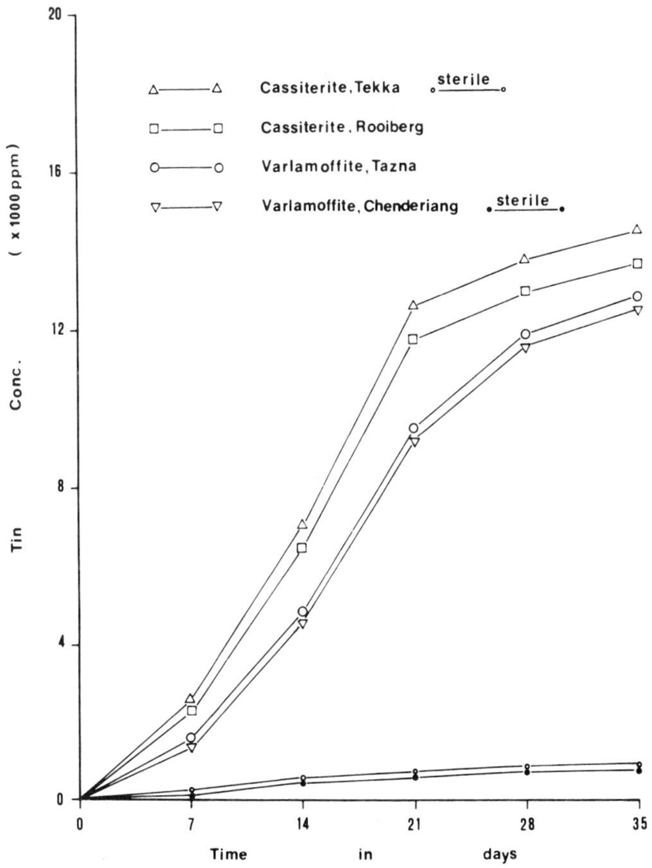

Fig. 6. Leaching of cassiterite and varlamoffite samples by *Thiobacillus ferrooxidans* at 32°C, 0.5% pulp density, initial pH 2.5, using minus 0.16 mm particle size

7.1 Leaching of Natural Tin Samples by *Thiobacillus ferrooxidans*

The amounts of tin extracted over a period of 35 days by *Thiobacillus ferrooxidans* at 32°C on leaching solutions of 0.5% pulp density and initial pH 2.5 from minus 0.16 mm particles of naturally occurring cassiterite, varlamoffite, and stannite samples are shown in Figs. 6 and 7 and Table 1.

The results show a significant extraction (about tenfold) for stannite compared with the cassiterites and varlamoffites (Figs. 6 and 7), confirming the preferential leaching of metallic sulphides by *Thiobacillus ferrooxidans.* Stannite from Hock Leong Mine gave a release of 31.30% Sn in 35 days with bacteria and 4.39% Sn in the absence of bacteria. This is approximately half that of the activity on synthetic stannite. The pH in the inoculated flask dropped from 2.5 to 2.2 during this period.

Cassiterite concentrates from Rooiberg and Tekka gave a more-or-less similar rate of dissolution over 35 days, releasing finally 2.88% and 2.77% Sn respectively.

Table 1. Leach rate and extraction (%) of tin from natural and synthetic samples by *Thiobacillus ferrooxidans*

Sample	Assay[a] % Sn	Leach rate ppm/h	Extraction[b] %
Natural			
1. TL23 Cassiterite (untreated concentrate) Rooiberg, S. Africa	47.5	27.98	2.88 (0.18)
2. TL24 Cassiterite (untreated concentrate) Tekka Malaysia	52.5	29.76	2.77 (0.18)
3. TL25 Stannite Ampang, Malaysia	30.0	178.57	31.30 (4.39)
4. TL26 Varlamoffite Tazna, Bolivia	52.0	23.51	2.47 (0.17)
5. TL27 Varlamoffite Chenderiang, Malaysia	46.8	23.21	2.67 (0.18)
Synthetic			
6. TL28 Stannite	27.5[c]	291.67	54.45 (4.60)
7. TL31 Kesterite	27.0[e]	366.07	72.66 (5.14)
8. TL32 Herzenbergite	78.0[d]	29.46	1.97 (0.15)
9. TL33 Ottemannite	71.1[e]	29.76	2.30 (0.17)
10. TL34 Berndtite	64.9[e]	29.17	2.47 (0.16)
11. TL35 Stannoidite	18.2[e]	303.57	97.13 (6.15)

a Using Stanton and McDonald's modified gallein method
b Extraction from minus 0.16 mm size fraction with 0.5% pulp density, initial pH 2.5 at 32°C (sterile controls in brackets)
c 99.60% extraction
d 99.07% extraction
e Theoretical value

The varlamoffites from Tazna, Bolivia, and Chenderiang, Malaysia, gave even better look-alike curves than the cassiterites, giving finally tin extractions of 2.47% and 2.67% respectively.

This varying response in leaching obtained for different samples of the same mineral is due, probably, to differences in microscopical intergrowths of minor accessory mineral species of the different localities.

With the cassiterites and varlamoffites, a pronounced decrease in the extraction rate was obvious after the 21st day of leaching. A probable significant accumulation of certain unresolved form of tin concentration appears to decrease the activity of the bacteria drastically.

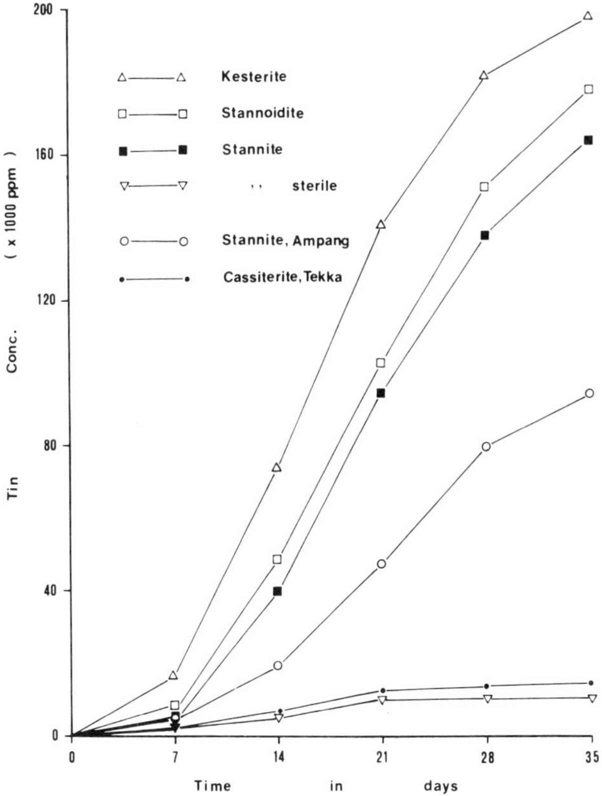

Fig. 7. Leaching of synthetic stannite, kesterite, and stannoidite by *Thiobacillus ferrooxidans*, using minus 0.16 mm size fraction, 0.5% pulp density, initial pH 2.5 at 32°C. The *curves* for natural stannite and cassiterite are for comparison

A notable result from this investigation of the tin dissolution activity of *Thiobacillus ferrooxidans* on cassiterite and varlamoffite is that the varlamoffite ($SnO_2 \cdot 2 H_2O$), which is believed to be of supergene origin (Russel and Vincent 1952; Antun, as reported by Singh and Bean 1968; Hosking 1973), does not appear to be more easily attacked by bacteria; in fact, it released even less tin than cassiterite. This confirms the fact that the response to leaching by the bacteria is not dependent on the varying resistance of the mineral to weathering but rather on its chemical composition and crystal structure.

7.2 Leaching of Synthetic Tin Minerals by *Thiobacillus ferrooxidans*

The plots showing the tin concentrations extracted from the synthetic minerals, kesterite, stannoidite, and stannite are presented in Fig. 7, while that for herzenbergite, ottemannite and berndtite are displayed in Fig. 8.

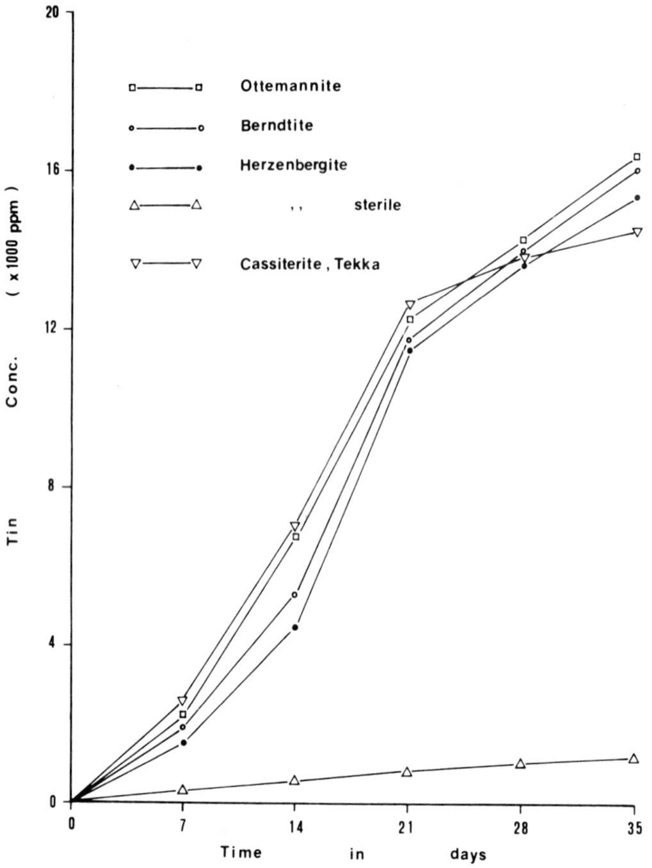

Fig. 8. Leaching of the binary tin sulphides, herzenbergite, ottemannite, and berndtite by *Thiobacillus ferrooxidans,* using minus 0.16 mm size fraction, 0.5% pulp density, initial pH 2.5 at 32°C. The *curve* for natural cassiterite is included for comparison

The group of quaternary sulphides, stannite, kesterite, and stannoidite, in general, show a distinctly higher tin extraction than the binary sulphides, herzenbergite, ottemannite, and berndtite (see also Table 1).

Most worthy of mention was the 97.13% tin extraction by *Thiobacillus ferrooxidans* from minus 0.16 mm stannoidite after 35 days of leaching at 32°C, 0.5% pulp density, and initial pH 2.5. Under the same conditions and parameters, kesterite and stannite yielded 72.66% and 54.45% Sn respectively. The pH, at the end of 35 days, was found to be 1.7 for the three quaternary tin sulphides. The sterile controls for stannoidite, kesterite, and stannite yielded 6.15%, 5.14%, and 4.60% Sn respectively.

On the other hand, the final tin extraction for herzenbergite, ottemannite, and berndtite in the presence of bacteria were 1.97%, 2.30%, and 2.47% Sn respectively. The rate of tin extraction for ottemannite was 30.06 ppm/h while that for herzenbergite and berndtite was about 29.46 ppm/h (from the 14th to 28th day of incubation).

The curve for cassiterite from Tekka, superimposed on those of the binary tin sulphides shows a more-or-less similar rate of extraction and yield for tin (Fig. 8). These binary oxides and sulphides of tin appear to be very resistant to biochemical leaching by *Thiobacillus ferrooxidans;* due, probably, to their high percentage of Sn or their tightly-bonded crystal lattices.

The high yield of tin from the quaternary tin sulphides, on the other hand, is probably due to the fact that besides Sn these sulphides have in their lattices Cu and Fe, two elements that are known to be readily extracted by *Thiobacillus ferrooxidans.* Excessive attack by the bacteria on Cu, Zn, and Fe resulted in weakening the lattice of these quaternary sulphides, thus exposing Sn to easier access and extraction. Another fact is the lower tin content in these sulphides, aptly demonstrated by stannoidite, which has the lowest tin content (about 18 wt.%), where 97.13%Sn extraction was achieved in 35 days.

7.3 Comparison for Leaching Natural and Synthetic Stannite by *Thiobacillus ferrooxidans*

In Fig. 7, the curve for natural stannite from Hock Leong Mine is superimposed on the plot for the synthetic quaternary tin sulphides, for comparison's sake, especially with synthetic stannite.

Although the natural stannite has a higher tin content (30.0% Sn), the synthetic stannite with 27.5% Sn gave a much higher yield on leaching by *Thiobacillus ferrooxidans.* From synthetic stannite 54.45% Sn was extracted, while the natural stannite sample only released 31.30% Sn. The leaching rate of synthetic stannite (291.67 ppm/ h) was also faster than that for natural stannite (178.57 ppm/h).

The lower tin extraction in natural stannite can be attributed to the fact that not the maximum surface area of the stannite is exposed to attack by bacteria, because of abundant microscopical intergrowths with other mineral species like sphalerite, galena, chalcopyrite, tennantite, stannoidite, and kesterite, and exsolution textures with sphalerite, kesterite, chalcopyrite, and stannoidite (Teh 1979).

On the other hand, the synthetic stannite sample is homogeneous, meaning that at the particular particle size the maximum stannite surface area is available to the bacteria. Furthermore, the fresh crystal surfaces of synthetic stannite are easily more "reactive" to the bacteria. Natural stannites may, normally, have a film of supergene material which may prevent the bacteria coming into direct contact with the stannite surfaces, hence cutting down on tin extraction.

8 Leaching with Organic Agents

Diagrams showing the effeciency of leaching or solubilizing of tin by various organic agents from the synthetic tin sulphides (stannite, kesterite, herzenbergite, ottemannite, and berndtite) are presented in Fig. 9, and those from the natural samples of cassiterite, varlamoffite, and stannite are plotted in Fig. 10.

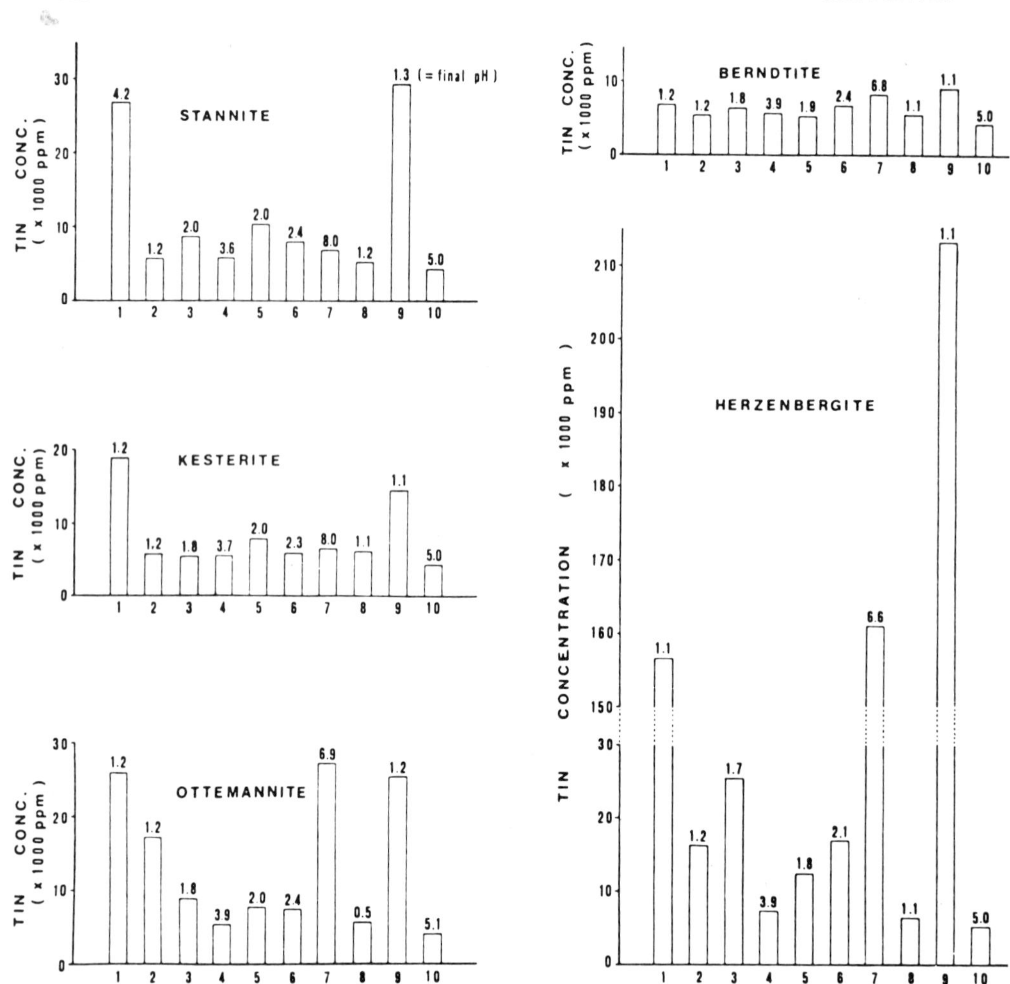

Fig. 9. Leaching of tin from synthetic tin sulphides, using organic agents. Diagrams derived on minus 0.16 mm size fraction, 1.33% initial pulp density solutions after 35 days of leaching at 32°C. Organic agents (0.05 M unless otherwise stated): *1* Oxalic acid (initial pH = 1), *2* Tannic acid (1.0), *3* Tartaric acid (2.0), *4* Acetic acid (3.0), *5* Citric acid (2.0), *6* Lactic acid (2.5), *7* EDTA (8.5), *8* H_2SO_4 (1.0), *9* Oxalic-citric acid 0.1 M (1.0), *10* H_2O (6.0)

8.1 Leaching Synthetic Tin Sulphides with Organic Agents

Generally, the organic agents oxalic acid and the oxalic-citric mixture, at their respective pH values, proved to be most suitable for the microbiological leaching of all the synthetic tin sulphides studied (Fig. 9).

In addition, citric acid, a strong complexing agent, was found to be also effective for the quaternary sulphides, stannite and kesterite.

On the other hand, the binary sulphides, in addition, undergo considerable dissolution, in fact more than citric acid with EDTA, also a strong complexing agent.

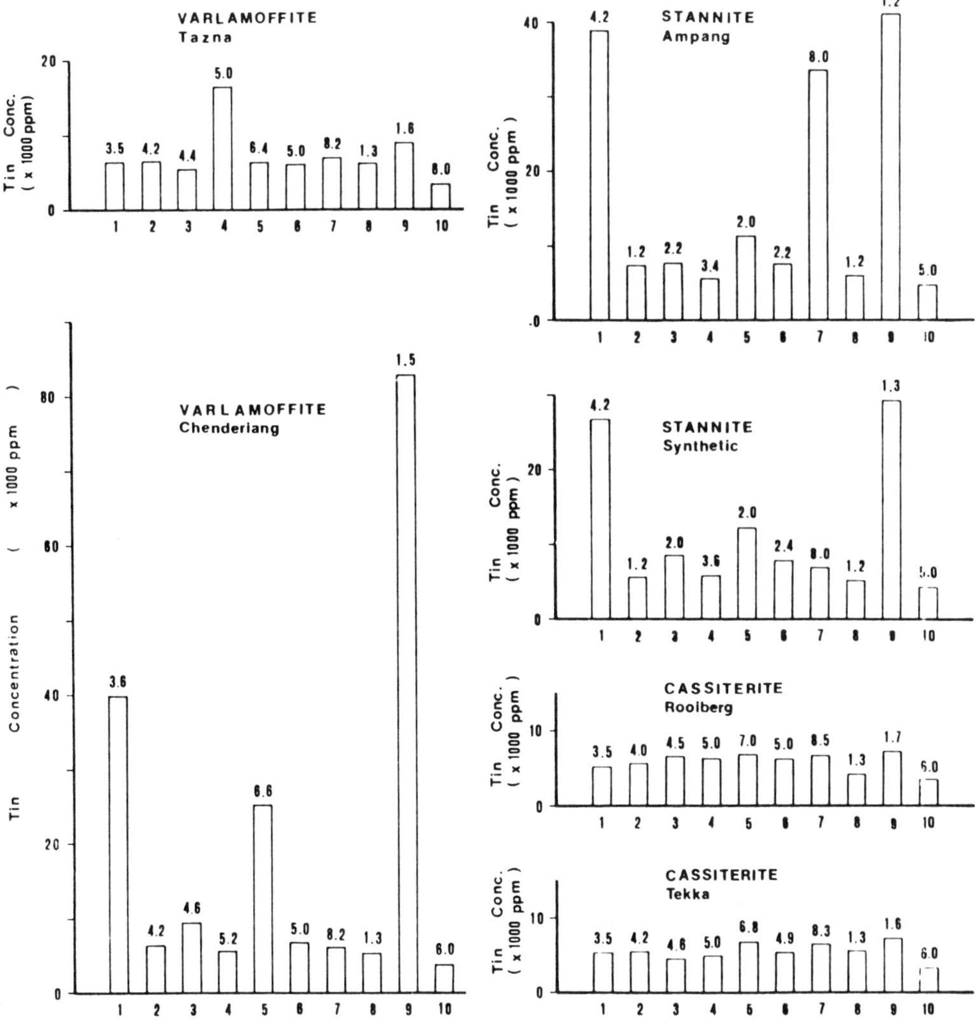

Fig. 10. Leaching of tin from natural tin sulphides using organic agents. Diagrams derived on minus 0.16 mm size fraction, 1.33% initial pulp density solutions after 35 days of leaching at 32°C. Organic agents (0.05 M unless otherwise stated): *1* Oxalic acid (Initial pH = 1), *2* Tannic acid (1.0), *3* Tartaric acid (2.0), *4* Acetic acid (3.0), *5* Citric acid (2.0), *6* Lactic acid (2.5), *7* EDTA (8.5), *8* H_2SO_4 (1.0), *9* Oxalic-citric acid 0.1 M (1.0), *10* H_2O (6.0)

Furtheron, tartaric acid and tannic acid (both strong complexing agents) are also effective for herzenbergite and ottemannite respectively.

The most striking result, however, is the exceptionally high tin yield from herzenbergite using the three most effective organic agents. The best extraction being 27.08% Sn from the oxalic-citric acid mixture after 35 days of leaching, with 1.33% initial pulp density and a final pH of 1.1, while the sulphuric acid and distilled water controls dissolved 0.80% Sn (final pH 1.1) and 0.65% Sn (final pH 5.0) respectively. With oxalic acid the yield was 19.91% Sn with a final pH of 1.1, while EDTA extracted 20.46% Sn with a final pH of 6.6.

Most effective for stannite and kesterite dissolubilization was also the oxalic-citric acid mixture, which yielded 10.61% Sn (final pH 1.3) and 7.44% Sn (final pH 1.1) respectively.

The organic agents, on the whole, appeared to have very little activity on berndtite.

8.2 Leaching Natural Tin Samples with Organic Agents

Diagrams presenting the leaching efficiency of the various organic agents on the natural tin samples of stannite, cassiterite, and varlamoffite are shown in Fig. 10.

The oxalic-citric acid mixture is also most effective on the natural samples except for varlamoffite from Tazna, Bolivia. It could be that the Bolivian varlamoffite has a matrix or a coating of supergene material on the varlamoffite grains which is not so reactive to the oxalic-citric acid mixture but is easier dissolubilized by acetic acid (3.16% Sn).

The best yield by the oxalic-citric mixture is on the varlamoffite from Chenderiang where 17.68% Sn were extracted after 35 days of leaching at 32°C with 1.33% pulp density and a final pH of 1.7. The sulphuric acid and distilled water controls yielded 1.13% and 0.79% Sn respectively.

With cassiterite, the oxalic-citric acid mixture is most effective for the two localities, releasing a more-or-less identical yield, 1.54% Sn (final pH 1.7) for the Tekka sample and 1.51% Sn (final pH 1.7) for the Rooiberg sample.

In addition, oxalic acid extracted 12.97% Sn from stannite from Hock Leong Mine and 8.50% Sn from varlamoffite from Chenderiang.

The plot of synthetic stannite superimposed on Fig. 10, shows that, in contrast to bacterial leaching, the natural stannite sample gave higher yields with organic agents like oxalic acid (12.96% vs 9.70%), citric acid (3.77% vs. 3.70%), EDTA (11.23% vs 2.51%) and the oxalic-citric acid mixture (13.68% vs 10.61%). This suggests that the biochemical processes involved in solubilizing tin are, in fact, more complex than homogeneity and availability of "reactive" surfaces.

9 Examination of Residue After Leaching

The flasks containing the various residues after leaching by *Thiobacillus ferrooxidans* and organic agents were allowed to dry up at room temperature. The contents were examined under the binoculars and, where necessary, polished sections and X-ray diffraction patterns were made.

The flasks involving leaching by *Thiobacillus ferrooxidans,* which initially contained stannite (both natural and synthetic), kesterite, stannoidite, herzenbergite, ottemannite, and berndtite, now show the presence of yellow, earthy, varlamoffite. The experiments on both synthetic and natural stannite with the bacteria alone without ferrous sulphate also yielded varlamoffite. This strongly substantiates the fact that the bacteria alone have been responsible for the varlamoffite formation and/or the bacteria have been responsible for the formation of ferric sulphate, which then oxidized the tin sulphides.

Further, the experiments with ferrous sulphate in the medium, but no bacteria, yielded no varlamoffite in all the tin sulphides. Again emphasizing the importance of the bacteria in catalyzing varlamoffite formation.

On the other hand, no varlamoffite was found in the tin sulphide residues leached by organic agents. This suggests that the tin is complexed in other more complicated forms by the organic agents.

10 Conclusions and Applications

This study confirms that it is possible to leach tin from both synthetic and natural tin minerals with the help of *Thiobacillus ferrooxidans* and organic agents, especially oxalic acid, EDTA, and the oxalic-citric acid mixture.

In other words, microorganisms which produce the organic agents, and *Thiobacillus ferrooxidans* are possible, effective participants involved in biochemical weathering processes leading to the oxidation and dissolution of stannite and related tin minerals in nature.

The results of this investigation show that a careful preliminary study of the mineralogical and petrographical aspects of tin deposits will help greatly in deciding on the correct method, conditions, and parameters for optimal leaching processes.

For cassiterite and varlamoffite, the search for an acid-producing, heterotrophic microorganism will appear to be the key to leaching processes on a large scale.

Thiobacillus ferrooxidans appears to be more applicable to sulphide-rich deposits. Leaching with this bacterium has also the advantage that the bacterium is able to reach very fine tin mineral grains which may be intimately intergrown with other mineral species. Further, the presence of copper, zinc, and iron, which are very common in most cassiterite- or stannite-bearing deposits, will help, indirectly, by making the tin grains more accessible to attack by the bacteria.

The varlamoffite found in the dried up, leached residues of the tin sulphides, namely stannite, kesterite, stannoidite, herzenbergite, ottemannite, and berndtite, provide the evidence that bacterial action (in this case demonstrated by *Thiobacillus ferrooxidans*) can be responsible for the genesis of supergene varlamoffite.

In this study it has been found that the optimum pulp density required by *Thiobacillus ferrooxidans* for effective leaching of tin is 0.1% (or less). Had this optimum pulp density been applied on the main leaching experiments, higher tin extractions would have resulted. This will make it feasible to apply such leaching processes in the many idle tin-tailing dumps in Malaysia (or other parts of the world), which may still contain "appreciable" quantities of tin for the bacteria, or in sulphide-rich deposits where tin sulphides with low Sn percentages (e.g. stannoidite and/or kesterite) occur.

References

Beck JV (1969) *Thiobacillus ferrooxidans* and its relation to the solubilization of ores of copper and iron. In: Perlman D (ed) Fermentation advances. Academic Press, London New York, 747 p

Beck JV, Brown DG (1968) Direct sulfide oxidation in the solubilization of sulfide ores by *Thiobacillus ferrooxidans.* J Bacteriol 96:1433–1434

Brockamp O (1974) Verwitterungsversuche mit organischen Säuren an einigen schwerlöslichen Fe-Pb-Zn und Cu Erzmineralien, angewendet auf die Genese des Kupferschiefers. Contrib Mineral Petrol 43:213–221

Bryner LC, Anderson R (1957) Microorganisms in leaching sulfide minerals. In Eng Chem 49(10): 1721–1724

Bryner LC, James AK (1958) Microorganisms in leaching sulfide minerals. Appl Microbiol 6:280–287

Colmer AR, Hinkle ME (1947) The role of microorganisms in acid mine drainage. A preliminary report. Science 106:253–256

Corrans IJ, Harris B, Ralph BJ (1972) Bacterial leaching: An introduction to its application and theory and a study on its mechanism of operation. J S Afr Inst Min Metall 72:221–230

Duncan DW (1967) Microbiological leaching of sulfide minerals. Aust Min 59:21–23

Duncan DW, Trussell PC (1964) Advances in the microbiological leaching of sulphide ores. Can Metall Q 3:43–55

Duncan DW, Walden CC (1972) Microbiological leaching in the presence of ferric iron. Dev Ind Microbiol 13:66–75

Duncan DW, Walden CC, Trussell PC (1966) Biological leaching of mill products. Can Min Metall Bull 59:1075–1079

Duncan DW, Landesman S, Walden CC (1967a) Role of *Thiobacillus ferrooxidans* in the oxidation of sulfide minerals. Can J Microbiol 13:397–403

Duncan DW, Walden CC Trussell PC, Lowe EA (1967b) Recent advances in the microbiological leaching of sulfide. Trans Soc Min Eng AIME 238(2):122–128

Ehrlich HL (1964) Bacterial oxidation of arsenopyrite and enargite. Econ Geol 59:1306–1312

Ehrlich HL, Fox SI (1967) Environmental effects on bacterial copper extraction from low-grade copper sulphide ores. Biotechnol Bioeng 9:471–485

Ginzburg II, Yashina RS, Matreyeva LA, Beletskiy VV, Nuzhdelouskaya TS (1963) Decomposition of certain minerals by organic substances. Chemistry of the earth's crust. I Trans Geochem Conf 100th Anniv Acadmician VI Vernadskiy, Izd An SSSR

Habashi F (1970) Die Auflösung von Sulfidmineralien. Ihre theoretische Grundlage und technische Anwendungen. Metall (Berlin) 24:1074

Hosking KFG (1973) Primary mineral deposits. In: Gobbett DJ, Hutchison CS (eds) Geology of the Malaya Peninsula (West Malaysia and Singapore). Wiley Interscience, New York, pp 335–390

Huang WH, Keller WD (1971) Dissolution of clay minerals in dilute organic acids at room temperature. Am Mineral 56:1082–1095

Huang WH, Keller WD (1972) Organic acids as agents of chemical weathering of silicate minerals. Nature (Lond) 239:149–151

Imai K, Sugio T, Tsuchida T, Tano T (1975) Effect of heavy metal ions on the growth and iron oxidizing activity of *Thiobacillus ferrooxidans.* Agric Biol Chem 39(7):1349–1354

Iskandar IK, Syers JK (1972) Metal-complex formation by lichen compounds. J Soil Sci 23:225–265

Kiel H (1977) Laugung von Kupferkarbonat- und Kupfersilikat-Erzen mit heterotrophen Mikroorganismen. In: Schwartz W (ed) Conference bacterial leaching. GBF Monogr Ser 4. Verlag Chemie, Weinheim New York, 270 pp

Kullerud G (1971) Experimental techniques in dry sulfide research. In: Ulmer GC (ed) Research techniques for high pressure and high temperature. Springer, Berlin Heidelberg New York, pp 289–315

Malouf EE, Prater JD (1961) Role of bacteria in the alteration of sulfide minerals. J Met 13:353–356

Marchlewitz B, Schwartz W (1961) Untersuchungen über die Mikrobenassoziationen saurer Grubenwasser. Z Allg Mikrobiol 1:100–114

McGoran CJM, Duncan DW, Walden CC (1969) Growth of *Thiobacillus ferrooxidans* on various substrates. Can J Microbiol 15:135–138

Muntz A (1890) Sur la decomposition des roches et al formation de la terre arable. CR Acad Sci Paris 110:1370–1372

Pings WB (1968) Bacterial leaching of minerals. Colo Sch Mines Miner Ind Bull 2(3):1–19

Razzell WE, Trussell PC (1963) Microbiological leaching of metallic sulfides. App Microbiol 11: 105

Rudolfs W (1922) Oxidation of pyrites by microorganisms and their use for making mineral phosphates available. Soil Sci 14:135

Rudolfs W, Helbronner A (1922) Oxidation of zinc sulphide by microorganisms. Soil Sci 14:459

Russel A, Vincent AE (1952) On the occurrence of varlamoffite (partly hydrated stannic oxide) in Cornwall. Mineral Mag XXIX:817–826

Scott TR, Dyson NF (1968) The catalyzed oxidation of zinc sulfide under acid pressure leaching conditions. Trans Metall Soc AIME 242:1915

Silver M (1970) Oxidation of elemental sulfur and sulfur compounds and CO_2 fixation by *Ferrobacillus ferrooxidans (Thiobacillus ferrooxidans)*. Can J Microbiol 16:845

Silverman MP (1967) Mechanism of bacterial pyrite oxidation. J Bacteriol 94:1046–1051

Silverman MP, Lundgren DG (1959) Studies on the chemoautotrophic iron bacterium *Ferrobacillus ferrooxidans*. I An improved medium and a harvesting procedure for securing high cell yields. J Bacteriol 77:642–647

Singh DS, Bean JH (1968) Some aspects of tin minerals in Malaysia. Pap Tech Conf Tin London 1967(2):457–478

Stanton RE, McDonald AJ (1961) Field determination of tin in geochemical soil and stream sediment surveys. Trans Inst Min Metall London 71:27–29

Teh GH (1979) Experimental, microbiological and mineralogical aspects of stannite and related tin-containing sulphides and their applications to selected stannite-bearing tin deposits of Peninsular Malaysia. Unpubl Dr rer nat Thesis, Univ Heidelberg, 144 pp

Temple KL, Colmer AR (1951) The autotrophic oxidation of iron by a new bacterium *Thiobacillus ferrooxidans*. J Bacteriol 62:605

Torma AE (1971) Microbiological oxidation of synthetic cobalt, nickel and zinc sulfides by *Thiobacillus ferrooxidans*. Rev Can Biol 30:209–216

Torma AE, Walden CC, Duncan DW, Branion RMR (1972) The effect of carbon dioxide and particle surface area on the microbiological leaching of a zinc sulfide concentrate. Biotechnol Bioeng 15:777–786

Tuovinen OH, Kelly DP (1972) Biology of *Thiobacillus ferrooxidans* in relation to the microbiological leaching of sulphide ores. Z Allg Mikrobiol 12:311–416

Tuovinen OH, Kelly DP (1974) Use of micro-organisms for the recovery of metals. Int Metall Rev 19:21–31

Wagner M, Schwartz W (1967) Geomikrobiologische Untersuchungen. VIII. Über das Verhalten von Bakterien auf der Oberfläche von Gesteinen und Mineralien und ihre Rolle bei der Verwitterung. Z Allg Mikrobiol 7:33–52

Woodcock JT (1967) Copper waste dump leaching. Proc Australas Inst Min Metall 224:47–66

A Contribution to the Stannite Problem

N. WANG[1]

Abstract

In an attempt to concretely define the various stannite-related minerals, part of the Cu-Sn-Fe-S system was experimentally investigated in the temperature range 425–600°C. Within the composition tetrahedron, two groups of mostly lens-like solid solution fields have been identified. The first group, which includes the phases $Cu_2Sn_{3.5}S_8$ and rhodostannite, is stable above the CuS-SnS-FeS reference plane with sulphur/metal ratio varying from 1.45 to 1.35 at 500°C. The second group found immediately below this reference plane consists of the stannite-related phases "isostannite", "intermediate stannite", Cu_2SnS_3, the "high stannoidite series" and the "lower series". The extensive stability ranges of some of the fields, particularly that of the high stannoidite series, partly explains the variable compositions and the observed exsolution textures of "stannites" found in nature.

All the fields are defined on the basis of characteristic superstructure diffraction patterns and optical homogeneity of synthetic materials.

1 Introduction

The classical "stannite problem" (Ramdohr 1944) which involves stannites and stannite-related species like stannoidite, mawsonite, kesterite, etc., is becoming a topic of increasing interest. A satisfactory solution of this problem requires an intensive exploration of the complicated 5-component system Cu-Sn-Fe-Zn-S over a wide temperature range. Recent publications involving both natural and synthetic materials are numerous and the data available cover an extensive composition range (Springer 1968, 1972; Oen 1970; Franz 1971; Bernhardt 1972; Harris and Owens 1972; Petruk 1973; Lee et al. 1975; Moh 1975; Hall et al. 1978; Kissin and Owens 1979; Teh 1979).

However, the solubility ranges of most of the synthetic phases, essential in defining the various "stannites", are still insufficiently known, not even in the Zn-free, 4-component subsystem. The frequently observed exsolution textures in stannites and related phases (Oen 1970; Moh 1975) suggest the preexistence of extensive solid solutions at elevated temperatures. With the exception of sulfphur-rich rhodostannite (Springer 1968; Wang 1975), these phases are all derivatives of the sphalerite structure. Their characteristic superstructure diffraction patterns permit an unequivocal identification of phases and phase boundaries at any specified temperature.

1 Mineralogisch-Petrographisches Institut, Universität Heidelberg, Im Neuenheimer Feld 236 6900 Heidelberg, FRG

The present work is part of an experimental approach to determine, within the Cu-Sn-Fe-S subsystem, the stability ranges of the various superstructure patterns under the experimentally permissible temperature conditions.

2 Methods of Investigation

Apart from a few runs with temperatures exceeding 720°C, most of the syntheses were performed at constant temperatures of 425°, 500°, and 600°C, using high-purity elements and presynthesized binary compounds as starting material, and standard silica tube technique as described elsewhere.

The various superstructure patterns of the mostly lens-like fields are extremely sensitive to the sulphur/metal ratio in the samples. By successsive addition of sulphur, this ratio was regulated from an initial value of 9/10 up to the final value of 10/10 where most of the s.s. fields of interest terminate. For the two fields above the CuS-SnS-FeS plane, this ratio attains a final value of up to 14.5/10.

A single run proceeds as follows: after the initial heating and regrinding, the reheated charge (2–3 days at 500°C or 1–2 weeks at 425°C) was quenched and examined by powder diffraction technique and by reflecting microscope. If found sulphur deficient (as indicated by the presence of the coexisting pair SnS-bornite, for instance), a calculated amount of sulphur (of the order of 1–2 mg per 1/2 g sample) was then added to the finely reground and reweighed charges for further runs. In places where the s.s. fields overlap or terminate, this process was repeated up to six times for a single metallic composition. The material loss per regrinding for polished section and for X-ray work was in most cases less than 5 mg (about 1 wt.%). The final sulphur/metal ratio of the homogeneous s.s. fields was rechecked at the key points by a number of new syntheses prepared directly from the elements.

The successive addition of sulphur not only appreciably reduces the number of samples in exploring the three-dimensional composition space, but also accelerates the reaction rate and prevents the metastable formation of certain "stubborn phases" like FeS_2.

Single crystals for the precession technique were prepared in 4–6 weeks in a salt flux of (2LiCl+3KCl) at temperatures ranging 500–580°C.

3 Experimental Results

Over the temperature ranges of 425–600°C, the salient features in the Cu-Sn-Fe-S composition tetrahedron are several subparallel, lens- to leaf-like solid solution fields near the CuS-SnS-FeS reference plane. Since the flat fields only slightly tilt to the plane, their maximum spatial extent can be represented without distortion by a projection from the sulphur apex onto this plane (Figs. 1 and 4). Details of the hitherto undefined fields are described in the following sections.

3.1 The High Stannoidite Series

Between 500 and 600°C, this lens-like field extends almost from the Cu-Sn-S ternary halfway across the tetrahedron to Fe-rich compositions close to stannoidite. The flat surface of this field dips slightly towards the chalcocite-bornite s.s. and strikes sub-parallel to the Cu_2SnFeS_4-Cu_2SnS_3 line. The sulphur/metal ratio varies along the long axis uniformly from 9.9/10 to 9.7/10 and approaches 10/10 at the Sn-rich flank of the field. Figure 1 shows, among others, a projection of this field at 500°C.

The characteristic superstructure pattern, which appears identical to kesterite (Cu_2ZnSnS_4) and to the Fe-free ternary phase $Cu_7Sn_3S_{10}$ (Wang 1977), remains essentially unchanged over the complete s.s. range (Fig. 2). The well-resolved reflections could be indexed, by reference to kesterite, on a tetragonal unit cell of a_o = 5.365Å, c_o = 10.73Å for the Sn-poor composition and a_o = 5.40Å, c_o = 10.80Å for the Sn-rich composition (Table 1). However, with identical powder pattern, single crystals of $Cu_7Sn_3S_{10}$ grown at 580°C indicate tetragonal symmetry with a unit cell a_o = 10.82Å and c_o = 10.82Å. This cell contains an undistorted, sphalerite-type sub-cell of a_s = 5.41Å. The minute departure from cubic symmetry involves only two very weak superstructure reflections indiscernible with the powder technique. Powder data of a member of the series, reindexed on this second possible unit cell, are given in Table 1 for comparison.

Table 1. Powder data of an intermediate member of the high stannoidite series quenched at 500°C. Diffractometer pattern with CuKα radiation

d_{obs} (Å)	I/I_1	hkl (I)[b]	hkl (II)[c]
5.37	5	002	020,002
4.80	12	011	021
3.802	3	012	022
3.103[a]	100	112	222
2.984	5	013	023
2.688[a]	25	020,004	040,004
2.406	4	014	024
2.349	6	211	124
2.000	5	015	025
1.902[a]	75	024,220	044
1.794	3	006	060,006
1.623[a]	45	132,116	262
1.555[a]	5	224	444

a Substructure reflections
b Referred to the unit cell of kesterite
c Referred to the unit cell of $Cu_7Sn_3S_{10}$ ($Cu_{22}Sn_{10}S_{32}$)

$Cu_7Sn_3S_{10}$ (cell formula $Cu_{22}Sn_{10}S_{32}$) is not stable as a phase at 500°C. Even at 600°C, the "high stannoidite" field does not seem to include this composition. On the other hand, a solid solution is to be expected at elevated temperatures between the "high stannoidite" and kesterite. The "unknown phase" of Petruk (1973), identified

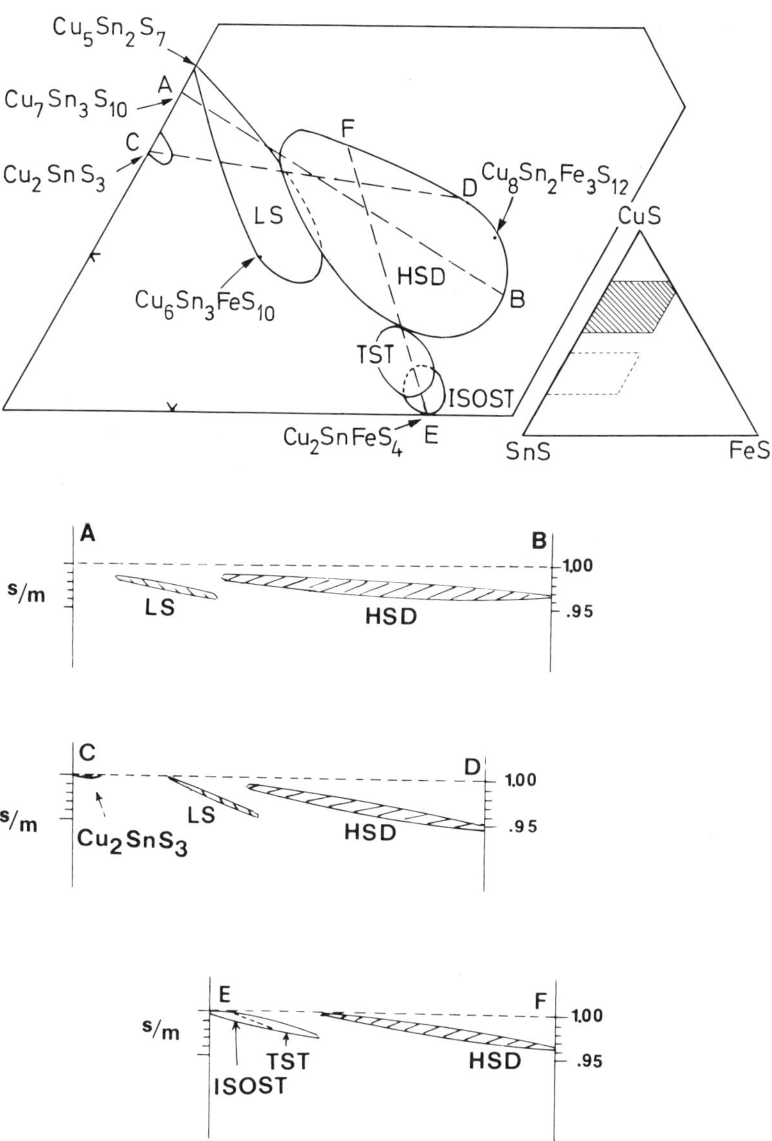

Fig. 1. Stannite-type superstructure fields at 500°C as projected on the CuS-SnS-FeS reference plane. The three sections *A–B*, *C–D*, *E–F* show the variation of s/m ratio over the s.s. range. *HSD* the "high stannoidite series"; *LS* the "lower series"; *TST* "intermediate stannite"; *ISOST* "isostannite"

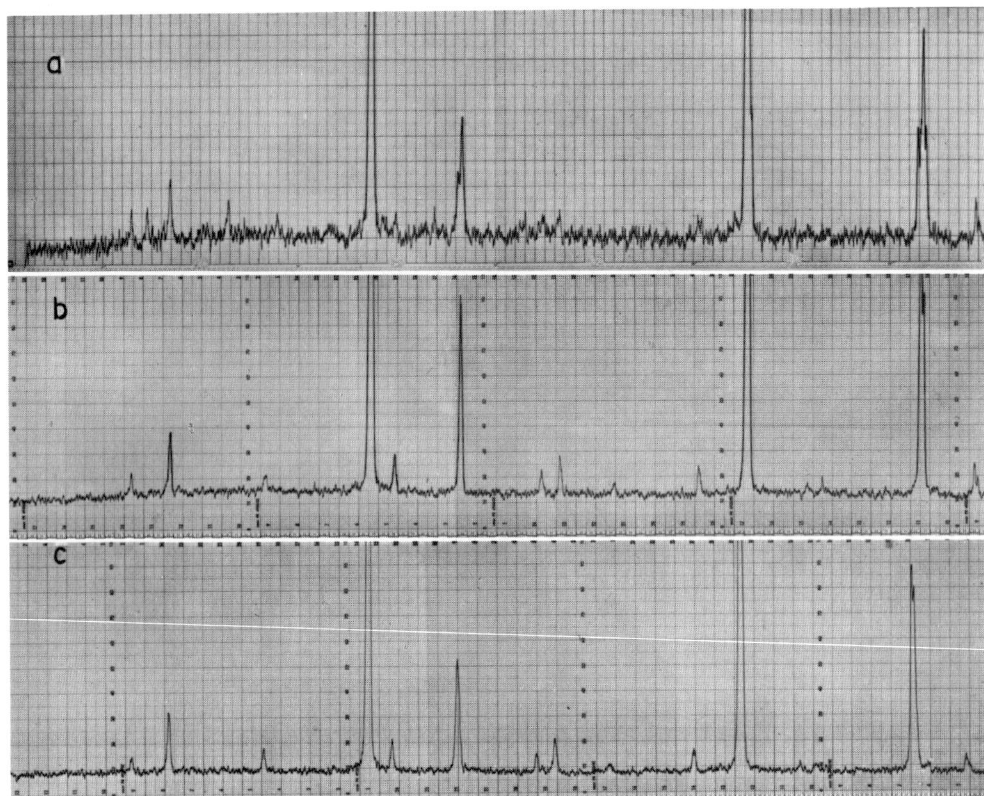

Fig. 2a–c. Diffractometer patterns of high stannoidite and related phases (CuKα radiation). **a** Stannoidite ($Cu_8Sn_2Fe_3S_{12}$), 412°C. **b** "High stannoidite" (an intermediate member), 424°C; **c** Kesterite (Cu_2SnZnS_4), 500°C

by Kissin and Owens (1979) as an intergrowth of two similar phases, possibly represents the exsolved equivalent of a member of this preexistant solid solution.

The Zn-free stannoidite ($Cu_8Sn_2Fe_3S_{12}$) transforms at 420°C to the "high stannoidite" (compare Lee et al. 1975). However, the extensive s.s. range even at 425°C and the variable s/m ratio ($\geqslant 9.4/10$) of the high temperature phase both indicate the nonpolymorphic nature of this transition. Stannoidite most probably relates to the "high stannoidite" as chalcopyrite does to the "intermediate solid solution".

3.2 The Lower Series

Near the high stannoidite field, a second elongated field, called the lower series, was identified between the Fe-free member $Cu_5Sn_2S_7$ and the Fe-rich member $Cu_{5.9}Sn_{2.7}Fe_{1.4}S_{9.8}$ from its characteristic superstructure pattern (Wang 1977). Over the complete s.s. range, this superstructure pattern remains unchanged where the intense substructure reflections tend to split with changing compositions. The supercell, presumably of very low symmetry, has not been determined. Attempts to grow single crystals in salt flux from homogeneous material were unsuccessful.

The two closely related fields are partially overlapped in projection. Similar to the high stannoidite series, the stability field of the lower series also dips towards the Cu_2S-Cu_9S_5 s.s with sulphur/metal ratio changing from 10/10 to 9.5/10. Part of its Sn-rich flank touches the CuS-SnS-FeS plane where it intercepts the Cu_2SnFeS_4-Cu_2SnS_3 join at the composition $Cu_6Sn_3FeS_{10}$. The trace of rhodostannite or pyrite found in samples with sulphur/metal ratio exceeding 10/10 indicates that the lower series field does not penetrate the reference plane at all temperatures (Fig. 1).

Cu_2SnS_3. In contrast to the other to fields, the stability range of the Cu_2SnS_3 field in the 4-component system is much restricted at 500°C. The particular Cu/Sn ratio appears to be essential for stabilizing the low symmetrical, triclinic superstructure (Wang 1974).

3.3 Stannites and Isostannite

Synthesis with the stoichiometric stannite composition Cu_2SnFeS_4 over the temperature range of 420–820°C yielded three completely different superstructure patterns. Apart from the high-temperature tetragonal pattern (Fig. 3a) obtained by quenching above 706°C (Moh 1975; Hall et al. 1978; Kissin and Owens 1979), two other patterns have been identified as temperature-dependent at this composition.

The cubic pattern stable at 500°C (Fig. 3b) corresponds to isostannite (Claringbull and Hey 1955; Franz 1971). The comparatively intense superstructure reflections indicate a highly ordered structure and a unit cell edge (a_o = 10.85Å) twice as large as the sphalerite-type subcell. Slight metal-rich charge as used by Franz at 420°C contains at 500°C besides isostannite a trace amount of coexisting SnS and a reddish brown phase.

In the temperature range of 580–600°C, isostannite transforms reversibly (sluggish and indetectable in the DTA pattern) to an intermediate modification of unknown symmetry. The superstructure pattern of this modification is characterized by a "twin peak" with d = 5.42Å, 5.32Å and by a slight splitting of the substructure reflections (Fig. 3c). At 500°C this pattern is stable only with Cu-enriched compositions and a sulphur/metal ratio $\leqslant 1$. The substructure peak splitting, as a measure of deviation of the subcell from cubic symmetry, increases with increasing Cu-content (Fig. 3d).

The stability ranges of the two phases extend from $Cu_5Sn_{2.5}Fe_{2.5}S_{10}$ (=Cu_2SnFeS_4) to $Cu_{5.3}Sn_{2.4}Fe_{2.3}S_{9.8}$ for "isostannite" and from $Cu_{5.2}Sn_{2.5}Fe_{2.3}S_{10}$ to $Cu_{5.6}Sn_{2.4}$-$Fe_2S_{9.7}$ for the "intermediate modification" at 500°C. The two lens-like, partially superposed fields dip, like the high stannoidite field, towards the bornite s.s., with the Sn-rich flanks touching the CuS-SnS-FeS reference plane (Fig. 1). In places where the two stannite fields overlap, a gradual change between the patterns was occasionally observed which makes the interpretation uncertain (Fig. 1, section E–F, dashed line).

A fourth pattern in the stannite group was obtained from the slight metal-rich composition $Cu_2SnFeS_{3.95}$ at 600°C (Fig. 3e). This pattern with the pronounced substructure splitting most probably corresponds to the "tetragonal stannite" described by Franz (1971). The small-angle superstructure reflections of this pattern, however, were incorrectly eliminated by Franz as part of isostannite reflections, leaving a pattern indexable on a tetragonal cell.

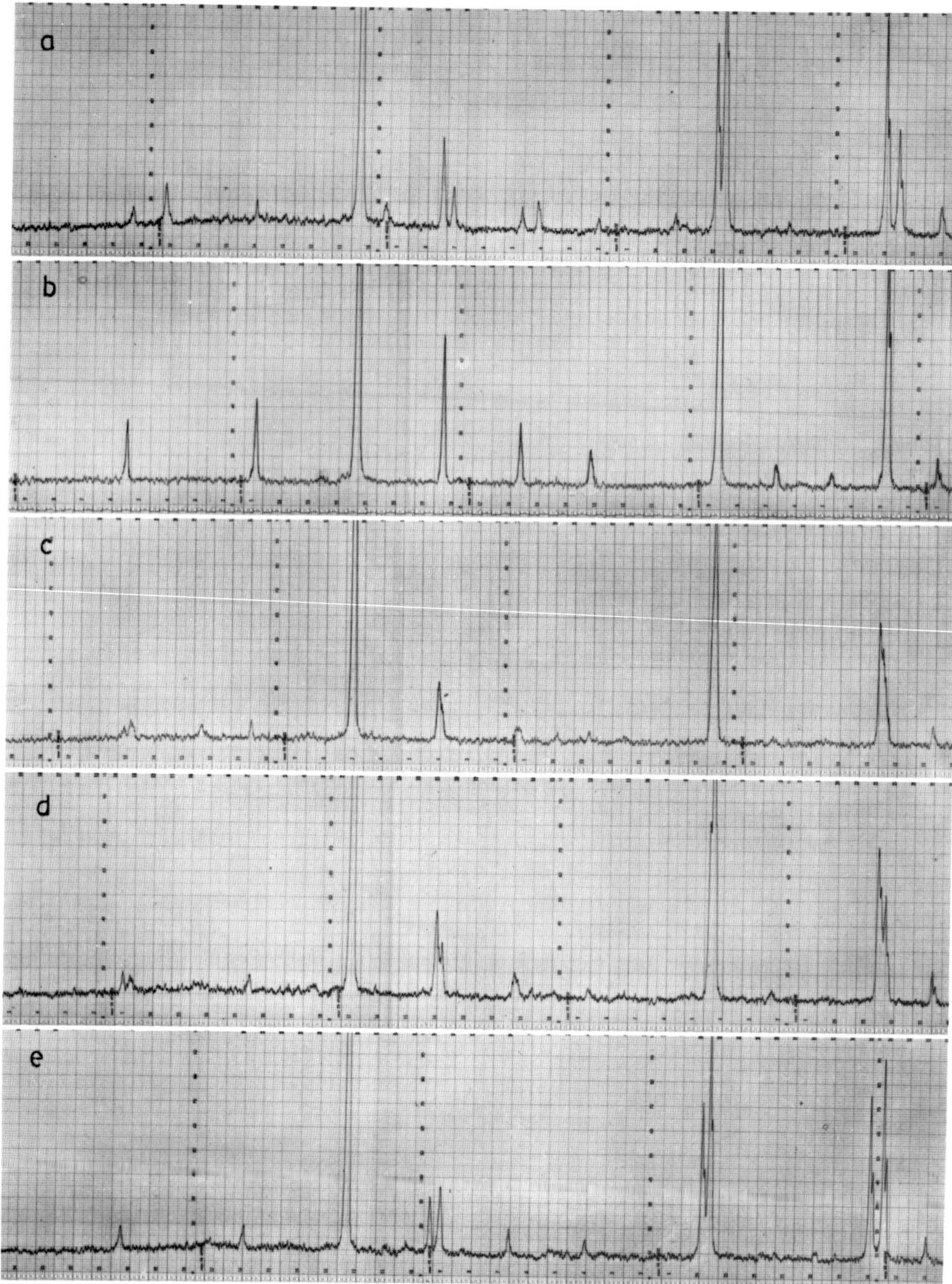

Fig. 3a–e. Diffractometer patterns of different stannite superstructures (CuKα radiation). **a** High stannite (Cu_2SnFeS_4), 720° C; **b** Isostannite (Cu_2SnFeS_4), 500° C; **c** "Intermediate stannite" (Cu_2SnFeS_4), 600° C; **d** "Intermediate stannite" (Cu-rich composition limit), 500° C; **e** "Tetragonal stannite" (Franz 1971) ($Cu_2SnFeS_{3.95}$), 600° C

3.4 Rhodostannite and $Cu_2Sn_{3.5}S_8$

Above the CuS-SnS-FeS reference plane, two other distinct fields with nearly identical powder patterns exist in the sulphur/metal range 1.35–1.45 (Fig. 4). The rhodostannite field, comparable both in size and shape to the lower series, has the limit compositions $Fe_{0.45}Cu_{1.91}Sn_{3.26}S_8$ and $Fe_{1.30}Cu_{1.72}Sn_{2.90}S_8$ at 500°C. The flat, lens-like field dips in the same way as the group of superstructure fields found below the reference plane, with sulphur/metal ratio decreasing towards Sn-poor compositions. In contrast, the $Cu_2Sn_{3.5}S_8$ field is much smaller. A two-phase field exists in between.

The successive addition of FeS in $Cu_2Sn_{3.5}S_8$ (Wang 1975) results in a decrease of s/m ratio in the bulk composition. Once this ratio reaches the lower boundary of the rhodostannite field, SnS starts to separate from the s.s. and coexists with rhodostannite in an assemblage "sulphur deficient" with respect to the s.s. field.

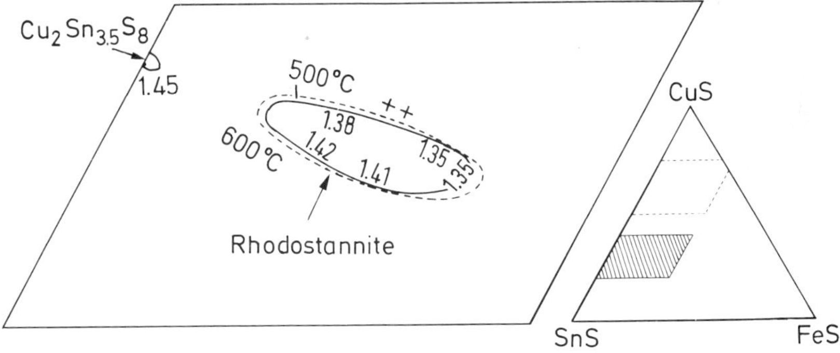

Fig. 4. Rhodostannite field at 500° and 600°C with varying s/m ratio as projected on the reference plane. The two *small crosses* correspond to analyses of natural rhodostannite

The two analyses of natural rhodostannite (Springer 1968) give compositions just outside the rhodostannite field at 500°C (Fig. 4, small crosses). However, with stannite as a coexisting phase, the Cu-rich boundary of this field is expected to pass through these two points at the temperature of formation of the natural rhodostannites.

4 Significance of the Experimental Results

Since Ramdohr (1944) published his data on the four "Zinnkies", the identity of the various "stannites" has always been controversial. There are two reasons for that. First, natural stannite-like phases cover a wide spectrum of compositions which includes zinc as an essential component. Variations of composition are usually accompanied by variations of optical behaviour, so that different names can be assigned to the same species by different authors. Second, the structure-dependent similarity of the diffraction patterns sometimes make it very difficult, if not impossible, to distinguish one species from the other.

The present results, though still incomplete, may provide a useful basis to tackle these problems. The two groups of s.s. fields identified above and immediately below the CuS-SnS-FeS reference plane not only set a "limit" to the compositions of "stannites" expected to occur in nature, they also concretely define the various "stannites" in terms of the characteristic superstructure patterns.

A complete solution of the "stannite problem", however, may not be reached until the low-temperature phase relations of the 5-component system become available, hopefully in the coming decade.

References

Bernhardt H-J (1972) Untersuchungen im pseudobinären System Stannin-Kupferkies. Neues Jahrb Mineral Monatsh 12:553–556

Claringbull GF, Hey MH (1955) Stannite and isostannite. Mineral Soc London Not 91(2) (Mineral Abstr 13:31, 1956)

Franz ED (1971) Kubischer Zinnkies und tetragonaler Zinnkies mit Kupferkies-Struktur. Neues Jahrb Mineral Monatsh 5:218–223

Hall SR, Szymanski JT, Stewart JM (1978) Kesterite $Cu_2(Zn,Fe)SnS_4$, and stannite $Cu_2(FeZn)SnS_4$, structurally similar but distinct minerals. Can Mineral 16:131–137

Harris DC, Owens DR (1972) A stannite-kesterite exsolution from British Columbia. Can Mineral 11:531–534

Kissin SA, Owens DR (1979) New data on stannite and related tin sulfide minerals. Can Mineral 17:125–135

Lee MS, Takenouchi S, Imai H (1975) Synthesis of stannoidite and mawsonite and their genesis in ore deposits. Econ Geol 70:834–843

Moh GH (1975) Tin-containing mineral systems. Part II. Phase relations and mineral assemblages in the Cu-Fe-Zn-Sn-S system. Chem Erde 34:1–61

Oen S (1970) Paragenetic relations of some Cu-Fe-Sn sulfides in the Mangualde pegmatite, North Portugal. Miner Deposita 5:59–84

Petruk W (1973) Tin sulfides from the deposit of Brunswick Tin Mines Limited. Can Mineral 12: 46–54

Ramdohr P (1944) Zum Zinnkiesproblem. Abh Preuss Akad Wiss 4:1–30

Springer G (1968) Electron probe analysis of stannite and related tin minerals. Mineral Mag 36: 1045–1051

Springer G (1972) The pseudobinary system Cu_2FeSnS_4-Cu_2ZnSnS_4 and its mineralogical significance. Can Mineral 11:535–541

Teh GH (1979) Experimental, microbiological and mineralogical aspects of stannite and related tin-containing sulfides and their application to selected stannite-bearing tin deposits of peninsular Malaysia. Unpubl Thesis, Univ Heidelberg, 144 pp

Wang N (1974) The three ternary phases in the system Cu-Sn-S. Neues Jahrb Mineral Monatsh 9: 424–431

Wang N (1975) Investigation of synthetic rhodostannite $Cu_2FeSn_3S_8$. Neues Jahrb Mineral Monatsh 4:166–171

Wang N (1977) Structural variations of non-stoichiometric Cu_2SnS_3. Neues Jahrb Mineral Abh 131:26–27

Mineralogical Studies

Crystallogenetic Disequilibrium in Ore Veins as Indicated by Hydrothermal Fluid Inclusions

A. BASSETT[1], G. DEICHA[2], and J. PROUVOST[3]

„Eine der ältesten Methoden der geologischen Temperaturmessungen überhaupt, bereits zurückgehend auf Sorby, die Untersuchung der Mutterlaugeneinschlüsse im Verhältnis zu ihren Libellen, hat heute auch für Erze Bedeutung gefunden."

Paul Ramdohr (1950)

Abstract

Since the publication of *Die Erzmineralien und ihre Verwachsungen* by Paul Ramdohr, mineralogical thermometry and geological barometry, founded on phase relationships inside of fluid inclusions, have found a number of applications, first of all as regards hydrothermal deposits. *Ore Genesis 1980* has to stress the new possibilities opened in the field of inclusion studies by recent progress in scanning electron microscopy, which enables a detailed insight of the inter- and intracrystalline cavities, for instance in gangue quartz. Such investigations lead us to pay more attention to the effects of the supersaturation degree of the hydrothermal solution, during the formation of the ore veins.

1 Introduction

Ever since Professor Paul Ramdohr wrote the above quoted lines, his encouragement has been of great value both to individual younger colleagues engaged in fluid inclusion studies, as well as to organizers and participants of colloquia devoted to this fast-expanding field of investigation. After the meeting on *Erz und Nebengestein* (Heidelberg, Febr. 20/21, 1970), we must recall his attendance and kind personal intervention at the symposium of the Commission on Ore-Forming Fluids in Inclusions (C.O.F.F.I.) during the Tokyo-Kyoto session (1970) of I.A.G.O.D. and at the session devoted to *Fluid Inclusions in Crystals,* during the Berlin meeting (1974) of I.M.A., which initiated the idea of a working group *Inclusions in Minerals,* created by the last General Assembly of I.M.A. at Novosibirsk (1978). This new group went into action at the I.M.A. meeting in Orleans held during the 1st week of July 1980.

1 Département de Minéralogie et de Cristallographie, Université Pierre et Marie Curie, Tour 16 (2 E), 4 place Jussieu, 75230 Paris Cedex 05, France

2 Département de Géologie Appliquée, Université Pierre et Marie Curie, Tour 16 (5 E), 4 place Jussieu, 75230 Paris Cedex 05, France

3 Laboratoire de Minéralogie, Université des Sciences et Techniques de Lille, 59650 Villeneuve d'Ascq, France

2 Growth Topography of Crystals and Fluid Inclusions

As thermobarogeochemistry (Ermakov 1979), with its heating, freezing and crushing stages as well as analytic equipment (Roedder 1972) now constitutes standard work in regional studies on ore genesis (Imai 1978), one of the new trends of fluid inclusion investigations is to pay more attention to the cavities that harbour bubbles and droplets of geochemical fluids. This approach is intended to obtain a more elaborate view of the complex systems which inclusions and related microscopic discontinuities constitute inside minerals and mineral aggregates. In our present short contribution we will consider only syngenetic structures. Syngenetic cavities, although sometimes perfectly isolated, appear to have been closely related in their genesis to the modifications of the growth topography of the external crystalline surface.

In crystallizing minerals, such outer surfaces lie in direct contact with either magma, pneumatolytic or hydrothermal fluids, evaporitic oceanic brine or any other mineral-forming medium, depending on the thermodynamic conditions. One of the most important and sometimes most ignored of these conditions is the crystallogenetic disequilibrium and the gradients in space and time (Raguin 1973).

Be it the undercooling of a silicate melt (Clocchiatti 1975) or the supersaturation of a salt solution in water (Sabouraud 1976), the crystallogenetic disequilibrium is the ruling factor of the variations in the intricacies of the growth topography of the deposited minerals, the microscopic topography upon which the morphology depends, the dimensions, the abundance and the mutual interconnections of the resulting lacunae of crystallization left inside the mineral (Fig. 1a).

As has often been stressed (Bruhns and Ramdohr 1939), especially in milky quartz from hydrothermal ore veins, the shape of the cavities appears to be highly irregular. Each inclusion displays its own geometrical outlines. Gangue quartz, at moderate magnification, shows lacunae which seem to be entirely bounded by curved surfaces. Such uneven surfaces reflect the initial character of the primary growth topography of the host mineral.

In fact such inner surfaces of the crystals are also subject to evolution, as long as they remain in contact with the mother solution trapped inside the inclusions. Such isolated droplets of mother solution allow some degree of recrystallization on the walls of the cavities.

Fig. 1a–f. Fractographic aspects (S.E.M.) of gangue quartz. Molybdenite-bearing vein from Fidmont (N.W. Quebec), Coll. Edmond Bruet, Lab. Struct. et Appl. Geol. Paris.

a Growth lacunae: intra-crystalline cavity of irregular shape and some other, smaller, inclusions.

b Microcrystalline grain arrangement: tight curved intergranular contacts.

c Formation of cavities at the meeting point of quartz individuals.

d Topography of grain surfaces near the boundary of two adjointing crystals.

e Detail of regular growth steps (in the submicron range) reflecting rhythmical layered crystallization

f Narrow gap between automorphous quartz crystals. N.B. Molybdenite, also of automorphous habit, occurs in microscopic "vugs" of the vein.

Scale shown under each figure in the dark rim: length of the white line corresponds to one micrometer. Scanning electron photographies taken by the authors at the Geological Laboratory of the Museum National d'Histoire Naturelle, Paris. Initial instr. magnification: **a** × 2200; **b** × 3600; **c** × 5400; **d–f** × 20000

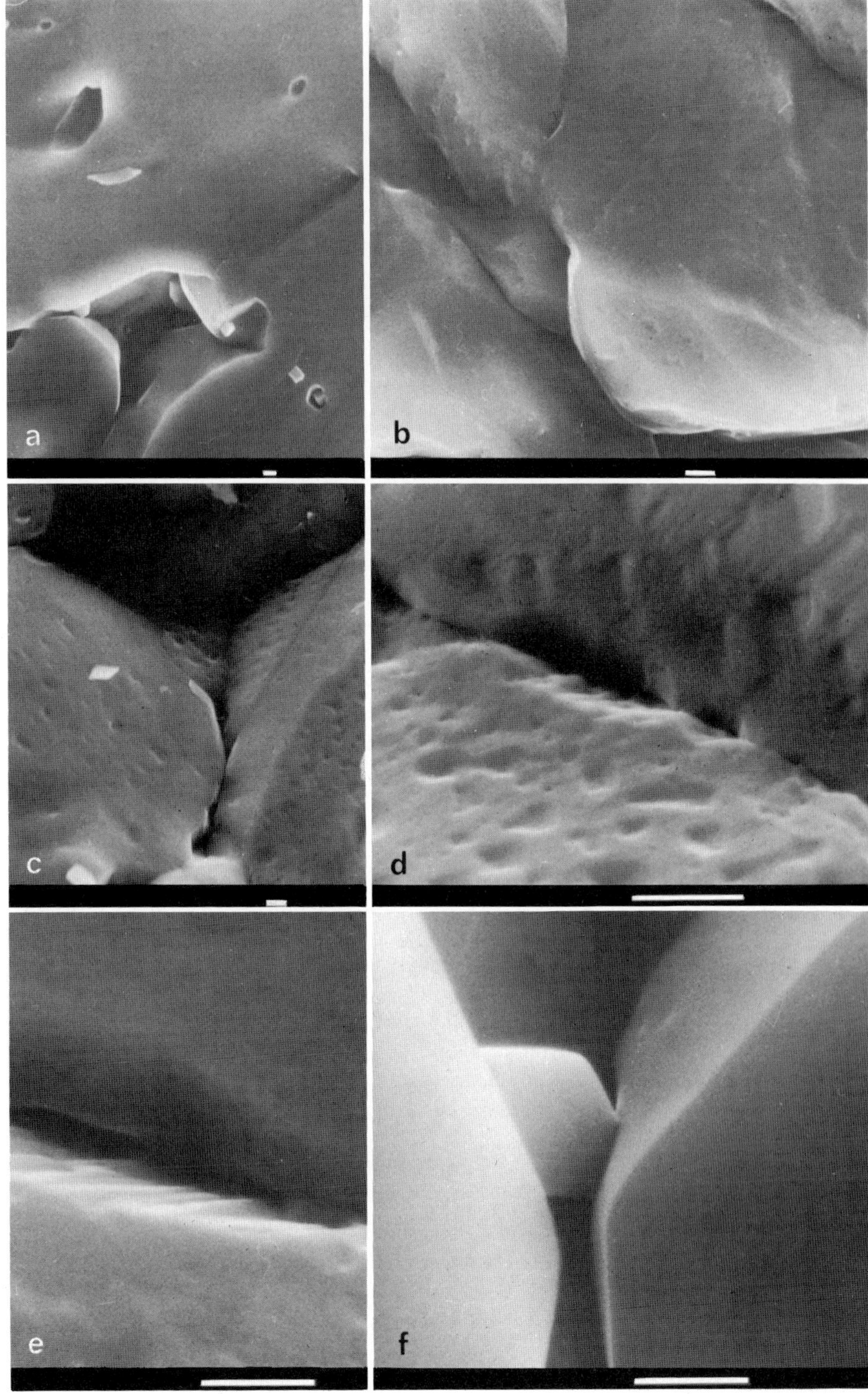
a
b
c
d
e
f

At higher magnifications, especially at those of scanning electron microscopy, we observe that the smallest cavities mostly display the morphology of nearly perfect "negative crystals".

The discrepancies between the crystallographic imperfection of the bigger syngenetic cavities and the ideal geometric habit of the smallest neighbouring inclusions give a hint as to the scope of disequilibrium at the time when these cavities were separated from the free surface of the growing mineral.

Experimental as well as industrial (Mullin 1972) hydrothermal synthesis of crystals teaches us that the equilibration of the process of crystal growth precludes the formation of inclusions. Direct observation of crystallization inside homogenized hydrothermal as well as hydromagmatic inclusions in gangue minerals has also taught us that the high degree of disequilibrium obtained in supersaturated and undercooled geochemical ore-bearing solutions leads to the formation of small and abundant cavities inside the obtained crystals.

3 Intergranular Contacts and Related Cavities

Electron fractography enables us to investigate the grain boundaries: for instance, inside milky quartz from an ore-bearing hydrothermal vein, the surface of an artificial fresh fracture shows, besides broken quartz grains, crystalline individuals that have been separated from their neighbours. The surface of such unbroken crystals offers a large choice of different kinds of intergranular topography, more or less elaborate (Fig. 1b). This induced topography depends upon the mutual crystallographic orientation of adjacent individuals: the simplest pattern occurs in the case of parallel intergrowth.

Sometimes a conspicuous step-structure appears at grain boundaries, which gives testimony to the rhythm of the layer growth during the deposition of quartz and thus, gives an idea of the magnitude of thickness of the involved crystallization layers (Fig. 1e). The already observed variations of this thickness are indeed very broad. As the height of the successive growth steps is linked to the degree of crystallogenetic disequilibrium, such observations give some indication of the conditions of formation.

The persistency, or the variation, in the conditions of growth are reflected in the regularity, or modification, of the height of the growth steps. Theoretically, the trend towards perfect equilibrium leading to the formation of smooth and plane boundaries of tight contacts is rare, most of grain boundaries presenting various kinds of discontinuities

In the same way that the supersaturation of the hydrothermal ore bearing solutions encourages the formation of growth lacunae inside the crystalline individuals, this same disequilibrium leads to the formation of intergranular cavities. The morphology, dimensions, abundance, and mutual connections of the microscopic hollows between adjoining quartz crystals also fluctuate in a very broad manner (Fig. 1c–f).

Sometimes such cavities harbour intergranular fluid inclusions, which can be confused with those trapped inside of the intragranular cavities. In the case of parallel-grown mosaic structures the differentiation cannot be asserted even by means of the polarizing microscope. Scanning electron fractography allows a better knowledge of the intricacies which can exist between growth lacunae and other discontinuities of the crystalline architecture: for instance, supersaturation twinnings also display cavities at their twinning boundaries which sometimes harbour inclusions.

4 Expected Further Developments

Taking into account the fact that the key word "inclusions" has been introduced in the programme of the 26th session of the International Geological Congress (Paris, 2nd and 3rd week of July 1980), it appears that the present year gives an excellent opportunity to evaluate the progress accomplished in the study of cavities observed in minerals since the publication of *Die Erzmineralien und ihre Verwachsungen* (Ramdohr 1950) and the printing of the first books entirely devoted to fluid inclusions in the Russian (Ermakov 1950), English (Smith 1953) and French (Deicha 1955) languages. It is hoped that in the next few years the contribution of scanning electron microscopy will increase our knowledge of the cavities containing fluids, especially in gangue and ore samples.

References

Bruhns W, Ramdohr P (1939) Petrographie (Gesteinskunde). Sammlung Göschen, Bd 173, 2. Aufl., Berlin, 117 pp

Clocchiatti R (1975) Les inclusions, vitreuses des cristaux de quartz. Mém Soc Geol Fr 122:1–113

Deicha G (1955) Les lacunes des cristaux et leurs inclusions fluides, signification dans la genèse des gîtes minéraux et des roches. Masson, Paris, 126 pp

Ermakov NP (1950) Research on mineral-forming solutions. Kharkov Univ Press, 460 pp (in Russian). Available in English translation in: Yermakov NP et al. (1965) Research on the nature of mineral-forming solutions, with special reference to the data from fluid inclusions. Int Ser Monogr in Earth Sciences, vol 22. Pergamon Press, Oxford, 743 pp

Ermakov NP (1979) Thermobarogeochemistry. Nedra Moscow

Imai H (1978) In: Imai H (ed) Geological studies of the mineral deposits in Japan and East Asia. University Press, Tokyo, pp 59–61

Mullin JW (1972) Crystallisation. Butterworth, London

Raguin E (1973) Les roches plutoniques dans leur rapports avec les gîtes minéraux. Colloque. Masson, Paris

Ramdohr P (1950) Die Erzmineralien und ihre Verwachsungen. Akademie Verlag, Berlin, 826 pp

Roedder E (1972) Composition of fluid inclusions. USGS Prof Pap 440-JJ. Waschington (DC), JJ1–JJ164

Sabouraud Ch (1976) Inclusions solides et liquides du gypse. Trav Lab Geol Ec Norm Super, Paris, 10

Smith FG (1953) Historical development of inclusion thermometry. University Press, Toronto, 149 pp

Fluid Inclusions and Trace Element Content of Quartz and Pyrite Pebbles from Witwatersrand Conglomerates: Their Significance with Respect to the Genesis of Primary Deposits

D.K. HALLBAUER and E.J.D. KABLE[1]

Abstract

Geological and mineralogical studies of quartz pebbles and detrital pyrite demonstrate the existence of differences in the provenance areas of the various Witwatersrand goldfields and provide information about the type of primary deposits.

The trace element content determined by INAA and AAS can be used to determine groupings and hence differences in provenance areas while fluid inclusion studies provide information on the type of primary deposits.

1 Introduction

As mining progresses on the Witwatersrand, the world's largest gold province, more knowledge is required about the distribution of the gold within each reef and hence the origin of the reef in order to develop new and more economic mining methods. A key to a better understanding of the distribution of gold may lie in a detailed study of the various components of the conglomerates in which it occurs. Because of the abundance and the variety of different types of quartz pebbles and pyrite within the gold-bearing conglomerate, they were studied to assess their relationship with respect to the distribution and origin of gold with a view to formulating a model of the distribution of gold within a reef.

This paper describes how differences in the provenances of the various Witwatersrand goldfields can be deduced from the chemical analysis of fluid inclusions and the trace element content of quartz pebbles and pyrite. Some conclusions are made about the type of deposits which supplied detritus to the Witwatersrand basin.

2 Pyrite

Pyrite is the most conspicuous mineral in Witwatersrand deposits. It occurs in the matrix of the conglomerate, as layers on foresets within the conglomerate, in the

1 Mining Technology Laboratory, Chamber of Mines of South Africa Research Organization, P.O. Box 91230, Auckland Park 2006, Johannesburg, South Africa

quartzitic rock above and below the reef in massive bands up to several centimetres thick and as pyrite stringers in many stratigraphic positions throughout the Witwatersrand sequence.

A major portion of the pyrite is present as round or rounded grains which are commonly termed "buckshot pyrite". It was this type of pyrite which led to an early postulation that sedimentary processes played a major role in the formation of the Witwatersrand deposits. Ramdohr (1955), Liebenberg (1955), Saager (1973) and others provided evidence of the sedimentary origin of "buckshot pyrite" as well as indications of its genetic relationship to the detrital gold.

Based on Ramdohr's (1955) classification of Witwatersrand pyrite, a genetic classification is suggested which discriminates between (a) allogenic detrital pyrite, (b) pyrite pseudomorphs and replacements, (c) synsedimentary pyrite, (d) authigenic pyrite, and (e) late hydrothermal and pseudohydrothermal formations.

As the allogenic detrital pyrite is assumed to be the most intimately related to the distribution and origin of the gold, it is dealt with exclusively in this paper.

In this type of pyrite [which includes Ramdohr's (1955) pebble pyrite, and Saager's (1973) and Utter's (1978) compact rounded pyrite] the inclusions and the trace element content are of primary nature.

The scanning electron microscope with an energy dispersive spectrometer was found to be best suited to investigate the morphology and composition of mineral inclusions as well as daughter minerals in fluid inclusions and fluid chemistry.

The results so far indicate that biotite and quartz are the most common silicate inclusions in allogenic detrital pyrite. While quartz is present as inclusions which are not related to fluid-filled voids (Plate 1.1), biotite occurs as genuine inclusions as well as a daughter mineral (Metzger et al. 1977) in fluid inclusions (Plate 1.2).

Daughter minerals usually contain those elements which are not soluble at lower temperatures. The presence of a daughter mineral unambiguously requires that its components are present as dissolved constituents in the inclusion fluid at the time of the original entrapment (Metzger et al. 1977).

Quartz is characteristically present as small elongated inclusions, often well crystallized and occasionally oriented parallel to the pyrite lattice directions but very rarely exceeding 10 μm in diameter (Plate 1.1). Quartz inclusions are accompanied by biotite, monazite and rutile in voids, probably as daughter minerals in fluid inclusions. Energy dispersive spectra obtained from biotite inclusions showed variations in the K and Mg contents, and in the case of a pyrite from the Basal Reef, small amounts of Ti and Cr were also measured (Plate 1.2).

Pyrite from the Ventersdorp Contact Reef (VCR) in the Carletonville area appears to contain orthoclase as the principal silicate inclusion, usually as a daughter mineral in fluid inclusions which contained (Na,K)Cl as the major dissolved species (Plate 1.3). Some large inclusions of molybdenite and tourmaline, so far not documented in pyrite from other localities, were observed in the detrital pyrite from the VCR, pointing to quartz-tourmaline-gold veins as a possible source.

Of particular interest is the occurrence of pyrophyllite, in allogenic detrital pyrite of the Vaal Reef. Large bodies of pyrophyllite which in places contain pyrite and cobaltite are situated in the possible hinterland of the Vaal Reef palaeo-placer within the Dominion Reef series. Since pyrophyllite was noted in the matrix of the Vaal Reef

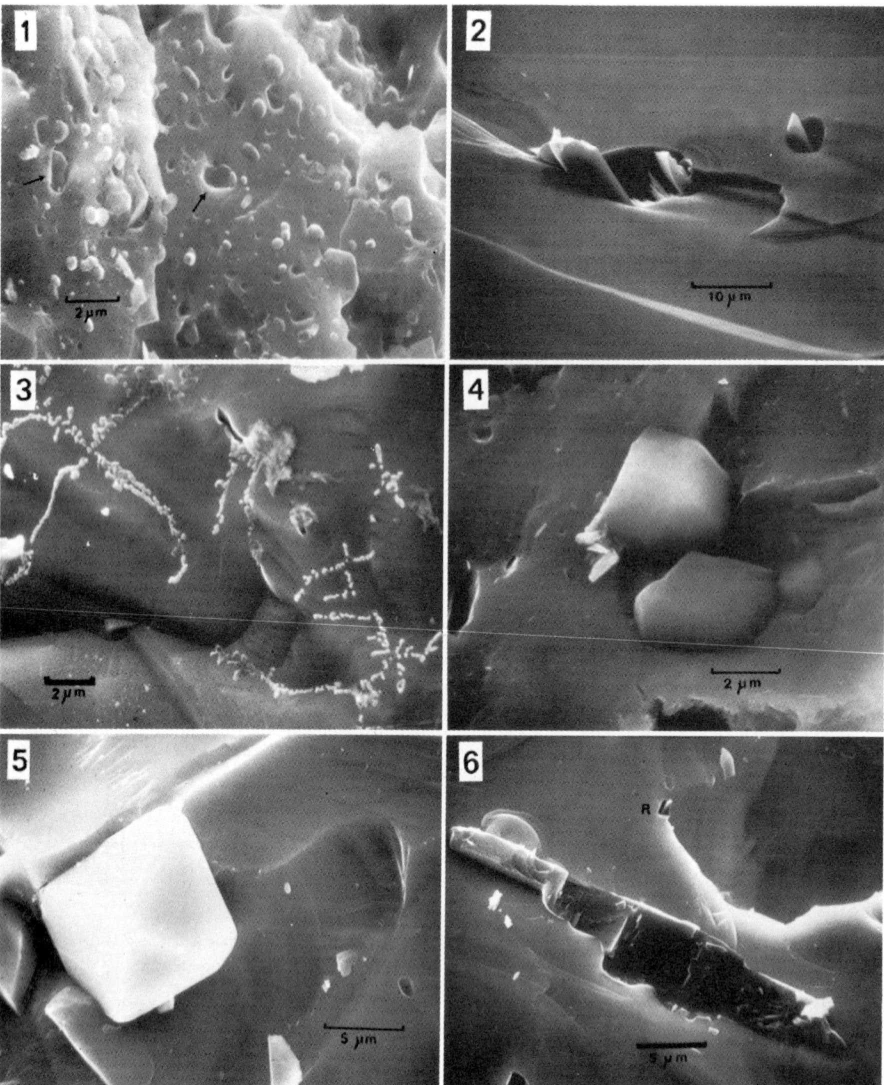

Plate 1. Inclusions in pyrite. **1** A multitude of quartz inclusions accompanied by biotite *(arrows)* in fluid inclusions. Basal Reef (Steyn Reef), Western Holdings gold mine. **2** Biotite as daughter mineral in fluid inclusions. Basal Reef, President Brand gold mine. **3** Dried fluid from fluid inclusions consisting of (Na,K)Cl. VCR, East Driefontein gold mine. **4** Spessartine inclusions in pyrite. "B" Reef, Loraine gold mine. **5** Magnetite inclusion in pyrite. Agnes gold mine, Barberton. **6** Large muscovite inclusion and rutile *(R)* in a fluid void. Fairview gold mine, Barberton

(von Rahden 1970) as well, the Dominion Reef series has to be considered as part of the Vaal Reef provenance area. Apart from the above-mentioned minerals others like anhydrite, sphene, apatite, spessartine (Plate 1.4) and brannerite have been observed as inclusions or daughter minerals.

Examples of pyrite from the Barberton Mountain Land, a model source area (Viljoen et al. 1970), have also been studied.

Pyrite from the Agnes and Fairview gold mines are different from the Witwatersrand detrital pyrite examined with regard to their dominant inclusions. The pyrite from the Agnes gold mine contains a large number of magnetite crystals (Plate 1.5) about 5 to 30 μm in diameter which are accompanied by oriented inclusions of rutile 0.1 to 1 μm in diameter. Small micaceous inclusions with a composition suggestive of biotite could occasionally be observed within the magnetite. Pyrite with octahedral habit which occur in the same samples, contain identical inclusions except that the micaceous inclusions occur in the pyrite and not in the magnetite. The characteristic inclusions in the pyrite from Fairview are large muscovite grains (up to 100 μm in diameter) accompanied by rutile (Plate 1.6), chalcopyrite, galena, quartz and zircon.

A suite of 16 elements including the transition elements, As, Sb, Au, Ag, La, Th and U have been determined routinely in pyrite concentrates by neutron activation and atomic absorption methods.

For purposes of comparing allogenic detrital pyrite from different reef horizons those elements which occupy lattice sites are considered to be the most meaningful. A comparison of Co-Ni relationship in pyrite from the Witwatersrand Goldfields with pyrite from the Archaean (Fig. 1), demonstrates that the Archaean pyrite (Saager 1973) tend to be depleted in Co relative to Ni, while detrital pyrite usually has higher Co contents. However, certain reef horizons, e.g., the Basal Reef, show very close similarities to the Archaean in terms of their Co and Ni contents while others like the Kimberley Reef are considerably different. It may be concluded from this information that a number of provenance areas must have been responsible for providing pyrite to the Witwatersrand goldfields.

Recent experiments conducted on single grains have indicated the presence of two geochemically different species of allogenic detrital pyrite in the VCR. Single grains of < 1 mm have higher concentrations of a number of elements including Ni and Co which are enriched by a factor of 10 compared to grains > 1 mm in size.

3 Quartz Pebbles

Although quartz pebbles make up the bulk of the Witwatersrand conglomerate, in the past mineralogists have paid little attention to them, apart from referring to their colour and texture.

In recent years quartz in mineral deposits and especially fluid inclusions in quartz have received increasing attention in an effort to gain a better understanding of the genesis of quartz and related ore deposits.

Shepherd (1977) distinguished between primary or pre-depositional inclusions in Witwatersrand pebbles formed at elevated temperatures and pressures, and secondary,

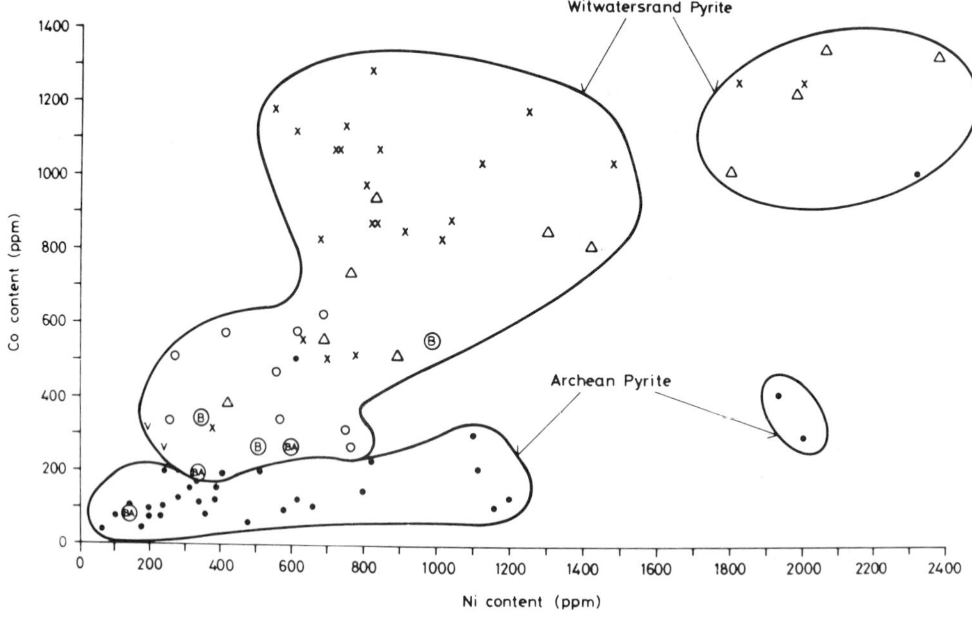

Fig. 1. Relationship between Co and Ni content for Archaean pyrite (data from Saager 1973) and Witwatersrand pyrite. *BA* Basal Reef; *B* "B" Reef; *V* Ventersdorp Contact Reef; *X* Kimberley Reef, Evander (Hirdes 1979) ; △ Kimberley Reef, Marievale (Hirdes 1979); ○ Vaal Reef, Klersdorp gold field

or post-depositional inclusions formed at lower temperatures. He also noted a possible correlation between the number of inclusions containing liquid CO_2 and the uranium content of some conglomerates, in particular the Black Reef, VCR and Vaal Reef.

Homogenization temperatures of the fluid inclusions in quartz pebbles indicate not only the temperature range at which the quartz was primarily formed, but, at the same time, indicate the geological processes after deposition.

Common to all fluid inclusions in quartz pebbles is a range of homogenization temperatures from about 110° to 180°C. Most of these inclusions can be classed as secondary or post-depositional. They provide evidence of a common thermal influence during the geological history and could possibly indicate the temperature range during metamorphism.

Homogenization temperatures above 250°C vary considerably from pebble to pebble. These fluid inclusions have the features of pre-depositional inclusions (Shepherd 1977) and characterize the variety of primary quartz pebbles in the Witwatersrand conglomerates.

The daughter minerals in primary fluid inclusions contain information about the composition of the original fluid inclusions which, according to Roedder (1967), are samples of the ore-forming fluids. Chlorides which are rich in base metals are present in some quartz pebbles, suggeesting that they have been derived from hydrothermal ore deposits.

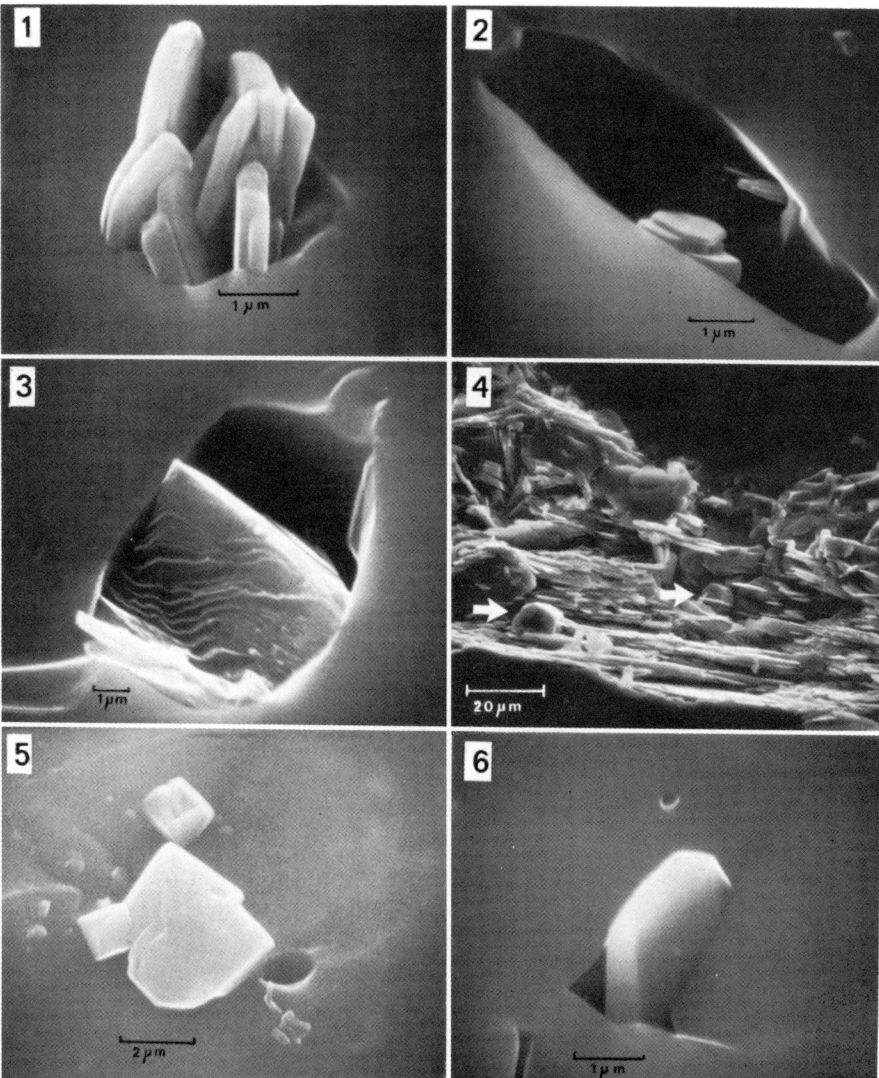

Plate 2. Inclusions in Witwatersrand quartz pebbles. **1** Orthoclase inclusions, Carbon Leader Reef, Blyvooruitzicht gold mine. **2** Muscovite daughter mineral in a fluid inclusion. Vaal Reef, Stilfontein gold mine. **3** Calcite daughter mineral in blue quartz. Carbon Leader Reef, Blyvooruitzicht gold mine. **4** Large fluid inclusion with apatite *(arrows)* and an iron-rich phyllosilicate. Carbon Leader Reef, Blyvooruitzicht. **5** Dried fluid containing NaCl with traces of Sn and Cs. "B" Reef, Free State Geduld gold mine. **6** Ba-feldspar as daughter mineral. "B" Reef, Free State Geduld gold mine

The daughter minerals observed in the fluid inclusions of Witwatersrand pebbles include orthoclase as the most common inclusion (Plate 2.1), muscovite (Plate 2.2), chlorite, rutile, apatite, corundum, anhydrite, various mixed chlorides such as NaCl, (Na,K)Cl, $CaCl_2$ and others. Of special interest is the abundance of calcium-bearing minerals: calcite (Plate 2.3), anhydrite, apatite (Plate 2.4) and calcium-bearing chloride in the blue opalescent quartz pebbles from many localities. These blue opalescent quartz pebbles are distinguishable by the absence of homogenization temperatures above 200°C which places them in a separate class from all other quartz pebbles.

Possible genetic links between quartz pebbles and older, primary deposits are being investigated. One striking example is a quartz pebble from the "B" Reef at Free State Geduld gold mine. This pebble, according to material which crystallized next to the voids, contained a NaCl-bearing fluid with traces of Sn and Cs (Plate 2.5). Some of the fluid inclusions contained cassiterite as a daughter mineral. The presence of these components indicates this quartz pebble was derived from a pegmatitic tin deposit similar to those which are known to occur in the Archean tin belt of Swaziland.

From the same locality, a quartz pebble of dark grey colour, known to contain exceptionally high trace amounts of REE, Cs and Ba was selected for inclusion studies. The dominant daughter mineral present in the fluid inclusions was identified as a barium feldspar (Plate 2.6). In some cases drops of dried fluid were analyzed which showed the presence of Al, K, Ca, Fe, Na, Ba, Mg, Mn with Cl and S as probable anions.

Barium-feldspar is known to occur in paragenesis with spessartine. This manganese garnet was found as inclusions in pyrite from the same locality. A genetic relationship between the quartz and pyrite is further substantiated by the presence of S in the fluid inclusions of the quartz. A possible common origin from a pegmatitic source is therefore suggested.

In addition to the analysis of the solid phases, the chemistry of fluid inclusions was investigated by the leaching of crushed and heated quartz pebbles with deionized water. The leach fluid was analyzed by atomic absorption spectrophotometry and other techniques. The result was recalculated for fluid content of the sample based on a mass loss at 700°C. Some of the results (Table 1) indicate that secondary quartz could be distinguished quite easily from a quartz pebble considered to be genetically related to a primary deposit.

As already mentioned, the geological history of the quartz pebbles in the Witwatersrand conglomerates is a very complex one. There are, however, no signs of substantial recrystallization or remobilization of quartz pebbles. It is therefore safe to assume that

Table 1. Analysis of leach solution of two quartz specimens (concentrations in ppm except where indicated)

	Cl	Na	K	Ca	Au	Mn	Co	Ni	Cu
Quartz pebble (VCR East Driefontein gold mine)	16%	3300	4400	1000	0.7	90	20	700	900
Secondary quartz (Libanon gold mine)	0.1%	1700	130	1100	0.1	1	1	1	5

most trace elements present in quartz pebbles still reflect the chemistry of the primary sources of the quartz.

Neutron activation techniques provide an ideal technique for the study of the chemistry of quartz pebbles to determine elements at the trace and ultra-trace levels. So far, samples from 12 different geological units have been investigated. Data were obtained for a wide range of elements including Na and K, the transition metals, several REE as well as Th, U, Au, As and Sb. The K content of the quartz samples is generally higher than the Na content, which ranged from 20–100 ppm. In some quartz pebbles values approach 7000 ppm. Quite a distinct group is represented by pebbles from the Steyn facies of the Basal Reef, with low K values ranging from 3–30 ppm.

A wide range of K values (20–6000 ppm) is common in the VCR and Kimberley Reef, while on average the highest content was determined in quartz pebbles from the Carbon Leader Reef (500–7000 ppm). This is similar to the values measured in quartz veins from basement granite west of Klerksdorp. However, quartz veins in Archean rocks from the Barberton district have comparatively low K contents (10–40 ppm).

In general, quartz pebbles show an enrichment of the light REE relative to the heavy REE which indicates a possible genetic relationship to granite and acidic primary rocks (Wedepohl 1978). Of special interest is the good positive correlation between high La content and high Au content within the various geological units.

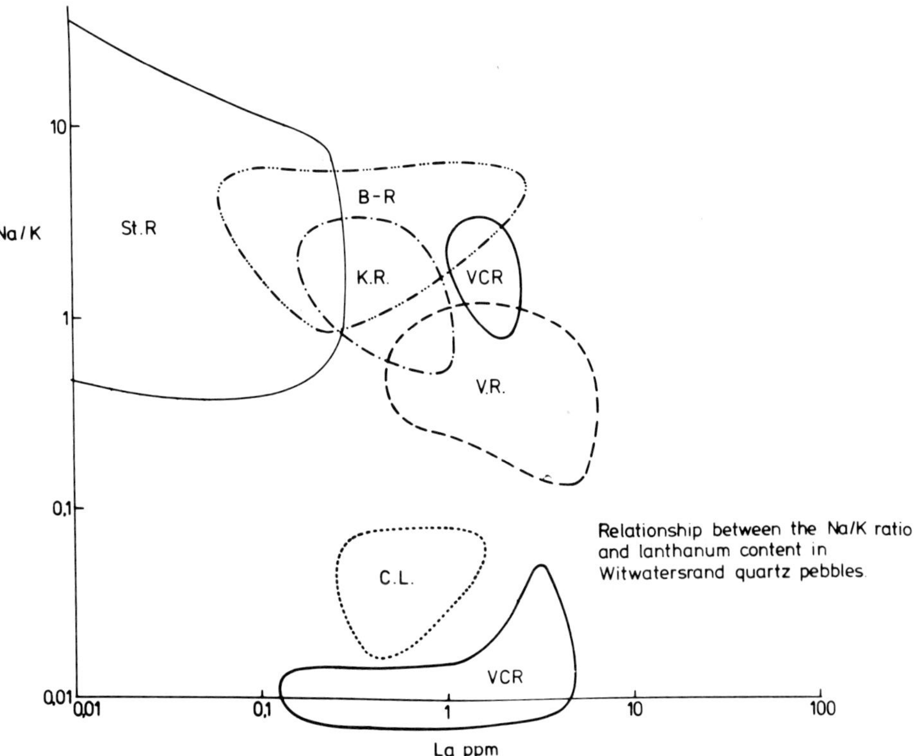

Fig. 2. Na:K and La distribution in quartz pebbles from the Steyn Reef *(St.R.)*, "B" Reef *(B-R)*, Kimberley Reef *(K.R.)*, Ventersdorp Contact Reef *(VCR)*, Vaal Reef *(V.R.)* and Carbon Leader Reef *(C.L.)*

Another interesting observation is that quartz specimens with high U contents (3–30 ppm) have high Au contents, while high Au contents are not necessarily associated with high U values. Generally, the Au content of quartz ranges from about 1 ppb (for secondary quartz veins older than the VCR) to about 8000 ppb in a quartz pebble from the Carbon Leader Reef. With the exception of a quartz vein from the basement granite west of Klerksdorp, all quartz specimens from secondary veins have very low Au contents.

The Th/U ratio in quartz pebbles is generally higher than the ratios in carbonaceous matter (0.09–0.27) and uraninite (0.17) and ranged from 0.5 to 20, with an average of 4. This is not only close to the average Th/U ratio in igneous rocks but indicates that most of the Th and U in quartz pebbles are genuine primary components and not later contaminants.

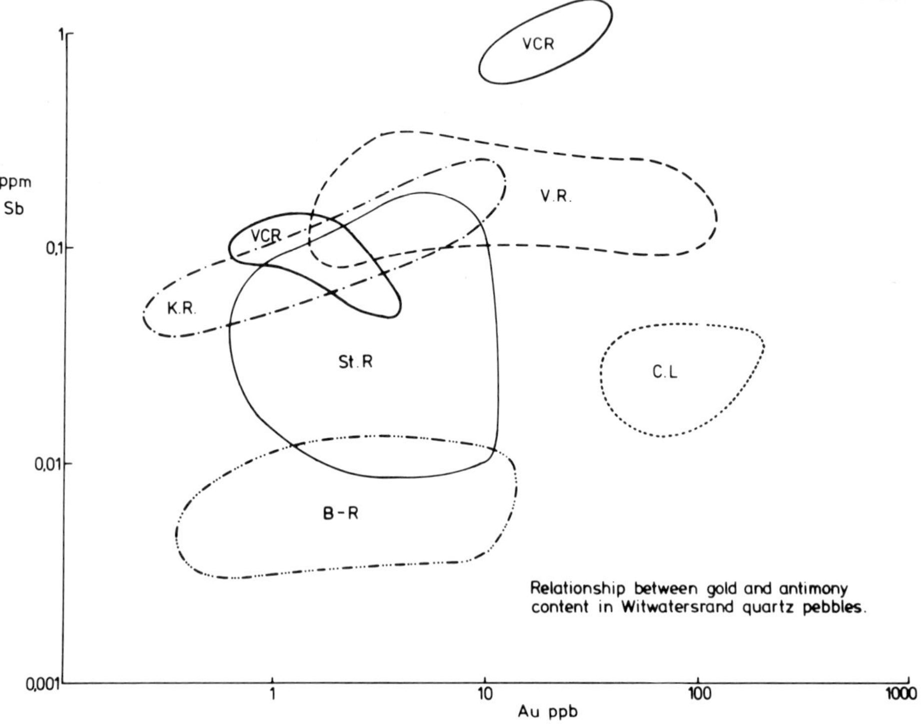

Fig. 3. Sb and Au distribution in quartz pebbles of same location as in Fig. 2

Scatter diagrams (Figs. 2 and 3) showing the relationship between Na/K ratio and the La content, and between the Sb and Au content in quartz pebbles, clearly indicate that the Carbon Leader specimens form a distinct group outside the range of other reef samples. Two distinct populations are apparent for the VCR in the Libanon and Kloof area. This suggests that two components are present in the VCR: one related to the underlying Kimberley Reef and the other component originating from a different source.

4 Conclusions

In this paper the authors have attempted to demonstrate that the Witwatersrand allogenic detrital pyrite and quartz pebbles provide information about their primary origin. Much of this information can be obtained from mineral inclusions, fluid inclusions and trace element contents.

Inclusions such as pyrophyllite, spessartine, orthoclase, molybdenite, Ti-Cr-biotite and others which were found only in pyrite of certain localities, and apatite, corundum, cassiterite and Ba-feldspar in quartz pebbles, indicate that the various goldfields of the Witwatersrand had different genetic types of primary deposits supplying detritus to the basin. Granitic pegmatites appear to play an important role as primary deposits for the Basal Reef and "B"-Reef, while rocks of the Dominion Reef volcanic sequence are possibly one of the sources of detritus for the Klerksdorp gold field. For portions of the Ventersdorp Contact Reef, quartz-tourmaline-gold-bearing veins are a likely source.

The investigations have not yet progressed far enough to formalize a relationship between gold and the various types of quartz and pyrite. This relationship requires further research.

Inclusion studies as well as the Co-Ni contents suggest that differences exist between pyrite from Barberton deposits and allogenic detrital pyrite from some Witwatersrand localities. Both inclusion and chemical studies demonstrate that single-grain pyrite analysis is preferable to the analysis of concentrates.

The trace element distribution in quartz pebbles provides evidence of geochemically different source areas. Attention has been drawn to the possibility of identifying reef horizons through inclusion studies and trace element analysis.

Th/U ratios in detrital pyrite, uraninite and carbonaceous matter are of the same magnitude (0.1–0.4) suggesting a similar type source. Quartz pebbles have similar ratios to those mentioned above as well as ratios representative of granitic rocks (av. 3.5–4), which emphasizes the presence of different genetic types of quartz in the Witwatersrand conglomerate.

References

Hirdes W (1979) The Proterozoic gold-uranium Kimberley Reef placers in the Evander and East Rand goldfields, Witwatersrand, South Africa: Different facies and their source area aspects. Unpubl PhD Thesis, Univ Heidelberg, 199 S

Liebenberg WR (1955) The occurrence and origin of gold and radioactive minerals in the Witwatersrand System. The Dominion Reef, the Ventersdorp Contact Reef and the Black Reef. Trans Geol Soc S Afr 58:101–254

Metzger FW, Kelly WC, Nesbitt BE, Essene EJ (1977) Scanning electron microscopy of daughter minerals in fluid inclusions. Econ Geol 72:141–152

Ramdohr P (1955) Neue Beobachtungen an Erzen des Witwatersrands in Südafrika und ihre genetische Bedeutung. Abh Dtsch Akad Wiss Berlin Kl Math Naturwiss 5:43

Roedder E (1967) Fluid inclusions as samples of ore fluids. In: Barnes HL (ed) Geochemistry of hydrothermal ore deposits. Holt, Rinehart and Winston, New York, pp 514–574

Saager R (1973) Geologische und geochemische Untersuchungen an primären und sekundären Goldvorkommen im frühen Präkambrium Süd-Afrikas: Ein Beitrag zur Deutung der primären Herkunft des Goldes in der Witwatersrand Lagerstätte. Unpubl Habil Univ Heidelberg, 150 S

Shepherd TJ (1977) Fluid inclusion study of the Witwatersrand gold-uranium ores. Philos Trans R Soc London Ser A 286:549–565

Utter T (1978) Morphology and geochemistry of different pyrite types from the Upper Witwatersrand System of the Klerksdorp goldfield, South Africa. Geol Rundsch 67(2):744–804

Viljoen RP, Saager R, Viljoen MJ (1970) Some thoughts on the origin and processes responsible for the concentration of gold in the early Precambrian of Southern Africa. Mineral Deposita 5:164–180

von Rahden HVR (1970) Mineralogical and geochemical studies of some Witwatersrand gold ores with special references to the nature of the phyllosilicates. Unpubl PhD Thesis, Univ Witwatersrand, p 89

Wedepohl KH (ed) (1978) Handbook of geochemistry, vol II/5 39, 57–E12. Springer, Berlin Heidelberg New York

The Nature and Significance of the Pleochroism of the Cassiterites of the Southeast Asian Tin Belt

K.F.G. HOSKING[1]

Abstract

The tin belt of southeast Asia is, in fact, composed of two parallel belts, an east and a west, and primary tin deposits of a number of different ages and types occur in each.

Certain types of primary tin deposits are restricted to one belt and the mineralogy of a given type in the one belt may differ significantly from the corresponding type in the other.

Intensely red-pale colour cassiterites containing Ta and possibly Nb in the lattice, are restricted to the west belt and are paramagnetic, whilst those that are brown-pale colour pleochroic, a phenomenon probably also due to Nb/Ta or W in the lattice, occur in both belts and may be ferromagnetic.

It is concluded that in both belts the sources of the tin were crustal ones, but that whilst in the west belt the sources were comparatively rich in Ta and Nb, those of the east belt were comparatively rich in iron.

Study of the pleochroic character of the cassiterites facilitates understanding of the genesis of the Southeast Asian tin deposits and can be profitable to the prospector and mineral engineer.

1 Introduction

The Southeast Asian tin belt, which is about 3900 km in length, stretches from Mainland Burma and Thailand to the Tin Islands of Indonesia (Fig. 1). The so-called belt is, in fact, composed essentially of an east and a west belt, which are approximately parallel, but locally fault-displaced.

Within these belts the tin deposits are associated with granites that are dominantly mesozonal in the west belt and essentially epizonal in the east one, although late epizonal elements occur in the west belt. Furthermore, there are distinct mineralogical differences between the granites of the two belts (Hutchison 1975) and also the trace-element pattern of the granites of the west belt differs significantly from that of the east belt (Yeap 1974). The granites of the two belts have a long and complex history and in each belt granites of distinctly different ages occur, sometimes in close association (Bignell 1972; Priem et al. 1975). Spatial and other considerations provide support for the view that within each belt primary tin deposits were developed on a number of distinctly separate occasions (Hosking 1977).

1 Camborne School of Mines, 1 B Penlu, Tuckingmill, Cornwall, TR 14 8 NL, England

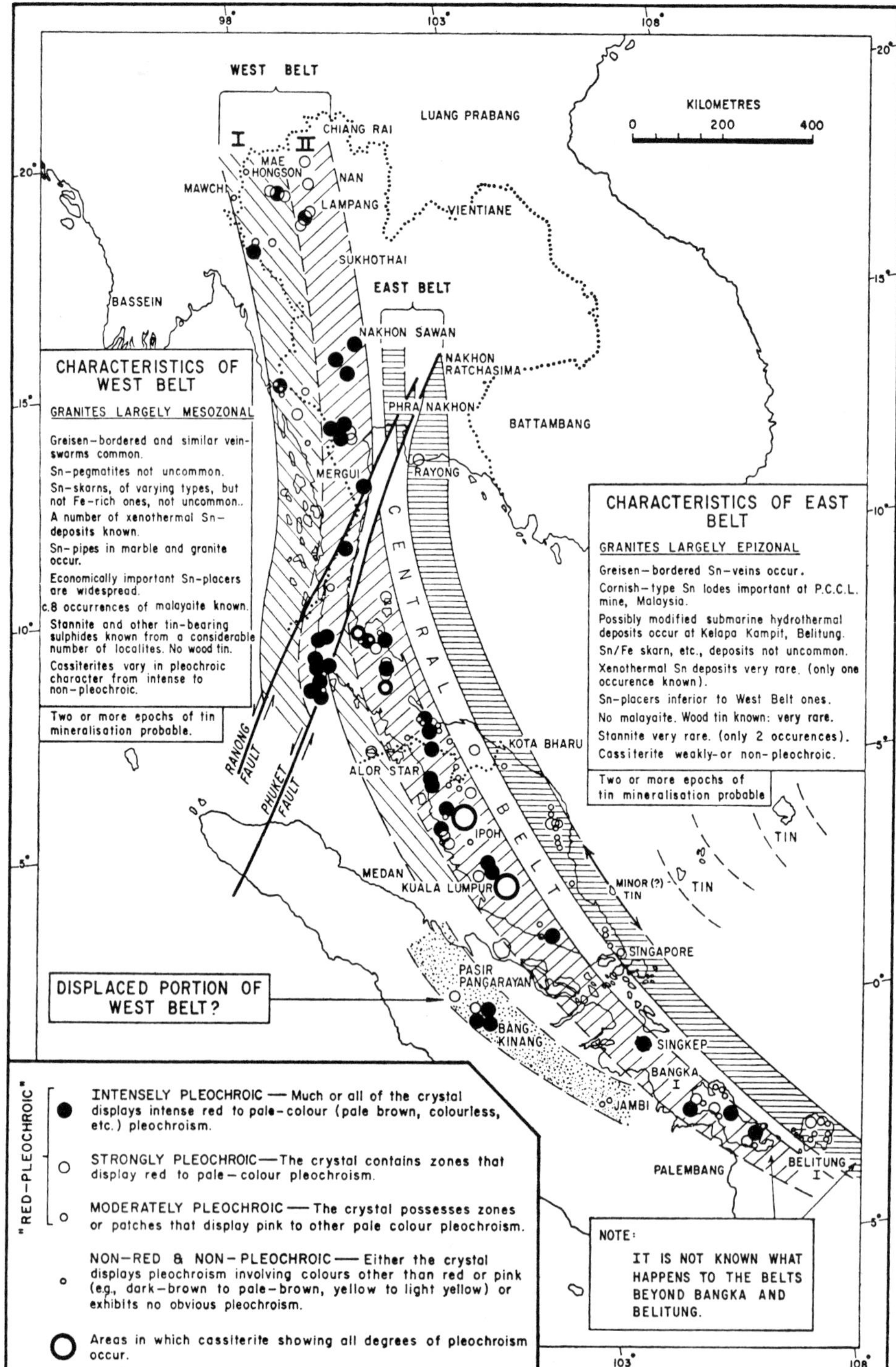

Fig. 1. Tin belts of Southeast Asia and the pleochroic character of the cassiterites within them. (After Hosking 1977)

Stanniferous vein-swarms, skarns and lodes are well represented in both belts, but stanniferous aplites and pegmatites, and pipe deposits appear to be confined to the west belt, while strata-bound deposits are only known to occur at Belitung (Billiton) and at Bukit Besi in the east belt.

It is also relevant to note that skarns bearing malayaite occur in the west belt and that such bodies containing this mineral appear to be absent from the east belt where iron-tin skarns are not uncommon. Also, only in the east belt are tin deposits, in addition to the skarn, commonly rich in magnetite, hematite and other iron-bearing species. In the west belt, also, no wood tin has been recorded, but tin-bearing sulphides are not uncommon and both beryllium and antimony species have been recorded from a number of primary tin deposits. In contrast, in the east belt wood tin has been recorded from two localities, but stannite and beryllium minerals are distinctly rare and no antimony species have been noted in any primary tin (or other) deposit (Hosking 1977).

2 Characteristics of the Pleochroic Cassiterites of the Tin Belt

Scrivenor (1928, p. 28) first studied the pleochroism exhibited by cassiterites from the tin belt but he confined his attention to specimens from Malaysia. He noted the following seven types:

Extraordinary ray	*Ordinary ray*
1. Carmine	Olive green
2. Carmine	Pale sepia
3. Pale carmine	Colourless
4. Orange-brown	Brown
5. Dark brown	Light brown
6. Mauve	Colourless
7. Yellow	Light yellow

Later van Overeem (1960, pp. 449–450) demonstrated that at Belitung cassiterites varied considerably in the nature of their pleochroism, and I have shown that such a variation is to be found throughout the whole length of the tin belt. However, it is probable that cassiterites displaying mauve/colourless pleochroism are very rare.

I have adopted a classification which divides the cassiterite into two groups, namely those that display a red to pale colour pleochroism (termed, for convenience, red pleochroism) and those that exhibit a non-red (usually brown to pale colour pleochroism) or are non-pleochroic. The red pleochroic group is further divided into intensely pleochroic, strongly pleochroic and moderately pleochroic sub-groups. The characteristics of the groups and sub-groups are to be found in Fig. 1.

Some of the cassiterites of the tin belt are magnetic, and although magnetic cassiterites from the tin belt have been investigated by a number of workers there is still much to be learned about them. It does appear, however, that red pleochroic cassiterites are often, perhaps always, paramagnetic, whilst brown pleochroic ones are ferromagnetic if they are magnetic at all.

3 Causes of the Pleochroism

There is considerable evidence in support of the view that substitution of pentavalent tantalum for quadrivalent tin in the lattice is largely, or possibly entirely responsible for the red pleochroism exhibited by some cassiterites from Southeast Asia and elsewhere. Perhaps, on occasion, niobium may be at least partly responsible for the red pleochroism under discussion. Gotman (1938) lends a little support to this view, in that having found that intensely red pleochroic cassiterites from Russian pegmatites contained more than 2% $(Nb,Ta)_2O_5$, suggested that the pleochroism was probably due, at least in part, to valency imbalance and lattice distortion resulting from the substitution of Sn^4 by Ta^5 and Nb^5.

Santokh Singh and Bean (1968) having investigated certain intensely red-pleochroic cassiterites from Semiling (Kedah) and Taiping (Perak) stated that tantalum was present in both as "discrete mineral inclusions and in diffuse solid solution showing zonal distribution", but that niobium was not detected in the Semiling material. They also concluded that the intense pleochroism (and unusual paramagnetic properties) of the cassiterites in question "may possibly be ascribed to a disturbance in the valancy balance, but there does not appear to be any lattice distortion in these samples." However, recent studies, involving more precise measurements, have shown that such cassiterites have distorted lattices (Hutchison, pers. comm.).

I think that in Southeast Asia the brown-pale colour pleochroism is generally due to Ta (and/or Nb), but particularly when the cassiterite has been derived from Sn/W deposits, the pleochroism under review may result from tungsten in the lattice.

In addition, some cassiterite crystals possess red pleochroic zones and other zones that are brown and/or orange and that may or may not be pleochroic: in part, their colour is a function of their iron content. Also, the variously coloured zones may be parallel to each other or a set of red parallel zones may be intersected by, or intersect a set of brown and/or orange parallel zones, so prompting questions concerning crystal development which have not yet been resolved.

Aranyakanon (1962, p. 33) observed a reddish-brown pleochroic halo around a grain of monazite that was protruding into a cassiterite crystal from Haad Som Pan (Thailand), and this suggested to him that red to pale colour pleochroism may, on occasion, be produced by radioactivity. My own observations suggest that in Southeast Asia this is a rather rare phenomenon but this may not be the case as Ramdohr (1961) has seen many examples of it from many fields.

4 The Distribution of the Cassiterites of the Tin Belt According to Their Pleochroism, and its Significance

That the cassiterites of Southeast Asia show great diversity in their pleochroic properties led me to investigate the distribution of the cassiterites of the tin belt according to their pleochroic character. Classification of the pleochroism was made by the system I have outlined earlier in this paper and the results were plotted on a map (Fig. 1). These emphasize that the so-called tin belt of Southeast Asia can be divided into two, approximately parallel, but in some respects dissimilar, tin-bearing belts, an east belt and a west belt, that in Malaysia are separated by an essentially tin-barren stretch. A case can also be made for further sub-dividing the west belt on the basis of the pleochroism of the cassiterites into two parallel members.

From the point of view of the pleochroism of the cassiterite, the west belt contains all the types mentioned in my classification, and the intensely red members are much in evidence. On the other hand, in the east belt there are only a few localities where even strongly red pleochroic cassiterites are to be found.

Generally, in tin/tantalum/niobium provinces during a single period of mineralization, tantalum/niobium species tend to be deposited largely early, as disseminations in granitoids, and/or within pegmatites/aplites. Any accompanying cassiterite, provided it is more or less coeval with the tantalum/niobium species, tends to be rich in Ta and possibly Nb and strongly or intensely pleochroic. In greisen-bordered and somewhat similar early deposits of such provinces, Ta/Nb species may be present, but almost invariably in quantities that are only of academic interest, and accompanying cassiterites that may be rich in Ta/Nb, and intensely or strongly red pleochroic. Stanniferous skarns in such provinces lack Ta/Nb species, as far as I am aware, but their cassiterites may display a faint or occasionally strong red pleochroism. The cassiterites of later hydrothermal veins in such provinces may provide a moderately red pleochroism, but often any pleochroism in evidence is of the non-red-pale colour variety. This general trend is particularly strongly supported by evidence from the west belt of the Southeast Asian tin province.

One might be tempted to argue that red pleochroic cassiterite is so little in evidence in the east belt because there the development of primary tin deposits was generally delayed until a rather late hydrothermal stage. However, there is more to it than this as cassiterite from the greisen-bordered veins of the east belt never displays the intensely red pleochroism so often exhibited by cassiterite from the same type of deposits in the west belt, nor do the cassiterites from the east belt skarns, unlike their west belt equivalents, ever present anything but a brown pleochroism or non-pleochroic character. The spectacularly zoned cassiterite from Pelepahkanon Mine, Johore, in the east belt, which occurs in what are probably early feldspathic veins, shows brown pleochroism. Thus one is led to the conclusion that in the west belt the tin always originated from sources rather rich in tantalum, whereas in the east belt the tin was derived from different sources that were comparatively poor in tantalum and, judging by the rich concentrations of iron species that commonly accompany the cassiterite in all the types of primary deposit but the greisen-bordered veins, these sources were also rich in iron.

Because there is evidence that there are at least two and possibly more important epochs of tin mineralization in the west belt (Hosking 1977) and on every occasion tantalum-rich cassiterite was deposited, I believe that the sources of tin and tantalum in the west belt were crustal ones that were tapped whenever circumstances permitted.

There are also reasons for believing that tin was deposited on more than one occasion in the east belt, and I think that there also the sources of the tin were crustal ones, but unrelated to those of the west belt. The iron of the stanniferous deposits of this belt were, I think, largely derived from iron-rich units that were invaded by the granitoids, although it is likely that some of this iron was originally derived from the mantle.

In addition to what has been written above, the disposition of the cassiterites according to their pleochroism also suggests that the east belt may continue from Malaysia under the Gulf of Siam and reappear at Rayong in Mainland Thailand. The disposition also provides strong reason for believing that whilst Bangka is a part of the west belt, its neighbour Belitung is largely, or entirely, a portion of the east belt. The pattern also suggests that the Sumatran tin province may be a displaced portion of the west belt, the displacement having been effected as a result of the opening of the Gulf of Malacca.

5 Practical Aspects

In addition to providing further thoughts concerning the sources of the tin in the stanniferous deposits of Southeast Asia the results of the study have also a practical value.

During exploration for tin and/or tantalum/niobium deposits in SE Asia (and elsewhere) in which the study of stream sediments plays a part, the pleochroism of any cassiterite present in the sample indicates the nature of the primary deposit or deposits from which the cassiterite originated. It may also serve to suggest whether or not it would be worthwhile to search for primary and/or secondary accumulations of tantalum/niobium species in the area.

The presence of an appreciable concentration of red pleochroic cassiterite in a given sample of tin ore would indicate to the smelters that the resulting slag would be likely to be a saleable product on account of its Ta content.

The pleochroic character of cassiterite may sometimes prompt thoughts concerning the tin potential of the body from which the cassiterite originated. Thus, the large lepidolite-bearing pegmatites of the Phangnga area of peninsular Thailand, which have been described by Garson and Bradshaw (1970) contain a number of tantalum species and cassiterite. However, the cassiterite does not display the red pleochroism that would be expected were it truly syngenetic with respect to the pegmatite, instead "there is a wide range . . . in pleochroism in green and brown shades." This suggests that the pegmatite behaved as a preferred host for late cassiterite and so it may well contain considerably more of this tin species than would have been expected were the cassiterite syngenetic with respect to the major species of the pegmatite.

Acknowledgment. This small contribution to tin mineralogy is dedicated to my friend Professor Dr. Paul Ramdohr who has made such major contributions to our knowledge of the tin species.

References

Aranyakanon P (1962) The cassiterite deposit of Haad Som Pan, Ranong Province, Thailand. Rep Invest R Dep Mines Bangkok 4

Bignell JD (1972) The geochronology of the Malayan granites. Unpubl PhD Thesis, Univ Oxford

Garson MS, Bradshaw N (1970) Lepidolite pegmatites in the Phangnga area of Peninsular Thailand. In: A second technical conference of tin, Bangkok 1969. ITC, London, pp 325–343

Gotman JD (1938) On the properties of tin in connection with conditions of its formation.Byull Mosk Ova Ispyt Prir Sect geol 16:130–160

Hosking KFG (1977) Known relationships between the 'hard-rock' tin deposits and the granites of Southeast Asia. Geol Soc Malaysia Bull 9:141–157

Hutchison CS (1975) Granites of the Southeast Asian Tin Province. In: Abstracts of papers presented at the 1975 regional conference on the geology and mineral resources of Southeast Asia, Jakarta, pp 23–25

Priem HNA, Boelrijk NAIM, Bon EH, Hebeda EH, Verdurmen EATh, Verschure RH (1975) Isotope geochronology in the Indonesian Tin Belt. Geol Mijnbouw 54:61–70

Ramdohr P (1961) Magnetische Cassiterite. Bull Commun Geol Finl 196:473–487

Santokh Singh D, Bean JH (1968) Some general aspects of tin minerals in Malaysia. In: A technical conference of tin, London 1967. ITC, London, pp 459–478

Scrivenor JB (1928) The geology of the Malayan ore-deposits. MacMillan, London

van Overeem AJA (1960) The geology of the cassiterite placers of Billiton (Indonesia). Geol Mijnbouw 22:444–457

Yeap CH (1974) Some trace element analyses of West Malaysia and Singapore granites. Geol Soc Malaysia Newslett 47:1–5

New Observations on Ore Formation and Weathering of the Kamariza Deposit, Laurion, SE Attica (Greece)

H. MEIXNER and W. PAAR [1]

Abstract

Ore-microscopic and X-ray investigations of hand specimens from the Kamariza deposit, Greece, resulted in the discovery of a complex paragenesis of silver minerals in the primary ores. This includes acanthite, andorite, argentopyrite, diaphorite, dyscrasite, miargyrite, "owyheeite", pyrargyrite, (?) pyrostilpnite, silver-bearing tetrahedrite and at least two phases which probably represent new minerals. The silver minerals are associated with native arsenic, stibarsenic and a variety of other components.

A detailed study of secondary mineralization has, for the first time, revealed the presence of dundasite in ores from Kamariza.

1 Introduction

It can be assumed that mining at Laurion was started as early as 1000 B.C. In the Phoenician era already ore was mined about 800 B.C.; as from 483 B.C. the wealth and prosperity of Athens was largely based on silver production from the Laurion deposit. Ardaillon (1897), Wilsdorf (1952), Marinos and Petrascheck (1956) and, in abstracts, Kohlberger (1976) and Grolig and Grolig (1978) reported on the history of the old mines. In recent times Putzer (1948), and Marinos and Petrascheck (1956) (in Greek) give a detailed description of the geology and ore deposits of the region.

For a long time fine-grained, silver-bearing galena was the main ore of the mine, but recently sphalerite, which is a minor constituent became of greater economic importance. Putzer (1948, p. 38, 39) quoted for lead ore in 1918 90–150 g Ag/t Pb, in 1934 160 g Ag/t Pb and in 1939 216 g Ag/t Pb, but galena with a silver content of up to 30 kg Ag/t PbS was reported from Ancient times. Samples of a pillar in the ancient "Satyri/Kamariza" pit showed a content of 20 kg Ag/t Pb (Cordella 1869). The content of silver was considered "an impurity in galena"; secondary native silver has been recorded as the only, and very rare, silver mineral.

The ore deposits of Laurion, especially those of Kamariza and Plaka, are world-famous for their great variety of minerals. Putzer (1948, p. 43, 44), Marinos and Petrascheck (1956, p. 200, 206), Kohlberger (1976, p. 116, 124), and Grolig and Grolig

1 Institut für Mineralogie und Petrographie der Universität Salzburg, Akademiestraße 26, A-5020 Salzburg, Austria

(1978, p. 20, 24) have given comprehensive references. Three groups of minerals of quite different genetic history can be clearly distinguished:

1. The primary mineralization. Only a surprisingly small number of mineral species, both ore and gangue, have been recorded so far.
2. The oxidation zones. Particularly well-developed. Various rare secondary minerals, including carbonates, sulphates, arsenates, and silicates occur; these have made Laurion a world-famous mineral locality.
3. Secondary minerals, mostly lead chlorides. These are reaction products of salt water and ancient lead slags.

One of the co-authors had the chance to visit the Laurion region and particularly the Kamariza deposit under the competent guidance of Dr. F. Kanaki-Mavridou (Athens) in April 1972. He was allowed to select, in the abandoned survey office, some mineral species, at that time not identified. They have come from one of the Kamariza deposits and have been investigated in the course of this study. As early as 1968 a sample containing only As, AsSb, and galena, has been provided by Strasser (Salzburg).

2 Notes on a Primary Mineralization from Kamariza, Laurion

The observations are based on the study of two ore samples from Kamariza. The exact locality (mine adit, etc.) is unknown. Therefore, no data on economic geology of this mineralization are available to the author. Both specimens have been investigated by ore microscopy and X-ray analysis.

The larger specimen (12 × 9 × 5 cm) consists of reniform aggregates of *native arsenic – stibarsenic* with distinctly concentric layering. Superficially, native arsenic has been altered to a crust of microcrystalline *arsenolite* octahedra. Sprays of tiny *stibnite* clusters (several millimetres in length), brownish *sphalerite* and *galena* are

Fig. 1. a Concentric shell structures of native arsenic *(as)* – stibarsenic *(sbas)*. The rhythmic pattern is enhanced by air etching. Stibnite *(st)* in leached cavities of native arsenic and as intercalated layer. Plane polarized light (PPL). Length of bar of Fig. a–i, m–p is 100 μm. **b** Gel-like intergrowths of native arsenic *(as)* – stibarsenic *(sbas)* – galena *(ga)* in galena. PPL. **c** "Atoll"-like structures of native arsenic, around As precipitations with nuclei of dyscrasite *(dy)*. PPL. **d** Safflorite *(sa)* "stars" in galena *(ga)*. PPL. **e** Galena *(ga)* with complex rimming of various Ag sulphosalts: "owyheeite" *(ow)*, miargyrite *(mi)* and andorite *(an)*. The latter occurs as lath-like inclusions in miargyrite. **f** Intergrowth of native arsenic *(as)* and galena *(ga)*; the latter is replaced by andorite *(an)* and "owyheeite" *(ow)*. Miargyrite *(mi)* as inclusions in calcite. PPL. **g** Galena *(ga)* with replacement by a complex intergrowth of andorite *(an)*, "owyheeite" *(ow)* and miargyrite *(mi)*. PPL. **h** Same as **g**, but crossed polars. **i** Inclusions of pyrargyrite *(py)* and miargyrite *(mi)* in galena *(ga)*. PPL. **j** Enlarged section of **i** ("miargyrite inclusion"), with rimming of "owyheeite" *(ow)* and inclusions of argentopyrite *(ar) (black outlined)*. PPL. **k** Argentopyrite (cf. **j**), exhibiting excellent twinning lamellation. Crossed polars. **l** Lamellar intergrowth of argentopyrite *(ar)* and pyrargyrite *(py)*. PPL. **m, n** Idiomorphic diaphorite *(di)* with distinct twinning lamellation. Partially crossed polars. **o** Radiated stibnite crystals growing around idiomorphic miargyrite *(mi)*. PPL. **p** Xenomorphic sphalerite grains and framboidal pyrite in calcite. PPL

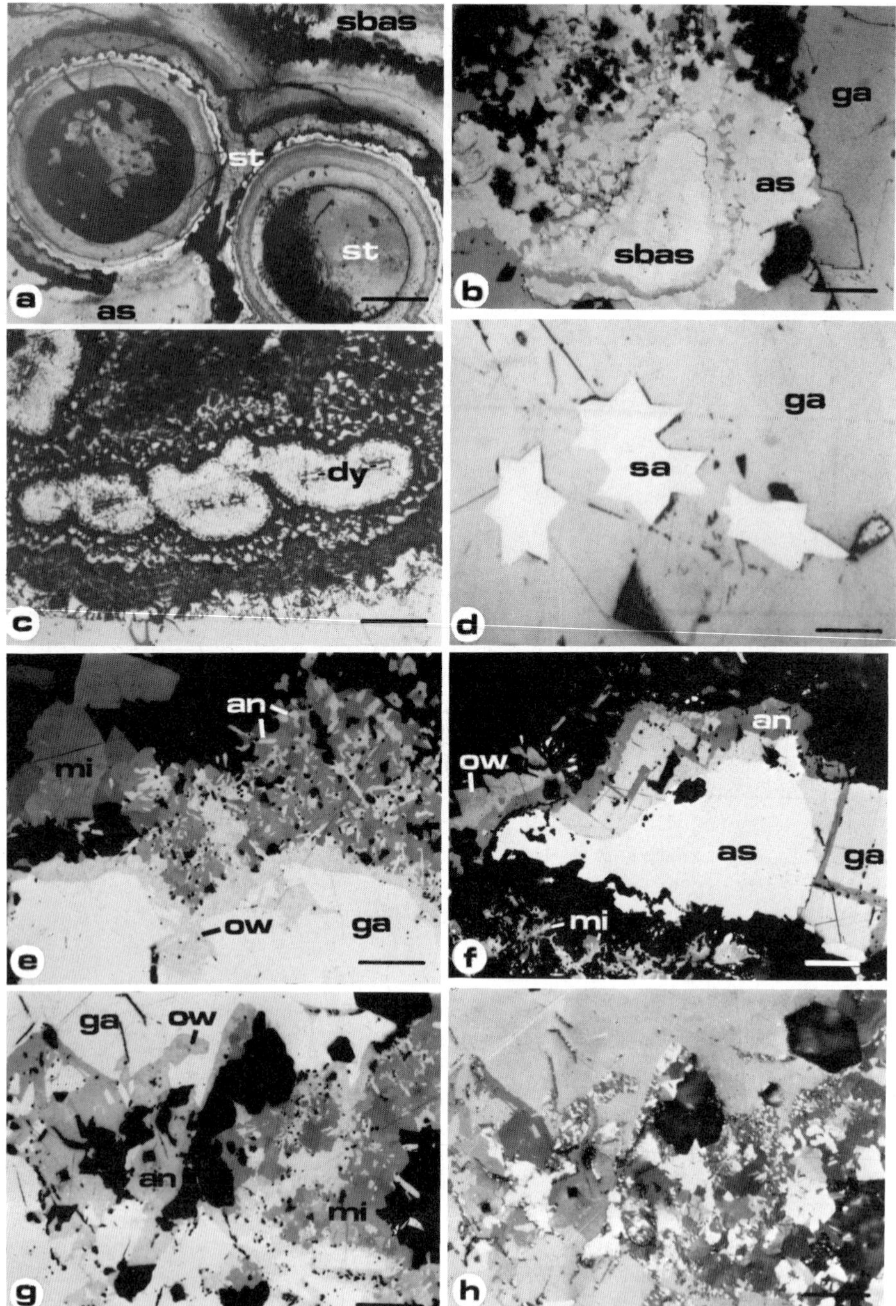

Fig. 1a–h (legend see p. 761)

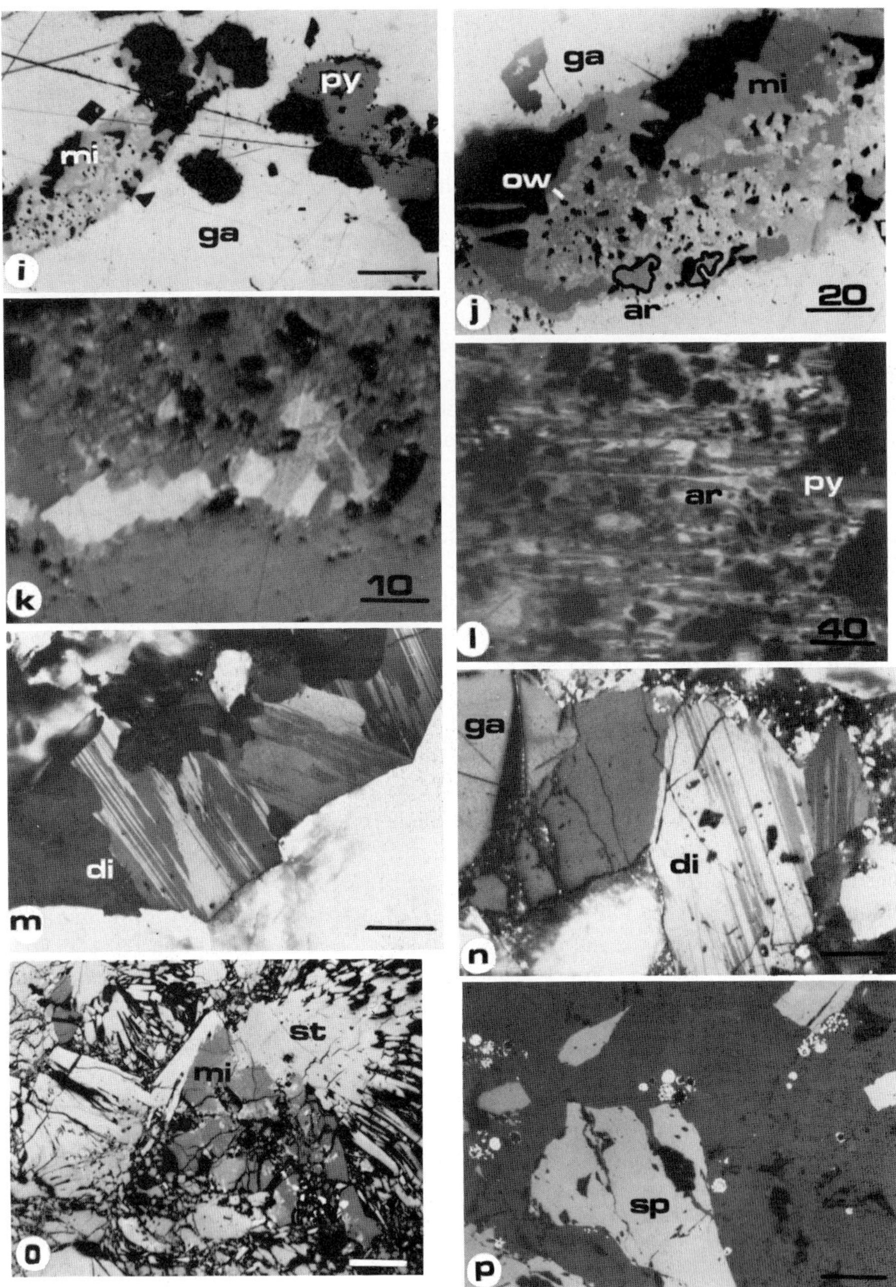

Fig. 1i–p (legend see p. 761)

present in minor amounts. Some of the stibnite crystals have been transformed to powdery *stibiconite.* Crystals of *calcite,* Fe-bearing *dolomite* and *quartz* are associated with the ores.

The following paragenesis has been observed microscopically (the phases are described in relation to the proposed crystallization sequence):

The earliest sulphides present in rather limited amounts are *Ag-bearing tetrahedrite, chalcopyrite* and *bournonite.*

Native arsenic, stibarsenic, and minor *antimony* (which together dominate all other components quantitatively) demonstrate in all details the typical structures and textures, so well documented by Ramdohr (1924, 1969), Van der Veen (1925), and others. Pronounced rhythmic structures (owing to the variable Sb-contents) are especially well visible by air etching (Fig. 1a). Zonally arranged xenomorphic grains of *galena* (Fig. 1b) and *safflorite* as starlike triplets (Fig. 1d) are the most frequently observed inclusions. *Dyscrasite* is rarely present as nuclei of roundish to elliptical crystal aggregates of *native arsenic* (Fig. 1c).

The crystallization of native arsenic and stibarsenic has been followed by major precipitation of sphalerite and galena, which usually contains relics of them. Galena of this stage carries numerous inclusions of Ag sulphosalts (Fig. 1g, i), which range in size from a few to 20–30 μm. The Ag-bearing assemblage consists, in order of decreasing abundance, of *miargyrite, diaphorite, andorite, "owyheeite", pyrargyrite, (?) pyrostilpnite, argentopyrite* and at least two phases, which do not correspond to known compositions. The larger portion of the Ag-sulphosalts occurs as intimate intergrowths rimming galena (Fig. 1e, f) and as isolated "islands" embedded in calcite (Fig. 1e).

Miargyrite is present as well developed idiomorphic crystals when associated with calcite (Fig. 1e). Very often it is completely surrounded by a narrow rim of a fine-grained 'owyheeite' like phase (Fig. 1e, j), which, in turn, is bordering galena. *Diaphorite* (Fig. 1 m, n) occurs as comparatively large crystals (± 1 mm) and exhibits excellent twinning lamellae after (?) (120). Reflectance (R_{589} = 35.7%–40.7%) and microhardness (VHN_{25} = 218–252 kp/mm^2; mean value = 241:10 indentation marks) are in good agreement with proved diaphorite from Příbram (C.S.S.R.). The powder diffraction diagram is identical (both intensities and d_{hkl}-values) with PDF 9–126 (diaphorite from Příbram). *Andorite* (Fig. 1e, g) and *pyrargyrite* (Fig. 1i) are observed as lath-like and xenomorphic grains, respectively. *Andorite* can be distinguished from diaphorite by different ore microscopic properties. Among others, stronger anisotropism (in greygreen-yellowish-brown colours) and a lower microhardness (VHN_{25} = 171 kp/mm^2; 10 i.marks) differ from diaphorite and thus are characteristic parameters. The powder diffraction diagram is very close to that of andorite from Takla Lake, British Colombia (PDF 10–499).

A mineral of the *"Silberkies"* group, probably *argentopyrite* (Quellmalz 1973) (brownish in air, distinctly birefringent, strong anisotropism: bluish and brownish colours; polishing hardness similar to galena), occurs rarely, associated with fine-grained brownish pyrite, miargyrite (Fig. 1j, k) and as lamellar intergrowths with pyrargyrite (Fig. 1 l). According to Ramdohr (1975) the latter texture can be interpreted as incipient decomposition of "Silberkies". Due to its small grain size it is impossible to isolate material for X-ray study.

Stibnite is one of the latest minerals crystallized in the sequence. Occasionally, euhedral inclusions of miargyrite (Fig. 1 o), safflorite, and rare overgrowths of *pyrostilpnite* are present.

3 A Dundasite Paragenesis from Kamariza

The sample (11 x 9 x 4 cm) consists largely of *limonite* (brown ochre), containing minor remains of unweathered *galena* (± 1–2 cm ϕ). Some *sphalerite, covellite* and *argentite* (in tiny grains and rimming galena) are visible microscopically, the latter contributing to the silver content of galena. Putzer (1948, p. 24) refers to ores such as galena, sphalerite, pyrite, and chalcopyrite. He points out that Au and Ag do not occur as sulphides.

Limonite is in places intensely overgrown by colourless, white and bluish, vitreous, globular and acicular secondary minerals. A very thin crust of globular, shell-like "Brauner Glaskopf" (goethite) can be noticed on limonite as substratum for other minerals. In polished sections it can be seen that the border zone of fissure fillings carries limonite intermingled with granular carbonate *(cerussite).* The vitreous crust is sometimes covered with white, cauliform *cerussite* crystals of only a few mm in size, but mostly with colourless, white to pale-bluish vitreous aggregates, dotted with radiated acicular clusters.

The vitreous globular parts are optically isotropic; they have N $\sim$ 1,490 (Na), contain major Al and can be termed *allophane,* $Al_2SiO_5 \cdot nH_2O$. The presence of this mineral was also verified for Kamariza by Daviot (1899).

Such allophane globules are often surrounded by radiating white acicular clusters. These globules, of about 1 to 2 mm in diameter, are surrounded by an acicular covering crust of 0.5 to 1.00 mm thickness. The crystals (0.003 x 0.01 – 0.02 x 1 mm) are lath-like, have straight extinction and carbonate-like birefringence. The optical parameters are $N_X \sim 1{,}600$, $1{,}740 > N_Z > 1{,}730$ (Na). The mineral effervesces, and is easily soluble, in dilute HCl. The above data suggest the mineral in question is *dundasite* $Pb_2Al_4(CO_3)_4(OH)_8 \cdot 3H_2O$, o'rhombic. X-ray analysis (Debye-Scherrer method), carried out by Dr. E. Kirchner, confirmed the determination. The powder-diffraction diagram is very close to that of dundasite from Dundas, Tasmania (PDF 21–936) with only minor deviations of the d-values (± 0.01–0.02 Å).

Dundasite is associated with small colourless grains of *cerussite.* According to Hintze (1930, p. 3530–3531) and Palache et al. (1951, p. 279–280) only few localities of dundasite have been recorded. Since then, a steady increase of new discoveries of dundasite has been noted. Characteristically, formation of dundasite seems to be restricted to environments where weathering products of galena (cerussite, etc.) occur with decomposed Al-bearing rocks (such as slates, etc.), which provided Al for the formation of allophane, $Al_2SiO_5 \cdot n\,H_2O$. Associations of dundasite and allophane are known from Wales, Cornwall, Zuckmantel/Zlaty Hora/Silesia, and now Kamariza/

Laurion.[2] The chemical composition of dundasite corresponds exactly to that recorded from these occurrences.

4 Conclusions

The preliminary results demonstrate that at least a considerable part of the Ag content of galena-sphalerite ores can be attributed to specific silver minerals. Precipitation of Ag took place at several stages: *Dyscrasite* was the first mineral to crystallize, followed by a variety of *(Pb)-Ag-Sb sulphosalts.* Though As had been present in comparatively large amounts (as As, SbAs), (Pb)-Ag-*Sb* sulphosalts have been formed exclusively. Deposition of Sb-bearing phases terminated with formation of *stibnite* during the waning stages of the mineralization. Secondary enrichment processes resulted in the formation of argentite-acanthite. All these silver minerals are younger than a sulphide association (chalcopyrite, tetrahedrite, bournonite, etc.), which is of very limited importance only in our samples. The colloidal textures of several phases (As, SbAs, particularly PbS) point to low temperature deposition with possibly a slight increase in temperature during crystallization of the various Ag sulphosalts. This is in accordance with the investigations by Wilke (1952) on the genetically similar mineralization of St. Andreasberg, West Germany. Intense weathering produced a great variety of secondary minerals at Laurion, such as arsenates, sulphates, etc. The number of these is now enlarged by the discovery of the rare new Pb-Al-hydrocarbonate, *dundasite!*

Acknowledgments. We are especially grateful to the "Fonds zur Förderung der wissenschaftlichen Forschung" (FFF), Vienna, which provided the instruments used in the ore-microscopy studies (grant No. 1062). Dr. E. Kirchner carried out the X-ray work, A. Geyer and I. Lueginger are responsible for the photography, Prof. E.F. Stumpfl revised the English text.

References

Ardaillon E (1897) Les mines du Laurium dans l'Antiquité. Bibl Ecol Fr Athenes Rome 77

Cordella A (1869–1871) Le Laurium. Marseille. (Ref Berg Hüttenmänn Ztg 1869, 209, see Z Prakt Geol 1896, 153)

Daviot (1899) Contribution à l'étude du Laurion. Bull Soc Hist Nat Autun 12:331–431

Grolig H, Grolig E (1978) Laurion in Attika. Lapis 3:16–25

Hintze C (1930) Handbuch der Mineralogie, Bd I/3/1. de Gruyter, Berlin Leipzig, pp 3530/3531

Kohlberger W (1976) Minerals of the Laurium Mines, Attica, Greece. Mineral Rec 7:114–125

Marinos GP, Petrascheck WE (1956) Laurium (Laurion). Geol Geophys Res 4(1):247 (in Greek with Engl. summary)

2 This paragenesis is verified by two occurrences just discovered in Austria: Dundasite crystals, one-tenth of the size of those at Kamariza, intermingled with cerussite and allophane have been found at Martinsbau near Guttaring and at Pleschitzen near Hirt, both in the province of Carinthia.

Palache Ch, Berman H, Frondel C (1951) The system of mineralogy, 7th edn, vol II. Wiley – Chapman and Hall, New York London, 1124 pp
Putzer H (1948) Die Erzlagerstätte von Laurion. Ann Geol Pays Hell 2:16–46
Quellmalz W (1973) Mineralogische Untersuchungen zum Problem der „Silberkiese". Abh Staatl Mus Mineral Geol Dresden 20:249
Ramdohr P (1924) Beobachtungen an opaken Erzen. Arch Lagerstaettenforsch 34:30
Ramdohr P (1969) The ore minerals and their intergrowths. Oxford London New York, 1174 pp
Ramdohr P (1975) Die Erzmineralien und ihre Verwachsungen. 4. Aufl. Akademie Verlag, Berlin, 1277 pp
Veen RW van der (1925) Mineragraphy and ore deposition, vol I. Naeff, The Hague, 167 pp
Wilke A (1952) Die Erzgänge von St. Andreasberg im Rahmen des Mittelharz-Ganggebietes. Beih Geol Jahrb 7:184
Wilsdorf H (1952) Bergleute und Hüttenmänner im Altertum. Akademie Verlag (Springer) Berlin, 284 pp

Bravoite and Co-Ni-Bearing Marcasite and Their Zoning Relationships

D. SCHACHNER-KORN[1]

Abstract

Bravoite and Ni-Co-bearing marcasite may display zoning relationships. Zoning in bravoite is due to changing contents of Ni and Co in the ore forming solutions. If the zoning is regular it is possible to determine the developed crystal forms by microscopic observation of polished sections. The shape of the crystals seen in the section may, however, be ambiguous with respect to the form (Fig. 1). It is possible to recognize habit changes developed during the growth of the crystal (Fig. 2).

Marcasite similar to pyrite incorporates Ni and Co in its crystal lattice. If it can develop idiomorphic faces, layers with varying contents of Ni and Co may be arranged parallel to the r and l faces which are commonly built up (Fig. 5). Where marcasite replaces zoned bravoite, the Ni and Co contents remain in the same places. Replacement of bravoite by marcasite follows the cube.

1 Zoned Bravoite

Zoning of bravoite has been observed long ago. It is described often on account of its varying appearance. It is mentioned usually in the titles of publications on ore deposits, even if it is only an accessory mineral. Marcasite with zones containing Ni and Co is, on the contrary, very rare (Schachner and Springer 1967).

Zoning of bravoite may be caused by changing contents of these two elements. That is manifested in varying reflectances and hues on microscopic examination. Much work has been done to correlate the Ni and Co contents with reflectances and colours.

The possible causes of zoning were explained on the basis of theoretical considerations. Disulphides with compositions and geometrical features similar to natural bravoites have also been synthesized (Clark and Kullerud 1959; Springer 1961).

Bravoite crystallizes in the form of cubes as well as pyritohedra. Combinations of these two forms are common. Bravoite is more brownish in comparison with pyrite and has no metallic lustre. In cavities, spherical aggregates are formed. They are built up of pyritohedra, which intergrow not exactly parallel (subparallel). Small cube faces flatten the edges. In numerous cases I observed the indices to be (320)[2] and not (210).

1 Institut für Mineralogie und Lagerstättenlehre der Technischen Hochschule Aachen, 5100 Aachen, FRG

2 Henglein assigns (320) already in 1914. Kalb and Meyer also in 1926 and, in detail, Kalb in 1952

In nearly all papers with figures of zoned bravoites, remarks are made about habit and habit changes. The difficulties in explaining the crystal forms from cross-sections are self-evident. In consideration of existing literature it is perhaps useful to give some examples showing this ambiguity of cross-sections (Korn 1933; Zimmermann and Amstutz 1973).

a) The cross-section perpendicular to a crystal axis of a pyritohedron (perpendicular to a twofold symmetry axis) can be a square or a rectangle if not traversing through the centre. The successive layer with compositional difference will be cut as a hexagon, provided its thickness is at least like the distance of the small pyritohedron on face (320) to the centre (Fig. 1). If the zone is smaller, the cross-section presents an octagon. Such a succession, square or rectangle inside and hexagon outside, does not necessarily indicate a habit change from cube inside to pyritohedron outside (see Kalb 1951, p. 560, Fig. 1b).

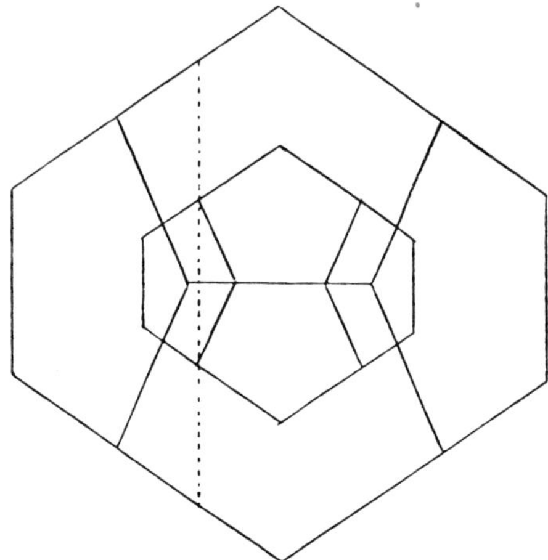

Fig. 1. A section vertical to the drawing and parallel to the finely dotted line is a square for the small pyritohedron, and a hexagon for the outer form

The cross-section of Fig. 1 is not a regular hexagon. It has two planes of symmetry normal to one another. Regular hexagons would arise in a cross-section through the centre of a pyritohedron (320) or (210) perpendicular to a threefold symmetry axis. A bravoite grain with a regular hexagon as cross-section can be just cut normal to a threefold symmetry axis of a cube or octahedron through the centre of the crystal. If not cut exactly through the centre the hexagon has also six 120° angles, but sides of two sorts of length, alternating.

b) An example for the possibility to find out the orientation of a section is given in Fig. 2. In this case only the change of habit provides the identification of the position.

Figure 2 demonstrates a section through a bravoite grain perpendicular to a threefold axis. The white zones are pyrite with small Ni content. The darker layers have varying contents of Ni and Co. The outer hexagon is regular and corresponds to the cube. The next successive dark layer displays additional three faces of the pyritohedron. The faces are only a little inclined to the cross-section and therefore seem broader and distinctly zoned. The next following layers show up distinctly a combination of cube/pyritohedron. In the centre no faces of the cube are seen. Hence, inside dominates the pyritohedron, outside the cube. The angles between the traces of the cube and the pyritohedron indicate that the pyritohedron has the indices (320) and not (210). The two angles between traces of (210) and (100) are 19° and 41°. The two angles between (320) and (100) are, in contrast, not so different, namely 26° and 34°. They are compatible with Fig. 2.

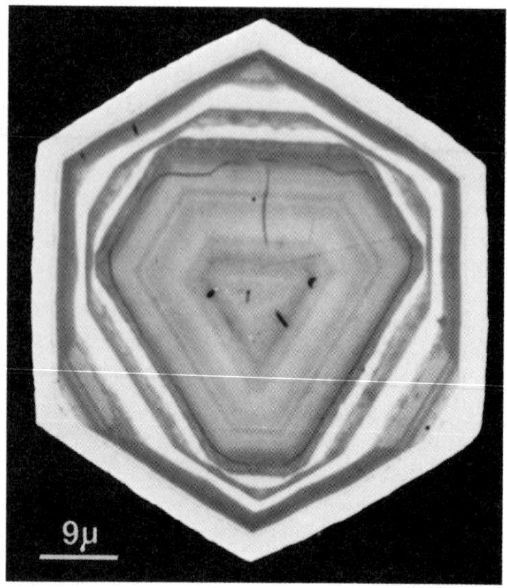

Fig. 2. Bravoite with habit change. Section vertical to the threefold axis. Combination: pyritohedron with cube. Outside cube only

2 Zoned Marcasite with Ni- and Co-Bearing Layers

In the strata-bound Pb-Zn ore deposits of Maubach and Mechernich bravoite is very common. It displays many types of habit. Sometimes bravoite forms large grains with broad zones. In this case it cements always several grains of quartz. In these so-called Bravoitknotten marcasite is almost constantly present. Its orientation is related to the shape and structure of bravoite. This shall be described in the following text.

a) Within the bravoite grain there exist irregularly confined parcels of marcasite. The zones of bravoite traverse these parcels of marcasite in the same direction and breadth. They are optically anisotropic and similarly oriented. Marcasite is here secondary, and replaces the bravoite respectively. The content of Ni and Co in the zones of lying in marcasite is small, up to 2.8% of Ni and 1.7% of Co. The amount in the corresponding zones of bravoite are nearly twice as high (Fig. 3).

b) Thin marcasite lamellae can replace the bravoite in two or three directions. These lamellae are parallel to the lines of triangular pits of cubic cleavage; the replacement is controlled by the cube, respectively. The zones are incorporated by growing marcasite lamellae (Fig. 4).

Fig. 3. Bravoite. The *light spot below left* is marcasite. It replaces zoned bravoite and preserves the zonation

c) Figure 5 demonstrates an unusual development of zoned marcasite. Layers containing Ni and Co are parallel to the idiomorphic boundary. The faces r and l are developed. [The cross-section is lying ± parallel to (100) because of the effect of anisotropy.] These idiomorphic zoned marcasites are primary and not the result of a replacement of bravoite as in examples (a) and (b). They are surrounded by "feldergeteiltem" (parquetted) bravoite, which is later in crystallization. Its varying reflectance (brownish and violet tints) is due to variation in composition. The contents of Co and Ni in bravoite are high. The measured concentrations are: Co 17%–21%, Ni 5%–7%. The layers in marcasite have only concentrations up to 1.7% of Co and 2.8% of Ni.

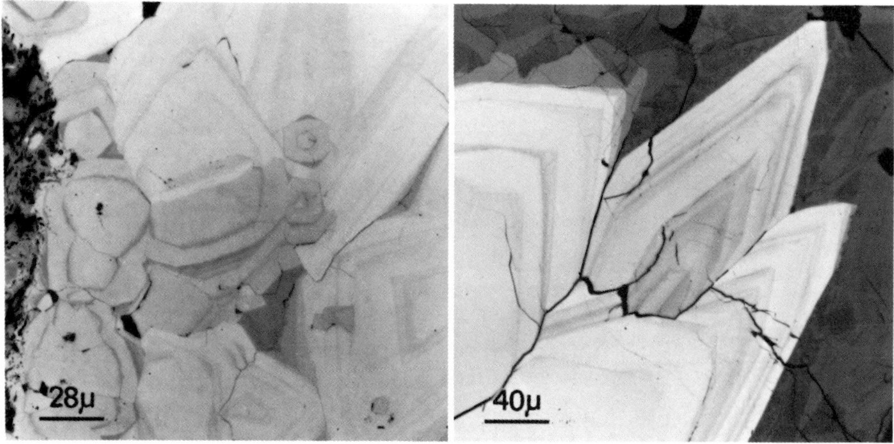

Fig. 4 **Fig. 5**

Fig. 4. Bravoite, *dark grey.* Marcasite, *white.* In the centre, marcasite replaces zoned bravoite

Fig. 5. Euhedral marcasites with layers of Ni and Co contents surrounded by bravoite with high Co and Ni contents

The analyses were made with an Electronmicroprobe GEOSCAN provided by the Deutsche Forschungsgemeinschaft.

References

Clark LA, Kullerud G (1959) The Fe-Ni-S-System, the phase relations between pyrite and vaesite in the presence of excess sulphur. Carnegie Inst Washington Yearb 58:142–145

Henglein M (1914) Über Kobaltnickelpyrit von Müsen im Siegenschen, ein neues Mineral der Kiesgruppe. Zentralbl Mineral 129–134

Kalb G (1951) Kristallmorphologie als Kriterium für Isomorphie. Naturwissenschaften 38:560–561

Kalb G (1952) Die Kristalltracht von Bravoit und Pyrit als Kriterium zur Beurteilung der Isomorphie. Neues Jahrb Mineral Monatsh 90–93

Kalb G, Meyer E (1926) Die Nickel- und Kobaltführung der Knottenerzlagerstätte von Mechernich. Zentralbl Mineral Abt A 26–28

Korn D (1933) Die Lagebestimmung opaker Minerale im Mikroskop. Lehrbuch der Erzmikroskopie. In: Schneiderhöhn H, Ramdohr P (eds) Lehrbuch der Erzmikroskopie, Bd I. Borntraeger, Berlin, S 265–296

Schachner D, Springer G (1967) Kobalt- und nickelhaltiger Markasit. Neues Jahrb Mineral Monatsh 152–154

Springer G (1961) Hydrothermalsynthesen im System Fe-Ni-Co-S. Dissertation, Aachen

Zimmermann RA, Amstutz GC (1973) Relationship of section of cube, octahedra and pyritohedra. Neues Jahrb Mineral Abh 120:15–30

Bismuth-Tellurium Associations: New Minerals of the Wehrlite-Pilsenite Assemblage from Hungary

K.I. SZTRÓKAY[1] and B. NAGY[2]

Abstract

The Te-Bi minerals associated with the hydrothermal-sulphide mineralization of the andesite range on the left bank of the Danube (Börzsöny Mts.) have in recent years been restudied and the individual mineral phases have been reclassed according to a more uniform approach. The constant silver content known even earlier is associated with the telluride phase, too. The ore nests or pockets rich in bismuth tellurides never represent homogeneous mineral phases. Because of the "hyperelectronic" nature of Bi and Te there arises a possibility for successive variations, a succession whose members are characterized by a gradual increase of S, appearing along with Te. And this holds true for all members including the final one, fitting in this succession even structurally, and constantly represented by Bi_2S_3. Relying on microprobe results, the authors recommend the materials of hitherto known bismuth telluric occurrences to be revised.

1 Introduction

Deposits of bismuth-tellurium compounds have for long been known to be confined to some ore parageneses associated with the volcanism in the inner zone of the Carpathians. It is also well known that, after samplings and discoveries by the Hungarian natural scientist, P. Kitaibel, and determination by M.H. Klaproth (1802), the discovery of the element Te was also connected with one of these deposits. Our investigations have concerned excavations, both old and new, of that deposit which yielded the first Te-containing mineral sample and which has been the single telluride ore mineralization ever known in Hungary. Out of the results here reported, primarily the latest microanalyses of Bi-Te-S compounds are worthy of attention. To present our relevant results would be timely also because, in spite of the rather abundant literature devoted to bismuth-tellurium mineral phases, there are scores of questions that are still to be cleared. This holds particularly true for the genetic circumstances of the deposits.

Obviously, this was the reason why P. Ramdohr (1975), in the 4th edition of his famous book, took an initiative in eliminating the controversies associated with the group of bismuth-tellurium compounds and in urging for measures to settle the problem, as pointed out, with keen insight, in the evaluation given by him.

1 Department of Mineralogy, Eötvös Lóránd University, Budapest 1088, Muzeum krt 4/a, Hungary

2 Hungarian Geological Institute, Budápest 1142, Népstadion ut 14, Hungary

The timeliness of the subject became evident when large-scale geological-geophysical research, using data of earlier ore mining, had been started in the Börzsöny Mountains. The results to be reported hereinafter were achieved under this research project.

2 Geological Sketch of the Börzsöny Mountains and Genetic Characterization of the Ore Mineralization

The geological setting of the Börzsöny is shown in Fig. 1.

It should be added to the above geological portrayal that beneath the Paleogene and Neogene volcano-sedimentary sequence the substratum of the mountains is constituted, in the north, by Lower Paleozoic crystalline granitoids and, in the south, by a Triassic carbonate (dolomite and limestone) complex. The two different basement complexes have their junction along a tectonic line running across the centre of the territory, and

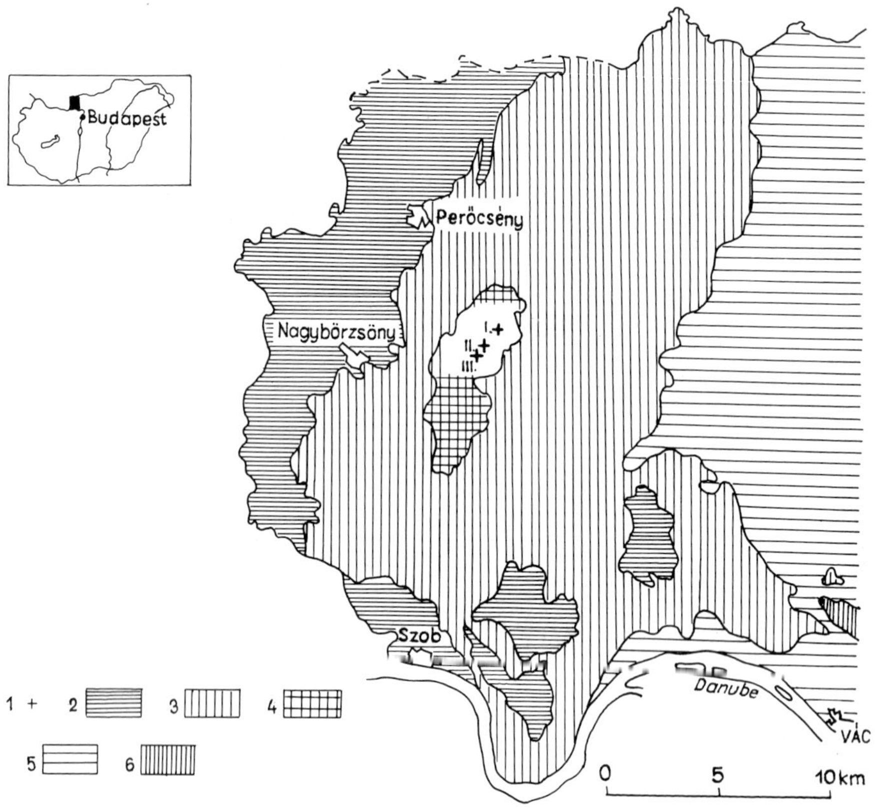

Fig. 1. Geological map of the Börzsöny's complex bismuth-telluride ore mineralization areas. *1* Ore mineralization areas, *2* Neogene (Upper Miocene) sediments, *3* Neogene (Miocene) stratovolcanic products, *4* Paleogene volcano-sedimentary products, *5* Oligocene sediments, *6* Triassic limestone

it is with this fault zone that the hydrothermal ore mineralization products within Paleogene volcanics are associated (Nagy 1978).

This ore-mineralized area is rimmed, in the west and south, by later (Neogene) andesite extrusion products and shallow-water marine sediments. Two distinct magmatic activities can be recognized as episodes provoked by tectonic deformation. Products of the earlier, Paleogene, volcanism are dacites to hornblende andesites and their pyroclastics as well as the associated sequence of argillaceous-marly sediments. The second paroxism, in the Neogene (Miocene), produced strictly stratovolcanic pyroxene andesite masses which are overlain by latest Miocene marine sediments. The products of this formation complex have obviously covered the whole study area. The final episode of the same tectonic phase appears to have brought about even a subvolcanic intrusion beneath the so-called Paleogene window. This process resulted in an upheaval of the Neogene "cover" and simultaneous metallogenic processes took place in the Paleogene sequence. In this area the heaviest erosion removed the Neogene, so that the Paleogene ore-bearing complex has been exposed to daylight (Czakó and Nagy 1976).

Earlier, there were more specialists dealing with ore formation in the area. While compiling this study, mainly the works of Koch and Grasselly (1952) and Pantó and Mikó (1964) were used by us.

3 Ore Parageneses

In the zone of the points on the chart in the Paleogene andesites and dacites, products of three ore mineralization phases of different character and temperature are known to occur (Nagy 1978).

1. Kuruc-patak. A copper ore paragenesis of porphyric habit occurring in hornblende andesites and biotitic hornblende dacites. Of the dispersed and stockwork-type ore minerals, magnetite, hematite, chalcopyrite, bornite, molybdenite and pyrite might be mentioned, being accompanied by quartz, calcite, anhydrite, fluorite and the zeolites, mordenite and desmine.

2. Rózsa-bánya. In the Rózsa-bánya area two assemblages of minerals of impregnation type characterized by stocks and "nests" have been discovered by detailed analytical studies: (a) The first ore generation is represented by galena, sphalerite (coarse-grained marmatite), chalcopyrite, pyrrhotite and pyrite. (b) The second ore mineralization phase is characterized by products of higher-temperature differentiation. Bi and Te compounds appear here, too. It is the arsenopyritic mineral assemblage that may be said to have played the principal role in the process associated, again, with basement rocks, i.e. characterized by a kind of rejuvenation. This ore mineralization came into existence partly as a result of the consumption, "replacement", of the earlier generation, so that the reactions of the two ore mineralization phases produced a rich mineral paragenesis (Pantó and Mikó 1964).

Consequently, the bismuth-tellurium minerals (tetradymite and its associates) appear within the assemblage of this second, rather peculiar, ore mineralization phase: hessite, plenty of arsenopyrite, cosalite, emplectite, gold, glaucodote, löllingite, cubanite, chalcopyrite-II, schapbachite, valleriite, sphalerite-II, stannite, galena-II, tetradymite, tennantite, bournonite, jamesonite, jordanite, meneghinite, pyrargyrite, proustite, sartorite, semseyite, stephanite, sternbergite, native As, marcasite-melnikovite. A finely fibrous tourmaline fabric in quartz and hydromica with tourmaline might be quoted as associates.

3. Fagyosasszony-bánya. The Fagyosasszony-bánya deposit is situated 550 m southwest of the previous ore deposit. Its veined hydrothermal ore paragenesis is substantially simpler: the ore mineral assemblage corresponds to the products of the previously outlined "first" ore mineralization: sphalerite (marmatite), galena, pyrrhotite, chalcopyrite, and pyrite. The genetic features suggest expressively meso- to epithermal circumstances. The associated (secondary) minerals are characterized by the predominance of clay minerals and carbonates.

Leaving aside the practical implications of the research and ore prospecting results achieved in the study area, the present paper has been intended to deal only with the latest observations concerning the Te-Bi compounds.

4 The Te-Bi Mineral Group

As emphasized by P. Ramdohr (1975), the information available on the Bi-Te-S system is insufficient. In his review he lists the phases hitherto figuring as minerals, and he suggests the idea of a solid solution formed at higher temperatures of which bismuthinite would differentiate at lower temperatures, so that the Bi_2Te_2S–Bi_2Te_3 mixture would be separated into tetradymite and tellurobismuthite. Later, in the chapter dealing with the "other" bismuth–tellurium compounds, he takes into consideration the observations and attempts at settling the problem made by Warren and Peacock (1945) and Sztrókay (1946). He believes that the phases poorest in S, wehrlite, and hedleyite are nearly identical ($\sim$ BiTe), while the additional phases: joseite Bi_3TeS, grünlingite Bi_4TeS_2, and oruetite Bi_8TeS_4, were earlier given somewhat arbitrary names. Se-containing ikunolite $Bi_4(S,Se)_3$ and laitakarite Bi_4Se_2S, phases distinguished later, might also be included in this category.

Besides the references quoted, a number of the publications were, in the meantime, devoted to the problem of these peculiar crystallization phases. Some workers (Glatz 1967) did even attempt to crystallize the elements Bi-Te synthetically. It should be mentioned here that the structural and physical properties of all these phases – e.g. low hardness, lamellar-layered habit, rhombohedral crystal structure, cleavage according to the basal plane – and even their mechanical translation characteristics, are nearly the same. Opaque surface reflexion white, with slightly yellow to light grey shades; reflectability strong, anisotropy also nearly identical, so that the individual members cannot be distinguished from one another, unless mounted in oil immersion. With a view to all these circumstances, we believe to have good reason to unite these compounds into one group and to examine them on the basis of identical crystal chemical principles (except for Ag_2Te) and, at the same time, to point out the cases in which it is justified to maintain the respective mineral name or, in case of identical element ratios, to rely on the principle of priority.

Last year, the staff of the Research Laboratory for Geochemistry, Hungarian Academy of Sciences, was so kind as to perform the microprobe analysis of telluride ore samples and of the two specimens of the original "wehrlite" sample of the deposit, together with a series of ore samples from the Börzsöny. The results are given in Table 1. Selected microanalyses are shown on Figs. 2 and 3.

Table 1. Electron microprobe analyses

	Phase 1 Wehrlite	Phase 2 Bismuth-telluride	Phase 3 Joseite	Phase 4 Tetradymite	Phase 5[a]	Phase 6 Bismuthinite	Phase 7 Hessite	Phase 8[a] a	b
	%	%	%	%	%	%	%	%	%
Bi	66.0	70.5	73.1	59.9	77.8	80.2	0.2	64.6	63.3
Ag	0.3	–	–	–	–	–	59.2	11.7	13.4
Te	34.0	28.7	22.2	35.6	14.2	0.1	38.9	0.4	0.5
S	–	–	2.9	4.2	6.1	18.0	–	17.2	16.0
Pb	–	–	–	–	–	–	–	4.4	4.7
Fe	–	–	–	–	–	–	–	–	0.5
	100.3	99.2	98.2	99.7	98.1	98.3	98.3	98.3	98.4

[a] New mineral phases now being elaborated in more detail. (The electron microprobe analyses were performed in 1979 at the Laboratory for Geochemical Research, Hungarian Academy of Sciences)

Listed according to the tabulation, the phases calculated from the analyses are the following:

1. BiTe wehrlite (pilsenite, tsumoite)
2. Bi_2Te_3 tellurobismuthite
3. Bi_4Te_2S joseite-B
4. Bi_2Te_2S tetradymite
5. Bi_3TeS_2 + New phase
6. Bi_2S_3 bismuthinite
7. Ag_2Te hessite

8a–b. 15 $Bi_2S_3 \cdot 5\ Ag_2S \cdot PbS$ ++ New (Te-free) phase

First of all, the investigation has corroborated the earlier statement (Sztrókay 1946) that the bismuth-tellurium compounds are in almost every case represented by two or more members in one and the same genesis. The phenomenon is mentioned in a number of references not quoted here. It was, however, not until the latest analyses (X-ray, electron microprobe) that these members could be separated or their relations to one another could be shown. Structure studies by Harker (1934) showed the rhombohedral symmetry of tetradymite and its layered pattern manifested in good cleavage. They also pointed out the sequence of atomic layers which, in turn, defined, in terms crystal physics, the sphere of telluride relations. L. Pauling's crystal chemical reasoning published recently (1975) is essential, as it may serve as a basis to substantiate the homeotypy inferred by Strunz (1963) from the subequal lattice constants (a_o, c_o, Z),

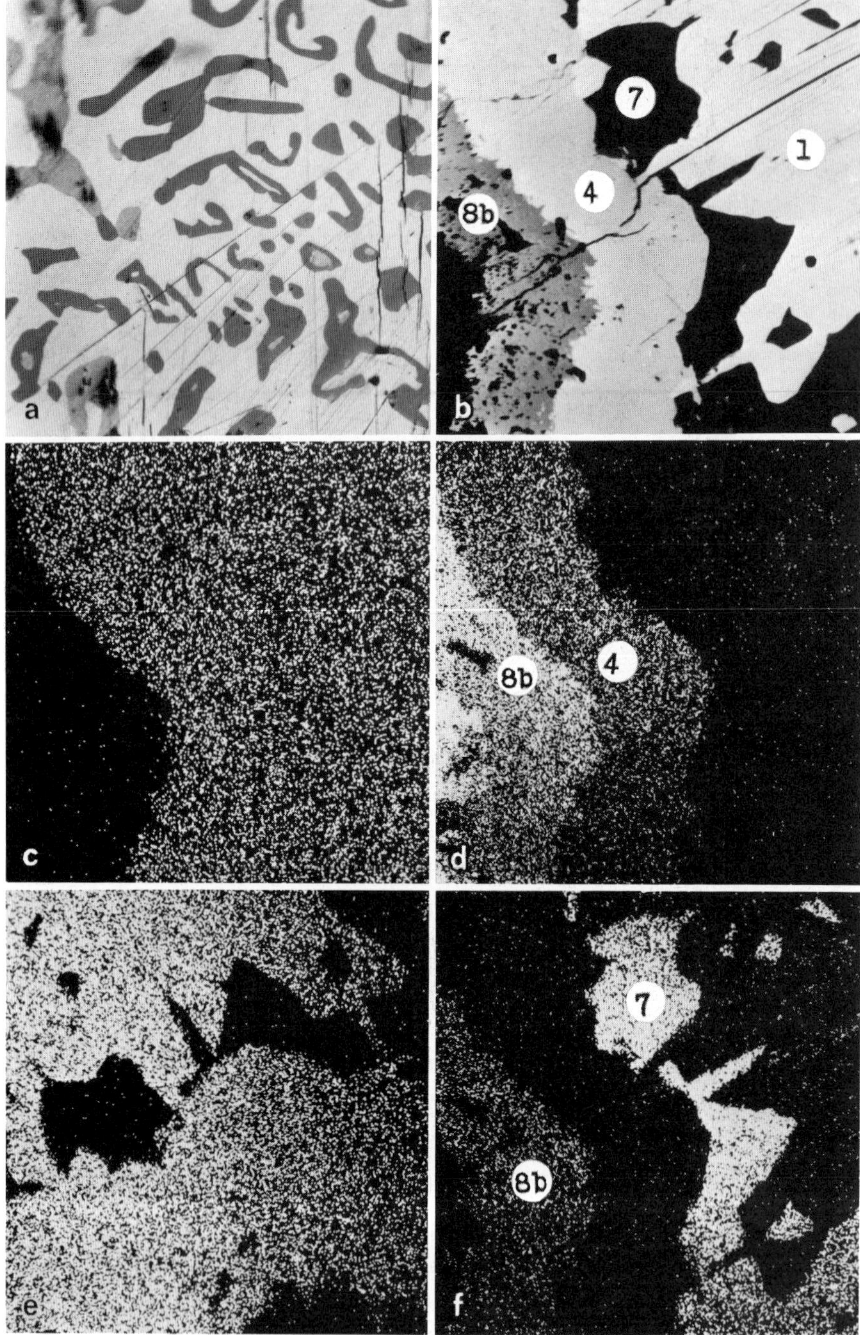

Fig. 2. **a** Differentiation of bismuthinite as viewed with the ore microscope. *On the right,* the lines of cleavage cross the Bi_2S_3 particles as well. Oil immersion. × 400. **b** Composition image. × 300. Phases: *1* wehrlite; *4* tetradymite; *7* hessite; *8b* new complex sulphide. Back-scattered electron photograph. **c** Scanning pattern of Te from **b**. **d** Scanning pattern of S from **b**. **e** Scanning pattern of Bi from **b**. **f** Scanning pattern of Ag from **b**

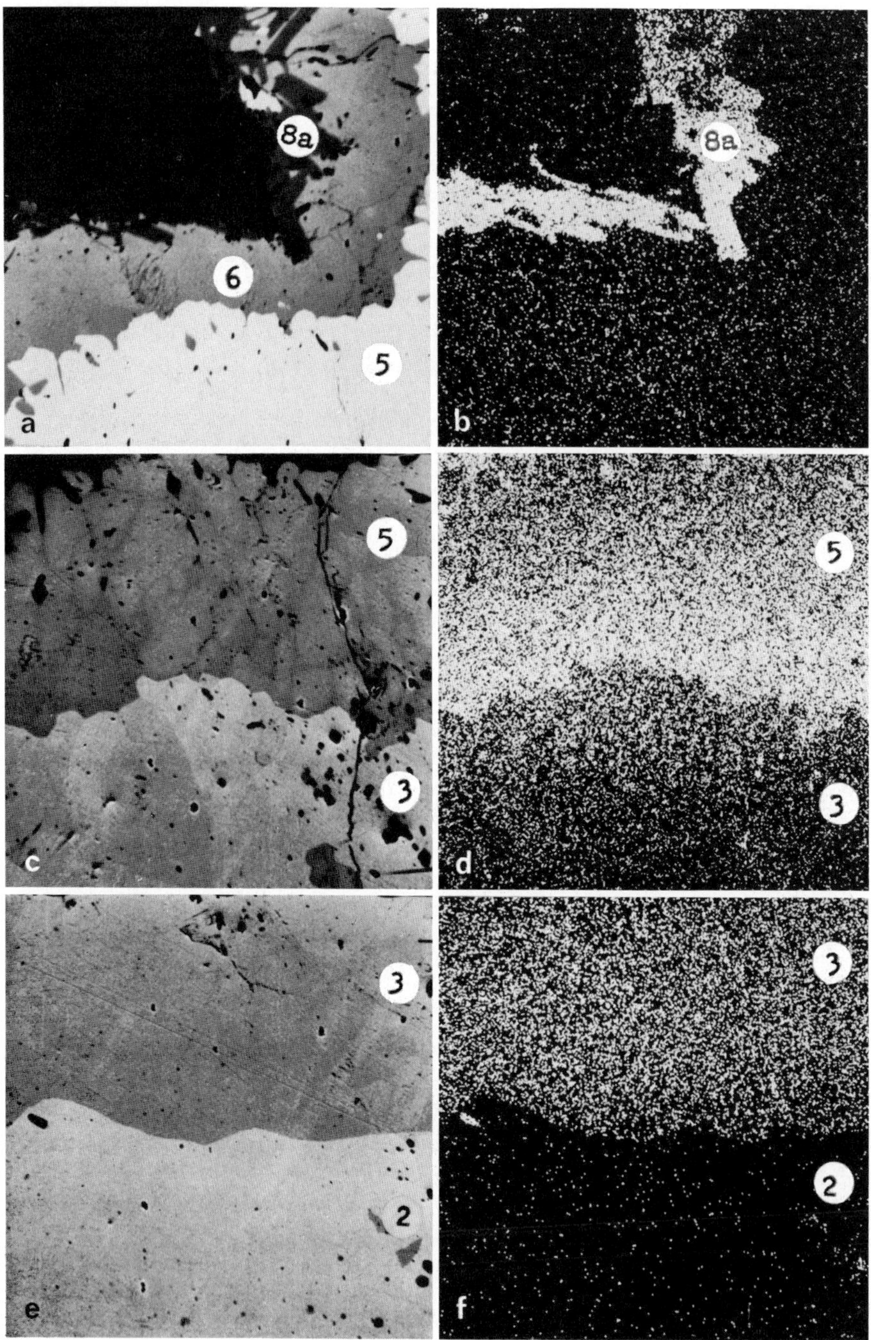

Fig. 3. a Composition image. × 300. Phases: *5* new S-containing bismuth telluride, *6* bismuthinite, *8*a new complex sulphide. **b** Scanning pattern of Ag from **a. c** Composition image. × 300. Phases: *5* new S-containing bismuth telluride; *3* joseite. **d** Scanning pattern of S from **c. e** Composition image. × 300. Phases: *3* joseite; *2* bismuth telluride. **f** Scanning pattern of S from **e**

and, in fact, to corroborate the existence of Bi-tellurides thus far considered stoichiometrically controversial.

The bond spacings observed in the tetradymite structure by Harker (1934) somewhat differ from the sum of radii of covalent single bonds. L. Pauling ascribes this to bond angle strain and sees in it an example of the metallic state. The interaction of adjacent elements in substances of this kind of metallic nature will provoke resonance bonds between alternative positions of elements. In tetradymite, according to Pauling, no transfer of electrons would take place between them either, owing to the hyperelectronic nature of the elements S, Te, and Bi. Their normal covalencies are 2 in the case of S and Te, and 3 in that of Bi when the bivalent S atom is bonded to 6 Bi atoms. Consequently, the two single covalent bonds resonate among the six positions, leading to bond number $n = \frac{1}{3}$. The Bi atom uses one of its three covalent bonds, while sulphur and the other two elements are bonded to the three adjacent telluric atoms, resulting in $n = \frac{2}{3}$. Bond angle strains will be brought about in the well-known five sheet Te-Be-S-Bi-Te sequence (Glatz 1967) which can be eliminated only by an arrangement in which each seventh atom is replaced by S in the two outer Te layers and Te is substituted for every seventh S atom in the middle S layer. In other words, the resulting sheet sequence will be Te_6S-Bi_7-TeS_6-Bi_7-Te_6S, hence the real formula, $Bi_{14}Te_{13}S_8$, giving composition of stable natural tetradymite. This being known, it is also essential that other variants may also satisfy the previously outlined conditions for a rhombohedral lattice cell pattern in the planes of sheet bonds of a lattice of $R\overline{3}$ symmetry and this will be really the prerequisite for the homeotypy of the elements Bi, Te, and S (Se), i.e. for their structural and genetic unity. At the same time, this provides a clue for understanding the opaque characteristics of the phases, e.g. their high reflectivity, etc.

5 Known and New Phases of the Paragenesis

What is worthy of attention in the microanalyses of members of the telluride assemblage formed in the II/a-b phase of the Börzsöny ore mineralization is the marked change compared with the information so far available. The higher temperature of the second of the two successive ore mineralization processes of the Rózsa-bánya deposit agrees with the genetic conditions of other known telluride-precious metal ore geneses (e.g. Sweden, Japan, Canada, etc.), and the results of Glatz's (1967) thermic (580–630°C) and phase equilibrium studies of the Bi_2Te_3-Bi_2S_3 system approximately coincide with the metallogenetic estimates.

As a new result, BiTe has been discovered among the earlier (Sztrókay 1946) ore mineralization phases (Fig. 2b) and Bi_2TeS_2, then defined as an unknown phase but later described as "csiklovaite" (Koch 1948), now can be replaced by other homeotypical compounds, as inferred from the microanalyses. Furthermore, a complex sulphide arises consistently from our measurements, representing the reaction product of ore margins.

1. BiTe, wehrlite figures for the first time in P. Ramdohr's list (1975), although he refers to it as pilsenite (= wehrlite). It should be pointed out that only the term wehrlite is correct. To use the name of the Pilsen community is erroneous and the first name must be used even by virtue of the principle of priority (wehrlite – Huot 1841; pilsenite – Kenngott 1853).

Lately, Japanese authors dealt with the compound (Shimazaki and Ozawa 1978) and, seemingly unaware of the relevant references, they referred to it as tsumoite. We need not say that this term is obsolete, more precisely, we have to recommend IMA's New Minerals Commission the tsumoite = wehrlite modification. It is more correct to write the formula in the form Bi_2Te_2. Consequently, if $a_o = 4.42$, $c_o = 24.05$ and $Z = 3$, then a_o will indicate a random Te substitution and its value will fit between Bi $a_o = 4.46$ – Bi_2Te_3 $a_o = 4.39$ Å. Thus, its lattice sheet number will be 12 and so $c_o/12 = 2.04$ Å, which fits excellently in the homeotypical series of trigonal Bi-Te compounds. The different value, $c_o = 30$ Å, of the Canadian wehrlite mentioned (Berry and Thompson 1962) in the excellent description of the mineral of Tsumo Mine as a difference of diagnostic value seems to be due to the disturbing effect of homeotypical associates.

2. Tellurobismuthite, Bi_2Te_3, is represented in a relatively low amount (Fig. 3e, f). Its structure was examined by several scientists. Lattice constants identical with those of the natural phase resulted even during the synthesis of the inclusion: $c_o = 30$ Å, its lattice sheet number being 15.

3. The formula Bi_4Te_2S corresponds completely to the joseite phase determined by Peacock (1941) which we distinguish from the earlier formula (Kenngott 1856) by using the symbol *B*; on the basis of $a_o = 4.34$, $c_o = 40.83$, $Z = 3$ and the (3×7=)21 atomic layers forming its lattice structure, it fits very well, with $c_o/21 = 1.95$ Å, in the succession. On Fig. 3 its compositional X-ray pattern scanning is marked with *3* as phase symbol, its SK radiograph being also indicated on Fig. 3. Adjacent to joseite-*B* is a (new) phase richer in sulphur, from which its distinction is a little obscure; here again, on the lower pair of images, the compound Bi_2Te_3 is the neighbour, the boundary of this being sharper.

4. Bi_2Te_2S, tetradymite (γ-phase). Its quantity within the assemblage is surprisingly poor, though it does not lack totally anywhere. Its position in the succession of differentiation is shown by Phase *4* on composition images of Fig. 2b.

5. Bi_3TeS_2: a new phase in the succession. With Pauling's substitution, its lattice consists of six atomic layers. Its cleavage, reflection, are completely similar to the other telluride members; a Te-containing phase most abundant in sulphur. In reality, it is the extreme member of the sulphur-saturation succession, which is reflected also by the microprobe analyses. The increasing abundance of sulphur is best indicated by the joseite → Bi_3TeS_2 transition (Fig. 3d).

6. Bi_2S_3, bismuthinite. A constant phase of bismuth-tellurium mineral assemblages. It is always associated with the S-richer members. Its fit with the telluride compounds is structurally oriented. Whenever there is an excess of Bi or a deficiency of Te in the Bi-Te-S system, the sulphide Bi_2S_3 will appear as a product of intergrowth, as already pointed out in an earlier paper (Sztrókay 1946). In the same paper the transformation $Bi_2S_3 \leftrightharpoons Bi_2Te_3$ (0001) was demonstrated in terms of lattice topology. When viewed with ore microscope, it can be clearly seen that the lines of cleavage continue in bis-

muthinite or they traverse this (Fig. 2a). In most cases, bismuthinite appears on the outer margin, poorest in tellurium, of the succession (Phase 5, a new phase) (Fig. 3a).

7. Ag_2Te, hessite. The ores of numerous Bi-telluride deposits contain varying quantities of Ag. The bond of silver entering the lattice of elements of hyperelectronic nature cannot be stable. Its only possible bond can be its being bonded, as type A_2B, to the more metallic tellurium. A peculiar, continuous zone of intergrowth will result between Te-rich wehrlite and tetradymite, showing the outward succession BiTe-Ag_2Te-Bi_2Te_2S (Fig. 2b–f).

8a–b. 15 $Bi_2S_3 \cdot 5\ Ag_2S \cdot PbS$. A new phase representing a sulphide of the elements Ag and Pb with predominance of Bi. A typical product of ore mineralization, it always appears on the margin bordering on the barren rock (Figs. 2b and 3a). Its optical reflection is lower even than that of bismuthinite; its inner structure is porous, with gas-filled cavities, and it contains barren grainlets as inclusions. At some points, the outlines of a lath-like lamella are recognizable (Fig. 3a, b).

The appearance of this new phase is further evidence corroborating the above argumentation. As a complex phase, it is the final episode of the genetic process characterizable in the succession of homeotypical phases by a gradual decrease of Te and an increase of S (and Bi). This latter ore mineral has incorporated Pb (and probably also Ag), as a higher-temperature (homogeneous) segregation derived from products of the first ore mineralization process.

6 Genetic Succession of Bismuth-Tellurium Minerals

In the foregoing those changes of the Bi-Te-S proportions were repeatedly referred to which, with preservation of the rhombohedral lattice frame and Pauling's (1975) bond forms, will lead to the formation of different mineral phases.

During the present study the following succession could be established within the mineral paragenesis of the Rózsa-bánya deposit in the Börzsöny Mountains:

a) Bi_2Te_3 – – Bi_4Te_2S – – Bi_3TeS_3 – – Bi_2S_3 – – Bi,Ag,Pb-sulphide
tellurobismuthite → joseite → new phase → bismuthinite → new phase

b) Another succession of the same trend:
BiTe – – (Ag_2Te) – – Bi_2Te_2S – – Bi,Ag,Pb-sulphide
wehrlite → hessite → tetradymite → new phase

The first succession (a) is represented by Phases 2, 3, 5, 6, and 8a on Fig. 3. On the left side, three composition images aligned vertically; on the right, the $K\alpha$ radiographs obtained for S and Ag in the same succession.

The second succession (b) is illustrated by the composition images of Fig. 2 and the relevant Te, Bi, Ag and S patterns.

7 Summary

In connection with the Paleogene and early Miocene andesite volcanism of the Börzsöny Mountains, Hungary, a succession of metallogenic processes took place, in one of which a Te-Bi ore mineral assemblage was formed at the subvolcanic level. Recurring, though at higher temperature now, at the location of an earlier Pb-Zn-(Cu) ore concentration, these processes produced a new ore mineral paragenesis with two new mineral phases identified within the resulting bismuth-tellurium succession. The ore-rich samples of the paragenesis exhibit a distinct grading in favour of the sulphur content. And this genetic picture is consistent with the physico-chemical control of the differentiation sequence.

References

Berry LG, Thompson RM (1962) X-ray powder data for ore minerals. The Peacock Atlas. Geol Soc Am Mem 85:281

Czakó T, Nagy B (1976) Correlation of phototectonic and metallogenic data from the Börzsöny Mountains (Hungarian). M All Foeldt Intez Evi Jel 1974 Evröl, pp 47–60

Glatz AG (1967) The Bi_2Te-Bi_2S_3 system and the synthesis of the mineral tetradymite. Am Mineral 52:161–170

Harker D (1934) The crystal structure of the mineral tetradymite. Z Kristallogr 89:175–181

Huot JIN (1841) Man Mineral Paris 1:188

Kenngott A (1853) Das Mohs'sche Mineralsystem. Carl Gerold und Sohn, Wien, S 121

Kenngott A (1856) Uebersicht der Resultate mineralogischer Forschungen im Jahr 1855. Leipzig, S 111/112

Klaproth MH (1802) Beiträge zur chemischen Kenntnis der Mineralkörper 1795. I:253, III:1

Koch S (1948) Bismuth minerals in the Carpathian Basin. Acta Mineral Petrogr 2:1–23

Koch S, Grasselly Gy (1952) The minerals of the sulfide ore deposit of Nagybörzsöny. Acta Mineral Petrogr 4:1–21

Nagy B (1978) Mineralogical, geochemical and metallogenetic study of ore mineralization types in the Börzsöny Mountains (Hungarian). M All Foeldt Int Evi Jelentése 1976 Evröl, pp 77–93

Pantó G, Miko L (1964) The Nagybörzsöny ore mineralization. M All Foeldt Int Évk 50:1–154 (Hungarian)

Pauling L (1975) The formula, structure, and chemical bonding of tetradymite, $Bi_{14}Te_{13}S_8$, and the phase $Bi_{14}Te_{15}S_6$. Am Mineral 60:994–997

Peacock MA (1941) On joseite, grünlingite, oruetite. Univ Toronto Stud Geol Ser 46:83–105

Ramdohr P (1975) Die Erzmineralien und ihre Verwachsungen. 4 Aufl. Akademie Verlag, Berlin

Shimazaki H, Ozawa T (1978) Tsumoite, BiTe, a new mineral from the Tsumo mine, Japan. Am Mineral 63:1162–1165

Strunz H (1963) Homöotypie Bi_2Se_2-Bi_2Se_3-Bi_3Se_4-Bi_4Se_5 usw. (Platynite, Ikunolite, Laitakarite). Neues Jahrb Mineral Monatsh 154–157

Sztrókay K (1946) Über den Wehrlit (Pilsenit). Ann Hist Nat Mus Hung 39:75–103

Warren HV, Peacock MA (1945) Hedleyite, a new bismuth telluride from British Columbia. Univ Toronto Stud Geol Ser 49:55–69

Rutile of Postmagmatic Mineral Formation

G. UDUBASA[1]

Abstract

This paper summarizes significant relationships of rutile-bearing assemblages in the late- and post-magmatic environments. The inherited nature, i.e., the derivation of TiO_2 from the pre-existing minerals, such as ilmenite, is assumed for the most rutile-bearing assemblages including various types of magmatic rocks, porphyry copper ores, rocks of igneous contacts, gold-bearing quartz ores of hydrothermal origin. The stability of rutile associated with pyrrhotite, pyrite, chalcopyrite, and sphene covers a relatively large PT field within a rather narrow range of sulphur fugacity.

1 Introduction

The stability relationships of the TiO_2-modifications including rutile, anatase, brookite, and TiO_2-II are not fully understood. Lindsley (1976, p. L-68) states that "the published 'phase boundaries' between rutile and anatase or brookite reflect kinetics rather than equilibrium". The rutile stability seems, however, to be superior under geologically significant conditions. Anatase is an apparent lower pressure and temperature phase (Dachille et al. 1968), although Jamieson and Olinger (1969) noted the anatase to be everywhere metastable with respect to rutile. The brookite formation has been proven to be strongly dependent on the stabilizing influence of sodium (Keesmann 1966, cited by Lerz 1968).

Ramdohr (1975) has suggested that the rutile represents an inherited mineral in the most natural occurrences. Lawrence and Savage (1975) conclude, however, that the rutile of the porphyry copper ores was deposited together with chalcopyrite.

Over 400 polished sections of rutile-bearing rocks and ores were microscopically studied. The samples include various types of magmatic rocks, porphyry copper ores, hornfelses, and near-contact igneous rocks, as well as gold- and base metal-bearing vein quartz of hydrothermal origin.

1 Institute of Geology and Geophysics, Bucharest–32, Romania

2 Rutile in Magmatic Rocks

Early, Newhouse (1936) and Ramdohr (1940) minutely described the accessory oxides and sulphides in igneous rocks. The more recent papers of Haggerty (1976a, b) provide an exhaustive examination of the oxides in magmatic rocks and of their transformations during subsequent oxidation. The oxide evolution usually ends with the apparition of rutile often associated with hematite. Elsdon (1975) assigned magnetite + rutile to the less common assemblages, although Rumble (1972, cited by Elsdon 1975) thermodynamically predicted that it must be more stable with respect to hematite + ilmenite. The stability of magnetite + rutile seems, however, to be greatly a function of the sulphur fugacity.

The primary magnetite + ilmenite intergrowths of some Alpine subvolcanic calc-alkaline rock series of north and west Romania exhibit gradual transformation by increasing intensity of deuteric and hydrothermal and supergene alterations (Fig. 1). Some differences do occur between Haggerty's (1976a) and this scheme of evolution of the oxides. Homogeneous titanohematite is shown to form firstly within the ilmenite ss (Fig. 1, II–IV). A similarly developed lamellar phase in the ilmenite was ascribed by Haggerty (1976a) to the ferrian rutile. In our rocks fine lamellae of titanohematite may occur without other phases. This mineral always shows a greenish tinge and the difference in the reflectivity with respect to the ilmenite host is greater than that between rutile and ilmenite. In addition, the titanohematite and the (ferrian) rutile rarely do occur together, rutile being as a rule a later phase.

The formation of rutile + hematite and/or rutile + sphene probably depends on the initial composition of the ilmenite ss and on the possibility of Ca extraction from the silicates. In case of sulphurization sphene becomes unstable with respect to rutile. Under supergene conditions rutile + sphene aggregates become more and more fine-grained (i.e. leucoxene), but by reheating they may recrystallize and the grain size especially of the sphene increases considerably (Fig. 1, IVc).

At very low temperature anatase and/or leucoxene form instead of rutile. The bleached leucoxene occurring in nonsulphide/sulphate environments may perhaps represent the naturally occurring alpha-titanic acid, i.e. the hydrated titanium oxide, $TiO_2 \cdot H_2O$. This compound forms from cold neutral solutions (Marcu 1979). The similar Sn-compound, the hydrocassiterite, has been proven to exist in nature (Moh and Hutchison 1977). In some oxidized rutile- and sulphide- (mainly pyrite) bearing rocks the stable assemblage seems to be anatase + goethite.

In the intensely sulphurized magnetite + ilmenite intergrowths the more stable phase is the ilmenite. Ramdohr (1975) shows beautiful examples of relict ilmenite lamellae in bornite from Ooikiep, Cape Province. The magnetite is completely (Fig. 248a) or partly (Fig. 248b) replaced by bornite. A similar feature exhibit some magnetite + ilmenite intergrowths from Sudbury. The magnetite was replaced both by sulphides (chalcopyrite and pyrrhotite) and silicates in which strictly oriented ilmenite lamellae appear (Fig. 2). The inherited nature of such ilmenite lamellae is beyond doubt.

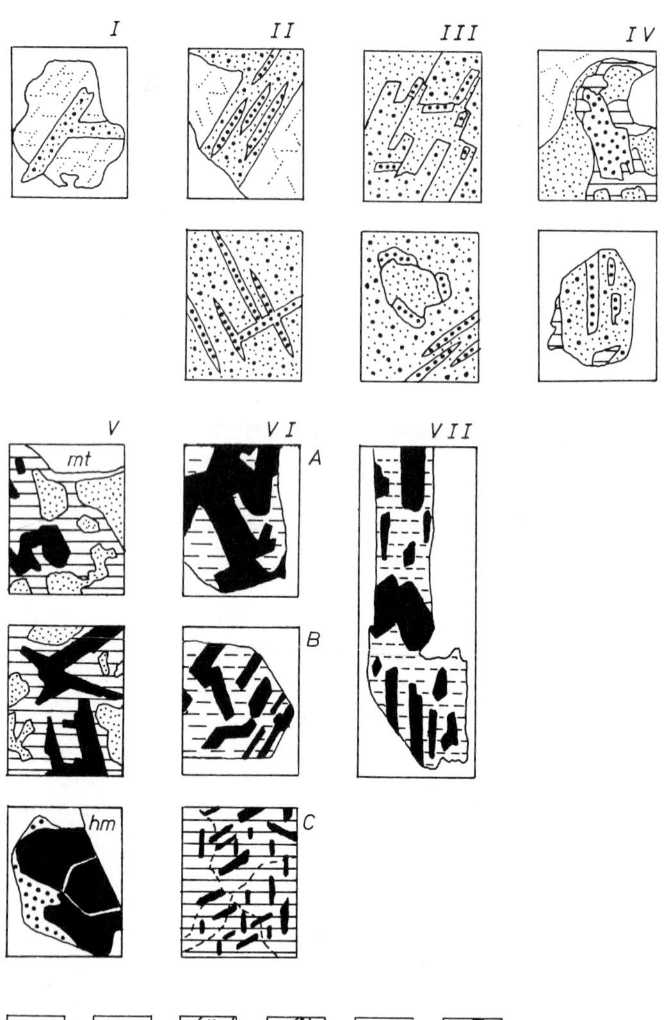

Fig. 1. Oxide transformation in subvolcanic rocks of andesitic composition (Tibles, Vlădeasa, and Metaliferi Mts., Romania). *I–VII* Stages of oxide evolution during deuteric *(II–III)*, autometamorphic *(IV–V)*, hydrothermal *(VI)*, and supergene *(VII)* alteration of the rocks. The evolution trends of both the sandwich-type lamellae of ilmenite and of the independent ilmenite grains are generally similar. Magnetite does not change until stage *VI* is attained. Under strongly oxidizing conditions magnetite transforms, however, into martite *(hm,* in *VC)*, which is probably coeval with titanohematite and ferrian rutile.

1 magnetite with cloth microtexture and without ilmenite fine lamellae *(mt)*; *2* high-temperature ilmenite solid solution series; *3* ilmenite, nearly pure end-member of the ilm_{SS}; *4* titanohematite (homogeneous hematite solid solution); *5* sphene (the length of the *dashed lines* in *VI* and *VII* reflects the decreasing grain size); *6* rutile (in *VII* it is very fine-grained and shows very peak bireflectance and anisotropic effects, i.e. leucoxene).

The enlargement of these microscopic drawings varies between 0.02 and 0.1 mm

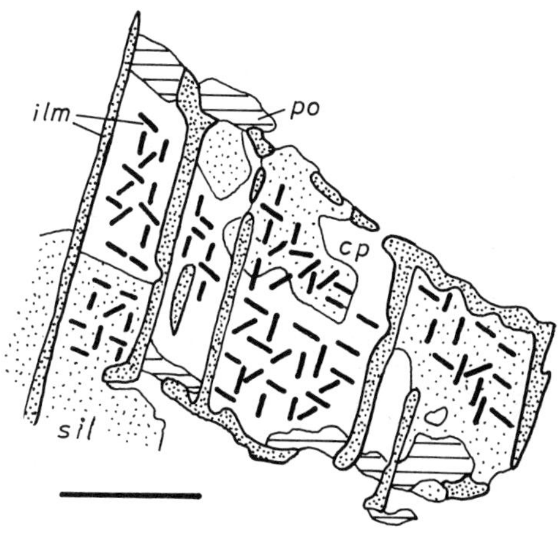

Fig. 2. Intergrowth of pyrrhotite *(po)*, ilmenite (*ilm*, both sandwich type and cloth microlamellae), silicates *(sil)* in chalcopyrite host *(cp)*. No magnetite present, being replaced both by cp+po and sil. Note the continuation of the *ilm* lamellae from *cp* into the *sil*. Sudbury, Clara Bella mine. Scale bar is 0.1 mm

3 Rutile in Porphyry Copper Systems

In order to explain the constant presence of rutile in some porphyry copper deposits Lawrence and Savage (1975) assume "a hydrothermal environment unusually rich in titanium" (p. 11). The rutile amount in such ores rarely exceeds 1.5%. This figure is of the same order with the TiO_2 content of the most frequent producers of the porphyry copper systems, i.e. granodiorites (0.48%), diorites (0.93%), and monzonites (1.11%) (Jackisch 1969). Therefore, there is no real titanium enrichment in the copper porphyries but the overall occurrence of rutile in such mineral assemblages may still point to a characteristic mineral feature of the porphyry type ores.

Microscopic examination carried out on polished sections of some porphyry copper ores (Butte, Montana, USA; Medet, Bulgaria; Recsk, Hungary; Deva, Roşia Poieni, and Bolcana-Troiţa, Romania) showed the persistent occurrence of rutile. It forms mainly intergrowths with sulphides and more seldom monomineralic fine-grained aggregates. The most impressive feature of these assemblages is the presence of oriented rutile lamella in pyrite and/or chalcopyrite (Fig. 3). Locally developed ilmenite remnants are indicative of rutile derivation from ilmenite of the intensely sulphurized primary intergrowths of magnetite + ilmenite. Intimate association of rutile and chalcopyrite, as shown by Lawrence and Savage (1975) and in Fig. 3c–d, suggests a subsequent recrystallization of the minerals accompanied by deferritization and partial replacement of rutile by chalcopyrite.

4 Rutile at Igneous Contacts

Around the Tibleş Neogene igneous complex, northern Romania, there exist Paleogene blackish shales containing carbonaceous matter, framboidal pyrite, and scattered anatase grains. At the contacts with the igneous rocks (monzodiorites, microdiorites, granodiorites, etc., Edelstein, Pop et al. 1979) the only opaque minerals observed are

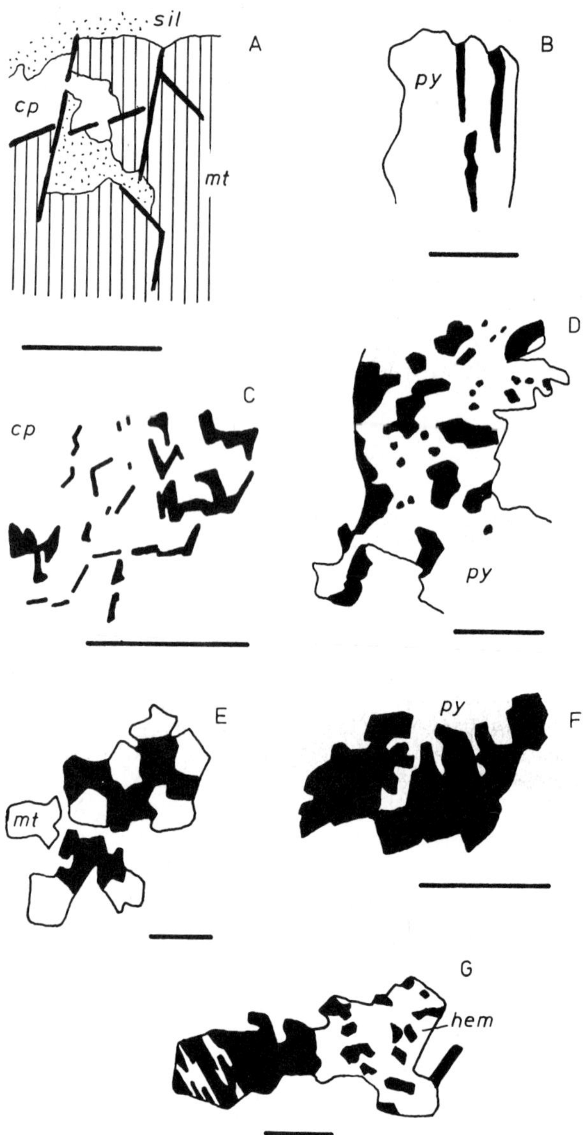

Fig. 3A–G. Rutile-bearing assemblages of the porphyry copper ores. **A** Magnetite *(mt)* partly replaced by chalcopyrite *(cp)* and silicates *(sil)*. The rutile lamellae *(black)* derive from the former ilmenite (sandwich type lamellae) in magnetite (compare Fig. 2). Bolcana-Troiţa, Romania. Scale bar is 0.05 mm. **B** Coarse irregular rutile lamellae in pyrite *(py)*. Butte, Montana. Scale bar is 0.025 mm. **C** Rutile inclusions in *cp*. The three directions of the ru lamellae are still recognizable although some replacement occurred. Bolcana-Troiţa. Scale bar is 0.015 mm. **D** Irregular rutile inclusions in pyrite. Roşia Poieni, Romania. Scale bar is 0.1 mm. **E** Rutile-magnetite assemblage. Bolcana-Troiţa. Scale bar is 0.01 mm. **F** Rutile inclusions in pyrite. Recsk, Hungary. Scale bar is 0.025 mm. **G** Rutile-hematite assemblage. Roşia Poieni. Scale bar is 0.02 mm

rutile and pyrrhotite. It is further interesting to note that these minerals appear both within the eruptive rock and the hornfelsized blackish shale, sometimes showing similar form and grain size (Fig. 4). The intense thermal recrystallization resulted thus in the homogenization of the ore minerals within the contacting rocks.

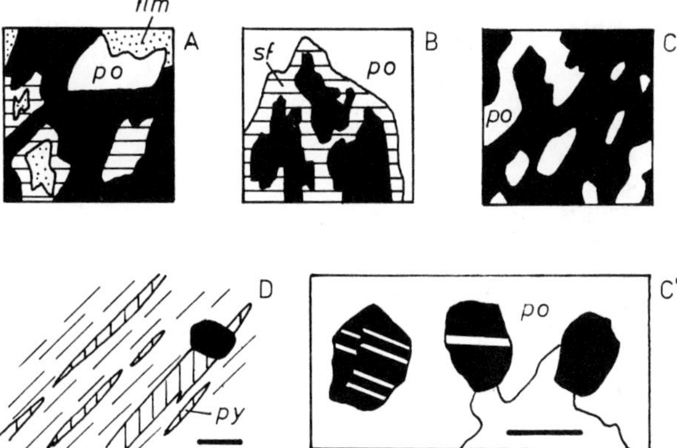

Fig. 4A–D. Rutile-bearing assemblages at igneous contacts, Tibles eruptive complex, Romania. **A–C** are stages of ilmenite and sphene replacement by pyrrhotite with increasing amount of rutile *(black)*. Near-contact monzodiorite (**A**), monzodiorite porphyry (**B**), and microdiorite (**C**). **C** is hornfelsized slaty marl with twinned rutile grains. Scale bar is 0.01 mm. **D** Elongate pyrite lenses *(py)* and rutile grains on the cleavage of an amphibole of a near-contact microdiorite porphyry. Scale bar is 0.01 mm. Size of **A, B** and **C** is about 0.025 × 0.025 mm

The rutile forms at the expense of the anatase of the shale and from the accessory titanomagnetite grains of the igneous rock. The relatively strong sulphurization of the contact zones of this eruptive complex results in the replacements of the magnetite by pyrrhotite. Gradual deferritization of the ilmenite lamellae or grains leads also to the formation of the rutile. The source of iron for pyrrhotite is the framboidal pyrite which gradually transforms towards the igneous contact. Further alteration of pyrrhotite to pyrite and, under supergene conditions, to goethite induces the topotaxic transformation of rutile into anatase. The stable assemblage is now anatase + goethite, which often appears also within the limonitic masses of the oxidized sulphide deposits.

5 Rutile in Hydrothermal Ores

"Rutile is rare in hydrothermal veins (proper in some places in gold-quartz veins), but forms frequently in the wall rocks from ilmenite, etc." (Ramdohr 1969, p. 985). The rutile occurrence at Barza-Carpen, Romania, a low temperature gold-quartz vein deposit of Miocene age, confirms this supposition (Udubasa 1978). Here the rutile formation was assigned to the hydrothermal process, according to the Lawrence and Savage (1975) hypothesis of the late ebullition of titanium into the hydrothermal systems.

Further examination of many polished sections of ore samples from either this or other gold- and base metal-bearing ore deposits of Miocene age in Romania shows the persistent appearance of the rutile in the gold ores, whereas very little or no rutile was observed in the base metal-bearing hydrothermal ones.

Rutile forms usually fine grains (up to 0.1 mm in size) showing normal bireflectance and anisotropic effects as well as simple twinning lamellae. The grains are embedded in quartz, and other ore minerals are generally lacking. Occasionally, the rutile grains show "neutral boundaries" with respect to pyrite. More seldom, needle-like rutile inclusions in pyrite were observed too (Fig. 5). The last feature reveals that the derivation of TiO_2 from the titanomagnetite grains of the host rocks forming xenoliths in the vein quartz cannot be excluded.

Fig. 5A, B. Subhedral pyrite grains in vein quartz with irregular inclusions of rutile (**A**) and with oriented rutile lamellae (**B**). **B** is supposed to represent a completely pyritized magnetite grain with ilmenite lamellae now transformed into rutile. Săsar ore deposit, North Romania. Scale bars are 0.1 (**A**) and 0.03 mm (**B**)

The selectivity of rutile occurrence in the gold ores may probably reflect a geochemical affinity (?) of gold and titanium under hydrothermal conditions. In addition, this fact represents a typomorphic feature of the gold-quartz ores, irrespective of whether the rutile is an inherited (and possibly a recrystallized) mineral or whether it has formed by hydrolysis of volatile compounds carried by hydrothermal solutions, as Lawrence and Savage (1975) stated.

6 Other Occurrences

In the metamorphic terrains rutile is a common mineral and it becomes more abundant in the higher grade rocks (Elsdon 1975; Ramdohr 1975). Novitzky (1973) suggested the following sequence of titanium minerals with increasing metamorphic grade:

anatase → leucoxene (rutile) → sphene → ilmenite → rutile. Rutile + magnetite assemblage is typical for lower grade metapelites. In the higher grade schists magnetite disappears and ilmenite + hematite occur instead (Rumble 1976).

In some stratabound sulphide deposits of different ages and compositions the rutile is a persistent presence in the enclosing rocks (Schulz 1972) or in the ores themselves (Ramdohr 1953; Udubasa 1972, 1980; Cailteux and Dimanche 1973). Ballintoni and Chitimus (1973) reported also rutile paramorphs after brookite from metapelitic rocks of a polymetamorphic ore-bearing series in the eastern Carpathians, Romania.

7 Discussion

The observations made on the appearance of rutile under different geological conditions can be used to help in elucidating the stability relationships of various rutile-bearing assemblages. Besides the known diagrams of the stability fields of the TiO_2 modifications as function of PT (Dachille et al. 1968) or PT and Na/Ti+Na ratio (Lerz 1968) an empirical diagram may be drawn based on the observed relationships of rutile and anatase with the iron minerals (Fig. 6). The narrow stability field of rutile + magnetite is limited especially with respect to the sulphur fugacity. This rare assemblage (Elsdon 1975) appears in such samples where no sphene, ilmenite, hematite, or pyrite were observed. The only sulphide present is chalcopyrite, cementing both magnetite and rutile. The equant grains of magnetite and rutile (Fig. 3E) strongly suggest the recrystallization of the primary intergrowths of magnetite and ilmenite under con-

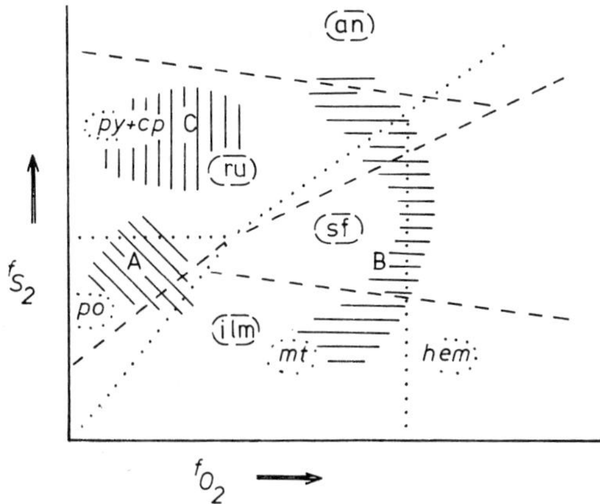

Fig. 6A–C. Empirical diagram of the stability fields of titanium compounds (*ilm* ilmenite; *sf* sphene; *ru* rutile; *an* anatase) as deduced from the observed relationships in natural occurrences. Stability fields of iron minerals after Raymahashay and Holland (1969, p. 291). **A** contact-metamorphic assemblages (*po* + *ru*); **B** porphyry copper systems (*ru* + *mt*, *ru* + *hem*, *ru* + *py*,*cp*); **C** field of hydrothermal gold-bearing quartz veins (*ru* + *py*)

ditions typical for porphyry copper environments. Similar morphological features exhibit some magnetite and ilmenite consisting of exolution-free equant grains derived from exsolution-rich titanomagnetite recrystallized during progressive metamorphism (Vertushkov et al. 1966).

The availability and the concentration of titanium appears to be the factor that controlled the occurrence of rutile in various assemblages. These results confirm Ramdohr's (1975) assumption on the inherited nature of rutile in most or even all natural occurrences. It forms therefore from ilmenite and anatase by increasing and decreasing sulphur fugacity, respectively.

Acknowledgments. The author expresses his thanks to the Mining Petroleum and Geology Ministry, Bucureşti, for permission to publish this paper. Thanks are also due to Mr. O. Edelstein and the geol. staff, Baia Mare, for furnishing some samples from the Tibles igneous complex, to Dr. G. Istrate and Mr. N. Pop for the samples from Medet, Butte, and Recsk.

Grateful acknowledgment is due to the Alexander von Humboldt-Stiftung, Bonn, for the Fellowship supported at Heidelberg University (1970–1972) where the writer has greatly benefited from the generous experience of Professor Ramdohr in the field of ore microscopy.

References

Balintoni I, Chitimus V (1973) On the presence of rutile paramorphs after brookite in the metamorphic rocks of the Tulghes Series, Eastern Carpathians. (Rumanian). Stud Cerc et Geol Geofiz Geogr Ser Geol 18:329–334

Cailteux J, Dimanche F (1973) Examen des oxydes de fer et titane dans l'environment du gisement de cuivre de Musoshi (Shaba, Republique du Zaire). Bull Soc Fr Minéral Cristallogr 96:378–382

Dachille F, Simons PY, Roy R (1968) Pressure-temperature studies of anatase, brookite, rutile and TiO_2-II. Am Mineral 53:1929–1939

Edelstein O, Pop N and geological staff, Baia Mare (1979) Unpublished report

Elsdon R (1975) Iron-titanium oxide minerals in igneous and metamorphic rocks. Min Sci Eng 7:48–70

Haggerty SE (1976a) Oxidation of opaque mineral oxides in basalts. In: Rumble D III (ed) Oxide minerals. Miner Soc Am, Short Course Not 3:Hg–1–Hg–98

Haggerty SE (1976b) Opaque mineral oxides in terrestrial igneous rocks. In: Rumble D III (ed) Oxide minerals. Miner Soc Am Short Course Not 3:Hg–101–Hg–300

Jackisch W (1969) Zur Geochemie und Lagerstättenkunde des Titans. Erzmetall 22:447–450, 498–504

Jamieson JC, Olinger B (1969) Pressure-temperature studies of anatase, brookite, rutile and TiO_2-II: A discussion. Am Mineral 54:1477–1481

Lawrence LJ, Savage EN (1975) Mineralogy of the titaniferous porphyry copper deposits of Melanesia. Australas Inst Min Metall Proc 256:1–14

Lerz H (1968) Über eine hydrothermale Paragenese von Anatas, Brookit und Rutil vom Dorfer Keesfleck, Prägraten, Osttirol. Neues Jahrb Mineral Monatsh 11:414–419

Lindsley DH (1976) Experimental studies of oxide minerals. In: Rumble D III (ed) Oxide minerals. Miner Soc Am Short Course Not 3:L–61–L–88

Marcu G (1979) Chemistry of metals (Rumanian). Edit Didactica si Pedagogica, Bucureşti, p 557

Moh GH, Hutchison ChS (1977) Hydrocassiterites. In: Ore minerals, G.H. Moh. Neues Jahrb Mineral Abh 131:13–17

Newhouse WH (1936) Opaque oxides and sulphides in common igneous rocks. Bull Geol Soc Am 47:1–52

Novitzky IP (1973) Sedimentary and sedimentary-metamorphic titanium concentrations of Rifeean Series in North-Western part of Kanin peninsula (Russian). In: Mineral deposits in sedimentary series. Nauka, Moscow, pp 9–15

Ramdohr P (1940) Die Erzmineralien in gewöhnlichen magmatischen Gesteinen. Abh Preuss Akad Wiss, Math-Naturwiss Kl 2:3–43

Ramdohr P (1953) Mineralbestand, Strukturen und Genesis der Rammelsberg-Lagerstätte. Geol Jahrb 67:367–494

Ramdohr P (1969) The ore minerals and their intergrowths. Pergamon Press, Oxford Braunschweig

Ramdohr P (1975) Die Erzmineralien und ihre Verwachsungen, 4 Aufl. Akademie Verlag, Berlin, 1277 pp

Raymahashay BC, Holland HD (1969) Redox relations accompanying hydrothermal wall rock alteration. Econ Geol 64:291–305

Rumble III D (1976) Oxide minerals in metamorphic rocks. In: Rumble D III (ed) Oxide minerals. Miner Soc Am Short Course Not 3:R–1–R–24

Schulz O (1972) Neue Ergebnisse über die Entstehung paläozoischer Erzlagerstätten am Beispiel der Nordtiroler Grauwackenzone. Proc 2nd Int Symp Mineral Deposits of the Alps, Ljubljana, pp 125–140

Udubasa G (1972) Syngenese und Epigenese in metamorphen und nichtmetamorphen Pb-Zn-Erzlagerstätten, aufgezeigt an den Beispielen Blazna-Tal (Ostkarpaten, Rumänien) und Ramsbeck (Westfalen, BRD). Inaug Dissertation, Univ Heidelberg, 145 S

Udubasa G (1978) Hydrothermal rutile in the Barza-Carpen gold-bearing ore deposits, Metaliferi Mountains. Dări Seamă Inst Geol Geofiz (Rom) LXIV/1:43–51

Udubasa G (1980) Lithology controls of ore localization: Blazna-Valley stratabound ore deposit, Eastern Carpathians, Romania (Rumanian). Dări Seamă Inst Geol Geofiz (Rom) Bucuresti (in press)

Vertushkov GN, Sokolov IUA, Lashkin VI (1966) Metamorphism of iron-titanium ore deposits of the Ufa group (Russian). Zap Vses Mineral Ova XCV:10–17

Subject Index

J. B. Dawson

Kimberlites and Their Xenoliths

1980. 84 figures, 35 tables. XII, 252 pages
(Minerals and Rocks, Volume 15)
ISBN 3-540-10208-6

In the past two decades there has been a rapid growth of interest in kimberlite, not only as a source of diamonds, but also as a rock type that has sampled the uppermantle more thoroughly than any other type of igneous activity. Whereas earlier books published on the subject in recent years have concentrated on specific topics such as the geology of diamond or upper-mantle mineralogy, the present volume provides on overall coverage of the geology of kimberlite from which inferences on the nature of the earth's upper-mantle can be drawn. This synthesis of up-to-date results and opinions from the recent geological literature will be highly appreciated by researchers in the field.

J. B. Gill

Orogenic Andesites and Plate Tectonics

1981. 109 figures. XIV, 390 pages
(Minerals and Rocks, Volume 16)
ISBN 3-540-10666-9

The solution to the questions of andesite genesis is a major, multidisciplinary undertaking facing geoscientists in the 1980's. **Orogenic Andesites and Plate Tectonic** was written in response to the growing need of researchers in this area for a clarification of the long-standing problem and identifications of profitable areas for future investigations.

This book critically summarizes the vast relevant literature on the tectonics, geophysics, volcanology, geology, geochemistry, and mineralogy of andesites and their volcanoes. The author cites over 1100 references in these specialties and include information on the location and rock composition of more than 300 recently active volcanoes. In addition, he provides a systematic and original evaluation of genetic hypotheses. Numerous cross references enhance the integrated subject development which consistently emphasizes the implications of data for theories of magma genesis. **Orogenic Andesites and Plate Tectonics** will prove an invaluable reference source to researchers and graduate students in the geosciences seeking a careful evaluation of genetic hypotheses and keydata, arguments or gaps in this promising field.

Springer-Verlag
Berlin
Heidelberg
New York

Ores in Sediments

VIII. International Sedimentological Congress Heidelberg,
August 31 – September 3, 1971
Editors: G. C. Amstutz, A. J. Bernard
Reprint 1976. 184 figures. VI, 350 pages
(International Union of Geological Sciences, Series A, Number 3)
ISBN 3-540-05712-9

From the reviews: "This book is the tangible result of the VIIIth International Sedimentological Congress ....
The editors deserve a special thanks from those of us to whom English is a useful language, ... since the book is published entirely in English. The printing and illustrations are good....
Topics range widely, from placers to metasomatic replacement, and from base and precious metals to iron, manganese, and uranium. The unifying theme is the association of metal concentrations with sediments. Four of the papers deal with elastic accumulations, placers, and iron ores. Four particularly informative papers describe the present-day accumulation of metals in sediments, three of the studies being from areas of active volcanism. The genetic relationship of mineralization to karst formation is interpreted by several authors, and the role of sedimentological facies in governing the localization of metal deposition is documented. ... This volume is a better source of interesting problem to be investigated than of definitive answers regarding diagenesis..." *"American Mineralogist"*

Time- and Strata-Bound Ore Deposits

Editors: D. D. Klemm, H.-J. Schneider

1977. 160 figures, 29 tables. XV, 444 pages (11 pages in German)
ISBN 3-540-08502-5

The problem of time- and strata-bound formation of ore deposits is an outstanding topic in actual international discussion. This book covers nearly all aspects of this very complex subject by original contributions. It integrates geochemical, physio-chemical, and microscopical results as well as field observations, partly in connection with plate tectonic aspects, reflecting recent progress in the fields from all continents.
The content of the book is subdivided into six main parts: Three of them deal with paramount ideas and general topics, as well as geochemical, physico-chemical, and mineralogical problems; three additional parts are, according to the leading theme, concerned with deposit-forming processes and indicative examples of metal enrichment in Precambrian. Paleozoic and Mesozoic. The different chapters contain results ranging from experimental and microscopical investigations up to field analyses in continental scale.
The various contributions by leading authorities are dedicated to Albert Maucher, an unflagging advocate of these ideas, on the occasion of his 70th anniversary.

Springer-Verlag
Berlin
Heidelberg
New York

Randall Library – UNCW
NXWW
QE390 .O 73 1982
Ore genesis, the state of the art : a volume in ho
3049002744695